Periodic Table of the Elements

Transition Elements

Key:

Symbol —	**Cl** 17 — Atomic Number
Atomic Mass	35.453
	$3p^5$ — Electron Configuration (outer shells only)

Group I	Group II											Group III	Group IV	Group V	Group VI	Group VII	Group VIII
H 1 1.0079 $1s^1$																	**He** 2 4.0026 $1s^2$
Li 3 6.941 $2s^1$	**Be** 4 9.0122 $2s^2$											**B** 5 10.811 $2p^1$	**C** 6 12.0107 $2p^2$	**N** 7 14.0067 $2p^3$	**O** 8 15.9994 $2p^4$	**F** 9 18.9984 $2p^5$	**Ne** 10 20.1797 $2p^6$
Na 11 22.9898 $3s^1$	**Mg** 12 24.3050 $3s^2$											**Al** 13 26.9815 $3p^1$	**Si** 14 28.0855 $3p^2$	**P** 15 30.9738 $3p^3$	**S** 16 32.065 $3p^4$	**Cl** 17 35.453 $3p^5$	**Ar** 18 39.948 $3p^6$
K 19 39.0983 $4s^1$	**Ca** 20 40.078 $4s^2$	**Sc** 21 44.9559 $3d^14s^2$	**Ti** 22 47.867 $3d^24s^2$	**V** 23 50.9415 $3d^34s^2$	**Cr** 24 51.9961 $3d^54s^1$	**Mn** 25 54.9380 $3d^54s^2$	**Fe** 26 55.845 $3d^64s^2$	**Co** 27 58.9332 $3d^74s^2$	**Ni** 28 58.6934 $3d^84s^2$	**Cu** 29 63.546 $3d^{10}4s^1$	**Zn** 30 65.409 $3d^{10}4s^2$	**Ga** 31 69.723 $4p^1$	**Ge** 32 72.64 $4p^2$	**As** 33 74.9216 $4p^3$	**Se** 34 78.96 $4p^4$	**Br** 35 79.904 $4p^5$	**Kr** 36 83.798 $4p^6$

LASER PHYSICS

LASER PHYSICS

PETER W. MILONNI
JOSEPH H. EBERLY

A JOHN WILEY & SONS, INC., PUBLICATION

Published by John Wiley & Sons, Inc., Hoboken, New Jersey
Published simultaneously in Canada

For general information on our other products and services or for technical support, please contact our Customer Care Department within the United States at (800) 762-2974, outside the United States at (317) 572-3993 or fax (317) 572-4002.

Wiley also publishes its books in a variety of electronic formats. Some content that appears in print may not be available in electronic formats. For more information about Wiley products, visit our web site at www.wiley.com.

Library of Congress Cataloging-in-Publication Data:

Milonni, Peter W.
    Laser physics / Peter W. Milonni, Joseph H. Eberly
        p. cm.
    Includes bibliographical references and index.
    ISBN 978-0-470-38771-9 (cloth)
    1. Lasers. 2. Nonlinear optics. 3. Physical optics. I. Eberly, J. H., 1935- II. Title.
    QC688.M55 2008
    621.36'6—dc22

                                                                2008026771

Printed in the United States of America
10 9 8 7 6 5 4 3 2

To our wives, Mei-Li and Shirley

CONTENTS

PREFACE

nano 10^{-9}
pico 10^{-12}
femto 10^{-15}
atto 10^{-18}
zepto 10^{-21}
yocto 10^{-24}

Judged by their economic impact and their role in everyday life, and also by the number of Nobel Prizes awarded, advances in laser science and engineering in the past quarter-century have been remarkable. Using lasers, scientists have produced what are believed to be the coldest temperatures in the universe, and energy densities greater than in the center of stars; have tested the foundations of quantum theory itself; and have controlled atomic, molecular, and photonic states with unprecedented precision.

Questions that previous generations of scientists could only contemplate in terms of thought experiments have been routinely addressed using lasers. Atomic clock frequencies can be measured to an accuracy exceeding that of any other physical quantity. The generation of femtosecond pulses has made it possible to follow chemical processes in action, and the recent availability of attosecond pulses is allowing the study of phenomena on the time scale of electron motion in atoms. Frequency stabilization and the frequency-comb spectra of mode-locked lasers have now made practical the measurement of absolute optical frequencies and promise ever greater precision in spectroscopy and other areas. Lasers are being used in adaptive optical systems to obtain image resolution with ground-based telescopes that is comparable to that of telescopes in space, and they have become indispensable in lidar and environmental studies. Together with optical fibers, diode lasers have fueled the explosive growth of optical networks and the Internet. In medicine, lasers are finding more and more uses in surgery and clinical procedures. Simply put, laser physics is an integral part of contemporary science and technology, and there is no foreseeable end to its progress and application.

The guiding theme of this book is lasers, and our intent is for the reader to arrive at more than a command of tables and formulas. Thus all of the chapters incorporate explanations of the central elements of optical engineering and physics that are required for a basic and detailed understanding of laser operation. Applications are important and we discuss how laser radiation interacts with matter, and how coherent and often very intense laser radiation is used in research and in the field. We presume that the reader has been exposed to classical electromagnetic theory and quantum mechanics at an undergraduate or beginning graduate level, but we take opportunities throughout to review parts of these subjects that are particularly important for laser physics.

The perceptive reader will notice that there is substantial overlap with a book we wrote 20 years ago called simply *Lasers*, also published by Wiley and still in print without revision or addition. Many readers and users of that book have told us that they particularly appreciated the frequent concentration on background optical physics as well as explanations of the physical basis for all aspects of laser operation. Naturally a book about lasers that is two decades old needs many new topics to be added to be even approximately current. However, while recognizing that additions are necessary, we also wanted to resist what is close to a law of nature, that a second book must weigh significantly more than its predecessor. We believe we have accomplished these goals by describing some of the most significant recent developments in laser physics together with an illustrative set of applications based on them.

The basic principles of lasers have not changed in the past twenty years, but there has been a shift in the kinds of lasers of greatest general interest. Considerable attention is devoted to semiconductor lasers and fiber lasers and amplifiers, and to considerations of noise and dispersion in fiber-optic communications. We also treat various aspects of chirping and its role in the generation of extremely short and intense pulses of radiation. Laser trapping and cooling are explained in some detail, as are most of the other applications mentioned above. We introduce the most important concepts needed to understand the propagation of laser radiation in the turbulent atmosphere; this is an important topic for free-space communication, for example, but it has usually been addressed only in more advanced and specialized books. We have attempted to present it in a way that might be helpful for students as well as laser scientists and engineers with no prior exposure to turbulence theory.

The book is designed as a textbook, but there is probably too much material here to be covered in a one-semester course. Chapters 1–7 could be used as a self-contained, elementary introduction to lasers and laser—matter interactions. In most respects the remaining chapters are self-contained, while using consistent notation and making reference to the same fundamentals. Chapters 9 and 10, for example, can serve as introductions to coherent propagation effects and nonlinear optics, respectively, and Chapters 12 and 13 can be read separately as introductions to photon detection, photon counting, and optical coherence. Chapters 14 and 15 describe some applications of lasers that will likely be of interest for many years to come.

We are grateful to A. Al-Qasimi, S. M. Barnett, P. R. Berman, R. W. Boyd, L. W. Casperson, C. A. Denman, R. Q. Fugate, J. W. Goodman, D. F. V. James, C. F. Maes, G. H. C. New, C. R. Stroud, Jr., J. M. Telle, I. A. Walmsley, and E. Wolf for comments on some of the chapters or for contributing in other ways to this effort.

1 INTRODUCTION TO LASER OPERATION

1.1 INTRODUCTION

The word *laser* is an acronym for the most significant feature of laser action: light amplification by stimulated emission of radiation. There are many different kinds of laser, but they all share a crucial element: Each contains material capable of amplifying radiation. This material is called the gain medium because radiation gains energy passing through it. The physical principle responsible for this amplification is called stimulated emission and was discovered by Albert Einstein in 1916. It was widely recognized that the laser would represent a scientific and technological step of the greatest magnitude, even before the first one was constructed in 1960 by T. H. Maiman. The award of the 1964 Nobel Prize in physics to C. H. Townes, N. G. Basov, and A. M. Prokhorov carried the citation "for fundamental work in the field of quantum electronics, which has led to the construction of oscillators and amplifiers based on the maser-laser principle." These oscillators and amplifiers have since motivated and aided the work of thousands of scientists and engineers.

In this chapter we will undertake a superficial introduction to lasers, cutting corners at every opportunity. We will present an overview of the properties of laser light, with the goal of understanding what a laser is, in the simplest terms. We will introduce the theory of light in cavities and of cavity modes, and we will describe an elementary theory of laser action.

We can begin our introduction with Fig. 1.1, which illustrates the four key elements of a laser. First, a collection of atoms or other material amplifies a light signal directed through it. This is shown in Fig. 1.1a. The amplifying material is usually enclosed by a highly reflecting cavity that will hold the amplified light, in effect redirecting it through the medium for repeated amplifications. This refinement is indicated in Fig. 1.1b. Some provision, as sketched in Fig. 1.1c, must be made for replenishing the energy of the amplifier that is being converted to light energy. And some means must be arranged for extracting in the form of a beam at least part of the light stored in the cavity, perhaps as shown in Fig. 1.1d. A schematic diagram of an operating laser embodying all these elements is shown in Fig. 1.2.

It is clear that a well-designed laser must carefully balance gains and losses. It can be anticipated with confidence that every potential laser system will present its designer with more sources of loss than gain. Lasers are subject to the basic laws of physics, and every stage of laser operation from the injection of energy into the amplifying medium to the extraction of light from the cavity is an opportunity for energy loss

Laser Physics. By Peter W. Milonni and Joseph H. Eberly
Copyright © 2010 John Wiley & Sons, Inc.

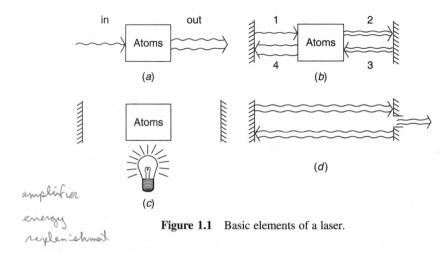

amplifier

energy
replenishment

Figure 1.1 Basic elements of a laser.

and entropy gain. One can say that the success of masers and lasers came only after physicists learned how atoms could be operated efficiently as thermodynamic engines.

One of the challenges in understanding the behavior of atoms in cavities arises from the strong feedback deliberately imposed by the cavity designer. This feedback means that a small input can be amplified in a straightforward way by the atoms, but not indefinitely. Simple amplification occurs only until the light field in the cavity is strong enough to affect the behavior of the atoms. Then the strength of the light as it acts on the amplifying atoms must be taken into account in determining the strength of the light itself. This sounds like circular reasoning and in a sense it is. The responses of the light and the atoms to each other can become so strongly interconnected that they cannot be determined independently but only self-consistently. Strong feedback also means that small perturbations can be rapidly magnified. Thus, it is accurate to anticipate that lasers are potentially highly erratic and unstable devices. In fact, lasers can provide dramatic exhibitions of truly chaotic behavior and have been the objects of fundamental study for this reason.

For our purposes lasers are principally interesting, however, when they operate stably, with well-determined output intensity and frequency as well as spatial mode structure.

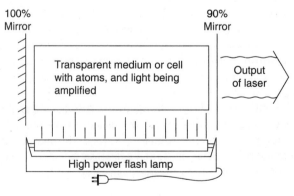

Figure 1.2 Complete laser system, showing elements responsible for energy input, amplification, and output.

The self-consistent interaction of light and atoms is important for these properties, and we will have to be concerned with concepts such as gain, loss, threshold, steady state, saturation, mode structure, frequency pulling, and linewidth.

In the next few sections we sketch properties of laser light, discuss modes in cavities, and give a theory of laser action. This theory is not really correct, but it is realistic within its own domain and has so many familiar features that it may be said to be "obvious." It is also significant to observe what is not explained by this theory and to observe the ways in which it is not fundamental but only empirical. These gaps and missing elements are an indication that the remaining chapters of the book may also be necessary.

1.2 LASERS AND LASER LIGHT

Many of the properties of laser light are special or extreme in one way or another. In this section we provide a brief overview of these properties, contrasting them with the properties of light from more ordinary sources when possible.

Wavelength

Laser light is available in all colors from red to violet and also far outside these conventional limits of the optical spectrum.[1] Over a wide portion of the available range laser light is "tunable." This means that some lasers (e.g., dye lasers) have the property of emitting light at any wavelength chosen within a range of wavelengths. The longest laser wavelength can be taken to be in the far infrared, in the neighborhood of 100–500 μm. Devices producing coherent light at much longer wavelengths by the "maser–laser principle" are usually thought of as masers. The search for lasers with ever shorter wavelengths is probably endless. Coherent stimulated emission in the XUV (extreme ultraviolet) or soft X-ray region (10–15 nm) has been reported. Appreciably shorter wavelengths, those characteristic of gamma rays, for example, may be quite difficult to reach.

Photon Energy

The energy of a laser photon is not different from the energy of an "ordinary" light photon of the same wavelength. A green–yellow photon, roughly in the middle of the optical spectrum, has an energy of about 2.5 eV (electron volts). This is the same as about 4×10^{-19} J (joules) $= 4 \times 10^{-12}$ erg. The large exponents in the last two numbers make it clear that electron volts are a much more convenient unit for laser photon energy than joules or ergs. From the infrared to the X-ray region photon energies vary from about 0.01 eV to about 100 eV. For contrast, at room temperature the thermal unit of energy is $kT \approx \frac{1}{40}$ eV $= 0.025$ eV. This is two orders of magnitude smaller than the typical optical photon energy just mentioned, and as a consequence thermal excitation plays only a very small role in the physics of nearly all lasers.

---

[1]A list of laser wavelengths may be found in M. J. Weber, *Handbook of Laser Wavelengths*, CRC, Boca Raton, FL, 1999.

Directionality

The output of a laser can consist of nearly ideal plane wavefronts. Only diffraction imposes a lower limit on the angular spread of a laser beam. The wavelength λ and the area A of the laser output aperture determine the order of magnitude of the beam's solid angle ($\Delta\Omega$) and vertex angle ($\Delta\theta$) of divergence (Fig. 1.3) through the relation

$$\Delta\Omega \approx \frac{\lambda^2}{A} \approx (\Delta\theta)^2. \qquad (1.2.1)$$

This represents a very small angular spread indeed if λ is in the optical range, say 500 nm, and A is macroscopic, say $(5 \text{ mm})^2$. In this example we compute $\Delta\Omega \approx (500)^2 \times 10^{-18} \text{ m}^2/(5^2 \times 10^{-6} \text{ m}^2) = 10^{-8}$ sr, or $\Delta\theta = 1/10$ mrad.

Monochromaticity

It is well known that lasers produce very pure colors. If they could produce exactly one wavelength, laser light would be fully monochromatic. This is not possible, in principle as well as for practical reasons. We will designate by $\Delta\lambda$ the range of wavelengths included in a laser beam of main wavelength λ. Similarly, the associated range of frequencies will be designated by $\Delta\nu$, the bandwidth. In the optical region of the spectrum we can take $\nu \approx 5 \times 10^{14}$ Hz (hertz, i.e., cycles per second). The bandwidth of sunlight is very broad, more than 10^{14} Hz. Of course, filtered sunlight is a different matter, and with sufficiently good filters $\Delta\nu$ could be reduced a great deal. However, the cost in lost intensity would usually be prohibitive. (See the discussion on spectral brightness below.) For lasers, a very low value of $\Delta\nu$ is 1 Hz, while a bandwidth around 100 Hz is spectroscopically practical in some cases (Fig. 1.4). For $\Delta\nu = 100$ Hz the relative spectral purity of a laser beam is quite impressive: $\Delta\nu/\nu \approx 100/(5 \times 10^{14}) = 2 \times 10^{-13}$.

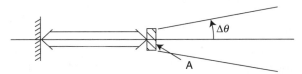

Figure 1.3 Sketch of a laser cavity showing angular beam divergence $\Delta\theta$ at the output mirror (area A).

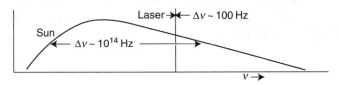

Figure 1.4 Spectral emission bands of the sun and of a representative laser, to indicate the much closer approach to monochromatic light achieved by the laser.

This exceeds the spectral purity (Q factor) achievable in conventional mechanical and electrical resonators by many orders of magnitude.

Coherence Time

The existence of a finite bandwidth $\Delta\nu$ means that the different frequencies present in a laser beam can eventually get out of phase with each other. The time required for two oscillations differing in frequency by $\Delta\nu$ to get out of phase by a full cycle is obviously $1/\Delta\nu$. After this amount of time the different frequency components in the beam can begin to interfere destructively, and the beam loses "coherence." Thus, $\Delta\tau = 1/\Delta\nu$ is called the beam's coherence time. This is a general definition, not restricted to laser light, but the extremely small values possible for $\Delta\nu$ in laser light make the coherence times of laser light extraordinarily long.

For example, even a "broadband" laser with $\Delta\nu \approx 1$ MHz has the coherence time $\Delta\tau \approx 1$ μs. This is enormously longer than most "typical" atomic fluorescence lifetimes, which are measured in nanoseconds (10^{-9} s). Thus even lasers that are not close to the limit of spectral purity are nevertheless effectively 100% pure on the relevant spectroscopic time scale. By way of contrast, sunlight has a bandwidth $\Delta\nu$ almost as great as its central frequency (yellow light, $\nu = 5 \times 10^{14}$ Hz). Thus, for sunlight the coherence time is $\Delta\tau \approx 2 \times 10^{-15}$ s, so short that unfiltered sunlight cannot be considered temporally coherent at all.

Coherence Length

The speed of light is so great that a light beam can travel a very great distance within even a short coherence time. For example, within $\Delta\tau \approx 1$ μs light travels $\Delta z \approx (3 \times 10^8 \text{ m/s}) \times (1 \text{ μs}) = 300$ m. The distance $\Delta z = c\,\Delta\tau$ is called the beam's coherence length. Only portions of the same beam that are separated by less than Δz are capable of interfering constructively with each other. No fringes will be recorded by the film in Fig. 1.5, for example, unless $2L < c\,\Delta\tau = \Delta z$.

Spectral Brightness

A light beam from a finite source can be characterized by its beam divergence $\Delta\Omega$, source size (usually surface area A), bandwidth $\Delta\nu$, and spectral power density P_ν (watts per hertz of bandwidth). From these parameters it is useful to determine the *spectral brightness* β_ν of the source, which is defined (Fig. 1.6) to be the power flow per unit

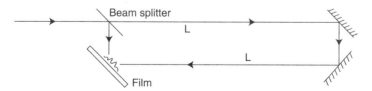

Figure 1.5   Two-beam interferometer showing interference fringes obtained at the recording plane if the coherence length of the light is great enough.

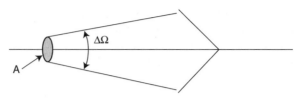

Figure 1.6 Geometrical construction showing source area and emission solid angle appropriate to discussion of spectral brightness.

area, unit bandwidth, and steradian, namely $\beta_\nu = P_\nu / A\,\Delta\Omega\,\Delta\nu$. Notice that $P_\nu / A\,\Delta\nu$ is the spectral intensity, so β_ν can also be thought of as the spectral intensity per steradian.

For an ordinary *nonlaser optical source*, brightness can be estimated directly from the blackbody formula for $\rho(\nu)$, the spectral energy density (J/m³-Hz):

$$\rho(\nu) = \frac{8\pi\nu^2}{c^3}\frac{h\nu}{e^{h\nu/k_B T}-1}. \tag{1.2.2}$$

The spectral intensity (W/m²-Hz) is thus $c\rho$, and $c\rho/\Delta\Omega$ is the desired spectral intensity per steradian. Taking $\Delta\Omega = 4\pi$ for a blackbody, we have

$$\beta_\nu = \frac{2\nu^2}{c^2}\frac{h\nu}{e^{h\nu/k_B T}-1}. \tag{1.2.3}$$

The temperature of the sun is about $T = 5800\text{K} \approx 20\times(300\text{K})$. Since the main solar emission is in the yellow portion of the spectrum, we can take $h\nu \approx 2.5\,\text{eV}$. We recall that $k_B T \approx \frac{1}{40}\,\text{eV}$ for $T = 300\text{K}$, so $h\nu/k_B T \approx 5$, giving $e^{h\nu/k_B T} \approx 150$ and finally

$$\beta_\nu \approx 1.5 \times 10^{-8} \text{ W/m}^2\text{-sr-Hz} \quad (\text{sun}). \tag{1.2.4}$$

Several different estimates can be made for laser radiation, depending on the type of laser considered. Consider first a *low-power He–Ne laser*. A power level of 1 mW is normal, with a bandwidth of around 10^4 Hz. From (1.2.1) we see that the product of beam cross-sectional area and solid angle is just λ^2, which for He–Ne light is $\lambda^2 \approx (6328\times 10^{-10}\text{ m})^2 \approx 4\times 10^{-13}\text{ m}^2$. Combining these, we find

$$\beta_\nu \approx 2.5 \times 10^5 \text{ W/m}^2\text{-sr-Hz} \quad (\text{He–Ne laser}). \tag{1.2.5}$$

Another common laser is the *mode-locked neodymium–glass laser*, which can easily reach power levels around 10^4 MW. The bandwidth of such a laser is limited by the pulse duration, say $\tau_p \approx 30$ ps (30×10^{-12} s), as follows. Since the laser's coherence time $\Delta\tau$ is equal to τ_p at most, its bandwidth is certainly greater than $1/\tau_p \approx 3.3\times 10^{10}\text{ s}^{-1}$. We convert from radians per second to cycles per second by dividing by 2π and get $\Delta\nu \approx 5\times 10^9$ Hz. The wavelength of a Nd:glass laser is 1.06 μm, so $\lambda^2 \approx 10^{-12}\text{ m}^2$. The result of combining these, again using $A\,\Delta\Omega = \lambda^2$, is

$$\beta_\nu \approx 2 \times 10^{12} \text{ W/m}^2\text{-sr-Hz} \quad (\text{Nd:glass laser}). \tag{1.2.6}$$

Recent developments have led to lasers with powers of terawatts (10^{12} W) and even petawatts (10^{15} W), so β_ν can be even orders of magnitude larger.

It is clear that in terms of brightness there is practically no comparison possible between lasers and thermal light. Our sun is 20 *orders of magnitude* less bright than a mode-locked laser. This raises an interesting question of principle. Let us imagine a thermal light source filtered and collimated to the bandwidth and directionality of a He−Ne laser, and the He−Ne laser attenuated to the brightness level of the thermal light. The question is: Could the two light beams with equal brightness, beam divergence, polarization, and bandwidth be distinguished in any way? The answer is that they could be distinguished, but not by any ordinary measurement of optics. Differences would show up only in the statistical fluctuations in the light beam. These fluctuations can reflect the quantum nature of the light source and are detected by photon counting, as discussed in Chapter 12.

Active Medium

The materials that can be used as the active medium of a laser are so varied that a listing is hardly possible. Gases, liquids, and solids of every sort have been made to *lase* (a verb contributed to science by the laser). The origin of laser photons, as shown in Fig. 1.7, is most often in a transition between discrete upper and lower energy states in the medium, regardless of its state of matter. He−Ne, ruby, CO_2, and dye lasers are familiar examples, but exceptions are easily found: The excimer laser has an unbound lower state, the semiconductor diode laser depends on transitions between electron bands rather than discrete states, and understanding the free-electron laser does not require quantum states at all.

Type of Laser Cavity

All laser cavities share two characteristics that complement each other: (1) They are basically linear devices with one relatively long optical axis, and (2) the sides parallel to this axis can be open, not enclosed by reflecting material as in a microwave cavity. There is no single best shape implied by these criteria, and in the case of ring lasers the long axis actually bends and closes on itself (Fig. 1.8). Despite what may seem

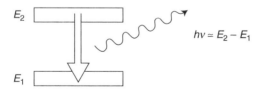

Figure 1.7 Photon emission accompanying a quantum jump from level 2 to level 1.

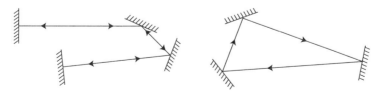

Figure 1.8 Two collections of mirrors making laser cavities, showing standing-wave and traveling-wave (ring) configurations on left and right, respectively.

obvious, it is not always best to design a cavity with the lowest loss. In the case of Q switching an extra loss is temporarily introduced into the cavity for the laser to overcome, and very high-power lasers sometimes use mirrors that are deliberately designed to deflect light out of the cavity rather than contain it.

Applications of Lasers

There is apparently no end of possible applications of lasers. Many of the uses of lasers are well known by now to most people, such as for various surgical procedures, for holography, in ultrasensitive gyroscopes, to provide straight lines for surveying, in supermarket checkout scanners and compact disc players, for welding, drilling, and scribing, in compact death-ray pistols, and so on. (The sophisticated student knows, even before reading this book, that one of these "well-known" applications has never been realized outside the movie theater.)

1.3 LIGHT IN CAVITIES

In laser technology the terms *cavity* and *resonator* are used interchangeably. The theory and design of the cavity are important enough for us to devote all of Chapter 7 to them. In this section we will consider only a simplified theory of resonators, a theory that is certain to be at least partly familiar to most readers. This simplification allows us to introduce the concept of cavity modes and to infer certain features of cavity modes that remain valid in more general circumstances. We also describe the great advantage of open, rather than closed, cavities for optical radiation.

We will consider only the case of a rectangular "empty cavity" containing radiation but no matter, as sketched in Fig. 1.9. The assumption that there is radiation but no matter inside the cavity is obviously an approximation if the cavity is part of a working laser. This approximation is used frequently in laser theory, and it is accurate enough for many purposes because laser media are usually only sparsely filled with active atoms or molecules.

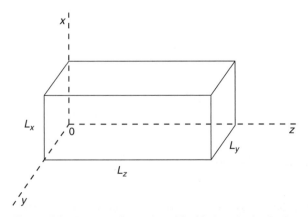

Figure 1.9 Rectangular cavity with side lengths L_x, L_y, L_z.

In Chapter 7 full solutions for the electric field in cavities of greatest interest are given. For example, the z dependence of the x component of the field takes the form

$$E_x(z) = E_0 \sin k_z z, \tag{1.3.1}$$

where E_0 is a constant. However, here we are interested only in the simplest features of the cavity field, and these can be obtained easily by physical reasoning.

The electric field should vanish at both ends of the cavity. It will do so if we fit exactly an integer number of half wavelengths into the cavity along each of its axes. This means, for example, that λ along the z axis is determined by the relation $L = n(\lambda/2)$, where $n = 1, 2, \ldots,$ is a positive integer and L is the cavity length. If we use the relation between wave vector and wavelength, $k = 2\pi/\lambda$, this is the same as

$$k_z = \frac{\pi}{L} n, \tag{1.3.2}$$

for the z component of the wave vector. By substitution into the solution (1.3.1) we see that (1.3.2) is sufficient to guarantee that the required boundary condition is met, i.e., that $E_x(z) = 0$ for both $z = 0$ and $z = L$.

If there were reflecting sides to a laser cavity, the same would apply to the x and y components of the wave vector. As we will show later, if the three dimensions are equivalent in this sense, the number of available modes grows extremely rapidly as a function of frequency. For example, a cubical three-dimensionally reflecting cavity 1 cm on a side has about 400 million resonant frequencies within the useful gain band of a He–Ne laser. Then lasing could occur across the whole band, eliminating any possibility of achieving the important narrow-band, nearly monochromatic character of laser light that we emphasized in the preceding sections.

The solution to this multimode dilemma was suggested independently in 1958 by Townes and A. L. Schawlow, R. H. Dicke, and Prokhorov. They recognized that a one-dimensional rather than a three-dimensional cavity was desirable, and that this could be achieved with an *open resonator* consisting of two parallel mirrors, as in Fig. 1.10. The difference in wave vector between two modes of a linear cavity, according to Eq. (1.3.2), is just π/L, so the mode spacing is given by $\Delta k = (2\pi/c)\,\Delta\nu$, or $\Delta\nu = c/2L$. For $L = 10$ cm we find

$$\Delta\nu = \frac{3 \times 10^8 \text{ m/s}}{(2)(0.10 \text{ m})} = 1500 \text{ MHz}, \tag{1.3.3}$$

for the separation in frequency of adjacent resonator modes. As indicated in Fig. 1.11, the number of possible modes that can lase is therefore at most

$$\frac{1500 \text{ MHz}}{1500 \text{ MHz}} = 1. \tag{1.3.4}$$

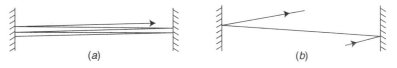

(a)   (b)

Figure 1.10 Sketch illustrating the advantage of a one-dimensional cavity. Stable modes are associated only with beams that are retroreflected many times.

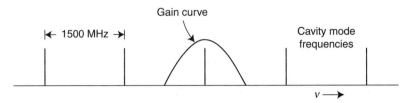

Figure 1.11   Mode frequencies separated by 1500 MHz, corresponding to a 10-cm one-dimensional cavity. A 1500-MHz gain curve overlaps only 1 mode.

The maximum number, including two choices of polarization, is therefore 2, considerably smaller than the estimate of 400 million obtained for three-dimensional cavities.

These results do not include the effects of diffraction of radiation at the mirror edges. Diffraction determines the x, y dependence of the field, which we have ignored completely. Accurate calculations of resonator modes, including diffraction, are often done with computers. Such calculations were first made in 1961 for the plane-parallel resonator of Fig. 1.10 with either rectangular or circular mirrors. Actually lasers are seldom designed with flat mirrors. Laser resonator mirrors are usually spherical surfaces, for reasons to be discussed in Chapter 7. A great deal about laser cavities can nevertheless be understood without worrying about diffraction or mirror shape. In particular, *for most practical purposes, the mode-frequency spacing is given accurately enough by* $\Delta v = c/2L$.

1.4   LIGHT EMISSION AND ABSORPTION IN QUANTUM THEORY

The modern interpretation of light emission and absorption was first proposed by Einstein in 1905 in his theory of the photoelectric effect. Einstein assumed the difference in energy of the electron before and after its photoejection to be equal to the energy hv of the photon absorbed in the process.

This picture of light absorption was extended in two ways by Bohr: to apply to atomic electrons that are not ejected during photon absorption but instead take on a higher energy within their atom, and to apply to the reverse process of photon emission, in which case the energy of the electron should decrease. These extensions of Einstein's idea fitted perfectly into Bohr's quantum mechanical model of an atom in 1913. This model, described in detail in Chapter 2, was the first to suggest that electrons are restricted to a certain fixed set of orbits around the atomic nucleus. This set of orbits was shown to correspond to a fixed set of allowed electron energies. The idea of a "quantum jump" was introduced to describe an electron's transition between two allowed orbits.

The amount of energy involved in a quantum jump depends on the quantum system. Atoms have quantum jumps whose energies are typically in the range $1-6$ eV, as long as an outer-shell electron is doing the jumping. This is the ordinary case, so atoms usually absorb and emit photons in or near the optical region of the spectrum. Jumps by inner-shell atomic electrons usually require much more energy and are associated with X-ray photons. On the other hand, quantum jumps among the so-called Rydberg energy levels, those outer-electron levels lying far from the ground level and near to the ionization limit, involve only a small amount of energy, corresponding to far-infrared or even microwave photons.

Molecules have vibrational and rotational degrees of freedom whose quantum jumps are smaller (perhaps much smaller) than the quantum jumps in free atoms, and the same is often true of jumps between conduction and valence bands in semiconductors. Many crystals are transparent in the optical region, which is a sign that they do not absorb or emit optical photons, because they do not have quantum energy levels that permit jumps in the optical range. However, colored crystals such as ruby have impurities that do absorb and emit optical photons. These impurities are frequently atomic ions, and they have both discrete energy levels and broad bands of levels that allow optical quantum jumps (ruby is a good absorber of green photons and so appears red).

1.5 EINSTEIN THEORY OF LIGHT–MATTER INTERACTIONS

The atoms of a laser undergo repeated quantum jumps and so act as microscopic transducers. That is, each atom accepts energy and jumps to a higher orbit as a result of some input or "pumping" process and converts it into other forms of energy—for example, into light energy (photons)—when it jumps to a lower orbit. At the same time, each atom must deal with the photons that have been emitted earlier and reflected back by the mirrors. These prior photons, already channeled along the cavity axis, are the origin of the stimulated component to the atom's emission of subsequent photons.

In Fig. 1.12 we indicate some ways in which energy conversion can occur. For simplicity we focus our attention on quantum jumps between two energy levels, 1 and 2, of an atom. The five distinct energy conversion diagrams of Fig. 1.12 are interpreted as follows:

(a) Absorption of an increment $\Delta E = E_2 - E_1$ of energy from the pump: The atom is raised from level 1 to level 2. In other words, an electron in the atom jumps from an inner orbit to an outer orbit.

(b) Spontaneous emission of a photon of energy $h\nu = E_2 - E_1$: The atom jumps down from level 2 to the lower level 1. The process occurs "spontaneously" without any external influence.

(c) Stimulated emission: The atom jumps down from energy level 2 to the lower level 1, and the emitted photon of energy $h\nu = E_2 - E_1$ is an exact replica of a photon already present. The process is induced, or stimulated, by the incident photon.

(d) Absorption of a photon of energy $h\nu = E_2 - E_1$: The atom jumps up from level 1 to the higher level 2. As in (c), the process is induced by an incident photon.

(e) Nonradiative deexcitation: The atom jumps down from level 2 to the lower level 1, but no photon is emitted so the energy $E_2 - E_1$ must appear in some other form [e.g., increased vibrational or rotational energy in the case of a molecule, or rearrangement ("shakeup") of other electrons in the atom].

All these processes occur in the gain medium of a laser. Lasers are often classified according to the nature of the pumping process (a) which is the source of energy for the output laser beam. In electric-discharge lasers, for instance, the pumping occurs

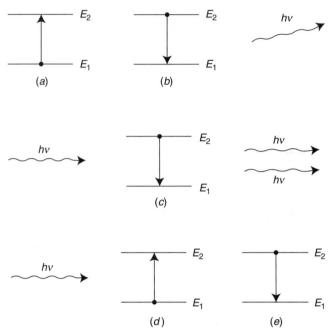

Figure 1.12   Energy conversion processes in a lasing atom or molecule: (*a*) absorption of energy $\Delta E = E_2 - E_1$ from the pump; (*b*) spontaneous emission of a photon of energy ΔE; (*c*) stimulated emission of a photon of energy ΔE; (*d*) absorption of a photon of energy ΔE; (*e*) nonradiative deexcitation.

as a result of collisions of electrons in a gaseous discharge with the atoms (or molecules) of the gain medium. In an optically pumped laser the pumping process is the same as the absorption process (d), except that the pumping photons are supplied by a lamp or perhaps another laser. In a diode laser an electric current at the junction of two different semiconductors produces electrons in excited energy states from which they can jump into lower energy states and emit photons.

This quantum picture is consistent with a highly simplified description of laser action. Suppose that lasing occurs on the transition defined by levels 1 and 2 of Fig. 1.12. In the most favorable situation the lower level (level 1) of the laser transition is empty. To maintain this situation a mechanism must exist to remove downward jumping electrons from level 1 to another level, say level 0. In this situation there can be no detrimental absorption of laser photons due to transitions upward from level 1 to level 2. In practice the number of electrons in level 1 cannot be exactly zero, but we will assume for simplicity that the rate of deexcitation of the lower level 1 is so large that the number of atoms remaining in that level is negligible compared to the number in level 2; this is a reasonably good approximation for many lasers. Under this approximation laser action can be described in terms of two "populations": the number n of atoms in the upper level 2 and the number q of photons in the laser cavity.

The number of laser photons in the cavity changes for two main reasons:

(i) Laser photons are continually being added because of stimulated emission.
(ii) Laser photons are continually being lost because of mirror transmission, scattering or absorption at the mirrors, etc.

Thus, we can write a (provisional) equation for the rate of change of the number of photons, incorporating the gain and loss described in (i) and (ii) as follows:

$$\frac{dq}{dt} = anq - bq. \tag{1.5.1}$$

That is, the rate at which the number of laser photons changes is the sum of two separate rates: the rate of increase (amplification or gain) due to stimulated emission, and the rate of decrease (loss) due to imperfect mirror reflectivity.

As Eq. (1.5.1) indicates, the gain of laser photons due to stimulated emission is not only proportional to the number n of atoms in level 2, but also to the number q of photons already in the cavity. The efficiency of the stimulated emission process depends on the type of atom used and other factors. These factors are reflected in the size of the amplification or *gain coefficient a*. The rate of loss of laser photons is simply proportional to the number of laser photons present.

We can also write a provisional equation for n. Both stimulated and spontaneous emission cause n to decrease (in the former case in proportion to q, in the latter case not), and the pump causes n to increase at some rate we denote by p. Thus, we write

$$\frac{dn}{dt} = -anq - fn + p. \tag{1.5.2}$$

Note that the first term appears in both equations, but with opposite signs. This reflects the central role of stimulated emission and shows that the decrease of n (excited atoms) due to stimulated emission corresponds precisely to the increase of q (photons).

Equations (1.5.1) and (1.5.2) describe laser action. They show how the numbers of lasing atoms and laser photons in the cavity are related to each other. They do not indicate what happens to the photons that leave the cavity, or what happens to the atoms when their electrons jump to some other level. Above all, they do not tell how to evaluate the coefficients a, b, f, p. They must be taken only as provisional equations, not well justified although intuitively reasonable.

It is important to note that neither Eq. (1.5.1) nor (1.5.2) can be solved independently of the other. That is, (1.5.1) and (1.5.2) are *coupled* equations. The coupling is due physically to stimulated emission: The lasing atoms of the gain medium can increase the number of photons via stimulated emission, but by the same process the presence of photons will also decrease the number of atoms in the upper laser level. This coupling between the atoms and the cavity photons is indicated schematically in Fig. 1.13.

We also note that Eqs. (1.5.1) and (1.5.2) are nonlinear. The nonlinearity (the product of the two variables nq) occurs in both equations and is another manifestation of

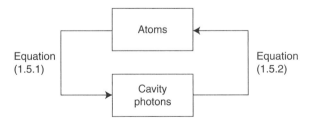

Figure 1.13 Self-consistent pair of laser equations.

stimulated emission. No established systematic methods exist for solving nonlinear differential equations, and there is no known general solution to these laser equations. However, they have a number of well-defined limiting cases of some practical importance, and some of these do have known solutions. The most important case is steady state.

In steady state we can put both dq/dt and dn/dt equal to zero. Then (1.5.1) reduces to

$$n = \frac{b}{a} \equiv n_t, \qquad (1.5.3)$$

which can be recognized as a *threshold* requirement on the number of upper-level atoms. That is, if $n < b/a$, then $dq/dt < 0$, and the number of photons in the cavity decreases, terminating laser action. The steady state of (1.5.2) also has a direct interpretation. From $dn/dt = 0$ and $n = n_t = b/a$ we find

$$q = \frac{p}{b} - \frac{f}{a}. \qquad (1.5.4)$$

This equation establishes a threshold for the pumping rate, since the number of photons q cannot be negative. Thus, the minimum or threshold value of p compatible with steady-state operation is found by putting $q = 0$:

$$p_t = \frac{fb}{a} = fn_t. \qquad (1.5.5)$$

In words, the threshold pumping rate just equals the loss rate per atom times the number of atoms present at threshold.

In Chapters 4–6 we will return to a discussion of laser equations. We will deal there with steady state as well as many other aspects of laser oscillation in two-level, three-level, and four-level quantum systems.

1.6 SUMMARY

The theory of laser action and the description of cavity modes presented in this chapter can be regarded only as caricatures. In common with all caricatures, they display outstanding features of their subject boldly and simply. All theories of laser action must address the questions of *gain*, *loss*, *steady state*, and *threshold*. The virtues of our caricatures in addressing these questions are limited. They do not even suggest matters such as *linewidth*, *saturation*, *output power*, *mode locking*, *tunability*, and *stability*.

Obviously, one must not accept a caricature as the truth. Concerning the many aspects of the truth that are distorted or omitted by these first discussions, it will take much of this book to get the facts straight. This is not only a matter of dealing with details within the caricatures, but also with concepts that are larger than the caricatures altogether.

One should ask whether lasers are better described by photons or electric fields. Also, is Einstein's theory always satisfactory, or does Schrödinger's wave equation play a role? Are Maxwell's equations for electromagnetic waves significant? The answer to these questions is no, yes, yes. Laser theory is usually based on Schrödinger's and Maxwell's equations, neither of which was needed in this chapter.

From a different point of view another kind of question is equally important in trying to understand what a laser is. For example, why were lasers not built before 1960? Are there any rules of thumb that can predict, approximately and without detailed calculation, how much one can increase the output power or change the operating frequency? What are the most sensitive design features of a gas laser? a chemical laser? a semiconductor laser? Is a laser essentially quantum mechanical, or can classical physics explain all the important features of laser operation?

It will not be possible to give detailed answers to all of these questions. However, these questions guide the organization of the book, and many of them are addressed individually. In the following chapters the reader should encounter the concepts of physics and engineering that are most important for understanding laser action in general and that provide the background for pursuing further questions of particular theoretical or practical interest.

2 ATOMS, MOLECULES, AND SOLIDS

2.1 INTRODUCTION

It is frequently said that quantum physics began with Max Planck's discovery of the correct blackbody radiation formula in 1900. But it was more than a quarter of a century before Planck's formula could be fully derived from a satisfactory theory of quantum mechanics. Nevertheless, once formulated, quantum mechanics answered so many questions that it was adopted and refined with remarkable speed between 1925 and 1930. By 1930 there were new and successful quantum theories of atomic and molecular structure, electromagnetic radiation, electron scattering, and thermal, optical, and magnetic properties of solids.

Lasers can be understood without a detailed knowledge of the quantum theory of matter. However, several consequences of the quantum theory are essential. This chapter provides a review of some results of quantum theory applied to simple models of atoms, molecules, and semiconductors.

2.2 ELECTRON ENERGY LEVELS IN ATOMS

In 1913 Niels Bohr discovered a way to use Planck's radiation constant h in a radically new, but still mostly classical, theory of the hydrogen atom. Bohr's theory was the first quantum theory of atoms. Its importance was recognized immediately, even though it raised as many questions as it answered.

One of the most important questions it answered had to do with the Balmer formula:

$$\lambda = \frac{bn^2}{n^2 - 4}, \tag{2.2.1}$$

where n denotes an integer. This relation had been found in 1885 by Johann Jacob Balmer, a Swiss school teacher. Balmer pointed out that if b were given the value 3645.6, then λ equaled the wavelength (measured in Ångstrom units, $1 \text{ Å} = 10^{-10} \text{ m}$) of a line in the hydrogen spectrum[1] for $n = 3$, 4, 5, and 6 (and possibly for higher integers as well, but no measurements existed to confirm or deny the possibility).

[1]Historically, the term spectral "line" arose because lines appeared as images of slits in spectrometers.

Laser Physics. By Peter W. Milonni and Joseph H. Eberly
Copyright © 2010 John Wiley & Sons, Inc.

For almost 30 years the Balmer formula was a small oasis of regularity in the field of spectroscopy—the science of measuring and cataloging the wavelengths of radiation emitted and absorbed by different elements and compounds. Unfortunately, the Balmer formula could not be explained, or applied to any other element, or even applied to other known wavelengths emitted by hydrogen atoms. It might well have been a mere coincidence, without any significance. Bohr's model of the hydrogen atom not only explained the Balmer formula, but also gave scientists their first glimpse of atomic structure. It still serves as the basis for most scientists' working picture of an atom.

Bohr adopted Rutherford's nuclear model that had been successful in explaining scattering experiments with alpha particles between 1910 and 1912. In other words, Bohr assumed that almost all the mass of a hydrogen atom is concentrated in a positively charged nucleus, allowing most of the atomic volume free for the motion of the much lighter electron. The electron was assumed attracted to the nucleus by the Coulomb force law governing opposite charges (Fig. 2.1). In magnitude this force is

$$F = \frac{1}{4\pi\epsilon_0}\frac{e^2}{r^2}. \tag{2.2.2}$$

Bohr also assumed that the electron travels in a circular orbit about the massive nucleus. Moreover, he assumed the validity of Newton's laws of motion for the orbit. Thus, in common with every planetary body in a circular orbit, the electron was assumed to experience an inward (centripetal) acceleration of magnitude

$$a = \frac{v^2}{r}. \tag{2.2.3}$$

Newton's second law of motion, $\mathbf{F} = m\mathbf{a}$, then gives

$$\frac{mv^2}{r} = \frac{1}{4\pi\epsilon_0}\frac{e^2}{r^2}, \tag{2.2.4}$$

which is the same as saying that the electron's kinetic energy, $T = \frac{1}{2}mv^2$, is half as great as the magnitude of its potential energy, $V = -e^2/4\pi\epsilon_0 r$. In the Coulomb field of the

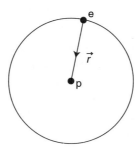

Figure 2.1 The electron in the Bohr model is attracted to the nucleus with a force of magnitude $F = e^2/4\pi\epsilon_0 r^2$.

nucleus the electron's total energy is therefore

$$E = T + V = -\frac{1}{4\pi\epsilon_0}\frac{e^2}{2r}.$$
(2.2.5)

These results are familiar consequences of Newton's laws. Bohr then introduced a single, radical, unexplained restriction on the electron's motion. He asserted that only certain circles are actually used by electrons as orbits. These orbits are the ones that permit the electron's angular momentum L to have one of the values

$$L = n\frac{h}{2\pi},$$
(2.2.6)

where n is an integer ($n = 1, 2, 3, \ldots$) and h is the constant of Planck's radiation formula:

$$h \approx 6.625 \times 10^{-34} \text{ J-s}.$$
(2.2.7)

With the definition of angular momentum for a circular orbit,

$$L = mvr,$$
(2.2.8)

it is easy to eliminate r between (2.2.4) and (2.2.8) and find

$$v = \frac{1}{4\pi\epsilon_0}\frac{e^2}{L} = \frac{1}{4\pi\epsilon_0}\frac{2\pi e^2}{nh}.$$
(2.2.9)

Then the combination of (2.2.4) and (2.2.5), namely

$$E = \frac{-mv^2}{2},$$
(2.2.10)

together with (2.2.9), gives the famous Bohr formula for the allowed energies of a hydrogen electron:

$$E_n = -\left(\frac{1}{4\pi\epsilon_0}\right)^2\frac{me^4}{2n^2\hbar^2}.$$
(2.2.11)

This can be seen, by comparison with (2.2.5), to be the same as a formula for the allowed values of electron orbital radius:

$$r_n = 4\pi\epsilon_0\frac{n^2\hbar^2}{me^2}.$$
(2.2.12)

In both (2.2.11) and (2.2.12) we have adopted the modern notation for Planck's constant:

$$\hbar = \frac{h}{2\pi} \approx 1.054 \times 10^{-34} \text{ J-s}.$$
(2.2.13)

The first thing to be said about Bohr's model and his unsupported assertion (2.2.6) is that they were not contradicted by known facts about atoms, and, for small values of n,

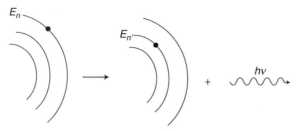

Figure 2.2   A radiative transition of an atomic electron in the Bohr model.

the allowed radii defined by (2.2.12) are numerically about right.[2] For example, the smallest of these radii (conventionally called "the Bohr radius" and denoted a_0) is

$$a_0 = r_1 = 4\pi\epsilon_0 \frac{\hbar^2}{me^2} \approx 0.53 \,\text{Å} \qquad (2.2.14)$$

This might have been an accident without further consequences. Since no way existed to measure such small distances with any precision, Bohr needed a connection between (2.2.12) and a possible laboratory experiment. A second unsupported assertion supplied the connection.

Bohr's second assertion was that the atom was stable when the electron was in one of the permitted orbits, but that jumps from one orbit to another were possible if accompanied by light emission or absorption. To be specific, Bohr combined earlier ideas of Planck and Einstein and stated that a jump from a higher to a lower orbit would find the decrease of the electron's energy transformed into a quantum of radiation that would be emitted in the process. In other words, Bohr postulated that

$$(\Delta E)_{n,n'} = h\nu = \text{energy of emitted photon.} \qquad (2.2.15)$$

Here $(\Delta E)_{n,n'}$ denotes the energy lost by the electron in switching orbits from r_n to $r_{n'}$ and ν is the frequency of the photon emitted in the process (Fig. 2.2).

The relation (2.2.15) led immediately to a connection between Bohr's theory and all the spectroscopic data known for atomic hydrogen. By using (2.2.11) for two different orbits, that is, for two different values of the integer n, we easily find for the energy decrement $(\Delta E)_{n,n'}$ the expression

$$E_n - E_{n'} = \left(\frac{1}{4\pi\epsilon_0}\right)^2 \frac{me^4}{2\hbar^2} \left(\frac{1}{n'^2} - \frac{1}{n^2}\right). \qquad (2.2.16)$$

Furthermore, the connection between the frequency and wavelength of a light wave is

$$\lambda = \frac{c}{\nu}. \qquad (2.2.17)$$

---

[2]Lord Rayleigh (1890) was able to estimate molecular dimensions by dropping olive oil onto a water surface. Assuming that an oil drop spreads until it forms a layer one molecule thick (a *molecular monolayer*), he could give a reasonable estimate of a molecular diameter from the area of the layer and the volume of the original drop. A century earlier, Benjamin Franklin tried the same oil-on-water experiment on a pond in London while on diplomatic assignment there.

Thus, Bohr's statement (2.2.15) and his energy formula (2.2.11) are actually equivalent to the postulate that *all* spectroscopic wavelengths of light associated with atomic hydrogen fit the formula

$$\lambda = \frac{hc}{\Delta E_{n,n'}} = (4\pi\epsilon_0)^2 \frac{4\pi\hbar^3 c}{me^4} \frac{n^2 n'^2}{n^2 - n'^2}, \tag{2.2.18}$$

where n and n' are integers to be chosen, but where all the other parameters are fixed.

It is obvious that if $n' = 2$ and $n > 2$, then (2.2.18) becomes

$$\lambda = (4\pi\epsilon_0)^2 \frac{16\pi\hbar^3 c}{me^4} \frac{n^2}{n^2 - 4}, \tag{2.2.19}$$

which is exactly the Balmer formula (2.2.1). The numerical value of the product of the coefficients in (2.2.19) is 3645.6 Å, just what Balmer had said the constant b was, 28 years earlier. Bohr's expression (2.2.18) was quickly found, for values of n' not equal to 2, to agree with other wavelengths associated with hydrogen but which had not fitted the Balmer formula (Problem 2.1).

Bohr's theory opened a new viewpoint on atomic spectroscopy. All observed spectroscopic wavelengths could be interpreted as evidence for the existence of certain allowed electron orbits in all atoms, even if formulas corresponding to Bohr's (2.2.18) were not known for any atom but hydrogen.

In Chapter 3 we will see that, for many aspects of the interaction of light and matter, atoms can be regarded as a set of electrons acting as harmonic oscillators. It might be supposed that the oscillation frequencies of the electrons in this classical electron oscillator model of an atom, which was used with considerable success before quantum theory, are associated somehow with the "transition frequencies" ν in Bohr's formula (2.2.15). In this way the most useful features of the classical oscillator model of an atom survived the quantum revolution unchanged. How this is possible, in view of the obvious fact that the assumed Coulomb force (2.2.2) between electron and nucleus is not a harmonic oscillator force, will be explained in Chapter 3.

It is easy to have second thoughts about Bohr's model, no matter how successful it is. For example, one can ask why (2.2.6) does not include the possibility $n = 0$. There is no apparent reason why zero angular momentum must be excluded, except that the energy formula (2.2.11) is not defined for $n = 0$. This point, whether physical significance can be assigned to the orbit with zero angular momentum, cannot be clarified within the Bohr theory, and it proved puzzling to physicists for more than a decade until quantum mechanics was developed. In a similar fashion, one can ask how Bohr's results are modified by relativity. The kinetic energy formula used above, $T = \frac{1}{2}mv^2$, is certainly nonrelativistic. Again, this point was not fully answered until the development of quantum mechanics.

• Relativistic corrections to Newtonian physics become important when particle velocities approach the velocity of light. If v is the velocity of a particle, then typically the first correction terms are found to be proportional to $(v/c)^2$, where c is the velocity of light. The value of $(v/c)^2$ can easily be estimated within the Bohr theory. It follows from (2.2.4) that

$$\left(\frac{v}{c}\right)^2 = \frac{1}{4\pi\epsilon_0}\frac{e^2}{rmc^2}. \tag{2.2.20}$$

By inserting r from (2.2.12) and taking the square root, the ratio v/c can be found for any of the allowed orbits:

$$\frac{v}{c} = \frac{1}{4\pi\epsilon_0}\frac{e^2}{n\hbar c}. \tag{2.2.21}$$

Equation (2.2.21) shows that the largest velocity to be expected in the Bohr atom is associated with the lowest orbit, $n = 1$. The ratio of this maximum velocity to the velocity of light is given by the remarkable (dimensionless) combination of electromagnetic and quantum mechanical constants, $e^2/4\pi\epsilon_0\hbar c$. The numerical value of this parameter is easily found:

$$\frac{1}{4\pi\epsilon_0}\frac{e^2}{\hbar c} = \frac{(1.602 \times 10^{-19}\,\text{C})^2}{(4\pi)(8.854 \times 10^{-12}\,\text{C/V-m})(1.054 \times 10^{-34}\,\text{J-s})(2.998 \times 10^8\,\text{m/s})}$$

$$= 0.007297 \quad (=1/137.04). \tag{2.2.22}$$

The value found in (2.2.22) is small enough that corrections to the Bohr model from relativistic effects are of the relative order of magnitude 10^{-4} or smaller, and thus negligible in most circumstances. Spectroscopic measurements, however, are commonly accurate to five significant figures. Arnold Sommerfeld, in the period 1915–1920, studied the relativistic corrections to Bohr's formulas and showed that they accounted accurately for some of the fine details or *fine structure* in observed spectra. For this reason the parameter $e^2/4\pi\epsilon_0\hbar c$ is called *Sommerfeld's fine-structure constant*.

The fine-structure constant appears so frequently in expressions of atomic radiation physics that it is very useful to remember its numerical value. Because the value given in (2.2.22) is very nearly equal to $1/137$, it is in this form that its value is memorized by physicists.   ●

Quantum States and Degeneracy

In the Bohr model a state of the electron is characterized by the *quantum number n*. Everything the model can say about the allowed states of the electron is given in terms of n.

The full quantum theory of the hydrogen atom also yields the allowed energies (2.2.11). However, in the quantum theory a state of the electron is characterized by other quantum numbers in addition to the *principal quantum number n* appearing in (2.2.11). The results of the quantum theory for the hydrogen atom, in addition to (2.2.11), are mainly the following:

(i) For each principal quantum number n ($=1, 2, 3, \ldots$) there are n possible values of the *orbital angular momentum quantum number* ℓ. The allowed values of ℓ are 0, 1, 2, \ldots, $n-1$. Thus, for $n = 1$ we can have only $\ell = 0$, whereas for $n = 2$ we can have $\ell = 0$ or 1, and so on.

(ii) For each ℓ there are $2\ell + 1$ possible values of the *magnetic quantum number* m. The possible values of magnetic quantum number m are $-\ell, -\ell+1, \ldots, -1, 0, 1, \ldots, \ell-1, \ell$.

(iii) In addition to orbital angular momentum, an electron also carries an intrinsic angular momentum, which is called simply *spin*. The spin of an electron always has magnitude $\frac{1}{2}$ (in units of $\hbar$). But in any given direction the electron spin can be either "up" or "down"; that is, quantum theory says that when the component of electron spin along any direction is measured, we will always find it to have one of two possible values.[3] Because of this, an electron state must also be labeled by an additional quantum number m_s, called the *spin magnetic quantum number*, whose only possible values are $\pm\frac{1}{2}$.

Thus, for a given n, there are n possible values of ℓ, and for each ℓ there are $2\ell + 1$ possible values of m, for a total of

$$\sum_{\ell=0}^{n-1} (2\ell + 1) = n^2, \tag{2.2.23}$$

states. And each of these states is characterized further by m_s, which may be $+\frac{1}{2}$ or $-\frac{1}{2}$. Therefore, there are $2n^2$ states associated with each principal quantum number n. In contrast to the Bohr model, in which an allowed state of the electron in the hydrogen atom is characterized by n, quantum theory characterizes each allowed state by the four quantum numbers n, ℓ, m, and m_s; and since the electron energy depends only on n [recall (2.2.11)], there are $2n^2$ states with the same energy for every value of n. These $2n^2$ states are called *degenerate states* or are said to be degenerate in energy.

Historical designations for the orbital angular momentum quantum numbers are still in use:

$\ell = 0$ designates the so-called *s* orbital
$\ell = 1$ *p* orbital
$\ell = 2$ *d* orbital
$\ell = 3$ *f* orbital
$\ell = 4$ *g* orbital

The first three letters came from the words *sharp*, *principal*, and *diffuse*, which described the character of atomic emission spectra in a qualitative way long before quantum theory showed that they could be associated systematically with different orbital angular momentum values for an electron in the atom.

The Periodic Table

Although hydrogen is the only atom for which explicit expressions such as (2.2.11) can be written down, we can nevertheless understand the gross features of the periodic table of the elements. That is, we can understand the chemical regularity, or periodicity, that occurs as the atomic number Z increases. The key to this understanding is the *exclusion principle* of Wolfgang Pauli (1925), which forbids two electrons from occupying the

---

[3]The magnitude of the spin angular momentum vector is (Section 2.4) $\sqrt{s(s+1)}\hbar = \sqrt{\left(\frac{1}{2}\right)\left(\frac{1}{2}+1\right)}\hbar = \sqrt{\frac{3}{4}}\hbar$, so that its two allowed components (spin "up" and spin "down") make angles $\cos^{-1}\left[\left(\pm\frac{1}{2}\right)/\sqrt{\frac{3}{4}}\right] = 54.74°$ and $(180 - 54.74) = 125.26°$ with the axis along which the spin is measured.

same quantum mechanical state. The Pauli exclusion principle may be proved only at an advanced level that is well beyond the scope of this book. We will simply accept it as a fundamental truth.

But the Pauli principle alone is not sufficient for an understanding of the periodic table. We must also deal with the electron–electron interactions in a multielectron atom. These interactions present us with an extremely complicated many-body problem that has never been solved. A useful approximation, however, is to assume that each electron moves independently of all the others; each electron is thought of as being in a spherically symmetric potential $V(r)$ due to the Coulomb field of the nucleus plus the $Z-1$ other electrons. In this independent-particle approximation an electron state is still characterized by the four quantum numbers (n, ℓ, m, m_s), as in the case of hydrogen. However, in this case the simple energy formula (2.2.11) does not apply, and in particular the energy depends on both n and ℓ (but not m or m_s) as sketched in Fig. 2.3.

The simplest multielectron atom, of course, is helium, in which there are $Z = 2$ electrons. The lowest energy state for each electron is characterized in the independent-particle approximation by the quantum numbers $n = 1$, $\ell = 0$, $m = 0$, and $m_s = \pm \frac{1}{2}$. Since the energy depends now on both n and ℓ, we can label this particular *electron configuration* as $1s$, a shorthand notation meaning $n = 1$ and $\ell = 0$. Both electrons are in the *shell $n = 1$*, one having spin up $\left(m_s = \frac{1}{2}\right)$, the other spin down $\left(m_s = -\frac{1}{2}\right)$. Since 2 is the maximum number of electrons allowed by the Pauli exclusion principle for the $1s$ configuration, we say that the $1s$ shell is completely filled in the helium atom.

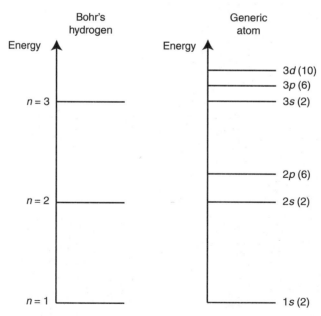

Figure 2.3   The main differences between Bohr's model for hydrogen and a generic many-electron atom arise from the Pauli exclusion principle and the dependence of level energies on both n and ℓ. In parentheses we show the number of different states, $(2\ell + 1) \times 2$, permitted by assignment of m and m_s values for each ℓ. Carbon's 6 electrons, for example, occupy the 2 states in each of the $1s$ and $2s$ levels and 2 of the $2p$ states. For clarity the energy separations are not properly scaled.

In the case $Z = 3$ (lithium), there is one electron left over after the $1s$ shell is filled. The next allowed electron configuration is $2s$ ($n = 2$, $\ell = 0$), and one of the electrons in lithium is assigned to this configuration. Since the $2s$ configuration can accommodate two electrons, the $2s$ *subshell* in lithium is only partially filled. The next element is beryllium, with $Z = 4$ electrons, and in this case the $2s$ subshell is completely filled, there being two electrons in this "slot."

For $Z = 5$ (boron), the added electron goes into the $2p$ configuration ($n = 2$, $\ell = 1$). This configuration can accommodate $2(2\ell + 1) = 6$ electrons. Thus, there are five other elements (C, N, O, F, and Ne) in which the outer subshell of electrons corresponds to the configuration $2p$. The eight elements lithium through neon, for which the outermost electrons belong to the $n = 2$ shell, constitute the first full row of the periodic table.

Inside the back cover of this book we list the first 36 elements and their electron configurations. The configurations are assigned in a similar manner as done above for $Z = 1$–10. Also listed is the ionization energy, defined as the energy required to remove one electron from the atom. For hydrogen the ionization energy W_I may be calculated from Eq. (2.2.11) with $n = 1$, that is, W_I is just the binding energy of the electron in ground-state hydrogen:

$$W_I = |E_1| = \left(\frac{1}{4\pi\epsilon_0}\right)^2 \frac{me^4}{2\hbar^2} = 2.17 \times 10^{-18} \text{ J}. \tag{2.2.24}$$

We already pointed out, in connection with photon energy in Chapter 1, that such small energies are usually expressed in units of electron volts, an electron volt being the energy acquired by an electron accelerated through a potential difference of 1 volt:

$$1 \text{ eV} = (1.602 \times 10^{-19} \text{ C})(1 \text{ V}) = 1.602 \times 10^{-19} \text{ J}. \tag{2.2.25}$$

The ionization energy of hydrogen is therefore

$$W_I = \frac{2.17 \times 10^{-18} \text{ J}}{1.602 \times 10^{-19} \text{ J/eV}} = 13.6 \text{ eV}. \tag{2.2.26}$$

The ionization energy of a hydrogen atom in any state (n, ℓ, m, m_s) is likewise $(13.6 \text{ eV})/n^2$.

The elements He, Ne, Ar, and Kr are chemically inactive. We note that each of these atoms has a completely filled outer shell. Evidently, an atom with a filled outer shell of electrons tends to be "satisfied" with itself, having very little proclivity to share its electrons with other atoms (i.e., to join in chemical bonds). However, a filled outer subshell does not necessarily mean chemical inertness. Beryllium, for instance, has a filled $2s$ subshell, but it is not inert. Furthermore, even some of the noble gases are not entirely inert.

The alkali metals Li, Na, and K have only one electron in an outer subshell, and their outer electrons are weakly bound, leading to low ionization energies of these elements. These elements are highly reactive; they will readily give up their "extra" electron. On the other hand, the halogens F, Cl, and Br are one electron short of a filled outer subshell. These atoms will readily take another electron, and so they too are quite reactive chemically and the halogens are sufficiently "eager" to combine with elements that can easily contribute an electron that they can form *negative* ions, stably but weakly binding an extra electron. This even includes H^-, the negative hydrogen ion. Hydrogen is in the odd position of having some properties in common with the alkali metals and some in common with the halogens.

The characterization of atomic electron states in terms of the four quantum numbers n, ℓ, m, and m_s, together with the Pauli exclusion principle, thus allows us to understand why Na is chemically similar to K, Mg is chemically similar to Ca, and so forth. These chemical periodicities, according to which the periodic table is arranged, are consequences of the way electrons fill in the allowed "slots" when they combine with nuclei to form atoms.

Of course, there is a great deal more that can be said about the periodic table. For a rigorous treatment of atomic structure, we must refer the reader to textbooks on atomic physics. As mentioned earlier, however, we can understand lasers without a more detailed understanding of atomic and molecular physics.

2.3   MOLECULAR VIBRATIONS

As in the case of atoms, there are only certain allowed energy levels for the electrons of a molecule. Quantum jumps of electrons in molecules are accompanied by the emission or absorption of photons that typically belong to the ultraviolet region of the electromagnetic spectrum. For our purposes the electronic energy levels of molecules are quite similar to those of atoms.

However, in contrast with atoms, molecules have vibrational and rotational as well as electronic energy. This is because the relative positions and orientations of the individual atomic nuclei in molecules are not absolutely fixed. The energies associated with molecular vibrations and rotations are also quantized, that is, restricted to certain allowed values. In this section and the next we will discuss the main features of molecular vibrational and rotational energy levels. Transitions between vibrational levels lie in the infrared portion of the electromagnetic spectrum, whereas rotational spectra are in the microwave region. Some of the most powerful lasers operate on molecular vibrational-rotational transitions.

Consider the simplest kind of molecule, namely a diatomic molecule such as O_2, N_2, or CO. There is a molecular binding force that is responsible for holding the two atoms together. To a first (and often very good) approximation the binding force is linear, so that the potential energy function is

$$V(x) = \tfrac{1}{2}(x_2 - x_1 - x_0)^2 = \tfrac{1}{2} k(x - x_0)^2, \tag{2.3.1}$$

where k is the "spring constant," $x = x_2 - x_1$ is the distance between the two nuclei, and x_0 is the internuclear separation for which the spring force

$$F = -k(x - x_0), \tag{2.3.2}$$

vanishes (Fig. 2.4). In other words, if the separation x is greater than x_0, the binding force is attractive and brings the nuclei closer; if x is less than x_0, the force is repulsive. The separation $x = x_0$ is therefore a point of stable equilibrium. The origin of the binding

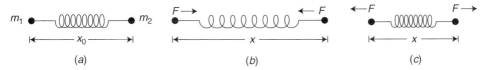

Figure 2.4 (*a*) When the two nuclei of a diatomic molecule are separated by the equilibrium distance x_0, there is no force between them. If their separation x is larger than x_0, there is an attractive force (*b*), whereas when x is less than x_0 the force is repulsive (*c*). The internuclear force is approximately harmonic, that is, springlike.

force is quantum mechanical; we will not attempt to explain it but will simply accept the result (2.3.1) and consider its consequences.

For simplicity, let us assume that the nuclei can move only in one dimension. The total energy of a diatomic system (i.e., the sum of kinetic and potential energies) is then

$$E = \tfrac{1}{2} m_1 \dot{x}_1^2 + \tfrac{1}{2} m_2 \dot{x}_2^2 + \tfrac{1}{2} k (x_2 - x_1 - x_0)^2, \qquad (2.3.3)$$

where the dots denote differentiation with respect to time, that is, $\dot{x} = dx/dt$. In terms of the reduced mass

$$m = \frac{m_1 m_2}{M}, \qquad (2.3.4)$$

where $M = m_1 + m_2$ is the total mass, and the center-of-mass coordinate

$$X = \frac{m_1 x_1 + m_2 x_2}{M}, \qquad (2.3.5)$$

we may write (2.3.3) as (Problem 2.2)

$$E = \tfrac{1}{2} M \dot{X}^2 + \tfrac{1}{2} m \dot{x}^2 + \tfrac{1}{2} k (x - x_0)^2. \qquad (2.3.6)$$

The first term is just the kinetic energy associated with the center-of-mass motion. We ignore it and focus our attention on the internal vibrational energy

$$E = \tfrac{1}{2} m \dot{x}^2 + \tfrac{1}{2} k (x - x_0)^2. \qquad (2.3.7)$$

The vibrational motion of a diatomic molecule must clearly be one dimensional, and so we lose nothing in the way of generality by restricting ourselves to one-dimensional vibrations from the start [Eq. (2.3.3)].

The quantum mechanics of the motion associated with the energy formula (2.3.7) has much in common with that for the hydrogen atom electron. The most important result is that the allowed energies E of the oscillator are also quantized. The quantized

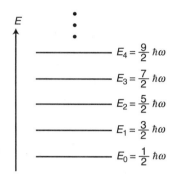

Figure 2.5   The energy levels of a harmonic oscillator form a ladder with rung spacing $\hbar\omega$.

energies are given by

$$E_n = \hbar\omega(n + \tfrac{1}{2}), \qquad n = 0, 1, 2, 3, \ldots, \tag{2.3.8}$$

with

$$\omega = \sqrt{k/m}. \tag{2.3.9}$$

This formula is clearly quite different from Bohr's formula for hydrogen. The quantum mechanical energy spectrum for a harmonic oscillator is simply a ladder of evenly spaced levels separated by $\hbar\omega$ (Fig. 2.5). The ground level of the oscillator corresponds to $n = 0$. However, an oscillator in its ground level is not at rest at its stable equilibrium point $x = x_0$. Even the lowest possible energy of a quantum mechanical oscillator has finite kinetic and potential energy contributions. At zero absolute temperature, where classically all motion ceases, the quantum mechanical oscillator still has a finite energy $\tfrac{1}{2}\hbar\omega$. For this reason the energy $\tfrac{1}{2}\hbar\omega$ is called the zero-point energy of the harmonic oscillator.

Of course, real diatomic molecules are not perfect harmonic oscillators, and their vibrational energies do not satisfy (2.3.8) precisely. Figure 2.6 shows the sort of potential

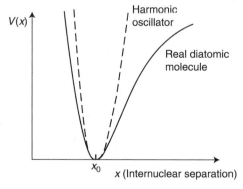

Figure 2.6   The potential energy function of a real diatomic molecule is approximately like that of a harmonic oscillator for values of x near x_0.

energy function $V(x)$ that describes the bonding of a real diatomic molecule. The Taylor series expansion of the function $V(x)$ about the equilibrium point x_0 is

$$V(x) = V(x_0) + (x - x_0)\left(\frac{dV}{dx}\right)_{x=x_0} + \frac{1}{2}(x - x_0)^2\left(\frac{d^2V}{dx^2}\right)_{x=x_0}$$

$$+ \frac{1}{6}(x - x_0)^3\left(\frac{d^3V}{dx^3}\right)_{x=x_0} + \cdots. \tag{2.3.10}$$

Here $V(x_0)$ is a constant, which we put equal to zero by shifting the origin of the energy scale. Also $(dV/dx)_{x=x_0} = 0$ because, by definition, $x = x_0$ is the equilibrium separation, at which the potential energy is a minimum. Furthermore (d^2V/dx^2) at $x = x_0$ is positive if x_0 is a point of stable equilibrium (Fig. 2.6). Thus, we can replace (2.3.10) by

$$V(x) = \tfrac{1}{2}k(x - x_0)^2 + A(x - x_0)^3 + B(x - x_0)^4 + \cdots, \tag{2.3.11}$$

where $A, B, \ldots$ are constants and $k = (d^2V/dx^2)_{x=x_0}$.

From (2.3.10) we can conclude that *any* potential energy function describing a stable equilibrium [i.e., $(dV/dx)_{x=x_0} = 0$, $(d^2V/dx^2)_{x=x_0} > 0$] can be approximated by the harmonic oscillator potential (2.3.1) for small enough displacements from equilibrium. Of course, what is "small" is determined by the constants $A, B, \ldots$ in (2.3.11), that is, by the shape of the potential function $V(x)$. If the terms involving third and/or higher powers of $x - x_0$ in (2.3.11) are not negligible, however, we have what is called an *anharmonic* potential. The energy levels of an anharmonic oscillator do not satisfy the simple formula (2.3.8).

Real diatomic molecules have vibrational spectra that are usually only slightly anharmonic. In conventional notation the vibrational energy levels of diatomic molecules are written in the form

$$E_v = hc\omega_e\left[\left(v + \tfrac{1}{2}\right) - x_e\left(v + \tfrac{1}{2}\right)^2 + y_e\left(v + \tfrac{1}{2}\right)^3 + \cdots\right], \tag{2.3.12}$$

where

$$v = 0, 1, 2, 3, \ldots, \tag{2.3.13}$$

and ω_e is in units of "wave numbers," i.e., cm^{-1}; $c\omega_e$ is the same as $\omega/2\pi = \nu$ in this notation. If the anharmonicity coefficients $x_e, y_e, \ldots$ are all zero, we recover the harmonic oscillator spectrum (2.3.8). Numerical values of $\omega_e, x_e, y_e, \ldots$ are tabulated in the literature.[4] Values of $\omega_e, x_e, y_e, \ldots$ are given for several diatomic molecules in Table 2.1. The deviations from perfect harmonicity are small until v becomes large, that is, until we climb fairly high up the vibrational ladder. The level spacing decreases as v increases, in contrast to the even spacing of the ideal harmonic oscillator (Fig. 2.7).

---

[4]A standard source is G. Herzberg, *Molecular Spectra and Molecular Structure. Volume I, Spectra of Diatomic Molecules*, Robert E. Krieger, Malabar, FL, 1989.

TABLE 2.1 Vibrational Constants of the Ground Electronic State for a Few Diatomic Molecules

Molecule	ω_e (cm^{-1})	x_e	y_e
H_2	4395.24	0.0268	6.67×10^{-5}
O_2	1580.36	0.00764	3.46×10^{-5}
CO	2170.21	0.00620	1.42×10^{-5}
HF	4138.52	0.0218	2.37×10^{-4}
HCl	2989.74	0.0174	1.87×10^{-5}

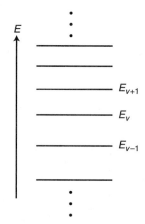

Figure 2.7 The vibrational energy level spacing of a real (anharmonic) diatomic molecule decreases with increasing vibrational energy.

• For a simple check on our theory, consider the two molecules hydrogen fluoride (HF) and deuterium fluoride (DF). These molecules differ only to the extent that D has a neutron and a proton in its nucleus and H has only a proton. Since neutrons have no effect on molecular bonding, we expect HF and DF to have the same potential function $V(x)$ and therefore the same "spring constant" k. According to (2.3.9), therefore, we should have

$$\frac{\omega_e^{HF}}{\omega_e^{DF}} = \left(\frac{m^{DF}}{m^{HF}}\right)^{1/2}, \qquad (2.3.14)$$

where m^{DF} and m^{HF} are the *reduced* masses of DF and HF, respectively, so that

$$\frac{m^{DF}}{m^{HF}} = \frac{m^D m^F}{m^D + m^F} \Big/ \frac{m^H m^F}{m^H + m^F} \approx \frac{(2)(19)}{2+19} \times \frac{1+19}{(1)(19)} \approx 1.90. \qquad (2.3.15)$$

From the value of ω_e for HF given in Table 2.1, therefore, we calculate

$$\omega_e^{DF} \approx (4138.52)(1.90)^{-1/2} = 2998.64 \text{ cm}^{-1}, \qquad (2.3.16)$$

and indeed this is very close to the tabulated value $\omega_e = 2998.25$ cm^{-1} for DF.[4]

Regarding the zero-point energy of molecular vibrations, consider a transition in which there is a change in both the electronic (e) and the vibrational (v) states of a diatomic molecule.

The transition energy is approximately

$$\Delta E_{e'v',ev} = E_{e'} - E_e + hc\left[\omega_{e'}\left(v' + \tfrac{1}{2}\right) - \omega_e\left(v + \tfrac{1}{2}\right)\right], \tag{2.3.17}$$

where the unprimed and primed labels refer to the initial and final states, respectively, and ω_e and $\omega_{e'}$ are the vibrational constants associated with the two electronic states. Now suppose that one of the nuclei of the diatomic molecule is replaced by a different isotope, for example, HF is replaced by DF. The electronic energy levels are approximately unchanged by this replacement, but the vibrational constants ω_e and $\omega_{e'}$ are changed to $\rho\omega_e$ and $\rho\omega_{e'}$, where ρ is the square root of the ratio of reduced masses of the two molecules, as in our example above comparing HF and DF. For the second, isotopically different molecule, then, the transition energy (2.3.17) is replaced by

$$\Delta E_{e'v',ev}^i = E_{e'} - E_e + hc\left[\rho\omega_{e'}\left(v' + \tfrac{1}{2}\right) - \rho\omega_e\left(v + \tfrac{1}{2}\right)\right]. \tag{2.3.18}$$

The vibrational spectra of the two isotopic molecules for the same electronic transition $E_e \to E_{e'}$ therefore differ by

$$\Delta E_{e'v',ev}^i - \Delta E_{e'v',ev} = hc(\rho - 1)\left[\omega_{e'}\left(v' + \tfrac{1}{2}\right) - \omega_e\left(v + \tfrac{1}{2}\right)\right], \tag{2.3.19}$$

and in particular, for $v = 0 \to v' = 0$,

$$\Delta E_{e'0,e0}^i - \Delta E_{e'0,e0} = \tfrac{1}{2}hc(\rho - 1)(\omega_{e'} - \omega_e). \tag{2.3.20}$$

This is nonzero because of zero-point energy, that is, it would vanish if the energy levels of a harmonic oscillator were given by $E_n = n\hbar\omega$ instead of $E_n = (n + \tfrac{1}{2})\hbar\omega$. The zero-point energy of molecular vibrations was confirmed in this way by R. S. Mulliken (1924), who compared the observed vibrational spectra of $B^{10}O^{16}$ and $B^{11}O^{16}$. This was before the quantum mechanical derivation by Heisenberg (1925) of the formula (2.3.8) for the energy levels of a harmonic oscillator. $\quad\bullet$

2.4   MOLECULAR ROTATIONS

The rotations of a diatomic molecule can be understood in two stages. First, we imagine the molecule to be a dumbbell consisting of two masses, m_1 and m_2, held together by a (massless) rigid rod of length x_0 (Fig. 2.8). The dumbbell can rotate about its center of mass. The moment of inertia I is mx_0^2, where m is the reduced mass (2.3.4) and x_0 is the distance separating the masses m_1 and m_2. If the angular velocity of rotation (radians per second) is ω_R, the angular momentum and kinetic energy are, respectively,

$$L = I\omega_R \quad \text{(magnitude of angular momentum vector)} \tag{2.4.1}$$

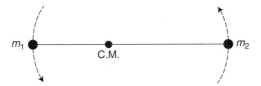

Figure 2.8   A dumbbell rotating about an axis through its center of mass serves as a classical model for the rotations of a diatomic molecule.

and

$$E = \tfrac{1}{2} I \omega_R^2 = \frac{L^2}{2I}.$$

(2.4.2)

These classical formulas are the starting point of a quantum-mechanical treatment of the rigid dumbbell, just as similar classical formulas underlie treatments of the hydrogen atom and the vibrations of molecules. It is found that the rotational energy (2.4.2) of the molecule has the allowed values

$$E_J = \frac{\hbar^2}{2I} J(J+1), \qquad J = 0, 1, 2, \ldots.$$

(2.4.3)

Actual diatomic molecules are, of course, not rigid dumbbells. In particular, the masses m_1 and m_2 do not stay a fixed distance x_0 apart. As the molecule rotates, the centrifugal force tends to increase the separation of the two masses, and therefore also the moment of inertia I. This decreases the rotational energy, the more so as the rate of rotation (i.e., J) increases. In the notation of molecular spectroscopy this is accounted for by writing

$$E_J = hcBJ(J+1) - hcDJ^2(J+1)^2,$$

(2.4.4)

where the J-independent quantities B and D have units of wave numbers.

The fact that the molecule can vibrate also tends to increase the effective moment of inertia, the more so as the vibrational quantum number v increases. This is accounted for by writing

$$B = B_e - \alpha_e \left(v + \tfrac{1}{2}\right),$$

(2.4.5)

where B_e and α_e (in cm^{-1}) are independent of v and J. The rotational energy levels associated with the vibrational level v of a diatomic molecule are therefore written as

$$E_J(v) = hc\left[B_e - \alpha_e\left(v + \tfrac{1}{2}\right)\right] J(J+1) - hcDJ^2(J+1)^2.$$

(2.4.6)

Higher-order corrections are necessary in general to explain the fine details of the rotational energy spectrum of a diatomic molecule. However, (2.4.6) is often accurate enough for practical purposes, and in fact the term involving D is often negligible.

The constants B_e, α_e, ... for different molecules are tabulated in the spectroscopic literature. The constants B_e and α_e for several molecules are given in Table 2.2. For our purposes it will suffice to make the rigid-dumbbell approximation and write

$$E_J(v) \approx E_J \approx hcB_eJ(J+1).$$

(2.4.7)

• Once again it is possible to check our theory with an example. A comparison of Eqs. (2.4.3) and (2.4.7) shows that the rotational constant B_e of a diatomic molecule should be inversely proportional to its moment of inertia $I = mx_0^2$. Since the equilibrium separation x_0 is determined primarily by chemical (i.e., electromagnetic) forces, we expect that it should be practically the same for the two molecules HF and DF. Thus, we expect

$$\frac{B_e^{\text{HF}}}{B_e^{\text{DF}}} = \frac{m^{\text{DF}}}{m^{\text{HF}}} = 1.90.$$

(2.4.8)

TABLE 2.2   Rotational Constants of the Ground Electronic State for the Molecules Listed in Table 4.1

Molecule	B_e (cm^{-1})	α_e (cm^{-1})
H_2	60.81	2.993
O_2	1.44567	0.01579
CO	1.9314	0.01749
HF	20.939	0.770
HCl	10.5909	0.3019

It follows from the data in Table 2.2, therefore, that the rotational constant for DF should be $B_e^{DF} \approx 11.02$ cm^{-1}, in excellent agreement with the value 11.007 cm^{-1} tabulated by Herzberg.[4]

Equations (2.4.2) and (2.4.3) imply that the square of the angular momentum in a state with angular momentum quantum number L is $L(L+1)\hbar^2$ rather than $L^2\hbar^2$. This is a general feature of the quantum theory of angular momentum; the square of the orbital angular momentum for the electron in a hydrogen atom in a state with orbital angular momentum quantum number ℓ, for example, is $\ell(\ell+1)\hbar^2$. It can be understood as a consequence of "space quantization," that is, the fact that the z component of angular momentum, L_z, has only the $2L+1$ allowed values $M = -L, -L+1, \ldots, L-1, L$. Since there is nothing special about the "z direction," and there are three space dimensions, the average $\langle L^2 \rangle$ of the square of the angular momentum must be three times the average of L_z^2:

$$\langle L^2 \rangle = 3\langle L_z^2 \rangle = 3\frac{1}{2L+1}\sum_{M=-L}^{L} M^2 = \frac{3}{2L+1}\frac{1}{3}L(L+1)(2L+1) = L(L+1), \quad (2.4.9)$$

where we have used the general identity

$$\sum_{M=-L}^{L} M^2 = \frac{1}{3}L(L+1)(2L+1). \quad (2.4.10)$$

In other words, once we accept space quantization as an experimental fact, we can understand why the square of the angular momentum must be $L(L+1)\hbar^2$ in a state with angular momentum quantum number L.   •

In summary, with every electronic state of a molecule there are associated vibrational constants $\omega_e, x_e, y_e, \ldots$ and rotational constants $B_e, \alpha_e, \ldots$. In Tables 2.1 and 2.2 the vibrational-rotational constants are given for the ground (lowest energy) electronic state.

2.5   EXAMPLE: CARBON DIOXIDE

In our treatment of molecular vibrations and rotations we have only considered the relatively simple case of diatomic molecules. Rather than now discussing general polyatomic molecules, which are more complicated but fundamentally much the same as diatomics, we will consider only the specific case of the carbon dioxide molecule. We choose this example because the CO_2 laser is one of the most important molecular lasers.

Carbon dioxide is a linear triatomic molecule (Fig. 2.9). Such a molecule has three so-called normal modes of vibration, shown in Fig. 2.9b. For obvious reasons these

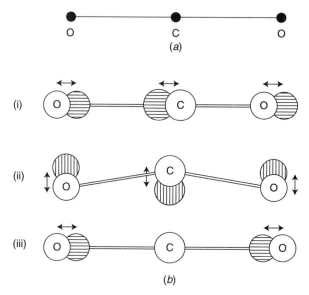

Figure 2.9   (*a*) Carbon dioxide (CO_2) is a linear triatomic molecule. (*b*) Normal modes of vibration of the CO_2 molecule: (i) the asymmetric stretch mode, (ii) the bending mode, and (iii) the symmetric stretch mode.

are called the asymmetric stretch, bending, and symmetric stretch modes. With each of these normal modes is associated a characteristic frequency of vibration. Each mode of vibration has a ladder of allowed energy levels associated with it, as in the case of a diatomic molecule (which has only one normal mode of vibration). The vibrational energy levels of the molecule may therefore be labeled by three integers v_1, v_2, and v_3 ($= 0, 1, 2, 3, \ldots$), and we have approximately

$$E(v_1, v_2, v_3) = hc\omega_e^{(1)}\left(v_1 + \tfrac{1}{2}\right) + hc\omega_e^{(2)}\left(v_2 + \tfrac{1}{2}\right) + hc\omega_e^{(3)}\left(v_3 + \tfrac{1}{2}\right), \qquad (2.5.1)$$

where $\omega_e^{(1)}$, $\omega_e^{(2)}$, and $\omega_e^{(3)}$ are the normal-mode frequencies in units of wave numbers. In reality each normal mode is slightly anharmonic, but the harmonic-oscillator approximation will suffice for our purposes. For CO_2 the normal-mode frequencies are

$$\omega_e^{(1)} = \omega(\text{symmetric stretch}) \approx 1388 \text{ cm}^{-1}, \qquad (2.5.2a)$$

$$\omega_e^{(2)} = \omega(\text{bending}) \approx 667 \text{ cm}^{-1}, \qquad (2.5.2b)$$

$$\omega_e^{(3)} = \omega(\text{asymmetric stretch}) \approx 2349 \text{ cm}^{-1}. \qquad (2.5.2c)$$

The first few vibrational energy levels ($v_1 v_2 v_3$) of the CO_2 molecule are indicated in Fig. 2.10.

Since CO_2 is a linear molecule, its rotational energy spectrum has the same character as that for diatomic molecules. The CO_2 rotational energy levels are thus given to a good approximation by Eq. (2.4.7):

$$E_J = hcB_eJ(J + 1), \qquad J = 0, 1, 2, \ldots, \qquad (2.5.3)$$

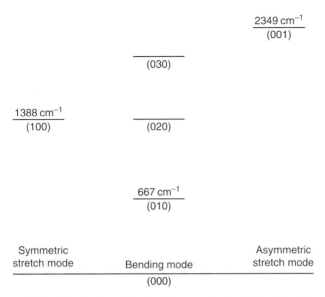

Figure 2.10   The first few vibrational energy levels of the CO_2 molecule.

where the rotational constant B_e for the CO_2 molecule is

$$B_e = 0.39 \text{ cm}^{-1}. \tag{2.5.4}$$

2.6   CONDUCTORS AND INSULATORS

In a gas the average distance between molecules (or atoms) is large compared to molecular dimensions. In liquids and solids, however, the intermolecular distance is comparable to a molecular diameter (Problem 2.3). Consequently the intermolecular forces are roughly comparable in strength to the interatomic bonding forces in the molecules. The molecules in liquids and solids are thus influenced very strongly by their neighbors.

What is generally called "solid-state physics" is mostly the study of crystalline solids, that is, solids in which the molecules are arranged in a regular pattern called a crystal lattice. The central fact of the theory of crystalline solids is that the discrete energy levels of the individual atoms are split into *energy bands*, each containing many closely spaced levels (Fig. 2.11). Between these allowed energy bands are gaps with no allowed energies. The way this happens is easy to explain with a simple example.

Imagine a sodium atom with its 11 electrons distributed according to the Pauli exclusion principle over its $1s$ level (2 electrons), $2s$ level (2 electrons), $2p$ level (6 electrons) and $3s$ level (1 electron). A second sodium atom has exactly the same energy levels occupied by 11 electrons in the same way. If the two sodium atoms are brought close together their two equal-energy $1s$ levels turn into two levels of "di-sodium," and these two levels of di-sodium have slightly different energies from their Na values and slightly different energies from each other. Similarly, the two $2s$ levels of Na become two slightly different levels of di-sodium, and so on for the higher levels.

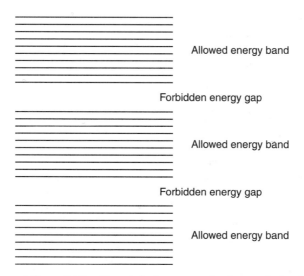

Figure 2.11   In a crystalline solid the allowed electron energy levels occur in *bands* of closely spaced levels. Between these allowed energy bands are forbidden gaps.

The same process occurs for three sodium atoms, in which case the $1s$ label applies to three slightly separated levels, the $2s$ label applies to three slightly separated levels, and so on. When the number of atoms N is as large as is appropriate to a macroscopic piece of sodium metal the N slightly separated levels are so closely bunched that they constitute an effectively continuous band of energies. The relatively large gap between the $1s$ and $2s$ levels in sodium atoms becomes the "forbidden gap" between $1s$ and $2s$ bands in sodium metal, where there are no longer any distinguishable levels. This is sketched in Fig. 2.12 for 1, 2, 6, and $N \sim 10^{23}$ atoms.

A one-dimensional model showing how such band structure arises in quantum theory is discussed in the Appendix to this chapter.

We can reach a crude understanding of the formation of energy bands by beginning with the case of two identical atoms. When the atoms are far apart and effectively non-interacting, their electron configurations and energy levels are identical. As they are brought closer together, however, the electrons of each atom begin to feel the presence

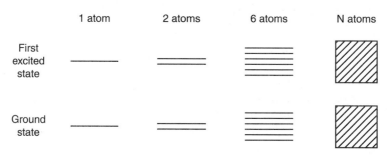

Figure 2.12   Sketch of the change in the energy values originally assigned to the ground and first excited states of an atom as more and more atoms are combined to form a solid. Note that the band gap energy can be identified with the original atomic level spacing, but is generally different in size.

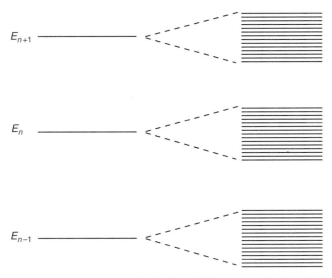

E_{n+1}

E_n

E_{n-1}

Figure 2.13 Crystalline solid energy bands formed from energy levels of isolated atoms.

of the other atom, and the energy levels become those of the two atoms as a whole.[5] The difference between these new energy levels depends upon the interatomic spacing (Fig. 2.12).

The difference between the highest and the lowest of these N levels depends on the interatomic distances, amounting typically to several electron volts for atomic spacings of a few angstroms, typical of solids. Now if we increase N, keeping the interatomic spacing fixed as in a crystalline solid, the total energy spread of the N levels stays about the same, but the levels become more densely spaced. For the large values of N typical of a solid (say, something like 10^{29} atoms$/$m^3), each set of N levels thus becomes in effect a continuous energy band (as in Fig. 2.13), which in some solids can be even wider then the original atomic level spacing.

The chemical and optical properties of atoms are determined primarily by their outer electrons. In solids, similarly, many important properties are determined by the electrons in the highest energy bands, the bands evolving out of the higher occupied states of the individual atoms. Consider, for instance, a solid in which the highest occupied energy band is only partially filled, as illustrated in Fig. 2.14a. In an applied electric field the electrons in this band can readily take up energy and move up within the band. A solid whose highest occupied band is only partially filled is a good *conductor* of electricity.

Now consider a solid whose highest occupied band is completely filled with electrons, as illustrated in Fig. 2.14b. In this case it is quite difficult for an electron to move because all the energetically allowed higher states in the band already have their full measure of electrons permitted by the Pauli principle. Therefore, a solid whose highest occupied energy band is filled will be an electrical *insulator*; in other words, its electrons will not flow freely when an electric field is applied. Implicit in

[5]These two energy levels correspond to symmetric and antisymmetric spatial wave functions for the electrons, with correspondingly antisymmetric and symmetric spin eigenfunctions. The twofold exchange degeneracy in the case of widely separated atoms is broken when their wave functions begin to overlap.

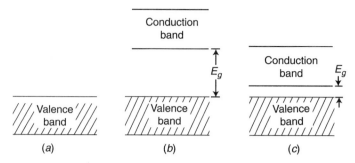

Figure 2.14   In a good conductor of electricity (*a*), the highest occupied band is only partially filled with electrons, whereas in a good insulator (*b*) it is filled. In (*b*) the energy gap E_g between the valence band and the conduction band is large. In the case (*c*) of a semiconductor, however, this gap is small, and electrons in the valence band can easily be promoted to the conduction band.

this definition of an insulator is the assumption that the forbidden energy gap between the highest filled band and the next allowed energy band, denoted E_g in Fig. 2.14, is large compared to the amount of energy an electron can pick up in the applied field.

Solids in which this band gap is not so large are called *semiconductors*. Their band structure is indicated in Fig. 2.14*c*. At the absolute zero of temperature the *valence* band of a semiconductor is completely filled, whereas the *conduction* band, the next allowed energy band, is empty. At room temperature, however, electrons in the valence band may have enough thermal energy to cross the narrow energy gap and go into the conduction band. Thus, diamond, which has a band gap of about 7 eV, is an insulator, whereas silicon, with a band gap of only about 1 eV, is a semiconductor. In a metallic conductor, by contrast, there is no band gap at all; the valence and conduction bands are effectively overlapped.

This characterization of solids as insulators, conductors, and semiconductors is obviously more descriptive than explanatory. To understand *why* a given solid is an insulator, conductor, or semiconductor, we must consider the nature of the forces binding the atoms (or molecules) together in the solid.

In covalent solids the atoms are bound by the sharing of outer electrons in partially filled configurations. In a true covalent solid, there are no free electrons, and so such solids do not conduct electricity very well. The covalent bonds tend to hold the atoms tightly together, thus causing covalent solids to be rather hard and have high melting points. A good example is diamond, in which carbon atoms are arranged in a lattice such that each atom is at the center of a tetrahedron formed by its four nearest neighbors.

In *ionic solids* such as NaCl, the binding is produced by electrostatic forces between oppositely charged ions. The reason for the binding is the same as in ionic molecules. In NaCl, for instance, the energy required to remove the 3*s* electron from Na and transfer it to Cl, to form Na^+ and Cl^-, is less than the electrostatic energy of attraction between the ions. Here again there are no free electrons available to conduct heat or electricity, and so ionic solids are not good conductors.

The so-called *molecular solids*, which include many organic compounds (e.g., teflon), are also poor conductors. In such solids the binding is due to the very weak van der Waals forces, which were originally postulated by J. D. van der Waals (1873) to explain deviations from the ideal gas law. The van der Waals energy of attraction between two molecules ordinarily varies as the inverse sixth power of the distance between the molecules. Because of the weakness of the van der Waals interaction, molecular solids are much easier to deform or compress than covalent or ionic solids.

Of course, electrical technology as we know it would be impossible without *metallic solids*, which are good conductors of electricity (copper, silver, etc.). In a metallic solid the electrons are not all tightly bound at crystal lattice sites. Some of the electrons are free to move over large distances in the metal, much as atoms move freely in a gas. This occurs because metals are formed from atoms in which there are one, two, or occasionally three outer electrons in unfilled configurations. The binding is associated with these weakly held electrons leaving their parent ions and being shared by all the ions, and so we can regard metallic binding as a kind of covalent binding. We can also think of the positive ions as being held in place because their attraction to the "electron gas" exceeds their mutual repulsion.

It is sometimes a useful approximation to regard the conduction electrons of a metal as completely free to move about. Of course, conduction electrons are not really completely free, as evidenced by the fact that even the very best conductors—copper, silver, and gold—have a finite resistance to the flow of electricity.

It should be emphasized that many solids do not fit so neatly into the covalent, ionic, molecular, or metallic categories. Furthermore many important properties of various solids are determined by imperfections such as impurities and dislocations in the crystal lattice. Steel, for instance, is much harder than pure iron because of the small amount of carbon that was mixed into the iron melt. Impurities can also determine the color of a crystal, as in the case of ruby (Section 3.1). We will shortly discuss how the addition of certain impurities in semiconductors is responsible for modern electronic technology.

2.7 SEMICONDUCTORS

A semiconductor is distinguished from an insulator by the fact that the band gap between the valence and conduction bands is small, about 2 eV or less. The important semiconductors silicon and germanium, for instance, are covalent solids with band gaps of about 1.12 and 0.67 eV, respectively, at 300 K. At very low temperatures they are insulators, but the conductivity increases rapidly with increasing temperature because valence-band electrons can be thermally excited into the conduction band. This increase of conductivity with increasing temperature is an important distinction between semiconductors and metals.[6] It was known to Faraday and other physicists early in the nineteenth century but was explained only when quantum mechanics and the band theory of solids were developed.

Another way to promote electrons into the conduction band of a semiconductor is by absorption of radiation, if the energy of an incident photon exceeds the gap energy E_g (i.e., $h\nu > E_g$, where $h\nu$ is the frequency of the radiation). This photoconductive effect is quite similar to the photoelectric effect, except that electrons are not actually released from the surface of the material. Photoconductive cells in which the current

---

[6]Perhaps it is worth emphasizing what a tremendous range of electrical conductivities is found in different materials. A good insulator might have a resistivity (the inverse of the conductivity) of 10^{20} Ω-m, whereas a metal may have a value 10^{-8} Ω-m. Room temperature resistivities of semiconductors, by contrast, are typically somewhere between 10^{-5} and 10^{7} Ω-m. This range of about 28 orders of magnitude is often cited as one of the broadest variations of any physical parameter.

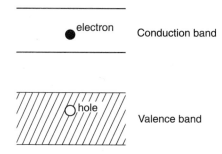

Figure 2.15 In going from the valence band to the conduction band, an electron leaves a hole in the valence band.

in an electric circuit is controlled by the intensity of incident light have applications similar to photoelectric cells (exposure meters in photography, automatic door openers, etc.), except that they require an auxiliary voltage supply to move electrons that have been put into the conduction band.

By far the most important means of producing conduction electrons in a semiconductor is by doping it with a certain type of impurity. Tiny junctions of differently doped semiconductors are the basis not only for transistors but also for light-emitting diodes and diode lasers. To understand the operation of such devices, however, we must first discuss the concept of a *hole* in the valence band of a semiconductor.

The basic idea is very simple. If an electron somehow goes from the valence band to the conduction band, it leaves a hole—the absence of an electron—in the valence band (Fig. 2.15). That is, a hole corresponds to the absence of an electron from an otherwise filled valence band. It turns out to be very useful to think of a hole as a particle like an electron. The removal of an electron increases the charge of the valence band, so clearly a hole must be a positively charged "particle," with charge opposite to that of an electron.

Consider a piece of a semiconductor in which electron–hole pairs have been created in some way (i.e., electrons have been put into the conduction band, leaving holes in the valence band). If we connect it with wires to the terminals of a battery, there will be a flow of current since there are "free" electrons in the conduction band ready to respond to an externally applied field. These electrons drift in the direction shown in Fig. 2.16, from the negative electrode to the positive. The net effect, as seen from the outside, is that electrons enter the semiconductor from the right and exit to the left. However, this is not the whole story, for electrons in the valence band are also affected by the potential difference. Specifically, an electron to the right of the hole indicated in Fig. 2.16 can fall into the hole; that is, it will go into the state previously occupied by another electron. In doing so, it leaves a hole at the site it left, which can now be filled by another electron.

This electron drift in the valence band constitutes a current in the same direction (left to right, by convention, in Fig. 2.16) as the current of the electrons in the conduction band. Equivalently, we can view the situation as one in which electrons in the conduction band are moving from right to left, while holes in the valence band are moving from left to right. In other words, we can describe the charge motion in the conduction band in terms of electrons, and that in the valence band in terms of holes, and *both electrons and holes contribute to the total current.*

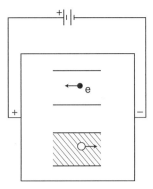

Figure 2.16 When a potential difference is applied to two ends of a semiconductor, electrons in the valence and conduction bands drift from the negative side to the positive. In the valence band, the effect is equivalent to the drift of positively charged holes from the positive side to the negative. The total current can therefore be attributed to electrons in the conduction band and holes in the valence band.

By doping a semiconductor with a certain kind of impurity, we can arrange for a current in the semiconductor to be due predominantly to either electrons or holes. In the former case the semiconductor is called *n type* (because electrons are negatively charged) and in the latter it is called *p type* (because holes have positive charge). To see how this works, we will consider the example of silicon doped with phosphorus.

In pure silicon each atom shares its four valence electrons in the unfilled $3s3p$ subshell (see inside cover) to form covalent bonds with its four nearest neighbors. Each silicon atom needs four more electrons to complete the sp configuration, and by sharing electrons in this way it comes closer to having a filled outer subshell. The crystal structure is that of diamond, with each silicon atom at the center of a regular tetrahedron (pyramid) and its four nearest neighbors at the vertices. This structure is a consequence of the fact that the bonds associated with shared electrons are spaced as far from each other as possible at equal angles from each atom. It is useful to represent the situation in the schematic, two-dimensional form of Fig. 2.17.

In its pure form silicon has a very low conductivity at room temperature because so few electrons can be thermally excited across the 1.12-eV energy gap $(e^{-E_g/kT} \approx e^{-1 \text{ eV}/1/40 \text{ eV}} = e^{-40})$. Under ordinary circumstances the current passed is so small as to be practically useless. To pass useful current, we must find a way to get more electrons into the conduction band or holes into the valence band.

Suppose that one of the silicon atoms in the crystal is replaced by an atom of phosphorus, which has *five* electrons in the unfilled $3s3p$ subshell. Four of these can contribute to the covalent bonding of the crystal, as indicated in Fig. 2.18, but there is one electron left over that cannot take part in the bonding. This fifth electron is very loosely bound, and so is free to move through the crystal when an electric field is applied. In other words, if we add a small amount of phosphorus to a silicon melt, the crystal that forms will be an *n*-type semiconductor.[7] We can also make an *n*-type semiconductor by doping silicon with other pentavalent elements such as arsenic and antimony.

---

[7]The proportion of dopant must be small in order to preserve the integrity of the host crystal lattice, since the dopant by itself forms its own crystal lattice structure.

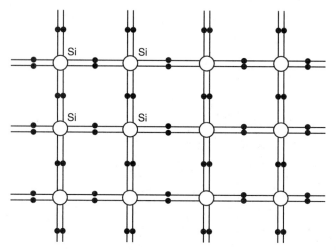

Figure 2.17   Schematic illustration of covalent bonding in silicon, in which each atom shares its four valence electrons with its four nearest neighbors.

Imagine instead that we replace a silicon atom by an atom of boron, which has *three* electrons in an unfilled outer shell. In this case there is one electron short of the four needed to join in complete covalent bonding in the host silicon lattice. Thus, if boron shares its three valence electrons with neighboring silicon atoms, there will be a missing bond in the crystal, as indicated in Fig. 2.19. This missing electron is a hole that can be filled by an electron that happens to be nearby. But when that electron fills the hole, it leaves another hole, which can be filled by another electron, and so in an electric field we get a migration of *holes* (or equivalently, of course, a migration of electrons in the opposite direction). In other words, by doping silicon with boron we can create

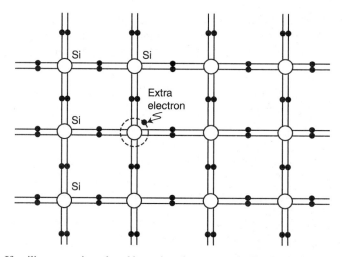

Figure 2.18   If a silicon atom is replaced by a phosphorus atom in Fig. 2.17, there is an extra electron left over that cannot take part in the covalent bonding. This electron is very loosely bound and therefore available for conduction of electric current.

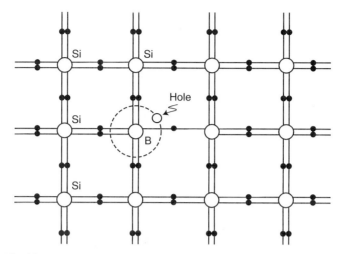

Figure 2.19 If a silicon atom in Fig. 2.17 is replaced by a boron atom, there will be a missing bond because there is one electron short of the number necessary for complete bonding. The missing electron is represented by a hole.

a p-type semiconductor. Other trivalent elements such as aluminum or gallium are also suitable dopants for this purpose.

As a matter of terminology, a dopant that produces an n-type semiconductor is called a *donor* because it donates electrons to the conduction band. A dopant that produces a p-type semiconductor is called an *acceptor* because it puts holes in the valence band, that is, it accepts electrons to fill the missing slots. In either case, of course, the crystal remains charge-neutral. Note also that the added impurities produce either electrons or holes, but no electron–hole pairs as in, for instance, photoconductivity. Thus, in an n-type semiconductor any current is due predominantly to electrons, whereas in p-type material it is due to holes.

Figure 2.20 shows an experiment that can distinguish between n-type and p-type semiconductors. Two ends of the material are connected to battery terminals to produce a current, and we also apply a magnetic field $\mathbf{B}$ at right angles to the current. The magnetic field exerts a force $q\mathbf{v} \times \mathbf{B}$ on particles of charge q moving with velocity $\mathbf{v}$. Regardless of the sign of q, this magnetic force is upward for the arrangement shown (Problem 2.4). Therefore, the top will become positively charged or negatively charged, depending on the sign of the charge carriers. This displacement of charge creates an electric force on the charge carriers, and this electric force opposes the magnetic force. This is called the *Hall effect*. In equilibrium these vertical electric and magnetic forces exactly cancel each other and the current flows horizontally. By measuring with a voltmeter the potential difference between the top and bottom of the sample, we can determine whether the charge carriers are positive or negative, and therefore whether the semiconductor is p type or n type.

Actually, some metals, such as beryllium, also exhibit an *anomalous Hall effect* in which the dominant charge carriers are positive. This is because beryllium has a filled $2s$ subshell in which the holes happen to be much more mobile than the $2p$ electrons. The important point, again, is that it is very convenient to think in terms of electrons and holes, even though the real charge carriers are, of course, the electrons.

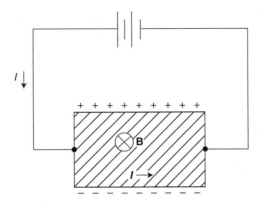

Figure 2.20 Experimental arrangement to determine the sign of the charge carriers (Hall effect). A magnetic field **B** is applied in the direction into the page. The top and bottom of the sample become charged $+$ and $-$, respectively, if the carriers are positive, and $-$ and $+$, respectively, if the carriers are negative.

- The extra electron indicated in Fig. 2.18 is not actually free but is attracted by the positively charged impurity ion. The electron–ion system is thus analogous to the hydrogen atom. It might be expected, therefore, that there are electron orbits about the ion with allowed energies given by the Bohr formula (2.2.11). In particular, the energy of the lowest energy state should be

$$E_1 = -\left(\frac{1}{4\pi\epsilon_0}\right)^2 \frac{me^4}{2\hbar^2},\qquad(2.7.1)$$

and an energy $|E_1| = 13.6$ eV should be required to free the extra electron and make it available for conduction.

This argument overestimates the binding energy of the extra electron for two reasons. First, the free-space permittivity ϵ_0 in (2.7.1) should be replaced by the material dielectric constant ϵ. Second, it turns out that the band theory ascribes to electrons (and holes) a certain *effective mass* m^*; because the electron is in the periodic potential of the crystal and not in free space, it acts *as if* its mass were smaller, say m^* (see the Appendix to this chapter). The value of m^* depends on the energy of the electron within an energy band and can also vary with direction within the crystal. Thus, we should replace ϵ_0 by ϵ and m by m^* in (2.7.1):

$$E_1 = -\left(\frac{\epsilon_0}{\epsilon}\right)^2 \left(\frac{m^*}{m}\right)\left(\frac{1}{4\pi\epsilon_0}\right)^2 \frac{me^4}{2\hbar^2} = -\left(\frac{\epsilon_0}{\epsilon}\right)^2 \left(\frac{m^*}{m}\right)(13.6 \text{ eV}).\qquad(2.7.2)$$

Using the values $\epsilon = 11.8\epsilon_0$ and $m^* = 0.26m$ for silicon, we obtain

$$E_1 = -0.025 \text{ eV}.\qquad(2.7.3)$$

Therefore, the extra electron is in fact bound very weakly, requiring only an energy of about 0.025 eV to put it into the conduction band. This small energy is about equal to $kT \approx \frac{1}{40}$ eV at room temperature, so that thermal excitation is enough to free some extra electrons in *n*-type doped silicon.

The Bohr levels of the extra electrons represent new energy levels not found in the pure semiconductor. These levels are called *donor levels*. Because they are small and negative, the donor levels lie just below the bottom of the conduction band.

The hole produced by an acceptor impurity as in Fig. 2.19 is likewise bound, and it also requires only a small amount of energy to be freed. Its Bohr energy levels, called *acceptor levels*, lie just above the top of the valence band.

In summary, the doping of a semiconductor with a donor or acceptor does not by itself produce conduction-band electrons or valence-band holes, as assumed in our discussion based on Figs. 2.18 and 2.19. However, the energy required to "ionize" these donors or acceptors is so small that thermal excitation at moderate temperatures will do the job.   •

Another important semiconductor, germanium, is similar to silicon in that it is tetravalent. It may, therefore, be doped in the same ways to produce *n*-type and *p*-type materials. In addition to such elemental semiconductors are "III–V" binary semi-conductors such as gallium arsenide or indium antimonide, in which trivalent and pentavalent atoms share in covalent bonding as a result of unfilled *sp* subshells (Problem 2.5).

2.8 SEMICONDUCTOR JUNCTIONS

Semiconductors of either *n* type or *p* type are not by themselves very useful. They are, after all, just second-class conductors. Their great utility is realized only when they are brought together to form junctions. All of semiconductor technology is based ultimately on the properties of the *pn* junction, the joining of an *n*-type material to a *p*-type material (Fig. 2.21).

We will imagine the junction to be abrupt, although, of course, it cannot be so on the atomic scale. Junctions can be made in practice from a single crystal, one region of which has been made *p* type whereas an adjoining region has been made *n* type. The boundary between the two regions can be made very narrow, typically less than a micron (1 μm = 10^{-6} m), and the sharp-boundary idealization is therefore a reasonable approximation.

In a *p*-type material the negative acceptor ions are fixed in position and the holes are mobile, whereas in an *n*-type material the positive donor ions are fixed and the electrons are mobile. Figure 2.22 illustrates what happens at a *pn* junction. Electrons from the *n* side are attracted by the positive holes at the boundary and drift over into the *p* side, while holes on the *p* side are attracted by electrons and drift across the boundary to the *n* side. When an electron meets a hole, it falls into it, becoming part of a covalent bond. This diffusion and annihilation of mobile charge carriers produces a *depletion region* at the boundary between the *p*- and *n*-type materials. The depletion region is short of both electrons and holes, consisting mainly of negative acceptor ions on the *p* side and positive donor ions on the *n* side (Fig. 2.22). Because of this charge separation, there is a static electric field pointing from the *n* side to the *p* side. This field

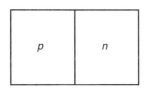

Figure 2.21   A *pn* junction is formed by joining *p*-type and *n*-type semiconductor materials.

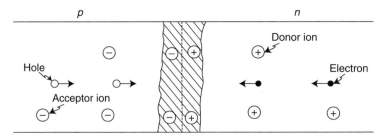

Figure 2.22   At a *pn* junction, electrons from the *n* side are attracted to the *p* side, and holes from the *p* side are attracted to the *n* side. When electron–hole pairs meet they are "annihilated" as the electron becomes part of a covalent bond. This results in a depletion region at the boundary, which is a region in which there are very few electrons or holes.

opposes further diffusion of electrons and holes and thus keeps the depletion region confined to a narrow layer at the boundary. In other words, there is some voltage drop, which we call V_0, in going from the *n* side to the *p* side (Fig. 2.23).

In thermal equilibrium there is a diffusion of individual electrons and holes across the junction, but no net flow of current. Holes on the *n* side, for instance, have no difficulty dropping down the potential-energy hill and going over to the *p* side. Holes on the *p* side, on the other hand, have to cross the potential-energy barrier eV_0 to get to the *n* side (here *e* is understood to be positive). According to statistical mechanics, the fraction of holes able to cross the barrier is given by the Boltzmann factor $\exp(-eV_0/k_BT)$, where k_B is Boltzmann's constant ($k_B \approx 1.38 \times 10^{-23}$ J/K $\approx \frac{1}{40}$ eV/300K) and *T* is the absolute temperature. If the current due to holes diffusing from the *n* side to the *p* side is to be exactly balanced by the hole current in the opposite direction, therefore, we must have

$$N_p(n \text{ side}) = N_p(p \text{ side})e^{-eV_0/k_BT}, \tag{2.8.1}$$

where N_p denotes the number of holes per unit volume. This equation shows, as expected, that there is a greater density of holes on the *p* side than on the *n* side. The same reasoning leads to the relation

$$N_n(p \text{ side}) = N_n(n \text{ side})e^{-eV_0/k_BT}, \tag{2.8.2}$$

for the number density N_n of electrons. According to these relations, the product N_nN_p is the same for the two sides of the junction.

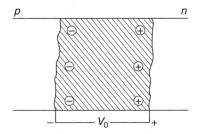

Figure 2.23   The charge separation due to donor and acceptor ions in the depletion layer of a *pn* junction results in a potential difference V_0.

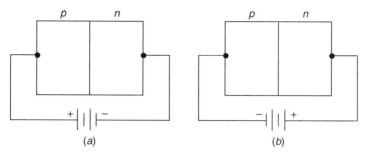

Figure 2.24 Forward-biased (*a*) and reverse-biased (*b*) *pn* junctions.

Now suppose that the *p* and *n* sides are connected to the terminals of a battery. If the *p* side is connected to the positive terminal and the *n* side to the negative terminal, the junction is said to be *forward biased* (Fig. 2.24a). Since the depletion layer is much more resistive to current than the bulk regions, most of the applied voltage *V* is dropped across the depletion layer. In other words, the applied voltage has the effect of lowering the barrier voltage V_0 to $V_0 - V$. We will assume for the present that *V* is small compared to V_0.

This lowering of the barrier potential by a forward-biased voltage does not affect very much the diffusion current of holes from the *n* side to the *p* side because with or without it they have no potential-energy barrier to cross. Their diffusion current, therefore, is unaffected by the applied voltage. The hole current from the *p* side to the *n* side, however, will increase with the applied voltage because now a fraction $e^{-(V_0-V)/k_BT}$ of them are able to cross over. For the net current of holes diffusing from the *p* side to the *n* side we have, therefore,

$$I_p \propto N_p(p \text{ side})e^{-e(V_0-V)/k_BT} - N_p(n \text{ side})$$

$$= N_p(n \text{ side})e^{eV/k_BT} - N_p(n \text{ side})$$

$$= N_p(n \text{ side})(e^{eV/k_BT} - 1), \tag{2.8.3}$$

where in the second line we have used (2.8.1). In other words, we have the relation

$$I_p = I_{p0}(e^{eV/k_BT} - 1), \tag{2.8.4}$$

for the net hole current flowing from the *p* side to the *n* side under forward bias, where I_{n0} is the hole diffusion current from the *n* side to the *p* side. A similar expression is obtained for the net electron current under forward biasing:

$$I_n = I_{n0}(e^{eV/k_BT} - 1), \tag{2.8.5}$$

where I_{n0} is the electron diffusion current flowing from the *p* side to the *n* side. The total current flowing from the *p* side to the *n* side of a forward-biased *pn* junction is, therefore,

$$I = I_p + I_n = I_0(e^{eV/k_BT} - 1) \quad \text{(forward biasing)}, \tag{2.8.6}$$

where $I_0 = I_{p0} + I_{n0}$ is called the *saturation current* of the junction.

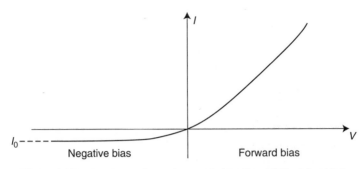

Figure 2.25   Current–voltage characteristics of an ideal *pn* junction.

Suppose instead that the leads are reversed, as in Fig. 2.24*b*. In this case the junction is said to be *reverse biased*. The barrier potential is now increased from V_0 to $V_0 + V$, and the net current is obtained simply by changing V to $-V$ in (2.8.6):

$$I = I_0(e^{-eV/k_BT} - 1) \quad \text{(reverse bias)}. \tag{2.8.7}$$

Equations (2.8.6) and (2.8.7) give the so-called current–voltage (*IV*) characteristics of an ideal *pn* junction (Fig. 2.25). Under forward biasing, the current increases rapidly (exponentially) with increasing voltage. Under reverse bias, the current *saturates* with increasing voltage to the value I_0. Typical barrier voltages V_0 in silicon and germanium diodes are roughly on the order of half a volt. Saturation current densities are extremely small, perhaps less than 10^{-10} A/cm^2; for a typical junction area of 10^{-2} cm^2, this amounts to a saturation current I_0 of less than 10^{-12} A. Formulas can be derived to estimate V_0 and I_0 as a function of carrier concentrations and other parameters, but we will not take the time to do so.

The key property of a *pn* junction, therefore, is that it will conduct current in one direction but not the other. That is, it can act as a diode, a sort of automatic switch that closes a circuit when voltage is applied in a forward sense, but blocks the flow of current otherwise. This diode can be used as a rectifier, converting ac current to dc current.

• Real semiconductor diodes do not display exactly the same *IV* characteristics as the idealized diode we have considered. For one thing, we have ignored electron–hole recombination within the depletion layer. This and other effects may be taken into account by replacing (2.8.6) by

$$I = I_0(e^{eV/\beta k_BT} - 1), \tag{2.8.8}$$

where the "ideality factor" β is a dimensionless parameter between 1 and 2, depending on T.

Furthermore, at large reverse-bias voltages a real diode no longer blocks the flow of current. At a certain "breakdown" voltage there is a sudden jump in the reverse current. One reason for this is that high-energy charge carriers colliding with atoms in the crystal lattice can ionize them, producing more charge carriers, which lead to further ionization and therefore increasing the current. This is called *avalanche breakdown*. Another mechanism for reverse-current generation is the *Zener effect* in which electrons undergo a quantum mechanical "tunneling" from the *p* side to the *n* side.   •

2.9   LIGHT-EMITTING DIODES

In our discussion of electrons and holes we have mentioned several times that an electron can "fall into" a hole, meaning that the electron can replace the missing electron represented by the hole. In doing so the electron becomes part of a covalent bond. This process is called *recombination*.

Now if an electron from the conduction band "recombines" with a hole in the valence band, it loses energy, having been free ($E > 0$) and then becoming bound ($E < 0$). There are two ways in which this energy can be discarded by the electron. One way is for it to appear as heat in the form of vibrations of the crystal atoms about their equilibrium positions. Another way is *radiative recombination* in which the electron transition is accompanied by the emission of a photon (Fig. 2.26). This is analogous to (spontaneous) emission by an atom. In fact the electron–hole pair forms an *exciton*, a mobile "quasi-particle" in which the electron and hole are bound by their Coulomb interaction and that decays with the emission of a photon in typically about a nanosecond. The emitted photon has a frequency ν satisfying $h\nu = E_f - E_i$, where E_i and E_f are the energies of the initial and final electron states, respectively. In a transition from the conduction band to the valence band, the minimum value of $E_f - E_i$ is clearly the gap energy. The maximum photon wavelength in interband radiative recombination is in turn given by

$$\lambda_{\max} = \frac{c}{\nu_{\min}} = \frac{hc}{E_g}. \tag{2.9.1}$$

For silicon, therefore, with a gap energy of 1.12 eV, $\lambda_{\max} = 1100$ nm. For germanium, with $E_g = 0.67$ eV, $\lambda_{\max} \cong 1900$ nm.

The light from a *light-emitting diode* (LED) is produced by radiative recombination of electrons and holes injected across the junction of a forward-biased *pn* diode. The electrons drifting from the *n* side to the *p* side recombine radiatively with holes, and

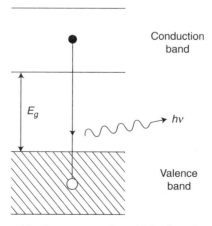

Figure 2.26   Radiative recombination process in which the electron–hole "annihilation" is accompanied by the emission of a photon.

the holes drifting from the p side to the n side recombine radiatively with electrons. On either side, of course, the emitted photons are produced when an electron makes a downward transition from a state of energy E_f to one of energy E_i. Not every recombination of an electron with a hole will be radiative because there are competing, nonradiative recombinations in which the energy lost by the electron appears in the form of crystal lattice vibrations. Although there are LEDs in which nearly every recombination process is radiative, the actual device efficiency is limited by other factors, as discussed below.

Interband recombination radiation has a distribution of wavelengths arising from the thermal distribution of electron energy within the conduction band. The maximum wavelength λ_{max} given by (2.9.1), however, provides a good estimate of the peak wavelength. Thus, we can deduce from it that LEDs made from Si or Ge junctions will not generate much visible radiation.

Actually there are other types of radiative recombination that produce longer wavelengths than the interband maximum λ_{max}. As noted in the preceding section, there are donor levels and acceptor levels associated with the impurities of a doped semiconductor, and these levels lie just below the bottom of the conduction band and just above the top of the valence band, respectively. Radiative recombination processes involving these impurity levels produce radiation of wavelengths $\lambda > \lambda_{max}$, as is clear from Fig. 2.27. In part (a) of the figure we indicate an interband radiative recombination transition, that is, a transition of an electron from the conduction band to the valence band. Part (b) shows a transition from a donor level to the valence band, while (c) shows a transition from the conduction band to an acceptor level. Finally, we show in (d) a transition from a donor level to an acceptor level. Processes (b)–(d) obviously lead to wavelengths greater than the interband process (a), and so LED wavelengths are often *greater* than the interband maximum (2.9.1). Because the differences $E_c - E_d$ and $E_a - E_v$ are small compared to E_g, however, (2.9.1) provides a good estimate of the sort of wavelength that can be expected with a given semiconductor.

The question of wavelength is obviously an important one if an LED is to be used for visual display purposes. Silicon and germanium, for instance, are eminently useful electronically because of the relative ease with which they can be doped and fabricated as diodes, but their band gaps are too small to make them useful as LEDs for visible radiation. Moreover, Si and Ge are radiatively too inefficient to be used in LEDs. This is because they are *indirect-band-gap* semiconductors, for which the

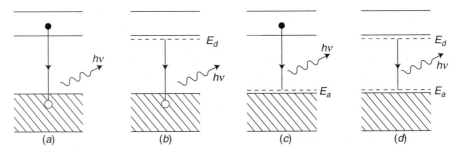

Figure 2.27   Radiative recombination involving a transition from (a) the conduction band to the valence band, (b) a donor level to valence band, (c) the conduction band to an acceptor level, and (d) a donor level to an acceptor level.

interband radiative recombination rates are very low. GaAs, by contrast, is a *direct-band-gap* semiconductor and is consequently a much more efficient radiator. Indirect-band-gap materials can be used in LEDs if there are efficient radiative pathways in addition to interband recombination.

- In the case of a direct band gap the minimum of the $E-k$ curve for the conduction band lies directly above the maximum of the $E-k$ curve for the valence band. (See the Appendix in this chapter for a discussion of a simplified one-dimensional model of electron wave functions and $E-k$ curves in a crystal, and also Fig. 15.2.) This means that a "vertical" transition can occur in which the electron energy (E) decreases by an amount equal to the energy of the emitted photon while there is no change in the $\mathbf{k}$ vector of the electron wave function. More precisely, the calculation of the transition rate leads to the $\mathbf{k}$ selection rule: the difference in the $\mathbf{k}$ vectors of the initial and final electron wave functions must be equal to the wave vector of the emitted photon; otherwise the transition is forbidden. But since the electron $\mathbf{k}$ vector has a much greater magnitude than the wave vector of the photon, this selection rule says that the electron $\mathbf{k}$ vector is approximately unchanged and momentum ($\hbar\mathbf{k}$) as well as energy is conserved in the transition.

  For an indirect band gap, however, the maximum of the $E-k$ curve for the valence band is offset from the minimum of the $E-k$ curve for the conduction band, and the wave vectors of the initial and final electron wave functions are not the same. To conserve momentum, therefore, a radiative transition in this case must be accompanied by a change in the momentum of the crystal lattice. Crystal lattice vibrations are characterized approximately in terms of the equally spaced energy levels of a harmonic oscillator, and the particle-like excitations associated with these energy levels are called *phonons*. Energy and momentum conservation in electron–hole recombination for an indirect band gap involves not only electrons and photons but also the phonons of the crystal lattice. As a consequence the rate and efficiency of the phonon-mediated, indirect-band-gap photon emission are generally much smaller than for direct-band-gap emission.   •

In addition to having a band gap large enough to produce visible radiation, a semiconductor to be used in an LED must, of course, have both *p*-type and *n*-type forms that can be made by suitable doping. As a rule of thumb, large-gap materials tend to have high melting points, making doping of a melt more difficult, and furthermore they tend to have low conductivities even when doped. Among the more commonly used LED materials is gallium arsenide (GaAs), with a band gap of 1.44 eV (and therefore $\lambda_{\mathrm{max}} \approx 861$ nm). Depending on the dopant, the dominant radiative recombination transition may be interband (Fig. 2.27a) or from the conduction band to an acceptor level (Fig. 2.27c).

Even if every charge carrier injected across the junction gave rise to an emitted photon, the efficiency of an LED would still be much less than 100%. An important reason for this is a phenomenon well known in classical optics: total internal reflection. For a quick review of this effect, recall that the refraction of light at an interface of two media is governed by **Snell's law**:

$$n_1 \sin \theta_1 = n_2 \sin \theta_2, \tag{2.9.2}$$

where n_1 and n_2 are the refractive indices on the two sides of the interface, and θ_1 and θ_2 are the corresponding angles of incidence, as in Fig. 2.28a. Now if $n_1 > n_2$, it is possible for light propagating from medium 1 to medium 2 to be reflected back into medium 1 instead of penetrating the interface and going into medium 2. This total internal

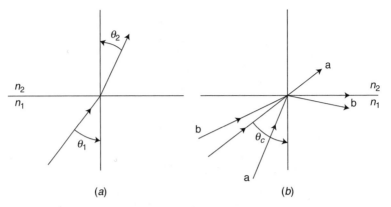

Figure 2.28   (a) Geometry for Snell's law if $n_1 < n_2$. (b) Total internal reflection occurs at the interface if $n_1 > n_2$ and the angle of incidence exceeds the critical angle θ_c given by (2.9.4).

reflection occurs at angles of incidence θ_1 greater than the critical value θ_c for which $\theta_2 = 90°$:

$$n_1 \sin \theta_c = n_2 \sin 90° = n_2, \qquad (2.9.3)$$

or

$$\theta_c = \sin^{-1}\left(\frac{n_2}{n_1}\right). \qquad (2.9.4)$$

This is illustrated in Fig. 2.28b.

Any light emerging from an LED and propagating into air is passing from a medium of higher index to a medium of lower index. This means that light approaching the LED–air interface at an angle greater than the critical angle θ_c will be reflected back into the LED instead of emerging as useful output radiation. In fact, the refractive indices of LED materials are often quite large, making the critical angle for total internal reflection rather small. In GaAs, for instance, $n \approx 3.6$, so that $\theta_c \approx 16°$ for the GaAs–air interface.

The deleterious effect of total internal reflection is minimized in the common LED design shown in Fig. 2.29. The junction is enclosed in a plastic case of refractive index $n \approx 1.5$. This reduces the effect of total internal reflection at the emitting surface because the critical angle for total internal reflection (2.9.4) is increased over that appropriate to a diode–air interface. Of course, there is still total internal reflection at the plastic–air interface, but this is minimized by shaping the plastic into the form of a hemispherical or similarly shaped dome. With this geometry, most of the light rays at the plastic–air interface have angles of incidence less than the critical angle for total internal reflection, and as a result the emission efficiency can typically be increased by ≈ 10 (Problem 2.6). The shape of the plastic enclosure also determines the extent to which the light emission is directional. A tubular shape, for example, can increase directivity as a result of side reflections. Alternatively, it is desirable in some applications to have a more diffuse emission, which can be accomplished by using a "diffusing lens" design in which tiny glass particles embedded in the plastic casing scatter the light from the *pn* junction and thereby produce a wider angular spread of radiation from the LED.

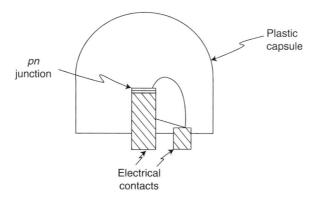

Figure 2.29 More light is extracted from an LED when there is a transparent plastic enclosure in the form of a hemispherical dome to reduce total internal reflection at the plastic–air interface.

There are two other effects that lower LED emission efficiency. One is simply the absorption of light, which can be significantly reduced by using a transparent material as the substrate for the *pn* junction. The other is the "Fresnel loss" due to reflection at the interface between the LED and the surrounding medium, which of course occurs even if total internal reflection is effectively eliminated. For light normally incident at an interface between the LED with refractive index n and a medium with refractive index n', for example, the power reflection coefficient is given by the Fresnel formula [Eq. (5.A.6)] $r = (n - n')^2/(n + n')^2$. Some of the reflected light can be retrieved by using a reflecting layer or cup at the "bottom" of the *pn* junction.

A wide range of LED colors has been realized by "band gap engineering" of mixed compound semiconductors such as gallium aluminum arsenide (GaAlAs) and gallium arsenide phosphide (GaAsP). For example, the band structure of $GaAs_{1-x}P_x$, where x is the mole fraction of P, is such that there is a direct band gap that monotonically increases with x from 1.44 eV when $x = 0$ (GaAs) to 2.1 eV when $x = 0.45$. $GaAs_{0.6}P_{0.4}$, for example, has a peak emission wavelength of 650 nm and is used in red LEDs. At $x = 0.45$ the compound has an *indirect* band gap and is nonradiative. It turns out, however, that doping the indirect-band-gap material with nitrogen allows direct-band-gap and therefore radiative electron–hole recombination. Thus, N-doped $GaAs_{0.15}P_{0.85}$, for example, has a peak emission wavelength of 589 nm and is used in yellow LEDs.

Blue LEDs based on gallium indium nitride (GaInN) became widely available in the 1990s. Together with red and green LEDs, they made it possible to produce any (visible) color by combining the light from three LEDs with appropriately adjusted currents and therefore output light intensities. White light is also produced using single, blue LEDs coated with wavelength-converting phosphors or "quantum dot" nanostructures that confine electrons to regions of linear dimension $\sim 2 - 10$ nm and act in some respects as "artificial atoms" with electron transition energies that vary with the size of the dot.

Although (red and green) LEDs became commercially available in the 1960s, research and development remained at a rather low level for a quarter-century. The recent advances have stemmed in considerable part from work on semiconductor lasers, including experimentation with different materials and dopants, progress in pure wafer fabrication and bonding techniques, and the development of suitable substrates for efficient extraction of radiation.

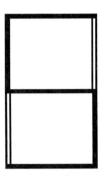

Figure 2.30   Seven-segment display format used with LEDs and LCDs. The ten digits 0 through 9 may be displayed by lighting selected segments.

Their use in cell phones resulted in a dramatic increase in the commercial production of light-emitting diodes in the 1990s and early 2000s. Infrared LEDs had already become ubiquitous in television remote controls and similar applications, while LEDs in the visible had replaced incandescent lamps in applications demanding compactness, low power consumption, and a high degree of reliability. For such purposes they are used either singly or in arrays. In the latter case a pattern or message can be conveyed when some of the LEDs are switched on. A simple and familiar example is the digital display used in clock radios and calculators. These commonly employ the seven-segment display shown in Fig. 2.30 in which each segment is an individual LED. The numerals 0–9 are displayed by turning on only certain of these LEDs at a time.

The energy efficiency and robustness of LEDs compared to incandescent or fluorescent glass lamps have made them increasingly important. It is estimated that in 2000 lighting accounted for 6–7% of the total power consumption in the United States. Incandescent lamps (like the everyday lightbulb) are notoriously inefficient: Only about 5% of the electrical power consumed is converted to light, with the rest wasted as heat. Efficiencies of fluorescent lights are 4–5 times greater but do not approach the 90% efficiencies possible with LEDs. Low power consumption, compactness, long life (tens of years or more), and resilience under jolts and vibrations make LED arrays ideal as light sources for traffic signals, for instance, and in the mid-1990s some cities in the United States began replacing incandescent traffic signal lights with LEDs. It is expected that in the near future most traffic lights will employ LED arrays. LED arrays were first used in automobile rear-center brake lights in the late 1980s, and front-end lights employing white-light LEDs were introduced as early as 2004. High-brightness LED arrays make possible the huge outdoor television screens that can be clearly seen even in daylight. Some industry analysts predict that by about 2015 most home lighting in the United States will employ LEDs.

•   In applications in which only a very small amount of power from a small battery is available, such as in digital wrist watches and many pocket calculators, the *liquid-crystal display*, or LCD, is used instead of the LED. LCDs consume less power because they do not generate any light of their own but use ambient light. Their operation is based on the properties of certain organic liquids of rod-shaped molecules. The molecules can take on certain organized relative alignments (hence the term *liquid "crystal"*) in such a way that the polarization of an incident light wave is rotated by 90° in passing through the LCD cell. The cell is a liquid-crystal layer, typically $\lesssim 10\,\mu$m thick, sandwiched between two clear plates whose inner surfaces are coated with a transparent conducting material arranged in a certain pattern. When a voltage is applied between the plates, the molecular alignment is altered and the polarization of incident light is no longer rotated by 90°. By using orthogonally oriented polarizing sheets in front of and behind the cell, and a mirror at the back, we can arrange for incident light to be reflected when there is no applied voltage, but for no light to be reflected from those areas where there is an applied voltage. Then we see the familiar black-and-white alphanumeric display patterns.

Liquid-crystal displays used in flat-screen televisions and laptop computers, for instance, are transmissive, or "backlit," rather than reflective, but the principle of operation is the same. The backlighting is done with very small fluorescent tubes or LEDs, together with a white panel behind the LCD that scatters light from the tubes to produce a uniform illumination of the LCD. The light from the screen is strongly polarized.   •

The light-emitting material in organic light-emitting diodes (OLEDs) consists of large organic molecules or polymers in a very thin layer, typically only a few hundred nanometers thick. The emitting layer is sandwiched between a cathode array and an anode array, with additional conductive layers serving to facilitate the injection of electrons and holes into the emitting layer. The color and brightness of the light produced by electron–hole recombination in the emitting layer depend on the type of organic molecules used and on the strength of the applied current. The electrode layers are anode and cathode strips, and the emitting pixels (picture elements) are at the intersections of these strips. The application of different current levels to different pixels determines which pixels are on or off for display or video. The emitting layers for OLEDs can be produced in large and flexible sheets, suggesting applications such as foldable electronic "newspapers" that can be updated minute by minute.

Transistors, consisting basically of two adjacent *pn* junctions (*pnp* or *npn*), are the most important application of semiconductor junctions, and their operation may be understood within the electron–hole framework we have used to discuss LEDs.

2.10 SUMMARY

In this chapter we have introduced some aspects of atoms, molecules, and solids that will be important for the remaining chapters. The most important aspect is the restriction of internal energies to a fixed set of allowed values.

We discussed the Bohr model of the hydrogen atom for three reasons. First, it was the first view of an atom that incorporated any quantum mechanical features (the postulated discrete values of orbital angular momentum). Second, it is still the model that most scientists and engineers use to *think* about atoms, although the mathematical machinery of quantum mechanics is needed to *calculate* about atoms. And third, its main results were correct. That is, it gives the right expression (2.2.11) for the energy levels of hydrogen, and Bohr's interpretation of atomic spectral lines on the basis of electron jumps between these levels was the key insight showing that atoms are not classical objects.

Obviously hydrogen is atypical in many ways. It is still the only element for which the exact values of the allowed energies can be written explicitly. Nevertheless, the results of the quantum theory of the hydrogen atom are useful in understanding the structure of other atoms. We have seen, for instance, how these results, together with the Pauli principle and the independent-particle approximation, explain the chemical regularities in the periodic table of the elements.

In much of laser physics it is sufficient to regard any atom as simply a "black box" for electrons, with the special property that the electrons inside can only be in certain energy slots. We can adopt a similar view for the electronic structure of molecules.

Of course, we cannot ignore the fact that molecules can also vibrate and rotate. However, the most important vibrational and rotational characteristics of molecules are, for us, very similar to their electronic characteristics. First and foremost among these similarities, of course, is the restriction of the vibrational and rotational energies to a fixed set of allowed values. Just as an electron can jump to a higher (or lower) energy level with the absorption (or emission) of a photon, so too can the molecule as a whole "jump" to a different vibrational or rotational state with the simultaneous absorption (or emission) of a photon. In fact, the electronic, vibrational, and rotational states of a molecule can *all* change as the molecule absorbs or emits a photon. Molecular spectra are more complicated than atomic spectra, but this simply means that the black box we call a molecule is more complicated on the inside than an atom.

The properties of solids are determined to a large extent by the outermost occupied electron orbitals of its constituent atoms or molecules. In crystalline solids the allowed electron energies are spread into energy bands as a consequence of the tight packing and periodic arrangement of atoms in a crystal lattice. The concept of energy bands provides a satisfactory interpretation of insulators, conductors, and semiconductors.

Semiconductor junctions are an especially interesting and important application of the quantum mechanics (band theory) of solids. In particular, the concept of a missing electron, or hole, as a sort of particle in its own right, greatly facilitates our understanding of semiconductor junctions. The basic *pn* junction acts as a diode, passing a current when it is forward biased but not when it is reverse biased. Light-emitting diodes are important not only in lighting and alphanumeric displays but also as the gain media of diode lasers.

The existence of atomic and subatomic particles as the basic building blocks of matter in all its forms is arguably the most basic and significant discovery of post-Newtonian science. A strong argument can also be made that the most far-reaching technological developments since the mid-20th century have involved the controlled manipulation of quantum states of these particles. Among these developments are nuclear power sources and transistor-based computer technology. The laser is another example. In this case populations of excited atomic and molecular states are created and controlled to generate light.

APPENDIX: ENERGY BANDS IN SOLIDS

In Section 2.6 we used the quantum mechanical result that the allowed electron energies in crystalline solids occur in bands, with forbidden energy gaps between these bands. We will now use a simple one-dimensional model of a solid to show how this band structure arises in quantum mechanics. This will also serve as an example of a full solution to the one-dimensional Schrödinger equation

$$\frac{d^2\psi}{dx^2} + \frac{2m}{\hbar^2}[E - V(x)]\psi = 0. \tag{2.A.1}$$

For the case of an electron in a periodic potential,

$$V(x + d) = V(x), \qquad (2.A.2)$$

where the distance d is the lattice spacing in our one-dimensional model. Note that (2.A.2) implies that $V(x + nd) = V(x)$, where n is any integer, so in our model the solid is infinitely long.

If $V(x)$ were identically zero, the reader could easily show by substitution that the solution of (2.A.1) is the free-particle plane-wave

$$\psi(x) = ue^{ikx}, \qquad (2.A.3)$$

with u some constant (complex) amplitude and k such that

$$E = \frac{\hbar^2 k^2}{2m}. \qquad (2.A.4)$$

For a potential that is not identically zero, and that satisfies (2.A.2), it is natural to try to satisfy the Schrödinger equation (2.A.1) with a wave function of the form

$$\psi(x) = u(x)e^{ikx}, \qquad (2.A.5a)$$

with k a real number and $u(x)$ now not a constant, but a function with the periodicity of the potential:

$$u(x + d) = u(x). \qquad (2.A.5b)$$

Indeed, it may be shown that a solution of the Schrödinger equation, with a potential satisfying (2.A.2), must be of the form (2.A.5). This statement is Floquet's theorem, and in solid-state physics it is called Bloch's theorem. We will use it in our treatment of the one-dimensional solid.

Different models of a one-dimensional solid are characterized by different choices of the potential $V(x)$ satisfying (2.A.2), but the most important results are insensitive to the specific $V(x)$ chosen. A particularly simple choice is the series of "square wells" shown in Fig. 2.31. This will serve as a crude idealization of the sort of potential encountered by an electron in a crystal lattice, each square well representing the effect of an atom at a lattice site. The lattice spacing in Fig. 2.31 is $a + b$, with a the width of each potential well.

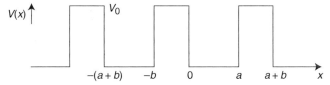

Figure 2.31   A model of the potential $V(x)$ encountered by an electron in a one-dimensional crystal lattice.

We consider first the solution in the "unit cell" $0 < x < a + b$. For $0 < x < a$, $V(x) = 0$, and

$$\psi(x) = Ae^{i\alpha x} + Be^{-i\alpha x}, \qquad (2.A.6)$$

where A and B are constants and

$$\alpha = \left(\frac{2mE}{\hbar^2}\right)^{1/2}. \qquad (2.A.7)$$

This solution is just a sum of free-particle, plane-wave solutions, one wave propagating to the right and the other to the left. This is the most general possible solution of (2.A.1) with $V = 0$. Similarly, for $a < x < a + b$, where $V(x) = V_0$, the general solution of (2.A.1) is[8]

$$\psi(x) = Ce^{\beta x} + De^{-\beta x}, \qquad (2.A.8)$$

where C and D are constants and

$$\beta = \left[\frac{2m}{\hbar^2}(V_0 - E)\right]^{1/2}. \qquad (2.A.9)$$

According to Bloch's theorem the wave function must have the form (2.A.5). We therefore use the solutions (2.A.6) and (2.A.8) to identify the function $u = \psi e^{-ikx}$:

$$u(x) = Ae^{i(\alpha-k)x} + Be^{-i(\alpha+k)x}, \qquad 0 < x < a, \qquad (2.A.10a)$$

$$u(x) = Ce^{(\beta-ik)x} + De^{-(\beta+ik)x}, \qquad a < x < a + b. \qquad (2.A.10b)$$

The wave equation (2.A.1) is a second-order differential equation. As such it demands that both ψ and $d\psi/dx$ be continuous functions of x because a function that is differentiable must be continuous. This means that u and du/dx must also be continuous functions of x. In particular, continuity of u and du/dx at $x = 0$ requires the following relations among A, B, C, and D in (2.A.10):

$$A + B = C + D, \qquad (2.A.11a)$$

$$i(\alpha - k)A - i(\alpha + k)B = (\beta - ik)C - (\beta + ik)D. \qquad (2.A.11b)$$

Now $u(x)$ must, according to Bloch's theorem, have the periodicity of the potential. Thus, u and du/dx must have the same values at $x = a$ as at $x = -b$. This condition of periodicity requires that

$$Ae^{i(\alpha-k)a} + Be^{-i(\alpha+k)a} = Ce^{-(\beta-ik)b} + De^{(\beta+ik)b}, \qquad (2.A.11c)$$

$$i(\alpha-k)Ae^{i(\alpha-k)a} - i(\alpha+k)Be^{-i(\alpha+k)a} = (\beta-ik)Ce^{-(\beta-ik)b} - (\beta+ik)De^{(\beta+ik)b}. \qquad (2.A.11d)$$

---

[8]Since we will be interested in the case $V_0 > E$, in which β is a real number, we have written (2.A.8) in terms of real exponentials. If $V_0 < E$, then β is purely imaginary and (2.A.8) is a sum of two plane waves, as in (2.A.6) for the case $V_0 = 0$.

The conditions (2.A.11) are four linear, homogeneous, algebraic equations for the four "unknowns" A, B, C, D. A trivial, uninteresting solution is $A = B = C = D = 0$. In order for a nontrivial solution to exist, the 4×4 determinant of the coefficients must vanish. After some algebra we find that this condition for a nontrivial solution takes the form

$$\frac{\beta^2 - \alpha^2}{2\alpha\beta} \sinh \beta b \sin \alpha a + \cosh \beta b \cos \alpha a = \cos k(a + b). \qquad (2.A.12)$$

Since a and b are fixed in our model of a one-dimensional solid, this equation imposes a relation among α, β, and k. As we now show, this relation gives rise to allowed energy bands separated by forbidden energy gaps.

It is convenient to consider a special case of (2.A.12) in which V_0 is very large and b is very small (Fig. 2.32). Specifically, we take $V_0 \to \infty$ and $b \to 0$ in such a way that $V_0 b$ remains a finite number. Since $\beta^2 \sim V_0$ for large V_0, this limit is such that $\beta^2 b$ has a finite limit as $\beta \to \infty$ and $b \to 0$. For convenience we denote this limiting value $2P/a$, which is the same as defining

$$P = \lim_{\beta \to \infty} \lim_{b \to 0} \left(\tfrac{1}{2} \beta^2 ab \right). \qquad (2.A.13)$$

Since $\beta b = (1/\beta)(\beta^2 b)$, it follows that $\beta b \to 0$ in this limit. Thus,

$$\lim_{\beta \to \infty} \lim_{b \to 0} \cosh \beta b = \lim_{x \to 0} \cosh x = 1, \qquad (2.A.14a)$$

and similarly

$$\lim_{\beta \to \infty} \lim_{b \to 0} \frac{\beta^2 - \alpha^2}{2\alpha\beta} \sinh \beta b = \frac{1}{\alpha a} \lim_{\beta \to \infty} \lim_{b \to 0} \frac{\beta^2 ab \sinh \beta b}{2} = \frac{1}{\alpha a} \lim_{\beta \to \infty} \lim_{b \to 0} \frac{\beta^2 ab}{2} = \frac{P}{\alpha a}, \qquad (2.A.14b)$$

since

$$\lim_{x \to 0} \frac{\sinh x}{x} = 1. \qquad (2.A.15)$$

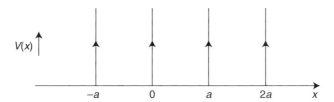

Figure 2.32 The Kronig–Penney model for the potential energy $V(x)$ for an electron in a one-dimensional crystal lattice. This is the limit of the potential of Fig. 2.31 for the case in which V_0 is very large and b is very small, such that $V_0 b$ is a finite number. In this limit the lattice spacing is a.

This limit of $\beta^2 \to \infty$ and $b \to 0$, such that $\beta^2 b$ stays finite, is called the *Kronig–Penney model*. It is useful as a simplification of (2.A.12). With (2.A.14), the condition (2.A.12) in this limit reduces to

$$P\frac{\sin \alpha a}{\alpha a} + \cos \alpha a = \cos ka. \tag{2.A.16}$$

If P is very small, the first term on the left may be neglected, and (2.A.16) becomes $\cos \alpha a = \cos ka$, or $\alpha = k$ (except possibly for a trivial shift of $2\pi/a$). Using the definition (2.A.7) of α, we see that $\alpha = k$ gives the free-particle $E-k$ relation (2.A.4).

If P is very large, on the other hand, then (2.A.16) can only make sense when $(\sin \alpha a)/\alpha a$ is very small. In the limit $P \to \infty$, then, we must have

$$\alpha a = n\pi, \qquad n = \pm 1, \pm 2, \pm 3, \ldots . \tag{2.A.17}$$

From the expression (2.A.7) for α, this condition is seen to restrict the electron energy to one of the values

$$E_n = \frac{n^2 \pi^2 \hbar^2}{2ma^2}, \qquad n = 1, 2, 3, \ldots . \tag{2.A.18}$$

These allowed energies are those for an electron in a single, infinitely deep ($V_0 \to \infty$) square well of width a (Problem 2.7). They may be regarded heuristically as the allowed levels of an electron in an isolated "atom" in the present model.

From the discussion in Section 2.6 we expect these discrete energy levels to broaden into bands when the atoms are brought together to form a crystal lattice. In Fig. 2.33 we plot the left-hand side of Eq. (2.A.16) for a case in which $P = 1$ has been arbitrarily chosen. Obviously, those values of αa for which this function exceeds unity do not allow (2.A.16) to be satisfied, since $|\cos ka| \le 1$ for all (real) values of ka. Those values of α for which $|(P \sin \alpha a)/\alpha a + \cos \alpha a| > 1$ define the *forbidden* values of E via the relation (2.A.4):

$$E = \frac{\hbar^2 \alpha^2}{2m} = \frac{\pi^2 \hbar^2}{2ma^2} \left(\frac{\alpha a}{\pi}\right)^2. \tag{2.A.19}$$

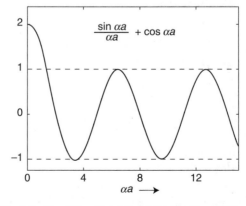

Figure 2.33   Plot of the left side of Eq. (2.A.16) for the Kronig–Penney model when $P = 1.0$.

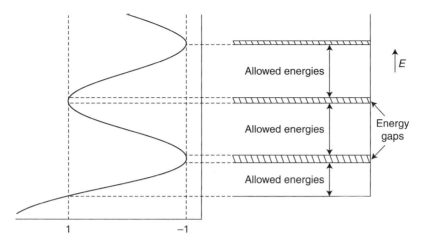

Figure 2.34   Allowed energies given by the Kronig–Penney model for $P = 1.0$. These energies are given by (2.A.19) for those values of $a\alpha$ for which Eq. (2.A.19) allows $|\cos ka| \leq 1$. The allowed energies appear in bands.

On the other hand, those values of αa for which $|P(\sin \alpha a)/\alpha a + \cos \alpha a| \leq 1$ permit (2.A.16) to be satisfied for real values of k, as required by Bloch's theorem, and the corresponding energies (2.A.19) are the *allowed* electron energies.

In Fig. 2.34 we plot the allowed energies, in units of $\pi^2 \hbar^2/2ma^2$, for the example $P = 1$ of Fig. 2.33. These are the results for a particularly simple one-dimensional model crystal. The theory for a three-dimensional crystal lattice leads similarly from the periodicity of the potential to a band structure for the allowed electron energies. This result is brought out by the one-dimensional model we have considered, and so we will not pursue a more complicated, albeit more realistic, three-dimensional model.

• Using certain physical assumptions, we can prove Bloch's theorem as follows. First, the periodicity of the potential in (2.A.1) suggests immediately that the probability distribution $|\psi(x)|^2$ for the electron should also be periodic:

$$|\psi(x + d)|^2 = |\psi(x)|^2, \tag{2.A.20}$$

which means that

$$\psi(x + d) = C\psi(x), \tag{2.A.21}$$

with

$$|C|^2 = 1. \tag{2.A.22}$$

Note that (2.A.21) implies that

$$\psi(x + nd) = C^n \psi(x). \tag{2.A.23}$$

Our one-dimensional crystal is assumed to be infinitely long; this is implicit in the assumption of the periodicity of the potential. The underlying assumption, of course, is that there are enough atoms in a real crystal to make the model of an infinite lattice a reasonable one. That is, "edge effects" in a real crystal are assumed to be very small. In this vein it is also reasonable to suppose there is some integer N, perhaps very large, such that

$$\psi(x + Nd) = C^N \psi(x) = \psi(x). \qquad (2.A.24)$$

It can be assumed that the distance Nd, after which the wave function repeats itself according to (2.A.24), is large enough that the assumption (2.A.24) will not affect any physical predictions of the model. In other words, edge effects associated with the artificial periodic boundary condition (2.A.24) do not have any real physical consequences.

Equation (2.A.24) implies that $C^N = 1$, which means that C must be one of the Nth roots of unity:

$$C = e^{2\pi iM/N}, \qquad M = 0, 1, 2, \ldots, N - 1. \qquad (2.A.25)$$

It then follows from (2.A.21) that $\psi(x)$ must have the form

$$\psi(x) = e^{2\pi iMx/Nd} u(x) = e^{ikx} u(x), \qquad (2.A.26a)$$

with $k = 2\pi M/Nd$ and

$$u(x + d) = u(x). \qquad (2.A.26b)$$

That is, Eqs. (2.A.21) and (2.A.25) are satisfied when $\psi(x)$ has the form (2.A.26). Bloch's theorem is easily extended to the case of a three-dimensional lattice.   •

It is instructive to plot E vs. k, as shown in Fig. 2.35 for the example $P = 1$. We also show for comparison the free-particle $E-k$ relation (2.A.4). In the $E-k$ curve the energy gaps occur at those values of k for which the right-hand side of (2.A.16) is $+1$, that is, for

$$k = \frac{n\pi}{a}, \qquad n = \pm 1, \pm 2, \pm 3, \ldots. \qquad (2.A.27)$$

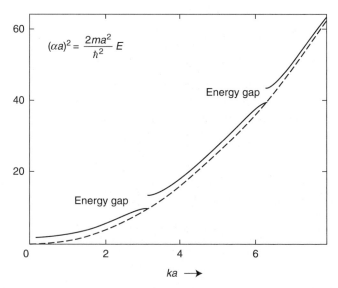

Figure 2.35   Plot of E vs. k for the Kronig–Penney model with $P = 1.0$.

This may be understood physically as follows. The wave function

$$\Psi(x, t) = \psi(x)e^{-iEt/\hbar} = u(x)e^{i(kx-Et/\hbar)}, \tag{2.A.28}$$

for an electron propagating down the lattice has wavelength $\lambda = 2\pi/|k|$ associated with the plane-wave factor

$$e^{i(kx-\omega t)} = e^{i(px-Et/\hbar)}. \tag{2.A.29}$$

If $2a = n\lambda$, where $n = 1, 2, 3, \ldots$, the spacing between the potential barriers in Fig. 2.35 is an integral number of half wavelengths. This means that the waves reflected from these barriers are all in phase and interfere constructively. In other words, when

$$|k| = \frac{2\pi}{\lambda} = \frac{n\pi}{a}, \qquad n = 1, 2, 3, \ldots, \tag{2.A.30}$$

the wave (2.A.28) is strongly reflected and forbidden from propagating unhindered down the lattice. This is why the energies associated with the k values (2.A.30), or equivalently (2.A.27), are forbidden.

• In Section 2.7 we invoked the concept of an *effective mass*, m^*, of an electron in a crystal lattice. To see how this concept arises, suppose that a force F acts on the electron. The rate of change of the electron energy as a result of this force is

$$\frac{dE}{dt} = Fv, \tag{2.A.31}$$

where v is the electron velocity. Now the force equals the rate of change of the momentum $p = \hbar k$:

$$F = \frac{dp}{dt} = \hbar \frac{dk}{dt}. \tag{2.A.32}$$

Thus

$$\frac{dE}{dt} = \hbar v \frac{dk}{dt}, \tag{2.A.33}$$

or

$$v = \frac{1}{\hbar} \frac{dE/dt}{dk/dt} = \frac{1}{\hbar} \frac{dE}{dk}, \tag{2.A.34}$$

which is the clear quantum analog of the well-known expression for the group velocity of a light pulse, $v_g = (dk/d\omega)^{-1}$ (Section 8.3). The acceleration of the electron is therefore

$$a = \frac{dv}{dt} = \frac{1}{\hbar} \frac{d}{dt}\left(\frac{dE}{dk}\right)$$

$$= \frac{1}{\hbar} \frac{dk}{dt} \frac{d}{dk}\left(\frac{dE}{dk}\right) = \frac{F}{\hbar^2} \frac{d^2E}{dk^2}. \tag{2.A.35}$$

This equation has the form $F = m^*a$, which defines the effective mass as

$$m^* = \hbar^2 \left(\frac{d^2E}{dk^2}\right)^{-1}. \tag{2.A.36}$$

The physical basis for effective mass can be understood as follows. The total force acting on the electron is the external force F_{ext} plus the force F_{crys} due to the atoms of the crystal lattice. This total force equals ma. In our derivation above, however, the force F is only F_{ext}; F_{crys} is accounted for only indirectly via the $E-k$ relation for the electron in the crystal. Thus, m^* arises from the proportionality of the electron acceleration to the *external* force.   •

PROBLEMS

2.1. **(a)** Equation (2.2.18) with $n' = 2$ and $n = 3, 4, 5, \ldots$ gives the *Balmer series* of the hydrogen spectrum. In what region of the electromagnetic spectrum (e.g., infrared, visible, ultraviolet) are the wavelengths of the Balmer series?

(b) Equation (2.2.18) with $n' = 1$ and $n = 2, 3, 4, \ldots$ gives the *Lyman series* of hydrogen. In what region of the spectrum are the wavelengths of the Lyman series?

(c) Equation (2.2.18) with $n' = 3$ and $n = 4, 5, 6, \ldots$ gives the *Paschen series* of hydrogen. In what region of the spectrum are these wavelengths?

2.2. Verify Eq. (2.3.6).

2.3. Given the fact that the molecular weight of water is 18, estimate the average distance between two water molecules in ice.

2.4. Show that the magnetic force acting on the charge carriers in the Hall-effect experiment of Fig. 2.20 is upward, regardless of whether the charges are positive or negative.

2.5. Assuming for GaAs a dielectric constant $\epsilon = 13.0\epsilon_0$, and an effective mass $m^* = 0.07m$, estimate the energy required to ionize donor impurities.

2.6. **(a)** Consider emission into air of an LED employing a *pn* junction *without* the dome indicated in Fig. 2.29. Show that total internal reflection at the interface with air reduces the emission efficiency by the factor $1 - \cos\theta_c$, where θ_c is the critical angle for total internal reflection. (Note: The solid angle of a cone with apex angle θ is $2\pi[1 - \cos(\theta/2)]$.)

(b) Show that the reduction factor calculated in part (a) is approximately equal to $1 - \sqrt{1 - 1/n^2}$, where n is the refractive index of the LED, and estimate this factor for GaAs. What is the expected increase in efficiency for a GaAs LED when it is designed with the plastic dome as in Fig. 2.29?

2.7. Consider a particle of mass m in an infinitely deep, one-dimensional square well of width a. Between the walls the particle is free ($V = 0$), but because it cannot penetrate the walls the wave function must vanish at $x = 0$ and $x = a$ and for all x outside those limits. Using the Schrödinger equation (2.A.1), show that the normalized stationary-state wave functions are given by

$$\psi_n(x) = \left(\frac{2}{a}\right)^{1/2} \sin\left(\frac{n\pi x}{a}\right), \qquad n = 1, 2, 3, \ldots,$$

with corresponding allowed energies

$$E_n = \frac{n^2 \pi^2 \hbar^2}{2ma^2}.$$

2.8. The binding energy of the ion H_2^+ (the energy required to separate to infinity the two protons and the electron) is -16.3 eV at the equilibrium separation 0.106 nm.

(a) What is the contribution to the energy from the Coulomb repulsion of the nuclei?

(b) What is the contribution to the energy from the Coulomb attraction of the electron to the nuclei?

(c) The *Hellman–Feynman theorem* says, in effect, that the force between the nuclei in a molecule can be calculated from the electrostatic repulsion between the nuclei and the electrostatic attraction of the nuclei to the electron distribution. According to this theorem, where must the squared modulus of the electron wave function in H_2^+ have its maximum value?

(d) Estimate the rotational constant B_e for H_2^+, and compare your result with the value 29.8 cm^{-1} tabulated in Herzberg's *Spectra of Diatomic Molecules*.[4]

3 ABSORPTION, EMISSION, AND DISPERSION OF LIGHT

3.1 INTRODUCTION

Most objects around us are not self-luminous but are neverthess visible because they scatter the light that falls upon them. Most objects are *colored*, however, because they absorb light, not simply because they scatter it. The colors of an object typically arise because materials selectively absorb light of certain frequencies, while freely scattering or transmitting light of other frequencies. Thus, if an object absorbs light of all visible frequencies, it is black. An object is red if it absorbs all (visible) frequencies except those our eyes perceive to be "red" (wavelengths roughly between about 630 and 680 nm), and so on.[1]

The physics of the absorption process is simplest in well-isolated atoms. These are found most commonly in gases. White light propagating through a gas is absorbed at the resonance frequencies of the atoms or molecules, so that one observes gaps in the wavelength distribution of the emerging light. On a spectrogram these gaps appear as bright lines on the dark, exposed background. The gaps, shown as lines in Fig. 3.1, correspond to the absorption of sunlight by the atmosphere *of the sun* before the light reaches Earth. The absorbed energy is partially converted into heat (translational kinetic energy of the atoms) when excited atoms (or molecules) that have absorbed radiation collide with other particles. The absorbed radiation is also partially reradiated in all directions at the frequency of the absorbed radiation. This is called resonance radiation, or resonance fluorescence. When the pressure of the gas is increased, collisions may rapidly convert the absorbed radiation into heat before it can be reradiated. In this case the resonance radiation is said to be quenched.

Most atoms have electronic resonance frequencies in the ultraviolet, although resonances in the visible and infrared are not uncommon. Sodium, for instance, has strong absorption lines in the yellow region at 589.0 and 589.6 nm, the Fraunhofer "D lines," and their position is indicated in Fig. 3.1.

Electronic resonances in molecules also tend to lie in the ultraviolet. We have "white" daylight because the atmosphere, consisting mostly of N_2 and O_2, does not absorb strongly at visible frequencies. As discussed in Chapter 2, the atoms of a molecule

---

[1]The principal features of the electromagnetic spectrum for our purposes are summarized inside the front cover of the book.

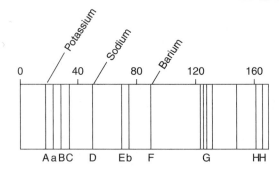

Figure 3.1   Absorption lines of the sun's atmosphere. The Fraunhofer D lines of sodium at 589.0 and 589.6 nm are not resolved in this sketch.

can vibrate back and forth, and the molecule as a whole can rotate. The vibrational and rotational resonances of molecules typically correspond to infrared and microwave frequencies, respectively, so that molecules typically have absorption resonances in the ultraviolet, infrared, and microwave regions of the spectrum.

Absorption in liquids and solids is much more complicated than in gases. In liquids and amorphous solids such as glass, the absorption lines have such large widths that they overlap. Water, for example, is obviously transparent in the visible but absorbs in the near infrared, that is, at infrared wavelengths not far removed from the visible. Its absorption curve is wide enough, in fact, that it extends into the red edge of the visible (Fig. 3.2). The weak absorption in the red portion of the visible spectrum explains why things appear green when one is sufficiently submerged under water.

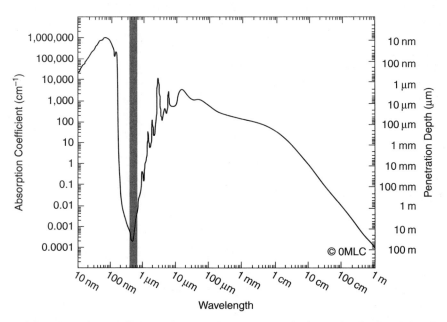

Figure 3.2   Absorption coefficient of water. (From D. Segelstein, M.S. Thesis, University of Missouri-Kansas City, 1981, as reproduced on the website of the Oregon Medical Laser Center (2007), http://omlc.ogi.edu/)

A broad absorption curve covering all visible wavelengths except those in a particular narrow band is characteristic of the molecules of a dye. The absorbed radiation is converted into heat before it can be reradiated. Such broad absorption curves and fast quenching rates require the high molecular number densities of liquids and solids.

The fact that metals contain approximately free electrons, as discussed in Chapter 2, makes them good reflectors of electromagnetic radiation of frequency less than the so-called *plasma frequency*, which for metals is usually in the ultraviolet. Thus, visible radiation is completely reflected by a metal, just as AM radio waves are reflected by the ionosphere. This strong reflection gives metals their shine. In a metal such as gold there is also absorption, associated with the electrons that remain bound to atoms, and it is this that gives the metal a characteristic color.

An insulator is usually transparent in the visible but opaque in the ultraviolet because its electrons are tightly bound and consequently give rise to absorption only at high frequencies, typically corresponding to wavelengths less than 400 nm. In semiconductors the absorption frequencies are smaller. Silicon, for example, absorbs visible wavelengths (it is black), but transmits radiation of wavelength greater than about 1 micron (1 μm).

Lattice defects (deviations from periodicity) can substantially modify the absorption spectra of crystalline solids. Ruby, for instance, is corundum (Al_2O_3) with an occasional (roughly 0.05% by weight) random substitution of Cr^{3+} ions in place of Al^{3+}. The chromium ions absorb green light and thus ruby is pink, in contrast to the transparency of pure corundum.

The variety of phenomena resulting from the selective absorption of certain wavelengths and the transmission of others is too broad to treat here. We mention only one important example, the "greenhouse effect."[2] Visible sunlight is transmitted by Earth's atmosphere and heats (by absorption) both land and water. The warmed Earth's surface is a source of thermal radiation, the dominant emission for ambient temperatures being in the infrared (Problem 3.1). This infrared radiation, however, is strongly absorbed by CO_2 and H_2O vapor in the atmosphere, preventing rapid escape into space. Without this effect Earth would be a much colder place. An increased burning of fossil fuels could conceivably enhance the greenhouse effect by increasing the level of CO_2 and other "greenhouse gases" in the atmosphere.

3.2 ELECTRON OSCILLATOR MODEL

In classical physics the motion of a particle is described by Newton's second law, $\mathbf{F} = m\mathbf{a}$. For a charged particle in an electromagnetic field $\mathbf{F}$ is the Lorentz force,

$$\mathbf{F} = e(\mathbf{E} + \mathbf{v} \times \mathbf{B}), \tag{3.2.1}$$

where e and $\mathbf{v}$ are the charge and velocity, respectively, of the particle.

---

[2]The term *greenhouse effect* is actually a misnomer, originating in the observation that the glass in a greenhouse, which is transparent in the visible but opaque to the infrared, plays an absorptive role similar to that of CO_2 and H_2O in Earth's atmosphere. This effect, however, does not contribute significantly to the warming of the air inside a greenhouse. A real greenhouse mainly prevents cooling by air currents. Although this point was demonstrated experimentally by R. W. Wood (1909), the contrary misperception persists even among scientists.

We assume that (3.2.1) applies to the individual protons and electrons in atoms. Although these particles and their interactions can be properly treated only using quantum theory, their interaction with light can be treated very accurately in most cases with classical laws and concepts. The quantum theoretical basis for our classical treatment is discussed in the Appendix to this chapter.

The electron has mass m_e and charge e (a negative number), and the oppositely charged core of the atom ("nucleus") has mass m_n and charge $-e$. The nucleus exerts a binding force $\mathbf{F}_{en}$ on the electron, depending on the relative separation $\mathbf{r}_{en} = \mathbf{r}_e - \mathbf{r}_n$, as shown in Fig. 3.3. The electron also exerts a force $\mathbf{F}_{ne}$ on the nucleus, and according to Newton's third law,

$$\mathbf{F}_{ne}(\mathbf{r}_{en}) = -\mathbf{F}_{en}(\mathbf{r}_{en}). \tag{3.2.2}$$

The Newton equations of motion for the electron and nucleus are therefore

$$m_e \frac{d^2 \mathbf{r}_e}{dt^2} = e\mathbf{E}(\mathbf{r}_e, t) + \mathbf{F}_{en}(\mathbf{r}_{en}), \tag{3.2.3a}$$

$$m_n \frac{d^2 \mathbf{r}_n}{dt^2} = -e\mathbf{E}(\mathbf{r}_n, t) + \mathbf{F}_{ne}(\mathbf{r}_{en}). \tag{3.2.3b}$$

In writing these equations we have dropped the magnetic contributions to the Lorentz force because optical phenomena do not normally involve relativistic particle velocities. We can safely disregard the magnetic force for our purposes here (Problem 3.2).

The interaction of electromagnetic fields with charges is mainly determined by the acceleration of the charges. The nucleus is so massive compared to an electron that its acceleration is generally negligible. In this case only the electron equation is needed. The binding force $\mathbf{F}_{en}$ is strong enough to restrict the atomic electrons to small excursions about the (approximately stationary) nucleus. Thus we can write $\mathbf{r}_e = \mathbf{r}_n + \mathbf{x}$, where $\mathbf{x}$ is a displacement of atomic dimension in size ($|\mathbf{x}| \lesssim 1\,\text{nm}$). The electric field varies spatially on the scale of an optical wavelength ($\lambda \approx 600\,\text{nm}$ for yellow light) and is not sensitive to variations as small as $|\mathbf{x}|$, so we have $\mathbf{E}(\mathbf{r}_e, t) \approx \mathbf{E}(\mathbf{r}_n, t)$.

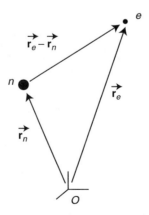

Figure 3.3   The position vectors $\mathbf{r}_e$ and $\mathbf{r}_n$ of the electron and nucleus, measured from some origin O. By Newton's third law the force $\mathbf{F}_{en}(\mathbf{r}_e - \mathbf{r}_n)$ exerted by the nucleus on the electron is equal in magnitude but opposite in direction to the force $\mathbf{F}_{ne}(\mathbf{r}_e - \mathbf{r}_n)$ exerted by the electron on the nucleus.

Within these approximations we can replace Eqs. (3.2.3) by

$$m \frac{d^2\mathbf{x}}{dt^2} = e\mathbf{E}(\mathbf{R}, t) + \mathbf{F}_{en}(\mathbf{x}), \tag{3.2.4}$$

as shown below. Here we have dropped the subscript e from the electron mass and have written $\mathbf{R}$ for the position of the stationary nucleus. Actually, $\mathbf{x}$ and $\mathbf{R}$ are the relative coordinate and center-of-mass coordinate of the electron–nucleus pair, and m is the associated reduced mass. These terms are defined in the black-dot section below. For our purposes it is accurate to continue to think of $\mathbf{R}$ as the position of the nucleus and m as the electron mass. Only in exceptional cases in which the two charges have nearly equal mass, such as the positronium atom (an atom in which the nucleus is a positron, i.e., an antielectron, rather than a proton), would significant corrections be required.

- The position of the center of mass of the electron–nucleus system is defined to be

$$\mathbf{R} = \frac{m_e \mathbf{r}_e + m_n \mathbf{r}_n}{M}, \tag{3.2.5}$$

where $M = m_e + m_n$ and $\mathbf{x}$ is the electron coordinate relative to the nucleus, $\mathbf{x} = \mathbf{r}_{en}$, in terms of which

$$\mathbf{r}_e = \mathbf{R} + \frac{m_n}{M}\mathbf{x}, \tag{3.2.6a}$$

$$\mathbf{r}_n = \mathbf{R} - \frac{m_e}{M}\mathbf{x}. \tag{3.2.6b}$$

Then Eqs. (3.2.3) may be written as

$$m_e \frac{d^2\mathbf{R}}{dt^2} + m \frac{d^2\mathbf{x}}{dt^2} = e\mathbf{E}\left(\mathbf{R} + \frac{m_n}{M}\mathbf{x}, t\right) + \mathbf{F}_{en}(\mathbf{x}), \tag{3.2.7a}$$

$$m_n \frac{d^2\mathbf{R}}{dt^2} - m \frac{d^2\mathbf{x}}{dt^2} = -e\mathbf{E}\left(\mathbf{R} - \frac{m_e}{M}\mathbf{x}, t\right) + \mathbf{F}_{ne}(\mathbf{x}), \tag{3.2.7b}$$

where

$$m = \frac{m_e m_n}{M} = \frac{m_e m_n}{m_e + m_n} \tag{3.2.8}$$

is the reduced mass of the electron–nucleus system.

By adding and subtracting Eqs. (3.2.7a) and (3.2.7b), and using (3.2.2), we obtain the equations of motion

$$M \frac{d^2\mathbf{R}}{dt^2} = e\left[\mathbf{E}\left(\mathbf{R} + \frac{m_n}{M}\mathbf{x}, t\right) - \mathbf{E}\left(\mathbf{R} - \frac{m_e}{M}\mathbf{x}, t\right)\right] \tag{3.2.9a}$$

and

$$m \frac{d^2\mathbf{x}}{dt^2} = \frac{e}{2}\left[\mathbf{E}\left(\mathbf{R} + \frac{m_n}{M}\mathbf{x}, t\right) + \mathbf{E}\left(\mathbf{R} - \frac{m_e}{M}\mathbf{x}, t\right)\right] + \mathbf{F}_{en}(\mathbf{x}) + \frac{1}{2}(m_n - m_e)\frac{d^2\mathbf{R}}{dt^2}. \tag{3.2.9b}$$

Equation (3.2.9a) describes the motion of the center of mass of the atom. In the absence of an external field **E**, the center of mass moves with constant velocity. Equation (3.2.9b) describes the motion of the relative coordinate **x** of the electron–nucleus system.

We have already remarked that optical radiation is characterized by wavelengths that are a few hundred nanometers or larger, and the electron–nucleus separations in atoms are typically only 0.1–1 nm in size. The extreme disparity of these sizes is the basis of a fundamental approximation called the *dipole approximation*. The dipole approximation arises from the leading terms of a Taylor series expansion of the type

$$F(X + \delta X) = F(X) + \delta X F'(X) + \tfrac{1}{2}(\delta X)^2 F''(X) + \cdots \tag{3.2.10}$$

applied to the electric field vectors in (3.2.9a) and (3.2.9b). The vector analog of the Taylor series is

$$\mathbf{E}\left(\mathbf{R} - \frac{m_e}{M}\mathbf{x}, t\right) = \mathbf{E}(\mathbf{R}, t) - \frac{m_e}{M}\mathbf{x} \cdot \nabla_\mathbf{R}\mathbf{E}(\mathbf{R}, t) + \cdots \tag{3.2.11a}$$

and

$$\mathbf{E}\left(\mathbf{R} + \frac{m_n}{M}\mathbf{x}, t\right) = \mathbf{E}(\mathbf{R}, t) + \frac{m_n}{M}\mathbf{x} \cdot \nabla_\mathbf{R}\mathbf{E}(\mathbf{R}, t) + \cdots, \tag{3.2.11b}$$

where $\nabla_\mathbf{R}$ is the gradient operation with respect to the coordinate **R**. If we retain only the first two terms in these Taylor series, Eqs. (3.2.9) become

$$M\frac{d^2\mathbf{R}}{dt^2} \approx e\mathbf{x} \cdot \nabla_\mathbf{R}\mathbf{E}(\mathbf{R}, t) \tag{3.2.12a}$$

$$m\frac{d^2\mathbf{x}}{dt^2} \approx e\mathbf{E}(\mathbf{R}, t) + \left(\frac{m_n - m_e}{M}\right)e\mathbf{x} \cdot \nabla_\mathbf{R}\mathbf{E}(\mathbf{R}, t) + \mathbf{F}_{en}(\mathbf{x}). \tag{3.2.12b}$$

The vector

$$\mathbf{d} = e\mathbf{x} \tag{3.2.13}$$

is the electric dipole moment of the electron–nucleus pair. In terms of **d** Eq. (3.2.12a) is

$$M\frac{d^2\mathbf{R}}{dt^2} = \mathbf{d} \cdot \nabla_\mathbf{R}\mathbf{E}(\mathbf{R}, t). \tag{3.2.14}$$

A more complete expression for the force on an electric dipole is given in Section 14.4.

Finally, we retain only the leading **E** term on the right-hand side of (3.2.12b) and obtain

$$m\frac{d^2\mathbf{x}}{dt^2} \approx e\mathbf{E}(\mathbf{R}, t) + \mathbf{F}_{en}(\mathbf{x}) \quad \text{(electric-dipole approximation)}, \tag{3.2.15}$$

which is Eq. (3.2.4) again, this time with m, **x**, and **R** more carefully defined.

For most of our purposes we can assume that the center-of-mass motion of the atom is unaffected by the field, so that we can ignore (3.2.14). However, this is possible only because we are interested mainly in effects associated with laser action, which depends mostly on internal transitions within atoms or molecules, transitions based on the relative coordinate **x**. For other purposes Eq. (3.2.12a) is essential. For example, the important topics of laser trapping and laser cooling (Section 14.4) depend directly on the effects produced by laser light on the atomic center of mass.

•

Note that with our approximations the force due to the electric field $\mathbf{E}(\mathbf{R}, t)$ in (3.2.4) can be written in terms of a potential

$$V(\mathbf{x}, \mathbf{R}, t) = -e\mathbf{x} \cdot \mathbf{E}(\mathbf{R}, t) \tag{3.2.16}$$

such that

$$e\mathbf{E}(\mathbf{R}, t) = -\nabla_{\mathbf{x}}[-e\mathbf{x} \cdot \mathbf{E}(\mathbf{R}, t)] = -\nabla_{\mathbf{x}}V(\mathbf{x}, \mathbf{R}, t), \tag{3.2.17}$$

where $\nabla_{\mathbf{x}}$ denotes the gradient with respect to the coordinate $\mathbf{x}$.

To proceed with (3.2.4) it is necessary to know $\mathbf{F}_{en}(\mathbf{x})$. For reasons that only quantum theory can explain (see the Appendix), the classical theory satisfactorily treats many important features of the interaction of light with matter by adopting an ad hoc hypothesis about $\mathbf{F}_{en}$ due to H. A. Lorentz (around 1900). This hypothesis states that an electron in an atom responds to light as if it were bound to its atom or molecule by a simple spring. As a consequence the electron can be imagined to oscillate about the nucleus.

This electron oscillator model, which was developed before atoms were understood to have massive nuclei, is not really a model of an atom as such, but rather a model of the way an atom responds to a perturbation. It simply asserts that each electron in an atom has a certain equilibrium position when there are no external forces. Under the influence of an electromagnetic field, the electron experiences the Lorentz force (3.2.1) and is displaced from its equilibrium position; according to Lorentz "the displacement will immediately give rise to a new force by which the particle is pulled back towards its original position, and which we may therefore appropriately distinguish by the name of elastic force."[3] Lorentz's assertion is equivalent to the replacement $\mathbf{F}_{en}(\mathbf{x}) \rightarrow -k_s\mathbf{x}$, where k_s is the "spring constant" associated with the hypothetical elastic force. This leads to the equation

$$m\frac{d^2\mathbf{x}}{dt^2} = e\mathbf{E}(\mathbf{R}, t) - k_s\mathbf{x}, \tag{3.2.18a}$$

or

$$\left(\frac{d^2}{dt^2} + \omega_0^2\right)\mathbf{x} = \frac{e}{m}\mathbf{E}(\mathbf{R}, t), \tag{3.2.18b}$$

where we have defined the electron's natural oscillation frequency $\omega_0 = \sqrt{k_s/m}$ (see Problem 3.3).

The reader who has even a slight familiarity with the quantum theory of atomic structure might well object that this is a hopelessly crude model of an atom. However, the Lorentz model is not intended to describe an atom as such, but only *how an atom interacts with light:*

> You may think that this is a funny model of an atom if you have heard about electrons whirling around in orbits. But that is just an oversimplified picture. The correct picture of an atom, which

[3]H. A. Lorentz, *The Theory of Electrons*, Dover, New York, 1952, p. 9.

is given by [quantum mechanics], says that, *so far as problems involving light are concerned, the electrons behave as though they were held by springs.*[4]

In fact, the electron oscillator does not correctly describe *all* aspects of the interaction of light with atoms, and in particular it does not describe some of the most important features of lasers. With some appropriate modifications, however, the electron oscillator model will allow us to proceed rather quickly and easily to a realistic theory of laser operation, and to do so using mainly physical rather than mathematical aspects of quantum theory. In the Appendix we show that the electron oscillator model can be regarded as a good approximation to the quantum theory of the interaction of an atom with light.

3.3 SPONTANEOUS EMISSION

In the case that there is no applied field the electron oscillator equation is

$$\frac{d^2\mathbf{x}}{dt^2} + \omega_0^2 \mathbf{x} = 0, \tag{3.3.1}$$

and the general solution of this equation is

$$\mathbf{x}(t) = \mathbf{x}_0 \cos \omega_0 t + \frac{\mathbf{v}_0}{\omega_0} \sin \omega_0 t, \tag{3.3.2}$$

where $\mathbf{x}_0$ and $\mathbf{v}_0$ are, respectively, the initial displacement and the initial velocity of the electron. Thus, provided $\mathbf{x}_0$ and $\mathbf{v}_0$ are not both zero, there is an oscillating electric dipole moment $\mathbf{d}(t) = e\mathbf{x}(t)$. According to the Larmor formula of classical electromagnetic theory, such an oscillating dipole radiates electromagnetic energy at the rate

$$\text{Pwr} = \left(\frac{1}{4\pi\epsilon_0}\right) \frac{2\ddot{\mathbf{d}}^2}{3c^3} = -\frac{dW}{dt}, \tag{3.3.3}$$

where W denotes the oscillator energy. Therefore, from (3.3.2), our electron oscillator should lose energy at the rate

$$\frac{dW}{dt} = -\left(\frac{1}{4\pi\epsilon_0}\right) \frac{2e^2}{3c^3} \left[\omega_0^4 \mathbf{x}_0^2 \cos^2 \omega_0 t + \omega_0^3 \mathbf{x}_0 \cdot \mathbf{v}_0 \sin 2\omega_0 t + \omega_0^2 \mathbf{v}_0^2 \sin^2 \omega_0 t\right], \tag{3.3.4}$$

where we have used the identity $2 \sin \omega_0 t \cos \omega_0 t = \sin 2\omega_0 t$.

As discussed below, we will be interested in frequencies ω_0 that are large, say $10^{15}\,\text{s}^{-1}$. Such rapid oscillations are not measured; what is measured is an average over times much larger than $1/\omega_0$. Therefore, we replace $\cos^2 \omega_0 t$, $\sin^2 \omega_0 t$, and $\sin 2\omega_0 t$ in (3.3.4) by their average values over such times, namely $\frac{1}{2}, \frac{1}{2}$, and 0. This "cycle-averaged" rate at which the electron oscillator's energy changes (decreases) as

---

[4]R. P. Feynman, R. B. Leighton, and M. Sands, *The Feynman Lectures on Physics*, Addison-Wesley, Reading, MA, 1963, Vol. I, p. 31–4.

electromagnetic energy is radiated is then

$$\frac{dW}{dt} = -\left(\frac{1}{4\pi\epsilon_0}\right)\frac{e^2}{3c^3}[\omega_0^4\mathbf{x}_0^2 + \omega_0^2\mathbf{v}_0^2] = -\left(\frac{1}{4\pi\epsilon_0}\right)\frac{2e^2\omega_0^2}{3mc^3}\left[\frac{1}{2}m\mathbf{v}_0^2 + \frac{1}{2}m\omega_0^2\mathbf{x}_0^2\right]$$

$$= -\left(\frac{1}{4\pi\epsilon_0}\right)\frac{2e^2\omega_0^2}{3mc^3}E, \tag{3.3.5}$$

where we have recognized the quantity in brackets in the second equality as the oscillator energy W.

This radiation by an electron oscillator that has been "excited," i.e., given a non-vanishing $d^2\mathbf{d}/dt^2$, corresponds to the spontaneous emission of radiation by an excited atom, which was mentioned briefly in Section 1.5. Since the frequency of the field radiated by the electron oscillator is the same as the oscillator frequency $\nu_0 = \omega_0/2\pi$, we associate the electron oscillator with an atomic transition of frequency ν_0 (Fig. 3.4). Thus, for an optical transition of wavelength $\lambda_0 = 500$ nm, $\omega_0 = 2\pi c/\lambda_0 \approx 3.8 \times 10^{15}$ s^{-1} and the rate of spontaneous emission predicted by the electron oscillator model is

$$\left(\frac{1}{4\pi\epsilon_0}\right)\left(\frac{2e^2\omega_0^2}{3mc^3}\right) \approx 9 \times 10^7 \text{ s}^{-1}. \tag{3.3.6}$$

This is a reasonable estimate for spontaneous emission rates of atomic transitions at optical wavelengths. However, spontaneous emission rates are not a quadratic function of transition frequency as predicted by (3.3.6). The 2p–1s transition of hydrogen at 121.6 nm, for instance, has a spontaneous emission rate 6.26×10^8 s^{-1}, whereas for the $3s$–$2p$ transition at 656.3 nm the rate is 6.31×10^6 s^{-1} (see Table 3.2). The ratio of these two rates is $(6.26 \times 10^8/6.31 \times 10^6) \sim 100$, whereas according to (3.3.6) this ratio should be $(6563/1216)^2 \sim 30$. To bring the classical radiation rate into numerical agreement with the rate at which excited atoms jump spontaneously from an energy level E_2 to a lower energy level E_1, with $E_2 - E_1 = h\nu_0$, we multiply (3.3.6) by a factor that, to conform to a notational convention, we write as $3f$. Thus, denoting the spontaneous emission rate for the quantum jump from energy level E_2 to energy level E_1

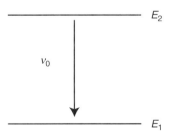

Figure 3.4 An atomic transition of frequency $\nu_0 = \omega_0/2\pi = (E_2 - E_1)/h$ and wavelength $\lambda_0 = c/\nu_0$.

by A_{21}, we write

$$A_{21} = \left(\frac{1}{4\pi\epsilon_0}\right)\frac{2e^2\omega_0^2}{mc^3}f. \tag{3.3.7}$$

It is evident that f must have different numerical values for different atomic transitions and provides a measure of the "strength" of the transition.

The factor f, called the *oscillator strength*, was introduced before the development of quantum theory in order to bring the electron oscillator model into numerical agreement with various spectroscopic data. It remains useful as a measure of the "strength" of a transition; numerical values of oscillator strengths are tabulated in various handbooks.[5] The quantum theoretical formula for the oscillator strength is derived in the Appendix.

A second modification of the classical oscillator model is required to describe spontaneous radiative transitions: We must take into account that atoms can only be found in certain states with "allowed" energies. Thus, the spontaneous emission rate A_{21} is the rate at which the number N_2 of atoms in the upper state of energy E_2 decreases and the number N_1 of atoms in the lower state of energy E_1 correspondingly increases (Fig. 3.4). The changes in the "populations" N_2 and N_1 due to spontaneous emission are described by the rate equations

$$\frac{dN_2}{dt} = -A_{21}N_2 \tag{3.3.8}$$

and

$$\frac{dN_1}{dt} = A_{21}N_2, \tag{3.3.9}$$

implying that $d(N_1 + N_2)/dt = 0$, i.e., the total number of atoms $N_1 + N_2$ in the upper and lower states of the transition stays the same.

As discussed in Section 3.6, most of the light around us is ultimately the result of spontaneous emission, and the phenomenon appears in many different contexts. The term *luminescence*, for instance, describes spontaneous emission from atoms or molecules excited by some means other than heating. If excitation occurs in an electric discharge such as a spark, the term *electroluminescence* is used. If the excited states are produced as a by-product of a chemical reaction, the emission is called *chemiluminescence*, or, if this occurs in a living organism (such as a firefly), *bioluminescence*. *Fluorescence* refers to spontaneous emission from an excited state produced by the absorption of light. *Phosphorescence* describes the situation in which the emission persists long after the exciting light is turned off and is associated with a metastable (long-lived) level, as illustrated in Fig. 3.5. Phosphorescent materials are used, for instance, in toy figurines that magically glow in the dark.

---

[5]A useful collection of atomic reference data is provided by the National Institute of Standards and Technology (NIST) and may be found on the Web as well as in a variety of published sources. See, for example, A. N. Cox, ed., *Allen's Astrophysical Quantities*, 4th ed., AIP Press, New York, 2000, or W. L. Wiese, M. W. Smith, and B. M. Glennon, *Atomic Transition Probabilities*, U.S. Government Printing Office, Washington, D.C., 1966. A more recent and readily available compendium of useful data on atomic transitions of interest has been prepared by D. A. Steck at http://steck.us/alkalidata/.

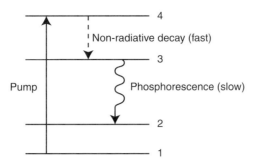

Figure 3.5 Model of phosphorescence. A molecule is pumped to level 4 by absorption of radiation, and then decays to level 3. Level 3 is metastable, i.e., it has a very small spontaneous emission rate. As a result the molecule continues to fluoresce long after the source of radiation has been shut off.

In most situations an excited level has several or many spontaneous decay channels, so that the general case is somewhat more complex than our notation A_{21} implies. For example, the solution of Eq. (3.3.8), the exponential decay law,

$$N_2(t) = N_2(0)e^{-A_{21}t}, \qquad (3.3.10)$$

indicates that the population of the upper level decays to zero with the characteristic time constant $\tau_2 = 1/A_{21}$. However, if level 2 has other decay channels open to it, they will obviously shorten the effective lifetime of level 2 and this expression for τ_2 will be incomplete.

According to quantum theory the spontaneous radiative lifetime of level n is determined by the sum of the rates for all possible radiative channels:

$$A_n = \sum_m A_{nm}, \qquad (3.3.11)$$

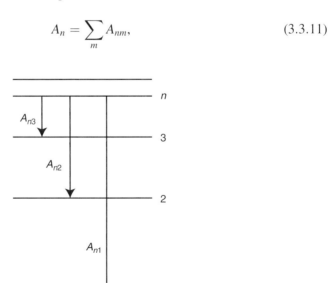

Figure 3.6 An atomic state n may make spontaneous transitions to lower states m with rates A_{nm}. The total spontaneous decay rate of state n is $A_n = \Sigma_m A_{nm}$.

and the correct expression for the upper-state lifetime is

$$\tau_n = \frac{1}{A_n} = \frac{1}{\sum_m A_{nm}}, \tag{3.3.12}$$

where the summation is over all states m with energy E_m lower than the energy level n (see Fig. 3.6). Numerical values of the "A coefficients" A_{nm} are usually included in tables of oscillator strengths. Radiative lifetimes of excited atomic states are typically on the order of 10–100 ns.

3.4   ABSORPTION

To account for the absorption of radiation in the electron oscillator model, it must be assumed that the oscillator is subject to a frictional force in addition to the applied field. Without a frictional force the electron oscillations induced by an applied field simply cause the radiation of electromagnetic energy according to the formula (3.3.3), with no *change* in the total electromagnetic energy. This corresponds to *scattering* of radiation rather than absorption. The actual origin of the friction-like force is itself a subject for discussion, which will be found in Section 3.8. For present purposes, however, we will take a frictional force for granted and explore its consequences.

We simply amend the Newton force law (3.2.18a) to read

$$m\frac{d^2\mathbf{x}}{dt^2} = e\mathbf{E}(\mathbf{R}, t) - k_s\mathbf{x} + \mathbf{F}_{\text{fric}}, \tag{3.4.1}$$

and make the simplest assumption compatible with the idea of frictional drag, namely that the frictional force is proportional to the velocity:

$$\mathbf{F}_{\text{fric}} = -b\mathbf{v} = -b\frac{d\mathbf{x}}{dt}. \tag{3.4.2}$$

Then the Newton equation of motion (3.2.18b) for an electron oscillator in a linearly polarized monochromatic plane wave takes the form

$$\frac{d^2\mathbf{x}}{dt^2} + 2\beta\frac{d\mathbf{x}}{dt} + \omega_0^2\mathbf{x} = \hat{\boldsymbol{\varepsilon}}\,\frac{e}{m}E_0\,\cos(\omega t - kz), \tag{3.4.3}$$

where for later convenience we have defined $\beta = b/2m$. The unit vector $\hat{\boldsymbol{\varepsilon}}$ defines the polarization of the applied field (see Problem 3.4).

Equation (3.4.3) for the electron oscillator with frictional damping is most easily solved by first writing it in complex form:

$$\frac{d^2\mathbf{x}}{dt^2} + 2\beta\frac{d\mathbf{x}}{dt} + \omega_0^2\mathbf{x} = \hat{\boldsymbol{\varepsilon}}\,\frac{e}{m}E_0 e^{-i(\omega t - kz)}, \tag{3.4.4}$$

where we follow the convention of writing $E_0 \cos(\omega t - kz)$ as $E_0 e^{-i(\omega t - kz)}$. This means that $\mathbf{x}(t)$ in (3.4.4) is also regarded mathematically as a complex quantity in our calculations, but only its real part is physically meaningful. In other words, we may defer the process of taking the real part of (3.4.4) until after our calculations, at which point the real part of our solution for $\mathbf{x}(t)$ is the (real) electron displacement. This approach is used frequently in solving linear equations such as (3.4.3). We solve (3.4.4) by writing

$$\mathbf{x}(t) = \mathbf{a}e^{-i(\omega t - kz)}, \tag{3.4.5}$$

and after inserting this in (3.4.4) we obtain

$$(-\omega^2 - 2i\beta\omega + \omega_0^2)\mathbf{a} = \hat{\boldsymbol{\varepsilon}}\frac{e}{m}E_0. \tag{3.4.6}$$

Therefore the assumed solution (3.4.5) satisfies Eq. (3.4.4) if

$$\mathbf{a} = \frac{-\hat{\boldsymbol{\varepsilon}}(e/m)E_0}{\omega^2 - \omega_0^2 + 2i\beta\omega}, \tag{3.4.7}$$

and the physically relevant solution is therefore

$$\mathbf{x}(t) = \mathrm{Re}\left[\frac{\hat{\boldsymbol{\varepsilon}}(e/m)E_0 e^{-i(\omega t - kz)}}{\omega_0^2 - \omega^2 - 2i\beta\omega}\right]. \tag{3.4.8}$$

Note that (3.4.8) actually gives only the steady-state solution of (3.4.3). Any solution of the homogeneous version of (3.4.3) can be added to (3.4.8), and the sum will still be a solution of (3.4.3). The homogeneous version is

$$\frac{d^2\mathbf{x}_{\mathrm{hom}}}{dt^2} + 2\beta\frac{d\mathbf{x}_{\mathrm{hom}}}{dt} + \omega_0^2\mathbf{x}_{\mathrm{hom}} = 0, \tag{3.4.9}$$

and its general solution is

$$\mathbf{x}_{\mathrm{hom}} = [\mathbf{A}\cos\omega_0't + \mathbf{B}\sin\omega_0't]e^{-\beta t}, \tag{3.4.10}$$

where underdamped oscillation ($\beta \ll \omega_0$) is by far the most common occurrence, so

$$\omega_0' = (\omega_0^2 - \beta^2)^{1/2} \approx \omega_0. \tag{3.4.11}$$

We will usually neglect the homogeneous part of the full solution to (3.4.3). This is obviously an approximation. The approximation is, however, an excellent one whenever

$$t \gg \frac{1}{\beta}. \tag{3.4.12}$$

Under this condition, $e^{-\beta t} \ll 1$, and we can safely neglect the homogeneous component (3.4.10) because it makes only a short-lived transient contribution to the solution.

Even though the damping time $1/\beta$ is very short, it is not the shortest time in the problem. Typically, the oscillation periods $T_0 = 2\pi/\omega_0$ and $T = 2\pi/\omega$ associated with the natural oscillation frequency ω_0 or the forcing frequency ω are very much shorter. In the case of ordinary optically transparent materials such as atomic vapors, glasses, and many crystals and liquids, both ω_0 and ω are typically in the neighborhood of 10^{15} s^{-1}, and β falls in a wide range of much smaller frequencies:

$$\beta \approx 10^6 \text{--} 10^{12} \text{ s}^{-1} \ll \omega_0, \omega. \tag{3.4.13}$$

Relations (3.4.12) and (3.4.13), taken together, imply that times of physical interest must be much longer than an optical period:

$$t \gg \beta^{-1} \gg \omega_0^{-1}, \, \omega^{-1}. \tag{3.4.14}$$

That is, steady-state solutions of (3.4.3) are valid for times that are many periods of oscillator vibration ($T_0 = 2\pi/\omega_0$) and forced vibration ($T = 2\pi/\omega$) removed from $t = 0$, but they cannot be used to predict the oscillator's response within the first few cycles after $t = 0$. This is, however, no real restriction in optical physics, as it is equivalent to

$$t \gg 10^{-15} \text{ s } (=10^{-3} \text{ ps} = 1 \text{ fs}). \tag{3.4.15}$$

One femtosecond (fs) is a time span one or two orders of magnitude smaller than can ordinarily be resolved optically.

To calculate the rate at which energy is absorbed from the field, we consider the rate at which work is done by the field on an oscillator at position z along the direction of propagation of the presumed plane-wave, monochromatic field:

$$\frac{dW}{dt} = \mathbf{F} \cdot \mathbf{v} = \mathbf{F} \cdot \frac{d\mathbf{x}}{dt} = \hat{\boldsymbol{\varepsilon}} e E_0 \cos(\omega t - kz) \cdot \left(\frac{d\mathbf{x}}{dt} \right). \tag{3.4.16}$$

The (steady-state) velocity $d\mathbf{x}/dt$ for the oscillator follows by differentiation of (3.4.8):

$$\frac{d\mathbf{x}}{dt} = \frac{\hat{\boldsymbol{\varepsilon}}(e/m)E_0}{(\omega_0^2 - \omega^2)^2 + 4\beta^2\omega^2} [2\beta\omega^2 \cos(\omega t - kz) - \omega(\omega_0^2 - \omega^2) \sin(\omega t - kz)]. \tag{3.4.17}$$

We now use this expression in (3.4.16) and average over times large compared to $1/\omega$ as in the preceding section. This amounts to the replacement of $\cos^2(\omega t - kz)$ by $\frac{1}{2}$ and $\sin(\omega t - kz)\cos(\omega t - kz)$ by 0, resulting in

$$\frac{dW}{dt} = \frac{e^2}{m}E_0^2 \frac{1}{\beta} \left[\frac{\beta^2\omega^2}{(\omega_0^2 - \omega^2)^2 + 4\beta^2\omega^2} \right] \tag{3.4.18}$$

for the cycle-averaged rate of work. We have used the fact that $\hat{\boldsymbol{\varepsilon}} \cdot \hat{\boldsymbol{\varepsilon}} = 1$, i.e., that $\hat{\boldsymbol{\varepsilon}}$ is a unit vector (see also Problem 3.5).

Since $\beta \ll \omega, \omega_0$, the dimensionless quantity in brackets in Eq. (3.4.18) will have a very small value unless the field frequency ω is near the oscillator resonance frequency ω_0. More precisely, frequency "detunings" $|\omega_0 - \omega|$ much larger than β result in very little absorption. Thus we make the approximation $\omega_0 + \omega \approx 2\omega_0$, or

$$(\omega_0^2 - \omega^2)^2 = (\omega_0 - \omega)^2(\omega_0 + \omega)^2 \approx 4\omega_0^2(\omega_0 - \omega)^2 \tag{3.4.19}$$

in (3.4.18), and likewise approximate ω^3 by ω_0^3 in the numerator and $4\beta^2 \omega^2$ by $4\beta^2 \omega_0^2$ in the denominator inside the brackets:

$$\frac{dW}{dt} \approx \frac{e^2}{m}E_0^2 \left[\frac{\beta\omega_0^2}{4\omega_0^2(\omega_0 - \omega)^2 + 4\beta^2\omega_0^2} \right] = \frac{\pi e^2}{4m}E_0^2 \left[\frac{(1/\pi)\beta}{(\omega_0 - \omega)^2 + \beta^2} \right]. \tag{3.4.20}$$

We can write the rate of absorption of energy by the oscillator, dW/dt, in terms of the circular frequency $\nu = \omega/2\pi = c/\lambda$ of the field:

$$\frac{dW}{dt} = \frac{e^2}{8m}E_0^2\left[\frac{(1/\pi)\delta\nu_0}{(\nu - \nu_0)^2 + \delta\nu_0^2}\right], \tag{3.4.21}$$

where

$$\nu_0 = \frac{\omega_0}{2\pi} \tag{3.4.22}$$

and

$$\delta\nu_0 = \frac{\beta}{2\pi}. \tag{3.4.23}$$

It is also convenient to write the absorption rate in terms of the field intensity

$$I_\nu = \tfrac{1}{2}c\epsilon_0 E_0^2, \tag{3.4.24}$$

where the subscript indicates that I_ν is the intensity of the assumed monochromatic field of frequency ν. Thus,

$$\frac{dW}{dt} = \frac{e^2}{4mc\epsilon_0}I_\nu L(\nu), \tag{3.4.25}$$

where the "lineshape function" $L(\nu)$, which determines the dependence of the absorption on the field frequency, is defined by

$$L(\nu) = \frac{\delta\nu_0/\pi}{(\nu - \nu_0)^2 + \delta\nu_0^2}. \tag{3.4.26}$$

This is called the *Lorentzian lineshape function*, or Lorentzian distribution, and is plotted in Fig. 3.7.

The Lorentzian function is a mathematically idealized lineshape in several respects. We have already shown that it is the near-resonance approximation to the more complicated function appearing in (3.4.18). The function is defined mathematically for negative frequencies, even though they have no physical significance. It is exactly normalized to unity when integrated over all frequencies, as is easily checked:

$$\int_{-\infty}^{\infty} d\nu\, L(\nu) = \frac{\delta\nu_0}{\pi}\int_{-\infty}^{\infty}\frac{d\nu}{(\nu - \nu_0)^2 + \delta\nu_0^2} = 1, \tag{3.4.27}$$

and the normalization is approximately the same when only the physical, positive frequencies are used. The approximation is excellent for $\delta\nu_0 \ll \nu_0$ [recall (3.4.13)]. In other words, the contribution of the unphysical negative frequencies is negligible

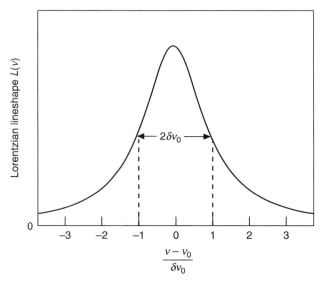

Figure 3.7   Lorentzian lineshape function.

because the linewidth is negligible compared to the resonance frequency, and in this sense $L(v)$ is physically as well as mathematically normalized to unity.

The maximum value of $L(v)$ occurs at the resonance frequency $v = v_0$:

$$L(v)_{max} = L(v_0) = \frac{1}{\pi \delta v_0}. \tag{3.4.28}$$

At $v = v_0 \pm \delta v_0$ we have

$$L(v_0 \pm \delta v_0) = \frac{1}{2\pi \delta v_0} = \frac{1}{2} L(v)_{max}. \tag{3.4.29}$$

Because of this property, $2\delta v_0$ is called the width of the Lorentzian function, or the *full width at half-maximum* (FWHM), and δv_0 is called the *half width at half-maximum* (HWHM). The Lorentzian function is fully specified by its width (FWHM or HWHM) and the frequency v_0 where it peaks. The peak value of the absorption rate is

$$\left(\frac{dW}{dt}\right)_{max} = \frac{e^2}{4mc\epsilon_0} I_v L(v_0) = \left(\frac{1}{4\pi\epsilon_0}\right) \frac{e^2}{mc\delta v_0} I_v, \tag{3.4.30}$$

and it decreases to half this resonance value when the field is "detuned" from resonance by the half width δv_0 of the Lorentzian function.

Our classical theory thus predicts that the absorption is strongest when the frequency of the light equals the oscillation frequency of the electron oscillator. Far out in the wings of the Lorentzian, where $|v - v_0| \gg \delta v_0$, there is very little absorption. A knowledge of the width δv_0 is therefore essential to a quantitative interpretation of absorption data.

To determine the numerical value of $\delta\nu_0$ in a given situation, we must consider in some detail the physical origin of this absorption width. This we do in Section 3.8.

We shall see that the absorption rate does not always have the Lorentzian form (3.4.25). However, we can in general write

$$\frac{dW}{dt} = \frac{e^2}{4mc\epsilon_0} I_\nu S(\nu), \tag{3.4.31}$$

where the lineshape function $S(\nu)$, whatever its form, is normalized to unity:

$$\int_0^\infty d\nu \, S(\nu) = 1. \tag{3.4.32}$$

As in the case of spontaneous emission, two changes to the classical oscillator theory of absorption are required to obtain quantitatively correct formulas. First, we introduce the oscillator strength f, in this case replacing (3.4.31) with

$$\frac{dW}{dt} = \frac{e^2 f}{4mc\epsilon_0} I_\nu S(\nu). \tag{3.4.33}$$

Second, we account for the fact that atoms are found only in states of allowed energy. That is, the absorption process proceeds from a state in which N_1 atoms are in a state of lower energy E_1 to a state in which N_2 atoms are in a state of higher energy $E_2 = E_1 + h\nu_0$, as indicated in Fig. 3.8. This suggests the replacement of dW/dt of the classical theory by $-\hbar\omega_0 dN_1/dt$ since the actual rate of energy absorption should be proportional to the number of atoms in the lower energy state from which the absorption proceeds. Thus, we relate the rate (3.4.31) at which energy is absorbed according to the classical theory to the rate of change of the "population" N_1 as follows:

$$\frac{dW}{dt} = -\frac{d}{dt}(\hbar\omega_0 N_1), \tag{3.4.34}$$

or

$$\frac{dN_1}{dt} = -\left(\frac{1}{4\pi\epsilon_0}\right)\left(\frac{\pi e^2 f}{mc\hbar\omega_0}\right) N_1 I_\nu S(\nu). \tag{3.4.35}$$

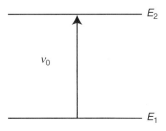

Figure 3.8 Atomic absorption transition of frequency $\nu_0 = (E_2 - E_1)/h$.

Similarly, because $dN_2/dt = -dN_1/dt$, the rate of change of the upper-level population N_2 due to absorption of light of frequency ν and intensity I_ν is

$$\frac{dN_2}{dt} = \left(\frac{1}{4\pi\epsilon_0}\right)\left(\frac{\pi e^2 f}{mc\hbar\omega_0}\right)N_1 I_\nu S(\nu). \tag{3.4.36}$$

It is convenient to use Eq. (3.3.7) to write the rate Eq. (3.4.35) for absorption in terms of the spontaneous emission rate A_{21} for the $E_2 \to E_1$ transition. Simple algebra yields

$$\frac{dN_1}{dt} = -\frac{1}{h\nu}\left[\frac{\lambda^2 A_{21}}{8\pi}\right]N_1 I_\nu S(\nu) = -\frac{dN_2}{dt}. \tag{3.4.37}$$

Because we are assuming that the applied field is close to the transition resonance frequency ν_0, we have used ν in place of ν_0 in this equation, and for later convenience we have also written the quantity in brackets in terms of the wavelength $\lambda = c/\nu$ rather than ν.

3.5   ABSORPTION OF BROADBAND LIGHT

Thus far, we have considered only absorption from a very narrowband, in fact perfectly monochromatic, field of frequency ν. In reality, of course, the applied field will not be perfectly monochromatic. For many purposes the rate of absorption of light having a distribution of frequencies can be obtained by simply summing the absorption rates associated with each frequency component:

$$\frac{dN_1}{dt} = -\frac{A_{21}}{8\pi h}N_1 \sum_\nu \frac{\lambda^2}{\nu} I_\nu S(\nu). \tag{3.5.1}$$

In many cases of interest the field is composed of a continuous range of frequencies, and the summation in (3.5.1) must be replaced by an integral:

$$\frac{dN_1}{dt} \longrightarrow -\frac{A_{21}}{8\pi h}N_1 \int_0^\infty \frac{c^2}{\nu^3} I(\nu)S(\nu)\, d\nu, \tag{3.5.2}$$

where $I(\nu)\, d\nu$ is the intensity of radiation in the frequency band from ν to $\nu + d\nu$.

It is convenient to define a spectral energy density $\rho(\nu)$, such that $\rho(\nu)\, d\nu$ is the electromagnetic energy per unit volume in the same frequency band (Fig. 3.9). The intensity, or energy flux, is the velocity of light times the energy density. Therefore $I(\nu) = c\rho(\nu)$ and

$$\frac{dN_1}{dt} = -\frac{A_{21}}{8\pi h}N_1 \int_0^\infty \frac{c^3}{\nu^3} \rho(\nu)S(\nu)\, d\nu. \tag{3.5.3}$$

We can now define "broadband light" as follows. Whenever the spectral energy density $\rho(\nu)$ is a broad, almost constant function of ν compared to the atomic lineshape

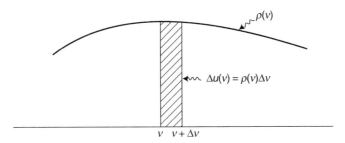

Figure 3.9   Spectral energy density $\rho(\nu)$ is defined such that $\Delta u(\nu) = \rho(\nu)\,\Delta\nu$ is the electromagnetic energy per unit volume in the narrow frequency interval from ν to $\nu + \Delta\nu$.

function $S(\nu)$, we can write

$$\int_0^\infty \frac{c^3}{\nu^3}\rho(\nu)S(\nu)\,d\nu \approx \frac{c^3}{\nu_0^3}\rho(\nu_0)\int_0^\infty S(\nu)\,d\nu = \frac{c^3}{\nu_0^3}\rho(\nu_0). \tag{3.5.4}$$

Whether $\rho(\nu)$ is flat enough in its variation to justify the approximation (3.5.4) depends on the lineshape function $S(\nu)$. The narrower the width of $S(\nu)$, the easier it is to satisfy (3.5.4). When this approximation is valid, we may say that we have broadband light and broadband absorption, as opposed to the opposite extreme of narrowband (i.e., nearly monochromatic) absorption. Both extremes are limiting cases of (3.5.3). The absorption rate for an atom exposed to broadband radiation is thus

$$\frac{dN_1}{dt} = -\frac{A_{21}}{8\pi h}\frac{c^3}{\nu_0^3}N_1\rho(\nu_0) = -\frac{dN_2}{dt}. \tag{3.5.5}$$

We see that for broadband absorption the rate at which the lower- and upper-state populations change is completely independent of the form of the lineshape function $S(\nu)$ and is simply proportional to the spectral energy density of the field at the atomic transition frequency ν_0. In the classical electron oscillator model, similarly, the rate of absorption of energy by an atom in a broadband field is (Problem 3.6)

$$\frac{dW}{dt} = \left(\frac{1}{4\pi\epsilon_0}\right)\frac{\pi e^2}{m}\rho(\nu_0). \tag{3.5.6}$$

3.6   THERMAL RADIATION

In thermal equilibrium the processes of absorption and emission balance each other in such a way that the spectral density of radiation is completely characterized by the temperature T. The spectral energy density of thermal radiation is the *Planck spectrum*:

$$\rho(\nu) = \frac{8\pi h\nu^3/c^3}{e^{h\nu/k_B T} - 1}, \tag{3.6.1}$$

where k_B ($= 1.380 \times 10^{-23}$ J/K $= 0.08614 \times 10^{-3}$ eV/K) is Boltzmann's constant. The historical significance of the Planck spectrum in the development of quantum theory is discussed in many textbooks. Our aim here is to see what the classical electron oscillator model and the quantum theory of emission and absorption, as formulated thus far, imply about thermal equilibrium radiation. Because the thermal spectral density $\rho(\nu)$ is certainly "broadband," we will use the absorption formulas appropriate to broadband radiation.

Consider first the classical electron oscillator model. The rates of emission and absorption are given by Eqs. (3.3.5) and (3.5.6), respectively. The equilibrium condition that emission and absorption are equal in the electron oscillator model is then

$$\left(\frac{1}{4\pi\epsilon_0}\right)\frac{2e^2\omega_0^2}{3mc^3}E = \left(\frac{1}{4\pi\epsilon_0}\right)\frac{\pi e^2}{m}\rho(\nu),\tag{3.6.2}$$

or, since $\omega_0 = 2\pi\nu_0$,

$$\rho(\nu_0) = \frac{8\pi\nu_0^2}{3c^3}E.\tag{3.6.3}$$

It is a well-known theorem of classical physics that, in thermal equilibrium at temperature T, the average energy E of an oscillator free to oscillate in three dimensions is $3k_BT$. Using this result in (3.6.3), we have

$$\rho(\nu) = \frac{8\pi\nu^2}{c^3}k_BT\tag{3.6.4}$$

for the spectral density of thermal radiation predicted by the classical electron oscillator model. We have written this result in terms of an arbitrary frequency ν rather than a single resonance frequency ν_0 in order to model an idealized blackbody that, as discussed below, absorbs radiation at *all* frequencies.

The spectrum (3.6.4) is the *Rayleigh–Jeans spectrum*, and it is an inexorable consequence of classical physics. It is an approximation to the Planck spectrum (3.6.1) when the quantum of energy $h\nu$ is small compared to k_BT, and so can be regarded as the "classical limit" of the Planck spectrum.

Let us now use the formulas (3.3.8), (3.3.9), and (3.5.5) of quantum theory for the emission and absorption. For dN_1/dt, for instance, we have

$$\frac{dN_1}{dt} = A_{21}N_2 - \frac{A_{21}}{8\pi h}\frac{c^3}{\nu_0^3}N_1\rho(\nu_0),\tag{3.6.5}$$

the first term being due to spontaneous emission and the second to absorption. Setting dN_1/dt (or dN_2/dt) equal to zero, since the populations of the atomic states must be constant when the atoms and the radiation are in equilibrium, we obtain the equilibrium radiation spectrum:

$$\rho(\nu_0) = \frac{8\pi h\nu_0^3}{c^3}\frac{N_2}{N_1}.\tag{3.6.6}$$

Now in thermal equilibrium at temperature T the ratio of the populations N_2 and N_1 must satisfy a general result of quantum statistical mechanics:[6]

$$\frac{N_2}{N_1} = e^{-(E_2 - E_1)/k_B T} = e^{-h\nu_0/k_B T}. \tag{3.6.7}$$

Therefore, the spectrum of thermal radiation predicted by this (not quite correct) argument is

$$\rho(\nu) = \frac{8\pi h \nu^3}{c^3} e^{-h\nu/k_B T}, \tag{3.6.8}$$

where, for the reason noted following Eq. (3.6.4), we have written the spectrum in terms of an arbitrary frequency ν rather than a specific frequency ν_0. The spectrum (3.6.8), which is called the *Wien spectrum*, is an approximation to the Planck spectrum when the quantum of energy $h\nu$ is large compared to $k_B T$.

To see why we have not obtained the correct spectrum of thermal equilibrium radiation using the (correct) results of the quantum theory of spontaneous emission and absorption, let us use (3.6.7) to write the Planck spectrum as

$$\rho(\nu_0) = \frac{8\pi h \nu_0^3}{c^3} \frac{1}{N_1/N_2 - 1} = \frac{8\pi h \nu_0^3}{c^3} \frac{N_2}{N_1 - N_2}, \tag{3.6.9}$$

or

$$\frac{c^3}{8\pi h \nu_0^3} \rho(\nu_0)(N_1 - N_2) = N_2, \tag{3.6.10}$$

and therefore

$$\frac{A_{21}}{8\pi h} \frac{c^3}{\nu_0^3} N_1 \rho(\nu_0) = A_{21} N_2 + \frac{A_{21}}{8\pi h} \frac{c^3}{\nu_0^3} N_2 \rho(\nu_0), \tag{3.6.11}$$

where we have multiplied through by the spontaneous emission rate A_{21}. This equation for $\rho(\nu_0)$, together with the Boltzmann condition (3.6.7), yields the Planck spectrum.

Without the second term on the right-hand side of (3.6.11) we would have Eq. (3.6.6), which was obtained by equating the rates of absorption [the left-hand side of (3.6.11)] and spontaneous emission (the first term on the right). To obtain the correct thermal radiation spectrum, therefore, we require another effect in addition to absorption and spontaneous emission. This "new" effect is described by the second term on the right-hand side of (3.6.11). Like spontaneous emission, the rate for this effect is proportional to the upper-state population N_2, so that it too is associated with the emission of radiation. Unlike spontaneous emission, however, the rate for this emission process is proportional

---

[6]We are ignoring any degeneracy of the energy levels E_2 and E_1, which does not affect the thermal radiation spectrum (Problem 3.8).

to the spectral density $\rho(\nu_0)$ of radiation already present. That is, the process described by the second term on the right-hand side of (3.6.11) is *stimulated emission.*

According to (3.6.11) the rate coefficient for stimulated emission, which we write as

$$\frac{A_{21}}{8\pi h} \frac{c^3}{\nu_0^3} \rho(\nu_0) = B_{21}\rho(\nu_0), \tag{3.6.12}$$

is identical to the rate coefficient for absorption, which we write as $B_{12}\rho(\nu_0)$. That is,[7]

$$B_{12} = B_{21}. \tag{3.6.13}$$

With this notation we can write Eq. (3.6.11) as

$$N_1 B_{12}\rho(\nu_0) = N_2 A_{21} + N_2 B_{21}\rho(\nu_0), \tag{3.6.14}$$

the left-hand side being the rate of absorption and the right-hand side the rate of spontaneous and stimulated emission.

• Equation (3.6.14) was first presented by Albert Einstein in 1916, more than a decade before what are now called the "Einstein A and B coefficients" could be derived from quantum theory. To obtain the Planck spectrum using discrete energy states and other aspects of the Bohr theory, Einstein postulated the processes of spontaneous emission, absorption, and stimulated emission, and that these processes could be characterized by rate coefficients as in (3.6.14). Arguing that (3.6.14) must be true for all temperatures and therefore for arbitrarily large spectral densities $\rho(\nu)$, Einstein concluded that $B_{12} = B_{21}$.

Equations (3.6.14) and (3.6.7) imply

$$\rho(\nu_0) = \frac{A_{21}/B_{21}}{e^{h\nu_0/k_B T} - 1}, \tag{3.6.15}$$

which in the "classical limit" $h\nu_0/k_B T \ll 1$ reduces to

$$\rho(\nu_0) = \frac{A_{21}}{B_{21}} \frac{k_B T}{h\nu_0}. \tag{3.6.16}$$

Reasoning that this limit should yield the Rayleigh–Jeans spectrum (3.6.4), Einstein deduced that

$$\frac{A_{21}}{B_{21}} = \frac{8\pi h\nu_0^3}{c^3}, \tag{3.6.17}$$

a result implied by our Eq. (3.5.5) when it is written as

$$\frac{dN_1}{dt} = -N_1 B_{12}\rho(\nu_0), \tag{3.6.18}$$

---

[7]Equation (3.6.13) and a few others are generalized in Section 3.7 when degenerate energy levels are considered.

where, using (3.3.7), we identify

$$B_{12} = \left(\frac{1}{4\pi\epsilon_0}\right)\frac{\pi e^2 f}{mh\nu_0}.$$

(3.6.19)

Einstein was thus able to derive essentially all the results for emission and absorption in broadband fields that we have obtained starting from the electron oscillator model and modifying it to incorporate results of quantum theory. A truly new insight achieved by Einstein was that the observed Planck spectrum implied that there must be stimulated emission in addition to spontaneous emission. The practical utilization of this new concept—the laser—was to occur more than 40 years later.　●

It is instructive to write the energy per unit volume of thermal radiation in the small frequency interval $\Delta\nu$ about ν, $\Delta u(\nu) = \rho(\nu)\Delta\nu$, in the form

$$\rho(\nu)\,\Delta\nu = h\nu\left(\frac{8\pi\nu^2}{c^3}\Delta\nu\right)\frac{1}{e^{h\nu/k_BT}-1}.$$

(3.6.20)

The first factor on the right is the energy of a photon of frequency ν. The quantity in parentheses is the number of electromagnetic field modes per unit volume in the small frequency interval $[\nu, \nu + \Delta\nu]$, assuming that any cavity containing the thermal radiation is large compared to the wavelength c/λ (Section 3.12). The last factor, $1/(e^{h\nu/k_BT}-1)$, can therefore be identified as the average number of photons of frequency ν in thermal equilibrium at temperature T.

This last quantity has a further significance that can be inferred by considering the ratio of the rate of stimulated emission in thermal equilibrium, $N_2 B_{21}\rho(\nu_0)$, to the rate of spontaneous emission, $N_2 A_{21}$, for a transition of frequency ν_0:

$$\frac{N_2 B_{21}\rho(\nu_0)}{N_2 A_{21}} = \frac{B_{21}}{A_{21}}\rho(\nu_0) = \frac{1}{e^{h\nu_0/k_BT}-1},$$

(3.6.21)

where we have used Eq. (3.6.15). In other words, the stimulated emission rate is equal to the spontaneous emission rate times the average number of photons at the transition frequency. Although this result has been inferred for the case of thermal radiation, it is more generally valid and may be stated as follows: *The rate of stimulated emission into any mode of the field is equal to the spontaneous emission rate into the mode, times the average number of photons already occupying that mode.* (See also Section 3.7.)

The Planck spectrum is independent of the atomic or molecular properties of the material in thermal equilibrium with radiation. According to (3.6.7), there will be more atoms in the lower level than the upper level of any transition. This means that any radiation incident on the material will lose energy. It is convenient to define an ideal *blackbody* as an object that absorbs *all* the radiation, of *any* frequency, incident upon it. In such a blackbody the absorption and emission of radiation are exactly balanced in a steady state of thermal equilibrium, and any radiation incident upon its surface would be completely absorbed.

Although no perfect blackbody is known to exist, it is possible to construct an excellent approximation to an ideal blackbody surface. Consider a cavity inside a metal block, with a small hole drilled through to provide an opening to the outside, as illustrated in Fig. 3.10. Any radiation incident on the hole from the outside is repeatedly reflected

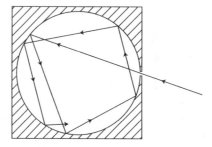

Figure 3.10   A cavity inside a metal block kept at constant temperature. A small hole allows radiation to enter the cavity, and the radiation is diffused by repeated internal scattering. The hole itself acts as the surface of a blackbody.

within the cavity, and eventually absorbed, so that the amount of incident radiation escaping back through the hole to the outside is negligibly small. The hole itself thus acts as the "surface" of a blackbody. Furthermore a small amount of equilibrium thermal radiation inside the cavity, produced by spontaneous and stimulated emission from the cavity walls, can escape through the hole to the outside. This radiation escaping through the hole is a sampling of the thermal radiation inside the cavity, and therefore the spectral density of this "cavity radiation" should satisfy the Planck formula. If the block is placed inside an oven, it can be kept in thermal equilibrium at some fixed temperature T. The earliest accurate measurements of such cavity radiation in the far-infrared spectral region from 12 to 18 μm, where the Wien law fails to agree with the data, were carried out by O. Lummer and E. Pringsheim in 1900. These and other measurements, particularly those of H. Rubens and F. Kurlbaum, motivated Planck to reconsider the existing theory of thermal radiation and led to his announcement of formula (3.6.1) at the October 19, 1900, session of the Prussian Academy of Science.

Many sources of radiation have spectral characteristics approximating those of an ideal blackbody. Stars, for instance, are certainly not perfect blackbodies, but they come sufficiently close to the ideal that we can estimate their surface temperatures by fitting their spectra to Planck's law (Fig. 3.11). In particular, the peak emission wavelength λ_{max} of a blackbody at temperature T (K) is given by (Problem 3.1)

$$\lambda_{max} = \frac{2.898 \times 10^6}{T} \text{ nm.} \tag{3.6.22}$$

Thus, the sun, which has a spectrum approaching that of a blackbody at 5800K, has a peak emission wavelength $\lambda_{max} \approx 500$ nm. Its total intensity at Earth's surface is about $0.14 \, \text{W/cm}^2 = 1.4 \, \text{kW/m}^2$. Equation (3.6.22) is consistent with the observation that the color of hot bodies shifts to shorter wavelengths with increasing temperature T. Thus, "white hot" is hotter than "red hot"; the filament of an incandescent lightbulb glows white, whereas a (cooler) toaster glows red. This shift of peak wavelength with temperature is evident in Fig. 3.12.

For wavelengths in the visible, and for temperatures less than several tens of thousands of kelvins, the ratio (3.6.21) is much less than unity. For the solar temperature $T = 5800$K, for instance, and $\lambda = 500$ nm, the ratio is about $1/142$. Thus, we can infer that more than 99% of the light from the sun is due to spontaneous rather than stimulated radiation processes.

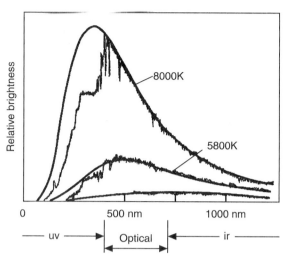

Figure 3.11 Comparison of blackbody emission (smooth curves) and stellar emission spectra for two temperatures, 8000 and 5800K. (After W. M. Protheroe, E. R. Capriotti, and G. H. Newsom, *Exploring the Universe*, 3rd ed., Merrill, Columbus, 1984.)

The total electromagnetic energy density of a blackbody is found by integrating (3.6.1), which leads to

$$\int_0^\infty \rho(\nu)\,d\nu = \frac{8\pi^5 k_B^4}{15c^3 h^3}\,T^4. \tag{3.6.23}$$

The total intensity, or power radiated per unit area, is then

$$I_{\text{total}} = \frac{c}{4}\int_0^\infty \rho(\nu)\,d\nu = \frac{2\pi^5 k_B^4}{15c^2 h^3}\,T^4 = \sigma T^4, \tag{3.6.24}$$

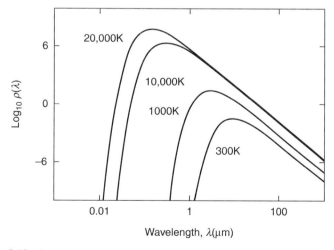

Figure 3.12 $\text{Log}_{10}\rho(\lambda)$ vs. λ for an ideal blackbody radiator at four temperatures.

where $\sigma = 5.67 \times 10^{-8}$ J-m^{-2}-s^{-1}-K^{-4} = 5.67×10^{-12} W/cm^2/K^4 is the *Stefan–Boltzmann constant*. Note that the intensity at each frequency in this formula is not simply the speed of light c times the spectral energy density at that frequency. The extra factor $\frac{1}{4}$ arises for two reasons. First, a factor $\frac{1}{2}$ arises because, along any axis through the blackbody, there are equal intensities of radiation propagating in opposite directions, but in (3.6.24) we are only interested in radiation propagating outward through the surface. A second factor of $\frac{1}{2}$ arises because the average component of light velocity normal to the surface is

$$(\cos \theta)_{\text{avg}} = \frac{1}{2\pi} \int_0^{2\pi} d\phi \int_0^{\pi/2} \cos \theta \sin \theta \, d\theta = \frac{1}{2}. \qquad (3.6.25)$$

Except at very high pressures, an atomic gas radiates only at certain discrete wavelengths corresponding to the spectral lines of the atoms. In solids, however, the interactions of the closely packed atoms cause the emitted radiation to have a continuous spectrum, which is approximately the Planck spectrum for the temperature T of the solid. The ratio of the power radiated by a given solid to the power radiated by a blackbody of equal area and temperature is called the *total emissivity*, ε, of the solid. The *spectral emissivity*, ε_λ, is defined similarly in terms of the power radiated within a narrow wavelength interval between λ and $\lambda + d\lambda$. The spectral emissivity and the reflectivity r_λ satisfy $\varepsilon_\lambda + r_\lambda = 1$, so that the spectral emissivity can be determined by measuring the reflectivity. For passive surfaces (not part of some laser device) we have $r_\lambda < 1$, so emissivities are always less than 1. Typically $0.2 < \varepsilon < 0.9$, but emissivities can be very small for highly reflecting surfaces.

The radiation of a perfect blackbody is isotropic, that is, the Planck spectrum does not depend on any direction of propagation. For real bodies, however, the spectral emissivity can depend not only on wavelength but also on direction and temperature. Emissivities of different materials are tabulated in handbooks.[8]

Figure 3.13 shows the radiation spectrum of tungsten, the filament material in household incandescent lightbulbs, compared with that of a blackbody for the temperature $T = 3000$K. In order for the filament to produce significant visible radiation, it must be heated to high temperatures; tungsten is used because, among other things, it has a high melting temperature (3655K) and proper working gives it the strength and ductility necessary for it to be formed in fine wire filaments.[9] As can be seen from Fig. 3.13, however, only a small part of the radiation from the heated filament lies in the visible.

• The limited lifetime (~ 1000 h) of tungsten filament lightbulbs is due to the evaporation of tungsten, which causes the blackening of a bulb over time. An inert fill gas (usually argon) is used to reduce the evaporation of tungsten particles, which are deposited on the upper part of the bulb as a result of convection currents that carry them upward. Thus, a bulb used "base up" on a ceiling will blacken near the stem of the bulb, whereas "base down" use results in blackening over the dome. The tungsten halogen lamps marketed in recent years employ a tungsten–halogen regenerative cycle in which the evaporated tungsten combines with a halogen (e.g., iodine) to form a compound, thus preventing the tungsten from being deposited on the glass

[8]See, for instance, R. C. Weast, ed., *CRC Handbook of Physics and Chemistry*, CRC, Boca Raton, FL, 1988, pp. E-390–E-392.
[9]This was one of the first successes of American industrially organized science. It came from the discovery in 1908 of how to make tungsten ductile by W. D. Coolidge in the General Electric Company laboratory in Schenectady, NY.

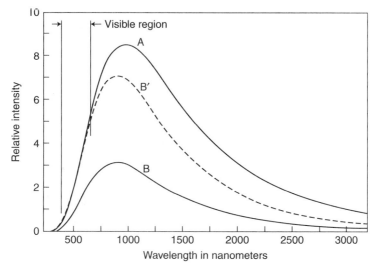

Figure 3.13 Radiation from tungsten at 3000K compared with a blackbody at the same temperature. Curves *A* and *B* give the relative intensities from 1 cm^2 of a blackbody and 1 cm^2 of tungsten, respectively, while curve *B'* is the relative intensity from 2.27 cm^2 of tungsten, which is seen to produce the same total intensity in the visible as the blackbody. (After J. E. Kaufman, *IES Lighting Handbook, Reference Volume*, Illuminating Engineering Society of North America, New York, 1984.)

bulb. When this compound comes in contact with the filament, however, the heat is sufficient to dissociate the compound into tungsten, which is redeposited on the filament, and the halogen, which is then available to continue the regenerative cycle.

The more efficient fluorescent tubes are, of course, not thermal light sources, but like thermal and all other nonlaser sources the light they generate derives from spontaneous emission. At each end of the tube is an electrode, one of which is a tungsten coil coated with a material that increases the efficiency with which electrons are ejected when the coil temperature exceeds about 1300K. The low-pressure ($\approx$ 0.008 Torr mercury and 1–3 Torr of other gases, 1 Torr = 1/760 atm of pressure) electric discharge along the axis of the tube causes the emission of 253.7 nm radiation from mercury atoms excited by collisions with electrons. The inner walls of the tube are coated with a "phosphor" that absorbs in the ultraviolet and emits in the visible. The operating lifetime in this case is determined primarily by the erosion of the electron emissive coating each time the lamp is turned on, so that the rated average life of fluorescent tubes is based on the number of starts, assuming 3 h of operation per start.

Lighting technology remains an active area of research and development, with particular focus in recent years on light-emitting diodes (LEDs). The reader wishing to pursue some of the many interesting aspects of the subject is referred not only to books and journals devoted to it but also to lighting and optical company catalogs or websites that sometimes contain tutorial information about lighting products.   •

3.7 EMISSION AND ABSORPTION OF NARROWBAND LIGHT

For lasers we are interested in stimulated emission in *narrowband* fields. Equation (3.4.37) gives the rate of change of level populations due to absorption in a narrowband field of intensity I_ν. As in the case of a broadband field, the stimulated emission rate in a narrowband field is the same as the absorption rate. Thus, the rate of change of N_2 due to both absorption and stimulated emission in a narrowband field is

$$\frac{dN_2}{dt} = \text{(absorption rate)} \times N_1 - \text{(stimulated emission rate)} \times N_2$$

$$= \text{(absorption rate)} \times (N_1 - N_2)$$

$$= -\frac{1}{h\nu}\frac{\lambda^2 A_{21}}{8\pi}(N_2 - N_1)I_\nu S(\nu). \tag{3.7.1}$$

Similarly, we must modify Eq. (3.4.37) for dN_1/dt to include the effect of stimulated emission:

$$\frac{dN_1}{dt} = \frac{1}{h\nu}\frac{\lambda^2 A_{21}}{8\pi}(N_2 - N_1)I_\nu S(\nu). \tag{3.7.2}$$

The absorption rate of narrowband light in these formulas was obtained assuming a linearly polarized, plane-wave, monochromatic field propagating in the z direction [recall Eq. (3.4.4)]. When absorption occurs, the field propagates in the same direction with the same polarization and frequency but with less intensity. In the case of stimulated emission the field also propagates in the same direction and with the same polarization and frequency as the incident field, but with *greater* intensity.

We have confined ourselves thus far to showing how absorption and stimulated emission affect the atomic level populations, without regard for how these processes affect the *field* except insofar as they decrease or increase the field energy. In Section 3.12 we derive an equation describing how the field intensity I_ν depends on the level populations N_1 and N_2.

The intensity I_ν can be expressed as the number of photons crossing a unit area per unit time, times the energy $h\nu$ of a single photon. Thus,

$$\text{Stimulated emission rate} = \frac{\lambda^2 A_{21}}{8\pi}S(\nu)$$

$$\times \text{ number of incident photons/(area-time).} \tag{3.7.3}$$

The quantity

$$\sigma(\nu) = \frac{\lambda^2 A_{21}}{8\pi}S(\nu), \tag{3.7.4}$$

which has the dimensions of an area, is the *cross section* for stimulated emission (and absorption). The relation (3.7.3) identifies the cross section as an effective area associated with the atomic transition, such that every photon intercepted by this area would induce an atom to undergo stimulated emission (or absorption). Needless to say there is no actual geometric object associated with this area; the cross section is nothing more than a conventional measure of the absorption strength of a transition. Note that it depends not only on the transition wavelength and *spontaneous* emission rate, but also on the lineshape function $S(\nu)$.

Of course, the level populations N_1 and N_2 can also change by spontaneous emission. The rate equations describing all three emission and absorption processes are obtained by adding the spontaneous emission terms (3.3.8) and (3.3.9) to (3.7.1) and (3.7.2):

$$\frac{dN_2}{dt} = -\frac{\sigma(\nu)}{h\nu} I_\nu (N_2 - N_1) - A_{21} N_2, \tag{3.7.5a}$$

$$\frac{dN_1}{dt} = \frac{\sigma(\nu)}{h\nu} I_\nu (N_2 - N_1) + A_{21} N_2. \tag{3.7.5b}$$

These are among the most important equations needed for a quantitative understanding of lasers. In general, it is necessary to modify them to include other processes, such as an external mechanism to supply and/or maintain atoms in the upper level, as well as collisions among atoms that can excite or deexcite the level populations. It is also necessary in general to account for the possibility that the two energy levels of the transition are degenerate and that the refractive index of the medium might differ significantly from unity. And in the case of semiconductor lasers, we must deal with a more complicated version of these rate equations because there the transitions are between *bands* of levels rather than between two discrete states. Nevertheless Eqs. (3.7.5) are the foundation for a considerable part of the theory of lasers.

- A well-known result of the quantum theory of radiation is that the stimulated emission rate into a single mode of the field is equal to q times the spontaneous emission rate into that mode, where q is the average number of photons initially occupying the mode. Thus, if R_{spon} is the rate for spontaneous emission into a certain mode, then the rate for stimulated emission into that mode is $R_{\text{stim}} = q R_{\text{spon}}$, or in other words the total emission rate into the mode is $(q + 1) R_{\text{spon}}$. We have already referred to this result following Eq. (3.6.21).

It is not immediately obvious that Eqs. (3.7.5) are consistent with this "$q + 1$" rule, one reason being that the stimulated emission rate $R_{\text{stim}} = \sigma(\nu) I_\nu / h\nu = (\lambda^2 A_{21}/8\pi)[I_\nu S(\nu)/h\nu]$ in these equations refers to a single field mode, that is, a field with a single frequency and polarization and propagating in a single direction, whereas A_{21} is the total rate for spontaneous emission in all possible directions of propagation and polarization. Moreover the stimulated emission rate depends on the lineshape function $S(\nu)$ at the applied field frequency ν, whereas A_{21} does not. It is therefore instructive to show that Eqs. (3.7.5) are in fact consistent with the $q + 1$ rule and can indeed be inferred from it.

As the remarks above suggest, $R_{\text{spon}} \neq A_{21}$. Let us write A_{21} [Eq. (3.3.7)] as

$$A_{21} = \left(\frac{1}{4\pi\epsilon_0}\right) \frac{\pi e^2 f}{m} \left[\frac{8\pi \nu_0^2}{c^3}\right]. \tag{3.7.6}$$

We recall that $(8\pi\nu^2/c^3)\,d\nu$ is the number of modes per unit volume in the frequency interval $[\nu, \nu + d\nu]$, so that the factor in square brackets in (3.7.6) is the density, in frequency, of modes per unit volume at the frequency ν_0 of the atomic transition. Thus, (3.7.6) expresses the spontaneous emission rate A_{21} as a sum of rates into all possible modes having the frequency ν_0.

But for a single field mode in a volume V—which for our purposes here might be imagined to be the volume of a laser cavity—the density in frequency of possible final states per unit volume is determined not by the field, which by assumption has only one frequency, ν, but by the atomic lineshape function $S(\nu)$. Therefore, to obtain the rate of spontaneous emission *into a single field mode* we replace the factor in square brackets in (3.7.6) by the frequency–space density $(1/V)S(\nu)$:

$$R_{\text{spon}} = \left(\frac{1}{4\pi\epsilon_0}\right)\frac{\pi e^2 f}{m}\frac{1}{V}S(\nu). \tag{3.7.7}$$

Comparison with (3.3.7) shows that

$$R_{\text{spon}} = \frac{\lambda^2 A_{21}}{8\pi}\frac{c}{V}S(\nu) \tag{3.7.8}$$

and therefore, when we multiply this by the photon number q, that the stimulated emission rate in Eqs. (3.7.5) is consistent with the $q + 1$ rule:

$$qR_{\text{spon}} = \frac{1}{h\nu}\frac{\lambda^2 A_{21}}{8\pi}\left[c\frac{qh\nu}{V}\right]S(\nu) = \frac{1}{h\nu}\frac{\lambda^2 A_{21}}{8\pi}I_\nu S(\nu) = \frac{\sigma(\nu)}{h\nu}I_\nu = R_{\text{stim}}, \tag{3.7.9}$$

where we have used $I_\nu = c(qh\nu/V)$, i.e., the intensity is the velocity of light times the field energy density $qh\nu/V$.   •

The relation $A_{21} = (8\pi h\nu_0^3/c^3)B_{21}$ [cf. Eq. (3.6.17)] can be used to write the stimulated emission rate as[10]

$$R_{\text{stim}} = \frac{\sigma(\nu)}{h\nu}I_\nu = \frac{1}{h\nu}\frac{\lambda^2 A_{21}}{8\pi}I_\nu S(\nu) = \frac{1}{c}B_{21}I_\nu S(\nu), \tag{3.7.10}$$

so that (3.7.5a), for instance, takes the form

$$\frac{dN_2}{dt} = -\frac{1}{c}(B_{21}N_2 - B_{12}N_1)I_\nu S(\nu) - A_{21}N_2. \tag{3.7.11}$$

The reader is encouraged to compare this equation with the simple model laser equation (1.5.2). N_2 here corresponds to n there, and N_1 is assumed zero there (complete inversion). What is the physical meaning of other differences? Can you identify f?

In our treatment of emission and absorption thus far we have not dealt explicitly with the possibility of level degeneracy, that is, that there might be more than one quantum state associated with each of the energy levels E_2 and E_1. The level populations N_2 and N_1 in Eq. (3.7.11), for example, are the total populations of the two energy levels, regardless of any degeneracies. Let us now label different possible states of energy E_2 and E_1 by m_2 and m_1, respectively. Let $\mathcal{N}_2(m_2)$ and $\mathcal{N}_1(m_1)$ denote the populations in these specific states, $\mathcal{A}(m_2, m_1)$ the rate of spontaneous emission from the upper state m_2 to the lower state m_1, and $\mathcal{R}(m_1, m_2)$ and $\mathcal{R}(m_2, m_1)$ the corresponding

---

[10]Recall that we are assuming $\nu \approx \nu_0$, so that we can replace ν by ν_0 in these formulas *except* in the lineshape function $S(\nu)$.

rates for absorption and stimulated emission, respectively. Thus

$$
\frac{d\mathcal{N}_2(m_2)}{dt} = -\sum_{m_1} [\mathcal{R}(m_2, m_1)\mathcal{N}_2(m_2) - \mathcal{R}(m_1, m_2)\mathcal{N}_1(m_1)]
$$

$$
-\sum_{m_1} \mathcal{A}(m_2, m_1)\mathcal{N}_2(m_2). \tag{3.7.12}
$$

Note that, for each of the three processes described by this rate equation, the total rate of change of $\mathcal{N}_2(m_2)$ is the sum of the rates for each possible channel $m_2 \leftrightarrow m_1$. We now sum both sides of (3.7.12) over the degenerate states m_2 and use the fact that $N_2 = \sum_{m_2} \mathcal{N}_2(m_2)$:

$$
\frac{dN_2}{dt} = -\sum_{m_1, m_2} [\mathcal{R}(m_2, m_1)\mathcal{N}_2(m_2) - \mathcal{R}(m_1, m_2)\mathcal{N}_1(m_1)]
$$

$$
-\sum_{m_1, m_2} \mathcal{A}(m_2, m_1)\mathcal{N}_2(m_2). \tag{3.7.13}
$$

This more general equation, and the one for N_1 that follows from $dN_1/dt = -dN_2/dt$, can sometimes be simplified. The most widely used simplification is to assume that all the states of a given level are equally populated, so that $\mathcal{N}_1(m_1)$ and $\mathcal{N}_2(m_2)$ are independent of m_1 and m_2 and equal to N_1/g_1 and N_2/g_2, respectively, where g_j is the degeneracy, or *statistical weight*, of level j. In this case

$$
\frac{dN_2}{dt} = -\sum_{m_1, m_2} \left[\frac{N_2}{g_2}\mathcal{R}(m_2, m_1) - \frac{N_1}{g_1}\mathcal{R}(m_1, m_2) \right] - \frac{N_2}{g_2}\sum_{m_1, m_2}\mathcal{A}(m_2, m_1)
$$

$$
= -\left[\frac{N_2}{g_2} - \frac{N_1}{g_1} \right] \sum_{m_1, m_2}\mathcal{R}(m_1, m_2) - \frac{N_2}{g_2}\sum_{m_1, m_2}\mathcal{A}(m_2, m_1). \tag{3.7.14}
$$

Comparing with (3.7.11), we make the identifications

$$
A_{21} = \frac{1}{g_2}\sum_{m_1, m_2}\mathcal{A}(m_2, m_1), \tag{3.7.15}
$$

$$
\frac{1}{c}I_\nu S(\nu)B_{21} = \frac{1}{g_2}\sum_{m_1, m_2}\mathcal{R}(m_1, m_2), \tag{3.7.16}
$$

$$
\frac{1}{c}I_\nu S(\nu)B_{12} = \frac{1}{g_1}\sum_{m_1, m_2}\mathcal{R}(m_1, m_2). \tag{3.7.17}
$$

Note therefore that

$$
g_2 B_{21} = g_1 B_{12}, \tag{3.7.18}
$$

which generalizes (3.6.13). Using (3.7.15)–(3.7.17) in (3.7.14), we have

$$\frac{dN_2}{dt} = -\frac{dN_1}{dt} = -\frac{1}{c}I_\nu S(\nu)[B_{21}N_2 - B_{12}N_1] - A_{21}N_2$$

$$= -\frac{1}{c}I_\nu S(\nu)B_{21}\left[N_2 - \frac{g_2}{g_1}N_1\right] - A_{21}N_2$$

$$= -\frac{1}{h\nu}\frac{\lambda^2 A_{21}}{8\pi}\left[N_2 - \frac{g_2}{g_1}N_1\right]I_\nu S(\nu) - A_{21}N_2, \qquad (3.7.19)$$

Situations where we cannot assume that degenerate states are equally populated can arise under excitation by polarized light, as discussed in Section 14.3.

We can use these relations to write the A and B coefficients in terms of the oscillator strength and level degeneracies. Equations (3.6.17) and (3.7.15)–(3.7.17) give

$$A_{21} = \frac{8\pi h\nu_0^3}{c^3}B_{21} = \frac{g_1}{g_2}\frac{8\pi h\nu_0^3}{c^3}B_{12} = \frac{g_1}{g_2}\frac{2\pi e^2 f}{\epsilon_0 mc\lambda_0^2}, \qquad (3.7.20)$$

where we have used Eq. (3.6.19) for B_{12}. In particular, if the transition wavelength λ_0 is expressed in meters,

$$A_{21} = (6.67 \times 10^{-5})\frac{g_1}{g_2}\frac{f}{\lambda_0^2} \text{ s}^{-1}. \qquad (3.7.21)$$

The appearance of the dimensionless factor $(g_1/g_2)f$ in this formula explains why the spontaneous emission rate is not simply a universal constant times λ_0^{-2}, as predicted by classical theory and discussed following Eq. (3.3.6).

- Denoting the nondegenerate states 1 and 2 in Eq. (3.A.26) by m_1 and m_2, we can write

$$\mathcal{A}(m_2, m_1) = \frac{e^2\omega_0^3}{3\pi\epsilon_0\hbar c^3}|\mathbf{x}_{m_1 m_2}|^2, \qquad (3.7.22)$$

where $\mathbf{x}_{m_1 m_2}$ is the matrix element between states m_1 and m_2 of the electron coordinate $\mathbf{x}$. (For a multielectron atom, $\mathbf{x}$ is the sum of the position vectors of all the electrons.) In the case of g_2 degenerate states m_2 associated with an atomic energy level E_2, each state m_2 has the same total radiative decay rate, namely $A_{21} = \sum_{m_1}\mathcal{A}(m_2, m_1)$, the sum of the spontaneous emission rates from m_2 to all possible lower states m_1 of the transition. This must be so because the states m_2 correspond simply to different "z components" of angular momentum, but the z direction is chosen arbitrarily; in other words, spherical symmetry requires that

$$A_{21} = \sum_{m_1}\frac{e^2\omega_0^3}{3\pi\epsilon_0\hbar c^3}|\mathbf{x}_{m_1 m_2}|^2 = \frac{1}{g_2}\sum_{m_1,m_2}\frac{e^2\omega_0^3}{3\pi\epsilon_0\hbar c^3}|\mathbf{x}_{m_1,m_2}|^2. \qquad (3.7.23)$$

Thus, the last equality in (3.7.20) implies that

$$f = \frac{1}{g_1}\frac{2m\omega_0}{3\hbar}\sum_{m_1,m_2}|\mathbf{x}_{m_1 m_2}|^2, \qquad (3.7.24)$$

which generalizes (3.A.25) to the case of degenerate levels.

Let us write Eq. (3.7.24) as

$$f_{12} = \frac{1}{g_1} \frac{2m\omega_{21}}{3\hbar} \sum_{m_1 m_2} |\mathbf{x}_{m_1 m_2}|^2 \qquad (3.7.25)$$

for the transition between levels 1 and 2, where $\omega_{21} = (E_2 - E_1)/\hbar \ (>0)$. Interchanging 1 and 2, we define

$$f_{21} = \frac{1}{g_2} \frac{2m\omega_{12}}{3\hbar} \sum_{m_1 m_2} |\mathbf{x}_{m_1 m_2}|^2 = -\frac{g_1}{g_2} f_{12}, \qquad (3.7.26)$$

which is negative; f_{12} and f_{21} are called oscillator strengths for absorption and emission, respectively. One motivation for introducing a separate oscillator strength for emission, which is not really necessary for our purposes, may be found at the end of Section 3.14.   ●

The degenerate states belonging to a given atomic energy level are those corresponding to different magnetic quantum numbers m. The application of a relatively weak magnetic field $\mathbf{B}$ establishes a preferred z direction and removes the degeneracy: Each of the states is shifted in energy by an amount proportional to $m|\mathbf{B}|$. This is the *Zeeman effect* discussed in many textbooks on quantum mechanics. The different values of m are defined with respect to some "z" direction, such as the direction of the magnetic field in the Zeeman effect. If the atom is exposed to isotropic (e.g., thermal) radiation, then by spherical symmetry the different magnetic substates must have equal populations, so that (3.7.19) is an exact consequence of the more general (3.7.12). In the case of atoms in unidirectional narrowband light, simplified rate equations such as (3.7.19) are often a good approximation if collisions between atoms or between atoms and container walls are effective in maintaining a nearly equal population distribution among degenerate magnetic substates, or if the intensity of the light is not large enough to produce a significant change in an initial thermal distribution of level populations. In the latter case the generalization of Eq. (3.6.7),

$$\frac{N_2}{N_1} = \frac{g_2 \mathcal{N}_2(m_2)}{g_1 \mathcal{N}_1(m_1)} = \frac{g_2}{g_1} e^{-(E_2 - E_1)/k_B T}, \qquad (3.7.27)$$

ensures the validity of (3.7.19). As discussed in Section 14.3, however, atoms can be "optically pumped" or "aligned" preferentially in certain magnetic substates.

3.8 COLLISION BROADENING

In Section 3.4 we showed that light is most strongly absorbed when it is nearly resonant with one of the natural oscillation frequencies of the atoms of a medium, and that absorption is due to "frictional" processes that damp out the electron oscillations. We have also shown that any frictional force in the Newton equation of an electron oscillator leads to a broadened absorption line, the lineshape being Lorentzian. We did not, however, give any fundamental explanation for the existence of frictional processes. We will now approach the question of absorption and lineshape from a more fundamental viewpoint, focusing our attention on "line-broadening" mechanisms in gases to answer the question of the origin of the frictional coefficient β.

It is a well-known result of experiment that, for sufficiently large pressures, the width of an absorption line in a gas increases as the pressure increases. This broadening is due to collisions of the atoms and is therefore called *collision broadening* or sometimes pressure broadening. Collision broadening is the most important line-broadening mechanism in gases at atmospheric pressures and is often dominant at much lower pressures as well. We will begin our study by considering the details of collision broadening.

Our treatment will follow the original approach of Lorentz. We will find, for instance, that a kind of frictional force arises naturally as a result of collisions, and that the damping rate β can be interpreted as simply the collision rate. We start with (3.4.1) *without* the frictional term and introduce the oscillator momentum $\mathbf{p}$ by writing

$$\frac{d\mathbf{x}}{dt} \equiv \frac{\mathbf{p}}{m}. \tag{3.8.1}$$

Then (3.4.1) can be rewritten in simple complex form as

$$\frac{d}{dt}\left(\mathbf{x} + i\frac{\mathbf{p}}{m\omega_0}\right) + i\omega_0\left(\mathbf{x} + i\frac{\mathbf{p}}{m\omega_0}\right) = i\hat{\boldsymbol{\varepsilon}}\,\frac{e}{m\omega_0}E(\mathbf{R}, t), \tag{3.8.2}$$

where one can easily check that the real part simply repeats the defining relation (3.8.1), and with the use of (3.8.1) the imaginary part is nothing other than (3.4.1) with the frictional force omitted. It will be convenient to have a shorthand form of this equation, so we define

$$\mathbf{x} + i\frac{\mathbf{p}}{m\omega_0} \equiv \mathbf{S} \tag{3.8.3}$$

and will examine the solution of the $\mathbf{S}$ equation,

$$\frac{d}{dt}\mathbf{S} + i\omega_0\mathbf{S} = i\hat{\boldsymbol{\varepsilon}}\,\frac{e}{m\omega_0}E(\mathbf{R}, t), \tag{3.8.4}$$

under the following interpretation of the effect of collisions.

We imagine collisions to occur in billiard-ball fashion, each collision lasting for a time that is very short compared to the time between collisions. We suppose that, immediately prior to a collision, the active electrons in an atom are oscillating along the axis defined by the field polarization, as indicated by (3.4.8). During a collision, the interaction between the two atoms causes a reorientation of the axes of oscillation. Since each atom in a gas may be bombarded by other atoms from any direction, we can assume that on the average all orientations of the displacements and momenta of the atomic electrons are equally probable following a collision, so after a collision the displacement and momentum both vanish on average. This is the assumption made by Lorentz. It is an assumption about the statistics of a large number of collisions rather than about the details of a single collision.

We will examine the consequences of this assumption as follows. We consider the evolution of the electron displacements of atoms that underwent their most recent collision at the representative earlier time t_1. The time of the last collision is known only in a statistical sense, and we will obtain a picture valid for the electrons in all the atoms by averaging over t_1. The simplest statistical model for the frequency of collisions is "Markovian," meaning memoryless. The fraction of atoms without a collision since t_1 must decrease during the waiting interval between t and $t + \delta t$, and in the memoryless model this change is directly proportional to the length of the interval δt but does not depend on t.

Specifically, if $f(t; t_1)$ is the fraction of dipoles at time t not having suffered a collision since t_1, then the fraction collision free at a time δt later is smaller by an amount proportional to the fraction "available" for collision, namely f itself, and to the time interval δt:

$$f(t + \delta t; t_1) = f(t; t_1) - \gamma_c f(t; t_1)\,\delta t, \tag{3.8.5}$$

where the proportionality constant γ_c is the rate at which collisions are occurring. In the limit of very small δt, this recipe for the surviving fraction becomes the simple equation

$$\frac{df}{dt} = -\gamma_c f, \tag{3.8.6}$$

and so long as γ_c is not dependent on t itself (the collisions are Markovian) the solution is

$$f(t; t_1) = e^{-\gamma_c(t - t_1)}. \tag{3.8.7}$$

Since the probability that a collision occurs in the time dt_1 is $\gamma_c f\,dt_1$, we can take account of all collisions by integrating over t_1. We will now indicate the time of latest collision explicitly and write the complex displacement as $\mathbf{S}(t, t_1)$, where we will enforce $\mathbf{S}(t_1, t_1) = 0$, corresponding to the fact that the starting displacement and momentum were zero following the collision. The collision-averaged complex displacement $\mathbf{S}$, which we will denote $\bar{\mathbf{S}}(t)$, is thus given by

$$\bar{\mathbf{S}}(t) \equiv \int_{-\infty}^{t} \mathbf{S}(t; t_1)\gamma_c f(t; t_1)\,dt_1. \tag{3.8.8}$$

By differentiating $\bar{\mathbf{S}}$ with respect to t we find that it satisfies a simple equation, one with an obvious physical interpretation. Using the evolution equation (3.8.4), and remembering that $\mathbf{S}(t_1, t_1) = 0$, we obtain

$$\frac{d}{dt}\bar{\mathbf{S}} + i(\omega_0 - i\gamma_c)\bar{\mathbf{S}} = i\hat{\boldsymbol{\varepsilon}}\,\frac{e}{m\omega_0}E(\mathbf{R}, t), \tag{3.8.9}$$

which we see to be exactly the same as Eq. (3.8.4) for the collisionless displacement, except for the appearance of the new γ_c term. Such an extra term is exactly what is needed to reproduce the β coefficient in the electron oscillator equation: writing $\bar{\mathbf{S}} = \bar{\mathbf{x}} + (i/m\omega_0)\bar{\mathbf{p}}$ in (3.8.9) and then taking the real and imaginary parts of that equation, we obtain

$$\frac{d^2\bar{\mathbf{x}}}{dt^2} + 2\gamma_c \frac{d\bar{\mathbf{x}}}{dt} + (\omega_0^2 + \gamma_c^2)\bar{\mathbf{x}} = \hat{\boldsymbol{\varepsilon}}\frac{e}{m}E(\mathbf{R}, t). \qquad (3.8.10)$$

While collisions occur frequently on a "normal" time scale, they are rare on the scale of an optical period (roughly 1 fs), so

$$\omega_0^2 \gg \gamma_c^2, \qquad (3.8.11)$$

and one safely ignores the contribution of γ_c^2 to the last term on the left side of (3.8.10). Then (3.8.10) is exactly the same as (3.4.4) when we equate β in (3.4.4) to the collision rate γ_c. In other words, an averaged treatment of collisions leads directly to "frictional" drag in the electron oscillation. The linewidth of the collision-broadened lineshape is [Eq. (3.4.23)]

$$\delta\nu_0 = \frac{\gamma_c}{2\pi} = \frac{1}{2\pi} \times \text{(collision rate)}. \qquad (3.8.12)$$

Collision broadening is often described in terms of a "dephasing" of the electron oscillators, as follows. Immediately after a collision the phase of the electron's oscillation has no correlation with the precollision phase. Collisions have the effect of "interrupting" the phase of oscillation, leading to an overall decay of the average electron displacement from equilibrium (Fig. 3.14). The damping rate γ_c is sometimes called a dephasing rate in order to distinguish it from an "energy decay" rate. The latter would appear as a frictional term in the equation of motion of each electron oscillator as well as in the average equation. In the absence of any inelastic collisions to decrease the energy of the electron oscillators, each oscillator would satisfy the Newton equation (3.2.18b) with no damping term. Due to elastic collisions, that is, collisions that only

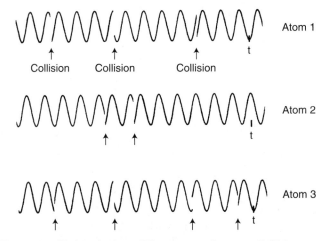

Figure 3.14   Electron oscillations in three different atoms in a gas. Collisions completely interrupt the phase of the oscillation. The average electron displacement associated with all the atoms in the gas therefore decays to zero at a rate given by γ_c, the inverse of the mean time between collisions.

interrupt the phase of the oscillation but do not produce any change in energy, the *average* electron displacement follows Eq. (3.4.4), which includes damping.

Collision Cross Sections

The collision rate γ_c may be expressed in terms of the number density N of atoms, the collision cross section σ between atoms, and the average relative velocity $\bar{v}$ of the atoms. Imagine some particular atom to be at rest and bombarded by a stream of identical atoms of velocity $\bar{v}$. If the number of atoms per unit volume in the stream is N, then the number of collisions per unit time undergone by the atom at rest is $N\sigma\bar{v}$, where the area σ is the *collision cross section* between the atom at rest and the atoms in the stream. The number of collisions per second is the same as if all the stream atoms within a cross-sectional area σ collide with the stationary atom. The idea here is the same as that used to define the absorption cross section for incident light.

According to the kinetic theory of gases, an atom of mass m_X has a mean velocity

$$v_{\text{rms}} = \left(\frac{8k_B T}{\pi m_X}\right)^{1/2} \tag{3.8.13}$$

in a gas in thermal equilibrium at temperature T. To obtain the average relative velocity $\bar{v}_{\text{rel}}$ of colliding atoms of masses m_X and m_Y in the gas, we replace m_X in (3.8.13) by the reduced mass

$$\mu_{X,Y} = \frac{m_X m_Y}{m_X + m_Y} = \left(\frac{1}{m_X} + \frac{1}{m_Y}\right)^{-1}. \tag{3.8.14}$$

Thus,

$$\bar{v}_{\text{rel}} = \left[\frac{8k_B T}{\pi}\left(\frac{1}{m_X} + \frac{1}{m_Y}\right)\right]^{1/2}. \tag{3.8.15}$$

It is convenient to express this in terms of the atomic (or molecular) weights M_X and M_Y:

$$\bar{v}_{\text{rel}} = \left[\frac{8RT}{\pi}\left(\frac{1}{M_X} + \frac{1}{M_Y}\right)\right]^{1/2}, \tag{3.8.16}$$

where R, the universal gas constant, is Boltzmann's constant times Avogadro's number. The collision rate for molecules of type X is therefore

$$\gamma_c = \sum_Y N(Y)\sigma(X, Y)\bar{v}_{\text{rel}}(X, Y)$$

$$= \sum_Y N(Y)\sigma(X, Y)\left[\frac{8RT}{\pi}\left(\frac{1}{M_X} + \frac{1}{M_Y}\right)\right]^{1/2}, \tag{3.8.17}$$

where the sum is over all species Y, including X.

The important "unknowns" in expression (3.8.17) are the collision cross sections $\sigma(X, Y)$, which often are not known very accurately. The simplest approximation to the cross section is the "hard-sphere" approximation. We write

$$\bar{\sigma}(X, Y) = \frac{\pi}{4}(d_X + d_Y)^2, \tag{3.8.18}$$

where d_X and d_Y are the hard-sphere molecular diameters, estimates of which can be made from measurements of various transport quantities such as thermal conductivities or diffusion constants; $\bar{\sigma}(X,Y)$ is the area of a circle of diameter $d_X + d_Y$, just what we would expect if the molecules acted like spheres of diameters d_X and d_Y. For CO_2, for example, the hard-sphere diameter is about 0.4 nm. From (3.8.18), therefore, the hard-sphere cross section for two CO_2 molecules is $\sigma(CO_2, CO_2) = 5.03 \times 10^{-19} \, \text{m}^2$. For a gas of pure CO_2 at $T = 300$K we find the average relative velocity of two colliding CO_2 molecules to be $\bar{v}_{\text{rel}} = 5.37 \times 10^2 \, \text{m/s}$. The collision rate (3.8.17) in the hard-sphere approximation is therefore

$$\gamma_c = N(5.03 \times 10^{-19} \, \text{m}^2)(5.37 \times 10^2 \, \text{m/s}) = 2.70 \times 10^{-16} N/s, \tag{3.8.19}$$

where N is the number of CO_2 molecules per cubic meter. For an ideal gas we calculate (Problem 3.10)

$$N = 9.65 \times 10^{24} \frac{P(\text{Torr})}{T}, \tag{3.8.20}$$

where $P(\text{Torr})$ is the pressure in Torr (1 atm = 760 Torr) and T is the temperature (K). From (3.8.17), finally, the collision rate for a gas of CO_2 at 300K is

$$\gamma_c = 8.69 \times 10^6 P(\text{Torr}) \, \text{s}^{-1}. \tag{3.8.21}$$

Thus, at a pressure of 1 atm we calculate the collision rate

$$\gamma_c = 6.60 \times 10^9 \, \text{s}^{-1}, \tag{3.8.22}$$

and from (3.8.12) the collision-broadened linewidth

$$\delta\nu_0 = 1.05 \times 10^9 \, \text{Hz}. \tag{3.8.23}$$

The actual collision-broadened linewidths can be larger, by as much as an order of magnitude or more, than those calculated in the hard-sphere approximation. The value calculated above, however, is reasonable, and it allows us to point out some general features of collision-broadened linewidths. First, we note that the collision rate (3.8.22) is very much smaller than an optical frequency, as assumed in (3.8.11). The linewidth $\delta\nu_0$ is thus also orders of magnitude less than an optical frequency. This explains why we can speak of absorption "lines" in a gas, even though the absorption occurs over a band of frequencies: the band has a width ($\sim 2\delta\nu_0$) that is very small compared to the resonance frequency ν_0.

From (3.8.21) we note that the linewidth is linearly proportional to the pressure. For this reason, experimental results for collision-broadened linewidths are often reported in units such as MHz-Torr^{-1}. The linewidth calculated above, for instance, may be expressed as 1.38 MHz-Torr^{-1} at 300K.

Our treatment of collision broadening only highlights some general features of a complex subject. In actual calculations we prefer always to use measured values of the collision-broadened linewidths. We note parenthetically that, for the 10.6-μm CO_2 laser line, the linewidth (1.38 MHz-Torr^{-1}) computed above is about three times smaller than the experimentally determined value. It is possible to calculate these widths more accurately, but this will not concern us. See also Problem 3.11.

3.9 DOPPLER BROADENING

The Doppler effect was demonstrated for sound waves in 1845 by C. H. D. Buys Ballot, who employed trumpeters performing in a moving train to demonstrate it. The mathematician C. J. Doppler had predicted the effect in 1842. His prediction applied also to light, although Maxwell's electromagnetic theory of light waves was still nearly a quarter of a century away.

Let us consider again a gaseous medium, this time only very weakly influenced by collisions (i.e., β is very small). Every electron oscillator will undergo practically undamped oscillation at the field frequency. Nevertheless, we will show that, because of the Doppler effect, an absorption line is broadened and its width can be much larger than β. We will find that the lineshape associated with the Doppler effect is not the Lorentzian function (3.4.26) but rather the Gaussian function given in Eq. (3.9.9) below.

To an atom moving with velocity $v \ll c$ away from a source of radiation of frequency ν, the frequency of the radiation appears to be shifted:

$$\nu' = \nu\left(1 - \frac{v}{c}\right). \tag{3.9.1}$$

This is the Doppler effect. It implies that a source of radiation (e.g., a laser) exactly resonant in frequency with an absorption line of a stationary atom will not be in resonance with the same absorption line in a moving atom, and the frequency offset is $\delta\nu = (v/c)\nu$. Similarly, a nonresonant absorption line of an atom may be brought into resonance with the field as a result of atomic motion. Since the atoms in a gas exhibit a wide variety of velocities, a broad range of different effective resonance frequencies will be associated with a given absorption line. In other words, the absorption line is broadened because of the Doppler effect, and is said to be Doppler-broadened.

For a gas in thermal equilibrium at the temperature T, the fraction $df(v)$ of atoms having velocities between v and $v + dv$ along any one axis is given by the (one-dimensional) Maxwell–Boltzmann distribution,

$$df(v) = \left(\frac{m_X}{2\pi k_B T}\right)^{1/2} e^{-m_X v^2/2k_B T}\, dv. \tag{3.9.2}$$

Here again k_B is the Boltzmann constant and m_X is the mass of an atom or molecule of species X. Because we have assumed that collisions are almost negligible, an atom with resonance frequency ν_0 and velocity v moving away from the source of radiation will only absorb radiation very near to (within $\Delta\nu = \beta/2\pi$ of) the frequency

$$\nu = \nu_0\left(1 + \frac{v}{c}\right). \tag{3.9.3}$$

The fraction of atoms absorbing within the frequency interval from ν to $\nu + d\nu$ is thus equal to the fraction of atoms with velocity in the interval from v to $v + dv$. From (3.9.3) we have (see also Problem 3.12)

$$v = \frac{c}{\nu_0}(\nu - \nu_0) \tag{3.9.4}$$

and $dv = (c/\nu_0)\,d\nu$. Using (3.9.2) we can determine that this fraction is

$$df_v(v) = \left(\frac{m_X}{2\pi k_B T}\right)^{1/2} e^{-m_X c^2 (\nu-\nu_0)^2 / 2k_B T\nu_0^2}\left(\frac{c}{\nu_0}\,d\nu\right). \tag{3.9.5}$$

Since the absorption rate at frequency ν must be proportional to $df_v(v)$, we may write the Doppler lineshape function as

$$S(\nu) = \left(\frac{m_X c^2}{2\pi k_B T\nu_0^2}\right)^{1/2} e^{-m_X c^2 (\nu-\nu_0)^2 / 2k_B T\nu_0^2}. \tag{3.9.6}$$

Because (3.9.2) was normalized to unity when integrated over velocity, (3.9.6) is normalized to unity with respect to the frequency offset (or "detuning") $\nu - \nu_0$, as required by the definition of a lineshape function. By direct calculation using (3.9.6) we find

$$\begin{aligned}
\int_0^\infty d\nu\, S(\nu) &= \left(\frac{m_X c^2}{2\pi k_B T\nu_0^2}\right)^{1/2}\int_0^\infty d\nu\, e^{-m_X c^2 (\nu-\nu_0)^2 / 2k_B T\nu_0^2}\\
&= \left(\frac{m_X c^2}{2\pi k_B T\nu_0^2}\right)^{1/2}\int_{-\nu_0}^\infty d\mu\, e^{-m_X c^2 \mu^2 / 2k_B T\nu_0^2}\\
&\approx \left(\frac{m_X c^2}{2\pi k_B T\nu_0^2}\right)^{1/2}\int_{-\infty}^\infty d\mu\, e^{-m_X c^2 \mu^2 / 2k_B T\nu_0^2}\\
&= \left(\frac{m_X c^2}{2\pi k_B T\nu_0^2}\right)^{1/2}\left[\frac{\nu_0}{c}\left(\frac{2\pi k_B T}{m_X}\right)^{1/2}\right] = 1. \tag{3.9.7}
\end{aligned}$$

We have used $m_X c^2/k_B \gg T$ to replace the lower limit of the integral by $-\infty$. (For the hydrogen atom, for example, $m_X c^2/k_B \approx 10^{13}$K.)

It is convenient to define

$$\delta v_D = 2\frac{v_0}{c}\left(\frac{2k_BT}{m_X}\ln 2\right)^{1/2},$$

(3.9.8)

in terms of which

$$S(v) = \frac{1}{\delta v_D}\left(\frac{4\ln 2}{\pi}\right)^{1/2} e^{-4(v-v_0)^2 \ln 2/\delta v_D^2},$$

(3.9.9)

and we recognize that δv_D is the width (FWHM) of the Doppler absorption curve, since (Fig. 3.15)

$$S\left(v_0 \pm \tfrac{1}{2}\delta v_D\right) = S(v_0)e^{-\ln 2} = \tfrac{1}{2}S(v_0).$$

(3.9.10)

δv_D is called the *Doppler width*.

The Doppler width is sometimes defined in terms of the $1/e$ point of the curve rather than the half-maximum point. Sometimes it is defined as the half width at half-maximum (HWHM) rather than the FWHM. Thus, one finds formulas in the literature differing by factors of 2, ln 2, and so forth. It is important to keep these possible differences in mind when comparing calculations.

In terms of the molecular weight M_X, and the wavelength $\lambda_0 = c/v_0$ of the absorption line, the Doppler width is

$$\delta v_D = \frac{2}{\lambda_0}\left(\frac{2RT}{M_X}\ln 2\right)^{1/2} = 2.15 \times 10^{11}\left[\frac{1}{\lambda_0}\left(\frac{T}{M_X}\right)^{1/2}\right] \text{Hz},$$

(3.9.11)

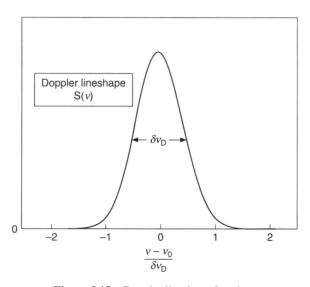

Figure 3.15 Doppler lineshape function.

where λ_0 is expressed in nanometers, M_X in grams, and T in kelvins. In these same units the formula

$$\frac{\delta\nu_D}{\nu_0} \approx 7.16 \times 10^{-7} \left(\frac{T}{M_X}\right)^{1/2} \tag{3.9.12}$$

for the ratio of the Doppler width to the resonance frequency is also useful.

The Doppler width depends only on the transition frequency, the gas temperature, and the molecular weight of the absorbing species. It is, therefore, much simpler to calculate than the collision-broadened width, which involves the collision cross section. As an example, consider the 632.8-nm line of Ne in the He–Ne laser. Since $M_{Ne} = 20.18$ g, we obtain from (3.9.11) the Doppler width

$$\delta\nu_D \approx 1500\,\text{MHz} \tag{3.9.13}$$

for $T = 400$K. For the 10.6-μm line of CO_2 and the same temperature, however, we find a much smaller Doppler width:

$$\delta\nu_D \approx 61\,\text{MHz}. \tag{3.9.14}$$

3.10 THE VOIGT PROFILE

Doppler broadening is an example of what is called *inhomogeneous* broadening. The term inhomogeneous means that individual atoms within a collection of otherwise identical atoms do not have the same resonant response frequencies. Thus, atoms in the collection can show resonant response over a range of frequencies. In the Doppler case this is because nominally identical individual atoms can have different velocities. These different velocities serve as tags or labels for the individual atoms, and any discussion of the behavior of a sample of such atoms must take account of all the velocity labels.

There are other possible inhomogeneities that have the same effect as the Doppler distribution of velocities. For example, impurity atoms embedded randomly in a crystal are subjected to different local crystal fields due to strains and defects. These have the effect of shifting the resonance frequency of each atom slightly differently. The distribution of such shifts acts very much like the Doppler distribution and gives rise to an inhomogeneous broadening of the absorption line associated with the nominally identical impurity atoms subjected to different local fields in the crystal. This type of random strain broadening is present in the resonance lines of impurity ions such as titanium or chromium in Ti:sapphire and ruby laser crystals, where the host is Al_2O_3 (corundum).

The line broadening associated with collisions is different, and is called *homogeneous*. This is because each atom can itself absorb light over a range of frequencies, due to the interruptions of its dipole oscillations by collisions. Since the collisional history of every atom is assumed to be the same, no greater collisional broadening is associated with the collection of atoms than is associated with an individual atom.

In general, we cannot characterize an absorption lineshape of a gas as a pure collision-broadened Lorentzian or a pure Doppler-broadened Gaussian. Both phase interrupting collisions and the Doppler effect may play a role in determining the lineshape.

We will now derive the absorption lineshape when both collision broadening and Doppler broadening must be taken into account.

Equation (3.4.26) gives the collision-broadened lineshape for each atom in the gas. If an atom has a velocity component v moving away from the source of light of frequency $\nu \approx \nu_0$, its absorption curve is Doppler shifted to

$$S(\nu, v) = \frac{(1/\pi)\delta\nu_0}{(\nu_0 - \nu + v\nu/c)^2 + \delta\nu_0^2}. \tag{3.10.1}$$

In other words, the peak absorption for this atom will occur at the field frequency ν such that (3.9.3) is satisfied:

$$\nu \approx \nu_0 + \frac{\nu_0 v}{c}. \tag{3.10.2}$$

The lineshape function for the gas is obtained by integrating over the velocity distribution (3.9.2):

$$
\begin{aligned}
S(\nu) &= \int_{-\infty}^{\infty} dv\, S(\nu, v) \left(\frac{M_X}{2\pi RT}\right)^{1/2} e^{-M_X v^2/2RT} \\
&= \left(\frac{M_X}{2\pi RT}\right)^{1/2} \frac{\delta\nu_0}{\pi} \int_{-\infty}^{\infty} \frac{dv\, e^{-M_X v^2/2RT}}{(\nu_0 - \nu + \nu_0 v/c)^2 + \delta\nu_0^2} \\
&= \frac{1}{\pi^{3/2}} \frac{b^2}{\delta\nu_0} \int_{-\infty}^{\infty} \frac{dy\, e^{-y^2}}{(y+x)^2 + b^2},
\end{aligned}
\tag{3.10.3}
$$

where we have made the change of variables

$$x = (4\ln 2)^{1/2} \left(\frac{\nu_0 - \nu}{\delta\nu_D}\right), \tag{3.10.4}$$

and we have defined

$$b = (4\ln 2)^{1/2} \frac{\delta\nu_0}{\delta\nu_D}. \tag{3.10.5}$$

The lineshape function (3.10.3) is called the *Voigt profile*.

In the case when the applied field is tuned exactly to the resonance frequency ν_0, we have $x = 0$ and therefore

$$S(\nu_0) = \frac{b^2}{\pi^{3/2}\delta\nu_0} \int_{-\infty}^{\infty} \frac{dy\, e^{-y^2}}{y^2 + b^2}. \tag{3.10.6}$$

The integral defines a known function:

$$\int_{-\infty}^{\infty} \frac{dy\, e^{-y^2}}{y^2 + b^2} = \frac{\pi}{b} e^{b^2} \operatorname{erfc}(b), \tag{3.10.7}$$

where

$$\text{erfc}(b) = \frac{2}{\pi^{1/2}} \int_b^\infty du\, e^{-u^2} \tag{3.10.8}$$

is the *complementary error function*. From (3.10.6) and (3.10.7), therefore, the lineshape function for the resonance frequency $\nu = \nu_0$ has the value

$$S(\nu_0) = \frac{b^2}{\pi^{3/2}\delta\nu_0} \frac{\pi}{b} e^{b^2} \text{erfc}(b) = \frac{b}{\pi^{1/2}\delta\nu_0} e^{b^2} \text{erfc}(b)$$

$$= \left(\frac{4\ln 2}{\pi}\right)^{1/2} \frac{1}{\delta\nu_D} e^{b^2} \text{erfc}(b). \tag{3.10.9}$$

This function is plotted versus the parameter b in Fig. 3.16.

$S(\nu_0)$ depends strongly on the ratio of the linewidths for collision and Doppler broadening. When the collision width $\delta\nu_0$ is much greater than the Doppler width $\delta\nu_D$, we have $b \gg 1$. For large values of b,

$$e^{b^2} \text{erfc}(b) \approx \frac{1}{\pi^{1/2}b} \quad (b \gg 1). \tag{3.10.10}$$

In this "collision-broadened limit," therefore, we have from (3.10.9) the result

$$S(\nu_0) \approx \frac{1}{\pi\delta\nu_0} \quad (b \gg 1), \tag{3.10.11}$$

which is exactly (3.4.28) for the case of pure collision broadening. In the limit in which the Doppler width is much greater than the collision-broadened width, on the other hand, we have $b \ll 1$, in which case the function

$$e^{b^2} \text{erfc}(b) \approx 1 \quad (b \ll 1). \tag{3.10.12}$$

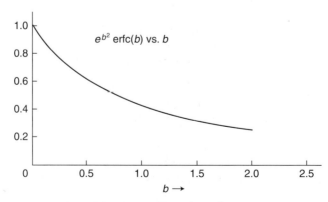

Figure 3.16   The function $e^{b^2}\text{erfc}(b)$.

Then from (3.10.9)

$$S(\nu_0) \approx \frac{1}{\delta\nu_D} \left(\frac{4\ln 2}{\pi}\right)^{1/2} \qquad (b \ll 1), \qquad (3.10.13)$$

which is the result (3.9.9) for pure Doppler broadening and $\nu = \nu_0$. The limits $\delta\nu_0 \gg \delta\nu_D$ and $\delta\nu_0 \ll \delta\nu_D$ thus reproduce the results for pure collision broadening and pure Doppler broadening, respectively. In general, for arbitrary values of b, $S(\nu_0)$, given by (3.10.9), must be evaluated using tables of erfc(b).

For the general case of arbitrary values of both the parameter b and the detuning parameter x, the lineshape function $S(\nu)$ given by Eq. (3.10.3) must be evaluated from tabulated values of the more complicated function

$$\int_{-\infty}^{\infty} \frac{dy\, e^{-y^2}}{(y+x)^2 + b^2} = \frac{\pi}{b}\mathrm{Re}\left(\frac{i}{\pi}\int_{-\infty}^{\infty}\frac{dy\, e^{-y^2}}{x+y+ib}\right) = \frac{\pi}{b}\mathrm{Re}[w(x+ib)], \qquad (3.10.14)$$

where w is the "error function of complex argument." Numerical values are tabulated in various mathematical handbooks.

In Table 3.1 we summarize our results for collision broadening and Doppler broadening, as well as the more general case of the Voigt profile.

TABLE 3.1   Collision, Doppler, and Voigt Lineshape Functions

Collision-Broadening Lineshape

$$S(\nu) = \frac{(1/\pi)\delta\nu_0}{(\nu - \nu_0)^2 + \delta\nu_0^2}$$

$$\delta\nu_0 = \frac{\text{collision rate}}{2\pi}$$

Doppler-Broadening Lineshape

$$S(\nu) = \frac{0.939}{\delta\nu_D} e^{-2.77(\nu-\nu_0)^2/\delta\nu_D^2}$$

$$\delta\nu_D = 2.15 \times 10^5 \left[\frac{1}{\lambda_0}\left(\frac{T}{M}\right)^{1/2}\right] \text{ MHz}$$

T = gas temperature (K)
M = molecular weight (g) of absorber
λ_0 = wavelength (nm) of absorption line

Voigt Lineshape

$$S(\nu) = \frac{0.939}{\delta\nu_D}\mathrm{Re}\, w(x+ib)$$

$$x = 1.67\frac{\nu_0 - \nu}{\delta\nu_D}$$

$$b = 1.67\frac{\delta\nu_0}{\delta\nu_D}$$

w = error function of complex argument

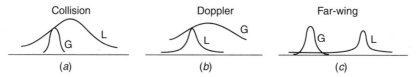

Figure 3.17 Sketch of factors in the integrand of (3.10.3) in three limiting cases: (*a*) collision-broadened limit, (*b*) Doppler-broadened limit, and (*c*) far-wing limit.

- Without going to numerical tables, and even without a study of the asymptotic properties of $w(x + ib)$, it is possible to evaluate the Voigt integral (3.10.3) in several limits because both factors in the integrand are normalized lineshapes themselves. There are three limits of interest, as shown in Fig. 3.17.

   Collisional Limit $(\delta v_0 \gg \delta v_D)$ In this case $S(v, v)$ is very broad and slowly varying compared to the narrow Gaussian velocity distribution (Fig. 3.17*a*). Since the Gaussian is normalized to unity, it acts like the delta function $\delta(v)$, and the Voigt integral reduces to $S(v) = S(v, v = 0)$, which is just the original collisional Lorentzian lineshape given in (3.4.26).

   Doppler Limit $(\delta v_D \gg \delta v_0)$ In this case the reverse is true (Fig. 3.17*b*), and the collisional function $S(v, v)$ acts like the delta function $\delta(v_0 - v + vv/c)$. Thus, the Voigt integral gives back the Gaussian function (3.9.9). Except at high pressures or in cases where the Doppler distribution is altered by atomic beam collimation it is usually valid to assume that the inequality $\delta v_D \gg \delta v_0$ is accurate and the Doppler limit applies.

   Far-Wing Limit $(|v - v_0| \gg \delta v_D, \delta v_0)$ This case refers to the spectral region far from line center, far outside the half widths of either the collisional or Doppler factors in the Voigt integrand. Thus, the integrand is the product of two peaked functions. Each peak falls in the remote wing of the other function (see Fig. 3.17*c*). Here the qualitative difference between Gaussian and Lorentzian functions is significant. The Gaussian is much more compact. It falls to zero much more rapidly than the Lorentzian. Because the Lorentzian's wings are falling relatively slowly, as $1/v^2$ for large v, it still has nonzero value at the position of the Gaussian peak. However, the value of the Gaussian function is effectively zero by comparison near the Lorentzian's peak. Thus, the contribution of the Gaussian function in the Lorentzian wing is much greater than that of the Lorentzian function in the Gaussian wing, and the Voigt integral can be replaced by (3.4.26) in its far wing:

$$S(v) \longrightarrow \frac{\delta v_0 / \pi}{(v - v_0)^2}. \tag{3.10.15}$$

This result, which can be derived more formally from the asymptotic behavior of the complementary error function, is anomalous in the sense that the lineshape behaves like a Lorentzian in the far wing even if the broadening is principally Doppler, not collisional $(\delta v_D \gg \delta v_0)$. •

3.11   RADIATIVE BROADENING

Under most circumstances in gaseous media the broadening of spectral lines is due mainly to collisions and atomic motion. However, even if these effects could be neglected, a spectral line would still have a nonvanishing "natural" width. This natural or *radiative* line broadening is associated with spontaneous emission.

If A_2 and A_1 are the spontaneous emission rates of the upper and lower states, respectively, of a transition, then the radiative lineshape given by quantum theory is the Lorentzian function

$$S(\nu) = \frac{\delta\nu_{\text{rad}}/\pi}{(\nu - \nu_0)^2 + \delta\nu_{\text{rad}}^2}, \tag{3.11.1}$$

where the radiative linewidth $\delta\nu_{\text{rad}}$ is

$$\delta\nu_{\text{rad}} = \frac{1}{4\pi}(A_2 + A_1). \tag{3.11.2}$$

The total spontaneous emission rates of the two states, A_2 and A_1, are each the sum of spontaneous emission rates to all possible lower states. In particular, A_2 includes A_{21}, the rate of spontaneous emission on the $2 \rightarrow 1$ transition. Thus, if state 1 is the ground state and state 2 the first excited state of an atom, then $A_2 = A_{21}$, $A_1 = 0$, and $\delta\nu_{\text{rad}} = A_{21}/4\pi$.

In a gas the total homogeneous linewidth of a transition is $\delta\nu_h = \delta\nu_0 + \delta\nu_{\text{rad}}$, where $\delta\nu_0$ is the width due to collision broadening. As the pressure goes to zero, $\delta\nu_0$ also goes to zero and the only contribution to the homogeneous linewidth is $\delta\nu_{\text{rad}}$. If furthermore the inhomogeneous (Doppler) broadening is negligible, then the line is homogeneously broadened with lineshape function (3.11.1). Unlike other sources of line broadening, spontaneous emission cannot ordinarily be reduced by changing certain variables such as pressure or temperature in an experiment. This is why the broadening due to spontaneous emission is called natural line broadening: the term "natural" is meant to imply that the linewidth $\delta\nu_{\text{rad}}$ is immutable, that is, that $\delta\nu_{\text{rad}}$ is fundamentally the smallest possible linewidth that can be realized in any experiment. This is not strictly true, however. We will see, for instance, that the spectrum of laser radiation can be much narrower than the natural linewidth of the laser transition.

• As remarked following Eq. (3.12.19), the spontaneous emission rate, and therefore $\delta\nu_{\text{rad}}$, is changed from its "free-space" value when the emission occurs in a host medium of refractive index $n(\nu_0) \neq 1$. The spontaneous emission rate can also be changed by placing the emitting atom inside a cavity in which only certain resonant frequencies are possible. If the transition frequency ν_0 is not an allowed cavity mode frequency, there is no spontaneous emission at ν_0. Similarly, if ν_0 is a cavity mode frequency, the spontaneous emission rate will vary with the position of the atom inside the cavity. The spontaneous emission rate also varies, for instance, with the distance of an atom from a reflecting surface. All of these effects have been observed and studied experimentally under the rubric of "cavity QED" (cavity quantum electrodynamics).

Radiative broadening can be understood as a consequence of the finite lifetime of the excited state. Any process that causes population to be removed from states 1 and 2 at the rates A_1 and A_2 will result in a (homogeneous) linewidth of the form (3.11.2). Note, however, that the collisional linewidth $\delta\nu_0$ does not involve population changes but only "elastic" collisions that, classically speaking, disrupt the electron oscillations without changing its energy (Section 3.8). The "lifetime" in this case is the average time between collisions. Population-changing collisions make a smaller contribution than $\delta\nu_0$ to the homogeneous linewidth because the probability of a population change in a collision is smaller (often much smaller) than one.

It follows from these remarks that stimulated emission and absorption should also produce a homogeneous line broadening. This *power broadening* is discussed in the following chapter.   •

3.12   ABSORPTION AND GAIN COEFFICIENTS

We now consider the propagation of narrowband radiation in a medium of atoms having a transition frequency equal, or nearly equal, to the frequency of the radiation. Our goal is to derive an equation describing the propagation of the intensity I_ν of the field. A more rigorous treatment is given in Chapter 9. While the treatment here is simplified, it yields a correct and very important formula for the *absorption coefficient* in the case of attenuation of light in an absorbing medium, and for the *gain coefficient* for the amplification of light that occurs in a laser.

The intensity I_ν is equal to the field energy density u_ν times the wave propagation velocity. The rate at which electromagnetic energy passes through a plane cross-sectional area A at z is $I_\nu(z)A$, and at an adjacent plane at $z + \Delta z$ this rate is $I_\nu(z + \Delta z)A$; the difference is

$$[I_\nu(z + \Delta z) - I_\nu(z)]A \cong \left[I_\nu(z) + \frac{\partial I_\nu}{\partial z}\Delta z - I_\nu(z)\right]A = \frac{\partial}{\partial z}(I_\nu A)\,\Delta z \qquad (3.12.1)$$

in the limit in which Δz is very small. This equation gives the rate at which electromagnetic energy leaves the volume $A\,\Delta z$, that is,

$$\frac{\partial}{\partial t}(u_\nu A\,\Delta z) = -\frac{\partial}{\partial z}(I_\nu A)\,\Delta z, \qquad (3.12.2)$$

where u_ν is the field energy density. Since A and Δz are constant, and $u_\nu = I_\nu/c$, we may write this equation in the form

$$\frac{1}{c}\frac{\partial I_\nu}{\partial t} + \frac{\partial I_\nu}{\partial z} = 0, \qquad (3.12.3)$$

the so-called *equation of continuity*. Equation (3.12.3) is an example of Poynting's theorem, in one space dimension, and is applicable to a plane wave propagating in vacuum.

If the wave propagates in a medium, however, we must replace the zero on the right-hand side of (3.12.3) by the rate per unit volume at which electromagnetic energy changes due to the medium. We can calculate this from the rate of change of the atomic level populations due to both absorption and stimulated emission. Letting N_1 be the number of atoms per unit volume in the lower energy level of the resonant transition, and N_2 the number per unit volume in the upper level, we can write the rate of change of the energy per unit volume in the medium due to stimulated emission and absorption as [recall Eq. (3.7.19)]

$$hv\frac{dN_2}{dt} = -\sigma(v)I_\nu\left(N_2 - \frac{g_2}{g_1}N_1\right) = -\frac{hv}{c}I_\nu S(v)B_{21}\left(N_2 - \frac{g_2}{g_1}N_1\right)$$

$$= -hvB_{21}u_\nu S(v)\left(N_2 - \frac{g_2}{g_1}N_1\right). \qquad (3.12.4)$$

Conservation of energy demands that the rate of change of field energy be minus the rate of change of the energy of the atoms of the medium. Thus, the change in intensity due to stimulated emission and absorption is described in the plane-wave approximation by the equation

$$\left(\frac{1}{c}\frac{\partial}{\partial t} + \frac{\partial}{\partial z}\right)I_\nu = \sigma(\nu)\left(N_2 - \frac{g_2}{g_1}N_1\right)I_\nu. \tag{3.12.5}$$

It is convenient to group the factors multiplying I_ν on the right-hand side into a single coefficient $g(\nu)$ having units of (length)$^{-1}$:

$$g(\nu) = \sigma(\nu)\left(N_2 - \frac{g_2}{g_1}N_1\right) = \frac{\lambda^2 A_{21}}{8\pi}\left(N_2 - \frac{g_2}{g_1}N_1\right)S(\nu). \tag{3.12.6}$$

Depending on whether $N_2 - (g_2/g_1)N_1$ is positive or negative, $g(\nu)$ is called the *gain coefficient* or the *absorption coefficient*, respectively. Thus, if $N_2 - (g_2/g_1)N_1 < 0$, we define the (positive) absorption coefficient as

$$a(\nu) = \frac{\lambda^2 A_{21}}{8\pi}\left(\frac{g_2}{g_1}N_1 - N_2\right)S(\nu) \quad \text{[absorption coefficient, } (g_2/g_1)N_1 > N_2]. \tag{3.12.7}$$

An important special case is that in which practically *all* the atoms are in their ground states, so that $N_2 \approx 0$ and $N_1 \approx N$, the total number of absorbing atoms per unit volume of the medium. In this case

$$a(\nu) \approx \frac{\lambda^2 A_{21}}{8\pi}\frac{g_2}{g_1}NS(\nu). \tag{3.12.8}$$

The terms *absorption coefficient* and *gain coefficient* are easily understood by considering the temporal steady state in which I_ν is independent of time and varies only with the distance z of propagation in the plane-wave approximation we are assuming. Then (3.12.5) simplifies to

$$\frac{dI_\nu}{dz} = g(\nu)I_\nu. \tag{3.12.9}$$

If furthermore the numbers of atoms per unit volume in the two levels of the resonant transition are independent of I_ν and z, so that $g(\nu)$ is independent of I_ν and z, then

$$I_\nu(z) = I_\nu(0)e^{g(\nu)z}. \tag{3.12.10}$$

Thus, the intensity decreases or increases exponentially with distance of propagation z in the medium, depending on whether the medium is absorbing [$a(\nu) > 0$] or amplifying [$g(\nu) > 0$], respectively. The exponential attenuation formula $I_\nu(z) = I_\nu(0)\,e^{-a(\nu)z}$ for an absorber is often called Lambert's law or Beer's law, and $a(\nu)^{-1}$ is called the Beer length (Problem 3.14).

The approximation of exponential attenuation of intensity in an absorber or exponential growth in an amplifier is a useful and often very accurate one. However, we cannot in general assume, as we have done in obtaining (3.12.10), that the atomic level populations are independent of the field intensity. In general, the intensity I_ν and the populations N_1 and N_2 *are not independent* but are determined by the *coupled* differential equations (3.12.5) and (3.7.5). These equations account not only for the change in intensity as the field propagates in the medium but also for the change in the level populations of the atoms due to absorption and emission induced by the field. That is, these equations determine the field intensity and the atomic level populations *self-consistently*.

More often than not we must account for changes in intensity and atomic level populations produced by effects other than absorption, stimulated and spontaneous emission. In other words, it is generally necessary to add more terms to Eqs. (3.12.5) and (3.7.5). For example, inelastic collisions of atoms will cause N_2 and N_1 to change in ways that are not accounted for by Eqs. (3.7.5), and these equations certainly do not account for the physical mechanisms responsible for creating "gain" [$g(\nu) > 0$] in a laser. And there may be scattering, diffraction, and other "loss" processes that cause the field intensity to change but are not included in Eq. (3.12.10). We shall deal with such effects in the following chapters.

Expression (3.12.6) for the gain coefficient may be generalized to include the refractive index of the host medium. This generalization, which is derived below, is[11]

$$g(\nu) = \frac{\lambda^2 A_{21}}{8\pi n^2} \left(N_2 - \frac{g_2}{g_1} N_1 \right) S(\nu), \tag{3.12.11}$$

where n is the refractive index at the frequency ν. This modification is significant in solid-state lasers, where n may differ appreciably from unity.

● To derive this result we first return to the thermal radiation energy density (3.6.1) and see how it is modified when the radiation is in a medium with refractive index $n(\nu)$. The denominator in (3.6.1) is unaffected by the refractive index, but the following argument shows that the numerator, the number of modes per unit volume in the frequency interval [$\nu, \nu + d\nu$], must depend upon $n(\nu)$.

Let the medium of refractive index $n(\nu)$ be a box with sides of length L_x, L_y, and L_z, such that radiation of frequency ν consists of standing waves inside the box. Along each of the x, y, and z directions there is one node of the field of frequency ν for each integral multiple of the wavelength. The number of nodes along the x direction, for instance, is $N_x = k_x L_x / 2\pi$, where k_x is the x component of the wave vector $\mathbf{k}$ and $|\mathbf{k}| = k = 2\pi n(\nu)\nu/c$ is the wave number (see Sec. 8.2). A small change Δk_x in k_x, therefore, implies a change $\Delta N_x = L_x \times \Delta k_x / 2\pi$ in the number of nodes along the x direction. (We assume $N_x \gg 1$.) Equating the change in the number of nodes with the change in the number of *modes*, we see that the number of modes of frequency ν in a volume $\Delta k_x \, \Delta k_y \, \Delta k_z \to d^3 k$ of "k space" must be $\Delta N_x \, \Delta N_y \, \Delta N_z = [L_x L_y L_z/(2\pi)^3] d^3 k = V d^3 k/(2\pi)^3$, where V is the volume of the box. Multiplying by 2 in order to account for the two independent (orthogonal) polarizations for each frequency and direction of propagation, we obtain the number of field modes of frequency ν per unit volume,

$$d\mathcal{N}_\nu = \frac{2}{(2\pi)^3} \, d^3 k = \frac{2}{(2\pi)^3} 4\pi k^2 \, dk, \tag{3.12.12}$$

in the volume element $d^3 k = 4\pi k^2 \, dk$ of k space.

---

[11]λ in this expression, and throughout this book, is the wavelength in *vacuum*.

If $n(v) = 1$, $k = 2\pi v/c$ and

$$d\mathcal{N}_v = \frac{8\pi v^2}{c^3} dv. \tag{3.12.13}$$

This result for the number of field modes per unit volume in the frequency interval $[v, v + dv]$ has already been used (Section 3.6), and we have now derived it.

If $n(v) \neq 1$,

$$d\mathcal{N}_v = \frac{2}{8\pi^3} 4\pi \left(\frac{2\pi nv}{c}\right)^2 \frac{2\pi}{c} \frac{d}{dv}[nv]\,dv, \tag{3.12.14}$$

where we have used the fact that $dk = d[2\pi nv/c] = (2\pi/c)d[nv]$.

Suppose that $d[nv]/dv = v\,dn/dv + n \approx n$ at the frequency v. Then

$$d\mathcal{N}_v \approx \frac{8\pi n^3 v^3}{c^3} dv. \tag{3.12.15}$$

In this approximation the spectral energy density of thermal radiation is [cf. (3.6.20)]

$$\rho(v) = \frac{hv(d\mathcal{N}_v/dv)}{e^{hv/k_BT} - 1} = \frac{8\pi hn^3 v^3/c^3}{e^{hv/k_BT} - 1}. \tag{3.12.16}$$

According to Eq. (3.6.15), therefore,

$$\frac{A_{21}}{B_{21}} = \frac{8\pi hn^3(v_0)v_0^3}{c^3}, \tag{3.12.17}$$

where A_{21} and B_{21} are now the Einstein A and B coefficients for the case where the atom is inside a host medium of refractive index $n(v)$.

Equation (3.12.4) implies that the stimulated emission cross section may be expressed as

$$\sigma(v) = hv \frac{B_{21}u_v}{I_v} S(v). \tag{3.12.18}$$

If we relate the intensity to the energy density by the formula $I_v = (c/n)u_v$, then (3.12.18) gives

$$\sigma(v) = hv \frac{n}{c} S(v) \frac{c^3 A_{21}}{8\pi n^3 v^3} = \frac{\lambda^2 A_{21}}{8\pi n^2} S(v), \tag{3.12.19}$$

and, using this result for the cross section in (3.12.6), we obtain (3.12.11).

The terms A_{21}, B_{21}, I_v, and u_v all depend on the refractive index. In the formula (3.12.11) for the gain coefficient, A_{21} is the spontaneous emission rate in the host medium; it may be shown that, aside from a possible Lorenz–Lorenz local field correction factor, the spontaneous emission rate in the medium is $n(v_0)$ times the rate in free space in the case of electric dipole transitions. In this case the gain (or absorption) coefficient is actually $1/n$ times its value when there is no "host" medium of refractive index n.

The generalization of (3.12.15) when $d[nv]/dv$ is not well approximated by n is

$$d\mathcal{N}_v = \frac{8\pi n^2 v^3}{c^3} \frac{d}{dv}[nv]\,dv = \frac{8\pi n^2 v^3}{c^2 v_g} dv, \tag{3.12.20}$$

where

$$v_g \equiv \frac{c}{d[nv]/dv} \qquad (3.12.21)$$

is the *group velocity* at frequency v (Section 8.3). Note that (3.12.15) differs from (3.12.20) by the replacement of the group velocity by the phase velocity, c/n. Equation (3.12.11) is valid even if the group velocity is not well approximated by the phase velocity. In this case (3.12.17) is replaced by $A_{21}/B_{21} = 8\pi h n^2(v_0)v_0^3/v_g c^2$, and the formula $I_v = (c/n)u_v$ by $I_v = v_g u_v$. This leads again to the cross section $\sigma(v) = (\lambda^2 A_{21}/8\pi n^2)S(v)$ and therefore to Eq. (3.12.11) for the gain coefficient. ●

3.13   EXAMPLE: SODIUM VAPOR

Application of the formulas derived in this chapter to specific absorbing or amplifying media requires some information about the absorbing or emitting species. In addition to the obvious requirement that we know at what wavelengths the atoms or molecules absorb or emit light, we must know the strength of the transitions as characterized, for instance, by a spontaneous emission rate or an oscillator strength. It is a highly nontrivial matter in general to calculate such quantities. Normally, we obtain such information from tabulations made by sophisticated computations or, more commonly, by laboratory measurements.

We will consider as an example the absorption of (yellow) light whose frequency is near the D lines of atomic sodium (Section 3.1). The spectroscopic features of the D lines are well known, and we begin by summarizing some of the most important of these features.

The single valence electron in the $(3s)$ ground level of sodium is characterized by the spin angular momentum quantum number $S = \frac{1}{2}$ as well as the orbital angular momentum quantum number $L = 0$. The orbital, spin, and total (orbital plus spin) angular momentum vectors are denoted by **L**, **S**, and **J**, respectively, with $\mathbf{J} = \mathbf{L} + \mathbf{S}$. According to quantum theory the $(2J + 1)$ allowed values of the angular momentum quantum number J are $|L - S|, |L - S| + 1, \ldots, L + S$. Since $L = 0$ and $S = \frac{1}{2}$, J has the single allowed value $\frac{1}{2}$. The ground state of sodium is labeled $3S_{1/2}$, the 3 corresponding to the principle quantum number n, S indicating an s orbital ($L = 0$), and the subscript $\frac{1}{2}$ being the value of J.

The first excited level is a $3p$ orbital ($L = 1$), so that the allowed values of J, according to the rule given in the preceding paragraph, are $\frac{1}{2}$ and $\frac{3}{2}$. The different states corresponding to these values of J are labeled $3P_{1/2}$ and $3P_{3/2}$, respectively, and have different energies as indicated in Fig. 3.18. This splitting of the $3p$ configuration into two energy levels is an example of *fine structure*. The 589.6-nm $3S_{1/2} \leftrightarrow 3P_{1/2}$ transition in sodium is called the D_1 line, while the 589.0-nm transition $3S_{1/2} \leftrightarrow 3P_{3/2}$ is referred to as the D_2 line. The splitting of the D lines is relatively small, corresponding to a frequency of about 520 GHz, and consequently its observation requires a spectrometer of moderately high resolution.

At a much finer scale of resolution it is observed that all three levels in Fig. 3.18 have a *hyperfine structure*, as shown in Fig. 3.19. Hyperfine structure arises from the fact that

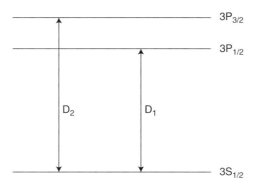

Figure 3.18 The sodium D lines associated with fine structure splitting.

there is an intrinsic (spin) angular momentum **I** of the nucleus in addition to the orbital and spin angular momenta of the electrons. The angular momentum **F** obtained by the addition of **I** and **J** has allowed quantum numbers found by applying again the quantum-mechanical rule for the addition of two angular momenta: $F = |J - I|, |J - I| + 1, \ldots,$ $J + I$. Since $I = \frac{3}{2}$ for the sodium nucleus, we can have either $F = \left(\frac{3}{2} - \frac{1}{2}\right)$ or $F = \left(\frac{3}{2} + \frac{1}{2}\right)$ for $3S_{1/2}$, and these states have an energy difference corresponding to about 1772 MHz. For $3P_{3/2}$, similarly, F can have the values 0, 1, 2, or 3 obtained from $J = I = \frac{3}{2}$, while for $3P_{1/2}$ F can be 1 or 2. The hyperfine splittings of the 3P levels are considerably smaller than the 1772-MHz hyperfine splitting of $3S_{1/2}$ (Fig. 3.19.)

In the absence of any magnetic field there is associated with each F level a set of $2F + 1$ degenerate substates, $M_F = -F, -F + 1, \ldots, F - 1, F$. The D_2 line, for instance, actually involves a total of 24 states (Fig. 3.19).

We will consider specifically the absorption of narrowband radiation with frequency ν near the D_2 resonance. For $T \sim 200K$ and vapor pressures $\lesssim 0.1$ Torr, the D_2 line is

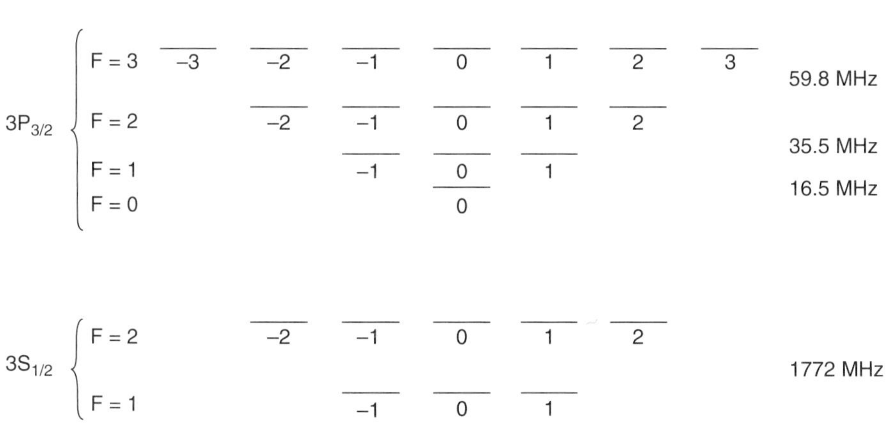

Figure 3.19 Hyperfine splittings of the sodium $3S_{1/2}$ and $3P_{3/2}$ levels. The energy differences indicated are not to scale.

Doppler broadened, with Doppler width [Eq. (3.9.11)]

$$\delta v_D = 2.15 \times 10^{11} \left[\frac{1}{589} \left(\frac{200}{23} \right)^{1/2} \right] \cong 1\,\text{GHz}. \tag{3.13.1}$$

This is less than the 1.772-GHz hyperfine separation of the $3S_{1/2}(F = 1)$ and $3S_{1/2}(F = 2)$ levels, but large compared to the separation of the different $3P_{3/2}$ levels. Therefore we will ignore the hyperfine splittings of the $3P_{3/2}$ level, giving a degeneracy $7 + 5 + 3 + 1 = 16$ for it. (See the black-dot section below.)

If the population of $3P_{3/2}$ is negligible, the rate of change due to absorption of the population of $3S_{1/2}(F = 1)$ is simply [cf. Eq. (3.7.13)]

$$\left[\frac{dN_1^{(1)}}{dt} \right]_{abs} = -\sum_{m_1=-1}^{1} \sum_{m_2} \mathcal{R}(m_1, m_2) \mathcal{N}_1^{(1)}(m_1), \tag{3.13.2}$$

where the superscript (1) is used to designate $3S_{1/2}(F = 1)$ states. Thus, $\mathcal{N}_1^{(1)}(m_1)$, with $m_1 = -1, 0, 1$, are the populations of atoms in the three $3S_{1/2}(F = 1)$ states. In thermal equilibrium at $T \sim 200\text{K}$ the $3S_{1/2}(F = 1) - 3S_{1/2}(F = 2)$ splitting is small compared to $k_B T$, so that each of the eight $(= g_1)$ $3S_{1/2}$ states has practically the same population, namely $\mathcal{N}_1^{(1)}(m_1) = N_1/g_1$, where N_1 is the total $3S_{1/2}$ population. Therefore, from (3.7.16),

$$\begin{aligned}
\left[\frac{dN_1^{(1)}}{dt} \right]_{abs} &= -\frac{N_1}{g_1} \sum_{m_1=-1}^{1} \sum_{m_2} \mathcal{R}(m_1, m_2) \\
&= -\frac{N_1}{g_1} g_2 \frac{1}{c} I_v S^{(1)}(v) B_{21} \\
&= -\frac{\lambda^2 A^{(1)}}{8\pi h v} \frac{g_2}{g_1} I_v S^{(1)}(v) N_1,
\end{aligned} \tag{3.13.3}$$

where

$$S^{(1)}(v) = \frac{1}{\delta v_D} \left(\frac{4 \ln 2}{\pi} \right)^{1/2} e^{-4(v - v_0^{(1)})^2 \ln 2/\delta v_D^2}, \tag{3.13.4}$$

$v_0^{(1)}$ is the $3S_{1/2}(F = 1) \rightarrow 3P_{3/2}$ transition frequency, and $A^{(1)}$ is the rate of spontaneous emission from $3P_{3/2}$ due to $3P_{3/2} \rightarrow 3S_{1/2}(F = 1)$ transitions. Similarly, using the superscript (2) to designate $3S_{1/2}(F = 2)$ states, we have

$$\left[\frac{dN_1^{(2)}}{dt} \right]_{abs} = -\frac{I_v}{h v} \frac{\lambda^2 A^{(2)}}{8\pi} \frac{g_2}{g_1} N_1 S^{(2)}(v), \tag{3.13.5}$$

where $A^{(2)}$ is the spontaneous emission rate of $3P_{3/2}$ due to $3P_{3/2} \rightarrow 3S_{1/2}(F = 2)$ transitions and

$$S^{(2)}(\nu) = \frac{1}{\delta\nu_D}\left(\frac{4\ln 2}{\pi}\right)^{1/2} e^{-4(\nu-\nu_0^{(2)})^2 \ln 2/\delta\nu_D^2}, \tag{3.13.6}$$

with $\nu_0^{(2)}$ the $3S_{1/2}(F = 2) \rightarrow 3P_{3/2}$ transition frequency, $\nu_0^{(2)} = \nu_0^{(1)} - 1772$ MHz. The total rate of change of $3S_{1/2}$ population due to absorption is $[dN_1/dt]_{abs} = [dN_1^{(1)}/dt]_{abs} + [dN_1^{(2)}/dt]_{abs}$, that is,

$$\left[\frac{dN_1}{dt}\right]_{abs} = -\frac{I_\nu}{h\nu}\frac{\lambda^2}{8\pi}\frac{g_2}{g_1}N_1\left[A^{(1)}S^{(1)}(\nu) + A^{(2)}S^{(2)}(\nu)\right]. \tag{3.13.7}$$

Equivalently, the absorption cross section for narrowband radiation with frequency near the D_2 line is

$$\sigma(\nu) = \frac{\lambda^2}{8\pi}\frac{g_2}{g_1}\left[A^{(1)}S^{(1)}(\nu) + A^{(2)}S^{(2)}(\nu)\right], \tag{3.13.8}$$

and the absorption coefficient is $a(\nu) = N_1\sigma(\nu)$.

The sodium $3P_{3/2}$ radiative lifetime is known experimentally to be about 16 ns, corresponding to a spontaneous emission rate $A_{21} = A^{(1)} + A^{(2)} = 1/(16 \text{ ns}) = 6.2 \times 10^7 \text{ s}^{-1}$. It might be expected, since there are five possible lower states in $3P_{3/2} \rightarrow 3S_{1/2}(F = 2)$ spontaneous emission, and three possible lower states for $3P_{3/2} \rightarrow 3S_{1/2}(F = 1)$, that $A^{(2)} = \left(\frac{5}{3}\right)A^{(1)}$ and therefore that $A_{21} = \left(\frac{8}{3}\right)A^{(1)}$, $A^{(1)} = \left(\frac{3}{8}\right)A_{21}$, and $A^{(2)} = \left(\frac{5}{8}\right)A_{21}$. These relations are in fact correct. They imply that

$$\sigma(\nu) = \frac{\lambda^2 A_{21}}{8\pi}\frac{g_2}{g_1}\left[\frac{3}{8}S^{(1)}(\nu) + \frac{5}{8}S^{(2)}(\nu)\right]. \tag{3.13.9}$$

For $T = 200$K and $\nu = \nu_0^{(2)}$,

$$\sigma(\nu) \cong \frac{\lambda^2 A_{21}}{8\pi}\frac{g_2}{g_1}\frac{5}{8}\frac{1}{\delta\nu_D}\left(\frac{4\ln 2}{\pi}\right)^{1/2}$$

$$= \frac{(5890 \times 10^{-10} \text{ m})^2}{8\pi}(6.2 \times 10^7 \text{ s}^{-1})\left(\frac{16}{8}\right)\left(\frac{5}{8}\right)$$

$$\times \frac{1}{1.08 \times 10^9 \text{ s}^{-1}}\left(\frac{4\ln 2}{\pi}\right)^{1/2}$$

$$= 9.3 \times 10^{-16} \text{ m}^2. \tag{3.13.10}$$

The cross section (3.13.9) is plotted in Fig. 3.20 for several different values of the temperature T, each of which implies a different Doppler width. As discussed in Section 14.1,

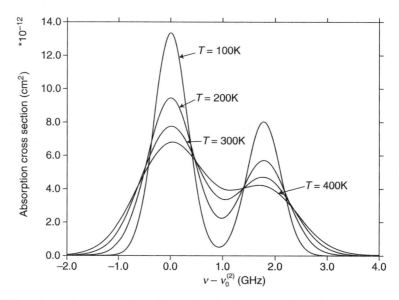

Figure 3.20   Absorption cross section of the sodium D_2 line [Eq. (3.13.9)] for different values of the temperature T. The hyperfine splitting shown is obviously unresolved at $T = 5800K$ and so inaccessible to Fraunhofer when he first resolved and named D_2 as distinct from D_1.

the dependence of the sodium absorption cross section on temperature has been used to measure the temperature variation in the mesosphere.

The transmission coefficient for narrowband radiation of frequency ν near the D_2 resonance is

$$\frac{I_\nu(z)}{I_\nu(0)} = e^{-a(\nu)z} = e^{-N_1\sigma(\nu)z}, \tag{3.13.11}$$

where z is the distance of propagation through the sodium vapor. The exponential dependence on $N_1\sigma(\nu)z$, together with the curves for $\sigma(\nu)$ shown in Fig. 3.20, indicates how strongly dependent on frequency the transmission coefficient can be (Problem 3.17).

These results are based on the assumption that the $3P_{3/2}$ population is very small, so that (3.13.2), (3.13.3), and (3.13.5) are valid. In other words, we have assumed that the effect of stimulated emission can be neglected in writing the rate equations for $N_1^{(1)}$ and $N_1^{(2)}$. For the population of $3P_{3/2}$ to remain small, its loss rate A_{21} due to spontaneous emission—which is assumed here to be larger than any collisional deexcitation rate—must be large compared to its growth rate $\sigma(\nu)I_\nu/h\nu$ due to absorption: $A_{21} \gg \sigma(\nu)I_\nu/h\nu$, or

$$I_\nu \ll \frac{h\nu A_{21}}{\sigma(\nu)} \approx 20\,\text{kW/m}^2 = 2\,\text{W/cm}^2 \tag{3.13.12}$$

if we assume $\nu = \nu_0^{(2)}$, in which case $\sigma(\nu)$ is given by (3.13.10). In the following chapter we discuss in more detail what it means for a field intensity to be large or small in its effect on level populations.

● Note that, if the $3S_{1/2}(F=1)-3S_{1/2}(F=2)$ splitting were very small compared to the Doppler width, we would have $S^{(1)}(\nu) \cong S^{(2)}(\nu) \equiv S(\nu)$ and

$$\sigma(\nu) \cong \frac{\lambda^2 A_{21}}{8\pi} \frac{g_2}{g_1} \left[\frac{3}{8} + \frac{5}{8} \right] S(\nu) = \frac{\lambda^2 A_{21}}{8\pi} \frac{g_2}{g_1} S(\nu). \tag{3.13.13}$$

We would in effect be justified in ignoring the energy difference between the $F=1$ and $F=2$ levels.

The same sort of arguments as used in Section 3.5 for broadband radiation can be made to show that, if the *radiation* has a spectral width $\delta\nu$ large compared to the $3S_{1/2}(F=1)-3S_{1/2}(F=2)$ separation, then the absorption is also given approximately by (3.13.13). In other words, if either the spectral width of the transition or the spectral width of the radiation is large compared to the $3S_{1/2}$ hyperfine splitting, we can treat $3S_{1/2}$ as a single "unresolved" level. In particular, this approximation can be made if the radiation is in the form of a pulse of duration $\tau_p(\sim 1/\delta\nu)$ that is short compared to $1/(2\pi \times 1772\,\text{MHz}) = 90\,\text{ps}$.

Such considerations can also be applied, of course, to excited states and can be used to justify our treatment of $3P_{3/2}$ as a single unresolved level in the calculation of $\sigma(\nu)$. ●

3.14 REFRACTIVE INDEX

Recall the classical oscillator formula (3.4.8) for the electron displacement due to an applied electric field $\hat{\boldsymbol{\varepsilon}} E_0 \cos(\omega t - kz)$:

$$\mathbf{x}(t) = \hat{\boldsymbol{\varepsilon}} \frac{e}{m} E_0 \left[\frac{\omega_0^2 - \omega^2}{(\omega_0^2 - \omega^2)^2 + 4\beta^2 \omega^2} \cos(\omega t - kz) + \frac{2\beta\omega}{(\omega_0^2 - \omega^2)^2 + 4\beta^2 \omega^2} \sin(\omega t - kz) \right]. \tag{3.14.1}$$

The first term in brackets is in phase with the electric field, whereas the second term is "in quadrature," that is, its phase differs by $\pi/2$ from that of the field. It is clear from the discussion leading to Eq. (3.4.20) that the in-quadrature part of the induced electric dipole moment $\mathbf{d} = e\mathbf{x}$ is responsible for absorption (or stimulated emission) of light. See also Problem 3.18.

The in-phase part of the induced dipole moment is responsible for the refractive index.[12] According to basic electromagnetic theory [see Eq. (8.2.21)], the refractive index at frequency ω of a medium of N atoms per unit volume is given by the formula $n^2(\omega) = 1 + N\alpha(\omega)/\epsilon_0$, where the *polarizability* $\alpha(\omega)$ is defined by writing the in-phase component of $\mathbf{d}$ as $\alpha\hat{\boldsymbol{\varepsilon}} E_0 \cos(\omega t - kz)$. Thus, from Eq. (3.14.1),

$$\alpha(\omega) = \frac{e^2}{m} \frac{\omega_0^2 - \omega^2}{(\omega_0^2 - \omega^2)^2 + 4\beta^2 \omega^2} \tag{3.14.2}$$

and

$$n^2(\omega) - 1 = \frac{Ne^2}{m\epsilon_0} \frac{\omega_0^2 - \omega^2}{(\omega_0^2 - \omega^2)^2 + 4\beta^2 \omega^2}. \tag{3.14.3}$$

---

[12]The theory of the propagation of light, including refractive effects, is treated in Chapters 8 and 9.

As in the case of spontaneous emission and absorption, this result of the classical oscillator model must be modified to include the oscillator strength f:

$$n^2(\omega) - 1 = \frac{Ne^2 f}{m\epsilon_0} \frac{\omega_0^2 - \omega^2}{(\omega_0^2 - \omega^2)^2 + 4\beta^2\omega^2}. \qquad (3.14.4)$$

Unlike absorption, the refractive index is usually attributable to *nonresonant* transitions, that is, transitions such that $|\omega_0^2 - \omega^2| \gg \beta\omega$. In this case

$$n^2(\omega) - 1 \approx \frac{Ne^2 f}{m\epsilon_0} \frac{1}{\omega_0^2 - \omega^2}. \qquad (3.14.5)$$

In this nonresonant situation, however, no one transition is necessarily dominant, and so we must add the contributions of all transitions connected to the ground state in which the atoms are presumed (for now) to reside. Thus, if the transitions from the ground state have oscillator strengths f_j and transition frequencies ω_j, the refractive index at the radiation frequency ω is given by the formula

$$n^2(\omega) - 1 = \frac{Ne^2}{m\epsilon_0} \sum_j \frac{f_j}{\omega_j^2 - \omega^2}. \qquad (3.14.6)$$

This result applies when there is one type of atom or molecule in the medium; more generally we simply add the contributions of the different species. In a gas, furthermore, the density N is generally sufficiently low that $n(\omega) \approx 1$ and therefore $n^2 - 1 = (n-1)(n+1) \approx 2(n-1)$. Thus, for a gas consisting of a single type of atom or molecule with number density N, the formula for the refractive index is approximately

$$n(\omega) = 1 + \frac{Ne^2}{2m\epsilon_0} \sum_j \frac{f_j}{\omega_j^2 - \omega^2}. \qquad (3.14.7)$$

It is interesting to relate this result to a formula that is often used in tabulations of the refractive index of gases. For this purpose we first rewrite (3.14.7) in terms of radiation wavelength $\lambda \ (= 2\pi c/\omega)$ and transition wavelengths $\lambda_j \ (= 2\pi c/\omega_j)$:

$$n(\lambda) = 1 + \frac{Ne^2}{8\pi^2\epsilon_0 mc^2} \sum_j \frac{\lambda_j^2 f_j}{1 - \lambda_j^2/\lambda^2}. \qquad (3.14.8)$$

As noted in the Introduction, electronic resonances in molecules (and in many atoms) tend to lie in the ultraviolet, in which case $\lambda_j \ll \lambda$ for optical wavelengths λ. In this case we can approximate $(1 - \lambda_j^2/\lambda^2)^{-1}$ by the first two terms of its binomial series expansion, $1 + \lambda_j^2/\lambda^2$:

$$n(\lambda) - 1 \approx A_1(1 + B_1/\lambda^2), \qquad (3.14.9)$$

where

$$A_1 = \frac{Ne^2}{8\pi^2 \epsilon_0 mc^2} \sum_j f_j \lambda_j^2 \qquad (3.14.10)$$

and

$$B_1 = \frac{\sum_j f_j \lambda_j^4}{\sum_j f_j \lambda_j^2}. \qquad (3.14.11)$$

An empirical relation of the form (3.14.9) was proposed by Cauchy in 1830, before the electromagnetic theory of light. Our derivation of *Cauchy's formula* gives explicit expressions for the coefficients A_1 and B_1. Unfortunately, it is difficult to calculate the numerical values of A_1 and B_1 for a given atom or molecule because we require the transition wavelengths and the oscillator strengths of all transitions connected to the ground state, including transitions to "continuum" states in which the electrons are unbound, that is, in which an atom is ionized.

For a gas at STP $[P = 760$ Torr, $T = 273$K, and, from Eq. (3.8.20), $N = 2.69 \times 10^{25}$ m$^{-3}]$,

$$A_1 = 1.2 \times 10^{10} \sum_j f_j \lambda_j^2. \qquad (3.14.12)$$

Consider as an example a gas of ground-state helium atoms, for which the 58.4-nm transition from the ground state to the first excited state has an oscillator strength of about 0.28. If we include only the contribution of this transition to the summations in (3.14.10) and (3.14.11), we obtain $A_1 = 1.2 \times 10^{-5}$ and $B_1 = 3.4 \times 10^{-15}$ m^2, in contrast to the tabulated (measured) values of 3.48×10^{-5} and 2.3×10^{-15} m^2, respectively.[13] Adding the contributions of transitions from the ground state to higher-energy bound states does not change these results very significantly because the oscillator strengths for these transitions are considerably smaller than that for the transition to the first excited state. We conclude that transitions to the continuum are mainly responsible for the discrepancy between our simple calculation and the measured values of the Cauchy constants for helium.

Cauchy's formula correctly accounts for the fact that most transparent materials we encounter daily (e.g., water, air, glass) have refractive indices greater than unity at visible wavelengths. According to our analysis, this is a consequence of these materials having resonance wavelengths λ_j that are small compared to optical wavelengths (which lie roughly between 400 and 700 nm). It also follows from (3.14.9) that $dn/d\lambda < 0$, which is also a familiar feature of refractive indices in the visible: A glass prism, for instance, causes violet to be dispersed more than red when it separates white light into its spectral components. In fact, the increase of $n(\lambda)$ with decreasing λ ($dn/d\lambda < 0$) is sufficiently ubiquitous that it is called "normal dispersion." An example of normal dispersion appears in Fig. 3.21.

[13]M. Born and E. Wolf, *Principles of Optics*, 7th ed., Cambridge University Press, Cambridge, 1999, p. 101.

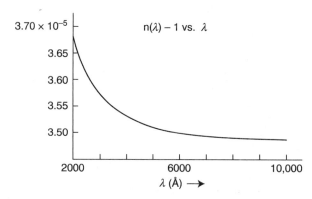

Figure 3.21   Refractive index of helium at standard temperature and pressure.

It is also interesting to consider the case in which the frequency of the radiation is much greater than the resonance frequencies of the medium. The simplest example occurs for free electrons, for which there is no binding force. The resonance frequencies ω_j are then zero, and the dispersion formula (3.14.6) reduces to [see Eq. (3.14.16)]

$$n(\omega) = \left(1 - \frac{Ne^2}{\epsilon_0 m \omega^2}\right)^{1/2} = \left(1 - \frac{\omega_p^2}{\omega^2}\right)^{1/2}, \qquad (3.14.13)$$

where N is now the density of free electrons (Problem 3.20) and

$$\omega_p = \left(\frac{Ne^2}{m\epsilon_0}\right)^{1/2} \qquad (3.14.14)$$

is called the *plasma frequency*. In some cases this result is known to be fairly accurate; it is applicable, for instance, to the upper atmosphere, where ultraviolet solar radiation produces free electrons by photoionization (Problem 3.20). Another example is the refraction of X rays by glass. In this case the resonance frequencies of the medium are not zero but are much less than X-ray frequencies. In both examples the refractive index is less than one.

If $\omega < \omega_p$, the refractive index (3.14.13) is a pure imaginary number. In this case the free-electron gas will not support a propagating electromagnetic wave, and an incident wave is instead reflected. This applies, for instance, to the propagation of radio waves in Earth's atmosphere. High-frequency FM radio waves are not reflected by the ionosphere ($\omega > \omega_p$), whereas the lower frequency AM waves are. AM radio broadcasts, therefore, reach more distant points on Earth's surface (Problem 3.20).

• Consider the limit of Eq. (3.14.6) in which the field frequency ω is very large compared to any of the transition frequencies ω_j:

$$n^2(\omega) - 1 = -\frac{Ne^2}{\epsilon_0 m \omega^2} \sum_j f_j. \qquad (3.14.15)$$

TABLE 3.2   Oscillator Strengths for Some Bound–Bound Transitions of Hydrogen

Initial state	1s	2s	2p		3s	3p		3d	
Final state	np	np	ns	nd	np	ns	nd	np	nf
$n = 1$	—	—	−0.139	—	—	−0.026	—	—	—
$n = 2$	0.4162	—	—	—	−0.041	−0.145	—	−0.417	—
$n = 3$	0.0791	0.435	0.0134	0.696	—	—	—	—	—
$n = 4$	0.0290	0.103	0.0030	0.122	0.484	0.032	0.619	0.011	1.018
$n = 5$	0.0139	0.042	0.0012	0.044	0.121	0.007	0.139	0.002	0.156

In this limit the transition frequencies ω_j are effectively zero, that is, the atom behaves as though its energy levels form a continuum, as is the case for unbound electrons. It is then plausible that in this limit the refractive index should be identical to that of N free electrons per unit volume, which is given by Eq. (3.14.13). This implies that the oscillator strengths f_j must obey the *electric dipole sum rule*

$$\sum_j f_j = 1, \tag{3.14.16}$$

which in fact may be derived using quantum mechanics. Our less rigorous "derivation" of this sum rule is based on the assumption that each of the N atoms per unit volume has one bound electron. In the case of Z electrons per atom Eq. (3.14.15) should reduce to $n^2(\omega) - 1$ for the case of NZ free electrons per unit volume, so that the sum rule for a Z-electron atom is

$$\sum_j f_j = Z. \tag{3.14.17}$$

Table 3.2 lists oscillator strengths for the hydrogen atom for "allowed" electric dipole transitions, i.e., transitions for which $\Delta\ell = \pm 1$.[14] The sum of the oscillator strengths of *all* the bound–bound $1s - np$ transitions is 0.565, so that the electric dipole sum rule implies that the sum of the oscillator strengths for transitions from $1s$ to continuum states is 0.435. For the $2s$ state the corresponding bound–bound and bound–free contributions to the f sums are 0.649 and 0.351. Table 3.2 exemplifies the fact, noted above for helium, that transitions to the first excited state tend to be stronger than transitions to higher-energy bound states. Note also that the transitions to continuum states contribute significantly to the sum of the oscillator strengths. For Z-electron atoms the transitions to continuum states make a contribution $\approx Z$ to the sum of the oscillator strengths for all the electrons. Bound–bound transitions of single electrons, such as those for the valence electrons of the alkali atoms, have oscillator strengths comparable in magnitude to those of hydrogen.   •

We have assumed in our discussion of the refractive index that the N atoms per unit volume are all in the ground state with high probability, but it is straightforward to deal with the more general situation where there are N_i atoms per unit volume in

---

[14]Selection rules and many other fundamental aspects of atomic theory are discussed in R. D. Cowan, *The Theory of Atomic Structure and Spectra* University of California Press, Berkeley, CA, 1981. Useful tabulations of oscillator strengths and formulas for atomic and molecular transitions may be found in *Allen's Astrophysical Quantities* and other sources (see footnote 5).

energy level E_i. Equation (3.14.7), for instance, generalizes to

$$n(\omega) = 1 + \frac{e^2}{2m\epsilon_0} \sum_i \sum_j \frac{N_i f_{ij}}{\omega_{ji}^2 - \omega^2}, \tag{3.14.18}$$

where $\omega_{ji} = (E_j - E_i)/\hbar$ and f_{ij} is the oscillator strength for the $i \to j$ transition. In particular, the contribution to the index from the $1 \to 2$ transition is

$$n(\omega)_{12} = 1 + \frac{e^2}{2m\epsilon_0} \left(\frac{N_1 f_{12}}{\omega_{21}^2 - \omega^2} + \frac{N_2 f_{21}}{\omega_{21}^2 - \omega^2} \right)$$

$$= 1 + \frac{e^2}{2m\epsilon_0} \frac{f_{12}}{\omega_{21}^2 - \omega^2} \left(N_1 - \frac{g_1}{g_2} N_2 \right), \tag{3.14.19}$$

where we have used Eq. (3.7.26) to relate f_{21} to f_{12}.

- The sum rule (3.14.17) may be written more generally as

$$\sum_j f_{ij} = \sum_{j>i} f_{ij} + \sum_{j<i} f_{ij} = \sum_{j>i} f_{ij} - \sum_{j<i} |f_{ij}| = Z \tag{3.14.20}$$

for any level i of an atom. In other words, the sum over oscillator strengths in the electric dipole sum rule must, in the case of excited states, include both the oscillator strengths for absorption (positive) and for emission (negative).

The sum rule for oscillator strengths played an important role in the formulation of quantum theory in the 1920s. It was already known, based on the physical argument we have used in going from (3.14.15) to (3.14.16), before some of the most important features of quantum mechanics (e.g., before the Schrödinger equation).

Historically, downward transitions associated with stimulated emission were referred to in terms of "negative oscillators." In the case of the refractive index, the term proportional to N_2 in Eq. (3.14.19) corresponds to such a negative oscillator. The contribution of negative oscillators to the refractive index was studied experimentally by R. Ladenburg and H. Kopfermann around 1928. They measured the variation of refractive index with electric current of a discharge tube filled with neon. According to our theory, for $\omega < \omega_{21}$ in (3.14.19), with neither 1 nor 2 the ground level, $n(\omega)_{12}$ should initially increase with increasing current because electron–atom collisions produce atoms in excited level 1. With further increase of the current, however, the rate of growth of $n(\omega)_{12}$ with current decreases because excited level 2 has appreciable population N_2 and acts as a negative oscillator. This sort of behavior was observed by Ladenburg and Kopfermann, thus confirming the role of negative oscillators.   ●

The atoms or molecules of a medium do not form a continuum but have empty space between them. As a result, there is a difference between the "mean" field and the actual field acting on a given atom. In many cases the only practical consequence of this difference is that the relation between the refractive index and the polarizability α becomes

$$\frac{n^2(\omega) - 1}{n^2(\omega) + 2} = \frac{N\alpha(\omega)}{3\epsilon_0}. \tag{3.14.21}$$

The origin of this "Lorentz–Lorenz relation" is discussed in many textbooks on electromagnetism. Note that when the refractive index is close to unity, so that $n^2(\omega) + 2 \approx 3$, the Lorentz–Lorenz relation reduces to the relation between n and α assumed in writing (3.14.3).

3.15 ANOMALOUS DISPERSION

In the preceding section we assumed that the radiation frequency ω is far removed from any absorption frequency of the medium. We now allow for the possibility that the radiation frequency is near an absorption resonance.

Equation (3.14.18) shows that $n^2(\omega) - 1$, and therefore the polarizability $\alpha(\omega)$, is additive over all the transition frequencies. Thus, we can write

$$\alpha(\omega) = \alpha_b(\omega) + \alpha_r(\omega), \tag{3.15.1}$$

where α_b and α_r are the contributions to the polarizability from nonresonant "background" transitions and resonant transitions, respectively. The nonresonant transitions may be associated with atoms in the host medium in which the absorbing atoms reside, or with the absorbing atoms themselves. In either case, from $n^2 = 1 + N\alpha/\epsilon_0$, we have

$$n^2(\omega) = 1 + \sum_j \frac{N_b}{\epsilon_0} \alpha_{bj}(\omega) + \frac{N_r}{\epsilon_0} \alpha_r(\omega), \tag{3.15.2}$$

where the sum is over all background species.

The first two terms in (3.15.2) determine $n_b(\omega)$, the index of refraction of the background or host material. Thus, we will write

$$n^2(\omega) = n_b^2(\omega) + \frac{N_r \alpha_r(\omega)}{\epsilon_0} = n_b^2(\omega)\left(1 + \frac{N_r \alpha_r(\omega)}{n_b^2(\omega)\epsilon_0}\right)$$

$$= n_b^2(\omega)\left(1 + \frac{N_r \alpha_r(\omega)}{\varepsilon_b(\omega)}\right), \tag{3.15.3}$$

where $\varepsilon_b = n_b^2 \epsilon_0$ is the dielectric permittivity of the background. If the resonant atoms are present in a monatomic beam, then the background material is vacuum or nearly so, and the background contributions can largely be ignored. Even in an atomic vapor n_b can be taken to be unity to three or four significant figures. However, in laser physics, the background material is frequently a solid or liquid. For example, the active atoms of a Ti : sapphire laser are titanium ions thinly dispersed throughout a solid lattice, and the molecules of a dye laser are dissolved in a liquid solvent. Then n_b is significantly different from unity, typically in the range 1.3–2.0. Because the resonances of the background are typically in the infrared or ultraviolet and n_b is effectively constant at optical frequencies, we will write n_b in place of $n_b(\omega)$ hereafter.

The resonant atoms do not make a correspondingly large contribution since they are usually present in such small concentrations. The concentration of the chromium ions in a ruby laser, for example, may be only 10^{25} per cubic meter or even less, much smaller than typical solid densities. As a consequence, the last term in (3.15.3) is typically much smaller than unity. Then, the total index of refraction can be expressed compactly as follows:

$$n(\omega) = n_b\left(1 + \frac{N_r \alpha_r(\omega)}{\varepsilon_b}\right)^{1/2} \approx n_b + \frac{N_r \alpha_r(\omega)}{2 n_b \epsilon_0}, \tag{3.15.4}$$

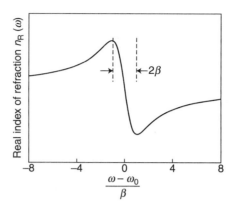

Figure 3.22   Anomalous dispersion curve for a collision-broadened absorption line.

where we have again used $\varepsilon_b = n_b^2 \epsilon_0$ after expanding the square root and keeping only the first term in the binomial series $(1 + x)^{1/2} = 1 + x/2 - x^2/8 + \cdots$.

Now $\alpha_r(\omega)$, the polarizability when ω is close to a resonant frequency ω_0, is given by (3.14.2) with $\omega \approx \omega_0$. Thus, writing $\omega_0^2 - \omega^2 = (\omega_0 + \omega)(\omega_0 - \omega) \approx 2\omega(\omega_0 - \omega)$, and introducing the oscillator strength f of the resonant transition, we have

$$\alpha \approx \frac{e^2 f}{m} \frac{2\omega(\omega_0 - \omega)}{4\omega^2(\omega_0 - \omega)^2 + 4\beta^2\omega^2} = \frac{e^2 f}{2m\omega} \frac{\omega_0 - \omega}{(\omega_0 - \omega)^2 + \beta^2} \qquad (3.15.5)$$

and therefore, from (3.15.4),

$$n(\omega) \approx n_b + \frac{Ne^2 f}{4n_b\epsilon_0 m\omega} \frac{\omega_0 - \omega}{(\omega_0 - \omega)^2 + \beta^2} , \qquad (3.15.6)$$

where $N = N_r$ is now the density of resonant atoms. This refractive index is plotted versus frequency in Fig. 3.22. On the low-frequency side of the resonance frequency, $n(\omega)$ increases with increasing frequency, that is, we have "normal dispersion." However, when ω gets within β of ω_0, $n(\omega)$ begins *decreasing* with increasing frequency. This decrease continues until ω is more than β from ω_0 on the high-frequency side, whereupon it again increases with increasing frequency. Because most media show normal dispersion at optical frequencies, the negative slope of the dispersion curve near an absorption line is called, for historical reasons, *anomalous dispersion*.

• Anomalous dispersion was observed by R. W. Wood in 1904. Wood studied the dispersion of light at frequencies near the sodium D lines. The basic idea of Wood's experiment is sketched in Fig. 3.23. Light enters a tube in which sodium vapor is produced by heating sodium. The vapor pressure decreases upward in the tube, so that for normal dispersion the light would be bent downward, in the direction of greater density and refractive index. The vapor thus acts as a kind of prism. The light emerging from the tube is focused onto the entrance slit of a spectroscope. Wood writes:

> On heating the tube, the sodium prism deviates the rays of different wave-length up or down by different amounts, curving the spectrum into two oppositely directed branches. The spectrum on the green side of the D lines will be found to bend down in the spectroscope, which means that the rays are

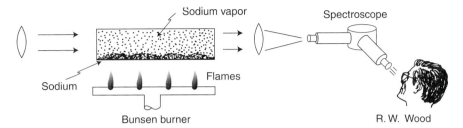

Figure 3.23 One of R. W. Wood's experiments on anomalous dispersion in sodium vapor.

deviated upwards in passing through the sodium tube, since the spectroscope inverts the image of its slit. This means that the phase velocity is greater in the sodium vapor than in vacuo, or the prism acts for these rays like an air prism immersed in water. The red and orange region is deviated in the opposite direction; these rays are therefore retarded by the vapor.

In other words, the refractive index on the low-frequency side of resonance was observed to be greater than unity, whereas on the high-frequency side it was less than unity. This is the behavior shown in Fig. 3.22.   •

Equation (3.15.6) assumes that all the resonant atoms are in the lower state of the transition. More generally, if N_1 and N_2 are the densities of atoms in the lower and upper states, respectively, of the resonant transition,

$$n(\omega) \approx 1 + \frac{e^2 f(N_1 - N_2)}{4\epsilon_0 m\omega} \frac{\omega_0 - \omega}{(\omega_0 - \omega)^2 + \beta^2} = 1 + \frac{e^2 f(N_1 - N_2)}{16\pi^2 \epsilon_0 mv} \frac{v_0 - v}{(v_0 - v)^2 + \delta v_0^2} \quad (3.15.7)$$

if, to simplify, we ignore level degeneracies and take the background refractive index n_b to be 1. Comparing this to the absorption coefficient (3.12.7) for a Lorentzian lineshape and no degeneracy,

$$a(v) = \frac{\lambda^2 A_{21}}{8\pi} (N_1 - N_2) \frac{(1/\pi)\delta v_0}{(v_0 - v)^2 + \delta v_0^2}, \quad (3.15.8)$$

we note that

$$n(v) - 1 = \frac{\lambda_0}{4\pi} \frac{v_0 - v}{\delta v_0} a(v), \quad (3.15.9)$$

where we have used (3.7.20) for the spontaneous emission rate A_{21}. Although this relation was obtained for a collision-broadened lineshape, and various simplifying assumptions were invoked, a similar relation holds more generally. Thus, in the case of a Voigt profile it can be shown that

$$n(v) - 1 = \frac{\lambda_0}{4\pi} \frac{\text{Im}[w(x+ib)]}{\text{Re}[w(x+ib)]} a(v), \quad (3.15.10)$$

where w, x, and b are defined in Section 3.10. The relation between the refractive index and the absorption coefficient is an example of a so-called *Kramers–Kronig relation* and may be derived on very general grounds based on "causality," or the condition that an effect cannot precede its cause. In the present context the "effect" and the "cause" are the induced electric dipole moment and the applied electric field, respectively.

Equation (3.15.9) exemplifies the general result that anomalous dispersion tends to be strongest in media with strong absorption (or gain) and narrow linewidth. As discussed in Section 5.9, the refractive index associated with the lasing transition can play an important part in determining the oscillation frequency of the laser, and the effect is largest in high-gain lasers with narrow gain profiles.

3.16   SUMMARY

Starting from the classical oscillator model of an atom, we have obtained some of the most important formulas used to describe the absorption and emission of light. To obtain these formulas we went beyond the classical oscillator model by introducing the oscillator strength and the fact that absorption and emission involve transitions between allowed energy levels.

We showed that the thermal radiation spectrum implies that there must be stimulated emission as well as spontaneous emission and absorption. Thermal radiation is broadband in the sense that its spectrum is much broader than the lineshape function of the atoms with which it is in equilibrium. In this case the atomic lineshape function does not appear in the stimulated emission and absorption rates, which are defined in terms of the Einstein B coefficients. In the opposite, narrowband limit of interest for lasers, the rates are proportional to the atomic lineshape function $S(\nu)$, where $S(\nu)$ is a Lorentzian function in the collision-broadening limit, for example, and a Gaussian in the Doppler-broadening limit. The formulas obtained for the absorption or gain coefficient will be central to much of our discussion of laser theory.

While our results have been cast in terms of absorption and emission by atoms, they apply more generally. The spectrum of thermal radiation, for instance, is independent of whether the absorbers and emitters are atoms or molecules, or whether the material is a gas, liquid, solid, or plasma. The formulas we have obtained will be found to be applicable to transitions between, say, two vibrational states of a molecule or two energy bands of a solid as well as between two states of an atom.

APPENDIX: THE OSCILLATOR MODEL AND QUANTUM THEORY

The electron oscillator model for the interaction of a bound electron with light can be regarded as an approximation to the quantum theory, as we now discuss starting from the time-dependent Schrödinger equation,

$$i\hbar \frac{\partial \psi}{\partial t} = H\psi, \tag{3.A.1}$$

where $\psi(\mathbf{x}, t)$ is the wave function for the electron at position $\mathbf{x}$ and H is the "Hamiltonian operator," which is the quantum mechanical expression for the energy of the electron.

The form of H depends on the energy we associate with the electron. For a bound electron in an electric field $\mathbf{E}(\mathbf{r}, t)$ we take $H = H_0 + H_I$, where H_0 accounts for the kinetic energy of the electron plus the potential energy associated with its binding to the nucleus, and H_I is the "interaction energy" arising from the applied electromagnetic field.

A standard approach to the solution of the Schrödinger equation (3.A.1) is based on the assumption that ψ can be written as a linear combination of the eigenfunctions of H_0: $\psi = \sum_n a_n \phi_n$, where the functions $\phi_n(\mathbf{x})$ are solutions of the time-*independent* Schrödinger equation that follows from (3.A.1) when we take $H = H_0$ and write $\psi(\mathbf{x}, t) = \phi_n(\mathbf{x}) \exp(-iE_n t/\hbar)$:

$$H_0 \phi_n(\mathbf{x}) = E_n \phi_n(\mathbf{x}). \tag{3.A.2}$$

Equation (2.A.1) is a special case of this equation with $H_0 = -(\hbar^2/2m)d^2/dx^2 + V(x)$. The eigenvalues E_n are the allowed energies of the electron when there is no applied field. For simplicity we will ignore the possibility of degeneracy, so that with each allowed energy there corresponds only one eigenfunction ϕ_n. In writing $\psi(\mathbf{x}, t)$ as a linear combination of the time-independent functions $\phi_n(\mathbf{x})$, it is evident that the coefficients a_n in this linear superposition must depend on the time t:

$$\psi(\mathbf{x}, t) = \sum_n a_n(t)\phi_n(\mathbf{x}). \tag{3.A.3}$$

The a_n are quantum mechanical "probability amplitudes" in the sense that $|a_n(t)|^2$ is the probability at time t that the state of the atomic electron is described by the eigenfunction $\phi_n(\mathbf{x})$. Loosely speaking, we say that $|a_n(t)|^2$ is the probability at time t that the electron is in the state $\phi_n(\mathbf{x})$.

We now use the expression (3.A.3) for ψ in Eq. (3.A.1) and obtain

$$i\hbar \sum_n \frac{da_n}{dt}\phi_n(\mathbf{x}) = \sum_n a_n(t)H_0\phi_n(\mathbf{x}) + \sum_n a_n(t)H_I\phi_n(\mathbf{x})$$

$$= \sum_n E_n a_n(t)\phi_n(\mathbf{x}) + \sum_n a_n(t)H_I\phi_n(\mathbf{x}), \tag{3.A.4}$$

where in the second line we have used (3.A.2). Next we use the fact that the eigenfunctions $\phi_n(\mathbf{x})$ of H_0 are orthogonal in the sense that

$$\int d^3\mathbf{x}\,\phi_m^*(\mathbf{x})\phi_n(\mathbf{x}) = 0 \quad \text{if } m \neq n. \tag{3.A.5}$$

As discussed in quantum mechanics textbooks, this orthogonality is a consequence of the fact that the Hamiltonian H_0 is a Hermitian operator (the electron energy is a real number). We assume furthermore that the ϕ_n are normalized such that

$$\int d^3\mathbf{x}\,\phi_n^*(\mathbf{x})\phi_n(\mathbf{x}) = 1, \tag{3.A.6}$$

which, as also discussed in quantum mechanics texts, allows us to interpret $\phi_n^*(\mathbf{x})\phi_n(\mathbf{x}) = |\phi_n(\mathbf{x})|^2$ as a normalized probability density. We now multiply both sides of (3.A.4) by $\phi_m^*(\mathbf{x})$, integrate both sides over all space, and use the properties

(3.A.5) and (3.A.6). The result is

$$
i\hbar \frac{da_m}{dt} = E_m a_m(t) + \sum_n a_n(t) \int d^3x \phi_m^*(\mathbf{x}) H_I \phi_n(\mathbf{x})
$$

$$
\equiv E_m a_m(t) + \sum_n V_{mn} a_n(t). \tag{3.A.7}
$$

This is a very general equation for the change in time of probability amplitudes. Note that in arriving at Eq. (3.A.7) we have not had to know the form of H_0, but only that it has eigenvalues E_n and eigenfunctions ϕ_n, and H_I, which is associated with the effect of the applied field on the electron, has also remained unspecified. To proceed further with the solution of the time-dependent Schrödinger equation we must know something about the V_{mn} coefficients, usually called "matrix elements":

$$
V_{mn} = \int d^3x \phi_m^*(\mathbf{x}) H_I \phi_n(\mathbf{x}). \tag{3.A.8}
$$

To this end we note that a displacement $\mathbf{x}$ of the electron from the nucleus implies an electric dipole moment $\mathbf{d} = e\mathbf{x}$, and therefore the interaction energy $-\mathbf{d} \cdot \mathbf{E} = -e\mathbf{x} \cdot \mathbf{E}$ in an electric field $\mathbf{E}$. Thus we assume that

$$
H_I = -e\mathbf{x} \cdot \mathbf{E}, \tag{3.A.9}
$$

and therefore that

$$
V_{mn} = \int d^3x \phi_m^*(\mathbf{x})[-e\mathbf{x} \cdot \mathbf{E}]\phi_n(\mathbf{x}). \tag{3.A.10}
$$

We assume furthermore that $\mathbf{E}$ is practically constant over the region of space for which $\phi_m^*(\mathbf{x})\phi_n(\mathbf{x})$ in (3.A.10) is not negligibly small. That is, as in Section 3.2, if $\mathbf{r}_e$ is the electron's position, we can write $\mathbf{E}(\mathbf{r}_e, t) = \mathbf{E}(\mathbf{R} + (m_n/M)\mathbf{x}, t) \approx \mathbf{E}(\mathbf{R}, t)$ whenever the electric field wavelength exceeds atomic dimensions. This is the electric dipole approximation discussed in Section 3.2; in this approximation H_I reduces to the potential energy (3.2.16), and $\mathbf{E}$ is independent of electron position $\mathbf{x}$ and can be removed from the integral in (3.A.10). The position $\mathbf{R}$ of the center of mass can be identified with the coordinate origin and we will drop it from the argument of $\mathbf{E}$. The matrix element (3.A.10) is then

$$
V_{mn}(t) = -e\mathbf{x}_{mn} \cdot \mathbf{E}(t), \tag{3.A.11}
$$

where

$$
\mathbf{x}_{mn} \equiv \int d^3x \phi_m^*(\mathbf{x})\mathbf{x}\phi_n(\mathbf{x}). \tag{3.A.12}
$$

According to quantum mechanics the average value (or "expectation value") of the electron displacement $\mathbf{x}$ at time t, denoted $\langle \mathbf{x}(t) \rangle$, is

$$\langle \mathbf{x}(t) \rangle = \int d^3\mathbf{x}\, \mathbf{x} |\psi(\mathbf{x},\, t)|^2 = \int d^3\mathbf{x}\, \psi^*(\mathbf{x},\, t)\mathbf{x}\psi(\mathbf{x},\, t)$$

$$= \int d^3x \left(\sum_m a_m^*(t)\phi_m^*(\mathbf{x}) \right) \mathbf{x} \left(\sum_n a_n(t)\phi_n(\mathbf{x}) \right)$$

$$= \sum_m \sum_n a_m^*(t)a_n(t) \int d^3x\, \phi_m^*(\mathbf{x})\mathbf{x}\phi_n(\mathbf{x})$$

$$= \sum_m \sum_n \mathbf{x}_{mn} a_m^*(t)a_n(t). \tag{3.A.13}$$

The variations in time of the coefficients $a_n(t)$ thus determine the variation in time of $\langle \mathbf{x}(t) \rangle$.

We want to compare $\langle \mathbf{x}(t) \rangle$ with the displacement $\mathbf{x}(t)$ of the classical electron oscillator model [Eq. (3.2.18b)]. For this purpose we now make an approximation that greatly simplifies the solution of the time-dependent Schrödinger equation: We assume that at any time t the electron has essentially zero probability of being found in any state other than the ground state or the first excited state of the atom. We denote these two states by subscripts 1 and 2, respectively. In this "two-state" approximation the set of Eqs. (3.A.7) reduces to the two equations

$$i\hbar \frac{da_1}{dt} = E_1 a_1 + V_{12}a_2 = E_1 a_1 - e\mathbf{x}_{12} \cdot \mathbf{E}a_2 \tag{3.A.14}$$

and

$$i\hbar \frac{da_2}{dt} = E_2 a_2 + V_{21}a_1 = E_2 a_2 - e\mathbf{x}_{12} \cdot \mathbf{E}a_1, \tag{3.A.15}$$

while the expectation value of the electron coordinate is

$$\langle \mathbf{x}(t) \rangle = [a_1^*(t)a_2(t) + a_2^*(t)a_1(t)]\mathbf{x}_{12}. \tag{3.A.16}$$

In writing these equations we have assumed that $\mathbf{x}_{21} = \mathbf{x}_{12}$ or, what is the same thing, that $\mathbf{x}_{12}$ is a real number (Problem 3.21).

It follows from Eqs. (3.A.14) and (3.A.15) that

$$\frac{d}{dt}(a_1^*a_2) = -i\omega_0(a_1^*a_2) + \frac{ie}{\hbar}\mathbf{x}_{12} \cdot \mathbf{E}\left[|a_1|^2 - |a_2|^2\right], \tag{3.A.17}$$

where

$$\omega_0 = \frac{E_2 - E_1}{\hbar} \tag{3.A.18}$$

is 2π times the Bohr frequency for transitions between the ground state and the first excited state of the atom. Differentiating both sides of (3.A.17) with respect to t, and adding the complex conjugate of the result, we obtain

$$\frac{d^2}{dt^2}(a_1^*a_2 + a_2^*a_1) = -\omega_0^2(a_1^*a_2 + a_2^*a_1) + \frac{2e}{\hbar}\omega_0 \mathbf{x}_{12} \cdot \mathbf{E}(|a_1|^2 - |a_2|^2) \qquad (3.A.19)$$

and consequently, from (3.A.16),

$$\frac{d^2}{dt^2}\langle\mathbf{x}\rangle + \omega_0^2\langle\mathbf{x}\rangle = \frac{2e\omega_0}{\hbar}\mathbf{x}_{12}(\mathbf{x}_{12} \cdot \mathbf{E})(|a_1|^2 - |a_2|^2). \qquad (3.A.20)$$

This result obviously resembles the classical oscillator equation (3.2.18b), the difference being in the right-hand side. To interpret this difference, we recall the circumstances for which the classical oscillator model was invented in the period around 1900. The phenomena that Lorentz and others sought to explain involved only natural light (from the sun) or light from man-made thermal sources (lamps). As discussed in Section 1.2, the spectral intensity of any such radiation is weak. This suggests that we focus our attention on the quantum mechanical equation (3.A.20) for the case in which the excited-state probability is close to zero, i.e., for $|a_2|^2 \ll 1$ and $|a_1|^2 \approx 1$. Then we can approximate (3.A.20) by

$$\frac{d^2}{dt^2}\langle\mathbf{x}\rangle + \omega_0^2\langle\mathbf{x}\rangle = \frac{2e\omega_0}{\hbar}\mathbf{x}_{12}(\mathbf{x}_{12} \cdot \mathbf{E}). \qquad (3.A.21)$$

This equation still differs from (3.2.18b), but only in the constants on the right-hand side. To proceed further, let's label the direction of the field as the z direction: $\mathbf{E} = E\hat{z}$, where $\hat{z}$ is the unit vector in the z direction. Then, taking the vector dot product of both sides of (3.A.21) with $\hat{z}$, we have

$$\frac{d^2}{dt^2}\langle z\rangle + \omega_0^2\langle z\rangle = \frac{2e\omega_0}{\hbar}z_{12}^2E, \qquad (3.A.22)$$

where $\langle z\rangle = \langle\mathbf{x}\rangle \cdot \hat{z}$ is the component of $\langle\mathbf{x}\rangle$ along the direction of the electric field. Note that z_{12}^2 means $(z_{12})^2$ and not $(z^2)_{12}$.

In the classical electron oscillator model an electric field pointing in the z direction induces an electron displacement in the z direction, and the Newton equation of motion for this displacement is

$$\frac{d^2z}{dt^2} + \omega_0^2z = \frac{e}{m}E. \qquad (3.A.23)$$

The approximate quantum mechanical equation of motion (3.A.22) for the expectation value of z is identical to the classical equation (3.A.23) if we replace e/m in the latter by (ef/m), where

$$f = \frac{2m\omega_0}{\hbar}z_{12}^2. \qquad (3.A.24)$$

As the notation suggests, f is the oscillator strength introduced in Sections 3.3 and 3.4 in order to bring results of the classical electron oscillator model for emission and absorption into numerical agreement with the results of quantum theory. Equation (3.A.24) can be used to calculate the numerical value of the oscillator strength of the atomic transition of frequency $\omega_0/2\pi$ if we know the wave functions $\phi_1(\mathbf{x})$ and $\phi_2(\mathbf{x})$ of the two states of the transition [see Eq. (3.A.12)]. Using the fact that $x_{12}^2 = y_{12}^2 = z_{12}^2$ for any atomic transition, and that $|\mathbf{x}_{12}|^2 = x_{12}^2 + y_{12}^2 + z_{12}^2$, we can write the expression for the oscillator strength more generally as

$$f = \frac{2m\omega_0}{3\hbar}|\mathbf{x}_{12}|^2. \tag{3.A.25}$$

Thus, for example, we can write the spontaneous emission rate (3.3.7) as

$$A_{21} = \frac{e^2|\mathbf{x}_{12}|^2\omega_0^3}{3\pi\epsilon_0\hbar c^3}. \tag{3.A.26}$$

Effects of level degeneracies on these expressions for f and A_{21} are derived in Section 3.7.

The quantum mechanical validation of the classical electron oscillator model is little short of wonderful. We have shown that, under conditions of low excitation probability, an atomic electron responds to an electric field exactly as if it were bound by a spring to the nucleus, with the natural oscillation frequency corresponding to the Bohr transition frequency. And to make the predictions of the electron-on-a-spring model agree quantitatively with quantum theory, we simply introduce the oscillator strength f. Thus, "*so far as problems involving light are concerned*, the electrons behave as though they were held by springs"[4]—provided that excited-state probabilities are small, which is certainly the case in practically all naturally occurring phenomena. We can also call attention to the $\hbar$ in the denominator of f, showing that f is truly quantum mechanical—there is no classical limit for it as $\hbar \rightarrow 0$.

We have justified the classical oscillator model using the approximation of including only two atomic states in our calculations. It is not difficult to justify the classical model without the two-state approximation; all that is really necessary is the approximation that the atom remains with high probability in the ground state. As a practical matter, however, it is usually not necessary, under conditions of low excitation probability, to include more than the ground and first excited levels of the atom. This is because atomic transitions between the ground state and the first excited state typically have a larger oscillator strength than other transitions involving the ground state, and therefore contribute most strongly to the expectation value $\langle\mathbf{x}\rangle$.

When excited-state probabilities are not small, we cannot make the approximation $|a_1|^2 \approx 1$. In particular, we cannot approximate $|a_1|^2 - |a_2|^2$ in the two-state model by 1. It is precisely this difference that gives rise to the "population difference" $(N_1 - N_2)$ appearing, for instance, in Eqs. (3.6.9) and (3.7.1), and which arises physically because of the possibility of stimulated emission as well as absorption.

PROBLEMS

3.1. (a) Show that the spectrum of thermal radiation for $T = 300\text{K}$ peaks at approximately 10 microns.

   (b) At what frequency ν does $\rho(\nu)$ have its maximum?

(c) Support or refute the statement (from S. Weinberg, *The First Three Minutes,* Bantam Books, New York, 1977, p. 57) that "the average distance between photons in black-body radiation is roughly equal to the typical photon wavelength."

3.2. Assuming the classical force $\mathbf{F} = e\mathbf{E} + e\mathbf{v} \times \mathbf{B}$ acting on a charge e in electric and magnetic fields $\mathbf{E}$ and $\mathbf{B}$, respectively, show that the magnetic force is small compared to the electric force when the charge has a velocity $|\mathbf{v}| \ll c$ in a plane-wave electromagnetic field.

3.3. Assume the "spring constants" k_s for the binding of electrons in atoms are approximately the same as those for the binding of atoms in molecules. If $\nu = 5 \times 10^{14}$ Hz is a typical electronic oscillation frequency, estimate the range of frequencies typical of atomic vibrations in molecules, given typical electron–atom mass differences. Does your estimate indicate that molecular vibrations lie in the infrared region of the spectrum?

3.4. Show that if the field polarization vector in Eq. (3.4.4) is taken to be complex: $\hat{\varepsilon} = \frac{1}{\sqrt{2}}(\hat{x} + i\hat{y})$, then the real part of the right-hand side of (3.4.4) represents a circularly polarized field with the same time-averaged intensity as the given linearly polarized field, that is, $\overline{\mathbf{E} \cdot \mathbf{E}} = \frac{1}{2}E_0^2$. Does this field vector rotate clockwise when viewed by an observer looking into the wave (i.e., looking back toward negative z)? If so, the wave is called right circularly polarized according to the optics convention for polarization.

3.5. Take the incident field to be circularly polarized (see Problem 3.4) and recalculate dW/dt to show that the result given in Eq. (3.4.18) remains unchanged.

3.6. Show that the rate of absorption of energy by an atom in a broadband field is given in the electron oscillator model by Eq. (3.5.6).

3.7. Estimate the temperature of a blacktop road on a sunny day. Assume the asphalt is a perfect blackbody.

3.8. Show that the spectrum of thermal radiation in a gas at temperature T is unaffected by degeneracies of the energy levels of the atoms.

3.9. Compare Eq. (3.7.11) with Eq. (1.5.2) obtained in the simplified laser model described in Chapter 1. What term in Eq. (3.7.11) corresponds to the parameter f in Eq. (1.5.2)? What is the physical meaning of the differences in the form of these two equations?

3.10. Show that the number of atoms (or molecules) per cubic meter of an ideal gas at pressure P and temperature T is given by (3.8.20).

3.11. The CO_2 molecule has strong absorption lines in the neighborhood of $\lambda = 10$ μm. Assuming that the cross sections of CO_2 molecules with N_2 and O_2 molecules are $\sigma(CO_2, N_2) = 1.20$ nm^2 and $\sigma(CO_2, O_2) = 0.95$ nm^2, estimate the collision-broadened linewidth for CO_2 in the atmosphere. (Note: Since the concentration of CO_2 is very small compared to N_2 and O_2 in air, you may assume that only N_2–CO_2 and O_2–CO_2 collisions contribute to the linewidth.) Compare this to the Doppler width.

3.12. Consider an atom of mass m with a resonance frequency v_0 and an initial velocity v in a direction away from a stationary source of radiation of frequency $v \approx v_0$. Assume that a photon of frequency v carries an energy hv and a linear momentum hv/c, that $v \ll c$, and that $hv/c \ll mv$. Using the conservation of energy and linear momentum, derive the formula (3.9.4) for the Doppler-shifted absorption frequency.

3.13. Consider a radiatively broadened transition of an atom. Assuming that the degenerate states of each energy level are equally populated, show that the stimulated emission cross section for narrowband radiation of wavelength $\lambda = c/v$ equal to the transition wavelength is simply $\sigma(v) = \lambda^2/2\pi$. What is the cross section for absorption? (Note: As discussed in Section 14.3, significantly different cross sections can result when the degenerate substates of each level are not equally populated, as occurs when there is "optical pumping.")

3.14. Beer's law in the study of dyes and optical filters states that, if T is the transmission coefficient at a given wavelength and a given dye concentration ρ, then an n-fold increase in ρ results in a transmission coefficient T^n. The equivalent statement in terms of the thickness z of a filter is called Bouguer's law. What functional dependence on ρ and z do these empirical laws imply for the transmitted intensity $I_v(z)$?

3.15. Consider the absorption coefficient $a(v_0)$ of a pure gas precisely at resonance. Show that $a(v_0)$ is proportional to the number density of atoms when the absorption line is Doppler broadened, but is independent of the number density when the pressure is sufficiently large that collision broadening is dominant.

3.16. Consider a cell of length L along the direction of propagation of collimated radiation of frequency v near that of an atomic line. Define the spectral brightness $\mathcal{I}_v(z)$ to be the radiant power per unit area, unit bandwidth, and steradian. Assume that the radiation is unpolarized and that spontaneously emitted radiation is isotropic.

(a) Derive the *equation of radiative transfer*,

$$\frac{d\mathcal{I}_v}{dz} = \frac{hv}{4\pi}A_{21}N_2 S(v) + \frac{hv}{c}B_{12}\left(N_1 - \frac{g_1}{g_2}N_2\right)S(v)\mathcal{I}_v$$

$$\equiv s_v - \kappa_v \mathcal{I}_v.$$

(b) Assuming that N_1 and N_2 are spatially uniform, show that

$$\mathcal{I}_v(L) = \mathcal{I}_v(0)e^{-\kappa_v L} + \frac{s_v}{\kappa_v}[1 - e^{-\kappa_v L}],$$

where $\mathcal{I}_v(0)$ is the spectral brightness of the radiation input to the cell.

(c) If the *optical depth* $\kappa_v L \ll 1$, the medium is said to be *optically thin*. Assuming that there is no radiation input to the cell [i.e., $\mathcal{I}_v(0) = 0$], show that for an optically thin medium

$$\mathcal{I}_\nu(L) \approx \frac{h\nu}{4\pi} A_{21} N_2 S(\nu) L.$$

[Note that the spectrum of the emitted radiation is identical to the absorption lineshape $S(\nu)$.] This result is the basis of one method of measuring oscillator strengths.

(d) If $\kappa_\nu L \gg 1$ the medium is said to be *optically thick*. Show that in this case [cf. Eq. (1.2.3)]

$$\mathcal{I}_\nu(L) \approx \frac{2h\nu^3/c^2}{e^{h\nu/k_B T} - 1},$$

where T is the temperature of the medium.

(e) Discuss the evolution of the spectrum of the emitted radiation as $\kappa_\nu L$ is increased from a very small number to a very large one.

3.17. (a) Estimate the absorption coefficient for 589.0 nm radiation in sodium vapor containing 2.7×10^{18} atoms/m^3 at 200°C. [See J. E. Bjorkholm and A. Ashkin, *Physical Review Letters* **32**, 129 (1973)].

(b) Assuming the same conditions as in (a), plot $I_\nu(z)/I_\nu(0)$ vs. z for $\nu = \nu_0^{(2)}$, $\nu = \nu_0^{(2)} \pm \delta\nu_D$, and $\nu = \nu_0^{(2)} \pm 2\delta\nu_D$.

3.18. Show that for circularly polarized light, for which $\hat{\boldsymbol{\varepsilon}} = \frac{1}{\sqrt{2}}(\hat{x} \pm i\hat{y})$, the remarks following Eq. (3.14.1) remain correct even though (3.14.1) itself applies only to linearly polarized light.

3.19. (a) What is the spontaneous emission rate for the helium $1S_0$–$2P_1$ transition at 58.4 nm?

(b) A cell is filled with helium at a temperature of 300K, and the density is sufficiently low that collision broadening is negligible. Calculate the absorption coefficient for the 58.4-nm transition.

3.20. (a) The position vector **x** for an electron moving with velocity much less than c in a plane monochromatic wave $\hat{\boldsymbol{\varepsilon}} E_0 \cos(\omega t - kz)$ is determined by the equation of motion $m d^2\mathbf{x}/dt^2 = e\hat{\boldsymbol{\varepsilon}} E_0 \cos(\omega t - kz)$. Show that the refractive index of an electron gas is given by Eq. (3.14.13).

(b) Assume that in the ionosphere the refractive index for 100-MHz radio waves is 0.90 and that the free electrons make the greatest contribution to the index. Estimate the number density of electrons.

(c) Why is the contribution of positively charged ions to the refractive index much smaller?

(d) Choose an AM and an FM radio station in your area and compare their frequencies to the plasma frequency of the ionosphere.

3.21. Show how Eqs. (3.A.14)–(3.A.24) are altered if we do not make the assumption that $\mathbf{x}_{21} = \mathbf{x}_{12}$.

4 LASER OSCILLATION: GAIN AND THRESHOLD

4.1 INTRODUCTION

In our superficial analysis of the laser in Chapter 1 we introduced certain concepts such as gain, threshold, and feedback and indicated their importance in our understanding of lasers. We also introduced certain coefficients (a, b, f, p), which we did not derive or explain very carefully. In this chapter we will begin a detailed description of laser oscillation.

The physical system we consider is a collection of atoms (or molecules) between two mirrors. By some pumping process, such as absorption of light from a flashlamp or electron-impact excitation in a gaseous discharge, some of these atoms are promoted to excited states. The excited atoms begin radiating spontaneously, as in an ordinary fluorescent lamp. A spontaneously emitted photon can induce an excited atom to emit another photon of the same frequency and direction as the first. The more such photons are produced by stimulated emission, the faster is the production of still more photons because the stimulated emission rate is proportional to the flux of photons already in the stimulating field. (Recall the discussion in Section 3.7.)

The mirrors of the laser keep photons from escaping completely, so that they can be redirected into the active laser medium to stimulate the emission of more photons. By making the mirrors partially transmitting, some of the photons are allowed to escape. They constitute the output laser beam. The intensity of the output laser beam is determined by the rate of production of excited atoms, the reflectivities of the mirrors, and certain properties of the active atoms. We will see, in this chapter and the next, exactly how the laser output depends on these quantities.

4.2 GAIN AND FEEDBACK

In Chapter 1 the growth rate of the number of laser photons in the cavity was described by an amplification coefficient a. It is closely related to the gain coefficient derived in Section 3.12.

Consider the propagation of narrowband radiation in a medium of atoms that have a transition frequency equal, or nearly equal, to the frequency of the radiation

Laser Physics. By Peter W. Milonni and Joseph H. Eberly
Copyright © 2010 John Wiley & Sons, Inc.

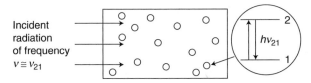

Figure 4.1 Propagation of radiation of frequency $\nu \approx \nu_{21}$ in a medium of atoms with a transition frequency ν_{21}.

(Fig. 4.1). If more atoms are in the upper level of the transition than the lower, there will be more stimulated emission than absorption, and the radiation can be amplified as it propagates. In such a case we say there is *gain* at the resonant frequency. The formula (3.12.6) for the gain coefficient shows that the same lineshape function for the medium applies for both absorption and stimulated emission. Our discussion of lineshape functions in Chapter 3 is therefore relevant to laser media as well as absorbing media. The only thing that distinguishes amplifying media from absorbing media is the sign of the *population inversion* $N_2 - N_1$ or, if the upper and lower levels have degeneracies g_2 and g_1, respectively, $N_2 - (g_2/g_1)N_1$.

Equation (3.12.10) gives an overly optimistic estimate of the growth of intensity in an amplifying ($g > 0$) medium, for it assumes that the gain is independent of intensity. As mentioned in Chapter 3 in connection with that equation, this is a valid assumption only for low intensities. In Sections 4.11 and 4.12 we will explain what it means to have a "high" intensity, and what are the implications of high intensity for the gain coefficient $g(\nu)$. For the present, however, let us accept the prediction of exponential growth as the first approximation to the actual behavior of light in an amplifier.

It is reasonable to expect that a laser can be built in the form of a pencil-shaped container of atoms for which $g > 0$ (Fig. 4.2). The consequences of such a geometry are easy to predict. Some photons are emitted along the axis of the container, where they can encounter other atoms and so induce the emission of more photons, propagating in the same direction and with the same frequency, by stimulated emission. As the number of such photons grows, the stimulated emission rate grows proportionately, so that we expect a burst of radiation to emerge from either end of the container. The direction and cross-sectional area of the beam of light so produced are determined by the container of the excited atoms.

As an example, suppose we have an amplifying medium with a gain coefficient $g = 0.01 \text{ cm}^{-1}$, an achievable gain in many laser media. With a length $L = 1$ m for the gain

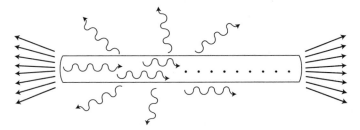

Figure 4.2 A mirrorless "laser." Photons emitted spontaneously along the axis of the tube of excited atoms are multiplied by stimulated emission, resulting in a burst of radiation.

cell, a spontaneously emitted photon at one end of the cell leads, according to Eq. (3.12.10), to an average total of

$$e^{(0.01\text{cm}^{-1})(100\text{cm})} = e^1 \approx 2.72 \qquad (4.2.1)$$

photons emerging at the other end. The output of such a "laser" is obviously not very impressive.

The way to increase the photon yield from such a device is to catch photons emerging from one end and repeatedly feed them back for more amplification. In this way we can, in effect, increase the length of the gain cell. The practical way to achieve this *feedback*, of course, is to have mirrors at the ends of the container.

● It is possible, in media with very high gain, to build mirrorless (sometimes called "superradiant") lasers. The light from such a device resembles that from a conventional laser insofar as it is bright, is quasi-monochromatic, and produces a small spot on a screen. However, it does not have the same degree of temporal and spatial coherence usually associated with lasers. We discuss these coherence properties in Chapter 13. ●

4.3 THRESHOLD

In a laser there is not only an increase in the number of cavity photons because of stimulated emission but also a decrease because of loss effects. These include scattering and absorption of radiation at the mirrors, as well as the "output coupling" of radiation in the form of the usable laser beam. To sustain laser oscillation the stimulated amplification must be sufficient to overcome these losses. This sets a lower limit on the gain coefficient $g(\nu)$, below which laser oscillation does not occur.

One thing we can do now is to predict, given the various losses that tend to diminish the intensity of radiation within the cavity, what minimum gain is necessary to achieve laser oscillation. The condition that the gain coefficient is greater than or equal to this lower limit is called the *threshold condition* for laser oscillation.

Ordinarily, the scattering and absorption of radiation within the gain medium of active atoms is quite small compared to the loss occurring at the mirrors of the laser. We will therefore consider in detail only the losses associated with the mirrors. Figure 4.3 shows a stylized version of a laser resonator, that is, an empty space bounded

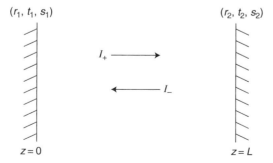

Figure 4.3   The two oppositely propagating beams in a laser cavity.

on two sides by highly reflecting mirrors. A beam of intensity I incident upon one of these mirrors is transformed into a reflected beam of intensity rI, where r is the *reflection coefficient* of the mirror. A beam of intensity tI, where t is the *transmission coefficient*, passes through the mirror. We might expect from the law of conservation of energy that

$$r + t = 1, \tag{4.3.1}$$

that is, the fraction of power reflected plus the fraction transmitted should be unity. Actually, however, some of the incident beam power may be absorbed by the mirror, tending to raise its temperature. Or some of the incident beam may be scattered away because the mirror surface is not perfectly smooth. Thus, the law of conservation of energy takes the form

$$r + t + s = 1, \tag{4.3.2}$$

where s represents the fraction of the incident beam power that is absorbed or scattered by the mirror.

Each of the mirrors of Fig. 4.3 is characterized by a set of coefficients r, t, and s. At the mirror at $z = L$ we have

$$I_\nu^{(-)}(L) = r_2 I_\nu^{(+)}(L), \tag{4.3.3a}$$

and similarly

$$I_\nu^{(+)}(0) = r_1 I_\nu^{(-)}(0) \tag{4.3.3b}$$

for the mirror at $z = 0$. Equations (4.3.3) are boundary conditions that must be satisfied by the solution of the equations describing the propagation of intensity inside the laser cavity.

What are these equations? We are now interested only in steady-state, or continuous-wave (cw), laser oscillation. Near the threshold of laser oscillation the intracavity intensity is very small, and therefore Eq. (3.12.10) is applicable. For light propagating in the positive z direction, therefore, we have

$$\frac{dI_\nu^{(+)}}{dz} = g(\nu)I_\nu^{(+)} \tag{4.3.4a}$$

near the threshold, where g may be taken to be constant. Light propagating in the negative z direction sees the same gain medium and so satisfies a similar equation (see Problem 4.1):

$$\frac{dI_\nu^{(-)}}{dz} = -g(\nu)I_\nu^{(-)}. \tag{4.3.4b}$$

The solutions of these equations are

$$I_\nu^{(+)}(z) = I_\nu^{(+)}(0)e^{g(\nu)z} \tag{4.3.5a}$$

and

$$I_\nu^{(-)}(z) = I_\nu^{(-)}(L)\exp[g(\nu)(L - z)]. \tag{4.3.5b}$$

From (4.3.5a) we see that

$$I_\nu^{(+)}(L) = I_\nu^{(+)}(0)e^{g(\nu)L} \tag{4.3.6}$$

at the right mirror ($z = L$), and the left-going beam has intensity

$$I_\nu^{(-)}(0) = I_\nu^{(-)}(L)e^{g(\nu)L} \tag{4.3.7}$$

at the left mirror ($z = 0$). In steady state the left-going beam has a fraction r_1 of itself reflected at the left mirror (at $z = 0$), and this fraction is just the right-going beam at $z = 0$. A similar consideration applies at the right mirror. Thus, we have

$$I_\nu^{(+)}(0) = r_1 I_\nu^{(-)}(0) = r_1\left[e^{g(\nu)L}I_\nu^{(-)}(L)\right] = r_1 e^{g(\nu)L}\left[r_2 I_\nu^{(+)}(L)\right]$$
$$= r_1 r_2 e^{g(\nu)L}\left[I_\nu^{(+)}(0)e^{g(\nu)L}\right] = \left[r_1 r_2 e^{2g(\nu)L}\right]I_\nu^{(+)}(0). \tag{4.3.8}$$

Similar manipulations, applied to any of the quantities $I_\nu^{(+)}(L), I_\nu^{(-)}(L)$, and $I_\nu^{(-)}(0)$ lead to the same result. Therefore, if $I_\nu^{(+)}(0)$ is not zero, we must have, at steady state,

$$r_1 r_2 e^{2gL} = 1. \tag{4.3.9}$$

The steady-state value of gain that allows (4.3.9) to be satisfied is also the value at which laser action begins. For smaller values there is net attenuation of I_ν in the cavity. Thus, the value of g that satisfies (4.3.9) is labeled g_t and called the *threshold gain*:

$$g_t = \frac{1}{2L}\ln\left(\frac{1}{r_1 r_2}\right) = -\frac{1}{2L}\ln\left(r_1 r_2\right). \tag{4.3.10}$$

This expression can be rewritten usefully in the common case that $r_1 r_2 \approx 1$. Then we define $r_1 r_2 = 1 - x$, or $x = 1 - r_1 r_2$, and use the first term in the Taylor series expansion $\ln(1 - x) \approx -x$, valid when $x \ll 1$, to obtain

$$g_t = \frac{1}{2L}(1 - r_1 r_2) \quad \text{(high reflectivities)}, \tag{4.3.11}$$

which is a satisfactory approximation to (4.3.10) if $r_1 r_2 > 0.90$. The difference between (4.3.11) and (4.3.10) is connected with the assumption that the intracavity field is spatially uniform: We will see in the next chapter that spatial uniformity is a good approximation when $1 - r_1 r_2$ is small, that is, when the mirrors are highly reflecting.

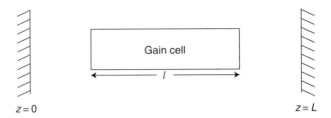

Figure 4.4   A laser in which the gain medium does not fill the entire distance L between the mirrors.

Note that if we are given the mirror reflectivities r_1 and r_2 and their separation L, and therefore the threshold gain, we can determine the population inversion necessary to achieve laser action from (3.12.6) and the atomic absorption (or stimulated emission) cross section.

Our derivation of (4.3.9) assumes that the gain medium fills the entire distance L between the mirrors. This assumption is valid for many solid-state lasers in which the ends of the gain medium are polished and coated with reflecting material. In gas and liquid lasers, however, the gain medium is usually contained in a cell of length $l < L$ that is not joined to the mirrors (Fig. 4.4). In this case the threshold condition is

$$ g_t = -\frac{1}{2l}\ln(r_1 r_2) \approx \frac{1}{2l}(1 - r_1 r_2) \quad \text{(high reflectivities)}. \qquad (4.3.12) $$

The threshold condition (4.3.12) [or (4.3.10)] assumes that "loss" occurs only at the mirrors. This loss is associated with transmission through the mirrors, absorption by the mirrors, and scattering off the mirrors into nonlasing modes. Absorption and scattering are minimized as much as possible by using mirrors of high optical quality. Transmission, of course, is necessary if there is to be any output from the laser.

Other losses might arise from scattering and absorption within the gain medium (from nearly resonant but nonlasing transitions). Such losses are usually small, but they are not difficult to account for in the threshold condition. If a is the effective loss per unit length associated with these additional losses, then the threshold condition (4.3.12) is modified as follows:

$$ g_t = -\frac{1}{2l}\ln(r_1 r_2) + a. \qquad (4.3.13) $$

For our purposes these "distributed losses" (i.e., losses not associated with the mirrors) may usually be ignored.

It is instructive at this point to consider an example. A typical 632.8-nm He–Ne laser might have a gain cell of length $l = 50$ cm and mirrors with reflectivities $r_1 = 0.998$ and $r_2 = 0.980$. Thus

$$ g_t = \frac{-1}{2(50)}\ln(0.998)(0.980)\,\text{cm}^{-1} = 2.2 \times 10^{-4}\,\text{cm}^{-1} \qquad (4.3.14) $$

is the threshold gain.

Using this value for g_t and Eq. (3.12.6), we may calculate the *threshold population inversion* necessary to achieve lasing:

$$\Delta N_t = \left(N_2 - \frac{g_2}{g_1}N_1\right)_t = \frac{8\pi g_t}{\lambda^2 A S(\nu)} = \frac{g_t}{\sigma(\nu)}. \tag{4.3.15}$$

The A coefficient for the 632.8-nm transition in Ne is

$$A \approx 1.4 \times 10^6\,\text{s}^{-1}. \tag{4.3.16}$$

For $T \approx 400\text{K}$ and the Ne atomic weight $M \approx 20$ g, we obtain from Table 3.1 the Doppler width $\delta\nu_D \approx 1500$ MHz. Thus,

$$S(\nu) \approx 6.3 \times 10^{-10}\,\text{s} \tag{4.3.17}$$

and

$$\Delta N_t \approx \frac{(8\pi)(2.2 \times 10^{-4}\,\text{cm}^{-1})}{(6328 \times 10^{-8}\,\text{cm})^2(1.4 \times 10^6\,\text{s}^{-1})(6.3 \times 10^{-10}\,\text{s})}$$
$$= 1.6 \times 10^9\,\text{atoms/cm}^3. \tag{4.3.18}$$

This is a lot of atoms, but it is nevertheless quite a small number compared to the total number of Ne atoms. For a (typical) Ne partial pressure of 0.2 Torr, the total

---

TABLE 4.1   Quantities and Formulas Related to Gain and Threshold

The Gain Coefficient

$$g(\nu) = \frac{\lambda^2 A}{8\pi n^2}\left(N_2 - \frac{g_2}{g_1}N_1\right)S(\nu)$$

$$= \sigma(\nu)\left(N_2 - \frac{g_2}{g_1}N_1\right)$$

$\lambda = \dfrac{c}{\nu}$ = wavelength of radiation

A = Einstein A coefficient for spontaneous emission on the $2 \rightarrow 1$ transition

n = refractive index at wavelength λ

N_2, N_1 = number of atoms per unit volume in levels 2 and 1

g_2, g_1 = degeneracies of levels 2 and 1

$S(\nu)$ = lineshape function (Table 3.1)

Threshold Gain

$$g_t = \frac{-1}{2l}\ln(r_1 r_2) + a \approx \frac{1}{2l}(1 - r_1 r_2) + a$$

l = length of gain medium

r_1, r_2 = mirror reflectivities

a = distributed loss per unit length

number of Ne atoms per cubic centimeter is [Eq. (3.8.20)] about 4.8×10^{15}. Thus, the ratio of the threshold population inversion to the total density of atoms of the lasing species is only

$$\frac{\Delta N_t}{N} = \frac{1.6 \times 10^9}{4.8 \times 10^{15}} = \frac{1}{3} \times 10^{-6}. \tag{4.3.19}$$

Sometimes the quantity e^{gl} is called the gain, and expressed in decibels, that is,

$$G_{dB} = 10 \log_{10}(e^{gl}) = 10 \log_{10}(10^{0.434gl}) = 4.34gl. \tag{4.3.20}$$

The threshold gain in our example is, thus,

$$(G_{dB})_t = (4.34)(2.2 \times 10^{-4}\,\mathrm{cm}^{-1})(50\,\mathrm{cm}) = 0.048\,\mathrm{dB}. \tag{4.3.21}$$

In the laser research literature gain is usually expressed in reciprocal centimeters, although the decibel is the preferred unit in fiber optics.

In Table 4.1 we collect the formulas and terms we have used in discussing gain and threshold.

4.4 PHOTON RATE EQUATIONS

To describe time-dependent phenomena, such as pulsed laser operation or the startup of continuous-wave lasing, we must include the time derivative $\partial I_\nu/\partial t$ in the propagation equation (3.12.5). For the right- and left-going waves in the laser resonator (Fig. 4.3), we write

$$\frac{\partial I_\nu^{(+)}}{\partial z} + \frac{1}{c}\frac{\partial I_\nu^{(+)}}{\partial t} = g(\nu)I_\nu^{(+)}, \tag{4.4.1a}$$

and

$$-\frac{\partial I_\nu^{(-)}}{\partial z} + \frac{1}{c}\frac{\partial I_\nu^{(-)}}{\partial t} = g(\nu)I_\nu^{(-)}, \tag{4.4.1b}$$

respectively. Addition of these equations gives

$$\frac{\partial}{\partial z}\left[I_\nu^{(+)} - I_\nu^{(-)}\right] + \frac{1}{c}\frac{\partial}{\partial t}\left[I_\nu^{(+)} + I_\nu^{(-)}\right] = g(\nu)(I_\nu^{(+)} + I_\nu^{(-)}). \tag{4.4.2}$$

We will see in the following chapter (Sections 5.2 and 5.5) that in many lasers there is very little gross variation of $I_\nu^{(+)} - I_\nu^{(-)}$ with z. Assuming this result, we approximate (4.4.2) by the ordinary differential equation

$$\frac{d}{dt}\left[I_\nu^{(+)} + I_\nu^{(-)}\right] = cg(\nu)\left[I_\nu^{(+)} + I_\nu^{(-)}\right]. \tag{4.4.3}$$

Note that here we are assuming spatial uniformity, just as we assumed temporal uniformity (steady state) in the preceding section. In Chapter 5 we will discuss the temporal steady-state rate equation that results from a more detailed consideration of the spatial boundary conditions.

If the gain medium does not completely fill the resonator (Fig. 4.4), then $g(v) = 0$ outside it. If we integrate both sides of (4.4.3) over z in the region $0 < z < L$, then the left side, which is independent of z, is simply multiplied by L. However, the right side is multiplied by l, which is less than L, since $g(v)$ is different from zero only inside the gain medium. Thus,

$$\frac{d}{dt}\left[I_v^{(+)} + I_v^{(-)}\right] = \frac{cl}{L}g(v)\left[I_v^{(+)} + I_v^{(-)}\right] \qquad (4.4.4)$$

is the generalization of (4.4.3) to the case $l < L$. Since the number of photons inside the cavity is proportional to the total intensity, we may also write

$$\frac{dq_v}{dt} = \frac{cl}{L}g(v)q_v, \qquad (4.4.5)$$

where q_v is the number of cavity photons associated with the frequency v.

Equation (4.4.5) describes the growth in time of the number of cavity photons as a result of the absorption and induced emission of photons by the gain medium. The factor $cg(v)l/L$ is the growth rate. Of course, we must also consider the loss of cavity photons due to output coupling, absorption and scattering at the mirrors, and the like. We can take account of the loss associated with the output coupling of laser radiation from the cavity, which is usually the most important loss mechanism, as follows.

Radiation reflected from the mirror at $z = L$ (Fig. 4.3) has an intensity that is r_2 times the incident intensity. After it is reflected from the mirror at $z = 0$, therefore, it has an intensity $r_1 r_2$ times its intensity before the round trip inside the resonator. In other words, a fraction $1 - r_1 r_2$ of intensity is lost. Since the time it takes to make a round trip is $2L/c$, the *rate* at which intensity is lost due to the imperfect reflectivity of the mirrors is $c(1 - r_1 r_2)/2L$. In terms of photons, this loss rate is

$$\left(\frac{dq_v}{dt}\right)_{\text{output coupling}} = -\frac{c}{2L}(1 - r_1 r_2)q_v. \qquad (4.4.6)$$

The total rate at which the number of cavity photons changes is therefore

$$\frac{dq_v}{dt} = \left(\frac{dq_v}{dt}\right)_{\text{gain}} + \left(\frac{dq_v}{dt}\right)_{\text{output coupling}} = \frac{cl}{L}g(v)q_v - \frac{c}{2L}(1 - r_1 r_2)q_v$$

$$= \frac{cl}{L}g(v)q_v - \frac{cl}{L}g_t q_v. \qquad (4.4.7)$$

If there are significant losses besides those occurring at the mirrors, they may be accounted for in a similar fashion. We will assume that these other losses are negligible, in which case (4.4.7) gives the rate of change with time of the number of cavity photons. Equivalently, we may write the rate equation

$$\frac{dI_v}{dt} = \frac{cl}{L}g(v)I_v - \frac{c}{2L}(1 - r_1 r_2)I_v \qquad (4.4.8)$$

for the total intensity $I_\nu = I_\nu^{(+)} + I_\nu^{(-)}$ inside the cavity. If we assume equal upper and lower level degeneracies, $g_1 = g_2$, then we may write (4.4.8) as

$$\frac{dI_\nu}{dt} = \frac{cl}{L}\frac{\lambda^2 A}{8\pi}(N_2 - N_1)S(\nu)I_\nu - \frac{c}{2L}(1 - r_1 r_2)I_\nu$$

$$= \frac{cl}{L}\sigma(\nu)(N_2 - N_1)I_\nu - \frac{c}{2L}(1 - r_1 r_2)I_\nu. \tag{4.4.9}$$

4.5   POPULATION RATE EQUATIONS

The population densities N_2 and N_1 also change in time, of course, due to stimulated emission, absorption, spontaneous emission, and collisions. The effects of the first three processes on the rate equations for N_2 and N_1 are given by Eqs. (3.7.5). Collisions affecting the upper- and lower-level populations are called "inelastic" in order to distinguish them from "elastic" collisions, which do not result in a change in the energy of the colliding atoms. To account for inelastic (population-changing) collisions, we simply assert that their effect is to knock population out of levels 1 and 2 into unspecified levels at the rates Γ_1 and Γ_2. Thus, we replace Eqs. (3.7.5) by

$$\frac{dN_2}{dt} = -\Gamma_2 N_2 - \frac{\sigma(\nu)}{h\nu}I_\nu(N_2 - N_1) - A_{21}N_2, \tag{4.5.1a}$$

$$\frac{dN_1}{dt} = -\Gamma_1 N_1 + \frac{\sigma(\nu)}{h\nu}I_\nu(N_2 - N_1) + A_{21}N_2. \tag{4.5.1b}$$

We must also account for the pumping process that produces the (positive) population inversion $(N_2 - N_1)$. To do this most simply we add a term K to the population equations and call it the pumping rate into the upper level. There are several methods of arranging pumping of this kind, as discussed in Chapter 11. For the time being we simply insert a term K. With this minor modification of the population equations (4.5.1), we have the following set of coupled equations for the light and the atoms in the laser cavity:

$$\frac{dN_1}{dt} = -\Gamma_1 N_1 + A_{21}N_2 + g(\nu)\Phi_\nu, \tag{4.5.2a}$$

$$\frac{dN_2}{dt} = -(\Gamma_2 + A_{21})N_2 - g(\nu)\Phi_\nu + K, \tag{4.5.2b}$$

$$\frac{d\Phi_\nu}{dt} = \frac{cl}{L}g(\nu)\Phi_\nu - \frac{c}{2L}(1 - r_1 r_2)\Phi_\nu. \tag{4.5.2c}$$

We have used

$$I_\nu = h\nu\Phi_\nu, \tag{4.5.3}$$

where Φ_ν is the photon flux, in rewriting (4.4.8) as (4.5.2c).

For some purposes it is useful to rewrite Eqs. (4.5.2) so that they refer to absolute numbers, rather than densities, of atoms and photons. This is easy to do. The total number of atoms in level 2 is $n_2 = N_2 V_g$, where V_g is the volume of the gain medium. Likewise the total number of atoms in the lower level of the laser transition is $n_1 = N_1 V_g$. The electromagnetic energy density u_ν in the cavity is related to intensity I_ν and photon flux Φ_ν by

$$u_\nu = \frac{I_\nu}{c} = \left(\frac{h\nu}{c}\right)\Phi_\nu, \tag{4.5.4}$$

and it is related to photon number q_ν by

$$u_\nu = \frac{h\nu q_\nu}{V},$$

where V is the cavity volume. These relations assume a uniform distribution of intensity within the cavity and a refractive index $n \approx 1$. Thus,

$$\Phi_\nu = \frac{cq_\nu}{V},$$

and Eqs. (4.5.2) may be rewritten in the form

$$\frac{dn_1}{dt} = -\Gamma_1 n_1 + A_{21} n_2 + \frac{cl}{L} g(\nu) q_\nu, \tag{4.5.5a}$$

$$\frac{dn_2}{dt} = -(\Gamma_2 + A_{21}) n_2 - \frac{cl}{L} g(\nu) q_\nu + p, \tag{4.5.5b}$$

$$\frac{dq_\nu}{dt} = \frac{cl}{L} g(\nu) q_\nu - \frac{c}{2L}(1 - r_1 r_2) q_\nu, \tag{4.5.5c}$$

where we have used the relations

$$\frac{V_g}{V} = \frac{l}{L} \qquad \text{and} \qquad K V_g = p. \tag{4.5.6}$$

Equations (4.5.5) imply that

$$\frac{d}{dt}(n_2 + q_\nu) = -(\Gamma_2 + A_{21}) n_2 + p - \frac{c}{2L}(1 - r_1 r_2) q_\nu. \tag{4.5.7}$$

This equation has an obvious interpretation. The left-hand side is the rate of change of the total number of *excitations*, that is, the number of atoms in the upper level 2 of the lasing transition plus the number of photons in the cavity. The first term on the right is the rate of decrease in the number of these excitations as a result of inelastic collisions and spontaneous emission from level 2. The second term is the rate of change associated with pumping of level 2. The last term is the rate at which excitation in the form of photons is lost from the cavity. Note that contributions from stimulated emission (or absorption) do not appear in (4.5.7) because they have canceled out: An increase in q_ν is always accompanied by an equal decrease in n_2. Further features of (4.5.7) are pointed out in Problem 4.2.

4.6   COMPARISON WITH CHAPTER 1

In Chapter 1, Section 1.5, we developed an intuitive quantum theory of the laser, introducing various rate constants in a largely ad hoc fashion. Now that we have developed rate equations for level populations and photons, it is interesting to return to this intuitive model and examine its validity.

First recall that $g(\nu) = \sigma(\nu)(N_2 - N_1)$ if the level degeneracies are equal. Thus,

$$g(\nu) \approx \sigma(\nu)N_2 = \frac{\sigma(\nu)n_2}{V_g} \tag{4.6.1}$$

if there is negligible occupation of level 1: $n_2 \gg n_1$. Now define two constant coefficients a and b:

$$a = \frac{c\sigma(\nu)l/L}{V_g} = \frac{c\sigma(\nu)}{V} \tag{4.6.2}$$

and

$$b = \frac{c}{2L}(1 - r_1 r_2). \tag{4.6.3}$$

Then Eq. (4.5.5c) for the photon number q_ν can be written in the compact form

$$\frac{dq_\nu}{dt} = an_2 q_\nu - bq_\nu, \tag{4.6.4}$$

which is exactly Eq. (1.5.1). Recall that in Chapter 1 we identified n as the number of atoms in level 2, here denoted n_2.

The equation for n_2 is easily obtained from (4.5.5b). We again invoke the assumption $n_2 \gg n_1$ to get

$$\frac{dn_2}{dt} = -an_2 q_\nu - fn_2 + p, \tag{4.6.5}$$

where

$$f = \Gamma_2 + A_{21}. \tag{4.6.6}$$

We see that Eq. (4.6.5) is the same as Eq. (1.5.2).

Thus, if the population inversion is large enough that N_1 is negligible compared to N_2, the theory developed in Chapter 1 agrees with our coupled photon-population rate equations. If N_2 is not much larger than N_1, the theory of Chapter 1 requires some minor modifications. What we were not able to do in Chapter 1, however, was to identify the constants a, b, f, and p in terms of fundamental atomic parameters like the Einstein A coefficient, the inelastic collision rate, the atomic absorption cross section, and mirror reflectivities. That has now been accomplished.

4.7   THREE-LEVEL LASER SCHEME

Thus far, we have not specified where levels 1 and 2 appear in the overall energy-level scheme of the lasing atoms. We might imagine that level 1 is the ground level and level 2 the first excited level of an atom (Fig. 4.5). When we attempt to achieve continuous laser oscillation using the two-level scheme of Fig. 4.5, however, we encounter a serious difficulty: The mechanism we use to excite atoms to level 2 can also deexcite them. For example, if we try to pump atoms from level 1 to level 2 by irradiating the medium, the radiation will induce both upward transitions $1 \rightarrow 2$ (absorption) and downward transitions $2 \rightarrow 1$ (stimulated emission).

As discussed in Section 4.11, the *best* we can do by this optical pumping process is to produce nearly the same number of atoms in level 2 as in level 1; we cannot obtain a positive steady-state population inversion using only two atomic levels in the pumping process.

One resolution of this difficulty is to make use of a third level, as in the *three-level laser* inversion scheme of Fig. 4.6. In such a laser, some pumping process acts between level 1 and level 3. An atom in level 3 cannot stay there forever. As a result of the pumping process, it may return to level 1, but for other reasons such as spontaneous emission or a collision with another particle, the atom may drop to a different level of lower energy. In the case of spontaneous emission the energy lost by the atom appears as radiation. In the case of collisional deexcitation, the energy lost by the atom may appear as internal excitation in a collision partner, or as an increase in the kinetic energy of the collision partners, or both. The key to the three-level inversion scheme of Fig. 4.6 is to have atoms in the pumping level 3 drop very rapidly to the upper laser level 2. This accomplishes two purposes. First, the pumping from level 1 is, in effect, directly

Figure 4.5   A two-level laser.

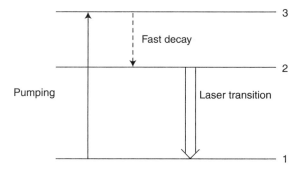

Figure 4.6   A three-level laser. Level 1 is the ground level, and laser oscillation occurs on the $2 \rightarrow 1$ transition.

from level 1 to the upper laser level 2, because every atom finding itself in level 3 converts quickly to an atom in level 2. Second, the rapid depletion of level 3 does not give the pumping process much chance to act in reverse and repopulate the ground level 1.

We will characterize the pumping process by a rate P, so that PN_1 is the number of atoms per cubic centimeter per second that are taken from ground level 1 to level 3. Thus, the rate of change of the population N_1 of atoms per cubic centimeter in level 1 is

$$\left(\frac{dN_1}{dt}\right)_{\text{pumping}} = -PN_1 \tag{4.7.1}$$

as a result of the pumping process. Since the pumping takes atoms from level 1 to level 3, and level 3 is assumed to decay very rapidly to level 2, we may also write (see Problem 4.3)

$$\left(\frac{dN_2}{dt}\right)_{\text{pumping}} \approx \left(\frac{dN_3}{dt}\right)_{\text{pumping}} = -\left(\frac{dN_1}{dt}\right)_{\text{pumping}} = PN_1 \tag{4.7.2}$$

for the rate of change of population of level 2 due to pumping.

Atoms in level 2 can decay, by spontaneous emission or via collisions, as indicated in population Eq. (4.5.2b) or (4.5.5b). For simplicity we will now assume that level 2 decays only into level 1 by these processes, and we will denote the rate by Γ_{21}. That is, we assume

$$\left(\frac{dN_2}{dt}\right)_{\text{decay}} = -\Gamma_{21}N_2, \qquad \left(\frac{dN_1}{dt}\right)_{\text{decay}} = \Gamma_{21}N_2, \tag{4.7.3}$$

for the population changes associated with the decay of level 2. The total rates of change of the populations of levels 1 and 2 are therefore

$$\frac{dN_1}{dt} = -PN_1 + \Gamma_{21}N_2 + \sigma\Phi_\nu(N_2 - N_1), \tag{4.7.4a}$$

$$\frac{dN_2}{dt} = PN_1 - \Gamma_{21}N_2 - \sigma\Phi_\nu(N_2 - N_1). \tag{4.7.4b}$$

Equations (4.7.4) imply the conservation law

$$\frac{d}{dt}(N_1 + N_2) = 0,$$

or

$$N_1 + N_2 = \text{const} = N_T. \tag{4.7.5}$$

By ignoring any other atomic energy levels, and assuming that level 3 decays practically instantaneously into level 2, we are assuming that each active atom of the gain medium must be either in level 1 or level 2. Therefore, the conserved quantity N_T is simply the total number of active atoms per unit volume.

We can now draw some important conclusions about the "threshold region" of steady-state (cw) laser oscillation. Near threshold the number of cavity photons is small enough that stimulated emission may be omitted from Eqs. (4.7.4). In particular, we can determine from these equations the threshold pumping rate necessary to achieve a population inversion, together with the threshold power expended in the process.

In the steady state N_1 and N_2 are not changing in time. The steady-state values $\overline{N}_1$ and $\overline{N}_2$, therefore, satisfy Eqs. (4.7.4) with $dN_1/dt = dN_2/dt = 0$. Thus, if Φ_ν is so small that the last terms in (4.7.4) are negligible, we find

$$\overline{N}_2 = \frac{P}{\Gamma_{21}}\overline{N}_1 \tag{4.7.6}$$

in the steady state. Since (4.7.5) must hold for all possible values of N_1 and N_2, including the steady-state values $\overline{N}_1$ and $\overline{N}_2$, we also have

$$\overline{N}_1 + \overline{N}_2 = N_T. \tag{4.7.7}$$

Equations (4.7.6) and (4.7.7) may be solved for $\overline{N}_1$ and $\overline{N}_2$ to obtain

$$\overline{N}_1 = \frac{\Gamma_{21}}{P + \Gamma_{21}}N_T \tag{4.7.8a}$$

and

$$\overline{N}_2 = \frac{P}{P + \Gamma_{21}}N_T. \tag{4.7.8b}$$

The steady-state threshold-region population inversion is therefore

$$\overline{N}_2 - \overline{N}_1 = \frac{P - \Gamma_{21}}{P + \Gamma_{21}}N_T. \tag{4.7.9}$$

To have a positive steady-state population inversion, and therefore a positive gain, we must obviously have

$$P > \Gamma_{21}, \tag{4.7.10}$$

which simply says that the pumping rate into the upper laser level must exceed the decay rate. The greater the pumping rate with respect to the decay rate, the greater the population inversion and gain.

The pumping of an atom from level 1 to level 3 requires an energy

$$E_3 - E_1 = h\nu_{31}. \tag{4.7.11}$$

The power per unit volume delivered to the active atoms in the pumping process is therefore

$$\frac{\text{Pwr}}{V} = h\nu_{31}P\overline{N}_1 \tag{4.7.12}$$

in the steady state. Using (4.7.8), we may write this as

$$\frac{\text{Pwr}}{V} = \frac{h\nu_{31}P\Gamma_{21}}{P + \Gamma_{21}}N_T. \tag{4.7.13}$$

Now from (4.7.10) we may regard

$$P_{\min} = \Gamma_{21} \tag{4.7.14}$$

as the minimum pumping rate necessary to reach positive gain. Substituting P_{min} for P in (4.7.13), we obtain

$$\left(\frac{\text{Pwr}}{V}\right)_{min} = \frac{1}{2}\Gamma_{21}N_T h\nu_{31} \tag{4.7.15}$$

as the minimum power per unit volume that must be exceeded to produce a positive gain. With this amount of pumping power delivered to the active medium, we see from (4.7.8) (with $P = P_{min} = \Gamma_{21}$) that half the active atoms are in the lower level of the laser transition and half are in the upper level. A pumping power density greater than (4.7.15) makes $\overline{N}_2 > \overline{N}_1$.

4.8   FOUR-LEVEL LASER SCHEME

Another useful model for achieving population inversion is the *four-level laser* scheme shown in Fig. 4.7. Pumping proceeds from the ground level 0 to the level 3, which, as in the three-level laser, decays rapidly into the upper laser level 2. In this model the lower laser level 1 is not the ground level, but an excited level that can itself decay into the ground level. This represents an advantage over the three-level laser, for the depletion of the lower laser level obviously enhances the population inversion on the laser transition. That is, a decrease in N_1 results in an increase in $N_2 - N_1$.

As in a three-level laser the decay from level 3 to level 2 is ideally instantaneous, that is, extremely rapid compared to any other rates in the population rate equations. Then we may take $N_3 \approx 0$, and the population rate equations for the four-level laser take the form

$$\frac{dN_0}{dt} = -PN_0 + \Gamma_{10}N_1, \tag{4.8.1a}$$

$$\frac{dN_1}{dt} = -\Gamma_{10}N_1 + \Gamma_{21}N_2 + \sigma(\nu)(N_2 - N_1)\Phi_\nu, \tag{4.8.1b}$$

$$\frac{dN_2}{dt} = PN_0 - \Gamma_{21}N_2 - \sigma(\nu)(N_2 - N_1)\Phi_\nu, \tag{4.8.1c}$$

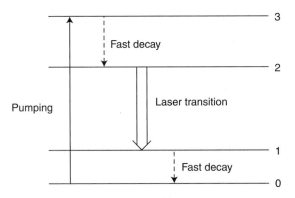

Figure 4.7   A four-level laser. The lower laser level 1 decays into the ground level 0.

where P is again the pumping rate out of the ground level 0, and PN_0 is the upper level pumping rate denoted by K in (4.5.2). Γ_{21} and Γ_{10} are the rates for the decay processes $2 \rightarrow 1$ and $1 \rightarrow 0$, respectively, and we have made the same approximation (4.7.2) for the pumping process as in the three-level case. Note that Eqs. (4.8.1) imply the conservation law

$$N_0 + N_1 + N_2 = \text{const} = N_T. \tag{4.8.2}$$

If the stimulated emission rate is very small compared to the pumping and decay rates, Eqs. (4.8.1) give the steady-state populations

$$\overline{N}_0 = \frac{\Gamma_{10}\Gamma_{21}}{\Gamma_{10}\Gamma_{21} + \Gamma_{10}P + \Gamma_{21}P} N_T, \tag{4.8.3a}$$

$$\overline{N}_1 = \frac{\Gamma_{21}P}{\Gamma_{10}\Gamma_{21} + \Gamma_{10}P + \Gamma_{21}P} N_T, \tag{4.8.3b}$$

$$\overline{N}_2 = \frac{\Gamma_{10}P}{\Gamma_{10}\Gamma_{21} + \Gamma_{10}P + \Gamma_{21}P} N_T. \tag{4.8.3c}$$

The steady-state population inversion of the laser transition is therefore (Problem 4.4)

$$\overline{N}_2 - \overline{N}_1 = \frac{P(\Gamma_{10} - \Gamma_{21})N_T}{\Gamma_{10}\Gamma_{21} + \Gamma_{10}P + \Gamma_{21}P}. \tag{4.8.4}$$

Thus, the pumped ($P \neq 0$) four-level system will always have a steady-state population inversion when

$$\Gamma_{10} > \Gamma_{21}, \tag{4.8.5}$$

that is, when the lower laser level decays more rapidly than the upper laser level. When we have

$$\Gamma_{10} \gg \Gamma_{21}, P, \tag{4.8.6}$$

then $N_0 \approx N_T$, $N_1 \approx 0$ and Eq. (4.8.4) reduces to

$$\overline{N}_2 - \overline{N}_1 \approx \overline{N}_2 \approx \frac{P}{P + \Gamma_{21}} N_T. \tag{4.8.7}$$

4.9 PUMPING THREE- AND FOUR-LEVEL LASERS

It is interesting to compare the pumping rates necessary for laser oscillation in the three- and four-level lasers. To achieve laser oscillation, we must produce a gain larger than the threshold value g_t, and therefore a population inversion greater than ΔN_t, where

$$g_t \equiv \sigma(\nu)\,\Delta N_t \tag{4.9.1}$$

is the threshold gain for frequency ν. Setting $\overline{N}_2 - \overline{N}_1$ equal to the threshold inversion in (4.7.9), we obtain

$$(P_t)_{\text{three-level laser}} = \frac{N_T + \Delta N_t}{N_T - \Delta N_t}\Gamma_{21} \tag{4.9.2}$$

for the threshold pumping rate for laser oscillation in the three-level laser. From (4.8.7), on the other hand, the pumping rate necessary for a four-level laser satisfying (4.8.6) is

$$(P_t)_{\text{four-level laser}} = \frac{\Delta N_t}{N_T - \Delta N_t} \Gamma_{21} \tag{4.9.3}$$

The ratio of (4.9.3) to (4.9.2) is

$$\frac{(P_t)_{\text{four-level laser}}}{(P_t)_{\text{three-level laser}}} = \frac{\Delta N_t}{N_T + \Delta N_t}. \tag{4.9.4}$$

Ordinarily, the population inversion (at threshold or otherwise) is very small compared to the total number of active atoms; recall our example near the end of Section 4.3. From (4.9.4), therefore, we see that

$$(P_t)_{\text{four-level laser}} \ll (P_t)_{\text{three-level laser}}. \tag{4.9.5}$$

Thus, a much larger pumping rate is necessary to achieve laser oscillation in a three-level laser than in a four-level laser.

In the three-level laser the pumping power density necessary to establish the threshold inversion ΔN_t is

$$\left(\frac{(\text{Pwr})_t}{V} \right)_{\text{three-level laser}} = h\nu_{31}(N_1)_t(P_t)_{\text{three-level laser}}, \tag{4.9.6}$$

where $(N_1)_t$ is given by (4.7.8a) with $P = (P_t)_{\text{three-level laser}}$. Assuming $\Delta N_t \ll N_T$, we conclude from (4.9.2) that

$$(P_t)_{\text{three-level laser}} \approx \Gamma_{21}, \tag{4.9.7}$$

and from (4.7.8) that

$$(N_1)_t \approx \frac{N_T}{2}. \tag{4.9.8}$$

Equation (4.9.6) therefore becomes

$$\left(\frac{(\text{Pwr})_t}{V} \right)_{\text{three-level laser}} \approx \frac{1}{2} h\nu_{31} N_T \Gamma_{21}, \tag{4.9.9}$$

which, because of the approximation $\Delta N_t \ll N_T$, is the same as (4.7.15). Thus, when $\Delta N_t \ll N_T$ the minimum pumping power density necessary to achieve positive gain is approximately the same as that necessary to reach threshold and laser oscillation. In the four-level case, on the other hand, we obtain

$$\left(\frac{(\text{Pwr})_t}{V} \right)_{\text{four-level laser}} \approx h\nu_{30} \Delta N_t \Gamma_{21} \tag{4.9.10}$$

for $\Delta N_t \ll N_T$. The ratio of (4.9.10) to (4.9.9) is

$$\frac{[(\text{Pwr})_t/V]_{\text{four-level laser}}}{[(\text{Pwr})_t/V]_{\text{three-level laser}}} \approx \frac{2\nu_{30}\,\Delta N_t}{\nu_{31}N_T} \tag{4.9.11}$$

if we take the upper-laser-level decay rate Γ_{21} to be the same in the two cases. This shows again that, other things being roughly commensurate, much less power is required to achieve laser oscillation in the four-level case.

4.10 EXAMPLES OF THREE- AND FOUR-LEVEL LASERS

Real lasers seldom fit very neatly into our three- and four-level categories. However, these idealizations sometimes provide a useful framework for rough estimates of pump power requirements. We will illustrate this for cw ruby and Nd : YAG solid-state lasers, which are approximately three-level and four-level systems, respectively.

Ruby was the gain medium for the very first laser, but the ruby laser has largely been replaced by other, more efficient solid-state lasers. It nevertheless illustrates nicely the concept of a three-level laser. As mentioned in Section 3.1, ruby is the crystal Al_2O_3 with chromium ions (Cr^{3+}) replacing some of the aluminum ions (Al^{3+}); the concentration of chromium is only about 0.05% by weight. The relevant energy levels for the ruby laser are those of the Cr^{3+} ion in the host crystal lattice. The laser is *optically pumped*, that is, a population inversion is obtained by the absorption of radiation from another laser or a lamp, typically a high-pressure Xe or Hg flashlamp (Section 11.12). It is approximately a three-level laser, although the third "level" really consists of two broad bands of energy, both decaying rapidly (rate $\approx 10^7$ s^{-1}) into the upper laser level 2. At room temperature the decay rate of the upper laser level is

$$\Gamma_{21} \approx \tfrac{1}{2} \times 10^3 \, \text{s}^{-1}. \tag{4.10.1}$$

The density of "active atoms" (i.e., Cr^{3+} ions) is (Problem 4.5)

$$N_T \approx 1.6 \times 10^{19} \, \text{cm}^{-3}. \tag{4.10.2}$$

The excitation energy required from the pump, the energy difference between the ground level and the lower pump band, is about 2.25 eV, corresponding to a wavelength of about 550 nm (green). From (4.9.9), therefore, the minimum pumping power density necessary to achieve nonnegative gain [and also, according to (4.7.15), approximately the pumping power density necessary for laser oscillation] is

$$\left(\frac{\text{Pwr}}{V}\right)_{\text{min}} \approx \frac{1}{2}\left(\frac{1}{2} \times 10^3 \, \text{s}^{-1}\right)(1.6 \times 10^{19} \, \text{cm}^{-3})(2.25 \, \text{eV}) \approx 1 \, \text{kW/cm}^3. \tag{4.10.3}$$

For a 5-cm-long ruby rod of radius 2 mm, the required pump power is

$$\text{Pwr} \approx \left(\frac{1 \, \text{kW}}{\text{cm}^3}\right) \pi \, (0.2 \, \text{cm})^2 (5 \, \text{cm}) = 600 \, \text{W}. \tag{4.10.4}$$

This is a rather large amount of power. In fact, much *more* power is actually required to operate a cw ruby laser because (4.10.4) only gives the power that must actually be absorbed by the Cr^{3+} ions. In reality only a small fraction (typically $\approx 0.1\%$) of the electrical power delivered to the lamp is converted to useful laser radiation.

It is also interesting to estimate the population inversion necessary for threshold gain in a typical ruby laser. At room temperature the 694.3-nm laser transition has a Lorentzian lineshape of width (HWHM)

$$\delta\nu_0 \approx 170\,\text{GHz}, \tag{4.10.5}$$

and the A coefficient for spontaneous emission is

$$A \approx 230\,\text{s}^{-1}. \tag{4.10.6}$$

The line-center cross section for stimulated emission is therefore

$$\sigma = \frac{\lambda^2 A}{8\pi n^2} \frac{1}{\pi\delta\nu_0} \approx 2.7 \times 10^{-20}\,\text{cm}^2, \tag{4.10.7}$$

where we have used the value $n = 1.76$ for the refractive index of ruby. If we assume a resonator with mirror reflectivities $r_1 = 1.0$ and $r_2 = 0.96$, and a scattering loss of 3% per round-trip pass through the gain cell, then the threshold gain for laser oscillation is [Eq. (4.3.13)]

$$g_t = \frac{-1}{2(5\,\text{cm})} \ln(0.96) + \frac{0.03}{2(5\,\text{cm})} = 7.1 \times 10^{-3}\,\text{cm}^{-1} \tag{4.10.8}$$

for a ruby rod 5 cm long. From (4.3.15), (4.10.7), and (4.10.8), therefore, we calculate a population inversion threshold

$$\Delta N_t \approx \frac{7.1 \times 10^{-3}\,\text{cm}^{-1}}{2.7 \times 10^{-20}\,\text{cm}^2} = 2.6 \times 10^{17}\,\text{cm}^{-3}. \tag{4.10.9}$$

This is much larger than the sort of population inversion necessary for a typical He–Ne laser [Eq. (4.3.18)]. The difference stems from the much larger stimulated emission cross section of the 632.8-nm laser transition of Ne, which in turn results from the much larger A coefficient and much smaller linewidth than in ruby. This illustrates an important point: Gas lasers obviously have a much smaller density of atoms than solid-state lasers, but this does not necessarily mean that they have smaller gains. In fact, many of the most powerful lasers are gas lasers. The reasons for this are discussed in Chapter 11.

In the 1.06-μm (1064-nm) Nd : YAG laser, the active atoms are also impurities in a crystal lattice, in this case Nd^{3+} ions in yttrium aluminum garnet ($Y_3Al_5O_{12}$, called YAG). The Nd : YAG laser is approximately a four-level system, with upper-level decay rate

$$\Gamma_{21} \approx 4400\,\text{s}^{-1} \tag{4.10.10}$$

and stimulated emission cross section[1]

$$\sigma \approx 3 \times 10^{-19} \, \text{cm}^2 \qquad (4.10.11)$$

at room temperature. If we assume the same threshold gain (4.10.8) as in our calculation for the ruby laser, we obtain a population inversion threshold

$$\Delta N_t \approx \frac{7.1 \times 10^{-3} \, \text{cm}^{-1}}{3 \times 10^{-19} \, \text{cm}^2} \approx 2 \times 10^{16} \, \text{cm}^{-3}, \qquad (4.10.12)$$

which, because of the relatively large stimulated emission cross section for Nd^{3+}, is considerably smaller than the value (4.10.9) for ruby.

The pump "level 3" for the Nd : YAG laser is actually a series of energy bands located between about 1.63 and 3.13 eV above the ground level. If we take the energy difference $E_3 - E_0$ in our four-level model (Fig. 4.7) to be the average value, 2.38 eV, we obtain from (4.9.10) the pumping power density for threshold:

$$\frac{(\text{Pwr})_{\min}}{V} \approx 30 \, \text{W/cm}^3. \qquad (4.10.13)$$

For a 5-cm Nd : YAG rod of radius 2 mm, therefore, the threshold pump power is

$$\text{Pwr} \approx 20 \, \text{W}, \qquad (4.10.14)$$

much smaller than the estimate (4.10.4) for ruby.

4.11 SATURATION

We remarked in Sections 3.12 and 4.2 that exponential growth of intensity in a gain medium is only an approximation, and that the approximation breaks down when the intensity is sufficiently large. Exponential attenuation in an absorbing medium is likewise a low-intensity approximation.

To understand this, let us return to the rate equations (3.7.5) for the populations of two nondegenerate levels. No upper-level pumping processes are included in these equations, only absorption and stimulated and spontaneous emission. We further assume that the intensity I_ν of the field is constant in time. The steady-state solutions $\overline{N}_2$ and $\overline{N}_1$ obtained by setting the derivatives equal to zero are easily found to be

$$\overline{N}_2 = \frac{\sigma(\nu) I_\nu / h\nu}{A_{21} + 2\sigma(\nu) I_\nu / h\nu} N = \frac{\frac{1}{2} I_\nu / I_\nu^{\text{sat}}}{1 + I_\nu / I_\nu^{\text{sat}}} N, \qquad (4.11.1a)$$

$$\overline{N}_1 = \frac{A_{21} + \sigma(\nu) I_\nu / h\nu}{A_{21} + 2\sigma(\nu) I_\nu / h\nu} N = \frac{1 + \frac{1}{2} I_\nu / I_\nu^{\text{sat}}}{1 + I_\nu / I_\nu^{\text{sat}}} N, \qquad (4.11.1b)$$

[1]There are significant differences in reported measurements of these parameters for Nd : YAG, and the estimates used here should be considered reliable only to within about a factor of 2.

where $N = N_1 + N_2 = \overline{N}_1 + \overline{N}_2$ and

$$I_\nu^{\text{sat}} = \frac{h\nu A_{21}}{2\sigma(\nu)} \tag{4.11.2}$$

is the *saturation intensity*. The absorption coefficient is then

$$a(\nu) = \sigma(\nu)(\overline{N}_1 - \overline{N}_2) = \frac{a_0(\nu)}{1 + I_\nu/I_\nu^{\text{sat}}}, \tag{4.11.3}$$

where

$$a_0(\nu) = \sigma(\nu)N \tag{4.11.4}$$

is the *small-signal absorption coefficient*, the absorption coefficient when the intensity I_ν is small compared to I_ν^{sat}. In this case $\overline{N}_1 \approx N$ and $\overline{N}_2 \approx 0$, that is, practically all the atoms are in the ground level 1.

As I_ν/I_ν^{sat} increases, the absorption coefficient "saturates," becoming smaller and smaller as I_ν increases. For $I_\nu \gg I_\nu^{\text{sat}}$, $\overline{N}_2 \approx \overline{N}_1 \approx N/2$. In this strongly saturated regime the (equal) rates of absorption and stimulated emission are so large that the atoms are equally likely to be found in the excited level as the ground level. The larger I_ν^{sat}, the larger the field intensity I_ν has to be to produce significant saturation of the transition. Saturation of an absorbing transition arises from the excitation of the upper level, which increases stimulated emission and reduces the absorption.

As discussed in the following section, the gain coefficient of an amplifying medium exhibits essentially the same saturation behavior. In this case the saturation arises from the growth due to stimulated emission of the lower-level population, which enhances absorption and thereby reduces the amplification of the field. The dependence of $g(\nu)$ on I_ν means that the solution of Eq. (3.12.9) is not the simple exponentially growing intensity (3.12.10). The correct solution, which is given in the following chapter, grows exponentially with z only as long as I_ν is small compared to I_ν^{sat}. The exponential attenuation in an absorber is likewise a valid approximation only for intensities small compared to I_ν^{sat}. The saturation intensity, whose numerical value is determined by the transition cross section and rates, thus provides the measure of whether a given field intensity is "large" or "small" in terms of its ability to saturate the transition.

A different, somewhat more restrictive interpretation of saturation is possible. Consider a homogeneously broadened transition having a Lorentzian lineshape of width $\delta\nu_0$. After some simple algebra, using Eqs. (4.11.2), (4.11.3), and (3.7.4), we find that

$$a(\nu) = \frac{\lambda^2 A_{21}}{8\pi} N \frac{(1/\pi)\delta\nu_0}{(\nu - \nu_0)^2 + \delta\nu_0'^2} = \frac{a_0(\nu_0)\delta\nu_0^2}{(\nu - \nu_0)^2 + \delta\nu_0'^2}, \tag{4.11.5}$$

where we define

$$\delta\nu_0' = \delta\nu_0\sqrt{1 + I_\nu/I_{\nu_0}^{\text{sat}}}. \tag{4.11.6}$$

We see that, in effect, the width $\delta\nu_0$ of the transition is increased by the factor $\sqrt{1 + I_\nu/I_{\nu_0}^{\text{sat}}}$. In other words, we can interpret the saturation of the transition with increasing intensity as an effective "power broadening" of the linewidth.

Saturation will always occur at sufficiently high intensities, regardless of whether the transition is homogeneously or inhomogeneously broadened. The saturation intensity (4.11.2), because of its dependence on $\sigma(\nu)$, will vary with the lineshape function $S(\nu)$. For a Doppler-broadened transition, for instance,

$$I_{\nu_0}^{\text{sat}} = \frac{h\nu A_{21}}{2(\lambda^2 A_{21}/8\pi)S(\nu_0)} = \frac{4\pi^2 hc}{\lambda^3}\delta\nu_D\sqrt{\frac{1}{4\pi\ln 2}} \qquad (4.11.7)$$

at line center ($\nu = \nu_0$), whereas for a transition with a Lorentzian lineshape (3.4.26),

$$I_{\nu_0}^{\text{sat}} = \frac{4\pi^2 hc}{\lambda^3}\delta\nu_0. \qquad (4.11.8)$$

These formulas are based on the assumption that spontaneous emission is the only (intensity-independent) decay process for the upper level of the transition, and that the lower level is the ground level of the atom. In general the saturation intensity will depend on both upper- and lower-level decay rates associated with collisional as well as radiative processes, and it can also depend on the level degeneracies. Here we are less interested in the detailed form of I_ν^{sat} as we are in the fact that the absorption and gain coefficients saturate, in many situations of practical interest, as $1/[1 + I_\nu/I_\nu^{\text{sat}}]$, whatever the form of I_ν^{sat}. In the following section we will derive saturation formulas specifically for the gain coefficient of our idealized three- and four-level lasers.

Although they account only for radiative excitation and deexcitation processes, the formulas obtained here for I_ν^{sat} are nevertheless useful in their own right. Consider as an example the absorption by sodium vapor of radiation resonant with the $3S_{1/2}(F = 2) \leftrightarrow 3P_{3/2}$ transition: $\nu = \nu_0^{(2)}$ in the notation of Section 3.13. For Doppler broadening at $T = 200$K, we calculated in Section 3.13 the Doppler width $\delta\nu_D = 1$ GHz. Then, from (4.11.7),

$$I_{\nu_0}^{\text{sat}} \approx 1.3\,\text{W/cm}^2. \qquad (4.11.9)$$

If instead we assume radiative broadening, for which $\delta\nu_0 = A_{21}/4\pi$ (Section 3.11), then

$$I_{\nu_0}^{\text{sat}} \approx \frac{\pi hc}{\lambda^3}A_{21} = 19\,\text{mW/cm}^2, \qquad (4.11.10)$$

where we have used $A_{21} = 6.2 \times 10^7\,\text{s}^{-1}$ for the spontaneous emission rate of the sodium D_2 line (Section 3.13). These results do not account for the level degeneracies and hyperfine structure, and thus do not include factors such as g_2/g_1 or $\frac{5}{8}$ or $\frac{3}{8}$ appearing in Eq. (3.13.9). However, because these omissions give rise only to factors of order unity, the numerical values for I_ν^{sat} are of the correct magnitude. The large disparity in these two saturation intensities is not unusual; saturation intensities can vary widely for the same absorbing or emitting atoms, depending on the physical situation.

• Saturation of an atomic transition has been observed rather directly in experiments using a sodium beam. Well-collimated atomic beams are formed by those atoms that have passed from an oven (used to produce a vapor) through two (or more) successive pinholes. Irradiation by a laser beam propagating at a right angle to the atomic beam nearly eliminates any Doppler broadening and results, typically, in purely radiative broadening of the resonant transition. By monitoring the intensity of the spontaneously emitted radiation, one can infer the dependence of the excited-state population on the laser intensity or frequency.

As discussed in Section 14.3, it is possible to "align" atoms by irradiating them with polarized light. For instance, if a sodium beam is irradiated with circularly polarized laser radiation, it can be "aligned" such that only transitions between the two states $3S_{1/2}(F = 2, M = 2)$ and $3P_{3/2}(F = 3, M = 3)$ are possible. For this transition the saturation intensity $I_{\nu_0}^{sat}$ can be shown to be $\pi h c A_{21}/3\lambda^3$, that is, 6.3 mW/cm^2, or one third the value given by Eq. (4.11.10), which assumes no alignment (Problem 14.8).

The FWHM radiative linewidth of the sodium D_2 line is [Eq. (3.11.2)] $2\delta\nu_0 = A_{21}/2\pi = 10$ MHz. According to (4.11.6), therefore, the power-broadened radiative linewidth (FWHM) of the $3S_{1/2}(F = 2, M = 2) \leftrightarrow 3P_{3/2}(F = 3, M = 3)$ transition should be

$$\delta\nu_0' \approx 10\sqrt{1 + I_\nu/6.3} \text{ MHz}, \tag{4.11.11}$$

where I_ν is the laser intensity in units of mW/cm^2. Measurements of $\delta\nu_0'$ for $I_\nu = 0.84, 3.5, 90,$ and 170 mW/cm^2 gave $\delta\nu_0' = 12.4 \pm 0.8, 13.8 \pm 0.9, 41.2 \pm 1.8,$ and 53.7 ± 2.8 MHz, respectively, in good agreement with the variation predicted by (4.11.11).[2] The dependence of the scattered intensity on the laser intensity at resonance, similarly, was found to be well described by the factor $1/[1 + I_\nu/I_{\nu_0}^{sat}]$. •

4.12 SMALL-SIGNAL GAIN AND SATURATION

Equation (4.7.9) gives the steady-state population inversion for a three-level laser when the stimulated emission rate is negligible. In general, of course, the stimulated emission rate is not negligible, and here we consider the steady-state population inversion in the more general case. For this we require the steady-state solutions of Eqs. (4.7.4). These may be obtained by noting that the following replacements for P and Γ_{21} (in the equations *without* stimulated emission),

$$P \longrightarrow P + \sigma\Phi_\nu, \tag{4.12.1a}$$

$$\Gamma_{21} \longrightarrow \Gamma_{21} + \sigma\Phi_\nu, \tag{4.12.1b}$$

are sufficient to reinstate all stimulated-emission terms. Here Φ_ν is the steady-state (i.e., time-independent) cavity photon flux. Thus, the steady-state solutions of (4.7.4) may be obtained by making the same replacements in the solutions (4.7.8). Likewise the steady-state population inversion $\overline{N}_2 - \overline{N}_1$ in the general case follows when the replacements (4.12.1) are made in (4.7.9):

$$\overline{N}_2 - \overline{N}_1 = \frac{(P - \Gamma_{21})N_T}{P + \Gamma_{21} + 2\sigma\Phi_\nu}. \tag{4.12.2}$$

This is the generalization of (4.7.9) to the case in which the stimulated emission rate is not negligible compared to P and Γ_{21}.

[2]M. L. Citron, H. R. Gray, C. W. Gabel, and C. R. Stroud, Jr., *Physical Review A* **16**, 1507 (1977).

The steady-state gain coefficient for a three-level laser follows from (3.12.6) and (4.12.2). Assuming $g_1 = g_2$, we have

$$g(\nu) = \frac{\sigma(\nu)(P - \Gamma_{21})N_T}{P + \Gamma_{21} + 2\sigma(\nu)\Phi_\nu} = \frac{\sigma(\nu)(P - \Gamma_{21})N_T}{P + \Gamma_{21}} \frac{1}{1 + [2\sigma(\nu)\Phi_\nu/(P + \Gamma_{21})]}$$

$$= \frac{g_0(\nu)}{1 + \Phi_\nu/\Phi_\nu^{\text{sat}}} = \frac{g_0(\nu)}{1 + I_\nu/I_\nu^{\text{sat}}}, \tag{4.12.3}$$

where we define the *small-signal gain*

$$g_0(\nu) = \frac{\sigma(\nu)(P - \Gamma_{21})N_T}{P + \Gamma_{21}} \tag{4.12.4}$$

and the *saturation flux*

$$\Phi_\nu^{\text{sat}} = \frac{P + \Gamma_{21}}{2\sigma(\nu)}. \tag{4.12.5}$$

The corresponding expressions for the saturation intensity and photon number are (see Problem 4.7)

$$I_\nu^{\text{sat}} = h\nu\Phi_\nu^{\text{sat}} = \frac{h\nu(P + \Gamma_{21})}{2\sigma(\nu)} \tag{4.12.6}$$

and

$$q_\nu^{\text{sat}} = \frac{V}{c}\Phi_\nu^{\text{sat}} = \frac{P + \Gamma_{21}}{2c\sigma(\nu)}V. \tag{4.12.7}$$

The gain coefficient for the three-level laser, therefore, saturates in the same way as the absorption coefficient (4.11.3) of a two-level transition. The saturation intensities in the two cases are different; in particular, I_ν^{sat} for the three-level laser depends not only on $h\nu/\sigma(\nu)$ but also on the pumping rate P and the decay rate Γ_{21}.

For $I_\nu \ll I_\nu^{\text{sat}}$, $g(\nu) \approx g_0(\nu)$, which, of course, is why $g_0(\nu)$ is called the "small-signal" gain coefficient. The maximum gain is $g_0(\nu_0)$, that is, the gain when $I_\nu \ll I_\nu^{\text{sat}}$ and the field frequency matches the line-center frequency ν_0, where $\sigma(\nu_0)$ has its maximum value. When the lineshape is Lorentzian, with HWHM width $\delta\nu_0$, we have

$$g(\nu) = g_0(\nu_0)\frac{1}{(\nu_0 - \nu)^2/\delta\nu_0^2 + 1 + (\Phi_\nu/\Phi_{\nu_0}^{\text{sat}})}. \tag{4.12.8}$$

The cavity frequencies at which there is small-signal gain sufficient to overcome loss in a laser are generally those within about $\delta\nu_0$ of line center ($\nu = \nu_0$); $\delta\nu_0$ can be called the small-signal *gain bandwidth*. In Section 1.3 we showed by way of an example how the gain bandwidth and the cavity mode spacing together determine the number of possible frequencies that can lase.

In Fig. 4.8 we plot $g(\nu)$ vs. ν as given in (4.12.8) for several values of Φ_ν and $g(\nu)$ vs. Φ_ν for several values of ν. Clearly, it is harder to saturate $g(\nu)$ away from line center. Alternatively, for higher fluxes the halfwidth of $g(\nu)$ is greater. This is exactly the

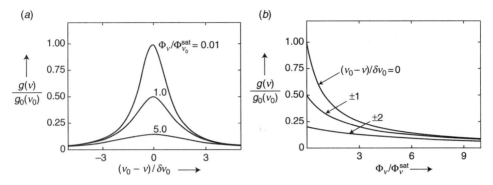

Figure 4.8   Saturated gain curves, according to Eq. (4.12.8).

same as the *power broadening* discussed in Section 4.11. The power-broadened gain bandwidth is the half width implied by (4.12.8), namely

$$\delta v_g = \delta v_0 \sqrt{1 + \Phi_v / \Phi_{v_0}^{sat}}. \tag{4.12.9}$$

It is greater than the small-signal gain bandwidth and is seen to be exactly the width $\delta v_0'$ given in (4.11.6), as it should be, if we put $P = 0$ and $\Gamma_{21} = A_{21}$ in the expression for Φ_v^{sat}.

In the case of a four-level laser we can obtain in a similar fashion a gain with the flux dependence (4.12.3), but with a saturation flux twice that given by (4.12.5) (Problem 4.8). The same basic flux dependence is also obtained when we include degeneracy and refractive index corrections to the gain coefficient (Section 3.12).

Three and four-level lasers are idealizations that are seldom fully realized in practice. The gain–saturation formulas (4.12.3) and (4.12.8) are, however, applicable to a wide variety of actual lasers. That is, although (4.12.3) may be derived from simple models, it often applies outside the range of validity of these models. It is the most commonly assumed formula for the intensity dependence of the gain on a homogeneously broadened laser transition. In Section 4.14 we will consider the case of an inhomogeneously broadened transition.

In Eq. (4.12.5) P and Γ_{21} are the "decay rates" of the lower and upper laser levels, respectively [cf. Eqs. (4.7.4)]. The larger the decay rates, the larger the saturation flux. This makes good sense physically, for the larger the decay rates, the larger must be the stimulated emission rate necessary to saturate the transition, that is, to equalize the population densities $\overline{N}_1$ and $\overline{N}_2$. In fact the saturation flux (4.12.5) for a three-level laser is just the intensity for which the stimulated emission rate is the average of the upper- and lower-level decay rates (Problem 4.8). In general, the larger these decay rates, the larger the saturation flux. Equation (4.12.5) is an example of this general result.

In most cases of practical interest the pump rate P is small compared to Γ_{21} in a three-level laser and to Γ_{21} and Γ_{10} in a four-level laser. Then

$$I_v^{sat} \simeq \frac{h v \Gamma_{21}}{2 \sigma(v)} \quad \text{(three-level laser)} \tag{4.12.10}$$

and (Problem 4.8)

$$I_v^{sat} \cong \frac{h\nu\Gamma_{21}}{\sigma(\nu)} \quad \text{(four-level laser).} \tag{4.12.11}$$

The relation $\sigma(\nu) \propto B_{21}$ [recall (3.12.18)] means that the saturation flux (4.12.5) is inversely proportional to the Einstein B coefficient for stimulated emission. This is another general result and is hardly surprising because the smaller B_{21} is, the greater the intensity necessary to achieve a given stimulated emission rate. For a Lorentzian line-shape function, (4.12.5) also predicts that the line-center saturation flux is directly proportional to the transition linewidth $\delta\nu_0$:

$$\Phi_{\nu_0}^{sat} = \frac{P + \Gamma_{21}}{2\sigma(\nu_0)} = \frac{4\pi^2 \delta\nu_0}{\lambda^2 A}(P + \Gamma_{21}). \tag{4.12.12}$$

This too is a general conclusion that is applicable beyond the three- and four-level models.

The most important results of this section are Eqs. (4.12.3) and (4.12.8). We have obtained these results for the specific case of an ideal three-level laser, but we have emphasized that they apply to a large variety of real lasers under conditions of homogeneous line broadening. Whereas the detailed equations for the small-signal gain and saturation intensity are specific to the particular laser under consideration, the expressions (4.12.3) and (4.12.8) are more generally applicable. Indeed, it will usually be difficult to *calculate* g_0 and I_v^{sat}, but we can often be confident nevertheless that the *form* of the intensity dependence of the gain described by (4.12.3) or (4.12.8) is correct.

We emphasize that these equations are applicable regardless of whether g_0 is positive (gain) or negative (absorption). That is, a medium may be saturated regardless of whether it is amplifying or absorbing. Thus, the absorption coefficient $a(\nu)$ of an absorbing medium will decrease as the intensity of the radiation is raised, as discussed in the preceding section. When the intensity is much larger than the line-center saturation intensity $I_{\nu_0}^{sat}$ characteristic of the medium, the absorption coefficient is very small [$a(\nu) \approx 0$], which means that the medium is practically transparent to high-intensity radiation. In this case the medium is sometimes said to be "bleached" because it no longer absorbs radiation that is resonant with one of its transition frequencies. What is happening in the case of such strong saturation is that the stimulated emission (and absorption) rate has become much greater than the decay rate of the *upper* level of the transition. An atom that has absorbed a photon will then be quickly induced to return to the lower level and give the photon back to the field by stimulated emission. This occurs, with high probability, before the absorbed energy can be dissipated as heat or fluorescence. Thus, no energy is lost by the incident field; the medium has been made effectively transparent ("bleached") by virtue of the high intensity of the field.

4.13 SPATIAL HOLE BURNING

In this section we will consider more carefully the meaning of the "intensity." Intensity refers to the electromagnetic energy flow per unit area per unit time, but in most lasers we have *standing* waves rather than traveling waves. The gain–saturation formulas (4.12.3) and (4.12.8) are often written with Φ_ν assumed to be the sum of the fluxes of the two traveling waves (Fig. 4.3):

$$\Phi_\nu \longrightarrow \Phi_\nu = \Phi_\nu^{(+)} + \Phi_\nu^{(-)}. \tag{4.13.1}$$

However, this is not quite correct, for it ignores the interference of the two traveling waves. The electromagnetic energy density u is proportional to the square of the electric field:[3]

$$u = \epsilon_0 \mathbf{E}^2(\mathbf{r}, t) = \epsilon_0 E_0^2 \cos^2 \omega t \sin^2 kz. \tag{4.13.2}$$

We replace $\cos^2 \omega t$ by $\frac{1}{2}$, its average value over times long compared to an optical period $2\pi/\omega \approx 10^{-14}$ s, and write

$$u = \frac{\epsilon_0}{2} E_0^2 \sin^2 kz. \tag{4.13.3}$$

Now a cavity standing-wave field is the sum of two oppositely propagating traveling-wave fields:

$$E(z, t) = E_0 \cos \omega t \sin kz = \tfrac{1}{2} E_0[\sin(kz - \omega t) + \sin(kz + \omega t)]$$
$$= E_+(z, t) + E_-(z, t), \tag{4.13.4}$$

where the two electric waves

$$E_\pm(z, t) = \tfrac{1}{2} E_0 \sin(kz \mp \omega t) \tag{4.13.5}$$

propagate in the positive $(+)$ and negative $(-)$ z directions. The time-averaged square of the electric field (4.13.5) gives a field energy density $u = u^{(+)} + u^{(-)}$, where

$$u^{(\pm)} = \frac{\epsilon_0}{8} E_0^2. \tag{4.13.6}$$

From (4.13.3) and (4.13.6), therefore, it follows that

$$\Phi_\nu \equiv \frac{c}{h\nu} u = 2\left[\Phi_\nu^{(+)} + \Phi_\nu^{(-)}\right] \sin^2 kz, \tag{4.13.7}$$

or, in terms of the intensity $I_\nu \equiv h\nu \Phi_\nu$,

$$I_\nu = 2\left[I_\nu^{(+)} + I_\nu^{(-)}\right] \sin^2 kz. \tag{4.13.8}$$

Thus, it is not correct to use (4.13.1) as the flux in the gain saturation formulas (4.12.3) and (4.12.8). We should use (4.13.7), which accounts properly for the

---

[3]For simplicity we assume in this section that the refractive index $n \approx 1$.

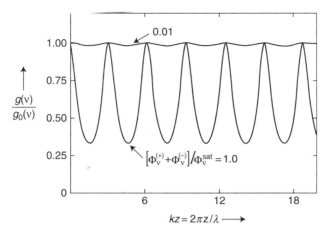

Figure 4.9   Spatial hole burning in the gain curve, according to Eq. (4.13.9).

interference of the two traveling-wave fields. Then the gain–saturation formula for a homogeneously broadened transition is

$$g(\nu) = \frac{g_0(\nu)}{1 + 2\left[\left(\Phi_\nu^{(+)} + \Phi_\nu^{(-)}\right)/\Phi_\nu^{\text{sat}}\right]\sin^2 kz}. \qquad (4.13.9)$$

This saturation formula replaces (4.12.3) when the standing-wave nature of the cavity field is properly accounted for.

The $\sin^2 kz$ term in Eq. (4.13.9) gives rise to what is called *spatial hole burning* in the gain coefficient $g(\nu)$. At points z for which $\sin^2 kz = 0$, $g(\nu)$ takes on its maximum value, namely the small-signal value $g_0(\nu)$. Where $\sin^2 kz = 1$, however, $g(\nu)$ has its minimum value, that is, it is most strongly saturated; a "hole" is "burned" in the curve of $g(\nu)$ vs. z (Fig. 4.9). The holes in this curve are separated by $\Delta z = \pi/k = \lambda/2$. Thus, $g(\nu)$ varies with z on the scale of the laser wavelength.

This rapid variation of $g(\nu)$ with z suggests the approximation of replacing $\sin^2 kz$ by its spatial average, $\frac{1}{2}$, in Eq. (4.13.9). In this approximation we take

$$g(\nu) = \frac{g_0(\nu)}{1 + \left(\Phi_\nu^{(+)} + \Phi_\nu^{(-)}\right)/\Phi_\nu^{\text{sat}}}, \qquad (4.13.10)$$

which is the result obtained by using (4.13.1) in the gain–saturation formula (4.12.3). This approximation ignores the spatial dependence of the intracavity field and, therefore, also the spatial hole burning of the gain coefficient. It is called the *uniform-field approximation* or the *spatial mean-field approximation*.

4.14   SPECTRAL HOLE BURNING

In an inhomogeneously broadened medium the different atoms have different central transition frequencies ν_0'. This may be due to their different velocities and the Doppler effect (Section 3.9), the presence of different isotopes having slightly different

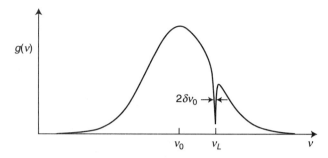

Figure 4.10   Spectral hole burning in an inhomogeneously broadened gain profile. Radiation of frequency ν_L saturates only a spectral packet of atoms with frequency $\nu \approx \nu_L$.

energy levels, spatially nonuniform electric and magnetic fields causing shifts in the energy levels, or a host of other effects.

Saturation of the absorption or gain coefficient is more complicated in the case of inhomogeneous broadening. Atoms with different central frequencies ν_0' will be saturated according to (4.11.5), but there is a *distribution* of resonance frequencies ν_0'. The absorption or gain coefficient is obtained by integrating the contributions from different frequency components, or *spectral packets*, each of which saturates to a different degree depending on its detuning from the cavity mode frequency ν.

Equation (4.11.5) implies that the absorption or gain is more strongly saturated for spectral packets with frequency $\nu_0' \approx \nu$; spectral packets with frequencies detuned from ν by much more than the homogeneous linewidth are hardly saturated at all. This is illustrated in Fig. 4.10. The selective saturation leads to *spectral hole burning* in the gain curve. The width of a hole is just the homogeneous linewidth $\delta\nu_0$ (if power broadening is small), while the depth is determined by the field intensity. When the field intensity is very large the hole "touches down," that is, the gain at the center of the hole is fully saturated.

Spectral hole burning is especially interesting in the case of a purely Doppler-broadened gain medium.[4] Suppose the cavity mode frequency ν is different from the center frequency ν_0 of the Doppler gain profile. Consider, for example, the traveling-wave field propagating in the positive z direction. This wave will strongly saturate the spectral packet of atoms with Doppler-shifted frequencies $\nu_0' = \nu$; the Doppler effect has brought these atoms into resonance with the wave. Therefore these atoms have the z component of velocity given by (cf. Section 3.9)

$$\nu = \nu_0' = \nu_0\left(1 + \frac{v}{c}\right),$$ (4.14.1a)

or

$$\frac{v}{c} = \frac{\nu - \nu_0}{\nu_0},$$ (4.14.1b)

where ν_0 is the resonance frequency of a stationary atom. If $\nu > \nu_0$, then v is positive. This means that the atoms that have been Doppler shifted into resonance are moving in

---

[4]Spectral hole burning is the origin of the *Lamb dip* in Doppler-broadened lasers, as discussed in Section 5.8.

the positive z direction, the same direction in which the traveling wave is propagating. If $v < v_0$, on the other hand, v/c is negative, or in other words the atoms must be moving in the negative z direction, opposite to the traveling wave, in order to be shifted into resonance.

Consider the saturation of the gain coefficient when there is Doppler broadening. For simplicity we will assume a single traveling wave of frequency ν. With atoms of velocity v along the z direction we associate the saturated gain coefficient

$$g(\nu, v) = \frac{\lambda^2 A_{21}}{8\pi} \frac{P - \Gamma_{21}}{P + \Gamma_{21}} N_T \frac{\delta\nu_0/\pi}{(\nu_0 - \nu + \nu_0 v/c)^2 + \delta\nu_g^2}$$

$$\equiv C \frac{\delta\nu_0/\pi}{(\nu_0 - \nu + \nu_0 v/c)^2 + \delta\nu_g^2} \qquad (4.14.2)$$

that follows from the three-level laser model (Section 4.12) with the Doppler shift included in the (power-broadened) Lorentzian lineshape function. We now calculate the total gain coefficient due to atoms of all velocities by integrating over the Maxwell–Boltzmann distribution for atoms of molecular weight M_X, exactly as in Section 3.9:

$$g(\nu) = \int_{-\infty}^{\infty} dv \left(\frac{M_X}{2\pi RT} \right)^{1/2} e^{-M_X v^2/2RT} g(\nu, v)$$

$$= C \left(\frac{M_X}{2\pi RT} \right)^{1/2} \frac{\delta\nu_0}{\pi} \int_{-\infty}^{\infty} \frac{dv \, e^{-M_X v^2/2RT}}{(\nu_0 - \nu + \nu_0 v/c)^2 + \delta\nu_g^2}$$

$$= C \frac{4\ln 2}{\pi^{3/2}} \frac{\delta\nu_0}{\delta\nu_D^2} \int_{-\infty}^{\infty} \frac{dy \, e^{-y^2}}{(x + y)^2 + b'^2}, \qquad (4.14.3)$$

where we have made the same change of variables as in Section 3.10 and defined x as in Eq. (3.10.4); b' is defined by replacing $\delta\nu_0$ by $\delta\nu_g$ in (3.10.5):

$$b' = (4\ln 2)^{1/2} \frac{\delta\nu_g}{\delta\nu_D}. \qquad (4.14.4)$$

As in Section 3.10 it is convenient to consider the case $x = 0$, the case where the field frequency ν exactly matches the central frequency ν_0 of the Doppler lineshape. Then, using (3.10.7),

$$g(\nu) = C \frac{4\ln 2}{\pi^{3/2}} \frac{\delta\nu_0}{\delta\nu_D^2} \frac{\pi}{b'} e^{b'^2} \text{erfc}(b') = C \left(\frac{4\ln 2}{\pi} \right)^{1/2} \frac{\delta\nu_0}{\delta\nu_D \delta\nu_g} e^{b'^2} \text{erfc}(b'). \qquad (4.14.5)$$

Suppose first that the transition is Doppler broadened and the intensity of the field is well below saturation, so that $\delta\nu_D \gg \delta\nu_0$, $\delta\nu_g \approx \delta\nu_0$, and $b' \approx (4\ln 2)^{1/2}\delta\nu_0/\delta\nu_D \ll 1$. Then, from (3.10.12), we obtain

$$g(\nu) \approx C \left(\frac{4\ln 2}{\pi} \right)^{1/2} \frac{1}{\delta\nu_D} = g_0^{(D)}(\nu_0), \qquad (4.14.6)$$

where $g_0^{(D)}(\nu_0)$ is the small-signal gain at $\nu = \nu_0$ for Doppler broadening. If $\delta\nu_D \gg \delta\nu_g$ but $\delta\nu_g$ differs significantly from $\delta\nu_0$, then $b' \ll 1$ and we can still use (3.10.13), but now

$$
g(\nu) \approx C\left(\frac{4\ln 2}{\pi}\right)^{1/2} \frac{\delta\nu_0}{\delta\nu_D \delta\nu_g} = C\left(\frac{4\ln 2}{\pi}\right)^{1/2} \frac{1}{\delta\nu_D} \frac{1}{\sqrt{1 + I_{\nu_0}/I_{\nu_0}^{\text{sat}}}}
$$

$$
= \frac{g_0^{(D)}(\nu_0)}{\sqrt{1 + I_{\nu_0}/I_{\nu_0}^{\text{sat}}}}. \tag{4.14.7}
$$

Thus, a Doppler-broadened transition saturates not as $1/[1 + I_{\nu_0}/I_{\nu_0}^{\text{sat}}]$ but as $1\big/\sqrt{1 + I_{\nu_0}/I_{\nu_0}^{\text{sat}}}$. This is true so long as $\delta\nu_g$ is not too large: if $\delta\nu_g \gg \delta\nu_D$, then the approximation (3.10.10) [with b' replacing b] is applicable and (4.14.5) becomes

$$
g(\nu) \approx C\left(\frac{4\ln 2}{\pi}\right)^{1/2} \frac{\delta\nu_0}{\delta\nu_D \delta\nu_g} \frac{1}{\pi^{1/2}} \frac{\delta\nu_D}{(4\ln 2)^{1/2}\delta\nu_g} = C\frac{1}{\pi}\frac{\delta\nu_0}{\delta\nu_g^2}
$$

$$
= C\frac{1/(\pi\delta\nu_0)}{1 + I_{\nu_0}/I_{\nu_0}^{\text{sat}}} = \frac{g_0^{(H)}(\nu_0)}{1 + I_{\nu_0}/I_{\nu_0}^{\text{sat}}}, \tag{4.14.8}
$$

where $g_0^{(H)}(\nu_0)$ is the line-center small-signal gain for the homogeneously broadened transition with Lorentzian HWHM $\delta\nu_0$. In this power-broadened limit we recover the factor $1/[1 + I_{\nu_0}/I_{\nu_0}^{\text{sat}}]$ characteristic of *homogeneous* broadening.

Inhomogeneously broadened lasers, and in particular Doppler-broadened lasers, are generally much more complicated to treat theoretically than homogeneously broadened lasers, especially if counterpropagating traveling waves and spatial hole burning are taken into account.

4.15 SUMMARY

In Chapter 1 we gave a very crude description of laser action and introduced some fundamental concepts such as gain and threshold. In the intervening chapters we have gone more deeply into the theory of the interaction of light and matter, and in the present chapter we have begun to apply what we have learned to laser theory.

The most important theoretical tools for our understanding of lasers are the rate equations for level populations and field intensities. These equations generally include effects of pumping, collisions, absorption, spontaneous and stimulated emission, field gain and loss, and other processes that may be pertinent for a particular laser. We have used such equations to discuss three- and four-level lasers, and in fact the use of rate equations will be a dominant theme of the following chapters. We have also discussed the concept of saturation which, as we will see in the next chapter, is a major consideration in determining how much output power can be obtained with a given laser.

PROBLEMS

4.1. Show how (3.12.5) and (4.3.4a) are modified if the light propagates toward $-z$ rather than $+z$, and derive (4.3.4b).

4.2. **(a)** Solve (4.5.7) as a function of time for the (unusual) case that the equation's loss parameters satisfy $\Gamma_{21} + A_{21} = (c/2L)(1 - r_1 r_2)$. Give the steady-state value of $n_2 + q_\nu$.

(b) Find the steady-state solution for q_ν in terms of p and n_2 for arbitrary loss parameters.

4.3. **(a)** Write the full set of equations for a three-level laser by modifying (4.7.4) and including the following equation for the third level (as shown in Fig. 4.6): $dN_3/dt = PN_1 - \Gamma_{32}N_3$, and show that the full set of equations satisfies $N_1 + N_2 + N_3 = \text{constant} = N_T$.

(b) Determine the steady-state values of the three level populations.

(c) Find the condition under which it is satisfactory to neglect level 3 [$N_3 \approx 0$] and to use Eqs. (4.7.2) and (4.7.4) as written in the text.

4.4. Solve Eqs. (4.8.1) for the steady-state value of $\overline{N}_2 - \overline{N}_1$, and show that (4.8.4) gives the limiting value as the stimulated emission rate decreases to zero.

4.5. Estimate the density of chromium ions in ruby, assuming that the concentration of chromium in ruby is about 0.05% by weight.

4.6. **(a)** Calculate the transition dipole moment $e|\mathbf{x}_{12}|$ for the $3S_{1/2}(F = 2, M = 2) \leftrightarrow 3P_{3/2}(F = 3, M = 3)$ transition of sodium.

(b) What is the oscillator strength of this transition?

4.7. Derive the formula analogous to (4.12.3) for a four-level laser, and write the expression for the saturation intensity.

4.8. Show that the saturation intensity I_ν^{sat} of a three-level laser [Eq. (4.12.6)] is the intensity for which the stimulated emission rate is the average of the upper- and lower-level decay rates of the laser transition. Find the corresponding expression that follows from laser equations (4.5.2).

4.9. **(a)** A gain cell of length 10 cm has a small-signal gain coefficient of 0.025 cm^{-1}. Two mirrors having the same reflectivity are placed at the ends of the cell. Assuming that scattering losses are negligible, calculate the reflectivity necessary for lasing.

(b) If the gain curve has a FWHM width of 1 GHz, what is the maximum number of modes that can lase?

5 LASER OSCILLATION: POWER AND FREQUENCY

5.1 INTRODUCTION

In this chapter we will consider mainly the output power and frequency of a continuous-wave (cw) laser, that is, a laser in the steady-state, time-independent mode of operation. Such a laser emits a steady, continuous beam of radiation. The other common mode of operation, in which the laser output is in the form of single or repeated pulses of radiation, will be discussed in the following chapter.

We will also restrict ourselves to the case of laser oscillation on a single resonator mode, postponing until the next chapter the case of multimode lasing. The assumption of single-mode oscillation allows us to focus on the essential ideas without undue complication.

5.2 UNIFORM-FIELD APPROXIMATION

We will now use the gain–saturation formula (4.12.3) to derive an expression for the output intensity of a cw laser oscillating on a single cavity mode. We continue to assume for now that the lasing transition is homogeneously broadened. In Section 5.4 we will consider the effect of spatial hole burning, but our discussion in the present section will be restricted to the mean-field (uniform-field) approximation.

In cw laser oscillation the cavity photon number is constant in time. This means that the field amplification due to stimulated emission exactly balances the attenuation due to output coupling, scattering, and other cavity loss processes. That is, the growth rate of the cavity photon number equals the decay rate.

We will assume all loss processes to be independent of the cavity intensity. This implies that the field attenuation rate in steady-state oscillation is no different from that at the threshold of oscillation, where the cavity intensity is practically zero. Thus, the growth rate of cavity photons in steady-state oscillation must also be the same as its threshold value. In other words, *in cw oscillation the gain is precisely equal to its threshold value g_t*. The gain is sometimes said to be "clamped" at its threshold value. Since the steady-state gain equals g_t, this clamping should determine the cavity intensity of the laser. We will now justify these assertions and examine their implications.

Laser Physics. By Peter W. Milonni and Joseph H. Eberly
Copyright © 2010 John Wiley & Sons, Inc.

From the photon rate Eq. (4.4.8) [which applies when $g(\nu)$ is independent of z, consistent with the uniform-field approximation] we infer that

$$\frac{cl}{L}g(\nu)I_\nu = \frac{c}{2L}(1 - r_1 r_2)I_\nu \qquad (5.2.1)$$

in steady-state oscillation ($dI_\nu/dt = 0$). Dividing through by I_ν, therefore, we have the condition

$$g(\nu) = \frac{1}{2l}(1 - r_1 r_2) \qquad (5.2.2)$$

for steady-state oscillation. But from (4.4.7) we recognize the right-hand side as the threshold gain necessary for oscillation. Equation (5.2.2), we recall, is applicable when $r_1 r_2 \approx 1$; our discussion in Section 4.4 implies that this case is consistent with the uniform-field approximation. Thus, the condition for steady-state oscillation in the uniform-field approximation is simply

$$g(\nu) = g_t \quad \text{(steady-state oscillation)}, \qquad (5.2.3)$$

as asserted above. Given the threshold gain g_t, which may be calculated from the length of the gain medium, the mirror reflectivities, and any significant scattering or other loss coefficients, we therefore have also the "clamped" gain for cw oscillation.

Now from the gain–saturation formula (4.12.3), and the cw oscillation condition (5.2.3), we must evidently have

$$g(\nu) = \frac{g_0(\nu)}{1 + (I_\nu^{(+)} + I_\nu^{(-)})/I_\nu^{\text{sat}}} = g_t, \qquad (5.2.4)$$

or

$$I_\nu^{(+)} + I_\nu^{(-)} = I_\nu^{\text{sat}}\left(\frac{g_0(\nu)}{g_t} - 1\right). \qquad (5.2.5)$$

This simple formula gives the total cw cavity intensity in terms of the saturation intensity I_ν^{sat}, the small-signal gain $g_0(\nu)$, and the threshold gain g_t.

Of course, the intensity on the left side of (5.2.5) is not the laser output intensity, but the intracavity intensity. The output intensity is

$$I_\nu^{\text{out}} = t_1 I_\nu^{(-)} + t_2 I_\nu^{(+)}, \qquad (5.2.6)$$

where t_1 and t_2 are the transmission coefficients of the mirrors at the laser frequency ν (Fig. 5.1a). Usually, only one of the mirrors is transmitting (Fig. 5.1b), in which case we write

$$I_\nu^{\text{out}} = t I_\nu^{(+)} \qquad (5.2.7)$$

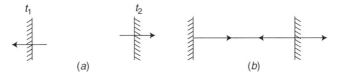

Figure 5.1   (*a*) A laser cavity with mirror transmission coefficients t_1 and t_2. (*b*) A laser cavity with one output mirror.

instead of (5.2.6). Here t is the transmission coefficient of the output mirror, and $I_\nu^{(+)}$ is the intensity of the traveling wave propagating toward the output mirror. For the cavity mode defined by (4.13.4) we have

$$I_\nu^{(+)} = I_\nu^{(-)}. \tag{5.2.8}$$

The equality of $I_\nu^{(+)}$ and $I_\nu^{(-)}$ holds whenever the cavity mirrors of Fig. 5.1 are perfectly reflecting, and so we might expect (5.2.8) to be a good approximation if the mirrors are highly reflecting, as is the case in many lasers. This expectation will be borne out in Section 5.5.

Combining the results (5.2.5), (5.2.7), and (5.2.8), we obtain

$$I_\nu^{\text{out}} = \frac{t}{2} I_\nu^{\text{sat}} \left(\frac{g_0(\nu)}{g_t} - 1 \right) \tag{5.2.9}$$

for the output intensity. Now according to (4.3.2), (5.2.2), and (5.2.3),

$$g_t = \frac{1}{2l}(1 - r) = \frac{1}{2l}(t + s), \tag{5.2.10}$$

so that

$$I_\nu^{\text{out}} = \frac{1}{2} t I_\nu^{\text{sat}} \left(\frac{2 g_0(\nu) l}{t + s} - 1 \right) \tag{5.2.11}$$

for "small" (and typical) output couplings, for which (5.2.2) describes the threshold gain.

It may be worthwhile to summarize the assumptions made in deriving the output intensity (5.2.11) of a single-mode, homogeneously broadened, cw laser. First of all, we have ignored any spatial variation of the intensity perpendicular to the cavity axis (the z direction). Furthermore, we have assumed that the intensity is also constant along the cavity axis; this is the uniform-field approximation, and it implies that the gain coefficient is likewise independent of z. Thus, we have assumed that the gain and intensity are constant *throughout* the laser cavity. In the derivation leading to (5.2.11) we have also assumed that one of the mirrors is essentially perfectly reflecting, while the output mirror's reflectivity is high enough that $I_\nu^{(+)} \approx I_\nu^{(-)}$.

Equation (5.2.11) gives the output intensity in terms of the small-signal gain $g_0(\nu)$, the saturation intensity I_ν^{sat}, and the length l of the gain medium, plus the coefficients t and s of the output mirror. Thus, a given gain medium, characterized by g_0, I_ν^{sat}, and l, can yield different laser beam intensities, depending on how the laser cavity (i.e., t and s) is chosen. We will next determine the largest possible output intensity that can be obtained from a given gain medium.

5.3 OPTIMAL OUTPUT COUPLING

It is a simple exercise to determine the optimum transmission coefficient t_{opt} of the output mirror, the value for t that maximizes the output intensity (5.2.11). We obtain (Problem 5.2)

$$t_{\text{opt}} = \sqrt{2g_0(\nu)ls} - s, \tag{5.3.1}$$

and therefore the maximum possible output intensity is

$$[I_\nu^{\text{out}}]_{\text{max}} = I_\nu^{\text{sat}} \left[\sqrt{g_0(\nu)l} - \sqrt{s/2} \right]^2. \tag{5.3.2}$$

When $t = t_{\text{opt}}$, the threshold gain, and therefore the gain in steady-state oscillation, is

$$(g_t)_{\text{opt}} = \frac{1}{2l}(1 - r)_{\text{opt}} = \frac{1}{2l}(t_{\text{opt}} + s) = \frac{1}{2l}\sqrt{2g_0(\nu)ls} = \sqrt{\frac{g_0(\nu)s}{2l}}. \tag{5.3.3}$$

The small-signal gain $g_0(\nu)$ must be greater than $s/2l$ in a laser, or else the threshold condition (5.2.10) could not be satisfied. If the scattering and absorption losses are small enough that $g_0(\nu) \gg s/2l$, then from (5.3.2)

$$[I_\nu^{\text{out}}]_{\text{max}} \approx g_0(\nu)I_\nu^{\text{sat}}l \qquad [g_0(\nu) \gg s/2l] \tag{5.3.4}$$

is the maximum possible intensity of radiation at frequency ν that can be extracted from the medium. That is, if the small-signal gain is much greater than the scattering loss coefficient $s/2l$, and we design the resonator to have the optimal output coupling (5.3.1), we will extract the maximum intensity (5.3.4) at frequency ν. Since the small-signal gain is generally greatest at line center ($\nu = \nu_0$), the maximum output intensity extractable from the medium is

$$[I_{\nu_0}^{\text{out}}]_{\text{max}} \approx g_0(\nu_0)I^{\text{sat}}l, \tag{5.3.5}$$

where $I^{\text{sat}} \equiv I_{\nu_0}^{\text{sat}}$.

The result (5.3.4) may perhaps be better appreciated by deriving it in a different way, using population rate equations. For this purpose we consider again the specific example of the ideal three-level laser, the population rate equations for which are given by (4.7.4). Since the ground level may be taken to have zero energy, the rate of change, due to stimulated emission, of the energy per unit volume stored in the atoms is

$$h\nu \left(\frac{dN_2}{dt} \right)_{\text{stimulated emission}} = -\sigma(\nu)(N_2 - N_1)I_\nu \tag{5.3.6}$$

in steady-state oscillation. Using Eq. (4.13.8) in the uniform-field approximation (replacing $\sin^2 kz$ by its average value $\frac{1}{2}$), we may write (5.3.6) as

$$hv\left(\frac{dN_2}{dt}\right)_{\text{stimulated emission}} = -g(v)(I_v^{(+)} + I_v^{(-)}). \tag{5.3.7}$$

This is the power per unit volume extracted from the active atoms by stimulated emission.

It follows that $g(I_v^{(+)} + I_v^{(-)})$ is the growth rate per unit volume of laser field energy in the active medium. But in steady-state oscillation this must equal the loss rate due to output coupling plus the loss rate associated with scattering and absorption processes; these are characterized by t and s, respectively. When scattering and absorption losses are small compared to output coupling, therefore, $g(I_v^{(+)} + I_v^{(-)})$ is approximately the power per unit volume lost by the active medium in the form of *output* laser radiation. From (5.2.5),

$$g(v)(I_v^{(+)} + I_v^{(-)}) = g(v)I_v^{\text{sat}}\left(\frac{g_0(v)}{g_t} - 1\right) = g_t I_v^{\text{sat}}\left(\frac{g_0(v)}{g_t} - 1\right)$$

$$= I_v^{\text{sat}}[g_0(v) - g_t]. \tag{5.3.8}$$

The maximum value of the power per unit volume that can be extracted as output laser radiation of frequency v is therefore

$$\left(\frac{\text{power}}{\text{volume}} \longrightarrow \text{laser radiation}\right)_{\text{max}} = g_0(v)I_v^{\text{sat}}, \tag{5.3.9}$$

and this is obtained when the small-signal gain g_0 is much larger than the threshold gain g_t. Thus, we can interpret the maximum possible intensity (5.3.4) that can be extracted from the gain medium as simply the maximum possible power per unit volume $g_0 I_v^{\text{sat}}$ that can be extracted from the medium, multiplied by the length l of the medium.

The theoretical upper limit (5.3.9) to the extracted power per unit volume of the gain medium is useful because it depends only on the properties of the active atoms and the pumping process. Consider as an example the ideal three-level laser. The input power per unit volume to the gain cell in steady-state oscillation is given by Eq. (4.7.12). For pump rate P, the theoretical upper limit of the input-to-output *power conversion efficiency* is therefore

$$e_{\text{max}} = \frac{g_0(v_0)I_v^{\text{sat}}}{hv_{31}P\bar{N}_1} \tag{5.3.10}$$

for line-center operation. $g_0(v_0)$ and I^{sat} for the three-level laser are given by (4.12.4) and (4.12.6), respectively, and some simple algebra yields

$$e_{\text{max}} = \frac{(P - \Gamma_{21})N_T hv_0}{2hv_{31}P\bar{N}_1}. \tag{5.3.11}$$

Since we are considering the case of optimal output coupling and $g_0 \gg s/2l$ for the purpose of obtaining a theoretical upper limit to the power conversion efficiency, we may take $I_{\nu_0}^{(+)} + I_{\nu_0}^{(-)} \gg I^{\text{sat}}$, for from (5.2.5) and (5.3.3),

$$\frac{I_{\nu_0}^{(+)} + I_{\nu_0}^{(-)}}{I^{\text{sat}}} = \frac{g_0(\nu_0)}{g_t} - 1 = \frac{g_0(\nu_0)}{\sqrt{g_0(\nu_0)s/2l}} - 1 = \sqrt{\frac{2g_0(\nu_0)l}{s}} - 1 \gg 1. \qquad (5.3.12)$$

In this case the laser transition is strongly saturated, that is, $\bar{N}_1 \approx \bar{N}_2 \approx \frac{1}{2}N_T$, and therefore

$$e_{\max} \approx \frac{P - \Gamma_{21}}{P} \frac{\nu_0}{\nu_{31}}. \qquad (5.3.13)$$

Finally, we assume $P \gg \Gamma_{21}$ in order to have a large small-signal gain [Eq. (4.12.4)]. Thus

$$e_{\max} \approx \frac{\nu_0}{\nu_{31}} = \frac{E_2 - E_1}{E_3 - E_1}. \qquad (5.3.14)$$

This is the theoretical upper limit to the power conversion efficiency. It is just the ratio of the quantum of energy $h\nu_0$ associated with the laser transition to the quantum of energy $h\nu_{31}$ associated with the pump transition of the three-level laser (Fig. 4.6). This ratio is called the *quantum efficiency* of the three-level laser. It is a property only of the energy-level structure of the active atoms. Similarly ν_0/ν_{30} is the quantum efficiency of the ideal four-level laser. In ruby and Nd : YAG lasers the pump level 3 is not a single, sharply defined level. Viewing them as approximately three- and four-level lasers, and using the numbers given in Section 4.10, we calculate that ruby and Nd : YAG lasers have quantum efficiencies $\leq 80\%$ and $\leq 50\%$, respectively (Problem 5.4).

Needless to say, the quantum efficiency is seldom approached in real lasers. First of all, the input-to-output power conversion efficiency, of which the quantum efficiency is the theoretical upper limit, does not give the actual overall efficiency of operation of the laser. It only gives the fraction of the power *actually delivered* to the active medium that is converted to laser output power. There is no account of the efficiency with which the pump power is generated and delivered.

In a carefully designed cw ruby laser, for example, about 25% of the electric power used by the lamp is actually converted to radiation with frequencies lying within the pump bands of the chromium ion, and, of course, not all of this radiation is actually incident on the ruby rod. The fraction of the incident radiation actually absorbed by the ruby is about 4%, and of this only the fraction equal to the quantum efficiency may be used for lasing. All things considered, the actual operating efficiency of a cw ruby laser system is on the order of a tenth of a percent. Although much higher efficiencies are available with modern lasers, the point is that the quantum efficiency defined by (5.3.14) usually has little bearing on the actual operating efficiency of the complete laser system consisting of the pump, the gain cell, and the laser resonator.

5.4   EFFECT OF SPATIAL HOLE BURNING

The effect of spatial hole burning is to reduce the output intensity. This can be understood as follows.

The gain saturates according to the formula (4.13.9), that is,

$$g(\nu) = \frac{g_0(\nu)}{1 + 4(I_\nu^{(+)}/I_\nu^{\text{sat}}) \sin^2 kz}, \tag{5.4.1}$$

where we have used Eq. (5.2.8) to write

$$I_\nu = I_\nu^{(+)} + I_\nu^{(-)} = 2I_\nu^{(+)}. \tag{5.4.2}$$

Our result (5.2.11) for the laser output intensity is based on the approximation of replacing $\sin^2 kz$ by its average value, that is, by ignoring the spatial dependence of the gain arising from the interference of the two traveling waves. We will now consider the effect of retaining the spatial variation (5.4.1) of the gain coefficient. In other words, we will now improve upon the uniform-field approximation by including the effect of spatial hole burning.

Equation (5.3.7) was written in the uniform-field approximation. Without this approximation we arrive at the expression

$$-h\nu \left(\frac{dN_2}{dt}\right)_{\text{stimulated emission}} = 2g(\nu)I_\nu \sin^2 kz = \frac{2g_0(\nu)I_\nu \sin^2 kz}{1 + 2(I_\nu/I_\nu^{\text{sat}}) \sin^2 kz}. \tag{5.4.3}$$

This is the power (at frequency ν) per unit volume, *at the point z*, extracted from the gain medium by stimulated emission. Equation (5.3.7) follows when $\sin^2 kz$ is replaced by $\frac{1}{2}$, its average value over distances large compared with a wavelength.

The gain "clamping" condition (5.2.3) does not apply in the "exact" theory in which the gain and intensity vary with z. In other words, if g is a function of z we can no longer say that the gain and loss coefficients *at every point* in the gain medium are equal in steady-state oscillation. It must still be true, however, that the rate at which the field gains energy equals the rate at which it loses energy. The former follows from the generalization (5.4.3) of (5.3.7):

$$\int_0^l gI \, dz = 2g_0 I \int_0^l \frac{dz \sin^2 kz}{1 + 2(I/I^{\text{sat}}) \sin^2 kz}, \tag{5.4.4}$$

where we have dropped subscript ν's to simplify the notation.

The rate of field intensity loss from the cavity is just

$$(t + s)I^{(+)} = \tfrac{1}{2}(t + s)I = g_t lI. \tag{5.4.5}$$

Note that the one-way intensity $I_\nu^{(+)} = I/2$ in the direction of the output mirror is independent of z. The right-hand sides of (5.4.4) and (5.4.5) must be equal in cw oscillation, and this equality determines I. From a table of integrals we find that, for $kl \gg 1$,

$$\int_0^l \frac{dz \sin^2 kz}{1 + 2(I/I^{\text{sat}}) \sin^2 kz} \cong \frac{l}{2} \frac{I^{\text{sat}}}{I} \left(1 - \frac{1}{\sqrt{1 + 2I/I^{\text{sat}}}}\right), \tag{5.4.6}$$

and, therefore, from the equality of (5.4.4) and (5.4.5),

$$1 - \frac{1}{\sqrt{1 + 2I/I^{\text{sat}}}} = \frac{g_t I}{g_0 I^{\text{sat}}}. \tag{5.4.7}$$

This expression can be written more simply:

$$\sqrt{x} = \frac{2g_0}{g_t} - x, \tag{5.4.8}$$

where

$$x = 1 + \frac{2I}{I^{\text{sat}}}. \tag{5.4.9}$$

Squaring both sides of (5.4.8), we obtain a quadratic equation for x, with the two solutions

$$x = 1 + \frac{2I}{I^{\text{sat}}} = \frac{2g_0}{g_t} + \frac{1}{2} \pm \sqrt{\frac{2g_0}{g_t} + \frac{1}{4}}. \tag{5.4.10}$$

Since x should be equal to 1 ($I = 0$) when $g_0/g_t = 1$, the desired solution is the one with the minus sign in the last term on the right:

$$1 + \frac{2I}{I^{\text{sat}}} = \frac{2g_0}{g_t} + \frac{1}{2} - \sqrt{\frac{2g_0}{g_t} + \frac{1}{4}}, \tag{5.4.11}$$

or

$$I = I^{\text{sat}} \left(\frac{g_0}{g_t} - \frac{1}{4} - \sqrt{\frac{g_0}{2g_t} + \frac{1}{16}} \right). \tag{5.4.12}$$

The output intensity is $I^{\text{out}} = tI^{(+)} = (t/2)I$, exactly as in the uniform-field approximation in which spatial hole burning is not included. This is because I^{out} is determined

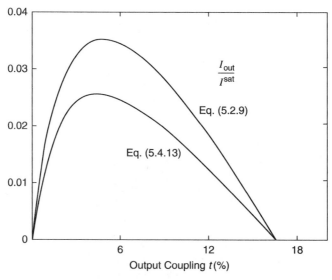

Figure 5.2   Effect of spatial hole burning on output intensity, assuming $g_0 l = 0.10$ and $s = 0.034$.

directly by the *one-way* intensity $I^{(+)}$ (Fig. 5.1), and there are no interference terms to worry about. Thus

$$I^{\text{out}} = \frac{t}{2} I^{\text{sat}} \left(\frac{g_0(\nu)}{g_t} - \frac{1}{4} - \sqrt{\frac{g_0(\nu)}{2g_t} + \frac{1}{16}} \right), \qquad (5.4.13)$$

which is different from the result (5.2.9) obtained when spatial hole burning is neglected.

Figure 5.2 shows the curve of output intensity vs. output coupling predicted by (5.2.11) for the example $g_0 l = 0.10$ and $s = 0.034$. Also shown is the curve predicted by the formula (5.4.13). The two predictions are seen to differ significantly, typically by about 30%. Thus, the effect of spatial hole burning is to *reduce* the output intensity, as already mentioned.

5.5 LARGE OUTPUT COUPLING

Our analysis of output power thus far has assumed that the output coupling is small and that the two traveling waves have equal intensities, $I^{(+)} = I^{(-)}$. We have also assumed that the time-averaged intensities $I^{(+)}$ and $I^{(-)}$ are independent of the axial coordinate z. We will now allow arbitrary output coupling and, therefore, allow the possibility that $I^{(+)}$ and $I^{(-)}$ may vary with z. We assume, however, that the variation of interest is much more gradual than the $\sin^2 kz$ variation due to spatial hole burning, and replace $\sin^2 kz$ by its average value $\frac{1}{2}$.

Thus, we work with the gain–saturation formula (5.2.4), which we now write in the form

$$g(z) = \frac{g_0}{1 + [I^{(+)}(z) + I^{(-)}(z)]/I^{\text{sat}}}. \qquad (5.5.1)$$

For notational simplicity we have again suppressed the ν dependence of the various terms in this equation, but we indicate explicitly the z dependence. In principle, g_0 could also depend on z, for example, if the pumping rate P depends on z, but here we assume that it does not. In steady-state oscillation the intensities $I^{(+)}$ and $I^{(-)}$ satisfy Eqs. (4.3.4):

$$\frac{dI^{(+)}}{dz} = g(z)I^{(+)}(z), \qquad (5.5.2a)$$

$$\frac{dI^{(-)}}{dz} = -g(z)I^{(-)}(z). \qquad (5.5.2b)$$

We will assume that all cavity loss processes (output coupling, scattering, absorption) occur at the mirrors. Thus, we will not include terms accounting for "distributed" loss within the cavity.

Equations (5.5.2) imply that

$$\frac{d}{dz}\left[I^{(+)}I^{(-)}\right] = I^{(+)}\frac{dI^{(-)}}{dz} + I^{(-)}\frac{dI^{(+)}}{dz} = 0, \tag{5.5.3}$$

or

$$I^{(+)}(z)I^{(-)}(z) = \text{const} = C, \tag{5.5.4}$$

where the constant C can be evaluated at either mirror:

$$I^{(+)}(0)I^{(-)}(0) = I^{(+)}(L)I^{(-)}(L) = C. \tag{5.5.5}$$

Let us now make use of (5.5.4) and (5.5.1) in (5.5.2):

$$\frac{1}{I^{(+)}}\frac{dI^{(+)}}{dz} = \frac{g_0}{1 + [I^{(+)} + C/I^{(+)}]/I^{\text{sat}}}, \tag{5.5.6a}$$

$$\frac{1}{I^{(-)}}\frac{dI^{(-)}}{dz} = \frac{-g_0}{1 + [I^{(-)} + C/I^{(-)}]/I^{\text{sat}}}. \tag{5.5.6b}$$

In this form the equations for $I^{(+)}(z)$ and $I^{(-)}(z)$ are uncoupled and can be solved separately.

Consider first the equation for $I^{(+)}(z)$. Writing (5.5.6a) in the form

$$g_0\,dz = \frac{1}{I^{(+)}}\left[1 + \frac{1}{I^{\text{sat}}}\left(I^{(+)} + \frac{C}{I^{(+)}}\right)\right]dI^{(+)}$$

$$= \frac{dI^{(+)}}{I^{(+)}} + \frac{1}{I^{\text{sat}}}dI^{(+)} + \frac{C}{I^{\text{sat}}}\frac{dI^{(+)}}{I^{(+)2}}, \tag{5.5.7}$$

and integrating from $z = 0$ to $z = L$ (Fig. 4.4), we have

$$\int_0^L g_0\,dz = \int_{I^{(+)}(0)}^{I^{(+)}(L)} \frac{dI^{(+)}}{I^{(+)}} + \frac{1}{I^{\text{sat}}}\int_{I^{(+)}(0)}^{I^{(+)}(L)} dI^{(+)} + \frac{C}{I^{\text{sat}}}\int_{I^{(+)}(0)}^{I^{(+)}(L)} \frac{dI^{(+)}}{I^{(+)2}},$$

or

$$g_0 l = \ln\frac{I^{(+)}(L)}{I^{(+)}(0)} + \frac{1}{I^{\text{sat}}}\left[I^{(+)}(L) - I^{(+)}(0)\right] - \frac{C}{I^{\text{sat}}}\left(\frac{1}{I^{(+)}(L)} - \frac{1}{I^{(+)}(0)}\right). \tag{5.5.8a}$$

Considering Eq. (5.5.6b), we obtain similarly

$$g_0 l = \ln\frac{I^{(-)}(0)}{I^{(-)}(L)} + \frac{1}{I^{\text{sat}}}\left[I^{(-)}(0) - I^{(-)}(L)\right] - \frac{C}{I^{\text{sat}}}\left(\frac{1}{I^{(-)}(0)} - \frac{1}{I^{(-)}(L)}\right). \tag{5.5.8b}$$

Since the small-signal gain g_0 is assumed to be constant inside the gain cell of length l (Fig. 4.4), but vanishes outside the gain cell, we have used, in both of Eqs. (5.5.8),

$$\int_0^L g_0 \, dz = g_0 l. \tag{5.5.9}$$

The boundary conditions (4.3.3), together with (5.5.4), may be used to express $I^{(-)}(0)$, $I^{(-)}(L)$, and $I^{(+)}(0)$ in terms of $I^{(+)}(L)$. Thus,

$$I^{(-)}(L) = \frac{C}{I^{(+)}(L)} = r_2 I^{(+)}(L), \tag{5.5.10}$$

so that

$$C = r_2 \left[I^{(+)}(L) \right]^2. \tag{5.5.11}$$

Furthermore,

$$I^{(+)}(0) = \frac{C}{I^{(-)}(0)} = r_1 I^{(-)}(0), \tag{5.5.12}$$

which means we may also write

$$I^{(-)}(0) = \sqrt{r_2/r_1} I^{(+)}(L), \tag{5.5.13}$$

and from (5.5.12),

$$I^{(+)}(0) = \sqrt{r_1 r_2} I^{(+)}(L). \tag{5.5.14}$$

We can use (5.5.11), (5.5.12), and (5.5.14) to express the right sides of (5.5.8) in terms of $I^{(+)}(L)$. From (5.5.8a),

$$g_0 l = \ln \frac{1}{\sqrt{r_1 r_2}} + \frac{I^{(+)}(L)}{I^{\text{sat}}} (1 - \sqrt{r_1 r_2}) - \frac{I^{(+)}(L)}{I^{\text{sat}}} \left(r_2 - \sqrt{\frac{r_2}{r_1}} \right), \tag{5.5.15a}$$

while from (5.5.8b),

$$g_0 l = \ln \frac{1}{\sqrt{r_1 r_2}} + \frac{I^{(+)}(L)}{I^{\text{sat}}} \left(\sqrt{\frac{r_2}{r_1}} - r_2 \right) - \frac{I^{(+)}(L)}{I^{\text{sat}}} (\sqrt{r_1 r_2} - 1). \tag{5.5.15b}$$

These equations are identical. We now solve for $I^{(+)}(L)$:

$$I^{(+)}(L) = \frac{I^{\text{sat}}(g_0 l + \ln \sqrt{r_1 r_2})}{1 + \sqrt{r_2/r_1} - r_2 - \sqrt{r_1 r_2}}$$

$$= \frac{\sqrt{r_1} I^{\text{sat}}}{(\sqrt{r_1} + \sqrt{r_2})(1 - \sqrt{r_1 r_2})} (g_0 l + \ln \sqrt{r_1 r_2}). \tag{5.5.16}$$

Finally, we can use (5.5.13) and (5.5.16) to obtain the output intensity:

$$I_{out} = t_1 I^{(-)}(0) + t_2 I^{(+)}(L)$$

$$= I^{sat}\left(t_2 + \sqrt{\frac{r_2}{r_1}}t_1\right)\frac{\sqrt{r_1}}{(\sqrt{r_1} + \sqrt{r_2})(1 - \sqrt{r_1 r_2})}(g_0 l + \ln\sqrt{r_1 r_2}). \qquad (5.5.17)$$

These results generalize our previous ones in that they apply to arbitrary values of output coupling. The principal assumptions in our derivation of (5.5.17) have been that (1) the gain medium is homogeneously broadened; (2) the gain saturates according to formula (5.5.1); (3) the small-signal gain g_0 and the saturation intensity I_{sat} are constant throughout the gain medium; (4) spatial variations of the cavity intensity transverse to the resonator axis can be neglected as a first approximation; (5) loss occurs only at the mirrors; and (6) spatial hole burning is averaged.

The analysis leading to (5.2.11) assumed all these things, and also that the output coupling (or other losses) is small. The analysis just given should therefore reproduce (5.2.11) in the limit of high mirror reflectivities.

To see that this is so, suppose that one of the mirrors is perfectly reflecting ($r_1 = 1$, $t_1 = s_1 = 0$). Then (5.5.17) becomes

$$I_{out} = tI^{sat}\frac{1}{(1 + \sqrt{r})(1 - \sqrt{r})}(g_0 l + \ln\sqrt{r}) = \frac{t}{1 - r}I^{sat}\left(g_0 l - \frac{1}{2}\ln\frac{1}{r}\right)$$

$$= \frac{t}{t + s}I^{sat}\left[g_0 l - \frac{1}{2}\ln(1 - t - s)^{-1}\right]$$

$$= \frac{(\frac{1}{2}t)\ln(1 - t - s)^{-1}}{t + s}I^{sat}\left(\frac{2g_0 l}{\ln(1 - t - s)^{-1}} - 1\right), \qquad (5.5.18)$$

where $r = r_2$, $t = t_2$, $s = s_2$. This is a generalization of (5.2.11) (Problem 5.5). When $t + s \ll 1$, we have

$$I_{out} \approx \frac{t}{2}I^{sat}\left(\frac{2g_0 l}{t + s} - 1\right), \qquad (5.5.19)$$

which is just (5.2.11). In going from (5.5.18) to (5.5.19) we have used the approximation

$$-\ln(1 - t - s) \approx t + s \qquad (t + s \ll 1). \qquad (5.5.20)$$

One interesting result of our analysis is that the total two-way intensity is relatively uniform in the z direction, even for moderately large output couplings. A simple way to see this is to use the boundary conditions (5.5.10), (5.5.13), and (5.5.14) to calculate the ratio of the total intensity at the output mirror to that at the other mirror, assumed again to be perfectly reflecting. With $r_1 = 1$ and $r_2 = r$, we obtain

$$\frac{I^{(+)}(L) + I^{(-)}(L)}{I^{(+)}(0) + I^{(-)}(0)} = \frac{1 + r}{2\sqrt{r}}. \qquad (5.5.21)$$

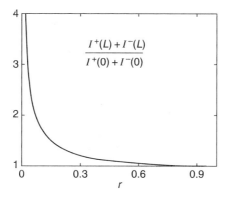

Figure 5.3   Ratio of total intensities at the two mirrors [Eq. (5.5.21)].

This is plotted in Fig. 5.3. It is seen that the total intensities are comparable at the two mirrors for reflectivities as low as 50%. This conclusion is consistent with a more detailed analysis in which $I^{(+)}(z)$ and $I^{(-)}(z)$ are calculated for $0 \leq z \leq L$. Equations (5.5.6) are the starting points for such an analysis. We can conclude that for most (but by no means all) lasers, $I^{(+)}$ and $I^{(-)}$ vary only mildly with z. This justifies the approximate theory of Section 5.4.

These results for arbitrarily large output couplings were first obtained by W. W. Rigrod.[1] The Rigrod analysis predicts an optimal output coupling that reduces to (5.3.1) when $t + s \ll 1$. We will not bother to show this. The reader is referred to Rigrod's study of graphs of optimal output coupling and output intensity as a function of small-signal gain g_0 and scattering loss coefficient s. Spatial hole burning in the case of arbitrary output couplings has subsequently been included in numerical computations. The effect is to reduce the output intensity calculated without spatial hole burning by about the same relative magnitude as in the case of small output coupling (cf. Fig. 5.2).

5.6   MEASURING GAIN AND OPTIMAL OUTPUT COUPLING

Equation (5.2.11) for output intensity, or its generalization (5.5.18), has been shown by experiment to be quite accurate, and it has been used extensively in laser design. This is so despite the fact that our formulas for laser output intensity were derived without taking atomic motion into account. Atomic motion tends to smear out the effect of spatial hole burning: An atom at a field nodal point does not stay there forever, as assumed in our simple analysis. The result of rapid atomic motion will be to average the field spatial variations, and so spatial hole burning is usually assumed to be negligible in gas lasers.

Figure 5.4 shows experimental results for the output power of a He–Ne laser as a function of mirror transmission. The three curves drawn through the experimental points are based on (5.2.11) and are seen to fit the data very nicely.

[1]W. W. Rigrod, *Journal of Applied Physics* **36**, 2487 (1965).

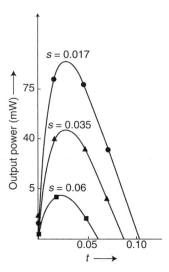

Figure 5.4   Experimental data on the output power of a 632.8-nm He–Ne laser. The solid curves are based on Eq. (5.2.11) for output intensity vs. output coupling. The three curves correspond to $s = 0.06$, 0.035, and 0.017. [After P. Laures, *Physics Letters* **10**, 61 (1964).]

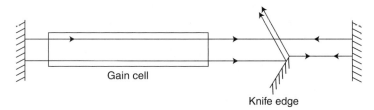

Figure 5.5   A knife edge may be used to determine the small-signal gain by the maximal-loss method. By a micrometer adjustment the knife edge is made to occlude an increasing fraction of the intracavity beam, until lasing ceases because the total loss exceeds the gain. This determines the maximal loss and, therefore, the small-signal gain for laser oscillation.

In general, the small-signal gain g_0 and the saturation intensity I^{sat} are difficult to calculate accurately because the pumping and decay rates of the relevant level populations may not be well known. One way to measure g_0 is the *maximal-loss method*. In this method the cavity loss is increased until the laser oscillation ceases. Since laser oscillation requires $g_0 > g_t$, and the cavity loss determines g_t, the maximal loss allowing laser oscillation is in fact just g_0. As illustrated in Fig. 5.5, the cavity loss may be varied by inserting a reflecting knife edge into the cavity. A micrometer adjustment determines the fraction of the intracavity intensity that is occluded by the knife edge, and thus the cavity loss is varied by turning the micrometer screw.

A variant of the maximal-loss method[2] may be used to determine not only the small-signal gain but also the optimal output coupling t_{opt} and the output power obtainable with $t = t_{opt}$. This method may be understood with reference to Fig. 5.6. The knife

---

[2]T. F. Johnston, *IEEE Journal of Quantum Electronics* **12**, 310 (1976).

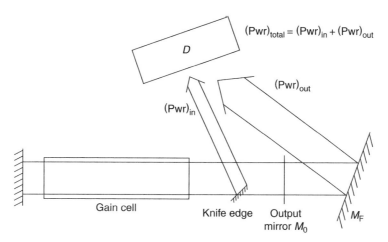

Figure 5.6 Experimental setup for determining the small-signal gain, the optimal output coupling, and the maximum possible output power. [After T. F. Johnston, *IEEE Journal of Quantum Electronics* **12**, 310 (1976).]

edge may be a small cube that has been coated on two adjacent faces with a reflecting material. The knife edge deflects part of the intracavity field out of the cavity and onto a power meter D; the amount of power deflected to D, $(\text{Pwr})_{\text{in}}$, may be varied by a micrometer adjustment. M_F is an extracavity folding mirror that directs the output power $(\text{Pwr})_{\text{out}}$ onto the same detector D.

With the knife edge inserted in the cavity there is a scattering coefficient

$$s = \frac{(\text{Pwr})_{\text{in}}}{\text{Pwr}^{(+)}}, \tag{5.6.1}$$

so that the effective output coupling is $t + s$, where t is the transmission coefficient of the output mirror M_0; here $\text{Pwr}^{(+)}$ is the power in the wave traveling toward M_0. Since $t = (\text{Pwr})_{\text{out}}/\text{Pwr}^{(+)}$, we have

$$s = \frac{(\text{Pwr})_{\text{in}}}{(\text{Pwr})_{\text{out}}} t. \tag{5.6.2}$$

We are assuming that the knife edge represents a small perturbation, so that $I^{(+)}$ is the intensity at both M_0 and the knife edge. Now the sum

$$(\text{Pwr})_{\text{total}} = (\text{Pwr})_{\text{in}} + (\text{Pwr})_{\text{out}} \tag{5.6.3}$$

represents the total output power of the laser. The micrometer setting can be varied until a value s_{opt} is obtained for which $(\text{Pwr})_{\text{total}}$ is a maximum. This gives the optimal output coupling as (if we assume $t < t_{\text{opt}}$)

$$t_{\text{opt}} = s_{\text{opt}} + t. \tag{5.6.4}$$

Furthermore, the output power at this optimal output coupling is just the maximum value obtained for $(\text{Pwr})_{\text{total}}$. The small-signal gain is determined by the value of s at which laser oscillation stops. In practice, this must be determined by extrapolation since

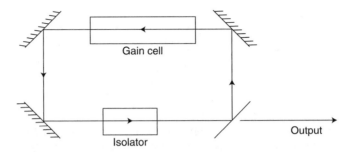

Figure 5.7   A unidirectional (traveling-wave) ring laser. An optical "isolator," which transmits only radiation propagating in the direction shown, is used to obtain traveling-wave rather than standing-wave laser oscillation.

both $(\text{Pwr})_{\text{in}}$ and $(\text{Pwr})_{\text{out}}$ go to zero as the laser threshold is approached, and (5.6.2) becomes indeterminate.

In lasers with liquid or solid gain media the spatial hole burning is not smoothed out by atomic motion to the same extent as in gas lasers, although other processes may tend to weaken its effect. Convincing evidence for spatial hole burning has been obtained with liquid dye lasers. The output power of a single-mode dye laser is found to increase significantly when operated as a ring laser instead of the two-mirror standing-wave configuration. In the ideal traveling-wave ring laser, sketched in Fig. 5.7, there is no standing-wave interference pattern and therefore no spatial hole burning.

We will see in the following chapter that spatial hole burning plays an important role in the multimode behavior of many lasers.

• When a linearly polarized field propagates a distance L through a material in which there is an applied dc magnetic induction field B along the direction of propagation, the direction of polarization is rotated by an angle

$$\theta = VBL, \tag{5.6.5}$$

where V, which depends on wavelength and temperature, is the *Verdet constant* of the material. This rotation of the plane of polarization is called the *Faraday effect*. The angle θ has the same magnitude and direction for waves propagating in opposite directions. Verdet constants of glasses used in commercial Faraday rotators are typically in the range $10\text{--}100$ rad Tesla^{-1} m^{-1}.

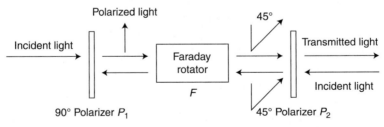

Figure 5.8   An optical isolator. A light beam incident from the left emerges from a polarizer P_1 with vertical polarization, has its polarization rotated $45°$ by a Faraday rotator F, and then passes through $45°$ polarizer P_2. Light incident on P_2 from the right will emerge from P_2 with the same $45°$ polarization but has its polarization rotated another $45°$ by F, so that it is then polarized horizontally and does not pass through P_1. The optical isolator therefore transmits light incident from the left but not from the right.

The Faraday effect is used to transmit light in one direction but not in the opposite direction, that is, to make an optical isolator (see Fig. 5.7). To see how this is done, suppose that light passes through a vertical polarizer P_1 and then through a Faraday rotator F that rotates the polarization by $45°$ (Fig. 5.8). A second polarizer P_2 oriented at the same $45°$ angle will then transmit the light. Light incident on P_2 from the opposite direction, however, will emerge from F with a polarization orthogonal to that transmitted by P_1, and will therefore not pass through P_1. In other words, light will propagate in the direction P_1FP_2 but not in the direction P_2FP_1.

In some applications it is necessary to avoid backreflections of laser radiation into the laser cavity. Such backreflections can have deleterious effects, in some cases causing the laser oscillation to become unstable. An optical isolator placed outside the laser cavity can substantially reduce backreflections. •

5.7 INHOMOGENEOUSLY BROADENED MEDIA

Recall that in an inhomogeneously broadened gain medium the different active atoms have different central transition frequencies ν_0, due to their different velocities and the Doppler effect (recall Section 3.9), spatially nonuniform electric and magnetic fields,[3] the presence of different isotopes having slightly different energy levels, and various other effects.

In the case of inhomogeneous broadening, the theory of laser oscillation can become enormously complex. For one thing, it becomes much more difficult to justify our assumption of single-mode oscillation, as we will see in Section 5.10. Here we will simply restrict ourselves to a few remarks.

As in the case of homogeneous broadening, we can define a small-signal gain and a saturation intensity for an inhomogeneously broadened gain medium. The small-signal gain is simply the gain when the field intensity is so small that it does not affect the population difference $N_2 - N_1$. In the case of Doppler broadening, for example, the small-signal gain is given by

$$g_0(\nu) = \frac{\lambda^2 A}{8\pi}(\Delta N)_0 S(\nu)$$

$$= \frac{\lambda^2 A}{8\pi}(\Delta N)_0 \frac{1}{\delta\nu_D}\left(\frac{4\ln 2}{\pi}\right)^{1/2}\exp\left[\frac{-4(\nu-\nu_0)^2(\ln 2)}{\delta\nu_D^2}\right], \qquad (5.7.1)$$

where $(\Delta N)_0 = (N_2 - N_1)_0$ is the small-signal (unsaturated) population difference and we have used Eq. (3.9.9) for the Doppler lineshape function. Thus,

$$g_0(\nu) = g_0(\nu_0)\exp\left[\frac{-4(\nu-\nu_0)^2(\ln 2)}{\delta\nu_D^2}\right], \qquad (5.7.2)$$

which replaces the formula (4.12.8) in the small-signal limit when the gain profile has the Doppler lineshape.

Real complications arise when we consider the saturation characteristics of an inhomogeneously broadened gain medium. Atoms with central transition frequency ν_0 will be saturated according to (4.12.8), but there is a *distribution* of resonance frequencies ν_0.

---

[3]The energy levels of atoms and molecules may be shifted by electric or magnetic fields. The energy-level shifts produced by electric and magnetic fields are called the Stark and Zeeman shifts, respectively. The relative shift is usually very small.

The gain coefficient is obtained by integrating the contributions from the different frequency components, each of which saturates to a different degree depending on its detuning from the cavity mode frequency ν. For many inhomogeneously broadened laser media (in particular, low-pressure gas lasers), this integration results in a gain–saturation formula of the form (4.14.7) if spatial hole burning and power broadening are ignored. Thus, (5.2.9) is replaced by

$$I_\nu^{\text{out}} = \frac{t}{2} I^{\text{sat}} \left[\left(\frac{g_0(\nu)}{g_t} \right)^2 - 1 \right] \tag{5.7.3}$$

in the case of inhomogeneous broadening (and small output coupling). However, these results apply to single-mode operation, and we will see in the next chapter that single-mode operation is seldom achieved with inhomogeneously broadened gain media.

In general, the gain profile is determined by both homogeneous and inhomogeneous broadening mechanisms. In a gaseous medium, for example, there is inhomogeneous broadening due to atomic motion. The lineshape function $S(\nu)$ entering the small-signal gain formula is then the Voigt profile, exactly as in the theory of absorption. The assumptions of pure homogeneous broadening or pure inhomogeneous broadening are, in general, *approximations* that apply when one type of broadening is dominant.

5.8   SPECTRAL HOLE BURNING AND THE LAMB DIP

We have noted that in the case of Doppler broadening the gains associated with different frequencies (spectral packets) are saturated to different degrees, depending on the detuning from the field frequency ν. A traveling wave propagating to the right (positive z direction) will saturate those atoms with the z component of velocity given by (Section 4.14)

$$\frac{v}{c} = \frac{\nu - \bar{\nu}}{\bar{\nu}}. \tag{5.8.1}$$

where $\bar{\nu}$ is the resonance frequency of a stationary atom. In the same way, a traveling wave propagating to the *left* will saturate those atoms with the z component of velocity given by

$$\frac{v}{c} = -\frac{\nu - \bar{\nu}}{\bar{\nu}}. \tag{5.8.2}$$

Therefore, a *standing-wave* cavity field will burn *two* holes in the Doppler line profile, as shown in Fig. 5.9.

If a laser is operating with its cavity mode frequency detuned from the resonance frequency $\bar{\nu}$ of a stationary atom (i.e., from the center of the Doppler line), it will burn a hole in the gain curve on either side of line center (Fig. 5.9). In other words, the cavity mode is "feeding" off two spectral packets. When the mode frequency is exactly at the center ($v = 0$) of the Doppler line, however, the two holes merge together because the field can now strongly saturate only those atoms having no z component of velocity. The gain when $\nu = \bar{\nu}$ comes predominantly from these atoms, which are saturated by both traveling waves. When $|\nu - \bar{\nu}|$ is larger than about a homogeneous

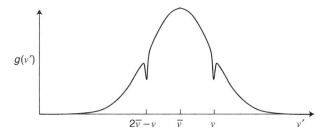

Figure 5.9   A standing-wave field burns two holes in the curve of gain vs. velocity. These holes are centered at velocities $v = \pm c(v - \bar{v})/\bar{v}$, where $\bar{v}$ is the Bohr frequency of a stationary atom. This results in the burning of two holes in the Doppler-broadened gain profile.

linewidth, however, the gain comes predominantly from the atoms with velocities given by (5.8.1) or (5.8.2), and these atoms are saturated by a single traveling wave. The gain saturation is therefore strongest at line center ($v = \bar{v}$), and consequently there is a dip in the laser output power at line center.

This dip in output power at line center was predicted by W. E. Lamb, Jr. in 1963, and is called the *Lamb dip*. Figure 5.10 shows very early experimental results confirming the prediction of the Lamb dip. The data also show, as expected, that the dip becomes more pronounced at higher power levels, where the degree of selective saturation (hole burning) is greatest.

The observation of the Lamb dip requires that the cavity mode frequency be swept across the gain profile. Since the cavity mode frequencies are given (approximately) by $v = mc/2L$, where m is an integer, this may be done by slowly varying the cavity length L, for example, by mounting one of the cavity mirrors on a piezoelectric crystal, which expands or contracts in certain directions when an electric field is applied.

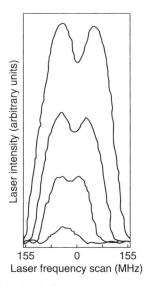

Figure 5.10   Early observation of the Lamb dip in the output power of a 1.15-nm He–Ne laser. [From A. Szöke and A. Javan, *Physical Review Letters* **10**, 521 (1963).]

- The fluctuations in the cavity length L due to mechanical vibrations and temperature variations cause the frequency of a single-mode laser to fluctuate. Various techniques are used when a high degree of frequency stability is required. The Lamb dip is the basis of one such method: The increase in laser power as the frequency drifts away from the center of the Doppler line serves as the "error signal" for a feedback circuit that adjusts the cavity length piezoelectrically to bring the output power back to its value at the center of the Doppler line. Frequency stability on the order of 1 part in 10^{10} can be obtained by this method.

The "inverse" Lamb dip can be used for frequency stabilization more generally when the gain medium may or may not exhibit a Lamb dip. The inverse Lamb dip occurs at the center of the Doppler line of an intracavity saturable *absorption* cell. The saturation of the absorption cell results in a *peak* of the laser output power instead of a dip, and a feedback system is used to fix the output power at this peak value. Because the absorption line can be made very narrow, the inverse Lamb dip can be used to stabilize the laser output frequency to 1 part in 10^{12} or better.

As implied by the preceding discussion, the width of the Lamb dip (or the inverse Lamb dip) is determined by the homogeneous linewidth of the gain (or absorption) line. This fact can be used to determine the ordinarily unobserved homogeneous linewidth of a Doppler-broadened transition.   •

5.9   FREQUENCY PULLING

We have mostly ignored the effect of the refractive index of the gain medium on laser oscillation, except insofar as it enters into the equations of Table 4.1 for gain and threshold. However, it turns out that the refractive index of the gain medium actually determines to some extent the laser oscillation frequency. We will now examine how this occurs.

A laser will oscillate at a frequency ν such that the optical length of the cavity is an integral number of half wavelengths. That is, $L = m\lambda/2$, or

$$\nu = \frac{mc}{2L} = \nu_m, \qquad (5.9.1)$$

where m is a positive integer. This applies to the *bare-cavity* case in which the gain and refractive index of the active medium are not taken into account.

In general, however, the effective *optical* length of a medium is not just its physical length but rather the product of its physical length and its refractive index $n(\nu)$. To account for the index of refraction of the active medium, therefore, we divide the cavity length into two parts: $L = l + (L - l)$, where l is the length of the gain cell and the remainder is empty cavity, as in Fig. 4.4. The *optical* length of the gain cell is $n(\nu)l$, so that (5.9.1) should be replaced by

$$\nu = \frac{mc/2}{n(\nu)l + (L - l)}, \qquad (5.9.2a)$$

or

$$\frac{l}{L}[n(\nu) - 1]\nu = \nu_m - \nu, \qquad (5.9.2b)$$

where $\nu_m = mc/2L$ is a bare-cavity mode frequency. Thus, *the laser oscillation frequency ν will be different from a bare-cavity mode frequency ν_m if $n(\nu) \neq 1$.*

Equation (5.9.2b) determining the laser oscillation frequency may be written in a different form. To do this, let us assume that $n(v)$ is determined primarily by the single nearly resonant, lasing atomic transition. In other words, we will assume that $n(v)$ is essentially the resonant (or "anomalous") refractive index (Section 3.15). Since other transitions contributing to the refractive index will usually be off resonance by many transition linewidths, this will often be an excellent approximation.

In the case of an absorbing medium the resonant refractive index is simply related to the absorption coefficient [recall (3.15.10)]. The same relation applies to an amplifying (gain) medium, simply by replacing the absorption coefficient by the negative of the gain coefficient. Thus, for a homogeneously broadened gain medium, we have

$$n(v) - 1 = -\frac{\lambda_0}{4\pi} \frac{v_0 - v}{\delta v_0} g(v), \tag{5.9.3}$$

where δv_0 is the homogeneous linewidth (HWHM). From (5.9.2) and (5.9.3), therefore, we obtain, using the fact that $v \approx c/\lambda_0$,

$$v = \frac{v_0[cg(v)l/4\pi L] + v_m \delta v_0}{[cg(v)l/4\pi L] + \delta v_0}. \tag{5.9.4}$$

The quantity

$$\delta v_c = \frac{cg(v)l}{4\pi L} \tag{5.9.5}$$

is related to the mirror reflectivities through the gain-clamping condition (5.2.3), and has the dimensions of frequency. It is called the *cavity bandwidth*. Thus, Eq. (5.9.4), the equation for the laser oscillation frequency v, is usually written in the form

$$v = \frac{v_0 \delta v_c + v_m \delta v_0}{\delta v_c + \delta v_0}, \tag{5.9.6a}$$

or alternatively

$$\delta v_c(v - v_0) = \delta v_0(v_m - v). \tag{5.9.6b}$$

Note that the second of these two expressions establishes that the lasing frequency v lies *between* v_0 and v_m, no matter which of them is larger. The actual frequency of laser radiation is therefore "pulled" toward the center of the gain profile and away from the bare-cavity frequency. This effect is called *frequency pulling*.

We will discuss the cavity bandwidth in some detail at the end of this section. For now it suffices to mention that $\delta v_0 \gg \delta v_c$ in most lasers. In most lasers, therefore,

$$v = \frac{v_0 \delta v_c/\delta v_0 + v_m}{1 + \delta v_c/\delta v_0} \approx \left(v_0 \frac{\delta v_c}{\delta v_0} + v_m \right) \left(1 - \frac{\delta v_c}{\delta v_0} \right)$$

$$\approx v_m + (v_0 - v_m)\frac{\delta v_c}{\delta v_0} \quad \text{(homogeneous broadening)}. \tag{5.9.7}$$

Now the starting point of our analysis leading to (5.9.7) was the relation (5.9.3) between the resonant refractive index and the gain. Although we have assumed a homogeneously broadened line, it is in fact *always* possible to relate the index and the gain (or absorption). [Eq. (3.15.10).] In the case of a Doppler-broadened medium, for example, we are led by analogous manipulations (for $|\nu_0 - \nu| \ll \delta\nu_D$) to the formula

$$\nu \approx \nu_m + (\nu_0 - \nu_m)\frac{\delta\nu_c}{\delta\nu_D}\sqrt{4\ln 2/\pi}$$

$$\approx \nu_m + 1.88(\nu_0 - \nu_m)\frac{\delta\nu_c}{\delta\nu_D} \quad \text{(Doppler broadening).} \tag{5.9.8}$$

If we use the gain-clamping condition (5.2.3), then (5.9.5) gives

$$\delta\nu_c = \frac{cl}{4\pi L}g_t, \tag{5.9.9}$$

and therefore

$$\frac{\delta\nu_c}{\delta\nu_0} = \frac{cl}{4\pi L}\frac{g_t}{\delta\nu_0}, \tag{5.9.10}$$

$$\frac{\delta\nu_c}{\delta\nu_D} = \frac{cl}{4\pi L}\frac{g_t}{\delta\nu_D}, \tag{5.9.11}$$

for homogeneous broadening and Doppler broadening, respectively. These results indicate that frequency pulling will be more readily observed for high-gain media with narrow-gain profiles. This prediction is in fact borne out experimentally.

For a 632.8-nm He−Ne laser with a threshold gain of 0.001 cm^{-1} and a Doppler width of 1500 MHz, for example, we have

$$\frac{\delta\nu_c}{\delta\nu_D} \approx 0.0016 \quad (l \approx L). \tag{5.9.12}$$

The 3.39-μm He−Ne laser, on the other hand, has a smaller Doppler width of about 280 MHz owing to the larger value of the transition wavelength. Furthermore, the gain at the 3.39-μm transition is typically much larger than at the 632.8-nm transition, so that lasing can be achieved with $g_t \approx 0.03$ cm^{-1}. In this case

$$\frac{\delta\nu_c}{\delta\nu_D} \approx 0.26 \quad (l \approx L). \tag{5.9.13}$$

Frequency pulling is therefore readily observed with a 3.39-μm He−Ne laser.

Frequency pulling is especially pronounced in the low-pressure He−Xe laser operating on the 3.51-μm transition of Xe. In this case the Doppler width is only about 100 MHz owing to the relatively large mass of Xe. Furthermore, the gain may be as high as 1 cm^{-1}. With $g_t = 0.5$ cm^{-1} we obtain for this laser the ratio

$$\frac{\delta\nu_c}{\delta\nu_D} \approx 12 \quad (l \approx L). \tag{5.9.14}$$

In this case the frequency pulling is so pronounced that the approximation (5.9.8) is inapplicable.

We can write (5.9.6) as

$$\nu^{(m)} = \frac{\nu_0 \delta\nu_c + \nu_m \delta\nu_0}{\delta\nu_c + \delta\nu_0}, \tag{5.9.15}$$

where $\nu^{(m)}$ denotes the "frequency-pulled" value of the bare-cavity mode frequency ν_m. The bare-cavity mode spacing $\nu_{m+1} - \nu_m = c/2L$ given in (5.9.1) is, therefore, "renormalized" by frequency pulling to the value

$$\nu^{(m+1)} - \nu^{(m)} = \frac{(\nu_{m+1} - \nu_m)\delta\nu_0}{\delta\nu_c + \delta\nu_0} = \frac{c}{2L}\frac{1}{1 + \delta\nu_c/\delta\nu_0}. \tag{5.9.16}$$

An analogous expression applies for an inhomogenously broadened gain medium.

If a laser is oscillating on several modes, therefore, the effect of frequency pulling is to reduce the mode spacing from $c/2L$ to the value (5.9.16) or the analogous expression for an inhomogeneous line. This effect of frequency pulling can be observed by looking at the beat (difference) frequency of the laser output with a sufficiently fast photodetector. In this way a reduction on the order of 2 has been observed in the mode spacing of a He–Xe laser.

When spectral hole burning is present in the case of inhomogeneous broadening, the analysis of frequency pulling becomes rather complicated, and our simple theory gives only a crude approximation to the actual situation. Since frequency pulling in many lasers is only a small effect, we will not discuss the complications due to hole burning.

It is not difficult to see why $\delta\nu_c$ is called the *cavity bandwidth*. According to Eq. (4.4.8), in the absence of an active medium in the cavity the cavity intensity satisfies the rate equation

$$\frac{dI_\nu}{dt} = -\frac{c}{2L}(1 - r_1 r_2)I_\nu = -\left(c\frac{l}{L}g_t\right)I_\nu. \tag{5.9.17}$$

In a bare cavity, therefore, the cavity intensity decays exponentially:

$$I_\nu(t) = I_\nu(0)\exp\left[-c\left(\frac{l}{L}\right)g_t t\right]. \tag{5.9.18}$$

Since the intensity is proportional to the square of the electric field amplitude E_ν, we may write

$$E_\nu(t) = E_\nu(0)\exp\left[-c\left(\frac{l}{2L}\right)g_t t\right]e^{-2\pi i \nu t} \tag{5.9.19}$$

for the decay of the electric field in a bare cavity. Equation (5.9.18) can be used to determine the *cavity lifetime* $L/(l\, cg_t)$ by the *cavity ring-down method*: A pulse is injected into the cavity and the intensity of light escaping from the cavity is measured as a function of time.

Now the fact that the field decays in time implies that it cannot be truly monochromatic. In fact, the frequency spectrum associated with the time-dependent field (5.9.19) is Lorentzian:

$$s(\nu') = \frac{\delta\nu_c/\pi}{(\nu' - \nu)^2 + \delta\nu_c^2}, \qquad (5.9.20)$$

where

$$\delta\nu_c = \frac{1}{2\pi}\frac{clg_t}{2L}, \qquad (5.9.21)$$

which, of course, is the same as (5.9.5) when the gain-clamping condition $g(\nu) = g_t$ is used. The step from (5.9.19) to (5.9.20) is a standard result of Fourier transform theory: The frequency spectrum of a quantity that decays exponentially in time is a Lorentzian function of frequency. We have already seen an example of this in Chapter 3, where we found that the exponential decay of the atomic dipole moments due to collisions led to a Lorentzian lineshape function.

Note that $\delta\nu_c$ depends only on properties of the bare cavity:

$$\delta\nu_c = \frac{1}{2\pi}\frac{cl}{2L}\left[\frac{1}{2l}\ln\left(\frac{1}{r_1 r_2}\right)\right] = \frac{1}{4\pi}\left[\frac{c}{2L}\ln\left(\frac{1}{r_1 r_2}\right)\right]. \qquad (5.9.22)$$

Frequently a laser cavity is characterized by the dimensionless *quality factor*[4]

$$Q = \frac{\nu}{2\delta\nu_c} = \frac{\text{mode frequency}}{\text{FWHM cavity bandwidth}}. \qquad (5.9.23)$$

A high-Q cavity is one with low loss, whereas a low-Q cavity has a high power loss rate. These results are easily generalized to include loss effects other than output coupling.

5.10 OBTAINING SINGLE-MODE OSCILLATION

In our discussion of laser theory we have thus far ignored any variations of the cavity intensity transverse to the cavity axis (in the xy plane). We will extend the theory beyond this approximation in Chapter 7, where we discuss the cavity modes of actual laser resonators.

Our assumption of single-mode oscillation in this chapter, therefore, has really been the assumption of a single *longitudinal* mode, whose frequency is given by (5.9.1): $\nu_m = mc/2L$. For many applications (e.g., holography) single-mode oscillation is highly desirable, and we now consider ways of obtaining it.

---

[4]A wide variety of oscillatory systems with damping can be characterized by a Q factor. Q was first introduced around 1920 by K. S. Johnson to characterize induction coils. Writing in *American Scientist* in 1955, E. I. Green noted that "[Johnson's] reason for choosing Q was quite simple. He says that it did not stand for 'quality factor' or anything else, but since the other letters of the alphabet had already been pre-empted for other purposes, Q was all he had left."

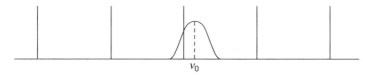

Figure 5.11   A case in which the cavity mode spacing $c/2L$ is larger than the width of the gain profile. Only a single longitudinal mode can lase.

The number of possible lasing modes can be estimated by counting the number of cavity modes, separated in frequency by $\Delta \nu = c/2L$, that lie within the gain bandwidth $\delta \nu_g$. This number is on the order of $\delta \nu_g / \Delta \nu$ for $\delta \nu_g > \Delta \nu$. If $\Delta \nu > \delta \nu_g$, however, the cavity mode frequencies are separated by more than the width of the gain profile, and at most only a single cavity frequency can have gain and lase (Fig. 5.11). In other words, *we can obtain single-mode oscillation by making the cavity short enough*. For a He–Ne laser with gain bandwidth $\delta \nu_g = \delta \nu_D \approx 1500$ MHz, for example, we require

$$\frac{c}{2L} = \frac{3 \times 10^{10} \text{ cm s}^{-1}}{2L} > 1.5 \times 10^9 \text{s}^{-1}, \tag{5.10.1}$$

or $L < 10$ cm. A disadvantage of this way of achieving single-mode oscillation is thus evident: It requires a small gain cell and therefore typically results in low output power.

The linewidths of liquid- and solid-laser transitions are usually much larger than in gases, 100 GHz being a typical order of magnitude. In such lasers the short-cavity approach to single-mode oscillation is not generally practical except in the case of diode lasers (Chapter 15).

Actually the gain-clamping condition (5.2.3) leads to a surprising conclusion: All lasers operating on a homogeneously broadened transition should oscillate on only a single mode, that for which the small-signal gain is greatest. This is illustrated in Fig. 5.12. In Fig. 5.12*a* we show a small-signal gain profile broad enough to allow five cavity modes to be above threshold. The *saturated* gain in steady-state oscillation must equal the threshold gain g_t (i.e., gain equals loss) according to the gain-clamping condition. The saturated-gain profile is shown in Fig. 5.12*b*. The saturated gain of one of

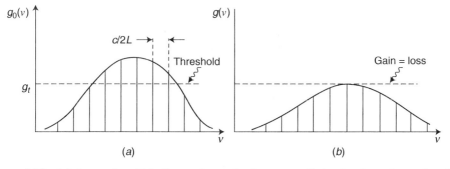

Figure 5.12   (*a*) A case in which five cavity modes have a small-signal gain g_0 larger than the threshold g_t for laser oscillation. (*b*) If the gain saturates homogeneously, only the mode with the largest small-signal gain is expected to lase. The others are saturated below the gain g_t necessary for laser oscillation.

the cavity modes—that with the largest small-signal gain—equals the threshold gain g_t. But the laser oscillation on this mode has saturated the gain to the extent that all other modes have gains *below* the threshold value; these other modes therefore cannot lase. Only the one mode can lase (Problem 5.7).

Many lasers with homogeneously broadened gain media do indeed oscillate on a single longitudinal mode. However, our argument for this assumes the validity of the gain-clamping condition (5.2.3) and ignores spatial hole burning. When spatial hole burning is present, the atoms of the gain medium are saturated to different degrees, depending on where they are within the standing-wave field, as indicated by Eq. (5.4.1). In this case the argument for single-mode oscillation illustrated in Fig. 5.12 does not apply. Multimode oscillation in a homogeneously broadened medium is therefore permitted.

In gas lasers for which the pressure is large enough to make the gain medium predominantly homogeneously (collision-) broadened, however, the output tends to be single-mode. In this case the effect of spatial hole burning is largely mitigated by atomic motion.

The argument illustrated in Fig. 5.12 is also inapplicable to inhomogeneously broadened laser media. In this case the presence of *spectral* hole burning means that different atoms saturate differently. The gain does not saturate to the same degree (i.e., homogeneously) for different spectral packets of atoms, in contrast to the homogeneous case, where there is in fact only a single spectral packet. One cavity mode might burn a deep hole in a particular spectral packet, without at all saturating other spectral packets on which other modes may lase. The He–Ne laser, for example, generally oscillates multimode.

Spatial and spectral hole burning thus invalidate the argument for single-mode oscillation illustrated in Fig. 5.12. If spatial hole burning is negligible, single-mode oscillation can be expected for a homogeneously broadened medium. Since spectral hole burning will always be present to some degree in an inhomogeneously broadened medium, however, we generally expect multimode oscillation in this case. These expectations are borne out experimentally.

This leads us to consider methods other than the short-cavity approach for achieving single-mode oscillation. These other methods have one feature in common: An additional loss mechanism is introduced to discriminate against all possible laser modes but one. That is, a situation is created in which all modes but one have a gain less than their loss. Note that this is just an extended application of the open-cavity principle, which is basic for optical-wave oscillation. An open cavity is extremely lossy except for axial waves. What we want now is a way to make the losses very large for most of these axial waves as well.

One important way to do this is to use a *Fabry-Pérot etalon* (Fig. 5.24). Consider the situation illustrated in Fig. 5.13a. An incident plane monochromatic wave is normally incident from a medium of refractive index n' onto a slab of material of index n and thickness d. Reflection and transmission occur at both interfaces. If $n > n'$, the reflected wave at the first interface is shifted in phase by π radians from the incident wave. This is a simple consequence of the Fresnel formulas for reflection and refraction (Section 5.12).

Now the Fabry-Pérot etalon can also produce a reflected wave as a result of reflection off the back face of the etalon and transmission again across the first interface (Fig. 5.13b). If $n > n'$, there is no phase change associated with reflection from the

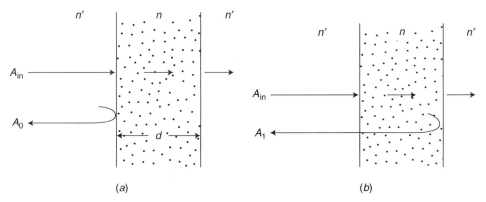

Figure 5.13   (*a*) A plane wave incident normally on a Fabry-Pérot etalon and the zeroth contribution to the reflected wave. (*b*) Another contribution (A_1) to the reflected wave in a Fabry-Pérot etalon.

internal back face of the etalon. The phase of this wave reflected from the etalon is thus shifted from that of the incident wave only because of propagation, that is, because it makes a round trip inside the etalon. This phase change is simply 2π times the optical length of propagation in wavelengths:

$$2\pi \frac{n(d+d)}{\lambda} = \frac{4\pi n d}{c} \nu. \tag{5.10.2}$$

The phase difference between the two reflected waves shown in Fig. 5.13 is thus

$$\frac{4\pi n d}{c}\nu - \pi = 2\pi\left(\frac{2\nu n d}{c} - \frac{1}{2}\right). \tag{5.10.3}$$

These two reflected waves will interfere destructively if their phase difference is equal to an odd integral multiple of π radians, that is, if

$$2\pi\left(\frac{2\nu n d}{c} - \frac{1}{2}\right) = (2m+1)\pi = 2\pi\left(m + \frac{1}{2}\right), \qquad m = 0, 1, 2,\ldots. \tag{5.10.4}$$

In other words, destructive interference occurs for frequencies

$$\nu_m = m\frac{c}{2nd}, \qquad m = 1, 2, 3,\ldots. \tag{5.10.5}$$

These are the "resonance frequencies" of the Fabry-Pérot etalon in the sense that they undergo minimal reflection and therefore maximal transmission. Actually, we should include the effect of multiple reflections within the etalon. As shown in the Appendix to this chapter, this leads to the same resonance frequencies (5.10.5) for maximal transmission. Furthermore, we can easily generalize to the case of an arbitrary angle of incidence (Problem 5.8), with the result that the resonance frequencies become

$$\nu_m = m\left(\frac{c}{2nd\cos\theta}\right), \qquad m = 1, 2, 3,\ldots. \tag{5.10.6}$$

If d is small enough, the spacing

$$\Delta v = v_{m+1} - v_m = \frac{c}{2nd\cos\theta} \tag{5.10.7}$$

between adjacent resonance frequencies of the etalon will be large compared to the width δv_g of the gain profile. By adjusting θ, a resonance frequency can be brought near the center of the gain profile, while the next resonance frequency lies outside the gain profile.[5]

The Fabry-Pérot etalon is widely used in spectroscopy. Because of its general importance in laser technology, we devote the Appendix to a more detailed discussion of its properties.

In referring to cavity modes in this section we have ignored polarization. A very common and convenient way of obtaining a linearly polarized output from a laser is to use *Brewster windows*. To understand this technique, it may be worthwhile to review briefly some results of electromagnetic theory.

For a plane wave incident upon a plane interface between two dielectric media, we define the *plane of incidence* as the plane formed by the propagation direction and a line perpendicular to the interface; this definition is unambiguous whenever the incident field is not normally incident. The polarization components parallel and perpendicular to the plane of incidence are referred to as p polarization and s polarization (s for the German word *senkrecht* for perpendicular), respectively, or alternatively as π polarization and σ polarization. For p polarization there is a particular angle of incidence θ_B, called *Brewster's angle*, for which there is no reflected wave (Section 5.12). If the wave is incident from a medium of index $n' \approx 1$ onto a medium of index n, the Brewster angle is given by

$$\theta_B = \tan^{-1} n. \tag{5.10.8}$$

For an air-to-glass interface ($n \sim 1.5$), θ_B is about $56°$.

Now if a plane wave of mixed polarization is incident at the angle θ_B, we can regard it as composed of p and s components. The p component will not be reflected. If a wave of arbitrary polarization is incident at Brewster's angle, the reflected field will therefore be completely polarized perpendicular to the plane of incidence. This is in fact a way of producing polarized light. Furthermore, the reflected wave will be partially polarized even if the angle is close but not quite equal to θ_B. This explains the success of Polaroid sunglasses in reducing the glare of reflected sunlight.

Now we can understand the use of *Brewster-angle windows* for obtaining linearly polarized laser radiation. Figure 5.14 illustrates a laser in which the ends of the gain cell are cut at the Brewster angle with respect to the cavity axis. The plane of incidence associated with the cavity field is obviously just the plane of the figure. Laser radiation that is linearly polarized in this plane will not suffer any reflection off the ends of the gain cell. Radiation polarized perpendicular to the plane of the figure, however, will have a greater loss coefficient because it is reflected at the windows. Lasing is therefore more favorable to linear polarization in the plane of incidence, as indicated in Fig. 5.14.

---

[5]In practice the tilt angle θ must not be too small, or else the etalon modifies the cavity frequencies and can no longer be regarded as a simple frequency filter.

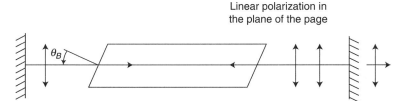

Figure 5.14   A laser with the gain-cell windows cut at the Brewster angle. The indicated polarization (parallel to the plane of incidence, which is the plane of the figure) will suffer no reflective loss at the windows, and therefore will lase preferentially. The orthogonal polarization will have a greater loss due to reflection at the windows.

5.11   THE LASER LINEWIDTH

The spectral width of laser radiation is in general different from the bare-cavity bandwidth. In steady-state laser oscillation the decay of the field amplitude described by (5.9.19) is exactly balanced by amplification due to stimulated emission. That is, the field amplitude does not actually decay. One might, therefore, expect that laser radiation should have a very narrow spectral width, and no natural minimum value for the width is readily apparent. Nevertheless, laser radiation, even from a cw single-mode laser, can never be made perfectly monochromatic.

The fundamental reason for this is spontaneous emission. An excited atom in the gain medium can drop spontaneously to the lower laser level, rather than be stimulated to do so by the cavity field. Whereas stimulated emission adds *coherently* to the stimulating field, that is, with a definite phase relationship, the spontaneously emitted radiation bears no phase relation to the cavity field. It adds incoherently to the cavity field. And the spontaneously emitted radiation has an inherent, Lorentzian distribution of frequencies (Section 3.11). Spontaneous emission, therefore, sets a fundamental lower limit on the laser linewidth. The actual laser linewidth will generally be much greater than this lower limit as a consequence of other causes of spectral broadening that could in principle be eliminated.

A proper treatment of spontaneous emission requires the quantum theory of radiation. Therefore, the problem of determining the fundamental lower limit to the spectral width of a laser can be solved rigorously only by using quantum electromagnetic theory. However, it is possible to give an argument that leads to the same answer given by the quantum theory of radiation.[6]

The argument is based on the heuristic energy–time uncertainty relation

$$\Delta E \, \Delta t > \hbar, \tag{5.11.1}$$

where we have multiplied the well-known Fourier result for root-mean-square deviations, $\Delta\omega \, \Delta t \geq 1$, by $\hbar$ on both sides, and defined ΔE from the Einstein relation $E = \hbar\omega$. Here ΔE is the uncertainty in the energy measured in a time interval Δt. If

---

[6]Two different derivations, which include the effects of the K and α parameters appearing in Eq. (5.11.16), are presented in Section 15.4.

an energy measurement process requires a time Δt, there is a lower limit $\Delta E = \hbar/\Delta t$ to the precision with which the energy can be determined.

In terms of the number of cavity photons q_v, the energy E in a lasing cavity mode is $E = h v q_v$, so that E may have uncertainties arising from both v and q_v:

$$\Delta E = h v \, \Delta q_v + h q_v \, \Delta v. \qquad (5.11.2)$$

Now the condition that the gain balances the loss in steady-state oscillation suggests that the amplitude of the cavity field should have a small relative uncertainty compared to the frequency, that is,

$$\frac{|\Delta q_v|}{q_v} \ll \frac{|\Delta v|}{v}. \qquad (5.11.3)$$

In other words, we suspect that the field amplitude should be relatively stable compared to the frequency; any small fluctuation of q_v from its steady-state value $\bar{q}_v$ should quickly relax to zero. Therefore, we assume

$$\Delta E \approx h \bar{q}_v \, \Delta v, \qquad (5.11.4)$$

or, using (5.11.1),

$$\Delta v \geq \frac{1}{2 \pi \bar{q}_v \, \Delta t}. \qquad (5.11.5)$$

The cavity-mode energy is determined both by stimulated emission and spontaneous emission into the cavity mode. The uncertainty ΔE, we argue, is due to spontaneous emission, which has not been accounted for in our laser intensity equations up to now. It is reasonable to assume in this heuristic approach that any sort of measurement of E requires a time interval Δt no larger than the spontaneous emission lifetime, i.e., $1/\Delta t$ must be at least as large as the spontaneous emission rate into the lasing cavity mode. Thus, we write (5.11.5) as

$$\Delta v \geq \frac{C}{2 \pi \bar{q}_v}, \qquad (5.11.6)$$

where C is the rate at which photons are emitted spontaneously into the single cavity mode.

Now we use the fact (recall Section 3.7) that the rate of spontaneous emission into any one field mode is equal to the rate of stimulated emission when there is one photon already in that mode. Therefore,

$$C = \text{(stimulated emission rate for one atom when } q_v = 1)$$
$$\times \text{(number of atoms in upper level)}. \qquad (5.11.7)$$

The first factor in parentheses is just $c \sigma(v)/V$ [recall (4.6.2) and (4.6.5) with $n_2 = q_v = 1$]. The second factor is $N_2 V_g$, where N_2 is the density of upper-state atoms and V_g is the gain volume. Therefore (5.11.7) is equivalent to

$$C = c\sigma(\nu)N_2 \frac{V_g}{V} = c\sigma(\nu)\frac{l}{L}\bar{N}_2 \qquad (5.11.8)$$

in steady-state oscillation, whence it follows from (5.11.6) that

$$\Delta\nu \geq \frac{c\sigma(\nu)\bar{N}_2 l/L}{2\pi\bar{q}_\nu}. \qquad (5.11.9)$$

This is our expression for the minimum possible linewidth of laser radiation.

It is convenient to rewrite this using

$$c\sigma(\nu) = \frac{cg(\nu)}{N_2 - N_1} = \frac{cg_t}{\Delta N_t}, \qquad (5.11.10)$$

where the second equality follows from the steady-state gain-clamping condition. As in Section 4.3, ΔN_t represents the threshold population inversion for laser oscillation. Furthermore $\bar{q}_\nu$ can be related to the output power $(\text{Pwr})_{\text{out}}$ of the laser, which is just $h\nu\bar{q}_\nu$ times the rate $cg_t l/L$ [cf. Eq. (4.5.5c)] at which photons are removed from the cavity, assuming that internal cavity losses are negligible:

$$\bar{q}_\nu = \frac{(\text{Pwr})_{\text{out}}}{h\nu c g_t}\frac{L}{l}. \qquad (5.11.11)$$

Using (5.11.10) and (5.11.11) in (5.11.9), therefore, we have

$$\Delta\nu \geq \frac{h\nu\bar{N}_2}{2\pi\,\Delta N_t}\left(\frac{cg_t l}{L}\right)^2 \frac{1}{(\text{Pwr})_{\text{out}}}. \qquad (5.11.12)$$

Finally, we use (5.9.9) to write cg_t in terms of the cavity bandwidth $\delta\nu_c$:

$$\Delta\nu \geq \frac{\bar{N}_2}{\Delta N_t}\frac{8\pi h\nu(\delta\nu_c)^2}{(\text{Pwr})_{\text{out}}} \equiv \Delta\nu_{\text{ST}}. \qquad (5.11.13)$$

A similar result was first obtained by Schawlow and Townes.[7] Note that $\Delta\nu_{\text{ST}}$ is the theoretical lower limit for the laser linewidth; the spectrum of single-mode laser radiation in this limit is predicted to be a Lorentzian with FWHM $\Delta\nu_{\text{ST}}$.

Consider, for example, a 632.8-nm He–Ne laser with mirror reflectivities $r_1 \approx 1.0$, $r_2 \approx 0.97$, a mirror separation of 30 cm, and an output power of 1 mW. Assuming that $\bar{N}_2/\Delta N_t$ is on the order of unity, we obtain a theoretical lower limit of 0.01 Hz for the linewidth. To observe such a tiny spectral width at 632.8 nm would require a cavity length that is fixed to an accuracy of $\Delta L = L\,\Delta\nu_{\text{ST}}/\nu \approx 6 \times 10^{-16}$ cm! It is nevertheless possible in some circumstances, when $\Delta\nu_{\text{ST}}$ is large enough and when frequency stabilization is employed, to approach a laser linewidth of $\Delta\nu_{\text{ST}}$.

[7]A. L. Schawlow and C. H. Townes, *Physical Review* **112**, 1940 (1958). This article is widely recognized as containing the first analysis of the conditions necessary to achieve laser oscillation at optical frequencies.

Examples like the one just given suggest that in most lasers the fundamental linewidth $\Delta\nu_{ST}$ is too small to be of practical interest, as the actual linewidth will be dominated by "technical noise" associated with mechanical vibrations and temperature variations that cause the cavity length to fluctuate. In semiconductor lasers, however, $\Delta\nu_{ST}$ is in fact of practical interest for certain applications, mainly because they have very small lengths compared to other lasers and because they often have relatively small mirror reflectivities; these characteristics imply relatively large cavity bandwidths (Chapter 15) and Schawlow–Townes linewidths that can easily dominate technical noise.

The Schawlow–Townes linewidth $\Delta\nu_{ST}$ is based on various approximations, one of which is the assumption that the laser output coupling is small, that is, that the mirror reflectivities are near unity. More generally the quantum lower limit to the laser linewidth can be written as

$$\Delta\nu = K\,\Delta\nu_{ST}, \tag{5.11.14}$$

where

$$K = \left[\frac{(\sqrt{r_1} + \sqrt{r_2})(1 - \sqrt{r_1 r_2})}{\sqrt{r_1 r_2}\,\ln(r_1 r_2)}\right]^2. \tag{5.11.15}$$

The factor K, which is ascribed to "excess spontaneous emission noise," approaches unity as the mirror reflectivities r_1, $r_2 \rightarrow 1$, and is typically between 1 and 2 for "stable" resonators of the type we have been assuming. For unstable resonators, however, K can be much greater than 1.[8]

Another approximation made in the derivation (5.11.13) is that intensity fluctuations, and in particular the coupling between intensity and phase fluctuations, can be ignored. The correction factor accounting for this coupling is written as $1 + \alpha^2$ and leads, together with the correction for excess spontaneous emission noise, to the expression

$$\Delta\nu = K(1 + \alpha^2)\,\Delta\nu_{ST} \tag{5.11.16}$$

for the fundamental laser linewidth. In semiconductor lasers $\alpha \approx 5 - 6$ is not unusual, so that the correction to the Schawlow–Townes linewidth associated with this parameter can be large. In general, $\Delta\nu_{ST}$ itself differs from (5.11.13) when internal cavity losses and dispersion are accounted for. Corrections to (5.11.13) are derived in Section 15.4.

•   Laser radiation typically has a linewidth much smaller than the "natural" linewidth due to radiative broadening of the lasing transition (Section 3.11). In the case of a maser ("microwave amplification by stimulated emission of radiation"), similarly, the linewidth can be much smaller than the "natural" width $1/t$ associated with the transit time of molecules in the microwave cavity. The fact that the radiation from a laser or maser could have a linewidth smaller than the "natural" linewidth was not immediately obvious. In light of our simple derivation of the Schawlow–Townes linewidth, it is interesting to note Charles Townes's recollection that[9]

> ... there was the uncertainty principle relating time and energy, a basic law for physicists. With lifetime t of molecules in the cavity limited (for the beam-type maser) by the time of transit, how could

---

[8]The stability of laser resonators is discussed in Chapter 7.
[9]C. H. Townes, *Making Waves*, AIP Press, New York, 1995, p. 27. The frequency width $1/t$ referred to here by Townes is due to "transit-time broadening," which is discussed in Section 9.11.

there be a frequency width much smaller than $1/t$? An electrical engineer accustomed to the almost monochromatic oscillation produced by an electron tube with positive feedback would perhaps not have given the problem a second thought. However, before oscillation was achieved I never succeeded in convincing two of my Columbia University colleagues, even after long discussion, that the frequency width could be very narrow. One insisted on betting me a bottle of Scotch that it would not. After successful oscillation, I remember interesting discussions on this point with Niels Bohr and [John] von Neumann. Each immediately questioned how such a narrow frequency could be allowed by the uncertainty principle. I was never sure that Bohr's immediate acceptance of my explanation based on a collection of molecules rather than a single one was because he was convinced, or was due simply to his kindness to a young scientist.   •

5.12 POLARIZATION AND MODULATION

The polarization of light plays a crucial role in the design of many optical components used in laser technology. We have already seen examples of this in our discussion of Faraday isolators (Section 5.6) and Brewster windows (Section 5.10), and in this section we will see how polarization can be used to modulate laser radiation. As discussed in the following section, such modulation is employed in one of the most effective methods of stabilizing the frequency of laser radiation. We begin by reviewing some basic aspects of polarization.

For a monochromatic plane wave propagating in the z direction in a medium of refractive index n we can write

$$\mathbf{E}(z,\,t) = \hat{x}E_x\,\cos(\omega t - kz) + \hat{y}E_y\,\cos(\omega t - kz + \phi), \qquad (5.12.1)$$

where $k = n\omega/c$, and the unit vectors $\hat{x}$ and $\hat{y}$ are orthogonal to each other and to the unit vector $\hat{z}$ pointing in the z direction ($\hat{x} \times \hat{y} = \hat{z}$). If $E_x = 0$ or $E_y = 0$ we have linear (or "plane") polarization. If $E_x = E_y = E_0$ and $\phi = \pm\pi/2$ we have circular polarization:

$$\mathbf{E}(z,\,t) = \frac{1}{\sqrt{2}}E_0[\hat{x}\,\cos(\omega t - kz) \mp \hat{y}\,\sin(\omega t - kz)]. \qquad (5.12.2)$$

The x and y components of this electric field in a plane of constant z trace out a circle, as indicated in Fig. 5.15. The field is said to be right-hand circularly polarized if $\mathbf{E}$ rotates

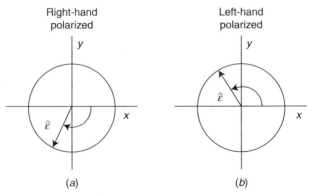

Figure 5.15   The x and y components of a circularly polarized field propagating in the z direction trace out a circle: (a) right-hand circular polarization; (b) left-hand circular polarization.

clockwise for an observer viewing the incoming light, and left-hand circularly polarized if **E** rotates counterclockwise.[10] Right- and left-hand circular polarization correspond to the $-$ and $+$ signs, respectively, in Eq. (5.12.2).

Linear and circular polarization are special cases of the general elliptical polarization described by the electric field (5.12.1). If $|E_x| \neq |E_y|$ but $\phi = \pm\pi/2$, $\mathbf{E} = \hat{x}E_x \times \cos(\omega t - kz) \mp \hat{y}E_y \sin(\omega t - kz)$ traces out an ellipse with major and minor axes parallel to $\hat{x}$ and $\hat{y}$; for arbitrary ϕ the major and minor axes are not parallel to $\hat{x}$ and $\hat{y}$.

The light from natural sources is generally unpolarized. The resultant electric field **E** from all the individual radiators of the source points in a single direction at a given point and at a given instant, but it varies so rapidly that as a practical matter it appears directionless. Linearly polarized light can be obtained by scattering, reflection, or transmission (Section 5.10), or by using polarizers consisting of "dichroic" materials that transmit light polarized in one direction but absorb light polarized in the perpendicular direction. If a wave linearly polarized in some direction $\hat{\boldsymbol{\epsilon}}$ is normally incident on a polarizer in which the transmitting direction is $\hat{x}$, the transmitted wave will be linearly polarized in the x direction and will be diminished in amplitude by the factor $\hat{\boldsymbol{\epsilon}} \cdot \hat{x}$. The intensity is therefore diminished by $(\hat{\boldsymbol{\epsilon}} \cdot \hat{x})^2 = \cos^2\theta$; this is the familiar Malus law. Similarly, unpolarized light incident on the polarizer will be reduced in intensity by 50%.

Circular polarization may be obtained using a quarter-wave plate consisting of a *birefringent* material, that is, a material in which the refractive index is different for different polarizations. All transparent crystals with noncubic lattice structure, such as ice and sugar, are birefringent. Such behavior can be understood using the electron oscillator model (Chapter 3), assuming the "spring constant" is different for different directions of the electron displacement. The model then predicts different refractive indices for different polarizations of the field propagating in the medium. It is easy to understand how such anisotropy might arise. In particular, imagine a long, rod-shaped molecule in which the elastic restoring force is different for electron displacements parallel and perpendicular to the axis. There will be different refractive indices for different directions of polarization whenever there is some degree of molecular alignment.

Consider a birefringent material in which the molecules are aligned along the y axis, as illustrated in Fig. 5.16. This preferred direction may be called the *optic axis*. Waves that are linearly polarized in the x or y directions will propagate with different refractive

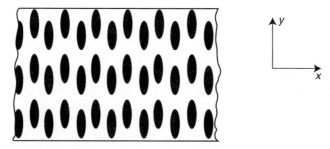

Figure 5.16   A birefringent material in which the molecules are aligned along the y axis.

---

[10]This is the convention in optics. It is unnatural in the sense that the right-hand rule applied to the wave vector $\mathbf{k} = k\hat{z}$ would suggest the opposite.

indices n_x or n_y. That is, a field

$$\mathbf{E}(z, t) = \hat{\boldsymbol{\varepsilon}}E_0 \cos(\omega t - kz) = \hat{\boldsymbol{\varepsilon}}E_0 \cos \omega\left(t - \frac{nz}{c}\right) \tag{5.12.3}$$

in the medium will propagate with refractive index $n = n_x$ if $\hat{\boldsymbol{\varepsilon}} = \hat{x}$, or $n = n_y$ if $\hat{\boldsymbol{\varepsilon}} = \hat{y}$. If the linearly polarized field

$$\mathbf{E}(0, t) = \varepsilon_x\hat{x}E_0 \cos \omega t + \varepsilon_y\hat{y}E_0 \cos \omega t \quad (\varepsilon_x^2 + \varepsilon_y^2 = 1) \tag{5.12.4}$$

is normally incident at the face $z = 0$ of the material, the field at $z = l$ in the material will be

$$\mathbf{E}(l, t) = \varepsilon_x\hat{x}E_0 \cos \omega\left(t - \frac{n_x l}{c}\right) + \varepsilon_y\hat{y}E_0 \cos \omega\left(t - \frac{n_y l}{c}\right). \tag{5.12.5}$$

Thus, the x and y components of the field, since they propagate with different phase velocities, develop a phase difference

$$\phi(l) = \frac{\omega l}{c}(n_x - n_y). \tag{5.12.6}$$

If the material has length l, the field at $z \geq l$ is

$$\mathbf{E}(z, t) = \varepsilon_x\hat{x}E_0 \cos[\omega t - k_x z] + \varepsilon_y\hat{y}E_0 \cos[\omega t - k_x z + \phi(l)], \tag{5.12.7}$$

where $k_x = n_x\omega/c$ and $\phi(l)$ [Eq. (5.12.6)] is the phase difference of the x and y components after a propagation distance l. The phase velocities for x and y polarizations are c/n_x and c/n_y, and the larger of these phase velocities defines the *fast axis* of the birefringent material.[11]

The field (5.12.7) has the form (5.12.1) with $k = k_x$ and $\phi = \phi(l)$. Thus, if $|\varepsilon_x| = |\varepsilon_y|$ and $\phi(l) = \pm \pi/2$, an incident linearly polarized field will emerge from the material of length l as a circularly polarized field. The first condition can be realized by orienting the fast axis at 45° with respect to the direction of linear polarization of the incident field. The second condition is satisfied when the length l is such that the optical path difference $l|n_x - n_y|$ is a quarter of the wavelength under consideration, that is, when we have a quarter-wave plate (QWP). When this condition is satisfied, an incident *elliptically* polarized field will result in a linearly polarized output field.

Consider, for example, the situation shown in Fig. 5.17. Unpolarized light is incident on a polarizing beam splitter (see the black-dot section below) that transmits linearly polarized light. This light is then incident on a QWP with its fast axis at 45° with respect to the polarization of the field. The circularly polarized light emerging from the QWP is then reflected off a mirror, which results in a field coming back through the QWP with linear polarization orthogonal to that of the field incident on the QWP from the polarizing beam splitter (Problem 5.9). This field is therefore polarized for maximum *reflection* from the polarizing beam splitter. Thus, if the QWP is aligned with its fast axis at 45° to the linear polarization of the field incident on it, the intensity at P_2 is maximized. This

---

[11]The propagation of light in birefringent media is discussed in more detail in Chapter 8.

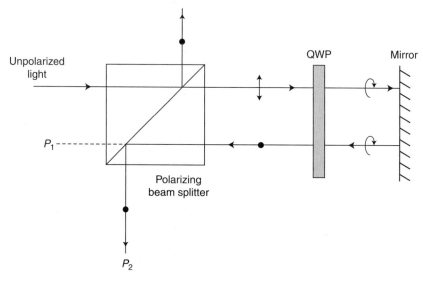

Figure 5.17   Arrangment for aligning a quarter-wave plate. When the fast axis of the QWP is at 45°
with respect to the polarization of the light incident upon it, the light intensity at point P_2 is maximized,
while that at P_1 is minimized.

arrangement is useful not only for aligning a QWP, but also in certain applications as an
optical isolator for "picking off" a reflected field (Section 5.13).

• The reflection and transmission of light at the interface between two dielectrics with refractive indices n_1 and n_2 are described by the Fresnel formulas.[12] These formulas are used in the
design of beam splitters and many other optical components.

If θ_i and θ_t are the angles of incidence and transmission, respectively, and the incident field is
linearly polarized parallel to the plane of incidence (Section 5.10), the Fresnel formulas relating
the reflected (r), transmitted (t), and incident (i) electric field amplitudes are (Fig. 5.18)

$$\frac{E_r^{\parallel}}{E_i^{\parallel}} = \frac{n_2 \cos \theta_i - n_1 \cos \theta_t}{n_2 \cos \theta_i + n_1 \cos \theta_t} = \frac{\tan(\theta_i - \theta_t)}{\tan(\theta_i + \theta_t)}, \tag{5.12.8a}$$

$$\frac{E_t^{\parallel}}{E_i^{\parallel}} = \frac{2n_1 \cos \theta_i}{n_2 \cos \theta_i + n_1 \cos \theta_t} = \frac{2 \cos \theta_i \sin \theta_t}{\sin(\theta_i + \theta_t) \cos(\theta_i - \theta_t)}. \tag{5.12.8b}$$

If the incident field is polarized perpendicular to the plane of incidence, then

$$\frac{E_r^{\perp}}{E_i^{\perp}} = \frac{n_1 \cos \theta_i - n_2 \cos \theta_t}{n_1 \cos \theta_i + n_2 \cos \theta_t} = -\frac{\sin(\theta_i - \theta_t)}{\sin(\theta_i + \theta_t)}, \tag{5.12.9a}$$

$$\frac{E_t^{\perp}}{E_i^{\perp}} = \frac{2n_1 \cos \theta_i}{n_1 \cos \theta_i + n_2 \cos \theta_t} = \frac{2 \cos \theta_i \sin \theta_t}{\sin(\theta_i + \theta_t)}. \tag{5.12.9b}$$

[12]See, for instance, M. Born and E. Wolf, *Principles of Optics*, 7th ed., Cambridge University Press,
Cambridge, 1999, Section 1.5.2, or J. D. Jackson, *Classical Electrodynamics*, 3rd ed., Wiley, New York,
1999, Section 7.3.

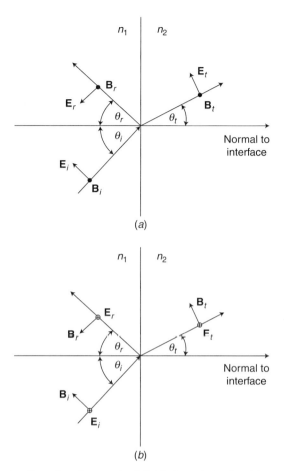

Figure 5.18   Incident, reflected, and transmitted fields for polarization (*a*) parallel and (*b*) perpendicular to the plane of incidence.

The angles θ_i and θ_t are related by Snell's law, $n_1 \sin \theta_i = n_2 \sin \theta_t$, which has been used in writing the second equality in each of these equations.[13] The angle of reflection, θ_r, is equal to the angle θ_i of incidence.

The Fresnel formula (5.12.8a) implies that polarized light can be obtained by reflection: if $\theta_i + \theta_t = \pi/2$, $E_r^{\parallel} = 0$ and the reflected field can only be *s* polarized. Using Snell's law, it follows that this condition is satisfied when $n_1 \sin \theta_i = n_2 \sin(\pi/2 - \theta_i) = n_2 \cos \theta_i$, that is, when $\theta_i = \theta_B \equiv \tan^{-1}(n_2/n_1)$, the Brewster angle.

A common type of beam splitter is a glass plate coated with a dielectric (or metallic) film chosen so as to give the desired relative intensities (e.g., a 50/50 ratio) of the reflected and transmitted beams at the design wavelength and angle. An antireflection coating (Problem 5.10) is applied to the back surface to eliminate "ghost" images. As implied by the Fresnel formulas, dielectric coatings result in polarization-dependent reflection/transmission ratios; this effect can be reduced, if necessary, using polarizers or by specially designed multilayered dielectric coatings. Coatings can be designed such that a polarized incident beam undergoes a 50/50 split into two beams having the same polarization as the incident beam.

---

[13]It is assumed in writing these equations that the magnetic permeabilities μ_1 and μ_2 are equal. This is generally an excellent approximation for optical frequencies.

Polarizing beam splitters are used to separate a beam into two beams propagating in different directions with different polarizations. Figure 5.17 indicates a polarizing beam splitter consisting of two matched right-angle prisms with a dielectric coating on the hypotenuse face of one of the prisms and with multilayer dielectric coatings to reduce reflections ("Fresnel losses") at the input and exit faces. The matched prisms result in equal path lengths for the reflected and transmitted beams and, unlike plate beam splitters, do not produce a deflection of the transmitted beam with respect to the incident beam. Other types of polarizing beam splitters employ birefringent material to separate an incident beam into two orthogonally polarized beams.   •

Many isotropic media can be made birefringent by the application of an electric field. Consider, for instance, a liquid consisting of long molecules having permanent electric dipole moments. The existence of a permanent dipole moment implies a certain asymmetry, namely a preponderance of positive charge at one end of the molecule and negative charge at the other. Because of collisions, the molecules are randomly oriented and the liquid will be macroscopically isotropic and will not exhibit birefringence. An applied electric field, however, will tend to align the molecules, creating anisotropy and making the liquid birefringent. This creation of birefringence by an applied electric field is called the *electro-optic effect*. In the *Kerr electro-optic effect* the induced optic axis is parallel to the applied field, and the difference in refractive indices for light polarized parallel and perpendicular to the optic axis is proportional to the square of the applied field.

In certain crystals an applied electric field creates an optic axis perpendicular to the field, and the difference in refractive indices for light polarized parallel and perpendicular to the optic axis is linearly proportional to the applied field. This is called the *Pockels electro-optic effect*.

The electro-optic effect provides one means of modulating laser radiation. In the case of the Pockels effect an electric field $\mathcal{E}$ in the z direction results in refractive indices (see Problem 5.11)

$$n_x = n_0 + \tfrac{1}{2}P\mathcal{E}, \tag{5.12.10a}$$

$$n_y = n_0 - \tfrac{1}{2}P\mathcal{E}, \tag{5.12.10b}$$

for specific x and y components of a field propagating in the z direction. The constant P characterizes the Pockels cell material at the optical frequency ω of interest, and n_0 is the refractive index at this frequency in the absence of the applied electric field $\mathcal{E}$. Suppose the Pockels cell is between two "crossed" polarizers (Fig. 5.19). The field emerging from the cell has the form [cf. Eq. (5.12.7)]

$$\mathbf{E}(z, t) = \varepsilon_x \hat{x} E_0 \cos\left(\omega t - k_0 z - \tfrac{1}{2}\phi\right) + \varepsilon_y \hat{y} E_0 \cos\left(\omega t - k_0 z + \tfrac{1}{2}\phi\right), \tag{5.12.11}$$

where $k = n_0 \omega / c$, $\phi = \omega l P \mathcal{E} / c$, and l is the length of the Pockels cell. The field transmitted by the second polarizer has the magnitude

$$E_{\text{out}}(z, t) = \varepsilon_x (\hat{x} \cdot \hat{a}) E_0 \cos\left(\omega t - k_0 z - \tfrac{1}{2}\phi\right) + \varepsilon_y (\hat{y} \cdot \hat{a}) E_0 \cos\left(\omega t - k_0 z + \tfrac{1}{2}\phi\right), \tag{5.12.12}$$

where the unit vector $\hat{a}$ defines the polarization direction transmitted by the second polarizer. Let us assume that $\hat{a}$ is at 45° with respect to $\hat{x}$ and $\hat{y}$, so that

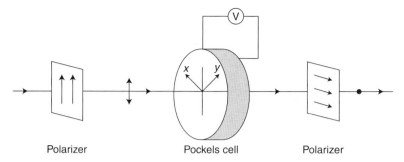

Figure 5.19 A voltage-controlled light modulator consisting of a Pockels cell and two crossed polarizers.

$\hat{x} \cdot \hat{a} = -\hat{y} \cdot \hat{a} = -1/\sqrt{2}$ and $\varepsilon_x = \varepsilon_y = 1/\sqrt{2}$ (Fig. 5.19). Then

$$E_{\text{out}}(z, t) = -\frac{1}{2}E_0[\cos(\omega t - k_0 z - \tfrac{1}{2}\phi)$$

$$-\cos(\omega t - k_0 z + \tfrac{1}{2}\phi)] = -E_0 \sin(\omega t - k_0 z) \sin\left(\frac{\phi}{2}\right) \qquad (5.12.13)$$

and

$$E_{\text{out}}^2(z, t) = E_0^2 \sin^2(\omega t - k_0 z) \sin^2\left(\frac{\phi}{2}\right). \qquad (5.12.14)$$

Averaging over a period $2\pi/\omega$ ($\approx 10^{-15}$ s) of the optical field gives

$$\overline{E_{\text{out}}^2}(z, t) = \tfrac{1}{2}E_0^2 \sin^2\left(\frac{\phi}{2}\right). \qquad (5.12.15)$$

The time-averaged square of the field incident on the Pockels cell is similarly $\tfrac{1}{2}E_0^2$, so that the ratio of the transmitted intensity and the intensity incident on the Pockels cell is

$$\frac{I_{\text{out}}}{I_{\text{in}}} = \sin^2\left(\frac{\phi}{2}\right) = \sin^2\left[\frac{\pi}{\lambda}PV\right], \qquad (5.12.16)$$

where we have expressed ϕ in terms of the wavelength λ and the voltage $V = \mathcal{E}\ell$ applied to the Pockels cell. Defining the *half-wave voltage* $V_0 = \lambda/2P$, we can write $\phi = \pi V/V_0$ and

$$\frac{I_{\text{out}}}{I_{\text{in}}} = \sin^2\left(\frac{\pi V}{2 V_0}\right). \qquad (5.12.17)$$

where $V = V_0$ implies that the Pockels cell produces a phase retardance of π, or half a wave. In this case maximum transmission is obtained. When $V = 2V_0$, the phase retardance is a full wave and the minimum transmission is obtained. An optical *chopper* in which the transmission varies in time between a minimum and a maximum can be made by varying the phase retardance (voltage) in time between full-wave and half-wave.

Crystals such as KDP (KH_2PO_4) or lithium niobate ($LiNO_3$) used to make Pockels cells typically have half-wave voltages in the kilovolt range. Kerr cells, in which the phase difference is proportional to the square of the voltage, have similar uses and typically require somewhat higher voltages than Pockels cells. Both Pockels and Kerr cells are very fast, with switching times (~ 1 ns or less) set practically by the speed with which the applied voltages can be varied. Liquid-crystal modulators require much smaller voltages but, because they involve the alignment of *molecules* by the applied field (rather than the motion of electrons), are fundamentally much slower, with switching times in the microsecond to millisecond range.

Equation (5.12.17) describes an amplitude modulation of a light wave. The electrooptic effect can also be used for *phase* modulation. Suppose, for instance, that a wave polarized in the x direction enters a Pockels cell in which the applied voltage varies sinusoidally such that $\phi = -2\beta \sin \Omega t$. Then [Eq. (5.12.11)]

$$\mathbf{E}(z, t) = \varepsilon_x \hat{x} E_0 \cos(\omega t - k_0 z + \beta \sin \Omega t). \tag{5.12.18}$$

It is convenient to write this in the complex form

$$\mathbf{E}(z, t) = \varepsilon_x \hat{x} E_0 e^{-i(\omega t - k_0 z)} e^{-i\beta \sin \Omega t} \tag{5.12.19}$$

where, as in Section 3.4, it is understood that we are to take the real part of the right-hand side. Assuming $\beta \ll 1$, we have

$$e^{-i\beta \sin \Omega t} = 1 - i\beta \sin \Omega t - \tfrac{1}{2}\beta^2 \sin^2 \Omega t + \cdots \approx 1 - i\beta \sin \Omega t$$
$$= 1 - \tfrac{1}{2}\beta \, (e^{i\Omega t} - e^{-i\Omega t}). \tag{5.12.20}$$

In this approximation

$$\mathbf{E}(z, t) = \varepsilon_x \hat{x} E_0 e^{-i(\omega t - k_0 z)} \big[1 - \tfrac{1}{2}\beta e^{i\Omega t} + \tfrac{1}{2}\beta e^{-i\Omega t} \big]$$
$$= \varepsilon_x \hat{x} E_0 \big[e^{-i\omega t} - \tfrac{1}{2}\beta e^{-i(\omega - \Omega)t} + \tfrac{1}{2}\beta e^{-i(\omega + \Omega)t} \big] e^{ik_0 z} \tag{5.12.21}$$

or, taking the real part of the right-hand side,

$$\mathbf{E}(z, t) = \varepsilon_x \hat{x} E_0 \big(\cos(\omega t - k_0 z) - \tfrac{1}{2}\beta \cos[(\omega - \Omega)t - k_0 z] + \tfrac{1}{2}\beta \cos[(\omega + \Omega)t - k_0 z] \big). \tag{5.12.22}$$

The (weak) sinusoidal modulation of the voltage across the Pockels cell, therefore, adds to the *carrier* wave of frequency ω two *sidebands* of frequency $\omega - \Omega$ and $\omega + \Omega$.

• More generally we can use the identity

$$e^{-i\beta \sin \Omega t} = J_0(\beta) - 2i \sum_{k=0}^{\infty} J_{2k+1}(\beta) \sin(2k + 1)\Omega t + 2 \sum_{k=1}^{\infty} J_{2k}(\beta) \cos 2k\Omega t, \tag{5.12.23}$$

where the J's are Bessel functions of the first kind. This implies sidebands at $\omega \pm m\,\Omega$, $m = 1$, 2, 3, ... For $\beta \ll 1$ the Bessel functions become increasingly small with increasing order: $J_0(\beta) \approx 1$, $J_1(\beta) \approx \beta/2$, $J_2(\beta) \approx \beta^2/8$, ... and, retaining only terms up to first order in β, we obtain (5.12.20). •

5.13 FREQUENCY STABILIZATION

The frequency at which a single-mode laser oscillates will fluctuate due to mirror vibrations and other sources of "noise." In applications requiring high-precision measurements, it is important that a laser oscillate at a well-defined and highly stable frequency. We have already mentioned how the Lamb dip can be used for frequency stabilization (Section 5.8).

Frequency stabilization, or frequency "locking," involves concepts such as error signals and feedback that are part of *control theory*. An example is provided by a thermostat that measures the temperature in a room and uses the "error signal," the difference between the measured temperature and the desired temperature, to turn a furnace on or off. In the case of Lamb-dip laser frequency stabilization, the error signal is the difference between the measured laser intensity and the intensity at the center of the Doppler line, which is a measure of the difference between the actual oscillation frequency and the frequency at which we want the laser to oscillate (Section 5.8).

There is another, widely used and very effective frequency stabilization method that makes use of a Fabry-Pérot etalon rather than an absorption resonance to produce an error signal. An advantage of the Fabry-Pérot is that it offers a wide range of possible resonance frequencies [recall Eq. (5.10.5)] to choose from, whereas Lamb-dip stabilization requires a saturable molecular transition close to the laser frequency. This wide range of possible "locking frequencies" is especially useful for the stabilization of tunable lasers.

When light from a laser is incident on a Fabry-Pérot etalon, the transmitted (or reflected) intensity can provide the error signal needed to stabilize the laser frequency. For instance, if the intensity reflected by the Fabry-Pérot is zero, the laser frequency must be a resonance frequency of the cavity. A deviation of the laser frequency from this cavity resonance will result in some reflected intensity that can serve as the error signal for a feedback loop that locks the laser frequency to the cavity resonance. An attractive feature of this approach is that the frequency locking is not affected by *intensity* fluctuations of the laser: as long as the reflected intensity is zero the laser frequency is locked to the cavity resonance.

Note, however, that the variation of the reflected (and transmitted) intensity is symmetric about a cavity resonance (cf. Fig. 5.23). This means that a nonzero reflected intensity (error signal) cannot tell us whether the laser frequency should be increased or decreased in order to bring it back to the locking frequency. The essence of the "Pound–Drever–Hall" frequency stabilization method is to get around this problem by producing an error signal that depends on the sign of the deviation of the laser frequency from the cavity resonance frequency. Such an error signal, which is obtained by modulating (or "dithering") the laser output to generate sideband frequencies, provides information as to whether the laser frequency is above or below the frequency we wish to lock it to.

Suppose that a Pockels cell is used to phase-modulate the laser radiation so that the electric field incident on the Fabry-Pérot is given approximately by [Eq. (5.12.21)]

$$E(t) = E_0 \left[e^{-i\omega t} - \tfrac{1}{2}\beta e^{-i(\omega-\Omega)t} + \tfrac{1}{2}\beta e^{-i(\omega+\Omega)t} \right]. \qquad (5.13.1)$$

The amplitude reflection coefficient of the Fabry-Pérot is defined by Eq. (5.A.12):

$$\mathcal{R}(\omega) \equiv \frac{A_R}{A_{\text{in}}} = \frac{e^{2i\omega nd/c} - 1}{1 - Re^{2i\omega nd/c}}\sqrt{R}, \tag{5.13.2}$$

and the reflected field is therefore

$$E_R(t) = E_0 \mathcal{R}(\omega)e^{-i\omega t} - \tfrac{1}{2}\beta E_0 \mathcal{R}(\omega - \Omega)e^{-i(\omega - \Omega)t} + \tfrac{1}{2}\beta E_0$$
$$\times \mathcal{R}(\omega + \Omega)e^{-i(\omega + \Omega)t}. \tag{5.13.3}$$

Let us assume for simplicity that the sidebands are sufficiently far from a cavity resonance that they are perfectly reflected, that is, for the sidebands, $R = 1$ and $\mathcal{R}(\omega \pm \Omega) \cong -1$. Then

$$E_R(t) \cong E_0 \mathcal{R}(\omega)e^{-i\omega t} + \tfrac{1}{2}\beta E_0 e^{-i(\omega - \Omega)t} - \tfrac{1}{2}\beta E_0 e^{-i(\omega + \Omega)t}. \tag{5.13.4}$$

The reflected intensity, $(c\epsilon_0/2)|E_R(t)|^2$, is then

$$I_R(t) \cong \tfrac{1}{2}c\epsilon_0 E_0^2\left[|\mathcal{R}(\omega)|^2 + \tfrac{1}{2}\beta^2 - \tfrac{1}{2}\beta^2 \cos 2\Omega t + 2\beta\, \text{Im}[\mathcal{R}(\omega)]\sin \Omega t\right]. \tag{5.13.5}$$

Since the (time-averaged) carrier and sideband intensities of the field incident on the Fabry-Pérot are $I_c = (c\epsilon_0/2)E_0^2$ and $I_s = (c\epsilon_0/2)\beta^2 E_0^2/4$, respectively, we can write the reflected intensity as

$$I_R(t) \cong |\mathcal{R}(\omega)|^2 I_c + 2I_s - 2I_s \cos 2\Omega t + 4\sqrt{I_c I_s}\, \text{Im}[\mathcal{R}(\omega)]\sin \Omega t, \tag{5.13.6}$$

and similarly, in our simplified model, for the reflected power:

$$(\text{Pwr})_R(t) \cong |\mathcal{R}(\omega)|^2(\text{Pwr})_c + 2(\text{Pwr})_s - 2(\text{Pwr})_s \cos 2\Omega t$$
$$+ 4\sqrt{(\text{Pwr})_c(\text{Pwr})_s}\, \text{Im}[\mathcal{R}(\omega)]\sin \Omega t. \tag{5.13.7}$$

It is the last term in the reflected power that is of interest in Pound–Drever–Hall stabilization. Consider $\text{Im}[\mathcal{R}(\omega)]$ when ω is close to a resonance of the etalon:

$$\omega = m\left(\frac{\pi c}{d}\right) + \delta\omega, \tag{5.13.8}$$

where we use Eq. (5.10.5) with $n = 1$ and assume that $|\delta\omega| \ll \pi c/d$. Then,

$$\text{Im}[\mathcal{R}(\omega)] = \text{Im}\left[\frac{(e^{2\pi im}e^{2i\delta\omega d/c} - 1)\sqrt{R}}{1 - Re^{2\pi im}e^{2i\delta\omega d/c}}\right] = \text{Im}\left[\frac{(e^{2i\delta\omega d/c} - 1)\sqrt{R}}{1 - Re^{2i\delta\omega d/c}}\right]$$
$$\cong \text{Im}\left[\frac{2i\delta\omega d/c}{1 - R}\right] = \frac{2d/c}{1 - R}\delta\omega \tag{5.13.9}$$

for $R \cong 1$, that is, for a high-finesse cavity (see the Appendix). Since it is proportional to $\delta\omega$, $\text{Im}[\mathcal{R}(\omega)]$ contains the information needed, for instance, for a servomechanism to

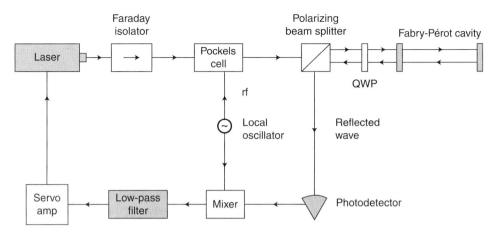

Figure 5.20 Simplified schematic for Pound–Drever–Hall frequency stabilization.

piezoelectrically adjust the spacing of the laser mirrors so as to increase (if $\delta\omega < 0$) or decrease (if $\delta\omega > 0$) the laser frequency to correct for the deviation $\delta\omega$ from the locking frequency. In terms of the frequency deviation $\delta\nu = \delta\omega/2\pi$ and the cavity bandwidth $\delta\nu_c$ [Eq. (5.9.22)] we have $\text{Im}[\mathcal{R}(\omega)] \simeq \delta\nu/\delta\nu_c$ and, therefore,

$$(\text{Pwr})_R(t) \cong |\mathcal{R}(\omega)|^2(\text{Pwr})_c + 2(\text{Pwr})_s - 2(\text{Pwr})_s \cos 2\Omega t$$

$$+ 4\sqrt{(\text{Pwr})_c(\text{Pwr})_s}\,\frac{\delta\nu}{\delta\nu_c}\sin\Omega t. \tag{5.13.10}$$

Figure 5.20 is a simplified schematic diagram for Pound–Drever–Hall frequency stabilization. The field from the laser passes through a Faraday isolator and is modulated by a Pockels cell driven by a radio-frequency (rf) "local oscillator." The resulting carrier and sideband waves are incident on a Fabry–Pérot cavity, and a polarizing beam splitter as described earlier (Fig. 5.17) directs the reflected field onto a photodetector. The output of the photodetector is combined with a portion of the rf modulation signal in a "mixer," a nonlinear device whose output is the product of two inputs. The part of the mixer output that is of interest comes from the product of the local oscillator field and the last term in (5.13.10); this part has a dc component arising from the product $\sin\Omega t\sin\Omega t = (\frac{1}{2})(1 - \cos\Omega t)$. The remaining terms in (5.13.10) lead to sinusoidal components at frequencies Ω and 3Ω in the output of the mixer. A low-pass filter picks out the dc component, which provides the error signal (voltage) for the "servo" amplifier that controls the laser frequency. Ideally, the error signal is then just

$$\varepsilon_s = 4\sqrt{(\text{Pwr})_c(\text{Pwr})_s}\,\frac{\delta\nu}{\delta\nu_c}. \tag{5.13.11}$$

In obtaining (5.13.10) we have assumed that $\mathcal{R}(\omega \pm \Omega) \cong -1$ and retained only the first-order dependence on the frequency deviation $\delta\nu$. It is straightforward to go beyond these approximations and obtain

$$C_{\sin}(\omega) = 2\sqrt{(\text{Pwr})_c(\text{Pwr})_s}\,\text{Im}[\mathcal{R}^*(\omega)\mathcal{R}(\omega - \Omega) - \mathcal{R}(\omega)\mathcal{R}^*(\omega + \Omega)] \tag{5.13.12}$$

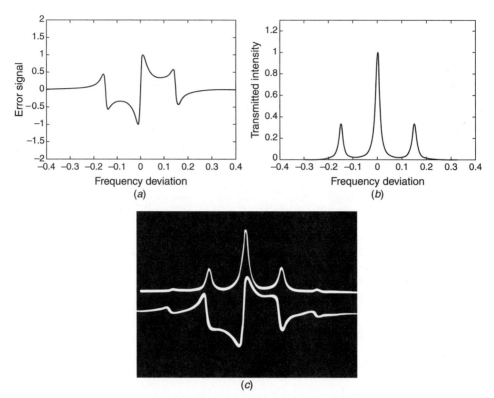

Figure 5.21 (a) $C_{\sin}(\omega)/\sqrt{(\mathrm{Pwr})_c(\mathrm{Pwr})_s}$ [Eq. (5.13.12)] for $R = 0.99$ and $\Omega = 0.15(c/2d)$. The frequency deviation $\delta\nu$ is in units of the free spectral range $\Delta\nu = c/2d$, of the Fabry-Pérot. (b) The normalized intensity transmitted by the Fabry-Pérot for $\beta = \sqrt{1/3}$. (c) Experimental results for the transmitted intensity (upper curve) and the error signal (lower curve). [From R. W. P. Drever, J. L. Hall, F. V. Kowalski, J. Hough, G. M. Ford, A. J. Munley, and H. Ward, *Applied Physics B* **31**, 97 (1983).]

for the coefficient multiplying $\sin \Omega t$ in $(\mathrm{Pwr})_R(t)$. This is plotted for $R = 0.99$ and $\Omega = 0.15(c/2d)$ in Fig. 5.21. We also plot the (normalized) intensity transmitted by the cavity as a function of the frequency deviation from a cavity resonance. Figure 5.21c shows experimental results obtained in the frequency stabilization of a dye laser ($\nu = 5.8 \times 10^{14}$ Hz) using a modulation frequency of 15 MHz and a cavity bandwidth $\delta\nu_c = 3$ MHz. Frequency stability of better than 100 Hz, or a relative frequency stability better than 1.7×10^{-13}, was obtained.

One of the practical complications is the inevitable phase delay between the two inputs to the mixer, so that the inputs are not simply two (in-phase) sine waves, as we have assumed. A "phase shifter" is used to introduce an adjustable phase shift between the rf and reflected signals such that, for laser frequencies near a resonance of the Fabry-Pérot, an error signal like that shown in Fig. 5.21 is obtained; then the inputs to the mixer are in phase.

Another advantage of Pound–Drever–Hall stabilization is that the Fabry-Pérot resonance, unlike a saturable atomic absorption resonance, is linear, so that an increase in laser power does not broaden the resonance.

As discussed below, the degree of frequency stability that is possible with the Pound–Drever–Hall method increases with increasing cavity finesse. Using Fabry-Pérot etalons with finesse exceeding 25,000, it has been demonstrated that two lasers can be locked to adjacent longitudinal modes of a single etalon with a relative frequency stability of better than 1 Hz. Diode laser linewidths ~ 10 Hz have been realized using etalons with finesse greater than 10^5.

• The method of laser frequency stabilization we have described is conceptually similar to a microwave stabilization method invented in the 1940s by R. V. Pound. Our description is obviously a greatly simplified one, as it is beyond our scope to delve into any details about the servo amplifier, the mixer, the phase shifter, or the low-pass filter, descriptions of which can best be found in the electronic engineering literature.

We have noted that the intensity of the field reflected off the Fabry-Pérot etalons varies symmetrically about a cavity resonance. In an older frequency stabilization technique the laser is locked to one side of a cavity resonance, so that a change in the reflected intensity provides information about the sign of the laser frequency fluctuation. The disadvantage of this "side-locking" technique, compared with Pound–Drever–Hall, is that a change in the reflected intensity can arise not only from a fluctuation in the laser frequency but also from a fluctuation in the laser intensity. Improving frequency stability by side locking, therefore, requires that the laser intensity be separately stabilized.

Another approach, which can be used to lock the laser frequency to the center of a cavity or atomic resonance, is to dither the laser frequency *slowly* compared to the linewidth of the resonance. In the case of a cavity resonance this means that Ω is small compared to $\delta\nu_c$ and that

$$\mathcal{R}^*(\omega)\mathcal{R}(\omega - \Omega) - \mathcal{R}(\omega)\mathcal{R}^*(\omega + \Omega) \cong -\Omega \frac{d}{d\omega}|\mathcal{R}(\omega)|^2. \qquad (5.13.13)$$

Thus, $C_{\sin}(\omega)$ [Eq. (5.13.12)] vanishes. However, the reflected power has a part that varies as $2\sqrt{(\text{Pwr})_c(\text{Pwr})_s}\cos\Omega t$ times the expression (5.13.13), the latter being *antisymmetric* in the deviation of the laser frequency from a cavity resonance. The phase shifter and mixer can, therefore, produce the error signal needed to lock the laser frequency to the peak of the resonance. A disadvantage of slow dithering is that the available servo bandwidth is limited by the (slow) modulation frequency. The Pound–Drever–Hall method, however, can be shown to allow for much larger modulation frequencies and servo bandwidths.

The cavity linewidth is related to the rate at which light escapes the cavity, or in other words to the inverse of the "storage time" of light in the cavity (cf. Section 5.9). If the laser frequency changes rapidly enough, the light inside the cavity may not be able to "readjust" fast enough to respond to these changes. The error signal, however, does respond to these rapid changes, so that in effect the laser is being locked to an average of the frequency changes over the storage time of the cavity. •

Single-frequency laser oscillation can also be obtained by *injection locking* with a "seed laser." In this technique the output of a single-mode, low-noise, usually low-power laser is injected through a resonator mirror of a "slave laser" having nearly the same oscillation frequency. Together with locking electronics controlling the length of the slave laser resonator, this acts to force the slave laser to oscillate at the seed laser frequency with much less noise than would be the case if it ran freely. Injection locking is used to reduce the noise of a high-power (slave) laser.

5.14 LASER AT THRESHOLD

We have mostly ignored the threshold regime except to identify it as the point beyond which useful laser output is obtained. We will now consider the threshold of laser oscillation more carefully. For this purpose we use the model of a three-level laser. In steady-state oscillation the population inversion for a three-level laser is given by the formula (4.12.2). When $P \gg \Gamma_{21}$ we have

$$\bar{N}_2 - \bar{N}_1 \approx \bar{N}_2 \approx \frac{PN_T}{P + 2\sigma(v)\Phi_v} = \frac{N_T}{1 + \bar{q}/q^{\text{sat}}}, \tag{5.14.1}$$

where q^{sat} is defined by Eq. (4.12.7) and $\bar{q}$ is the intracavity steady-state photon number, and once again we suppress the subscript v, with the understanding that we are considering a single field of frequency v.

Near threshold, spontaneous emission may not be negligible compared with stimulated emission, contrary to what we have assumed in writing the photon rate Eq. (4.4.7). Therefore, let us amend (4.4.7) to include a term giving the growth of q due to spontaneous emission. For this purpose we use again the fact that the rate of spontaneous emission into a single field mode is just the rate of stimulated emission with one incident photon; recall Section 3.7. Thus, we may simply replace q by $q + 1$ in the first term on the right side of (4.4.7):

$$\frac{dq}{dt} = c\sigma(v)\frac{l}{L}\bar{N}_2(q + 1) - \frac{l}{L}cg_t q. \tag{5.14.2}$$

The steady-state solution of this equation is obtained by setting the right-hand side to zero, which gives

$$\bar{q} = \frac{\sigma(v)\bar{N}_2}{g_t - \sigma(v)\bar{N}_2}. \tag{5.14.3}$$

Now

$$cg_t = c\sigma(v)\,\Delta N_t, \tag{5.14.4}$$

and if we define two convenient dimensionless parameters

$$x = \frac{\bar{N}_2}{\Delta N_t} \quad \text{and} \quad y = \frac{N_T}{\Delta N_t}, \tag{5.14.5}$$

we may write (5.14.3) in the simplified form

$$\bar{q} = \frac{x}{1 - x}, \tag{5.14.6}$$

and we may similarly write (5.14.1) as

$$x = \frac{N_T/\Delta N_t}{1 + \bar{q}/q^{\text{sat}}} = \frac{y}{1 + \bar{q}/q^{\text{sat}}}. \tag{5.14.7}$$

Equations (5.14.6) and (5.14.7) may be solved simultaneously for $\bar{q}$. Some simple algebra yields the quadratic equation

$$\bar{q}^2 + q^{\text{sat}}(1 - y)\bar{q} - q^{\text{sat}}y = 0, \tag{5.14.8}$$

which has the solution

$$\frac{\bar{q}}{q^{\text{sat}}} = \tfrac{1}{2}(y - 1) + \tfrac{1}{2}\sqrt{(y - 1)^2 + \frac{4y}{q^{\text{sat}}}}. \tag{5.14.9}$$

To obtain a nonnegative value for the cavity photon number $\bar{q}$ we have taken in (5.14.9) the positive-definite root of (5.14.8).

Equation (5.14.1) shows that, for an ideal three-level laser, $\bar{N}_2 \approx N_T$ when $\bar{q} \to 0$. That is, N_T is equal to the small-signal population inversion. It follows that y is equal to the small-signal gain divided by the threshold gain:

$$y = \frac{g_0}{g_t}. \tag{5.14.10}$$

If $y < 1$, the device is below the threshold for laser oscillation; if $y > 1$, it is above threshold. Far above threshold ($y \gg 1$) we have from (5.14.9) the result

$$\bar{q} \approx q^{\text{sat}}\left(\frac{g_0}{g_t} - 1\right), \tag{5.14.11}$$

which may be used to obtain our previous expression (5.2.9) for the laser output intensity.

To study the threshold region $y \approx 1$, we need an estimate of the number q^{sat} appearing in Eq. (5.14.9) for the cavity photon number; q^{sat} is defined by (4.12.7). Assuming again that $P \gg \Gamma_{21}$, we have

$$q^{\text{sat}} \approx \left(2\pi\frac{\epsilon_0 m}{e^2}\right)\left(\frac{1}{f}\right)\delta\nu_0(PV), \tag{5.14.12}$$

where f is the oscillator strength of the laser transition. In many lasers PV is roughly on the order of 10^3 m^3 s^{-1}. Taking $f \approx 1$ and $\delta\nu_0 \approx 10$ GHz, therefore, we have the reasonable estimate $q^{\text{sat}} \approx 10^{10}$ for the saturation photon number. We will therefore study the properties of (5.14.9) near the threshold region $y \approx 1$ by *defining* $q^{\text{sat}} = 10^{10}$ as a reasonable value for typical lasers.

Exactly at threshold the cavity photon number is given by (5.14.9) with $y = 1$:

$$(\bar{q})_{\text{threshold}} = \tfrac{1}{2}q^{\text{sat}}\sqrt{4/q^{\text{sat}}} = \sqrt{q^{\text{sat}}} = 10^5. \tag{5.14.13}$$

This is, to be sure, a small number of photons compared with what (5.14.11) predicts well above threshold. However, it is much larger than the average photon number per mode of frequency ν of a thermal field,

$$(q)_{\text{thermal}} = \left(e^{h\nu/kT} - 1\right)^{-1}, \tag{5.14.14}$$

for any realistic value of temperature. For ν corresponding to the He–Ne 632.8-nm line, for example,

$$(q)_{\text{thermal}} \approx 0.023 \tag{5.14.15}$$

for the solar temperature $T \approx 6000K$. The conclusion is that a laser exactly at threshold channels many more photons into the "lasing" mode than it would if it were an ordinary thermal source. We will see that the same thing happens even fairly far *below* threshold.

It is interesting to consider the rate of change of $\bar{q}$ with y in the vicinity of threshold. From (5.14.9) we calculate

$$\frac{d\bar{q}_\nu}{dy} = \tfrac{1}{2}q^{\text{sat}}\left(1 + \frac{y - 1 + 2/q^{\text{sat}}}{\sqrt{(y-1)^2 + 4y/q^{\text{sat}}}}\right), \tag{5.14.16}$$

so that exactly at threshold we have

$$\left(\frac{d\bar{q}}{dy}\right)_{\text{threshold}} \approx \tfrac{1}{2}q^{\text{sat}} = \tfrac{1}{2} \times 10^{10}. \tag{5.14.17}$$

Thus, the curve of $\bar{q}$ vs. y has an extremely large, positive slope at $y = 1$. This is evident in Fig. 5.22, which plots $\bar{q}$ vs. y from Eq. (5.14.9) for $q^{\text{sat}} = 10^{10}$. We conclude that there is an extremely rapid rise in the cavity photon number at the point where the medium is pumped just above threshold. We note also that $\bar{q}$ is much larger than the typical "thermal" value given in Eq. (5.14.15) even for $y = 0.1$.

The sudden transition at threshold that occurs in a laser is so abrupt that it is not usually observed unless an experiment is designed specifically for its observation. An even more profound transition occurs at threshold in the *photon-statistical* properties of the laser output radiation, as discussed in Section 13.14.

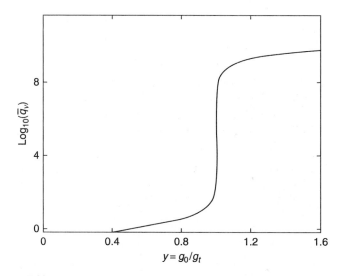

Figure 5.22   The remarkable jump in the cavity photon number at threshold.

APPENDIX: THE FABRY-PÉROT ETALON

Consider again the case of a monochromatic plane wave incident normally upon a Fabry-Pérot etalon (Fig. 5.13). With the first interface ($n' \to n$) between the two media we associate power reflection and transmission coefficients r and t, respectively. We similarly denote by r' and t' the reflection and transmission coefficients for the second interface ($n \to n'$). The coefficients r, r', t, and t' are given in terms of n and n' by the Fresnel formulas.

If A_{in} is the (complex) amplitude of the incident wave, then

$$A_0 = \sqrt{r}\, A_{in} e^{i\pi} = -\sqrt{r}\, A_{in} \tag{5.A.1}$$

is the amplitude of the wave reflected from the first interface. The phase shift of π introduced in (5.A.1) is again a consequence of our assumption that $n > n'$. (The reflected *intensity* is r times the incident intensity and is independent of whether $n > n'$ or $n < n'$.)

The etalon can also produce a reflected wave as a result of transmission at the first interface, reflection at the second interface, and finally transmission through the first interface. The amplitude of the reflected wave is $\sqrt{tr't'}A_{in}$, and it also has a phase shift

$$\frac{2\pi}{\lambda}(2dn) = \frac{4\pi \nu n d}{c} = \Phi \tag{5.A.2}$$

with respect to the incident wave. Because $n > n'$, there are no π phase shifts associated with reflections off the inner faces of the etalon. Therefore,

$$A_1 = \left(\sqrt{r'tt'}\, e^{i\Phi} \right) A_{in}. \tag{5.A.3}$$

A contribution to the total reflected wave also arises from transmission across the first interface, *two* round-trip passes through the etalon as a result of three reflections, and finally transmission across the first interface. For this contribution we have

$$A_2 = \sqrt{tr'r'r't'}\, e^{2i\Phi} A_{in}. \tag{5.A.4}$$

Continuing in this manner, it is easy to see that the total reflected field due to multiple reflections inside the etalon has an amplitude

$$A_R = A_0 + A_1 + A_2 + A_3 + \cdots$$
$$= \left[-\sqrt{r} + \sqrt{r'}\sqrt{tt'}\, e^{i\Phi} \left(1 + \sqrt{r'r'}\, e^{i\Phi} + \sqrt{r'r'r'r'}\, e^{2i\Phi} + \cdots \right) \right] A_{in}. \tag{5.A.5}$$

From the Fresnel formulas (5.12.8a) and (5.12.9a) it follows that, for normal incidence,

$$r = \left(\frac{n - n'}{n + n'} \right)^2 = r'. \tag{5.A.6}$$

It is convenient to define

$$R = r = r' \qquad \text{and} \qquad T = \sqrt{tt'}, \tag{5.A.7}$$

and it follows from energy conservation (and the Fresnel formulas) that

$$R + T = 1. \tag{5.A.8}$$

Then we may write (5.A.5) as

$$\frac{A_R}{A_{in}} = -\sqrt{R}\left[1 - Te^{i\Phi}\left(1 + Re^{i\Phi} + R^2 e^{2i\Phi} + \cdots\right)\right]. \tag{5.A.9}$$

The infinite series can be summed easily:

$$1 + x + x^2 + x^3 + \cdots = \frac{1}{1 - x} \qquad (|x| < 1), \tag{5.A.10}$$

and so we have

$$1 + Re^{i\Phi} + R^2 r^{2i\Phi} + \cdots = \frac{1}{1 - Re^{i\Phi}} \tag{5.A.11}$$

in (5.A.9). Thus,

$$\frac{A_R}{A_{in}} = -\sqrt{R}\left(1 - \frac{Te^{i\Phi}}{1 - Re^{i\phi}}\right) = -\sqrt{R}\frac{1 - (R + T)e^{i\Phi}}{1 - Re^{i\Phi}} = -\frac{1 - e^{i\Phi}}{1 - Re^{i\Phi}}\sqrt{R}, \tag{5.A.12}$$

where the last step follows from (5.A.8). The fraction of the incident *intensity* reflected by the etalon is therefore given by the Airy formula:

$$\frac{I_R}{I_{in}} = \left|\frac{A_R}{A_{in}}\right|^2 = \left|\frac{1 - e^{i\Phi}}{1 - Re^{i\Phi}}\right|^2 R = \frac{4R\sin^2(\Phi/2)}{(1 - R)^2 + 4R\sin^2(\Phi/2)}. \tag{5.A.13}$$

Similarly, the fraction of the transmitted intensity is

$$\frac{I_T}{I_{in}} = 1 - \frac{I_R}{I_{in}} = \frac{(1 - R)^2}{(1 - R)^2 + 4R\sin^2(\Phi/2)}. \tag{5.A.14}$$

It may be shown that for nonnormal incidence we simply replace (5.A.2) by

$$\Phi = \frac{4\pi \nu n d}{c}\cos\theta. \tag{5.A.15}$$

From (5.A.13) and (5.A.14) it follows that the Fabry-Pérot etalon is perfectly transmitting for ν and θ such that

$$\frac{\Phi}{2} = m\pi, \qquad m = 1, 2, 3, \ldots. \tag{5.A.16}$$

This is precisely equivalent to the condition (5.10.6) obtained by considering only a single reflection inside the etalon.

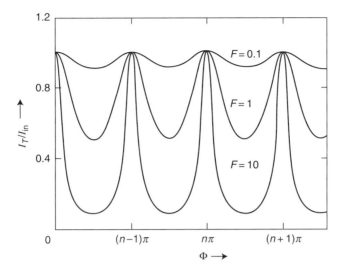

Figure 5.23   Transmission function of the Fabry-Pérot etalon for three values of F.

The multiple-reflection analysis leading to (5.A.11) and (5.A.14) allows us to investigate the bandpass characteristics of the Fabry-Pérot etalon. That is, we can determine how the transmission through the etalon falls off as the frequency ν is displaced from a resonance frequency (5.10.6). In Fig. 5.23 we plot the transmission function (5.A.14) versus Φ for several values of the parameter

$$F = \frac{4R}{(1-R)^2}. \tag{5.A.17}$$

In terms of this parameter the transmitted fraction is

$$\frac{I_R}{I_{\text{in}}} = \frac{1}{1 + F\sin^2(\Phi/2)}. \tag{5.A.18}$$

For small values of F there is considerable transmission of *all* frequencies. For large values of F ($R \rightarrow 1$), however, only a narrow band of frequencies centered at each resonance frequency (5.10.6) is transmitted. In spectroscopic applications this results in a trade-off between the "throughput," or amount of intensity transmitted, and the "resolution," or narrowness of the bandwidths of transmitted frequencies. Figure 5.24 shows how an intracavity Fabry-Pérot etalon can be used to realize single-mode oscillation in a laser.

The resonance frequency spacing (5.10.7) is called the *free spectral range* of the Fabry-Pérot etalon. The ratio of the free spectral range to the half-width of the frequency band centered on a resonance frequency may be shown to be

$$\mathcal{F} = \frac{\pi}{2}\sqrt{F} = \frac{\pi\sqrt{R}}{1-R}. \tag{5.A.19}$$

$\mathcal{F}$ is called the *finesse* of the Fabry-Pérot etalon. The greater the finesse, the sharper the bands of transmitted frequencies relative to their separation. $\mathcal{F}$ is typically around

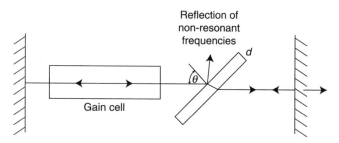

Figure 5.24   An intracavity Fabry-Pérot etalon can be used to filter out all cavity mode frequencies except those satisfying the condition (5.A.16) for transmission without reflection. If the "free spectral range" $c/2nd \cos \theta$ is large compared with the width of the gain profile, only a single longitudinal mode can lase.

30–100, but values $\approx 10^5$ can be achieved under special circumstances (see Problem 5.10).

PROBLEMS

5.1. A cavity for a 632.8-nm He–Ne laser is 50 cm long with reflection coefficient $r_1 = 1.0$ for one mirror and $r_2 = 0.98$ for the other. Losses other than output coupling are very small and may be ignored. The output power of the laser is measured to be about 10 mW on a single mode.

   (a) What is the cavity photon number and what is the photon output rate?

   (b) It has been written that "if the light from a thousand suns were to shine in the sky, that would be the glory of the Mighty One." Assume that the sun is an ideal blackbody radiator at $T = 6000$K. Estimate the flux of photons in a frequency band of width $\delta\nu \approx 10$ MHz centered at 632.8 nm that can be obtain from 1000 suns. How does this compare to the photon flux that can be obtained from He–Ne or other lasers?

5.2. Show that the output coupling that maximizes the output intensity (5.2.11) is given by (5.3.1), and determine the output maximum.

5.3. A high-power CO_2 laser has a small-signal gain $g_0(\nu_0) \approx 0.005 \, \text{cm}^{-1}$ at line center. The laser transition is homogeneously broadened with a Lorentzian linewidth (HWHM) $\delta\nu_0 \approx 1$ GHz. The gain medium fills nearly the entire 50 cm between the cavity mirrors. One of the mirrors is nominally perfectly reflecting, while the output mirror is characterized by a scattering–absorption coefficient $s = 2\%$.

   (a) Determine the output mirror transmission coefficient t that will produce the greatest amount of output power from this laser.

   (b) The saturation intensity I^{sat} for this laser is estimated to be about 100 kW/cm^2. What is the output intensity if the cavity is designed to have the maximal output power?

   (c) Estimate the intracavity intensity. Why might such a laser be designed to have water-cooled mirrors?

5.4. Using numerical values given in Section 4.10, estimate the quantum efficiencies of the ruby and Nd : YAG lasers.

5.5. Assuming $s = 0.04$, plot Eq. (5.2.11) for the output intensity as a function of the output coupling t for the three cases $g_0 l = 0.1, 1.0, 30.0$. Compare these results to those based on Eq. (5.5.18).

5.6. Should the Lamb dip occur with any inhomogeneously broadened gain medium, or only the specific case of Doppler broadening?

5.7. Do you think that most lasers have a cavity bandwidth much larger or smaller than the linewidth of the gain profile? Is any implicit assumption about this made in our discussion related to Fig. 5.12?

5.8. Derive Eq. (5.10.6) for the resonance frequencies of a Fabry-Pérot etalon for an arbitrary angle of incidence.

5.9. (a) Show that the light propagating from the quarter-wave plate to the polarizing beam splitter in Fig. 5.17 is polarized orthogonally to the light propagating from the polarizing beam splitter to the quarter-wave plate. (Note: It is important for this and various other applications to bear in mind that circular polarization changes upon reflection from a mirror: right-hand circularly polarized light becomes left-handed circularly polarized and vice versa. The sense in which the electric field vector rotates is preserved in the reflection, but the propagation direction is reversed.)

(b) Consider the arrangement shown in Fig. 5.25. Show that, with proper alignment of the quarter-wave plate, this arrangment can be used as an optical isolator that minimizes backreflection.

5.10. Let a wave of amplitude A_{in} and wavelength λ be normally incident from a medium of refractive index n_1 onto a layer of index n_2 and thickness d followed by a medium of index n_3 (Fig. 5.26).

(a) Show that the amplitude reflection and transmission coefficients for the $n_i \rightarrow n_j$ interface are

$$\mathcal{R}_{ij} = \frac{n_i - n_j}{n_i + n_j}, \qquad \mathcal{T}_{ij} = \frac{2n_i}{n_i + n_j}, \tag{1}$$

regardless of the polarization of the incident field.

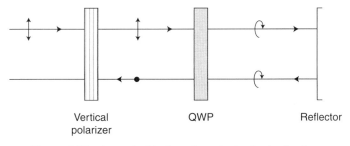

Vertical polarizer   QWP   Reflector

Figure 5.25 An optical isolator for reducing backreflection.

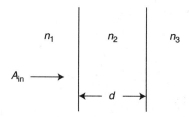

Figure 5.26 Plane wave of amplitude A_{in} incident normally on a layered dielectric. The layer of refractive index n_2 can be chosen so as to reduce or enhance the reflectivity that would be obtained without it.

(b) Show that Eqs. (1) are consistent with energy conservation. (Recall that the Poynting vector for a plane wave in a medium of refractive index n has magnitude $\frac{1}{2}n\epsilon_0 cE_0^2$, where E_0 is the electric field amplitude.)

(c) Show that the reflected wave amplitude A_R is given by

$$\frac{A_R}{A_{in}} = \mathcal{R}_{12} + \frac{\mathcal{T}_{12}\mathcal{T}_{21}\mathcal{R}_{23}e^{i\Phi}}{1 - \mathcal{R}_{21}\mathcal{R}_{23}e^{i\Phi}}, \tag{2}$$

where $\Phi = 4\pi dn_2/\lambda$.

(d) Suppose that the layer in Fig. 5.26 has an optical path length of a quarter of a wavelength, that is, $n_2 d = \lambda/4$ and therefore $\Phi = \pi$. Show that $A_R = 0$ if $n_2 = \sqrt{n_1 n_3}$. In this case the layer makes a perfect *antireflection coating*.

(e) Consider a quarter-wave layer of MgF_2 ($n_2 = 1.38$) deposited on crown glass ($n_3 = 1.52$) in air. Show that the power reflection coefficients with and without the MgF_2 layer are 1 and 4%, respectively. (Multiple layers can be used to nearly eliminate any reflected field.)

(f) Suppose the MgF_2 layer is replaced by a quarter-wave layer of ZnS ($n_2 = 2.3$). What is the power reflection coefficient in this case? (Multilayered dielectrics can have reflectivities greater than 0.99999.)

5.11. The constant P in Eq. (5.12.10) is usually written as rn_0^3, where r is the "linear electro-optic coefficient." For KDP $r = 10.6$ pm/V (1 pm = 1 picometer = 10^{-12} m) and $n_0 = 1.51$ for $\lambda \sim 600$ nm. What is the half-wave voltage of a Pockels cell that uses KDP as the electro-optic medium?

6 MULTIMODE AND PULSED LASING

6.1 INTRODUCTION

Thus far we have restricted our study of the laser to the case of continuous-wave, single-mode operation. In this chapter we will consider time-dependent, transient effects, including relaxation oscillations and Q switching. We will also extend our single-mode theory to the case in which several or many cavity modes can oscillate simultaneously. This allows us, in particular, to understand mode locking, a way to obtain ultrashort pulses of light.

6.2 RATE EQUATIONS FOR INTENSITIES AND POPULATIONS

In the preceding two chapters we found it convenient and instructive to describe the strength of the cavity field either in terms of intensity I_ν or photon number q_ν. In the present chapter it will be convenient to use the intensity description. We will therefore begin with a brief review of the appropriate equations coupling the intensity and the laser level population densities N_2 and N_1.

In general, the cavity intensity will vary both in time and space. We will continue in this chapter to make the plane-wave approximation in which the intensity is assumed to be uniform in any plane perpendicular to the cavity axis. Furthermore we showed in Section 5.5 that, for the most common situation in which the mirror reflectivities are large (say, $>50\%$), the cavity intensity is approximately uniform along the cavity axis if we ignore the rapidly varying $\sin^2 kz$ interference term. So it is useful again to make the uniform-field approximation, but now to include the time dependence of the cavity intensity. First, we recall Eq. (4.4.8):

$$\frac{dI_\nu}{dt} = \frac{cl}{L}\left[g(\nu)I_\nu - \frac{1}{2l}(1 - r_1 r_2)I_\nu\right] = \frac{cl}{L}[g(\nu) - g_t]I_\nu. \qquad (6.2.1)$$

For simplicity, we will assume that the gain cell fills the entire space between the mirrors. Then $l = L$ and

$$\frac{dI_\nu}{dt} = c[g(\nu) - g_t]I_\nu. \qquad (6.2.2)$$

Recall that $I_\nu = I_\nu^{(+)} + I_\nu^{(-)}$ is the sum of the two traveling-wave intensities; in the case of high mirror reflectivities, the two are approximately equal.

In terms of the cavity intensity we can write the population rate equations [cf. (4.5.2) with $\Gamma_1 = \Gamma_2 = 0$ and $A \rightarrow \Gamma_{21}$]:

$$\frac{dN_2}{dt} = -\frac{g(\nu)I_\nu}{h\nu} - \Gamma_{21}N_2 + K_2, \tag{6.2.3a}$$

$$\frac{dN_1}{dt} = \frac{g(\nu)I_\nu}{h\nu} + \Gamma_{21}N_2 + K_1, \tag{6.2.3b}$$

where the rates Γ_{21}, K_2, and K_1 are, again, level decay and pumping rates. Since N_2 and N_1 are populations per unit volume, the pumping rates have units of $(\text{volume})^{-1}$ $(\text{time})^{-1}$. Equations (6.2.2) and (6.2.3) are coupled rate equations for I_ν, N_2, and N_1. The coupling is through the gain coefficient

$$g(\nu) = \frac{\lambda^2 A}{8\pi}(N_2 - N_1)S(\nu) = \sigma(\nu)(N_2 - N_1), \tag{6.2.4}$$

where we assume for simplicity that $g_2/g_1 = 1$.

The population rate equations (6.2.3) are easily modified to suit a particular laser medium. We have already described such modifications in the case of the stylized three- and four-level models. Further modifications are described in Chapter 10, where we consider specific population inversion mechanisms. Since we will be describing in this chapter some rather general phenomena that transcend specific inversion schemes, it will be adequate to use the simple rate equations (6.2.3) for the laser level population densities.

For many purposes the rate equations (6.2.2) and (6.2.3) may be simplified somewhat. One simplifying assumption is that $N_1 \ll N_2$, that is, that the lower laser level population is negligible compared to the upper laser level population. This would be the case in a four-level laser, where the lower level decays very rapidly compared to the stimulated emission (absorption) rate. Then $g(\nu) = \sigma(\nu)N_2$, and (6.2.2) and (6.2.3a) become

$$\frac{dI_\nu}{dt} = c\sigma(\nu)N_2 I_\nu - cg_t I_\nu, \tag{6.2.5a}$$

$$\frac{dN_2}{dt} = -\frac{\sigma(\nu)}{h\nu}N_2 I_\nu - \Gamma_{21}N_2 + K_2. \tag{6.2.5b}$$

6.3   RELAXATION OSCILLATIONS

The coupled equations (6.2.5) for I_ν and N_2 are simple in appearance, but they have no known general solution. However, it is easy to find the *steady-state* solutions, which we denote $\bar{I}_\nu$ and $\bar{N}_2$. These are obtained simply by replacing the left sides of (6.2.5) by zero and solving the resulting algebraic equations, with the result

$$\bar{I}_\nu = h\nu\left(\frac{K_2}{g_t} - \frac{\Gamma_{21}}{\sigma(\nu)}\right) \quad \text{and} \quad \bar{N}_2 = \frac{g_t}{\sigma(\nu)}. \tag{6.3.1}$$

These solutions may also be written in a different form to show explicitly how $\bar{N}_2$ saturates with increasing $\bar{I}_\nu$ (Problem 6.1).

It is possible to solve these equations approximately if the laser is operating very near to steady state. In this case we write

$$I_\nu = \bar{I}_\nu + \varepsilon, \tag{6.3.2a}$$

$$N_2 = \bar{N}_2 + \eta, \tag{6.3.2b}$$

and assume

$$|\varepsilon| \ll \bar{I}_\nu, \tag{6.3.3a}$$

$$|\eta| \ll \bar{N}_2. \tag{6.3.3b}$$

This approximation allows the equations (6.2.5) to be linearized and solved, as follows. Using (6.3.2) in (6.2.5a), we have

$$\frac{d}{dt}(\bar{I}_\nu + \varepsilon) = c\sigma(\bar{N}_2 + \eta)(\bar{I}_\nu + \varepsilon) - cg_t(\bar{I}_\nu + \varepsilon), \tag{6.3.4}$$

which is the same (since $d\bar{I}_\nu/dt = 0$) as

$$\frac{d\varepsilon}{dt} = c\sigma(\bar{N}_2 + \eta)(\bar{I}_\nu + \varepsilon) - cg_t(\bar{I}_\nu + \varepsilon)$$

$$= c\sigma(\bar{N}_2\bar{I}_\nu + \bar{N}_2\varepsilon + \eta\bar{I}_\nu + \eta\varepsilon) - cg_t(\bar{I}_\nu + \varepsilon). \tag{6.3.5}$$

Now $\bar{I}_\nu$ and $\bar{N}_2$ are such as to make the right sides of (6.2.5) vanish. In particular,

$$c\sigma\bar{N}_2\bar{I}_\nu - cg_t\bar{I}_\nu = 0. \tag{6.3.6}$$

Using this relation in (6.3.5), we obtain the much simpler equation

$$\frac{d\varepsilon}{dt} = c\sigma\eta\bar{I}_\nu + c\sigma\eta\varepsilon. \tag{6.3.7}$$

This is still nonlinear (because of the term $\eta\varepsilon$), but now the nonlinearity is very small because it involves the product of the small quantities η and ε. Near enough to steady state [recall (6.3.3)], such second-order small terms can be dropped altogether without significant error. Thus, we obtain the following linear equation for the time dependence of the departure of the cavity intensity from its steady-state value:

$$\frac{d\varepsilon}{dt} = (c\sigma\bar{I}_\nu)\eta, \tag{6.3.8}$$

where the factor in parentheses is constant in time. The same procedure can be applied to (6.2.5b). Again the product $\eta\varepsilon$ is very small and can be dropped, and again the definitions of $\bar{I}_\nu$ and $\bar{N}_2$ can be used to cancel some terms. The result is

$$\frac{d\eta}{dt} = -\frac{g_t}{h\nu}\varepsilon - \frac{\sigma K_2}{g_t}\eta. \tag{6.3.9}$$

Equations (6.3.8) and (6.3.9) are still coupled to each other, but they are now linear and easily solved. We use (6.3.8) to replace η in (6.3.9) by $(c\bar{\sigma}\bar{I}_v)^{-1}\, d\varepsilon/dt$ to get

$$\frac{d^2\varepsilon}{dt^2} + \gamma\frac{d\varepsilon}{dt} + \omega_0^2\varepsilon = 0, \tag{6.3.10}$$

where we define

$$\gamma = \frac{\sigma K_2}{g_t} \tag{6.3.11}$$

and

$$\omega_0^2 = \frac{c\sigma g_t}{h\nu}\bar{I}_v. \tag{6.3.12}$$

The solution to (6.3.10) is easily found to be

$$\varepsilon(t) = Ae^{-\gamma t/2}\cos(\omega t + \phi), \tag{6.3.13}$$

where A and ϕ are constants and the frequency of oscillation is

$$\omega = \sqrt{\omega_0^2 - \gamma^2/4}. \tag{6.3.14}$$

For definiteness we assume $\omega_0 > \gamma/2$, making ω real. Thus, near to the steady state, the cavity intensity oscillates about the steady-state value $\bar{I}_v$, and gradually approaches $\bar{I}_v$ at the (exponential) rate $\gamma/2$:

$$I_v = \bar{I}_v + Ae^{-\gamma t/2}\cos(\omega t + \phi). \tag{6.3.15}$$

This is called a *relaxation oscillation*. Similar behavior is observed in a wide variety of nonlinear systems.

Although the relaxation–oscillation solution (6.3.13) is valid only if $|\varepsilon| \ll \bar{I}_v$ [recall (6.3.3)], the nature of the solution is of general importance. The critical feature of the solution is that γ is positive. This guarantees that the steady-state solution $\bar{I}_v$ is a *stable* solution. That is, if some outside agent slightly disturbs the laser while it is running in steady state, the effect of the disturbance decays to zero, thus returning the laser to steady state again. If γ were negative, a small disturbance would grow, and the steady state would therefore be unstable and of little practical significance.

We may write the period T_r and lifetime τ_r of the relaxation oscillations as (Problem 6.1)

$$T_r = \frac{2\pi}{\omega_0} = \frac{2\pi}{\sqrt{(c/\tau_{21})(g_0 - g_t)}} \tag{6.3.16}$$

and

$$\tau_r = \frac{1}{\gamma} = \frac{g_t}{g_0}\tau_{21}, \tag{6.3.17}$$

where g_0 is the small-signal gain and $\tau_{21} = \Gamma_{21}^{-1}$ is the lifetime of the upper laser level. From (6.3.17) and (6.3.11) we see that the duration of the relaxation oscillations decreases with increasing pumping rate K_2 of the gain medium. Likewise the period T_r of the relaxation oscillations should decrease with increased g_0. These predicted trends are consistent with many experimental observations.

It is possible to observe relaxation oscillations in the output intensity of a laser after it is turned on and approaches a steady-state operation. Perturbations in the pumping power can also cause relaxation oscillations to appear spontaneously. In some cases, especially in solid-state lasers, the relaxation time τ_r may be relatively large, making relaxation oscillations readily apparent on an oscilloscope trace of the laser output intensity.

As an example, consider a ruby laser with mirror reflectivities $r_1 \approx 1.0$, $r_2 \approx 0.94$, and a ruby rod of length $l = 5.0$ cm. For such a laser $g_t \approx (1/2l)(1 - r_2) = 0.006$ cm^{-1}, so that $g_t \approx 1.8 \times 10^8$ s^{-1}. For ruby the upper-level lifetime $\tau_{21} \approx 2 \times 10^{-3}$ s. Assuming a pumping level such that $g_0/g_t = 2.0$, we compute from (6.3.16) and (6.3.17) the period and lifetime of relaxation oscillations:

$$T_r \approx 21 \ \mu s, \tag{6.3.18}$$

$$\tau_r \approx 2 \ ms. \tag{6.3.19}$$

Relaxation–oscillation periods are often in the microsecond range, as in this example. The damping time τ_r is particularly large in ruby because of its unusually long upper-level lifetime τ_{21}. Relaxation oscillations are therefore particularly pronounced in ruby. The output of a continuously pumped ruby laser typically consists of a series of irregular spikes, and this spiking behavior is usually attributed to relaxation oscillations being continuously excited by various mechanical and thermal perturbations.

6.4   *Q* SWITCHING

Q switching is a way of obtaining short, powerful pulses of laser radiation. *Q* refers to the quality factor of the laser resonator, as discussed in Section 5.9; recall that a high-*Q* cavity is one with low loss, whereas a lossy cavity will have a low *Q*. The term *Q* switching therefore refers to an abrupt change in the cavity loss. Specifically, it is a sudden switching from a low value to a high value, that is, a sudden lowering of the cavity loss. In this section we will describe how *Q* switching works, and in the following section how it is achieved in actual lasers.

Suppose we pump a laser medium inside a very lossy cavity. Because the loss is so large, laser action is precluded even if the upper-level population N_2 is pumped to a very high value. No field builds up by stimulated emission in the gain cell, and, if pumping is very strong, the gain can grow to a large, small-signal value without any laser oscillation to deplete or saturate it. Suddenly we lower the loss to a value permitting laser oscillation. We now have a small-signal gain much larger than the threshold gain for oscillation.

What happens in this situation, of course, is that there is a rapid growth of intensity inside the cavity. The intensity builds up quickly to a large value, resulting in a large stimulated emission rate and therefore a rapid extraction of energy from the gain cell. The result of the *Q* switching is therefore a short, intense pulse of laser radiation. Pulses as short as 10^{-8}–10^{-7} s are routinely obtained by *Q* switching.

This qualitative explanation of Q switching may be substantiated by solving the rate equations (6.2.5). For this purpose it is convenient to define the dimensionless quantities

$$x = \frac{I_v}{ch\nu N_t} \quad \text{and} \quad y = \frac{N_2}{N_t}, \qquad (6.4.1)$$

where N_t is the threshold population inversion density. The threshold gain is $g_t = \sigma(\nu)N_t$. y is the ratio of the population inversion to the threshold inversion under our assumption $N_1 \ll N_2$; equivalently, it is the ratio of gain g to threshold gain g_t. x is easily seen to be the ratio of the cavity photon density to the threshold population inversion density (Problem 6.2).

In terms of x, y, and the dimensionless time variable

$$\tau = cg_t t, \qquad (6.4.2)$$

equation (6.2.5a) for the cavity intensity takes the form (Problem 6.2)

$$\frac{dx}{d\tau} = (y - 1)x. \qquad (6.4.3a)$$

We will assume that the duration of the Q-switched pulse is short enough that pumping and spontaneous decay of N_2 during this interval is negligible, and only stimulated decay due to the intense pulse occurs. This assumption allows us to ignore the second and third terms on the right-hand side of the rate equation (6.2.5b) and to write the simpler equation (Problem 6.2)

$$\frac{dy}{d\tau} = -xy. \qquad (6.4.3b)$$

The validity of this assumption can always be checked after a solution of Eqs. (6.4.3) has been obtained.

The result of a numerical integration of Eq. (6.4.3) is shown in Fig. 6.1. The pumping level prior to Q switching is assumed to be such that $y(0) = 2$. We observe that the

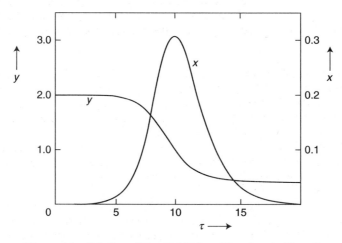

Figure 6.1   Solution of Eq. (6.4.3) for $y(0) = 2$ and $x(0) \approx 0$.

normalized intensity x grows until the population inversion drops below threshold, at which point the intensity begins to decrease.

As a specific example, consider the case of a 694.3-nm ruby laser in which $\sigma \approx 2.7 \times 10^{-20}$ cm^2. Suppose one of the mirrors is highly reflecting ($r_1 \approx 1.0$), the other has a reflectivity $r \approx 0.90$, and the ruby rod has a length $l = 5$ cm. Then

$$g_t \approx \frac{1}{2l}(1 - r) = 0.01 \text{ cm}^{-1} \tag{6.4.4}$$

and $cg_t \approx 3 \times 10^8$ s^{-1}. From (6.4.2), therefore,

$$\tau = 3 \times 10^8 t \quad (t \text{ measured in seconds}). \tag{6.4.5}$$

The Q-switched pulse of Fig. 6.1 has a width of about $\tau = 4$, corresponding to an actual pulse duration of

$$t_p = (3 \times 10^8)^{-1}(4) = 13 \text{ ns.} \tag{6.4.6}$$

The variable x in Fig. 6.1 has a peak value of about 0.3, corresponding to a peak intensity of [Eq. (6.4.1)]

$$I_{\text{peak}} = (ch\nu)N_t(0.3) = \left(\frac{c^2 h}{\lambda}\right)\left(\frac{g_t}{\sigma}\right)(0.3) \approx 10^9 \text{ W/cm}^2, \tag{6.4.7}$$

as the reader may easily verify. This is a very large amount of power—much larger than would be obtainable if the same laser were operated as a continuous-wave device (Problem 6.3). For a beam cross-sectional area of 0.1 cm^2 the total energy in the Q-switched pulse is

$$\text{Energy} \approx (I_{\text{peak}})(t_p)(0.1 \text{ cm}^2) \approx 1 \text{ J.} \tag{6.4.8}$$

• Equations (6.4.3) imply that, if $x(0) = 0$, then x and y remain fixed at their initial values. Physically, this is incorrect and occurs only because in writing (6.4.3) we left out the effect of spontaneous emission. Spontaneous emission has the effect of giving x a small but nonzero initial value, allowing it to grow from this initial value. In other words, spontaneous emission provides the first few "seed" photons needed to initiate the growth of laser intensity by stimulated emission.

In obtaining the numerical results shown in Fig. 6.1 a fourth-order Runge–Kutta integration algorithm was used, with a step size $\Delta\tau = 0.01$. An initial value of 10^{-4} was assumed for $x(0)$. The numerical results for the pulse shape, duration, and peak value are insensitive to the (small) initial value assumed for x. This is because x grows to values large compared to its initial value. Then the number of cavity photons becomes so large that spontaneous emission is negligible compared to stimulated emission.

The value of τ at which the pulse intensity reaches its peak, however, does depend upon the choice of $x(0)$. If this aspect of the problem is of concern, therefore, one should include properly the effect of spontaneous emission in the rate equations (6.4.3) as well as various details of the Q switching.

The reader may wish to experiment with Eqs. (6.4.3) or various other differential equations that appear in this book and in the laser research literature. We include at the end of the book (Section 16.A) a FORTRAN listing of the Runge–Kutta algorithm used to obtain the results in Fig. 6.1. •

6.5   METHODS OF Q SWITCHING

There are various ways to Q-switch a laser. The most popular ones allow the gain to build up to a large value and then switch the cavity Q factor within a time interval that is short compared to the photon lifetime $(cg_t)^{-1}$ before the onset of laser oscillation. We will discuss three common methods of Q switching.

Rotating Mirrors

One way to Q-switch is to have one of the cavity mirrors rotating about an axis perpendicular to the cavity axis (Fig. 6.2). The loss is then very large except during the brief period when the mirrors are nearly parallel. A typical angular velocity of the rotating mirror is about 10,000 revolutions per minute (rpm).

A similar mechanical method of Q switching involves a rotating chopper wheel. In this method, however, the Q switching is effected relatively slowly, even for a wheel velocity of 10,000 rpm. This is because lasing can begin before the shutter fully exposes the gain cell to the cavity mirrors.

Electro-optical Switches

The most common methods of Q switching employ electro-optical or acousto-optical switches. Electro-optical (Kerr or Pockels) shutters control the cavity Q by means of an applied voltage, as indicated in Fig. 6.3. The voltage and orientation of the Kerr cell are such that the (linearly polarized) light passing through the polarizer is converted to circularly polarized light. After reflection off the cavity mirror this circularly polarized light is converted by the Kerr cell to light linearly polarized orthogonally to the polarizer axis. The presence of the Kerr cell thus prevents feedback, and the cavity is in effect a very

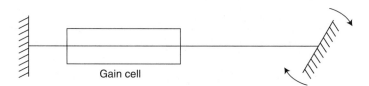

Gain cell

Figure 6.2   A laser cavity with a rotating mirror for Q switching.

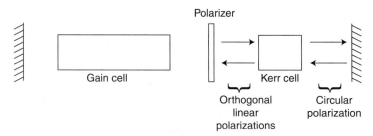

Figure 6.3 Q switching with a Kerr cell. With the Kerr cell on the cavity loss is large but is suddenly lowered when the Kerr cell is switched off.

lossy one. If the voltage across the Kerr cell is switched off, however, the cell is no longer birefringent, and a laser pulse is produced with the sudden increase in the cavity Q.

Saturable Absorbers

Another way of Q switching is to place in the laser cavity a "shutter" consisting of a cell of absorbing material whose absorption coefficient can be saturated (or "bleached") by the laser radiation. Saturation can occur in both absorbing and amplifying media. The absorption (or gain) coefficient decreases with increasing intensity of resonant radiation, becoming nearly zero when the intensity is much larger than a characteristic saturation intensity I_ν^{sat} of the medium.

When the gain medium is first pumped, the gain threshold is very high. The cavity loss is too large because of the absorbing cell to allow laser oscillation. The medium can therefore be pumped to a high gain without generating significant light intensity. Once the gain is high enough to overcome the loss, however, the cavity intensity grows rapidly. This in turn rapidly saturates the absorption cell, and the effective cavity loss drops abruptly. The whole process leads to an output pulse in a manner similar to that with a mechanical Q switch (Problem 6.4).

The use of a saturable absorber for Q switching is often called *passive* Q switching, in contrast to the *active* Q switching achieved mechanically or electro-optically as described above.

The passive Q switch is obviously simpler in terms of the necessary auxiliary equipment than the two active Q switches we have described. It enjoys an additional advantage: A passive Q switch will often give an output pulse concentrated mostly in a single mode. The reason for this is that it takes a finite time for the saturable absorber to become highly saturated and thus to raise the cavity Q. In the meantime the cavity intensity originating from spontaneous emission noise builds up on different modes, and it grows to a greater degree on those modes with the lowest loss per pass. Since a photon can typically make several thousand round trips between the cavity mirrors before the absorber saturates, even small differences in the losses of different modes become significant. The result is that only the lowest-loss mode (or modes) appear in the Q-switched pulse.

In the active Q switches, the switch to high Q is much more rapid, typically occurring during only several tens of photon round trips in the cavity. Small differences in mode losses per pass may then not be sufficient to discriminate among different modes. The output frequency spectrum of a Q-switched ruby laser has long been known to be narrower if a passive Q switch is used instead of a rotating mirror or a Kerr (or Pockels) cell.

6.6 MULTIMODE LASER OSCILLATION

In Section 5.10 we noted that a laser with a homogeneously broadened gain medium tends to oscillate on a single longitudinal mode if the effect of spatial hole burning is small. This expectation is borne out in collision-broadened gas lasers, where atomic motion tends to smear out the effect of spatial hole burning. A similar effect can occur in solid-state lasers in which there is a diffusion of excitation among the atoms. In general, however, oscillation will occur on many longitudinal modes, especially when the gain medium is pumped far above threshold, allowing many modes under the gain curve to meet the threshold condition.

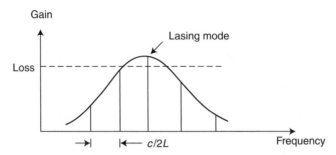

Figure 6.4   A case in which several modes lie under the gain curve, but only one can lase.

Single-longitudinal-mode oscillation is generally precluded by spectral hole burning in inhomogeneously broadened lasers, unless the laser cavity is very short (Fig. 5.11) or the pumping rate permits only one mode to reach threshold (Fig. 6.4). He–Ne lasers, for example, usually oscillate on several longitudinal modes in the absence of any mode selection mechanism (e.g., an etalon). Figure 6.5 shows the output spectrum of a typical, low-pressure 632.8-nm He–Ne laser having a mirror separation $L = 1$ m. Figure 6.5a is the result obtained at a relatively low pumping level. Only one longitudinal mode is above threshold. As the pumping level is raised by increasing the discharge current, however, several modes under the 1700-MHz Doppler profile can oscillate (Fig. 6.5b), and their frequency spacing is near $c/2L = 150$ MHz, as expected.

A rigorous theory of laser oscillation must therefore describe the case in which several or many modes oscillate simultaneously. In this case we cannot formulate the theory in terms of a single cavity photon number or intensity. Instead we must specify the photon number or intensity *for each mode*. The analysis is especially complicated by spectral hole burning in the case of inhomogeneous broadening.

The rate equations (4.5.2), or their simplified version (6.2.5), describe the rate of change of the total number of atoms per unit volume in a particular atomic level. They do not take account of the fact that different atoms may have different line-center frequencies and therefore different stimulated emission cross sections $\sigma(v)$ for radiation of frequency v. That is, there is no account of inhomogeneous line broadening. If there is inhomogeneous broadening, we must write separate rate equations for different spectral packets of atoms (Section 4.14). Different spectral packets will then saturate to different degrees (spectral hole burning), and the complications can be enormous in the multimode case. A proper description of this case would, for our purposes, be inordinately lengthy.

In spite of these complexities, there are situations where the gain of a multimode, inhomogeneously broadened laser saturates homogeneously in the sense that every

Figure 6.5   Typical output spectra of a 1-m-long He–Ne laser for (a) low and (b) high pumping (discharge current) levels.

spectral packet saturates in the same manner. In this case a saturation formula such as (4.12.3) is applicable, and the total output power on all modes is well described by the Rigrod-type analysis discussed in the preceding chapter. One situation in which this is realized approximately is when the longitudinal mode spacing $c/2L$ is small compared to the homogeneous linewidth $\delta\nu_0$, that is, when there are many longitudinal modes lying within the frequency interval $\delta\nu_0$. Evidence for the validity of this approximation may be found in the results of experiments with a low-pressure 3.51-μm He–Xe laser, which is highly inhomogeneously (Doppler) broadened.[1] The cavity mirrors were separated by over 10 m in order to permit the oscillation of a large number of longitudinal modes. The total output power was described quite well by the Rigrod theory for a homogeneously broadened laser (Section 5.5).

There are other effects tending to "homogenize" the gain saturation. In a gas laser, for instance, collisions will change the z component of an atom's velocity and tend to "fill in" a spectral hole. In other words, collisions act to preserve the Maxwell–Boltzmann velocity distribution and, therefore, the Doppler gain profile. Collisions thus act in opposition to the spectral hole-burning effect of the field. At high intensity levels the effective homogeneous linewidth is also increased due to power broadening (Section 4.12).

6.7 PHASE-LOCKED OSCILLATORS

In a Q-switched laser the light pulse must make several passes through the gain medium after the cavity Q is switched. Feedback is necessary in order to build up a large field amplitude by stimulated emission. For some applications it is desirable to have pulses of light even shorter than can be achieved by Q switching. Such powerful, ultrashort pulses of light can be obtained by the technique called *mode locking*.

Whereas Q switching may involve either a single mode or many modes, mode locking is a fundamentally multimode phenomenon. Specifically, mode locking involves the "locking" together of the phases of many cavity longitudinal modes. The purpose of this section is to consider a simple analog of a mode-locked laser. We will consider the problem of adding the displacements of N harmonic oscillators with equally spaced frequencies. That is, we consider the sum of

$$x_n(t) = x_0 \sin(\omega_n t + \phi_0), \tag{6.7.1}$$

where

$$\omega_n = \omega_0 + n\Delta, \qquad n = -\frac{N-1}{2},\ -\frac{N-1}{2}+1,\ -\frac{N-1}{2}+2,\ldots,\frac{N-1}{2}. \tag{6.7.2}$$

In other words, the amplitudes x_0 and phases ϕ_0 of the oscillators are identical, and their frequencies ω_n are equally spaced by Δ and centered at ω_0, as shown in Fig. 6.6. The sum of the displacements is

$$X(t) = \sum_n x_n(t) = -\sum_{-(N-1)/2}^{(N-1)/2} x_0 \sin(\omega_n t + \phi_0). \tag{6.7.3}$$

---

[1] L. W. Casperson, *IEEE Journal of Quantum Electronics* **QE-9**, 250 (1973).

Figure 6.6 A collection of N frequencies running from $\omega_0 - \frac{1}{2}(N-1)\Delta$ to $\omega_0 + \frac{1}{2}(N-1)\Delta$ as in Eq. (6.7.2).

Since $\sin x$ is the imaginary part of e^{ix}, we may write this as

$$X(t) = x_0 \mathrm{Im}\left(\sum_n e^{i(\omega_0 t + \phi_0 + n\,\Delta t)}\right) = x_0 \mathrm{Im}\left(e^{i(\omega_0 t + \phi_0)}\sum_n e^{in\Delta t}\right). \qquad (6.7.4)$$

The general identity

$$\sum_{-(N-1)/2}^{(N-1)/2} e^{iny} = \frac{\sin(Ny/2)}{\sin(y/2)} \qquad (6.7.5)$$

proved below allows us to write (6.7.4) as

$$X(t) = x_0 \mathrm{Im}\left[e^{i(\omega_0 t + \phi_0)}\frac{\sin(N\Delta t/2)}{\sin(\Delta t/2)}\right] = x_0 \sin(\omega_0 t + \phi_0)\left[\frac{\sin(N\Delta t/2)}{\sin(\Delta t/2)}\right]$$

$$= A_N(t)x_0 \sin(\omega_0 t + \phi_0). \qquad (6.7.6)$$

The function $A_N(t)$ is plotted in Fig. 6.7 for $N=3$ and $N=7$. In general $A_N(t)$ has equal maxima

$$A_N(t)_{\mathrm{max}} = N \qquad (6.7.7)$$

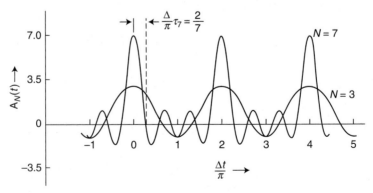

Figure 6.7 The function $A_N(t) = \sin(\frac{1}{2}N\Delta t)/\sin(\frac{1}{2}\Delta t)$ vs. $\Delta t/\pi$.

at values of t given by

$$t_m = m\left(\frac{2\pi}{\Delta}\right) \equiv mT, \qquad m = 0, \pm1, \pm2, \ldots. \tag{6.7.8}$$

As N increases, the maxima of $A_N(t)$ become larger. They also become more sharply peaked. A measure of their width is the time interval τ_N indicated in Fig. 6.7 for $N = 7$:

$$\tau_N = \frac{2\pi}{N\Delta} = \frac{T}{N}. \tag{6.7.9}$$

We have thus shown that the addition of N oscillators of equal amplitudes and phases, and equally spaced frequencies (6.7.2), gives maximum total oscillation amplitudes equal to N times the amplitude of a single oscillator. These maximum amplitudes occur at intervals of time T [Eq. (6.7.8)]. For large N we have, loosely speaking, a series of large-amplitude "spikes." The smaller the frequency spacing Δ between the individual oscillators, the larger the time interval $T = 2\pi/\Delta$ between spikes, and conversely. The temporal duration of each spike is $\tau_N = T/N$, so the spikes get sharper as N is increased.

We have assumed for simplicity that each oscillator has the same phase ϕ_0 [Eq. (6.7.1)]. A more general kind of phase locking occurs when the phase differences of the oscillators are constant but not necessarily zero:

$$\phi_n = \phi_0 + n\alpha, \tag{6.7.10a}$$

or

$$\phi_{n+1} - \phi_n = \alpha. \tag{6.7.10b}$$

In this case the sum of the oscillator displacements (6.7.3) is replaced by

$$X(t) = \sum_n x_0 \sin(\omega_n t + \phi_n) = x_0 \mathrm{Im}\left(e^{i(\omega_0 t + \phi_0)} \sum_{-(N-1)/2}^{(N-1)/2} e^{in(\Delta t + \alpha)}\right), \tag{6.7.11}$$

and this may be evaluated to give the total displacement

$$X(t) = x_0 \sin(\omega_0 t + \phi_0)\left[\frac{\sin N(\Delta t + \alpha)/2)}{\sin(\Delta t + \alpha/2)}\right], \tag{6.7.12}$$

having basically the same properties as (6.7.6) obtained with $\alpha = 0$. (See also Problem 6.5.)

• We prove (6.7.5) as follows. Let the sum be denoted S_N. For convenience we will first evaluate

$$S_{N+1} = \sum_{n=-N/2}^{+N/2} e^{iny}. \tag{6.7.13}$$

The first step is to shift the summation label by introducing

$$m = n + \frac{N}{2},$$
(6.7.14)

so that

$$S_{N+1} = \sum_{m=0}^{N} e^{i(m-N/2)y} = e^{-iNy/2} \sum_{m=0}^{N} e^{imy} = e^{-iNy/2} \sum_{m=0}^{N} (e^{iy})^m.$$
(6.7.15)

The second step is to make use of the identity

$$\sum_{m=0}^{N} x^m = \frac{1 - x^{N+1}}{1 - x}.$$
(6.7.16)

Then we can write

$$S_{N+1} = e^{-iNy/2} \frac{1 - e^{i(N+1)y}}{1 - e^{iy}}$$
$$= e^{-iNy/2} \frac{e^{i(N+1)y/2}}{e^{iy/2}} \frac{e^{-i(N+1)y/2} - e^{i(N+1)y/2}}{e^{-iy/2} - e^{iy/2}} = \frac{\sin(N+1)y/2}{\sin y/2},$$
(6.7.17)

and so we have proved that

$$S_N = \frac{\sin Ny/2}{\sin y/2},$$
(6.7.18)

as claimed in (6.7.5).   •

6.8 MODE LOCKING

What we have in the example of the preceding section is a simple model of a mode-locked laser. The individual oscillators in the model play the role of individual longitudinal-mode fields, while their frequency spacing Δ represents the mode (angular) frequency separation $2\pi(c/2L) = \pi c/L$. The assumption of equal oscillator phase differences α ("phase locking") in the model corresponds to the locking together of the phases of the different cavity modes.

Our oscillator model suggests that, if we can somehow manage to lock together the phases of N longitudinal modes of a laser, then the output light of the laser will consist of a *train of pulses* separated in time by $T = 2\pi/\Delta = 2L/c$. The temporal duration of each pulse in the train will be $\tau_N = T/N = 2L/cN$. The larger the number N of phase-locked modes, the greater the amplitude, and the shorter the duration, of each individual pulse in the train. As we will see, this is indeed the essence of the mode-locking technique for obtaining very short, powerful laser pulses.

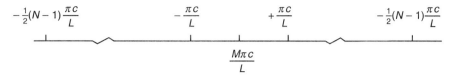

Figure 6.8 The distribution of N cavity mode frequencies as given by Eq. (6.8.7b). The situation is exactly the same as in Fig. 6.7 for the case of N phase-locked oscillators.

The number of longitudinal modes that can simultaneously lase is determined by the gain linewidth (FWHM) $\Delta\nu_g$ and the frequency separation $c/2L$ between modes (cf. Fig. 6.8). Under sufficiently strong pumping of the gain medium we expect that approximately

$$M = \frac{\Delta\nu_g}{c/2L} = \frac{2L}{c}\Delta\nu_g \qquad (6.8.1)$$

longitudinal modes can oscillate simultaneously. The shortest pulse length we expect to achieve by mode locking is therefore

$$\tau_{\min} = \tau_M = \frac{2L}{cM} = \frac{1}{\Delta\nu_g}. \qquad (6.8.2)$$

That is, *the shortest pulse duration we can achieve by mode locking is (approximately) the reciprocal of the gain linewidth.* As an example, consider the 632.8-nm He–Ne laser with a gain linewidth $\Delta\nu_g = \delta\nu_D = 1700$ MHz. For such a laser the shortest pulses obtainable by mode locking are of duration

$$\frac{1}{\delta\nu_D} = \frac{1}{1700 \times 10^6}\ \mathrm{s}^{-1} = 1\ \mathrm{ns}. \qquad (6.8.3)$$

In other words, for this laser, mode locking is not much of an improvement over Q switching for the production of short pulses. This is often true of gas lasers. Their gain linewidths are so narrow that very short (say, picosecond, 10^{-12} s duration) pulses cannot be obtained by mode locking.

On the other hand, consider a 693.4-nm ruby laser with $\Delta\nu_g \approx 10^{11}\ \mathrm{s}^{-1}$. For this laser mode-locked pulses of 10^{-11} s may be obtained.

Liquid dye lasers typically have broad gain profiles, with $\Delta\nu_g \approx 10^{12}\ \mathrm{s}^{-1}$ or more. With such lasers mode-locked pulses in the picosecond range are routinely obtained.

A basic understanding of mode-locked laser oscillation may be reached by extending only slightly our analysis of phase-locked oscillators. We associate with the mth longitudinal mode an electric field

$$E(z, t) = \hat{\boldsymbol{\varepsilon}}_m \mathcal{E}_m(z) \sin(\omega_m t + \phi_m) = \hat{\boldsymbol{\varepsilon}}_m \mathcal{E}_m \sin k_m z \sin(\omega_m t + \phi_m), \qquad (6.8.4)$$

where

$$k_m = m\frac{\pi}{L}, \qquad m = 1, 2, 3, \ldots, \qquad (6.8.5a)$$

and

$$\omega_m = k_m c = m\frac{\pi c}{L}, \qquad m = 1, 2, 3, \ldots . \tag{6.8.5b}$$

For simplicity let us assume that the mode fields all have the same magnitude ($\mathcal{E}_0$) and polarization, so that we can do our calculations with scalar quantities. Furthermore let us consider, without much loss of generality, the simplest example of phase locking, in which all $\phi_m = 0$. Then the total electric field in the cavity is

$$E(z, t) = \sum_m E_m(z, t) = \mathcal{E}_0 \sum_m \sin k_m z \sin \omega_m t, \tag{6.8.6}$$

where the summation is over all oscillating modes.

For a cavity 1 m long, $\pi c/L \approx 9 \times 10^8$ Hz. For near-optical frequencies, of course, the lasing frequencies ω_m will be much larger; at a wavelength of 600 nm, $\omega = 2\pi c/\lambda = 3 \times 10^{15}$ Hz. The integer m in (6.8.5) will therefore typically be in the millions. So let us take $m \rightarrow M + n$ and write equations (6.8.5) as

$$k_m = \frac{(M + n)\pi}{L}, \tag{6.8.7a}$$

$$\omega_m = \frac{(M + n)\pi c}{L}, \tag{6.8.7b}$$

where M is a very large positive integer ($M \approx 10^6$) and n runs from $-\frac{1}{2}(N - 1)$ to $+\frac{1}{2}(N - 1)$, corresponding to a total of $N(\ll M)$ modes centered at the frequency $M\pi c/L$ (Fig. 6.8). Then (6.8.6) becomes

$$E(z, t) = \mathcal{E}_0 \sum_{-(N-1)/2}^{(N-1)/2} \sin\frac{(M + n)\pi z}{L} \sin\frac{(M + n)\pi ct}{L}$$

$$= \frac{1}{2}\mathcal{E}_0 \sum_n \left[\cos\frac{(M + n)\pi(z - ct)}{L} - \cos\frac{(M + n)\pi(z + ct)}{L}\right] \tag{6.8.8}$$

for the total electric field in the laser cavity.

Now we proceed as in the preceding section. The sum

$$\sum_{-(N-1)/2}^{(N-1)/2} \cos\frac{(M + n)\pi(z - ct)}{L} = \text{Re} \sum_{-(N-1)/2}^{(N-1)/2} e^{i(M+n)\pi(z-ct)/L}$$

$$= \text{Re}\left\{e^{iM\pi(z-ct)/L}\frac{\sin[\pi N(z - ct)/2L]}{\sin[\pi(z - ct)/2L]}\right\}$$

$$= \left[\cos\frac{M\pi(z - ct)}{L}\right]\frac{\sin[\pi N(z - ct)/2L]}{\sin[\pi(z - ct)/2L]}, \tag{6.8.9}$$

where we have again used the identity (6.7.5). Similarly

$$\sum_n \cos \frac{(M+n)\pi(z+ct)}{L} = \left\{ \cos \frac{M\pi(z+ct)}{L} \frac{\sin[\pi N(z+ct)/2L]}{\sin[\pi(z+ct)/2L]} \right\}. \tag{6.8.10}$$

From (6.8.8), then,

$$E(z,t) = \frac{\mathcal{E}_0}{2} \left\{ \cos k_0(z-ct) \frac{\sin[\pi N(z-ct)/2L]}{\sin[\pi(z-ct)/2L]} - \cos k_0(z+ct) \frac{\sin[\pi N(z+ct)/2L]}{\sin[\pi(z+ct)/2L]} \right\}, \tag{6.8.11}$$

where $k_0 = \pi M/L$.

The functions

$$A_N^{(\pm)}(z,t) = \frac{\sin[\pi N(z \pm ct)/2L]}{\sin[\pi(z \pm ct)/2L]} \tag{6.8.12}$$

appearing in (6.8.11) have basically the same form—and effect—as the function $A_N(t)$ appearing in Eq. (6.7.6) for the phase-locked oscillator model. In particular, $A_N^{(\pm)}(z,t)$ has maxima occurring at

$$z \pm ct = m(2L), \qquad m = 0, \pm 1, \pm 2, \ldots. \tag{6.8.13}$$

If we put our attention on a fixed value of z inside the cavity, for instance, there are pulses of peak amplitude $N\mathcal{E}_0/2$ appearing at time intervals of $2L/c$, each pulse having a duration T/N (Fig. 6.9). If we fix our attention on the spatial distribution of $E(z,t)$ at a fixed time t, we find pulses of amplitude $N\mathcal{E}_0/2$ with spatial separation $2L$, each pulse having a spatial extent of $2L/N$ (Fig. 6.10).

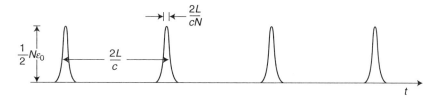

Figure 6.9 A mode-locked pulse train as a function of time, observed at a fixed position z.

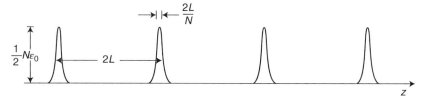

Figure 6.10 A mode-locked pulse train as a function of coordinate z, observed at a fixed instant of time.

In other words, the field (6.8.11) represents two trains of pulses, one moving in the positive z direction and the other in the negative z direction. In the usual situation in which output is obtained through one of the cavity mirrors, the laser radiation appears as a single train of pulses of temporal separation and duration $2L/c$ and $2L/cN$, respectively. All this confirms our conclusions deduced from the phase-locked oscillator model.

The fact that the pulses of a mode-locked train are separated in time by the round-trip cavity transit time $2L/c$ suggests a "bouncing-ball" picture of a mode-locked laser: We can regard the mode locking as generating a pulse of duration $2L/cN$, and this pulse keeps bouncing back and forth between the cavity mirrors. Focusing our attention on a particular plane of constant z in the resonator, we observe a train of identical pulses moving in either direction.

In most lasers the phases ϕ_n of the different modes will undergo random and uncorrelated variations in time. In this case the total intensity is the sum of the individual mode intensities. In mode-locked lasers, however, the mode phases are correlated and the total intensity is not simply the sum of the individual mode intensities. In fact, the individual pulses in the mode-locked train have an intensity N times larger than the sum of the individual mode intensities. The average power, however, is essentially unaltered by mode locking the laser (Problem 6.6).

• Before discussing how mode locking can be accomplished, it is worth noting that "phase locking" or "synchronization" phenomena occur in many *nonlinear* oscillatory systems besides lasers, and indeed these phenomena have been known for a very long time. C. Huygens (1629–1695), for instance, observed that two pendulum clocks hung a few feet apart on a thin wall tend to have their periods synchronized as a result of their small coupling via the vibrations of the wall. Near the end of the 19th century, Lord Rayleigh found that two organ pipes of slightly different resonance frequencies will vibrate at the same frequency when they are sufficiently close together. The contractive pulsations of the heart's muscle cells become phase-locked during the development of the fetus. Fibrillation of the heart occurs when they get out of phase for some reason and results in death unless the heart can be shocked back into the normal condition of cell synchronization. There are other biological examples of phase locking, but detailed theoretical analyses are obviously extremely difficult or impossible for such complex systems. Modern applications of synchronization principles are made in high-precision motors and control systems.   •

6.9   AMPLITUDE-MODULATED MODE LOCKING

The process by which phase or mode locking is forced upon a laser is fundamentally a nonlinear one, and a rigorous analysis of it is complicated. We will therefore rely largely on semiquantitative explanations.

Consider again the scalar electric field

$$E_m(z, t) = \mathcal{E}_m \sin k_m z \sin(\omega_m t + \phi_m) \tag{6.9.1}$$

associated with a longitudinal mode. Suppose that the amplitude $\mathcal{E}_m$ is not constant but rather is modulated periodically in time according to the formula

$$\mathcal{E}_m = \mathcal{E}_0(1 + \varepsilon \cos \Omega t), \tag{6.9.2}$$

where Ω is the modulation frequency and $\mathcal{E}_0$ and ε are constants. Thus, we have an amplitude-modulated field:

$$E_m(z, t) = \mathcal{E}_0(1 + \varepsilon \cos \Omega t) \sin(\omega_m t + \phi_m) \sin k_m z. \qquad (6.9.3)$$

Since

$$\cos \Omega t \sin(\omega_m t + \phi_m) = \tfrac{1}{2} \sin(\omega_m t + \phi_m + \Omega t) + \tfrac{1}{2} \sin(\omega_m t + \phi_m - \Omega t), \quad (6.9.4)$$

we can write the field (6.9.3) as a sum of harmonically varying parts:

$$E_m(z, t) = \mathcal{E}_0\Big\{ \sin(\omega_m t + \phi_m) + \frac{\varepsilon}{2} \sin[(\omega_m + \Omega)t + \phi_m]$$
$$+ \frac{\varepsilon}{2} \sin[(\omega_m - \Omega) + \phi_m] \Big\} \sin k_m z. \qquad (6.9.5)$$

The frequency spectrum of the field (6.9.5) is shown in Fig. 6.11. The amplitude modulation of the field (6.9.1) of frequency ω_m has generated *sidebands* of frequency $\omega_m \pm \Omega$. These sidebands are displaced from the *carrier frequency* ω_m by precisely the modulation frequency Ω. Sideband generation is a well-known consequence of amplitude modulation.

In a laser the mode amplitudes $\mathcal{E}_m$ are determined by the condition that the gain equals the loss. If the loss (or gain) is periodically modulated at a frequency Ω, we expect the fields $E_m(z, t)$ associated with the various modes to be amplitude modulated (AM) with this frequency. In other words, we expect sidebands to be generated about each mode frequency ω_m, as in (6.9.5). In particular, if the modulation frequency Ω is equal to the mode frequency spacing

$$\Delta = \omega_{m+1} - \omega_m = \frac{\pi c}{L}, \qquad (6.9.6)$$

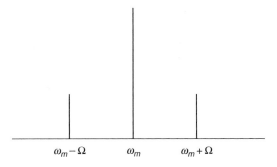

Figure 6.11   Frequency spectrum of the amplitude-modulated field (6.9.5). The sidebands at $\omega_m \pm \Omega$ have amplitudes $\varepsilon/2$ times as large as the carrier amplitude at ω_m. In this case $\varepsilon/2 < 1$.

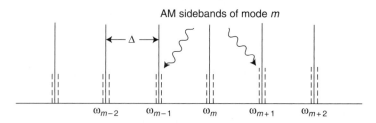

Figure 6.12   Longitudinal modes amplitude-modulated at the frequency Δ equal to their spacing. For clarity the AM sidebands are indicated as dashed lines slightly displaced from the mode frequencies ω_m.

the sidebands associated with each mode match exactly the frequencies of the two adjacent modes (Fig. 6.12). In this case each mode becomes strongly coupled to its nearest-neighbor modes, and it turns out that there is a tendency for the modes to lock together in phase. Loss or gain modulation at the mode separation frequency Δ is therefore one way of mode locking. Borrowing terminology from radio engineering, we call this *AM mode locking*.

The dimensionless factor ε appearing in (6.9.2) is called the modulation index. It is usually small, but it must be large enough to couple the different modes sufficiently strongly. This is analogous to the synchronization phenomenon observed in the 17th century by Huygens with pendulum clocks. Their frequencies were locked together when the clocks were mounted just a meter or so apart, but larger separations weakened their coupling and destroyed the locking effect. If ε is too large, on the other hand, the locking effect is also weakened. This is analogous to the distortion arising in AM radio when the carrier wave is "overmodulated," that is, when $\varepsilon > 1$.

A heuristic way to understand why AM mode locking occurs in lasers is first to suppose that lasing can occur only in brief intervals when the periodically modulated loss is at a minimum. These minima occur in time intervals of $T = 2\pi/\Delta = 2L/c$ if the modulation frequency $\Omega = \Delta$. Between these times of minimum loss the loss is too large for laser oscillation. Thus, we can have laser oscillation only if it is possible to generate a train of short pulses separated in time by T. This is possible if the modes lock together and act in unison, for then we generate a mode-locked train of pulses separated by time T. Thus, mode locking has been described as a kind of "survival of the fittest" phenomenon.

6.10 FREQUENCY-MODULATED MODE LOCKING

We will now consider the case where the *phase* of the field (6.9.1) is periodically modulated rather than the amplitude:

$$E_m(z, t) = \mathcal{E}_m \sin k_m z \sin(\omega_m t + \phi_m + \delta \cos \Omega t). \qquad (6.10.1)$$

The dimensionless constant δ gives the amplitude of the modulation of frequency Ω. As in the case of amplitude modulation, this phase modulation gives rise to sideband frequencies about the carrier frequency ω_m. As we will now see, however, phase modulation produces a whole series of sidebands.

The time-dependent part of (6.10.1) may be written as

$$\sin(\omega_m t + \phi_m + \delta \cos \Omega t) = \sin(\omega_m t + \phi_m) \cos(\delta \cos \Omega t)$$
$$+ \cos(\omega_m t + \phi_m) \sin(\delta \cos \Omega t) \qquad (6.10.2)$$

Now we make use of two mathematical identities:

$$\cos(x \cos \theta) = J_0(x) + 2 \sum_{k=1}^{\infty} (-1)^k J_{2k}(x) \cos(2k\theta), \qquad (6.10.3a)$$

$$\sin(x \cos \theta) = 2 \sum_{k=0}^{\infty} (-1)^k J_{2k+1}(x) \cos[(2k + 1)\theta], \qquad (6.10.3b)$$

where $J_n(x)$ is the Bessel function of the first kind of order n. The first few lowest-order Bessel functions are plotted in Fig. 6.13. These plots are all we will need to know about them. The functions (6.10.3) appear in (6.10.2) with $x = \delta$ and $\theta = \Omega t$. Thus

$$\sin(\omega_m t + \phi_m + \delta \cos \Omega t) = \sin(\omega_m t + \phi_m)\left[J_0(\delta) + 2\sum_{k=1}^{\infty} (-1)^k J_{2k}(\delta)\cos(2k\Omega t)\right]$$
$$+ 2\cos(\omega_m t + \phi_m)\sum_{k=0}^{\infty} (-1)^k J_{2k+1}(\delta)\cos[(2k+1)\Omega t]$$
$$= \sin(\omega_m t + \phi_m)[J_0(\delta) - 2J_2(\delta)\cos 2\Omega t$$
$$+ 2J_4(\delta)\cos 4\Omega t - 2J_6(\delta)\cos 6\Omega t + \cdots]$$
$$+ 2\cos(\omega_m t + \phi_m)[J_1(\delta)\cos \Omega t - J_3(\delta)\cos 3\Omega t$$
$$+ J_5(\delta)\cos 5\Omega t - \cdots]. \qquad (6.10.4)$$

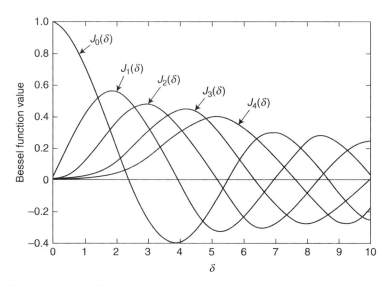

Figure 6.13 The first few lowest-order Bessel functions of the first kind, $J_n(\delta)$.

Using the identities $\sin x \cos y = \frac{1}{2}[\sin(x+y) + \sin(x-y)]$ and $\cos x \cos y = \frac{1}{2}[\cos(x+y) + \cos(x-y)]$, therefore, we have

$$\sin(\omega_m t + \phi_m + \delta \cos \Omega t) = J_0(\delta)\sin(\omega_m t + \phi_m)$$

$$+ J_1(\delta)\{\cos[(\omega_m + \Omega)t + \phi_m] + \cos[(\omega_m - \Omega)t + \phi_m]\}$$

$$- J_2(\delta)\{\sin[(\omega_m + 2\Omega)t + \phi_m] + \sin[(\omega_m - 2\Omega)t + \phi_m]\}$$

$$- J_3(\delta)\{\cos[(\omega_m + 3\Omega)t + \phi_m] + \cos[(\omega_m - 3\Omega)t + \phi_m]\}$$

$$+ J_4(\delta)\{\sin[(\omega_m + 4\Omega)t + \phi_m] + \sin[(\omega_m - 4\Omega)t + \phi_m]\}$$

$$+ J_5(\delta)\{\cos[(\omega_m + 5\Omega)t + \phi_m] + \cos[(\omega_m - 5\Omega)t + \phi_m]\}$$

$$- \cdots \tag{6.10.5}$$

after a simple rearrangement of terms in (6.10.4). Whereas amplitude modulation produces one sideband on either side of the carrier frequency ω_m, phase modulation in general produces a whole series of pairs of sidebands. If the "modulation index" δ is somewhat less than unity, however, we observe from (6.10.5) and Fig. 6.13 that the first pair of sidebands at $\omega_m \pm \Omega$ is strongest. As the strength of the modulation increases, that is, as δ increases, more sideband pairs become important. Figure 6.14 shows the frequency spectrum of the function (6.10.5) for $\delta = 1$ and $\delta = 5$.

Again borrowing the terminology of radio engineering, we refer to this type of modulation as *frequency modulation* (FM). As in the AM case, frequency modulation at the mode separation frequency $\Omega = \Delta = \pi c/L$ causes the sidebands associated with each mode to be in resonance with the carrier frequencies of other modes. This results in a strong coupling of these modes and a tendency for them to lock together and produce a mode-locked train of pulses. This is called *FM mode locking*.

• Information cannot be transmitted with a purely monochromatic wave. The basic idea of radio communication is to modulate a monochromatic (carrier) wave in some way (AM or FM), transmit it, then demodulate it at a receiver to recover the information or message contained in the original modulation. Because radiation from extraneous sources such as lightning, electric

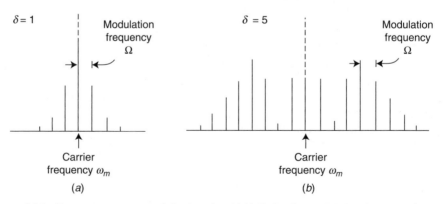

Figure 6.14   Frequency spectrum of the function (6.10.5) for the modulation index (*a*) $\delta = 1$ and (*b*) $\delta = 5$.

power generators, and car ignition systems can "modulate" the amplitude but not the frequency of a radio transmission, FM is much less susceptible than AM to the interference ("static") such sources produce at the receiver. The much larger bandwidth of FM radio compared to AM gives it much greater fidelity. Each FM station is allowed a certain bandwidth (200 kHz in the United States, 100 kHz in Europe) that is more than enough to cover the full musical range of the human ear.

In radio communication, in contrast to our discussion, it is customary to distinguish between *phase* modulation and *frequency* modulation. A phase-modulated wave has the form

$$A(t) = A_0 \sin[\omega_m t + \phi_M(t)]. \tag{6.10.6}$$

The modulation, or "message," is defined by $\phi_M(t)$. A frequency-modulated wave has the form

$$A(t) = A_0 \sin\left[\omega_m t + \int_0^t \Phi_M(t')\, dt'\right], \tag{6.10.7}$$

where now the "message" $\Phi_M(t)$ appears in the instantaneous frequency defined by

$$\omega_i(t) \equiv \frac{d}{dt}\left[\omega_m t + \int_0^t \Phi_M(t')\, dt'\right] = \omega_m + \Phi_M(t). \tag{6.10.8}$$

Obviously, the forms (6.10.6) and (6.10.7) are special cases of each other. The difference between the two in communications relates to how the *demodulation* is performed to recover the transmitted message. The demodulation of a frequency-modulated wave involves electronics that produces a voltage proportional to the instantaneous frequency, whereas a demodulator of a phase-modulated wave produces a voltage proportional to the instantaneous phase. Our discussion of FM mode locking does not require consideration of any demodulation, and therefore it is not necessary for our purposes to distinguish between phase modulation and frequency modulation. ●

6.11 METHODS OF MODE LOCKING

Lasers can be mode locked in a variety of ways. We will focus our attention on three common and illustrative techniques.

Acoustic Loss Modulation

This method is based on the diffraction of light by sound waves, that is, on Brillouin scattering. A sound wave is basically a wave of density variation—and therefore refractive index variation—in a material medium. As discussed in the Appendix to this chapter, a sound wave can therefore act as a "diffraction grating" for light. A sound wave of wavelength λ_S diffracts light of wavelength λ with diffraction angle θ (Fig. 6.18) satisfying [Eq. (6.A.3)]

$$\sin\theta = \frac{\lambda}{2n\lambda_S}, \tag{6.11.1}$$

where n is the refractive index of the medium.

A standing sound wave in a medium may be represented by a refractive index variation of the form

$$\Delta n(x, t) = a \sin(\omega_S t + \theta) \sin k_S x. \tag{6.11.2}$$

The periodic spatial modulation $\sin k_S x$ of the refractive index gives rise to diffraction at the angle θ given by (6.11.1) with $\lambda_S = 2\pi/k_S$. The temporal oscillation at frequency ω_S means that the diffraction is most effective at times t such that $\sin(\omega_S t + \theta) = \pm 1$, for at these times the "diffraction grating" represented by $\sin k_S x$ has its largest amplitude $(\pm a)$. Thus, the diffracting strength of the standing acoustic wave varies harmonically in time with frequency $2\omega_S$.

We can now understand how the diffraction of light by sound can be used to periodically modulate the cavity loss in a laser, and thereby to achieve AM mode locking. If a block of material having a standing acoustic wave is inside the cavity, the diffractive loss associated with it will oscillate with frequency $2\omega_S$. If $2\omega_S = \Delta = \pi c/L$, the cavity loss is modulated at the mode frequency separation, as desired for mode locking. Since audible sound waves have frequencies roughly from 20 Hz to 2×10^4 Hz, while the mode separations in a laser are typically much larger, it is clear that ultrasonic acoustic modulation is required for mode locking. This may be done by driving a block of quartz with a piezoelectric crystal.

Electro-optical Phase Modulation

This method is based on the electro-optical effect. Consider a linearly polarized monochromatic wave propagating in the z direction in a medium with refractive index n:

$$\mathbf{E}(z, t) = \hat{x}\mathcal{E}_0 \cos(\omega t - kz) = \hat{x}\mathcal{E}_0 \cos \omega\left(t - \frac{n}{c}z\right). \tag{6.11.3}$$

Suppose we have a Pockels-type electro-optical medium in which the refractive index for light polarized in the x direction is linearly proportional to an applied electric field E_a:

$$n = n_0 + \beta E_a. \tag{6.11.4}$$

Therefore, the electric field (6.11.3) in such an electro-optical medium in which an external field E_a is applied is

$$\mathbf{E}(z, t) = \hat{x}\mathcal{E}_0 \left(\omega t - \frac{n_0\omega}{c}z - \frac{\beta\omega}{c}E_a z\right) = \hat{x}\mathcal{E}_0 \cos\left[\omega\left(t - \frac{n_0}{c}z\right) - \phi\right], \tag{6.11.5}$$

where

$$\phi = \frac{\beta\omega}{c}E_a z. \tag{6.11.6}$$

After a distance l of propagation in the medium, the field has the phase

$$\phi = \frac{\beta\omega}{c}V, \tag{6.11.7}$$

where $V = E_a l$ is the potential difference due to the field E_a. Thus, if an electro-optical cell is inserted in a laser cavity, the laser can be FM mode locked by varying the applied voltage V sinusoidally at the mode separation frequency Δ.

In general, a linearly polarized electric field entering an electro-optical medium can be decomposed into two orthogonally polarized components, each of which has a different refractive index. The two orthogonal polarization directions are determined by the orientation of the cell and the applied field E_a. In deriving (6.11.7), we have assumed that the incident field is linearly polarized along one of these directions. In the general case the field will have components in both directions, and in a Pockels cell the two components will have different values of β. This results in a phase difference between the two field components. If the cell produces a total phase change of $90°$, for example, the incident linearly polarized field will be converted to a circularly polarized field, as in the case illustrated in Fig. 6.3 for a Kerr cell. That is, a cell containing an electro-optical material can act as a quarter-wave plate. The advantage of using electro-optical media rather than naturally birefringent materials, of course, is the switching and control capabilities one has through the adjustment of the bias voltage.

Saturable Absorbers

As in the case of Q switching, a "passive" AM mode locking may be achieved through the use of a saturable absorber. Assume for simplicity that the absorption coefficient a of the absorption cell saturates according to the formula

$$a = \frac{a_0}{1 + I/I^{\text{sat}}} \tag{6.11.8}$$

for a homogeneously broadened line. It is also convenient (but not necessary) to assume that the saturation intensity I^{sat} of the absorption line is very large compared to the laser intensity I. Then (6.11.8) is approximated by

$$a \approx a_0 - \frac{a_0 I}{I^{\text{sat}}}. \tag{6.11.9}$$

Suppose first that there are two oscillating cavity modes, so that the total cavity electric field is

$$E(z, t) = \mathcal{E}_1 \sin k_1 z \sin(\omega_1 t + \phi_1) + \mathcal{E}_2 \sin k_2 z \sin(\omega_2 t + \phi_2), \tag{6.11.10}$$

and the cavity intensity is

$$\begin{aligned}
I(z, t) = c\epsilon_0 E^2(z, t) = c\epsilon_0 [\mathcal{E}_1^2 \sin^2 k_1 z \sin^2(\omega_1 t + \phi_1) + \mathcal{E}_2^2 \sin^2 k_2 z \sin^2(\omega_2 t + \phi_2) \\
+ 2\mathcal{E}_1 \mathcal{E}_2 \sin k_1 z \sin k_2 z \sin(\omega_1 t + \phi_1) \sin(\omega_2 t + \phi_2)].
\end{aligned} \tag{6.11.11}$$

Now the last term can be rewritten using the identity

$$\begin{aligned}
2 \sin(\omega_1 t + \phi_1) \sin(\omega_2 t + \phi_2) = \cos[(\omega_1 - \omega_2)t + \phi_1 - \phi_2] \\
- \cos[(\omega_1 + \omega_2)t + \phi_1 + \phi_2].
\end{aligned} \tag{6.11.12}$$

The frequencies ω_1, ω_2, and $\omega_1 + \omega_2$ are very large compared to the mode separation frequency $\omega_1 - \omega_2 = \Delta$. If we average the intensity (6.11.11) over a few optical periods, therefore, we obtain

$$\bar{I}(z,t) = \frac{c\epsilon_0}{2}\left[\mathcal{E}_1^2 \sin^2 k_1 z + \mathcal{E}_2^2 \sin^2 k_2 z + 2\mathcal{E}_1\mathcal{E}_2 \sin k_1 z \sin k_2 z \cos(\Delta t + \phi_1 - \phi_2)\right].$$

$$(6.11.13)$$

The intensity $\bar{I}$ has a time dependence that is simply a sinusoidal oscillation at the mode beat frequency Δ. The absorption coefficient (6.11.9) averaged over a few optical periods will therefore have this same time dependence. In other words, if a saturable absorber described by (6.11.8) is placed inside the laser cavity, it results in a cavity loss modulated at the mode separation frequency, and therefore acts to mode-lock the laser. The argument may be extended to the case of N cavity modes, and we conclude that mode-locked operation may be achieved by placing a cell containing a saturable absorber inside the cavity.

This technique is commonly used in mode-locked solid-state and dye lasers, which, as discussed in Section 6.8, are especially attractive in this regard.

• Although both Q switching and mode locking may be accomplished by inserting a cell containing a saturable absorber into the laser cavity, there are somewhat different requirements for the absorber in the two cases.

In the case of mode locking the absorber should respond very quickly to any changes in the cavity intensity. This was implied in our discussion above, where it was assumed that the saturation behavior of the absorber is fixed according to (6.11.8); there are no transient terms showing how a changes from $a_0(1 + I_1/I^{\text{sat}})^{-1}$ to $a_0 (1 + I_2/I^{\text{sat}})^{-1}$ as I changes from I_1 to I_2. Rather, it was assumed that a reacts instantaneously to variations in I, or at least with a response time shorter than $2L/c$. This requires the absorber to have a short relaxation time, whereas a longer one would be tolerable for Q switching.

Similarly, it is desirable for Q switching that the saturation intensity of the absorber be considerably smaller than that of the laser gain medium. This ensures a large, unsaturated gain

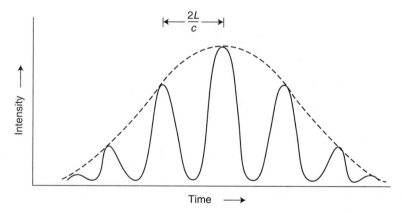

Figure 6.15 Output intensity vs. time of a Q-switched, mode-locked laser. The dashed curve is the envelope of the mode-locked pulse train. Each contributing mode viewed individually has the time dependence of the Q-switched envelope, but because the modes are locked the total output is in the form of a group of pulses separated in time by $2L/c$.

after the loss associated with the absorption cell is fully saturated, and allows the giant pulse to build up. In the case of mode locking, however, a relatively large absorber saturation intensity can still give rise to the required modulation. Our assumption $I \ll I^{\text{sat}}$ in (6.11.9) was made for convenience, not necessity.

Thus, it is possible to Q-switch a laser with one absorption cell and mode-lock it with another. Frequently, however, both effects are present with a saturable absorber, and a Q-switched laser will show signs of mode locking. The output of such a Q-switched, mode-locked laser is indicated in Fig. 6.15.   •

In Section 11.13 we discuss *Kerr lens mode locking*, which is currently used to generate femtosecond laser pulses.

6.12 AMPLIFICATION OF SHORT PULSES

In many applications it is necessary to amplify laser radiation by passing it through a medium with a population inversion on a resonant transition. The amplifier is often made of the same material, and pumped in the same way, as the gain cell of the laser. The most important difference between the laser and the amplifier is simply that the amplifier does not have a resonator with mirrors for feedback. Radiation incident on the amplifier undergoes amplification by stimulated emission and emerges at the other end with greater energy. A series of amplifiers may be employed in tandem, and mirrors may be used to allow the beam to make several passes through a single amplifier. In this section we will consider a pulse of radiation making a single pass through an amplifier.

We will assume that the duration of the laser pulse is short compared to any pumping or relaxation times, so that the changes in level populations in the amplifier are due mainly to stimulated emission and absorption. The rate equations for the level population densities of the amplifying transition are then simply

$$\frac{\partial N_2}{\partial t} = -\frac{\sigma}{h\nu}(N_2 - N_1)I, \tag{6.12.1a}$$

$$\frac{\partial N_1}{\partial t} = \frac{\sigma}{h\nu}(N_2 - N_1)I, \tag{6.12.1b}$$

since pumping and relaxation processes do not affect N_2 and N_1 significantly during the pulse; this condition for a "short" pulse typically requires pulse lengths shorter than about a nanosecond. Equations (6.12.1) may be combined to form a single equation for the population difference $N \equiv N_2 - N_1$:

$$\frac{\partial N}{\partial t} = -\frac{2\sigma}{h\nu}NI. \tag{6.12.2}$$

We also write the (plane-wave) equation for the variation in space and time of the intensity:

$$\frac{\partial I}{\partial z} + \frac{1}{c}\frac{\partial I}{\partial t} = \sigma NI. \tag{6.12.3}$$

Let us integrate both sides of (6.12.3) over time:

$$\int_{-\infty}^{\infty} \left(\frac{\partial I}{\partial z} + \frac{1}{c} \frac{\partial I}{\partial t} \right) dt = \int_{-\infty}^{\infty} \sigma N I \, dt. \tag{6.12.4}$$

Here $t = -\infty$ and $t = +\infty$ denote times long before and after the pulse has "turned on" at z, so that $I(z, t = -\infty) = I(z, t = +\infty) = 0$. Thus,

$$\int_{-\infty}^{\infty} \frac{\partial I}{\partial t} \, dt = I(z, t = +\infty) - I(z, t = -\infty) = 0 \tag{6.12.5}$$

and so (6.12.4) becomes

$$\frac{d\phi}{dz} = \sigma \int_{-\infty}^{\infty} N(z, t) I(z, t) \, dt, \tag{6.12.6}$$

where

$$\phi(z) = \int_{-\infty}^{\infty} I(z, t) \, dt \tag{6.12.7}$$

is called the *fluence* and is a measure of the total energy content of the pulse. Note that the fluence has units of energy per unit area and should not be confused with the photon flux Φ (number of photons per unit area and time) that was introduced in Chapter 4 and used extensively in Chapter 5.

Equation (6.12.6) may be simplified by solving (6.12.2) for $N(z, t)$:

$$N(z, t) = N(z, -\infty) \exp\left[-\frac{2\sigma}{h\nu} \int_{-\infty}^{t} I(z, t') \, dt' \right], \tag{6.12.8}$$

where $N(z, -\infty)$ is the population inversion at z before the pulse has arrived. Thus, we find

$$
\begin{aligned}
\frac{d\phi}{dz} &= \sigma N(z, -\infty) \int_{-\infty}^{\infty} I(z, t) \exp\left[-\frac{2\sigma}{h\nu} \int_{-\infty}^{t} I(z, t') \, dt' \right] dt \\
&= -\sigma N(z, -\infty) \int_{-\infty}^{\infty} \frac{h\nu}{2\sigma} \frac{\partial}{\partial t} \left\{ \exp\left[-\frac{2\sigma}{h\nu} \int_{-\infty}^{t} I(z, t') \, dt' \right] \right\} dt \\
&= -\frac{h\nu}{2} N(z, -\infty) \left\{ \exp\left[-\frac{2\sigma}{h\nu} \int_{-\infty}^{\infty} I(z, t) \, dt \right] - 1 \right\}.
\end{aligned}
\tag{6.12.9}
$$

Then (6.12.7) allows us to write

$$
\begin{aligned}
\frac{d\phi}{dz} &= \frac{h\nu}{2} N(z, -\infty) \left[1 - \exp\left(-\frac{2\sigma\phi(z)}{h\nu} \right) \right] \\
&= \frac{h\nu}{2} N(z, -\infty) \left[1 - \exp\left(-\frac{\phi(z)}{\phi_{\text{sat}}} \right) \right],
\end{aligned}
\tag{6.12.10}
$$

where we define the *saturation fluence*

$$\phi_{\text{sat}} \equiv \frac{h\nu}{2\sigma}. \tag{6.12.11}$$

This is just the photon energy divided by twice the stimulated emission cross section.

The differential equation (6.12.10) may be written in terms of the ratio $\theta\,(z) = \phi(z)/\phi_{\text{sat}}$:

$$\frac{d\theta}{dz} = g_0(1 - e^{-\theta}), \tag{6.12.12}$$

where we have used the fact that the small-signal gain coefficient $g_0(z, -\infty) = \sigma N(z, -\infty)$. In many cases of interest the spatial variations of g_0 are small, and we can take g_0 to be a constant in the differential equation (6.12.12) for the fluence. Then this equation has the solution

$$\theta(z) = \ln[1 + e^{g_0 z}(e^{\theta(0)} - 1)], \tag{6.12.13}$$

or

$$\phi_{\text{out}} = \phi_{\text{sat}} \ln\{1 + G_0[\exp(\phi_{\text{in}}/\phi_{\text{sat}}) - 1]\}, \tag{6.12.14}$$

where $\phi_{\text{out}} \equiv \phi(L)$ is the output fluence of an amplifier of length L with small-signal gain factor $G_0 = e^{g_0 L}$, given the input fluence $\phi_{\text{in}} = \phi(0)$ to the amplifier. We can also write (6.12.14) in terms of the total gain $G \equiv \phi_{\text{out}}/\phi_{\text{in}}$:

$$G = \frac{\phi_{\text{out}}}{\phi_{\text{in}}} = X^{-1} \ln[1 + G_0(e^X - 1)], \qquad X \equiv \frac{\phi_{\text{in}}}{\phi_{\text{sat}}}. \tag{6.12.15}$$

It is important to note that this solution for the output fluence is independent of the shape of the pulse as a function of time. As long as the pulse is confined, to a good approximation, to a finite duration, and this duration is short compared to any pumping and relaxation times, Eq. (6.12.15) gives us the output fluence as a function of the small-signal gain, length, and saturation fluence of the amplifier. In Fig. 6.16 we plot G as a function of X, assuming a small-signal gain factor $G_0 = 5000$.

If $X = \phi_{\text{in}}/\phi_{\text{sat}} \ll 1$, then (6.12.15) becomes

$$G \approx X^{-1} \ln(1 + G_0 X). \tag{6.12.16}$$

If furthermore $G_0 X \ll 1$, then $\ln(1 + G_0 X) \approx G_0 X$ and we have

$$G \approx G_0 = e^{g_0 L} \tag{6.12.17}$$

for the small-signal total gain. If $e^X = \exp(\phi_{\text{in}}/\phi_{\text{sat}}) \gg 1$, on the other hand, then

$$G \approx X^{-1} \ln(G_0 e^X) = X^{-1} \ln[\exp(g_0 L + X)] = X^{-1}(g_0 L + X)$$

$$= 1 + \frac{g_0 L}{X}, \tag{6.12.18a}$$

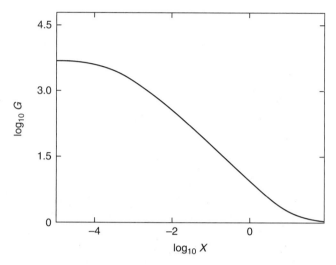

Figure 6.16   Gain $G(X)$ [Eq. (6.12.15)] for $G_0 = 5000$.

or

$$\phi_{out} \approx \phi_{in} + (g_0 \phi_{sat})L. \tag{6.12.18b}$$

This result identifies $g_0 \phi_{sat}$ as the largest energy per unit volume that can be extracted from the amplifier when ϕ_{in} is large compared to ϕ_{sat}. This is analogous to the result (5.3.9) for a cw laser, where $g_0 I^{sat}$ is the largest possible rate of energy extraction per unit volume. Using the fact that $g_0 = \sigma N$ and $\phi_{sat} = h\nu/2\sigma$, we can write (6.12.18) in the form

$$\phi_{out} \approx \phi_{in} + \frac{Nh\nu}{2} L. \tag{6.12.19}$$

This says that *the largest extractable energy density corresponds to taking half a photon, on average, from each excited atom of the amplifier*. The reason for the factor $\frac{1}{2}$ is simply that in the limit of large ϕ_{in}/ϕ_{sat} under consideration, the amplifier is well saturated, with the upper- and lower-level populations having equal probabilities.

This theory of short-pulse amplification is referred to as the *Frantz–Nodvik model*[2] and is useful in the design and interpretation of short-pulse amplification experiments. More generally for short pulses it is necessary to allow for coherent propagation effects (Chapter 9) that are not included in the Franz–Nodvik model.

6.13   AMPLIFIED SPONTANEOUS EMISSION

The theory of pulse propagation presented in the preceding section is used frequently in the analysis of laser amplifiers. However, for high-gain systems this theory ignores an important phenomenon: The amplifier can amplify not only the input field from a

[2]L. M. Frantz and J. S. Nodvik, *Journal of Applied Physics* **24**, 2346 (1963).

laser oscillator (or another amplifier), but also the spontaneous radiation emitted by the excited molecules of the amplifier itself. It is easy to see that spontaneously emitted photons at one end of an amplifier, which happen to be directed along the amplifier axis, or close to that direction, can *stimulate* the emission of more photons and lead to substantial output radiation at the other end of the amplifier. This radiation, which appears regardless of whether there is any input radiation, is called *amplified spontaneous emission.*

It is clear that amplified spontaneous emission (ASE) will have at least some properties resembling laser radiation. In particular, it will be narrowband in frequency and it will also be highly directional, simply because the amplifier is long and thin. For these reasons high-gain systems emitting ASE are often referred to as "mirrorless lasers." Such mirrorless lasing, also called "superradiance," is well known in the 3.39 μm He−Ne laser, and in high-gain excimer, dye, and semiconductor laser media.

For a simple estimate of ASE, let us consider the steady-state equation for the propagation of intensity in an amplifying medium characterized by a spatially uniform gain coefficient g:

$$\frac{dI}{dz} = gI. \tag{6.13.1}$$

If $I(0) = 0$, this equation predicts that $I(z) = 0$ for all z. In other words, this equation does not account for ASE. To include the possibility of ASE we add to (6.13.1) the effect of spontaneous emission:

$$\frac{dI}{dz} = gI + (A_{21}N_2 h\nu)\left(\frac{\Omega}{4\pi}\right). \tag{6.13.2}$$

The added term is the contribution to dI/dz from spontaneous emission of photons of energy $h\nu$ by N_2 excited molecules per unit volume with spontaneous emission rate A_{21}. Since spontaneous emission may be assumed here to be (statistically) isotropic, we have included a factor $\Omega/4\pi$, where Ω is an appropriate solid angle; this factor accounts for the fact that only a fraction $\Omega/4\pi$ of spontaneously emitted photons are emitted in directions for which amplification can occur. In the simplest approximation Ω is taken to be A/L^2, where A is the cross-sectional area of the amplifier and L is its length.

In the small-signal regime in which g and N_2 are independent of I, we have the following solution of Eq. (6.13.2):

$$I(z) = \left(\frac{A_{21}h\nu\Omega}{4\pi}\right)\left(\frac{N_2}{g_0}\right)(e^{g_0 z} - 1) = \left(\frac{A_{21}h\nu\Omega}{4\pi}\right)\left(\frac{N_2}{g_0}\right)[G(z) - 1], \tag{6.13.3}$$

where g_0 is the small-signal gain at frequency ν, and we define the gain factor $G(z) = \exp(g_0 z)$. For simplicity we will ignore the possibility of level degeneracies and write $g_0 = \sigma(N_2 - N_1)$, where σ is the stimulated emission cross section at frequency ν. For a homogeneously broadened transition having a Lorentzian lineshape of full width at half-maximum $\Delta\nu$ we have

$$\sigma = \left(\frac{\lambda^2 A_{21}}{8\pi}\right)\left(\frac{2}{\pi\,\Delta\nu}\right) = \frac{\lambda^4 A_{21}}{4\pi^2 c\,\Delta\lambda}, \tag{6.13.4}$$

where $\Delta\lambda = (c/v^2)\,\Delta v$ is the width in the emission wavelength $\lambda = c/v$. In this case (6.13.3) becomes

$$
\begin{aligned}
I(z) &= \left(\frac{\pi h v^3 \Omega}{c^2}\right)\Delta v \frac{N_2}{N_2 - N_1}[G(z) - 1]\\
&= \left(\frac{\pi h c^2 \Omega}{\lambda^5}\right)\Delta\lambda \frac{N_2}{N_2 - N_1}[G(z) - 1]
\end{aligned}
\tag{6.13.5}
$$

for the growth of ASE intensity in the amplifier.

- Amplified spontaneous emission is sometimes described from the perspective of an "effective noise input." This approach is based on (6.13.1) rather than (6.13.2). It begins by noting that, since Eq. (6.13.1) gives $I(z) = 0$ for all z if $I(0) = 0$, some effective input $I(0) = I_{\text{eff}} \neq 0$ is necessary in order to obtain a nonvanishing $I(z)$. An expression for I_{eff} is obtained by recalling that in a cavity of dimensions large compared to a wavelength there are $(8\pi v^2/c^3)\,\Delta v$ modes of the field per unit volume in the frequency interval $[v, v + \Delta v]$ (Section 3.12). And according to the quantum theory of radiation, each of these modes has a zero-point energy $\frac{1}{2}hv$. There is therefore a zero-point field energy per unit volume

$$
\rho_0(v) = \left(\frac{8\pi v^2}{c^3}\right)\frac{1}{2}hv\,\Delta v = \left(\frac{4\pi h v^3}{c^3}\right)\frac{c\,\Delta\lambda}{\lambda^2}
\tag{6.13.6}
$$

in the wavelength interval $[\lambda, \lambda + \Delta\lambda]$. Clearly we should take $\Delta\lambda$ to be on the order of the spectral width of the laser transition. For our purposes we replace $\Delta\lambda$ in (6.13.6) by $\pi\,\Delta\lambda$, where $\Delta\lambda$ is the transition linewidth (FWHM). Then the "quantum noise" intensity at the laser transition wavelength λ is given by

$$
I_{\text{eff}} = \frac{c\rho_0(v)\Omega}{4\pi} = \left(\frac{\pi h c^2 \Omega}{\lambda^5}\right)\Delta\lambda,
\tag{6.13.7}
$$

where we have inserted the factor $\Omega/4\pi$ to account for the fact that only those modes within the solid angle Ω appropriate to the amplifier can act as effective noise sources. Thus, from (6.13.1),

$$
I(z) = I_{\text{eff}}e^{g_0 z} = \left(\frac{\pi h c^2 \Omega}{\lambda^5}\right)\Delta\lambda e^{g_0 z},
\tag{6.13.8}
$$

which reproduces (6.13.5) when $G \gg 1$ and there is complete inversion ($N_1 = 0$). The concept of effective noise input can also be used for arbitrary population inversion and gain, but this requires a more rigorous analysis employing the quantum theory of radiation and detection. •

In most cases of practical interest the solid angle Ω is very small. For a cylindrical gain volume of cross-sectional diameter D the divergence angle due to diffraction is $\theta \approx \lambda/D$ (Section 7.11), and so $\Omega \approx \lambda^2/S$, $S = \pi D^2/4$ being the cross-sectional area. Then

$$
I(z) \approx \frac{\pi h v^3}{c^2}\frac{\lambda^2}{S}\Delta v \frac{N_2}{N_2 - N_1}[G(z) - 1] = \pi\frac{hv}{S}\Delta v \frac{N_2}{N_2 - N_1}[G(z) - 1].
\tag{6.13.9}
$$

Assuming the ASE is unpolarized, we obtain the "noise" power $(\text{Pwr})_N$ due to ASE at a single polarization by multiplying (6.13.9) by $\frac{1}{2}$ and by the area S:

$$(\text{Pwr})_N \approx \frac{N_2}{N_2 - N_1} h\nu\, \Delta\nu(G - 1) = n_{\text{sp}} h\nu\, \Delta\nu(G - 1), \qquad (6.13.10)$$

where

$$n_{\text{sp}} \equiv \frac{N_2}{N_2 - N_1} \qquad (6.13.11)$$

is called the *spontaneous emission factor*; $(\text{Pwr})_N$ is usually defined as

$$(\text{Pwr})_N = n_{\text{sp}} h\nu B(G - 1), \qquad (6.13.12)$$

where B is the amplifier bandwidth, or more generally the bandwidth of a detector that responds to a limited range of frequencies within the amplifier bandwidth $\Delta\nu$.

• Since the so-called noise per mode formula (6.13.12) is a fundamental equation that appears in various contexts, it is appropriate to present a more rigorous derivation. Consider a polarized field propagating in the z direction. Such a field of frequency ν defines a single mode, and the rate of spontaneous emission into such a mode is given by Eq. (3.7.8):

$$R_{\text{spon}}(\nu) = \frac{A_{21} c^2}{8\pi \nu^2} \frac{c}{V} S(\nu), \qquad (6.13.13)$$

where $S(\nu)$ is the gain lineshape function, A_{21} is the rate of spontaneous emission into *all* modes, and V is the mode volume. Now consider modes with a given polarization and propagation direction and having a continuous distribution of frequencies within a bandwidth B. If B is small relative to ν and the gain bandwidth defined by $S(\nu)$, the change in the spontaneous emission power over a small distance Δz due to atoms radiating spontaneously into the frequency bandwidth B is approximately

$$\Delta\text{Pwr} = h\nu N_2 V R_{\text{spon}} \mathcal{N}(B)\, \Delta z. \qquad (6.13.14)$$

$\mathcal{N}(B)$ is the number of longitudinal modes per unit length within the frequency range B. As discussed in Section 3.7, $\mathcal{N}(B) = \Delta k/2\pi = B/c$. Then, for $\Delta z \to 0$,

$$\frac{d\text{Pwr}}{dz} = h\nu N_2 V R_{\text{spon}} \frac{B}{c} = h\nu N_2 \frac{A_{21} \lambda^2}{8\pi} S(\nu) B. \qquad (6.13.15)$$

The spontaneously emitted radiation is amplified as it propagates in the medium with (small-signal) gain coefficient g_0, and therefore the ASE noise power propagates according to the equation

$$\frac{d(\text{Pwr})_N}{dz} = g_0 (\text{Pwr})_N + h\nu N_2 \frac{A_{21} \lambda^2}{8\pi} S(\nu) B; \qquad (6.13.16)$$

the solution of this equation with $(\text{Pwr})_N(0) = 0$ is

$$(\text{Pwr})_N(z) = h\nu \frac{N_2}{N_2 - N_1} B[e^{g_0 z} - 1] = n_{\text{sp}} h\nu B[G(z) - 1], \qquad (6.13.17)$$

where we have used the formula $g_0(\nu) = (\lambda^2 A_{21}/8\pi)(N_2 - N_1)S(\nu)$. •

The expression for $(Pwr)_N$ also applies to an *absorbing* medium when we replace $G = \exp(g_0 z)$ by $\exp(-az)$, where a is the absorption coefficient:

$$(Pwr)_N = n_{sp}h\nu B(e^{-az} - 1) = \frac{N_2}{N_1 - N_2}h\nu B(1 - e^{-az}). \qquad (6.13.18)$$

In thermal equilibrium at temperature T, $N_2/N_1 = \exp(-h\nu/k_B T)$ and

$$(Pwr)_N = \frac{h\nu}{e^{h\nu/k_B T} - 1}B(1 - e^{-az}). \qquad (6.13.19)$$

Thus, an absorber also generates spontaneous emission noise. In the case of large loss $(e^{-az} \to 0)$ it acts as a blackbody, and $(Pwr)_N$ is just the power within a bandwidth B of traveling-wave thermal radiation. When $e^{-az} \to 0$ and $k_B T \gg h\nu$ Eq. (6.13.19) gives

$$(Pwr)_N = k_B TB, \qquad (6.13.20)$$

a formula well known in radio engineering.

Amplified spontaneous emission radiation can have spatial coherence comparable to true laser radiation, but it generally lacks the same degree of temporal coherence. (See Chapter 13 for a discussion of spatial and temporal coherence.) The latter property is understandable from the fact that ASE is basically amplified "noise." Equation (6.13.17) implies that the spectrum of ASE is narrower than the gain bandwidth: $\exp[g(\nu)z] - 1$ has a narrower distribution in frequency than $g(\nu)$. The bandwidth of ASE is typically a few times smaller than the gain bandwidth.

Amplified spontaneous emission can limit the amplification of an input signal if it becomes strong enough to saturate the gain, or if the input signal power is small compared to the "effective noise input" for ASE. While our derivation of (6.13.12) does not account for saturation, it allows us to obtain a simple estimate of the importance of ASE "noise per mode" in high-gain amplifiers (Section 11.14).

• Because it accounts in an approximate way for ASE into a solid angle Ω, Eq. (6.13.3) can be used to estimate ASE intensity when there is no restriction on the number of transverse field modes. Comparison of (6.13.16) and (6.13.2) shows that the latter does not account for the lineshape function $S(\nu)$ that determines the contribution of spontaneous emission to the intensity at frequency ν. The approximation (6.13.3) can therefore be expected to overestimate the total ASE intensity. To obtain a better approximation we replace the second term on the right-hand side of (6.13.2) by $A_{21}N_2 h\nu(\Omega/4\pi)S(\nu)\,d\nu$, as in (6.13.16), and then integrate over z and ν to obtain the total noise intensity:

$$I(z) = A_{21}h\nu_0\frac{\Omega}{4\pi}N_2\int_{-\infty}^{\infty}\frac{S(\nu)}{g_0(\nu)}\left[e^{g_0(\nu)z} - 1\right]d\nu. \qquad (6.13.21)$$

We have replaced $h\nu$ by $h\nu_0$ and extended the lower limit of integration from 0 to $-\infty$; both approximations introduce negligible error because the integrand is sharply peaked at $\nu = \nu_0$, which is much greater than the width of the gain lineshape function $S(\nu)$. For definiteness we will assume a Lorentzian form for the latter:

$$g_0(\nu) = \frac{\lambda^2 A_{21}}{8\pi}(N_2 - N_1)\frac{\delta\nu_0/\pi}{(\nu - \nu_0)^2 + \delta\nu_0^2} = \frac{\lambda^2 A_{21}}{8\pi}(N_2 - N_1)\frac{1}{\pi\delta\nu_0}\frac{\delta\nu_0^2}{(\nu - \nu_0)^2 + \delta\nu_0^2}$$

$$= \sigma(\nu_0)(N_2 - N_1)\frac{\delta\nu_0^2}{(\nu - \nu_0)^2 + \delta\nu_0^2}, \tag{6.13.22}$$

and therefore $S(\nu)/g_0(\nu) = [\sigma(\nu_0)(N_2 - N_1)\pi\delta\nu_0]^{-1}$. Next we introduce the new variable $x = (\nu - \nu_0)/\delta\nu_0$ in (6.13.21):

$$I(z) = A_{21}h\nu_0\frac{\Omega}{4\pi}n_{\rm sp}\frac{1/\pi}{\sigma(\nu_0)}\int_{-\infty}^{\infty}\left[e^{g_0(\nu_0)z/(x^2+1)} - 1\right]dx. \tag{6.13.23}$$

The integral can be approximated using the "method of steepest descents." We write

$$e^{g_0(\nu_0)z/(x^2+1)} - 1 = e^{F(x)}, \qquad F(x) = \ln\left[e^{g_0(\nu_0)z/(x^2+1)} - 1\right], \tag{6.13.24}$$

and expand $F(x)$ in a Taylor series about the point $x = 0$ where it has its maximum value: $F(x) = F(0) + \frac{1}{2}F''(0)x^2 + \cdots = F(0) - Kx^2 + \cdots$, where $F'(0) = 0$ and $K > 0$ since $F(0)$ is a maximum. Keeping only the first two nonvanishing terms in the Taylor series, we obtain

$$\int_{-\infty}^{\infty}\left[e^{g_0(\nu_0)z/(x^2+1)} - 1\right]dx \cong e^{F(0)}\int_{-\infty}^{\infty}e^{-Kx^2}dx = e^{F(0)}\sqrt{\frac{\pi}{K}}. \tag{6.13.25}$$

From the definition of $F(x)$, it follows that $F(0) = \exp[g_0(\nu_0)z] - 1 = G - 1$, $K = G\ln G/(G - 1)$, and therefore

$$\int_{-\infty}^{\infty}\left[e^{g_0(\nu_0)z/(x^2+1)} - 1\right]dx \cong \sqrt{\pi}\frac{(G - 1)^{3/2}}{(G\ln G)^{1/2}} \tag{6.13.26}$$

and

$$I(z) \cong A_{21}h\nu_0\frac{\Omega}{4\pi}n_{\rm sp}\frac{1}{\sigma(\nu_0)}\frac{(G - 1)^{3/2}}{(\pi G\ln G)^{1/2}} \tag{6.13.27}$$

Finally, to put this result into another form that appears often in the research literature, we use the formula $I_{\nu_0}^{\rm sat} = h\nu_0\Gamma_{21}/\sigma(\nu_0)$ [Eq. (4.12.11)] for the line-center saturation intensity of a laser described by the four-level model to write (6.13.27) as

$$I(z) \cong \frac{A_{21}}{\Gamma_{21}}\frac{\Omega}{4\pi}n_{\rm sp}I_{\nu_0}^{\rm sat}\frac{(G - 1)^{3/2}}{(\pi G\ln G)^{1/2}}. \tag{6.13.28}$$

This can be compared to the expression (6.13.3), which can be written similarly as

$$I(z) \cong \frac{A_{21}}{\Gamma_{21}}\frac{\Omega}{4\pi}n_{\rm sp}I_{\nu_0}^{\rm sat}(G - 1). \tag{6.13.29}$$

As expected, this approximation predicts a somewhat larger ASE intensity than (6.13.28) in cases of practical interest. It should also be noted that (6.13.28) depends on the assumption of a Lorentzian lineshape; different lineshapes will lead to modifications of (6.13.28) by numerical factors ~ 2.[3] •

[3]O. Svelto, S. Taccheo, and C. Svelto, *Optics Communications* **149**, 277 (1998).

6.14   ULTRASHORT LIGHT PULSES

With mode-locked lasers it is possible to produce ultrashort, extremely intense pulses of radiation. Mode-locked lasers using saturable absorbers are used to produce, rather routinely, picosecond pulses with peak powers in some cases exceeding 10^{11} W (100 GW). In Chapter 11 we will discuss techniques for producing even shorter and more powerful light pulses; here we make some qualitative remarks about ultrashort pulses and ways to generate them.

There are many scientific and technological applications of these ultrashort light pulses. For instance, they can be used to study extremely fast (femtosecond time scale) photoprocesses in molecules and semiconductors (Section 14.7). Especially promising applications may be possible in biological systems, where it has already been determined that certain fundamental chemical reactions occur on picosecond time scales. Intense laser pulses have also been of interest in connection with laser isotope separation and controlled thermonuclear fusion, and it seems safe to say that newer applications will continue to be developed. Not surprisingly, therefore, an entire field of research has grown up around the generation of ultrashort light pulses.

One way to generate ultrashort pulses is with a *colliding-pulse laser*. This is a mode-locked, three-mirror ring laser with two countercirculating pulse trains (Fig. 6.17). Pulses in each direction pass through a very thin (about 10 μm) jet of a saturable absorbing dye. The absorption coefficient of the absorber is smallest when the intensity is largest. Therefore, the cavity loss is least when the countercirculating pulses collide and overlap within the thin dye jet. The thinness of the absorber forces the pulses to overlap within a very short distance and thus over a very short time interval ($\tau \sim 10$ μm/$c \sim 3 \times 10^{-14}$ s).

Another technique involves *chirping*, by which an ultrashort pulse from a laser can be further compressed. A *chirped* pulse is one in which the carrier frequency ω has a small time dependence, typically a linear time dependence of the form $\omega = \omega_0 + \beta t$. A pulse can be deliberately chirped by passing it through a medium with a nonlinear refractive index, that is, a medium in which the refractive index depends upon the electric field. The chirping results in a spectral broadening of the pulse, that is, it extends the range of frequency components contained in the pulse. A chirped pulse can be compressed by passing it through a dispersive (i.e., frequency-dependent) delay line. If the higher frequency components of the pulse travel more slowly than the lower frequency components, the delay line is designed to make them catch up with the lower frequencies

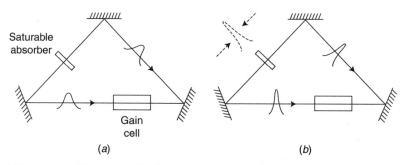

Figure 6.17   (*a*) A colliding-pulse ring laser with countercirculating pulses. (*b*) The lowest-loss condition is for the colliding pulses to synchronize and overlap inside the thin saturable absorber.

on the leading temporal edge of the pulse. By broadening the spectral width of the pulse by chirping, therefore, it is possible to narrow it in time. Optical pulses shorter than 10 fs (1 femtosecond $= 10^{-15}$ s) have been achieved by compressing in this way ultrashort pulses from a colliding-pulse dye laser. The most common way of generating such short pulses uses Kerr lens mode locking of Ti : sapphire lasers (Section 11.13).

APPENDIX: DIFFRACTION OF LIGHT BY SOUND

The diffraction of light by sound waves can be understood by analogy with the diffraction of X rays by crystals. The atoms of a crystal are spaced in a regular pattern, and consequently they scatter radiation cooperatively, with well-defined phase relations between the fields scattered by different atoms. This results in scattering only in certain well-defined directions, and the process is usually called "diffraction" instead of "scattering." Figure 6.18 shows a wave incident upon a stack of crystal planes separated by a distance d. The allowed diffraction angles are determined by the condition of constructive interference of the fields "reflected" from different planes. As shown in the figure, these diffraction angles satisfy the Bragg diffraction formula

$$2d \sin \theta = m\lambda, \qquad m = 1, 2, 3, \ldots, \qquad (6.A.1)$$

where λ is the wavelength of radiation and d is the separation distance between adjacent crystal planes.

- Since d is on the order of 1 Å in actual crystals, only wavelengths in the X-ray region can satisfy (6.A.1) and the requirement $|\sin \theta| \leq 1$. The measurement of X-ray diffraction angles thus provides information about crystal structure. Indeed the use of crystals as "diffraction gratings" for X rays has been one of the most important techniques of modern science for probing the structure of matter (e.g., in the discovery of the double-helix structure of DNA). This technique was originally suggested by Max von Laue. The idea arose in connection with the question whether X rays were particles or waves. L. Brillouin predicted the diffraction of radiation by sound waves in 1922, and it was first observed 10 years later by P. W. Debye and F. W. Sears. •

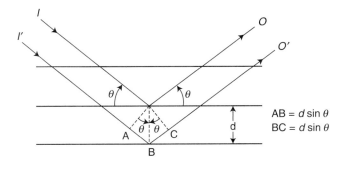

Figure 6.18   Diffraction of a plane wave by a stack of crystal planes. Constructive interference of the two waves IO and $I'O'$ occurs when their path difference $AB + BC$ is equal to an integral multiple m of the wavelength λ. This gives the Bragg diffraction formula $2d \sin \theta = m\lambda$ for the allowed diffraction angles θ. A more complete analysis shows that these angles give the only directions in which scattering occurs.

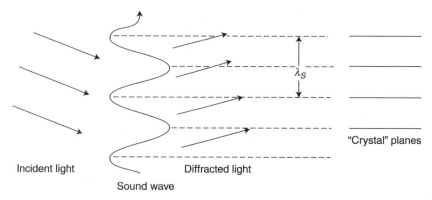

Figure 6.19 Intuitive picture of diffraction of light by sound as diffraction from a fictitious set of "crystal" planes defined by the intensity maxima of the sound waves.

The refractive-index variation associated with a sound wave of wavelength λ_S has a spatial dependence of the form

$$\Delta n(x) = a \sin k_S x, \tag{6.A.2}$$

where $k_S = 2\pi/\lambda_S$ and a depends on material constants of the medium and the intensity of the sound wave. Equation (6.A.2) arises from the fact that a sound wave is basically a wave of density variation. Figure 6.19 shows an intuitive way of understanding the diffraction of light by a sound wave. We regard the planes of constant x where (6.A.2) is a maximum as "crystal" planes that, because of their regular spacing λ_S, will diffract light only in certain well-defined directions. Indeed it turns out that the diffraction angles are given by the Bragg formula (6.A.1) with $d = \lambda_S$ and $m = 1$:

$$2\lambda_S \sin \theta = \frac{\lambda}{n}. \tag{6.A.3}$$

We have included the effect of the refractive index n of the medium.

The important difference between (6.A.1) and (6.A.3) is that there are no higher-order diffraction angles corresponding to $m > 1$ in (6.A.3). This difference arises from the fact that the "diffraction grating" associated with the sound wave indicated in Fig. 6.19 is really a spatially continuous one, not a discrete set of crystal planes with nothing in between.

Equation (6.A.3) gives only the diffraction angle. It does not tell us the strength with which the sound wave diffracts light, that is, the fraction of light intensity diffracted after a given distance of propagation. This is determined by a and the wavelength of the light.

PROBLEMS

6.1. (a) Write the steady-state solutions of Eq. (6.2.5) in such a way as to show the saturation of $\bar{N}_2$ with increasing $\bar{I}_\nu$.

    (b) Verify Eq. (6.3.16) and (6.3.17) for the period and lifetime of relaxation oscillations.

6.2. (a) Show that the quantity x defined in Eq. (6.4.1) is the cavity photon number density divided by the threshold population inversion density.

(b) Show that Eqs. (6.2.5) may be written as (6.4.3) when the change of variables (6.4.1)–(6.4.2) is made and pumping and relaxation of N_2 is ignored.

6.3. (a) Why is it that so much more power can be obtained from a Q-switched laser than in ordinary continuous-wave operation?

(b) Suppose a Q-switched laser using a rotating mirror or a saturable absorber is pumped continuously. How do you expect the laser to behave?

6.4. Set up the equations for x and y [Eqs. (6.4.1)] in the case where Q switching is done with a saturable absorber with absorption coefficient $a = a_0(1 + I/I_\nu^{sat})^{-1}$. Solve these equations numerically using, for example, the Runge–Kutta algorithm of Section 16.A. You will have to assume values of the small-signal absorption coefficient a_0 and the saturation intensity I_ν^{sat} of the absorber. Determine how x and y depend on the choice of a_0 and I_ν^{sat}.

6.5. (a) Suppose we have N oscillators whose frequencies are given by (6.7.2) and whose phases ϕ_n are fixed but not "locked" according to (6.7.10). Discuss the properties of the sum $X(t)$ of the oscillator displacements in this case. Can the maximum value of $X(t)$ be as large as in the phase-locked case?

(b) Suppose that the phases ϕ_n are randomly chosen from an ensemble and are completely uncorrelated, so that $\langle e^{i(\phi_m - \phi_n)} \rangle = \delta_{mn}$, where $\langle \ldots \rangle$ indicates an ensemble average. Compute $\langle X(t) \rangle$ and $\langle X^2(t) \rangle$.

6.6. (a) Show that each pulse of a mode-locked pulse train has an intensity N times larger than the sum of the intensities of the individual modes constituting it.

(b) Show that the average intensity of a mode-locked pulse train is equal to the sum of the intensities of the individual modes constituting it.

6.7. Make a plot of the time-dependent factor in the curly brackets in Eq. (6.9.5), choosing $\phi_m = 0$, $\Omega = \omega_m/10$, and (a) $\varepsilon = \frac{1}{2}$, (b) $\varepsilon = 1$, and (c) $\varepsilon = 5$.

6.8. Consider the 632.8-nm He–Ne laser.

(a) Estimate the shortest pulse that can be obtained by mode-locking such a laser.

(b) What is the duration of each pulse of the mode-locked train if the gain tube has length $l = 10$ cm and the mirror separation $L = 40$ cm?

(c) What is the separation between the mode-locked pulses in part (b)?

(d) Why do liquid dye and solid-state lasers produce much shorter mode-locked pulses than typical gas lasers?

6.9. (a) Estimate the average power, in watts, expended by a normal human adult. Assume that a "normal human adult" consumes 2500 dietitian's calories (2500×4185 J) per day, and that his output energy just balances his input energy.

(b) Estimate the intensity at a distance of 1 m from a 60-W light bulb.

(c) Estimate the average electrical power used to operate a typical house in your area.

6.10. It is possible to effectively "switch off" the output mirror of a laser. What is the advantage of *cavity dumping* in this way? Can you think of a way to do this?

6.11. Our discussion of amplified spontaneous emission in Section 6.13 assumes that the amplifier is continuously pumped. Discuss the modifications necessary to treat the case in which all the atoms of the amplifier are excited at $t = 0$ and emit spontaneously thereafter without any pumping.

7 LASER RESONATORS AND GAUSSIAN BEAMS

7.1 INTRODUCTION

Until now we have supposed a laser resonator to consist of two highly reflecting, flat, parallel mirrors separated by some distance L. The only important property of such a resonator for our purposes thus far is that it has "longitudinal modes" separated in frequency by $c/2L$. We have not concerned ourselves with how the field inside the resonator varies in directions transverse to the line joining the centers of the mirrors. In fact, we have assumed the field to be uniform in any plane perpendicular to this so-called optical axis. In this chapter we will consider laser resonators more realistically. We will consider some of the important characteristics of actual laser resonators, beginning with a rather simple approach based on geometrical optics, and gradually working our way up to a description based on Maxwell's equations.

Most of our treatment of laser resonators will assume that the laser medium is passive. That is, the electromagnetic modes of the laser resonator will be assumed to be the same as the modes of an empty resonator having no gain medium. This is a good approximation if the gain coefficient and refractive index of the medium are fairly uniform throughout the medium. This is obviously a useful approximation because it allows us to consider laser resonators independently of the laser medium. Fortunately, it is often an accurate approximation. It is not, however, applicable to fiber lasers, where the gain and index are tightly confined near the fiber axis. However, even in the case of fibers the fields can often be approximated by Gaussian beams, which we treat in detail in this chapter. Modes of optical fibers are discussed in the following chapter.

Figure 7.1a shows a light ray normal to the mirrors of a resonator with flat, parallel mirrors. The ray keeps retracing its path on successive reflections from the mirrors. If the mirrors are not perfectly parallel, however, the ray will eventually escape from the resonator, as indicated in Fig. 7.1b. The misaligned resonator of Fig. 7.1b requires greater gain for laser oscillation than the resonator of Fig. 7.1a. We might find, for instance, that a laser with flat mirrors turns off (i.e., laser action ceases) at the slightest misalignment of the mirrors. Obviously, this is undesirable if we wish to construct a practical and durable laser. Figure 7.2 shows a much more commonly used type of laser resonator, consisting of mirrors with spherical surfaces. This is the type of resonator used in most gas lasers, for instance. In Section 7.3 we will see why.

Laser Physics. By Peter W. Milonni and Joseph H. Eberly
Copyright © 2010 John Wiley & Sons, Inc.

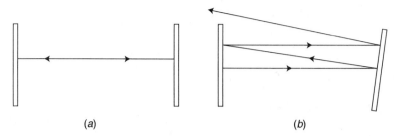

(a)   *(b)*

Figure 7.1   A laser resonator with flat, parallel mirrors. A light ray parallel to the optical axis remains inside the resonator if the mirrors are perfectly parallel (*a*). Otherwise it eventually escapes (*b*).

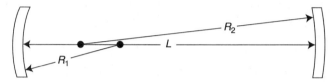

Figure 7.2   A laser resonator with mirrors that are spherical surfaces with radii of curvature R_1 and R_2.

7.2   THE RAY MATRIX

In geometrical optics light propagation is described in terms of rays. We may define a ray at each point on a wave as an arrow drawn normal to the wavefront (Fig. 7.3). We will assume that the direction of a ray is the direction of energy flow. There is no physical significance to the "length" of a ray; a ray merely represents a direction of propagation at a given point. When we adopt this ray picture, we are ignoring the polarization of the light waves. Our ray picture is a crude but useful representation of the actual physical situation.

In this section we will develop a convenient formalism for ray propagation. This formalism will turn out to be useful for the description of Gaussian laser beams, which are discussed in Section 7.5.

In situations of practical interest we are dealing with light waves traveling more or less in a single direction, which we will call the z direction. The rays we envision point almost parallel to the z axis. At any point on the wave we imagine a ray having a lateral displacement $r(z)$, measured from the z axis, and a slope (Fig. 7.4)

$$r'(z) = \frac{dr}{dz}. \tag{7.2.1}$$

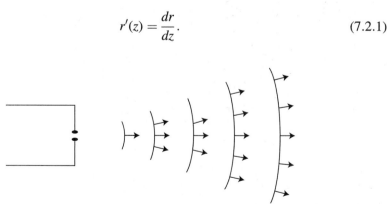

Figure 7.3   Rays drawn on a wave represent the direction of propagation.

Figure 7.4   A ray is characterized by its displacement r and slope r' measured from some z axis.

Because of our assumption of nearly unidirectional propagation along z, the slope $r'(z)$ of a ray will be very small, so that (Fig. 7.4)

$$r'(z) = \tan \theta \approx \sin \theta \approx \theta. \tag{7.2.2}$$

Such rays are called *paraxial rays*. We will assume, as is implicit in our definition of the ray displacement r and slope r', that we have cylindrical symmetry about the z axis. The slope of a ray is taken to be positive or negative depending on whether the displacement r is increasing or decreasing in the direction of propagation.

We would like to relate the displacement and slope of a ray at a point z to the displacement and slope at a point z'. Consider, for example, the simple case of vacuum propagation from z_1 to z_2. In vacuum there is nothing to change the direction of a ray, so we have (Fig. 7.5)

$$r(z_2) = r(z_1) + r'(z_1)(z_2 - z_1), \tag{7.2.3}$$

and

$$r'(z_2) = r'(z_1). \tag{7.2.4}$$

In matrix notation we may write Eqs. (7.2.3) and (7.2.4) as

$$\begin{bmatrix} r(z_2) \\ r'(z_2) \end{bmatrix} = \begin{bmatrix} 1 & z_2 - z_1 \\ 0 & 1 \end{bmatrix} \begin{bmatrix} r(z_1) \\ r'(z_1) \end{bmatrix}. \tag{7.2.5}$$

A ray is completely characterized by the 2×1 matrix, or *column vector*,

$$\begin{bmatrix} r \\ r' \end{bmatrix}, \tag{7.2.6}$$

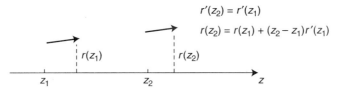

Figure 7.5   The transformation of a ray as a result of free propagation over a distance $z_2 - z_1$.

and Eq. (7.2.5) relates the *final ray*

$$\begin{bmatrix} r_f \\ r_f' \end{bmatrix} = \begin{bmatrix} r(z_2) \\ r'(z_2) \end{bmatrix}, \tag{7.2.7}$$

to the *initial ray*

$$\begin{bmatrix} r_i \\ r_i' \end{bmatrix} = \begin{bmatrix} r(z_1) \\ r'(z_1) \end{bmatrix}. \tag{7.2.8}$$

Thus, according to Eq. (7.2.5), the vacuum propagation of a ray through a distance $d = z_2 - z_1$ is described by the matrix equation

$$\begin{bmatrix} r_f \\ r_f' \end{bmatrix} = \begin{bmatrix} 1 & d \\ 0 & 1 \end{bmatrix} \begin{bmatrix} r_i \\ r_i' \end{bmatrix}. \tag{7.2.9}$$

Given the initial ray with displacement r_i and slope r_i', this equation tells us how that ray is modified by propagation through a distance d.

Consider next the more interesting example of the transformation of a (paraxial) ray by a thin lens of focal length f (Fig. 7.6). Immediately to the right of the lens the ray's lateral displacement r_f is the same as the initial displacement r_i immediately to the left:

$$r_f = r_i. \tag{7.2.10}$$

The slope of the ray, however, is changed by the lens. From the thin lens equation relating the object and image distances with the focal length of the lens, we obtain (Fig. 7.6)

$$r_f' = -\frac{r_i}{d_i} = \frac{r_i}{d_o} - \frac{r_i}{f} = r_i' - \frac{r_i}{f}. \tag{7.2.11}$$

In matrix notation Eqs. (7.2.10) and (7.2.11) take the form

$$\begin{bmatrix} r_f \\ r_f' \end{bmatrix} = \begin{bmatrix} 1 & 0 \\ -1/f & 1 \end{bmatrix} \begin{bmatrix} r_i \\ r_i' \end{bmatrix}. \tag{7.2.12}$$

One more example will be of interest to us, namely the case of a spherical mirror with radius of curvature R. The displacement of the ray is the same immediately before and after reflection from the mirror, that is, $r_f = r_i$. The slope of the ray after reflection,

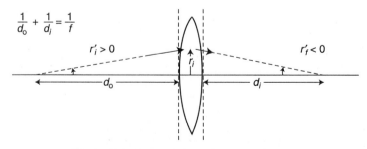

Figure 7.6 Ray transformation by a thin lens.

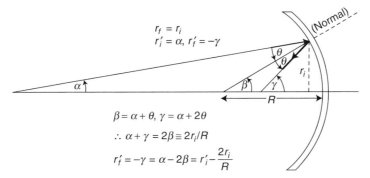

$$r_f = r_i$$
$$r'_i = \alpha, \; r'_f = -\gamma$$

$$\beta = \alpha + \theta, \; \gamma = \alpha + 2\theta$$

$$\therefore \; \alpha + \gamma = 2\beta \cong 2r_i/R$$

$$r'_f = -\gamma = \alpha - 2\beta = r'_i - \frac{2r_i}{R}$$

Figure 7.7 Paraxial ray transformation by a spherical mirror surface with radius of curvature R. The relation between r'_f and r'_i is obtained by applying the trigonometric theorem that an exterior angle of a triangle equals the sum of the two opposite interior angles, and the approximation $\beta \approx 2r_i/R$ that holds for paraxial rays.

however, is (Fig. 7.7)

$$r'_f = r'_i - \frac{2r_i}{R}. \tag{7.2.13}$$

In matrix notation, therefore, the ray transformation by the spherical mirror is given by the equation

$$\begin{bmatrix} r_f \\ r'_f \end{bmatrix} = \begin{bmatrix} 1 & 0 \\ -2/R & 1 \end{bmatrix} \begin{bmatrix} r_i \\ r'_i \end{bmatrix}. \tag{7.2.14}$$

There is a sign convention for r', namely $r' > 0$ if r is increasing with propagation, $r' < 0$ otherwise. With this in mind, our sign convention for the radius of curvature R of a spherical mirror is easily checked: R is positive for a concave mirror (Fig. 7.7) and negative for a convex mirror. Similarly, the focal length f of a lens is positive for a converging lens (Fig. 7.6) and negative for a diverging lens. These statements may be verified by making sketches like those in Figs. 7.6 and 7.7. Thus, Eqs. (7.2.12) and (7.2.14) apply also to diverging lenses and convex mirrors, respectively, provided f and R are taken to be negative in those cases.

We have considered thus far the transformation of a ray by three different "optical elements"—empty space of length d, a thin lens of focal length f, and a spherical mirror of radius of curvature R. In general, an optical element will transform a ray according to the matrix equation

$$\begin{bmatrix} r_f \\ r'_f \end{bmatrix} = \begin{bmatrix} A & B \\ C & D \end{bmatrix} \begin{bmatrix} r_i \\ r'_i \end{bmatrix}. \tag{7.2.15}$$

The 2×2 matrix on the right-hand side of this equation is called the *ray matrix*, or *ABCD* matrix, for the optical element. Equations (7.2.9), (7.2.12), and (7.2.14) give the ray matrices for a straight section of length d, a thin lens of focal length f, and a spherical mirror of radius of curvature R, respectively.

Let us consider the effect on a ray of an open path section of length d followed by a thin lens of focal length f. If a ray has displacement r_i and slope r_i' initially, then after the open section of propagation it has displacement r and slope r' given by Eq. (7.2.9):

$$\begin{bmatrix} r \\ r' \end{bmatrix} = \begin{bmatrix} 1 & d \\ 0 & 1 \end{bmatrix} \begin{bmatrix} r_i \\ r_i' \end{bmatrix}. \tag{7.2.16}$$

This gives the "initial" ray displacement and slope immediately before passage through the lens. The "final" ray displacement and slope are therefore given by Eq. (7.2.12):

$$\begin{bmatrix} r_f \\ r_f' \end{bmatrix} = \begin{bmatrix} 1 & 0 \\ -1/f & 1 \end{bmatrix} \begin{bmatrix} r \\ r' \end{bmatrix} = \begin{bmatrix} 1 & 0 \\ -1/f & 1 \end{bmatrix} \begin{bmatrix} 1 & d \\ 0 & 1 \end{bmatrix} \begin{bmatrix} r_i \\ r_i' \end{bmatrix}$$

$$= \begin{bmatrix} 1 & d \\ -1/f & 1 - d/f \end{bmatrix} \begin{bmatrix} r_i \\ r_i' \end{bmatrix}. \tag{7.2.17}$$

The matrix

$$\begin{bmatrix} 1 & d \\ -1/f & 1 - d/f \end{bmatrix} = \begin{bmatrix} 1 & 0 \\ -1/f & 1 \end{bmatrix} \begin{bmatrix} 1 & d \\ 0 & 1 \end{bmatrix} \tag{7.2.18}$$

is therefore the ray matrix for the combined optical system consisting of an open section of length d followed by a thin lens of focal length f. It is the product of the ray matrices for an open section and a lens. It follows that if we have any number of optical elements in some sequence, then the ray matrix for the system comprising all these elements is the matrix product of the ray matrices of the individual elements. Since the matrix product M_1M_2 is in general not the same as M_2M_1, the order of the matrices in the product is important. Thus, the system ray matrix is the ray matrix of the first optical element encountered, multiplied on the left by the ray matrix of the second optical element, multiplied on the left by the ray matrix of the third element, and so forth. The reader may easily show, for instance, that the ray matrix for the system consisting of an open section followed by a thin lens is different from the ray matrix for a thin lens followed by an open section (Problem 7.1). This means, of course, that the effects of the two systems on a ray are different.

7.3 RESONATOR STABILITY

One of the simplest but most important questions concerning a laser resonator is whether it is stable. To see what this means, consider an arbitrary (paraxial) ray bouncing back and forth between the mirrors of a resonator. If the ray remains within the resonator, the resonator is said to be stable. If, however, the ray escapes from the resonator after a number of reflections, the resonator is unstable. Figure 7.1b, for example, shows that a misaligned flat-mirror resonator is unstable. In general, a stability criterion for a laser resonator can be expressed in terms of the radii of curvature of the mirrors and the distance separating the mirrors. We will now derive this stability criterion with the aid of the *ABCD* matrix.

Consider the resonator sketched in Fig. 7.2, consisting of mirrors of radii of curvature R_1 and R_2, separated by a distance L. As drawn, the mirrors are concave. Our analysis, however, will apply also to the case of convex mirrors if we recall that a convex mirror by convention has a negative radius of curvature. We note also that a flat mirror may be regarded as a spherical mirror surface with an infinite radius of curvature.

Imagine a ray starting at the left mirror of Fig. 7.2. After a round trip through the resonator, this ray will have been transformed by a straight section of length L, a spherical mirror of radius of curvature R_2, another straight section of length L, and finally a spherical mirror of radius of curvature R_1. The ray matrix describing the ray transformation by a round trip through the resonator is

$$
\begin{bmatrix} A & B \\ C & D \end{bmatrix} = \begin{bmatrix} 1 & 0 \\ -\dfrac{2}{R_1} & 1 \end{bmatrix} \begin{bmatrix} 1 & L \\ 0 & 1 \end{bmatrix} \begin{bmatrix} 1 & 0 \\ -\dfrac{2}{R_2} & 1 \end{bmatrix} \begin{bmatrix} 1 & L \\ 0 & 1 \end{bmatrix}
$$

$$
= \begin{bmatrix} 1 - \dfrac{2L}{R_2} & 2L - \dfrac{2L^2}{R_2} \\ \dfrac{4L}{R_1 R_2} - \dfrac{2}{R_1} - \dfrac{2}{R_2} & 1 - \dfrac{2L}{R_2} - \dfrac{4L}{R_1} + \dfrac{4L^2}{R_1 R_2} \end{bmatrix}. \tag{7.3.1}
$$

After N round trips through the resonator, therefore, the initial ray with displacement r_i and slope r_i' is transformed to the ray with displacement r_N and slope r_N' given by

$$
\begin{bmatrix} r_N \\ r_N' \end{bmatrix} = \begin{bmatrix} A & B \\ C & D \end{bmatrix}^N \begin{bmatrix} r_i \\ r_i' \end{bmatrix}, \tag{7.3.2}
$$

where the ray (*ABCD*) matrix is defined by (7.3.1). This ray matrix has determinant[1]

$$
AD - BC = 1. \tag{7.3.3}
$$

Using this fact, and defining an angle θ by

$$
\cos \theta = \tfrac{1}{2}(A + D), \tag{7.3.4}
$$

it may be shown (see below) that

$$
\begin{bmatrix} A & B \\ C & D \end{bmatrix}^N = \frac{1}{\sin \theta} \begin{bmatrix} A \sin N\theta - \sin (N-1)\theta & B \sin N\theta \\ C \sin N\theta & D \sin N\theta - \sin (N-1)\theta \end{bmatrix}. \tag{7.3.5}
$$

[1]The simplest way to check this is to note that the ray matrix (7.3.1) is a product of four matrices, each having determinant equal to one. Since the determinant of the product of matrices is equal to the product of the determinants, (7.3.3) follows.

• The result (7.3.5) for a 2×2 matrix satisfying (7.3.3) is sometimes called "Sylvester's theorem." It may be proved by induction: It obviously holds for the case $N = 1$, and so we try to show that if it holds for a single given (but arbitrary) N, it must hold also for $N + 1$. If we can show this, Sylvester's theorem is proved.

Thus, let us assume that (7.3.5) holds, so that

$$
\begin{bmatrix} A & B \\ C & D \end{bmatrix}^{N+1} = \begin{bmatrix} A & B \\ C & D \end{bmatrix} \begin{bmatrix} A & B \\ C & D \end{bmatrix}^{N}
$$

$$
= \frac{1}{\sin \theta} \begin{bmatrix} A & B \\ C & D \end{bmatrix} \begin{bmatrix} A \sin N\theta - \sin(N-1)\theta & B \sin N\theta \\ C \sin N\theta & D \sin N\theta - \sin(N-1)\theta \end{bmatrix}
$$

$$
= \frac{1}{\sin \theta} \begin{bmatrix} (A^2 + BC) \sin N\theta - A \sin(N-1)\theta & B(A+D) \sin N\theta - B \sin(N-1)\theta \\ C(A+D) \sin N\theta - C \sin(N-1)\theta & (BC + D^2) \sin N\theta - D \sin(N-1)\theta \end{bmatrix}.
$$

(7.3.6)

Using (7.3.3) and (7.3.4), we see that the $(1, 1)$ element of this matrix is

$$
\begin{aligned}
(A^2 + BC) \sin N\theta - A \sin(N-1)\theta &= (A^2 + AD - 1) \sin N\theta - A \sin(N-1)\theta \\
&= A(A+D) \sin N\theta - \sin N\theta - A \sin(N-1)\theta \\
&= 2A \sin N\theta \cos \theta - \sin N\theta - A \sin(N-1)\theta \\
&= 2A \left[\tfrac{1}{2} \sin(N+1)\theta + \tfrac{1}{2} \sin(N-1)\theta \right] \\
&\quad - \sin N\theta - A \sin(N-1)\theta \\
&= A \sin(N+1)\theta - \sin N\theta.
\end{aligned}
$$

(7.3.7)

The remaining three matrix elements of (7.3.6) may be evaluated similarly. We obtain

$$
\begin{bmatrix} A & B \\ C & D \end{bmatrix}^{N+1} = \frac{1}{\sin \theta} \begin{bmatrix} A \sin(N+1)\theta - \sin N\theta & B \sin(N+1)\theta \\ C \sin(N+1)\theta & D \sin(N+1)\theta - \sin N\theta \end{bmatrix}.
$$

(7.3.8)

But this is just Eq. (7.3.5) with N replaced by $N + 1$. Thus, (7.3.5) is true for $N = 1$, and we have just shown that if it is true for any N, then it must be true also for $N + 1$. This proves Sylvester's theorem.

It now follows from Eq. (7.3.2) that

$$
\begin{bmatrix} r_N \\ r'_N \end{bmatrix} = \frac{1}{\sin \theta} \begin{bmatrix} A \sin N\theta - \sin(N-1)\theta & B \sin N\theta \\ C \sin N\theta & D \sin N\theta - \sin(N-1)\theta \end{bmatrix} \begin{bmatrix} r_i \\ r'_i \end{bmatrix},
$$

(7.3.9)

where, from (7.3.4) and (7.3.1),

$$
\cos \theta = \frac{1}{2} \left(1 - \frac{2L}{R_2} + 1 - \frac{2L}{R_2} - \frac{4L}{R_1} + \frac{4L^2}{R_1 R_2} \right) = 1 - \frac{2L}{R_1} - \frac{2L}{R_2} + \frac{2L^2}{R_1 R_2}.
$$

(7.3.10)

Equation (7.3.9) gives the ray displacement and slope after N round trips through the resonator. We observe that r_N (and r'_N) stays finite as long as θ is real. If θ is a complex number, however, then $\sin N\theta = (e^{iN\theta} - e^{-iN\theta})/2i$ can be very large for large N, and in fact diverges as $N \to \infty$. In other words, if θ is not purely real, r_N itself will diverge, that is, the ray will escape from the confines of the resonator. Thus, the condition for resonator stability is for θ to be real, which means that $|\cos \theta| \leq 1$, or, from (7.3.10),

$$-1 \leq 1 - \frac{2L}{R_1} - \frac{2L}{R_2} + \frac{2L^2}{R_1 R_2} \leq 1,$$

$$0 \leq 1 - \frac{L}{R_1} - \frac{L}{R_2} + \frac{L^2}{R_1 R_2} \leq 1. \tag{7.3.11}$$

This stability condition is usually written in the laser literature as

$$0 \leq g_1 g_2 \leq 1, \tag{7.3.12}$$

where

$$g_1 = 1 - \frac{L}{R_1}, \tag{7.3.13a}$$

Plane-parallel resonator
$R_1 = R_2 = \infty$
$g_1 g_2 = 1$

Spherical resonator (concentric)
$R_1 = R_2 = L/2$
$g_1 g_2 = 1$

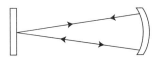

Hemispherical resonator
$R_1 = \infty$, $R_2 = L$
$g_1 g_2 = 0$

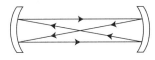

Confocal resonator
$R_1 = R_2 = L$
$g_1 g_2 = 0$

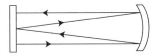

Hemiconfocal resonator
$R_1 = \infty$, $R_2 = 2L$
$g_1 g_2 = \frac{1}{2}$

Figure 7.8 Examples of stable resonators.

and

$$g_2 = 1 - \frac{L}{R_2}. \tag{7.3.13b}$$

These are called the g *parameters* of the resonator. If the g parameters are such that (7.3.12) is satisfied, the resonator is stable. If $g_1 g_2 < 0$ or $g_1 g_2 > 1$, however, the resonator is unstable.

The ray-matrix approach allows us to check immediately whether a given resonator is stable, without having to perform a ray trace such as that shown in Fig. 7.1*b*. Whether a given resonator is stable or unstable depends only on the radii of curvature of the mirrors and the distance separating them. Figures 7.8 and 7.9 show examples of stable and unstable resonators, respectively. The reader may easily check in each case whether the resonator is stable or unstable (Problem 7.2).

Our stability analysis has assumed perfect mirror reflectivities. In reality, of course, some energy will be taken from the intraresonator laser field because of imperfect mirror reflectivities. We have already noted (Chapter 4) that transmissive output coupling through one (or both) of the mirrors is one such loss mechanism. In addition to such loss mechanisms as output coupling, scattering, or absorption, a laser with an unstable resonator will have a large loss associated with the escape of radiation past the mirrors, as indicated by ray tracing as in Figs. 7.1*b* and 7.9. Because of this additional loss factor, unstable resonators typically require media with higher gain to sustain laser oscillation. This is not to say that unstable resonators should always be avoided. On the contrary, unstable resonators offer several advantages for certain high-power

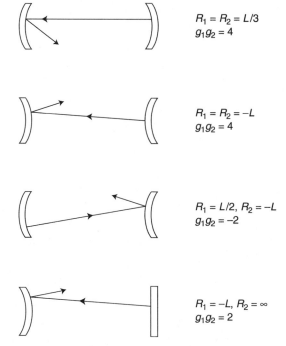

Figure 7.9   Examples of unstable resonators.

lasers (Section 7.14). In more familiar devices such as commercial He–Ne lasers, however, stable resonators are usually employed.

The plane-parallel resonator of Fig. 7.1 is not used for practical lasers because it becomes unstable with only slight misalignment of the mirrors. The resonators of most lasers have at least one spherical mirror surface. The hemispherical resonator of Fig. 7.8, for instance, is perhaps the most commonly used design for He–Ne lasers.

7.4 THE PARAXIAL WAVE EQUATION

Many important properties of laser resonators are consequences of the wave nature of light. A complete understanding of laser resonators, therefore, demands a treatment based on Maxwell's equations rather than geometrical rays. In this section we will examine an approximate solution of the Maxwell wave equation that is very important for laser resonators.

Let us first recall the wave equation (see Chapter 8) for the electric field in vacuum:

$$\nabla^2 E(\mathbf{r}, t) - \frac{1}{c^2} \frac{\partial^2}{\partial t^2} E(\mathbf{r}, t) = 0. \tag{7.4.1}$$

We have written the scalar wave equation instead of the full vector equation. Our treatment will, therefore, account for diffraction and interference of the radiation inside a resonator, but not for polarization effects. A fully vectorial treatment of laser resonators is complicated, but fortunately the scalar theory is quite adequate for our purposes. We will be interested in solutions of (7.4.1) of the form

$$E(\mathbf{r}, t) = \mathcal{E}(\mathbf{r})e^{-i\omega t}, \tag{7.4.2}$$

that is, monochromatic fields. When this expression is used in the wave equation (7.4.1), we obtain the *Helmholtz equation* for $\mathcal{E}(\mathbf{r})$:

$$\nabla^2 \mathcal{E}(\mathbf{r}) + k^2 \mathcal{E}(\mathbf{r}) = 0, \tag{7.4.3}$$

where

$$k^2 = \frac{\omega^2}{c^2}. \tag{7.4.4}$$

A solution of the Helmholtz equation for $\mathcal{E}(\mathbf{r})$ will provide a monochromatic solution (7.4.2) of the wave equation.

One solution of (7.4.3) is

$$\mathcal{E}(\mathbf{r}) = \mathcal{E}_0 e^{i\mathbf{k} \cdot \mathbf{r}}, \tag{7.4.5}$$

where $\mathcal{E}_0$ is a constant and $\mathbf{k}$ is a vector whose squared magnitude is given by (7.4.4). Such a plane-wave solution has the same value for all points in any plane normal to $\mathbf{k}$. If we take $\mathbf{k}$ to point in the z direction, for instance, the solution (7.4.5) has the same value ($\mathcal{E}_0 e^{ikz}$) in any plane defined by a constant value of z. Another solution of

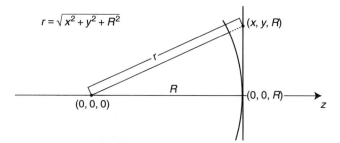

Figure 7.10   Geometry for Eq. (7.4.7).

(7.4.3), valid for all $r \neq 0$, is (Problem 7.3)

$$\mathcal{E}(\mathbf{r}) = \frac{A}{r} e^{ikr}, \tag{7.4.6}$$

where A is an arbitrary number. This solution has a constant value on any sphere centered at the origin and is therefore called a *spherical wave*. This is the form of solution we would associate with a point source at the origin, with the intensity of the wave (square of $|\mathcal{E}|$) decreasing with distance r according to the inverse square law.

Consider the plane $z = R$ (Fig. 7.10). In this plane

$$r = (x^2 + y^2 + R^2)^{1/2} = R \left(1 + \frac{x^2 + y^2}{R^2} \right)^{1/2}. \tag{7.4.7}$$

If we restrict ourselves to a small "patch" of observation about the point ($x = 0$, $y = 0$, $z = R$), so that $x^2 + y^2$ is small compared to R^2, then

$$\left(1 + \frac{x^2 + y^2}{R^2} \right)^{1/2} = 1 + \frac{x^2 + y^2}{2R^2} + \cdots, \tag{7.4.8}$$

according to the binomial expansion. Thus, we use the approximation

$$kr \approx kR + \frac{k(x^2 + y^2)}{2R}. \tag{7.4.9}$$

The field of the spherical-wave solution (7.4.6) on the plane $z = R$ in the vicinity of ($x = 0$, $y = 0$) is therefore

$$\mathcal{E}(\mathbf{r}) = \frac{A}{R} e^{ikR} e^{ik(x^2+y^2)/2R}, \tag{7.4.10}$$

at points for which (7.4.8) is satisfied. Note that we have simply replaced r by R for the approximate evaluation of the factor A/r in (7.4.6). In the factor e^{ikr}, however, we have retained in (7.4.10) both terms on the right-hand side of (7.4.9). This is necessary because, although $(x^2 + y^2)/2R$ may be very small compared to R, it need not be very small compared to a wavelength ($\lambda = 2\pi/k$), and so the second term in (7.4.9) cannot be neglected in e^{ikr}. In order for (7.4.10) to be a good approximation, the next term in the binomial expansion of r must be very small compared to a wavelength. This condition is (Problem 7.3)

$$\frac{a^2}{\lambda R} \ll \left(\frac{R}{a}\right)^2, \tag{7.4.11}$$

where

$$a^2 = x^2 + y^2. \tag{7.4.12}$$

The field (7.4.10) is an accurate approximation to the spherical wave (7.4.6) when (7.4.8) or (7.4.11) is satisfied, that is, when we consider a small enough radius a of observation about the point $(0, 0, R)$ in the plane $z = R$. This approximation is used frequently in physical optics.

Now a laser beam propagates as a nearly unidirectional wave with some finite cross-sectional area. Plane and spherical waves are obviously not beams. A spherical wave is not unidirectional, and a plane wave has an infinite cross-sectional area. We therefore seek solutions of (7.4.3) that look more like beams. To this end we try a solution of the form

$$\mathcal{E}(\mathbf{r}) = \mathcal{E}_0(\mathbf{r})e^{ikz}, \tag{7.4.13}$$

which differs from the plane wave (7.4.5) by the fact that its amplitude is not a constant. In writing (7.4.13) we are seeking a solution that has nearly the unidirectionality of a plane wave, without having an infinite beam cross section. We assume that the variations of $\mathcal{E}_0(\mathbf{r})$ and $\partial\mathcal{E}_0(\mathbf{r})/\partial z$ within a distance of the order of a wavelength in the z direction are negligible, that is,

$$\lambda\left|\frac{\partial\mathcal{E}_0}{\partial z}\right| \ll |\mathcal{E}_0|, \qquad \lambda\left|\frac{\partial^2\mathcal{E}_0}{\partial z^2}\right| \ll \left|\frac{\partial\mathcal{E}_0}{\partial z}\right|, \tag{7.4.14}$$

or, since $k = 2\pi/\lambda$,

$$\left|\frac{\partial\mathcal{E}_0}{\partial z}\right| \ll k|\mathcal{E}_0|, \qquad \left|\frac{\partial^2\mathcal{E}_0}{\partial z^2}\right| \ll k\left|\frac{\partial\mathcal{E}_0}{\partial z}\right|. \tag{7.4.15}$$

Thus, we are assuming that $\mathcal{E}(\mathbf{r})$ varies approximately as e^{ikz} over distances z on the order of several wavelengths.

The field (7.4.13) must satisfy the Helmholtz equation:

$$\left(\frac{\partial^2}{\partial x^2} + \frac{\partial^2}{\partial y^2} + \frac{\partial^2}{\partial z^2}\right)\mathcal{E}_0(\mathbf{r})e^{ikz} + k^2\mathcal{E}_0(\mathbf{r})e^{ikz} = 0. \tag{7.4.16}$$

Now

$$\frac{\partial^2}{\partial z^2}\mathcal{E}_0(\mathbf{r})e^{ikz} = \left(\frac{\partial^2\mathcal{E}_0}{\partial z^2} + 2ik\frac{\partial\mathcal{E}_0}{\partial z} - k^2\mathcal{E}_0\right)e^{ikz} \approx \left(2ik\frac{\partial\mathcal{E}_0}{\partial z} - k^2\mathcal{E}_0\right)e^{ikz}, \quad (7.4.17)$$

in the approximation (7.4.15), and thus from (7.4.16) we have

$$\left(\frac{\partial^2}{\partial x^2} + \frac{\partial^2}{\partial y^2} + 2ik\frac{\partial}{\partial z}\right)\mathcal{E}_0(\mathbf{r}) \approx 0. \quad (7.4.18)$$

We therefore seek solutions to this equation, that is, fields of the form (7.4.13) satisfying (7.4.15). The equation

$$\nabla_T^2\mathcal{E}_0 + 2ik\frac{\partial\mathcal{E}_0}{\partial z} = 0, \quad (7.4.19)$$

is called the *paraxial wave equation*. Here the *transverse Laplacian* is

$$\nabla_T^2 = \frac{\partial^2}{\partial x^2} + \frac{\partial^2}{\partial y^2}. \quad (7.4.20)$$

We will now consider important solutions of this partial differential equation.

7.5   GAUSSIAN BEAMS

The intensity of a "beamlike" wave propagating in the z direction is negligible at points sufficiently far from the z axis. A Gaussian beam intensity profile, for example, has the form

$$I(x, y, z) \sim |\mathcal{E}_0|^2 e^{-2(x^2+y^2)/w^2}, \quad (7.5.1)$$

in a plane normal to the direction (z) of propagation. At a lateral distance w from the z axis, the intensity is a factor e^{-2} ($=0.135$) smaller than its value on axis. If the beam were projected onto a screen we would see a spot of radius $\sim w$, and so w is called the spot size of the Gaussian beam.

Laser beams are frequently observed to have an intensity profile like (7.5.1). With this in mind, we try to construct a solution of (7.4.19) having the form

$$\mathcal{E}_0(\mathbf{r}) = Ae^{ik(x^2+y^2)/2q(z)}e^{ip(z)}, \quad (7.5.2)$$

where A is a constant and $q(z)$ and $p(z)$ are to be determined. Note that if

$$\frac{1}{q} = \frac{2i}{kw^2} = \frac{i\lambda}{\pi w^2}, \quad (7.5.3)$$

then (7.5.2) gives the Gaussian intensity profile (7.5.1). In assuming a solution of the form (7.5.2), in which q depends upon z, we are allowing for the possibility that the spot size of a Gaussian beam can vary with distance of propagation, which is in fact known to occur in laser beams.

For the function (7.5.2) we have

$$\frac{\partial \mathcal{E}_0}{\partial z} = iA \left[\frac{dp}{dz} - \frac{k}{2}(x^2 + y^2) \frac{1}{q^2} \frac{dq}{dz} \right] e^{ik(x^2+y^2)/2q(z)} e^{ip(z)}, \tag{7.5.4}$$

and

$$\nabla_T^2 \mathcal{E}_0 = A \left[\frac{2ik}{q} - \frac{k^2}{q^2}(x^2 + y^2) \right] e^{ik(x^2+y^2)/2q(z)} e^{ip(z)}, \tag{7.5.5}$$

so that

$$\nabla_T^2 \mathcal{E}_0 + 2ik \frac{\partial \mathcal{E}_0}{\partial z} = A \left[\frac{k^2}{q^2}(x^2 + y^2) \left(\frac{dq}{dz} - 1 \right) - 2k \left(\frac{dp}{dz} - \frac{i}{q} \right) \right] e^{ik(x^2+y^2)/2q(z)} e^{ip(z)}. \tag{7.5.6}$$

Therefore, the form (7.5.2) is indeed a solution to Eq. (7.4.19) if $p(z)$ and $q(z)$ satisfy

$$\frac{dq}{dz} = 1, \tag{7.5.7}$$

and

$$\frac{dp}{dz} = \frac{i}{q}. \tag{7.5.8}$$

These equations have the solutions

$$q(z) = q_0 + z, \tag{7.5.9}$$

and

$$p(z) = i \ln \frac{q_0 + z}{q_0}, \tag{7.5.10}$$

where $q_0 = q(0)$ and we assume $p(0) = 0$.

Since q may be complex, we write

$$\frac{1}{q(z)} = \frac{1}{R(z)} + \frac{i\lambda}{\pi w^2(z)}, \tag{7.5.11}$$

with R and w real. This way of writing $1/q$ is suggested by (7.5.3), to which (7.5.11) reduces when $R \to \infty$, that is, when q is purely imaginary. With $1/q$ written this way, we have

$$e^{ik(x^2+y^2)/2q(z)} = e^{ik(x^2+y^2)/2R(z)} e^{-(x^2+y^2)/w^2(z)}. \tag{7.5.12}$$

Using (7.5.10) and (7.5.11), we also have

$$e^{ip(z)} = \exp\left(-\ln\frac{q_0 + z}{q_0}\right) = \frac{q_0}{q_0 + z} = \frac{1}{1 + z/q_0} = \frac{1}{1 + z/R_0 + i\lambda z/\pi w_0^2}, \quad (7.5.13)$$

where R_0 and w_0 denote the values of R and w at $z = 0$.

If R_0 and w_0 are known, Eqs. (7.5.9) and (7.5.11) give $R(z)$ and $w(z)$ for all values of z. Since the designation $z = 0$ is arbitrary, let us choose the plane $z = 0$ to be that for which R_0 is infinitely large, that is,

$$R_0 = \infty, \quad (7.5.14)$$

and

$$\frac{1}{q_0} = \frac{i\lambda}{\pi w_0^2}. \quad (7.5.15)$$

It then follows from (7.5.9) that

$$\frac{1}{q(z)} = \frac{1}{q_0 + z} = \frac{1/q_0}{1 + z(1/q_0)} = \frac{i\lambda/\pi w_0^2}{1 + iz\lambda/\pi w_0^2} = \frac{i\lambda/\pi w_0^2 + (1/z)(\lambda z/\pi w_0^2)^2}{1 + (\lambda z/\pi w_0^2)^2}$$

$$= \frac{1}{R(z)} + \frac{i\lambda}{\pi w^2(z)}. \quad (7.5.16)$$

Equating separately the real and imaginary parts, we have

$$R(z) = z + \frac{z_0^2}{z}, \quad (7.5.17)$$

$$w(z) = w_0\sqrt{1 + z^2/z_0^2}, \quad (7.5.18)$$

where we have defined z_0 by

$$z_0 = \frac{\pi w_0^2}{\lambda}. \quad (7.5.19)$$

This new parameter is known as the *Rayleigh range*, and is discussed below. The alternative term *confocal parameter* (exactly twice the Rayleigh range) is also used to characterize Gaussian beams.

Finally, let us note that (7.5.14) allows us to write (7.5.13) as

$$e^{ip(z)} = \frac{1}{1 + iz/z_0} = \frac{1}{\sqrt{1 + z^2/z_0^2}}e^{-i\phi(z)}, \quad (7.5.20)$$

where

$$\phi(z) = \tan^{-1}\left(\frac{z}{z_0}\right). \quad (7.5.21)$$

With this result and Eq. (7.5.12), we can rewrite our solution (7.5.2) in terms of $R(z)$, $w(z)$, and z_0 satisfying (7.5.17)–(7.5.19). Then the complete paraxial wave solution takes the form

$$\mathcal{E}(\mathbf{r}) = \frac{Ae^{ikz}e^{-i\phi(z)}}{\sqrt{1+z^2/z_0^2}}e^{ik(x^2+y^2)/2R(z)}e^{-(x^2+y^2)/w^2(z)}, \qquad (7.5.22)$$

with A a constant. Thus we obtain the full expression (7.4.13) for a "beamlike" solution to the wave equation, or at least one that is valid within the approximation (7.4.15). The solution (7.5.22) has the Gaussian intensity profile (7.5.1) in any plane $z = $ constant. The spot size $w(z)$ has a minimum value w_0 in some plane $z = 0$ and grows with distance from this plane according to the formula (7.5.18). This behavior is sketched in Fig. 7.11. Note that we do not have an expression for the minimum spot size w_0. The solution (7.5.22) is characterized, except for the trivial constant A, by w_0 and the wavelength λ. For obvious reasons (Fig. 7.11) the plane $z = 0$ is called the *beam waist*.

The distance z_0 defined by (7.5.19) is such that

$$w(z_0) = w_0\sqrt{2}. \qquad (7.5.23)$$

The Rayleigh range z_0 is thus a measure of the length of the waist region, where the spot size is smallest. The smaller the spot size w_0 at the beam waist, the smaller the Rayleigh range, and thus the greater the rate of growth with z of the spot size from the waist. This result is similar to what happens when a plane wave is diffracted by a circular aperture in an opaque screen. The smaller the aperture diameter D, the greater the diffraction. In fact the far-field divergence angle of the diffracted beam (Fig. 7.12) may be defined by

$$\theta = 1.22\frac{\lambda}{D}. \qquad (7.5.24)$$

This result is derived in Section 7.11. We may define the divergence angle of our Gaussian beam similarly, as (Fig. 7.11)

$$\theta \approx \frac{w(z)}{z} \approx \frac{w_0}{z_0} = \frac{\lambda}{\pi w_0} \qquad (z \gg z_0). \qquad (7.5.25)$$

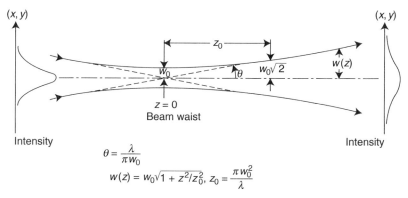

Figure 7.11 Variation of spot size $w(z)$ of a Gaussian beam.

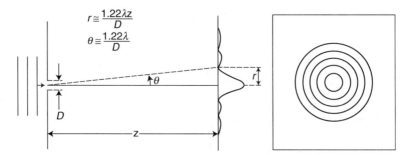

Figure 7.12 Diffraction of a monochromatic plane wave of wavelength λ by a circular aperture of diameter D. The diffraction intensity pattern in the far field consists of a central bright spot of radius r surrounded by faint concentric rings. The rings are not truly equally spaced, and the intensities shown for the rings (side lobes in the left sketch) are exaggerated. See Fig. 7.29.

Thus, the divergence angle of a Gaussian beam is of the same order as that associated with the diffraction of a plane wave by an aperture of diameter $D \sim w_0$ located at the beam waist.

The intensity of the field (7.5.22) averaged over an optical period is (Problem 7.3)

$$I(x, y, z) = \frac{c\epsilon_0}{2} |\mathcal{E}(x, y, z)|^2 = \frac{(c\epsilon_0/2)|A|^2}{1 + z^2/z_0^2} e^{-2(x^2+y^2)/w^2(z)}. \qquad (7.5.26)$$

The rate at which energy crosses any plane defined by a constant value of z is therefore

$$\int_{-\infty}^{\infty} \int_{-\infty}^{\infty} dx\, dy\, I(x, y, z) = \frac{(c\epsilon_0/2)|A|^2}{1 + z^2/z_0^2} \int_{-\infty}^{\infty} \int_{-\infty}^{\infty} dx\, dy\, e^{-2(x^2+y^2)/w^2(z)}$$

$$= \frac{c\epsilon_0}{4} |A|^2 (\pi w_0^2). \qquad (7.5.27)$$

The fact that this expression is independent of z is, of course, consistent with the conservation of energy. For $z \gg z_0$ the beam intensity (7.5.26) has an inverse-square dependence on the distance z from the waist:

$$I(x, y, z) \approx \frac{I_0}{z^2} z_0^2 e^{-2(x^2+y^2)/w^2(z)} \qquad (z \gg z_0), \qquad (7.5.28)$$

where

$$I_0 = \frac{c\epsilon_0}{2} |A|^2. \qquad (7.5.29)$$

The spot size at large distances from the beam waist is

$$w(z) \approx \frac{w_0 z}{z_0} = \frac{\lambda z}{\pi w_0} \qquad (z \gg z_0), \qquad (7.5.30)$$

and is seen to grow linearly with distance z from the beam waist.

It is also interesting to consider the electric field at large distances from the beam waist:

$$\mathcal{E}(\mathbf{r}) \approx \frac{Az_0}{z} e^{i(kz-\pi/2)} e^{ik(x^2+y^2)/2R(z)} e^{-(x^2+y^2)/w^2(z)}, \tag{7.5.31}$$

for $z \gg z_0$. In this limit Eq. (7.5.17) gives

$$R(z) \approx z \qquad (z \gg z_0), \tag{7.5.32}$$

so that

$$\mathcal{E}(\mathbf{r}) \approx -iAz_0 \left[\frac{1}{z} e^{ikz} e^{ik(x^2+y^2)/2z}\right] e^{-(x^2+y^2)/w^2(z)}. \tag{7.5.33}$$

The factor in brackets has exactly the form (7.4.10) of a spherical wave with its center of curvature located at the beam waist ($z = 0$) (Fig. 7.13). The field (7.5.33) in fact has exactly the form of a spherical wave for points close enough to the beam axis that

$$e^{-(x^2+y^2)/w^2(z)} \approx 1. \tag{7.5.34}$$

The beamlike fields of the type (7.5.22) are sometimes called *Gaussian spherical waves*.

The properties of the Gaussian beam field (7.5.22) are collected in Table 7.1. These properties are illustrated in Figs. 7.11 and 7.13. Note that to the left of the beam waist the radius of curvature is negative, corresponding to concave surfaces of constant phase in the direction of propagation (Fig. 7.13). The field (7.5.22), except for the "strength" A, is completely specified by the wavelength λ and the spot size w_0 at the beam waist. Given λ and w_0, we can determine the spot size and radius of curvature everywhere. A monochromatic Gaussian beam is therefore fully characterized by three parameters: w_0, the location of its waist, and the field amplitude.

Actually, there is one restriction on w_0, namely w_0 must be large compared to the wavelength λ. This restriction is found by requiring the field (7.5.22) to satisfy (7.4.15), which was assumed in the derivation of the paraxial wave equation (7.4.19). In other words, our solution (7.5.22) must be consistent with the approximations used in obtaining it. This restriction implies that the beam divergence angle (7.5.25) must be small, that is, that our Gaussian beams are paraxial.

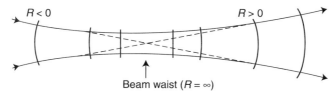

Figure 7.13   Variation of the radius of curvature of a Gaussian beam. At large distances from the beam waist the surfaces of constant phase are spheres centered on the waist.

TABLE 7.1   Gaussian Beam Solutions of the Paraxial Wave Equation

$E(\mathbf{r}) = \mathcal{E}(\mathbf{r})e^{-i\omega t}$ (electric field)

$\mathcal{E}(\mathbf{r}) = A\dfrac{w_0}{w(z)}e^{i[kz - \tan^{-1}(z/z_0)]}e^{ik(x^2+y^2)/2R(z)}e^{-(x^2+y^2)/w^2(z)}$

$I(\mathbf{r}) = \dfrac{c\epsilon_0}{2}|A|^2 e^{-2(x^2+y^2)/w^2(z)}$ (intensity)

$w(z) = w_0\sqrt{1 + \dfrac{z^2}{z_0^2}}$ (spot size)

$R(z) = z + \dfrac{z_0^2}{z}$ (radius of curvature)

$z_0 = \pi w_0^2/\lambda$ (Rayleigh range)

$\theta = \lambda/\pi w_0$ (divergence angle)

7.6   THE *ABCD* LAW FOR GAUSSIAN BEAMS

We have found that a Gaussian beam remains a Gaussian beam as it propagates in vacuum. The beam spot size and radius of curvature change with propagation, but the basic Gaussian spherical wave form (7.5.22) is always maintained. Equations (7.5.17) and (7.5.18) for $R(z)$ and $w(z)$ follow from the simple propagation law (7.5.9) for the q parameter of a Gaussian beam. This propagation law describes the effect on a Gaussian beam of propagation in empty space, that is, propagation through an "optical element" consisting of a straight section of empty space of length z. If the q parameter has the value q_i in the plane $z = z_i$, then its value in the plane $z = z_f$ is

$$q_f = q_i + d, \tag{7.6.1}$$

where $d = z_f - z_i$. This result, which follows trivially from (7.5.7), is the generalization of (7.5.9) to the case in which z_i is not necessarily zero. We will now consider the effect on a Gaussian beam of other optical elements, such as lenses and spherical mirrors.

Let us consider first the effect of a thin lens of local length f on a Gaussian beam. Suppose the radius of curvature and spot size immediately before the lens are R_1 and w_1, respectively. Immediately to the right of the lens the spot size should also be w_1 since the lens is not expected to alter the transverse intensity distribution of the beam. The beam curvature, however, will be changed by the lens. We can determine the beam curvature immediately to the right of the lens using the thin-lens equation (Fig. 7.14). The spherical wavefront immediately to the left of the lens has the same phase distribution as it would have if there were a point source on the lens axis at a distance $d_0 = R_1$ to the left of the lens. Such a point object will focus to a point image at a distance d_i to the right of the lens. The phase distribution of the beam immediately to the right of the lens should therefore correspond to a spherical wave of radius of curvature R_2 converging to a point at d_i (Fig. 7.14). In our sign convention, a spherical wavefront propagating in the positive z direction has a positive curvature if it is convex when viewed from $z = \infty$, and negative curvature if it is concave. This is consistent with our convention that concave spherical mirrors have positive curvature and convex mirrors have negative curvature. If a plane wave is incident on a concave mirror, for example, the reflected wave will have a negative radius of curvature, consistent with Eq. (7.2.14)

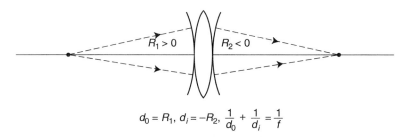

$$d_0 = R_1, \; d_i = -R_2, \; \frac{1}{d_0} + \frac{1}{d_i} = \frac{1}{f}$$

Figure 7.14 The transformation of a diverging spherical wavefront of curvature R_1 to a converging spherical wavefront of curvature R_2 by a thin lens of focal length f.

with $R > 0$. Thus, R_1 is positive and R_2 negative for the case in Fig. 7.14, or $d_0 = R_1$, $d_i = -R_2$, with d_0 and d_i both positive for the converging lens shown. The thin-lens equation, that is,

$$\frac{1}{d_0} + \frac{1}{d_i} = \frac{1}{f}, \tag{7.6.2}$$

then gives

$$\frac{1}{R_2} = \frac{1}{R_1} - \frac{1}{f}, \tag{7.6.3}$$

which is the desired relation between the radii of curvature of the spherical wavefront immediately before and after transformation by the lens. Equation (7.6.3) shows that R_2 must be *positive* if $0 < R_1 < f$. In this case the lens is too weak to focus the spherical wave to a point behind the lens. Equation (7.6.3) also applies to the case of a diverging lens, for which $f < 0$.

Using (7.6.3), we may relate the *final q* parameter q_f of a Gaussian spherical wave, just after passage through the lens, to the *initial* value q_i just before the lens. From the definitions

$$\frac{1}{q_i} = \frac{1}{R_1} + \frac{i\lambda}{\pi w_1^2}, \tag{7.6.4}$$

and

$$\frac{1}{q_f} = \frac{1}{R_2} + \frac{i\lambda}{\pi w_2^2}, \tag{7.6.5}$$

we have, since $w_1 = w_2$,

$$\frac{1}{q_f} = \frac{1}{q_i} - \frac{1}{f}, \tag{7.6.6}$$

or

$$q_f = \frac{1}{1/q_i - 1/f} = \frac{q_i}{-q_i/f + 1}. \tag{7.6.7}$$

Thus, whereas (7.6.1) gives the transformation of the q parameter of a Gaussian beam arising from free propagation through a distance d, (7.6.7) is the transformation effected by a thin lens of focal length f.

By similar reasoning, we can determine the transformation of the q parameter effected by a straight section of propagation of length d, followed by a lens of focal length f. The q transformation due to this "optical system" is found to be

$$q_f = \frac{q_i + d}{-q_i/f + 1 - d/f}. \tag{7.6.8}$$

In each of these three examples, the q parameter of a Gaussian beam is transformed according to an equation of the form

$$q_f = \frac{Aq_i + B}{Cq_i + D}. \tag{7.6.9}$$

The coefficients A, B, C, and D in the three cases (7.6.1), (7.6.7), and (7.6.8) may be read off from the matrices

$$\begin{bmatrix} A & B \\ C & D \end{bmatrix} = \begin{bmatrix} 1 & d \\ 0 & 1 \end{bmatrix}, \tag{7.6.10}$$

$$\begin{bmatrix} A & B \\ C & D \end{bmatrix} = \begin{bmatrix} 1 & 0 \\ -1/f & 1 \end{bmatrix}, \tag{7.6.11}$$

and

$$\begin{bmatrix} A & B \\ C & D \end{bmatrix} = \begin{bmatrix} 1 & d \\ -1/f & 1 - d/f \end{bmatrix}. \tag{7.6.12}$$

Now we recall from Section 7.2 that (7.6.10) is just the ray matrix associated with a straight section of propagation of length d [Eq. (7.2.9)]. Similarly (7.6.11) is the ray matrix for a thin lens of focal length f [Eq. (7.2.11)]. And finally (7.6.12) is precisely the ray matrix for a straight section of length d followed by a thin lens of focal length f [Eq. (7.2.17)], i.e., it is the matrix (7.6.10) multiplied from the left by the matrix (7.6.11).

These results are special cases of the *ABCD law for Gaussian beams*: The transformation of a Gaussian beam by an optical system may be obtained from Eq. (7.6.9), where the coefficients A, B, C, and D are given by the ray matrix of the optical system.

The advantage of the *ABCD* law is that it allows us to evaluate the transformation of a Gaussian beam using the ray matrix of *geometrical* optics. The transformation by an optical system of the Gaussian beam (7.5.22), which is a solution of the paraxial wave equation, may be inferred from the way the system transforms geometrical, paraxial rays. If we know the ray matrix for the optical system, we can predict how it modifies the q parameter, and therefore the spot size and radius of curvature, of a Gaussian beam. This remarkable property of Gaussian beams proves very useful in tracing the behavior of a (Gaussian) laser beam through various optical systems of lenses and mirrors.

As an example of the application of the *ABCD* law, consider the arrangement shown in Fig. 7.15. A Gaussian beam is incident upon a lens located at the waist of the Gaussian

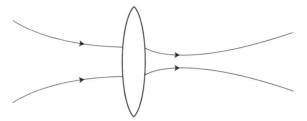

Figure 7.15   A Gaussian beam incident on a lens at its waist is focused to a new waist.

beam, where $R = \infty$ and the q parameter is

$$q_i = -\frac{i\pi w_0^2}{\lambda}, \tag{7.6.13}$$

with w_0 again the spot size at the waist. The beam passes through the lens, and we wish to determine the distance d behind the lens at which the transmitted beam has its waist. First, we calculate the ray matrix for the optical system consisting of a thin lens of focal length f followed by a straight section of propagation of length d:

$$\begin{bmatrix} A & B \\ C & D \end{bmatrix} = \begin{bmatrix} 1 & d \\ 0 & 1 \end{bmatrix} \begin{bmatrix} 1 & 0 \\ -1/f & 1 \end{bmatrix} = \begin{bmatrix} 1 - d/f & d \\ -1/f & 1 \end{bmatrix}. \tag{7.6.14}$$

The q parameter of the Gaussian beam is transformed by this optical system to

$$q_f = \left[\frac{1}{R(d)} + \frac{i\lambda}{\pi w^2(d)}\right]^{-1} = \frac{-(i\pi w_0^2/\lambda)(1 - d/f) + d}{(i\pi w_0^2/\lambda f) + 1}, \tag{7.6.15}$$

where we have used the *ABCD* law (7.6.9) with the ray matrix (7.6.14) and the initial q parameter (7.6.13). By equating the real and imaginary parts of both sides of (7.6.15), we obtain after some straightforward algebra the expressions

$$R(d) = \frac{(d/z_0)^2 + (1 - d/f)^2}{d/z_0^2 - (1/f)(1 - d/f)}, \tag{7.6.16}$$

and

$$w^2(d) = w_0^2 \left(1 - \frac{d}{f}\right)^2 + w_0^2 \left(\frac{d}{z_0}\right)^2, \tag{7.6.17}$$

for the radius of curvature $R(d)$ and the spot size $w(d)$ of the beam at any distance d behind the lens. Here z_0 is the Rayleigh range of the beam incident upon the lens.

The waist of the transmitted beam occurs at the distance d behind the lens that minimizes w_0, or equivalently the distance d for which $R(d) = \infty$. From (7.6.16), therefore, we see that d must satisfy the equation

$$\frac{d}{z_0^2} - \frac{1}{f}\left(1 - \frac{d}{f}\right) = 0, \tag{7.6.18}$$

so that

$$d = \frac{f}{1 + f^2/z_0^2}, \tag{7.6.19}$$

is the distance behind the lens where the new waist is located. Using this value of d in Eq. (7.6.17), we obtain the spot size w_0' at the new waist:

$$w_0' = \frac{\lambda f}{\pi w_0} \frac{1}{\sqrt{1 + f^2/z_0^2}} \cong \frac{\lambda f}{\pi w_0} \quad (z \gg z_0). \tag{7.6.20}$$

Equation (7.6.20) is an important result, for it indicates that *a Gaussian beam can be focused to a very small spot*. This property of Gaussian laser beams is important in many applications in which it is necessary to focus radiation onto a small area.

In the following section we will see how to compute the spot size of a laser beam emerging from a given resonator. For a He–Ne laser the spot size w_0 at the waist is typically on the order of 1 mm. From (7.6.20) with $\lambda = 632.8$ nm, therefore, we obtain (for $f \ll z_0$)

$$w_0' \approx 2 \times 10^{-4} f. \tag{7.6.21}$$

For lens focal lengths f on the order of centimeters, therefore, we see that the new spot size at the (new) waist is considerably smaller than the unfocused value, $w_0 \sim 1$ mm.

7.7 GAUSSIAN BEAM MODES

The mirrors of a laser force radiation to pass through the gain cell repeatedly, thereby enhancing its amplification by stimulated emission. The time taken by light to traverse the distance L (measured in centimeters) between the mirrors is

$$T = \frac{L}{c} \approx \frac{1}{3} \times 10^{-10} L \text{ s.} \tag{7.7.1}$$

If we measure the transverse intensity profile of the radiation from a laser, we normally find a steady profile that does not change in a time corresponding to successive reflections off the mirrors. Such a steady spatial pattern of intensity implies a steady spatial pattern of the field inside the resonator too. This is called a *mode* of the resonator.

To bring out more clearly the idea of a resonator mode, consider a plane normal to the optical axis of the resonator (Fig. 7.16). Radiation passing through this plane from the left will propagate to the mirror on the right, be reflected, and then pass through the plane from the right. After reflection from the mirror on the left, it propagates back to our imaginary plane, thus completing a round trip through the resonator. The key point is that, if the field is a mode of the resonator, it must have exactly the same value on the imaginary

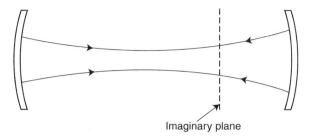

Figure 7.16 Laser resonator with imaginary plane drawn through the optical axis.

plane after a round trip as before. And this must be true regardless of where the plane is chosen inside the resonator.

If the Gaussian field of Table 7.1 is to be a mode of a resonator, then evidently its q parameter must not change in a round trip through the resonator. This condition may be examined using the *ABCD* law (7.6.9) for Gaussian beams. The condition for a Gaussian beam mode of a resonator is thus

$$q(z) = \frac{Aq(z) + B}{Cq(z) + D}, \tag{7.7.2}$$

where A, B, C, and D are the elements of the ray matrix for the optical system defined by a round trip through the resonator; these matrix elements will generally depend upon z. To have a Gaussian beam mode, Eq. (7.7.2) must hold for all values of z between the mirrors.

The condition (7.7.2) for a Gaussian beam mode is general. It can be applied to more complicated resonators than we are considering. For instance, it can be used for the case in which a lens is placed between the mirrors. For our purposes, however, the algebra is simpler if we follow a more intuitive approach. In this approach we consider the propagation of the field from one mirror to the other, as sketched in Fig. 7.17.

If a Gaussian beam is to be a mode of a resonator with spherical mirrors, *its radius of curvature at each mirror must be equal in magnitude to that of the mirror*. If this were not true, the mirror would change the magnitude of the beam radius upon reflection, and we would therefore not have a mode of the resonator. This physically reasonable result may be proved formally using the *ABCD* law for Gaussian beams (Problem 7.4). The radius of curvature of the Gaussian beam at the left-hand mirror of Fig. 7.17 is

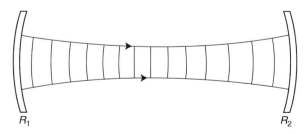

Figure 7.17 A Gaussian beam with radii of curvature at the mirrors equal in magnitude to those of the mirrors. A mode of the resonator is formed by the right-going Gaussian beam and the reflected, left-going beam.

[recall Eq. (7.5.17)]

$$R(z_1) = z_1 + \frac{z_0^2}{z_1}, \tag{7.7.3}$$

where z_0 is the Rayleigh range of the beam and z_1 gives the location of the mirror as measured from the beam waist. The beam at z_1 in Fig. 7.17 is a converging Gaussian spherical wave, and therefore has a negative radius of curvature according to our sign convention. On the other hand, the (concave) mirror at z_1 has a positive radius of curvature, R_1. Therefore

$$R(z_1) = z_1 + \frac{z_0^2}{z_1} = -R_1, \tag{7.7.4}$$

if the radius of curvature of the Gaussian beam is to have the same magnitude as that of the mirror. Similarly,

$$R(z_2) = z_2 + \frac{z_0^2}{z_2} = R_2, \tag{7.7.5}$$

where R_2 is the radius of curvature of the right-hand mirror of Fig. 7.17. The Gaussian beam in this case is diverging and thus has a positive radius of curvature, equal in magnitude and sign to that of the (concave) mirror drawn. It is easily checked that (7.7.4) and (7.7.5) apply regardless of whether the mirrors are concave or convex.

If the mirror separation is L, then we may also write the equation

$$z_2 - z_1 = L. \tag{7.7.6}$$

Equations (7.7.4)–(7.7.6) are three equations for the three "unknowns" z_1, z_2, and z_0^2. Their solution gives us z_1, z_2, and z_0^2 in terms of the mirror radii (R_1 and R_2) and mirror separation L. After straightforward algebra we obtain

$$z_1 = \frac{-L(R_2 - L)}{R_1 + R_2 - 2L}, \tag{7.7.7}$$

$$z_2 = \frac{L(R_1 - L)}{R_1 + R_2 - 2L}, \tag{7.7.8}$$

$$z_0^2 = \frac{L(R_1 - L)(R_2 - L)(R_1 + R_2 - L)}{(R_1 + R_2 - 2L)^2}. \tag{7.7.9}$$

In Section 7.4, where we discussed Gaussian beams as solutions of the paraxial wave equation in free space, the location of the beam waist was arbitrary. There was nothing to tell us its location. Equation (7.7.7) [or (7.7.8)], however, locates the beam waist of a Gaussian beam mode with respect to a mirror of the resonator. Similarly, the spot size w_0 at the waist was essentially a free parameter in Section 7.4, whereas (7.7.9) allows us to express w_0 explicitly in terms of the wavelength, the mirror radii of curvature, and their separation. Specifically, we obtain from (7.5.19) and (7.7.9) the minimum beam spot size

$$w_0 = \frac{(\lambda/\pi)^{1/2}[L(R_1 - L)(R_2 - L)(R_1 + R_2 - L)]^{1/4}}{(R_1 + R_2 - 2L)^{1/2}}. \tag{7.7.10}$$

In Table 7.2 we express z_1, z_2, and w_0 in terms of the resonator g parameters (7.3.13). From these expressions we obtain the beam spot sizes at the mirrors using (7.5.18):

$$w_1 = w_0 \sqrt{1 + \frac{z_1^2}{z_0^2}}, \tag{7.7.11a}$$

and

$$w_2 = w_0 \sqrt{1 + \frac{z_2^2}{z_0^2}}, \tag{7.7.11b}$$

TABLE 7.2   Properties of Lowest-Order Gaussian Modes (TEM$_{00}$) of Stable Resonators

Definitions and Conventions

$$g_i = 1 - \frac{L}{R_i}$$

$R_i > 0$ for concave mirrors, $R_i < 0$ for convex mirrors.

Location of Mirrors with Respect to Beam Waist

$$z_1 = \frac{-Lg_2(1 - g_1)}{g_1 + g_2 - 2g_1g_2}, \qquad z_2 = z_1 + L$$

Spot Sizes at Mirrors

$$w_1 = \left(\frac{\lambda L}{\pi}\right)^{1/2}\left[\frac{g_2}{g_1(1 - g_1g_2)}\right]^{1/4}$$

$$w_2 = \left(\frac{\lambda L}{\pi}\right)^{1/2}\left[\frac{g_1}{g_2(1 - g_1g_2)}\right]^{1/4}$$

$$w_0 = \left(\frac{\lambda L}{\pi}\right)^{1/2}\left[\frac{g_1g_2(1 - g_1g_2)}{(g_1 + g_2 - 2g_1g_2)^2}\right]^{1/4}$$

$ = $ spot size at beam waist

Resonance Frequencies

$$\nu_m = \frac{c}{2L}\left(m + \frac{1}{\pi}\cos^{-1}\sqrt{g_1g_2}\right)$$

which are given in terms of g_1 and g_2 in Table 7.2. The spot size for $z_1 < z < z_2$ may also be obtained from the propagation law (7.5.18). Similarly, the beam radius of curvature for any z may be determined using (7.5.17).

The results listed in Table 7.2 follow from the requirement that the radii of curvature of the Gaussian beam at the mirrors coincide in magnitude with those of the mirrors. The field reflected from the right-hand mirror of Fig. 7.17 is also a Gaussian beam whose radii of curvature at the mirrors match (in magnitude) those of the mirrors. The standing wave formed by the left- and right-going Gaussian beams thus form a mode of the resonator: They are not changed on successive reflections, and consequently the standing-wave pattern stays fixed.

The phase of our Gaussian-mode field along the optical axis ($r = 0$) of the resonator is (Table 7.1)

$$\theta(z) = kz - \tan^{-1}\left(\frac{z}{z_0}\right). \tag{7.7.12}$$

(It is convenient to consider the phase change along the optical axis because there the mirror surfaces are separated by exactly L.) The condition for a mode is that the field does not change in a round trip through the resonator. This means that the phase change of the field in a round trip should be an integral multiple of 2π, or that the one-way phase change is an integral multiple of π:

$$\theta(z_2) - \theta(z_1) = k(z_2 - z_1) - \left[\tan^{-1}\frac{z_2}{z_0} - \tan^{-1}\frac{z_1}{z_0}\right] = m\pi, \quad m = 0, 1, 2, \ldots. \tag{7.7.13}$$

This expression gives the allowed values of k, and therefore the resonance frequencies $\nu = kc/2\pi$. Using Eqs. (7.7.6)–(7.7.8), we find after some algebra (Problem 7.5) the allowed mode frequencies

$$\nu_m = \frac{c}{2L}\left(m + \frac{1}{\pi}\cos^{-1}\sqrt{g_1 g_2}\right), \tag{7.7.14}$$

where the sign of the square root is understood to be the same as the sign of g_1 (equal to the sign of g_2 for a stable resonator).

A resonator with plane-parallel mirrors has $g_1 g_2 = 1$, and so (7.7.14) gives back the result found earlier for this case [recall (5.9.1)]:

$$\nu_m = m\frac{c}{2L}. \tag{7.7.15}$$

In the case of spherical but nearly flat mirrors we obtain (Problem 7.5)

$$\nu_m \approx \frac{c}{2L}\left[q + \frac{1}{\pi}\left(\frac{L}{R_1} + \frac{L}{R_2}\right)^{1/2}\right], \quad L \ll R_1, R_2. \tag{7.7.16}$$

In all cases for which our Gaussian mode applies we have the frequency spacing

$$\Delta\nu = \nu_m - \nu_{m-1} = \frac{c}{2L},$$ (7.7.17)

between different modes, *exactly as for plane-parallel resonator mirrors*.

We note from Table 7.2 that all the properties of our Gaussian modes—the radius of curvature and spot size as a function of z inside the resonator, the location of the beam waist in terms of the mirror locations, and the allowed mode frequencies—follow from the resonator g parameters, the wavelength λ, and the mirror separation L. Actually, the mode with the properties listed in Table 7.2 is a special type of *zero-order* Gaussian mode. We will understand this when we consider high-order Gaussian modes in the following section. We must first, however, add three caveats.

First, we note from Table 7.2 that w_0^4 is negative if

$$g_1 g_2 (1 - g_1 g_2) \leq 0.$$ (7.7.18)

When this occurs we do not have a beamlike solution with a finite cross section. Thus, our Gaussian mode can only be valid for values of $g_1 g_2$ not satisfying (7.7.18), that is, for

$$0 \leq g_1 g_2 \leq 1.$$ (7.7.19)

This is precisely the condition (7.3.12) for resonator stability. Therefore, *Gaussian beam modes apply only to stable resonators*. In fact our analysis also generally breaks down if either $g_1 g_2 = 1$ or $g_1 g_2 = 0$, for then the spot size becomes infinite on at least one of the mirrors (Table 7.2). The cases $g_1 g_2 = 0$ and $g_1 g_2 = 1$ are the boundaries between stability and instability, the region of *marginal stability*.

The second restriction on the validity of our Gaussian beam modes is that the mirrors must be large enough to intercept the beam without any spillover. Otherwise the beam is not simply reflected at a mirror, and a more complicated, diffraction analysis is required. The transverse dimensions of the mirrors when viewed along the optical axis may be characterized by some effective radius a if the x and y dimensions are not too disparate. In the case of identical, flat, circular mirrors, a will be just the mirror radius. For the mirror to reflect a Gaussian beam without appreciable spillover, we require that

$$a \gg w_1, w_2,$$ (7.7.20)

where the spot sizes w_1 and w_2 at the mirrors are given in Table 7.2. These expressions for w_1 and w_2 are of the form

$$w_1 = \left(\frac{\lambda L}{\pi}\right)^{1/2} F_1(g_1, g_2),$$ (7.7.21a)

$$w_2 = \left(\frac{\lambda L}{\pi}\right)^{1/2} F_2(g_1, g_2),$$ (7.7.21b)

where $F_1(g_1, g_2)$ and $F_2(g_1, g_2)$ are typically of order unity. Thus (7.7.20) requires in this case $a \gg (\lambda L/\pi)^{1/2}$, or

$$N_F = \frac{a^2}{\lambda L} \gg 1, \tag{7.7.22}$$

where N_F is called the *Fresnel number* of the resonator.

Condition (7.7.22) is normally not difficult to meet. Consider, for example, a He–Ne laser wavelength of 632.8 nm and a mirror separation L of 50 cm. Condition (7.7.22) in this example becomes

$$a \gg 0.56 \text{ mm}, \tag{7.7.23}$$

which will obviously be satisfied for reasonably designed mirrors. For a 10.6-μm CO_2 laser with the same mirror separation we require

$$a \gg 2.3 \text{ mm}, \tag{7.7.24}$$

which again is a reasonable condition.

The third restriction on the validity of the Gaussian mode analysis is that

$$N_F \ll \left(\frac{L}{a}\right)^2, \tag{7.7.25}$$

must be satisfied. This condition may be understood from essentially the same condition (7.4.11) for the accurate approximation of the spherical wave (7.4.6) by (7.4.10). In the present context (7.7.25) ensures that the Gaussian beam field (7.5.22) has approximately spherical wavefronts that match the spherical mirror surfaces. Condition (7.7.25) is well satisfied in most practical resonators.

7.8   HERMITE–GAUSSIAN AND LAGUERRE–GAUSSIAN BEAMS

The Gaussian beam of Table 7.1 results from the solution (7.5.22) of the paraxial wave equation (7.4.19). However, this is not the only solution. In this section we will consider a more general type of Gaussian beam solution of the paraxial wave equation. The Gaussian beam of Table 7.1 will emerge as a special case of this more general solution.

We arrived at the solution (7.5.22) of the paraxial wave equation by guessing a solution of the form (7.5.2). In attempting to obtain other solutions, we will proceed in a similar fashion, assuming a solution of the form

$$\mathcal{E}_0(\mathbf{r}) = A g\left[\frac{x}{w(z)}\right] h\left[\frac{y}{w(z)}\right] e^{iP(z)} e^{ik(x^2+y^2)/2q(z)}. \tag{7.8.1}$$

We assume $w(z)$ and $q(z)$ are the same as before, i.e., that the spot size and radius of curvature of our more general Gaussian beam are given in Table 7.1. In fact if

$$P(z) = p(z),$$

and

$$g\left[\frac{x}{w(z)}\right] = h\left[\frac{y}{w(z)}\right] = 1,$$

then the trial solution (7.8.1) reduces exactly to (7.5.2). The fact that g and h are functions of $[x/w(z)]$ and $[y/w(z)]$, respectively, means that the intensity pattern associated with (7.8.1) will scale according to the spot size $w(z)$. This intensity pattern will be a function of $[x/w(z)]$ and $[y/w(z)]$, as is the intensity pattern given in Table 7.1. Our task is to find g, h, and P such that (7.8.1) satisfies the paraxial wave equation.

Using our trial solution (7.8.1) in the paraxial wave equation (7.4.19), we obtain differential equations for g, h, and P. Since the algebra is straightforward but rather tedious, we will omit the details of the derivation and give only the main steps. First, we use the fact that g and h are functions of the independent variables

$$\xi = \frac{x}{w(z)} \qquad \text{and} \qquad \eta = \frac{y}{w(z)}, \tag{7.8.2}$$

respectively, to write

$$\frac{\partial g}{\partial x} = \frac{dg}{d\xi}\frac{\partial \xi}{\partial x} = \frac{1}{w(z)}\frac{dg}{d\xi}, \tag{7.8.3a}$$

$$\frac{\partial^2 g}{\partial x^2} = \frac{1}{w^2(z)}\frac{d^2 g}{d\xi^2}, \tag{7.8.3b}$$

$$\frac{\partial g}{\partial z} = \frac{dg}{d\xi}\frac{\partial \xi}{\partial z} = -\frac{x}{w^2(z)}\frac{dw}{dz}\frac{dg}{d\xi}, \tag{7.8.3c}$$

with analogous results for the partial derivatives of h. We then use these results, together with Eqs. (7.5.11), (7.5.17), and (7.5.18), in the paraxial wave equation (7.4.19). We obtain

$$\frac{1}{g(\xi)}\left(\frac{d^2 g}{d\xi^2} - 4\xi\frac{dg}{d\xi}\right) + \frac{1}{h(\eta)}\left(\frac{d^2 h}{d\eta^2} - 4\eta\frac{dh}{d\eta}\right) + \left(\frac{2ik}{q(z)} - 2k\frac{dP}{dz}\right)w^2(z) = 0, \tag{7.8.4}$$

after division by $g(\xi)h(\eta)$. The functions $g(\xi)$, $h(\eta)$, and $P(z)$ must satisfy this equation in order for (7.8.1) to satisfy the paraxial wave equation.

Now the first term on the left-hand side of (7.8.4) is a function only of the independent variable ξ, the second term is a function only of the independent variable η, and the third term is a function only of the independent variable z. Thus, equation (7.8.4) cannot hold for all values of the *independent* variables ξ, η, and z unless each of these terms is separately constant. Therefore, we write

$$\frac{1}{g(\xi)}\left(\frac{d^2 g}{d\xi^2} - 4\xi\frac{dg}{d\xi}\right) = -a_1, \tag{7.8.5}$$

$$\frac{1}{h(\eta)}\left(\frac{d^2 h}{d\eta^2} - 4\eta\frac{dh}{d\eta}\right) = -a_2, \tag{7.8.6}$$

and

$$\left[\frac{2ik}{q(z)} - 2k\frac{dP}{dz}\right]w^2(z) = a_1 + a_2, \tag{7.8.7}$$

where a_1 and a_2 are constants. Thus, we have reduced the problem of solving the partial differential equation (7.4.19) in three independent variables to the problem of solving the three ordinary differential equations (7.8.5)–(7.8.7). This is an example of the method of "separation of variables."

It is convenient to write (7.8.5) in a slightly different form by defining the new variable

$$u = \sqrt{2}\xi. \tag{7.8.8}$$

Since

$$\frac{dg}{d\xi} = \frac{dg}{du}\frac{du}{d\xi} = \sqrt{2}\frac{dg}{du}, \tag{7.8.9a}$$

and

$$\frac{d^2 g}{d\xi^2} = 2\frac{d^2 g}{du^2}, \tag{7.8.9b}$$

we have

$$\frac{d^2 g}{du^2} - 2u\frac{dg}{du} + \frac{a_1}{2}g = 0. \tag{7.8.10}$$

The reason we have chosen to write (7.8.5) in this form is that Eq. (7.8.10) arises in many different problems. It appears, for example, in the quantum mechanics of the harmonic oscillator. A solution of (7.8.10) stays finite as $u \to \infty$ only if the constant a_1 satisfies

$$\frac{a_1}{2} = 2m, \qquad m = 0, 1, 2, \ldots. \tag{7.8.11}$$

The allowed (finite) solutions of (7.8.10) are the Hermite polynomials, the first few of which are

$$H_0(u) = 1,$$
$$H_1(u) = 2u,$$
$$H_2(u) = 4u^2 - 2,$$
$$H_3(u) = 8u^3 - 12u,$$
$$H_4(u) = 16u^4 - 48u^2 + 12.$$

The allowed solutions for the function g in our trial solution (7.8.1) are thus

$$g\left[\frac{x}{w(z)}\right] = H_m\left[\sqrt{2}\frac{x}{w(z)}\right], \qquad m = 0, 1, 2, \ldots. \tag{7.8.12}$$

In a similar fashion we obtain the allowed solutions

$$h\left[\frac{y}{w(z)}\right] = H_n\left[\sqrt{2}\frac{y}{w(z)}\right], \qquad n = 0, 1, 2, \ldots, \tag{7.8.13}$$

for the function h.

It remains to determine $P(z)$. Using Eqs. (7.5.11), (7.5.17), and (7.5.18), we obtain from (7.8.7) the differential equation

$$\frac{dP}{dz} = \frac{iz}{z^2 + z_0^2} - \frac{z_0(m+n+1)}{z^2 + z_0^2}, \tag{7.8.14}$$

which may be integrated to give

$$P(z) = i \ln \sqrt{1 + \frac{z^2}{z_0^2}} - (m+n+1)\phi(z), \tag{7.8.15}$$

or

$$e^{iP(z)} = \frac{e^{-i(m+n+1)\phi(z)}}{\sqrt{1 + z^2/z_0^2}} = \frac{w_0}{w(z)}e^{-i(m+n+1)\phi(z)}, \tag{7.8.16}$$

where $\phi(z) = \tan^{-1}(z/z_0)$, as in (7.5.21). Collecting the results (7.8.12), (7.8.13), and (7.8.16), we have a solution (7.8.1) to the paraxial wave equation. The electric field is thus

$$\mathcal{E}_{mn}(x, y, z) = \frac{Aw_0}{w(z)}H_m\left[\sqrt{2}\frac{x}{w(z)}\right]H_n\left[\sqrt{2}\frac{y}{w(z)}\right]e^{i[kz-(m+n+1)\tan^{-1} z/z_0]}$$

$$\times e^{ik(x^2+y^2)/2R(z)}e^{-(x^2+y^2)/w^2(z)}. \tag{7.8.17}$$

Note that when $m = n = 0$ we recover the solution (7.5.22), so our previous Gaussian beam solution is therefore the "lowest-order" or "zero-order" case of (7.8.17). Another important point is that $R(z)$ and $w(z)$ are independent of m and n; all higher-order Gaussian beams are characterized by the same functions $R(z)$ and $w(z)$ as the lowest-order one. Furthermore, all higher-order Gaussian beams satisfy the same *ABCD* law as the lowest-order one: a Gaussian beam of order (m, n) remains a Gaussian beam of the same order after propagation in free space or transformation by a thin lens or a spherical mirror, but its q parameter is changed according to the *ABCD* law.

Because they have the same $w(z)$ and $R(z)$ as the lowest-order beam, the higher-order Gaussian beams also form modes of stable resonators satisfying (7.7.20). All the properties of Table 7.2 apply as well to such higher-order modes, *except* for the resonance frequencies. Following exactly the same approach that led to (7.7.14), we obtain for a Gaussian mode of order (m, n) the allowed mode frequencies

$$\nu_{qmn} = \frac{c}{2L}\left[q + \frac{1}{\pi}(m + n + 1)\cos^{-1}\sqrt{g_1 g_2}\right], \qquad (7.8.18)$$

with q a positive integer or zero, and with the sign convention of (7.7.14).

Gaussian modes characterized by different values of m and n are said to be different *transverse modes* because their intensity patterns transverse to the optical axis are different. Modes associated with different values of q are said to be different *longitudinal modes*. Thus, a given transverse mode (m, n) may be associated with different longitudinal modes (q), and vice versa. A Gaussian mode is specified by the three integers (q, m, n), that is, by its longitudinal and transverse mode character. The transverse character of a Gaussian mode is conventionally designated TEM_{mn}, meaning "transverse electromagnetic of order (m, n)."

There is a wealth of experimental evidence to corroborate our Gaussian mode analysis for stable resonators satisfying conditions (7.7.20) and (7.7.25). One way to record the intensity pattern of a laser beam is illustrated in Fig. 7.18, while Fig. 7.19 shows a "power-in-the-bucket" method of measuring the spot size of a lowest order (TEM_{00}) Gaussian beam. A direct recording of the intensity pattern is usually made with a

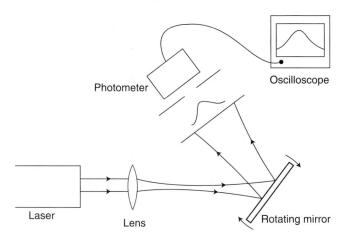

Figure 7.18   One way of recording the intensity pattern of a laser beam. The lens is used to expand the laser beam, which normally has a very small diameter (~ 1 mm). The mirror rotates at an angular velocity of a few revolutions per second.

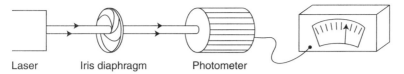

Figure 7.19 Determining the spot size of a TEM$_{00}$ Gaussian beam. The power is first measured with the iris diaphragm fully open. The iris aperture is then reduced until the measured power is 86.5% of the initial value. Then the aperture radius is equal to the spot size. (See Problem 7.6.)

charge-coupled device (CCD) camera, which measures the intensity distribution over a two-dimensional array of pixels, each of which is typically of width ≈ 15–40 μm. CCD cameras are useful for pulsed lasers, and they allow three-dimensional contour plots as well as two-dimensional distributions to be displayed. CCD cameras are easily saturated, and so the beam must usually be attenuated before entering the camera. The attenuation can introduce beam distortions, but otherwise the principal limiting factor on the accuracy of the beam intensity measurement is the pixel size.

A high-resolution measurement of an intensity profile can be made by scanning a pinhole across the beam and electronically correlating the measured intensity with the position of the pinhole to produce a display of the intensity distribution on a computer screen. Beam profiling instruments of various types are commercially available.

Figure 7.20 shows actual intensity patterns recorded with a He–Ne laser operating at 632.8 nm. The reader may easily show that these intensity patterns may be associated with the various low-order transverse Gaussian modes indicated (Problem 7.7). It should be noted that the intensity patterns tend to be larger for the higher-order modes. For such modes condition (7.7.20) may not hold as well as for lower-order modes. Consequently, the higher-order modes tend to suffer greater loss due to diffractive spillover at the mirrors.

In general, a laser is able to oscillate simultaneously on a number of transverse (and longitudinal) modes. To achieve oscillation on a single transverse mode, as in the intensity patterns of Fig. 7.20, it is necessary to have some sort of *mode discrimination*. That is, it is necessary to have high losses for all transverse modes except one. A laser in which the gain is concentrated near the optical axis, for instance, will tend to oscillate on lower-order modes. In many applications a TEM$_{00}$ Gaussian mode is desired. To discriminate against the higher-order modes in this case, a circular aperture may be inserted into the resonator to produce high losses on all but the lowest-order mode. The Hermite–Gaussian modes we have found have rectangular symmetry and thus would appear to be inapplicable in the case of mirrors with a circular cross section. In practice, however, slight mirror misalignments or other "perturbations" will result in rectangularly symmetric rather than circularly symmetric modes. One such perturbation is the use of Brewster-angle windows. In this case our scalar electric field may be assumed to be the field component having the favored linear polarization. The directions of the electric field vectors for our Gaussian modes in this case are shown in Fig. 7.21.

A different type of solution to the paraxial wave equation (7.4.19) is obtained by expressing the latter in terms of cylindrical coordinates (r, ϕ, z),

$$\left(\frac{\partial^2}{\partial r^2} + \frac{1}{r}\frac{\partial}{\partial r} + \frac{1}{r^2}\frac{\partial^2}{\partial \phi^2} + 2ik\frac{\partial}{\partial z} \right)\mathcal{E}(r, \phi, z) = 0, \qquad (7.8.19)$$

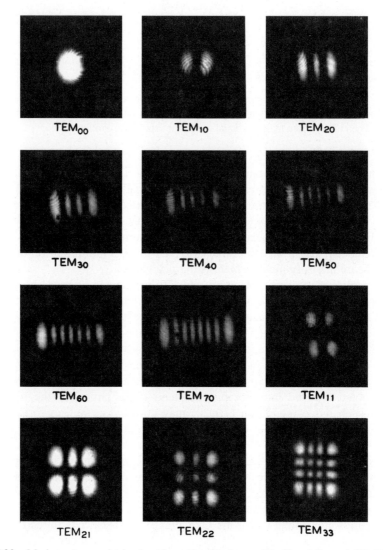

Figure 7.20 Mode patterns obtained with a He–Ne laser. [H. Kogelnik and W. W. Rigrod, *Proceedings of the IRE (Correspondence)* **50**, 220 (1962). © IEEE]

and assuming solutions with azimuthal variation $\exp(i\ell\phi)$, with $\ell = 0, \pm 1, \pm 2, \ldots$ in order to have $\mathcal{E}(r, \phi + 2\pi, z) = \mathcal{E}(r, \phi, z)$. It is straightforward to show that these solutions have the form

$$\mathcal{E}_{p\ell}(r, \phi, z) = A \frac{w_0}{w(z)} e^{i\ell\phi} \left[\frac{r\sqrt{2}}{w(z)}\right]^{|\ell|} L_p^{|\ell|}\left(\frac{2r^2}{w^2(z)}\right) e^{-r^2/w^2(z)} e^{ikr^2/2R(z)} e^{-i[(2p+\ell+1)\tan^{-1}(z/z_0)]}.$$

$$(7.8.20)$$

Here $p = 0, 1, 2, \ldots$, w_0, $w(z)$ and $R(z)$ have exactly the same form as in the case of Hermite–Gaussian modes, and the functions L_p^ℓ are associated Laguerre polynomials,

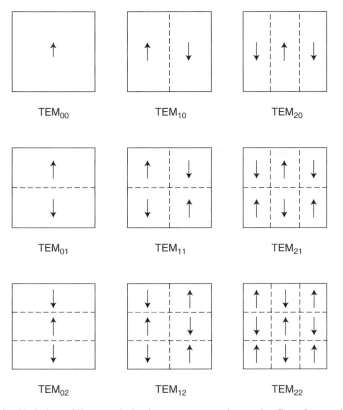

Figure 7.21 Variation of linear polarization across some low-order Gaussian mode patterns.

the lowest-order ones defined by

$$L_0^\ell(x) = 1,$$
$$L_1^\ell(x) = (\ell + 1) - x,$$
$$L_2^\ell(x) = \frac{(\ell + 1)(\ell + 2)}{2!} - (\ell + 2)x + \frac{x^2}{2!}.$$

Note that the lowest-order ($p = \ell = 0$) Laguerre–Gaussian beam is identical to the lowest-order ($m = n = 0$) Hermite–Gaussian beam, which is just the familiar Gaussian beam (Section 7.5).

The mode $\mathcal{E}_{01}(r, \phi, z)$ has an intensity profile proportional to

$$|\mathcal{E}_{01}(r, \phi, z)|^2 = |A|^2 \frac{2w_0^2 r^2}{w^4(z)} e^{-2r^2/w^2(z)}. \tag{7.8.21}$$

The intensity vanishes on axis ($r = 0$) and grows to a maximum at $r = w(z)/2$ before diminishing rapidly as r increases further, and so $\mathcal{E}_{01}(r, \phi, z)$ is often called a "doughnut" mode. Expressing this mode in rectangular coordinates ($x = r \cos \phi$,

$y = r \sin \phi$) as

$$\mathcal{E}_{01}(x, y, z) = A \frac{w_0 \sqrt{2}}{w^2(z)} e^{-r^2/w^2(z)}(x + iy)e^{ikr^2/2R(z)}e^{-2i\tan^{-1}(z/z_0)}, \qquad (7.8.22)$$

it is seen that it is just a superposition of the ($m = 1$, $n = 0$) Hermite–Gaussian mode with i times the ($m = 0$, $n = 1$) Hermite–Gaussian mode. In fact any Laguerre–Gaussian mode can be expressed as a linear combination of Hermite–Gaussian modes and vice versa; both types of modes form a complete set.

But Laguerre–Gaussian beams with $\ell \neq 0$ have a property that Hermite–Gaussian modes do not: they have nonvanishing orbital angular momentum. The Laguerre–Gaussian mode described by the electric field $\mathcal{E}_{p\ell}(r, \phi, z)$ has an orbital angular momentum about the z axis that amounts to $\ell \hbar$ per photon.[2] Such a beam incident on an absorbing dielectric particle can therefore cause the particle to rotate. This consequence of the conservation of angular momentum has been demonstrated with optical tweezers (Section 14.5). It has also been demonstrated that Laguerre–Gaussian beams can be used to produce vortices in a Bose-Einstein condensate (Section 14.6). Because of their orbital angular momentum, Laguerre–Gaussian modes are often called OAM (orbital angular momentum) states. For reasons of symmetry mentioned earlier in connection with Hermite–Gaussian modes, Laguerre–Gaussian beams are not ordinarily produced by lasers, although it has been possible to generate such modes in lasers with very accurately aligned mirrors.[3]

7.9   RESONATORS FOR HE–NE LASERS

Many factors are involved in the design of laser resonators, including, of course, the intended applications of the laser. Such things as mechanical stability and thermal expansion coefficients of Brewster windows and mirrors must also be considered. A detailed technical discussion of resonator construction is inappropriate here. Instead, we will briefly apply some of the results obtained in the preceding sections to the design of resonators for commercially available He–Ne lasers.

He–Ne lasers usually have hemispherical resonators (Fig. 7.8). This type of resonator has low sensitivity to mirror misalignments and is easily adjusted. Unlike the confocal resonator, for example, it is not difficult to obtain TEM_{00} oscillation with a hemispherical resonator. This mode is desirable for applications such as alignment and holography. Higher-order Gaussian modes do not offer the same low beam divergence or the same ability to be focused down to a tiny spot.

For the hemispherical resonator we have $g_1 = 1$, $g_2 = 0$, and $g_1 g_2 = 0$. It is therefore on the border between stability and instability, as indicated by the fact that the spot size

---

[2]This may be understood from a quantum-mechanical perspective, in that $\mathcal{E}_{p\ell}(r, \phi, z)$ is an eigenfunction of the orbital angular momentum operator $L_z = i\hbar \, d/d\phi$. The orbital angular momentum is distinct from the *spin* or intrinsic angular momentum of light, which is determined by the polarization and has a maximum magnitude per photon of $\hbar$.

[3]For further reading on the orbital angular momentum of light see, for instance, the special issue on "Atoms and the Orbital Angular Momentum of Light," *Journal of Optics B: Quantum and Semiclassical Optics* **4** (April 2002).

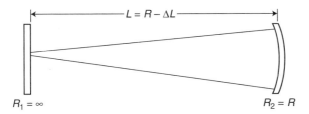

Figure 7.22   Quasi-hemispherical resonator used in many commercial lasers.

w_2 in Table 7.2 is infinite, whereas w_1 vanishes. Actual "hemispherical" resonators are used with the mirror separation L slightly less than the radius of curvature R_2, so that $g_2 > 0$ and $g_1 g_2 > 0$ (Fig. 7.22). The resonator is then stable.

Taking $R_1 = \infty$ (flat mirror), $R_2 = R$, and

$$L = R - \Delta L, \qquad R \gg \Delta L > 0, \tag{7.9.1}$$

we can use the results of Table 7.2 to obtain approximate expressions for the spot sizes in terms of L and ΔL. At the flat mirror we have the spot size (Problem 7.8)

$$w_1 \approx \left(\frac{\lambda^2 L \, \Delta L}{\pi^2} \right)^{1/4}, \tag{7.9.2}$$

and at the spherical mirror

$$w_2 \approx \left(\frac{\lambda^2 L^3}{\pi^2 \, \Delta L} \right)^{1/4}. \tag{7.9.3}$$

Thus, the Gaussian mode spot size is much larger at the spherical mirror than at the flat mirror, as indicated in Fig. 7.22. In fact, it follows from Table 7.2 that the beam waist occurs at the flat mirror.

The spot sizes of Gaussian beam modes are usually very small. This is inefficient in the sense that the laser beam intersects only a small fraction of the total volume of the gain medium. Equation (7.9.3) indicates that w_2 can be made large by making ΔL smaller. By using a micrometer adjustment screw to vary ΔL, w_2 can be made to cover a significant portion of the output coupling mirror of Fig. 7.22. In this way the size of the output beam can be varied while the laser is on (Problem 7.8). In practice, however, the spot sizes of commercial He−Ne lasers are not adjustable. In fact, many such lasers are of the "hard-seal" type in which the mirror spacing is permanently fixed.

Many commercial gas lasers have *collimating mirrors* at the output port. Figure 7.23 illustrates a design of such a mirror. The surface with radius of curvature R_1 is highly reflecting, while the bulk of the "mirror" consists of quartz or glass of refractive index n. Curvatures R_1 and R_2 and refractive index n are chosen in such a way that the output beam is collimated, that is, there is a new waist near the out-coupling mirror (Problem 7.9). On the outer surface with radius of curvature R_2 in Fig. 7.23 is an anti-reflective coating to minimize the power in any secondary beam. (For a small additional

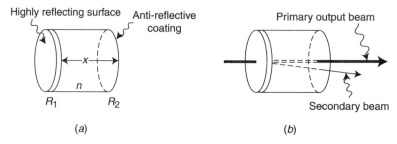

Figure 7.23   (*a*) Design of collimating mirror and (*b*) formation of a secondary beam.

cost, one manufacturer of He–Ne lasers has guaranteed that the secondary beam will be no brighter than $1/500$ of the primary beam.)

Equation (7.5.25) shows that the larger the spot size at the waist of a Gaussian beam, the smaller the beam divergence angle. In many applications it is important to minimize the spot size over large distances of propagation. This can be accomplished by mounting a "beam expander" to the output port of the laser. A beam expander is basically a telescope in reverse. Figure 7.24 shows the basic principle of operation of Galilean and Keplerian beam-expanding telescopes. A typical commercially available beam expander magnifies the beam waist by a factor of 10. Figure 7.25 shows the spot size as a function of distance for a He–Ne laser ($\lambda = 632.8$ nm) with and without such a beam expander. Figure 7.26 shows a way of measuring the divergence angle of the beam (see Problem 7.11).

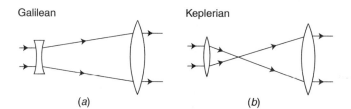

Figure 7.24   Beam expanders of the (*a*) Galilean and (*b*) Keplerian type.

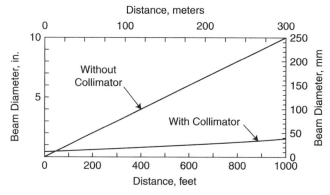

Figure 7.25   Spot size as a function of distance for a He–Ne laser beam with and without a beam collimator. (Adapted from an old Hughes Aircraft Company catalog.)

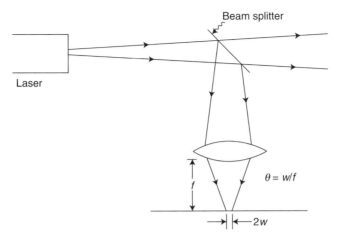

Figure 7.26   Setup for measuring the divergence angle θ of a Gaussian beam.

The reader interested in the construction details of laser resonators is advised to consult the websites and brochures provided by various manufacturers. These often include useful summaries of the properties of Gaussian beam modes. This chapter should provide the reader with sufficient background to read these publications as well as the research literature on resonators.

7.10   DIFFRACTION

We define diffraction, in broad terms, as the bending of light around some obstacle. As such, diffraction is a distinctly wavelike phenomenon, inexplicable from the viewpoint of geometrical rays of light. Some appreciation of diffraction is necessary for a more complete understanding of laser resonators.

Let us start by mentioning *Huygens' principle*. Huygens' principle says that *we can imagine every point on a wavefront to be a point source for a spherical wave* (Fig. 7.27). This way of thinking about waves allows us to make accurate estimates about how

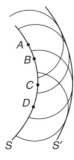

Figure 7.27   Huygens' principle says we can imagine each point (such as A, B, C, D) on a wavefront S to be a source of spherical waves. The superposition of all these spherical waves gives the wavefront (S') elsewhere in space.

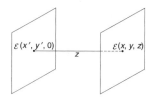

Figure 7.28   Given the field distribution of a monochromatic wave in the plane $z = 0$, what is the field in a plane $z > 0$? We can answer this question using Huygens' principle.

they propagate, either in free space or in a medium or around obstacles. The first thing we will do is to express Huygens' principle in quantitative terms, so that we will be able not only to understand why diffraction occurs but also to calculate diffraction patterns for different obstacles. As in our discussion of Gaussian beams, we will confine ourselves to a scalar-wave approach, ignoring for simplicity the polarization of electromagnetic fields.

Consider the situation indicated in Fig. 7.28. We have a monochromatic wave propagating in the z direction, and we are given the field values $\mathcal{E}(x, y, 0)$ in the xy plane. We want to know what the field is on some observation plane a distance z away. In other words, what is $\mathcal{E}(x, y, z)$, given $\mathcal{E}(x, y, 0)$? Huygens' principle tells us how to calculate the field on the observation plane: We do it by regarding every point in the plane $(x, y, z = 0)$ as a point source of spherical waves of the type (7.4.6). Every point $(x', y', 0)$ in the "source plane" acts as a source for a spherical wave that contributes to the point (x, y, z) on the observation plane a field

$$\Delta\mathcal{E}(x, y, z) = -\frac{i}{\lambda}\frac{e^{ikr}}{r}\mathcal{E}(x', y', 0)\,\Delta a', \tag{7.10.1}$$

where

$$r = [(x - x')^2 + (y - y')^2 + z^2]^{1/2}, \tag{7.10.2}$$

$k = 2\pi/\lambda$, and $\Delta a'$ is a tiny element of area surrounding the point $(x', y', 0)$ on the source plane, where the field has the value $\mathcal{E}(x', y', 0)$. We will not attempt to explain the origin of the factor $-i/\lambda$ in (7.10.1); it arises in a slightly more mathematical version of Huygens' principle. The factor i will not be very important to us anyway because it will not affect the calculation of measurable intensities. The appearance of the wavelength λ in (7.10.1) ensures that the two sides of the equation are dimensionally consistent.

Following exactly the same approach as that leading from (7.4.6) to (7.4.10), we replace the Huygens spherical wave (7.10.1) by

$$\Delta\mathcal{E}(x, y, z) \approx -\frac{i}{\lambda z}e^{ikz}e^{ik[(x-x')^2+(y-y')^2]/2z}\mathcal{E}(x', y', 0)\,\Delta a'. \tag{7.10.3}$$

The theory of diffraction based on the approximation (7.10.3) is called the Fresnel approximation, or simply *Fresnel diffraction*. Finally, we calculate the complete field at the point (x, y, z) on the observation plane by integrating over all the point sources on the source plane, that is, by summing up the contributions from all the Huygens spherical waves:

$$\mathcal{E}(x, y, z) \approx -\frac{ie^{ikz}}{\lambda z} \int \int \mathcal{E}(x', y', 0)e^{ik[(x-x')^2+(y-y')^2]/2z}dx' \, dy'. \tag{7.10.4}$$

The exponential in the integrand of (7.10.4) can be rewritten

$$e^{ik(x^2+y^2)/2z}e^{ik(x'^2+y'^2)/2z}e^{-ik(xx'+yy')/z}. \tag{7.10.5}$$

The first factor on the right is independent of the integration variables x', y' and may be pulled outside the integral in (7.10.4). The important special case of *Fraunhofer diffraction* occurs if the inequality

$$z \gg k(x'^2 + y'^2), \tag{7.10.6}$$

holds for all points (x', y') on the source plane aperture. In this case,

$$e^{ik(x'^2+y'^2)/2z} \approx 1, \tag{7.10.7}$$

and

$$\mathcal{E}(x, y, z) \approx -\frac{ie^{ikz}}{\lambda z}e^{ik(x^2+y^2)/2z} \int \int \mathcal{E}(x', y', 0)e^{-ik(xx'+yy')/z} \, dx' \, dy'. \tag{7.10.8}$$

This can be called the *Fraunhofer diffraction integral and is basically just the two-dimensional Fourier transform of $\mathcal{E}(x', y', 0)$.*

• The Fresnel approximation retains only the first two terms of the expansion

$$kr = k[(x - x')^2 + (y - y')^2 + z^2]^{1/2}$$

$$= kz + \frac{k}{2z}[(x - x')^2 + (y - y')^2] - \frac{k}{8z^3}[(x - x')^2 + (y - y')^2]^2 + \cdots, \tag{7.10.9}$$

in e^{ikr}. As such it is a valid approximation if the third term is small, that is, if

$$z^3 \gg \frac{\pi}{4\lambda}[(x - x')^2 + (y - y')^2]_{max}^2 \quad \text{(Fresnel approximation)}, \tag{7.10.10}$$

where $[\ldots]_{max}$ denotes the largest value of interest of the quantity in brackets. The Fraunhofer approximation, on the other hand, assumes that

$$z \gg \frac{\pi}{\lambda}[x'^2 + y'^2]_{max} \quad \text{(Fraunhofer approximation)}, \tag{7.10.11}$$

which is less often satisfied in the sense that larger values of z are required than in the Fresnel approximation. An example of (7.10.11) will be given in the following section. ●

7.11 DIFFRACTION BY AN APERTURE

In this section we will use the Fraunhofer diffraction formula to treat the diffraction of light by an aperture. This example is useful for a qualitative understanding of other diffraction problems, as we will see. The Fraunhofer diffraction formula is very important, and it is generally much easier to work with than the Fresnel formula (i.e., the integrals are easier to work out). In the following sections we will relate the Fresnel diffraction formula (7.10.4) to the paraxial wave equation, Gaussian beams, and more general types of laser resonators.

We will consider a circular aperture of diameter $D = 2a$, upon which is incident a uniform monochromatic plane wave. We have already indicated in Fig. 7.12 the solution of this diffraction problem, and now we will derive the solution using the Fraunhofer diffraction formula (7.10.8).

Since we are considering a uniform plane wave incident upon the aperture, we have $\mathcal{E}(x', y', 0) = \mathcal{E}_0$ everywhere on the aperture and $\mathcal{E}(x', y', 0) = 0$ everywhere on the opaque screen. That is, the aperture is uniformly illuminated with a field of amplitude $\mathcal{E}_0$, whereas the screen is perfectly absorbing and as such is not a source of any Huygens wavelets. The integral in (7.10.8), therefore, becomes an integral over the $x'y'$ coordinates of the aperture alone:

$$\int\int \mathcal{E}(x', y', 0)e^{-ik(xx'+yy')/z} \, dx' \, dy' = \mathcal{E}_0 \int\int e^{-ik(xx'+yy')/z} \, dx' \, dy'. \tag{7.11.1}$$

Because the aperture is circular, it is convenient to use circular coordinates for both the source and observation planes:

$$x = r\cos\theta, \qquad y = r\sin\theta, \tag{7.11.2a}$$

$$x' = r'\cos\theta', \qquad y' = r'\sin\theta', \tag{7.11.2b}$$

$$xx' + yy' = rr'(\cos\theta\,\cos\theta' + \sin\theta\,\sin\theta') = rr'(\cos\theta' - \theta), \tag{7.11.2c}$$

$$dx' \, dy' = r' \, dr' \, d\theta'. \tag{7.11.2d}$$

In terms of these coordinates we have

$$\int\int_{\text{aperture}} e^{-ik(xx'+yy')/z} \, dx' \, dy' = \int_0^a r' \, dr' \int_0^{2\pi} e^{-ikrr'\cos(\theta'-\theta)/z} \, d\theta'$$

$$= 2\pi \int_0^a J_0\left(\frac{krr'}{z}\right) r' \, dr', \tag{7.11.3}$$

where we have written the integral over θ' as a zeroth-order Bessel function of the first kind, J_0. This integral representation of J_0 is a standard one that may be found in integral tables. Furthermore

$$\int_0^a J_0\left(\frac{krr'}{z}\right) r' \, dr' = \frac{a^2}{2} \frac{2J_1(kar/z)}{kar/z}, \tag{7.11.4}$$

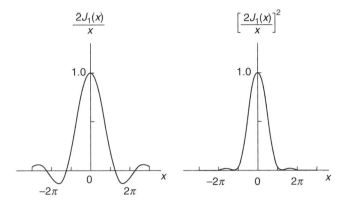

Figure 7.29   The functions $2J_1(x)/x$ and $[2J_1(x)/x]^2$.

where J_1 is the first-order Bessel function of the first kind, a graph of which is given in Fig. 6.13. In Fig. 7.29 we plot the function $2J_1(x)/x$ appearing in Eq. (7.11.4).

Using these results in (7.10.8), we have

$$\mathcal{E}(r, z) = (-ie^{ikz}e^{ikr^2/2z})\frac{\pi a^2}{\lambda z}\frac{2J_1(kar/z)}{(kar/z)}, \tag{7.11.5}$$

for the field at the observation plane. Note that $\mathcal{E}$ is independent of θ, compatible with the circular symmetry in our example. The intensity distribution over the observation plane is determined by $|\mathcal{E}|^2$ and is therefore independent of the first factor in (7.11.5), since it has unit modulus. Thus, after writing $2\pi/\lambda$ for k everywhere, we obtain the intensity distribution corresponding to (7.11.5):

$$I(r, z) = I_0\left(\frac{\pi a^2}{\lambda z}\right)^2\frac{[2J_1(2\pi ar/\lambda z)]^2}{(2\pi ar/\lambda z)^2}, \tag{7.11.6}$$

where I_0 is the intensity of the incident (uniform) plane wave. This is the Fraunhofer diffraction pattern for a circular aperture. Its general properties may be inferred from Fig. 7.29.

For a given distance z of the observation plane from the plane containing the aperture, the intensity has its maximum at $r = 0$, that is, "on axis," where the intensity is

$$I(0, z) = \left(\frac{\pi a^2}{\lambda z}\right)^2 I_0. \tag{7.11.7}$$

We may write (7.11.6) as

$$I(r, z) = I(0, z)\frac{[2J_1(2\pi ar/\lambda z)]^2}{(2\pi ar/\lambda z)^2}. \tag{7.11.8}$$

The factor multiplying $I(0, z)$ determines the variation of intensity with r in the observation plane. The intensity distribution (7.11.8) is called the *Airy pattern*, after the British astronomer G. B. Airy,[4] who derived it in 1835.

Away from $r = 0$ in the observation plane the intensity decreases, reaching zero at the value of r satisfying $J_1(2\pi ar/\lambda z) = 0$. Since $J_1(x) = 0$ at $x = 1.22\pi$ (Fig. 7.29), there is a zero in the Airy pattern at

$$r_0 \cong 1.22\frac{\lambda z}{D}. \tag{7.11.9}$$

As indicated in Fig. 7.12, this radius of the central bright spot, or *Airy disk*, increases linearly with z. We may define the divergence angle

$$\theta = \tan^{-1}\frac{r_0}{z} \approx \frac{r_0}{z} = 1.22\frac{\lambda}{D}, \tag{7.11.10}$$

which is Eq. (7.5.24). The small-angle approximation in (7.11.10) is made because we are considering apertures large compared to a wavelength, that is, λ/D is small.

As we increase r beyond r_0, the intensity rises above zero again, but it is much smaller than in the central bright region. There is a pattern of successively dimmer concentric rings, as indicated in Fig. 7.12. The intensity in the observation plane does not cut off sharply at any radius r, and the light spreads out beyond the geometrical size of the aperture. Of course, this is just what we mean by "diffraction."

Another feature of the Airy pattern is that the radius r_0 of the central bright spot is directly proportional to the wavelength and inversely proportional to the diameter of the aperture. This feature is characteristic of the diffraction of light by apertures of arbitrary shapes: *The larger the wavelength and the smaller the aperture, the more pronounced will be the diffractive "spreading" of the field.*

• To check condition (7.10.11) for the validity of the Fraunhofer approximation, consider the following example. Let $\lambda = 632.8$ nm and $a = 1$ cm. In this case $(x'^2 + y'^2)_{max} = a^2$ and (7.10.11) gives the condition $z \gg 496$ m for the validity of the Fraunhofer approximation. [The reader may easily convince himself that the condition (7.10.10) for the validity of the Fresnel approximation can be satisfied with much smaller values of z.] This makes Fraunhofer diffraction seem of doubtful relevance in the laboratory. However, it turns out that the use of lenses restores the relevance of Fraunhofer diffraction even for small values of z (Problem 7.12). Indeed, a whole branch of optics called *Fourier optics* is based essentially on this property of lenses.[5]

It may be shown, for instance, that the image of a point source in the focal plane of a circular lens is not a point, as predicated by geometrical ray optics, but an Airy pattern. Thus, the image of a star formed in a telescope is an Airy pattern. If D is the diameter of the lens, and f its focal length, this Airy pattern is given by (7.11.8) with $a = D/2$ and $z = f$. The diameter of the Airy disk is

$$d_0 = 2r_0 = 2.44\frac{f\lambda}{D}. \tag{7.11.11}$$

---

[4]Airy is believed to be the first person (in 1827) to correct astigmatism in the eye (his own) by using cylindrical eyeglass lenses.

[5]See, for example, J. W. Goodman, *Introduction to Fourier Optics*, 2nd ed. McGraw-Hill, New York, 1996. Fourier optics is so named because it deals extensively with Fourier transforms such as, for instance, Eq. (7.10.8).

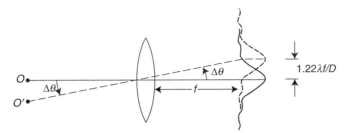

Figure 7.30   According to the Rayleigh criterion, two points O and O' are just resolved in the focal plane of a lens when the first minimum of the image of one coincides with the central maximum of the image of the other.

The fact that the image of a point is an Airy pattern rather than a point is essential to the understanding of *resolution limits* of optical instruments such as the human eye. If two distant objects subtend a very small angle, it is hard to distinguish (or "resolve") them if their diffraction patterns in the focal plane of a lens overlap. A useful measure of resolution is provided by the *Rayleigh criterion*: We say that two points are "just resolved" if the central maximum of the Airy pattern of the image of one coincides with the first zero in the Airy pattern of the other (Fig. 7.30). If the point sources were brought closer together, we would say (somewhat arbitrarily) that they were no longer resolvable. Thus, two point sources are just resolved if their angular separation ($\Delta\theta$) is equal to the angular radius of the Airy disk of the image of either source (Fig. 7.30):

$$(\Delta\theta)_{min} = \frac{r_0}{f} = 1.22\frac{\lambda}{D}. \tag{7.11.12}$$

As an example, consider the human eye. Assuming a lens aperture (i.e., pupil) diameter $D = 2.5$ mm, and a mean optical wavelength $\lambda = 550$ nm, we have a resolution limit of

$$(\Delta\theta)_{min} = 2.7 \times 10^{-4} \text{ rad.} \tag{7.11.13}$$

If we are to distinguish the headlights of a car as two separate sources, their angular separation subtended at the eye must, according to the Rayleigh criterion, be no smaller than $(\Delta\theta)_{min}$, that is,

$$\frac{h}{d} \geq (\Delta\theta)_{min}, \tag{7.11.14}$$

where h is the distance between the headlights and d is the distance from the observer to the car. Thus, the two headlights are resolvable if the car is not farther than a distance $d_{max} = h/(\Delta\theta)_{min}$ away. Taking $h = 1.2$ m, this distance is

$$d_{max} = \frac{1.2 \text{ m}}{2.7 \times 10^{-4}} = 4.4 \text{ km.} \tag{7.11.15}$$

If the car is farther away, the headlights will be blurred together and not resolvable. (Actually, we are ignoring the refractive index of the vitreous humor, and the fact that the pupil diameter of the dark-adapted eye is dilated.)

Equation (7.11.11) for the diameter of the Airy disk associated with a point source is frequently written as

$$d_0 = 2.44\lambda f\#, \qquad (7.11.16)$$

where the dimensionless number $f\# = f/D$ is called the *f number*. The intensity of light in the focal plane of the lens may be shown to be inversely proportional to the square of the *f* number; this is probably intuitively reasonable to anyone who has ever tried to burn things using sunlight and a magnifying glass.

The "stops" marked on manual-focus cameras reflect this dependence of focal-plane intensity (or, optically speaking, "illumination") on *f* number. Sequences of numbers such as 2, 2.8, 4, 5.6, 8, 11, 16, 22 on the diaphragm (aperture) setting for the camera lens are *f* numbers such that the square of each $f\#$ is approximately twice the square of the preceding one. A shift from one $f\#$ to the next one requires that the exposure time be doubled in order to get the same film exposure ("exposure" equals illumination times exposure time). An $f/2.8$ setting with an exposure time of $1/250$ s, for instance, will give the same exposure as an $f/4$ setting with a $1/125$ s exposure time. The minimum $f\#$ of a given lens is obtained with the diaphragm wide open. The smaller this minimum $f\#$, the "faster" the lens is said to be (and usually the more expensive).

Telescopes are also characterized by their $f\#$, in this case defined as the effective focal length divided by the diameter of the primary mirror or lens. Whereas photographers change the $f\#$ by changing the diaphragm setting, astronomers usually do it by changing the focal length.

The field radiated, reflected, or transmitted by a two-dimensional object can be written as a Fourier integral with "spatial frequencies" k_x, k_y, k_z such that $k^2 = k_x^2 + k_y^2 + k_z^2 = n^2\omega^2/c^2$, where n is the refractive index at wavelength λ. Information about the smallest spatial variations on the object is contained in the largest spatial frequencies k_x and k_y. It might be argued that the largest value of $k_x^2 + k_y^2$ is given by $(k_x^2 + k_y^2)_{max} = k^2$, and therefore that resolution of spatial structure on an object is limited to scales

$$\Delta \sim \frac{2\pi}{k} = \frac{2\pi c}{n\omega} = \frac{\lambda}{n}, \qquad (7.11.17)$$

and larger. Indeed it is often said that the resolution possible with a lens, for example, is limited by the wavelength. Note, however, that in writing (7.11.17) we implicitly assumed that $k_z^2 \geq 0$, implying that $(k_x^2 + k_y^2)_{max} = k^2$. This assumption ignores so-called *evanescent waves* for which $k_z^2 = n^2\omega^2/c^2 - k_x^2 - k_y^2 \leq 0$; evanescent waves decay with distance z from an object as

$$\exp(-|k_z|z) = \exp\left(-\sqrt{k_x^2 + k_y^2 - n^2\omega^2/c^2}\,z\right). \qquad (7.11.18)$$

They are present whenever the field in some plane $z = 0$ has spatial frequency components k_x, k_y such that $k_x^2 + k_y^2 > n^2\omega^2/c^2$. In total internal reflection at a glass–air interface, for example, $k_x^2 + k_y^2 < n_g^2\omega^2/c^2$ on the glass side, whereas $k_x^2 + k_y^2 > n_a^2\omega^2/c^2$ on the air side. (n_g and n_a are, respectively, the refractive indices of glass and air.) The evanescent wave on the air side decreases with distance z as $\exp(-az)$, where $a = (4\pi n_a/\lambda)\sqrt{(n_g^2/n_a^2)\sin^2\theta_i - 1}$ and θ_i is the angle of incidence. Evanescent waves are therefore necessary in general to characterize completely the field. Since they decay exponentially with distance, they can usually be ignored as a practical matter, in which case (7.11.17) provides a valid practical limit to resolution. But this is not a fundamental limit, and in fact techniques have been developed to "capture" the evanescent field components and thereby to resolve spatial variations on an object on a scale smaller than the wavelength of the light from the object.[6] The fact that subwavelength resolution is possible can be proved more generally based on the analytic properties of the Fourier-integral representation of the object field.[5]   •

7.12   DIFFRACTION THEORY OF RESONATORS

The Fresnel diffraction formula (7.10.4) is intimately related to our theory of Gaussian beams based on the paraxial wave equation. If $\mathcal{E}_0(x, y, z)$ satisfies (7.4.19), and is specified in the plane $(x, y, z = 0)$, then

$$\mathcal{E}_0(x, y, z) = -\frac{i}{\lambda z} \int \int \mathcal{E}_0(x', y', 0) e^{ik[(x-x')^2 + (y-y')^2]/2z} \, dx' \, dy', \tag{7.12.1}$$

and therefore [recall Eq. (7.4.13)]

$$\mathcal{E}(x, y, z) = -\frac{ie^{ikz}}{\lambda z} \int \int \mathcal{E}(x', y', 0) e^{ik[(x-x')^2 + (y-y')^2]/2z} \, dx' \, dy', \tag{7.12.2}$$

which is precisely (7.10.4). In other words, the Fresnel diffraction formula (7.10.4) is the solution of the paraxial wave equation when the field is specified in a plane $(x, y, z = 0)$ transverse to the propagation direction.

•   We will not take the time to derive (7.12.1) from (7.4.19). For the reader familiar with such things, we note that the function

$$K(x, y; x', y'; z) = -\frac{ie^{ikz}}{\lambda z} e^{ik[(x-x')^2 + (y-y')^2]/2z}, \tag{7.12.3}$$

is a Green function, or *propagator*, of the paraxial wave equation.

Our Gaussian beam solutions of the paraxial wave equation therefore satisfy (7.12.2). This is most easily seen in the Fraunhofer limit (7.10.8). In this case $\mathcal{E}(x, y, z)$ and $\mathcal{E}(x', y', 0)$ are related by a Fourier transform, and since the Fourier transform of a Gaussian function is again a Gaussian function, it follows that a Gaussian beam remains a Gaussian beam as it propagates.   •

Suppose we have a laser resonator and we know the field $\mathcal{E}_0(x, y, 0)$ on the mirror at $z = 0$. Then from (7.12.1) we know that the field at the mirror at $z = L$ is[7]

$$\mathcal{E}_0(x, y, L) = -\frac{i}{\lambda L} \int \int \mathcal{E}_0(x', y', 0) e^{ik[(x-x')^2 + (y-y')^2]/2L} \, dx' \, dy'$$

$$= \int \int K(x, y; x', y') \mathcal{E}_0(x', y', 0) \, dx' \, dy'. \tag{7.12.4}$$

---

[6]See, for example, J. B. Pendry, *Physical Review Letters* **85**, 3966 (2000) for a discussion of the possibility of realizing a "perfect lens" with *metamaterials* in which the refractive index is negative.

[7]The implicit assumption of planar mirrors here is of no real consequence. The plane $z = 0$ could be *any* plane between the mirrors, and the condition for a resonator mode [Eq. (7.12.7)] is that the field spatial pattern is reproduced in a round trip that begins and ends on this plane.

Similarly, we can use (7.12.4) in (7.12.1) to obtain the field on the first mirror *after one round trip through the resonator:*

$$\tilde{\mathcal{E}}_0(x, y, 0) = \int\int K(x, y; x', y')\mathcal{E}_0(x', y', L)\, dx'\, dy'$$

$$= \int\int K(x, y; x', y')\, dx'\, dy' \int\int K(x', y'; x'', y'')\mathcal{E}_0(x'', y'', 0)\, dx''\, dy''$$

$$= \int\int\left[\int\int K(x, y; x', y')K(x', y'; x'', y'')\, dx'\, dy'\right]\mathcal{E}_0(x'', y'', 0)\, dx''\, dy''$$

$$= \int\int \tilde{K}(x, y; x'', y'')\mathcal{E}_0(x'', y'', 0)\, dx''\, dy'', \tag{7.12.5}$$

where

$$\tilde{K}(x, y; x'', y'') = \int\int K(x, y; x', y')K(x', y'; x'', y'')\, dx'\, dy'. \tag{7.12.6}$$

Continuing in this manner, we obtain the field at each mirror after any arbitrary number of round trips (or "bounces") through the resonator. Now according to our discussion in Section 7.7, a *mode* of the resonator is defined as a field distribution that does not change on successive bounces inside the resonator. Because we are dealing with the modes of an empty resonator with no gain medium, the mode amplitudes will decrease on successive bounces due to diffraction at the mirrors, but the field distribution, or spatial pattern, will not. That is, a mode of the field is such that $\tilde{\mathcal{E}}_0(x, y, 0)$ is simply $\mathcal{E}_0(x, y, 0)$ times some (complex) number:

$$\tilde{\mathcal{E}}_0(x, y) = \gamma\mathcal{E}_0(x, y), \tag{7.12.7a}$$

or

$$\gamma\mathcal{E}_0(x, y) = \int\int \tilde{K}(x, y; x', y')\mathcal{E}_0(x', y')\, dx'\, dy', \tag{7.12.7b}$$

where for simplicity we drop the explicit reference to the z dependence.

Since diffractive losses at the mirrors can only diminish the total field energy inside the resonator, we must have $|\gamma| < 1$. In an actual "loaded" laser resonator, stimulated emission in the gain medium will compensate for this diffractive loss and all other losses. Equation (7.12.7) is simply a condition that the empty-cavity field distribution must satisfy if steady-state mode patterns are to exist.

For resonators with Gaussian beam modes, the integral equation (7.12.7) does not give any new results. However, the formulation of the mode problem based on (7.12.7) is useful when the spot sizes w_1, w_2 given in Table 7.2 are comparable to or larger than the mirror radii. As discussed in Section 7.7, our Gaussian beam mode analysis in this case is inapplicable. From Table 7.2 it is seen that this happens when $g_1 g_2$ is close to 0 or 1, or is such that the stability condition (7.7.19) is violated.

When the Gaussian beam mode analysis breaks down, the resonator modes may be found by solving (7.12.7) numerically. This is usually done according to the method developed by Fox and Li in 1961: One starts by *assuming* some field $\mathcal{E}_0(x, y)$ on a mirror, usually just $\mathcal{E}_0(x, y) = $ constant. The field is then "propagated" to the other mirror by doing the integral (7.12.4) numerically. The field obtained on the second

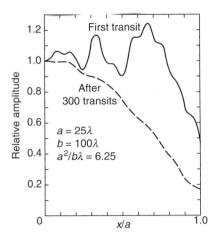

Figure 7.31   Field amplitude distribution computed by Fox and Li for a resonator with flat rectangular mirrors. The parameters are such that the Fresnel number $a^2/b\lambda = 6.25$. The iteration process was begun assuming a uniform plane-wave field on one of the mirrors. [From A. G. Fox and T. Li, *Bell System Technical Journal* **40**, 453 (1961).]

mirror is then propagated back to the first mirror by another numerical computation based on (7.12.4). This procedure is then iterated until the field on the mirrors is unchanged (within some prescribed numerical error) on successive iterations, except for a constant factor γ. The field so obtained is then a solution of (7.12.7), that is, it is a mode of the resonator. In practice, this method will yield straightforwardly only one mode, that of lowest round-trip loss, but certain numerical "tricks" can be employed to obtain higher-loss modes.

Figures 7.31 and 7.32 are reproduced from the original Fox–Li study. For the example shown, about 300 iterations were necessary for the iterative procedure to converge on a mode of the resonator.

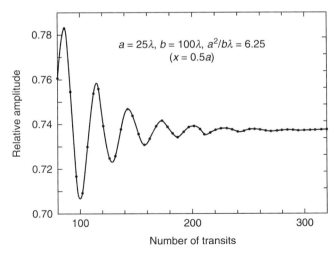

Figure 7.32   Dependence on iteration number of the amplitude at a fixed point on a mirror, as computed by Fox and Li for the case of Fig. 7.31.

- Equation (7.12.7) may be written in the *operator form*

$$\hat{\Gamma}\mathcal{E}_0 = \gamma\mathcal{E}_0, \tag{7.12.8}$$

where $\hat{\Gamma}$ is the operator corresponding to a round trip through the resonator, that is, $\hat{\Gamma}$ is *defined* by its effect on functions $f(x, y)$:

$$\hat{\Gamma}f(x, y) \equiv \iint \tilde{K}(x, y; x', y')f(x', y')\,dx'\,dy'. \tag{7.12.9}$$

According to (7.12.8), the modes of a resonator are the eigenfunctions of the operator $\hat{\Gamma}$ for the resonator. The number γ is the eigenvalue corresponding to the eigenfunction $\mathcal{E}_0(x, y)$. This sounds a bit like the mathematics of quantum mechanics. Here, however, $\hat{\Gamma}$ is generally not a Hermitian operator; for instance, the eigenvalues γ are complex, whereas the eigenvalues of a Hermitian operator are always real.   •

7.13 BEAM QUALITY

Most applications of lasers are based at least in part on the collimation and focusing properties of laser beams. In the ideal case a laser beam is *diffraction limited*, that is, its divergence is completely determined by diffraction, as we have assumed in our discussion of Gaussian beams. In reality the output from a laser may not be a TEM$_{00}$ Gaussian beam. Imperfections such as mirror misalignments will cause part of the output to belong to higher-order modes, or cause random phase variations across the output wavefront, resulting in "off-axis" radiation and a beam divergence greater than that associated solely with the diffraction of the TEM$_{00}$ mode. This could result in a significant increase in the spot size of the focused beam and a decrease in the focused intensity.

The diffraction-limited divergence angle of the idealized lowest-order Gaussian beam is given by Eq. (7.5.25):

$$\theta = \frac{\lambda}{\pi w_0}. \tag{7.13.1}$$

Based on this equation, it is common to write the divergence angle of a laser beam as

$$\theta_M = \frac{M^2\lambda}{\pi w_0}, \tag{7.13.2}$$

which defines M^2, a commonly used "times diffraction limit" measure of *beam quality*. The spot size of the beam after focusing with a lens of focal length f is

$$w'_M = f\theta_M = \frac{M^2\lambda f}{\pi w_0}, \tag{7.13.3}$$

as compared to (7.6.20). M^2 is one of the specifications typically provided by laser manufacturers. For a typical He–Ne laser operating on the TEM$_{00}$ mode, $M^2 < 1.1$, whereas for high-power multimode lasers M^2 might be ≈ 10 or even larger. Formula (7.13.3) is the basis for the measurement of M^2.

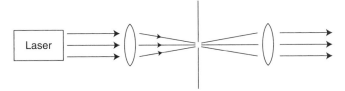

Figure 7.33 Beam cleanup by spatial filtering.

Another frequently used measure of beam quality is the *Strehl ratio*, which is defined as the ratio of the peak intensity in the focal plane to the diffraction-limited peak intensity. The Strehl ratio so defined is $1/M^2$.

We have assumed that the beam divergence can be characterized by a single M, but more generally it is necessary to specify values of M along each of two orthogonal directions in the plane perpendicular to the propagation direction.

Beam quality can be improved ("beam cleanup") using the *spatial filter* sketched in Fig. 7.33. The first lens focuses the laser output onto the pinhole, which has a diameter slightly larger than the diameter of the focused beam. Off-axis rays are focused at points away from the focal spot of the main beam, and therefore do not pass through the pinhole. The light from the pinhole then passes through the second lens, resulting in a well-collimated, "high-quality" beam. Note that beam cleanup by spatial filtering can be done simultaneously with beam expansion (see Fig. 7.24).

7.14 UNSTABLE RESONATORS FOR HIGH-POWER LASERS

Our emphasis on stable laser resonators should not be taken to imply that *unstable* resonators have no practical applications. On the contrary, unstable resonators enjoy certain advantages, and they are essential to the design of some important high-power lasers.

Stable resonators have some drawbacks if one wants to build a high-power device. A major disadvantage is that the modes of stable resonators tend to be concentrated in very thin, needlelike regions within the resonator. Therefore, they do not overlap a very large portion of the gain medium, and this obviously presents a problem if high-power extraction from the medium is desired. A Gaussian beam mode of a stable resonator, for instance, has a spot size on the order of $(\lambda L/\pi)^{1/2}$ [see Eq. (7.7.21)]. For a CO_2 laser with $\lambda = 10.6$ μm and $L = 1$ m,

$$\left(\frac{\lambda L}{\pi}\right)^{1/2} = 1.8 \text{ mm}, \tag{7.14.1}$$

a typical sort of beam "size" for Gaussian beam modes of stable resonators.

Unstable resonators, however, typically have much larger mode volumes and can therefore make better use of the available gain region. Figure 7.34 shows an important practical example of an unstable resonator, the so-called positive-branch (because $g_1 g_2 > 1$) confocal resonator. As indicated, the intracavity field fills a large portion of the cavity and can be made larger simply by using larger mirrors. The "magnification" m is a function only of the g parameters of the mirrors.

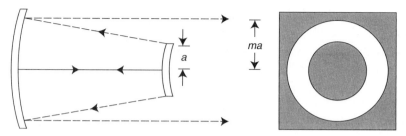

Figure 7.34   A positive-branch ($g_1 g_2 > 1$) confocal unstable resonator. The near-field output is a collimated, annular beam.

Iterative computations of the Fox–Li type reveal that the modes of unstable resonators like that shown in Fig. 7.34 are distinctly non-Gaussian. To a first approximation the lowest-loss mode has a nearly uniform intensity profile on the mirrors. The output beam for the resonator shown in Fig. 7.34 is a collimated annular (doughnut-shaped) beam in the *near field* close to the resonator. In the far field this output beam has a central bright spot on axis. In the limit of large magnification this far field approaches an Airy pattern, with most of the intensity concentrated in the central bright spot.

Unstable resonators offer other advantages in addition to their large mode volumes. For instance, they tend to yield higher output powers when operating on the lowest-loss transverse mode rather than on several (or many) modes. This property is not generally shared by stable-resonator lasers, and it is an important advantage in many applications. In addition, unstable-resonator lasers use *all-reflective optics*. That is, the output does not pass through any mirrors but simply spills around the mirror edges. At high power levels, where mirror damage is an important consideration, the mirrors can often be water cooled without much difficulty. Obviously, the problem of mirror damage and thermal distortion is not so easily surmountable in stable laser resonators employing transmissive output coupling.

The theory of unstable-resonator lasers does not differ in any fundamental way from that of stable-resonator lasers. For this reason, and because stable resonators are more common, we will not consider in any detail the mode characteristics of unstable resonators.

7.15   BESSEL BEAMS

In discussing Gaussian beams in Section 7.5 we introduced both the paraxial factorization (7.4.13),

$$\mathcal{E}(\mathbf{r}) = \mathcal{E}_0(\mathbf{r})e^{ikz}, \tag{7.15.1}$$

and several approximations [recall Eqs. (7.4.14) and (7.4.15) based on the presumed slow variation of the plane-wave envelope $\mathcal{E}_0$]. These are natural steps to take when dealing with beams that spread very little. Surprisingly, they do not lead to a description of the beams that spread the least of all. There is a set of ideally nonspreading beams that are described by Bessel functions rather than by Hermite–Gaussian or Laguerre–Gaussian functions.

In contrast to the "weak" paraxial factorization (7.4.13), there is also a "strong" factorization.[8] It is introduced by requiring the envelope function $\mathcal{E}_0$ to be completely independent of z. That is, in place of (7.15.1) we write

$$\mathcal{E}(\mathbf{r}) = \mathcal{E}_0(x, y)e^{i\beta z}, \tag{7.15.2}$$

where we have indicated explicitly the absence of z dependence in the envelope function. In this case, in place of (7.4.19) we find the equation

$$[\nabla_T^2 + \beta^2]\mathcal{E}_0(x, y) = 0, \tag{7.15.3}$$

where ∇_T^2 is the transverse Laplacian defined in (7.4.20). It is easy to check that Eq. (7.15.3), which is the Helmholtz equation in two dimensions rather than three, is an *exact* consequence of the strong paraxial factorization (7.15.2). There are no leftover terms required to be negligible, as there were in the transition from (7.4.16) to (7.4.18). The solution to the two-dimensional Helmholtz equation was known to be given in terms of Bessel functions at least 50 years before the time of Helmholtz. The solutions are most conveniently expressed in cylindrical coordinates ρ and ϕ:

$$x = \rho \cos \phi \qquad \text{and} \qquad y = \rho \sin \phi. \tag{7.15.4}$$

The solution that is finite at the origin is given by

$$\mathcal{E}_0 \longrightarrow \mathcal{E}_m(x, y) = AJ_m(\alpha\rho)e^{im\phi}, \tag{7.15.5}$$

where A is a constant and $J_m(x)$ is the mth Bessel function, the same functions introduced in our discussion of FM mode locking in Section 6.10 and shown in Fig. 6.13. In order for (7.15.5) to satisfy the two-dimensional Helmholtz equation, and for (7.15.2) to satisfy the full three-dimensional Helmholtz equation (7.4.3), it is only necessary that α and β be connected by the frequency of the light:

$$\alpha^2 + \beta^2 = \left(\frac{\omega}{c}\right)^2. \tag{7.15.6}$$

It is clear from (7.15.5) that only the lowest-order solution, the one with $m = 0$, is cylindrically symmetric (independent of ϕ). This is analogous to the situation found earlier with Hermite–Gaussian modes in Section 7.8. Inspection of Fig. 6.13 shows that the lowest-order mode also gives the most intense beam near the axis (near $\rho = 0$).

The most remarkable feature of the Bessel mode solutions described here is that they are, in the ideal case, *completely nondiffracting*. To show precisely what this statement means, let us compute the intensity of radiation associated with the general Bessel solution (7.15.5). The physical electric field is the real part of $\mathcal{E}(\mathbf{r})e^{-i\omega t}$, which in this case is given by

$$E_m(\mathbf{r}, t) = AJ_m(\alpha\rho) \cos(\omega t - kz + m\phi). \tag{7.15.7}$$

---

[8]J. Durnin, *Journal of the Optical Society of America A* **4**, 651 (1987).

The cycle-averaged power flow of the mth Bessel beam mode is then easily seen to be given by

$$I_m(\mathbf{r}, t) = \frac{c\epsilon_0}{2} A^2 J_m^2(\alpha\rho), \qquad (7.15.8)$$

and this function has the surprising property that it is completely independent of z.

That is, at every value of z (every distance of propagation) the intensity of an ideal Bessel beam has exactly the same x, y dependence. By contrast, Gaussian beams are characterized by their waist function $w(z)$, which grows from a minimum value w_0 in the course of propagation, and it is the growth of the beam waist that determines a Gaussian beam's far field (recall Fig. 7.11). By contrast, Bessel beams are characterized by the same transverse distribution of intensity (the same "waist") at every value of z and so do not diverge at all. In this sense they can be said to constitute perfectly nondiffracting beams.

One may recall that the intensity distribution of light propagated through a circular aperture, graphed in Fig. 7.29, is expressed in (7.11.6) in terms of the Bessel function J_1. The argument of the Bessel function in that application, however, depends explicitly on z and so is quite different from that derived here for a Bessel beam. This comparison of (7.11.6) and (7.15.5) shows that nonspreading Bessel beams will not be created by transmitting a plane wave through a circular aperture. That raises the question how Bessel beams can be realized in practice. One might think that any practical realization would involve apertures that would inevitably prevent the development of the strong paraxial character expressed in (7.15.2), and thus lead back to a Gaussian beam. However, this is not the case, and the remarkable properties of Bessel beams can actually be realized over substantial regions of space.

To explain one method for creating a Bessel beam in practice, we will concentrate on the most important case, the cylindrically symmetric lowest-order Bessel beam. The zero-order Bessel function $J_0(\alpha\rho)$ can be represented by an integral:

$$J_0(\alpha\rho) = \frac{1}{2\pi} \int_0^{2\pi} e^{i\alpha(x\cos\phi + y\sin\phi)} \, d\phi. \qquad (7.15.9)$$

The full x, y, z dependence of the solution is then obtained by combining (7.15.2) with (7.15.5) for $m = 0$, and using the integral expression for J_0 to get

$$\mathcal{E}_0(\mathbf{r}) = \frac{A}{2\pi} \int_0^{2\pi} e^{i[\alpha(x\cos\phi + y\sin\phi) + \beta z]} \, d\phi. \qquad (7.15.10)$$

This can be rewritten and interpreted directly in physical terms by defining a "wave vector" $\mathbf{q}$ whose components are given by

$$\mathbf{q} = (\alpha\cos\phi, \ \alpha\sin\phi, \ \beta). \qquad (7.15.11)$$

Then we have

$$\mathcal{E}_0(\mathbf{r}) = \frac{A}{2\pi} \int_0^{2\pi} e^{i\mathbf{q}\cdot\mathbf{r}} \, d\phi. \qquad (7.15.12)$$

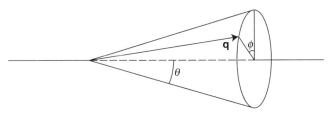

Figure 7.35 The cone of wave vectors making up the Bessel beam defined by (7.15.11) and (7.15.12).

The advantage of this form is that it can be interpreted physically. It says that a zero-order Bessel mode consists of all possible plane waves with wave vectors **q** whose length is restricted by $\alpha^2 + \beta^2 = (\omega/c)^2$, whose polar angle of inclination to the z axis is fixed at the value $\tan \theta = \sqrt{q_x^2 + q_y^2}/q_z = \alpha/\beta$, and whose azimuthal angles ϕ are completely unrestricted. In other words, these are all wave vectors of length $q = \omega/c$ lying on the surface of a cone with opening angle θ, as shown in Fig. 7.35.

This mathematical description suggests a method of creating a Bessel beam in practice. A very narrow annular aperture normal to the z axis can be illuminated from one side. If the aperture slit is narrow enough, it acts as a circular line source of light. A lens placed with the aperture in its focal plane will then transmit a cone of light just as (7.15.12) requires. In the space beyond the lens the optical field will be given by $J_0(\alpha\rho)$. Of course, the inevitably finite size of the circular aperture and of the lens will tend to destroy some of the ideal features of the Bessel beam. In particular, as Fig. 7.36 shows, the cone of light will be only finitely wide and its elements will not overlap on the z axis beyond a certain point.

We denote by Z_{max} the point on the z axis where the geometrical rays shown in the figure no longer overlap. It is found by simple geometrical arguments to be given by

$$Z_{max} = \frac{r}{\tan \theta},\qquad(7.15.13)$$

where $\tan \theta = \alpha/\beta$ and r is the radius of the lens. However, the Bessel solution is more than just a geometrical ray property of the light field, and experiments reported in 1987

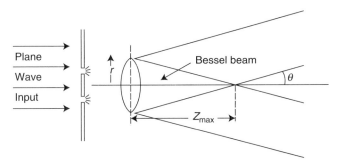

Figure 7.36 Cross section of an arrangement for creating the Bessel beam (7.15.10). Illumination of a narrow annular slit from the left by a plane wave creates a circle of coherent point sources. These give rise to the required cone of rays on the right side of the lens. Here $Z_{max} = r/\tan \theta = r\beta/\alpha$. [From J. Durnin, J. J. Miceli, Jr., and J. H. Eberly, *Physical Review Letters* **58**, 1499 (1987).]

showed that the nondiffracting nature of the Bessel mode persists even beyond the point $z = Z_{max}$, although the intensity is much diminished.

As an example of some of the striking differences between Bessel and Gaussian beams, let us consider the propagation of a very narrow Bessel beam of wavelength $\lambda = 500$ nm in a region empty of lenses or other media or boundaries. Suppose that at $z = 0$ the aperture (lens diameter in Fig. 7.36) has area $\pi r^2 = (5 \text{ mm})^2$ and that the central Bessel spot is rather small with area $\pi a^2 = (50 \text{ μm})^2$. It can easily be determined geometrically that $(Z_{max})^2 = (\pi r^2/\lambda)(\pi a^2/\lambda)$, or $Z_{max} = \pi r a/\lambda$, which in this example means $Z_{max} \approx 0.5$ m. For comparison, a Gaussian beam with its waist in the aperture plane and as small as the central Bessel spot [namely, a beam area of $(50 \text{ μm})^2$] will have a Rayleigh range [recall Eq. (7.5.19)] given by $z_0 \approx 1$ cm, much smaller than Z_{max}. This illustration is designed to be "favorable" for a Bessel beam, to show that it can propagate without spreading and with significant intensity much farther than a Gaussian beam with an equally small central spot. However, the Gaussian beam would not have to have its waist in the aperture plane, as we assumed here. With a sufficiently large lens at $z = 0$, we could arrange for the Gaussian waist (focal spot) to occur at any distance down the z axis. Even then, wherever the $(50\text{-μm})^2$ Gaussian focal spot is placed, it will have a depth of field of only ≈ 1 cm. The significant difference between Bessel and Gaussian beams is therefore the much greater depth of field of the Bessel beam's "focal region."

Bessel mode beams are of course not ideal in every respect, and the lack of beam divergence is achieved at some cost, both theoretically and practically. In principle, the Bessel solution is valid only in infinite transverse x–y space. This is true also of the Gaussian beam solution. However, the Gaussian beam solution falls to zero so rapidly with increasing x and y (away from the z axis) that a Gaussian beam does not "notice" that infinite x–y space is actually not available to it in any real laboratory. This is less true of the Bessel beam, which falls to zero in the transverse directions rather slowly:

$$I_m(\alpha\rho) = \frac{c\epsilon_0}{2}A^2 J_m^2(\alpha\rho) \approx \frac{1}{\alpha\rho}\cos^2(\alpha\rho + \delta), \qquad (7.15.14)$$

for large values of x and y. An important consequence of the cosine-squared character of (7.15.14) is that the energy of a Bessel beam is contained in concentric rings of width given by $\alpha\rho = \pi$, and the intensity is approximately equal in each ring:

$$\int_{\text{one ring}} I_m(\alpha\rho)\rho\,d\rho\,d\phi \approx \frac{2\pi}{\alpha}\int \cos^2(\alpha\rho + \delta)\,d\rho = \frac{\pi}{\alpha^2}, \qquad (7.15.15)$$

where the transverse integration is carried out over one period of J_m. Thus, we see that an ideal Bessel mode beam must carry an infinite amount of energy in its transverse skirt. This is also true (obviously) of plane waves, but is in strong contrast to a Gaussian beam, for which only a very small fraction of energy will be in the skirt outside the central beam spot.

Finally, we should mention that Eq. (7.15.14) explains the apparent conflict between the existence of nondiffracting beams and the familiar Fourier principle (or, in quantum theory, the Heisenberg uncertainty principle), which requires

$$\Delta k_\perp \Delta \rho > (2\pi)^2, \qquad\qquad (7.15.16)$$

where $\Delta k_\perp$ is the dispersion of the transverse component of the beam wave vector ($\hbar k_\perp$ is the transverse photon momentum in quantum theory) and $\Delta \rho$ is the transverse dispersion in the beam wave solution. As it happens, because of the slow falloff of the intensity of the Bessel beam's skirt, as given in (7.15.14), the dispersion $\Delta \rho$ actually diverges. Because $\Delta \rho$ is infinite, $\Delta k_\perp$ can be zero (no diffraction at all) without violating (7.15.16).

PROBLEMS

7.1. **(a)** Prove that the ray matrix for the optical system consisting of a straight section followed by a thin lens is different from that for a thin lens followed by a straight section.

(b) Verify Eq. (7.3.3).

(c) Using ray matrix multiplication, show that two thin lenses of focal lengths f_1 and f_2 placed in contact are equivalent to a single thin lens of focal length

$$f = \frac{f_1 f_2}{f_1 + f_2} \qquad \text{or} \qquad \frac{1}{f} = \frac{1}{f_1} + \frac{1}{f_2}.$$

7.2. Show that the resonators sketched in Figs. 7.8 and 7.9 are stable and unstable, respectively.

7.3. **(a)** Prove that the spherical wave (7.4.6) satisfies the Helmholtz equation (7.4.3).

(b) Verify condition (7.4.11) for the validity of (7.4.10).

(c) Verify Eq. (7.5.26) for the intensity of a Gaussian beam.

(d) Show that the intensity (7.5.26) of a Gaussian beam may be written in the form

$$I(r, z) = \frac{2P}{\pi w^2(z)} e^{-2r^2/w^2(z)},$$

where P is the total beam power:

$$P = 2\pi \int_0^\infty I(r, z) r \, dr.$$

7.4. Show that the magnitude of the radius of curvature of a Gaussian beam is changed upon reflection from a spherical mirror, unless (a) the mirror has infinite radius of curvature (flat mirror), or (b) the radius of curvature of the mirror equals that of the Gaussian beam.

7.5. Verify Eqs. (7.7.14) and (7.7.16) for the resonance frequencies of TEM_{00} Gaussian modes of resonators with nearly flat mirrors.

7.6. Consider the arrangement sketched in Fig. 7.19 for the measurement of the spot size of a TEM$_{00}$ Gaussian beam. Show that the aperture passes about 86.5% of the total beam power when its radius equals the spot size.

7.7. (a) Show that the intensity patterns of Fig. 7.20 may be associated with the transverse Gaussian modes indicated.

   (b) Show that the low-order, linearly polarized Gaussian modes have the polarization patterns indicated in Fig. 7.21.

7.8. (a) Derive expressions (7.9.2) and (7.9.3) for the Gaussian mode spot sizes on the mirror of a quasi-hemispherical resonator.

   (b) A 632.8-nm He–Ne laser with a mirror separation of 50 cm has a micrometer adjustment to vary the separation between a flat mirror and an output mirror of radius of curvature $R = 50$ cm. If spot sizes between 1 and 2 mm are desired for the output beam, over what range must the mirror separation vary? Over what range will the spot size at the flat mirror vary?

7.9. Consider the optical element sketched in Fig. 7.23. Take the refractive indices inside and outside the element to be n and 1.0, respectively.

   (a) Show that the ray matrices associated with the first and second spherical interfaces are

$$\begin{bmatrix} 1 & 0 \\ \dfrac{n-1}{nR_1} & \dfrac{1}{n} \end{bmatrix} \quad \text{and} \quad \begin{bmatrix} 1 & 0 \\ \dfrac{1-n}{R_2} & n \end{bmatrix},$$

   respectively.

   (b) Determine the ray matrix for the optical element in terms of n, R_1, R_2, and the thickness x.

   (c) A particular He–Ne laser (Spectra-Physics Model 124) employed a collimating mirror with $R_1 = 2$ m, $R_2 = 64$ cm, and $x = 4$ mm. The waist of the output beam occurred at a distance of 28 cm from the collimating mirror. Determine the refractive index n.

7.10. Figure 7.25 plots the output beam spot size vs. distance for a 632.8-nm He–Ne laser, both with and without a beam collimator with a magnification of 10. Derive the equations satisfied by these two performance curves.

7.11. Figure 7.26 shows a possible experimental setup for measuring the divergence angle θ of a Gaussian beam. Show that

$$\theta = \frac{w}{f},$$

   where w is the beam spot size in the focal plane of the lens of focal length f. Does it matter where we intercept the beam from the laser?

7.12. In the paraxial approximation a thin lens converts an incident wave with a constant phase in the (x, y) plane to a converging spherical wave with phase $\exp[-ik(x^2 + y^2)/2f]$, where $f (>0)$ is the focal length [cf. Fig. 7.14 with $R_1 = \infty$ and $R_2 = f$].

Show that, for the purpose of calculating the intensity in the focal plane ($z = f$), the effect of the lens is to replace the Fresnel diffraction formula (7.10.4) by the Fraunhofer formula (7.10.8). (Note: Departures from paraxiality are responsible for *aberrations* that reduce image quality.) Show more generally that the lens law,

$$\frac{1}{d_0} + \frac{1}{d_i} = \frac{1}{f},$$

is satisfied, where d_0 and d_i are defined in Fig. 7.14.

7.13. A company that builds pulsed Nd : glass lasers used in making razor blades and in other laser welding applications provides a focus spot estimator. A "multiplier value," such that the focal length in millimeters times the multiplier value gives the focus spot size in microns, is given for each of a set of beam diameters. For a 350-W laser the multiplier values for beam diameters of 10, 15, and 20 mm are 1.5, 1.0, and 0.75, respectively. For a 550-W laser the corresponding multiplier values are 3.5, 2.3, and 1.75.

(a) Explain how these multiplier values might have been calculated.

(b) Does the beam quality increase or decrease with laser power?

7.14. A laser is to be used to make cuts in steel. You are asked to choose between two 10.6-μm CO_2 lasers, one with an output power of 300 W and a beam quality characterized by $M^2 = 1.2$ and the other with an output of 400 W and $M^2 = 2$. Both lasers have a beam size of 10 mm before focusing. Which is the better choice in terms of maximizing the focused intensity?

8 PROPAGATION OF LASER RADIATION

8.1 INTRODUCTION

It is important for a wide variety of applications to have a basic understanding of how laser radiation propagates in various media under various conditions depending on the laser wavelength, intensity, pulse duration, and other factors. In our treatment of propagation thus far we have made some simplifying assumptions, such as the plane-wave approximation, to describe how the radiation intensity propagates in an absorbing or amplifying medium, or we have gone beyond the plane-wave approximation but restricted ourselves to idealized monochromatic fields. We have also largely ignored the possibility of spatial variations of the refractive index.

The utility of these simplifications should not be underestimated. The (monochromatic) Gaussian beams discussed in the preceding chapter, for instance, do in fact provide an accurate description of the output of many lasers. However, it is also true that there are important features of the propagation of laser radiation that are not amenable to treatments based on plane waves, monochromaticity, or a constant refractive index. We will now consider the propagation of both monochromatic and nonmonochromatic laser beams in media other than vacuum or the absorbing or amplifying media considered thus far. We begin, in the following section, by deriving the wave equation (8.2.13), which is the basis for the analysis of nearly every propagation problem of interest and was the starting point for our derivation of the paraxial wave equation for monochromatic fields in Chapter 7.

The most important examples of nonmonochromatic radiation are laser pulses produced, for instance, by Q switching or mode locking. In this chapter we introduce some basic concepts of laser pulse propagation, such as group velocity and group velocity dispersion, and we describe methods for mitigating the deleterious effects of group velocity dispersion. These concepts are especially important in the theory of ultrashort pulse generation and in fiber-optic communication systems (Section 15.6). We treat the propagation of radiation in optical fibers as an important example of propagation in the case of a spatially varying refractive index. We then discuss two aspects of electromagnetic wave propagation—Rayleigh scattering and birefringence—that are important not only for applications but also for an understanding of many natural optical phenomena. This is followed by a discussion of some of the concepts involved in the propagation of laser radiation in Earth's turbulent atmosphere, where the refractive index undergoes random fluctuations.

Laser Physics. By Peter W. Milonni and Joseph H. Eberly
Copyright © 2010 John Wiley & Sons, Inc.

We restrict ourselves in this chapter to media characterized by a *linear* refractive index, that is, a refractive index that is independent of the strength of the field; the propagation is then described by linear partial differential equations [e.g., Eq. (8.2.20)]. Nonlinear propagation effects are considered in Chapter 10.

8.2 THE WAVE EQUATION FOR THE ELECTRIC FIELD

Our treatment of absorption and gain in Chapter 3 was based on the rate at which an atom in an electric field gains or loses energy. Equations such as (3.12.5) were derived on the basis of energy conservation rather than the fundamental Maxwell equations for the electromagnetic field. Similarly, our derivation of the paraxial wave equation in Chapter 7 proceeded not from Maxwell's equations but rather from a consequence of these equations, namely the wave equation (7.4.1) for the electric field in vacuum. We will now describe in much more detail the propagation of light, and in particular the highly directional (and mostly paraxial) light from a laser, starting from Maxwell's equations. We presume that the reader has some familiarity with Maxwell's equations and understands that they are the basis for the most fundamental description of electric and magnetic fields and therefore of light.[1] For a dielectric medium with no free electric charges or currents, Maxwell's equations are

$$\nabla \cdot \mathbf{D} = 0, \tag{8.2.1}$$

$$\nabla \cdot \mathbf{B} = 0, \tag{8.2.2}$$

$$\nabla \times \mathbf{E} = -\frac{\partial \mathbf{B}}{\partial t}, \tag{8.2.3}$$

$$\nabla \times \mathbf{H} = \frac{\partial \mathbf{D}}{\partial t}. \tag{8.2.4}$$

The electric displacement $\mathbf{D}$ is defined by

$$\mathbf{D} = \epsilon_0 \mathbf{E} + \mathbf{P}, \tag{8.2.5}$$

where $\mathbf{P}$ is the polarization density, that is, the electric dipole moment per unit volume, and the "permittivity of free space," ϵ_0, is given by $1/4\pi\epsilon_0 = 8.98755 \times 10^9$ N-m^2/C^2. The magnetic intensity vector $\mathbf{H}$ is defined by

$$\mathbf{B} = \mu_0(\mathbf{H} + \mathbf{M}), \tag{8.2.6}$$

where $\mathbf{M}$ is the magnetic dipole moment per unit volume and $\mu_0 = 4\pi \times 10^{-7}$ N/A^2 is the permeability of free space.

---

[1] This is true even in the quantum theory of light in which the electric and magnetic fields are quantum mechanical *operators* satisfying Maxwell's equations.

The magnetic susceptibility χ_m of a medium is defined by writing $\mathbf{M} = \chi_m \mathbf{H}$, in which case (8.2.6) is usually written as $\mathbf{B} = \mu \mathbf{H}$, where $\mu = \mu_0(1 + \chi_m)$ is the magnetic permeability of the medium. In the case of a time-varying electromagnetic field μ is a function of the field frequency, but at or near optical frequencies this dependence is negligible and in fact for practical purposes we can take $\mu = \mu_0$, that is, $\mathbf{B} = \mu_0 \mathbf{H}$. Then, applying the curl operation to both sides of (8.2.3), we obtain

$$\nabla \times (\nabla \times \mathbf{E}) = -\nabla \times \frac{\partial \mathbf{B}}{\partial t} = -\mu_0 \frac{\partial}{\partial t}(\nabla \times \mathbf{H}) = -\mu_0 \frac{\partial^2 \mathbf{D}}{\partial t^2}. \tag{8.2.7}$$

Now we use the general identity

$$\nabla \times (\nabla \times \mathbf{E}) = \nabla(\nabla \cdot \mathbf{E}) - \nabla^2 \mathbf{E} \tag{8.2.8}$$

of vector calculus to write (8.2.7) as

$$\nabla(\nabla \cdot \mathbf{E}) - \nabla^2 \mathbf{E} = -\mu_0 \frac{\partial^2 \mathbf{D}}{\partial t^2}. \tag{8.2.9}$$

Finally, we use the definition (8.2.5) of $\mathbf{D}$ and rearrange terms:

$$\nabla^2 \mathbf{E} - \nabla(\nabla \cdot \mathbf{E}) - \frac{1}{c^2}\frac{\partial^2 \mathbf{E}}{\partial t^2} = \frac{1}{\epsilon_0 c^2}\frac{\partial^2 \mathbf{P}}{\partial t^2}. \tag{8.2.10}$$

Here we have used the fact that

$$\epsilon_0 \mu_0 = \frac{1}{c^2}, \tag{8.2.11}$$

where $c = 2.99792458 \times 10^8$ m/s is the velocity of light in vacuum.

Equation (8.2.10) is a partial differential equation with independent variables x, y, z, and t. It tells us how the electric field $\mathbf{E}(x, y, z, t)$ depends on the electric dipole moment density $\mathbf{P}$ of the medium. We will be particularly interested in *transverse* fields (also called *solenoidal* or *radiation* fields), which satisfy (see below)

$$\nabla \cdot \mathbf{E} = 0. \tag{8.2.12}$$

Transverse fields therefore satisfy the wave equation

$$\nabla^2 \mathbf{E} - \frac{1}{c^2}\frac{\partial^2 \mathbf{E}}{\partial t^2} = \frac{1}{\epsilon_0 c^2}\frac{\partial^2 \mathbf{P}}{\partial t^2}. \tag{8.2.13}$$

This is the fundamental electromagnetic field equation for most purposes.[2] In the absence of any medium the right-hand side is zero and we have the "homogeneous" wave equation

$$\nabla^2 \mathbf{E} - \frac{1}{c^2}\frac{\partial^2 \mathbf{E}}{\partial t^2} = 0, \tag{8.2.14}$$

which was the basis for our derivation of the paraxial wave equation in Section 7.4.

---

[2]As noted in Section 3.2 (see also Problem 3.2), light interacts with matter primarily through its electric field, the effect of the magnetic field being very small. For this reason we always write propagation equations for the electric field.

To make use of (8.2.13) we must specify the polarization **P**. This cannot be done within the framework of Maxwell's equations alone, for **P** is a property of the material medium in which the field **E** propagates; we need to know how an electric dipole moment density **P** is produced in the medium.

Let us for simplicity consider here a single component of the electric field and polarization vectors, and replace (8.2.13) by the scalar wave equation,

$$\nabla^2 E - \frac{1}{c^2}\frac{\partial^2 E}{\partial t^2} = \frac{1}{\epsilon_0 c^2}\frac{\partial^2 P}{\partial t^2}. \tag{8.2.15}$$

Let us furthermore consider a monochromatic field of frequency $\nu = \omega/2\pi$, and write

$$E(\mathbf{r}, t) = \mathcal{E}(\mathbf{r}, \omega)e^{-i\omega t}, \tag{8.2.16a}$$

and similarly

$$P(\mathbf{r}, t) = \mathcal{P}(\mathbf{r}, \omega)e^{-i\omega t}. \tag{8.2.16b}$$

In this case (8.2.13) reduces to

$$\nabla^2 \mathcal{E}(\mathbf{r}, \omega) + \frac{\omega^2}{c^2}\mathcal{E}(\mathbf{r}, \omega) = -\frac{\omega^2}{\epsilon_0 c^2}\mathcal{P}(\mathbf{r}, \omega). \tag{8.2.17}$$

If the medium consists of N particles per unit volume, each of which has a dipole moment **p**, the polarization density is $N\mathbf{p}$. We assume that the medium has no polarization in the absence of a field, so that the only dipole moments **p** in the medium are induced by the field, that is, the electric field causes charge displacements that result in an induced dipole moment

$$p(\mathbf{r}, \omega) = \alpha(\omega)\mathcal{E}(\mathbf{r}, \omega) \tag{8.2.18}$$

in an atom at **r**. The polarizability $\alpha(\omega)$, which we have already introduced in Section 3.14, depends not only on ω but also on the transition frequencies and oscillator strengths of the atoms of the medium. In terms of the atom density and the polarizability, Eq. (8.2.17) becomes

$$\nabla^2 \mathcal{E}(\mathbf{r}, \omega) + \frac{\omega^2}{c^2}\mathcal{E}(\mathbf{r}, \omega) = -\frac{\omega^2}{\epsilon_0 c^2}N\alpha(\omega)\mathcal{E}(\mathbf{r}, \omega), \tag{8.2.19}$$

or

$$\nabla^2 \mathcal{E}(\mathbf{r}, \omega) + n^2(\omega)\frac{\omega^2}{c^2}\mathcal{E}(\mathbf{r}, \omega) = 0, \tag{8.2.20}$$

where

$$n^2(\omega) = 1 + \frac{1}{\epsilon_0}N\alpha(\omega). \tag{8.2.21}$$

Equation (8.2.20) implies that when a field of frequency ω propagates in a medium, the velocity of light—which in the absence of any medium has the same value, c, for all frequencies—is $c/n(\omega)$. In other words, $n(\omega)$ is the refractive index. More precisely, $c/n(\omega)$ is the phase velocity of light of frequency ω in a medium of refractive index $n(\omega)$. To see this, consider a monochromatic plane wave propagating in the z direction:

$$E(z, t) = \mathcal{E}_0 e^{-i(\omega t - kz)}, \tag{8.2.22}$$

or

$$\mathcal{E}(\mathbf{r}, \omega) = \mathcal{E}_0 e^{ikz}. \tag{8.2.23}$$

In this case (8.2.20) implies $k^2 = n^2(\omega)\omega^2/c^2$, or $k = \pm n(\omega)\omega/c$, and

$$E(z, t) = \mathcal{E}_0 e^{-i\omega[t \mp n(\omega)z/c]}, \tag{8.2.24}$$

which shows that $c/n(\omega)$ is indeed the phase velocity of light of frequency ω. Explicit formulas for the refractive index in terms of transition wavelengths and oscillator strengths have been given in Section 3.14.

Note that, for a field of frequency ω,

$$\mathbf{D}(\mathbf{r}, \omega) = \epsilon_0 \mathbf{E}(\mathbf{r}, \omega) + \mathbf{P}(\mathbf{r}, \omega) = \epsilon_0 \mathbf{E}(\mathbf{r}, \omega) + N\alpha(\omega)\mathbf{E}(\mathbf{r}, \omega)$$

$$= \varepsilon(\omega)\mathbf{E}(\mathbf{r}, \omega), \tag{8.2.25}$$

where $\varepsilon(\omega)$ is the permittivity of the medium. We have assumed that $\mathbf{p}$ (and therefore $\mathbf{D}$) is linearly proportional to $\mathbf{E}$, that $\varepsilon(\omega)$ is a scalar, i.e., that $\mathbf{D}$ points in the same direction as $\mathbf{E}$, and that $\varepsilon(\omega)$ does not depend on $\mathbf{r}$. In this case $\nabla \cdot \mathbf{D}(\mathbf{r}, \omega) = 0 = \varepsilon(\omega)\nabla \cdot \mathbf{E}(\mathbf{r}, \omega)$ in a charge-neutral medium. Then, since $\nabla \cdot \mathbf{E}(\mathbf{r}, \omega) = 0$ for every frequency ω, the transversality condition (8.2.12) for $\mathbf{E}(\mathbf{r}, t)$ must be satisfied. As discussed in Section 8.8, $\mathbf{E}$ and $\mathbf{D}$ do not always point in the same direction.

• The fact that the velocity c appearing in the wave equation (8.2.13) is numerically equal to the speed of light in vacuum led Maxwell to conclude that light is an electromagnetic phenomenon, or, in his words, "the luminiferous ether and the electromagnetic medium are one." The first conclusive experimental corroboration of Maxwell's theory was provided by Hertz in 1887, 8 years after Maxwell's death. Hertz observed that sparks between metal spheres produced sparks across a second pair of spheres and showed that the "disturbance" (radiation) transmitted between the two pairs of spheres could be reflected, focused, and refracted. He produced oscillatory sparking at 3×10^7 cycles per second with an induction coil and, by determining the nodes of a standing-wave pattern, inferred a wavelength of 9.6 m and, therefore, a propagation velocity of about 3×10^8 m/s for the electromagnetic disturbance.

The first evidence for a *finite* velocity of light was obtained by Olaf Roemer (1676) from his observations of the eclipses of Jupiter's moon Io. Jupiter's moons orbit the planet in nearly the same plane in which Jupiter and Earth orbit the sun and, as seen from Earth, they are periodically eclipsed by Jupiter. Roemer observed that successive eclipses occurred more frequently when Earth was moving toward Jupiter than when it was moving away—a consequence of what we would today call the Doppler effect—and attributed the difference to a finite velocity of light.[3] In 1725 Bradley deduced from the aberration of starlight, that is, the changes in the apparent position of a star due to the motion of Earth, that $c \sim 3 \times 10^8$ m/s.

---

[3] A detailed discussion of Roemer's work is given by J. H. Shea, *American Journal of Physics* **66**, 561 (1998).

Fizeau (1849) made the first terrestrial determination of c using a rotating toothed wheel such that light passing through a gap between the "teeth" and reflected off a mirror 8.6 km away could either pass through a gap or be blocked by the rotating wheel. From the gap spacing and the angular velocity of the wheel, Fizeau obtained $c = 3.15 \times 10^8$ m/s. Using a rotating mirror instead of a wheel, Foucault obtained $c = 2.986 \times 10^8$ m/s. In a long series of experiments Michelson used a rotating mirror to obtain, in 1935, $c = (2.99774 \pm 1.1) \times 10^8$ m/s. Actually these experiments measure the *group* velocity of light, which under the conditions of the experiments is very nearly equal to c, the velocity of light in vacuum (Section 8.3).

Similar measurements of c can be made using a laser and a rotating mirror or by sweeping a laser beam across a small hole in an opaque screen to produce a "pulse" incident on a beam splitter near the hole. Part of the pulse reflected off the beam splitter is incident on a photodiode that triggers an oscilloscope, and the other part travels some distance before being reflected onto a second photodiode that produces a second pulse profile on the oscilloscope. Comparison of the two pulse traces yields, from the sweep speed of the oscilloscope, the propagation time of the second pulse and therefore its velocity. The accuracy of such "direct" measurements of c is limited by the accuracy ($\sim 1\%$) with which the propagation time can be inferred from the oscilloscope trace.

Because of its fundamental significance, dozens of careful measurements of c have been carried out since Michelson's experiments. The most accurate measurements by far have employed frequency-stabilized lasers (Section 5.13) and the relation $c = \lambda \nu$, where λ and ν are the wavelength and frequency, respectively. Whereas optical wavelengths can be measured very accurately, the measurement of optical frequencies ($\sim 10^{14} - 10^{15}$ Hz) is much more challenging. Optical frequencies have been measured using techniques for frequency multiplication of nearly monochromatic millimeter waves produced by klystrons (frequencies ~ 70 GHz) and subsequent measurement of a beat frequency between the frequency-multiplied radiation and the frequency-stabilized laser. In this way laser frequencies have been measured to an accuracy of a few parts in 10^{-10} or better, resulting in determinations of c with uncertainties < 1 m/s, an improvement by a factor ~ 100 over previous (prelaser) measurements.[4] The redefinition of the meter by the International Committee on Weights and Measurements in 1983 resulted in the value $c = 299792458$ m/s for the velocity of light in vacuum.   ●

8.3 GROUP VELOCITY

The phase velocity $c/n(\omega)$ characterizes an idealized monochromatic wave, which has no beginning or end and obviously cannot be realized. Any electromagnetic wave of finite duration has a distribution of frequencies, and consequently when it propagates in a dielectric medium it has a distribution of phase velocities. But pulses often propagate with a well-defined velocity, called the *group velocity*, which we now discuss.

The derivation of the formula for the group velocity is somewhat complicated, and so we begin with a trivial "derivation" that leads to the correct result. We consider a field that is the superposition of two plane waves with the same polarization and amplitude but differing slightly in frequency ω and "wave number" $k = n\omega/c$. The sum of the two waves is proportional to

$$E(z, t) = \cos[(\omega + \Delta\omega)t - (k + \Delta k)z] + \cos[(\omega - \Delta\omega)t - (k - \Delta k)z]$$

$$= 2\cos(\omega t - kz)\cos\Delta\omega\left(t - \frac{\Delta k}{\Delta\omega}z\right). \tag{8.3.1}$$

---

[4]Even more accurate measurements of optical frequencies are discussed in Section 14.7.

The factor $\cos(\omega t - kz)$ is a "carrier" wave with phase velocity $v_p = \omega/k = c/n(\omega)$. The second cosine factor gives the wave modulation, or "envelope." If $\Delta\omega$ and Δk are small in magnitude compared to ω and k, the envelope is slowly varying in space and time compared to the carrier and propagates with the velocity $\Delta\omega/\Delta k$, the velocity of points z such that $t - (\Delta k/\Delta\omega)z$ is constant. From this we might guess that if we add waves with a small and *continuous* spread in frequencies and wave numbers around ω and k, we will obtain similarly a carrier wave with phase velocity $c/n(\omega)$ and an envelope that propagates with the velocity

$$v_g = \frac{d\omega}{dk}. \tag{8.3.2}$$

This is the correct formula for the group velocity. Writing $k = n(\omega)\omega/c$, we have $1/v_g = dk/d\omega = [n(\omega) + \omega\, dn/d\omega]/c$, or

$$v_g = \frac{c}{n + \dfrac{\omega\, dn}{d\omega}}. \tag{8.3.3}$$

The group velocity is commonly written as c/n_g, where the *group index*

$$n_g \equiv n + \frac{\omega\, dn}{d\omega}. \tag{8.3.4}$$

We will now derive this formula in a manner that is not only more rigorous but that also allows us to understand better the physical significance of group velocity. It will suffice to make the plane-wave approximation for the electric field propagating in the z direction, writing it as a continuous superposition of plane waves with different frequencies and guided by the monochromatic solution given in Eq. (8.2.24):

$$E(z, t) = \int_{-\infty}^{\infty} d\omega'\tilde{E}(\omega')e^{-i[\omega't - k(\omega')z]}, \tag{8.3.5}$$

where again $k(\omega') = n(\omega')\omega'/c$. We assume that the field oscillates primarily at a carrier frequency, which we denote by ω, and write $\omega' = \omega + \Delta$ in (8.3.5), so that we can write this equation equivalently as

$$\begin{aligned}
E(z, t) &= \int_{-\infty}^{\infty} d\Delta\, \tilde{E}(\omega + \Delta)e^{-i(\omega+\Delta)t}e^{ik(\omega+\Delta)z} \\
&= e^{-i[\omega t - k(\omega)z]}\int_{-\infty}^{\infty} d\Delta\, \tilde{E}(\omega + \Delta)e^{-i\Delta t}e^{i[k(\omega+\Delta)-k(\omega)]z} \\
&\equiv \mathcal{E}(z, t)e^{-i[\omega t - k(\omega)z]}.
\end{aligned} \tag{8.3.6}$$

The Taylor series expansion of $k(\omega + \Delta)$ gives

$$k(\omega + \Delta) - k(\omega) = \Delta\left(\frac{dk}{d\omega}\right)_\omega + \frac{1}{2}\Delta^2\left(\frac{d^2k}{d\omega^2}\right)_\omega + \frac{1}{6}\Delta^3\left(\frac{d^3k}{d\omega^3}\right)_\omega + \cdots, \tag{8.3.7}$$

so that differentiation of $\mathcal{E}(z, t)$ with respect to z results in

$$
\frac{\partial \mathcal{E}}{\partial z} = i \int_{-\infty}^{\infty} d\Delta \, \tilde{E}(\omega + \Delta) \left[\Delta \left(\frac{dk}{d\omega} \right)_{\omega} + \frac{1}{2} \Delta^2 \left(\frac{d^2 k}{d\omega^2} \right)_{\omega} + \frac{1}{6} \Delta^3 \left(\frac{d^3 k}{d\omega^3} \right)_{\omega} + \cdots \right]
$$
$$
\times e^{-i\Delta t} e^{i[k(\omega + \Delta) - k(\omega)]z}. \tag{8.3.8}
$$

Noting that

$$
\frac{\partial \mathcal{E}}{\partial t} = -i \int_{-\infty}^{\infty} d\Delta \, \Delta \tilde{E}(\omega + \Delta) e^{-i\Delta t} e^{i[k(\omega + \Delta) - k(\omega)]z},
$$
$$
\frac{\partial^2 \mathcal{E}}{\partial t^2} = - \int_{-\infty}^{\infty} d\Delta \, \Delta^2 \tilde{E}(\omega + \Delta) e^{-i\Delta t} e^{i[k(\omega + \Delta) - k(\omega)]z}, \tag{8.3.9}
$$
$$
\frac{\partial^3 \mathcal{E}}{\partial t^3} = i \int_{-\infty}^{\infty} d\Delta \, \Delta^3 \tilde{E}(\omega + \Delta) e^{-i\Delta t} e^{i[k(\omega + \Delta) - k(\omega)]z},
$$

we can write (8.3.8) as

$$
\frac{\partial \mathcal{E}}{\partial z} = - \left(\frac{dk}{d\omega} \right)_{\omega} \frac{\partial \mathcal{E}}{\partial t} - \frac{i}{2} \left(\frac{d^2 k}{d\omega^2} \right)_{\omega} \frac{\partial^2 \mathcal{E}}{\partial t^2} + \frac{1}{6} \left(\frac{d^3 k}{d\omega^3} \right)_{\omega} \frac{\partial^3 \mathcal{E}}{\partial t^3} + \cdots,
$$

or

$$
\frac{\partial \mathcal{E}}{\partial z} + \frac{1}{v_g} \frac{\partial \mathcal{E}}{\partial t} + \frac{i}{2} \left(\frac{d^2 k}{d\omega^2} \right)_{\omega} \frac{\partial^2 \mathcal{E}}{\partial t^2} - \frac{1}{6} \left(\frac{d^3 k}{d\omega^3} \right)_{\omega} \frac{\partial^3 \mathcal{E}}{\partial t^3} + \cdots = 0, \tag{8.3.10}
$$

which is sometimes written as

$$
\frac{\partial \mathcal{E}}{\partial z} + \frac{1}{v_g} \frac{\partial \mathcal{E}}{\partial t} + \frac{i}{2} \beta \frac{\partial^2 \mathcal{E}}{\partial t^2} - \frac{1}{6} \beta_3 \frac{\partial^3 \mathcal{E}}{\partial t^3} + \cdots = 0, \tag{8.3.11}
$$

where, for $n \geq 3$,

$$
\beta_n \equiv \left(\frac{d^n k}{d\omega^n} \right)_{\omega}. \tag{8.3.12}
$$

This equation for the envelope function $\mathcal{E}(z, t)$ determines the evolution of the field $E(z, t) = \mathcal{E}(z, t) \exp(-i[\omega t - k(\omega)z])$.

So far there are no approximations. If we now recall the assumption that $E(z, t)$ oscillates primarily at frequency ω, which was used to motivate the Taylor series expansion in Eq. (8.3.7), then it is clear that $\mathcal{E}(z, t)$ must be slowly varying in time compared to $\exp(-i\omega t)$. Another way to say this is that $\tilde{E}(\omega + \Delta)$ in Eq. (8.3.6) is negligible unless $|\Delta| \ll \omega$, so that terms of higher powers in Δ in (8.3.8) must be small. Whether the term that goes as Δ^n is small enough to be neglected depends, of course, on the value of the coefficient $(d^n k / d\omega^n)_{\omega}$ multiplying it. It is often the case that the dispersion of the medium, that is, the dependence of $k = n(\omega)\omega/c$ on ω, is such that the terms with $n \geq 2$

are very small compared to $(1/v_g)\partial\mathcal{E}/\partial t$. Then (8.3.10) can be replaced by

$$\frac{\partial\mathcal{E}}{\partial z} + \frac{1}{v_g}\frac{\partial\mathcal{E}}{\partial t} = 0. \tag{8.3.13}$$

The approximation that $E(z, t)$ not only varies slowly in t compared to $\exp(-i\omega t)$ but also slowly in z compared to $\exp[ik(\omega)z])$, similar to the approximation made in Section 7.4 [recall Eqs. (7.4.14) and (7.4.15)], is called the *slowly varying envelope approximation*. The replacement of (8.3.10) by (8.3.13) is an additional approximation, based on the assumption that the dispersion of the medium and the variation of $\mathcal{E}$ with t are sufficiently small that we can ignore all the terms in (8.3.10) involving second and higher derivatives of $\mathcal{E}$ with respect to t.

Equation (8.3.13) has solutions of the form[5]

$$\mathcal{E}(z, t) = F\left(t - \frac{z}{v_g}\right) = F(\tau) \qquad \left(\tau = t - \frac{z}{v_g}\right), \tag{8.3.14}$$

implying *distortionless* propagation at the velocity v_g of the field envelope. For example, if the field envelope at $z = 0$ corresponds to a Gaussian pulse of light,

$$\mathcal{E}(0, t) = \mathcal{E}_0 e^{-t^2/2\tau_p^2}, \tag{8.3.15}$$

it retains the same Gaussian form at all propagation distances z:

$$\mathcal{E}(z, t) = \mathcal{E}_0 e^{-(t-z/v_g)^2/2\tau_p^2}. \tag{8.3.16}$$

Note that the time τ_p characterizes the pulse duration at all points z, that is, the pulse not only remains Gaussian, it also maintains the same Gaussian width (and the same peak amplitude, $\mathcal{E}_0$).

- The phase velocity, $c/n(\omega)$, exceeds the speed of light in vacuum when $n(\omega) < 1$. This occurs, for instance, in a plasma when the field frequency ω is greater than the plasma frequency (Section 3.14). A phase velocity greater than c is not in conflict with the theory of special relativity, which requires that the velocity of a *signal* not exceed c. A signal in this sense represents information; phase velocity is simply the velocity of points of constant phase of a *monochromatic* wave, which cannot carry information because it is simply a sinusoidal oscillation of the field for all times. To encode information on a wave we must turn it on and off, for example, and such a wave form cannot be monochromatic.

The formula (8.3.3) indicates that the group velocity of a pulse can exceed c when $(dn/d\omega)_\omega$ is negative, as occurs, for instance, in the case of anomalous dispersion (Section 3.15). Again there is no conflict with special relativity because group velocity, like phase velocity, is not the velocity with which a signal, or information, is transmitted. The reasons for this are rather subtle in the case of group velocity, and here we will only touch on the basic idea.

If we have a field like the Gaussian (8.3.16), which is smoothly varying and has no beginning or end, we can determine the field at a time $t + \Delta t$ at any point z from the field at

---

[5]To see this, note that, with $\mathcal{E}(z, t)$ of the form (8.3.14), $\partial\mathcal{E}/\partial z = [dF/d\tau][\partial\tau/\partial z] = -(1/v_g)\,dF/d\tau$ and $\partial\mathcal{E}/\partial t = [dF/d\tau][\partial\tau/\partial t] = dF/d\tau$, so that (8.3.13) is satisfied. Note also that the function $F(\tau)$ is arbitrary; for example, $F(\tau) = \exp(-\tau^2/\tau_p^2)$ and $F(\tau) = \tau^2$ are both solutions of (8.3.13). What determines the actual form of $F(\tau)$?

time t by a Taylor expansion:

$$\mathcal{E}(z, t + \Delta t) = \mathcal{E}(z, t) + \left(\frac{\partial \mathcal{E}}{\partial t}\right)\Delta t + \frac{1}{2}\left(\frac{\partial^2 \mathcal{E}}{\partial t^2}\right)(\Delta t)^2 + \cdots. \tag{8.3.17}$$

Therefore, the field at a given time is fully predictable from the field at earlier times and so does not provide any information not already contained in the field at earlier times. While $\mathcal{E}$ may propagate with a group velocity greater than c, no new *information* is being propagated with a velocity greater than c. New information is only transmitted when the field varies in such a way that it is not a simple "analytic continuation" like (8.3.17) of its value at an earlier time. In other words, new information is associated with a *discontinuity* in the field or one of its derivatives, such that a Taylor expansion is inapplicable. It is such a point of discontinuity that represents information that, according to special relativity, cannot propagate with a velocity exceeding c.

Group velocities exceeding c have in fact been measured in various experiments. In one type of experiment the repetition frequency $v_g/2L$ of pulses in a resonant absorber has been found to exceed the value $c/2L$ characteristic of a mode-locked laser of length L (see Problem 15.4).

Equation (8.3.3) suggests that the group velocity of a pulse can even be infinite or negative if $dn/d\omega$ is negative and sufficiently large in magnitude. An infinite group velocity occurs when the peak of a pulse like (8.3.16) at the end of the medium occurs at the same time as the peak of the pulse at the entrance to the medium. A negative group velocity means that the peak of the exiting pulse occurs *before* the pulse at the entrance to the medium reaches its peak. Infinite and negative group velocities have also been observed experimentally.

According to Eq. (8.3.3) the group velocity can be very *small* compared to c if $dn/d\omega$ is large and positive. It has been demonstrated experimentally that pulses of light can propagate with group velocities on the order of a few meters per second or less; in fact it has been shown that the group velocity can be controlled to such an extent that a pulse of light can be brought to a complete stop and then regenerated. Such effects are considered in Section 9.10.

While group velocities that are very large or very small are fascinating, they occur under circumstances that are sufficiently unusual that they have not yet appeared in practical applications, and therefore we will not consider them further here.   •

8.4 GROUP VELOCITY DISPERSION

Distortionless pulse propagation at the group velocity v_g is often—but not always—an excellent approximation to what is observed. Deviations from this idealization are due to the terms neglected in going from Eq. (8.3.10) to the approximation (8.3.13).

Corrections to the approximation (8.3.13) involve second and higher derivatives of $\mathcal{E}$ with respect to time. In other words, corrections to (8.3.13) should be most significant when $\mathcal{E}$ varies rapidly in time, as occurs in the case of very short pulses. Consider the first correction to (8.3.13) obtained by including the term proportional to $\partial^2 \mathcal{E}/\partial t^2$ in (8.3.10):

$$\frac{\partial \mathcal{E}}{\partial z} + \frac{1}{v_g}\frac{\partial \mathcal{E}}{\partial t} + \frac{i}{2}\beta\frac{\partial^2 \mathcal{E}}{\partial t^2} = 0, \tag{8.4.1}$$

where we recall

$$\beta = \frac{d^2 k}{d\omega^2} = \frac{d}{d\omega}\left(\frac{1}{v_g}\right) = \frac{1}{c}\left(2\frac{dn}{d\omega} + \omega\frac{d^2 n}{d\omega^2}\right). \tag{8.4.2}$$

β, which is evaluated at the carrier frequency ω, is a measure of *group velocity dispersion* (GVD), that is, the variation of the group velocity with frequency. In terms of the wavelength $\lambda = 2\pi c/\omega$,

$$\beta = \frac{\lambda^3}{2\pi c^2}\frac{d^2 n}{d\lambda^2}. \tag{8.4.3}$$

If there is no group velocity dispersion ($\beta = 0$), a pulse will propagate at the group velocity v_g without any change in its temporal shape; for example, the Gaussian pulse (8.3.16) will retain exactly its Gaussian form.[6] To solve Eq. (8.4.1) with $\beta \neq 0$ it is convenient to introduce the new independent variables

$$\eta = z \quad \text{and} \quad \tau = t - z/v_g,$$

in terms of which

$$\frac{\partial \mathcal{E}}{\partial z} = \frac{\partial \mathcal{E}}{\partial \eta}\frac{\partial \eta}{\partial z} + \frac{\partial \mathcal{E}}{\partial \tau}\frac{\partial \tau}{\partial z} = \frac{\partial \mathcal{E}}{\partial \eta} - \frac{1}{v_g}\frac{\partial \mathcal{E}}{\partial \tau},$$

$$\frac{\partial \mathcal{E}}{\partial t} = \frac{\partial \mathcal{E}}{\partial \eta}\frac{\partial \eta}{\partial t} + \frac{\partial \mathcal{E}}{\partial \tau}\frac{\partial \tau}{\partial t} = 0 + \frac{\partial \mathcal{E}}{\partial \tau},$$

$$\frac{\partial^2 \mathcal{E}}{\partial t^2} = \frac{\partial^2 \mathcal{E}}{\partial \tau^2},$$

and (8.4.1) becomes

$$\frac{\partial \mathcal{E}}{\partial \eta} + \frac{i}{2}\beta\frac{\partial^2 \mathcal{E}}{\partial \tau^2} = 0 \qquad (\tau = t - z/v_g). \tag{8.4.4}$$

This equation describes the propagation of the field envelope $\mathcal{E}$ in a reference frame moving with the group velocity v_g. It has the same form as the paraxial wave equation (7.4.18) for a monochromatic field when in that equation we ignore variations along the y direction:

$$\frac{\partial \mathcal{E}}{\partial \eta} - \frac{i}{2k}\frac{\partial^2 \mathcal{E}}{\partial x^2} = 0. \tag{8.4.5}$$

That is, Eq. (8.4.4) is identical to (8.4.5) when we replace τ by x and β by $-1/k$.[7]

Equation (8.4.5) describes the propagation of a monochromatic wave of wavelength $\lambda = 2\pi/k$ when we account for variations along only one direction (x) tranverse to the direction (z) of propagation. If the wave has a transverse spread $\Delta x \sim a$ at $z = 0$, Eq. (8.4.5) tells us that after a propagation distance z the transverse diffractive spread will be $\Delta x \sim z/ka$. By analogy, therefore, Eq. (8.4.4) implies that a plane-wave

[6]We are assuming there is no absorption or amplification that might change the pulse shape.
[7]These equations have the same form as the Schrödinger equation $i\hbar\partial\psi/\partial t = -(\hbar^2/2m)\partial^2\psi/\partial x^2$ for a free particle of mass m. See Problem 8.4.

pulse of duration τ_p will, after propagating a distance L, have a duration

$$\Delta t \sim \frac{L}{(1/|\beta|)\tau_p} = \frac{L|\beta|}{\tau_p}. \tag{8.4.6}$$

In other words, GVD should cause a pulse to spread in time, the spread relative to the initial pulse duration τ_p being

$$\frac{\Delta t}{\tau_p} \sim L \frac{|\beta|}{\tau_p^2}, \tag{8.4.7}$$

which will be small if the propagation distance L is small compared to

$$L_{\text{GVD}} \equiv \frac{\tau_p^2}{|\beta|}. \tag{8.4.8}$$

L_{GVD} decreases—pulse distortion due to group velocity dispersion increases—as the pulse duration decreases or as $|\beta|$ increases. For propagation distances small compared to L_{GVD}, we can expect a pulse to propagate without much change in its shape.

As this argument suggests, the analogy between the propagation of a plane-wave pulse with group velocity dispersion [Eq. (8.4.4)] and the paraxial propagation of a monochromatic wave [Eq. (8.4.5)] is very useful. In the case of paraxial wave propagation, Gaussian beams play a special role: An initially Gaussian beam remains Gaussian upon propagation in free space (Section 7.6). Likewise, laser pulses with Gaussian *temporal* profiles play a special role in the theory of pulse propagation in dispersive media. By analogy with (7.5.2), let us assume a solution of Eq. (8.4.4) of the form

$$\mathcal{E}(z, \tau) = A e^{-i\tau^2/2\beta q(z)} e^{ip(z)}, \tag{8.4.9}$$

where A is a constant and the functions $q(z)$ and $p(z)$ are solutions of the equations

$$\frac{dq}{dz} = 1, \qquad \frac{dp}{dz} = \frac{i}{2q} \tag{8.4.10}$$

that follow upon substitution of (8.4.9) in (8.4.4). Thus

$$q(z) = z + q(0) = z + q_0,$$
$$p(z) = \frac{i}{2}\ln\left(1 + \frac{z}{q_0}\right), \tag{8.4.11}$$

where we have taken $p(0) = 0$; therefore

$$\mathcal{E}(z, \tau) = A e^{-(1/2)\ln(1+z/q_0)} e^{-i\tau^2/2\beta(z+q_0)} = \frac{A}{\sqrt{1+z/q_0}} e^{-i\tau^2/2\beta(z+q_0)}. \tag{8.4.12}$$

For a Gaussian pulse at $z = 0$,

$$\mathcal{E}(0, \tau) = Ae^{-\tau^2/2\tau_p^2} = Ae^{-i\tau^2/2\beta q_0}, \tag{8.4.13}$$

implying $q_0 = i\tau_p^2/\beta$. After some straightforward algebra we obtain

$$\mathcal{E}(z, \tau) = \frac{A}{\sqrt{1 - i\beta z/\tau_p^2}} e^{-\tau^2/2\tau_p^2(z)} e^{-i\Theta(z,\tau)} \tag{8.4.14}$$

and

$$|\mathcal{E}(z, \tau)|^2 = \frac{\tau_p}{\tau_p(z)} |A|^2 e^{-\tau^2/\tau_p^2(z)}, \tag{8.4.15}$$

where

$$\tau_p(z) = \sqrt{\tau_p^2 + (\beta z/\tau_p)^2} = \tau_p\sqrt{1 + z^2/L_{\text{GVD}}^2}, \tag{8.4.16}$$

$$\Theta(z, \tau) = \frac{\beta z \tau^2}{2\tau_p^2 \tau_p^2(z)}. \tag{8.4.17}$$

Therefore, a Gaussian pulse of duration τ_p at $z = 0$ remains Gaussian as it propagates in a medium with GVD parameter β, but it spreads in time according to (8.4.16). It also acquires a phase $\Theta(z, \tau)$ that varies with both z and τ. For a propagation distance z large enough that $|\beta|z \gg \tau_p^2$

$$\frac{\tau_p(z)}{\tau_p} \cong \frac{|\beta|z}{\tau_p^2}, \tag{8.4.18}$$

which is equivalent to (8.4.6), and

$$\Theta(z, \tau) \cong \frac{\tau^2}{2\beta z}. \tag{8.4.19}$$

Comparing Eq. (8.4.16) and (7.5.18), we see that τ_p and L_{GVD} for a plane-wave pulse with a Gaussian temporal shape correspond, respectively, to w_0 and z_0 for a monochromatic beam with a Gaussian transverse spatial profile.

The temporal broadening of a pulse due to GVD occurs simply because different frequency components of the pulse propagate with different velocities:

$$\Delta t = \left| \Delta \left(\frac{z}{v_g} \right) \right| \cong z \left| \frac{d}{d\omega} \left(\frac{1}{v_g} \right) \right| \Delta \omega = |\beta|z\,\Delta\omega \approx \tau_p(z) \tag{8.4.20}$$

for $\tau_p(z) \gg \tau_p$, where $\Delta\omega$ characterizes the spectral width of the pulse. We can also obtain this result from (8.4.18) by writing $\tau_p = 1/\Delta\omega$:

$$\tau_p(z) = \frac{|\beta|z}{\tau_p} = |\beta|z\,\Delta\omega. \tag{8.4.21}$$

Using $\Delta\omega = -(2\pi c/\lambda^2)\Delta\lambda$ to relate the spread $\Delta\lambda$ in wavelength to the spread $\Delta\omega$ in frequency, we have

$$\tau_p(z) = |D|z\,\Delta\lambda, \tag{8.4.22}$$

where the *dispersion parameter D* is defined by

$$D = -\frac{2\pi c}{\lambda^2}\beta = -\frac{\lambda}{c}\frac{d^2 n}{d\lambda^2}. \tag{8.4.23}$$

β is conventionally expressed in ps^2/km (picoseconds^2/kilometer) and D in ps/(km-nm) (Problem 8.5); β (and D) can be positive or negative. Since

$$\beta = \left(\frac{d}{d\omega}\right)\left(\frac{1}{v_g}\right) = -\left(\frac{1}{v_g^2}\right)\frac{dv_g}{d\omega}, \tag{8.4.24}$$

a positive β means that group velocity decreases with increasing frequency, while a negative β means that group velocity increases with increasing frequency.

- From Eq. (8.4.4) it can be shown that (Problem 8.6)

$$\mathcal{E}(z,\tau) = \left(\frac{i}{2\pi|\beta|z}\right)^{1/2}\int_{-\infty}^{\infty} d\tau'\,\mathcal{E}(0,\tau')e^{-i(\tau-\tau')^2/2\beta z}, \tag{8.4.25}$$

which gives the field at any z and τ in terms of the temporal profile $[\mathcal{E}(0,\tau)]$ of the field at the input plane $z = 0$. If $|\beta| \gg \tau'^2/z$ for all τ' for which $\mathcal{E}(0,\tau')$ is nonnegligible, we can approximate (8.4.25) by

$$\mathcal{E}(z,\tau) = \left(\frac{i}{2\pi|\beta|z}\right)^{1/2}e^{-i\tau^2/2\beta z}\int_{-\infty}^{\infty} d\tau'\,\mathcal{E}(0,\tau')e^{i\tau\tau'/\beta z}, \tag{8.4.26}$$

Equations (8.4.25) and (8.4.26) are analogous to formulas (7.10.4) and (7.10.8) for Fresnel and Fraunhofer diffraction, respectively.

From (8.4.26),

$$|\mathcal{E}(z,\tau)|^2 = \frac{1}{2\pi|\beta|z}\left|\int_{-\infty}^{\infty} d\tau'\,\mathcal{E}(0,\tau')e^{i\Omega_0\tau'}\right|^2, \tag{8.4.27}$$

where the frequency $\Omega_0 = \tau/\beta z$. In other words, if the propagation distance z is large enough that the "Fraunhofer limit" $|\beta| \gg \tau'^2/z$ is realized, the measured intensity of a pulse ($\propto |\mathcal{E}(z,\tau)|^2$) is directly proportional to

$$S(0,\Omega_0) = \left|\frac{1}{2\pi}\int_{-\infty}^{\infty} d\tau\,\mathcal{E}(0,\tau)e^{i\Omega_0\tau}\right|^2. \tag{8.4.28}$$

This is the basis for one method for measuring β and the dispersion parameter D of an optical fiber: $S(0,\Omega_0)$, and the intensity profile of a pulse after it propagates in a fiber of sufficient length, can be used to infer the values of β and D.

There are other interesting consequences of the analogy between the paraxial propagation of a monochromatic wave and the propagation of a plane-wave pulse in a dispersive medium. For example, two pulses separated in time will spread and overlap after propagating a sufficiently large distance in a medium with group velocity dispersion, and the temporal profile of the intensity of the total field will have the same form as the two-slit Young interference pattern for a monochromatic field. This has been observed in the propagation of pulses in fibers.   •

Any material medium can in principle exhibit group velocity dispersion. As already noted, and as can be seen directly from Eqs. (8.4.4) and (8.4.16), GVD is most significant for short pulses. Consider, for example, the propagation in a sapphire rod of a pulse with a wavelength of 800 nm, for which $\beta \cong 58 \text{ ps}^2/\text{km} = 58 \text{ fs}^2/\text{mm}$. From (8.4.16),

$$\tau_p(z) = \tau_p \sqrt{1 + (5.8 \times 10^{-5} z)^2 / \tau_p^4}, \tag{8.4.29}$$

where $\tau_p(z)$ and τ_p are to be expressed in picoseconds and z in millimeters. For an initial pulse duration of 1 ps there is practically no change in the pulse duration after propagation in a 20-mm rod. A 10-fs pulse incident on the same rod, however, is stretched to 116 fs after propagation through the rod. This example illustrates why *compensation for group velocity dispersion* has been the most crucial part of the technology for the generation of pulses as short as $\sim$50 fs or less (Section 11.13).

Reducing the effects of group velocity dispersion has also been very important in fiber-optical communications because of the long propagation paths involved. The GVD pulse spreading can cause two pulses to overlap as they propagate in a fiber, thus blurring their separate identities and the information they are meant to convey. For standard fibers used in telecommunications, $\beta \cong 20 \text{ ps}^2/\text{km}$ at 1.55 μm. For an initial 10-ps pulse the pulse duration after a propagation distance of 15 km is, according to Eq. (8.4.16), about 32 ps. If each pulse represents 1 bit of information, the maximum transmission rate before there is significant pulse overlap is roughly $1/32 \text{ ps} \approx 30 \text{ Gb/s}$, about three times slower than could be achieved without GVD.

Various methods have been developed to reduce or eliminate the effect of group velocity dispersion by passing pulses through an optical element that acts to *shorten* them. To understand such methods, it is useful first to briefly review some basic concepts from the theory of Fourier transforms.

We define the Fourier transform $\tilde{\mathcal{E}}(z, \Omega)$ of $\mathcal{E}(z, \tau)$ by writing

$$\mathcal{E}(z, \tau) = \int_{-\infty}^{\infty} d\Omega \, \tilde{\mathcal{E}}(z, \Omega) e^{-i\Omega\tau} \tag{8.4.30}$$

or the inverse relation

$$\tilde{\mathcal{E}}(z, \Omega) = \frac{1}{2\pi} \int_{-\infty}^{\infty} d\tau \, \mathcal{E}(z, \tau) e^{i\Omega\tau}. \tag{8.4.31}$$

The *spectrum* of $\mathcal{E}(z, \tau)$ may be defined by [cf. (8.4.28)]

$$S(z, \Omega) = |\tilde{\mathcal{E}}(z, \Omega)|^2, \tag{8.4.32}$$

and it follows that[8]

$$\int_{-\infty}^{\infty} d\Omega \, S(z, \Omega) = \frac{1}{2\pi} \int_{-\infty}^{\infty} d\tau |\mathcal{E}(z, \tau)|^2. \tag{8.4.33}$$

---

[8]Readers familiar with the Dirac delta function $\delta(x)$ will recognize that (8.4.33) and the equivalence of (8.4.30) and (8.4.31) follow from the property
$$\frac{1}{2\pi} \int_{-\infty}^{\infty} du \, e^{ixu} = \delta(x).$$

● The spectrum of the electric field is easily related to the spectrum of the pulse envelope. The electric field is

$$E(z, t) = \mathcal{E}(z, \tau)e^{-i\omega t}e^{ik(\omega)z} = e^{-i\omega t}e^{ik(\omega)z}\int_{-\infty}^{\infty}d\Omega\,\tilde{\mathcal{E}}(z, \Omega)e^{-i\Omega t}e^{i\Omega z/v_g}$$

$$= e^{ik(\omega)z}\int_{-\infty}^{\infty}d\Omega\,\tilde{\mathcal{E}}(z, \Omega)e^{-i(\Omega+\omega)t}e^{i\Omega z/v_g} = e^{ik(\omega)z}\int_{-\infty}^{\infty}d\omega'\,\tilde{\mathcal{E}}(z, \omega'-\omega)e^{-i\omega't}e^{i(\omega'-\omega)z/v_g}$$

$$= e^{i\omega(1/v_p-1/v_g)z}\int_{-\infty}^{\infty}d\omega'\,\tilde{\mathcal{E}}(z, \omega'-\omega)e^{-i\omega'(t-z/v_g)}, \tag{8.4.34}$$

where we have used the expression $v_p = \omega/k(\omega) = c/n(\omega)$ for the phase velocity. Thus, we can identify

$$\tilde{E}(z, \omega') = \tilde{\mathcal{E}}(z, \omega'-\omega)e^{i\omega(1/v_p-1/v_g)z}e^{i\omega'z/v_g} \tag{8.4.35}$$

as the Fourier transform of the electric field and

$$|\tilde{E}(z, \omega')|^2 = |\tilde{\mathcal{E}}(z, \omega'-\omega)|^2 \tag{8.4.36}$$

as its spectrum. If $|\tilde{E}(z, \omega')|^2$ is described by a bell-shaped curve that has its peak at the carrier frequency ω, then the spectrum $|\tilde{\mathcal{E}}(z, \Omega)|^2$ of the pulse envelope has the same bell shape and has its peak at $\Omega = 0$. ●

Equation (8.4.4) implies that

$$\frac{\partial\mathcal{E}}{\partial z} + \frac{i}{2}\beta\frac{\partial^2\mathcal{E}}{\partial\tau^2} = \int_{-\infty}^{\infty}d\Omega\left(\frac{\partial\tilde{\mathcal{E}}}{\partial z} - \frac{i}{2}\beta\Omega^2\tilde{\mathcal{E}}\right)e^{-i\Omega\tau} = 0, \tag{8.4.37}$$

or $\partial\tilde{\mathcal{E}}/\partial z = (i\beta\Omega^2/2)\tilde{\mathcal{E}}$ and therefore

$$\tilde{\mathcal{E}}(z, \Omega) = \tilde{\mathcal{E}}(z_0, \Omega)e^{i\beta\Omega^2(z-z_0)/2}, \tag{8.4.38}$$

where $\tilde{\mathcal{E}}(z_0, \Omega)$ is the Fourier transform of the pulse envelope $\mathcal{E}(z_0, \tau)$ at some input plane defined by z_0. In the case of an initial Gaussian pulse $\mathcal{E}(z_0, \tau) = A\exp(-\tau^2/2\tau_p^2)$, for example,

$$\tilde{\mathcal{E}}(z_0, \Omega) = \frac{A}{2\pi}\int_{-\infty}^{\infty}d\tau\,e^{-\tau^2/2\tau_p^2}e^{i\Omega\tau} = \frac{A}{\pi}\int_0^{\infty}d\tau\,e^{-\tau^2/2\tau_p^2}\cos\Omega\tau$$

$$= \frac{A\tau_p}{\sqrt{2\pi}}e^{-\Omega^2\tau_p^2/2}. \tag{8.4.39}$$

More generally, from (8.4.31) and (8.4.38),

$$\tilde{\mathcal{E}}(z, \Omega) = \frac{A\tau_p}{\sqrt{2\pi}}e^{-\Omega^2\tau_p^2/2}e^{i\beta\Omega^2(z-z_0)/2} \tag{8.4.40}$$

for an initially Gaussian pulse. It can be verified that (8.4.30) and (8.4.40) with $z_0 = 0$ reproduce Eq. (8.4.15).

Note that, for any pulse shape,

$$\int_{-\infty}^{\infty} d\tau |\mathcal{E}(z, \tau)|^2 = \int_{-\infty}^{\infty} d\tau |\mathcal{E}(z_0, \tau)|^2, \tag{8.4.41}$$

which means that the total energy of the pulse is unchanged by propagation in a medium whose only effect on propagation is dispersion. This result follows from (8.4.38) in the case of group velocity dispersion, but it holds no matter what the nature of the dispersion, so long as field attenuation or amplification are negligible.

Consider now the propagation of a pulse in a medium of length L_1 with $\beta = \beta_1$ and $v_g = v_{g1}$, followed by propagation in a medium of length L_2 with $\beta = \beta_2$ and $v_g = v_{g2}$. The field at the end of the first medium follows from (8.4.30) and (8.4.38):

$$\mathcal{E}(L_1, t) = \int_{-\infty}^{\infty} d\Omega\, \tilde{\mathcal{E}}(0, \Omega) e^{i\beta_1 \Omega^2 L_1/2} e^{-i\Omega t} e^{i\Omega L_1/v_{g1}} \equiv \int_{-\infty}^{\infty} d\Omega\, \tilde{\mathcal{E}}(L_1, \Omega) e^{-i\Omega t}, \tag{8.4.42}$$

where for simplicity we have defined the input plane of the first medium by setting $z_0 = 0$. The field after propagation through the second medium is similarly

$$\mathcal{E}(L_1 + L_2, t) = \int_{-\infty}^{\infty} d\Omega\, \tilde{\mathcal{E}}(L_1, \Omega) e^{i\beta_2 \Omega^2 L_2/2} e^{-i\Omega t} e^{i\Omega L_2/v_{g2}}$$

$$= \int_{-\infty}^{\infty} d\Omega\, \tilde{\mathcal{E}}(0, \Omega) e^{i[\beta_2 L_2 + \beta_1 L_1]\Omega^2/2} e^{-i\Omega[t - L_1/v_{g1} - L_2/v_{g2}]}. \tag{8.4.43}$$

If the second medium is such that

$$\beta_2 L_2 = -\beta_1 L_1, \tag{8.4.44}$$

therefore,

$$\mathcal{E}(L_1 + L_2, t) = \int_{-\infty}^{\infty} d\Omega\, \tilde{\mathcal{E}}(0, \Omega) e^{-i\Omega[t - L_1/v_{g1} - L_2/v_{g2}]}, \tag{8.4.45}$$

or, according to (8.4.30),

$$\mathcal{E}(L_1 + L_2, t) = \mathcal{E}\left(0, t - \frac{L_1}{v_{g1}} - \frac{L_2}{v_{g2}}\right). \tag{8.4.46}$$

For a Gaussian pulse $A \exp(-t^2/2\tau_p^2)$ incident on the first medium, for example,

$$\mathcal{E}(L_1 + L_2, t) = A e^{-(t - t_d)^2/2\tau_p^2}, \tag{8.4.47}$$

where $t_d = L_1/v_{g1} + L_2/v_{g2}$ is the group delay time for propagation through both media. Thus, if (8.4.44) is satisfied, *the incident pulse retains its temporal shape and there is no pulse spreading.* Such compensation for group velocity dispersion is realized when the second medium has opposite GVD from the first medium ($\beta_2/\beta_1 < 0$) and its length is chosen in accord with (8.4.44).

Most materials are positively dispersive ($\beta > 0$) in the visible and near-infrared. To compensate for group velocity dispersion in a positively dispersive medium, we require a medium with negative dispersion ($\beta < 0$). Negative dispersion means that the group velocity increases with increasing frequency, or in other words that the "red" components of a pulse have smaller group velocities than the "blue" components and therefore propagate from one point to another with larger delay times. Negative dispersion needed to compensate for group velocity dispersion is often realized not with negatively dispersive materials as such but with optical elements such as prisms or gratings in which the red components are delayed with respect to the blue components. In glass, for example, the red components of a pulse travel faster than the blue components and therefore undergo less refraction, but the *angular* dispersion (i.e., the "bending" of different frequencies by different amounts) of a glass prism is such that negative dispersion can in effect be obtained by delaying the red components with respect to the blue. Figure 8.1 shows a two-prism arrangement commonly employed for this purpose. The red components are delayed with respect to the blue components by having them traverse more glass (i.e., more optical path length = distance $\times$ refractive index). Writing (8.4.24) as $\beta = L^{-1} d(L/v_g)/d\omega = L^{-1} dt_d/d\omega$, where t_d is the group delay time for a propagation distance L, suggests that negative dispersion can in effect be obtained by introducing delay times that decrease with increasing frequency, and this is what is accomplished by the two-prism arrangement shown.

A rather involved Fourier analysis shows in fact that this arrangement produces a phase shift with a contribution that is quadratic in frequency and negative and depends, among other things, on the separation of the prisms; according to the preceding analysis, this is exactly what is required for compensation of group velocity dispersion. Moving the prisms farther apart (increasing L in Fig. 8.1) increases the optical path difference between the red and blue frequency components and therefore increases the negative dispersion. The negative dispersion is also increased by increasing the distance d. The degree of negative dispersion can therefore be "tuned" by changing the positions of the prisms. A mirror placed after the second prism results in a double pass through the prism pair and a phase-shifted pulse propagating in the direction opposite to that of the pulse incident on the first prism. Alternatively, a second pair of prisms can be used to produce a phase-shifted pulse propagating in the same direction as the input pulse and aligned with it [cf. Fig. 8.3].

Group velocity dispersion compensation is also commonly achieved with a pair of diffraction gratings (Fig. 8.2). Here again angular dispersion results in time delays that increase with decreasing frequency, resulting in a contribution to the phase shift that is negative and varies quadratically with frequency. The degree of negative dispersion can be controlled by changing the positions of the gratings. As with the use of prisms, a mirror can be used to produce a phase-shifted pulse propagating in the

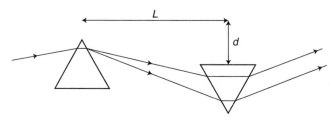

Figure 8.1 Two-prism arrangement for introducing delay times that increase with decreasing frequency as a result of angular dispersion.

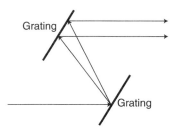

Figure 8.2 Two-grating arrangement for compensation of group velocity dispersion.

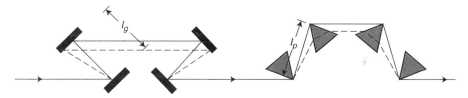

Figure 8.3 Arrangement of four diffraction gratings and four prisms to compensate for both second- and third-order dispersion. [After R. L. Fork, C. H. Brito Cruz, P. C. Becker, and C. V. Shank, *Optics Letters* **12**, 483 (1987)].

direction opposite to the incoming pulse, or a second pair of gratings can be used to produce a phase-shifted pulse propagating in the same direction and aligned with the incoming pulse. Compensation for group velocity dispersion with prisms or gratings can be accomplished inside or outside a laser resonator.

• Gratings have two advantages over prisms for GVD compensation: They typically produce a larger degree of negative dispersion, and material dispersion, which causes a positive dispersion ($\beta > 0$), plays no role because the pulses do not pass through any material. On the other hand they have greater "insertion loss" than (Brewster-angled) prisms and when used in resonators are consequently most advantageous for high-gain lasers, particularly fiber lasers.

We have ignored third and higher derivatives with respect to time in the partial differential equation (8.3.11), as group velocity and group velocity dispersion are associated with first and second derivatives, respectively, with respect to time [Eq. (8.4.1)]. In the case of very short pulses, say $\tau_p < 10\,\text{fs}$, third-order effects depending on $d^3k/d\omega^3 = z^{-1}d^3\phi/d\omega^3$ must also be accounted for. Prism and grating pulse compressors introduce third-order phase shifts of opposite sign, and when used in tandem can produce a range of both second- and third-order phase shifts. As early as 1987 the design shown in Fig. 8.3 was employed to compress 50-fs pulses from a colliding-pulse dye laser and an amplifier to 6 fs. The grating and prism separations shown in the figure were $\ell_g = 0.5\,\text{cm}$ and $\ell_p = 71\,\text{cm}$, respectively. Before compression, the pulses were passed through an optical fiber that caused spectral broadening due to self-phase modulation (Section 10.4). It was estimated based on theoretical considerations that a pulse emerging from the fiber had a second-order phase shift with $d^2\phi_f/d\omega^2 \sim 700\,\text{fs}^2$ and that for the grating and prism pairs $d^2\phi_g/d\omega^2 \sim -1820\,\text{fs}^2$ and $d^2\phi_p/d\omega^2 \sim -1528\,\text{fs}^2$, respectively, at the 620-nm carrier wavelength.[9] For material dispersion

---

[9]Approximate formulas for $\phi_g(\omega)$ and $\phi_p(\omega)$ are given by C. H. Brito Cruz, P. C. Becker, R. L. Fork, and C. V. Shank, *Optics Letters* **13**, 123 (1988).

due mainly to lenses in the system it was estimated that $d^2\phi_m/d\omega^2 \sim 2900 \text{ fs}^2$, so that $d^2\phi_g/d\omega^2 + d^2\phi_p/d\omega^2 + d^2\phi_m/d\omega^2 \sim -d^2\phi_f/d\omega^2$. The third-order contributions were estimated to be $d^3\phi_g/d\omega^3 \sim 1560 \text{ fs}^3$, $d^3\phi_p/d\omega^3 \sim -3055 \text{ fs}^3$, and $d^3\phi_m/d\omega^3 \sim 1620 \text{ fs}^3$, and the sum of these third-order contributions was therefore small and consistent with experimental observations.

As noted earlier, any material medium can be a source of GVD as well as higher-order dispersion. In a laser, for example, GVD can result not only from propagation in the gain medium but also from a saturable absorber in the case of passive mode locking, or from prisms and other intracavity components that may be present, or even from the frequency dependence of the mirror reflectivities. The latter can be a major source of GVD.

In fact any optical element for which transmission or reflection are frequency dependent—a prism, a grating, a mirror, an interferometer, etc.—can be a source of GVD: an incident field of frequency ω results in a transmitted or reflected field of the form $A(\omega)\exp(-i[\omega t - \phi(\omega)])$, where $A(\omega)$ and $\phi(\omega)$ are real. For an incident pulse there is a group delay due to $d\phi/d\omega$ and a second-order dispersion due to $d^2\phi/d\omega^2$ upon reflection or transmission. If parameters such as a mirror spacing can be chosen such that the second-order dispersion is negative, the optical element can be used for GVD compensation.

The prism and grating compensators of Figs. 8.1 and 8.2 are very useful in practice because they produce relatively large negative dispersion with little change in pulse energy. Also very useful in practice are *chirped Bragg mirrors*. One type of Bragg mirror consists of alternating layers of two materials with different refractive indices, each layer having an optical thickness d. The optical path difference in a Fresnel reflection off each layer is $2d$, and successive layers have amplitude reflection coefficients of

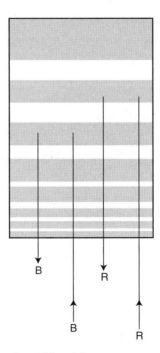

Figure 8.4 Chirped Bragg mirror in which red frequency components (R) of an incident field are reflected with a greater time delay than blue components (B).

opposite sign. There is, therefore, constructive interference of reflected waves when a field incident normally to the layers, for example, has a wavelength $\lambda = 4d$. In other words, Bragg mirrors are designed to strongly reflect only certain wavelengths. Their reflectivity depends on the number of layers and the difference in the two refractive indices, the latter being the primary determinant of the reflectivity bandwidth. Strongly wavelength-dependent reflectivity is also obtained when there is a continuous (smooth) periodic variation of the refractive index, as in fiber Bragg gratings (Section 11.14). In a "chirped" Bragg mirror the refractive index is not periodic but varies such that the condition for strong reflectivity is satisfied at different penetration distances for different wavelengths. Figure 8.4 indicates how lower frequency components of a pulse incident on a chirped Bragg mirror can suffer greater time delays for reflection as a result of the greater penetration depths required for strong reflectivity.   ●

8.5   CHIRPING

In addition to the temporal spreading of a pulse, GVD results in a time-dependent frequency. This is easily understood: If $\beta > 0$, for example, the higher-frequency Fourier components of a pulse travel with a smaller velocity than the lower-frequency components ($dv_g/d\omega < 0$) and therefore arrive at z later. The frequency at z therefore increases with time. By analogy with the sounds made by birds, such a time-dependent frequency is called a chirp (Section 6.14). In this section we will discuss chirping and related matters in a bit more detail.

The expression for the electric field given by Eqs. (8.3.6), (8.4.14), and (8.4.19) is

$$E(z, t) = \frac{A}{\sqrt{1 - i\gamma}} e^{-i[\omega t - k(\omega)z]} e^{-\tau^2/2\tau_p^2(z)} e^{-i\gamma\tau^2/2\tau_p^2(z)}, \tag{8.5.1}$$

where we have defined the dimensionless parameter

$$\gamma = \frac{\beta z}{\tau_p^2}. \tag{8.5.2}$$

The phase of the electric field is time dependent; we can define an "instantaneous frequency"

$$\omega_{\text{inst}}(t) = \frac{\partial}{\partial t}\left[\omega t + \frac{\gamma\tau^2}{2\tau_p^2(z)}\right] = \omega + \frac{\gamma\tau}{\tau_p^2(z)} = \omega + \frac{\gamma(t - z/v_g)}{\tau_p^2(z)}. \tag{8.5.3}$$

In this case there is a linear dependence of the instantaneous frequency on time, that is, a *linear chirp*.

From the discussion of GVD compensation in the preceding section, it should be clear that a negative chirp ($\gamma < 0$—the instantaneous frequency decreases with time) can result in the temporal compression of a pulse. Consider the propagation in a GVD medium of a Gaussian pulse incident at $z = 0$ with a negative linear chirp:

$$\mathcal{E}(0, t) = Ae^{-t^2/2\tau_p^2} e^{i\alpha_c t^2/2\tau_p^2}, \tag{8.5.4}$$

where $\alpha_c > 0$ specifies the magnitude of the chirp. The Fourier transform of this pulse envelope is

$$
\tilde{\mathcal{E}}(0, \Omega) = \frac{A}{2\pi} \int_{-\infty}^{\infty} dt \, e^{-t^2/2\tau_p^2} e^{i\alpha_c t^2/2\tau_p^2} e^{i\Omega t}
$$

$$
= \frac{A\tau_p}{\sqrt{2\pi}(1 + \alpha_c^2)^{1/4}} e^{(i/2)\tan^{-1}\alpha_c} e^{-(1/2)\Omega^2 \tau_p^2/(1+\alpha_c^2)} e^{-(i/2)\alpha_c \Omega^2 \tau_p^2/(1+\alpha_c^2)}. \quad (8.5.5)
$$

The propagation of such a chirped pulse by a distance z in a medium characterized by a group velocity v_g and a GVD dispersion parameter β results in the field [recall (8.4.38)]

$$
\mathcal{E}(z, \tau) = \int_{-\infty}^{\infty} d\Omega \, \tilde{\mathcal{E}}(0, \Omega) e^{i\beta \Omega^2 z/2} e^{-i\Omega\tau}
$$

$$
= \frac{A\tau_p}{\sqrt{2\pi}(1 + \alpha_c^2)^{1/4}} e^{(i/2)\tan^{-1}\alpha_c}
$$

$$
\times \int_{-\infty}^{\infty} d\Omega \, e^{i[\beta z - \alpha_c \tau_p^2/(1+\alpha_c^2)]\Omega^2/2} e^{-(1/2)\Omega^2 \tau_p^2/(1+\alpha_c^2)} e^{-i\Omega\tau}, \quad (8.5.6)
$$

with $\tau = t - z/v_g$. The evaluation of the integral yields a rather complicated expression; to simplify matters we write explicitly only the quantity proportional to the pulse intensity:

$$
|\mathcal{E}(z, \tau)|^2 = |A|^2 \sqrt{\frac{1 + \alpha_c^2}{F}} e^{-(1+\alpha_c^2)\tau^2/F\tau_p^2}, \quad (8.5.7)
$$

in which we define

$$
F = 1 + \left[\alpha_c - \frac{\beta z}{\tau_p^2}(1 + \alpha_c^2) \right]^2. \quad (8.5.8)
$$

Equation (8.5.7) implies the pulse duration

$$
\tau_p(z) = \frac{\tau_p}{\sqrt{1 + \alpha_c^2}} \sqrt{F} = \frac{\tau_p}{\sqrt{1 + \alpha_c^2}} \sqrt{1 + \left[\alpha_c - \frac{\beta z}{\tau_p^2}(1 + \alpha_c^2) \right]^2}, \quad (8.5.9)
$$

which reduces to (8.4.16) when there is no chirp ($\alpha_c = 0$).

This expression shows that a pulse propagating in a dispersive medium can be temporally compressed if it has an initial, negative linear chirp. At a propagation distance z_s such that

$$
\left[\alpha_c - \frac{\beta z_s}{\tau_p^2}(1 + \alpha_c^2) \right]^2 \ll 1, \quad (8.5.10)
$$

for example, the pulse duration is reduced from its initial value τ_p to

$$\tau_p(z_s) = \frac{\tau_p}{\sqrt{1 + \alpha_c^2}}. \qquad (8.5.11)$$

It is easily shown from (8.5.9) that this is the shortest pulse duration achievable in the case of a chirped Gaussian pulse with initial duration τ_p, and that it occurs at the propagation distance

$$z_s = \frac{\alpha_c}{1 + \alpha_c^2} \frac{\tau_p^2}{\beta}. \qquad (8.5.12)$$

Beyond this distance the pulse duration increases and is given approximately by (8.4.16) when $z \gg z_s$.

This compression of a chirped pulse can be understood from the mathematical correspondence between the plane-wave propagation of a pulse in a dispersive medium and the paraxial propagation of a monochromatic wave. In particular, a negative chirp plays the role in this correspondence to a thin lens: A lens of focal length f imposes a spatial phase factor $\exp[-ik(x^2 + y^2)/2f]$ on an incident monochromatic wave (Problem 7.12), and the chirping of a pulse introduces a temporal phase factor of the same form [Eq. (8.5.4)]. Thus, the temporal compression of a pulse by negative chirping is analogous to the transverse spatial compression (focusing) produced by a lens, and the subsequent increase in the pulse duration with propagation distance as described by (8.5.9) is analogous to the diffractive spreading of a beam that propagates past the distance at which its spot size is smallest [cf. Fig. 7.15].

We saw in the preceding section how chirping produced by prisms, gratings, and other optical elements can be used to compress pulses and in particular to compensate for the group velocity dispersion in a medium through which a pulse has propagated and incurred a phase distortion that causes it to broaden in time. Equation (8.5.9) shows that chirping a pulse before it enters a dispersive medium—by passing it through some dispersive optical element, for example—can also shorten its duration. It is not surprising that pulse chirping is one of the most widely used techniques in applications requiring ultrashort laser pulses (Section 14.7). In the remainder of this section we will focus on a few other basic concepts relevant to the description of chirped optical pulses.

The spectrum of the chirped pulse (8.5.4) is

$$S(0, \Omega) = |\tilde{\mathcal{E}}(0, \Omega)|^2 = \frac{|A|^2 \tau_p^2}{2\pi\sqrt{1 + \alpha_c^2}} e^{-\Omega^2 \tau_p^2/(1+\alpha_c^2)}. \qquad (8.5.13)$$

The FWHM spectral width $\Delta\Omega$ of the pulse at z is defined by $S(z, \pm\Delta\Omega/2) = S(z, 0)/2$, that is,

$$\Delta\Omega = \frac{\sqrt{4\ln 2}}{\tau_p}\sqrt{1 + \alpha_c^2}, \qquad (8.5.14)$$

which is larger by the factor $\sqrt{1 + \alpha_c^2}$ than the spectral width of an unchirped ($\alpha_c = 0$) Gaussian pulse. Using

$$|\mathcal{E}(0, t)|^2 = |A|^2 e^{-t^2/\tau_p^2}, \tag{8.5.15}$$

we can define similarly a FWHM pulse width:

$$\Delta\tau = \tau_p\sqrt{4\ln 2}, \tag{8.5.16}$$

and the *time–bandwidth product* with the spectral and temporal widths defined in this way for a Gaussian pulse is[10]

$$\Delta\Omega\,\Delta\tau = 4\ln 2\sqrt{1 + \alpha_c^2}. \tag{8.5.17}$$

This is larger than the time–bandwidth product $4\ln 2$ of an unchirped Gaussian pulse. Chirping increases the time–bandwidth product by broadening the pulse spectrum while preserving the pulse width. As is clear from (8.5.4), it does not change the pulse energy.

After a chirped pulse propagates in a medium with GVD, it has a spectrum that can be read off from (8.5.6):

$$S(z, \Omega) = \frac{|A|^2 \tau_p^2}{2\pi\sqrt{1 + \alpha_c^2}} e^{-\Omega^2 \tau_p^2/(1+\alpha_c^2)}. \tag{8.5.18}$$

The spectrum is seen to be independent of z and therefore unchanged by propagation, as must be the case for propagation in any linearly dispersive medium without attenuation or amplification; the pulse energy is likewise unchanged. The (FWHM) spectral width is

$$\Delta\Omega = \frac{\sqrt{4\ln 2}}{\tau_p}\sqrt{1 + \alpha_c^2}, \tag{8.5.19}$$

and the pulse width implied by (8.5.7) is

$$\Delta\tau = \tau_p\sqrt{4\ln 2}\,\frac{\sqrt{F}}{\sqrt{1 + \alpha_c^2}} \tag{8.5.20}$$

after a propagation distance z. Thus,

$$\Delta\Omega\,\Delta\tau = 4\ln 2\sqrt{F} = 4\ln 2\sqrt{1 + \left[\alpha_c - \frac{\beta z}{\tau_p^2}(1 + \alpha_c^2)\right]^2}. \tag{8.5.21}$$

The pulse is spectrally broadened, and its width $\Delta\tau$ varies as it propagates and can be larger or smaller than the width (8.5.16) of an initial (unchirped) pulse. At $z = z_s$, $\tau_p\sqrt{4\ln 2}/\sqrt{1 + \alpha_c^2}$ $\tau_p\sqrt{4\ln 2}/\sqrt{1 + \alpha_c^2}$ and the time–bandwidth product is identical

---

[10]The time–bandwidth product is often defined as $\Delta\nu\,\Delta\tau = (2\ln 2/\pi)\sqrt{1 + \alpha_c^2}$, where $\Delta\nu = \Delta\Omega/2\pi$ is the spectral width expressed in terms of the circular frequency ν rather than the angular frequency Ω.

to that of an unchirped Gaussian pulse in the absence of GVD:

$$\Delta\Omega\,\Delta\tau = 4\ln 2. \tag{8.5.22}$$

$\Delta\Omega\,\Delta\tau = 4\ln 2$ is the smallest time–bandwidth product for a Gaussian pulse. A pulse with the smallest width $\Delta\tau$ for a given spectral width $\Delta\Omega$ is said to be *transform limited* (or *bandwidth limited*). Thus, a chirped pulse that has propagated a distance z_s in a medium with GVD is transform limited, whereas at $z = 0$—where it has the same spectral width—it is not. Ultrashort pulses obtained by chirped compression are usually transform limited or nearly so.

The spectrum $|\tilde{\mathcal{E}}(\Omega)|^2$ cannot in general be uniquely related to $|\mathcal{E}(t)|^2$ because there is no phase information in either of these quantitites. As in the example of a chirped Gaussian pulse, a pulse with $\tilde{\mathcal{E}}(\Omega) = |\tilde{\mathcal{E}}(\Omega)|\exp[i\Theta(\Omega)]$ will not be transform-limited if the phase Θ is a nonlinear function of Ω. The time–bandwidth product in any case satisfies

$$\Delta\Omega\,\Delta\tau \geq K, \tag{8.5.23}$$

where the constant K depends on the pulse shape and is $4\ln 2$ in the case of a Gaussian pulse.[11] The pulse width $\Delta\tau = K/\Delta\Omega$ for a transform-limited pulse. It should be noted, however, that K also depends on how we choose to define the spectral and temporal widths. We have chosen to define these as full widths at half maxima, but obviously we are free to choose other definitions. For example, the rms pulse width is defined by

$$\Delta\tau_{\mathrm{rms}} = \left[\frac{\int_{-\infty}^{\infty}dt\ t^2|\mathcal{E}(z,\,t)|^2}{\int_{-\infty}^{\infty}dt|\mathcal{E}(z,\,t)|^2}\right]^{1/2}, \tag{8.5.24}$$

and the rms spectral width $\Delta\Omega_{\mathrm{rms}}$ is defined similarly.

- Instantaneous frequencies of a pulse are in general distinct from the frequency components of its spectrum: $\omega_{\mathrm{inst}}(t)$ can have values Ω such that $S(z,\Omega) = 0$, and conversely $S(z,\Omega)$ can be non-zero at frequencies Ω not appearing at any time in $\omega_{\mathrm{inst}}(t)$. In other words, instantaneous frequencies and frequencies for which $S(z,\Omega) \neq 0$ are in general physically distinct properties of a pulse. The frequencies appearing in $S(z,\Omega)$ are those inferred, for example, when an interferometer is used to analyze the spectral content of light. Instantaneous frequencies, on the other hand, better describe measurements of the time-dependent phase difference of two interfering pulses.[12]

Pulse compression by chirping has long been employed in pulsed radar systems, where shorter pulses obviously allow greater range resolution. Lord Rayleigh wrote that a flight of stairs at his estate in Terling returned "an echo of the clap of the hands as a note resembling the chirp of a sparrow," and chirped handclap echoes can also be heard near corrugated walls acting as diffraction gratings for sound waves.[13]   •

8.6 PROPAGATION MODES IN FIBERS

The theory of the guided propagation of light in optical fibers is straightforward conceptually but somewhat complicated in its algebraic details, which we will largely skip over;

---

[11]Pulse shaping and measurement are discussed in Section 11.13.
[12]For examples and a discussion see L. Mandel, *American Journal of Physics* **42**, 840 (1974).
[13]F. S. Crawford, Jr., *American Journal of Physics* **38**, 378 (1970).

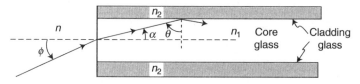

Figure 8.5   An optical fiber, viewed along a direction perpendicular to the fiber axis.

the interested reader will not find it difficult to fill in at least some of the steps, or to find comprehensive discussions and references in more specialized books.[14]

Optical fibers guide light by total internal reflection, which was briefly reviewed in Section 2.9. Figure 8.5 is an enlarged view of a segment of an optical fiber; the core diameter may be as small as a few microns, as explained below. The critical angle for total internal reflection is [Eq. (2.9.4)]

$$\theta_c = \sin^{-1}\left(\frac{n_2}{n_1}\right), \tag{8.6.1}$$

where n_2 and n_1 are the refractive indices of the cladding and the core, respectively (Fig. 8.5). Total internal reflection occurs for angles of incidence $\theta \geq \theta_c$. This implies a maximum "acceptance angle" for which light injected into the fiber will undergo total internal reflection. Applying Snell's law to the dielectric interface at the entrance to the fiber in Fig. 8.5, we have

$$n \sin \phi = n_1 \sin \alpha = n_1 \sin\left(\frac{\pi}{2} - \theta\right) = n_1 \cos \theta = n_1 \sqrt{1 - \sin^2 \theta}. \tag{8.6.2}$$

For $\theta = \theta_c$,

$$n \sin \phi = n_1 \sqrt{1 - \frac{n_2^2}{n_1^2}} = \sqrt{n_1^2 - n_2^2} \equiv \text{NA}, \tag{8.6.3}$$

where the number NA is called the *numerical aperture* of the fiber. According to these equations the angle

$$\phi_{\max} = \sin^{-1}\left(\frac{\text{NA}}{n}\right) \tag{8.6.4}$$

is the maximum acceptance angle at which there is total internal reflection. For a fiber in air ($n \cong 1$) with a core refractive index $n_1 = 1.53$ and a cladding index $n_2 = 1.50$, NA = 0.3 and the maximum acceptance angle is $\phi_{\max} \cong 18°$. As in this example, the difference between n_1 and n_2 is typically only a few percent, and so one conventionally introduces the small parameter $\Delta = (n_1 - n_2)/n_1$, in terms of which

$$\text{NA} = \sqrt{n_1 \Delta(n_1 + n_2)} \cong n_1 \sqrt{2\Delta}. \tag{8.6.5}$$

---

[14]See, for instance, G. P. Agrawal, *Fiber-Optic Communication Systems*, 3rd ed., Wiley, New York, 2002, and references therein.

The numerical aperture is obviously a measure of the amount of light that can be taken in and guided by the fiber. However, fibers with large numerical apertures have disadvantages for communication purposes because they admit a large number of propagation modes and therefore suffer from an effect known as *intermodal dispersion*. We discussed in the preceding section the *material dispersion* associated with the frequency dependence of the refractive index, but in fibers there is also a pulse-broadening effect associated with different angles of incidence θ in Fig. 8.5. Since different angles are associated with different modes of propagation, this dispersive effect is called *intermodal*. To estimate the pulse broadening due to intermodal dispersion, consider the propagation paths for two pulses, one propagating along the core axis and the other having an angle of incidence θ at the core–cladding interface (Fig. 8.5). For a fiber length L the off-axis pulse has a total propagation length $L/\cos \alpha$, whereas the propagation length for the on-axis pulse is simply L. These different propagation paths imply a difference ΔT in the propagation times for pulses with group velocity v_g to reach the end of the fiber. For the lowest-order modes of a fiber it is found that $v_g \cong c/n_1$, the phase velocity in the core. Thus,

$$\Delta T \cong L \frac{(1/\cos \alpha) - 1}{c/n_1} \cong \frac{n_1 \alpha^2 L}{2c}, \tag{8.6.6}$$

where the angle α is assumed to be very small. For the maximum acceptance angle defined by (8.6.4), it follows from (8.6.2) that $n_1 \sin \alpha = \text{NA}$, or $\alpha = \text{NA}/n_1$ in the small-angle approximation. Then (8.6.6) becomes

$$\Delta T \cong \frac{(\text{NA})^2}{2n_1 c} L \tag{8.6.7}$$

for these two modes. A multimode pulse will therefore undergo a temporal broadening.

Intermodal dispersion is reduced when the fiber is of the *graded-index* type rather than the *step-index* type illustrated in Fig. 8.5. In a graded-index fiber the refractive index does not have a sharp, steplike decrease from n_1 to n_2. Instead the index decreases more smoothly from the center of the fiber. An index distribution that is frequently used in practice is described by the formula

$$n^2 = n_c^2 (1 - a_2^2 r^2), \tag{8.6.8}$$

where n_c is the refractive index at the center, r is the distance from the center, and a_2^2 is a constant. The advantage of a graded-index fiber is a consequence of the following result, which we will not take the time to derive: The temporal spread ΔT for a graded-index fiber is proportional to $(\text{NA})^4$ rather than to $(\text{NA})^2$ as in the step-index case. Thus, a small numerical aperture implies smaller intermodal disperion in a graded-index fiber than in a step-index fiber. It is easy to understand physically why this is so. In the graded-index case the light rays along the axis of the fiber travel a shorter path than off-axis rays but have a smaller phase velocity because of the larger index on-axis [Eq. (8.6.8)]. The graded index therefore reduces the difference in propagation times of different modes.

Intermodal dispersion is completely absent in a *single-mode fiber*. We now consider in more detail the propagation modes of an optical fiber.

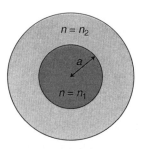

Figure 8.6   Cross-sectional view of a step-index fiber.

For laser resonators we defined a mode as a field distribution that does not change in form upon back-and-forth propagation in the resonator. In the case of an optical fiber, similarly, we define a mode as a field distribution that retains its form during propagation in the fiber. Thus, we require that the electric field satisfy the Helmholtz equation (8.2.20), with the refractive index having the spatial distribution appropriate to the fiber under consideration. In addition to satisfying (8.2.20), the field must, of course, satisfy the appropriate boundary conditions. We will consider a step-index fiber with $n = n_1$ for $r \leq a$ and $n = n_2$ for $r > a$ (Fig. 8.6).

The fiber geometry obviously suggests the use of cylindrical coordinates (r, ϕ, z), in terms of which Eq. (8.2.20) takes the form

$$\frac{\partial^2 \mathcal{E}}{\partial r^2} + \frac{1}{r}\frac{\partial \mathcal{E}}{\partial r} + \frac{1}{r^2}\frac{\partial^2 \mathcal{E}}{\partial \phi^2} + \frac{\partial^2 \mathcal{E}}{\partial z^2} + n^2\frac{\omega^2}{c^2}\mathcal{E} = 0. \tag{8.6.9}$$

Since a rotation by 2π about the fiber axis cannot affect the field, a solution of (8.6.9) must not change when 2π is added to ϕ. Thus, $\mathcal{E}$ must vary with ϕ as $\exp(im\phi)$, where $m = 0, \pm 1, \pm 2, \dots$. We seek solutions describing propagation along the z axis, and therefore write[15]

$$\mathcal{E}(r, \phi, z) = F(r)e^{im\phi}e^{i\beta z}, \tag{8.6.10}$$

where the propagation constant β is at this point unspecified. Such a field retains its form except for a phase factor $[\exp(i\beta z)]$, and therefore defines a mode of the fiber. Using this form in (8.6.9), we obtain for the radial function $F(r)$ the ordinary differential equation

$$\frac{d^2 F}{dr^2} + \frac{1}{r}\frac{dF}{dr} + \left(n^2\frac{\omega^2}{c^2} - \beta^2 - \frac{m^2}{r^2}\right)F = 0. \tag{8.6.11}$$

Thus, in the core region,

$$\frac{d^2 F}{dr^2} + \frac{1}{r}\frac{dF}{dr} + \left(\kappa^2 - \frac{m^2}{r^2}\right)F = 0 \qquad (r \leq a), \tag{8.6.12}$$

---

[15]We are employing here the method of separation of variables, as discussed in Section 7.8.

where

$$\kappa^2 = n_1^2 \frac{\omega^2}{c^2} - \beta^2 \equiv n_1^2 k_0^2 - \beta^2. \tag{8.6.13}$$

Equation (8.6.12) has the form of the Bessel differential equation. The solutions that remain finite as $r \to 0$ are the Bessel functions $J_m(\kappa r)$ of the first kind, which we have already used in the preceding chapter (Sections 7.11 and 7.15):

$$F(r) = A J_m(\kappa r) \qquad (r \le a), \tag{8.6.14}$$

where A is a constant.

In the cladding region we write

$$\frac{d^2 F}{dr^2} + \frac{1}{r}\frac{dF}{dr} - \left(\gamma^2 + \frac{m^2}{r^2}\right)F = 0 \qquad (r > a), \tag{8.6.15}$$

where

$$\gamma^2 = \beta^2 - n_2^2 \frac{\omega^2}{c^2} \equiv \beta^2 - n_2^2 k_0^2. \tag{8.6.16}$$

We assume that γ^2 is positive, that is, that γ is real, in order to have solutions for $r > a$ that go to 0 as $r \to \infty$. These solutions are of the form

$$F(r) = B K_m(\gamma r) \qquad (r > a), \tag{8.6.17}$$

where K_m is a modified Bessel function of the second kind. Plots of $J_m(x)$ and $K_m(x)$ are readily found in various handbooks or on the Web. For our purposes at this point we need only know that $J_m(\kappa r)$ is finite at $r = 0$ and that $K_m(\gamma r) \to 0$ as $r \to \infty$, which are necessary conditions if the solutions (8.6.15) and (8.6.17) are to be applicable in the core and cladding regions, respectively.

Unlike our approach in Chapter 7 to obtain the modes of laser resonators, we have not invoked here the paraxial approximation. In fact, the solutions given by (8.6.10) and (8.6.14) for the field in the core are of the same form as the (nonparaxial) Bessel beam modes of Section 7.15, except that here the propagation is in a medium with refractive index n_1 rather than free space, and the propagation constant β is fixed by the fact that the tangential components of the field must be continuous at the core–cladding interface. In the case of a fiber the paraxial approximation may not be a good one because the field is guided by total internal reflection and, depending on the difference $n_1 - n_2$, the angles that rays make with respect to the fiber axis are not necessarily small.

The Helmholtz equation (8.2.20) applies to a single component of the electric field envelope $\mathcal{E}$, and also to a single component of the slowly varying magnetic field envelope $\mathcal{H}$. Given $\mathcal{E}_z$ and $\mathcal{H}_z$, for instance, we can obtain $\mathcal{E}_x$, $\mathcal{E}_y$, $\mathcal{H}_x$, and $\mathcal{H}_y$ from the Maxwell equations

$$\nabla \times \mathbf{E} = i\omega\mu_0 \mathbf{H}, \tag{8.6.18}$$

$$\nabla \times \mathbf{H} = -i\omega\varepsilon\mathbf{E} = -i\omega n^2 \epsilon_0 \mathbf{E}, \tag{8.6.19}$$

for a field that varies with time as $\exp(-i\omega t)$ and with z as $\exp(i\beta z)$. Consider, for example, the component H_x of $\mathbf{H}$. From (8.6.18),

$$i\omega\mu_0 H_x = \frac{\partial E_z}{\partial y} - \frac{\partial E_y}{\partial z} = \frac{\partial E_z}{\partial y} - i\beta E_y \tag{8.6.20}$$

and, from (8.6.19),

$$E_y = \frac{i}{\omega n^2 \epsilon_0}\left(-\frac{\partial H_z}{\partial x} + \frac{\partial H_x}{\partial z}\right). \tag{8.6.21}$$

Using (8.6.21) in (8.6.20), and $\epsilon_0\mu_0 = 1/c^2$, we obtain for the core region ($n^2 = n_1^2$)

$$H_x = -\frac{i}{\kappa^2}\left(\omega n_1^2 \epsilon_0 \frac{\partial E_z}{\partial y} - \beta \frac{\partial H_z}{\partial x}\right). \tag{8.6.22}$$

In the same fashion we obtain, in both the core and cladding regions, E_x, E_y, and H_x, H_y in terms of E_z and H_z. Of course, we can express this as well in terms of cylindrical components of the slowly varying envelope functions: We can express $\mathcal{E}_r$, $\mathcal{E}_\phi$, $\mathcal{H}_r$, and $\mathcal{H}_\phi$ in terms of $\mathcal{E}_z$ and $\mathcal{H}_z$ satisfying

$$\mathcal{E}_z(r, \phi, z) = AJ_m(\kappa r)e^{im\phi}e^{i\beta z}, \tag{8.6.23a}$$

$$\mathcal{H}_z(r, \phi, z) = BJ_m(\kappa r)e^{im\phi}e^{i\beta z}, \tag{8.6.23b}$$

for $r \leq a$ and

$$\mathcal{E}_z(r, \phi, z) = CK_m(\gamma r)e^{im\phi}e^{i\beta z}, \tag{8.6.23c}$$

$$\mathcal{H}_z(r, \phi, z) = DK_m(\gamma r)e^{im\phi}e^{i\beta z}, \tag{8.6.23d}$$

for $r > a$.

Maxwell's equations require that the tangential components of $\mathbf{E}$ and $\mathbf{H}$ be continuous at the core–cladding interface at $r = a$. That is, $\mathcal{E}_z$, $\mathcal{E}_\phi$, $\mathcal{H}_z$, and $\mathcal{H}_\phi$ must be continuous at $r = a$. Requiring this continuity leads to four homogeneous linear algebraic equations for the constants A, B, C, and D appearing in Eqs. (8.6.23), i.e., equations of the form $a_{1j}A + a_{2j}B + a_{3j}C + a_{4j}D = 0$, $j = 1, 2, 3, 4$. In order for these equations to have nonvanishing solutions for A, B, C, and D, the determinant of the coefficient matrix (a_{ij}), $i, j = 1, 2, 3, 4$, must vanish. This requirement takes the form of a complicated equation involving $J_m(\kappa a)$, $J_m'(\kappa a)$, $K_m(\gamma a)$, and $K_m'(\gamma a)$, where the primes denote derivatives. This "characteristic equation," which must be solved numerically, determines the propagation constant β for given values of ω, a, n_1, and n_2, that is, for a given frequency ω and for a given core radius a and core and cladding refractive indices n_1 and n_2, respectively.

For given values of ω, a, and n_1, n_2, the values of β determined by the numerical solution of the characteristic equation will depend on the integer m. For each m there is in general more than one solution for β; these different solutions can be denoted $\beta_{mj}, j = 1, 2, 3, \ldots$, and each β_{mj} defines a mode of the fiber. That is, a mode is defined by the pair of integers m and j that specify the spatial dependence of the electric and magnetic fields. The electric and magnetic fields for each mode are defined by Eqs. (8.6.23) and the equations relating the other field components to $\mathcal{E}_z$ and $\mathcal{H}_z$, with κ and γ depending on β_{mj} [Eq. (8.6.13) and (8.6.16)].

We are interested in *guided modes* in which the electric and magnetic fields fall off with radial distance from the fiber. Consider the fields (8.6.23) for $\gamma r \gg 1$. In this limit

$$K_m(\gamma r) \approx \left(\frac{\pi}{2\gamma r}\right)^{1/2} \left(1 - \frac{4m^2 - 1}{8\gamma r}\right) e^{-\gamma r}, \tag{8.6.24}$$

and the electric and magnetic fields (8.6.23) for a mode characterized by this radial dependence decay exponentially with distance from the fiber if γ is real ($\gamma^2 > 0$). If γ is purely imaginary ($\gamma^2 < 0$), however, the mode is not "guided"; the exponential decay of (8.6.24) is replaced by $\exp(-i|\gamma|r) = \cos|\gamma|r - i \sin|\gamma|r$ for $\gamma = i|\gamma|$. Therefore, $\gamma^2 = 0$ defines the "cut-off" between guided and unguided modes: $\gamma^2 > 0$ implies a guided mode, whereas $\gamma^2 < 0$ implies an unguided mode.

From (8.6.13) and (8.6.16) we see that $\gamma^2 = 0$ implies that $\kappa = k_0\sqrt{n_1^2 - n_2^2}$. The dimensionless "V parameter,"

$$V = k_0 a \sqrt{n_1^2 - n_2^2} = \frac{\omega}{c} a \sqrt{n_1^2 - n_2^2} = \frac{\omega a}{c} \mathrm{NA}, \tag{8.6.25}$$

determines the number of modes: fibers with large V parameters have many modes as determined by numerical solutions of the characteristic equation. The number of modes is found to be approximately $V^2/2$ for $V \gg 1$. But if V is made small enough, it is found that only the *fundamental* mode with $m = 0$ is guided by the fiber. Such single-mode fibers are of special interest for communication systems, and we will therefore devote the following section to them.

8.7 SINGLE-MODE FIBERS

In the preceding section we noted that the requirement that $\mathcal{E}_z$, $\mathcal{E}_\phi$, $\mathcal{H}_z$, and $\mathcal{H}_\phi$ be continuous at the core–cladding interface leads to a complicated characteristic equation involving $J_m(\kappa a)$, $J'_m(\kappa a)$, $K_m(\gamma a)$, and $K'_m(\gamma a)$. Numerical solutions of this equation determine the guided modes of the fiber for real values of the parameter γ defined by (8.6.16). Analysis of the characteristic equation shows that when $V < V_c$, where V_c is defined as the smallest value of V satisfying $J_0(V) = 0$, the fiber supports only the single mode with $m = 0$; there are no other guided modes for $V < V_c$.

The smallest "zero" of $J_0(x)$, that is, the smallest x such that $J_0(x) = 0$, is approximately 2.405. Thus, a step-index fiber will support only a single mode when $V_c < 2.405$, or in other words when

$$\frac{\omega}{c} a \sqrt{n_1^2 - n_2^2} = \frac{2\pi a}{\lambda} \mathrm{NA} < 2.405. \tag{8.7.1}$$

For $\lambda = 2\pi c/\omega = 1.3$ μm, $n_1 = 1.450$, and $n_2 = 1.443$, this single-mode condition is satisfied if the core radius $a < 3.5$ μm. These values are in the range characteristic of the single-mode fibers used in communication systems (Section 15.6). Obviously, the single-mode condition can be satisfied if the wavelength is large enough or if the core diameter and the numerical aperture are small enough.

In order to realize the single-mode condition (8.7.1) for wavelengths of interest and for core diameters that are not unreasonably small, the numerical aperture $NA = \sqrt{n_1^2 - n_2^2}$ must be small. Fibers typically have values of $\Delta = (n_1 - n_2)/n_1 \approx$ 0.01 and, as noted in the preceding section, the guided modes in this case are approximately paraxial, with z components of the field small compared to the transverse (x and y) components. That is, the guided modes are approximately transverse, and a linearly polarized mode has an electric field component of the form

$$\mathcal{E}_x(r, \phi, z) = \mathcal{E}_x(r, z) = \mathcal{E}_0 \frac{J_0(\kappa r)}{J_0(\kappa a)} e^{i\beta z} \qquad (r \leq a)$$

$$= \mathcal{E}_0 \frac{K_0(\gamma r)}{K_0(\gamma a)} e^{i\beta z} \qquad (r > a), \qquad (8.7.2)$$

where $\mathcal{E}_0$ is a constant specifying the amplitude of the field at $r = a$. The function $J_0(x)$ peaks at $x = 0$ ($J_0(0) = 1$) and its falloff to 0 at $x = 2.405\ldots$ follows roughly a bell-shaped curve, while the variation of $K_0(x)$ for large values of x is given by (8.6.24). The field (8.7.2) for a single-mode fiber is therefore often approximated by a Gaussian function:

$$\mathcal{E}_x(r, z) \approx \mathcal{E}_0 e^{-r^2/w^2} e^{i\beta z}, \qquad (8.7.3)$$

where the spot size w depends on the V parameter of the fiber and is $\cong a$ for $V \cong 2$.[14] Thus, *the single guided modes of fibers of interest for optical communication systems are approximately paraxial, transverse, and Gaussian, with a spot size on the order of the core diameter.*

● The calculation and characterization of the modes of an optical fiber are obviously rather complicated, and it is beyond our scope to delve much further into the subject. A few more general remarks, however, are appropriate.

The astute reader will have noticed that we have in effect assumed an infinite cladding region. The justification for this assumption is the exponential decay of the electric and magnetic fields of the guided modes outside the core [Eq. (8.6.24)]. Optical fibers are in fact designed so that the fields are negligibly small at the outer surface of the cladding. If this were not the case, light would be lost due to scattering from surface irregularities on the outer surface of the fiber.

As already noted, the core and cladding refractive indices in optical fibers typically differ by only a few percent. The critical angle for total internal reflection is therefore relatively large, making the guided modes approximately paraxial and the z components of the electric and magnetic fields small in magnitude compared to the transverse components. Each (m, j) mode is then approximately transverse and we can associate with it two "degenerate" orthogonal linear polarizations having the same (r, ϕ, z) dependence. If the fiber cross section were perfectly circular, a linearly polarized field would maintain its polarization, but in reality there are always slight imperfections in the core diameter, for instance, that cause the fiber to be birefringent in the sense that the *mode index* β/k_0 is different for the two orthogonally polarized modes. A "single-mode" fiber will have two mode indices, $\bar{n}_x$ and $\bar{n}_y$, and this causes the two orthogonal polarizations to exchange power. In practice, the injection of a linearly polarized field into the fiber results in an output field whose polarization is unpredictable as a consequence of random fluctuations of the birefringence. In *polarization preserving fibers* a relatively large and deterministic birefringence is introduced to overcome the random birefringence.   ●

The major breakthroughs that led to the widespread use of optical fibers in communication systems were the development of fibers with low attenuation and of compact (diode) lasers for efficiently coupling light into fibers (Section 15.6). In the early 1970s fibers were developed at Corning Glass Works with attenuations $\mathcal{A} \sim 20$ dB/km at wavelengths around 1 μm, compared to attenuations ~ 1000 dB/ km characteristic of the fibers manufactured earlier. The fused silica currently used to make optical fibers absorbs in the ultraviolet as a consequence of electronic resonances of the SiO_2 molecules and in the infrared as a consequence of molecular vibrations. The ultraviolet and infrared absorption together produce a broad absorption spectrum with $\mathcal{A} < 0.03$ dB/km in the wavelength range $1.3-1.6$ μm used in fiber-optic communications, and with an absorption minimum at 1.55 μm. Water vapor and, to a lesser extent, metallic impurities, are the dominant sources of absorption losses in silica fibers, and these losses, together with the loss due to Rayleigh scattering from local density fluctuations, exceed the "intrinsic" absorption loss of pure silica. All the sources of power loss in currently manufactured telecommunication fibers combine to produce an attenuation minimum of about 0.2 dB/km at 1.55 μm.

- In fiber optics the attenuation is commonly expressed in decibels per kilometer (dB/km). If $(\text{Pwr})_{\text{in}}$ and $(\text{Pwr})_{\text{out}}$ are the input and output powers, the attenuation in decibels is defined by [recall the definition (4.3.20)]

$$\mathcal{A} = 10 \log_{10} \frac{(\text{Pwr})_{\text{in}}}{(\text{Pwr})_{\text{out}}}. \tag{8.7.4}$$

A 3-dB attenuation means that the output power is half the input power. In terms of an attenuation coefficient a_0 per unit length, $(\text{Pwr})_{\text{in}}/(\text{Pwr})_{\text{out}} = \exp(a_0 L)$, where L is the length of the fiber. a_0 and $\mathcal{A}$ are related by $\mathcal{A}\,(\text{dB}) = 10 \log_{10} \exp(a_0 L)$, or

$$e^{a_0 L} = (10^{0.434})^{a_0 L} = 10^{\mathcal{A}(\text{dB})/10} \tag{8.7.5}$$

and therefore $a_0 = (0.23/L)\mathcal{A}\,(\text{dB})$ and

$$a_0\,(\text{cm}^{-1}) = 2.3 \times 10^{-6} \mathcal{A}\,(\text{dB}/\text{km}). \tag{8.7.6}$$

Decibel units are sometimes convenient simply because of the fact that the logarithm of a product of two numbers is equal to the sum of the two logarithms. For example, when a fiber with an attenuation (gain) of 10 dB is followed by a fiber with an attenuation (gain) of 20 dB, the overall attenuation (gain) is 30 dB.

The remarkable transmission capabilities of glass telecom fibers can be appreciated by a comparison with ordinary window glass, which has an optical attenuation coefficient $a_0 \sim 0.05$ cm^{-1}, about 100,000 times that of a fiber with $\mathcal{A} = 0.2$ dB/km. The small attenuation of transatlantic fiber cable allows repeaters (amplifiers) to be placed ~ 70 km apart.

The bending flexibility of fibers compared to the brittleness of bulk glass is mainly a consequence of their small surface areas. Fracture in glass and many other materials arises from voids that act to concentrate the effect of an applied stress. In glass, the voids are associated with tiny surface cracks that can grow under an applied stress and lead to fracture. The theory suggesting that the brittleness of glass is a surface effect, and therefore should be reduced when the surface area is decreased, was developed in the early 1920s by A. A. Griffith, who showed that "hot-drawing" glass into fibers dramatically increased its strength.

Fibers for guiding light had been proposed and tested in the 1920s and 1930s, but the fibers at the time were unclad and inefficient transmitters of light. The development of *fiber bundles* for "fiberscopes," the precursors of modern endoscopes, spurred renewed interest in optical fibers in the 1950s; these are also of interest for generating high powers (albeit with generally poor beam quality) by combining the outputs of single-fiber lasers (Section 11.14). Their invention was spurred by the need in many applications to guide light around obstacles without the usual methods based on lenses and mirrors. In the first publication on fiber bundles [*Nature* **173**, 39 (1954)], and on the use of a lower-index sheath around a single fiber, A. C. S. van Heel wrote that:

> Consideration of the construction of the eye of some insects suggested another approach. If a bundle or sheaf of thin transparent fibres is cut off perpendicularly at both ends and an optical image is formed on one end, it will be seen at the other end, as the light entering one fibre can only leave this at the other end, provided leakage of light from one fibre to another of the bundle is prevented. Moreover, the cylindrical wall of each fibre must reflect the light as nearly completely as possible, because of the numerous reflexions occurring when the fibres are thin compared to their length. Preliminary experiments ... have shown that coating the fibres with silver or any other metal yields an unsatisfactory transmission. A much better result was obtained when the fibres were coated with a layer of lower refractive index, which ensured total reflexion. This coating was isolated from the neighboring fibres by a thin coat of black paint. In this way, flexible 'image rods' have been obtained with satisfactory transmission, a very good contrast in the end image, and with the possibility of using forms bent in any direction (up to at least 360°). •

Light can escape a bent fiber: Rays incident on the core–cladding interface with an angle of incidence greater than the critical angle for total internal reflection can have an angle of incidence *smaller* than the critical angle when they encounter a bend. Bending loss in a fiber is characterized by an attenuation coefficient α_B such that after a propagation distance ℓ the light inside the fiber diminishes in power by the factor $\exp(-\alpha_B\ell)$; the fraction $1 - \exp(-\alpha_B\ell)$ of the power at $\ell = 0$ is radiated out of the fiber. Approximate calculations yield the result that α_B for a fiber mode depends on the radius of curvature R of a bend primarily through an exponential factor $\exp(-2\gamma^3 R/3\beta^2) \equiv \exp(-R/R_c)$, where β is the propagation constant for the mode and γ is defined by (8.5.16). Bending radii much smaller than R_c will result in significant loss of power in the fiber due to radiation from the fiber. Small values of R_c make the fiber less susceptible to bending loss; R_c is a function of the core and cladding radii and refractive indices that is not in general amenable to a simple analytical form (Problem 8.7). It increases with decreasing numerical aperture and with mode order, that is, higher-order modes have greater loss for a given bending radius than the lowest-order mode. Experiments generally support the predictions of the theory, although data analyses must also account for losses associated with the tensile strength and other characteristics of a particular fiber. Rough rules of thumb are that bending radii greater than about 10 times the fiber diameter result in acceptably small radiation loss and that fibers with numerical apertures smaller than 0.06 are too sensitive to bending to be practical. In addition to "macrobending" loss, there can also be significant "microbending" loss due to small, random bending radii along the fiber.

We have already mentioned intermodal dispersion, which can cause different pulses in a fiber to overlap and thereby limit the rate at which information in the form of "0" and "1" pulses can be transmitted. While single-mode fibers do not suffer from intermodal dispersion, there are nevertheless other types of dispersion that can limit their information transmission rate. One of these, of course, is group velocity dispersion. Another is *polarization-mode dispersion* arising from the fact that two orthogonal polarization components can have different group velocities as a consequence of the random birefringence effect described above. In Section 15.6 we will discuss further the effects of dispersion in fiber-optic communications.

8.8   BIREFRINGENCE

In Section 5.12 we discussed applications of *birefringent* materials in which the refractive index is different for different linear polarizations of the field. Like Rayleigh scattering, birefringence is a ubiquitous phenomenon of interest beyond the physics and applications of lasers. However, like Rayleigh scattering, it is sufficiently important in various laser applications to warrant more than the phenomenological discussion of Section 5.12. We will now discuss birefringence in more detail, starting from the wave equation (8.2.10) for the electric field.

Writing the Cartesian components $\mathbf{E}_i$ and $\mathbf{P}_i$ of $\mathbf{E}$ and $\mathbf{P}$ as in (8.2.16) for a field of frequency ω, Eq. (8.2.10) becomes

$$\nabla^2 \mathcal{E}_i - \frac{\partial}{\partial x_i} \sum_{j=1}^{3} \frac{\partial \mathcal{E}_j}{\partial x_j} + \frac{\omega^2}{c^2} \mathcal{E}_i = -\frac{\omega^2}{\epsilon_0 c^2} \mathcal{P}_i, \tag{8.8.1}$$

where x_i is the ith Cartesian component of $\mathbf{r}$, e.g., $x_1 = x$, $x_2 = y$, and $x_3 = z$. Now in general $\mathbf{P}$ does not point in the same direction as $\mathbf{E}$. In other words, the medium can in general be *anisotropic*. For instance, we could have

$$\mathcal{P}_x = \epsilon_0(\chi_{xx}\mathcal{E}_x + \chi_{xy}\mathcal{E}_y + \chi_{xz}\mathcal{E}_z) \tag{8.8.2}$$

and likewise for the y and z components of $\mathbf{P}$. The coefficients χ_{xx}, χ_{xy}, and χ_{xz} are components of the electric susceptibility *tensor*. As discussed in Section 5.12, the fact that these components can be different can be understood physically in terms of a restoring force that can be different for different directions in which an electron in the material is displaced from its equilibrium position. In the notation employed in (8.8.1), we can write

$$\mathcal{P}_i = \epsilon_0 \sum_{j=1}^{3} \chi_{ij}(\omega)\mathcal{E}_j. \tag{8.8.3}$$

Then (8.8.1) takes the form

$$\nabla^2 \mathcal{E}_i + \frac{\omega^2}{\epsilon_0 c^2} \sum_{j=1}^{3} \epsilon_{ij}(\omega)\mathcal{E}_j - \frac{\partial}{\partial x_i} \sum_{j=1}^{3} \frac{\partial \mathcal{E}_j}{\partial x_j} = 0. \tag{8.8.4}$$

Here the 3×3 matrix

$$\epsilon_{ij}(\omega) = \epsilon_0[\delta_{ij} + \chi_{ij}(\omega)] \tag{8.8.5}$$

is the dielectric *tensor* and we employ the "Kronecker delta" δ_{ij}, defined to be 1 if $i = j$ and 0 if $i \neq j$. Lest there be any confusion about the notation, we write Eq. (8.8.4) explicitly for the component $\mathcal{E}_x$ of the electric field amplitude:

$$\left(\frac{\partial^2}{\partial x^2} + \frac{\partial^2}{\partial y^2} + \frac{\partial^2}{\partial z^2}\right)\mathcal{E}_x + \frac{\omega^2}{\epsilon_0 c^2}\left(\epsilon_{xx}\mathcal{E}_x + \epsilon_{xy}\mathcal{E}_y + \epsilon_{xz}\mathcal{E}_z\right) - \frac{\partial}{\partial x}\left(\frac{\partial \mathcal{E}_x}{\partial x} + \frac{\partial \mathcal{E}_y}{\partial y} + \frac{\partial \mathcal{E}_z}{\partial z}\right) = 0, \tag{8.8.6}$$

where $\epsilon_{xx} = \epsilon_0(1 + \chi_{xx})$, $\epsilon_{xy} = \epsilon_0\chi_{xy}$, and $\epsilon_{xz} = \epsilon_0\chi_{xz}$.

Similar equations can, of course, be written in any Cartesian coordinate system (X, Y, Z), where X, Y, and Z are each a linear combination of whatever Cartesian coordinates x, y, and z were used in writing (8.8.6). In particular, it is convenient to write these equations in the particular coordinate system (X, Y, Z) in which the matrix ϵ_{ij} is diagonal:

$$\epsilon_{ij}(\omega) = \epsilon_{ii}(\omega)\delta_{ij}, \tag{8.8.7}$$

for example, $\epsilon_{XY} = \epsilon_{XZ} = 0$. The coordinate system in which ϵ_{ij} is diagonal defines the *principal dielectric axes* of the material. In this coordinate system we have, for example,

$$\left(\frac{\partial^2}{\partial X^2} + \frac{\partial^2}{\partial Y^2} + \frac{\partial^2}{\partial Z^2}\right)\mathcal{E}_X + \frac{\omega^2}{\epsilon_0 c^2}\epsilon_{XX}\mathcal{E}_X - \frac{\partial}{\partial X}\left(\frac{\partial\mathcal{E}_X}{\partial X} + \frac{\partial\mathcal{E}_Y}{\partial Y} + \frac{\partial\mathcal{E}_Z}{\partial Z}\right) = 0. \tag{8.8.8}$$

The condition $\nabla \cdot \mathbf{D} = 0$ for a charge-neutral medium implies

$$\frac{\partial D_X}{\partial X} + \frac{\partial D_Y}{\partial Y} + \frac{\partial D_Z}{\partial Z} = \epsilon_{XX}\frac{\partial\mathcal{E}_X}{\partial X} + \epsilon_{YY}\frac{\partial\mathcal{E}_Y}{\partial Y} + \epsilon_{ZZ}\frac{\partial\mathcal{E}_Z}{\partial Z} = 0. \tag{8.8.9}$$

We will restrict ourselves to the important case of *uniaxial* birefringent crystals in which two of the three ϵ's along the principal dielectric axes, say ϵ_{XX} and ϵ_{YY}, are equal; the Z axis then defines the optic axis of the crystal (Section 5.12), and we will see how the polarization and direction of propagation with respect to this axis determine the refractive index.[16] With (8.8.9) and the definitions

$$n_o^2(\omega) = \frac{\epsilon_{XX}(\omega)}{\epsilon_0} = \frac{\epsilon_{YY}(\omega)}{\epsilon_0} \quad \text{and} \quad n_e^2(\omega) = \frac{\epsilon_{ZZ}(\omega)}{\epsilon_0}, \tag{8.8.10}$$

Eq. (8.8.8) becomes

$$\left(\frac{\partial^2}{\partial X^2} + \frac{\partial^2}{\partial Y^2} + \frac{\partial^2}{\partial Z^2}\right)\mathcal{E}_X + \frac{\omega^2}{c^2}n_o^2(\omega)\mathcal{E}_X - \left[1 - \frac{n_e^2(\omega)}{n_o^2(\omega)}\right]\frac{\partial^2\mathcal{E}_Z}{\partial X\partial Z} = 0, \tag{8.8.11}$$

and similarly

$$\left(\frac{\partial^2}{\partial X^2} + \frac{\partial^2}{\partial Y^2} + \frac{\partial^2}{\partial Z^2}\right)\mathcal{E}_Y + \frac{\omega^2}{c^2}n_o^2(\omega)\mathcal{E}_Y - \left[1 - \frac{n_e^2(\omega)}{n_o^2(\omega)}\right]\frac{\partial^2\mathcal{E}_Z}{\partial Y\partial Z} = 0 \tag{8.8.12}$$

and

$$\left(\frac{\partial^2}{\partial X^2} + \frac{\partial^2}{\partial Y^2} + \frac{\partial^2}{\partial Z^2}\right)\mathcal{E}_Z + \frac{\omega^2}{c^2}n_e^2(\omega)\mathcal{E}_Z - \left[1 - \frac{n_e^2(\omega)}{n_o^2(\omega)}\right]\frac{\partial^2\mathcal{E}_Z}{\partial Z^2} = 0. \tag{8.8.13}$$

Equations (8.8.11)–(8.8.13) are expressed in terms of the coordinates (X, Y, Z) along the principal dielectric axes of the crystal. Of more direct interest, of course, are the wave

[16]An anisotropic crystal may have *two* optic axes, in which case it is called *biaxial*.

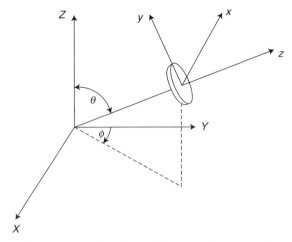

Figure 8.7 Principal dielectric axes (X, Y, Z) and "laboratory" axes (x, y, z). $z =$ direction of propagation, $Z =$ optic axis.

equations expressed in terms of the "laboratory" coordinates (x, y, z). These equations may be derived from (8.8.11)–(8.8.13) as follows:

The (x, y, z) coordinates of a point in the crystal may be written as linear combinations of its (X, Y, Z) coordinates. From the angles θ and ϕ shown in Fig. 8.7 it can be seen that

$$
\begin{pmatrix} x \\ y \\ z \end{pmatrix} = \begin{pmatrix} -\cos\phi & \sin\phi & 0 \\ -\cos\theta\sin\phi & -\cos\theta\cos\phi & \sin\theta \\ \sin\theta\sin\phi & \sin\theta\cos\phi & \cos\theta \end{pmatrix} \begin{pmatrix} X \\ Y \\ Z \end{pmatrix}. \tag{8.8.14}
$$

Thus,

$$
\begin{aligned}
\frac{\partial}{\partial X} &= \frac{\partial x}{\partial X}\frac{\partial}{\partial x} + \frac{\partial y}{\partial X}\frac{\partial}{\partial y} + \frac{\partial z}{\partial X}\frac{\partial}{\partial z} \\
&= -\cos\phi\frac{\partial}{\partial x} - \sin\phi\cos\theta\frac{\partial}{\partial y} + \sin\phi\sin\theta\frac{\partial}{\partial z}
\end{aligned} \tag{8.8.15}
$$

and likewise

$$
\frac{\partial}{\partial Y} = \sin\phi\frac{\partial}{\partial x} - \cos\phi\cos\theta\frac{\partial}{\partial y} + \cos\phi\sin\theta\frac{\partial}{\partial z} \tag{8.8.16}
$$

and

$$
\frac{\partial}{\partial Z} = \sin\theta\frac{\partial}{\partial y} + \cos\theta\frac{\partial}{\partial z}. \tag{8.8.17}
$$

Using these relations we can rewrite Eqs. (8.8.11)–(8.8.13) in terms of the laboratory coordinates (x, y, z). Consider first a wave polarized along the X direction $(\mathcal{E}_Y = \mathcal{E}_Z = 0)$. The Laplacian operator has the same form in different Cartesian coordinate systems, that is, $\partial^2/\partial X^2 + \partial^2/\partial Y^2 + \partial^2/\partial Z^2 = \partial^2/\partial x^2 + \partial^2/\partial y^2 + \partial^2/\partial z^2$, as may be shown straightforwardly using Eqs. (8.8.15)–(8.8.17). Therefore (8.8.11) has the form

$$\left(\frac{\partial^2}{\partial x^2} + \frac{\partial^2}{\partial y^2} + \frac{\partial^2}{\partial z^2}\right)\mathcal{E}_X + \frac{\omega^2}{c^2}n_o^2(\omega)\mathcal{E}_X = 0 \tag{8.8.18}$$

in the (x, y, z) coordinate system. Similarly, for a wave polarized along the Y direction $(\mathcal{E}_X = \mathcal{E}_Z = 0)$,

$$\left(\frac{\partial^2}{\partial x^2} + \frac{\partial^2}{\partial y^2} + \frac{\partial^2}{\partial z^2}\right)\mathcal{E}_Y + \frac{\omega^2}{c^2}n_o^2(\omega)\mathcal{E}_Y = 0. \tag{8.8.19}$$

We see therefore that $n_o(\omega)$ *is the index of refraction* for waves with polarization perpendicular to the optic axis of the crystal. Waves with polarization perpendicular to the optic axis are called *ordinary waves*.

For a field polarized *parallel* to the optic axis $(\mathcal{E}_X = \mathcal{E}_Y = 0)$, it follows from Eqs. (8.8.13) and (8.8.15)–(8.8.17) that

$$\left(\frac{\partial^2}{\partial x^2} + \frac{\partial^2}{\partial y^2} + \frac{\partial^2}{\partial z^2}\right)\mathcal{E}_Z + \frac{\omega^2}{c^2}n_e^2(\omega)\mathcal{E}_Z$$

$$- \left[1 - \frac{n_e^2(\omega)}{n_o^2(\omega)}\right]\left(\sin^2\theta\frac{\partial^2}{\partial y^2} + \cos^2\theta\frac{\partial^2}{\partial z^2} + \sin 2\theta\frac{\partial^2}{\partial y\partial z}\right)\mathcal{E}_Z = 0. \tag{8.8.20}$$

In particular, for a plane wave propagating in the z direction, $\mathcal{E}_Z = \mathcal{E}_Z(z)$ and therefore

$$\frac{\partial^2\mathcal{E}_Z}{\partial z^2} + \frac{\omega^2}{c^2}n_e^2(\omega)\mathcal{E}_Z - \left[1 - \frac{n_e^2(\omega)}{n_o^2(\omega)}\right]\cos^2\theta\frac{\partial^2\mathcal{E}_Z}{\partial z^2} = 0, \tag{8.8.21}$$

or

$$\frac{\partial^2\mathcal{E}_Z}{\partial z^2} + \frac{\omega^2}{c^2}n_e^2(\omega, \theta)\mathcal{E}_Z = 0, \tag{8.8.22}$$

where $n_e^2(\omega, \theta)$ is defined by the equation

$$\frac{1}{n_e^2(\omega,\theta)} = \frac{\cos^2\theta}{n_o^2(\omega)} + \frac{\sin^2\theta}{n_e^2(\omega)}. \tag{8.8.23}$$

It follows from (8.8.22) that $n_e(\omega, \theta)$ *is the index of refraction for waves polarized parallel to the optic axis and propagating in a direction z making an angle θ with respect to the optic axis.* Waves with polarization parallel to the optic axis are called *extraordinary waves*. Equation (8.8.23) shows that $n_e(\omega)$ is the refractive index for an extraordinary wave propagating in a direction perpendicular to the optic axis ($\theta = \pi/2$, $\sin\theta = 1$, $\cos\theta = 0$). Extraordinary waves propagate with a refractive index $n_e(\omega, \theta)$ that depends on their direction of propagation with respect to the optic axis, whereas ordinary waves propagate with a refractive index $n_o(\omega)$ regardless of their direction of propagation. The refractive index for *any* wave propagating in a direction parallel to the optic axis will be $n_o(\omega)$, as can be seen from (8.8.23) with $\theta = 0$. $n_o(\omega)$ and $n_e(\omega)$ are

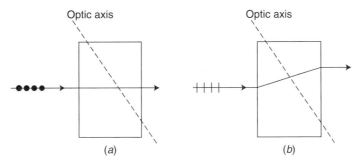

Figure 8.8   (*a*) Ordinary and (*b*) extraordinary waves in a uniaxial birefringent crystal.

called the *principal refractive indices* of the crystal and are generally temperature-as well as wavelength-dependent.

In Section 5.12 we assumed that the optic axis—which is actually a *direction* in the crystal rather than a single axis—was parallel to the crystal surface onto which a field is normally incident. Equation (8.8.23) gives the refractive index for extraordinary waves when the field is not necessarily propagating orthogonally to the optic axis, i.e., when θ may be different from $\pi/2$.

Figure 8.8 illustrates what is "ordinary" and "extraordinary" about the two types of wave that can propagate in a uniaxial crystal. Figure 8.8*a* shows a normally incident field that is linearly polarized in a direction perpendicular to the plane of incidence formed by the optic axis and the propagation direction. In this case the field simply propagates through the crystal in the expected or "ordinary" way, i.e., according to Snell's law. In Fig. 8.8*b*, however, the field is linearly polarized parallel to the plane of incidence. In this case something "extraordinary" happens: The wave is deflected at the boundaries and the rays emergent at the exit face are displaced with respect to the incident rays. Note that there is a displacement even though the field is normally incident on the crystal surface; no such displacement would be expected from Snell's law.

This displacement means that if a birefringent crystal such as Iceland spar (calcite, $CaCO_3$) is laid over a small dot on a piece of paper, we will see a double image of the spot. This phenomenon of *double refraction* (sometimes called *anomalous refraction*) was noted over 400 years ago by European sailors visiting Iceland.

If unpolarized light is incident on a doubly refracting crystal, it is separated into ordinary and extraordinary ways that are linearly polarized orthogonally to each other. This splitting of a light beam into two orthogonally polarized beams in calcite was observed by Arago and Fresnel early in the nineteenth century. Since their polarizations are orthogonal, the two beams do not interfere. This led Thomas Young (1817), and later Fresnel, to propose that light waves are transverse, for the absence of interference could not be explained if light waves are longitudinal, like sound waves in air.

Separation of the two orthogonally polarized waves resulting from double refraction can be used to construct polarizers such as the Nicol prism. The more familiar polarizers used in sunglasses, for example, are based on the fact that in certain crystals the molecules are aligned in such a way that light polarized in one direction is transmitted, whereas the light polarized in the perpendicular direction is strongly absorbed. Such materials are said to be *dichroic*. (Some such materials appear colored in white light because the effect is wavelength dependent, whence the term dichroic.) The most common types of polarizer are Polaroid filters, invented around 1926 by Edwin Land.

One type of Polaroid filter consists of a plastic sheet in which are embedded needlelike crystals of dichroic herapathite (quinine sulfate periodide). The most common Polaroid is made by dipping a plastic (whose long molecules have been aligned by stretching) in iodine, which makes the plastic dichroic.

• The dichroic property of herapathite was known long before Land, but the crystals were very fragile and difficult to grow in sizes large enough to be useful. The essence of Land's idea was to embed the tiny crystals in a plastic that was stretched while soft to align them. Land wrote as follows about the discovery of herapathite:[17] "In the literature are a few pertinent high spots in the development of polarizers, particularly the work of William Bird Herapath, a physician in Bristol, England, whose pupil, a Mr. Phelps, had found that when he dropped iodine into the urine of a dog that had been fed quinine, little scintillating green crystals formed in the reaction fluid. Phelps went to his teacher, and Herapath then did something which I think was curious under the circumstances; he looked at the crystals under a microscope and noticed that in some places they were light where they were overlapped and in some places they were dark. He was shrewd enough to recognize that here was a remarkable phenomenon, a new polarizing material." •

It is often appropriate to replace Eq. (8.8.18)–(8.8.20) by their paraxial approximations. For an ordinary wave propagating in the z direction we write

$$\mathcal{E}_X(x,\ y,\ z) = \mathcal{E}_0^{(o)}(x,\ y,\ z)e^{in_o(\omega)\omega z/c}, \tag{8.8.24}$$

and likewise for $\mathcal{E}_Y(x,\ y,\ z)$, and we assume that $\mathcal{E}_0^{(o)}$ is slowly varying in z compared to $\exp[in_o(\omega)\omega z/c]$. Then (8.8.18) is replaced by the paraxial wave equation (Section 7.4)

$$\left(\frac{\partial^2}{\partial x^2} + \frac{\partial^2}{\partial x^2}\right)\mathcal{E}_0^{(o)} + 2in_o(\omega)\frac{\omega}{c}\frac{\partial \mathcal{E}_0^{(o)}}{\partial z} = 0. \tag{8.8.25}$$

The paraxial approximation for extraordinary waves is a bit more complicated. In this case we write

$$\mathcal{E}_Z(x,\ y,\ z) = \mathcal{E}_0^{(e)}(x,\ y,\ z)e^{in_e(\omega,\theta)\omega z/c} \tag{8.8.26}$$

for an extraordinary wave propagating in the direction z making an angle θ with respect to the optic axis. Then Eq. (8.8.20) is replaced by

$$\frac{\partial^2 \mathcal{E}_0^{(e)}}{\partial x^2} + [1 - r(\omega)\sin^2\theta]\frac{\partial^2 \mathcal{E}_0^{(e)}}{\partial y^2} - ir(\omega)n_e(\omega,\ \theta)\frac{\omega}{c}\frac{\partial \mathcal{E}_0^{(e)}}{\partial y}$$
$$+ 2in_e(\omega,\ \theta)\frac{\omega}{c}[1 - r(\omega)\cos^2\theta]\frac{\partial \mathcal{E}_0^{(e)}}{\partial z} = 0, \tag{8.8.27}$$

[17]E. H. Land, *Journal of the Optical Society of America* **41**, 957 (1951). Land is also recognized among other things for his contributions to reconnaissance, three-dimensional movies, the theory of color vision, and for his invention of the Polaroid camera (1947), which was stimulated by his daughter's asking why it took so long to develop photographs from a family vacation in Santa Fe, New Mexico. The student might wish to consider what Land once said to an interviewer: "My whole life has been spent trying to teach people that intense concentration for hour after hour can bring out in people resources they didn't know they had."

where we have defined

$$r(\omega) = 1 - \frac{n_e^2(\omega)}{n_o^2(\omega)} \tag{8.8.28}$$

and used the paraxial approximation in which $\partial^2 \mathcal{E}_0^{(e)}/\partial z^2$ is negligible compared to the last term on the left-hand side of (8.8.27), and in which we take

$$\frac{\partial^2}{\partial y \, \partial z} \left[\mathcal{E}_0^{(e)} e^{in_e(\omega,\theta)\omega z/c} \right] \cong in_e(\omega, \theta) \frac{\omega}{c} \frac{\partial \mathcal{E}_0^{(e)}}{\partial y} e^{in_e(\omega,\theta)\omega z/c}. \tag{8.8.29}$$

Using an identity that follows from (8.8.23):

$$n_e(\omega, \theta)[1 - r(\omega)\cos^2 \theta] = n_e(\omega)[1 - r(\omega)\cos^2 \theta]^{1/2} \tag{8.8.30}$$

and the definition

$$\rho(\theta) = \frac{-\frac{1}{2} r(\omega) \sin 2\theta}{1 - r(\omega)\cos^2 \theta} = \frac{1}{n_e(\omega, \theta)} \frac{\partial n_e(\omega, \theta)}{\partial \theta}, \tag{8.8.31}$$

we write (8.8.27) as

$$\frac{\partial^2 \mathcal{E}_0^{(e)}}{\partial x^2} + [1 - r(\omega)\sin^2 \theta] \frac{\partial^2 \mathcal{E}_0^{(e)}}{\partial y^2} + 2in_e(\omega)\frac{\omega}{c}[1 - r(\omega)\cos^2 \theta]^{1/2} \left[\frac{\partial \mathcal{E}_0^{(e)}}{\partial z} + \rho(\theta)\frac{\partial \mathcal{E}_0^{(e)}}{\partial y} \right] = 0. \tag{8.8.32}$$

The factors $[1 - r(\omega)\sin^2 \theta]$ and $[1 - r(\omega)\cos^2 \theta]$ are typically ≈ 1; the propagation equation (8.8.32) for extraordinary waves then differs from Eq. (8.8.25) for ordinary waves mainly because of the term $\rho(\theta)\partial \mathcal{E}_0^{(e)}/\partial y$. To appreciate the physical significance of this term, let us ignore diffraction, which is accounted for by the first two terms on the left-hand side of (8.8.32) (Problem 8.9). Then we have the propagation equation

$$\frac{\partial \mathcal{E}_0^{(e)}}{\partial z} + \rho(\theta)\frac{\partial \mathcal{E}_0^{(e)}}{\partial y} = 0. \tag{8.8.33}$$

Solutions of this equation are of the form

$$\mathcal{E}_0^{(e)}(y, z) = \mathcal{E}_0^{(e)}(y - \rho(\theta)z), \tag{8.8.34}$$

which implies that, after a distance z of propagation, the field is displaced by $\rho(\theta)z$ along the y direction. In other words, as z increases, the field "walks off" the z direction by $\rho(\theta)z$, corresponding to the *walk-off angle* $\tan^{-1} \rho(\theta)$. This is the physical significance of $\rho(\theta)$: it is the tangent of the walk-off angle between ordinary and extraordinary waves in the "sensitive direction" y along which the extraordinary wave is polarized (Fig. 8.7). Thus, $\rho(\theta)$ provides a quantitative measure of the walk-off illustrated in Fig. 8.8b.

8.9 RAYLEIGH SCATTERING

A laser beam will generally be attenuated as it propagates. This occurs because of absorption, as discussed in Chapter 1, or because of scattering of radiation out of the beam. Rayleigh scattering—the scattering of light by particles small compared to a wavelength—is important for the understanding of many natural optical phenomena and must often be considered in applications involving the propagation of laser radiation. Among the propagation effects discussed so far, it is the first to repay attention to the particulate nature of media, i.e., to differences between the individual, nominally identical, constituents of any medium.

To understand Rayleigh scattering we begin with the electric field scattered by a particle considered as a dipole $\hat{\mathbf{x}}p(t)$ oscillating under the influence of an incident field. In the radiation zone the scattered field is

$$\mathbf{E}(\mathbf{r}, t) = \frac{1}{4\pi\epsilon_0} [(\hat{\mathbf{x}} \cdot \hat{\mathbf{r}})\hat{\mathbf{r}} - \hat{\mathbf{x}}] \frac{1}{c^2 r} \frac{d^2}{dt^2} p(t - r/c), \tag{8.9.1}$$

where $\hat{\mathbf{x}}$ is the unit vector in the direction of the electric field inducing the dipole moment and $\hat{\mathbf{r}}$ is the unit vector pointing from the dipole to the point of observation ($\mathbf{r} = r\hat{\mathbf{r}}$). The attentuation coefficient a_R associated with this scattering can be derived as follows.

The relation

$$p(t) = \alpha(\omega)E_0 \cos \omega t \tag{8.9.2}$$

gives the dipole moment that is induced in a particle with polarizability $\alpha(\omega)$ in an electric field $\hat{\mathbf{x}}E_0 \cos \omega t$. We have already presented formula (3.3.3) for the power radiated in all directions by such a dipole moment $p(t)$:

$$\mathrm{Pwr} - \left(\frac{1}{4\pi\epsilon_0}\right) \frac{2}{3c^3} (-\omega^2 \alpha(\omega)E_0 \cos \omega t)^2. \tag{8.9.3}$$

We can relate the electric field to the corresponding intensity I in a medium with refractive index $n(\omega) \approx 1$ by the formula $I = c\epsilon_0 E_0^2 \cos^2 \omega t$, and can write this as

$$\mathrm{Pwr} = \left(\frac{1}{4\pi\epsilon_0}\right) \frac{2\omega^4}{3c^3} \frac{\alpha^2(\omega)}{c\epsilon_0} I \equiv \sigma_R(\omega)I. \tag{8.9.4}$$

Here the cross section for Rayleigh scattering is given by

$$\sigma_R(\omega) = \frac{8\pi\omega^4}{3} \left(\frac{\alpha(\omega)}{4\pi\epsilon_0 c^2}\right)^2. \tag{8.9.5}$$

The Rayleigh attentuation coefficient $a_R(\omega) = N\sigma_R(\omega)$ is then obtained for a dilute medium of N such particles per unit volume, via the formula $n^2(\omega) - 1 = N\alpha(\omega)/\epsilon_0$:

$$a_R(\omega) = N\sigma_R(\omega) = \left(\frac{\omega}{c}\right)^4 \frac{[n^2(\omega) - 1]^2}{6\pi N}. \tag{8.9.6}$$

The attenuation is due to the fact that a dipole induced by an incident wave radiates its own electromagnetic field, causing a spatial redistribution of the field of the incident wave. In other words, radiation is scattered out of the incident wave, causing its intensity to diminish with distance z as $\exp(-a_R z)$: $I(z) = I(0)\exp(-a_R z)$.

The ω^4 (or $1/\lambda^4$) dependence of a_R means that the amount of scattering increases sharply with increasing frequency. (The refractive index n generally varies much more slowly with ω than ω^4.) Rayleigh used this dependence to explain why the sky is blue and the sunset red. When we look at the sky away from the sun on a sunny day, we see light that has been scattered by air molecules exposed to sunlight. This scattered light is predominantly blue because the high-frequency components of the visible solar radiation are scattered more strongly than the low-frequency components. The sunset, however, is reddish because the sunlight has traveled a sufficient distance through Earth's atmosphere that much of the high-frequency components have been scattered away.

Consider the Rayleigh scattering of visible radiation by molecules in Earth's atmosphere. Taking $\lambda = 600$ nm, and $n \cong 1.0003$ for the refractive index of air at optical frequencies, we find from (8.9.6) that $a_R^{-1} \approx 4.4 \times 10^{-21} N$ m, where N is the number of molecules per cubic meter. Assuming an ideal gas at standard temperature and pressure, we calculate $N \approx 2.69 \times 10^{25}$ and therefore $a_R^{-1} \approx 118$ km for the distance in which 600-nm radiation is attenuated by a factor $e^{-1} \approx 37\%$. Rayleigh compared such calculations with astronomers' estimates for the transmission of stellar radiation through Earth's atmosphere. He drew the important conclusion that the scattering of light by molecules alone, without suspended particles (dust), is strong enough to cause the blue sky, which he poetically called the "heavenly azure." This explanation of the blue sky suggests, in fact, that the sky should be violet since violet light should be scattered more strongly than blue. One reason the sky appears blue rather than violet is that the eye is more sensitive to blue. Furthermore the solar spectrum is not uniform but has somewhat less radiation at the shorter visible wavelengths.

There is an interesting conceptual inconsistency here between our derivation of a_R and the exact solution of Maxwell's equations for the electric field propagating in a continuous, uniform medium of dipoles characterized by a real index of refraction $n(\omega)$. The exact solution for such a field in (8.2.24) shows no evidence of either scattered light or of attenuation. This is correct. There is no attenuation when a light beam passes through such an idealized collection of dipoles.

One can see that our derivation here amounts to assigning the same scattered power to every dipole in the medium. i.e., $a_R(\omega)$ is obtained by an addition of intensities whereas, according to first principles, addition of fields not intensities is correct. An important characteristic of atmospheric light propagation resolves this apparent contradiction. The exact (non-scattering) solution of Maxwell's equations is based on the assumption that the density N describes a uniform continuum. The addition of intensities in the case of Rayleigh scattering is correctly a consequence of density *fluctuations* that must be accounted for.

• The important role of density fluctuations in Rayleigh scattering is indicated by the following argument that takes account of different positions for different particles. The electric dipole moment induced in a particle at $\mathbf{r}_i$ by an incident field $\mathbf{E}_0 \cos(\omega t - \mathbf{k}_0 \cdot \mathbf{r})$ is

$$\mathbf{p}_i(t) = \alpha(\omega)\mathbf{E}_0 e^{-i\omega t} e^{i\mathbf{k}_0 \cdot \mathbf{r}_i}, \tag{8.9.7}$$

where as usual it is implicit that we should take the real part of the right-hand side. The electric field from this dipole at a point $\mathbf{r}$ in the "radiation zone," i.e., the part of the electric field that varies with distance $|\mathbf{r} - \mathbf{r}_i|$ from the dipole as $1/|\mathbf{r} - \mathbf{r}_i|$, is proportional to [cf. (8.9.1) and (8.9.2)]

$$\ddot{p}(t - |\mathbf{r} - \mathbf{r}_i|/c) = -\omega^2 \alpha(\omega) E_0 e^{-i\omega(t - |\mathbf{r} - \mathbf{r}_i|)} e^{i\mathbf{k}_0 \cdot \mathbf{r}_i}, \tag{8.9.8}$$

where again $k_0 = \omega/c$. Now

$$k_0 |\mathbf{r} - \mathbf{r}_i| = k_0 \sqrt{\{r^2 - 2\mathbf{r} \cdot \mathbf{r}_i + r_i^2\}} \cong k_0 r - k_0 \hat{\mathbf{r}} \cdot \mathbf{r}_i \tag{8.9.9}$$

for large distances from the dipole ($r \gg r_i$), and therefore

$$\begin{aligned} \ddot{p}(t - |\mathbf{r} - \mathbf{r}_i|/c) &\cong -\omega^2 \alpha(\omega) E_0 e^{-i\omega(t - r/c)} e^{i\mathbf{k}_0 \cdot \mathbf{r}_i} e^{-ik_0 \hat{\mathbf{r}} \cdot \mathbf{r}_i} \\ &= -\omega^2 \alpha(\omega) E_0 e^{-i\omega(t - r/c)} e^{i\mathbf{K} \cdot \mathbf{r}_i}, \end{aligned} \tag{8.9.10}$$

where $\mathbf{K} = \mathbf{k}_0 - k_0 \hat{\mathbf{r}}$ is the difference between the $\mathbf{k}$ vectors of the incident ($\mathbf{k}_0$) and scattered ($k_0 \hat{\mathbf{r}}$) plane waves. For $\mathcal{N}$ dipoles at positions $\mathbf{r}_1, \mathbf{r}_2, \ldots, \mathbf{r}_{\mathcal{N}}$, the total scattered field at large distances from the dipoles is proportional to[18]

$$\begin{aligned} -\omega^2 \alpha(\omega) E_0 e^{-i\omega(t - r/c)} \sum_{i=1}^{\mathcal{N}} e^{i\mathbf{K} \cdot \mathbf{r}_i} &= -\omega^2 \alpha(\omega) E_0 e^{-i\omega(t - r/c)} \sum_{i=1}^{\mathcal{N}} e^{i\Phi_i} \\ &= -\omega^2 \alpha(\omega) E_0 e^{-i\omega(t - r/c)} F(\mathbf{K}), \end{aligned} \tag{8.9.11}$$

where we have defined the phase $\Phi_i = \mathbf{K} \cdot \mathbf{r}_i$ and the "structure factor"

$$F(\mathbf{K}) = \sum_{i=1}^{\mathcal{N}} e^{i\Phi_i}. \tag{8.9.12}$$

If the scatterers have a continuous and uniform distribution, then $F(\mathbf{K}) = 0$ except for the forward scattering direction for which $\mathbf{K} = 0$; this follows from the fact that, if $\mathbf{K} \neq 0$, there will for every Φ_i be a Φ_j such that $\exp(i\Phi_j) = -\exp(i\Phi_i)$. In other words, if the scatterers are densely and uniformly distributed, there is no side scattering.

The scattered power contains the factor

$$|F(\mathbf{K})|^2 = \left| \sum_{i=1}^{\mathcal{N}} e^{i\Phi_i} \right|^2 = \mathcal{N} + \sum_{i \neq j}^{\mathcal{N}} \sum_{j=1}^{\mathcal{N}} e^{i(\Phi_i - \Phi_j)}, \tag{8.9.13}$$

which for a medium with a randomly fluctuating density of dipoles must be treated by averaging over dipole positions. The average of the second term on the right-hand side vanishes if the particle positions are uncorrelated, as in an ideal gas and approximately the case in a gas of weakly interacting particles, and the result is proportional to the number $\mathcal{N}$ of scatterers, just as if individual scattered intensities (powers) had been added, as we did in the first place. As we derived in (8.9.6), the Rayleigh attenuation coefficient is proportional to N.

A rigorous treatment of the role of density fluctuations in light scattering is a significant problem of statistical physics, first solved by both Smoluchowski and Einstein in the early 1900s. ●

---

[18]This assumes that multiple scattering is negligible, i.e., that the field from any dipole is not scattered by any other dipole. This is a good approximation at the low densities of interest for Rayleigh scattering in the atmosphere.

A less obvious characteristic of skylight is that it is polarized. This effect, which was discovered in 1811 by Arago, is easily observed with polarized sunglasses. The extent of polarization appears to be strongest from directions near $90°$ to the direction of the sun from the observer. It is known that bees are sensitive to the polarization of light and use it for navigation. Human eyes, of course, are not directly sensitive to polarization.

To understand the polarization of light by Rayleigh scattering, consider again the electric field (8.9.1) in the radiation zone of an oscillating dipole. If we observe the scattered field at right angles to the plane defined by the directions of polarization and propagation, we see from (8.9.1) that it will be polarized in the x direction, since

$$(\hat{\mathbf{x}} \cdot \hat{\mathbf{r}})\hat{\mathbf{r}} - \hat{\mathbf{x}} = (\hat{\mathbf{x}} \cdot \hat{\mathbf{y}})\hat{\mathbf{y}} - \hat{\mathbf{x}} = -\hat{\mathbf{x}} \tag{8.9.14}$$

when $\hat{\mathbf{r}} = \hat{\mathbf{y}}$ (Fig. 8.9). If instead the incident field inducing the field is polarized in the y direction, there is no scattered field in the y direction since

$$(\hat{\mathbf{y}} \cdot \hat{\mathbf{r}})\hat{\mathbf{r}} - \hat{\mathbf{y}} = (\hat{\mathbf{y}} \cdot \hat{\mathbf{y}})\hat{\mathbf{y}} - \hat{\mathbf{y}} = 0. \tag{8.9.15}$$

The direction of the dipole moment induced by an unpolarized wave propagating in the z direction (Fig. 8.9) will be rapidly varying in the xy plane. (Recall the discussion in Section 5.12.) Equations (8.9.14) and (8.9.15) show that the dipole radiates in the y direction only when its oscillation has a nonzero component in the x direction (Fig. 8.9), and in that case the radiation is polarized in the x direction. Rayleigh scattering thus produces polarized light. This explains the polarization of skylight produced by scattering of the (unpolarized) light from the sun.

The theory of light scattering becomes much more complicated when the particle dimensions are not negligible compared to the wavelength. In this case the light scattered at $90°$ is not completely polarized. And as the particle size increases, the scattering cross section becomes less sensitive to the wavelength; the radiation scattered from white light becomes "whiter" as the particle size increases. This explains why cirrus clouds, consisting of water droplets suspended in air, are white.

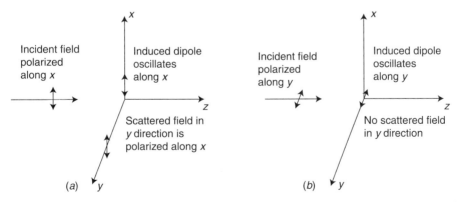

Figure 8.9   An incident field propagating in the z direction with polarization (*a*) along x, in which case the field scattered in the y direction is also polarized along x, and (*b*) along y, in which case there is no scattering in the y direction.

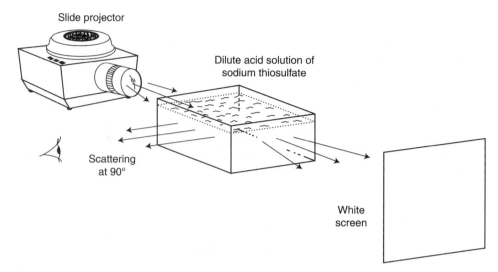

Slide projector

Dilute acid solution of
sodium thiosulfate

Scattering
at 90°

White
screen

Figure 8.10   Experiment demonstrating the effect of particle size in light scattering. In the initial stage of precipitation the sulfur particles suspended in sodium thiosulfate solution are small compared to an optical wavelength. The tank takes on a blue color, the light scattered at 90° is strongly polarized, and on the white screen one observes the "red sunset." After a few minutes the particles have grown larger than a wavelength of visible light. Then the tank has a cloudy appearance and the light scattered at 90° is no longer strongly polarized.

•   All these features of light scattering may be observed beautifully in a simple experiment that the reader can perform with readily available materials (Fig. 8.10). A few spoonfuls of photographic fixing powder (sodium thiosulfate) are dissolved in a small tank of water. The addition of about 100 mL of dilute sulfuric acid causes small grains of sulfur to precipitate out of solution after a few minutes. In the initial stages of the precipitation these grains are very small and the scattered light has a faintly bluish hue (the "blue sky"). The scattered light viewed at 90° (Fig. 8.10) is observed with a pair of polarized sunglasses to be strongly polarized. The light transmitted through the tank has a yellowish and eventually a strongly reddish hue (the "sunset"). After several more minutes the scattered light is not blue but white ("clouds"), and it is no longer strongly polarized. At this stage we are observing light scattered from sulfur grains that have grown to a size comparable to or larger than optical wavelengths.

Such an experiment was performed by John Tyndall in 1869. Rayleigh (1881) found that the "hypo" (sodium thiosulfate) solution demonstrates the scattering effects especially well.

It should be borne in mind, of course, that the notion of "color" is a subjective one, and that there is not a strictly one-to-one correspondence between wavelength and color: Different combinations of wavelengths can produce the same perceived color while certain colors, such as brown, are not associated with any single wavelength. The theory of color vision has attracted the attention of many physicists, including Newton, Young, Helmholtz, Maxwell, Rayleigh, and Land.[19] Even now, however, color vision is not completely understood. Although it would be inappropriate for us to explore the subject here, it may be worthwhile to mention that in dim light we see nearly in "black and white" (Section 12.7). This explains, for example, why we do not see the brilliant colors of the Crab nebula that are apparent in photographs taken with long exposure times. Another example is the night sky itself. It looks black instead of blue because of its low intensity; a long-exposure photograph shows it to be "really" blue.   •

[19]See E. H. Land, "Experiments in Color Vision," *Scientific American* **200**, May 1959, p. 84.

8.10   ATMOSPHERIC TURBULENCE

A laser beam propagating in air undergoes scattering, absorption, and diffractive spreading. But there is another aspect of propagation in air that we have not yet considered: *atmospheric turbulence* causes the refractive index to fluctuate randomly. These fluctuations are responsible for the twinkling of starlight and for the limited resolution of ground-based telescopes. They can also cause a laser beam to spread more than would be expected from diffraction in vacuum, and to "wander" and "scintillate." To describe these effects we must have a theory for the statistics of the refractive index fluctuations.

A useful measure of these fluctuations is the refractive index *structure function*,

$$\mathcal{D}_n(\mathbf{r}_1, \mathbf{r}_2) = \langle [n(\mathbf{r}_1) - n(\mathbf{r}_2)]^2 \rangle = \langle [\tilde{n}(\mathbf{r}_1) - \tilde{n}(\mathbf{r}_2)]^2 \rangle, \qquad (8.10.1)$$

where $\langle .. \rangle$ denotes an averaging over the fluctuations, $\mathbf{r}_1$ and $\mathbf{r}_2$ are two points in space, and

$$\tilde{n}(\mathbf{r}) = n(\mathbf{r}) - \langle n(\mathbf{r}) \rangle \qquad (8.10.2)$$

is the deviation of $n(\mathbf{r})$ from its average value. Another important quantity is the *covariance*:

$$B_n(\mathbf{r}_1, \mathbf{r}_2) = \langle \tilde{n}(\mathbf{r}_1)\tilde{n}(\mathbf{r}_2) \rangle = \langle n(\mathbf{r}_1)n(\mathbf{r}_2) \rangle - \langle n(\mathbf{r}_1) \rangle \langle n(\mathbf{r}_2) \rangle. \qquad (8.10.3)$$

We note for later use that

$$\mathcal{D}_n(\mathbf{r}_1, \mathbf{r}_2) = B_n(\mathbf{r}_1, \mathbf{r}_1) + B_n(\mathbf{r}_2, \mathbf{r}_2) - 2B_n(\mathbf{r}_1, \mathbf{r}_2). \qquad (8.10.4)$$

The refractive index for air is given approximately by the Cauchy formula [Eq. (3.14.9)]

$$n(\lambda) - 1 = \left(7.76 \times 10^{-5} + \frac{0.584}{\lambda^2} \right) \frac{P}{T}, \qquad (8.10.5)$$

where λ is the wavelength (in nanometers) and P and T are the pressure (in millibars) and the temperature (in kelvins), respectively. Fluctuations in pressure and temperature cause the refractive index to fluctuate. For our purposes we may assume that pressure fluctuations are quickly washed out by sound (pressure) waves, whereas temperature fluctuations are equilibrated by conduction and persist much longer. The refractive index fluctuations are therefore due mainly to temperature fluctuations; from (8.10.5),

$$\tilde{n}(\mathbf{r}_1) = \Delta n \cong -7.9 \times 10^{-5} P \frac{\Delta T}{T^2} = -\left[7.9 \times 10^{-5} \frac{P}{T^2} \right] \tilde{T} \qquad (8.10.6)$$

for $\lambda = 550$ nm. With this expression we can relate the refractive index structure function $\mathcal{D}_n(\mathbf{r}_1, \mathbf{r}_2)$ to the temperature structure function $\mathcal{D}_T(\mathbf{r}_1, \mathbf{r}_2) = \langle [\tilde{T}(\mathbf{r}_1) - \tilde{T}(\mathbf{r}_2)]^2 \rangle$:

$$\mathcal{D}_n(\mathbf{r}_1, \mathbf{r}_2) = \left[7.9 \times 10^{-5} \frac{P}{T} \right]^2 \mathcal{D}_T(\mathbf{r}_1, \mathbf{r}_2). \tag{8.10.7}$$

One goal of the theory of atmospheric turbulence is to derive formulas for structure functions such as $\mathcal{D}_T(\mathbf{r}_1, \mathbf{r}_2)$.

Atmospheric turbulence refers to the irregular, apparently random fluctuations in space and time of the speed and direction of air currents. It is responsible for the shimmering appearance of distant objects down a road on a hot day and for the diffusion of smoke, water vapor, and aerosols. Turbulence in liquids is observed, for example, in the form of swirls or "eddies" around rocks in the otherwise steady flow of a stream. Such swirls are also observed in winds (e.g., the swirling motion of small pieces of paper on the ground) or in sketches by Leonardo da Vinci, who was among the first to study turbulence and who named it. Blood flow in arteries is an example of turbulent flow. The turbulence experienced on airplanes depends upon the size of the dominant eddies; eddies smaller than the size of the plane make for a choppy ride, while larger eddies produce sudden upward and downward jolts, or climbing and dropping in the case of the largest eddies.

Turbulence requires a source of energy. In the case of atmospheric turbulence this source of energy is solar radiation. Temperature inhomogeneities result from the different degrees of solar heating of different portions of Earth's surface. The scale of the inhomogeneities is reduced by wind and convection, and on a sufficiently small "outer scale" L_0 the atmosphere under "fully developed turbulence" may be regarded on average as a mixture of eddies of all scale sizes. Larger eddies break up into smaller eddies until an "inner scale" ℓ_0 is reached at which eddies are stable against further breakup. This occurs when the rate of kinetic energy transfer to smaller eddies equals the energy dissipation rate due to viscosity and heating. Earth's atmosphere is typically a good example of such fully developed turbulence. In the troposphere $\ell_0 \sim 1-10$ mm, the smallest values occurring near ground, while $L_0 \sim 1-100$ m, the smallest values again occurring near ground. These numbers vary considerably with atmospheric conditions, terrain, altitude, and other factors.

Turbulence is much too complicated to formulate in a rigorous, "first-principles" way, and in fact it is not very well understood. The theory relies in part on phenomenology and scaling arguments, and one must generally invoke various simplifying assumptions, which may apply only approximately to the "real world," in order to make progress. One such assumption is that the turbulence is *homogeneous*, so that $\mathcal{D}_T(\mathbf{r}_1, \mathbf{r}_2)$ depends only on the difference $\mathbf{r}_1 - \mathbf{r}_2$. Another assumption is that the turbulence is *isotropic*, meaning that $\mathcal{D}_T$ does not depend on the direction of $\mathbf{r}_1 - \mathbf{r}_2$. The assumptions of homogeneous and isotropic turbulence imply that $\mathcal{D}_T(\mathbf{r}_1 - \mathbf{r}_2)$ depends only on $r = |\mathbf{r}_1 - \mathbf{r}_2|$: $\mathcal{D}_T(\mathbf{r}_1, \mathbf{r}_2) = \mathcal{D}_T(r)$.

The most widely applied theory of turbulence is based on the work of the Russian mathematician A. N. Kolmogorov in the 1940s. The most important result of the theory for our purposes is that the temperature structure function $\mathcal{D}_T(r)$ scales as $r^{2/3}$ for $\ell_0 < r < L_0$:

$$\mathcal{D}_T(r) = C_T^2 r^{2/3} \qquad (\ell_0 < r < L_0), \tag{8.10.8}$$

and, therefore, according to (8.10.7), so does the refractive index structure function:

$$\mathcal{D}_n(r) = \left[7.9 \times 10^{-5} \frac{P}{T}\right]^2 C_T^2 r^{2/3} = C_n^2 r^{2/3} \qquad (\ell_0 < r < L_0), \qquad (8.10.9)$$

where C_n^2 is called the refractive index *structure constant*.

- The theory of atmospheric turbulence is replete with powers of $\frac{2}{3}, \frac{5}{3}, \frac{11}{3}, \ldots$ originating in scaling arguments. Consider, for example, the scaling of velocity v with scale length ℓ. The rate of change of kinetic energy per unit mass per unit time is proportional to $v^2/(\ell/v)$. Kinetic energy transfer to smaller and smaller scales continues until this rate equals the energy dissipation rate ε: $v^3/\ell \propto \varepsilon$, or $v^2 \propto \ell^{2/3}$. In fact, according to the Kolmogorov theory, the velocity structure function $\mathcal{D}_v(r) = C_v^2 r^{2/3}$ for $\ell_0 < r < L_0$.

Kolmogorov obtained a "power spectral density" $\Phi_n(\mathbf{K})$ that gives an average distribution of eddy dimensions L_x, L_y, and L_z, where $L_x = 2\pi/K_x$, $L_y = 2\pi/K_y$, and $L_z = 2\pi/K_z$: $\Phi_n(\mathbf{K}) = 0.033 C_n^2 K^{-11/3}$ for $\ell_0 \to 0$ and $L_0 \to \infty$. For our purposes $\Phi_n(\mathbf{K})$ may be defined as the Fourier transform of the refractive index covariance (8.10.3):

$$B_n(\mathbf{r}) = \int \Phi_n(\mathbf{K}) e^{-i\mathbf{K}\cdot\mathbf{r}} d^3 K. \qquad (8.10.10)$$

Various modifications of this formula have been proposed for finite values of ℓ_0 and L_0, the most commonly used one being the von Karman form,

$$\Phi_n(\mathbf{K}) = \frac{0.033 C_n^2 e^{-(K\ell_0)^2}}{[K^2 + 1/L_0^2]^{11/6}}. \qquad (8.10.11)$$

Atmospheric turbulence may be thought of in terms of eddies having a *continuous* distribution of sizes determined by $\Phi_n(\mathbf{K})$. Our considerations presume that the wavelength λ of light propagating in the turbulent atmosphere is much smaller than the inner scale ℓ_0 of turbulence. This makes the results inapplicable to propagation through clouds, which are said to be examples of "turbid" media.

Experimental tests of the theory of atmospheric turbulence leading to the expression (8.10.9) for the refractive index structure function are complicated by the variablity of atmospheric conditions, but the theory is fairly well supported by experiment. The temperature structure constant C_T^2 has been measured by attaching thermometers to tiny balloons; such measurements, together with the relation between C_T^2 and C_n^2 given by (8.10.9), allow the numerical value of C_n^2 to be inferred. It depends on atmospheric conditions and altitude as well as the time of year and the time of day. Near ground level $C_n^2 \approx 10^{-17}$ m$^{-2/3}$ when atmospheric turbulence is weak, whereas $C_n^2 \approx 10^{-13}$ m$^{-2/3}$ under conditions of strong turbulence. Measurements indicate that C_n^2 is typically smallest an hour or so before sunrise, a preferred time for hot air balloon launches, and after sunset, and largest around noon on clear days. It decreases rapidly with altitude, and at about 3 km it is typically about 1000 times smaller than it is near ground level.

8.11 THE COHERENCE DIAMETER

To include the effect on propagation of a refractive index that fluctuates randomly in time is a very complicated problem. The difficulty of the subject is compounded by the variety of assumptions and approximations in the literature. Our principal goal in this section is to introduce the coherence diameter, one of the most important concepts in the theory of optical propagation in the turbulent atmosphere. To this end we will simplify the theory as much as seems possible without losing sight of the basic physics.

We will begin by considering the simpler situation in which the refractive index varies in space but not in time; even this is difficult to treat in general. There are, of course, special cases, such as a lens or an optical fiber, where the effects of spatial variations of the refractive index are well understood and obviously beneficial. But arbitrary and uncontrolled variations of the refractive index are generally detrimental because, among other things, they limit the focusing and imaging capabilities of optical systems.

Consider the propagation in a medium with refractive index $n(x, y, z)$ of a monochromatic wave with complex electric field amplitude $\mathcal{E}(x, y, 0)$ in a plane $z = 0$. For simplicity we consider a single polarization component and relate the field at $z > 0$ to the field at $z = 0$ by the Fraunhofer formula (7.10.8):

$$\mathcal{E}(x, y, z) = -\frac{ie^{ikz}}{\lambda z} e^{ik(x^2+y^2)/2z} \iint \mathcal{E}(x', y', 0) e^{-ik(xx'+yy')/z} \, dx' \, dy', \qquad (8.11.1)$$

where $k = 2\pi/\lambda$ and the integration is over an aperture centered at $x = y = 0$ in the plane $z = 0$. This equation describes propagation from $z = 0$ *in vacuum*, whereas we are interested in propagation in a medium with refractive index $n(x, y, z)$. To make the general problem tractable, various approximations must be invoked, one of the simplest of which is to replace (8.11.1) by

$$\mathcal{E}(x, y, z) = -i\frac{1}{\lambda z} e^{ik(x^2+y^2)/2z} \iint \mathcal{E}(x', y', 0) e^{-ik(xx'+yy')/z} e^{i\phi(x',y',z)} \, dx' \, dy' \qquad (8.11.2)$$

for $n \neq 1$, where

$$\phi(x', y', z) = k \int_0^z n(x', y', z') \, dz'. \qquad (8.11.3)$$

In this "phase screen approximation" each point (x', y') on the wave in the plane $z = 0$ is simply multiplied by the phase factor $\exp[i\phi(x', y', z)]$. Since the direction of ray propagation is normal to surfaces of constant phase, this approximation ignores the bending of rays and diffraction effects associated with it. Obviously, this approximation will be a poor one if the variations of $\mathcal{E}(x', y', 0)$ are sufficiently great and if z is sufficiently large; conditions for the accuracy of the phase screen approximation can be derived, but we will just assume here that z is sufficiently small that (8.11.2) is accurate enough for our purposes. In numerical computations one can model the propagation using a sequence of phase screens, each for a propagation distance small enough that (8.11.2) is accurate. For simplicity we consider only a single phase screen. We also resort to a more compact notation, replacing (8.11.2) and (8.11.3) by

$$\mathcal{E}(\mathbf{R}, z) = -i\frac{1}{\lambda z} e^{ikR^2/2z} \int \mathcal{E}(\mathbf{R}', 0) e^{-ik\mathbf{R} \cdot \mathbf{R}'/z} e^{i\phi(\mathbf{R}',z)} \, d^2R' \qquad (8.11.4)$$

and

$$\phi(\mathbf{R}', z) = k \int_0^z n(\mathbf{R}', z) \, dz'. \qquad (8.11.5)$$

Suppose the field has its maximum intensity at $\mathbf{R} = 0$ in the plane $z = $ constant. This maximum intensity is proportional to

$$|\mathcal{E}(0, z)|^2 = \frac{1}{\lambda^2 z^2} \left| \int \mathcal{E}(\mathbf{R}', 0) e^{i\phi(\mathbf{R}', z)} d^2 R' \right|^2. \tag{8.11.6}$$

If the phase is sufficiently small that we can accurately replace $\exp(i\phi)$ by $1 + i\phi - \phi^2/2$, then

$$|\mathcal{E}(0, z)|^2 \cong \frac{1}{\lambda^2 z^2} \left| \int \mathcal{E}(\mathbf{R}', 0) \left[1 + i\phi(\mathbf{R}', z) - \frac{1}{2} \phi^2(\mathbf{R}', z) \right] d^2 R' \right|^2$$

$$\cong \frac{1}{\lambda^2 z^2} \int \int \mathcal{E}(\mathbf{R}', 0) \mathcal{E}^*(\mathbf{R}'', 0) \left(1 - \frac{1}{2} [\phi(\mathbf{R}', z) - \phi(\mathbf{R}'', z)]^2 \right) d^2 R' \, d^2 R'', \tag{8.11.7}$$

and if we take $\mathcal{E}(\mathbf{R}, 0) = $ const. $\equiv C$ over the aperture in the plane $z = 0$, as in the case of an incident plane wave, then

$$|\mathcal{E}(0, z)|^2 \cong \frac{|C|^2}{\lambda^2 z^2} \left| \int \left[1 + i\phi(\mathbf{R}', z) - \frac{1}{2} \phi^2(\mathbf{R}', z) \right] d^2 R' \right|^2$$

$$\cong \frac{|C|^2 S^2}{\lambda^2 z^2} [1 - (\overline{\phi^2} - \overline{\phi}^2)] = \frac{|C|^2 S^2}{\lambda^2 z^2} [1 - (\Delta\phi)^2_{\mathrm{rms}}], \tag{8.11.8}$$

where S is the aperture area in the plane $z = 0$, and we define the average of ϕ over the aperture as

$$\overline{\phi} = \frac{1}{S} \int \phi(\mathbf{R}', z) \, d^2 R' \tag{8.11.9}$$

and likewise the average of the square of ϕ over the aperture as

$$\overline{\phi^2} = \frac{1}{S} \int \phi^2(\mathbf{R}', z) \, d^2 R'. \tag{8.11.10}$$

The Strehl ratio is the ratio of the peak intensity defined by (8.11.6) to the peak intensity in the "diffraction-limited" case in which there are no phase variations (Section 7.13). It is a simple and convenient measure of the departure from the ideal situation in which only diffraction acts to limit the peak intensity. In our example, and with our small-phase approximation, the Strehl ratio (SR) is simply[20]

$$\mathrm{SR} \cong 1 - (\Delta\phi)^2_{\mathrm{rms}} = 1 - \left(\frac{2\pi}{\lambda} \right)^2 (\Delta\Phi)^2_{\mathrm{rms}}, \tag{8.11.11}$$

where $\Phi(x', y', z) = (\lambda/2\pi)\phi(x', y', z) = \int_0^z n(x', y', z') \, dz'$ is the optical path length. Thus, a $(\Delta\Phi)_{\mathrm{rms}}$ of only $\lambda/10$ produces a Strehl ratio of 0.6. That is, root-mean-square variations of the optical path length on the order of only a tenth of a wavelength

---

[20]A more widely used form of the Strehl ratio is $\exp[-(\Delta\phi)^2_{\mathrm{rms}}]$, which reduces to (8.11.11) when $(\Delta\phi)^2_{\mathrm{rms}} \ll 1$.

cause the peak intensity to drop to 60% of the diffraction-limited peak intensity—*relatively small phase variations significantly reduce the peak intensity*. The diminution of peak intensity implies, from energy conservation, that the beam is spread out by the phase variations. Since the Fraunhofer formula we have used describes the field in the focal plane of a lens (Problem 7.12) as well as the free-space propagation to the far field, it is clear that the spreading out of intensity due to refractive index variations can result in image blurring. In this sense the Strehl ratio is a useful, quantitative measure of image resolution.

In the case of the atmosphere, the *random fluctuations in time* of the refractive index will likewise be detrimental to imaging with a telescope or to the focusing of a laser beam to a small spot. If the refractive index undergoes rapid fluctuations, so will the phase ϕ, and the *observed* peak intensity will be proportional to the average, $\langle |\mathcal{E}(0, z)|^2 \rangle$, over the phase fluctuations. In the approximation that the phase is small we have, from (8.11.7),

$$\langle |\mathcal{E}(0, z)|^2 \rangle = \frac{1}{\lambda^2 z^2} \int \int \mathcal{E}(\mathbf{R}', 0)\mathcal{E}^*(\mathbf{R}'', 0)\left(1 - \frac{1}{2}\langle [\phi(\mathbf{R}', z) - \phi(\mathbf{R}'', z)]^2 \rangle \right) d^2R' \, d^2R''.$$

$$(8.11.12)$$

The phase structure function $\mathcal{D}_s(\mathbf{r}', \mathbf{r}'') = \langle [\phi(\mathbf{R}', z) - \phi(\mathbf{R}'', z)]^2 \rangle = \langle [\phi(\mathbf{r}') - \phi(\mathbf{r}'')]^2 \rangle$, like the refractive index structure function, is generally assumed to depend only on the difference $|\mathbf{r}' - \mathbf{r}''|$. Using this assumption, we define

$$\mathcal{D}_s(|\mathbf{R}' - \mathbf{R}''|) = \langle [\phi(\mathbf{R}', z) - \phi(\mathbf{R}'', z)]^2 \rangle \qquad (8.11.13)$$

and write (8.11.12) as

$$\langle |\mathcal{E}(0, z)|^2 \rangle = \frac{1}{\lambda^2 z^2} \int \int \mathcal{E}(\mathbf{R}', 0)\mathcal{E}^*(\mathbf{R}'', 0)\left[1 - \frac{1}{2}\mathcal{D}_s(|\mathbf{R}' - \mathbf{R}''|)\right] d^2R' \, d^2R''. \quad (8.11.14)$$

Another generally assumed property of the phase fluctuations is that they are governed by Gaussian statistics, that is, that the probability distribution of the phase is Gaussian. With this assumption it can be shown (Problem 8.15) that (8.11.14) is an approximation, for small phase fluctuations, to the more general expression

$$\langle |\mathcal{E}(0, z)|^2 \rangle = \frac{1}{\lambda^2 z^2} \int \int \mathcal{E}(\mathbf{R}', 0)\mathcal{E}^*(\mathbf{R}'', 0)e^{-(1/2)\mathcal{D}_s(|\mathbf{R}'-\mathbf{R}''|)} \, d^2R' \, d^2R''. \qquad (8.11.15)$$

Defining $\mathbf{K} = (\mathbf{R}' - \mathbf{R}'')/\lambda z$, we can write this equivalently as[21]

$$\langle |\mathcal{E}(0, z)|^2 \rangle = \int \int \mathcal{E}(\mathbf{R} + \lambda z\mathbf{K}, 0)\mathcal{E}^*(\mathbf{R}, 0)e^{-(1/2)\mathcal{D}_s(\lambda z K)} \, d^2R \, d^2K. \qquad (8.11.16)$$

---

[21]The introduction here of *spatial* frequencies (K_x, K_y) leads to expressions such as (8.11.26) that are identical to those obtained from a much more general approach based on the optical transfer function (OTF) of an optical imaging system. For the theory and application of the OTF to atmospheric turbulence and other optical problems requiring a statistical description see, for example, J. W. Goodman, *Statistical Optics*, Wiley, New York, 2000. See also the remarks near the end of this section.

• The justification for the assumption of Gaussian statistics comes from the central limit theorem, which, loosely speaking, says that the probability distribution for the sum of a large number of independent random variables with arbitrary probability distributions is Gaussian. Equation (8.11.3) expresses the phase $\phi(x', y', z)$ as the sum of a large number of random variables that, because they are associated with different parts of the medium, may be presumed to be approximately independent.

The refractive index in the atmosphere fluctuates randomly in *time*, whereas the probability distributions and the averages $\langle \ldots \rangle$ we work with refer to an *ensemble* of possible values of the refractive index (or phase). Implicit in our discussion is the assumption of *ergodicity* which, again speaking loosely, means that the average over the ensemble of possible values of a fluctuating quantity $x(t)$ is equivalent to an average over time of that quantity. A necessary but not sufficient condition for ergodicity of a "random process" $x(t)$ is *stationarity* in the sense that the ensemble averages $\langle x(t) \rangle$ and $\langle x(t)x(t + \tau) \rangle$ do not depend on t, or in other words these averages are independent of the origin of time. For the Gaussian random processes of interest here, this "wide-sense" stationarity is often, but not always, sufficient to guarantee ergodicity, which also requires that correlations of the random process decay rapidly enough in time.[22]

Another implicit assumption in our discussion is that the "exposure time" is long enough that the phase varies in time over a substantial portion of its ensemble of possible values. Since, as discussed below, the relevant time scale for atmospheric fluctuations is typically on the order of a millisecond, it is generally assumed that exposure times much greater than about 10 ms are "long" in this sense. Such exposure times are usually realized when photographs are taken of faint astronomical objects, for example. In the case of a short exposure time, in which case the atmospheric fluctuations are approximately "frozen," such a photograph is typically a speckled interference pattern, whereas a long exposure time washes out the speckles. •

Our considerations thus far reveal the importance of the phase structure function: Subject to our assumptions and approximations, $\mathcal{D}_s(r)$ fully characterizes the phase fluctuations. We will now show how $\mathcal{D}_s(r)$ can be related to the refractive index structure function $\mathcal{D}_n(r)$, the form of which was discussed in Section 8.10.

From (8.10.2), (8.11.5), and (8.11.13) it follows that

$$\mathcal{D}_s(|\mathbf{R}' - \mathbf{R}''|) = k^2 \left\langle \left(\int_0^z [\tilde{n}(\mathbf{R}', z') - \tilde{n}(\mathbf{R}'', z')]\, dz' \right)^2 \right\rangle, \qquad (8.11.17)$$

or, since this depends only on the distance r between the points $(\mathbf{R}', z')$ and $(\mathbf{R}'', z')$,

$$\mathcal{D}_s(r) = k^2 \int_0^z \int_0^z \left[2B_n(z' - z'') - 2B_n\left(\sqrt{(z' - z'')^2 + r^2} \right) \right] dz'\, dz'', \qquad (8.11.18)$$

where r is the radial variable along the direction perpendicular to the direction (z) of propagation and B_n is the refractive index covariance defined by (8.10.3) and having the properties

$$B_n(z' - z'') = \langle \tilde{n}(r, z')\tilde{n}(r, z'') \rangle = \langle \tilde{n}(0, z')\tilde{n}(0, z'') \rangle,$$

$$B_n\left(\sqrt{(z' - z'')^2 + r^2} \right) = \langle \tilde{n}(r, z')\tilde{n}(0, z'') \rangle = \langle \tilde{n}(r, z'')\tilde{n}(0, z') \rangle. \qquad (8.11.19)$$

---

[22]For further discussion of stationary and ergodic random processes see, for instance, L. Mandel and E. Wolf, *Optical Coherence and Quantum Optics*, Cambridge University Press, New York, 1995, Chapter 2.

The identity

$$2B_n(z' - z'') - 2B_n\left(\sqrt{(z' - z'')^2 + r^2}\right) = \left[2B_n(0) - 2B_n\left(\sqrt{(z' - z'')^2 + r^2}\right)\right]$$
$$- [2B_n(0) - 2B_n(z' - z'')] \qquad (8.11.20)$$

and the relation (8.10.4) imply that

$$2B_n(z' - z'') - 2B_n\left(\sqrt{(z' - z'')^2 + r^2}\right) = D_n\left(\sqrt{(z' - z'')^2 + r^2}\right) - D_n(z' - z''), \quad (8.11.21)$$

which allows us to rewrite Eq. (8.11.18) in terms of the refractive index structure function:

$$D_s(r) = k^2 \int_0^z \int_0^z \left[D_n\left(\sqrt{(z' - z'')^2 + r^2}\right) - D_n(z' - z'')\right] dz' dz''. \qquad (8.11.22)$$

Next we use the formula (8.10.9) for the refractive index structure function to evaluate the integral (8.11.22). We omit the details and just state the result:

$$D_s(r) = k^2 C_n^2 z r^{5/3} \int_{-\infty}^{\infty} [(x^2 + 1)^{1/3} - x^{2/3}] dx = 2.91 k^2 C_n^2 z r^{5/3}. \qquad (8.11.23)$$

The *coherence diameter* r_0 is defined by writing

$$D_s(r) = 6.88 \left(\frac{r}{r_0}\right)^{5/3}. \qquad (8.11.24)$$

The reason for the factor 6.88 will become clear shortly. r_0 is given explicitly by the formula

$$r_0 = \left[\frac{6.88}{2.91(4\pi^2)}\right]^{3/5} \left(\frac{\lambda^2}{C_n^2 z}\right)^{3/5} = 0.185 \left(\frac{\lambda^2}{C_n^2 z}\right)^{3/5}, \qquad (8.11.25)$$

and its physical significance is brought out by the following considerations.

With the phase structure function (8.11.24), Eq. (8.11.16) for the "on-axis" peak intensity becomes

$$\langle |\mathcal{E}(0, z)|^2 \rangle = \iint \mathcal{E}(\mathbf{R} + \lambda z \mathbf{K}, 0) \mathcal{E}^*(\mathbf{R}, 0) e^{-3.44(\lambda z K / r_0)^{5/3}} d^2 R \, d^2 K. \qquad (8.11.26)$$

Suppose the field $\mathcal{E}(\mathbf{R}, 0)$ in the plane $z = 0$ is a constant C over a circular aperture of diameter D centered at $\mathbf{R} = 0$. Then the integral over R is just the overlap area of two circles of diameter D with centers separated by $\lambda z K$, and can be evaluated using simple geometrical considerations:

$$\int \mathcal{E}(\mathbf{R} + \lambda z \mathbf{K}, 0) \mathcal{E}^*(\mathbf{R}, 0) d^2 R = |C|^2 \frac{D^2}{2} \left[\cos^{-1}\left(\frac{\lambda z K}{D}\right) - \left(\frac{\lambda z K}{D}\right)\left(1 - \sqrt{\frac{\lambda z K}{D}}\right)\right]$$
$$(8.11.27)$$

for $\lambda z K / D \leq 1$ and 0 otherwise. Therefore, since $\int (\ldots) d^2 K = 2\pi \int (\ldots) K \, dK$ when the integrand $(\ldots)$ depends only on the magnitude K of $\mathbf{K}$, Eq. (8.11.26) becomes

$$\langle |\mathcal{E}(0, z)|^2 \rangle = |C|^2 \pi D^2 \int_0^{D/\lambda z} \left[\cos^{-1}\left(\frac{\lambda z K}{D} \right) - \left(\frac{\lambda z K}{D} \right) \left(1 - \sqrt{\frac{\lambda z K}{D}} \right) \right]$$

$$\times e^{-3.44(\lambda z K / r_0)^{5/3}} K \, dK$$

$$= \pi |C|^2 \frac{D^4}{\lambda^2 z^2} \int_0^1 \left[\cos^{-1} u - u\sqrt{1 - u^2} \right] e^{-3.44(Du/r_0)^{5/3}} u \, du, \qquad (8.11.28)$$

where we have introduced the dimensionless integration variable $u = \lambda z K / D$.

Equation (8.11.28) gives the peak intensity and, like the Strehl ratio, serves as a measure of the achievable "resolution" at z. We define a dimensionless resolution $\mathcal{R}$ by dividing $\langle |\mathcal{E}(0, z)|^2 \rangle$ by $(\pi D^2 / 4)(|C|^2 / \lambda z)$:[23]

$$\mathcal{R} = \frac{4D^2}{\lambda z} \int_0^1 \left[\cos^{-1} u - u\sqrt{1 - u^2} \right] e^{-3.44(Du/r_0)^{5/3}} u \, du. \qquad (8.11.29)$$

The resolution so defined increases with the aperture diameter D, simply because a larger illuminated aperture results in a larger peak intensity in the observation plane. The maximum value of $\mathcal{R}$ is realized for $D \to \infty$:

$$\mathcal{R}_{\text{max}} = \frac{\pi}{4} \left(\frac{r_0}{\lambda z} \right)^2, \qquad (8.11.30)$$

as shown below. Thus,

$$\frac{\mathcal{R}}{\mathcal{R}_{\text{max}}} = \frac{16}{\pi} \left(\frac{D}{r_0} \right)^2 \int_0^1 \left[\cos^{-1} u - u\sqrt{1 - u^2} \right] e^{-3.44(Du/r_0)^{5/3}} u \, du. \qquad (8.11.31)$$

- Let $x = \alpha u^{5/3}$, $\alpha = 3.44(D/r_0)^{5/3}$. Then

$$\mathcal{R} = \frac{4D^2}{\lambda z} \alpha^{-6/5} \frac{3}{5} \int_0^\alpha \left[\cos^{-1}\left(\frac{x}{\alpha} \right)^{3/5} - \left(\frac{x}{\alpha} \right)^{3/5} \sqrt{1 - \left(\frac{x}{\alpha} \right)^{6/5}} \right] x^{1/5} e^{-x} \, dx. \qquad (8.11.32)$$

As $D \to \infty$, $\alpha \to \infty$ and the integral becomes $\int_0^\infty \cos^{-1}(0) x^{1/5} e^{-x} \, dx = (\pi/2)(0.918)$. Thus,

$$\mathcal{R}_{\text{max}} = \lim_{D \to \infty} \mathcal{R} = \frac{4D^2}{\lambda z} \left(\frac{1}{3.44} \right)^{6/5} \left(\frac{r_0}{D} \right)^2 \frac{3}{5} \frac{\pi}{2} (0.918) = \frac{\pi}{4} \frac{r_0^2}{\lambda z}. \qquad (8.11.33)$$

The simple multiplier $\pi/4$ results from the choice of the factor $6.88 \, (= 2 \times 3.44)$ used in the definition (8.11.25) of r_0. •

---

[23]The normalization used to define $\mathcal{R}$ has no particular significance and is done in accordance with the definition of $\mathcal{R}$ employed in the first paper in which the coherence diameter r_0 associated with atmospheric turbulence was introduced: D. L. Fried, *Journal of the Optical Society of America* **56**, 1372 (1966). Equations (8.11.29) and (8.11.31) are Eqs. (5.11) and (5.15), respectively, of Fried's paper.

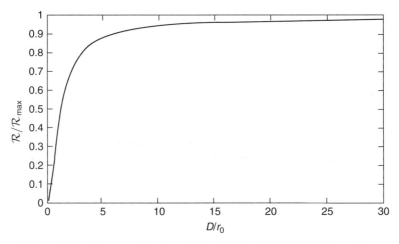

Figure 8.11 The normalized resolution $\mathcal{R}/\mathcal{R}_{\mathrm{max}}$ [Eq. (8.11.31)] vs. D/r_0.

In Fig. 8.11 $\mathcal{R}/\mathcal{R}_{\mathrm{max}}$ is plotted vs. D/r_0. For small values of D/r_0 the resolution increases rapidly with increasing aperture diameter D, but as D/r_0 is made larger, the resolution slowly approaches its maximum value given by (8.11.30): *increasing D much beyond r_0 does not significantly improve resolution*. In other words, an imaging system with an aperture size much larger than r_0 will not have significantly greater resolution than a system with aperture size r_0.

Consider the example of a laser beam with wavelength λ and initial diameter $\sim D$. If it propagates a large distance z in vacuum the peak intensity at z will be proportional to D^2, and the spot size will be proportional to $\lambda z/D$ (Chapter 7). If it propagates in the atmosphere with coherence diameter $r_0 \ll D$, however, the peak intensity will be proportional to r_0^2 [Eq. (8.11.30)], and the spot size will be proportional to $\lambda z/r_0$, that is, the peak intensity is smaller and the spot size larger than when the beam propagates the same distance in vacuum.

Our analysis based on the Fraunhofer formula also applies to the case of a lens or focusing mirror of diameter D. A large telescope with primary mirror diameter $D \gg r_0$, for example, will have no better resolution than a diffraction-limited telescope of diameter r_0. As discussed below, $r_0 \approx 20$ cm under excellent atmospheric seeing conditions (weak turbulence). Thus, a large ground-based telescope, in spite of its greater light-gathering ability, will have no greater resolution than a good 8-inch telescope.

Another measure of resolution or imaging quality is the angular resolution or *seeing* angle θ, smaller values of θ implying better seeing conditions. For a small diffraction-limited telescope of diameter D, $\theta \sim \lambda/D$. (Recall from Section 7.11 that, according to the Rayleigh criterion, two points are "just resolved" by a lens of diameter D if their angular separation is $1.22\lambda/D$.) For a large telescope whose resolution is limited by atmospheric turbulence, however, the seeing angle is $\theta \sim \lambda/r_0$. Ground-based telescopes at the best sites (i.e., the mountain-top sites where C_n^2 is small and atmospheric turbulence is least detrimental) have a seeing angle $\theta \sim (550 \text{ nm})/(10 \text{ cm}) \sim 1$ arcsec in the visible under good seeing conditions ($r_0 \sim 10$ cm).[24] For the 2.4-m Hubble space telescope, by

[24]1 degree = 60 arcminutes = 3600 arcseconds, 1 arcsec = (1/3600) degree = 4.85×10^{-6} radian. The moon subtends an angle of about $\frac{1}{2}$ degree as seen from Earth, while the width of a thumb at a distance of 3 miles subtends an angle of about 1 arcsec.

contrast, $\theta \sim (550\,\mathrm{nm})/(2.4\,\mathrm{m}) = 0.05$ arcsec. If the effects of atmospheric turbulence could be eliminated, a 10-m ground-based telescope could ideally have an angular resolution $\theta \sim (550\,\mathrm{nm})/(10\,\mathrm{m}) = 0.01$ arcsec in the visible. Techniques for compensating for the effects of atmospheric turbulence are discussed in Section 14.2.

Since r_0 varies with wavelength approximately as $\lambda^{6/5}$, the effects of turbulence are greater at smaller wavelengths. In particular, the seeing angle (λ/r_0) due to turbulence varies as $\lambda^{-1/5}$, implying that seeing is better at larger wavelengths for a given level of turbulence (C_n^2). The $\lambda^{-1/5}$ dependence of seeing has been verified experimentally,[25] as have various other scaling relations predicted by the Kolmogorov theory of turbulence.

As discussed at the end of the preceding section, C_n^2 varies considerably, and in particular it decreases rapidly with altitude, so a more appropriate definition than (8.11.25) of r_0 is

$$r_0 = 0.185 \left[\frac{\lambda^2}{\int_0^L C_n^2(z)\,dz} \right]^{3/5}, \qquad (8.11.34)$$

where the integration is over the propagation path of total length L. The integral of C_n^2 from a mountain-top observatory upward is such that $r_0 \approx 20$ cm when the seeing is excellent, whereas $r_0 \approx 5$ cm under poor seeing conditions. In the case of a horizontal line of sight along which C_n^2 is assumed to have the constant value 10^{-15} m$^{-2/3}$ near ground, r_0 is calculated from (8.11.25) to be 5 cm for $\lambda = 550$ nm and $z = 2.7$ km.

From (8.11.24) and the definition of the phase structure function $\mathcal{D}_s(r)$, it is clear that r_0 is the distance over which the phase fluctuations are well correlated. At any given time the transverse phase profile of a wave can be imagined to be divided into patches of size r_0, which depends on C_n^2, the wavelength, and the distance of propagation through turbulence. These patches move about because of local air motion. A *wavefront coherence time* τ_c is defined as r_0/V, where V is a mean wind velocity (typically ~ 10 m/s). Typical values of τ_c are in the $1-10$ ms range. Adaptive-optical systems that improve the image resolution of telescopes by correcting for turbulence-induced phase distortions must, therefore, have response times shorter than about a millisecond, or bandwidths greater than about 1 kHz (Section 14.2).

- The derivation of Eqs. (8.11.23)–(8.11.25) is based on the phase screen approximation (8.11.2). However, these equations can be shown by a more lengthy analysis to be accurate when the bending of rays and diffraction effects are properly accounted for, and when amplitude fluctuations arising from the refractive index fluctuations are also taken into account; in other words, their accuracy is not restricted to short propagation paths. The analysis shows that $\mathcal{D}_s(r)$ as given by (8.11.23) and (8.11.24) is actually the sum of phase and amplitude structure functions.

We have also assumed that the incident field at $z = 0$ is uniform over a circular aperture. In the case of noncircular apertures, or fields that are not uniform over an aperture, Eq. (8.11.27) must be modified accordingly.

In order to bring out some of the important features of propagation through turbulence, and to introduce the coherence diameter r_0, we have restricted our considerations to the on-axis $(\mathbf{R} = 0)$ intensity. More generally we can define the Fourier transform

$$\tau(\mathbf{K}) = \int \langle |\mathcal{E}(\mathbf{R}, z)|^2 \rangle e^{2\pi i \mathbf{K} \cdot \mathbf{R}}\, d^2 R, \qquad (8.11.35)$$

[25]R. W. Boyd, *Journal of the Optical Society of America* **68**, 877 (1978).

in terms of which $\langle|\mathcal{E}(\mathbf{R}, z)|^2\rangle$ is given by the inverse Fourier transform

$$\langle|\mathcal{E}(\mathbf{R}, z)|^2\rangle = \int \tau(\mathbf{K})e^{-2\pi i\mathbf{K}\cdot\mathbf{R}} \, d^2K. \tag{8.11.36}$$

If $\mathcal{E}(\mathbf{R}, 0) = A(\mathbf{R}) \exp i\phi(\mathbf{R})$, and if we treat phase fluctuations as described in this section, then it can be shown straightforwardly, using the Fraunhofer formula to express $\mathcal{E}(\mathbf{R}, z)$ in terms of $\mathcal{E}(\mathbf{R}, 0)$, that

$$\tau(\mathbf{K}) = \left[\int A(\mathbf{R} + \lambda z\mathbf{K}, 0)A^*(\mathbf{R}, 0)d^2R\right] e^{-(1/2)\mathcal{D}_s(\lambda zK)} \equiv \tau_0(\mathbf{K})e^{-(1/2)\mathcal{D}_s(\lambda zK)}. \tag{8.11.37}$$

Knowing $\tau(\mathbf{K})$, therefore, we can use (8.11.36) to calculate the intensity over the entire observation plane at z, not just the on-axis intensity which is proportional to $\langle|\mathcal{E}(0, z)|^2\rangle$ $= \int \tau(\mathbf{K})d^2K$. $\tau(\mathbf{K})$ is closely related to the *optical transfer function* (OTF), or more precisely the average OTF defined in this context as

$$\overline{\mathcal{H}}(\mathbf{K}) = \frac{\int \langle|\mathcal{E}(\mathbf{R}, z)|^2\rangle e^{2\pi i\mathbf{K}\cdot\mathbf{R}} \, d^2R}{\int \langle|\mathcal{E}(\mathbf{R}, z)|^2\rangle \, d^2R}. \tag{8.11.38}$$

As discussed following Eq. (8.11.16), we have assumed a long "exposure time" in which averages are taken over times long compared to the characteristic atmospheric coherence time. Now a general phase aberration can be expressed as a sum of components giving rise to different effects. (This is done formally in terms of so-called Zernike polynomials.) One component, called *tilt*, produces lateral displacements of a laser beam or an image in the focal plane of a lens. In the long exposure limit, these displacements are averaged out and contribute to the observed beam spreading or image blurring. In the case of short exposure times, however, tilt results in only a beam deflection or "wander" and not a beam spreading or image blurring. The calculation of the short-exposure OTF, therefore, requires removal of the tilt component before averaging, with the result that the structure function (8.11.24) in the two limiting cases $z \ll D^2/\lambda$ and $z \gg D^2/\lambda$ is replaced by

$$\mathcal{D}_s(r) = 6.88\left(\frac{r}{r_0}\right)^{5/3}\left[1 - \alpha\left(\frac{r}{D}\right)^{1/3}\right], \tag{8.11.39}$$

where $\alpha = 1$ for $z \ll D^2/\lambda$ and $\alpha = \frac{1}{2}$ for $z \gg D^2/\lambda$. The definition of the coherence diameter r_0 remains the same in the short-exposure case, and calculations show that $\mathcal{R}/\mathcal{R}_{max}$ is essentially the same as for long exposure times for $D/r_0 < \frac{1}{2}$ and $D/r_0 > 100$, while significantly higher resolutions are obtained in the short-exposure case for intermediate values of D/r_0. ●

8.12   BEAM WANDER AND SPREAD

We have discussed atmospheric turbulence in terms of the fluctuations of the *real* refractive index. A field propagating in such a medium does not lose energy. Of course, depending on the wavelength, there may well be absorption and loss of field energy arising from the imaginary part of the refractive index; CO_2 laser radiation, for example, will be absorbed by CO_2 molecules in the atmosphere. There will also be Rayleigh scattering.

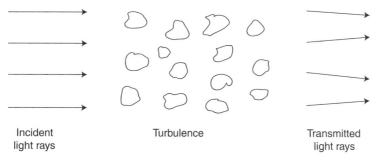

Incident   Turbulence   Transmitted
light rays   light rays

Figure 8.12   A turbulent medium viewed in terms of cells of different sizes and refractive indices.

Because absorption and Rayleigh scattering result mainly in attenuation of a propagating beam, we will ignore them and concentrate on the effects of refractive turbulence.

The discussion in Section 8.10 suggests that we can describe the turbulent atmosphere in terms of eddies or "cells" (Fig. 8.12) having a statistical distribution of diameters and refractive indices. Each eddy can be thought of as a weak lens that is moved erratically about with local wind currents. Eddies of different sizes compared to the diameter of a laser beam will have different effects on the beam, as we now discuss.

One effect of turbulence on a laser beam is *beam wander*, or the random motion of the beam spot over a receiver. This effect of refractive index fluctuations is well known to astronomers as the image dancing or "jitter" in a telescope. Beam wander results from turbulent eddies larger than the beam diameter, as shown in Fig. 8.13. These large eddies cause the beam centroid to deflect without much change in the beam diameter, and the random motions of the eddies result in beam wander.

We can obtain results of more rigorous analyses for the mean-square lateral beam displacement using the following crude argument, which assumes that geometrical optics is adequate. Consider a point on a surface of constant phase of a laser beam of initial diameter D that has propagated a distance z through turbulence. We define an angle α that represents the direction of ray propagation, as in Fig. 8.14. If we ignore the possibility that a ray can bend when it traverses the distance z, then a ray at angle α corresponds to a transverse phase change $\Delta\phi = (k \sin \alpha)D \cong kD\alpha$ for small α, or a root-mean-square (rms) lateral displacement $\langle\alpha^2\rangle z^2 \cong z^2\langle(\Delta\phi)^2\rangle/k^2D^2$. Now, if we take $\Delta\phi$ to be the phase change $\phi(D, z) - \phi(0, z)$ over the full diameter D, we will overestimate the rms lateral beam displacement $\langle\rho^2\rangle$; good agreement with more detailed calculations is obtained if we take $\langle(\Delta\phi)^2\rangle \approx (\frac{1}{3})\langle[\phi(D, z) - \phi(0, z)]^2\rangle$, or

$$\langle\rho^2\rangle \approx \frac{1}{3}\frac{z^2}{k^2D^2}\langle[\phi(D, z) - \phi(0, z)]^2\rangle = \frac{1}{3}\frac{z^2}{k^2D^2}\mathcal{D}_s(D) = \frac{1}{3}\frac{6.88}{k^2D^2}z^2\left(\frac{D}{r_0}\right)^{5/3}, \quad (8.12.1)$$

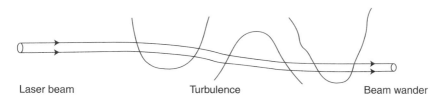

Laser beam   Turbulence   Beam wander

Figure 8.13   Deflection of a laser beam by turbulent eddies large compared to the beam diameter.

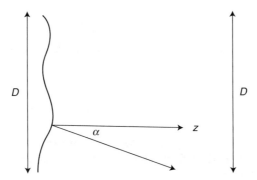

Figure 8.14   At a given point on a surface of constant phase, α is the angle between the normal to the surface and the direction z of beam propagation. According to geometrical optics, α gives the direction of ray propagation.

where we have used Eq. (8.11.24) for the phase structure function. Using the definition (8.11.25) for r_0, we have, for a horizontal propagation path over which C_n^2 may be assumed to be constant,

$$\langle \rho^2 \rangle \approx C_n^2 D^{-1/3} z^3. \tag{8.12.2}$$

Except for numerical factors of order unity, this is the result obtained by more rigorous analyses.[26] Likewise it applies, with $z = f$, to the beam spot jitter in the image plane when a beam is incident on a lens of focal length f before passing through turbulence. In the case of a Gaussian beam, we can use (8.12.2) with $D = 2w_0$, where w_0 is the initial spot size, to estimate the beam wander. Note that, according to (8.12.2), beam wander is (approximately) independent of wavelength.

It is instructive to consider the rms wander of a beam relative to the diameter of the same beam that has propagated the same distance in vacuum. In the near field the beam in vacuum retains approximately its initial diameter D, and the ratio of the rms beam wander to this diameter is

$$\frac{\sqrt{\langle \rho^2 \rangle}}{D} \approx \frac{\sqrt{C_n^2 D^{-1/3} z^3}}{D} = \sqrt{C_n^2 D^{-7/6} z^{3/2}}, \tag{8.12.3}$$

whereas in the far field the beam diameter $D(z)$ in vacuum is $\approx \lambda z / D$ and

$$\frac{\sqrt{\langle \rho^2 \rangle}}{D(z)} \approx \frac{\sqrt{C_n^2 z} D^{5/6}}{\lambda}. \tag{8.12.4}$$

[26]See, for instance, the review by R. L. Fante, *Proceedings of the IEEE* **63**, 1669 (1975). Equations (8.12.2) and (8.12.5) are consistent with Eqs. (40) and (37), respectively, of that article. An introduction to various aspects of the propagation of laser radiation in the atmosphere is given by H. Weichel, *Laser Beam Propagation in the Atmosphere* (SPIE, Bellingham, WA, 1990). See also L. C. Andrews and R. L. Phillips, *Laser Beam Propagation through Random Media*, second edition (SPIE, Bellingham, WA, 2005). As these reviews make clear, atmospheric turbulence is an extremely complicated phenomenon, with many poorly understood features. Caution must be exercised in applying propagation formulas beyond their range of validity.

For $C_n^2 = 10^{-15}$ m$^{-2/3}$, $\lambda = 550$ nm, and $D = 2$ mm, $\sqrt{\langle \rho^2 \rangle}/D(z) < 1$ when $z < 10$ km. This illustrates the point that in many situations the beam wander is relatively small. Even if it is small, however, beam wander can cause a beam to miss a small target or not to stay on target for a sufficiently long time. The simplest solution for beam wander, of course, is to have a sufficiently large receiver or target.

The motion of the large eddies that cause beam deflection results in the beam spot being continuously deflected in the observation plane, the time scale of the jitter being D/V, where V characterizes the flow velocity of the eddies transverse to the direction of propagation. A photograph taken with an exposure time shorter than this time scale would show a single deflected beam spot, and a sequence of such photographs would show that the spot dances randomly about. The spot is also broadened compared to the diffraction-limited spot size, as discussed below. But a single photograph with a *long* exposure time would smear out the deflected spots and show a single spot that has spread compared to a spot observed with a short exposure time. In other words, "short-term" beam wander described by (8.12.2) contributes to the "long-term" *beam spread*.

But there is an additional contribution to the beam spread that does not depend on the exposure time and in particular is responsible for the broadening of a deflected spot seen with a short exposure time. Like beam wander, this contribution to beam spread is well known to astronomers—it causes the spread in the telescope image of a star. In the case of laser beam propagation, it arises from eddies that are small compared to the beam diameter (Fig. 8.15). These smaller eddies introduce phase fluctuations that result in the peak intensity being smaller than its diffraction-limited value and the intensity distribution being broadened. They act in effect as weak lenses with different refractive indices, and their net effect is to limit the lateral distance over which the phase fluctuations on the transmitted wave are well correlated. The characteristic correlation diameter depends on the strength of the turbulence (C_n^2), the wavelength, and the distance of propagation through turbulence. We have already defined and calculated such a diameter, namely the coherence diameter r_0 [Eq. (8.11.25)]. If r_0 is smaller than the beam diameter D, the transverse phase fluctuations are correlated over areas ($\pi r_0^2/4$) smaller than the beam area ($\pi D^2/4$), and consequently we expect that the far-field beam diameter is determined not by D but by r_0. In other words, the far-field peak intensity should be proportional to r_0^2 rather than D^2 and the divergence angle should be $\approx \lambda/r_0$ rather than $\approx \lambda/D$. The beam spread after a propagation distance z should be $\approx \lambda z/r_0$. These

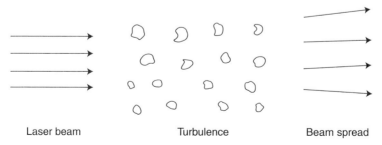

Laser beam   Turbulence   Beam spread

Figure 8.15 Spreading of a laser beam caused by turbulence cells small compared to the beam diameter.

expectations are borne out by the calculation leading to (8.11.31) and the discussion following it.

Detailed calculations for beam spread in various cases of interest where the beam at $z = 0$ is not uniform over a circular aperture of diameter D are in accord with the calculation leading to (8.11.31). In the case of a focused Gaussian beam having its waist at $z = 0$ with initial spot size w_0, for example, the beam spread is calculated to be

$$\langle \rho_L^2 \rangle \approx \frac{\lambda^2 z^2}{2\pi^2} \left(\frac{1}{r_0^2} + \frac{1}{w_0^2} \right) + \frac{w_0^2}{2} \left(1 - \frac{z}{f} \right)^2 \tag{8.12.5}$$

where f is the focal length. $\sqrt{\langle \rho_L^2 \rangle}$ is calculated as the distance from the z axis where the average intensity is reduced by a factor $e^{-1} \cong 0.37$ from its on-axis value. The approximation (8.12.5) is valid for propagation distances $z \ll \lambda^2/[4\pi^2 C_n^2 \ell_0^{5/3}]$, which is the limit of greatest practical interest. For $\lambda = 550$ nm, $C_n^2 = 10^{-15}$ m$^{-2/3}$, and the turbulence inner scale $\ell_0 = 2$ mm, for example, (8.12.5) is valid for propagation distances $z \ll 240$ km. Equation (8.12.5) provides an estimate for the long-exposure beam spread including the contribution from beam wander.

In the focal plane $z = f$, or in the case of far-field propagation, (8.12.5) gives a "beam diameter"

$$2\sqrt{\langle \rho_L^2 \rangle} \approx \frac{\sqrt{2}}{\pi} \frac{\lambda z}{r_0} \approx \frac{1}{2} \frac{\lambda z}{r_0} \tag{8.12.6}$$

for $w_0 \gg r_0$, which is consistent with the discussion following Eq. (8.11.31).

Our description of beam wander and spread is somewhat oversimplified. In particular, the motion of the turbulent eddies or "weak lenses" across a beam can lead, as a consequence of interference among different parts of the beam, to regions of high intensity, or "hot spots," interspersed with regions of low intensity in the observation plane. The pattern of hot spots changes on a time scale dictated by the motion of the eddies, and is typically on the order of milliseconds. This *beam breakup*, which is typically observed under conditions of strong turbulence, results in multiple spots even in the case of short exposure times; a long exposure time blurs these multiple spots into a single patch, and $\sqrt{\langle \rho_L^2 \rangle}$ in this case gives approximately the mean radius of this patch.

8.13   INTENSITY SCINTILLATIONS

The best known example of intensity fluctuations, or scintillations, is the "twinkling" of starlight. In this section we will consider the intensity probability distribution associated with the scintillation of a laser beam (or the light from a star) propagating through the turbulent atmosphere. In Section 13.14 we will discuss photon counting experiments that confirm this "log-normal distribution" for starlight as well as for laser radiation that has propagated in the atmosphere.

Consider the intensity I received at a point in the plane $z = $ constant. The field reaches this point after propagating a distance z through turbulence. We imagine the propagation path through turbulence to be divided into a large number $\mathcal{N}$ of sections of length Δz, where $z = \mathcal{N} \Delta z$ is the total propagation distance. Because of the random deflections caused by the refractive index fluctuations, the intensity I at the observation point fluctuates, and so does the field incident on any point on each of our imaginary sections along the propagation path. Suppose that after propagating through the jth section the intensity along the propagation path to the observation point on the detector changes by a factor T_j. If I_0 is the intensity at $z = 0$, then the intensity after the propagation distance z is

$$I = I_0 \times T_1 T_2 \ldots T_{\mathcal{N}} \tag{8.13.1}$$

at the observation point. Since the logarithm of a product is the sum of the logarithms,

$$\ln \left[\frac{I}{I_0} \right] = \sum_{j=1}^{\mathcal{N}} \ln T_j \equiv \chi \quad \text{and} \quad I = I_0 e^\chi. \tag{8.13.2}$$

It is easy to include the effect of Rayleigh scattering and absorption in the factors T_j, but we will not do so. The T_j's therefore account here only for the fluctuations in the intensity at points along the propagation path to the detector; the total power in the field is conserved, but the intensity at individual points in a plane of constant z fluctuates.

Because of turbulence the T_j fluctuate randomly, and we will take them to be independent random variables, that is, we assume that the effects of the different sections on the intensity are uncorrelated. This is the crucial assumption in our heuristic derivation; with this assumption, χ is the sum of a large number of independent random variables $\ln T_j$ and, therefore, according to the central limit theorem, has a Gaussian probability distribution:

$$P(\chi) = \frac{1}{\sigma_\chi \sqrt{2\pi}} e^{-[\chi - \langle \chi \rangle]^2 / 2\sigma_\chi^2}, \tag{8.13.3}$$

where $\langle \chi \rangle$ is the average value of χ, and $\sigma_\chi^2 = \langle [\chi - \langle \chi \rangle]^2 \rangle$ is the variance; neither $\langle \chi \rangle$ nor σ^2 are specified in our model.

To find the probability distribution of the intensity I, we use the fact that $P(\chi) d\chi = P(I) dI$, that is, the probability of finding χ in the interval $[\chi, \chi + d\chi]$ is the same as the probability $P(I) dI$ of finding the corresponding intensity $I = I(\chi)$ in the interval $[I, I + dI]$. Thus, $P(I) = P(\chi) d\chi / dI = (I_0/I)P(\chi)$, or

$$P(I) = \frac{I_0}{\sigma I \sqrt{2\pi}} e^{-(\chi - \langle \chi \rangle)^2 / 2\sigma^2} \qquad [\chi = \ln (I/I_0)]. \tag{8.13.4}$$

Now

$$\langle e^\chi \rangle = \int_{-\infty}^{\infty} P(\chi) e^\chi \, d\chi = \frac{1}{\sigma \sqrt{2\pi}} \int_{-\infty}^{\infty} e^{-(\chi - \langle \chi \rangle)^2 / 2\sigma^2} e^\chi \, d\chi = e^{\langle \chi \rangle + \sigma^2/2}, \tag{8.13.5}$$

that is,

$$\langle \chi \rangle = \ln\langle e^{\chi} \rangle - \frac{\sigma^2}{2} = \ln\left(\frac{\langle I \rangle}{I_0}\right) - \frac{\sigma^2}{2}. \tag{8.13.6}$$

Using this result in (8.13.4), we have

$$P(I) = \frac{I_0}{\sigma I \sqrt{2\pi}} e^{-[\ln(I/I_0) - \ln(\langle I \rangle/I_0) + \sigma^2/2]^2/2\sigma^2} = \frac{I_0}{\sigma I \sqrt{2\pi}} e^{-[\ln(I/\langle I \rangle) + \sigma^2/2]^2/2\sigma^2}, \tag{8.13.7}$$

where σ^2 is the variance of χ. To relate it to the variance of I we note that $\langle I^2 \rangle = I_0^2 \langle e^{2\chi} \rangle$ and that $\langle e^{2\chi} \rangle = e^{2\langle \chi \rangle + 2\sigma^2}$, which is derived in the same manner as (8.13.5). Then, using (8.13.6), we obtain the "normalized" intensity variance

$$\sigma_I^2 \equiv \frac{\langle I^2 \rangle - \langle I \rangle^2}{\langle I \rangle^2} = \frac{I_0^2}{\langle I \rangle^2}\langle e^{2\chi} \rangle - 1 = \frac{I_0^2}{\langle I \rangle^2} e^{2\ln(\langle I \rangle/I_0) + \sigma^2} - 1 = e^{\sigma^2} - 1. \tag{8.13.8}$$

Note that $\langle I \rangle$ is the mean intensity at the detector, which in general would include the effects of Rayleigh scattering or absorption in the propagation to z. We have ignored these effects, but in fact they do not change the log-normal distribution, even though they obviously play a role in determining what the mean intensity at the detector is. This mean intensity, together with σ_I, fully characterizes the distribution (8.13.7).

The actual theory of intensity scintillations is much more complicated than our heuristic model and, subject to the assumption that the turbulence is sufficiently weak, leads to an intensity probability distribution of the form (8.13.7) with σ^2 related to σ_I^2 by Eq. (8.13.8) and

$$\sigma^2 = 1.23 C_n^2 k^{7/6} z^{11/6} \tag{8.13.9}$$

for plane waves and a horizontal propagation path of length z. A more complicated expression for σ^2 can be derived for the case of a Gaussian beam, for instance. In practice C_n^2 is not known without additional measurements, and *measured* values of $\langle I \rangle$ and σ_I can be used to obtain a fit to the distribution (8.13.7).

The probability distribution (8.13.7) is called the *log-normal distribution*. Figure 8.16 is a plot of $P(I)$ for $C_n^2 = 10^{-15}$ m$^{-2/3}$, $\lambda = 550$ nm, and $z = 2$ km, in which case (8.13.9) and (8.13.8) give $\sigma_I^2 = 0.27$ and $\sigma^2 = 0.24$. Obviously, $P(I)$ is not symmetric about $P(\langle I \rangle)$ but is skewed toward values of I less than $\langle I \rangle$. We can define a *fade probability* as the probability that I will be found to be less than its mean value:

$$P_{\text{fade}} \equiv \int_0^{\langle I \rangle} P(I)\, dI. \tag{8.13.10}$$

For the distribution $P(I)$ of Fig. 8.16 it is found that $P_{\text{fade}} = 0.59$. Similarly, we can define a *surge probability* $P_{\text{surge}} = 1 - P_{\text{fade}} = 0.41$ for the example of Fig. 8.16.

As in our treatment of beam wander, our model for intensity scintillations deals with time averages and does not address the time scale of the fluctuations, which are typically in the 1–10 ms range. It should also be noted that the results of this section apply under

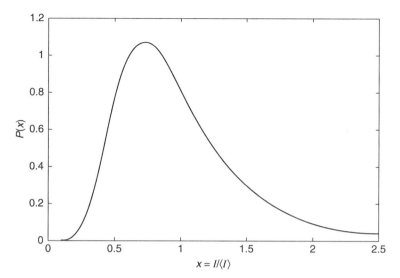

Figure 8.16   The log-normal distribution $P(I)$ [Eq. (8.13.7)] vs. $I/\langle I \rangle$ for $\sigma^2 = 0.24$.

conditions of *weak turbulence*, which in the research literature is conventionally defined by $\sigma^2 < 0.3$. It is important to note, however, that the results of weak turbulence theory are often found both theoretically and experimentally to be fairly accurate also for strong turbulence ($\sigma^2 > 0.3$).

An effect that is not explained by weak turbulence theory is "saturation of scintillations": the measured variance of the intensity can saturate and even decrease with increasing propagation distance (typically >1 km for horizontal propagation paths near ground). This effect is attributed to multiple scattering, that is, the rescattering of light that has already been scattered or deflected by refractive index fluctuations.

Scintillations can be reduced by *aperture averaging*. Our model leading to the log-normal distribution for $P(I)$ applies for a pointlike detector, or in practice to a detector, like the human eye, with a small aperture. More generally we must take into account the fact that the intensity, and therefore the variance of the intensity, varies across the detector. If the detector aperture is large enough, we are effectively averaging over intensity fluctuations that are statistically independent, so that the fluctuations in the measured power are smoothed out. In practice, aperture smoothing is often accounted for by modifying (8.13.8) to read

$$\sigma_I^2 = A(e^{\sigma^2} - 1), \tag{8.13.11}$$

where A is a function of $D^2/\lambda z$ that decreases with increasing detector aperture diameter D.

8.14   REMARKS

The wave equation (8.2.10) is the basis for our understanding of an enormous variety of optical phenomena. We have restricted ourselves in this chapter to the case of linear media in which the polarization density **P** at each field frequency is simply proportional to the electric field at that frequency, that is, to media characterized by a refractive index n that can vary with frequency or with position, or can undergo random fluctuations, but is independent of the electric (or magnetic) field. The theory of Gaussian beams and diffraction in Chapter 7 was based on the simplest case, $n \cong 1$.

The generation and application of pulses of laser radiation require consideration of the variation of the refractive index with frequency. We have introduced two concepts associated with the frequency dependence of the refractive index, namely group velocity and group velocity dispersion, both of which are particularly important for the understanding of ultrashort pulse generation and for fiber-optic communication systems (Section 15.6). Group velocity dispersion can cause laser pulses to broaden in time and can limit the rate at which information can be transmitted by fibers; we have described in this chapter how chirping can be used to avert such broadening. We have discussed the guiding of light by optical fibers as an example of the application of the wave equation to the case where n varies spatially. Group velocity and group velocity dispersion can be understood within the plane-wave approximation to the wave equation, neglecting any spatial variations of n, while essential features of the propagation modes of fibers can be understood without dealing with frequency variations of n.

How laser radiation propagates depends on the wavelength, beam diameter, and duration as well the characteristics of the propagation medium and involves consideration of diffraction, absorption or amplification, scattering, pulse distortion and spreading, and birefringence, among other things. While all these effects are described at once by the wave equation (8.2.10), we have generally discussed each of them separately, based on assumptions and approximations that render other effects negligible. We have not always spelled out in detail the specific assumptions or approximations made; the reader may find it interesting to think about what approximations to Maxwell's equations and the wave equation are being made in our analyses of different propagation effects throughout this book.

It should be evident that the propagation of laser radiation and, of course, electromagnetic radiation in general, is too broad a subject to treat in full detail in a few chapters—or even a few books! Propagation in the turbulent atmosphere, for example, is itself a subject about which books and many research papers continue to be written. In our discussion we ignored attenuation due to Rayleigh scattering or absorption in order to focus attention on concepts such as the refractive index structure function and the coherence diameter that characterize the turbulent medium. In a given application it might well be necessary, depending on the wavelength and the propagation distance, to take attenuation into account. For intense pulses of radiation it might also be necessary to account for the ionization of the air by the laser, or for Raman scattering, self-focusing, and other nonlinear effects, in addition to temporal distortions of the pulse. But the example of propagation in the turbulent atmosphere illustrates how useful it can be to isolate and study a single effect at a time. In Section 14.2, for instance, we will see how the concept of a coherence diameter, which derives from considerations of turbulence alone, is used to design adaptive optical systems that correct for image degradation due to atmospheric turbulence and in particular to improve the resolution of telescopes; this is true even though attenuation and other propagation effects come into play when the light from an astronomical object propagates through Earth's atmosphere.

PROBLEMS

8.1. Is it surprising that Gaussian beams do not satisfy the condition $\nabla \cdot \mathbf{E} = 0$? Show in what sense the non-transverse character is "small."

8.2. (a) The absorption coefficient $a(\nu)$, defined such that the intensity of a beam of radiation in a medium attenuates with distance z of propagation as $\exp[-a(\nu)z]$, is often expressed in terms of the real and imaginary parts (ϵ_R and ϵ_I) of the electric permittivity ϵ. Show that

$$a(\nu) = \frac{2\pi\nu}{c} \frac{\epsilon_I/\epsilon_0}{\sqrt{\epsilon_R/\epsilon_0}}$$

if ϵ_I is sufficiently small compared to ϵ_R, which is very often the case.

(b) Figure 8.17 shows $\kappa_R = \epsilon_R/\epsilon_0$ and $\kappa_I = \epsilon_I/\epsilon_0$ at microwave frequencies for water at various temperatures. Estimate the "penetration depth" $1/a(\nu)$ at 2.45 GHz, the operating frequency of most microwave ovens. Why are microwave ovens not made to operate at the frequency at which ϵ_I is largest for water? [See C. F. Bohren, *American Journal of Physics* **65**, 12 (1997).]

8.3. The relation $v_p v_g = c^2$ between the phase velocity v_p and the group velocity v_g appears in the theory of the propagation of electromagnetic waves in waveguides. Does this relation apply for propagation in a dispersive medium? What dependence of the refractive index on frequency is implied by this relation?

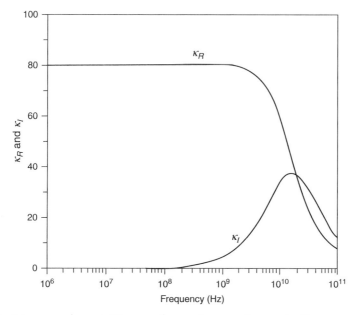

Figure 8.17 (*a*) $\kappa_R = \epsilon_R/\epsilon_0$ and (*b*) $\kappa_I = \epsilon_I/\epsilon_0$ at microwave frequencies for water at 293K. After M. Vollmer, *Physics Education* **39**, 74 (2004).

8.4. The Schrödinger equation for a particle of mass m in a potential $V(x, y)$ in two dimensions is

$$i\hbar \frac{\partial \psi}{\partial t} = -\frac{\hbar^2}{2m}\left(\frac{\partial^2}{\partial x^2} + \frac{\partial^2}{\partial y^2}\right)\psi + V\psi.$$

(a) What assumptions and substitutions make Eq. (8.4.4) for a plane-wave pulse identical to this Schrödinger equation?

(b) Let $\mathcal{E}(\mathbf{r}, \omega) = \mathcal{E}_0(\mathbf{r}, \omega)e^{i\omega z/c}$ in the Helmholtz equation (8.2.20) and assume that $\mathcal{E}_0$ is slowly varying in z compared to $e^{i\omega z/c}$. Show that

$$\left(\frac{\partial^2}{\partial x^2} + \frac{\partial^2}{\partial y^2}\right)\mathcal{E}_0 + 2i\frac{\omega}{c}\frac{\partial \mathcal{E}_0}{\partial z} + \frac{\omega^2}{c^2}(n^2 - 1)\mathcal{E}_0 = 0$$

in the "slowly varying envelope approximation." Is this equation valid if the refractive index n depends on x, y, and z? What substitutions make this equation identical to the Schrödinger equation above? What form of n puts it into the form of the Schrödinger equation for a two-dimensional harmonic oscillator?

8.5. An optical fiber has a dispersion parameter $D = 16$ ps/(km-nm) at a wavelength of 1.55 μm. Assuming an input pulse at this wavelength and with a duration $\tau_p = 100$ ps, estimate the propagation distance in the fiber for which the pulse duration doubles. What is L_{GVD} in this example?

8.6. Derive Eq. (8.4.25).

8.7. Consider a step-index fiber with core radius a and a bending radius R, as shown in Fig. 8.18. A ray associated with a low-order mode of the fiber propagates along the fiber axis when there is no bending, but due to the bending makes an angle of incidence θ at the core–cladding interface. If $\theta < \theta_c$, where θ_c is the critical angle for total internal reflection, then, according to geometrical optics, the mode will be unguided. Show that this condition for the loss of guiding due to bending may be expressed approximately as $R < a/\Delta$, where Δ is defined by (8.6.5).

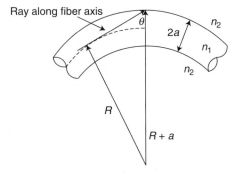

Figure 8.18 Ray associated with a low-order mode of a fiber with core radius a and bending radius R.

8.8. The uniaxial crystal KDP (potassium disodium phosphate) has refractive indices $n_o = 1.512$ and $n_e = 1.470$ at a wavelength of 546 nm. (When $n_e < n_o$, as for KDP, the material is said to be *negative uniaxial*; when $n_e > n_o$, it is called *positive uniaxial*.) Two 546-nm beams, one polarized as an ordinary wave and the other as an extraordinary wave, propagate in the same direction, at an angle of 60° with respect to the optic axis of a KDP crystal.

    (a) What is the refractive index of the extraordinary wave?

    (b) Calculate the walk-off between the two beams after a propagation distance of 1 cm. Would the result change if the crystal were positive uniaxial?

8.9. For an extraordinary wave having an initial beam diameter D and propagating a distance L, what conditions must be met in order that we can ignore diffraction and describe the propagation by Eq. (8.8.33)?

8.10. Referring to Fig. 8.5, we can write the propagation constant β for an optical fiber as $k_0 n_1 \sin\theta$. Show that the condition $\gamma^2 > 0$ for a guided mode is equivalent to the condition $\theta > \theta_c$ for total internal reflection. (Actually, this argument is strictly correct for a slab waveguide but not for a *rounded* optical fiber.)

8.11. An optical fiber is to be manufactured such as to allow a single propagation mode at a wavelength of 1.55 μm. The refractive indices of the core and cladding are 1.50 and 1.49, respectively. What is the maximum allowable core radius?

8.12. A typical He–Ne laser operating at 632.8 nm contains about five times as much He as Ne, with a total pressure of about 1 Torr. Assuming that the length of the gain tube is about 50 cm, estimate the fraction of laser radiation intensity lost due to Rayleigh scattering in passing a billion times through the gain tube. (Note: For Ne at standard temperature and pressure the constants A_1 and B_1 in the Cauchy formula (3.14.9) are $A_1 = 6.66 \times 10^{-5}$ and $B_1 = 2.4 \times 10^{-15}\, m^2$, respectively.) This illustrates the fact that Rayleigh scattering is usually very weak in gas laser media.

8.13. Liquid water has ~ 1000 more molecules per unit volume than air. It is observed, however, that the attenuation coefficient due to Rayleigh scattering for light is only about 200 times greater for water than for air. Why is the attenuation coefficient due to Rayleigh scattering not more like 1000 times greater for water than for air?

8.14. **(a)** What is the magnitude of the refractive index fluctuation corresponding to a temperature fluctuation of 1K in the atmosphere at standard temperature and pressure ($P = 760$ Torr $= 1013.2$ mbar, $T = 293$K)?

    (b) What is the coherence diameter r_0 for 1.06 μm laser radiation propagating a distance of 1 km parallel to the ground? Assume a refractive index structure constant $C_n^2 = 10^{-17}\, m^{-2/3}$.

8.15. **(a)** Using Eq. (8.11.4) and the assumptions leading to (8.11.14), show that

$$\langle |\mathcal{E}(0, z)|^2 \rangle = \frac{1}{\lambda^2 z^2} \int\int \mathcal{E}^*(\mathbf{R}'', 0)\mathcal{E}(\mathbf{R}', 0)\langle e^{i[\phi(\mathbf{R}',z) - \phi(\mathbf{R}'',z)]} \rangle\, d^2R'\, d^2R''.$$

(b) If $\phi(\mathbf{R}, z)$ has a Gaussian probability distribution with zero mean, then so does $\psi \equiv \phi(\mathbf{R}', z) - \phi(\mathbf{R}'', z)$. Show therefore that

$$\langle e^{i\psi} \rangle = e^{-\langle \psi^2 \rangle/2} = e^{-(1/2)\mathcal{D}_s(|\mathbf{R}' - \mathbf{R}''|)},$$

which then implies (8.11.15).

8.16. **(a)** A 10.6-μm Gaussian laser beam with an initial spot size $w_0 = 3$ mm propagates 1 km in air from its waist to a receiver. Assuming that the air is characterized by a C_n^2 of 10^{-15} m$^{-2/3}$, what is the diameter of the receiving aperture required to receive essentially all of the laser power?

(b) Estimate the degree of attenuation due to Rayleigh scattering as the beam propagates to the receiver.

(c) Estimate the normalized intensity variance $(\langle I^2 \rangle - \langle I \rangle^2)/\langle I \rangle^2$ at a point on the receiver.

9 COHERENCE IN ATOM-FIELD INTERACTIONS

9.1 INTRODUCTION

In Chapter 3 we briefly discussed the influence of the relaxation rate denoted β, associated with generalized "frictional" effects of various kinds. Their lifetime $1/\beta$ fell into the range of times $t \gg 1/\beta \gg 1/\omega$, where t indicated a generic time of interest. Throughout that chapter and implicitly in all later discussions until now, this allowed the neglect of terms whose temporal variation was sensitive to "friction" via the exponential damping factor $e^{-\beta t}$. However, in optical physics there are many interesting and important phenomena associated with short pulses that have durations on the order of $1/\beta$ or even much shorter, for which $\beta t \gg 1$ cannot be assumed. These phenomena are not damped by the "frictional" influences in the way we have considered. Consequently, they exhibit a type of "coherence" that has been missing from our discussions in the preceding chapters. We will now begin to examine the consequences of short-pulse interactions, and at the same time develop a clear picture of the meaning of the term coherence in interaction physics.

As a first-stage bonus of a more general treatment of the interaction of laser light with matter we will be able to explain how the loss of coherence is, somewhat counterintuitively, in fact necessary for the validity of the laser rate equations that were semi-intuitively presented in prior chapters. Additional consequences of coherence include such unusual effects as periodic population oscillations, atoms behaving as psuedo-spins and, in a highly absorptive medium, lossless transparency that is transient and "self-induced" by a laser pulse travelling through it.

A fundamental approach to any discussion of optical radiation must be based on Maxwell's equations for the field and on the time-dependent Schrödinger equation for the atoms of the propagation medium. In this chapter the time-dependent Schrödinger equation is used to obtain the "density matrix" equations for the atoms. The rate equations used in the preceding chapters to treat the effects of light on atoms and molecules will be shown to be *approximations* to these density matrix equations, which are the basis for the explanation of coherent and nonlinear propagation effects.

As discussed later, in Chapter 13, the coherence of a radiation field can refer either to its temporal or spatial characteristics. In this chapter we will assume that the field incident on the propagation medium is fully coherent, which for our purposes means simply that the amplitude and phase of the electric field have no random temporal or spatial

Laser Physics. By Peter W. Milonni and Joseph H. Eberly
Copyright © 2010 John Wiley & Sons, Inc.

fluctuations. Such a field exhibits interference effects that cannot be understood in terms of field intensity alone, and requires reference to the electric field.

Just as the radiation must, in general, be characterized by the electric field rather than intensity, the atoms of the propagation medium must, in general, be characterized in terms of probability *amplitudes* for different states rather than the probabilities themselves. Conditions under which this applies are considered in this chapter. Under these conditions the propagation medium is said to have "atomic coherence," and the propagation of light in the medium is said to be coherent in the sense that it is correctly described by equations for the electric field and the density matrix, but not by rate equations for the field intensity and the atomic-level populations.

9.2   TIME-DEPENDENT SCHRÖDINGER EQUATION

Recall that the probability amplitudes $a_n(t)$ were introduced in Section 3.A by writing the time-dependent wave function for an atomic electron as

$$\psi(\mathbf{x},\, t) = \sum_n a_n(t)\phi_n(\mathbf{x}), \tag{9.2.1}$$

where the $\phi_n(\mathbf{x})$ are the (time-independent) wave functions defined by Eq. (3.A.2). In terms of the probability amplitudes $a_n(t)$ the time-dependent Schrödinger equation, $i\hbar\, \partial\psi/\partial t = H\psi$, takes the form

$$i\hbar\, \frac{da_m}{dt} = E_m a_m(t) + \sum_n V_{mn} a_n(t). \tag{9.2.2}$$

We used this equation in the Appendix to Chapter 3 to explain why the classical electron oscillator model so often provides an accurate description of the response of atoms to light. The "orbitals" $\phi_n(\mathbf{x})$ form a complete set in terms of which $\psi(\mathbf{x}, t)$ can be expanded as in (9.2.1). From the orthogonality and normalization properties (3.A.5) and (3.A.6), respectively, of these functions, and the normalization

$$\int_{\text{all space}} \psi^*(\mathbf{x},\, t)\psi(\mathbf{x},\, t)\, d^3x = 1, \tag{9.2.3}$$

it follows from (9.2.1) that

$$\int \left(\sum_m a_m\phi_m\right)^* \left(\sum_n a_n\phi_n\right) d^3x = \sum_m a_m^* a_n \int \phi_m^* \phi_n\, d^3x$$

$$= \sum_n |a_n|^2 = 1. \tag{9.2.4}$$

$|a_n(t)|^2$ is the probability at time t that the atomic electron is in its nth allowed orbital.

There is a significant shift in viewpoint between (9.2.3) and (9.2.4), even though they both express the same normalization. Recall that $|\psi(\mathbf{x}, t)|^2\, d^3x$ is the electron probability assigned to the differential volume element d^3x. There is no reference to orbitals in this assignment, and indeed many or all of the orbitals may make a contribution to the probability within d^3x. On the other hand, $|a_n|^2$ plays the opposite role. It is the probability that the electron is in the nth orbital, without any reference to the spatial location of the electron in that orbital. In laser physics information about orbital number for an atomic electron is usually much more useful than information about its spatial location, because often only a few orbitals have significant occupation probabilities. For this reason the form (9.2.2) of the time-dependent Schrödinger equation is usually much more useful than the (equivalent) partial differential equation $i\hbar \, \partial\psi/\partial t = H\psi$.

• Let us write out the equations for the probability amplitudes a_n in (9.2.2) in order:

$$i\hbar\dot{a}_1 = E_1a_1 + V_{11}a_1 + V_{12}a_2 + V_{13}a_3 + \cdots$$
$$i\hbar\dot{a}_2 = E_2a_2 + V_{21}a_1 + V_{22}a_2 + V_{23}a_3 + \cdots$$
$$i\hbar\dot{a}_3 = E_3a_3 + V_{31}a_1 + V_{32}a_2 + V_{33}a_3 + \cdots$$

$$\vdots \qquad\qquad (9.2.5)$$

They can also be written as a single matrix equation:

$$i\hbar\dot{\boldsymbol{\psi}} = \mathbf{H}\boldsymbol{\psi}, \qquad\qquad (9.2.6)$$

where

$$\boldsymbol{\psi} = \begin{bmatrix} a_1 \\ a_2 \\ a_3 \\ \cdot \\ \cdot \\ \cdot \end{bmatrix} \qquad\qquad (9.2.7)$$

and

$$\mathbf{H} = \begin{bmatrix} E_1 + V_{11} & V_{12} & V_{13} & \cdots \\ V_{21} & E_2 + V_{22} & V_{23} & \cdots \\ V_{31} & V_{32} & E_3 + V_{33} & \cdots \\ \cdot & \cdot & \cdot & \cdots \\ \cdot & \cdot & \cdot & \cdots \\ \cdot & \cdot & \cdot & \cdots \end{bmatrix}. \qquad\qquad (9.2.8)$$

This matrix form of the Schrödinger equation is the origin of the term "matrix element" for V_{nm}. In this form $\mathbf{H}$ is called the *Hamiltonian matrix* and $\boldsymbol{\psi}$ the *state vector*. Werner Heisenberg's original approach to quantum mechanics (1925) was through such matrices. Physical observables were represented by Hermitian matrices with matrix elements satisfying the relation $V_{nm}^* = V_{mn}$. It was not immediately appreciated that Heisenberg's "matrix mechanics" is equivalent to Schrödinger's "wave mechanics."   •

9.3 TWO-STATE ATOMS IN SINUSOIDAL FIELDS

According to Bohr's description of quantum jumps, an atom can increase its energy by jumping from an orbit with energy E to one with higher energy E' if a photon of frequency $\omega = (E' - E)/\hbar$ is simultaneously absorbed. The reverse process is associated with the emission of a photon of frequency ω.

We associate photons of frequency ω with an electromagnetic wave of the same frequency. According to our analysis in Section 3.2, an external electromagnetic field interacts with an electron via the time-dependent potential

$$V(\mathbf{x}, \mathbf{R}, t) = -e\mathbf{x} \cdot \mathbf{E}(\mathbf{R}, t) \tag{9.3.1}$$

in the dipole approximation, where $\mathbf{x}$ is the electron-nuclear distance and $\mathbf{R}$ is the center of atomic mass. We consider now a monochromatic plane wave for $\mathbf{E}$:

$$\mathbf{E}(\mathbf{R}, t) = \tfrac{1}{2}\hat{\boldsymbol{\varepsilon}}E_0 e^{i(\mathbf{k}\cdot\mathbf{R}-\omega t)} + \text{c.c.} \longrightarrow \tfrac{1}{2}\hat{\boldsymbol{\varepsilon}}E_0 e^{-i\omega t} + \text{c.c.}, \tag{9.3.2}$$

where the unit vector $\hat{\boldsymbol{\varepsilon}}$ defines the field polarization, c.c. means complex conjugate, and for convenience $\mathbf{R}$ has been put at the origin. Note that the polarization vector $\hat{\boldsymbol{\varepsilon}}$ can be complex, as in the case of circular polarization; recall Problem 3.4.

The implication of Bohr's rule for quantum jumps is that only pairs of energy levels in the atom that are separated by $\Delta E = \hbar\omega$ are affected by radiation present at frequency ω. We will therefore begin our study by restricting our attention to just two of the electronic energy levels. These are shown in Fig. 9.1 and designated 1 and 2, with energies E_1 and E_2, such that $\Delta E = E_2 - E_1 = \hbar\omega$.

For such a two-state system the expression (9.2.1) is simply

$$\psi(\mathbf{x}, t) = a_1(t)\phi_1(\mathbf{x}) + a_2(t)\phi_2(\mathbf{x}), \tag{9.3.3}$$

and the corresponding Schrödinger equation (9.2.2) reduces to

$$i\hbar\dot{a}_1(t) = E_1 a_1(t) + V_{11}a_1(t) + V_{12}a_2(t), \tag{9.3.4a}$$

$$i\hbar\dot{a}_2(t) = E_2 a_2(t) + V_{21}a_1(t) + V_{22}a_2(t). \tag{9.3.4b}$$

Level 1 may be the ground level but need not be. In most cases of interest the parity selection rule (see Problem 9.1) requires the diagonal matrix elements V_{11} and V_{22} of

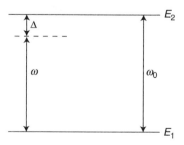

Figure 9.1   Energy levels of a hypothetical "two-state atom." The "detuning" $\Delta = \omega_0 - \omega$ is the difference between $(E_2 - E_1)/\hbar$ and ω. Radiation of angular frequency ω is nearly resonant with the $E_1 \rightarrow E_2$ transition when $\Delta \ll \omega_0 - \omega$.

the atom–field interaction to be zero. Then

$$ i\hbar\dot{a}_1(t) = E_1 a_1(t) + V_{12} a_2(t), \tag{9.3.5a} $$

$$ i\hbar\dot{a}_2(t) = E_2 a_2(t) + V_{21} a_1(t). \tag{9.3.5b} $$

Equations (9.3.5) give the time variation of the probability amplitudes a_1 and a_2 for the two-state system. If the two-state model is a reasonable approximation, we can assume that the atom has negligible probability of being in any state other than ϕ_1 and ϕ_2. In other words, the probability that the atom will be found in one or the other of these two states is unity at any time:

$$ |a_1(t)|^2 + |a_2(t)|^2 = 1. \tag{9.3.6} $$

This is the two-state version of (9.2.4).

Equations (9.3.5) show how the 1-2 and 2-1 matrix elements of V are involved in changes in the amplitudes $a_1(t)$ and $a_2(t)$. From (9.3.1) and (9.3.2) we can express these matrix elements more explicitly as

$$ V_{12}(t) = -e\mathbf{x}_{12} \cdot \tfrac{1}{2}(\hat{\boldsymbol{\varepsilon}} E_0 e^{-i\omega t} + \text{c.c.}), \tag{9.3.7a} $$

$$ V_{21}(t) = -e\mathbf{x}_{21} \cdot \tfrac{1}{2}(\hat{\boldsymbol{\varepsilon}} E_0 e^{-i\omega t} + \text{c.c.}), \tag{9.3.7b} $$

where $\mathbf{x}_{12}$ is the 1-2 matrix element of $\mathbf{x}$ defined by

$$ \mathbf{x}_{12} \equiv \int \phi_1^*(\mathbf{x})\, \mathbf{x}\, \phi_2(\mathbf{x})\, d^3x \tag{9.3.8} $$

and $\mathbf{x}_{21}(=\mathbf{x}_{12}^*)$ is defined by switching the subscripts 1 and 2. Note that $\mathbf{x}_{12}$ is generally a complex-valued vector because the ϕ's may be complex. The numerical value of $\mathbf{x}_{12}$ depends on the wave functions ϕ_1 and ϕ_2, so the size of the matrix elements V_{12} and V_{21} must be expected to vary from atom to atom. As a typical magnitude (associated with an optical transition to or from an atomic ground state), one can expect $|\mathbf{x}_{12}|$ to differ from the Bohr radius $a_0 \approx 0.05$ nm by less than a factor of 10.

With $V_{12}(t)$ and $V_{21}(t)$ given by Eqs. (9.3.7) we can insert them in (9.3.5). It is convenient to adopt several conventions at the same time. We will work with frequencies instead of energies, so we divide through by $\hbar$ and define

$$ \omega_0 = \frac{E_2 - E_1}{\hbar}, \tag{9.3.9} $$

$$ \chi_{21} = e(\mathbf{x}_{21} \cdot \hat{\boldsymbol{\varepsilon}})\frac{E_0}{\hbar}, \tag{9.3.10} $$

$$ \chi_{12} = e(\mathbf{x}_{12} \cdot \hat{\boldsymbol{\varepsilon}})\frac{E_0}{\hbar}. \tag{9.3.11} $$

Also, we now set the arbitrary zero of energy at E_1, so $E_2 \to E_2 - E_1 = \hbar\omega_0$. Then, Eqs. (9.3.5) become

$$ i\dot{a}_1(t) = -\tfrac{1}{2}(\chi_{12}e^{-i\omega t} + \chi_{21}^* e^{i\omega t})a_2, \tag{9.3.12a} $$

$$ i\dot{a}_2(t) = \omega_0 a_2(t) - \tfrac{1}{2}(\chi_{21}e^{-i\omega t} + \chi_{12}^* e^{i\omega t})a_1. \tag{9.3.12b} $$

In the absence of any radiation field ($\chi_{12} = \chi_{21} = 0$) we find $a_1(t) = a_1(0)$ from (9.3.12a) and $a_2(t) = a_2(0) \exp(-i\omega_0 t)$ from (9.3.12b). In the presence of a nearly resonant field (9.3.2) oscillating at frequency $\omega \approx \omega_0$, we adopt similar trial solutions:

$$a_1(t) = c_1(t), \tag{9.3.13a}$$

$$a_2(t) = c_2(t)e^{-i\omega t}, \tag{9.3.13b}$$

and find these equations for $c_1(t)$ and $c_2(t)$:

$$i\dot{c}_1(t) = -\tfrac{1}{2}(\chi_{12}e^{-2i\omega t} + \chi_{21}^*)c_2, \tag{9.3.14a}$$

$$i\dot{c}_2(t) = (\omega_0 - \omega)c_2 - \tfrac{1}{2}(\chi_{21} + \chi_{12}^* e^{2i\omega t})c_1. \tag{9.3.14b}$$

Equations (9.3.14) are more useful because of their isolation of the $\exp(\pm 2i\omega t)$ terms. For optical frequencies ω, these terms oscillate so rapidly compared to every other time variation in the equations that they can be assumed to average to zero over any realistic time interval. In this way it is argued that they can simply be discarded. This is known as the *rotating-wave approximation* (abbreviated RWA in the literature on optical resonance phenomena). An essential ingredient is the near-resonance assumption $\omega \approx \omega_0$, or $\Delta = \omega_0 - \omega \ll \omega_0$, ω, in magnitude. It leads to these elementary working equations:

$$i\dot{c}_1(t) = -\tfrac{1}{2}\chi^* c_2, \tag{9.3.15a}$$

$$i\dot{c}_2(t) = \Delta c_2 - \tfrac{1}{2}\chi c_1, \tag{9.3.15b}$$

where we have dropped the subscript 21 from χ_{21} and have introduced Δ to stand for the atom–field frequency offset, or *detuning*, as already shown in Fig. 9.1:

$$\chi = \chi_{21} = (e\mathbf{x}_{21} \cdot \hat{\boldsymbol{\varepsilon}})\frac{E_0}{\hbar}, \tag{9.3.16}$$

$$\Delta = \omega_0 - \omega. \tag{9.3.17}$$

Evidently χ is the field–atom interaction energy in frequency units. It is known as the *Rabi frequency*, as discussed below.

If $\hat{\boldsymbol{\varepsilon}}E_0$ is a constant vector, χ can be taken to be a purely real number. This can be arranged by the right choice of phases of the wave functions ϕ_1 and ϕ_2 (see Problem 9.2). Unless the context indicates otherwise, we will assume this has been done.

The great advantage of Eqs. (9.3.15) is their relative simplicity. The smallness of the coefficients Δ and χ (compared to ω and ω_0) shows that the c's are "slow" variables (compared to the a's). They contain the essential physics once the rapid oscillations associated with the frequencies ω and ω_0 are removed by the RWA. The solutions for the c's are easily found (see Problem 9.3):

$$c_1(t) = \left(\cos\frac{\Omega t}{2} + i\frac{\Delta}{\Omega}\sin\frac{\Omega t}{2}\right)e^{-i\Delta t/2}, \tag{9.3.18a}$$

$$c_2(t) = \left(i\frac{\chi}{\Omega}\sin\frac{\Omega t}{2}\right)e^{-i\Delta t/2}. \tag{9.3.18b}$$

We have assumed the atom to be in state 1 initially: $c_1(0) = 1$, $c_2(0) = 0$, and we have introduced the generalized Rabi frequency

$$\Omega = (\chi^2 + \Delta^2)^{1/2}, \tag{9.3.19}$$

which reduces to the ordinary Rabi frequency χ at exact resonance ($\Delta = 0$). For this reason χ is sometimes called the resonance Rabi frequency.

The corresponding probabilities $P_1(t) = |a_1(t)|^2$ and $P_2(t) = |a_2(t)|^2$ are

$$P_1(t) = \frac{1}{2}\left[1 + \left(\frac{\Delta}{\Omega}\right)^2\right] + \frac{1}{2}\left(\frac{\chi}{\Omega}\right)^2 \cos \Omega t, \tag{9.3.20a}$$

$$P_2(t) = \frac{1}{2}\left(\frac{\chi}{\Omega}\right)^2 [1 - \cos \Omega t]. \tag{9.3.20b}$$

The justification for defining Ω and χ exactly as we have and calling them (instead of $\Omega/2$ or 2Ω) "the" Rabi frequencies is evident in Fig. 9.2, where $P_2(t)$ is plotted. It is clear that Ω is precisely the frequency at which probability oscillates between levels 1 and 2. It is easy to check that $P_1(t) + P_2(t) = 1$ for all t, so $P_1(t)$ simply oscillates at the same frequency with the opposite phase from $P_2(t)$. It is also easy to see that, for a field detuning Δ large compared to χ, $P_2(t) \cong 0$ and $P_1(t) \cong 1$, that is, the probability is close to 1 that the atom stays in the lower state.

The "Rabi oscillations" shown in Fig. 9.2 depend explicitly on the electric field strength E_0 and cannot be described by population rate equations such as (3.7.11) in which only the field *intensity* appears. Rabi oscillations are a direct consequence of the time-dependent Schrödinger equation, whereas, as already noted, the population rate equations provide an approximate—but often very accurate—description of the

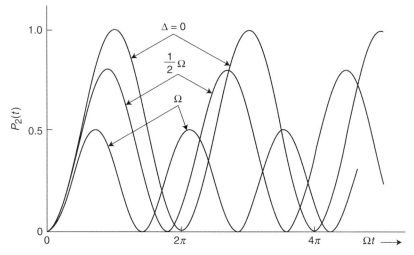

Figure 9.2 Plot of the upper-state probability $P_2(t)$ given by Eq. (9.3.20b). Note that larger detuning Δ corresponds to higher frequency of Rabi oscillations but lower amplitude.

response of a medium to an applied field. This is explained in Section 9.6. Of course we can use Eq. (9.3.16) to relate the Rabi frequency to the intensity (see Problem 9.4):

$$\chi^2 = \frac{e^2 |\mathbf{x}_{21} \cdot \hat{\boldsymbol{\epsilon}}|^2}{\hbar^2} E_0^2 = \frac{2e^2 |\mathbf{x}_{21} \cdot \hat{\boldsymbol{\epsilon}}|^2}{\epsilon_0 \hbar^2 c} \, I. \tag{9.3.21}$$

9.4 DENSITY MATRIX AND COLLISIONAL RELAXATION

The identification of the expectation value $\langle \mathbf{x} \rangle$ of the electron coordinate in the two-state model of an atom allows us to put the classical electron oscillator in its correct perspective, as discussed in Chapter 3. The fact that $\langle \mathbf{x} \rangle$ is determined by $a_1^* a_2$ and $a_1 a_2^*$ [recall Eq. (3.A.16)] suggests, correctly, that these combinations might be more useful than either a_1 or a_2 alone in our development of a general quantum theory of absorption and emission.

We will pursue this approach by obtaining the equations of motion for these combination variables, except that we will focus on the related but simpler quantities c_1 and c_2 defined in (9.3.13). First, we adopt a conventional notation and use the Greek letter ρ (rho) to define

$$\rho_{12} \equiv c_1 c_2^*, \tag{9.4.1a}$$

$$\rho_{21} \equiv c_2 c_1^*, \tag{9.4.1b}$$

$$\rho_{11} \equiv c_1 c_1^* = |c_1|^2, \tag{9.4.1c}$$

$$\rho_{22} \equiv c_2 c_2^* = |c_2|^2. \tag{9.4.1d}$$

The ρ's are elements of the so-called density matrix of the atom, as we explain briefly at the end of this section. However, independent of this terminology, it is clear that ρ_{11} and ρ_{22} are just new ways to write the levels' occupation probabilities. The physical meanings of ρ_{12} and ρ_{21} are related to the electron displacement [cf. Eq. (3.A.16)], so we can think of ρ_{21} as the complex amplitude of the electron's displacement $\mathbf{x}$.

By using Eq. (9.3.15) repeatedly, we can easily derive the following equations for the ρ's (Problem 9.5):

$$\dot{\rho}_{12} = i \Delta \rho_{12} + i \frac{\chi^*}{2} (\rho_{22} - \rho_{11}), \tag{9.4.2a}$$

$$\dot{\rho}_{21} = -i \Delta \rho_{21} - i \frac{\chi}{2} (\rho_{22} - \rho_{11}), \tag{9.4.2b}$$

$$\dot{\rho}_{11} = -\frac{i}{2} (\chi \rho_{12} - \chi^* \rho_{21}), \tag{9.4.2c}$$

$$\dot{\rho}_{22} = \frac{i}{2} (\chi \rho_{12} - \chi^* \rho_{21}). \tag{9.4.2d}$$

The solutions of these equations can be constructed from the solutions for $c_1(t)$ and $c_2(t)$ given in (9.3.18).

However, the equations themselves are not yet in their most useful form. This is because they do not reflect the existence of relaxation, arising from quasi-random processes such as collisions. The same statistical principles employed to treat collisions in Chapter 3 will be used again here. There will be one added complication compared to the classical case, originating with the population variables ρ_{22} and ρ_{11}, which have no classical counterparts.

First, we will concentrate on the electron's complex displacement variable ρ_{21} and on one type of collision, namely purely elastic collisions, which do not affect the populations ρ_{11} and ρ_{22}. If the radiation field present is steady, then $\chi = $ constant. We require the solution for $\rho_{21}(t)$ that vanishes at an earlier time t_1, which (as in Chapter 3) we associate with the atom's most recent collision. We assume that collisions are frequent, so that $t - t_1$ is short enough to neglect changes in $\rho_{22} - \rho_{11}$. The required solution is

$$\rho_{21}(t)_{t_1} = -\frac{\chi(\rho_{22} - \rho_{11})}{2\Delta}(1 - e^{-i\Delta(t - t_1)}). \tag{9.4.3}$$

This can be checked by substitution in (9.4.2b), remembering to hold χ and $\rho_{22} - \rho_{11}$ constant. Next we average this solution over all possible earlier times t_1 at which a collision might have occurred, using the expression [recall (3.8.7)]

$$df(t; t_1) = \gamma_c e^{-\gamma_c(t - t_1)} dt_1 \tag{9.4.4}$$

for the probability that a collision occurs in the time dt_1, γ_c being the collision rate. The result for this average is

$$\bar{\rho}_{21}(t) = -\frac{\gamma_c}{2\Delta}\chi(\rho_{22} - \rho_{11})\int_{-\infty}^{t} e^{-\gamma_c(t - t_1)}(1 - e^{-i\Delta(t - t_1)})\, dt_1$$

$$= -\frac{\chi}{2}(\rho_{22} - \rho_{11})\frac{1}{\Delta - i\gamma_c}. \tag{9.4.5}$$

This same result, obtained by a collision average, can also be reached by a simple modification of the original equation of motion. It can be checked (Problem 9.6) that collisions are already included if we rewrite Eqs. (9.4.2a) and (9.4.2b) for ρ_{21} and ρ_{12} as follows:

$$\dot{\rho}_{12} = -(\gamma_c - i\Delta)\rho_{12} + i\frac{\chi^*}{2}(\rho_{22} - \rho_{11}), \tag{9.4.6a}$$

$$\dot{\rho}_{21} = -(\gamma_c + i\Delta)\rho_{21} - i\frac{\chi}{2}(\rho_{22} - \rho_{11}). \tag{9.4.6b}$$

As in the classical electron oscillator model, we cannot apply these equations any longer to an individual atom. Instead they represent an "average" atom in the sense of the collision average in (9.4.5). For notational convenience we have omitted the overbar indicating the collision average.

Note that Eqs. (9.4.6) can be read as if the average atom's ρ_{12} and ρ_{21} variables undergo change for two reasons. That is, we can interpret (9.4.6b) as the result of adding two independent rates of change:

$$\dot{\rho}_{21} = (\dot{\rho}_{21})_{\text{elastic collisions}} + (\dot{\rho}_{21})_{\text{Schr. equation}}, \tag{9.4.7}$$

where

$$(\dot{\rho}_{21})_{\text{elastic collisions}} = -\gamma_c \rho_{21} \qquad (9.4.8a)$$

and

$$(\dot{\rho}_{21})_{\text{Schr. equation}} = -i\,\Delta\rho_{21} - i\frac{\chi}{2}(\rho_{22} - \rho_{11}). \qquad (9.4.8b)$$

Such an interpretation will be helpful in dealing with the effect of collisions on the level populations ρ_{22} and ρ_{11}, for which there are no classical analogs.

The elastic collision rate γ_c appearing in (9.4.6) is often referred to as the atomic dipole's "decoherence" (or dephasing) rate, as discussed for the classical electron oscillator model in Section 3.2, following Eq. (3.8.12); in the quantum mechanical description, elastic collisions are responsible for a damping of the average electron displacement (or atomic dipole moment). However, inelastic collisions, in which an electron can change its energy level, can also occur.

To account for inelastic collisions, we simply assert that their effect is to knock populations out of levels 1 and 2 into other unspecified levels of the atom at the fixed rates Γ_1 and Γ_2. At the same time we can include the effect of spontaneous photon emission as a special type of "collision" that transfers population between the two specified levels, from 2 to 1. Following Einstein's notation we will denote the spontaneous emission rate by A_{21}. Then we write, in analogy to (9.4.7),

$$\dot{\rho}_{22} = (\dot{\rho}_{22})_{\text{collisions}} + (\dot{\rho}_{22})_{\text{spont. emission}} + (\dot{\rho}_{22})_{\text{Schr. equation}}, \qquad (9.4.9)$$

where

$$(\dot{\rho}_{22})_{\text{collisions}} = -\Gamma_2 \rho_{22}, \qquad (9.4.10a)$$

$$(\dot{\rho}_{22})_{\text{spont. emission}} = -A_{21}\rho_{22}, \qquad (9.4.10b)$$

$$(\dot{\rho}_{22})_{\text{Schr. equation}} = \frac{i}{2}(\chi\rho_{12} - \chi^*\rho_{21}). \qquad (9.4.10c)$$

In a similar vein we write the separate contributions to $\dot{\rho}_{11}$:

$$(\dot{\rho}_{11})_{\text{collisions}} = -\Gamma_1 \rho_{11}, \qquad (9.4.11a)$$

$$(\dot{\rho}_{11})_{\text{spont. emission}} = +A_{21}\rho_{22}, \qquad (9.4.11b)$$

$$(\dot{\rho}_{11})_{\text{Schr. equation}} = -\frac{i}{2}(\chi\rho_{12} - \chi^*\rho_{21}). \qquad (9.4.11c)$$

Note that the contribution from spontaneous emission to $\dot{\rho}_{11}$ is positive, and just equal to the negative contribution to $\dot{\rho}_{22}$, on the assumption that the atom makes a jump from level 2 to level 1 while emitting a photon spontaneously.

As a result of these contributions from collisions and spontaneous emission, we obtain the following equations for the level populations:

$$\dot{\rho}_{11} = -\Gamma_1 \rho_{11} + A_{21}\rho_{22} - \frac{i}{2}(\chi \rho_{12} - \chi^* \rho_{21}), \tag{9.4.12a}$$

$$\dot{\rho}_{22} = -(\Gamma_2 + A_{21})\rho_{22} + \frac{i}{2}(\chi \rho_{12} - \chi^* \rho_{21}). \tag{9.4.12b}$$

Again, for notational convenience, we do not include overbars indicating collision averages. However, because of the collisions, these equations apply only in an average sense to the atoms under consideration.

Finally, we must return to the elastic-collision-averaged ρ_{12} and ρ_{21} equations. What is the effect of inelastic collisions on them? A simple answer is based on the obvious relation $|\rho_{12}| = (\rho_{11}\rho_{22})^{1/2}$, which holds as a direct consequence of the definitions (9.4.1). This relation says that the effect of collisions on the *magnitude* of ρ_{12}, as distinct from the effect on its *phase*, is directly related to the effect on the level populations in a specific way. That is, if inelastic collisions alone cause ρ_{11} and ρ_{22} to decay, i.e.,

$$\rho_{11}(t)|_{\text{collisions}} = \rho_{11}(0)e^{-\Gamma_1 t}, \tag{9.4.13a}$$

$$\rho_{22}(t)|_{\text{collisions}} = \rho_{22}(0)e^{-\Gamma_2 t}, \tag{9.4.13b}$$

which are the solutions to (9.4.10a) and (9.4.11a), then inelastic collisions alone cause $|\rho_{12}(t)|$ to decay as

$$|\rho_{12}(t)| = [\rho_{11}(t)\rho_{22}(t)]^{1/2} = \{\rho_{11}(0)\rho_{22}(0) \exp[-(\Gamma_1 + \Gamma_2)t]\}^{1/2}$$
$$= |\rho_{12}(0)| \exp\left(-\frac{\Gamma_1 + \Gamma_2}{2}t\right). \tag{9.4.14}$$

In words, the effect on ρ_{12} of inelastic collisions alone is to add an extra decay rate to the elastic collision decay rate γ_c. This added rate is just $(\Gamma_1 + \Gamma_2)/2$, one-half the sum of the population decay rates for ρ_{11} and ρ_{22}.

Thus, we write our final equations for ρ_{12} and ρ_{21} averaged over both elastic and inelastic collisions (and including spontaneous emission) in the form

$$\dot{\rho}_{12} = -(\beta - i\Delta)\rho_{12} + i\frac{\chi^*}{2}(\rho_{22} - \rho_{11}), \tag{9.4.15a}$$

$$\dot{\rho}_{21} = -(\beta + i\Delta)\rho_{21} - i\frac{\chi}{2}(\rho_{22} - \rho_{11}), \tag{9.4.15b}$$

where β is the total relaxation rate:

$$\beta = \gamma_c + \tfrac{1}{2}(\Gamma_1 + \Gamma_2 + A_{21}). \tag{9.4.16}$$

Only the γ_c term in β refers to elastic ("soft" or "dephasing") collisions, but it is often dominant. It is usually likely that an atom suffers many distant soft dephasing collisions

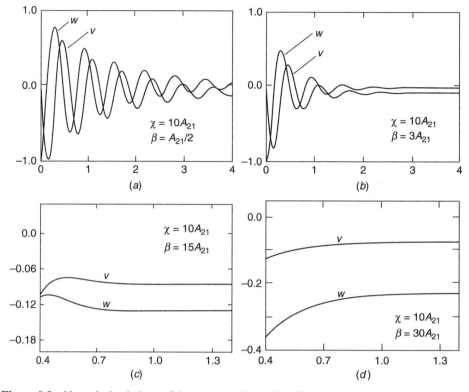

Figure 9.3 Numerical solutions of the v, w equations (9.4.19) for a range of collisional damping rates. Note scale changes. Times is in units of A_{21}^{-1}.

for every close collision that is hard enough to cause population changes. Thus, to a good approximation in many cases,

$$\beta \cong \gamma_c \gg \tfrac{1}{2}(\Gamma_1 + \Gamma_2 + A_{21}). \tag{9.4.17}$$

To a surprising degree, laser action of the usual kind depends very strongly on this inequality. We will require (9.4.17) in the following section.

The effects of collisional dephasing relaxation can be illustrated in detail by integrating the coupled equations for the ρ's (see Problem 9.7). In Fig. 9.3 we show the solutions for a wide range of parameters. We have chosen a special case that is free of complications. We take $\Gamma_1 = \Gamma_2 = 0$ (no transfer of probability to levels other than 1 and 2), and we take $\Delta = 0$ (exact resonance). Since $\Gamma_1 = \Gamma_2 = 0$, we have $d\rho_{11}/dt + d\rho_{22}/dt = 0$. Thus, $\rho_{11} + \rho_{22} = 1$ (conservation of probability), and it is enough to determine either ρ_{11} or ρ_{22}. Actually, it is most convenient to deal with the inversion $\rho_{22} - \rho_{11}$ since it enters Eqs. (9.4.15) naturally. Furthermore, Eqs. (9.4.15) show that at exact resonance $\rho_{21} + \rho_{12}$ is coupled only to itself (for real χ) and plays no role in the dynamics, so we can pay attention solely to the difference, $\rho_{12} - \rho_{21}$, which in any event is the variable that couples directly to ρ_{11} and ρ_{22}, as Eqs. (9.4.12) make clear.

Thus, we can focus on the two real variables:

$$v = i(\rho_{21} - \rho_{12}), \tag{9.4.18a}$$

$$w = \rho_{22} - \rho_{11}, \tag{9.4.18b}$$

which obey the equations (at resonance, and in the absence of the Γ_1 and Γ_2 collisions and for real χ)

$$\dot{v} = -\beta v + \chi w, \tag{9.4.19a}$$

$$\dot{w} = -A_{21}(1 + w) - \chi v. \tag{9.4.19b}$$

These equations follow directly from (9.4.12) and (9.4.15) and the definitions (9.4.18). They are discussed further in the following section. Of course, from (9.4.16) and the absence of Γ_1 and Γ_2, we have $\beta = \gamma_c + A_{21}/2$.

The solutions shown in Fig. 9.3 are chosen to illustrate the influence of elastic collisions. As the elastic collision rate γ_c increases from zero, the damping parameter β also increases and the oscillatory ("coherent") response of the atom to the applied radiation changes to nonoscillatory ("incoherent") relaxation. Note the changes in the scale needed in the figure to make evident the different types of response.

• The notation used for the ρ's suggests that they are the elements of a 2×2 matrix:

$$\mathbf{\rho} = \begin{bmatrix} \rho_{11} & \rho_{12} \\ \rho_{21} & \rho_{22} \end{bmatrix}. \tag{9.4.20}$$

This is indeed the case, and quantum statistical mechanics is devoted in large part to the study of such matrices. They were introduced into quantum theory independently by L. D. Landau and J. von Neumann before 1930. For historical reasons $\mathbf{\rho}$ is called the *density matrix* of the system, and in this case $\mathbf{\rho}$ is the density matrix of a two-state atom.

The density matrix is a generalization of a related 2×2 matrix:

$$\begin{bmatrix} c_1 c_1^* & c_1 c_2^* \\ c_2 c_1^* & c_2 c_2^* \end{bmatrix}, \tag{9.4.21}$$

and the two are occasionally confused. Note that they are *not* the same matrix, despite the original definition in (9.4.1): $\rho_{11} = c_1 c_1^*$, $\rho_{12} = c_1 c_2^*$, and so on. This is because the ρ's are now understood to refer to collision averages of $c_1 c_1^*$, etc. Thus, Eqs. (9.4.12) and (9.4.15) for the elements of the density matrix cannot be obtained from simpler equations for c_1 and c_2 separately. [The reader is challenged to try to construct equations for c_1 and c_2 that can be used to obtain (9.4.12) and (9.4.15).] This is the most important sense in which the cc^* combinations are more physical than c's and c^*'s alone.

The existence of the matrix (9.4.20) establishes a definite meaning to the terms "diagonal" and "off-diagonal." Obviously, ρ_{11} and ρ_{22} are the elements on the diagonal, and ρ_{12} and ρ_{21} are the off-diagonal elements associated with "atomic coherence" effects such as Rabi oscillations. This terminology is frequently applied to the damping rates. Referring to Eqs. (9.4.12) we see that Γ_1 and $\Gamma_2 + A_{21}$ can be called the diagonal damping rates, and from Eqs. (9.4.15) we see that β is the off-diagonal damping rate. A fundamental relation, obtained from (9.4.16), is illustrated by the inequality

$$\beta \geq \tfrac{1}{2}(\Gamma_1 + \Gamma_2 + A_{21}). \tag{9.4.22}$$

As we have seen, because the off-diagonal elements ρ_{12} and ρ_{21} have a complex phase as well as a magnitude, they are susceptible to purely phase-destructive as well as population-changing relaxation. ●

9.5   OPTICAL BLOCH EQUATIONS

Equations (9.4.19) were derived under the assumptions that $\Delta = \Gamma_1 = \Gamma_2 = 0$ and $\chi = \chi^*$. In the more general case in which Δ is arbitrary, we can write the density matrix equations in the form (Problem 9.8)

$$\frac{du}{dt} = -\beta u - \Delta v, \tag{9.5.1a}$$

$$\frac{dv}{dt} = -\beta v + \Delta u + \chi w, \tag{9.5.1b}$$

$$\frac{dw}{dt} = -A_{21}(w + 1) - \chi v. \tag{9.5.1c}$$

These equations follow from (9.4.12), (9.4.15), (9.4.18), and the notation

$$u = \rho_{21} + \rho_{12} \tag{9.5.2}$$

for the sum of the off-diagonal elements of the density matrix.

Equations (9.5.1) are easily modified to allow for collisional processes that transfer population from the upper state to the lower state at the rate Γ_{21}; such processes have the same effect on the density matrix elements as the rate A_{21} at which population is transferred from the upper state to the lower state when a photon is emitted spontaneously. Thus, to include the rate Γ_{21} in the density matrix Eqs. (9.5.1) we need only replace A_{21} by the total rate $A_{21} + \Gamma_{21}$ at which population is transferred from the upper state to the lower state by spontaneous emission and collisions. The resulting equations are conventionally written as

$$\frac{du}{dt} = -\frac{1}{T_2'} u - \Delta v, \tag{9.5.3a}$$

$$\frac{dv}{dt} = -\frac{1}{T_2'} v + \Delta u + \chi w, \tag{9.5.3b}$$

$$\frac{dw}{dt} = -\frac{1}{T_1}(w - w_0) - \chi v, \tag{9.5.3c}$$

where

$$\frac{1}{T_1} \equiv A_{21} + \Gamma_{21}, \tag{9.5.4a}$$

$$\frac{1}{T_2'} \equiv \gamma_c + \frac{1}{2}(A_{21} + \Gamma_{21}) = \gamma_c + \frac{1}{2T_1}. \tag{9.5.4b}$$

We have written w_0 in Eq. (9.5.3c) instead of the -1 appearing in (9.5.1c); w_0 is the steady-state value of the population difference when $\chi = 0$ (no applied field) and is normally taken to be equal to -1 (atom in the lower state). In thermal equilibrium at temperature T, for example, the probability that the two-state atom is in the lower state is

$$P_1 = \frac{e^{-E_1/k_BT}}{e^{-E_1/k_BT} + e^{-E_2/k_BT}} = \frac{1}{1 + e^{-(E_2-E_1)/k_BT}} = \frac{1}{1 + e^{-\hbar\omega_0/k_BT}}, \qquad (9.5.5)$$

where k_B in this equation is Boltzmann's constant. For optical frequencies and room temperature $\hbar\omega_0 \approx 100k_BT$, so $P_1 \cong 1$, and $w_0 \cong -1$. But w_0 can more generally differ significantly from -1; in a medium with a population inversion, for example, $w_0 > 0$.

Because they have the same form as a set of equations used by F. Bloch and others in the 1940s to describe magnetic resonance phenomena, these equations for an atom in a field that is near resonance with a particular transition are called the *optical Bloch equations*. The density matrix equations for magnetic resonance and for optical resonance have the same form because they both describe two-state systems: the "spin-up" and "spin-down" states of a spin-$\frac{1}{2}$ system in the former case and the two states of an atomic transition in the latter.

When the time over which the field applied to the two-state system is short compared to T_1 and T_2', we can ignore the damping terms in the equations for u, v, and w describing the response to the field and replace (9.5.3) by

$$\frac{du}{dt} = -\Delta v, \qquad (9.5.6a)$$

$$\frac{dv}{dt} = \Delta u + \chi w, \qquad (9.5.6b)$$

$$\frac{dw}{dt} = -\chi v. \qquad (9.5.6c)$$

In optical resonance theory these equations have a useful geometrical interpretation in terms of a "Bloch vector" (or "pseudospin vector") $\mathbf{S}$, which is analogous to the spin vector in magnetic resonance theory. In this geometrical interpretation, which was first proposed by R. P. Feynman, F. L. Vernon, Jr., and R. W. Hellwarth in 1957, we define

$$\mathbf{S} = \hat{\mathbf{1}}u + \hat{\mathbf{2}}v + \hat{\mathbf{3}}w, \qquad (9.5.7)$$

where the unit vectors $\hat{\mathbf{1}}$, $\hat{\mathbf{2}}$, $\hat{\mathbf{3}}$ (analogous to $\hat{\mathbf{x}}$, $\hat{\mathbf{y}}$, and $\hat{\mathbf{z}}$) are mutually orthogonal in a fictitious three-dimensional space. We also define a "torque" vector or "axis" vector $\mathbf{Q}$:

$$\mathbf{Q} = -\hat{\mathbf{1}}\chi + \hat{\mathbf{3}}\Delta. \qquad (9.5.8)$$

Then it can be shown by considering each component separately that the vector equation

$$\frac{d\mathbf{S}}{dt} = \mathbf{Q} \times \mathbf{S} \qquad (9.5.9)$$

is exactly equivalent to the three Eqs. (9.5.6). For example, the $\hat{2}$ (analogous to $\hat{y}$) component of (9.5.9) is

$$\frac{dv}{dt} = (\mathbf{Q} \times \mathbf{S})_2 = Q_3 S_1 - Q_1 S_3 = \Delta u + \chi w, \tag{9.5.10}$$

which is the same as (9.5.6b).

The role of $\mathbf{Q}$ as a torque or axis vector follows from the vector form of (9.5.9). Since $\mathbf{Q} \times \mathbf{S}$ is perpendicular to $\mathbf{S}$ (by the definition of the cross product), the effect of $\mathbf{Q}$ is only to rotate $\mathbf{S}$ about the direction of $\mathbf{Q}$. It cannot lengthen or shorten $\mathbf{S}$. It is easy to confirm that the magnitude $S^2 = \mathbf{S} \cdot \mathbf{S}$ is constant:

$$\frac{dS^2}{dt} = 2\mathbf{S} \cdot \frac{d\mathbf{S}}{dt} = 2\mathbf{S} \cdot (\mathbf{Q} \times \mathbf{S}) = 0. \tag{9.5.11}$$

- The constant magnitude of $\mathbf{S}$ has a physical meaning. From (9.5.7) and the definitions of u, v, and w,

$$S^2 = u^2 + v^2 + w^2 = (\rho_{21} + \rho_{12})^2 - (\rho_{21} - \rho_{12})^2 + (\rho_{22} - \rho_{11})^2$$
$$= 4\rho_{21}\rho_{12} + \rho_{22}^2 - 2\rho_{22}\rho_{11} + \rho_{11}^2. \tag{9.5.12}$$

This expression can be reduced further by using the original definitions of the ρ's in (9.4.1): $\rho_{12} = c_1 c_2^*$, and so forth. We find

$$S^2 = 4c_2 c_1^* c_1 c_2^* + (c_2 c_2^*)^2 - 2c_2 c_2^* c_1 c_1^* + (c_1 c_1^*)^2 = (c_2 c_2^* + c_1 c_1^*)^2 = 1^2 = 1. \tag{9.5.13}$$

The unit length of the Bloch vector $\mathbf{S}$ is thus seen to be equivalent (in the absence of collisions and spontaneous emission) to the conservation of probability in the two-state atom.   •

Whenever $S^2 = 1$, the tip of the Bloch vector lies on the surface of a unit sphere. Only the angles of the vector change with time. In effect, we have now shown that the time evolution of the two-state density matrix (with elements ρ_{21}, etc.) is equivalent to changes in the orientation of $\mathbf{S}$, as sketched in Fig. 9.4.

Let us consider the orientation of the Bloch vector in the fictitious $\hat{1}$-$\hat{2}$-$\hat{3}$ space. First, we note from (9.4.18b) that increasing or decreasing the degrees of inversion $\rho_{22} - \rho_{11}$ corresponds to moving up or down the "vertical" $\hat{3}$ axis. When the Bloch vector points straight up and $w = 1$, then obviously $\rho_{22} = 1$ and $\rho_{11} = 0$, and the atomic population is entirely in the upper level. Thus, the north pole of the unit sphere ("Bloch sphere") corresponds to a fully inverted atom. By the same reasoning, at the south pole $w = -1$ and the atom is in its lower energy level.

The rotation angles of the Bloch vector also have direct physical interpretations. At resonance ($\Delta = 0$) the dynamical evolution consists entirely of rotation about the $\hat{1}$ axis, since we have

$$\mathbf{Q} \longrightarrow -\hat{1}\chi \quad \text{(on resonance)}. \tag{9.5.14}$$

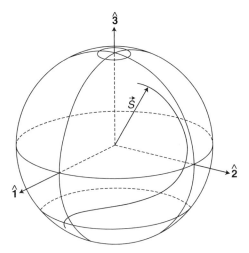

Figure 9.4 Bloch vector and its trajectory on the Bloch sphere.

Then we can characterize the Bloch vector by a single parameter, namely Θ, the rotation angle about the $\hat{1}$ axis:

$$v = -\sin\Theta, \tag{9.5.15a}$$

$$w = -\cos\Theta, \tag{9.5.15b}$$

where Θ is measured from the *south* pole since that is the normal initial state, corresponding to the atom in its lower energy level (Fig. 9.5).

Substitution of (9.5.15) into (9.5.6c) gives an equation for Θ:

$$\frac{d\Theta}{dt} = \chi, \tag{9.5.16}$$

which leads to the obvious solution

$$\Theta = \chi t \tag{9.5.17}$$

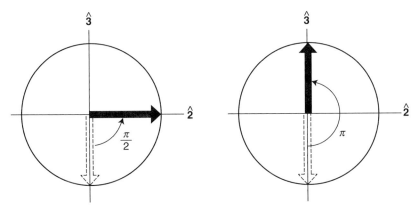

Figure 9.5 Effect of $\pi/2$ and π pulses on the Bloch vector for an on-resonance atom.

and

$$v = -\sin \chi t, \tag{9.5.18a}$$

$$w = -\cos \chi t. \tag{9.5.18b}$$

Agreement with the solutions found earlier in Eqs. (9.3.20) is worth checking in the same limit $\Delta = 0$. For example, since $\rho_{11} + \rho_{22} = 1$, we can write

$$\rho_{11} = \frac{1 - w}{2}, \tag{9.5.19a}$$

$$\rho_{22} = \frac{1 + w}{2}, \tag{9.5.19b}$$

and combine this with (9.5.18b) to obtain

$$\rho_{11} = \tfrac{1}{2}(1 + \cos \chi t), \tag{9.5.20a}$$
$$\rho_{22} = \tfrac{1}{2}(1 - \cos \chi t), \tag{9.5.20b}$$

which are the same as (9.3.20a) and (9.3.20b) on resonance.

If the Rabi frequency χ is real but not constant in time, solutions (9.5.15) are the same, but Θ is then given by

$$\Theta(t) = \int_0^t \chi(t') \, dt', \tag{9.5.21}$$

where

$$\chi(t) = \frac{e(\mathbf{x}_{21} \cdot \hat{\boldsymbol{\varepsilon}})E_0(t)}{\hbar} \equiv \frac{\mu E_0(t)}{\hbar}. \tag{9.5.22}$$

In this way the Bloch vector formalism connects a property of the atoms, namely the Bloch vector rotation angle on resonance, directly with a property of the incident radiation field, namely the time integral of the field amplitude.

The integral (9.5.21) is called the *area* of the pulse. This name derives from the fact that an integral can be viewed as an area, as in Fig. 9.6. The solutions (9.5.15) make it clear that pulses with areas equal to certain multiples of π are special. For example, in Fig. 9.5 a π pulse turns Θ through $180°$ about the $\hat{\mathbf{1}}$ axis, and thus inverts the atomic population from the lower level to the upper level. Much more surprising is the effect of a 2π pulse. It rotates the resonant Bloch vector through $360°$ and thus returns the

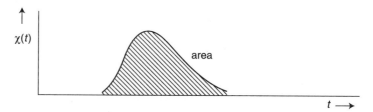

Figure 9.6   Area of a laser pulse according to (9.5.21).

atom exactly to its initial state. The same is true of $4\pi, 6\pi, \ldots$ pulses. There is therefore a set of pulses that *have no net effect* on the resonant atoms when propagating in an absorbing medium at resonance. Area is a pulse parameter completely overlooked by "conventional" optical spectroscopy based on rate equations for the upper- and lower-state populations, but one with fundamental significance.

If E_0 is allowed to be time dependent, we gain the ability to treat the interaction of atoms with light *pulses* as well as with steady light beams. At the same time two working assumptions must be modified. First, the argument employed to justify the rotating-wave approximation in going from (9.3.14) to (9.3.15) fails if χ itself contributes rapid temporal variation. Thus, we must assume that $E_0(t)$ is a slowly varying function in a sense made precise in (9.6.2) below. Second, if not just the amplitude but the phase of E_0 changes in time we can no longer assume that an adjustment of the wave function phase will make $\chi(t)$ real. In order to recognize this important shift in assumptions we now change our symbol for the electric field's complex amplitude: $E_0(t) \rightarrow \mathcal{E}(t)$, and the Rabi frequency on resonance becomes

$$\chi(t) = \frac{\mu \mathcal{E}(t)}{\hbar}. \tag{9.5.23}$$

For complex $\chi(t)$ the Bloch equations take a slightly more general form that the reader may easily verify:

$$\frac{d}{dt}(u - iv) = -(\beta + i\Delta)(u - iv) - i\chi w, \tag{9.5.24a}$$

$$\frac{dw}{dt} = -\frac{1}{T_1}(1 + w) + \frac{i}{2}[\chi(u + iv) - \chi^*(u - iv)]. \tag{9.5.24b}$$

• Equations (9.5.6) and (9.5.9) are no more than Schrödinger's equation written in another convenient form. Bloch's name should actually be associated with these equations only after relaxation processes are taken into account in a particular way. As already noted, the Bloch equations, including relaxation, were originally written in the context of spin resonance. In that case the Bloch vector **S** is actually the magnetic spin vector. Relaxation processes for spin commonly include phase-interrupting collisions of the kind we have considered in Section 9.4, but not inelastic decay to other levels such as we have associated with the rates Γ_1 and Γ_2.

The notation T_1 and T_2' for the diagonal and off-diagonal relaxation times was used in the original magnetic resonance context to designate the "longitudinal" and "transverse" lifetimes of the spin. The terms longitudinal and transverse refer to polar and equatorial directions on the Bloch sphere, directions in magnetic resonance experiments either along, or transverse to, the static magnetic field (conventionally the z axis). In view of (9.5.4) one has the frequently quoted inequality

$$T_2' \leq 2T_1 \tag{9.5.25}$$

between the transverse and longitudinal lifetimes.   •

As discussed in Section 9.4, the "coherent" response of an atom to a resonant field is most pronounced when the transverse (off-diagonal) relaxation rate is small. When β is sufficiently large, the density matrix elements do not undergo Rabi oscillations at a frequency depending on the electric field strength, but simply relax "incoherently" to their steady-state values in the case of a constant-amplitude field (Fig. 9.3). This is discussed in more detail in the next section, where it is shown that, when the off-diagonal relaxation of the density matrix elements is sufficiently rapid, the atom responds to the field exactly as described by the rate equations for the upper and lower state populations. In this incoherent or rate-equation limit the response to the field depends on the field intensity $I(t)$ rather than the electric field amplitude $\mathcal{E}(t)$; in particular, in the case of an applied pulse, the response is independent of the pulse area.

9.6   MAXWELL–BLOCH EQUATIONS

The optical Bloch equations describe a wide variety of coherent atom–field interactions. Here we are interested specifically in developing a formalism able to deal with situations not encountered in our discussions of propagation in Chapters 7 and 8. These are coherent *propagation* effects described by the coupling of the optical Bloch equations to Maxwell's equations in which the field is strong enough to stimulate substantial population exchange between levels 1 and 2. We will use the approximation in which the field is taken to be a plane wave propagating in the z direction:

$$\mathbf{E}(\mathbf{r},\,t) = \hat{\boldsymbol{\epsilon}}\mathcal{E}(z,\,t)e^{-i\omega(t-z/c)}, \tag{9.6.1}$$

where $\mathcal{E}(z,\,t)$ is the unknown complex amplitude to be determined and, as usual, the physical electric field is understood to be the real part of this expression. The distinction between the envelopes $\mathcal{E}(z,\,t)$ in Eqs. (8.3.6) and (9.6.1) will be clarified later in this section. A natural assumption implied by the form (9.6.1) is that the amplitude $\mathcal{E}(z,\,t)$ varies slowly compared to the carrier wave $e^{-i\omega(t-z/c)}$. This justifies inequalities such as

$$\left|\frac{\partial\mathcal{E}}{\partial z}\right| \ll \frac{\omega}{c}|\mathcal{E}|, \qquad \left|\frac{\partial^2\mathcal{E}}{\partial z^2}\right| \ll \frac{\omega}{c}\left|\frac{\partial\mathcal{E}}{\partial z}\right|, \qquad \left|\frac{\partial\mathcal{E}}{\partial t}\right| \ll \omega|\mathcal{E}|. \tag{9.6.2}$$

In physical terms these inequalities state that $\mathcal{E}(z,\,t)$ represents a smooth enough envelope in both space and time. We have already employed slowly varying envelope approximations in Sections 7.4, 8.3, and 8.8, for instance. In the vast majority of cases of practical interest these approximations are excellent since they would be violated only if $\mathcal{E}(z,\,t)$ represented a pulse shorter than a few optical periods ($\sim 10^{-15}$ s) in time or a few wavelengths (~ 1 μm) in space.

In the plane-wave and slowly varying envelope approximations for $\mathbf{E}(\mathbf{r},\,t)$ we have, for the left-hand side of the wave equation (8.2.13),

$$\nabla^2\mathbf{E} - \frac{1}{c^2}\frac{\partial^2\mathbf{E}}{\partial t^2} = \left[\frac{\partial^2\mathcal{E}}{\partial z^2} + 2i\frac{\omega}{c}\frac{\partial\mathcal{E}}{\partial z} - \frac{\omega^2}{c^2}\mathcal{E} - \frac{1}{c^2}\left(\frac{\partial^2\mathcal{E}}{\partial t^2} - 2i\omega\frac{\partial\mathcal{E}}{\partial t} - \omega^2\mathcal{E}\right)\right]\hat{\boldsymbol{\epsilon}}e^{-i\omega(t-z/c)}$$

$$\cong 2i\frac{\omega}{c}\left(\frac{\partial\mathcal{E}}{\partial z} + \frac{1}{c}\frac{\partial\mathcal{E}}{\partial t}\right)\hat{\boldsymbol{\epsilon}}e^{-i\omega(t-z/c)}. \tag{9.6.3}$$

On the right-hand side of (8.2.13) we require the polarization $\mathbf{P}$, the electric dipole moment per unit volume of the medium, which for our purposes here consists of two-state atoms, each having a transition frequency $\omega_0 \approx \omega$. Each atom is assumed to have an electric dipole moment expectation value $\mathbf{p}$ given by the electron charge e times the expectation value $\langle \mathbf{x} \rangle$ [recall Eq. (3.A.16)]:

$$\mathbf{p} = e\langle \mathbf{x} \rangle = e\left(a_1^* a_2 \mathbf{x}_{12} + a_2^* a_1 \mathbf{x}_{21}\right) = e\left(c_1^* c_2 \mathbf{x}_{12} e^{-i\omega t} + c_2^* c_1 \mathbf{x}_{12}^* e^{i\omega t}\right), \qquad (9.6.4)$$

where we have used (9.3.13). In deriving the equations for c_1 and c_2 in Section 9.2 we ignored any spatial variations of the field, which of course cannot be done when we want to describe how the field propagates. To account for the spatial variations of a plane-wave field acting on an atom with coordinate z along the direction of propagation, we simply replace $e^{-i\omega t}$ by $e^{-i\omega(t-z/c)}$ in (9.6.4). Thus, for an atom at z,

$$\mathbf{p}(z, t) = \rho_{21}(z, t)e\mathbf{x}_{12}e^{-i\omega(t-z/c)} + \rho_{21}^*(z, t)e\mathbf{x}_{12}^* e^{i\omega(t-z/c)}. \qquad (9.6.5)$$

Note that the ρ_{ij}, like $\mathcal{E}$, are assumed in their definitions to be slowly varying compared to $e^{\pm i\omega(t-z/c)}$; this assumption is seen to be consistent with Eqs. (9.4.2).

If there is a uniform distribution of N atoms per unit volume, the polarization $\mathbf{P}$ is

$$\mathbf{P}(z, t) = 2Ne\mathbf{x}_{12}\rho_{21}(z, t)e^{-i\omega(t-z/c)}, \qquad (9.6.6)$$

where, as in (9.6.1), the real part of the right-hand side is the (real) physical polarization $N\mathbf{p}(z, t)$. Since $\rho_{21}(z, t)$ varies much more slowly in time than $e^{-i\omega t}$,

$$\frac{\partial^2 \mathbf{P}}{\partial t^2} \cong -2N\omega^2 e\mathbf{x}_{12}\rho_{21}(z, t)e^{-i\omega(t-z/c)} \qquad (9.6.7)$$

and therefore, from (8.2.13) and (9.6.3),

$$2i\frac{\omega}{c}\left(\frac{\partial \mathcal{E}}{\partial z} + \frac{1}{c}\frac{\partial \mathcal{E}}{\partial t}\right)\hat{\boldsymbol{\varepsilon}}e^{-i\omega(t-z/c)} \cong -\frac{1}{\epsilon_0 c^2}2N\omega^2 e\mathbf{x}_{12}\rho_{21}e^{-i\omega(t-z/c)}. \qquad (9.6.8)$$

The approximate equation for the propagation of the electric field is then

$$\frac{\partial \mathcal{E}}{\partial z} + \frac{1}{c}\frac{\partial \mathcal{E}}{\partial t} = \frac{i\omega}{\epsilon_0 c}N\mu^* \rho_{21}, \qquad (9.6.9)$$

where we have taken the dot product of both sides of (9.6.8) with $\hat{\boldsymbol{\varepsilon}}^*(\hat{\boldsymbol{\varepsilon}} \cdot \hat{\boldsymbol{\varepsilon}}^* = 1)$ and used the abbreviation [Eq. (9.5.22)]

$$\mu^* = e(\mathbf{x}_{12} \cdot \hat{\boldsymbol{\varepsilon}}^*) \qquad (9.6.10)$$

for the projection of the transition dipole moment on the direction of the field polarization.

Using the relations (9.4.18) and (9.5.2), we can write (9.6.9) as

$$\frac{\partial \mathcal{E}}{\partial z} + \frac{1}{c}\frac{\partial \mathcal{E}}{\partial t} = \frac{i\omega}{2\epsilon_0 c}N\mu^*(u - iv). \tag{9.6.11}$$

The variables u and v on the right-hand side of this equation satisfy the Bloch equations (9.5.24), with d/dt in those equations replaced by $\partial/\partial t$ because u, v, and w are now functions of z as well as t:

$$\frac{\partial}{\partial t}(u - iv) = -(\beta + i\Delta)(u - iv) - i\chi w, \tag{9.6.12a}$$

$$\frac{\partial w}{\partial t} = -\frac{1}{T_1}(1 + w) + \frac{i}{2}[\chi(u + iv) - \chi^*(u - iv)], \tag{9.6.12b}$$

with $\chi = \mu\mathcal{E}/\hbar$. The coupled Eqs. (9.6.11) and (9.6.12) are called the *Maxwell–Bloch equations*.[1] In this form they describe the coupling between the field and a collection of two-state atoms, each of which has a transition frequency ω_0 that differs from the field carrier frequency ω by an amount characterized by the detuning $\Delta = \omega_0 - \omega$. The generalization to the case of inhomogeneous broadening, where there is a distribution of detunings, is straightforward and is considered in Section 9.8.

Note that Eq. (9.6.11) assumes that the field is coupled only to the two-state atoms in our model of resonant pulse propagation. Any significant absorption or scattering processes due to other matter in the medium will require a modification of (9.6.11). For example, we could account for an absorption coefficient a_b due to "background" atoms in the medium by adding the term $a_b\mathcal{E}$ to the left-hand side of (9.6.11).

With these equations we have a self-consistent formulation of the propagation of a field in a resonant medium described as a collection of two-state atoms. That is, the coupled Maxwell–Bloch equations allow the atoms and the field to influence each other mutually and at a fundamental level (Fig. 9.7). We already saw in Chapter 1 an example of such a mutual interaction (recall Fig. 1.13), but there the theory was completely empirical. We now reexamine some earlier results, including those of Chapter 1, from our present, more satisfactory perspective.

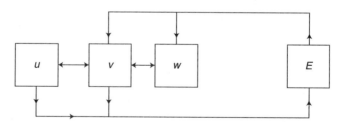

Figure 9.7   The mutual interactions embodied in the Maxwell–Bloch equations. The coupling is much more intricate than in the conventional rate-equation theory illustrated in Fig. 1.13.

[1]Because of the approximations made in deriving these equations—in particular the slowly varying envelope approximation that reduces the wave equation for the electric field to the first-order partial differential equation (9.6.11)—these equations are sometimes called the *reduced Maxwell–Bloch equations*.

Rate-Equation Approximation

The formal solution of Eq. (9.6.12a) is

$$u(t) - iv(t) = [u(0) - iv(0)]e^{-(\beta+i\Delta)t} - i\int_0^t \chi(t')w(t')e^{(\beta+i\Delta)(t'-t)}\,dt'. \qquad (9.6.13)$$

If the atom is initially (at $t = 0$) in the lower or upper state, then $u(0) = v(0) = 0$ and the first term on the right vanishes. We can also ignore this term for times $t \gg \beta^{-1}$. In either case we write

$$u(t) - iv(t) = -i\int_0^t \chi(t')w(t')e^{(\beta+i\Delta)(t'-t)}\,dt'. \qquad (9.6.14)$$

In many cases of practical interest, and in most lasers, the off-diagonal damping rate β is so large that χ and w are nearly constant over time intervals $\sim 1/\beta$. Then

$$u(t) - iv(t) = 2\rho_{21} \cong -i\chi(t)w(t)\int_0^t e^{(\beta+i\Delta)(t'-t)}\,dt' = \frac{-i\chi(t)w(t)}{\beta+i\Delta}[1 - e^{-(\beta+i\Delta)t}]$$

$$= \frac{-i\chi(t)w(t)}{\beta+i\Delta} \qquad (9.6.15)$$

for times $t \gg \beta^{-1}$. In this approximation u and v track the variations in time of $w(t)$; an example of this "adiabatic following" is seen in Fig. 9.3. For this approximation to be valid $\chi(t)$ and $w(t)$ must vary slowly compared to $e^{-\beta t}$, and $w(t)$ in particular can be expected to be slowly varying in this sense when the inequality (9.4.17) is satisfied. When these conditions are fulfilled, the off-diagonal density matrix elements can be completely eliminated, and the so-called rate equations are obtained for the diagonal elements ρ_{11} and ρ_{22} alone, as follows.

Using the approximation (9.6.15) for $\rho_{21}(t)$ and $\rho_{12}(t) = \rho_{21}^*(t)$ in the population equations (9.4.12),[2] we obtain

$$\frac{\partial\rho_{11}}{\partial t} = -\Gamma_1\rho_{11} + A_{21}\rho_{22} + \frac{|\chi|^2\beta/2}{\Delta^2 + \beta^2}(\rho_{22} - \rho_{11}), \qquad (9.6.16a)$$

$$\frac{\partial\rho_{22}}{\partial t} = -(\Gamma_2 + A_{21})\rho_{22} - \frac{|\chi|^2\beta/2}{\Delta^2 + \beta^2}(\rho_{22} - \rho_{11}). \qquad (9.6.16b)$$

Now all references to the off-diagonal variables ρ_{21} and ρ_{12} have been eliminated, and the two populations (or population probabilities) ρ_{11} and ρ_{22} are coupled only to each other.

---

[2]Of course, we can also use (9.6.15) in Eq. (9.6.12b) for w. Equations (9.4.12) are more general than (9.6.12b), however, because they allow for collisional processes that transfer population into states other than the states 1 and 2 of the resonant transition.

Equations (9.6.16) apply to a single atom in the average sense mentioned already. By multiplying both sides by N, the density of resonant (two-state) atoms, and defining the population densities $N_1 = N\rho_{11}$ and $N_2 = N\rho_{22}$, we obtain

$$\frac{\partial N_1}{\partial t} = -\Gamma_1 N_1 + A_{21} N_2 + \frac{|\chi|^2 \beta/2}{\Delta^2 + \beta^2}(N_2 - N_1), \tag{9.6.17a}$$

$$\frac{\partial N_2}{\partial t} = -(\Gamma_2 + A_{21})N_2 - \frac{|\chi|^2 \beta/2}{\Delta^2 + \beta^2}(N_2 - N_1). \tag{9.6.17b}$$

When the right-hand sides are expressed in terms of the intensity $I = (c\epsilon_0/2)|\mathcal{E}|^2$, these equations are seen to have the same form as the population rate equations (4.5.1). Note that, in order for $w = (N_2 - N_1)/N$ to be slowly varying compared to β, as we have assumed, the stimulated emission rate $\frac{1}{2}|\chi|^2\beta/(\Delta^2 + \beta^2)$ should be small compared to β. For $\Delta = 0$, this requires that the Rabi frequency should be small compared to β, a condition that is almost always satisfied in lasers.

Similarly the "adiabatic approximation" (9.6.15), when applied to (9.6.9), gives

$$\frac{\partial \mathcal{E}}{\partial z} + \frac{1}{c}\frac{\partial \mathcal{E}}{\partial t} = \frac{i\omega}{\epsilon_0 c}N\mu^*\left(\frac{-i\chi/2}{\beta + i\Delta}\right)(\rho_{22} - \rho_{11}) = \frac{\omega|\mu|^2}{2\epsilon_0 \hbar c}\frac{1}{\beta + i\Delta}\mathcal{E}(N_2 - N_1) \tag{9.6.18}$$

and therefore

$$\mathcal{E}^*\frac{\partial \mathcal{E}}{\partial z} + \frac{1}{c}\mathcal{E}^*\frac{\partial \mathcal{E}}{\partial t} = \frac{\omega|\mu|^2}{2\epsilon_0 \hbar c}\frac{1}{\beta + i\Delta}|\mathcal{E}|^2(N_2 - N_1). \tag{9.6.19}$$

Adding the complex conjugates to both sides of this equation, and using $\mathcal{E}^*\partial\mathcal{E}/\partial z + \mathcal{E}\,\partial\mathcal{E}^*/\partial z = \partial|\mathcal{E}|^2/\partial z$ and likewise for the derivatives with respect to t, we have

$$\frac{\partial|\mathcal{E}|^2}{\partial z} + \frac{1}{c}\frac{\partial|\mathcal{E}|^2}{\partial t} = \frac{\omega|\mu|^2}{2\epsilon_0 \hbar c}\frac{2\beta}{\Delta^2 + \beta^2}|\mathcal{E}|^2(N_2 - N_1), \tag{9.6.20}$$

or, in terms of the intensity,

$$\frac{\partial I}{\partial z} + \frac{1}{c}\frac{\partial I}{\partial t} = \frac{\omega|\mu|^2}{\epsilon_0 \hbar c}\frac{\beta}{\Delta^2 + \beta^2}I(N_2 - N_1) = gI, \tag{9.6.21}$$

which has the same form as Eq. (3.12.5) [or (4.4.1a)] for the propagation of the intensity of a plane wave.

For pulses of duration much shorter than the relaxation times $1/\Gamma_1$, $1/\Gamma_2$, and $1/A_{21}$, we can approximate the rate equations (9.6.17) by Eqs. (6.12.1). Then the coupled Eqs. (9.6.17) and (9.6.21) have the same form as the coupled atom–field Eqs. (6.12.2) and (6.12.3) used in Section 6.12 to describe the amplification of short pulses in a gain medium.

We have shown that when the off-diagonal density matrix elements relax quickly compared to the temporal variations of the population difference or the electric field amplitude, the Maxwell–Bloch equations may be approximated by rate equations that do not involve off-diagonal density matrix elements. These rate equations for populations and intensity have the same form as the rate equations we have used in earlier chapters to describe light amplification and laser oscillation, and are valid approximations for most laser media.

The inequality (9.4.17) is necessary for the validity of the rate equation approximation. In more physical terms, this condition requires that the homogeneous linewidth of the transition be much larger than the sum of the population decay rates. Similarly the condition that the field is not too strong means that the Rabi frequency should be small compared to the homogeneous linewidth. Even if these conditions are satisfied, however, we still require for the validity of the rate-equation approximation that the field amplitude $\mathcal{E}$ varies slowly on the time scale of the inverse homogeneous linewidth. This last condition can be violated, for instance, in the case of very short pulses of light, which, as discussed in Section 9.9, can propagate in ways that find no explanation in terms of coupled rate equations for populations and intensities.

- Equations (9.6.17) and (9.6.21) involve $|\mu|^2 = e^2|\mathbf{x}_{12} \cdot \hat{\boldsymbol{\varepsilon}}|^2$. For various practical reasons—unpolarized radiation, rotational or collisional disorientation, etc.—it is often the orientational average of this quantity that is relevant. It is not difficult to calculate the required average and show that (Problem 9.9)

$$e^2\langle|\mathbf{x}_{12} \cdot \hat{\boldsymbol{\varepsilon}}|^2\rangle_{\text{orientation}} = \langle|\mu|^2\rangle_{\text{orientation}} = \frac{1}{3}e^2|\mathbf{x}_{12}|^2 = \frac{\pi\epsilon_0\hbar c^3}{\omega_0^3}A_{21}, \qquad (9.6.22)$$

where in the last equality we have used formula (3.A.26) for the spontaneous emission rate A_{21}. Thus,

$$\langle|\chi|^2\rangle_{\text{orientation}} = \frac{1}{\hbar^2}\langle|\mu|^2\rangle_{\text{orientation}}|\mathcal{E}|^2 = \frac{\pi\epsilon_0 c^3}{\hbar\omega_0^3}A_{21}|\mathcal{E}|^2 = \frac{2\pi c^2}{\hbar\omega_0^3}A_{21}I = \frac{1}{\hbar\omega_0}\frac{\lambda_0^2}{2\pi}A_{21}I, \quad (9.6.23)$$

where $\lambda_0 (= c/\nu_0 = 2\pi c/\omega_0)$ is the transition wavelength. Therefore,

$$\frac{\langle|\chi|^2\rangle_{\text{orientation}}\beta/2}{\Delta^2 + \beta^2} = \frac{1}{\hbar\omega_0}\frac{\lambda_0^2}{2\pi}\frac{\beta/2}{\Delta^2 + \beta^2}A_{21}I \cong \frac{1}{h\nu}\frac{\lambda^2 A_{21}}{8\pi}L(\nu), \qquad (9.6.24)$$

where we have used (3.4.23) and (3.4.26) for the Lorentzian lineshape function $L(\nu)$ and have approximated ν_0 and λ_0 by the field frequency ν and wavelength λ, respectively.

More generally $L(\nu)$ is replaced by whatever lineshape function $S(\nu)$ is appropriate, and (9.6.24) is seen to be just $\sigma(\nu)/h\nu$, where $\sigma(\nu)$ is the cross section for stimulated emission [Eqs. (3.7.4) and (3.7.5b)]. Thus, the orientationally averaged field-dependent term in Eqs. (9.6.17) is the familiar [Eq. (3.7.5b)]

$$\frac{\sigma(\nu)}{h\nu}I(N_2 - N_1).$$

Similarly, the orientational average of the right-hand side of (9.6.21) is

$$\sigma(\nu)(N_2 - N_1)I = g(\nu)I, \qquad (9.6.25)$$

where $g(\nu)$ is the gain coefficient.

Equations (9.6.17) and (9.6.21), in other words, become exactly the same population and photon rate equations employed earlier when we replace $|\mu|^2$ by its orientational average. As discussed in Section 14.3, this orientational averaging is very often—but certainly not always—appropriate.   •

Refractive Index

The absorption coefficient

$$a(\nu) = -g(\nu) = \frac{\omega|\mu|^2}{\epsilon_0 \hbar c} \frac{\beta}{\Delta^2 + \beta^2} (N_1 - N_2) \qquad \left(\nu = \frac{\omega}{2\pi}\right) \tag{9.6.26}$$

appearing in (9.6.21) can be seen from our derivation of the rate equations to be an *approximate* measure of how the field intensity changes with propagation in a resonant medium. The characterization of the medium by a refractive index also follows from this rate-equation approximation to the Maxwell–Bloch equations. To see this, write (9.6.18) as

$$\frac{\partial \mathcal{E}}{\partial z} + \frac{1}{c}\frac{\partial \mathcal{E}}{\partial t} = -\tfrac{1}{2}a(\omega)\mathcal{E} + iX\mathcal{E}, \tag{9.6.27}$$

where

$$X = i\frac{\omega|\mu|^2}{2\epsilon_0 \hbar c}\frac{\Delta}{\Delta^2 + \beta^2}(N_1 - N_2). \tag{9.6.28}$$

The physical significance of X is most easily seen by assuming a monochromatic field; if we write the complex field amplitude for the field in this case as

$$\mathcal{E}(z) = \mathcal{E}_0 e^{-a(\nu)z/2} e^{i[n(\omega)-1]\omega z/c}, \tag{9.6.29}$$

we obtain from (9.6.27) the relation

$$n(\omega) - 1 = \frac{c}{\omega}X = \frac{|\mu|^2}{2\epsilon_0 \hbar}\frac{\Delta}{\Delta^2 + \beta^2}(N_1 - N_2) = \frac{\Delta}{\beta}\frac{c}{2\omega}a(\nu) = \frac{2\pi(\nu_0 - \nu)}{2\pi\delta\nu_0}\frac{c}{4\pi\nu}a(\nu)$$

$$= \frac{\lambda_0}{4\pi}\frac{\nu_0 - \nu}{\delta\nu_0}a(\nu) \tag{9.6.30}$$

when we approximate the field wavelength λ by the transition wavelength λ_0. $n(\omega)$ is the refractive index at frequency ω, as is seen from the expression [Eq. (9.6.1)]

$$E(z) = \mathcal{E}(z)e^{-i\omega(t-z/c)} = \mathcal{E}_0 e^{-a(\nu)z/2}e^{-i\omega(t-n(\omega)z/c)} \tag{9.6.31}$$

for the assumed monochromatic plane wave, and indeed expression (9.6.30) merely reproduces our earlier result (3.15.9) for the refractive index associated with a resonant transition.

•   For another example of how known results follow as approximations to the Maxwell–Bloch equations, let us consider a field that is sufficiently far from resonance that the atoms of the medium remain, with high probability, in their lower states. Then $w(t) \cong -1$, or $N_1 - N_2 \cong N_1 \cong N$, the density of resonant atoms, and (9.6.14) can be approximated by

$$u(t) - iv(t) = i \int_0^t \chi(t') e^{(\beta + i\Delta)(t' - t)} \, dt', \tag{9.6.32}$$

or, after successive integrations by parts,

$$u(t) - iv(t) = \frac{i}{\beta + i\Delta} \chi(t) - \frac{i}{(\beta + i\Delta)^2} \frac{\partial \chi}{\partial t} + \frac{i}{(\beta + i\Delta)^3} \frac{\partial^2 \chi}{\partial t^2} + \cdots \tag{9.6.33}$$

for times $t \gg \beta^{-1}$. Consistent with our assumption that the field is far from resonance, let us replace $\beta + i\Delta$ by $i\Delta$, which is equivalent to ignoring absorption:

$$u(t) - iv(t) \cong \frac{1}{\Delta} \chi(t) + \frac{i}{\Delta^2} \frac{\partial \chi}{\partial t} - \frac{1}{\Delta^3} \frac{\partial^2 \chi}{\partial t^2} \tag{9.6.34}$$

if $|\Delta|$ is sufficiently large that third- and higher-order derivatives of χ with respect to t can be neglected. With these approximations (9.6.11) takes the form

$$\frac{\partial \mathcal{E}}{\partial z} + \frac{1}{c} \frac{\partial \mathcal{E}}{\partial t} = \frac{iN\omega|\mu|^2}{2\epsilon_0\hbar c} \left[\frac{1}{\Delta} \mathcal{E} + \frac{i}{\Delta^2} \frac{\partial \mathcal{E}}{\partial t} - \frac{1}{\Delta^3} \frac{\partial^2 \mathcal{E}}{\partial t^2} \right]. \tag{9.6.35}$$

Under our assumption of large detuning from resonance ($\Delta^2 \gg \beta^2$), $N\omega|\mu|^2/2\epsilon_0\hbar c\Delta$ is just $(\omega/c)[n(\omega) - 1]$ [recall (9.6.30)]. Thus, we can replace (9.6.35) by

$$\frac{\partial \mathcal{E}}{\partial z} + \frac{1}{c} \left[1 + \frac{N\omega|\mu|^2}{2\epsilon_0\hbar\Delta^2} \right] \frac{\partial \mathcal{E}}{\partial t} = i\frac{\omega}{c}[n(\omega) - 1]\mathcal{E} - i\frac{N\omega|\mu|^2}{2\epsilon_0\hbar c\Delta^3} \frac{\partial^2 \mathcal{E}}{\partial t^2}. \tag{9.6.36}$$

Now $n(\omega) - 1 \cong N|\mu|^2/2\epsilon_0\hbar\Delta$ implies

$$\frac{dn}{d\omega} = -\frac{dn}{d\Delta} = \frac{N|\mu|^2}{2\epsilon_0\hbar\Delta^2}, \tag{9.6.37}$$

so that the term in brackets on the left-hand side of (9.6.36) is $1 + \omega \, dn/d\omega$, which is approximately $d(n\omega)/d\omega$ for $n \cong 1$.[3] Then, since $c[d(n\omega)/d\omega]^{-1}$ is the group velocity v_g [Eq. (8.3.3)], we write (9.6.36) as

$$\frac{\partial \mathcal{E}}{\partial z} + \frac{1}{v_g} \frac{\partial \mathcal{E}}{\partial t} = i\frac{\omega}{c}[n(\omega) - 1]\mathcal{E} - i\frac{N\omega|\mu|^2}{2\epsilon_0\hbar c\Delta^3} \frac{\partial^2 \mathcal{E}}{\partial t^2}. \tag{9.6.38}$$

In similar fashion, using again the approximation $n(\omega) - 1 = N|\mu|^2/2\epsilon_0\hbar\Delta$, we obtain, after some elementary algebra,

$$2\frac{dn}{d\omega} + \omega\frac{d^2n}{d\omega^2} = \frac{N\omega_0|\mu|^2}{\epsilon_0\hbar\Delta^3} \cong \frac{N\omega|\mu|^2}{\epsilon_0\hbar\Delta^3} = c\frac{d^2k}{d\omega^2}, \tag{9.6.39}$$

---

[3]The approximation $n \cong 1$ is consistent with our assumption that $\Delta^2 \gg \beta^2$. Of course, $\Delta^2 \gg \beta^2$ does not always imply that $n \cong 1$. If n is much different from 1, however, other effects, such as Lorentz–Lorenz local field corrections, must be taken into account. Our treatment here is most directly applicable to gaseous media.

where $k = n(\omega)\omega/c$; $d^2k/d\omega^2$ characterizes the group velocity dispersion [recall Eq. (8.4.2)]. We can therefore express (9.6.38) in the form

$$\frac{\partial \mathcal{E}'}{\partial z} + \frac{1}{v_g}\frac{\partial \mathcal{E}'}{\partial t} + \frac{i}{2}\frac{d^2k}{d\omega^2}\frac{\partial^2 \mathcal{E}'}{\partial t^2} = 0, \qquad (9.6.40)$$

where we have removed a phase factor by defining $\mathcal{E}'(z, t) = \mathcal{E}(z, t)e^{i[n(\omega)-1]\omega z/c}$. Equation (9.6.40) is identical to the Eq. (8.4.1) obtained in Chapter 8 by a different approach. ●

9.7   SEMICLASSICAL LASER THEORY

Our derivation of the Maxwell–Bloch equations was based on the so-called *semiclassical* theory of the interaction of light with matter. The semiclassical theory ignores the quantum mechanical nature of the electromagnetic field while using the full regalia of the Schrödinger equation to determine the behavior of the matter. The result is a theory that is ultimately inconsistent because quantum fluctuations associated with the electromagnetic field do affect atomic behavior, but the theory does not recognize such effects except in isolated instances. For example, semiclassical theory cannot account for all aspects of spontaneous emission (see Chapter 12), and in particular it cannot account for the fundamental limit to the laser linewidth. Except for our derivation of the fundamental laser linewidth in Section 5.11, and an occasional, usually unnecessary reference to photons, our entire presentation has been based on semiclassical theory.

The justification for the semiclassical theory is in fact extremely strong in a wide domain. This domain embraces almost all of radiation theory where the numbers of photons are much larger than unity. In most laser modes the number of photons is practically unbounded, and effects due to field quantization are insignificant so long as attention remains on stimulated processes. On the other hand, it is quite possible to use laser fields to probe the subtle correlations and fluctuations inherent in quantum theory, but in order to do so an observation must be undertaken of a quantity that is sensitive to single-photon differences. Some such observations are discussed in Chapters 12 and 13. They are rarely directly important for an understanding of laser operation.

The Maxwell–Bloch equations provide the basis for the most rigorous semiclassical theory of laser operation. In this theory the atomic dipoles, which are determined by off-diagonal density matrix elements, serve as sources of radiation, and this radiation in turn drives these dipoles into oscillation and causes them to radiate. The self-consistent solution of the coupled atom–field equations determines the laser power and frequency.

To simplify the discussion, and to focus on the self-consistent solution of the Maxwell–Bloch equations, we will assume a laser cavity consisting of two plane, parallel, and nearly perfectly reflecting mirrors. The electric field corresponding to the mth mode of the cavity then has the form

$$\mathbf{E}_m(z, t) = \hat{\boldsymbol{\epsilon}}_m \mathcal{E}_m(t) \sin k_m z e^{-i\omega t}, \qquad (9.7.1)$$

where $k_m = m\pi/L$ and L is the cavity length. The intracavity field can be written as a superposition of all possible cavity modes: $\mathbf{E}(z, t) = \sum_m \mathbf{E}_m(z, t)$. Note that the complex mode amplitude $\mathcal{E}_m(t)$ does not depend on z, since $\sin k_m z$ is assumed to fully express the z dependence of the mth mode. The frequency ω of laser oscillation is not known a priori, but can be expected to be close to one of the cavity mode frequencies ω_m.

It is convenient to express the polarization's z dependence in terms of cavity mode functions as well. That is, we write $\mathbf{P}(z, t) = \sum_m \mathbf{P}_m(z, t)$, where [recall (9.6.6)]

$$\mathbf{P}_m(z, t) = 2Ne\mathbf{x}_{12}\rho_{21}^{(m)}(z, t) \sin k_m z e^{-i\omega t}. \tag{9.7.2}$$

Then from the wave equation (8.2.13) we obtain, in the slowly varying envelope approximation,

$$\sum_m \left[\frac{\partial}{\partial t} - i(\omega - \omega_m)\right]\mathcal{E}_m(t) \sin k_m z = \sum_m \frac{i\omega}{\epsilon_0}N\mu^*\rho_{21}^{(m)}(z, t) \sin k_m z, \tag{9.7.3}$$

where we have defined $\omega_m = k_m c = m\pi c/L$ and made the approximation $\omega^2 - k_m^2 c^2 = (\omega + k_m c)(\omega - k_m c) \cong 2\omega(\omega - \omega_m)$. Equation (9.7.3) implies that different cavity modes are coupled through the z dependence of $\rho_{21}^{(m)}$. In many lasers this coupling is not very important, and for this reason we will replace $\rho_{21}^{(m)}(z, t)$ by its cavity-average value and then simply cancel the two $\sin k_m z$ factors in (9.7.3). For notational convenience we also drop the superscripts (m) to obtain the single-mode, approximate Maxwell equation

$$\frac{d\mathcal{E}}{dt} - i(\omega - \omega_m)\mathcal{E} = \frac{i\omega}{\epsilon_0}N\mu^*\rho_{21}(t). \tag{9.7.4}$$

We have not accounted for the imperfect cavity mirror reflectivities and other effects that attenuate the intracavity field. To account for such effects we add a term $\frac{1}{2}b\mathcal{E}_m$ to the left-hand side of (9.7.4):

$$\frac{d\mathcal{E}}{dt} + \frac{1}{2}b\mathcal{E} - i(\omega - \omega_m)\mathcal{E} = \frac{i\omega}{\epsilon_0}N\mu^*\rho_{21}(t). \tag{9.7.5}$$

In the case of an empty cavity, the right-hand side vanishes and we can assume that $\omega = \omega_m$. In that case (9.7.5) implies that the field amplitude decays exponentially as $e^{-bt/2}$, that is, the field intensity decays as e^{-bt}. In other words, b is the rate at which the field intensity decreases in the absence of any medium inside the laser cavity.

To bring out the essential features of semiclassical laser theory based on the Maxwell–Bloch equations, we will treat only steady-state laser operation here, and use the adiabatic approximation (9.6.15) for ρ_{21}:

$$\rho_{21} = \frac{-i\chi/2}{\beta + i\Delta}(\rho_{22} - \rho_{11}), \tag{9.7.6}$$

where again $\chi = \mu\mathcal{E}/\hbar$. Equation (9.7.5) becomes

$$\frac{d\mathcal{E}}{dt} + \frac{1}{2}b\mathcal{E} - i(\omega - \omega_m)\mathcal{E} = \frac{i\omega}{\epsilon_0}N\mu^*\frac{-i\chi/2}{\beta + i\Delta}(\rho_{22} - \rho_{11})$$

$$= \frac{|\mu|^2\omega}{2\epsilon_0\hbar}\frac{\beta - i\Delta}{\Delta^2 + \beta^2}\mathcal{E}(N_2 - N_1), \tag{9.7.7}$$

or

$$\frac{d\mathcal{E}}{dt} = \frac{1}{2}[-b + 2i(\omega - \omega_m) + c(g - i\delta)]\mathcal{E}, \tag{9.7.8}$$

where

$$g = \frac{|\mu|^2 \omega}{\epsilon_0 c \hbar} \frac{\beta}{\Delta^2 + \beta^2}(N_2 - N_1) \tag{9.7.9}$$

is the gain coefficient and we define

$$\delta = \frac{\Delta}{\beta}g = \frac{\omega_0 - \omega}{\beta}g. \tag{9.7.10}$$

Equation (9.7.8) may be regarded as the fundamental equation of semiclassical laser theory for the laser field. It is coupled to equations for the medium through the dependence of g and δ on $N_2 - N_1$.

Most of the essential aspects of laser behavior are associated with steady-state or cw (continuous wave) operation. In steady state the terms on the right-hand side of (9.7.8) must cancel: $-2i(\omega - \omega_m) + b = c(g - i\delta)$, or

$$g = \frac{b}{c}, \tag{9.7.11}$$

$$\omega_m - \omega = \frac{cg}{2\beta}(\omega - \omega_0). \tag{9.7.12}$$

Equation (9.7.11) is just the gain clamping condition (Section 5.2) for steady-state laser operation. Writing (9.7.12) as [recall (9.6.30)]

$$\omega_m - \omega = \omega[n(\omega) - 1], \tag{9.7.13}$$

we see that it is exactly the Eq. (5.9.2) determining the laser oscillation frequency ω when, as assumed here, the gain medium fills the entire laser cavity.

There is, of course, much more to semiclassical laser theory than the derivation of the steady-state conditions for the gain and the laser oscillation frequency. The most obvious extension of the theory presented in this section is to include multimode lasing and mode coupling effects. This extension is straightforward but cumbersome, especially for inhomogeneously broadened gain media and when hole burning effects are included. Detailed discussions of the semiclassical theory of these effects can be found in the research literature and in books. Our treatment of them based on rate equations in Chapters 4–6 is quite adequate for most practical purposes, and consequently we will not discuss here the more general semiclassical theory of the laser. The principal message of this section is simply that the coupled Maxwell–Bloch equations contain the basic elements of the most rigorous semiclassical theory of the laser.

• As already mentioned, and as discussed further in the following section, the Maxwell–Bloch equations describe various effects that are not accounted for by coupled rate equations for atomic populations and field intensities. We will consider briefly here, in the context of our simplified semiclassical theory of a single-mode laser with a homogeneously broadened gain medium, a rather dramatic example of the difference between Maxwell–Bloch equations and the rate equations for populations and intensities.

To simplify things we will assume that the transition frequency ω_0 coincides with a cavity mode frequency ω_m. Then, according to (5.9.6), we can take $\omega = \omega_m = \omega_0$ and therefore $\Delta = \delta = 0$. Equation (9.5.3a) then implies that $u(t) = 0$ if $u(0) = 0$, which will be the case if the atoms are in either the upper state or the lower state of the lasing transition at time $t = 0$. Then the Maxwell–Bloch equations (9.5.3) and (9.7.5) take the simpler form:

$$\frac{dv}{dt} = -\frac{1}{T_2'} v + \chi w, \tag{9.7.14a}$$

$$\frac{dw}{dt} = -\frac{1}{T_1}(w - w_0) - \chi v, \tag{9.7.14b}$$

$$\frac{d\mathcal{E}}{dt} = -\frac{1}{2} b\mathcal{E} + \frac{\omega}{2\epsilon_0} N\mu^* v. \tag{9.7.14c}$$

For our purposes it is useful to define the new independent variable $t' = t/T_2'$, the new dependent variables $x = \sqrt{B\Lambda}\mathcal{E}/\mathcal{E}_s$, $y = \sqrt{B\Lambda}v/v_s$, and $z = (w_0 - w)/w_s$, where v_s, w_s, and $\mathcal{E}_s$ are the steady-state solutions of Eqs. (9.7.14) for $\mathcal{E}$, v, and w (Problem 9.10), and the new parameters $B = T_2'/T_1$, $\Lambda = w_0/w_s - 1$, $r = w_0/w_s$, and $\sigma = \frac{1}{2}bT_2'$. With these definitions Eqs. (9.7.14) can be written as

$$\frac{dx}{dt'} = -\sigma(x - y), \tag{9.7.15a}$$

$$\frac{dy}{dt'} = -y - xz + rx, \tag{9.7.15b}$$

$$\frac{dz}{dt'} = xy - Bz. \tag{9.7.15c}$$

In this form these are exactly the equations introduced by E. N. Lorenz in 1963 and are now well known in chaos theory as the *Lorenz model*. For certain values of the parameters σ, r, and B, solutions of the Lorenz model equations exhibit "very sensitive dependence on initial conditions," the hallmark of deterministic chaos—the effectively *random* evolution of a system defined by a perfectly *deterministic* set of equations such as (9.7.15). Lorenz offered the now-famous butterfly metaphor for deterministic chaos, which we can paraphrase as follows: Suppose we have a set of equations that we use to make weather forecasts, and that all the initial conditions required to solve these equations (atmospheric pressure, temperature, . . .) are known. We feed all the required data into a computer program that solves our equations, but neglect to account for the fluttering of a butterfly somewhere in Bermuda. If our system of equations is chaotic, the tiny change in initial conditions caused by the butterfly will make detailed long-term weather prediction impossible; whatever the butterfly does might eventually affect, unpredictably, the weather in Beijing!

The (nonzero) steady-state solutions are unstable (Problem 9.10) when the two conditions

$$\sigma > B + 1, \tag{9.7.16a}$$

and

$$r > \frac{\sigma(\sigma + B + 3)}{\sigma - B - 1} \tag{9.7.16b}$$

are satisfied, in which case x, y, and z can evolve chaotically in time. For the single-mode laser model these conditions for chaos become

$$b > \frac{2}{T_2'} + \frac{2}{T_1}, \tag{9.7.17a}$$

$$\frac{w_0}{w_s} > \frac{b\left(\frac{1}{2}bT_2' + T_2'/T_1 + 3\right)}{b - 2/T_1 - 2/T_2'}. \tag{9.7.17b}$$

The first condition requires that the homogeneous linewidth $(1/T_2')$ of the lasing transition is smaller than the cavity bandwidth (b) and, as noted in Section 5.9, this "bad-cavity condition" is *not* satisfied in typical lasers. The second condition requires that the laser be pumped well above threshold, and in fact (9.7.17b) may be shown to require that it be pumped *at least nine times above threshold* (Problem 9.10), a condition that, again, is not realized in most lasers. But lasers have in fact been observed to exhibit unstable and chaotic behavior, and by varying a parameter such as the cavity loss, various *routes to chaos* typical of chaotic systems have been observed in lasers having either homogeneously or inhomogeneously broadened gain media. Since such behavior is deleterious in most applications, and can usually be avoided or controlled, we will not pursue further here the subject of chaotic lasers, except to note the following: If we make the rate-equation approximation in Eqs. (9.7.14) by replacing v by its quasi-steady value $T_2'\chi w$, the resulting two (rate) equations coupling w and $\mathcal{E}$ *cannot* exhibit chaos for any values of the (time-independent) parameters b, T_2', etc. appearing in these equations. In other words, the Maxwell–Bloch equations in this example allow the possibility of chaos, whereas the rate equations predict only nonchaotic, stable laser oscillation.   ●

9.8   RESONANT PULSE PROPAGATION

In Section 8.3 we considered the propagation of an optical pulse in a dispersive medium that does not significantly absorb (or amplify) the pulse, and in Section 6.12 we described the amplification of a pulse in an inverted medium whose population relaxation times are much longer than the pulse duration. These analyses represent approximations to the more general Maxwell–Bloch equations, which allow not only for absorption or amplification but also for pulse durations much shorter than *both* population and dipole ("off-diagonal") relaxation times. For such "ultrashort" pulses we can neglect atomic relaxation processes over times on the order of the pulse duration and replace the optical Bloch equations (9.6.12) by

$$\frac{\partial}{\partial t}(u - iv) = -i\Delta(u - iv) - i\chi w, \tag{9.8.1a}$$

$$\frac{\partial w}{\partial t} = \frac{i}{2}[\chi(u + iv) - \chi^*(u - iv)]. \tag{9.8.1b}$$

In general, both the amplitude and phase of the electric field envelope function $\mathcal{E}(z, t)$ will vary with z and t. In order to focus on a few of the more interesting predictions of the Maxwell–Bloch equations, we will assume here that there is no temporal phase modulation of the field, in which case we can write

$$\frac{\mu}{\hbar}\mathcal{E}(z, t) = A(z, t)e^{i\phi(z)}, \tag{9.8.2}$$

where $A(z, t)$ and $\phi(z)$ are real. Using this form of the field in Eq. (9.6.11), we obtain

$$\frac{\partial A}{\partial z} + \frac{1}{c}\frac{\partial A}{\partial t} + iA\frac{d\phi}{dz} = \frac{iN\omega|\mu|^2}{2\epsilon_0 c\hbar}(u - iv)e^{-i\phi} \equiv \frac{iN\omega|\mu|^2}{2\epsilon_0 c\hbar}(U - iV), \qquad (9.8.3)$$

where U and V are taken to be real. Similarly, from (9.8.1) we obtain

$$\frac{\partial U}{\partial t} = -\Delta V, \qquad (9.8.4a)$$

$$\frac{\partial V}{\partial t} = \Delta U + Aw, \qquad (9.8.4b)$$

$$\frac{\partial w}{\partial t} = -AV. \qquad (9.8.4c)$$

For an atom exactly on resonance ($\Delta = 0$) and in the lower state before the arrival of the pulse, the solutions of these equations are $U(z, t; \Delta = 0) = 0$ and [cf. Eqs. (9.5.15)]

$$V(z, t; \Delta = 0) = -\sin\Theta(z, t), \qquad (9.8.5a)$$

$$w(z, t; \Delta = 0) = -\cos\Theta(z, t), \qquad (9.8.5b)$$

where the pulse area

$$\Theta(z, t) = \int_{-\infty}^{t} A(z, t')\, dt' \qquad (9.8.6)$$

has the same form as (9.5.21) but with two differences: (1) $\Theta(z, t)$ is now z dependent, and (2) the "initial" time $t = -\infty$ refers here to a time long before the arrival at z of the pulse.

A solution of the optical Bloch equations for $\Delta \neq 0$ is obtained straightforwardly if we assume, following McCall and Hahn,[4] that

$$V(z, t; \Delta) = F(\Delta)V(z, t; \Delta = 0) = -F(\Delta)\sin\Theta(z, t), \qquad (9.8.7)$$

where $F(\Delta)$, to be determined, characterizes the response of off-resonant atoms. With this assumption

$$\frac{\partial}{\partial t}w(z, t; \Delta) = -A(z, t)V(z, t; \Delta) = -\frac{\partial\Theta(z, t)}{\partial t}F(\Delta)V(z, t; \Delta = 0)$$

$$= \frac{\partial\Theta(z, t)}{\partial t}\sin\Theta(z, t)F(\Delta), \qquad (9.8.8)$$

---

[4]S. L. McCall and E. L. Hahn, *Physical Review* **183**, 457 (1969).

which we can easily integrate:

$$w(z, t; \Delta) = -1 + F(\Delta) \int_{-\infty}^{t} \frac{\partial \Theta(z, t')}{\partial t'} \sin \Theta(z, t') \, dt'$$
$$= -1 + F(\Delta) - F(\Delta) \cos \Theta(z, t), \tag{9.8.9}$$

where we have again assumed that each atom is in the lower state ($w = -1$) before the arrival of the pulse.

We obtain an equation for $\Theta(z, t)$ as follows. From (9.8.4a) and (9.8.7),

$$\Delta \frac{\partial U}{\partial t} = \Delta^2 F(\Delta) \sin \Theta, \tag{9.8.10}$$

while from (9.8.4b) we obtain, after some straightforward algebra,

$$\frac{\partial^2 \Theta}{\partial t^2} [1 - F(\Delta)] = \Delta \frac{\partial U}{\partial t}. \tag{9.8.11}$$

Equating these two expressions for $\Delta \, \partial U / \partial t$ yields the equation for $\Theta(z, t)$:

$$\frac{\partial^2 \Theta}{\partial t^2} = \frac{\Delta^2 F(\Delta)}{1 - F(\Delta)} \sin \Theta. \tag{9.8.12}$$

Now since Θ is independent of Δ, this equation can only hold if the factor multiplying $\sin \Theta$ is independent of Δ. For reasons that will soon be clear, we denote this factor by $1/\tau_p^2$, i.e.,

$$F(\Delta) = \frac{1}{1 + \Delta^2 \tau_p^2}, \tag{9.8.13}$$

and rewrite (9.8.12) as

$$\frac{\partial^2 \Theta}{\partial t^2} - \frac{1}{\tau_p^2} \sin \Theta = 0. \tag{9.8.14}$$

The solution of this equation satisfying the initial conditions $A = \partial A / \partial t = 0$ at $t = \pm \infty$—that is, the condition that the field consists of a single pulse that vanishes at $t = \pm \infty$—is easily verified to be

$$\Theta(z, t) = 4 \tan^{-1} \left[e^{(t - t_0)/\tau_p} \right], \tag{9.8.15}$$

where t_0 depends on z but not t. Therefore

$$A = \frac{\partial \Theta}{\partial t} = \frac{2}{\tau_p} \operatorname{sech} \left(\frac{t - t_0}{\tau_p} \right), \tag{9.8.16}$$

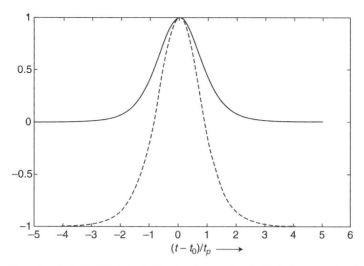

Figure 9.8   The electric field (9.8.17), divided by the amplitude $2\hbar/|\mu|\tau_p$, as a function of time in units of τ_p. Also shown (dashed curve) is the population inversion $w(z, t; \Delta = 0)$ for an atom exactly at resonance.

or, from (9.8.2),

$$|\mathcal{E}(z, t)| = \frac{2\hbar}{|\mu|\tau_p} \, \mathrm{sech}\left(\frac{t - t_0}{\tau_p}\right). \tag{9.8.17}$$

This hyperbolic secant solution for the electric field strength is shown in Fig. 9.8, where it is seen that the time τ_p can be identified with the pulse duration. It is rather remarkable that this form for the electric field strength of a pulse propagating in a medium taken to consist of initially unexcited two-state atoms is determined solely by the optical Bloch equations: *We have not used the field equation (9.6.11) or any other aspect of Maxwell's equations in deducing the expression (9.8.17) for the electric field strength.* Solutions for U, V, and w follow likewise from the Bloch equations and the assumption (9.8.7):

$$U = \frac{2\Delta\tau_p}{1 + \Delta^2\tau_p^2} \, \mathrm{sech}\left(\frac{t - t_0}{\tau_p}\right), \tag{9.8.18a}$$

$$V = \frac{2}{1 + \Delta^2\tau_p^2} \, \mathrm{sech}\left(\frac{t - t_0}{\tau_p}\right) \tanh\left(\frac{t - t_0}{\tau_p}\right), \tag{9.8.18b}$$

$$w = -1 + \frac{2}{1 + \Delta^2\tau_p^2} \, \mathrm{sech}^2\left(\frac{t - t_0}{\tau_p}\right). \tag{9.8.18c}$$

These solutions satisfy the "conservation of probability" condition $U^2 + V^2 + w^2 = 1$. The solution (9.8.18c) for the population inversion is shown in Fig. 9.8. Note that an atom at z is in the lower state ($w = -1$) before and after the arrival of the pulse at z, but that its population inversion is $-1 + 2/(1 + \Delta^2\tau_p^2)$ at $t = t_0$, at which time the pulse at z has its greatest amplitude. In particular, an on-resonance atom is in the upper state ($w = 1$) at $t = t_0$.

As noted following Eqs. (9.6.12), the field equation (9.6.11), or similarly (9.8.3), assumes that the field sources are two-state atoms with the single detuning Δ. More generally, as in a Doppler-broadened gas, for example, there is a distribution of detunings characterized by a lineshape function $G(\Delta)$, $\int_{-\infty}^{\infty} G(\Delta)\,d\Delta = 1$, and (9.8.3) is replaced by the more general form

$$\frac{\partial A}{\partial z} + \frac{1}{c}\frac{\partial A}{\partial t} + iA\frac{d\phi}{dz} = \frac{iN\omega|\mu|^2}{2\epsilon_0 c\hbar}\int_{-\infty}^{\infty} G(\Delta)[U(z, t; \Delta) - iV(z, t; \Delta)]\,d\Delta. \quad (9.8.19)$$

It is shown below that the total pulse area

$$\theta(z) = \Theta(z, \infty) = \int_{-\infty}^{\infty} A(z, t)\,dt \quad (9.8.20)$$

obeys the equation

$$\frac{d\theta}{dz} = -\frac{a}{2}\sin\theta, \quad (9.8.21)$$

with a the absorption coefficient for light with frequency ω equal to the Bohr transition frequency ω_0. We can adduce some important consequences of this equation without even writing its solution. First, it is obvious that pulses with θ equal to an integral multiple of π will not change in area as they propagate, since $d\theta/dz = 0$. Second, by writing $\theta = n\pi + \varepsilon$ in Eq. (9.8.21), where n is a positive or negative integer and $|\varepsilon| \ll 1$, we conclude that pulses with θ equal to odd integral multiples of π are unstable, whereas pulses with θ equal to even integral multiples of π are stable (Problem 9.14). The total pulse area θ approaches an even integral multiple of π after a distance of propagation z large compared to the absorption length a^{-1}. If $0 < \theta < \pi$ initially, the pulse area will diminish to 0 with propagation, whereas if $\theta > \pi$ initially it will tend, according to (9.8.21), to an even integral multiple of π. Note that the solution (9.8.16) deduced from the Bloch equations alone corresponds to an area

$$\theta = \frac{2}{\tau_p}\int_{-\infty}^{\infty} \text{sech}\left(\frac{t - t_0}{\tau_p}\right)dt = 2\pi. \quad (9.8.22)$$

Pulses with initial areas between π and 3π will reshape upon propagation into such 2π hyperbolic secant pulses, whereas solutions of the Maxwell–Bloch equations for an absorbing medium indicate that a single pulse with initial area greater than 4π will split into distinct 2π pulses having different amplitudes, durations, and group velocities.

These consequences of the Maxwell–Bloch equations for "ultrashort" pulse propagation in a resonant medium have been deduced under various approximations, for example, the medium is modeled as an idealized collection of two-state atoms, the field propagates as a plane wave, and the atom–field interaction is treated in the rotating-wave approximation. Nevertheless, as discussed in the following section, these predictions of the Maxwell–Bloch equations have in fact been confirmed in the laboratory.

• To prove the area theorem[4] (9.8.21), we first take the real part of each side of (9.8.19) and integrate from $t = -\infty$ to a time T long after a pulse has passed the point z:

$$\frac{\partial}{\partial z}\Theta(z, T) = \frac{N\omega|\mu|^2}{2\epsilon_0 c\hbar}\int_{-\infty}^{\infty} d\,\Delta G(\Delta)\int_{\infty}^{T} V(z, t; \Delta)\,dt$$

$$= -\frac{N\omega|\mu|^2}{2\epsilon_0 c\hbar}\int_{-\infty}^{\infty} d\Delta\,\frac{G(\Delta)}{\Delta}\int_{-\infty}^{T}\frac{\partial U(z, t; \Delta)}{\partial t}\,dt$$

$$= -\frac{N\omega|\mu|^2}{2\epsilon_0 c\hbar}\int_{-\infty}^{\infty} d\Delta\,\frac{G(\Delta)}{\Delta}U(z, T; \Delta). \qquad (9.8.23)$$

Let $T_0 < T$ be some other time after which a pulse has passed the point z. Since from time T_0 to time T an atom at z is not affected by any field, we can solve Eqs. (9.8.4a) and (9.8.4b) with $A = 0$:

$$U(z, T; \Delta) = U(z, T_0; \Delta)\cos\Delta(T - T_0) - V(z, T_0; \Delta)\sin\Delta(T - T_0) \qquad (9.8.24)$$

and therefore

$$\frac{\partial\Theta(z, T)}{\partial z} = -\frac{N\omega|\mu|^2}{2\epsilon_0 c\hbar}\left[\int_{-\infty}^{\infty} d\Delta\,\frac{G(\Delta)}{\Delta}U(z, T_0; \Delta)\cos\Delta(T - T_0)\right.$$

$$\left. -\int_{-\infty}^{\infty} d\Delta\,\frac{G(\Delta)}{\Delta}V(z, T_0; \Delta)\sin\Delta(T - T_0)\right]. \qquad (9.8.25)$$

The first integral in brackets oscillates rapidly for large $T - T_0$ and can be assumed to average to zero, except possibly near $\Delta = 0$. But Eqs. (9.8.4) imply that U is an odd function of Δ, and therefore that $[G(\Delta)/\Delta]U(z, T_0; \Delta)$ remains finite as $\Delta \rightarrow 0$. We can therefore assume that the integral makes no contribution.

On the other hand, $V(z, T_0; \Delta)$ is inferred from (9.8.4) to be an even function of Δ. The dominant contribution to the second integral is therefore

$$V(z, T_0; 0)G(0)\int_{-\infty}^{\infty} d\Delta\,\frac{\sin\Delta(T - T_0)}{\Delta} = \pi G(0)V(z, T_0; 0). \qquad (9.8.26)$$

Then, taking $T \rightarrow \infty$ in (9.8.25), we have

$$\frac{d\theta}{dz} = -\left[\frac{N\omega|\mu|^2}{2\epsilon_0 c\hbar}\pi G(0)\right]\sin\theta, \qquad (9.8.27)$$

where we have used (9.8.5a) for an absorbing medium [$w(z, \pm\infty) = -1$] as well as $\theta(z) = \Theta(z, \infty)$. The factor in brackets is exactly half the absorption coefficient $[\lambda^2 A_{21}/8\pi]NS(\nu_0)$, given by (3.12.8) for the case assumed here of nondegenerate energy levels, when we identify $G(0)$ with the lineshape function $S(\omega)$ evaluated at $\omega = \omega_0$.

The proof of the area theorem assumes that there is a distribution $G(\Delta)$ of detunings, or in other words that the medium is inhomogeneously broadened, and that every atom of the medium is in the lower state when the field acting on it vanishes. In the case of an amplifier with every atom excited when there is no field, the area theorem takes the same form but with $-a/2$ in (9.8.21) replaced by $a/2$. For an amplifier the stable pulse areas are *odd* integral muliples of π. •

9.9 SELF-INDUCED TRANSPARENCY

One of the most remarkable phenomena predicted by the Maxwell–Bloch equations is the absorptionless propagation of short light pulses in a resonant medium consisting, initially, of entirely unexcited, effectively two-state atoms. Such a medium would, of course, be expected on the basis of the rate equations for the atomic level populations and the field intensity to be strongly absorbant.

This effect—*self-induced transparency*—was first observed experimentally by S. L. McCall and E. L. Hahn after they discovered it in numerical solutions of the Maxwell–Bloch equations.[4] What they found was that, after a few absorption lengths, a short pulse not only propagated with constant area 2π but also that it did so without a change in its shape: $\mathcal{E}(z, t) = \mathcal{E}(t - z/v_g)$, where v_g is the (constant) group velocity. The original McCall–Hahn experiments with pulses from a Q-switched ruby laser and a ruby absorber were quickly followed by observations of self-induced transparency (SIT) in Doppler-broadened gases.

It is easy to see that the Maxwell–Bloch equations admit solutions of this form for a short pulse, that is, a pulse of duration short compared to the relaxation times T_1 and T'_2 of the atoms. As in Section 8.4 we introduce the new independent variable $\tau = t - z/v_g$ and assume, based on the known properties of the McCall–Hahn solutions, that U, V, w, and A are functions of τ rather than of z and t separately. Then, since $\partial/\partial t = d/d\tau$, solutions of Eqs. (9.8.4) have the form (9.8.18) when we again make the assumption (9.8.7):

$$U = \frac{2\Delta\tau_p}{1 + \Delta^2\tau_p^2} \operatorname{sech} \frac{\tau}{\tau_p}, \tag{9.9.1a}$$

$$V = \frac{2}{1 + \Delta^2\tau_p^2} \operatorname{sech} \frac{\tau}{\tau_p} \tanh \frac{\tau}{\tau_p}, \tag{9.9.1b}$$

$$w = -1 + \frac{2}{1 + \Delta^2\tau_p^2} \operatorname{sech}^2 \left(\frac{\tau}{\tau_p}\right). \tag{9.9.1c}$$

Likewise $A(z, t)$ has the hyperbolic secant form (9.8.16):

$$A(z, t) = \frac{2}{\tau_p} \operatorname{sech} \frac{\tau}{\tau_p}. \tag{9.9.2}$$

Since $\partial A/\partial z + (1/c)\partial A/\partial t = (1/c - 1/v_g)\, dA/d\tau$, Eq. (9.8.19) implies that

$$\left(\frac{1}{c} - \frac{1}{v_g}\right) \frac{dA}{d\tau} = \frac{N\omega|\mu|^2}{2\epsilon_0 c\hbar} \int_{-\infty}^{\infty} G(\Delta)V(\tau; \Delta)\, d\Delta. \tag{9.9.3}$$

Therefore, from (9.9.1b) and (9.9.2) and $(d/d\tau)\operatorname{sech}(\tau/\tau_p) = -(1/\tau_p)\operatorname{sech}(\tau/\tau_p)$ $\tanh(\tau/\tau_p)$, we have the following equation for the group velocity:

$$\frac{1}{v_g} = \frac{1}{c} + \frac{N\omega|\mu|^2}{2\epsilon_0 c\hbar} \int_{-\infty}^{\infty} \frac{G(\Delta)\, d\Delta}{\Delta^2 + 1/\tau_p^2}. \tag{9.9.4}$$

These solutions for SIT have an interesting physical interpretation. According to Eq. (9.9.2), an atom at z experiences a field that starts from 0 and builds to a peak amplitude of $2\hbar/|\mu|\tau_p$. During the time it takes for the field to reach this peak amplitude, the probability that the atom is in the upper state of the resonant transition goes from 0 to $\frac{1}{2}(w+1) = [1 + \Delta^2\tau_p^2]^{-1}$ according to (9.9.1c); in particular, an atom with $\Delta = 0$ reaches the upper state when the pulse amplitude is greatest. Then the pulse amplitude and the probability that the atom is excited both decrease to zero. These features are evident in Fig. 9.8. The hyperbolic secant pulse has just the right form that allows it to propagate without changing its shape and without losing energy to the absorbing medium; any pulse reshaping occurs over a few absorption lengths, as the pulse area approaches the stable value required by the area theorem. The pulse "induces" its transparency through its nonlinear interaction with the atoms, giving energy to the atoms during the first "half" of the pulse and taking it back during the second "half" (Fig. 9.8).

This continual transfer of energy from its leading part to its trailing part suggests that the SIT pulse propagates with a group velocity smaller than c. That this is the case is clear from (9.9.4) and the fact that $G(\Delta) \geq 0$. This expression for the group velocity can be simplified when the distribution of atomic frequencies is broad compared to the width $\Delta\omega_p \sim 1/\tau_p$ of the pulse spectrum. In this case the integral in (9.9.4) can be approximated by replacing $G(\Delta)$ in the integrand by $G(0)$:

$$\frac{1}{v_g} \cong \frac{1}{c} + \frac{N\omega|\mu|^2}{2\epsilon_0 c\hbar} G(0) \int_{-\infty}^{\infty} \frac{d\Delta}{\Delta^2 + 1/\tau_p^2} = \frac{N\omega|\mu|^2}{2\epsilon_0 c\hbar} G(0)\pi\tau_p, \tag{9.9.5}$$

or

$$\frac{c}{v_g} \cong 1 + \frac{1}{2}ac\tau_p, \tag{9.9.6}$$

where again a is the on-resonance absorption coefficient. Thus, if the spatial length $c\tau_p$ of the SIT pulse is much greater than the absorption length $1/a$ of the medium, the group velocity can be much less than c.

The phase velocity is determined by $\phi(z)$ [Eq. (9.8.2)]. The imaginary part of (9.8.19) gives an equation for $d\phi/dz$ which, for SIT [$A = A(\tau)$ and $U = U(\tau; \Delta)$] must be independent of z: $\phi(z) = a_0 + a_1 z$ with a_0 some constant phase and

$$a_1 = \frac{N\omega|\mu|^2}{2\epsilon_0 c\hbar} \int_{-\infty}^{\infty} \frac{\Delta G(\Delta)\, d\Delta}{\Delta^2 + 1/\tau_p^2}. \tag{9.9.7}$$

Then Eqs. (9.6.1), (9.8.2), and (9.9.2) give

$$\mathbf{E}(z, t) = \hat{\boldsymbol{\epsilon}}\mathcal{E}(\tau)e^{i\phi(z)}e^{-i\omega t}e^{i\omega z/c} = \hat{\boldsymbol{\epsilon}}\,\frac{\hbar}{\mu}\left[\frac{2}{\tau_p}\text{sech}\,\frac{\tau}{\tau_p}\right]e^{-i\omega t}e^{i(a_0 + a_1 z)}e^{i\omega z/c}$$

$$= \hat{\boldsymbol{\epsilon}}\left[\frac{2\hbar}{\mu\tau_p}\text{sech}\,\frac{\tau}{\tau_p}\right]e^{ia_0}e^{-i\omega t}e^{i\omega(1 + ca_1)z/c}, \tag{9.9.8}$$

which identifies $1 + ca_1$ as the refractive index:

$$n(\omega) = 1 + \frac{N\omega|\mu|^2}{2\epsilon_0 \hbar} \int_{-\infty}^{\infty} \frac{\Delta G(\Delta)\, d\Delta}{\Delta^2 + 1/\tau_p^2}. \tag{9.9.9}$$

When $G(\Delta)$ is the Doppler lineshape function, this expression is easily shown to reduce to (3.15.10) with the homogeneous linewidth $\delta\nu_0$ in that equation replaced by $1/2\pi\tau_p$. In other words, $1/2\pi\tau_p$ plays the role here of an effective homogeneous linewidth. Our short-pulse assumption, $\tau_p \ll T_2'$, is equivalent to the assumption that the homogeneous linewidth is negligible compared to $1/2\pi\tau_p$, which may be taken to define the spectral width of the SIT pulse. The inhomogeneous linewidth, on the other hand, can be large compared to the spectral width of the pulse, as assumed in writing (9.9.6).

Note that the SIT pulse duration τ_p is not fixed by the McCall–Hahn solutions of the Maxwell–Bloch equations. The particular value that τ_p assumes is determined by the input pulse and the pulse reshaping that occurs as the pulse area approaches 2π. In other words, the duration of a pulse will vary as it evolves into a 2π SIT pulse (Problem 9.16).

There is a more obvious type of "induced transparency" than SIT, namely that associated with the saturation of an absorber by sufficiently intense pulses: The leading edge of the pulse can saturate an absorbing transition, so that the lower and upper state probabilities are nearly equal and absorption and amplification (stimulated emission) become about equally likely for the rest of the pulse. This effect, unlike SIT, is incoherent in that it can be described by population rate equations and does not require any off-diagonal coherence of the density matrix. A major difference between SIT and incoherent saturation is the much larger pulse delay, or small group velocity, that is typical of SIT pulses. Group velocities $\sim c/1000$, consistent with the prediction (9.9.6), have been measured in SIT experiments.

- SIT pulses are examples of "solitary waves," waves that are localized in space and in time and that propagate without a change in shape. But they are also examples of very special solitary waves called *solitons*. What distinguishes solitons from other solitary waves is that two "colliding" solitons emerge essentially unchanged, somewhat like the elastic collision of two particles; the term "soliton" reflects this particle-like property. Solitons occur only in nonlinear systems and are often regarded qualitatively in terms of a balancing between effects that cause a wave to spread or disperse and effects that cause it to focus or collapse. The fact that two solitons can collide "elastically" is remarkable in that one would ordinarily expect them to be greatly distorted due to their nonlinear interaction, when in fact the only remnant of the collision is a phase shift of each soliton.

  The name soliton was introduced in the 1960s, but the first observation of such a wave was made by John Scott Russell in 1834. Russell observed a water wave traveling "without change of form or diminution of speed" on a canal after a barge drawn along a narrow channel by two horses came to a sudden stop. He "followed it on horseback and overtook it still rolling on at a rate of some eight or nine miles an hour, preserving its original figure some thirty feet long and a foot to a foot and a half in height."

  Self-induced transparency is but one example of how solitary waves and solitons can appear in the nonlinear propagation of laser radiation. Applications of solitons in optical communications are discussed briefly in Chapter 15. (See also Section 10.4). •

9.10   ELECTROMAGNETICALLY INDUCED TRANSPARENCY

A coherent atom-field interaction such as SIT usually, but not always, requires short pulses of radiation. Another "induced transparency" effect occurs when there are two applied fields, each near resonance with an atomic transition, and the atoms can be regarded as *three*-state systems. In the case indicated by the energy-level diagram of Fig. 9.9, a "coupling" field of frequency ω_c can lead to the *electromagnetically induced transparency* (EIT) of the probe field of frequency ω_p. Neither the probe nor the coupling field in EIT is short compared to typical atomic relaxation times. Another interesting feature of EIT is that the group velocity of the probe field can be just a few meters per second—or even zero! In this section we briefly discuss EIT not only to provide another example of a coherent propagation effect but also to show how density matrix equations are derived when more than two atomic states must be accounted for. In other words, we will show how to generalize the optical Bloch equations for a case where more than two atomic states play a significant role.

In terms of the probability amplitudes a_1, a_2, and a_3 for the three states of Fig. 9.9, the time-dependent Schrödinger equation takes the form [Eqs. (9.2.5)]

$$i\hbar\dot{a}_1 = E_1 a_1 + V_{13} a_3, \tag{9.10.1a}$$

$$i\hbar\dot{a}_2 = E_2 a_2 + V_{23} a_3, \tag{9.10.1b}$$

$$i\hbar\dot{a}_3 = E_3 a_3 + V_{31} a_1 + V_{32} a_2. \tag{9.10.1c}$$

Here $V_{ij} = -e\mathbf{x}_{ij} \cdot \mathbf{E}(t)$ vanishes for $ij = 12$ or 21 due to our assumption that the $1 \leftrightarrow 2$ transition is not allowed, and we assume again that the diagonal matrix elements vanish: $V_{ii} = 0$. Taking E_1 as the zero of energy, and defining $\omega_{ji} = (E_j - E_i)/\hbar$, we rewrite (9.10.1) as

$$\dot{a}_1 = -\frac{i}{\hbar} V_{13} a_3, \tag{9.10.2a}$$

$$\dot{a}_2 = -i\omega_{21} a_2 - \frac{i}{\hbar} V_{23} a_3, \tag{9.10.2b}$$

$$\dot{a}_3 = -i\omega_{31} a_3 - \frac{i}{\hbar} V_{31} a_1 - \frac{i}{\hbar} V_{32} a_2. \tag{9.10.2c}$$

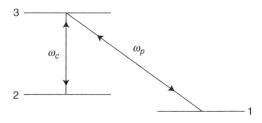

Figure 9.9   Three-state model in which transitions are allowed between states 1 and 3 and between states 2 and 3 but not between states 1 and 2. In the presence of a coupling field of frequency $\omega_c \approx (E_3 - E_2)/\hbar$, a probe field of frequency $\omega_p \approx (E_3 - E_1)/\hbar$ can propagate without absorption.

The electric field $\mathbf{E}(t)$ is the sum of the coupling and probe fields:

$$\mathbf{E}(t) = \tfrac{1}{2}\hat{\boldsymbol{\varepsilon}}_c \mathcal{E}_c e^{-i\omega_c t} + \tfrac{1}{2}\hat{\boldsymbol{\varepsilon}}_p \mathcal{E}_p e^{-i\omega_p t} + \text{c.c.} \qquad (9.10.3)$$

For simplicity we do not explicitly indicate the spatial dependence of the fields.

Based on the assumptions that $\omega_c \approx \omega_{32}$ and $\omega_p \approx \omega_{31}$, we can make an approximation similar to the rotating-wave approximation made in Section 9.2 for a two-state atom. Consider, for example, Eq. (9.10.2c) for $a_3(t)$. The dominant contributions from the second and third terms on the right-hand side will come from terms oscillating approximately as $e^{-i\omega_{31} t}$, which is how $a_3(t)$ oscillates in the absence of any coupling and probe fields. It is similarly the part of the second term on the right-hand side of (9.10.2b) that oscillates approximately as $e^{-i\omega_{21} t}$ ($\omega_{21} \approx \omega_p - \omega_c$) that will most strongly determine the time evolution of $a_2(t)$, while the most significant contribution to the evolution of $a_1(t)$ comes from the part of the right-hand side of (9.10.2a) that oscillates very slowly. Thus, retaining only these resonant contributions to the time evolution of a_1, a_2, and a_3, we approximate Eqs. (9.10.2) by

$$\dot{a}_1 = i\chi_{31}^{(p)*} e^{i\omega_p t} a_3, \qquad (9.10.4a)$$

$$\dot{a}_2 = -i\omega_{21} a_2 + i\chi_{32}^{(c)*} e^{i\omega_c t} a_3, \qquad (9.10.4b)$$

$$\dot{a}_3 = -i\omega_{31} a_3 + i\chi_{31}^{(p)} e^{-i\omega_p t} a_1 + i\chi_{32}^{(c)} e^{-i\omega_c t} a_2, \qquad (9.10.4c)$$

where we define

$$\chi_{ij}^{(p)} = \frac{e\mathbf{x}_{ij} \cdot \hat{\boldsymbol{\varepsilon}}_p \mathcal{E}_p}{2\hbar}, \qquad (9.10.5a)$$

$$\chi_{ij}^{(c)} = \frac{e\mathbf{x}_{ij} \cdot \hat{\boldsymbol{\varepsilon}}_c \mathcal{E}_c}{2\hbar}. \qquad (9.10.5b)$$

Following again the approach used in the case of the two-state atom, we introduce new, slowly varying probability amplitudes c_1, c_2, and c_3 by writing [cf. (9.3.13)]

$$a_1 = c_1(t), \qquad (9.10.6a)$$

$$a_2 = c_2(t)e^{-i(\omega_p - \omega_c)t}, \qquad (9.10.6b)$$

$$a_3 = c_3(t)e^{-i\omega_p t}, \qquad (9.10.6c)$$

in terms of which Eqs. (9.10.4) become

$$\dot{c}_1 = i\chi_{31}^{(p)*} c_3, \qquad (9.10.7a)$$

$$\dot{c}_2 = -i\Delta c_2 + i\chi_{32}^{(c)*} c_3, \qquad (9.10.7b)$$

$$\dot{c}_3 = -i\Delta c_3 + i\chi_{31}^{(p)} c_1 + i\chi_{32}^{(c)} c_2, \qquad (9.10.7c)$$

with a detuning Δ now defined as the difference between the $1 \leftrightarrow 3$ transition frequency and the probe frequency ω_p:

$$\Delta = \omega_{31} - \omega_p. \qquad (9.10.8)$$

We have assumed for simplicity that $\omega_c = \omega_{32}$, and in writing (9.10.7b) we have used the relation $\omega_{21} - (\omega_p - \omega_c) = \omega_{21} - \omega_p + \omega_{32} = -(\omega_p - \omega_{31}) = -\Delta$.

It is now straightforward to derive the equations for the density matrix elements $\rho_{ij} = c_i c_j^*$ defined as in (9.4.1). If the probe field is sufficiently weak, we can assume that the probability amplitudes c_2 and c_3 for the two excited states in Fig. 9.9 are small and that $|c_1(t)|^2 \cong 1$. Then we obtain in particular the following approximate equations for the density matrix elements ρ_{12} and ρ_{13} (Problem 9.18):

$$\dot{\rho}_{12} = i\,\Delta\rho_{12} - i\chi_{32}^{(c)}\rho_{13}, \tag{9.10.9a}$$

$$\dot{\rho}_{13} = i\,\Delta\rho_{13} - i\chi_{31}^{(p)*} - i\chi_{32}^{(c)*}\rho_{12}. \tag{9.10.9b}$$

Under our assumption of weak excitation we do not require equations for any of the other density matrix elements, since we already know their (approximate) values: $\rho_{11}(t) = 1$, $\rho_{23}(t) = \rho_{32}(t) = \rho_{22}(t) = \rho_{33}(t) = 0$. Damping due to collisions and spontaneous emission is assumed to be described by relaxation rates γ_{12} and γ_{13}, and we account for this damping by replacing (9.10.9) by

$$\dot{\rho}_{12} = i(\Delta + i\gamma_{12})\rho_{12} - i\chi_{32}^{(c)}\rho_{13}, \tag{9.10.10a}$$

$$\dot{\rho}_{13} = i(\Delta + i\gamma_{13})\rho_{13} - i\chi_{31}^{(p)*} - i\chi_{32}^{(c)*}\rho_{12}. \tag{9.10.10b}$$

The steady-state solutions of Eqs. (9.10.10), i.e., the solutions for times $t \gg \gamma_{12}^{-1},\ \gamma_{13}^{-1}$, are obtained by setting the derivatives equal to zero on the left-hand sides. For ρ_{13} we obtain the steady-state solution

$$\rho_{13} = \frac{\chi_{31}^{(p)*}(\Delta + i\gamma_{12})}{(\Delta + i\gamma_{12})(\Delta + i\gamma_{13}) - |\chi_{32}^{(c)}|^2}. \tag{9.10.11}$$

The steady-state electric dipole moment at the probe field frequency is therefore [cf. Eq. (9.6.5)]

$$\mathbf{p} = e\mathbf{x}_{13}\rho_{31}e^{-i\omega_p t} + \text{c.c.} = \frac{e\mathbf{x}_{13}\chi_{31}^{(p)}(\Delta - i\gamma_{12})}{(\Delta - i\gamma_{12})(\Delta - i\gamma_{13}) - |\chi_{32}^{(c)}|^2}e^{-i\omega_p t} + \text{c.c.}, \tag{9.10.12}$$

which we can write as

$$\mathbf{p} = \tfrac{1}{2}\alpha(\omega_p)\hat{\boldsymbol{\varepsilon}}_p \mathcal{E}_p e^{-i\omega_p t} + \text{c.c.}, \tag{9.10.13}$$

where $\alpha(\omega_p)$ is the (complex) polarizability at the probe frequency. Using the definitions $\chi_{31}^{(p)} = e\mathbf{x}_{31} \cdot \hat{\boldsymbol{\varepsilon}}_p \mathcal{E}_p / 2\hbar$ [Eq. (9.10.5b)], $\mu_{13} = e\mathbf{x}_{13} \cdot \hat{\boldsymbol{\varepsilon}}_p^*$ [Eq. (9.6.10)], and the relation $\hat{\boldsymbol{\varepsilon}}_p^* \cdot \hat{\boldsymbol{\varepsilon}}_p = 1$, we deduce from (9.10.12) that

$$\alpha(\omega_p) = \frac{|\mu_{13}|^2}{\hbar}\frac{\Delta - i\gamma_{12}}{(\Delta - i\gamma_{12})(\Delta - i\gamma_{13}) - |\chi_{32}^{(c)}|^2}. \tag{9.10.14}$$

We define the complex refractive index $n(\omega_p)$ by the familiar formula $n^2(\omega_p) = 1 + N\alpha(\omega_p)/\epsilon_0$, where N is the number density of atoms, and assume that the medium is sufficiently dilute that we can write

$$n(\omega_p) = 1 + \frac{N}{2\epsilon_0}\alpha(\omega_p) = n_R(\omega_p) + in_I(\omega_p), \tag{9.10.15}$$

where n_R and n_I are the real and imaginary parts of n.

Plane-wave propagation in the z direction implies the spatial dependence

$$e^{in(\omega_p)\omega_p z/c} = e^{-\omega_p n_I(\omega_p)z/c}e^{i\omega_p n_R(\omega_p)z/c} \tag{9.10.16}$$

for a plane-wave probe field. In other words, the (power) attenuation coefficient and the (real) refractive index at the probe frequency are, respectively,

$$
\begin{aligned}
a(\omega_p) &= \frac{\omega_p}{c\epsilon_0}N\alpha_I(\omega_p) \\
&= \frac{N}{\epsilon_0}\frac{\omega_p}{c}\frac{|\mu_{13}|^2}{\hbar}\frac{\gamma_{12}(\gamma_{12}\gamma_{13} + |\chi_{32}^{(c)}|^2) + \Delta^2\gamma_{13}}{[\Delta^2 - \gamma_{12}\gamma_{13} - |\chi_{32}^{(c)}|^2]^2 + \Delta^2(\gamma_{12} + \gamma_{13})^2}
\end{aligned}
\tag{9.10.17}
$$

and

$$
\begin{aligned}
n_R(\omega_p) &= 1 + \frac{N}{2\epsilon_0}\alpha_R(\omega_p) \\
&= 1 + \frac{N}{2\epsilon_0}\frac{|\mu_{13}|^2}{\hbar}\frac{\Delta(\Delta^2 + \gamma_{12}^2 - |\chi_{32}^{(c)}|^2)}{[\Delta^2 - \gamma_{12}\gamma_{13} - |\chi_{32}^{(c)}|^2]^2 + \Delta^2(\gamma_{12} + \gamma_{13})^2}.
\end{aligned}
\tag{9.10.18}
$$

If $\Delta = 0$, i.e., if the probe frequency is tuned exactly to the $1 \leftrightarrow 3$ transition, $n(\omega_p) = 1$ and

$$a(\omega_{31}) = \frac{N\omega_p|\mu_{13}|^2}{\epsilon_0\hbar c}\frac{\gamma_{12}}{\gamma_{12}\gamma_{13} + |\chi_{32}^{(c)}|^2}, \tag{9.10.19}$$

which for zero coupling field ($\chi_{32}^{(c)} = 0$) reduces to the familiar line-center absorption coefficient due to the resonant, homogeneously broadened $1 \leftrightarrow 3$ transition. When $\chi_{32}^{(c)} \neq 0$, however, $a(\omega_{31})$ can be very small if $|\chi_{32}^{(c)}|^2 \gg \gamma_{12}\gamma_{13}$. In other words, a sufficiently intense coupling field can "induce transparency" at the probe frequency.

The dephasing rate γ_{12} of the *nonallowed* transition $1 \leftrightarrow 2$ can be very small compared to the dephasing rate γ_{13} of the (allowed) transition $1 \leftrightarrow 3$. In Fig. 9.10 we plot $a(\omega_p)$ vs. Δ for $\chi_{32}^{(c)} = 0$ and $\chi_{32}^{(c)} = 3\gamma_{13}$, assuming $\gamma_{12} = 0.02\gamma_{13}$. For large enough values of the Rabi frequency $|\chi_{32}^{(c)}|$ the absorption spectrum of the probe field has two peaks separated by $|\chi_{32}^{(c)}|$ and is nearly zero, as in Fig. 9.10b, at the resonance frequency $\omega_p = \omega_{31}$.

EIT arises from the coupling of ρ_{13}, which determines the dipole moment at the probe frequency [Eqs. (9.10.12) and (9.10.13)], to the off-diagonal density matrix element ρ_{12};

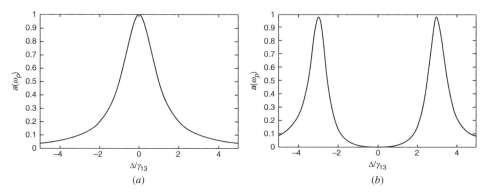

Figure 9.10 $a(\omega_p)$ (in units of $N\omega_p|\mu_{13}|^2/\epsilon_0 c\hbar\,\gamma_{13}$) vs. Δ/γ_{13} as given by Eq. (9.10.17) for $\gamma_{12}/\gamma_{13} = 0.02$ and (a) $\chi_{32}^{(c)} = 0$, (b) $|\chi_{32}^{(c)}| = 3\gamma_{13}$.

this coupling is described by Eqs. (9.10.10). EIT can therefore be attributed to the $1 \leftrightarrow 2$ atomic coherence, that is, to the fact that ρ_{12} is not zero. In a different but equivalent formulation it can also be interpreted as a quantum interference effect in which probability amplitudes interfere destructively to give a net absorption probability of zero. For the realization of EIT it is generally necessary that the decoherence rate γ_{12} be very small, and consequently EIT has been observed primarily in dilute gases.[5] For stationary atoms γ_{12} can be made as small as a few Hz. The larger values in gases are a consequence of so-called transit-time broadening (Section 9.11) arising from the fact that the atoms are in the field for a finite period of time; γ_{12} in this case can then be reduced by adding a noble buffer gas.

Let us now consider the refractive index given by Eq. (9.10.18). For $|\chi_{32}^{(c)}|^2 \gg \Delta^2$, γ_{12}^2, $\gamma_{12}\gamma_{13}$, $|\Delta|\gamma_{13}$,

$$n(\omega_p) \cong 1 - \frac{N}{2\epsilon_0\hbar}\frac{|\mu_{13}|^2}{|\chi_{32}^{(c)}|^2}\Delta \tag{9.10.20}$$

and the group velocity $v_g = c/[n + \omega\,dn/d\omega]$ at the probe frequency is (Problem 9.19)

$$v_g(\omega_p) = \frac{c|\chi_{32}^{(c)}|^2}{|\chi_{32}^{(c)}|^2 + N\omega_p|\mu_{13}|^2/2\epsilon_0\hbar}, \tag{9.10.21}$$

which is approximately $c\left[2\hbar\epsilon_0|\chi_{32}^{(c)}|^2/N\omega_p|\mu_{13}|^2\right]$ for $N\omega_p|\mu_{13}|^2/2\hbar\epsilon_0|\chi_{32}^{(c)}|^2 \gg 1$. From this formula we conclude that the group velocity in EIT can be extremely small—on the order of a few meters per second—compared to c (Problem 9.19).

Such group velocities were first reported in 1999. It has also been observed that EIT pulses can propagate with very little distortion, implying that the distance L_{GVD} defined

[5]A review of experimental and theoretical aspects of EIT is given by S. E. Harris, *Physics Today* (July, 1997), pp. 36–42.

by Eq. (8.4.8) is larger than the propagation length. Thus, an initial Gaussian pulse

$$\mathcal{E}_0(z = 0, t) = Ae^{-t^2/2\tau_p^2} \quad (A = \text{constant}) \tag{9.10.22}$$

will propagate approximately as

$$\mathcal{E}_0(z, t) = Ae^{-(t-z/v_g)^2/2\tau_p^2} \tag{9.10.23}$$

in the EIT medium, compared to the field

$$\mathcal{E}_0(z, t) = Ae^{-(t-z/c)^2/2\tau_p 2} \tag{9.10.24}$$

that would propagate in the absence of the EIT medium. Comparison of (9.10.23) and (9.10.24) shows that the EIT pulse, while retaining its temporal duration as it propagates, undergoes a *spatial* compression by the factor c/v_g, which, as we have seen, can be very large (Problem 9.19).

• Equation (9.10.21) implies that the group velocity can be made very small but not zero since there is no EIT when there is no coupling field. However, if $|\chi_{32}^{(c)}|^2$ is made to vary slowly in time, and if the propagation length of the probe field is not too large, then Eq. (9.10.21) remains applicable and implies a time-dependent group velocity that vanishes when $|\chi_{32}^{(c)}(t)|^2 = 0$. In other words, by varying the coupling field the probe pulse can be brought to a complete stop ($v_g = 0$). In fact an initial probe pulse can be stopped and then fully recovered, with the same amplitude and phase as the original pulse, by appropriately varying the coupling field. The EIT medium can thus serve as a "memory" for the amplitude and phase information of the probe pulse; this information can be stored over times shorter than the decoherence time γ_{12}^{-1}, which can be on the order of a millisecond or longer, depending on the nature of the dephasing collisions in the EIT medium.[6]   •

9.11 TRANSIT-TIME BROADENING AND THE RAMSEY EFFECT

Coherent effects such as Rabi oscillations, self-induced transparency, and electromagnetically induced transparency require that an atomic transition have a long dephasing time and therefore a small homogeneous linewidth. This is usually achieved in impurity-doped crystals at very low-temperatures or by using dilute gases to reduce collisional broadening, together with pulse durations much shorter than radiative lifetimes. But even without collisions and spontaneous emission, the response of an atom to a short pulse exhibits a spectral width just because the atom–field interaction occurs over a finite time t. We noted following Eq. (9.9.9), for example, that the inverse of the pulse duration in SIT can act as an effective homogeneous linewidth. Arguments based on the energy–time uncertainty relation (or Fourier transform theory) suggest that a spectral width $\approx 1/t$ can be associated with an atom–field interaction time t. We can verify this using Eqs. (9.3.15) for the probability amplitudes of the upper and lower states of

---

[6]For a review of "abnormal" group velocities see, for instance, P. W. Milonni, *Fast Light, Slow Light, and Left-Handed Light*, Institute of Physics, Bristol, UK, 2005.

a transition. If the Rabi frequency χ is constant in time and real, the excited-state probability $|c_2(t)|^2$ at time t is given by Eq. (9.3.20b), which for $\chi^2 \ll \Delta^2$ simplifies to

$$P_2(t) \cong \chi^2 \frac{\sin^2 \frac{1}{2}\Delta t}{\Delta^2} = \frac{1}{4}\chi^2 t^2 \frac{\sin^2 \frac{1}{2}\Delta t}{\left(\frac{1}{2}\Delta t\right)^2} \qquad (9.11.1)$$

for atom–field interaction times t that are short compared to any decoherence time. The function $\sin^2 x / x^2$ is plotted in Fig. 10.3. In the present context it implies that the excitation probability $P_2(t)$ drops to about 40% of its peak value, which occurs at $\Delta = 0$, when the field frequency ω is such that $|\Delta|t = \pi$, or $\omega = \omega_0 \pm \pi/t$. We therefore define a nominal resonance width $\Delta\nu = (2\pi)(2\pi/t) = 1/t$.

This effect of a finite atom–field interaction time is important in the case of atomic beams, where collisions are negligible and the transitions probed by an applied field typically have radiative lifetimes much longer than the time an atom spends in the field. An atom with velocity v interacts with the field for a time a/v, where a is the spatial extent of the field. For an atomic beam obtained from a vapor near room temperature, $v \sim 5 \times 10^4$ cm/s and $\Delta\nu \sim v/a \sim 0.5$ MHz if we assume a laser beam of width $a = 1$ mm. This example indicates that *transit-time broadening* is usually small, but it can nevertheless be the largest contributor to the resonance width of an atomic beam. In this example it exceeds radiative broadening if the radiative lifetime is larger than the transit time $a/v = 2$ μs.

- The connection between our derivation of the transit-time spectral width and that based on Fourier transform theory may be seen from the formal solution of Eq. (9.3.15b) with $c_1 \cong 1$ and a time-dependent Rabi frequency $\chi(t) = (e\mathbf{x}_{12} \cdot \hat{\boldsymbol{\epsilon}}/\hbar)E_0(t)$:

$$c_2(t) = \frac{i}{2}\int_{-\infty}^t dt' \chi(t')e^{i\Delta(t'-t)} = \frac{i}{2}\left(\frac{e\mathbf{x}_{12} \cdot \hat{\boldsymbol{\epsilon}}}{\hbar}\right)\int_{-\infty}^t dt\, E_0(t')e^{i\Delta(t'-t)}. \qquad (9.11.2)$$

It is assumed that c_2 is zero at $t = -\infty$, i.e., at times before the field is applied. For a field that is zero except for times between 0 and t, during which it has a constant amplitude E_0,

$$c_2(t) = \frac{i}{2}\left(\frac{e\mathbf{x}_{12} \cdot \hat{\boldsymbol{\epsilon}}}{\hbar}\right)E_0\int_0^t dt' e^{i\Delta(t'-t)} = \frac{i}{2}\left(\frac{e\mathbf{x}_{12} \cdot \hat{\boldsymbol{\epsilon}}}{\hbar}\right)E_0 e^{-i\Delta t/2}\frac{\sin\frac{1}{2}\Delta t}{\frac{1}{2}\Delta}, \qquad (9.11.3)$$

or

$$P_2(t) = |c_2(t)|^2 = \frac{1}{4}\chi^2 t^2 \frac{\sin^2 \frac{1}{2}\Delta t}{\left(\frac{1}{2}\Delta t\right)^2}, \qquad (9.11.4)$$

which, of course, is the same as (9.11.1), obtained now from the squared modulus of the Fourier transform of $E_0(t)$:

$$\left|\int_{-\infty}^\infty dt'\, E_0(t')e^{i\Delta t'}\right|^2 = E_0^2 \left|\int_0^t dt'\, e^{i\Delta t'}\right|^2. \qquad (9.11.5)$$

We can easily generalize to the more realistic case of, say, a monochromatic field with a Gaussian spatial profile. If the field along the direction x in which an atom moves varies as $\exp(-x^2/w^2)$, the atom experiences a time-dependent field $E_0 \exp(-v^2t^2/w^2)$ whose Fourier transform has squared modulus

$$\left| E_0 \int_{-\infty}^{\infty} dt\, e^{i\Delta t} e^{-v^2 t^2/w^2} \right|^2 = E_0^2 \frac{\pi w^2}{v^2} e^{-\Delta^2 w^2/2v^2}. \qquad (9.11.6)$$

This implies a resonance with a maximum at $\Delta = 0$ ($\omega = \omega_0$) and a FWHM width $\Delta\nu = (2v/\pi w)\sqrt{2\ln 2}$, consistent with the width $\Delta\nu \approx v/a$ for the case of a field with constant amplitude over the distance a traversed by the atom.   ●

In some applications, most notably in atomic frequency standards and clocks (Section 14.3), transit-time broadening must be kept as small as possible. It is obviously reduced if atoms are made to move very slowly across a field with a large beam diameter. The large mass of the cesium atom (133 amu) helps in achieving small velocities in atomic beam clocks. Atomic "fountains" employing laser cooling of atoms lead to further improvements (Section 14.4). But it is also necessary to minimize line-broadening effects arising from spatial inhomogeneities of fields through which the atoms move. Since atoms typically traverse a distance ~ 0.5 m in atomic clocks, the field inducing a "clock transition" would have to be largely free of gradients over this distance. Sufficient field uniformity is extremely difficult to realize over such a large distance, and in practice a different approach is followed in which atoms interact with two separate fields. In this approach, the advantages of which are discussed below, an atom is irradiated over a time τ with a resonant field, left free of any oscillatory field for a time T, and then irradiated again for a time τ with a field essentially identical to the first field. In the application to atomic beams, atoms move through a field for a time τ, drift freely for a time T, and then traverse a second field for a time τ.

To understand this method of *separated oscillatory fields* we again consider solutions of the time-dependent Schrödinger equation in the form (9.3.15). For a field with a duration τ and a constant Rabi frequency χ, the general solution of Eqs. (9.3.15) is

$$c_1(\tau) = \left[e^{-i\Delta\tau/2} \left(\cos \tfrac{1}{2}\Omega\tau + \frac{i\Delta}{\Omega} \sin \tfrac{1}{2}\Omega\tau \right) \right] c_1(0)$$

$$+ \left[e^{-i\Delta\tau/2} \frac{i\chi^*}{\Omega} \sin \tfrac{1}{2}\Omega\tau \right] c_2(0), \qquad (9.11.7a)$$

$$c_2(\tau) = \left[e^{-i\Delta\tau/2} \frac{i\chi}{\Omega} \sin \tfrac{1}{2}\Omega\tau \right] c_1(0)$$

$$+ \left[e^{-i\Delta\tau/2} \left(\cos \tfrac{1}{2}\Omega\tau - \frac{i\Delta}{\Omega} \sin \tfrac{1}{2}\Omega\tau \right) \right] c_2(0). \qquad (9.11.7b)$$

We have allowed for χ to be complex in order to account for any phase difference between the two "separated" fields, and have defined $\Omega = \sqrt{|\chi|^2 + \Delta^2}$. During the time interval from τ to $\tau + T$, there is no field ($\chi = 0$), and we can obtain $c_1(\tau + T)$ and $c_2(\tau + T)$ by using (9.11.7) with $\chi = 0$ and replacing τ by T and $c_1(0)$ and $c_2(0)$ by $c_1(\tau)$ and $c_2(\tau)$, respectively:

$$c_1(\tau + T) = e^{-i\Delta T/2} \left(\cos \tfrac{1}{2}\Delta T + i \sin \tfrac{1}{2}\Delta T \right) c_1(\tau) = c_1(\tau), \qquad (9.11.8a)$$

$$c_2(\tau + T) = e^{-i\Delta T/2} \left(\cos \tfrac{1}{2}\Delta T - i \sin \tfrac{1}{2}\Delta T \right) c_2(\tau) = e^{-i\Delta T} c_2(\tau). \qquad (9.11.8b)$$

We can obtain the probability amplitudes $c_1(\tau + T + \tau)$ and $c_2(\tau + T + \tau)$ after the second field of duration τ by using (9.11.7) with $c_1(0)$ and $c_2(0)$ replaced by $c_1(\tau + T)$ and $c_2(\tau + T)$, respectively. We assume that $\chi = |\chi|e^{-i\phi_1}$ for the first field (applied between $t = 0$ and $t = \tau$) and $\chi = |\chi|e^{-i\phi_2}$ for the second field (applied between $t = \tau + T$ and $t = \tau + T + \tau$). The amplitudes $c_1(2\tau + T)$ and $c_2(2\tau + T)$ after the second field are obtained straightforwardly: the upper state probability $P_2(2\tau + T) = |c_2(2\tau + T)|^2$ for $|\chi|^2 \gg \Delta^2$ and $c_1(0) = 1$, $c_2(0) = 0$ is[7]

$$P_2(2\tau + T) = \sin^2 |\chi| \tau \, \cos^2 \tfrac{1}{2}(\Delta T + \phi_1 - \phi_2)$$

$$= 2 \sin^2 \tfrac{1}{2}|\chi|\tau \cos^2 \tfrac{1}{2}|\chi|\tau\{1 + \cos\left[(\omega_0 - \omega)T + \phi_1 - \phi_2\right]\}. \qquad (9.11.9)$$

In the opposite limit $|\chi|^2 \ll \Delta^2$ we obtain

$$P_2(2\tau + T) = 4\frac{|\chi|^2}{\Delta^2} \sin^2 \tfrac{1}{2}\Delta\tau \cos^2 \tfrac{1}{2}(\Delta(T + \tau) + \phi_1 - \phi_2) \qquad (9.11.10)$$

and $c_1(0) = 1$, $c_2(0) = 0$. This generalizes (9.11.1) to the case of two fields separated by a "dark period" T.

The factor $\cos^2 \tfrac{1}{2}(\Delta T + \phi_1 - \phi_2)$ appears in both (9.11.9) and (9.11.10) and is the essential feature of the method of separated oscillatory fields (see Problem 9.20). The oscillation period of this factor can be made very small just by making T longer. Stretching T in the case of an atomic beam is much easier than making τ longer since it only involves increasing the length over which the atoms drift "in the dark" in the space between the two oscillatory fields. From (9.11.10), for example, we see that the transit-broadened lineshape is the same as in (9.11.1), except that it is sharply modulated with a period $2\pi/T$. This is shown in Fig. 9.11. The rapid modulations are known as *Ramsey fringes*. Note that the central peak within the modulated envelope in Fig. 9.11 is much narrower than the envelope, with width approximately $2\pi/T$ rather than the $2\pi/\tau$ implied by (9.11.1).

The phase difference $\phi_1 - \phi_2$ between the two fields has interesting consequences. If $\phi_1 - \phi_2 = \pi$, for instance, $P_2(2\tau + T)$ is *zero* at the resonance frequency $\omega = \omega_0$. If $\phi_1 - \phi_2 = \pi/2$, the resonance curve describing the variation of $P_2(2\tau + T)$ with Δ has a "dispersion" form (cf. Fig. 3.22). Such effects of this phase difference have been employed in experimental studies of resonance curves.[7]

The narrow widths of Ramsey fringes have inspired some highly sensitive spectroscopic techniques ("Ramsey spectroscopy"). But the most important and widespread application of the method of separated oscillatory fields has been in the area of atomic clocks and frequency standards (Section 14.3), where it circumvents the problem, noted earlier, of realizing fields that are sufficiently homogeneous over the distances traversed by the atomic beams; this is discussed further below. The Ramsey method has other advantages, one being that the average over transit times of atoms in the beam results in a resonance curve that is about 40% narrower than that obtained with a single field irradiating the atoms over the same average transit time.[7]

---

[7]The second equality in (9.11.9) puts $P_2(2\tau+T)$ in the form of Eq. (9) of N. F. Ramsey, *Applied Physics* B**60**, 85 (1995), to which the reader is referred for further background and discussion by the inventor of the method of separated oscillatory fields.

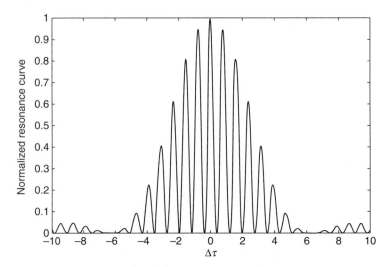

Figure 9.11   The normalized resonance curve (9.11.10) showing Ramsey fringes. In this example $\phi_1 = \phi_2$ and $T = 8\tau$, which means that the time an atom spends in the dark region between fields is 8 times longer than the time τ it spends in each of the fields.

The method of separated oscillatory fields, as we have described it, is difficult to apply in the case of optical transitions. At optical frequencies the atoms pass through many wavelengths as they traverse a distance ~ 1 cm through each of the two separated fields,[8] and the velocity-dependent phase shifts they experience act to wash out the factor $\cos^2 \frac{1}{2}(\Delta T + \phi_1 - \phi_2)$ when it is averaged over atomic velocities. In the case of the microwave fields used in atomic clocks, by contrast, the phase shifts are small compared to the wavelength, so that these Doppler phase shifts are absent. Various techniques involving multiple separated fields and nonlinear effects have been developed to extend the method of separated fields to the optical domain.

- To see why the method of separated fields overcomes the problem of field inhomogeneities, let us assume that the only inhomogeneities are those due to a static magnetic field that an atom traverses for a time T between the two oscillatory fields; this is in fact the situation of interest in atomic clocks. The static magnetic field induces Zeeman shifts in the energy levels of the two states of the atomic transition of interest, and because of its motion through this field an atom experiences a time-dependent frequency shift. Thus, we replace Δ in (9.3.15b) by $\Delta + f(t)$, where $f(t)$ accounts for the frequency shift due to the inhomogeneous magnetic field. In the "dark" region free of any oscillatory field ($\chi = 0$) we therefore write $\dot{c}_2 = -i[\Delta + f(t)]c_2(t)$, or, for $\tau < t < \tau + T$,

$$c_2(t) = c_2(\tau) \exp\left[-i\Delta(t - \tau) - i\int_\tau^t dt' f(t')\right] = e^{-i\overline{\Delta}(t-\tau)}, \tag{9.11.11}$$

where

$$\overline{\Delta} = \Delta + \frac{1}{t - \tau}\int_\tau^t dt' f(t') \tag{9.11.12}$$

[8]The Ramsey method in atomic beam clocks is implemented with a U-shaped cavity, as shown in Fig. 14.13.

is the *average* detuning of the oscillatory field from the atomic resonance frequency. Equation (9.11.11) generalizes (9.11.8b) to the present case in which the atom experiences time-dependent frequency shifts due to an inhomogeneous, static magnetic field in the dark region. Expressions (9.11.9) and (9.11.10) in this case are unchanged except for the replacement of Δ by $\overline{\Delta}$. This replacement implies a possible *displacement* of the resonance peak (if $\overline{\Delta} \neq \Delta$), but no *broadening* of the resonance curve due to the magnetic field inhomogeneities. This is in contrast to the line broadening that results when an atom moves through an inhomogeneous static magnetic field *and* an oscillatory field, and explains the advantage of using separated oscillatory fields.

Norman F. Ramsey was awarded the Nobel Prize in Physics in 1989 for his "invention [in 1949] of the separated oscillatory fields method and its use in the hydrogen maser and other atomic clocks." He has given an interesting account of how he arrived at the idea while he was working on an atomic beam apparatus and worried that he "was probably not going to succeed in making my field uniform enough":[9]

> The idea [of the separated oscillatory fields method] got stimulated by my giving a course in optics where there is a device known as the Michelson Stellar interferometer, which one of my professors at Cambridge University described in a rather dramatic fashion by saying, "If you had a telescope and were looking at a star that is very bright, so you don't have to worry about light gathering and if you didn't have quite enough resolution to tell whether it is a singular or double star. If you take a can of black paint and paint over the middle of the telescope, get twice the resolution. You could then tell whether it was single or double." I was actually giving this somewhat rather dramatic discussion of it to my class, and it suddenly occurred to me that although I was happy to get twice the resolution, I wasn't very much concerned about that. But it occurred to me that if you paint over the middle of the telescope with a can of black paint, it must not depend very much on the quality of the glass underneath the paint. Maybe I could do something analogous that wouldn't depend upon the quality of the magnetic field, if I put radio frequency fields only at the ends. I wasn't clear what the application would be, but it started me thinking in that direction. It's miscellaneous things that make you think in a new direction.   •

The Ramsey effect is a consequence of atomic coherence, as our explanation of it based on probability *amplitudes* rather than populations makes clear. It can be interpreted in various ways (Problem 9.20), all invoking this coherence in one way or another.

9.12 SUMMARY

The main message of this chapter is that it is sometimes inadequate to describe matter interacting with light solely in terms of energy-level occupation probabilities and field intensity. A more general description must be based on density matrices that account not only for energy-level probabilities but also for probability *amplitudes*. And laser radiation, not surprisingly, must in general be described in terms of an electric field rather than just an intensity.

Starting from the time-dependent Schrödinger equation for an atom in a monochromatic field, we obtained differential equations for the evolution in time of the probability amplitudes for different atomic states. For the important case in which only two atomic states have significant occupation probabilities we replaced these equations by those for a "two-state atom" and discussed the Rabi oscillations of the diagonal and off-diagonal

---

[9]From *Norman Ramsey, Electrical Engineer*, an oral history conducted in 1995 by Andrew Goldstein, IEEE History Center, Rutgers University, New Brunswick, NJ.

density-matrix elements for an atom in a resonant field. We then modified these equations to account for collisions and other effects and obtained the optical Bloch equations and, for the propagation of a resonant light pulse in a medium consisting of two-state atoms, the Maxwell–Bloch equations. Aside from effects that require a fully quantum electrodynamical treatment of the field—which seldom need to be considered in applications—these equations provide the framework for the most rigorous approach to laser theory, laser–matter interactions, and the propagation of laser radiation. We found that the Maxwell–Bloch equations reduce to the familiar rate equations for atomic-level populations and field intensity under certain approximations, and that they predict phenomena, such as self-induced transparency, that cannot be explained by the rate equations. We showed how these equations can be generalized to cases in which it is necessary to account for more than two atomic states and for more than a single monochromatic field by considering a "three-state model" and the phenomenon of electromagnetically induced transparency. Finally, we considered transit-time broadening and the Ramsey effect, a consequence of atomic coherence that is especially important in the physics of atomic and molecular beams.

PROBLEMS

9.1. (a) Show that $x_{11} = 0$ [see Eq. (9.3.8)]. To obtain this result you must assume that $|\phi_1(\mathbf{x})|^2$ is an even function of $\mathbf{x}$. More precisely, $\phi_1(\mathbf{x})$ must have a definite *parity*, i.e., $\phi_1(-\mathbf{x})$ is identically the same as either $\phi_1(\mathbf{x})$ (even parity) or $-\phi_1(\mathbf{x})$ (odd parity).

   (b) Show that x_{12} as defined in Eq. (9.3.8) must vanish if the wave functions $\phi_1(\mathbf{x})$ and $\phi_2(\mathbf{x})$ have the same definite parity.

9.2. Every solution of the Schrödinger equation (3.A.2) remains a solution when multiplied by a constant K, and it remains normalized according to (3.A.6) if K is a pure phasor: $K = e^{i\gamma}$. In this sense every $\phi(\mathbf{x})$ has an arbitrary constant phase that can be adjusted for convenience. Assume that an initial phase choice for the wave functions $\phi_1(\mathbf{x})$ and $\phi_2(\mathbf{x})$ leads to the complex matrix element $\chi_{12} = \alpha - i\beta$ (where α and β are real).

   (a) Replace ϕ_1 by $K\phi_1$. Find the value of K that makes χ_{12} real.

   (b) What is the new, purely real value of χ_{12}?

9.3. (a) Find the second-order differential equation satisfied by the probability amplitudes c_1 and c_2 by differentiation and substitution between Eqs. (9.3.15).

   (b) Write the general solution for $c_2(t)$ in terms of $\sin(\Omega t/2)$ and $\cos(\Omega t/2)$, and fix the coefficients to fit the initial condition $c_1(0) = 0$, $c_2(0) = 1$.

   (c) The initial condition specified in (b) is opposite to the one used to obtain the solutions (9.3.18) in the text. Comment on the differences (if any) between (9.3.18) and the solutions obtained in (b).

9.4. (a) For dipole matrix elements with the magnitude ea_0, list the Rabi frequencies that are obtained with these intensities (i) $I = 1$ mW/cm^2, (ii) $I = 100$ W/cm^2, (iii) $I = 1$ MW/cm^2. (b) Find the intensity for which the Rabi frequency is large enough to invalidate the rotating wave approximation.

9.5. Derive Eqs. (9.4.2a)–(9.4.2d) from Eqs. (9.3.15).

9.6. Solve Eq. (9.4.6b) by assuming that χ and $\rho_{22} - \rho_{11}$ are constant. Show that the result approaches (9.4.5) in the limit $t \gg 1/\gamma_c$.

9.7. Find the steady-state values of ρ_{21} and ρ_{22} predicted by Eqs. (9.4.12) and (9.4.15). Note that the true steady-state values must be zero, due to "leakage" from levels 1 and 2 into other atomic levels via the collision rates Γ_1 and Γ_2. Assume that this leakage occurs very slowly, and thus put $\Gamma_1 = \Gamma_2 = 0$. Assume $\rho_{11} + \rho_{22} = 1$ initially, and note that $\rho_{11} + \rho_{22}$ is a constant [add Eqs. (9.4.12) to check].

9.8. Derive Eqs. (9.5.1).

9.9. Let $\mathbf{x}_{21}$ and $\hat{\boldsymbol{\varepsilon}}$ be complex vectors, and let $\hat{\boldsymbol{\varepsilon}}$ be a unit vector in the complex sense: $\hat{\boldsymbol{\varepsilon}} \cdot \hat{\boldsymbol{\varepsilon}}^* = 1$. Designate $\mathbf{x}_{21} = \mathbf{p} + i\mathbf{q}$ and $\hat{\boldsymbol{\varepsilon}} = \boldsymbol{\alpha} + i\boldsymbol{\beta}$, where $\mathbf{p}$, $\mathbf{q}$, $\boldsymbol{\alpha}$, and $\boldsymbol{\beta}$ are purely real vectors.
  (a) Show that $\alpha^2 + \beta^2 = 1$.
  (b) Write $|\mathbf{x}_{21} \cdot \hat{\boldsymbol{\varepsilon}}|^2$ in terms of $\mathbf{p}$, $\mathbf{q}$, $\boldsymbol{\alpha}$, and $\boldsymbol{\beta}$.
  (c) Compute the average of $(\mathbf{p} \cdot \boldsymbol{\alpha})^2$ over all relative $\mathbf{p} - \alpha$ angles. Note that the spherical average of $\cos^2 \theta$ is *not* $\frac{1}{2}$.
  (d) Extrapolate from (c), using (b) as well, to evaluate the average of $|\mathbf{r}_{12} \cdot \hat{\boldsymbol{\varepsilon}}|^2$.

9.10. (a) Find the steady-state solutions v_s, w_s, and $\mathcal{E}_s$ of Eqs. (9.7.14).
  (b) Write $v = v_s + \varepsilon_1$, $w = w_s + \varepsilon_2$, and $\mathcal{E} = \mathcal{E}_s + \varepsilon_3$ in Eqs. (9.7.14) and find the differential equations satisfied by ε_1, ε_2, and ε_3. Assuming that ε_1, ε_2, and ε_3 are small perturbations of the steady-state solutions of the Lorenz model, retain only terms up to first order in these perturbations and write the resulting linear differential equations satisfied by them. Show that these differential equations have solutions that grow exponentially in time when conditions (9.7.17) are met, that is, that the steady-state solutions of the Lorenz model are unstable under these conditions.
  (c) Solve the Lorenz model equations numerically for parameters σ, B, and r satisfying (9.7.17). Consider two slightly different initial conditions and compute the "distance" $d(t) = [(x_1 - x_2)^2 + (y_1 - y_2)^2 + (z_1 - z_2)^2]^{1/2}$ between the two "trajectories" $x_1(t)$, $y_1(t)$, $z_1(t)$ and $x_2(t)$, $y_2(t)$, $z_2(t)$ in the xyz "phase space." This is a simple measure of sensitivity to initial conditions. Do you think that $d(t)$ can be made arbitrarily small for arbitrarily long times t for any two arbitrarily close initial conditions by using a sufficiently accurate algorithm for the numerical solution of the Lorenz-model equations? [A quantitative measure of sensitivity to initial conditions is provided by the number $\lambda \equiv \lim_{t \to \infty} (1/t) \log d(t)$ computed for two initially close trajectories. λ is the largest *Lyapunov exponent* of the dynamical system [e.g., the Lorenz model equations (9.7.14)], and if $\lambda > 0$, the system is said to exhibit *very sensitive dependence on initial conditions*, or *chaos*.]

9.11. Show that the Maxwell–Bloch Equations (9.6.11) and (9.6.12) imply that

$$\epsilon_0 c \left(\frac{\partial}{\partial z} + \frac{1}{c} \frac{\partial}{\partial t} \right) |\mathcal{E}|^2 + N\hbar\omega \frac{\partial w}{\partial t} = -N\hbar\omega \frac{1}{T_1}(w + 1).$$

Determine the physical dimension of the terms and give a physical interpretation of this equation.

9.12. **(a)** Solve Eqs. (16.B.1) numerically for a Gaussian pulse having a duration much shorter than the relaxation time $1/A_{21}$ of the amplifying medium, assuming that the length L of the medium is 10 times the inverse of the small-signal gain coefficient g_0. Verify that the pulse fluence at the end of the medium satisfies the formula (6.12.14).

   (b) Now take the pulse duration to be 5 times the medium relaxation time. Compare the output pulse fluence in this case with that obtained in part (a).

9.13. Write a computer program to solve the Maxwell–Bloch equations (9.6.11) and (9.6.12) numerically for a Gaussian pulse incident on an absorbing medium with $\beta = 2/T_1$. Assume that the length of the medium is 5 times the inverse of the small-signal absorption coefficient, that the detuning $\Delta = \beta$, and that the peak intensity of the incident pulse is equal to the saturation intensity of the medium.

   (a) Compare the time variation at the end of the medium of the intensity and the Bloch variables u, v, and w for incident pulse durations equal to $1/10\beta$, $1/\beta$, and $10/\beta$.

   (b) Compare the propagation velocity of the peak of the pulse with the group velocity for each of the three cases in part (a).

   (c) Repeat part (b) for $\Delta = -\beta$.

   (d) Repeat parts (a)–(c) for the case of an amplifying medium.

9.14. **(a)** Show that the steady-state solutions $\theta = n\pi$ given by the area theorem [Eq. (9.8.21)] are stable if n is an even integer but unstable if n is an odd integer.

   (b) Do you think that it is possible to realize 0π pulses? Draw the envelope of one.

   (c) Is self-induced transparency possible in a homogeneously broadened absorber?

   (d) The SIT hyperbolic secant pulse is "exponentially small" at any point z at times $t \to -\infty$, that is, at times much earlier than the arrival of the pulse at z. However, it is never *identically* zero. Do you think that the hyperbolic secant pulse is therefore unphysical? Can you identify what approximation in the derivation of the hyperbolic secant solution is responsible for this?

9.15. In connection with optical solitons one asks how intense is the field. If $T_2' \sim 1$ ns and the pulse duration is $\tau_p \sim 1$ ps, to satisfy $\tau_p \ll T_2'$, then what peak intensity is required to make a 2π pulse in a resonant medium with $|\mu| \simeq ea_0/10$?

9.16. Consider a short pulse propagating in an absorber with initial area slightly less than 3π. To appreciate how the amplitude and duration of the pulse can change as it evolves into a 2π pulse, make the simplifying assumption of a "square pulse" with amplitude and duration $\mathcal{E}_{3\pi}$ and $\tau_{3\pi}$ initially and $\mathcal{E}_{2\pi}$ and $\tau_{2\pi}$ after the pulse area becomes 2π. In this approximation $\mathcal{E}_{3\pi}\tau_{3\pi} = 3\pi$ and $\mathcal{E}_{2\pi}\tau_{2\pi} = 2\pi$.

   (a) Using the result of Problem 9.11, and assuming that we can effectively take $T_1 \to \infty$, show that $\mathcal{E}_{3\pi}^2\tau_{3\pi} = \mathcal{E}_{2\pi}^2\tau_{2\pi}$ in the square-pulse approximation.

   (b) Show that, as the area approaches 2π, the pulse increases in amplitude and decreases in duration.

9.17. Using the square-pulse model of Problem 9.16, obtain approximate expressions, in terms of τ_p, T_1, and the Doppler width $\delta\nu_D$ of a Doppler-broadened gas, for the intensities necessary for self-induced transparency and for incoherent saturation. Provide some numerical estimates for the required intensities for the two types of "induced transparency."

9.18. Derive the approximate density matrix Eqs. (9.10.9).

9.19. (a) Derive Eq. (9.10.21) for the group velocity at the probe frequency in electro-magnetically induced transparency.

　　　(b) Estimate this group velocity for a probe field at the sodium D_2 line frequency ($\lambda = 589$ nm), an atomic density $N = 3 \times 10^{12}$ cm^{-3}, and a coupling field intensity of 12 mW/cm^2. Assume $|\mu_{23}|^2/|\mu_{13}|^2 \approx 1$. [See L. V. Hau et al., *Nature* **297**, 594 (1999).]

9.20. The two interaction events of the Ramsey effect are often interpreted as two slits in a Young two-slit experiment. Try to explain why this makes sense. To begin your explanation, derive or look up the formula for the light intensity transmitted to the screen of a two-slit experiment, *allowing for each of the slits to have a finite width*. You should be able to make use of the fact that the slit spacing and width parameters have a correspondence with the T and τ parameters in the Ramsey effect.

10 INTRODUCTION TO NONLINEAR OPTICS

Because lasers can produce such high intensities, the propagation of laser radiation often exhibits nonlinear features, such as saturated absorption, that are not normally observed in the propagation of light from conventional sources. The variation of the intracavity field in a laser, where the gain coefficient varies nonlinearly with the intracavity intensity, is itself a consequence of nonlinear propagation. Self-induced transparency is another example of nonlinear (and coherent) propagation.

When *nonlinear optics* was originally explored as a new field made possible by lasers, it consisted of phenomena in which the polarization **P** and therefore the refractive index can be expressed as a low-order polynomial in the electric field. This typically happens when the field is sufficiently intense that its interaction with the medium is nonlinear, but sufficiently far from any resonances that the medium does not exhibit any significant saturation. Whether the study of coherent nonlinear effects such as self-induced transparency, which cannot be described by expressing **P** as a low-order polynomial in **E**, should be considered as part of nonlinear optics is obviously just a question of semantics. We now turn our attention mainly to some "low-order" nonlinear optical effects in which we can characterize the polarization using only the first few terms of a power series in the electric field.

10.1 MODEL FOR NONLINEAR POLARIZATION

To see how such a power series can arise, let us consider again a two-state atom in a monochromatic field. To simplify matters we will take $\chi_{12} = \chi_{21}^* = \mu E_0/\hbar$, where $\mu = e x_{12}$. From Eqs. (9.3.12) it follows identically that

$$\frac{d^2p}{dt^2} + \omega_0^2 p = -\frac{2\mu^2\omega_0}{\hbar} E_0 w(t) \cos \omega t, \qquad (10.1.1a)$$

$$\frac{dw}{dt} = \frac{2E_0}{\hbar\omega_0} \frac{dp}{dt} \cos \omega t, \qquad (10.1.1b)$$

Laser Physics. By Peter W. Milonni and Joseph H. Eberly
Copyright © 2010 John Wiley & Sons, Inc.

where $p = \mu(a_1^* a_2 + a_1 a_2^*)$ is the electric dipole moment expectation value [cf. (9.6.4)] and $w(t) = |a_2|^2 - |a_1|^2$ is again the population difference between the upper and lower states of our two-state atom. [Recall also Eq. (3.A.20) and our discussion of the classical electron oscillator model.] Here we are employing the two-state model not as an approximation to an atom in the case where the applied field is near a resonance and other states of the atom have negligible population, but only as a *model* for how the atom–field interaction can give rise to a nonlinear polarization.

To this end we proceed in a perturbative fashion, denoting by $p^{(n)}(t)$ and $w^{(n)}(t)$ the approximations to p and w valid up to the nth power in the electric field amplitude E_0. If $E_0 = 0$ the atom remains in the lower state, so that $p^{(0)}(t) = 0$ and $w^{(0)}(t) = -1$; $p^{(1)}(t)$ is the solution of the differential equation

$$\frac{d^2 p^{(1)}}{dt^2} + \omega_0^2 p^{(1)} = -\frac{2\mu^2 \omega_0}{\hbar} E_0 w^{(0)}(t) \cos \omega t = \frac{2\mu^2 \omega_0}{\hbar} E_0 \cos \omega t. \tag{10.1.2}$$

Ignoring the homogeneous solution of this equation in order to concentrate on the response to the applied field (and recognizing that the homogeneous solution damps to zero when we include relaxation), we have

$$p^{(1)}(t) = \frac{2\mu^2 \omega_0 E_0/\hbar}{\omega_0^2 - \omega^2} \cos \omega t. \tag{10.1.3}$$

Since $p^{(0)}(t) = 0$, it is clear from (10.1.1b) that $w^{(1)}(t) = 0$. The next approximation to $w(t)$ is $w^{(2)}(t)$, which satisfies

$$\frac{dw^{(2)}}{dt} = \frac{2E_0}{\hbar \omega_0} \frac{dp^{(1)}}{dt} \cos \omega t = -\frac{2E_0}{\hbar} \left(\frac{2\mu^2 E_0/\hbar}{\omega_0^2 - \omega^2} \right) \omega \sin \omega t \cos \omega t$$

$$= -\left[\frac{2\mu^2 \omega E_0^2/\hbar^2}{\omega_0^2 - \omega^2} \right] \sin 2\omega t, \tag{10.1.4}$$

and therefore

$$w^{(2)}(t) = w^{(0)}(t) + \left[\frac{2\mu^2 \omega E_0^2/\hbar^2}{\omega_0^2 - \omega^2} \right] \frac{1}{2\omega} (\cos 2\omega t - 1)$$

$$= -\left[1 + \frac{\mu^2 E_0^2/\hbar^2}{\omega_0^2 - \omega^2} \right] + \left[\frac{\mu^2 E_0^2/\hbar^2}{\omega_0^2 - \omega^2} \right] \cos 2\omega t. \tag{10.1.5}$$

Then for the next-order approximation to $p(t)$ we have

$$\frac{d^2 p^{(3)}}{dt^2} + \omega_0^2 p^{(3)} = -\frac{2\mu^2 \omega_0}{\hbar} E_0 w^{(2)}(t) \cos \omega t$$

$$= \frac{2\mu^2 \omega_0 E_0}{\hbar} \left[1 + \frac{\mu^2 E_0^2/2\hbar^2}{\omega_0^2 - \omega^2} \right] \cos \omega t$$

$$- \frac{\mu^2 \omega_0 E_0}{\hbar} \frac{\mu^2 E_0^2/\hbar^2}{\omega_0^2 - \omega^2} \cos 3\omega t, \tag{10.1.6}$$

where we have used the identity $\cos \omega t \cos 2\omega t = \frac{1}{2}[\cos \omega t + \cos 3\omega t]$. The solution of (10.1.6) is

$$p^{(3)}(t) = \frac{2\mu^2 \omega_0 E_0/\hbar}{\omega_0^2 - \omega^2}\left[1 + \frac{\mu^2 E_0^2/2\hbar^2}{\omega_0^2 - \omega^2}\right]\cos \omega t$$

$$-\frac{\mu^4 \omega_0/\hbar^3}{(\omega_0^2 - \omega^2)(\omega_0^2 - 9\omega^2)}E_0^3 \cos 3\omega t. \qquad (10.1.7)$$

Then the nonlinear polarization at the *third-harmonic* frequency 3ω in the case of N atoms per unit volume is

$$\mathcal{P}^{(\mathrm{NL})}(3\omega) = -N\frac{\mu^4 \omega_0}{\hbar^3}\frac{1}{(\omega_0^2 - \omega^2)(\omega_0^2 - 9\omega^2)}\mathcal{E}^3(\omega) \qquad [\mathcal{E}(\omega) \equiv E_0], \qquad (10.1.8)$$

while the second term in brackets multiplying $\cos \omega t$ in (10.1.7) gives rise to a nonlinear polarization at the applied field frequency ω. We have obtained this result by a perturbative solution, without the rotating-wave approximation, of Eqs. (10.1.1) for a two-state atom in a monochromatic field. We have not included damping terms as in the optical Bloch equations; doing so would add transition linewidths to the denominator in (10.1.8). These are needed whenever there is any possibility of a resonance enhancement of a nonlinear polarization, as occurs, for instance, for $\omega \approx \omega_0$ or $\omega \approx \omega_0/3$ in our two-state model.

A similar perturbative solution including all states of an atom leads likewise to an expression for the polarization as a power series expansion in the applied electric field, or in other words to expressions for the *nonlinear* susceptibilities. We now consider some general properties of nonlinear susceptibilities for different media.

10.2 NONLINEAR SUSCEPTIBILITIES

Let us first recall that the ith component of the *linear* polarizability has the general form [Eq. (8.8.3)]

$$\mathcal{P}_i^{(\mathrm{L})}(\omega) = \epsilon_0 \sum_{j=1}^{3}\chi_{ij}(\omega)\mathcal{E}_j(\omega) \quad \left(P_i = \tfrac{1}{2}[\mathcal{P}_i(\omega)e^{-i\omega t} + \text{c.c.}], \; E_j = \tfrac{1}{2}[\mathcal{E}_j(\omega)e^{-i\omega t} + \text{c.c.}]\right),$$

$$(10.2.1)$$

where $\chi_{ij}(\omega)$ is the ij matrix element of the linear susceptibility tensor (Section 8.8). The total polarization at a frequency ω is the sum of the linear polarization and the nonlinear polarization at frequency ω; the latter can result from electric fields at *different* frequencies, as in the third-harmonic nonlinear polarization given by Eq. (10.1.8) in the two-state model. For the lowest order nonlinear polarization at a frequency ω_3 we write

$$\mathcal{P}_i^{(\mathrm{NL})}(\omega_3) = \epsilon_0 \sum_{j=1}^{3}\sum_{k=1}^{3}\chi_{ijk}(-\omega_3, \omega_1, \omega_2)\mathcal{E}_j(\omega_1)\mathcal{E}_k(\omega_2), \qquad (10.2.2)$$

where $\chi_{ijk}(-\omega_3, \omega_1, \omega_2)$ is the nonlinear susceptibility tensor that characterizes the "mixing" of fields at ω_1 and ω_2 to produce a polarization at frequency $\omega_3 = \omega_1 + \omega_2$. We allow for the fact that electric fields in two different directions (j and k) can generate a polarization in yet another direction (i) in an anisotropic medium. To conform with one of several different notational conventions in nonlinear optics, we have introduced a minus sign in front of ω_3 in the definition of the nonlinear susceptibility; with this choice the sum of the three arguments of $\chi_{ijk}(-\omega_3, \omega_1, \omega_2)$ vanishes. We will also employ the definition

$$\mathcal{E}_i(-\omega) \equiv \mathcal{E}_i^*(\omega). \tag{10.2.3}$$

Finally, it is convenient to abbreviate (10.2.2) as

$$\mathcal{P}_i^{(NL)}(\omega_3) = \epsilon_0 \chi_{ijk}(-\omega_3, \omega_1, \omega_2)\mathcal{E}_j(\omega_1)\mathcal{E}_k(\omega_2), \tag{10.2.4}$$

where it is understood that we must sum over the *repeated indices* (they appear twice) j and k on the right. This convention of summing over repeated indices is called the *Einstein summation convention*; it allows us to dispense with the summation symbols Σ_j and Σ_k. Using (10.2.3) we can similarly write, for instance,

$$\mathcal{P}_i^{(NL)}(\omega_2) = \epsilon_0 \chi_{ijk}(-\omega_2, -\omega_1, \omega_3)\mathcal{E}_j(-\omega_1)\mathcal{E}_k(\omega_3). \tag{10.2.5}$$

The χ_{ijk} may be taken to be purely real in lossless media, i.e., when ω_1, ω_2, and ω_3 are far from any absorption resonances. In this case χ_{ijk} satisfies *overall permutation symmetry*, which simply means that subscripts and frequencies together may be freely permuted:

$$\chi_{ijk}(-\omega_1, \omega_2, \omega_3) = \chi_{jik}(\omega_2, -\omega_1, \omega_3) = \chi_{kji}(\omega_3, \omega_2, -\omega_1), \tag{10.2.6}$$

where $\chi_{jik}(\omega_2, -\omega_1, \omega_3)$, for instance, is defined by

$$\mathcal{P}_j^{(NL)}(-\omega_2) \equiv \mathcal{P}_j^*(\omega_2) = \epsilon_0 \chi_{jik}(\omega_2, -\omega_1, \omega_3)\mathcal{E}_i(-\omega_1)\mathcal{E}_k(\omega_3). \tag{10.2.7}$$

Comparing this with

$$\mathcal{P}_i^{(NL)}(\omega_1) = \epsilon_0 \chi_{ijk}(-\omega_1, \omega_2, \omega_3)\mathcal{E}_j(\omega_2)\mathcal{E}_k(\omega_3) \tag{10.2.8}$$

and (10.2.6), we see that, loosely speaking, permutation symmetry means that, in a "three-wave mixing" process involving fields of frequency ω_1, ω_2, and ω_3, *the nonlinear susceptibility for the process is the same regardless of which field is being generated and which fields are doing the generating*. Equation (10.2.7), for example, describes the generation of ω_2 by the mixing of ω_1 and ω_3, whereas (10.2.8) describes the generation of ω_1 by the mixing of ω_2 and ω_3. From (10.2.6) it follows that the nonlinear susceptibilities for these two processes are the same.

Another symmetry property is known to hold approximately in lossless media. Namely, the subscripts i, j, and k may be freely permuted:

$$\chi_{ijk} = \chi_{jik} = \chi_{kji} = \chi_{ikj}. \tag{10.2.9}$$

This is called *Kleinman's symmetry conjecture*. Combined with overall permutation symmetry, it states basically that $\chi_{ijk}(-\omega_1, \omega_2, \omega_3)$ is insensitive to the values of ω_1, ω_2, and ω_3.

The nonlinear susceptibility tensor allows us to describe various effects within the same framework. For instance, $\chi_{ijk}(0, -\omega, \omega)$ describes the generation of a static (dc) polarization due to a field at ω:

$$\mathcal{P}_i^{(\text{NL})}(0) = \epsilon_0 \chi_{ijk}(0, -\omega, \omega)\mathcal{E}_j^*(\omega)\mathcal{E}_k(\omega). \tag{10.2.10}$$

This gives rise to a static electric field inside the medium. Thus, an optical field incident on the material produces a (dc) voltage; this is called optical rectification. Similarly, $\chi_{ijk}(-\omega, \omega, 0)$ describes the generation of a polarization at ω due to the mixing of a field at ω with a static field ($\omega = 0$):

$$\mathcal{P}_i^{(\text{NL})}(\omega) = \epsilon_0 \chi_{ijk}(-\omega, \omega, 0)\mathcal{E}_j(\omega)\mathcal{E}_k(0) = \epsilon_0 A_{ij}(\omega)\mathcal{E}_j(\omega), \tag{10.2.11}$$

where $A_{ij}(\omega) \equiv \chi_{ijk}(-\omega, \omega, 0)\mathcal{E}_k(0)$. Comparing with (10.2.1), we see that $A_{ij}(\omega)$ acts in effect as a contribution to the *linear* susceptibility tensor at frequency ω. This contribution is linearly proportional to the strength of the applied static field and is responsible for the linear electro-optic effect, or Pockels effect, described in Section 6.5.

Quantum theory provides expressions for both linear and nonlinear susceptibilities. Consider, for example, the nonlinear susceptibility $\chi_{ijk}(-2\omega, \omega, \omega)$ for second-harmonic generation, the "degenerate" case ($\omega_1 = \omega_2$, $\omega_3 = 2\omega$) of three-wave mixing (Section 10.5). Suppose for simplicity that the pump field at ω is linearly polarized along the x direction and we are interested in the polarization 2ω in the x direction. Then the nonlinear susceptibility of interest is $\chi_{111}(-2\omega, \omega, \omega)$, and a perturbative solution for the density matrix gives

$$\chi_{111}(-2\omega, \omega, \omega) = \frac{N}{\epsilon_0 \hbar^2} \sum_{m,n} \mu_{gn}\mu_{nm}\mu_{mg} \left[\frac{1}{(\omega - \omega_{ng})(2\omega - \omega_{mg})} - \frac{1}{(\omega + \omega_{ng})(2\omega + \omega_{nm})} \right.$$
$$\left. + \frac{1}{(\omega + \omega_{ng})(2\omega + \omega_{mg})} - \frac{1}{(\omega - \omega_{mg})(2\omega + \omega_{nm})} \right], \tag{10.2.12}$$

where N is the number of molecules per unit volume and the subscript g refers to the ground state. In writing this formula, it is assumed that all the atoms or molecules remain in their ground states. $\omega_{nm} = (E_n - E_m)/\hbar$ is the transition frequency between states n and m, and $\mu_{nm} = ex_{nm}$ is the x component of the transition electric dipole moment between states n and m.

This formula predicts a "resonance enhancement" when $\omega \approx \omega_{mg}$ or $\omega \approx \omega_{mg}/2$ for some transition $g \to m$ of the molecule. Near such a "one-photon" or "two-photon" resonance we must include the effect of a transition linewidth, which has not been done in writing (10.2.12). If the field at such a resonance frequency is a sufficiently short pulse, it may be necessary to account for off-diagonal coherence of the resonant transition, using optical Bloch equations in the case of a one-photon resonance or their generalizations in the case of multiphoton resonances. Or, if the field at such a resonance frequency has a duration long compared to the off-diagonal coherence time of the resonant transition and is sufficiently intense, it may be necessary to account for saturation of the transition. In practical applications involving nonlinear optical effects such as second-harmonic generation or parametric amplification, the resonances occur only for frequencies in the ultraviolet, and linewidths, coherent effects, and saturation may be safely ignored. Such effects are well described in terms of low-order expansions of the polarization in powers of the electric field, or in other words by nonlinear susceptibilities such as (10.2.12).

Let us recall a *selection rule* for "allowed" (electric dipole) transitions between states m and n of definite (odd or even) parity: For μ_{mn} to be nonzero the states m and n must have different parity (Problem 9.1). Then, referring to formula (10.2.12), if $\mu_{gn} \neq 0$, the states g and n must have different parity; if $\mu_{nm} \neq 0$, then n and m have different parity, which means therefore that m and g must have the same parity. But then μ_{mg} must be zero, and so the right-hand side of (10.2.12) must vanish. In fact it follows that *the non-linear susceptibility* $\chi_{ijk}(-\omega_3, \omega_1, \omega_2)$ *vanishes whenever the medium has states of definite parity.* In other words, three-wave mixing processes can occur only when the quantum states of a medium do not have definite parity.

Materials in which the quantum states have definite parity are said to be "centrosymmetric," or to have inversion symmetry. It might be thought that all materials are centrosymmetric. However, the wave functions of molecules at the lattice sites of a crystal, for instance, are modified by neighboring molecules and as a consequence can lose their parity. Then the preceding argument based on definite parity does not apply, and three-wave mixing can occur. In fact three-wave mixing processes can also occur near the surface of a centrosymmetric material, where the inversion symmetry associated with the bulk of the material is broken. It is also worth noting that three-wave mixing can occur in materials that are not birefringent.

In the case of a single atom, a reversal of the direction of an applied electric field results in a reversal in the direction of the induced electric dipole moment. In a noncentrosymmetric material, however, a reversal in the direction of the electric fields does not result in a reversal in the direction of the induced dipole moment density. That is, changing the signs of $\mathcal{E}_j$ and $\mathcal{E}_k$ in (10.2.5) does not change the sign of $\mathcal{P}_i^{(NL)}$; if it did, (10.2.5) would imply that $\mathcal{P}_i^{(NL)}(\omega_3) = -\mathcal{P}_i^{(NL)}(\omega_3)$, or $\mathcal{P}_i^{(NL)}(\omega_3) = 0$, as in the case of a centrosymmetric material.

It should be noted that the relations between nonlinear susceptibility, nonlinear polarization, and electric fields require some additional consideration in the case of degenerate and higher-order wave-mixing processes. Consider the example of second-harmonic generation (Section 10.5) in which an applied field at frequency ω produces a nonlinear polarization and therefore a field at the second-harmonic frequency 2ω, and assume for simplicity that the electric fields and nonlinear polarization have one and the same Cartesian component. For this three-wave process it is the square of the electric field at the fundamental frequency ω that determines the nonlinear polarization:

$$\left[\tfrac{1}{2}\mathcal{E}(\omega)e^{-i\omega t} + \tfrac{1}{2}\mathcal{E}(-\omega)e^{i\omega t}\right]^2 = \tfrac{1}{4}\mathcal{E}^2(\omega)e^{-2i\omega t} + \tfrac{1}{2}\mathcal{E}(\omega)\mathcal{E}(-\omega) + \tfrac{1}{4}\mathcal{E}^2(-\omega)e^{2i\omega t}, \quad (10.2.13)$$

and the nonlinear polarization at the second-harmonic frequency is

$$P(2\omega) = \tfrac{1}{2}\mathcal{P}^{(L)}e^{-2i\omega t} + \tfrac{1}{2}\mathcal{P}^{(NL)}(2\omega)e^{-2i\omega t} + \text{c.c.} \qquad (10.2.14)$$

The nonlinear susceptibility $\chi(-2\omega,\,\omega,\,\omega)$ is *defined* by relating the coefficients of $e^{-2i\omega t}$ in (10.2.13) and (10.2.14) as follows:

$$\tfrac{1}{2}\mathcal{P}^{(NL)}(2\omega) = \epsilon_0 \chi(-2\omega,\,\omega,\,\omega)\tfrac{1}{4}\mathcal{E}^2(\omega), \qquad (10.2.15)$$

or

$$\mathcal{P}^{(NL)}(2\omega) = \tfrac{1}{2}\epsilon_0 \chi(-2\omega,\,\omega,\,\omega)\mathcal{E}^2(\omega), \qquad (10.2.16)$$

which differs from the definition (10.2.4) that applies for the nondegenerate ($\omega_1 \neq \omega_2$) case. In that case the square of the total electric field acting to produce the field at ω_3 is

$$\left[\tfrac{1}{2}\mathcal{E}(\omega_1)e^{-i\omega_1 t} + \tfrac{1}{2}\mathcal{E}(\omega_2)e^{-i\omega_2 t} + \text{c.c.}\right]^2 = \tfrac{1}{2}\mathcal{E}(\omega_1)\mathcal{E}(\omega_2)e^{-i(\omega_1+\omega_2)t} + \cdots, \qquad (10.2.17)$$

and the polarization at $\omega_3 = \omega_1 + \omega_2$ is given by (10.2.14) with 2ω replaced by ω_3. We define $\epsilon_0\chi(-\omega_3,\,\omega_1,\,\omega_2)$ analogously to $\epsilon_0\chi(-2\omega,\,\omega,\,\omega)$ above, as the ratio between $\tfrac{1}{2}\mathcal{P}^{(NL)}(\omega_3)$ and the factor multiplying $e^{-i\omega_3 t}$ in (10.2.17):

$$\tfrac{1}{2}\mathcal{P}^{(NL)}(\omega_3) = \epsilon_0 \chi(-\omega_3,\,\omega_1,\,\omega_2)\tfrac{1}{2}\mathcal{E}(\omega_1)\mathcal{E}(\omega_2), \qquad (10.2.18)$$

which is equivalent to (10.2.4).

We have considered only the lowest-order optical nonlinearity, a quadratic dependence of the polarization on the electric field associated with three-wave mixing processes. There is also *four*-wave mixing in which three electric fields with frequencies ω_1, ω_2, and ω_3 generate a fourth field at a frequency $\omega_4 = \omega_1 + \omega_2 + \omega_3$. Unlike three-wave mixing, four-wave mixing can occur in centrosymmetric media. The degenerate case $\omega_1 = \omega_2 = \omega_3 = \omega$ and $\omega_4 = 3\omega$ is called *third-harmonic generation*, or frequency tripling. Four-wave mixing processes are characterized by a nonlinear susceptibility $\chi(-\omega_4,\,\omega_1,\,\omega_2,\,\omega_3)$ such that the nonlinear polarization at frequency ω_4 is proportional to $\chi(-\omega_4,\,\omega_1,\,\omega_2,\,\omega_3)\mathcal{E}(\omega_1)\mathcal{E}(\omega_2)\mathcal{E}(\omega_3)$.

An expression analogous to (10.2.12) may be derived for $\chi(-3\omega,\,\omega,\,\omega,\,\omega)$, for instance:

$$\chi(-3\omega,\,\omega,\,\omega,\,\omega) = \frac{N}{\epsilon_0\hbar^3}\sum_{\ell,m,n}\mu_{g\ell}\mu_{\ell m}\mu_{mn}\mu_{ng}A_{\ell mn}, \qquad (10.2.19a)$$

$$A_{\ell mn} = \frac{1}{(\omega_{\ell g} - 3\omega)(\omega_{mg} - 2\omega)(\omega_{ng} - \omega)} + \frac{1}{(\omega_{\ell g} + \omega)(\omega_{mg} + 2\omega)(\omega_{ng} + 3\omega)}$$

$$+ \frac{1}{(\omega_{\ell g} + \omega)(\omega_{mg} + 2\omega)(\omega_{ng} - \omega)} + \frac{1}{(\omega_{\ell g} + \omega)(\omega_{mg} - 2\omega)(\omega_{ng} - \omega)}. \qquad (10.2.19b)$$

As in (10.2.12), it is assumed that the electric fields are all linearly polarized in the same direction and that the molecules remain in their ground states g with probability $\cong 1$. The expression for the nonlinear polarization in this degenerate case is obtained from the same kind of considerations leading to (10.2.16):

$$\mathcal{P}^{(\mathrm{NL})}(3\omega) = \tfrac{1}{4}\epsilon_0 \chi(-3\omega, \omega, \omega, \omega)\mathcal{E}^3(\omega). \tag{10.2.20}$$

Compare this with (10.2.19) when it is specialized to a two-state atom (Problem 10.1):

$$\chi(-3\omega, \omega, \omega, \omega) = -N\frac{4\mu^4\omega_0}{\epsilon_0\hbar^3}\frac{1}{(\omega_0^2 - \omega^2)(\omega_0^2 - 9\omega^2)}. \tag{10.2.21}$$

Together with (10.2.20), this implies the nonlinear polarization (10.1.8). The term multiplying $\cos \omega t$ and proportional to E_0^3 in (10.1.7) is similarly related to the nonlinear susceptibility $\chi(-\omega, \omega, \omega, -\omega)$ in the two-state model. Because the energy eigenstates of an isolated atom have definite parity, the nonlinear polarization involves an odd number of factors of the electric field. Similar calculations, without the assumption of definite-parity states, lead to a nonlinear polarization involving an even number of factors of the electric field, as in three-wave mixing.

10.3 SELF-FOCUSING

Consider a monochromatic field $E = \mathcal{E}(\omega)e^{-i\omega t}$ in a centrosymmetric, isotropic medium. It follows from the discussion in the preceding section that the lowest-order nonlinear polarization at the field frequency is

$$\mathcal{P}^{\mathrm{NL}}(\omega) = \tfrac{3}{4}\epsilon_0\chi(-\omega, \omega, \omega, -\omega)\mathcal{E}(\omega)\mathcal{E}(\omega)\mathcal{E}(-\omega) = \tfrac{3}{4}\epsilon_0\chi(-\omega, \omega, \omega, -\omega)\mathcal{E}(\omega)|\mathcal{E}(\omega)|^2, \tag{10.3.1}$$

which, to simplify the notation, we write as

$$\mathcal{P}^{\mathrm{NL}}(\omega) = \epsilon_0\chi_3(\omega)\mathcal{E}(\omega)|\mathcal{E}(\omega)|^2. \tag{10.3.2}$$

The linear polarization at the field frequency is

$$\mathcal{P}^{\mathrm{L}}(\omega) = \epsilon_0\chi(\omega)\mathcal{E}(\omega) \equiv \epsilon_0\chi_1(\omega)\mathcal{E}(\omega), \tag{10.3.3}$$

and so the polarization up to third order in the electric field strength is

$$\mathcal{P}(\omega) = \mathcal{P}^{\mathrm{L}}(\omega) + \mathcal{P}^{\mathrm{NL}}(\omega) = \epsilon_0\chi_1(\omega)\mathcal{E}(\omega) + \epsilon_0\chi_3(\omega)|\mathcal{E}(\omega)|^2\mathcal{E}(\omega), \tag{10.3.4}$$

and Eq. (8.2.17) for the complex electric field amplitude becomes

$$\nabla^2\mathcal{E} + \frac{\omega^2}{c^2}\mathcal{E} = -\chi_1\frac{\omega^2}{c^2}\mathcal{E} - \chi_3\frac{\omega^2}{c^2}|\mathcal{E}|^2\mathcal{E} \equiv -\chi\frac{\omega^2}{c^2}\mathcal{E} \tag{10.3.5}$$

if the field intensity is not so large as to require that we include contributions to the polarization higher than third order in the electric field.

The fact that $\mathcal{P}$ depends on $|\mathcal{E}|^2$ means that the refractive index depends on $|\mathcal{E}|^2$:

$$n^2 = 1 + \chi = 1 + \chi_1 + \chi_3|\mathcal{E}|^2 \equiv n_0^2 + \chi_3|\mathcal{E}|^2. \tag{10.3.6}$$

Therefore

$$n = n_0 \left(1 + \frac{\chi_3}{n_0^2}|\mathcal{E}|^2\right)^{1/2} \cong n_0 + \frac{\chi_3}{2n_0}|\mathcal{E}|^2 \equiv n_0 + \frac{n_2}{2}|\mathcal{E}|^2, \tag{10.3.7}$$

where $n_0(\omega)$ is the usual, *linear* refractive index of the material. The factor $\frac{1}{2}$ is introduced in defining n_2 in order to conform with a notational convention (among several) in which n is expressed as

$$n = n_0 + n_2 E^2, \qquad E = \tfrac{1}{2}[\mathcal{E}(\omega)e^{-i\omega t} + \text{c.c.}]. \tag{10.3.8}$$

Since the average of E^2 over an optical period is just $|\mathcal{E}|^2/2$, (10.3.7) and (10.3.8) are effectively equivalent. More generally we can write

$$n = n_0 + n_2 E^2 + n_4 E^4 + \cdots, \tag{10.3.9}$$

but the "Kerr nonlinearity" (10.3.8) suffices for our purposes, and indeed it is very often an excellent approximation. It is commonly written in terms of the time-averaged intensity I:

$$n = n_0 + n_{2I}I = n_0 + n_{2I}\frac{c\epsilon_0 n_0}{2}|\mathcal{E}|^2, \tag{10.3.10}$$

where n_{2I}, equal to $n_2/c\epsilon_0 n_0$ in our notation, has dimensions cm^2/W. When the field frequency is far below resonance frequencies of electronic transitions, n_{2I} is positive, which is generally the case at optical frequencies.

One consequence of (10.3.10) is immediately obvious. If $n_2 > 0$, the refractive index is largest where the intensity is largest. A Gaussian laser beam, for example, experiences the largest refractive index at the center of the beam, and the index monotonically decreases away from the beam axis (Fig. 10.1). In effect, then, the medium acts as a (positive) lens tending to focus the beam to a small spot size. Since the beam itself induces the nonlinear polarization and therefore the focusing, this effect is called *self-focusing*.

Figure 10.1 Self-focusing. A Gaussian beam in a medium with refractive index $n = n_0 + (n_2/2)|\mathcal{E}|^2$, $n_2 > 0$, leads to a refractive-index variation as shown. This index acts to focus the beam.

To attain a more quantitative understanding of self-focusing, we consider the nonlinear wave equation

$$\nabla_T^2 \mathcal{E}_0 + 2ik \frac{\partial \mathcal{E}_0}{\partial z} + \frac{k^2 n_2}{n_0} |\mathcal{E}_0|^2 \mathcal{E}_0 = 0 \qquad (10.3.11)$$

that follows from (10.3.5)–(10.3.7) when we define

$$\mathcal{E}(x, y, z) = \mathcal{E}_0(x, y, z)e^{ikz}, \qquad k = n_0\omega/c, \qquad (10.3.12)$$

and make the familiar approximation that the variation with respect to z of $\mathcal{E}_0$ is slow compared to that of e^{ikz}. The term $\nabla_T^2 \mathcal{E}_0$ accounts for diffraction; without it (10.3.11) describes only plane-wave propagation and cannot account for the decrease in beam diameter due to self-focusing. If the beam cross section is characterized by some radius a_0, then

$$\nabla_T^2 \mathcal{E}_0 \sim a_0^{-2} \mathcal{E}_0. \qquad (10.3.13)$$

We expect that self-focusing can compete with diffraction if the last term on the left-hand side of (10.3.11) is comparable in magnitude to the first, that is, if

$$\frac{k^2 n_2}{n_0} |\mathcal{E}_0|^2 \mathcal{E}_0 \sim a_0^{-2} \mathcal{E}_0 \qquad (10.3.14)$$

or

$$a_0^2 |\mathcal{E}_0|^2 \sim \frac{n_0}{k^2 n_2}. \qquad (10.3.15)$$

Since the beam intensity $I = (n_0 c \epsilon_0/2)|\mathcal{E}_0|^2$, we expect, based on (10.3.15), that a critical beam power on the order of

$$(\text{Pwr})_{\text{cr}} \sim (\pi a_0^2)I = \frac{\pi n_0 c \epsilon_0}{2} a_0^2 |\mathcal{E}_0|^2 = \frac{\pi n_0 c \epsilon_0}{2} \frac{n_0}{k^2 n_2} = \frac{c \epsilon_0 \lambda^2}{8 \pi n_2} = \frac{\lambda^2}{8 \pi n_0 n_2 I} \qquad (10.3.16)$$

is required for self-focusing to overcome the diffractive spreading of the beam. This result for $(\text{Pwr})_{\text{cr}}$ is in reasonably good agreement with results obtained by numerical integration of the nonlinear partial differential equation (10.3.11).

Consider as an example the liquid CS_2, which has a rather large n_2 value around 3.2×10^{-14} cm^2/W and a linear refractive index $n_0 \cong 1.6$ at a wavelength of 600 nm.[1] We estimate from (10.3.16) a critical power $(\text{Pwr})_{\text{cr}} \approx 3$ kW. This shows that self-focusing can occur even at relatively modest beam powers in strongly nonlinear materials. For fused silica, by contrast, $n_0 \cong 1.5$, $n_2 \cong 3.2 \times 10^{-16}$ cm^2/W, and $(\text{Pwr})_{\text{cr}} \approx 0.3$ MW, while for air, $n_0 \cong 1$, $n_2 \cong 5.0 \times 10^{-19}$ cm^2/W, and $(\text{Pwr})_{\text{cr}} \approx 300$ MW.

[1]R. W. Boyd, *Nonlinear Optics*, 3rd ed., Academic Press, San Diego, 2008, p. 212.

Note that it is the beam *power* that must exceed a certain threshold for self-focusing, not the beam intensity. Thus, if a beam of a certain power Pwr $<$ (Pwr)$_{cr}$ is focused with a lens to create a very large intensity, self-focusing will not occur, simply because the reduction in beam diameter increases the diffractive spreading that must be overcome to realize net focusing. When self-focusing does occur, it is typical for the beam to break up into several focal spots or "filaments." Filamentation is also observed when there are small-scale transverse intensity ripples on a beam. The ripples produce small-scale transverse refractive-index variations, or effectively an index "diffraction grating" $n_{2I}I(x, y)$, that enhances filamentation.

In many instances self-focusing does not profoundly alter a particular phenomenon of interest, but it does significantly modify the predictions of theoretical analyses that ignore it. For example, self-focusing in liquids can reduce the incident laser intensity required for stimulated Raman scattering by a factor ~ 100. The large power densities resulting from self-focusing can also cause optical breakdown and material damage at lower incident intensities than might otherwise be expected. Self-focusing therefore sets limitations on the design of various high-power laser systems.

• We now outline an approximate approach to the solution of (10.3.11) that supports the estimate (10.3.16) for the critical power for self-focusing and also provides an estimate of the focal length. We begin by writing

$$\mathcal{E}_0(\mathbf{r}) = A(\mathbf{r})e^{ikS(\mathbf{r})}, \tag{10.3.17}$$

where A and S are real functions of $\mathbf{r}$. This approach is used frequently in classical optics; the function $S(\mathbf{r})$ is called the *eikonal*, after the Greek word for "image." Using (10.3.17) in (10.3.11), we obtain

$$\frac{\partial A^2}{\partial z} + \nabla_T \cdot (A^2 \nabla_T S) = 0, \tag{10.3.18a}$$

$$2\frac{\partial S}{\partial z} + (\nabla_T S)^2 = \frac{\nabla_T^2 A}{k^2 A} + \frac{n_2 A^2}{n_0}. \tag{10.3.18b}$$

Assume for A a Gaussian beam form:

$$A(\mathbf{r}) = \frac{A_0 w_0}{w(z)} e^{-r^2/w^2(z)}. \tag{10.3.19}$$

Using this assumption in (10.3.18a), and the replacements $\nabla_T \to \hat{\mathbf{r}}\partial/\partial r$, $\nabla_T^2 \to \partial^2/\partial r^2 + (1/r)\partial/\partial r$ in the case of cylindrical symmetry, we obtain for S the equation

$$\frac{\partial^2 S}{\partial r^2} + \frac{1}{r}\left(1 - \frac{4r^2}{w^2}\right)\frac{\partial S}{\partial r} + \frac{2}{w}\left(\frac{2r^2}{w^2} - 1\right)\frac{dw}{dz} = 0. \tag{10.3.20}$$

It is easily checked that this equation has a solution of the form

$$S = \frac{r^2}{2w}\frac{dw}{dz}. \tag{10.3.21}$$

[Actually we can add to (10.3.21) any function of z and still satisfy (10.3.20), but (10.3.21) is adequate for our purposes here.] Next we use (10.3.21), together with (10.3.19), in (10.3.18b), and the result is the following equation for $w(z)$:

$$r^2 \frac{d^2 w}{dz^2} = \frac{2}{k^2 w} \left(\frac{2r^2}{w^2} - 1 \right) + \frac{n_2 A_0^2 w_0^2}{n_0 w} e^{-2r^2/w^2}. \tag{10.3.22}$$

This equation reveals a flaw in the presumption (10.3.19) that a Gaussian wave form is maintained during propagation in the nonlinear medium: w is supposed to depend only on z, but (10.3.22) implies it varies also with r. It turns out that the "aberrationless approximation" (10.3.19) is valid only near the z axis. Based on this consideration, we write

$$e^{-2r^2/w^2} \approx 1 - \frac{2r^2}{w^2} \tag{10.3.23}$$

in (10.3.22), and then equate coefficients of r^2 on the two sides of the resulting equation. This gives the following equation for w:

$$\frac{d^2 w}{dz^2} = \frac{4/k^2 - 2n_2 A_0^2 w_0^2/n_0}{w^3}, \tag{10.3.24}$$

which is consistent with the assumption that w is independent of r. Assuming $w = w_0$ and $dw/dz = 0$ at $r = 0$, we have the following solution to (10.3.20):

$$w(z) = w_0 \left[1 - \left(\frac{P_0}{P_{\text{cr}}} - 1 \right) \frac{z^2}{z_0^2} \right]^{1/2}, \tag{10.3.25}$$

where

$$(\text{Pwr})_0 = \tfrac{1}{4} \pi n_0 c \epsilon_0 w_0^2 A_0^2 \tag{10.3.26}$$

is the beam power,

$$(\text{Pwr})_{\text{cr}} = \frac{c \epsilon_0 \lambda^2}{8 \pi n_2}, \tag{10.3.27}$$

and

$$z_0 = \frac{k w_0^2}{2} = \frac{n_0 \pi w_0^2}{\lambda}. \tag{10.3.28}$$

From (10.3.25) we see that, for $(\text{Pwr})_0 \ll (\text{Pwr})_{\text{cr}}$,

$$w(z) \approx w_0 \left(1 + \frac{z^2}{z_0^2} \right)^{1/2}, \tag{10.3.29}$$

which corresponds to the growth in spot size of a Gaussian beam, as in Chapter 7. The critical power (10.3.27) agrees with our rough estimate (10.3.16). For a beam power $(\text{Pwr})_0 = (\text{Pwr})_{\text{cr}}$ we see that $w(z) = w_0$, so that diffraction and self-focusing cancel each other and the beam neither spreads nor focuses as it propagates. This is called *self-trapping*.

At a focusing distance

$$z_f \equiv z_0 \left(\frac{(\mathrm{Pwr})_0}{(\mathrm{Pwr})_{\mathrm{cr}}} - 1 \right)^{1/2}$$ (10.3.30)

the beam spot size vanishes, $w(z) = 0$. Of course this cannot occur, for it corresponds to infinite intensity. In practice, various other nonlinear effects, which are left out of our analysis here, come into play to prevent the avalanching of self-focusing. These include, for instance, optical breakdown and stimulated Raman and Brillouin scattering. The distance z_f is nevertheless a useful gauge of the distance required for substantial self-focusing.   •

A nonlinear refractive index can occur simply because of saturation. Equations (3.15.9) and (4.11.5), for example, imply that the contribution of an absorption line to the refractive index is

$$n(\nu) \cong 1 - \frac{\lambda_0}{4\pi} \frac{\nu - \nu_0}{\delta\nu_0} \frac{a_0(\nu_0)}{1 + [(\nu - \nu_0)/\delta\nu_0]^2}$$

$$+ \frac{\lambda_0}{4\pi} \frac{\nu - \nu_0}{\delta\nu_0} a_0(\nu_0) \frac{I_\nu/I_{\nu_0}^{\mathrm{sat}}}{1 + [(\nu - \nu_0)/\delta\nu_0]^2}$$ (10.3.31)

to first order in the intensity. Of course, the complete refractive index is obtained by including the "background" contribution, but it is clear from (10.3.31) that the index will be of the form (10.3.10) with $n_{2I} > 0$ for frequencies on the "blue" side ($\nu > \nu_0$) of the absorption resonance. In other words, saturation of anomalous dispersion can give rise to self-focusing—or self-*defocusing* if $\nu < \nu_0$.

10.4 SELF-PHASE MODULATION

The nonlinear refractive index (10.3.7) has another important consequence. To describe it we consider Eq. (10.3.11) in the plane-wave approximation ($\nabla_T^2 \mathcal{E} \to 0$) but include temporal variations in order to treat the propagation of a pulsed electric field $E(z, t) = \mathcal{E}(z, t)\exp[-i\omega t - k(\omega)z]$:

$$\frac{\partial \mathcal{E}}{\partial z} + \frac{1}{c} \frac{\partial \mathcal{E}}{\partial t} = -\frac{1}{2ik} \left(\frac{k^2 n_2}{n_0} \right) |\mathcal{E}|^2 \mathcal{E}.$$ (10.4.1)

By writing c instead of the group velocity v_g, and neglecting any higher-order derivatives of $\mathcal{E}(z, t)$ with respect to time, we are making the simplifying assumption that the only significant dispersive effect of the medium arises from the nonlinear part of its refractive index. In terms of the (time-averaged) intensity I and the nonlinear index n_{2I}, Eq. (10.4.1) is

$$\frac{\partial \mathcal{E}}{\partial z} + \frac{1}{c} \frac{\partial \mathcal{E}}{\partial t} - i\frac{\omega}{c} n_{2I} I \mathcal{E} = 0,$$ (10.4.2)

or, in terms of $\eta = z$ and $\tau = t - z/c$ [cf. (8.4.4)],

$$\frac{\partial \mathcal{E}}{\partial \eta} - i\frac{\omega}{c}n_{2I}I\mathcal{E}_0 = 0. \tag{10.4.3}$$

For our purposes here an approximation to the physics described by (10.4.3) will suffice. We will assume that the intensity incident on the medium at $z = 0$ is Gaussian in time and retains its form for sufficiently small propagation distances in the medium:

$$I(\eta, \tau) \cong I(0, \tau) = I_0 e^{-\tau^2/\tau_p^2}. \tag{10.4.4}$$

We will assume furthermore that $n_{2I}I(0, \tau)$ is negligible except near the peak value of the pulse intensity and approximate (10.4.3) by

$$\frac{\partial \mathcal{E}}{\partial \eta} - \frac{i\omega}{c}n_{2I}I_0\left(1 - \frac{\tau^2}{\tau_p^2}\right)\mathcal{E} \cong 0. \tag{10.4.5}$$

Then

$$\mathcal{E}(\eta, \tau) \cong \mathcal{E}_0(0, \tau)e^{i\theta(\eta)}e^{-i\phi(\eta,\tau)}, \tag{10.4.6}$$

where $\theta(\eta) = \omega n_{2I}I_0\eta/c$ and

$$\phi(\eta, \tau) = \frac{\omega n_{2I}I_0\tau^2\eta}{c\tau_p^2}. \tag{10.4.7}$$

The instantaneous frequency at z of the pulse is

$$\omega_{\text{inst}}(t) = \frac{\partial}{\partial t}[\omega\tau + \phi(\eta, \tau)] = \omega + \frac{2(\omega z/c)(n_{2I}I_0)(t - z/c)}{\tau_p^2}, \tag{10.4.8}$$

which corresponds to a linear chirp as in (8.5.3). In other words, a nonlinear refractive index $n = n_0 + n_{2I}I$ causes chirping. Because this temporal modulation of the phase depends on the intensity of the pulse itself, it is called *self-phase modulation* (or sometimes "self-chirping"), and it is frequently employed to chirp a pulse in order to subsequently compress it by means of gratings or prisms as discussed in Section 8.4.

Our greatly simplified analysis of self-phase modulation does not by any means capture all the effects that are possible when a pulse propagates in a material. More realistic analyses—often based on numerical methods such as those described in the Appendices—proceed from nonlinear partial differential equations that are more complicated than (10.4.3). If, for example, we allow for transverse spatial variations of a pulse in the paraxial approximation, and for group velocity dispersion, the appropriate equation for $\mathcal{E}(x, y, z, t)$ is (Problem 10.2)

$$\frac{1}{2ik}\nabla_T^2\mathcal{E} + \frac{\partial \mathcal{E}}{\partial \eta} + \frac{i}{2}\beta\frac{\partial^2 \mathcal{E}}{\partial \tau^2} - \frac{i\omega}{2c}n_2|\mathcal{E}|^2\mathcal{E} = 0, \tag{10.4.9}$$

where $\eta = z$, $\tau = t - z/v_g$, and $\beta - (d/d\omega)(1/v_g)$. Analyses of this equation predict *temporal solitons* such that the effects of group velocity dispersion and self-phase modulation cancel each other, and *spatial solitons* in which self-focusing cancels the effect of diffraction. Both types of soliton have been observed.

The dimensionless *B parameter* (or "*B* integral"),

$$B = \frac{2\pi}{\lambda} \int n_{2I} I \, dz, \tag{10.4.10}$$

is often used as a measure of the strength of nonlinear effects due to n_{2I}. Field intensities, propagation distances, and values of n_{2I} such that $B \gtrsim 1$ are generally found to result in significant nonlinear effects, including self-focusing and self-phase modulation.

10.5   SECOND-HARMONIC GENERATION

As we have seen in the example of self-focusing, it is straightforward to obtain a wave equation for the electric field produced by a nonlinear polarization. We will now do this for second-harmonic generation. In this example a field of frequency ω is incident on a (noncentrosymmetric) medium and produces a nonlinear polarization at the frequency 2ω, which in turn acts as the source of a field at 2ω.

We start with the scalar wave equation (8.2.15) for a single component E of the electric field:

$$\nabla^2 E - \frac{1}{c^2} \frac{\partial^2 E}{\partial t^2} = \mu_0 \frac{\partial^2 P}{\partial t^2}. \tag{10.5.1}$$

For simplicity, we will make the plane-wave and monochromatic approximations and write the second-harmonic field as

$$E = \frac{1}{2} \left[\mathcal{E}_{2\omega}(z) e^{-i(2\omega t - k_{2\omega} z)} + \mathcal{E}_{2\omega}^*(z) e^{i(2\omega t - k_{2\omega} z)} \right], \tag{10.5.2}$$

where $k_{2\omega} = n(2\omega)(2\omega/c)$, $n(2\omega)$ being the refractive index of the medium for radiation of frequency 2ω. Assuming $\mathcal{E}_{2\omega}(z)$ to be slowly varying in z compared to $\exp(ik_{2\omega}z)$, we make the familiar approximation

$$\nabla^2 E = \frac{d^2 E}{dz^2} \approx \frac{1}{2} \left(2ik_{2\omega} \frac{d\mathcal{E}_{2\omega}}{dz} - k_{2\omega}^2 \mathcal{E}_{2\omega} \right) e^{-i(2\omega t - k_{2\omega} z)} + \text{c.c.}, \tag{10.5.3}$$

which, together with

$$\frac{1}{c^2} \frac{\partial^2 E}{\partial t^2} = -\frac{2\omega^2}{c^2} [\mathcal{E}_{2\omega} e^{-i(2\omega t - k_{2\omega} z)} + \text{c.c.}], \tag{10.5.4}$$

gives the left-hand side of (10.5.1):

$$\nabla^2 E - \frac{1}{c^2} \frac{\partial^2 E}{\partial t^2} \approx \left[ik_{2\omega} \frac{d\mathcal{E}_{2\omega}}{dz} - \frac{1}{2} \left(k_{2\omega}^2 - \frac{4\omega^2}{c^2} \right) \mathcal{E}_{2\omega} \right] e^{-i(2\omega t - k_{2\omega} z)} + \text{c.c.} \tag{10.5.5}$$

for the second-harmonic field (10.5.2).

The right-hand side of (10.5.1) has both linear and nonlinear contributions at frequency 2ω:

$$P = \tfrac{1}{2}\left[\mathcal{P}_{2\omega}^{(L)}e^{-i(2\omega t - k_{2\omega}z)} + \mathcal{P}_{2\omega}^{(NL)}e^{-2i(\omega t - k_\omega z)} + \text{c.c.}\right],\tag{10.5.6}$$

so that

$$\mu_0\frac{\partial^2 P}{\partial t^2} = -2\mu_0\omega^2\left[\mathcal{P}_{2\omega}^{(L)}e^{-i(2\omega t - k_{2\omega}z)} + \mathcal{P}_{2\omega}^{(NL)}e^{-2i(\omega t - k_\omega z)} + \text{c.c.}\right].\tag{10.5.7}$$

Combining this with (10.5.5) and (10.5.1), we obtain

$$\left[ik_{2\omega}\frac{d\mathcal{E}_{2\omega}}{dz} - \frac{1}{2}\left(k_{2\omega}^2 - \frac{4\omega^2}{c^2}\right)\mathcal{E}_{2\omega}\right]e^{-i(2\omega t - k_{2\omega}z)} = -2\mu_0\omega^2\mathcal{P}_{2\omega}^{(L)}e^{-i(2\omega t - k_{2\omega}z)}$$

$$-2\mu_0\omega^2\mathcal{P}_{2\omega}^{(NL)}e^{-2i(\omega t - k_\omega z)}.\tag{10.5.8}$$

Now $\mathcal{P}_{2\omega}^{(L)} = \epsilon_0\chi(2\omega)\mathcal{E}_{2\omega}$ and

$$k_{2\omega}^2 = (2\omega)^2\epsilon_{2\omega}\mu_0 = \left(\frac{4\omega^2}{c^2}\right)n^2(2\omega) = \frac{4\omega^2}{c^2}[1 + \chi(2\omega)],\tag{10.5.9}$$

where $\chi(2\omega)$ is the linear susceptibility for frequency 2ω, taken to be real because we are assuming that absorption and other loss processes are negligible. The use of these relations in (10.5.8) results in the equation

$$ik_{2\omega}\frac{d\mathcal{E}_{2\omega}}{dz}e^{-i(2\omega t - k_{2\omega}z)} = -2\mu_0\omega^2\mathcal{P}_{2\omega}^{(NL)}e^{-2i(\omega t - k_\omega z)},\tag{10.5.10}$$

or

$$\frac{d\mathcal{E}_{2\omega}}{dz} = i\omega\sqrt{\frac{\mu_0}{\epsilon_{2\omega}}}\mathcal{P}_{2\omega}^{(NL)}e^{i(2k_\omega - k_{2\omega})z}\tag{10.5.11}$$

relating the second-harmonic field to the nonlinear polarization. Finally, it is convenient to define a quantity $\bar{d}$ by writing [(10.2.16)]^2

$$\mathcal{P}^{(NL)}(2\omega) = \tfrac{1}{2}\epsilon_0\chi(-2\omega,\ \omega,\ \omega)\mathcal{E}^2(\omega) \equiv \bar{d}\mathcal{E}_\omega^2(z).\tag{10.5.12}$$

Then (10.5.11) becomes

$$\frac{d\mathcal{E}_{2\omega}}{dz} = i\omega\sqrt{\frac{\mu_0}{\epsilon_{2\omega}}}\,\bar{d}\mathcal{E}_\omega^2(z)e^{i\Delta kz},\tag{10.5.13}$$

---

^{2}Sometimes $\bar{d}$ is defined by writing $\mathcal{P}^{(NL)}(2\omega) = \epsilon_0\bar{d}\mathcal{E}_\omega^2$; then $\bar{d}$ is typically expressed in pm/V (picometers per volt). This definition is employed, for instance, in F. Zernike and J. E. Midwinter, *Applied Nonlinear Optics*, Wiley, New York, 1973. Our definition follows that of A. Yariv, *Quantum Electronics*, 2nd ed., Wiley, New York, 1975. Further discussion of properties of the nonlinear susceptibility, and tabulations of $\bar{d}$ for various crystals, may be found in these books.

where

$$\Delta k = 2k_\omega - k_{2\omega} = \frac{2\omega}{c}[n(\omega) - n(2\omega)]. \qquad (10.5.14)$$

To solve (10.5.13) for the second-harmonic field amplitude $\mathcal{E}_{2\omega}(z)$, we must know $\mathcal{E}_\omega(z)$. In the simplest situation there is little attenuation of the fundamental (frequency ω) wave, and we make the approximation that $\mathcal{E}_\omega$ is a constant:

$$\mathcal{E}_\omega(z) \approx \mathcal{E}_\omega(0). \qquad (10.5.15)$$

In this approximation

$$\mathcal{E}_{2\omega}(z) \approx i\omega\sqrt{\frac{\mu_0}{\epsilon_{2\omega}}}\,\bar{d}\mathcal{E}_\omega^2(0)\int_0^z e^{i\Delta k z'}\,dz' = i\omega\sqrt{\frac{\mu_0}{\epsilon_{2\omega}}}\,\bar{d}\mathcal{E}_\omega^2(0)\left(\frac{1}{i\Delta k}[e^{i\Delta k z} - 1]\right). \quad (10.5.16)$$

Using

$$\frac{1}{i\Delta k}[e^{i\Delta k z} - 1] = \frac{e^{i\Delta k z/2}}{i\Delta k}\left[e^{i\Delta k z/2} - e^{-i\Delta k z/2}\right] = z e^{i\Delta k z/2}\left(\frac{\sin\frac{1}{2}\Delta k z}{\frac{1}{2}\Delta k z}\right), \qquad (10.5.17)$$

we write the solution (10.5.16) for the second-harmonic field, in the approximation of an unattenuated fundamental wave, as

$$\mathcal{E}_{2\omega}(z) \approx i\omega\sqrt{\frac{\mu_0}{\epsilon_{2\omega}}}\,\bar{d}\mathcal{E}_\omega^2(0)z e^{i\Delta k z/2}\left(\frac{\sin\frac{1}{2}\Delta k z}{\frac{1}{2}\Delta k z}\right). \qquad (10.5.18)$$

The first observation in the 1960s of various nonlinear optical effects followed quickly after the first successful laser operation; second-harmonic generation was reported in 1961. Light from a ruby laser ($\lambda = 694.3$ nm) incident on a quartz crystal caused ultraviolet light at half the wavelength of the laser radiation to be generated (Fig. 10.2). *Frequency doublers*, acting like this quartz crystal, are now widely used.

Suppose the nonlinear crystal indicated in Fig. 10.2 has length L. The second-harmonic electric field at the exit face of the crystal is then given by (10.5.18) with

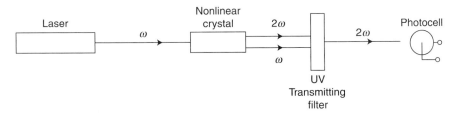

Figure 10.2 Schematic experimental arrangement for detection of second-harmonic generation.

$z = L$, and so

$$|\mathcal{E}_{2\omega}(L)|^2 = \frac{\mu_0 \omega^2 \bar{d}^2}{\epsilon_{2\omega}} |\mathcal{E}_\omega(0)|^4 L^2 \left(\frac{\sin \frac{1}{2} \Delta k L}{\frac{1}{2} \Delta k L} \right)^2. \tag{10.5.19}$$

The (time-averaged) intensities of the fields at ω and 2ω are

$$I_\omega = \frac{1}{2} \sqrt{\frac{\epsilon_\omega}{\mu_0}} |\mathcal{E}_\omega|^2, \qquad I_{2\omega} = \frac{1}{2} \sqrt{\frac{\epsilon_{2\omega}}{\mu_0}} |\mathcal{E}_{2\omega}|^2. \tag{10.5.20}$$

It follows from (10.5.19) that

$$I_{2\omega}(L) = \frac{2\mu_0^{3/2}}{\epsilon_\omega \sqrt{\epsilon_{2\omega}}} \omega^2 \bar{d}^2 I_\omega^2(0) L^2 \left(\frac{\sin \frac{1}{2} \Delta k L}{\frac{1}{2} \Delta k L} \right)^2$$

$$= 2 \left(\frac{\mu_0}{\epsilon_0} \right)^{3/2} \frac{\omega^2 \bar{d}^2}{n^2(\omega) n(2\omega)} I_\omega^2(0) L^2 \left(\frac{\sin \frac{1}{2} \Delta k L}{\frac{1}{2} \Delta k L} \right)^2. \tag{10.5.21}$$

From this we have the *power conversion efficiency* e_{SHG} for second-harmonic generation:

$$e_{\text{SHG}} = \frac{I_{2\omega}(L)}{I_\omega(0)} = 2 \left(\frac{\mu_0}{\epsilon_0} \right)^{3/2} \frac{\omega^2 \bar{d}^2}{n^2(\omega) n(2\omega)} I_\omega(0) L^2 \left(\frac{\sin \frac{1}{2} \Delta k L}{\frac{1}{2} \Delta k L} \right)^2. \tag{10.5.22}$$

If it happens that $n(\omega) = n(2\omega)$, then $\Delta k = 0$ [Eq. (10.5.14)] and therefore

$$\left(\frac{\sin \frac{1}{2} \Delta k L}{\frac{1}{2} \Delta k L} \right)^2 \longrightarrow \lim_{x \to 0} \frac{\sin^2 x}{x^2} = 1 \tag{10.5.23}$$

and

$$e_{\text{SHG}} = 2 \left(\frac{\mu_0}{\epsilon_0} \right)^{3/2} \frac{\omega^2 \bar{d}^2}{n^3} I_\omega(0) L^2, \tag{10.5.24}$$

where $n = n(\omega) = n(2\omega)$.

As an example, consider second-harmonic generation of 347.1-nm radiation in a quartz crystal irradiated by a Q-switched ruby laser. For quartz $\bar{d} \approx 4 \times 10^{-24}$ (mks units), and $n \approx 1.5$. Assuming $I_\omega \approx 10^8 \, \text{W/cm}^2$ and $L = 1$ cm, we compute from (10.5.24) the power conversion efficiency (Problem 10.3)

$$e_{\text{SHG}} \approx 37\%. \tag{10.5.25}$$

This rough estimate is misleading in two respects. First, the second-harmonic generation of 347.1-nm radiation occurs at the expense of the ruby laser radiation. That is, the laser radiation is converted to second-harmonic radiation, and for a conversion efficiency as large as (10.5.25) the approximation of no conversion of the fundamental wave is a poor one. Indeed, by taking L large enough in (10.5.24), we would predict $e_{SHG} > 100\%$, in violation of energy conservation. In other words, the formula (10.5.24) applies only if e_{SHG} is small; otherwise we have to go beyond the approximation (10.5.15) and include the depletion of the pump radiation as it is converted to the second harmonic. In such a more accurate analysis there are *two* equations of the type (10.5.13) coupling $\mathcal{E}_\omega(z)$ and $\mathcal{E}_{2\omega}(z)$ (Section 10.7).

But there is a more serious shortcoming of the computation leading to (10.5.25), namely, the assumption that $\Delta k = 0$. For quartz we have $n(694 \text{ nm}) \approx 1.54$ and $n(347 \text{ nm}) \approx 1.57$. Then, from (10.5.14) (Problem 10.3),

$$\Delta k \approx 5.4 \times 10^5 \, \text{m}^{-1} \qquad (10.5.26)$$

and

$$\Delta k L \approx 5.4 \times 10^3 \qquad (10.5.27)$$

for $L = 1$ cm. Therefore, the last factor in (10.5.22) is

$$\frac{\sin^2 \frac{1}{2}\Delta k L}{(\frac{1}{2}\Delta k L)^2} \approx \frac{\sin^2 2700}{(2700)^2} \sim 10^{-7}. \qquad (10.5.28)$$

The conversion efficiency (10.5.25) obtained using (10.5.23) is replaced by

$$e_{SHG} \sim (37\%)(10^{-7}) \sim 4 \times 10^{-8}, \qquad (10.5.29)$$

a far cry from (10.5.25)! The principle at work here is phase matching, or *mis*matching in this case. Phase matching is extremely important in nonlinear optics, and we devote the next section to it.

10.6 PHASE MATCHING

The dimensionless number ΔkL in (10.5.27), for example, determines the phase mismatching factor $\exp(2ik_\omega L - ik_{2\omega} L)$ between the fundamental and second-harmonic waves over the distance $z = L$. It comes from a difference in the indices of refraction for these two frequencies. Our rough estimate of the power conversion efficiency e_{SHG} for ruby laser radiation in quartz indicates that a phase mismatch can be a strongly negative effect in harmonic generation. Indeed, our estimate of e_{SHG} is consistent with the very weak second-harmonic signals observed in early experimental studies of second-harmonic generation. Efficient second-harmonic generation requires the pump and second-harmonic fields to somehow be *phase matched*.

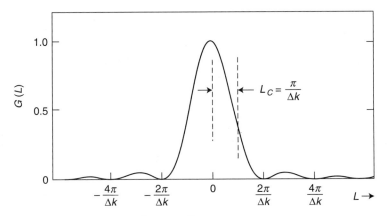

Figure 10.3   The function $G(L) = \sin^2 \alpha L/(\alpha L)^2$, where $\alpha = \Delta k/2$. The phase-matching coherence length L_c is indicated.

The reason for the phase difference $\Delta k\, L$ between the pump and second-harmonic fields is simple: the second-harmonic field propagates with the phase velocity $c/n(2\omega)$, whereas (10.5.6) shows that its nonlinear polarization source has phase velocity $c/n(\omega)$. Because of this difference, the two fields get out of step with each other. Since the medium will only allow the second-harmonic field to have phase velocity $c/n(2\omega)$, there is a reduction in second-harmonic generation determined by $n(2\omega) - n(\omega)$, and this reduction is expressed quantitatively by the factor $(\sin^2 \frac{1}{2}\Delta k\, L)/(\frac{1}{2}\Delta k\, L)^2$.

The function $\sin^2 x/x^2$ is plotted in Fig. 10.3. At $x = \pi/2$ this function drops to about 40% of its peak value at $x = 0$; we define a distance L_c, called the *coherence length*,[3] by $\frac{1}{2}|\Delta k|L_c = \pi/2$, or

$$L_c = \left| \frac{\pi}{\Delta k} \right|. \tag{10.6.1}$$

L_c is the distance over which there is significant generation of the second harmonic. If there is perfect phase matching, L_c is effectively infinite, and e_{SHG} increases as the square of the length L of the crystal, as indicated in Eq. (10.5.24), until the depletion of the pump field becomes important. When phase matching is not realized, however, only the coherence length L_c determines the conversion efficiency—increasing L beyond L_c does no good. Without some method of phase matching the pump and second-harmonic fields, the coherence length is usually small, typically on the order of 10 μm. Figure 10.4 shows experimental data verifying the variation of e_{SHG} with $\Delta k\, L$ according to the function $\sin^2 (\frac{1}{2}\Delta k\, L)/(\frac{1}{2}\Delta k\, L)^2$.

There are various techniques for phase matching the pump and second-harmonic fields, i.e., for making $|\Delta k|$ as small as possible. One common method employs the birefringence of uniaxial crystals used for second-harmonic generation. We will focus on this method, which is sometimes called "angle phase matching" for reasons that will become clear.

---

[3]There should be no confusion between this coherence length and the coherence length introduced in Chapter 13.

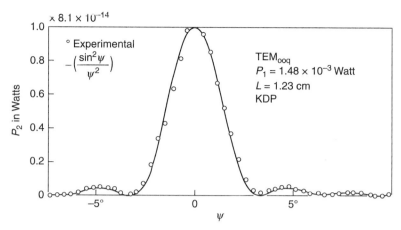

Figure 10.4   Variation of second-harmonic power with $\psi = \frac{1}{2}\Delta k L$, as measured for 1.15-$\mu$m pump radiation in the crystal KDP. [From A. Ashkin, G. D. Boyd, and J. M. Dziedzic, *Physical Review Letters* **11**, 14 (1963).]

As discussed in Sections 5.12 and 8.8, a birefringent, or doubly refracting material is one in which the refractive index depends upon the direction of polarization of a light wave. Only two types of wave can propagate in a uniaxial birefringent crystal: (a) "ordinary" waves plane polarized perpendicular to the plane formed by the optic axis and the direction of propagation and (b) "extraordinary" waves polarized parallel to this plane. The crystal has different refractive indices for ordinary and extraordinary waves.

The refractive index for ordinary waves is denoted n_o. The refractive index n_e of an extraordinary wave depends upon its angle of propagation θ relative to the optic axis. Recall Eq. (8.8.23):

$$\frac{1}{n_e^2(\omega,\,\theta)} = \frac{\cos^2\theta}{n_o^2(\omega)} + \frac{\sin^2\theta}{n_e^2(\omega)}, \tag{10.6.2}$$

where $n_e(\omega) = n_e(\omega,\,\theta = \pi/2)$. This angular dependence of the extraordinary index is the key to angle phase matching.

Consider a positive uniaxial crystal, that is, a birefringent crystal for which $n_e(\omega) > n_o(\omega)$. (Recall Problem 8.8.) For such a crystal we can phase-match the pump and second-harmonic waves by letting the pump wave of frequency ω be an extraordinary wave propagating at an angle θ_p to the optic axis, while the second-harmonic wave of frequency 2ω is an ordinary wave propagating in this direction. Phase matching is achieved if θ_p is chosen such that

$$n_o(2\omega) = n_e(\omega,\,\theta_p), \tag{10.6.3}$$

or

$$\frac{1}{n_o^2(2\omega)} = \frac{1}{n_e^2(\omega,\,\theta_p)} = \frac{\cos^2\theta_p}{n_o^2(\omega)} + \frac{\sin^2\theta_p}{n_e^2(\omega)}. \tag{10.6.4}$$

The solution of this equation for $\sin^2 \theta_p$ is

$$\sin^2 \theta_p = \frac{n_o(\omega)^{-2} - n_o(2\omega)^{-2}}{n_o(\omega)^{-2} - n_e(\omega)^{-2}}. \tag{10.6.5}$$

Assuming the crystal to be normally dispersive [i.e., $n_o(2\omega) > n_o(\omega)$ and therefore $n_o(\omega)^{-2} - n_o(2\omega)^{-2} > 0$], we see that $n_e(\omega) > n_o(\omega)$ is necessary in order to satisfy the requirement that $\sin^2 \theta_p > 0$, and furthermore

$$n_o(\omega)^{-2} - n_o(2\omega)^{-2} < n_o(\omega)^{-2} - n_e(\omega)^{-2}, \tag{10.6.6}$$

or

$$n_e(\omega) > n_o(2\omega) \quad \text{(positive uniaxial crystal)} \tag{10.6.7}$$

is necessary for $\sin^2 \theta_p < 1$. In other words, this method of phase matching can be used in a positive uniaxial crystal when (10.6.7) is satisfied, since $n_o(2\omega) > n_o(\omega)$ and (10.6.7) together imply $n_e(\omega) > n_o(\omega)$, as required above. Problem 10.4 concerns the possibility of angle phase-matched second-harmonic generation when 694.3-nm ruby laser radiation is incident on a quartz crystal.

In practice, the pump wave is sent into the crystal as an extraordinary wave at the angle θ_p to the optic axis. Then the second-harmonic wave is automatically generated as an ordinary wave propagating in the direction θ_p because this is the only direction and polarization for which there is phase matching and substantial second-harmonic generation.

For a negative uniaxial crystal [$n_e(\omega) < n_o(\omega)$] the situation is reversed: The angle between the pump and second-harmonic waves is (Problem 10.5)

$$\sin^2 \theta_p = \frac{n_o(\omega)^{-2} - n_o(2\omega)^{-2}}{n_e(2\omega)^{-2} - n_o(2\omega)^{-2}}, \tag{10.6.8}$$

and phase matching can be achieved if

$$n_e(2\omega) < n_o(\omega) \quad \text{(negative uniaxial crystal).} \tag{10.6.9}$$

Consider the example of second-harmonic generation when 694.3-nm radiation is incident on the negative uniaxial crystal KDP (potassium dihydrogen phosphate, KH_2PO_4). For this crystal $n_o(\omega) = 1.505$, $n_o(2\omega) = 1.534$, $n_e(\omega) = 1.465$, and $n_2(2\omega) = 1.487$. Thus (10.6.9) is satisfied and angle phase matching is achieved with the incident laser beam linearly polarized as an ordinary wave incident at the angle θ_p given by (10.6.8):

$$\sin^2 \theta_p = 0.606, \qquad \theta_p = 51°. \tag{10.6.10}$$

Phase-matched second-harmonic generation can also be realized using two incident fields of frequency ω, one polarized as an ordinary wave and the other as an

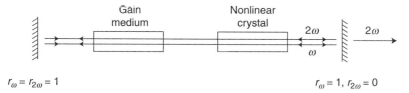

Figure 10.5   Intracavity second-harmonic generation.

extraordinary wave. When the two incident fields have the same polarization, the phase matching is said to be *type I phase matching*; when one is ordinary and the other extraordinary, the phase matching is said to be *type II*.

As we have seen, very high pump-to-second-harmonic conversion efficiencies can be predicted under phase-matched conditions. Under such conditions pump depletion accompanying harmonic generation must be included in a detailed analysis.

• Equation (10.5.24) indicates that the power conversion efficiency for second-harmonic generation is proportional to the pump intensity. High conversion efficiencies in most instances require the high peak intensities available only from pulsed lasers. However, it is desirable for some purposes to generate a cw second-harmonic field. One way to do this is by intracavity second-harmonic generation, that is, by inserting a nonlinear crystal inside the cavity of a cw laser (Fig. 10.5). The basic advantage of this technique is that it exposes the crystal to a cw pump field of much greater intensity than can be obtained outside the laser in ordinary (extracavity) second-harmonic generation. We recall from Chapter 5 that the output intensity of a laser is typically only a small fraction of the intracavity intensity.

In Fig. 10.5 the output mirror of the laser is replaced by one that is perfectly reflecting at the laser frequency ω, but perfectly transmitting at the second-harmonic frequency 2ω. The principal loss mechanism for the intracavity laser field is then its conversion to the second harmonic inside the nonlinear crystal. The optic axis of the crystal is oriented at the appropriate phase-matching angle θ_p.

Suppose that the power conversion efficiency for second-harmonic generation inside the crystal can be made equal to the optimal output coupling for the laser without the intracavity crystal. With the crystal in place the loss—due now to second-harmonic generation—is still optimal for the laser. In this case we still extract the maximum possible power from the laser, but now this power is at the second-harmonic frequency.

In practice, however, other effects conspire to reduce the second-harmonic output power. The insertion of the nonlinear crystal in the laser represents an additional loss mechanism because of scattering, and so crystals of high optical quality are needed to keep this loss small. Furthermore the crystal may be slightly absorbing at ω and 2ω, thus raising its temperature. The refractive indices n_o and n_e of the crystal often vary appreciably with temperature, so that even a small temperature rise might seriously reduce the phase matching. Commercial cw intracavity doubling systems, such as a Nd–YAG laser with an insertable lithium iodate (LiIO$_3$) crystal, can nevertheless have conversion efficiencies of 10% or more.   •

Another common way of phase matching is based on "periodically poled" structures in which the orientation of a crystal axis is varied periodically in such a way that the sign of the coefficient $\bar{d}$ changes periodically with position. The essence of this *quasi-phase-matching* in the case of second-harmonic generation, for example, can be understood from the Eq. (10.5.13) describing the generation of the second-harmonic field. Suppose we make the approximation (10.5.16) of no pump depletion and assume for definiteness that $\Delta k > 0$, so that the coherence length $L_c = \pi/\Delta k$. The second-harmonic field at $z = 2L_c$, for instance, is found from (10.5.13) to be

$$\mathcal{E}_{2\omega}(2L_c) = i\omega\sqrt{\frac{\mu_0}{\epsilon_{2\omega}}}\, \bar{d}\mathcal{E}_\omega^2(0)\int_0^{2L_c} dz\, e^{i\pi z/L_c} = \frac{\omega}{\pi}\sqrt{\frac{\mu_0}{\epsilon_{2\omega}}}\, \bar{d}\mathcal{E}_\omega^2(0)L_c(e^{2\pi i} - 1)$$

$$= 0, \tag{10.6.11}$$

whereas at $z = L_c$,

$$\mathcal{E}_{2\omega}(L_c) = i\omega\sqrt{\frac{\mu_0}{\epsilon_{2\omega}}}\, \bar{d}\mathcal{E}_\omega^2(0)\int_0^{L_c} dz\, e^{i\pi z/L_c} = \frac{\omega}{\pi}\sqrt{\frac{\mu_0}{\epsilon_{2\omega}}}\, \bar{d}\mathcal{E}_\omega^2(0)L_c(e^{i\pi} - 1)$$

$$= -\frac{2\omega}{\pi}\sqrt{\frac{\mu_0}{\epsilon_{2\omega}}}\, \bar{d}\mathcal{E}_\omega^2(0)L_c. \tag{10.6.12}$$

We noted earlier that, without phase matching ($\Delta k \neq 0$), increasing the crystal length beyond L_c does not increase the second-harmonic conversion efficiency; (10.6.11) and (10.6.12) show in particular that $\mathcal{E}_{2\omega}(z)$ decreases from a finite value at $z = L_c$ to zero at $z = 2L_c$. But suppose now that the nonlinear coefficient changes from $\bar{d}$ for $0 \leq z \leq L_c$ to a different value $\bar{d}'$ for $L_c < z \leq 2L_c$. In this case (10.5.13) in the approximation of no pump depletion gives

$$\mathcal{E}_{2\omega}(2L_c) = -\frac{2\omega}{\pi}\sqrt{\frac{\mu_0}{\epsilon_{2\omega}}}(\bar{d} - \bar{d}')\mathcal{E}_\omega^2(0)L_c, \tag{10.6.13}$$

and, if $\bar{d}' = -\bar{d}$, $\mathcal{E}_{2\omega}(2L_c) = 2\mathcal{E}_{2\omega}(L_c)$. If the sign of $\bar{d}$ is periodically switched in this manner the second-harmonic field *increases monotonically* with z; the period for switching the sign of $\bar{d}$ for this quasi-phase-matching is twice the coherence length L_c.

Periodically poled lithium niobate (PPLN) has been found to be very useful for quasi-phase-matching. $LiNO_3$ is ferroelectric: each unit cell of the crystal has a small electric dipole moment, and an applied electric field can change the orientation of these dipole moments. A very strong field (~ 22 kV/mm) applied for a few milliseconds can invert the dipole moments, and if such fields are applied in equally spaced sections along some (z) direction they result in a periodically "poled" material in which the electric dipole moment and therefore the sign of the nonlinear coefficient varies periodically. The required electric fields are produced by applying appropriate voltages to a periodic electrode structure. As noted earlier, coherence lengths are usually very small, and therefore the spatial periodicity in PPLN is small. For frequency doubling of 1.06-μm radiation, for example, the poling period required for quasi-phase-matched generation of 0.53 μm radiation is only about 6.6 μm at room temperature.

10.7 THREE-WAVE MIXING

In the more general case of three-wave mixing we can derive coupled wave equations for the complex amplitudes $\mathcal{E}_1$, $\mathcal{E}_2$, and $\mathcal{E}_3$ of the waves at ω_1, ω_2, and ω_3, respectively. Since the derivation is straightforward we simply write these equations:

$$\frac{d\mathcal{E}_1}{dz} = i\omega_1 \sqrt{\frac{\mu_0}{\epsilon_1}}\, \bar{d}\mathcal{E}_2^* \mathcal{E}_3 e^{-i\Delta kz}, \tag{10.7.1a}$$

$$\frac{d\mathcal{E}_2}{dz} = i\omega_2 \sqrt{\frac{\mu_0}{\epsilon_2}}\, \bar{d}\mathcal{E}_1^* \mathcal{E}_3 e^{-i\Delta kz}, \tag{10.7.1b}$$

$$\frac{d\mathcal{E}_3}{dz} = i\omega_3 \sqrt{\frac{\mu_0}{\epsilon_3}}\, \bar{d}\mathcal{E}_1 \mathcal{E}_2 e^{i\Delta kz}, \tag{10.7.1c}$$

with $\omega_3 = \omega_1 + \omega_2$ and

$$\Delta k = k_1 + k_2 - k_3 \qquad \left[k_j = n(\omega_j)\frac{\omega_j}{c} \right], \tag{10.7.2}$$

$$\epsilon_j = \epsilon_0 n^2(\omega_j) = \epsilon_0 n_j^2. \tag{10.7.3}$$

Equations (10.7.1) couple the three wave amplitudes $\mathcal{E}_1$, $\mathcal{E}_2$, and $\mathcal{E}_3$. That is, they describe the coupling, or mixing, of three waves. Nonlinear optical processes described by equations like (10.7.1) are thus examples of three-wave mixing, second-harmonic generation being a special example. Note that $\bar{d} = \frac{1}{2}\epsilon_0 \chi(-\omega_1, -\omega_2, \omega_3)$ in (10.7.1a), $\bar{d} = \frac{1}{2}\epsilon_0 \chi(-\omega_2, -\omega_1, \omega_3)$ in (10.7.1b), and $\bar{d} = \frac{1}{2}\epsilon_0 \chi(-\omega_3, \omega_1, \omega_2)$ in (10.7.1c); in writing Eqs. (10.7.1) we have used the assumption, discussed in Section 10.2, that the nonlinear susceptibilities are independent of the frequencies ω_1, ω_2, and ω_3.

Difference-frequency generation, in which two waves mix to produce radiation at their difference frequency, is also described by the coupled wave equations (10.7.1). For instance, according to (10.7.1b) we can use *pump* radiation at ω_3 and *signal* radiation at ω_1 to generate an *idler* field at ω_2. Or we can mix a field at ω_2 with the pump at ω_3 to generate a field at ω_1; in this case the field at ω_2 is called the signal and that at ω_1 the idler. What is conventionally called the idler or the signal wave depends on the initial conditions.

From (10.7.1) it follows that (Problem 10.6)

$$\frac{1}{\omega_1}\frac{d}{dz}\left(\sqrt{\frac{\epsilon_1}{\mu_0}}|\mathcal{E}_1|^2 \right) = \frac{1}{\omega_2}\frac{d}{dz}\left(\sqrt{\frac{\epsilon_2}{\mu_0}}|\mathcal{E}_2|^2 \right) = -\frac{1}{\omega_3}\frac{d}{dz}\left(\sqrt{\frac{\epsilon_3}{\mu_0}}|\mathcal{E}_3|^2 \right). \tag{10.7.4a}$$

In other words, for any three-wave mixing process with $\omega_3 = \omega_1 + \omega_2$ we have [recall (10.5.20)]

$$\frac{1}{\omega_1}(\text{rate of change of energy at }\omega_1) = \frac{1}{\omega_2}(\text{rate of change of energy at }\omega_2)$$

$$= -\frac{1}{\omega_3}(\text{rate of change of energy at }\omega_3). \tag{10.7.4b}$$

Such equations are called *Manley–Rowe relations*, and they have a remarkably simple interpretation in terms of photons: Each of the equal terms in these equations represents the rate of change of the *number of photons* at the corresponding frequency, so that the creation of a photon at ω_3 in sum-frequency generation, for instance, is accompanied by the annihilation of a photon at ω_1 and a photon at ω_2.

The coupled wave equations for second-harmonic generation are found similarly to be

$$\frac{d\mathcal{E}_\omega}{dz} = i\omega\sqrt{\frac{\mu_0}{\epsilon_\omega}}\, \bar{d}\mathcal{E}_\omega^* \mathcal{E}_{2\omega} e^{i\Delta kz}, \tag{10.7.5a}$$

$$\frac{d\mathcal{E}_{2\omega}}{dz} = i\omega\sqrt{\frac{\mu_0}{\epsilon_{2\omega}}}\, \bar{d}\mathcal{E}_\omega^2 e^{-i\Delta kz}, \tag{10.7.5b}$$

where Δk is defined by (10.5.14). When we neglect pump depletion in second-harmonic generation, we are simply ignoring (10.7.5a) and replacing $\mathcal{E}_\omega$ by a constant value in (10.7.5b). The analysis reduces in this case to that following (10.5.15).

Second-harmonic generation is, of course, just the degenerate case of sum-frequency generation with $\omega_1 = \omega_2 = \omega$ and $\omega_3 = 2\omega$. Note, however, that (10.7.5) does *not* follow from (10.7.1) by simply letting $\omega_1 = \omega_2 = \omega$, $\mathcal{E}_1 = \mathcal{E}_2 = \mathcal{E}_\omega$, and $\mathcal{E}_3 = \mathcal{E}_{2\omega}$ in the latter. In particular, (10.7.5b) differs from (10.7.1c), when these (improper) substitutions are made, by a factor $\frac{1}{2}$. This difference has already been discussed in Section 10.2. It can also be appreciated from a different point of view: from (10.7.5) we obtain the Manley–Rowe relation

$$\frac{1}{\omega}\frac{d}{dz}\left(\sqrt{\frac{\epsilon_\omega}{\mu_0}}|\mathcal{E}_\omega|^2\right) = -2\frac{1}{2\omega}\frac{d}{dz}\left(\sqrt{\frac{\epsilon_{2\omega}}{\mu_0}}|\mathcal{E}_{2\omega}|^2\right). \tag{10.7.6}$$

We can interpret this as saying that *two* photons from the field at ω are annihilated to produce *one* photon at 2ω in second-harmonic generation. This correct interpretation, however, does not follow from the Manley–Rowe relation (10.7.4) with $\omega_1 = \omega_2 = \omega$, $\omega_3 = 2\omega$.

10.8 PARAMETRIC AMPLIFICATION AND OSCILLATION

Consider the implication of the Manley–Rowe relation (10.7.4) for difference-frequency generation of light at $\omega_3 - \omega_1 = \omega_2$. In this process waves of frequency ω_3 (the pump) and ω_1 (the signal) mix to produce a wave at the idler frequency ω_2. According to (10.7.4), the decrease in power at ω_3 is accompanied by an increase in power at *both* ω_1 and ω_2. This amplification of light at ω_1 and ω_2 in the presence of light at ω_3 is called *parametric amplification*. In the "degenerate" case in which $\omega_1 = \omega_2 = \omega_3/2$, the process is essentially just the inverse of second-harmonic oscillation: Instead of generating the second-harmonic frequency 2ω at the expense of a pump field at the fundamental frequency ω, the frequency ω is generated at the expense of a pump field at frequency 2ω.

Parametric amplification is described by the coupled wave equations (10.7.1). Let us assume that the pump wave at ω_3 is approximately undepleted, so that $\mathcal{E}_3(0) \, \mathcal{E}_3(0)$; this is a good approximation under the common circumstance that the relative power converted to ω_1 and ω_2 is small. Then (10.7.1a) and (10.7.1b) give

$$\frac{d\mathcal{E}_1}{dz} = i \left[\omega_1 \sqrt{\frac{\mu_0}{\epsilon_1}} \, \bar{d}\mathcal{E}_3(0) \right] \mathcal{E}_2^*(z) = i \sqrt{\frac{\omega_1}{\omega_2}} \, b_1 \mathcal{E}_2^*(z), \qquad (10.8.1a)$$

$$\frac{d\mathcal{E}_2^*}{dz} = -i \left[\omega_2 \sqrt{\frac{\mu_0}{\epsilon_2}} \, \bar{d}\mathcal{E}_3^*(0) \right] \mathcal{E}_1(z) = -i \sqrt{\frac{\omega_2}{\omega_1}} \, b_2^* \mathcal{E}_1(z), \qquad (10.8.1b)$$

where

$$b_i = \left[\omega_1 \omega_2 \left(\frac{\mu_0}{\epsilon_i} \right) \right]^{1/2} \bar{d}\mathcal{E}_3(0), \qquad i = 1, 2. \qquad (10.8.2)$$

In writing (10.8.1) we have assumed perfect phase matching, $\Delta k = 0$, for simplicity; (10.8.1b) is obtained from the complex conjugate of (10.7.1b). We now differentiate (10.8.1a) and use (10.8.1b):

$$\frac{d^2\mathcal{E}_1}{dz^2} = i \sqrt{\frac{\omega_1}{\omega_2}} b_1 \frac{d\mathcal{E}_2^*}{dz} = i \sqrt{\frac{\omega_1}{\omega_2}} b_1 \left(-i \sqrt{\frac{\omega_2}{\omega_1}} b_2^* \mathcal{E}_1 \right) = K^2 \mathcal{E}_1, \qquad (10.8.3a)$$

where

$$K = \left(\frac{\omega_1 \omega_2}{n_1 n_2} \frac{\mu_0}{\epsilon_0} \right)^{1/2} \bar{d} |\mathcal{E}_3(0)|. \qquad (10.8.3b)$$

Similarly we obtain

$$\frac{d^2\mathcal{E}_2}{dz^2} = K^2 \mathcal{E}_2. \qquad (10.8.4)$$

The uncoupled equations (10.8.3) and (10.8.4) may be solved in terms of the fields $\mathcal{E}_1(0)$ and $\mathcal{E}_2(0)$ at the input face $z = 0$ of the nonlinear medium:

$$\mathcal{E}_1(z) = \mathcal{E}_1(0) \cosh Kz + i \sqrt{\frac{\omega_1}{\omega_2}} \mathcal{E}_2^*(0) \sinh Kz, \qquad (10.8.5a)$$

$$\mathcal{E}_2(z) = \mathcal{E}_2(0) \cosh Kz + i \sqrt{\frac{\omega_2}{\omega_1}} \mathcal{E}_1^*(0) \sinh Kz, \qquad (10.8.5b)$$

and we recall that

$$\cosh x = \tfrac{1}{2}(e^x + e^{-x}), \qquad \sinh x = \tfrac{1}{2}(e^x - e^{-x}). \qquad (10.8.5c)$$

If we imagine injecting into the nonlinear medium a pump wave at ω_3 and a signal wave at ω_1, then $\mathcal{E}_2(0) = 0$, i.e., there is no idler wave at the input face of the medium. In this case the solutions (10.8.5) imply

$$|\mathcal{E}_1(z)|^2 = |\mathcal{E}_1(0)|^2 \cosh^2 Kz, \tag{10.8.6a}$$

$$|\mathcal{E}_2(z)|^2 = \frac{\omega_2}{\omega_1}|\mathcal{E}_1(0)|^2 \sinh^2 Kz. \tag{10.8.6b}$$

In the limit $Kz \ll 1$ these solutions reduce to

$$|\mathcal{E}_1(z)|^2 \approx |\mathcal{E}_1(0)|^2(1 + K^2 z^2), \tag{10.8.7a}$$

$$|\mathcal{E}_2(z)|^2 \approx \frac{\omega_2}{\omega_1}|\mathcal{E}_1(0)|^2 K^2 z^2. \tag{10.8.7b}$$

Let us consider a numerical example to see what to expect for Kz. For the crystal LiNbO$_3$ (lithium niobate), we assume $\bar{d} \approx 4 \times 10^{-23}$ (mks units) and $n_1 \approx n_2 \approx n_3 \approx 1.5$ when the pump is an Nd–YAG laser ($\lambda = 1.06\,\mu$m) and the signal and idler are each at half the pump frequency. We compute (Problem 10.6)

$$K \approx 2 \times 10^{-4}\sqrt{I_3(0)}\ \text{cm}^{-1}, \tag{10.8.8}$$

where $I_3(0)$ is the pump intensity in units of W/cm^2. If, for example, $I_3(0) = 1$ MW/cm^2 and $z = 1$ cm, $Kz \approx 0.2$. This example suggests that the limiting forms (10.8.7) will be applicable in many circumstances.

Thus far we have assumed perfect phase matching. Without phase matching, and assuming again that $\mathcal{E}_3(z) \approx \mathcal{E}_3(0)$ and $Kz \ll 1$ and also that $b_1 \approx b_2$, we obtain

$$|\mathcal{E}_1(z)|^2 \approx |\mathcal{E}_1(0)|^2\left(1 + (Kz)^2\frac{\sin^2\frac{1}{2}\Delta k\,z}{\left(\frac{1}{2}\Delta k\,z\right)^2}\right), \tag{10.8.9a}$$

$$|\mathcal{E}_2(z)|^2 \approx \frac{\omega_2}{\omega_1}(Kz)^2|\mathcal{E}_1(0)|^2\frac{\sin^2\frac{1}{2}\Delta k\,z}{\left(\frac{1}{2}\Delta k\,z\right)^2}, \tag{10.8.9b}$$

where Δk is defined by (10.7.2). For $\Delta k \to 0$, these results reduce, as they should, to Eqs. (10.8.7), obtained under the assumption of perfect phase matching. The effect of phase matching is thus to introduce a factor $\left(\sin^2\frac{1}{2}\Delta k\,z\right)/\left(\frac{1}{2}\Delta k\,z\right)^2$, just as in second-harmonic generation. Now, however, Δk is given by (10.7.2) rather than (10.5.14). When $\omega_1 = \omega_2 = \omega_3/2$, as in second-harmonic generation or degenerate parametric amplification, the two expressions for Δk are identical.

The problem of phase matching in parametric amplification is basically the same as in second-harmonic generation. For instance, we can have angle phase matching in parametric amplification in a negative uniaxial crystal by propagating the pump as an ordinary wave and the signal and idler as extraordinary waves.

• Parametric amplification is a general phenomenon. A mechanical example of degenerate parametric amplification is provided by the playground swing. A swinger pumps the swing by raising and lowering her center of gravity as she tucks her legs in or extends them. The swing is optimally pumped when the pumping frequency is twice the swing oscillation frequency. By parametric amplification, energy is being fed from the pump at frequency 2ω to the swing oscillation at ω.

The parametric resonance principle was well known to 19th-century physicists. Lord Rayleigh, for instance, noted several examples, including a pendulum whose point of support vibrates vertically at twice the natural pendulum frequency. Such parametric resonance phenomena are often described by an equation of the type

$$\ddot{x} + \omega^2(t)x = 0, \tag{10.8.10}$$

where the frequency $\omega(t)$ varies in time according to the formula

$$\omega^2(t) = \omega_0^2(1 + \varepsilon \cos \omega't), \tag{10.8.11}$$

with ε small compared to 1. Parametric amplification occurs for $\omega' = 2\omega_0$, as may be shown either by perturbation theory or by a numerical solution of the differential equation (10.8.10). Similar equations are encountered in electronic parametric processes when a circuit parameter (e.g., capacitance) is made to vary sinusoidally.   •

Suppose the signal and idler waves in parametric amplification are propagating back and forth within a resonator containing the nonlinear medium. Because of this feedback, parametric *oscillation* is possible by balancing the amplification against transmission loss and whatever other attenuation processes are at work. Figure 10.6 illustrates the design of an *optical parametric oscillator* (OPO). The laser oscillator puts out a beam at frequency ω_3, which is focused onto a nonlinear crystal. The crystal is contained in a cavity with mirrors that are transparent to radiation of frequency ω_3, but one of the mirrors allows a small fraction of light at ω_1 and ω_2 incident upon it to be transmitted as the output of the parametric oscillator. The waves at ω_1 and ω_2 bounce back and forth inside this cavity, undergoing parametric amplification in the crystal. A variation of the crystal orientation and therefore the phase-matching angle will result in a change in the frequencies ω_1 and ω_2 at which parametric oscillation occurs. In this way the output wavelengths of parametric oscillators can be continuously varied or "tuned" over a fairly large range to obtain some particular wavelength.

Parametric oscillation is similar to laser oscillation. In a medium with population inversion and gain over some range of frequencies, there is amplification of an injected signal. If the signal is continually fed back into the gain medium by using a resonator supporting modes within the gain bandwidth, sustained laser oscillation is possible when the small-signal gain exceeds the loss. In the parametric oscillator, however, no population inversion is needed: The signal and idler are amplified at the expense of the energy in the pump wave rather than energy stored in the form of molecular

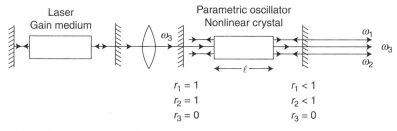

Figure 10.6   Schematic layout of a parametric oscillator with pump frequency $\omega_3 = \omega_1 + \omega_2$.

excitation in the nonlinear medium; the medium serves only to mix the pump, signal, and idler waves. Nevertheless, as in the laser, there is a threshold condition for oscillation. This threshold condition for parametric oscillation is for the pump intensity to exceed a certain level (Problem 10.7).

In parametric amplification we require, in addition to the pump wave, either the signal or idler (or both) to have some initial energy. This is evident, for instance, in (10.8.5). Where does the initial signal or idler energy come from in a parametric oscillator such as that sketched in Fig. 10.6? This is quite analogous to asking where the "initial photon" comes from in a laser oscillator, and the answer is the same: The initial radiation triggering the parametric oscillation comes from spontaneous emission. This is not "ordinary" spontaneous emission, however, in which a molecule drops from an excited energy level to a lower one with the emission of a photon. Rather, it is *parametric fluorescence*, in which a nonlinear crystal exposed to radiation of frequency ω_3 can emit *two* photons at ω_1 and ω_2, such that $\omega_1 + \omega_2 = \omega_3$. The crystal has, loosely speaking, acted to split the incident photon into two outgoing photons; this is discussed further in the following section. Parametric amplification is basically just a *stimulated* emission of the two photons in the presence of a signal photon.

• Simultaneous parametric oscillation at ω_1 and ω_2 requires that both frequencies be resonant frequencies of the OPO cavity:

$$\omega_1 = m_1 \frac{\pi c}{n(\omega_1)L} \quad \text{and} \quad \omega_2 = m_2 \frac{\pi c}{n(\omega_2)L}, \tag{10.8.12}$$

where m_1 and m_2 are integers and L is the mirror separation of the cavity containing the nonlinear crystal. In general, however, it is impossible to satisfy both these conditions as well as the phase-matching condition $n(\omega_1)\omega_1 + n(\omega_2)\omega_2 = n(\omega_3)\omega_3$ with $\omega_3 = \omega_1 + \omega_2$. Furthermore other effects that are difficult to control will cause ω_1 and ω_2 to drift randomly. Temperature variations, for instance, will change the refractive indices, while mechanical vibrations cause L to vary. Slight frequency variations in the pump laser will also cause variations in ω_1 and ω_2. For these reasons it is often preferable to operate an OPO so that only one of the two frequencies ω_1 and ω_2 can oscillate; this can be done by introducing a frequency-selective element, such as a material that absorbs at one of the frequencies but not the other. Then the OPO is said to be *singly resonant* rather than *doubly resonant*. Most commercial OPOs are singly resonant and pumped by pulsed lasers; because of their relative complexity and the need to have a pump laser and temperature control, OPOs have not been widely used commercially. However, their tunability and the fact that they can access wavelengths that might not be available from lasers has made them increasingly important in lidar applications.   •

10.9   TWO-PHOTON DOWNCONVERSION

Parametric fluorescence in which a pump wave at ω_3 incident on a nonlinear crystal results in the spontaneous generation of signal and idler photons at ω_1 and ω_2 may be called spontaneous two-photon downconversion, or simply *two-photon downconversion*. The downconversion efficiency is often so small that the signal and idler are well described as single-photon fields. The two downconverted photons are correlated not only in frequency ($\omega_1 + \omega_2 = \omega_3$) and wave vector ($\mathbf{k}_1 + \mathbf{k}_2 = \mathbf{k}_3$ in the case of

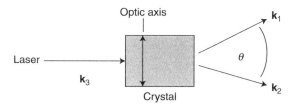

Figure 10.7   Two-photon downconversion in which the pump field with frequency ω_3 and wave vector $\mathbf{k}_3$ is polarized as an extraordinary wave incident at $90°$ to the optic axis of a negative uniaxial crystal. The single-photon downconverted fields are polarized as ordinary waves and the angle between their wave vectors ($\mathbf{k}_1$ and $\mathbf{k}_2$) is θ.

perfect phase matching) but also, when angle phase matching is employed, in polarization. As we shall see, these correlations have interesting consequences.

Figure 10.7 shows a typical arrangement for realizing two-photon downconversion. A pump wave is polarized as an extraordinary wave incident on a nonlinear crystal with its wave vector $\mathbf{k}_3$ perpendicular to the optic axis, which is oriented such that there is phase matching with the signal and idler waves polarized as ordinary waves. Then the frequencies and wave vectors satisfy $\omega_3 = \omega_1 + \omega_2$ and $\mathbf{k}_3 = \mathbf{k}_1 + \mathbf{k}_2$. Squaring both sides of the latter equation, we have $k_3^2 = k_1^2 + k_2^2 + 2\mathbf{k}_1 \cdot \mathbf{k}_2$, or

$$n_e^2(\omega_3)\omega_3^2 = n_o^2(\omega_1)\omega_1^2 + n_o^2(\omega_2)\omega_2^2 + 2n_o(\omega_1)n_o(\omega_2)\omega_1\omega_2 \cos\theta, \qquad (10.9.1)$$

where θ is the angle between $\mathbf{k}_1$ and $\mathbf{k}_2$. The downconverted radiation for a given pair of frequencies ω_1 and ω_2 satisfying these equations is emitted in a cone with opening angle θ. In the setup of Fig. 10.7 one can use filters to select a certain pair of downconverted frequencies ω_1 and ω_2, or two apertures separated by some angle θ to determine a particular pair of frequencies passing through the apertures. In the case of apertures the two "unknowns" ω_1 and ω_2 are determined by ω_3, the refractive indices $n_e(\omega_3)$, $n_o(\omega_1)$, $n_o(\omega_2)$, the angle θ, and Eqs. (10.9.1) and $\omega_3 = \omega_1 + \omega_2$.

Of course the finite diameter of the apertures results in a small range of angles $\Delta\theta$ about θ, which contributes to deviations $\Delta\omega_1$ and $\Delta\omega_2$ of the downconverted frequencies from ω_1 and ω_2, respectively. Assuming the pump frequency ω_3 is fixed, we have $\Delta\omega_3 = \Delta\omega_1 + \Delta\omega_2 = 0$, or $\Delta\omega_2 = -\Delta\omega_1$. Then, from (10.9.1) (Problem 10.8),

$$\Delta\omega_1 = \frac{n_o(\omega_1)n_o(\omega_2)\omega_1\omega_2\Delta\theta\sin\theta}{n_o^2(\omega_1)\omega_1 - n_o^2(\omega_2)\omega_2 + n_o(\omega_1)n_o(\omega_2)(\omega_2 - \omega_1)\cos\theta} \qquad (10.9.2)$$

is the deviation from ω_1 due to a deviation $\Delta\theta$ in the angle θ for which there is phase matching for the downconversion of photons of frequency ω_1 and ω_2.

- In an experiment employing the negative uniaxial crystal $LiIO_3$, for example, the pump (ω_3) wave from a single-mode argon–ion laser had a wavelength of 351.1 nm while the signal (ω_1) and idler (ω_2) photons had wavelengths of 632.8 nm and 788.7 nm, respectively.[4] Based on empirical

[4]T. J. Herzog, J. G. Rarity, H. Weinfurter, and A. Zeilinger, *Physical Review Letters* **72**, 629 (1994).

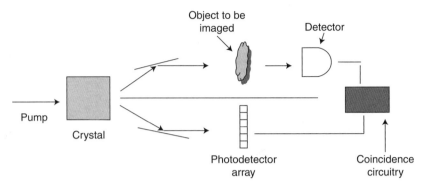

Figure 10.8   One photon of a downconverted pair is incident on some object to be imaged, while its correlated partner is incident on a photodetector array. Coincidence counting of photons passing through the object and their correlated partners at the photodetector array provides a "ghost" image of the object, i.e., an image formed by light that has not passed through the object.

formulas for the refractive indices of various crystals of interest for nonlinear optics,[5] we deduce the refractive indices $n_e(\omega_3) = 1.7197$, $n_o(\omega_1) = 1.8810$, and $n_o(\omega_2) = 1.8657$ for LiIO$_3$ and calculate $\theta = 47.2°$. In the experiment there were two 0.8-mm-diameter apertures, 90 cm apart, in the path of both the signal and idler fields. There was therefore an effective *spread* $\Delta\theta \approx \frac{1}{2}(0.08 \text{ cm})/(90 \text{ cm}) \approx 4.45 \times 10^{-4}$ rad in θ and, from (10.9.2) and the values above for the wavelengths, refractive indices, and phase-matching angle, a *width* (Problem 10.8)

$$\Delta\omega_1 \approx 10^{13} \text{ s}^{-1} \tag{10.9.3}$$

in the downconverted frequency ω_1. This corresponds to a "coherence length" (recall the definition given in Chapter 1) $2\pi c/\Delta\omega_1 \approx 190 \ \mu$m for the signal and idler fields, in approximate accord with the 260 μm measured in the experiments.   ●

As already noted, the frequencies of the downconverted photons are correlated: If one of the downconverted photons has a frequency ω_1, the other must have a frequency $\omega_2 = \omega_3 - \omega_1$. The downconverted photons are similarly correlated in their wave vectors. In Fig. 10.8 we show an example of how these two-photon correlations can be put to use.[6] One of two downconverted photons is incident on an object consisting of a two-dimensional transmission pattern, while its **k**-correlated partner is incident on a photodetector array that serves as an imaging detector. A photon passing through the object can be registered by a detector, while its partner can be registered somewhere on the photodetector array. Electronic circuitry is used to determine when the correlated photon pairs are counted in coincidence, and therefore when the output from the photodetector array should be recorded. If, for instance, there is complete absorption over some section of the object, a photon at the corresponding element on the photodetector array will never be counted in coincidence with its partner photon, and the complete absence of any coincidence signal in this case provides the information that the corresponding element on the object is dark. A high rate of photon-counting coincidences

[5]D. Eimerl, S. Velsko, L. Davis, and F. Wang, *Progress in Crystal Growth and Characterization of Materials* **20**, 59 (1990).

[6]T. B. Pittman, Y. H. Shih, D. V. Strekalov, and A. V. Sergienko, *Physical Review* A**52**, R3429 (1995).

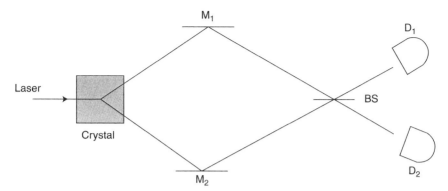

Figure 10.9 Photons generated by two-photon downconversion are directed by mirrors M_1 and M_2 to opposite sides of a lossless beam splitter BS, followed by reflection or transmission from which they can be counted at detectors D_1 and D_2.

for some other element on the photodetector array, on the other hand, implies that the corresponding element at the object is bright. In this way a two-dimensional "ghost" image of the object is recorded as a pattern of bright and dark pixels.

Coincidence imaging can also be performed with laser pulses that are angularly correlated in a "classical" way simply by the use of rotating mirrors and beam splitters, for instance. However, as discussed below, two-photon downconversion can result in distinctly quantum mechanical correlations that cannot be realized with laser light. It has been demonstrated that these correlations can be used to realize better image resolution than is possible by classical optical techniques.[7]

The photon pairs in two-photon downconversion are also correlated in time in that they are created essentially simultaneously. An interesting consequence of this simultaneity and the single-photon nature of the downconverted fields is revealed by the experiment indicated in Fig. 10.9, which shows the two photons directed to opposite sides of a lossless beam splitter. After propagation by equal distances and reflection or transmission at the beam splitter the photons can be counted by photodetectors D_1 and D_2, and a determination is made of the probability of counting photons at *both* D_1 and D_2. The remarkable result (see below) is that, for a $50/50$ beam splitter, this *coincidence* probability is zero: *either both photons are counted at D_1 or both photons are counted at D_2.*

This result is found when the paths traversed by the downconverted fields are of equal length, or when the path difference is small compared to the coherence length of the signal and idler fields. If the path difference is much larger than the coherence length, photon coincidences at D_1 and D_2 occur with probability $\frac{1}{2}$ while there is also a probability of $\frac{1}{2}$ that two photons arrive at one or the other of the two detectors. In other words, the coincidence probability vanishes at zero path difference and approaches $\frac{1}{2}$ for large path differences. The *Hong-Ou-Mandel interferometer* of Fig. 10.9 can therefore be used to determine single-photon propagation times with extremely high

[7]R. S. Bennink, S. J. Bentley, R. W. Boyd, and J. C. Howell, *Physical Review Letters* **92**, 033601(2004).

resolution.[8] Experiments have demonstrated temporal resolution of a few femto-seconds—about a million times shorter than that of conventional photon detection systems.

• The zero coincidence result can be explained by invoking a few results from a more advanced, fully quantum theoretical analysis. Photons can be detected at D_1 *and* D_2 (Fig. 10.9) in two ways: (i) both photons are reflected by BS or (ii) both photons are transmitted by BS. In stating these possibilities, we are acknowledging that we cannot split a photon and have it be partially reflected and partially transmitted; a photon is completely transmitted or completely reflected by the beam splitter BS.

Now according to quantum theory the probability *amplitude* for process (i) is $\mathcal{R}^2$, where $\mathcal{R}$ is the field (amplitude) reflection coefficient; we assume that the reflection coefficient is the same for both the signal and idler fields. Similarly, the probability amplitude for process (ii) is $\mathcal{T}^2$, where $\mathcal{T}$ is the amplitude transmission coefficient. The total probability amplitude for having photons arrive at both D_1 and D_2 is then $\mathcal{R}^2 + \mathcal{T}^2$, and the total probability P_{12} for having photons arrive at both D_1 and D_2 is the absolute square of the total probability amplitude:

$$P_{12} = |\mathcal{R}^2 + \mathcal{T}^2|^2. \tag{10.9.4}$$

Now we write $\mathcal{R} = |\mathcal{R}|e^{i\phi_R}$ and $\mathcal{T} = |\mathcal{T}|e^{i\phi_T}$ and use the result, which is valid in classical as well as quantum optics, that $\cos(\phi_R - \phi_T) = 0$, that is, that there is a $\pi/2$ phase shift in the reflected field compared to the transmitted field in the case of a lossless beam splitter (Problem 10.9). Then

$$P_{12} = \left| |\mathcal{R}|^2 e^{2i\phi_R} + |\mathcal{T}|^2 e^{2i\phi_T} \right|^2 = \left| |\mathcal{R}|^2 e^{2i(\phi_R - \phi_T)} + |\mathcal{T}|^2 \right|^2 = \left| |\mathcal{R}|^2 e^{i\pi} + |\mathcal{T}|^2 \right|^2$$
$$= \left| |\mathcal{R}|^2 - |\mathcal{T}|^2 \right|^2, \tag{10.9.5}$$

since $e^{i\pi} = -1$, and therefore $P_{12} = 0$ for a 50/50 beam splitter ($|\mathcal{R}|^2 = |\mathcal{T}|^2$).

The reader who has studied advanced quantum theory will recognize that what we have done here is to add probability amplitudes for the two *indistinguishable* processes (i) and (ii); it is a general feature of quantum theory that the total probability amplitude for a process that can occur in different but indistinguishable ways is the sum of the probability amplitudes for the different ways. The processes (i) and (ii) here are indistinguishable so long as the path difference for the two photons is small compared to the coherence time of the signal and idler fields. When this is not the case, (i) and (ii) can no longer be regarded as indistinguishable, and it is found in a more general calculation that $P_{12} \neq 0$.

The distinctly quantum feature of our derivation of (10.9.4) is the assumption that a *single photon* cannot be "split" by a beam splitter. If instead of a single photon we have a pulse of light from a laser or a thermal source, what happens, of course, is that the pulse is partly transmitted and partly reflected by a beam splitter. If we think of the pulse as consisting of a huge number of photons, we conclude that each photon has some probability of being completely transmitted or completely reflected, and the net effect is that a fraction $|\mathcal{T}|^2$ of the total number of photons is transmitted while a fraction $1 - |\mathcal{T}|^2 = |\mathcal{R}|^2$ is reflected. The observed net effect of the beam splitter on the pulse is then exactly what is expected from classical optics, and we do not see any quantum (photon) features. Such considerations are the subject of Chapter 12. •

[8]C. K. Hong, Z. Y. Ou, and L. Mandel, *Physical Review Letters* **59**, 2044 (1987); A. M. Steinberg, P. G. Kwiat, and R. Y. Chiao, *Physical Review Letters* **68**, 2421 (1992).

We have already made note of the polarization correlations of the downconverted photons. In the example of Fig. 10.7 the downconverted photons have the same horizontal linear polarization (H), perpendicular to the optic axis, while the pump has vertical linear polarization (V), parallel to the optic axis. The polarization state of the downconverted photons is written as $|H\rangle_1|H\rangle_2$, where a single-particle state is denoted $|\ldots\rangle$ according to the "Dirac notation" used in quantum theory. Thus, $|\;\rangle_1$ is a state for photon 1 and H designates horizontal polarization, $|H\rangle_2$ is the state in which photon 2 has horizontal polarization, and the two-photon "product" state $|H\rangle_1|H\rangle_2$ is that for which both photons have horizontal polarization. The notation may seem strange to readers who have not encountered it before, but the physics it represents is simple: downconversion in our example results in polarization states that are correlated in that the two photons have the same polarization H. $|H\rangle_1|H\rangle_2$ is simply a conventional way of denoting the polarization state of the two photons together. We could write similar expressions for the frequency and wavevector correlations of the photons, but for simplicity we will consider only polarization. We will use the example of polarization to show that there is a far more interesting type of correlation than that described by the product state $|H\rangle_1|H\rangle_2$, and that this type of correlation can be realized in two-photon downconversion.

Suppose that a second crystal is placed just to the right of the crystal of Fig. 10.7, and that it is identical to the first crystal but has its optic axis oriented horizontally, perpendicular to the optic axis of the first crystal.[9] A vertically polarized pump photon can be downconverted into two H photons in the first crystal, but it will not downconvert in the second crystal, where there is no phase matching for a vertically polarized pump. Similarly, a horizontally polarized pump photon will pass through the first crystal without downconversion but can downconvert into two V photons in the second crystal. Now suppose that a pump photon linearly polarized at $45°$ to the optic axes is incident on this two-crystal system. In this case there are two equally likely possibilities: two H photons are produced in the first crystal or two V photons in the second.

The two-photon polarization correlation now is of a different character than that described by a product state $|H\rangle_1|H\rangle_2$ or $|V\rangle_1|V\rangle_2$. In this case photon 1 is equally likely to be H or V polarized. If it is found to be H polarized, then photon 2 must be H polarized. If it is found to be V polarized, photon 2 must also be V polarized. Similarly, if the polarization of photon 2 is measured and found to be $H(V)$, then the polarization of photon 1 must be $H(V)$. In other words, a measurement of the polarization of one photon allows us to predict *with certainty* the polarization of its partner. The two-photon polarization state according to quantum theory is a linear combination of the two product states $|H\rangle_1|H\rangle_2$ and $|V\rangle_1|V\rangle_2$, a so-called *entangled state* such as[10]

$$\frac{1}{\sqrt{2}}(|H\rangle_1|H\rangle_2 + |V\rangle_1|V\rangle_2). \qquad (10.9.6)$$

Entangled quantum states of multiparticle systems exhibit some of the most profound differences between classical and quantum physics, and they are nearly always invoked in studies of the conceptual foundations of quantum theory. The polarization-entangled

[9]P. G. Kwiat, E. Waks, A. G. White, I. Appelbaum, and P. H. Eberhard, *Physical Review* A**60**, R773 (1999).
[10]The $1/\sqrt{2}$ is a normalization factor that is of no concern for our purposes. More generally a two-photon entangled state has the form $(1/\sqrt{2})(|H\rangle_1|H\rangle_2 + e^{i\phi}|V\rangle_1|V\rangle_2)$. In the two-crystal experiment the (real) phase ϕ depends on crystal thickness and other factors.

states produced in two-photon downconversion have been particularly important in experimental research in this area.

10.10 DISCUSSION

In this brief introduction to nonlinear optics we have restricted ourselves to examples of three- and four-wave mixing processes. Three-wave mixing in a medium with a nonlinear susceptibility $\chi(-\omega_3, \omega_2, \omega_1)$ involves the mixing of waves of frequency ω_1 and ω_2 to generate a wave of frequency $\omega_3 = \omega_1 + \omega_2$. Examples of three-wave mixing include second-harmonic generation, optical rectification, parametic amplification and oscillation, and two-photon downconversion. Four-wave mixing in a medium with a nonlinear susceptibility $\chi(-\omega_4, \omega_3, \omega_2, \omega_1)$ involves the mixing of waves of frequency ω_1, ω_2, and ω_3 to generate a wave of frequency $\omega_4 = \omega_1 + \omega_2 + \omega_3$. We considered only two examples of four-wave mixing, namely self-focusing and self-phase modulation arising from the nonlinear susceptibility $\chi(-\omega, \omega, \omega, -\omega)$.

It should be clear that there is a wide variety of three- and four-wave mixing processes as well as many other nonlinear optical processes described by higher-order nonlinear susceptibilities. *Third-harmonic generation*, for example, is a four-wave mixing process described by a nonlinear susceptibility $\chi(-3\omega, \omega, \omega, \omega)$: A sufficiently intense field of frequency ω produces a nonlinear polarization and therefore a field at frequency 3ω. It can be described by coupled wave equations similar to those we used in our discussions of second-harmonic generation and three-wave mixing. As in those examples, there is a phase-matching condition for third-harmonic generation, namely $n(3\omega) = n(\omega)$ in the simplest case of colinear fundamental and third-harmonic fields. Because it is a four-wave mixing process, third-harmonic generation is not restricted, like second-harmonic generation generally is, to noncentrosymmetric crystals and can occur in essentially any medium. Second- and third-harmonic generation are widely used as sources of radiation at frequencies for which lasers are not readily available.

We have also restricted ourselves to phenomena associated with the real parts of nonlinear susceptibilities. Recall that the imaginary part of the linear susceptibility gives rise to absorption—a molecular transition is accompanied by the absorption of a photon. Similarly, imaginary parts of nonlinear susceptibilities imply multiphoton absorption, such as two-photon absorption in which a molecular transition is accompanied by the absorption of two photons. Whereas the one-photon absorption process due to the imaginary part of the linear susceptibility is described in simplest terms by the equation $dI/dz = -a_1 I$, where I is the field intensity and a_1 is the absorption coefficient, the effect of n-photon absorption on the intensity is described by $dI/dz = -a_n I^n$, where a_n is the n-photon absorption coefficient. In the case of two-photon absorption, for example,

$$\frac{dI}{dz} = -a_2 I^2, \tag{10.10.1}$$

with the solution

$$I(z) = \frac{I(0)}{1 + a_2 z I(0)}, \tag{10.10.2}$$

which contrasts with the exponential (Beer law) attenuation of intensity in the case of a one-photon absorption process. As in the case of one-photon processes, spontaneous and stimulated multiphoton processes are also possible. For example, experiments have demonstrated lasing based on two-photon stimulated emission. It is also possible to have coherent multiphoton effects analogous to the Rabi oscillations in one-photon atom–field interactions (Chapter 9).

In addition to multiphoton absorption processes involving bound electron states, there are multiphoton *ionization* transitions in which one or more electrons are freed from an atom or molecule. Multiphoton ionization will occur in any atom or molecule when the laser radiation is sufficiently intense. Thus, any transparent medium under sufficiently intense irradiation will suffer *optical breakdown*, the formation of a (typically) high-temperature plasma due to ionization of the medium.[11] Usually, there is a visible spark, which can be quite large. The plasma can be highly opaque, resulting in a blockage of the laser beam. In pure air at standard temperature and pressure, for example, the threshold intensity for optical breakdown (BD) by laser radiation of wavelength λ is

$$I_{BD} \sim \frac{3 \times 10^{11}}{\lambda^2} \; W/cm^2, \tag{10.10.3}$$

for laser pulses longer than about 1 μs. For much shorter pulses the threshold condition for breakdown involves the pulse fluence (the time integral of the pulse intensity).

Optical breakdown is an avalanche process in which electrons already present in the medium take up energy from the field and so become able to produce more electrons by ionizing molecules with which they collide. Avalanche ionization and breakdown occur if, among other things, the rate of free-electron production exceeds the rate of loss due to attachment, recombination, and diffusion of the electrons out of the interaction region (e.g., the focal volume of laser radiation). The avalanche process requires some primary electrons to be present initially. Multiphoton ionization is believed to be important in creating the primary electrons, and therefore setting the stage for avalanche ionization and breakdown. The presence of particulate matter (aerosols) such as dust results in a lowering of the threshold intensity for optical breakdown of gases. Dirty air, for instance, can have breakdown threshold intensities several orders of magnitude smaller than that given by (10.10.3).

The extremely high intensities possible with laser pulses can lead to propagation effects in which different nonlinear processes play important roles and cannot be treated independently. For example, a laser pulse in air can undergo self-focusing while also producing a plasma. This can result in the propagation of a "self-channeling" pulse that retains a steady radial shape over a distance several times the Rayleigh range. The theory of such effects generally requires numerical solutions of coupled, nonlinear partial differential equations, using techniques such as those described in Section 16.C.

The interaction of intense laser pulses with matter cannot generally be understood in a perturbative fashion based on nonlinear susceptibilities and power series in the electric field strength, as we have done for second-harmonic generation, for instance. This is true not only for pulses propagating in a macroscopic medium, but also for the effect of an

---

[11]This is distinct from *thermal breakdown*, which arises from absorption of laser radiation and a consequent heating of the medium. In a gas the temperature rise can lead to collisional ionization and plasma formation, whereas in a solid it can lead to melting, vaporization, and surface plasma formation.

intense pulse on a single atom. Under extremely intense laser irradiation atoms have been found to produce a large number of odd harmonics of the fundamental laser frequency (Section 14.7); in some cases harmonics higher than a few hundred have been observed. The fact that only odd harmonics are generated is a consequence of the inversion symmetry of an atom, just as in the case of our two-state model in Section 10.1.

PROBLEMS

10.1. (a) Show that (10.2.21) follows from (10.2.19) in the limiting case of a two-state atom.

(b) Now consider, instead of the two-state model, a nonlinear electron oscillator model in which (3.2.18b) is replaced by

$$\frac{d^2x}{dt^2} + \omega_0^2 x + ax^n = \frac{e}{m} E_0 \cos \omega t$$

in the case of an applied field of frequency ω. Find, for $n = 2$, the solution for the induced dipole moment ex including terms up to second order in the field amplitude E_0.

(c) For this nonlinear oscillator model with $n = 3$, find the solution for the induced dipole moment including terms up to third order in E_0.

10.2. Derive Eq. (10.4.9) for the paraxial propagation of a pulse in a medium with group velocity dispersion and a Kerr-type nonlinear refractive index.

10.3. (a) Derive Eq. (10.5.25).

(b) Derive Eq. (10.5.26).

10.4. For quartz the refractive indices for the frequency corresponding to $\lambda = 694.3$ nm are $n_o(\omega) \approx 1.5408$ and $n_e(\omega) \approx 1.5498$, while at the second-harmonic frequency $n_o(2\omega) \approx 1.5664$ and $n_e(2\omega) \approx 1.5774$. Discuss the possibility of angle phase matching for second-harmonic generation with ruby laser radiation in

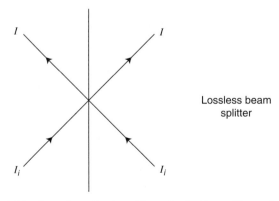

Figure 10.10   Two fields of equal amplitude and intensity incident with equal angles of incidence on opposite sides of a lossless beam splitter.

quartz. What is the angle θ_p between the pump and second-harmonic waves for angle phase matching?

10.5. Verify expression (10.6.8) for angle phase matching in a negative uniaxial crystal. Does the second-harmonic field propagate as an ordinary or an extraordinary wave?

10.6. (a) Derive the Manley–Rowe relations (10.7.4).

(b) Derive Eq. (10.8.8).

10.7. Modify Eqs. (10.8.1) to include loss coefficients for the fields at ω_1 and ω_2. Derive the threshold condition for parametric oscillation.

10.8. (a) Derive Eq. (10.9.2) and state the assumptions used in your derivation.

(b) Derive Eq. (10.9.3).

10.9. Consider two fields, each of amplitude E_i and intensity I_i, incident with equal angles of incidence on opposite sides of a lossless beam splitter as shown in Fig. 10.10. The "output" fields will have the same intensity $I \propto |\mathcal{R}E_i + \mathcal{T}E_i|^2 = |\mathcal{R} + \mathcal{T}|^2|E_i|^2$, where $\mathcal{R}$ and $\mathcal{T}$ are the (complex) amplitude reflection and transmission coefficients, respectively. Since the beam splitter is lossless, we must have $|\mathcal{R} + \mathcal{T}|^2 = 1$. Writing $\mathcal{R} = |\mathcal{R}|e^{i\phi_R}$ and $\mathcal{T} = |\mathcal{T}|e^{i\phi_T}$, show that $\cos(\phi_R - \phi_T) = 0$.

10.10. In deriving Eq. (10.9.4) we assumed that a photon cannot be "split" by a beam splitter, that is, that a photon is an indivisible unit of energy. But two-photon downconversion is sometimes described as a splitting of a photon of frequency ω_3 into photons of frequency ω_1 and ω_2. Discuss this contradiction.

11 SOME SPECIFIC LASERS AND AMPLIFIERS

11.1 INTRODUCTION

In thermal equilibrium the ratio of (nondegenerate) upper- and lower-state populations of an atomic or molecular transition is

$$\frac{N_2}{N_1} = \frac{e^{-E_2/k_B T}}{e^{-E_1/k_B T}} = e^{-(E_2 - E_1)/k_B T} = e^{-h\nu/k_B T}, \tag{11.1.1}$$

where ν is the transition frequency. This ratio is always less than one. This means that a medium in complete thermal equilibrium is always an absorber rather than an amplifier of radiation, regardless of how hot it is.

In order to have a population inversion on a transition, therefore, the level populations N_2 and N_1 must have a nonthermal distribution. If we insist on thinking in terms of a temperature, we see from (11.1.1) that a population inversion ($N_2 > N_1$) is associated with a "negative (absolute) temperature." The concept of negative absolute temperature was sometimes used in the early days of maser and laser research, but it can be misleading because it applies only to the lasing levels, while the rest of the atom or molecule exists at an entirely different and generally positive temperature.

The amplification of radiation by stimulated emission requires a population inversion, and in order to have a population inversion we must "pump" the medium to overcome its natural tendency to reach a thermal equilibrium. The suitability of a material as a laser medium thus depends, among other things, on how readily we can force it away from thermal equilibrium and establish a population inversion.

Arthur Schawlow joked that "anything will lase if you hit it hard enough."[1] Lasers are often classified according to how the gain medium is "hit," that is, according to the method used to obtain population inversion and gain. Thus, we speak of optically pumped lasers, electric-discharge lasers, chemical lasers, gas-dynamic lasers, etc. We will now discuss some of the methods for realizing population inversion and gain and also for generating extremely short and extremely intense pulses of light; each of these

---

[1] In public lectures Schawlow used an "edible laser" in which the gain medium was a gelatin dessert spiked with sodium fluorescein dye.

Laser Physics. By Peter W. Milonni and Joseph H. Eberly
Copyright © 2010 John Wiley & Sons, Inc.

methods draws on principles associated with different, broad areas of research and development in chemistry, physics, and engineering.

11.2   ELECTRON-IMPACT EXCITATION

A common means of pumping gas lasers is an electric discharge, which may be produced in a gas contained inside a glass tube by applying a high voltage to electrodes on either side of the tube. Electrons are ejected from the negative electrode (the cathode) and drift toward the positive electrode (the anode). When an electron collides with an atom (or molecule), there is some probability that the atom makes a transition to a higher energy state. This process of *electron-impact excitation* occurs in neon lamps, in which neon atoms excited by collisions with electrons undergo spontaneous emission and emit the deep red light so familiar from neon sign advertisements. Electron-impact excitation is also the microscopic basis of fluorescent lamps. In this case the electrons excite mercury atoms, which emit strongly in the ultraviolet; the tube is coated with a material that absorbs in the ultraviolet and emits in the visible.

We will discuss in Sections 11.4 and 11.7 the electrical excitation of He–Ne and CO_2 lasers. First, however, it will be worthwhile to discuss some general aspects of electron–atom (–molecule) collisions.

The simplest electron–atom process is an elastic collision in which the kinetic energy of the electron–atom system is conserved in the collision. In other words, none of the initial kinetic energy in an elastic collision is converted to internal energy of the atom. Furthermore there is relatively little exchange of kinetic energy between the electron and the atom in such an elastic collision; this is a simple consequence of the much smaller mass of the electron compared to the mass of the atom (Problem 11.2).

In an inelastic collision with an atom, the kinetic energies of the electron before and after the collision are different. In an inelastic *collision of the first kind*, the electron loses kinetic energy. Energy lost by the electron is converted to internal excitation energy of the atom and, of course, the total energies (kinetic plus internal) before and after the collision are the same. These collisions of the first kind are what we normally refer to as electron-impact excitation processes. In electron-impact excitation of molecules, the internal energy added to the molecule may be in the form of vibrational and rotational energy as well as electronic energy.

If an electron collides with an already excited atom or molecule, it can cause the atom or molecule to drop to a lower level of excitation, the energy difference now going into an increase in the kinetic energy of the system. This type of inelastic collision is called a *collision of the second kind* (or a "superelastic" collision).

A gas laser may be pumped *directly* by electron-impact excitation, in the sense that collisions of the active atoms with the electrons are the sole source of the population inversion. In this case the rates for the various excitation (collisions of the first kind) and deexcitation (collisions of the second kind) processes enter into the population rate equations as pumping and decay rates, respectively. Frequently, however, electron-impact excitation produces a population inversion *indirectly*, in the sense that it sets the stage for another process that acts more directly to produce a positive gain. The most important of these other processes is excitation transfer from one atom (or molecule) to another, which we discuss in the following section.

The rates at which electrons excite atoms and molecules in collisions of the first kind are determined by the collision cross sections σ. These in turn depend upon the relative velocity of the electron–atom (–molecule) pair. Thus, if a mono-energetic beam of N_e electrons per unit volume is incident upon an atom, the rate at which the atom is raised from level i to level $f(E_f > E_i)$ is

$$R_{if} = N_e \sigma_{if}(v)v, \tag{11.2.1}$$

where v is the relative velocity prior to collision and $\sigma_{if}(v)$ is the cross section for the process. The determination of the electron-impact excitation cross sections $\sigma_{if}(v)$ for electron–atom, electron–molecule, and electron–ion collisions is an old and active branch of atomic and molecular physics. These cross sections are important not only for the understanding and design of electric-discharge lasers, but also for a great many other phenomena, including such things as lightning and the aurora borealis.

In an electric discharge the electrons do not all have the same kinetic energy; there is a distribution $f(E)$ of electron energies, defined such that $f(E)\,dE$ is the fraction of electrons with energy in the interval $[E, E + dE]$. In this case R_{if} is obtained by averaging over the electron energy distribution:

$$R_{if} = N_e \left(\frac{2}{m}\right)^{1/2} \int_0^\infty \sigma_{if}(E)E^{1/2}f(E)\,dE, \tag{11.2.2}$$

where we have used $E = \frac{1}{2}mv^2$ to relate the electron energy and velocity. Collisions with electrons can also ionize atoms and molecules and break apart (dissociate) molecules. Equation (11.2.2) applies to any of these processes, each characterized by a cross section $\sigma(E)$. The rate for a collision of the second kind may also be expressed in the form (11.2.2).[2]

Electron energy distribution functions $f(E)$ in electric-discharge lasers are often not well described by the Boltzmann distribution $f(E) \sim \exp(-E/k_BT)$. They are frequently approximated by such a distribution, however, in which case the "electron temperature" is much greater than the temperature of the atomic (or molecular) gas in the discharge, typically being measured in thousands of degrees. The distribution $f(E)$ is sometimes obtained by direct numerical solution of the Boltzmann transport equation. In this case the cross sections for all important electron collision processes are essential input data for the computations.

11.3 EXCITATION TRANSFER

One way for an excited atom to transfer energy to another atom is by photon transfer: The photon spontaneously emitted by one atom is absorbed by the other. In this way the first atom drops to a lower level and the second atom is raised to a higher level, i.e., there is an excitation transfer between the two atoms. This process has a negligible probability of occurrence unless the photon emitted by the donor atom is within the absorption

---

[2]The cross section $\sigma_{fi}(E)$ for a collision of the second kind may be related to the cross section $\sigma_{if}(E)$ for the reverse process (collision of the first kind) by using the principle of detailed balance.

linewidth of the acceptor atom, that is, there must be a resonance (or near-resonance) of the atomic transitions.

Actually, the process of excitation transfer via spontaneous emission is quite negligible compared to other transfer processes that result from a direct nonradiative (e.g., collisional) interaction between two atoms. The calculation of excitation transfer rates between atoms (and molecules) is usually very complicated, and experimental determinations of transfer rates are essential; such studies form an entire field of research.

It would take us too far afield to discuss experimental and theoretical techniques used to obtain excitation transfer cross sections and rates. We will only describe some salient results of such studies. In the following sections we discuss the essential role of excitation transfer in two important lasers, He–Ne and CO_2.

The most important fact about excitation transfer is that the transfer cross section is large when the corresponding atomic or molecular transition frequencies are approximately equal. However, excitation transfer can occur between two species A and B even if the transitions are not precisely resonant. The *energy defect* ΔE (Fig. 11.1) can be made up by translational degrees of freedom, so there is no contradiction with the law of conservation of energy.

In the case of a positive energy defect, for instance, the "extra" energy ΔE appears as additional kinetic energy of A and B after the excitation transfer. In other words, A and B have more translational energy after the collision than before. Since the temperature of a gas is a measure of its translational energy content, this means that an *exothermic* ($\Delta E > 0$) process raises the temperature of the A–B system.

Similarly, in an *endothermic* ($\Delta E < 0$) process the "defect" in energy is made up at the expense of the kinetic energy of the collision partners. The kinetic energy of A and B after the excitation transfer is less than that before the transfer. Therefore, exothermic excitation transfer processes tend to lower the temperature of the system. It may be shown from the *principle of detailed balance* that the rate for an exothermic process is a factor $\exp(\Delta E/k_B T)$ greater than the reverse, endothermic process. Thus, if R is the rate for an exothermic process, then $R\exp(-\Delta E/k_B T)$ is the rate for the reverse process.

It is conventional to designate an excited atom or molecule by an asterisk. The exothermic process indicated in Fig. 11.1*a* is written out as a reaction as follows:

$$A^* + B \longrightarrow A + B^* + \Delta E. \tag{11.3.1}$$

Likewise the endothermic process of Fig. 11.1*b* is written symbolically as

$$A^* + B \longrightarrow A + B^* - \Delta E. \tag{11.3.2}$$

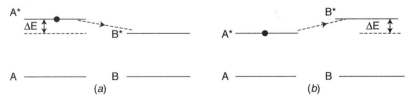

Figure 11.1   (*a*) Exothermic and (*b*) endothermic excitation from atom A to atom B.

We can write the following rate equation for the number density N_{A^*} of excited atoms A^* (or molecules) of species A due to the process (11.3.1):

$$\frac{dN_{A^*}}{dt} = -R(N_{A^*}N_B - e^{-\Delta E/k_B T}N_A N_{B^*}).$$

(11.3.3)

The first term on the right is associated with the process (11.1a), described by the rate constant R. Since this process occurs only if an A atom is excited *and* a B atom is not, the rate of decrease of excited A atoms is proportional to the product of the number densities of excited A atoms (N_{A^*}) and unexcited B atoms (N_B).[3] Similarly, the second term, associated with the process of Fig. 11.1b, is proportional to N_{B^*} times N_A because the process occurs only when a B atom is excited and an A atom is unexcited. The principle of detailed balance, which is discussed below, has been used to relate the forward and reverse rates in (11.3.3). In addition to (11.3.3) we may write rate equations for N_A, N_{B^*}, and N_B for the processes indicated in Fig. 11.1 (Problem 11.3).

• In one form the principle of detailed balance is the requirement that in thermodynamic equilibrium the rate of *any* process must be exactly balanced by the rate associated with the reverse of that process. In thermodynamic equilibrium, therefore, the right side of (11.3.3) must vanish. This means that

$$\frac{N_{A^*}}{N_A} = e^{-\Delta E/k_B T}\frac{N_{B^*}}{N_B}$$

(11.3.4)

for thermal equilibrium. But since

$$\Delta E = (E_{A^*} - E_A) - (E_{B^*} - E_B)$$

(11.3.5)

(recall Fig. 11.1a), we can write (11.3.4) as

$$\frac{N_{A^*}}{N_A} = e^{-(E_{A^*}-E_A)/k_B T}\left(\frac{N_{B^*}}{N_B}e^{(E_{B^*}-E_B)/k_B T}\right).$$

(11.3.6)

However, it is always true in thermal equilibrium that

$$\frac{N_{A^*}}{N_A} = e^{-(E_{A^*}-E_A)/k_B T}$$

(11.3.7a)

and

$$\frac{N_{B^*}}{N_B} = e^{-(E_{B^*}-E_B)/k_B T}$$

(11.3.7b)

according to the Boltzmann law. Therefore, we see that (11.3.3) is consistent with the steady-state Boltzmann law (for which $dN_{A^*}/dt = 0$). However, it is important to recognize that (11.3.3) applies regardless of whether the level populations are actually in thermal equilibrium, as long as the *translational* degrees of freedom of the gas are in thermal equilibrium at the temperature T.

---

[3]The product is a consequence of the fact that the *probability* of A being excited and B being unexcited is the probability of A being excited times the probability of B being unexcited. That is, the two "events"—A excited and B unexcited—are assumed to be statistically independent.

Thus, the principle of detailed balance relates the forward and reverse rates for a process in just such a way as to be consistent with the equilibrium distribution of states that the process would establish were it acting alone. In the present example of collisional excitation transfer in a gas whose translational degrees of freedom are characterized by a temperature T, the appropriate distribution is that of Boltzmann. In the case of electron-impact excitation, in which the electrons might not be well described by a Boltzmann distribution, the principle of detailed balance relates the forward and reverse rates in a way that would bring the atomic states into equilibrium with whatever the electron energy distribution happens to be. If both electron-impact excitation and collisional excitation transfer are occurring, as in many gaseous laser media, the steady-state distribution of the active atomic levels will not in general be a Boltzmann distribution. That is, the active levels will not be populated according to the statistical distribution (11.3.7). If they were, collisional excitation would obviously be unattractive as a laser pumping mechanism.   •

It is important to recognize that these energy transfer processes are not the only ones that can occur in a collision involving an excited atom or molecule. In a molecular gas, for example, collisions between molecules result in vibration-to-vibration (VV) energy transfer between the molecules. But there is also a probability that in a collision a vibrationally excited molecule will jump to a lower vibrational level, with the difference in energy between the two levels appearing as an increase in the translational kinetic energy of the colliding molecules. The latter process is called vibration-to-translation (VT) energy transfer.

11.4   He–Ne LASERS

The He–Ne electric-discharge laser, with its red output beam at 632.8 nm, is one of the most familiar gas lasers. He–Ne lasers can be made to operate at many other (mainly infrared) wavelengths. Whatever the operating wavelength, the active lasing species in He–Ne lasers is the Ne atom. It is excited by the transfer of excitation from He atoms, which in turn are excited by collisions with electrons. The population inversion mechanism in He–Ne lasers thus involves a combination of electron-impact excitation (of He) and excitation transfer (from He to Ne).

Figure 11.2 shows simplified energy-level diagrams for the neon and helium atoms. The 3.39-μm, 1.15-μm, and 632.8-nm lines of neon, which are the strongest lasing transitions in He–Ne lasers, are indicated. The common upper level of the 3.39-μm and 632.8-nm transitions, designated $3s_2$, is populated by excitation transfer from nearly resonant He atoms excited by electron impact to the 2^1S level. The upper level of the 1.15-μm transition is nearly resonant with the 2^3S level of He, and is populated by excitation transfer from He atoms in that excited state. Actually, Ne is also pumped directly into excited states by electron impacts, but the excitation transfer from He is the dominant pumping mechanism. The excited levels 2^1S_0 and 2^3S_1 of He, in addition to being nearly resonant with levels of Ne (and therefore allowing strong collisional excitation transfer) have the advantage of being forbidden by a selection rule to de-excite by spontaneous emission. This allows these levels to "hold" energy for delivery to Ne during collisions. (The total decay rates of He 2^1S and 2^3S levels due to collisions with Ne atoms are about 2×10^5 s^{-1} and 10^4 s^{-1}, respectively, per Torr of Ne.)

The partial pressures of Ne and He in typical He–Ne lasers are roughly 0.1 and 1 Torr, respectively. At these low pressures the upper-state lifetimes are determined

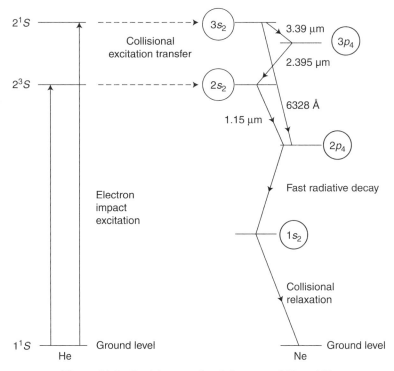

Figure 11.2 Partial energy-level diagrams of He and Ne.

predominantly by spontaneous emission rather than collisional deexcitation. The $3s_2$ and $2s_2$ levels[4] of Ne have short radiative (i.e., spontaneous emission) lifetimes, roughly 10–20 ns, due to the strong allowed ultraviolet transitions to the ground state. For Ne pressures typical of He–Ne lasers, however, these radiative lifetimes are actually about 10^{-7} s because of *radiative trapping*. This occurs when the spontaneously emitted photons are reabsorbed by atoms in the ground state, thereby effectively increasing the lifetime of the emitting level. Since the ground state is generally the most highly populated level even when there is population inversion [recall the numerical estimates in Section 4.3], radiative trapping is significant only from levels connected to the ground level by an allowed transition. Thus, the Ne $3p_4$ and $2p_4$ levels, which are forbidden by a selection rule from decaying spontaneously to the ground level, are not radiatively trapped and have lifetimes of about 10^{-8} s, roughly 10 times shorter than the $3s_2$ and $2s_2$ levels.

This means that the $s \rightarrow p$ transitions indicated in Fig. 11.2 have favorable lifetime ratios for lasing, that is, their lower (p) levels decay more quickly than their upper (s) levels, making it easier to establish a population inversion. The integrated absorption coefficients of the 632.8-nm and 3.39-μm lines have roughly the same magnitude, but the 3.39-μm line has a Doppler width about 5.4 times smaller than the 632.8-nm line, and consequently a considerably larger line-center gain. Without some mechanism

---

[4]We follow here the Paschen notation for the Ne levels. A more modern and systematic notation, based on Racah symbols, is not often encountered in the laser literature.

for suppressing oscillation on the 3.39-μm line, therefore, the familiar 632.8-nm line would not lase (Problem 11.4).

The 632.8-nm and 1.15-μm transitions have a common lower level, $2p_4$, which decays rapidly into the $1s$ level. The latter is forbidden by a selection rule from decaying radiatively into the ground level, and is therefore relatively long-lived. This is bad for laser oscillation on the 632.8-nm and 1.15-μm lines because electron-impact excitation can pump Ne atoms from $1s$ to $2p_4$, thereby reducing the population inversion on these lines. However, Ne atoms in the $1s$ level can decay to the ground level when they collide with the walls of the gain tube. In fact, it is found that the gain on the 632.8-nm and 1.15-μm lines increases when the tube diameter is decreased; this is attributed to an increase in the atom–wall collision rate with decreasing tube diameter.

The first gas laser, which was also the first cw laser, was a 1.15-μm He–Ne laser constructed in 1960 by A. Javan, W. R. Bennet, Jr., and D. R. Herriott. Inexpensive He–Ne lasers have long since been available commercially and have many practical applications. In most of these applications a *visible* laser beam is desired, and it is therefore necessary to suppress the infrared lines in order to obtain oscillation at 632.8 nm. This is done by discriminating against the infrared lines by using a cavity in which these lines have greater loss than the 632.8-nm line, and therefore a higher threshold for oscillation. Typically, this is accomplished by coating the cavity mirrors with dielectric materials that reflect in the visible but transmit in the infrared.

- Laser mirrors must be of much finer optical quality than the lenses and mirrors used in many other optical instruments. For most industrial applications, laser mirrors that are flat to within a quarter of a wavelength ($\lambda/4$) are adequate, but for interferometric applications the degree of surface flatness required is $\sim\lambda/20$. The reflecting surface can be either a metallic or dielectric coating, or some combination of the two. The highest-reflectivity mirrors currently available are made from a few tens of dielectric layers having a width of $\lambda/4$ and alternating between two different refractive indices (cf. Problem 5.10 and also the discussion of Bragg mirrors in Section 8.4). Such "quarter-wave stacks" can have reflectivities exceeding 99.999% or more in a narrow band around a specified wavelength. Dielectric layers can also be designed, for instance, to reflect some fraction f_1 of radiation of wavelength λ_1 while reflecting a fraction f_2 of radiation of wavelength λ_2.

Figure 11.3 indicates the general structure of a commercial He–Ne laser. For reasons discussed in Chapter 7, one of the mirrors is flat whereas the other is curved. The gain tube, which is typically 10–30 cm long, is shown with Brewster-angle windows to give a linearly polarized output (Section 5.10). The mirrors can be attached to the glass tube with an epoxy

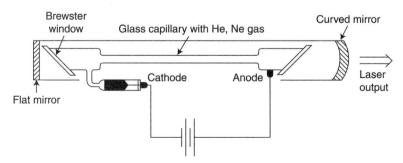

Figure 11.3   Basic structure of a He–Ne laser.

resin. However, epoxy eventually deteriorates and gas leaks out of the tube. "Hard-seal mirrors" have been developed to increase the life of the gain tube; the mirrors are bonded directly to the metal housing of the laser to form a tight seal without any epoxy. These lasers are low-power devices, producing outputs typically measured in milliwatts. ●

11.5 RATE EQUATION MODEL OF POPULATION INVERSION IN He–Ne LASERS

We will consider a rate equation model of population inversion in a 3.39-μm He–Ne laser. Our goal is to account for some observed results on the variation of small-signal gain with the electric current i through the gain tube.

Figure 11.4 summarizes our notation for the pertinent energy-level population densities. N_2 and N_1 denote the population densities (atoms/cm^3) of the upper and lower levels of the 3.39-μm laser transition. The ground-level population densities of He and Ne are denoted by $\tilde{N}_0$ and N_0, respectively, and $\tilde{N}_2$ denotes the population density of the excited level 2^1S of He. In our simplified model we will ignore all other levels of the He and Ne atoms. In other words, we assume that other levels are required only to explain some finer details that are not presently of interest to us.

The rate of change of N_2 due to excitation transfer with He is given by

$$\left(\frac{dN_2}{dt}\right)_{\substack{\text{excitation} \\ \text{transfer}}} = R\tilde{N}_2 N_0 - Re^{-\Delta E/k_B T}\tilde{N}_0 N_2, \tag{11.5.1}$$

where R is the rate constant for the excitation transfer collisions and

$$\Delta E = E[\text{He}(2^1S)] - E[\text{Ne}(3s_2)] \tag{11.5.2}$$

is the energy defect of the $\text{He}(2^1S) \rightarrow \text{Ne}(3s_2)$ inelastic collision. This energy defect is quite small, about 0.04 eV, and so we will take $\exp(-\Delta E/k_B T) \sim 1$ as an approximation in (11.5.1) (Problem 11.5). We therefore write

$$\frac{dN_2}{dt} = \left(\frac{dN_2}{dt}\right)_{\substack{\text{excitation} \\ \text{transfer}}} - R_2 N_2 = R\tilde{N}_2 N_0 - R\tilde{N}_0 N_2 - R_2 N_2, \tag{11.5.3}$$

where the decay rate R_2 is the rate of decrease of N_2 due to spontaneous emission and other deexcitation processes.

Figure 11.4 Simplified version of Fig. 11.2 for a model of population inversion on the 3.39-μm transition of Ne.

Similarly,

$$\frac{d\tilde{N}_2}{dt} = \left(\frac{d\tilde{N}_2}{dt}\right)_{\substack{\text{excitation} \\ \text{transfer}}} + \left(\frac{d\tilde{N}_2}{dt}\right)_{\substack{\text{electron} \\ \text{impact}}} + \left(\frac{d\tilde{N}_2}{dt}\right)_{\substack{\text{decay} \\ \text{processes}}}. \tag{11.5.4}$$

The first term is just the negative of (11.5.1) (Problem 11.3). The second term is the rate of change of $\tilde{N}_2$ due to electron-impact excitation (and deexcitation) of He 2^1S, which we write as

$$\left(\frac{d\tilde{N}_2}{dt}\right)_{\substack{\text{electron} \\ \text{impact}}} = K_1\tilde{N}_0 - K_2\tilde{N}_2, \tag{11.5.5}$$

K_1 is the rate for the electron-impact excitation of He 2^1S from the ground level of He, i.e., for the process

$$\text{He}(1^1S) + e \longrightarrow \text{He}(2^1S) + e, \tag{11.5.6}$$

whereas K_2 is the rate for the reverse, "superelastic" collision. Finally, the third term on the right side of (11.5.4) is the rate of decrease of $\tilde{N}_2$ due to other deexcitation processes, and is assumed to be characterized by some constant $\tilde{R}_2$. Thus,

$$\frac{d\tilde{N}_2}{dt} = -R\tilde{N}_2N_0 + RN_2\tilde{N}_0 + K_1\tilde{N}_0 - K_2\tilde{N}_2 - \tilde{R}_2\tilde{N}_2. \tag{11.5.7}$$

In a similar fashion we can write rate equations for $\tilde{N}_0$ and N_0. However, as we noted in the preceding section, these ground-state populations are very large compared to excited-state populations, and remain relatively unchanged by the pumping process. We will therefore make the approximation that $\tilde{N}_0$ and N_0 are constants. This reduces (11.5.3) and (11.5.7) to linear differential equations with constant coefficients. In particular, the steady-state values of N_2 and $\tilde{N}_2$, denoted $\overline{N}_2$ and $\overline{\tilde{N}}_2$, are easily obtained by setting the derivatives in these equations to zero:

$$\overline{\tilde{N}}_2 = \frac{K_1 A \tilde{N}_0}{1 + K_2 A}, \tag{11.5.8a}$$

$$\overline{N}_2 = \frac{RN_0}{R\tilde{N}_0 + R_2}\frac{K_1 A \tilde{N}_0}{1 + K_2 A}, \tag{11.5.8b}$$

where A is a constant in our simple-minded model:

$$A = \frac{R_2 + \tilde{N}_0}{\tilde{R}_2(R_2 + R\tilde{N}_0) + RR_2N_0}. \tag{11.5.9}$$

Now the electron-impact rates K_1 and K_2 are directly proportional to the number density (N_e) of electrons and therefore to the electric current i in the discharge tube, as is evident from Eq. (11.2.2). We can, therefore, write Eqs. (11.5.8) in terms of the current:

$$\overline{N}_2 = \frac{ai}{1+bi}, \tag{11.5.10a}$$

$$\overline{\tilde{N}}_2 = \frac{ci}{1+bi}, \tag{11.5.10b}$$

where a, b, and c are constants.

In order to express the gain in terms of the current, we need an expression for $\overline{N}_1$, the (steady-state) population density of the lower laser level. We will assume that N_1 varies according to the rate equation

$$\frac{dN_1}{dt} = K_3 N_0 - R_1 N_1, \tag{11.5.11}$$

where K_3 is the rate at which electron impacts pump ground-state Ne atoms up to the $3p_4$ level, and R_1 is the total decay rate of that level.[5] The steady-state solution of this equation is obtained as usual by setting $dN_1/dt = 0$, and is given by

$$\overline{N}_1 = \frac{K_3 N_0}{R_1} \propto \text{current.} \tag{11.5.12}$$

From the expression for the gain coefficient g in Table 4.1, therefore, we obtain from (11.5.10) and (11.5.12) the gain–current relation

$$g = \frac{\alpha i}{1+bi} - \beta i, \tag{11.5.13}$$

where α and β are constants.

Note that, since no stimulated emission terms were included in our model, Eq. (11.5.13) is actually a prediction about the variation of *small-signal* gain with current. This variation was studied in the early days of He–Ne laser research and, as indicated in Fig. 11.5, the functional form (11.5.13), with certain values of the constants α, β, and b, can be used to fit very well the measured data on small-signal gain vs. current.

One can also deduce relative values of various level population densities by measuring the intensities of fluorescent "sidelight" radiation at different wavelengths. That is, by measuring the strength of the spontaneous emission from a side of the laser tube, one can estimate the relative population of the upper level from which this emission proceeds (Problem 11.6). Such results are shown in Fig. 11.6, and provide evidence for the approximate proportionality of the He 2^1S and Ne $3s_2$ populations, as predicted by (11.5.10). They show furthermore that the Ne $3p_4$ (and $2p_4$) population density is very nearly proportional to the current, as predicted by (11.5.12).

---

[5]It is not known precisely how the lower levels, such as $3p_4$, are populated in He–Ne lasers. There is apparently some electron-impact excitation from ground-state Ne atoms, as evidenced by the fact that pure Ne can be made to lase (weakly). The p levels may also be populated via the decay of higher-lying Ne levels that are populated by electron-impact excitation from ground-state or excited-state Ne atoms. In neglecting the deexcitation of $3p_4$ by electron impact, we are assuming that the corresponding rate is much smaller than the decay rate R_1 due to spontaneous emission, etc.

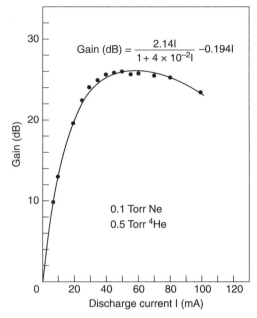

Figure 11.5   Measured variation of gain with current in a 3.39-μm He–Ne laser. [From A. D. White and E. I. Gordon, *Applied Physics Letters* **3**, 197 (1963).]

For small currents the small-signal gain is proportional to the current. As the current increases, however, two effects work against a further increase of the gain. First, there is the deexcitation of He 2^1S by electron impact; this is described by the rate constant $K_2 \propto b$, and is thus associated with the denominator $1 + bi$ in (11.5.13). Because of this denominator, the first term in (11.5.13) does not increase linearly with i, but rather saturates to the constant value α/b for large currents ($bi \gg 1$). Second, there is the proportionality of the

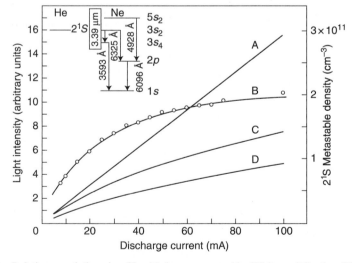

Figure 11.6   Relative populations in a He–Ne laser measured by White and Gordon. The four curves refer to: A, 492.8-nm line; B, 632.8-nm line; C, 609.6-nm line; D, 359.3-nm line, and the circles on curve B are measurements of the 2^1S metastable population density.

lower-laser-level population Ne $3p_4$ to the current, which is associated with the term βi in (11.5.13). Beyond a certain range of current values, therefore, the output power of a He–Ne laser will decrease with increasing current, and lasing will eventually cease altogether.

It should be emphasized that we have not formulated a first-principles theory of population inversion on the 3.39-μm line of a He–Ne laser. For instance, we have not specified the values of the electron–atom collision rates K_1, K_2, and K_3. This approach to understanding population inversion processes—that is, trying to understand general trends rather than obtaining detailed quantitative predictions—is more often the rule than the exception in laser theory and design. This is partly because many of the rates for the processes determining population inversion are often not well known and, because of the complexity of the processes, theoretical analyses are highly involved and not always highly reliable. Furthermore, in a given device various quantities, such as number densities of atoms and electrons, may be hard to specify accurately, so that even a trustworthy theory might be of only limited value. In spite of these disclaimers, however, some level of theoretical analysis is indispensable to the building of new lasers and the improvement of old ones.

11.6 RADIAL GAIN VARIATION IN He–Ne LASER TUBES

It is usually assumed that the small-signal gain g_0 is approximately constant throughout the gain medium. For many purposes this is a reasonable approximation, but it is generally not strictly true. We will now describe how the small-signal gain in a He–Ne laser varies with the radial distance r from the axis of the gain tube, which is assumed to have a circular cross section.

We will show below that the (free-) electron number density N_e varies with r according to the formula

$$N_e(r) = N_e(0)J_0\left(\frac{2.405r}{R}\right), \tag{11.6.1}$$

where R is the tube radius, J_0 is the zeroth-order Bessel function,[6] and $N_e(0)$ is the electron number density at the tube axis, where $r = 0$ $[J_0(x) = 1$ for $x = 0]$. Since $J_0(2.405\ldots) = 0$, it follows that N_e is zero on the tube wall, as expected.

The current density j is proportional to N_e, so that

$$j(r) = j(0)J_0\left(\frac{2.405r}{R}\right). \tag{11.6.2}$$

The current i is the integral of j over the cross-sectional area of the tube:

$$i = 2\pi\int_0^R j(r)r\,dr = 2\pi j(0)\int_0^R J_0\left(\frac{2.405r}{R}\right)r\,dr \approx 1.36j(0)R^2, \tag{11.6.3}$$

where we have used the properties $\int_0^z J_0(x)\,x\,dx = zJ_1(z)$ and $J_1(2.405) \approx 0.519$.

In the preceding section we derived the formula (11.5.13) for the current dependence of the gain, neglecting any r variation of the electron number density. In other words, we

---

[6]A graph of the function $J_0(x)$ is given in Fig. 6.13.

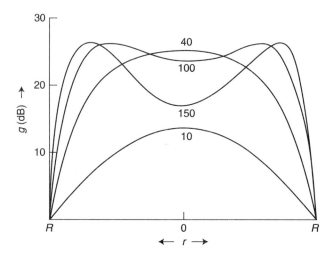

Figure 11.7   Radial dependence of small-signal gain, obtained by plotting the function (11.6.4) for the parameters α, b, β given in Fig. 11.5. The four curves correspond to currents $i = 10$, 40, 100, and 150 mA, as indicated.

took N_e and j as simply proportional to the current i, with no r dependence. Let us assume that (11.5.13) actually gives the small-signal gain at the *center* of the gain tube, i.e., that the left side of (11.5.13) is really $g(r = 0)$. Then, since N_e (and j) varies radially as $J_0(2.405r/R)$, we have for $g(r) \neq 0$ the formula (11.5.13) with i replaced by $iJ_0(2.405r/R)$:

$$g(r) = \frac{\alpha i J_0(2.405r/R)}{1 + b i J_0(2.405r/R)} - \beta i J_0\left(\frac{2.405r}{R}\right). \tag{11.6.4}$$

A graph of $g(r)$ is shown in Fig. 11.7 for several values of i, using the parameters α, b, and β of Fig. 11.5. For small currents $g(r)$ goes as $J_0(2.405r/R)$, having its maximum on axis and falling off to zero at $r = R$. As the current is raised $g(r)$ becomes flatter near the axis, and with higher currents has a dip on axis. That is, as the current is raised, a stage is reached where the small-signal gain has a local minimum along the axis of the tube.

Similar results for the radial variation of the small-signal gain are obtained whenever the on-axis gain-vs.-current curve has a maximum and then "turns over" with increasing current, as in Fig. 11.5. Therefore radial variations of the form shown in Fig. 11.7 are expected not only in He–Ne lasers but in a wide variety of other electric-discharge lasers. Although these lasers employ somewhat different population inversion processes, they all follow the general trends predicted in Fig. 11.7. Experimental results for He–Ne and CO_2 lasers are shown in Fig. 11.8. We refer the reader to the papers cited in the figure for details of the different measurements.

• The basic starting point of the analysis above is the formula (11.6.1) for the electron density in the (cylindrical) discharge tube. This formula may be understood from the following argument.

If the electron density has some nonvanishing gradient, the electrons will tend to redistribute, or *diffuse*, just as a gradient in the temperature of a gas gives rise to a flow of heat. We will assume that the diffusion of electrons follows the *Fick law*, i.e., that the electron diffusion rate across a

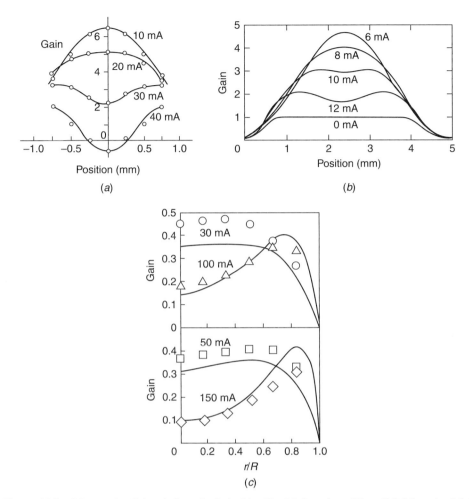

Figure 11.8 Measured radial variation of gain in (*a*) a He–Ne laser laser [From I. P. Mazanko, M. I. Molchanov, N.-D. D. Ogurok, and M. V. Sviridov, *Optics and Spectroscopy* **30**, 495 (1971).]; (*b*) a He–Xe laser [From P. A. Wolff, N. B. Abraham, and S. R. Smith, *IEEE Journal of Quantum Electronics* **QE-13**, 400 (1977).]; (*c*) a CO_2 laser [From W. J. Wiegand, M. C. Fowler, and J. A. Benda, *Applied Physics Letters* **18**, 365 (1971).]

given area is directly proportional, but opposite in direction, to the gradient. In other words, the flux **f** (particles per second per unit area) of electrons is given by

$$\mathbf{f} = -D\nabla N_e, \tag{11.6.5}$$

where the constant of proportionality D is the *diffusion coefficient*. If the electron density is increasing in some direction (positive gradient), Eq. (11.6.5) says there will be a compensating diffusion of electrons in the opposite direction. We will not attempt to justify (11.6.5), but it is worth mentioning that this Fick law accurately describes many diffusion processes, such as the diffusion of gas particles due to a density or thermal gradient, or the diffusion of neutrons in a nuclear reactor.

The rate at which electrons diffuse out of some closed surface S is given by the surface integral of the flux:

$$\text{Rate of diffusion out of volume } V = \int_S \mathbf{f} \cdot \hat{\mathbf{n}} \, dA, \qquad (11.6.6)$$

where V is the volume enclosed by S and $\hat{\mathbf{n}}$ is the unit vector outwardly normal to S. From the divergence theorem (i.e., Gauss's law) and (11.6.5) we have

$$\text{Rate of diffusion out of volume } V = \int_V \nabla \cdot \mathbf{f} \, dV = -D \int_V \nabla^2 N_e \, dV. \qquad (11.6.7)$$

Now in a steady-state discharge the rate of diffusion of electrons out of V must be exactly balanced by the rate at which free electrons are produced inside V, in order that the total number of electrons within V be constant. Free electrons are produced when an electron collides with an atom (or ion) and ionizes it, leaving another free electron plus a positive ion. This electron-impact ionization is the electron production process that must balance the diffusive loss (11.6.7). Letting Q_i denote the ionization rate, we have

$$\text{Rate of production of free electrons inside } V = \int_V Q_i N_e \, dV. \qquad (11.6.8)$$

Equating (11.6.7) and (11.6.8), we obtain

$$\nabla^2 N_e + \frac{Q_i}{D} N_e = 0. \qquad (11.6.9)$$

This equation, subject to whatever boundary conditions are to be imposed, determines the electron density N_e.

For a cylindrical discharge tube it is convenient to write out the Laplacian ∇^2 in terms of the cylindrical coordinates r, θ, z:

$$\nabla^2 N_e = \left(\frac{\partial^2 N_e}{\partial r^2} + \frac{1}{r} \frac{\partial N_e}{\partial r} + \frac{1}{r^2} \frac{\partial^2 N_e}{\partial \theta^2} + \frac{\partial^2 N_e}{\partial z^2} \right). \qquad (11.6.10)$$

We will assume circular symmetry (no θ dependence of N_e), and that the z dependence of N_e can be ignored to a good approximation. Then the last two terms in $\nabla^2 N_e$ above may be dropped, and (11.6.9) becomes

$$\frac{d^2 N_e}{dr^2} + \frac{1}{r} \frac{dN_e}{dr} + \frac{Q_i}{D} N_e = 0. \qquad (11.6.11)$$

This differential equation has the solution

$$N_e(r) = N_e(0) J_0(r\sqrt{Q_i/D}), \qquad (11.6.12)$$

where the constant $N_e(0)$ is the value of N_e at the tube axis, $r = 0$.

To satisfy the boundary condition that the electron density vanishes on the wall of the tube, i.e., that $N_e(R) = 0$, we require that

$$J_0(R\sqrt{Q_i/D}) = 0. \qquad (11.6.13)$$

In other words, $R\sqrt{Q_i/D}$ must be a zero of the zeroth-order Bessel function J_0. In order to ensure that $N_e(r)$ given by (11.6.12) is positive-definite for all values of $r \leq R$, furthermore, we require

$R\sqrt{Q_i/D}$ be the *first* zero of J_0, which is about 2.405. Thus

$$\sqrt{Q_i/D} = 2.405/R, \tag{11.6.14}$$

which, together with (11.6.12), gives (11.6.1).

The electrons in the discharge, because of their much higher average velocity, might be expected to diffuse to the walls much more quickly than the positive ions, producing an excess of negative charge near the walls. However, an electric field is set up by the charges (a "space charge" field) in such a way as to retard the diffusion of electrons and effectively "drag along" the positive ions. In this *ambipolar diffusion* both the positive and negative charge carriers have the same diffusion constant. Our simplified derivation of (11.6.1) assumes ambipolar diffusion, and D in our analysis is in fact the ambipolar diffusion coefficient.   •

11.7 CO_2 ELECTRIC-DISCHARGE LASERS

The electric-discharge carbon dioxide laser has a population inversion mechanism similar in some respects to the He–Ne laser: the upper CO_2 laser level is pumped by excitation transfer from the nitrogen molecule, with N_2 itself excited by electron impact.

The relevant energy levels of the CO_2 and N_2 molecules are vibrational-rotational levels of their electronic ground states. We discussed the vibrational-rotational characteristics of the CO_2 molecule in Section 2.5, and indicated in Fig. 2.10 the relative energy scales of the three normal modes of vibration, the so-called symmetric stretch, bending, and asymmetric stretch modes (Fig. 2.9). Like all diatomic molecules, N_2 has a single "ladder" of vibrational levels corresponding to a single mode of vibration (Fig. 2.7). In Fig. 11.9 we show the CO_2 and N_2 vibrational energy level diagrams side by side.

Figure 11.9 shows that the first excited vibrational level ($v = 1$) of the N_2 molecule lies close to the level (001) of CO_2. Because of this near resonance, there is a rapid excitation transfer between $N_2(v = 1)$ and $CO_2(001)$, the upper laser level. $N_2(v = 1)$ is itself a long-lived (metastable) level, so it effectively stores energy for eventual transfer to

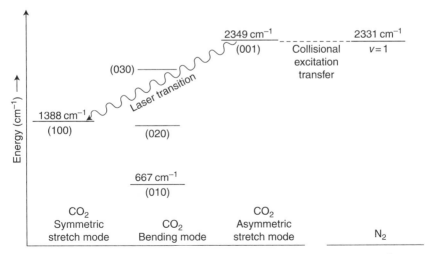

Figure 11.9 Vibrational energy levels of CO_2 and N_2. The energies are given in cm^{-1}, a unit corresponding to a frequency $(c)(1 \text{ cm}^{-1}) \approx 3.0 \times 10^{10}$ Hz, or an energy $h\nu \approx 1.2 \times 10^{-4}$ eV.

$CO_2(001)$; it is also efficiently pumped by electron-impact excitation. As in the case of the He−Ne laser, therefore, advantage is taken of a fortuitous near resonance between an excited state of the lasing species and an excited, long-lived collision partner.

Laser action in CO_2 lasers occurs on the vibrational transition $(001) \rightarrow (100)$ of CO_2. This transition has a wave number around $(2349 - 1388)$ cm^{-1} = 961 cm^{-1} (Fig. 11.9), or a wavelength around $(961$ cm$^{-1})^{-1}$ = 10.4 μm in the infrared. The laser wavelength depends also on the rotational quantum numbers of the upper and lower laser levels. For the case in which the upper and lower levels are characterized by $J = 19$ and 20, respectively, the wavelength is about 10.6 μm, the most common CO_2 laser wavelength.

The (100) and (020) vibrational levels of CO_2 are essentially resonant. This "accidental degeneracy" results in a strong quantum mechanical coupling in which states in effect lose their separate identities.[7] Furthermore the (010) and (020) levels undergo a very rapid vibration-to-vibration (VV) energy transfer:

$$CO_2(020) + CO_2(000) \quad \longrightarrow \quad CO_2(010) + CO_2(010). \qquad (11.7.1)$$

For practical purposes, then, the stimulated emission on the $(001) \rightarrow (100)$ vibrational band takes CO_2 molecules from (001) to (010). The (010) level thus acts in effect like a lower laser level that must be rapidly "knocked out" in order to avoid a bottleneck in the population inversion.

Fortunately, it is relatively easy to deexcite the (010) level by vibration-to-translation (VT) processes:

$$CO_2(010) + A \quad \longrightarrow \quad CO_2(000) + A, \qquad (11.7.2)$$

where A represents some collision partner. The VT deexcitation of (010) effectively depopulates the lower laser level and also puts CO_2 molecules in the ground level, where they can be pumpd to the upper laser level by the VV excitation transfer

$$N_2(v = 1) + CO_2(000) \quad \longrightarrow \quad N_2(v = 0) + CO_2(001). \qquad (11.7.3)$$

In high-power CO_2 lasers the lifetime of the $CO_2(010)$ level may be on the order of 1 μs due to collisions of CO_2 with He, N_2, and CO_2 itself. Of course, the VT process (11.7.2) is exothermic and results in a heating of the laser medium; some of the other VT and VV processes in the CO_2 laser have the same effect. This heating of the laser medium is a very serious problem in high-power lasers. In the next section we will see how it may be overcome.

Electron impacts excite CO_2 as well as N_2 vibrations. Furthermore there are various other processes that have to be accounted for in an accurately predictive rate-equation model of a CO_2 laser. Because of the many applications of high-power CO_2 lasers, such models have been developed and are often quite accurate. These models are computer programs that numerically integrate rate equations for the various level populations and the intensity. They also compute the electron energy distribution function, and from this the electron-impact excitation rates (Section 11.2). Our discussion captures only the bare essence of the population inversion process but is sufficient for a qualitative understanding of CO_2 lasers.

---

[7]For a discussion of this *Fermi resonance* effect see, for instance, G. Herzberg, *Molecular Spectra and Molecular Structure: Infrared and Raman Spectra of Polyatomic Molecules*, Robert E. Krieger Publishing, Malabar, FL, 1990.

In the laser research literature one finds expressions for the gain coefficients of CO_2 and other infrared molecular-vibration lasers; at first glance these expressions often do not resemble the "standard" formula for the gain coefficient given in Table 4.1. In the Appendix to this chapter we carry out the steps leading from the formula in Table 4.1 to an expression that appears frequently in the literature.[8] As an example of the use of this expression we estimate the absorption coefficient for 10.6-μm CO_2 laser radiation propagating in air at sea level.

11.8 GAS-DYNAMIC LASERS

We mentioned in Section 4.10 that many of the most powerful lasers employ gaseous gain media. Although solid laser media have much higher molecular densities than gases, and therefore a potential for higher gains, they also are more susceptible to heating that can damage the gain medium or induce distortions in it that can degrade the spatial coherence of the laser radiation, especially in the case of high-power, cw or long-pulse operation. The major damage mechanism for gaseous media, however, is photoionization, and this is usually not a concern except at extremely high intensities, perhaps $10^{10} - 10^{14}\,\text{W}/\text{cm}^2$ or higher, depending on the circumstances.

The pumping and lasing of a gaseous medium also generates waste heat, and this is deleterious to the scaling of a gas laser up to very high powers. Various factors, such as an increase in collisional deexcitation rates, contrive to reduce the power and coherence properties of the laser radiation when the gas gets too hot. By the late 1960s the highest-power lasers were CO_2 lasers generating several kilowatts of power. This certainly represents a good deal of radiation intensity when it is concentrated in a narrow laser beam; such a beam can drill holes in quarter-inch steel in a matter of seconds. But these lasers typically had "folded" resonator designs that made the gain medium effectively a hundred meters or more long in some cases. Because of the heat generated as an unavoidable product of lasing, they seemed at the time to be approaching some sort of practical upper limit in the quest for higher and higher powers.

In 1968, however, a CO_2 laser was developed that produced more than 60 kW of continuous-wave output power. The new idea was to remove the waste heat by using a laser medium consisting of a gas *flowing* through the laser cavity (Fig. 11.10). In this way the hot gas is expelled while fresh, cooler gas is continually flowing in and lasing. High-power lasers operated in this way are called *gas-dynamic lasers* when the gas flow velocity is supersonic.

It should be evident by now that laser engineering involves many diverse areas of atomic, molecular, electronic, solid-state, and optical physics. The gas-dynamic laser spawned a new branch of laser research connected with aerodynamic effects. We will not be able to discuss any technical aspects of this interplay of radiation physics and aerodynamics; instead we will describe qualitatively the population inversion mechanism in gas-dynamic lasers.

In the CO_2 gas-dynamic laser a gas mixture containing CO_2 and N_2 is heated in a high-pressure container, or plenum. The temperature in the plenum may be 1500–2000K, with a pressure on the order of several tens of atmospheres. The translational,

---

[8]Similar expressions for absorption coefficients associated with infrared molecular vibrational-rotational transitions are used in studies of the atmospheric "greenhouse effect."

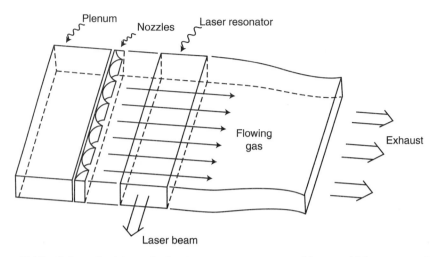

Figure 11.10 Schematic picture of a flowing-gas arrangement used in some high-power gas lasers.

rotational, and vibrational degrees of freedom of the CO_2 and N_2 molecules in the plenum are in thermal equilibrium, and the population densities satisfy a (high-temperature) Boltzmann distribution.

The gas is then suddenly allowed to leave the plenum through an array of nozzles. A supersonic expansion results in which the gas (translational) temperature and pressure are drastically reduced, to about 300–400K and 50 Torr, respectively. The rotational degrees of freedom of the molecules also relax quickly to the new, much cooler thermal equilibrium. The key point, however, is that the *vibrational* relaxation rates (VV and VT) are much lower; the vibrational degrees of freedom are thus temporarily "frozen" near the original, high-temperature Boltzmann distribution. On the other hand, the gas temperature itself as measured by the average translational energy $\frac{3}{2}k_B T$ of the molecules, is greatly decreased by the expansion. We then have a *nonequilibrium flow* in which different degrees of freedom in the gas are characterized by very different temperatures.

Population inversion in the nonequilibrium flow results for two reasons. First, the VT collision rates at the gas translational temperature (300–400K) are such that the decay of the lower level of a CO_2 vibrational-rotational transition near 10.6 μm is rapid, giving a favorable lifetime ratio of upper and lower levels. Second, the N_2 vibrational decay rates (due to spontaneous emission and VT collisions) are very low, so that the N_2 molecules store energy for excitation transfer to $CO_2(001)$.

Note that the population inversion achieved in this manner does not require any "external" process such as optical pumping or an electric discharge. In this sense the gas-dynamic laser is thermodynamically similar to classical power generators (e.g., steam engines) in that, using hot and cold reservoirs, it converts thermal energy into a more useful form of energy.

11.9 CHEMICAL LASERS

Chemical reactions can produce excited-state species and the population inversions necessary for *chemical lasers*. In other words, chemical energy of molecular bonding can be converted into electromagnetic energy in the form of laser radiation.

We will consider briefly one example of a chemical laser, the hydrogen fluoride (HF) laser. HF lasers operate on several (sometimes many) HF vibrational-rotational transitions around 2.6–2.8 μm. Vibrationally excited HF molecules are produced as a result of two exothermic chemical reactions:

$$F + H_2 \longrightarrow HF^* + \Delta H_1, \tag{11.9.1}$$

$$H + F_2 \longrightarrow HF^* + \Delta H_2. \tag{11.9.2}$$

The heats of reaction of these two processes are $\Delta H_1 \approx 31.6$ kcal/mol and $\Delta H_2 \approx 98.0$ kcal/mol, and they are therefore referred to as the "cold" and "hot" reactions, respectively. HF^* denotes a vibrationally excited HF molecule. The cold reaction (11.9.1) produces HF molecules in excited vibrational levels up to $v \approx 3$, whereas the hot reaction (11.9.2) results in significant population up to $v \approx 10$.

The overall result of the cold and hot reactions is summarized by writing

$$H_2 + F_2 \longrightarrow 2HF^* + \Delta H_1 + \Delta H_2. \tag{11.9.3}$$

But this disguises the role of atomic fluorine (F) in the cold reaction (11.9.1); without F, the production rate of HF^* is too slow. Since F atoms bond to form F_2, the F_2 molecules must somehow be dissociated into two F atoms. There are several ways of doing this. One utilizes the collisional dissociation of F_2 by a collision partner A:

$$F_2 + A \longrightarrow 2F + A \tag{11.9.4}$$

in a high-temperature chamber. Another involves the use of radiation to free F atoms from chemical bonding; this photochemical process is called *photolysis*, or, if pulsed radiation is used, *flash photolysis*.[9] Another common means of getting F atoms for

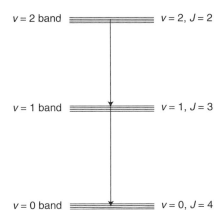

Figure 11.11 HF and other molecular lasers can lase simultaneously on two or more coupled transitions, as indicated here for a $(v = 2, J = 2) \rightarrow (v = 1, J = 3) \rightarrow (v = 0, J = 4)$ cascade.

---

[9]An important example of photolysis occurs in the upper atmosphere. O_2 molecules are dissociated by ultraviolet solar radiation, and the freed O atoms react with O_2 to produce ozone, O_3. The "ozone layer" absorbs far-ultraviolet solar radiation that is harmful to living organisms.

HF chemical lasers is by electron-impact dissociation of F_2 or another molecule bonding F atoms, such as sulfur hexafluoride:

$$SF_6 + e \quad \longrightarrow \quad SF_5 + F + e. \tag{11.9.5}$$

Typically HF lases on low-lying vibrational-rotational transitions such as ($v = 1$, $J = 3$) $\rightarrow$ ($v = 0$, $J = 4$). Lasing tends to occur on two or more vibrational-rotational transitions simultaneously, as indicated in Fig. 11.11. The various chemical and VT processes that occur in HF lasers result in a considerable heating of the gain medium. High-power HF chemical lasers are therefore frequently of the flowing-gas type. This and other chemical lasers, such as the 1.315-μm chemical oxygen iodine laser (COIL) oscillating on a magnetic dipole transition of atomic iodine, are among the most powerful lasers, producing megawatts of continuous-wave infrared radiation.

11.10 EXCIMER LASERS

There are molecules that can exist only in excited electronic levels, the ground level being dissociative. In such a molecule the potential energy curve for the ground level has no local minimum, and so there is no stable ground level (Fig. 11.12). A molecule of this sort is called an *excimer*, a contraction for "excited dimer." In the transition indicated in Fig. 11.12 the lower level very quickly dissociates into two unbound atoms. The dissociation time is on the order of a vibrational period, around 10^{-13} s. This effective absence of any lower-level population is the most significant feature of an excimer laser operating on such a bound-free transition. Obviously, such a laser has a very favorable lifetime ratio of upper and lower levels.

Another attractive feature of excimer lasers is their wavelength, which extends from the visible to the ultraviolet, depending on the particular excimer. Moreover the

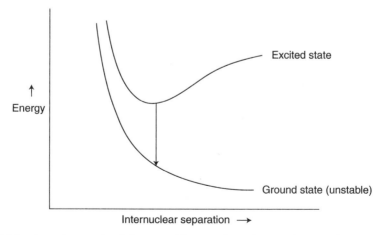

Figure 11.12   An excimer molecule has no stable ground state because the potential energy curve has no local minimum.

bound-free nature of the laser transition allows for tunability over a considerable range of wavelengths (≈ 5 nm), since well-defined vibrational-rotational transitions do not occur. Such a tunable source of coherent ultraviolet radiation is very useful for a variety of applications, such as high-resolution studies of molecular electronic spectra.

The KrF excimer laser at around 248 nm is one of the most efficient high-power ultraviolet lasers. The population inversion process for this and other rare-gas monohalide lasers (e.g., ArF, XeF, XeCl) involves a relatively large number of reactions; computer models of such lasers sometimes include ~ 100 rate equations for different processes. These lasers are pumped either by an electric discharge or by an electron beam. Of the processes leading directly to excited XY molecules, where X and Y refer to a rare-gas atom and a halogen, respectively (e.g., X = Kr, Y = F), two are especially important. One is the ion–ion recombination process in which ions X^+ and Y^- combine in the presence of a third body to produce an excited XY molecule:

$$X^+ + Y^- \longrightarrow XY^*. \tag{11.10.1}$$

The other is the "harpooning reaction"

$$X^* + YR \longrightarrow XY^* + R, \tag{11.10.2}$$

where R represents some radical attached to the halogen Y. The rate for such a process tends to be largest when $R = Y$, as in the reaction

$$Kr^* + F_2 \longrightarrow KrF^* + F \tag{11.10.3}$$

in the KrF laser. Ion–ion recombination processes like

$$Kr^+ + F^- \longrightarrow KrF^* \tag{11.10.4}$$

are very important in the monofluoride excimer lasers because of the rapid production of F^- ions by the dissociative electron attachment reaction

$$F_2 + e \longrightarrow F^- + F. \tag{11.10.5}$$

High-power KrF lasers typically contain a gas mix of around 90% Ar, less than 10% Kr, and about 0.5% F_2. An electron beam produces electron–ion pairs, and the "secondary" electrons so generated take part in processes such as (11.10.5). For pressures less than about 1 atm, Ar^+ and F^- undergo ion–ion recombination to form ArF^*, which reacts with Kr to form KrF^*. At higher pressures the charge-transfer reactions

$$Ar^+ + 2Ar \longrightarrow Ar_2^+ + Ar \tag{11.10.6}$$

and

$$Ar_2^+ + Kr \longrightarrow Kr^+ + 2Ar \tag{11.10.7}$$

provide Kr^+ ions for the reaction (11.10.4). Kr_2^+ ions can also be formed by reactions such as

$$Kr^+ + 2Kr \longrightarrow Kr_2^+ + Kr, \qquad (11.10.8)$$

and the molecular krypton ions can then react with F^- in an ion–ion recombination reaction to form KrF^*.

•   An approximate expression for the gain coefficient for an excimer laser such as KrF may be derived as follows from the general expression for $g(\nu)$ given in Table 4.1. For a transition from a bound upper state to a lower, unbound state, we assume that the lower-level population has in effect a very large decay rate, so that $N_1 \approx 0$ and

$$g(\nu) \approx \frac{\lambda^2 A}{8\pi} N S(\nu), \qquad (11.10.9)$$

where N is the upper-level population density and we assume $n(\nu) \cong 1$. Since $NS(\nu)\, d\nu \equiv (\partial N/\partial \nu)\, d\nu$ is effectively the number of excited molecules per unit volume for which there is gain in the frequency interval $[\nu, \nu + d\nu]$, we rewrite (11.10.9) as

$$g(\nu) \approx \frac{\lambda^2 A}{8\pi} \frac{\partial N}{\partial \nu}. \qquad (11.10.10)$$

Let R denote the internuclear separation and consider a transition from a vibrational level of energy E_v of the excited, bound electronic state to an unbound state of energy $E(R)$; $E(R)$ is the potential energy curve for unbound ground states (Fig. 11.12). The transition frequency $\nu = [E_v - E(R)]/h$, and therefore

$$\frac{\partial N}{\partial \nu} = \frac{\partial N}{\partial R}\left(\frac{\partial \nu}{\partial R}\right)^{-1} = -h\frac{\partial N}{\partial R}\left(\frac{\partial E}{\partial R}\right)^{-1} = h\frac{\partial N}{\partial R}\left|\frac{\partial E}{\partial R}\right|^{-1} \qquad (11.10.11)$$

and

$$g(\nu) \approx \frac{h\lambda^2 A}{8\pi} \frac{\partial N}{\partial R}\left|\frac{\partial E}{\partial R}\right|^{-1}, \qquad (11.10.12)$$

where we have used $dE/dR < 0$ for the unbound lower state (Fig. 11.12).

To obtain an expression for $\partial N/\partial R$ we make the simplifying approximation that all the molecules in the excited electronic state are in the vibrational ground level ($v = 0$). Then the average population density of molecules within the internuclear separation interval $[R, R + \Delta R]$ is $\Delta N = N(v = 0)|\psi_0(x)|^2\Delta R$, where $N(v = 0)$ is the total population density of the $v = 0$ vibrational level of the excited electronic state; $x = R - R_0$, where R_0 is the internuclear separation at which the potential energy curve for the excited electronic state has its minimum; $\psi_0(x) = (\alpha/\pi)^{1/4} \exp(-\alpha x^2/2)$ is the (harmonic oscillator) wave function for the $v = 0$ vibrational state; and $\alpha = \mu\omega_0/\hbar$, where μ and ω_0 are, respectively, the reduced mass and the angular vibrational frequency. Thus, $\partial N/\partial R = N(v = 0)|\psi_0(x)|^2$ and

$$g(\nu) \approx \frac{h\lambda^2 A}{8\pi}\left(\frac{\alpha}{\pi}\right)^{1/2} N(v = 0)e^{-\alpha x^2}\left|\frac{dE}{dR}\right|^{-1}. \qquad (11.10.13)$$

where dE/dR is evaluated at $R = R_0$, which is generally a good approximation because it is roughly constant near $R = R_0$ (Fig. 11.12). (For KrF, for example, quantum mechanical computations for the Kr–F interaction yield $|dE/dR| \approx 3.6$ eV/nm.) This also implies that

$\nu = [E_v - E(R)]/h$ is a linear function of R and therefore x, so that the exponent in (11.10.13) varies quadratically with ν. In other words, $g(\nu)$ is a Gaussian function of ν, a consequence of the assumption that the upper, bound vibrational state is the ground state of a harmonic oscillator. This Gaussian form of the gain (and fluorescence) spectrum is consistent with experimental data for rare-gas monohalide excimer lasers.   •

11.11 DYE LASERS

Liquid dye lasers provide especially interesting examples of optical pumping, i.e., population inversion by absorption of radiation illuminating the laser medium. The active molecules in these lasers are large organic molecules in a solvent such as alcohol or water. The most useful feature of dye lasers is their tunability: By means of some adjustment, a dye laser can be made to oscillate over a wide range of optical wavelengths. This tunability has been extremely useful in atomic spectroscopy.

Figure 11.13 is an energy-level diagram typical of dye molecules. With each electronic level of the molecule is associated a set of vibrational and rotational energy levels, which are spaced very closely compared to electronic energy-level spacings. The vibrational energy levels are typically separated by $1200-1700$ cm^{-1}, whereas the rotational level spacings are roughly two orders of magnitude smaller. The symbols S_0, S_1, S_2, T_1, T_2 in Fig. 11.13 label electronic energy levels and the associated manifold of vibrational and rotational levels. The symbols S and T stand for singlet and triplet electronic levels, respectively. In a singlet level the total electron spin quantum number is zero ($S = 0$), whereas in a triplet level $S = 1$. A level having a total spin quantum number S is $(2S + 1)$-fold degenerate, whence the names singlet ($2S + 1 = 1$) and triplet ($2S + 1 = 3$). For our purposes only one thing about these S and T levels is important: $S \longleftrightarrow S$ and $T \longleftrightarrow T$ transitions are radiatively allowed, whereas $S \longleftrightarrow T$ transitions are

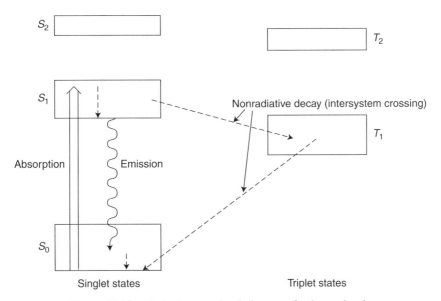

Figure 11.13 Typical energy-level diagram of a dye molecule.

radiatively forbidden. That is, the oscillator strengths are zero for $S \longleftrightarrow T$ transitions. This is a consequence of the dipole selection rule $\Delta S = 0$ for electron spin.

As indicated in Fig. 11.13, the ground level S_0 is a singlet state. In thermal equilibrium practically all the dye molecules are in the S_0 level. Because $S \longleftrightarrow T$ radiative transitions are forbidden, optical pumping can promote the molecule from the ground level S_0 to the higher-energy singlet states (S_1, S_2, etc.) but not to any of the triplet states. To begin with, therefore, let us focus our attention only on the allowed singlet–singlet transitions.

Optical pumping takes the molecule from one of the vibrational-rotational levels of the ground electronic state (S_0) to one of the vibrational-rotational levels of the first excited singlet electronic state (S_1). The $S_0 \rightarrow S_1$ transition is possible over a broad frequency range because there is a broad range of vibrational-rotational levels associated with both S_0 and S_1. The transition frequencies typically lie in the visible or near visible, and so the dye gives the solution a certain color because of selective absorption. Furthermore the oscillator strengths for allowed transitions in dye molecules are usually quite large. This can be understood from the large size of organic dye molecules. In the hydrogen atom, for instance, the transition electric dipole moments are on the order of ea_0, where $a_0 = 0.053$ nm is the Bohr radius, roughly the "size" of the hydrogen atom. Dye molecules are much larger, and consequently have much larger dipole moments and oscillator strengths. Dye molecules therefore have strong absorption bands.

An excited dye molecule tends to decay very quickly to the lowest lying vibrational level of a given electronic state. The decay process is nonradiative, and typical lifetimes are in the picosecond range. A dye molecule in electronic state S_1, for example, will quickly decay to the "bottom" of the S_1 manifold, as indicated in Fig. 11.13. A crude description of the pumping of a dye laser therefore follows the four-level scheme of Section 4.8 (Fig. 11.14). Absorption of radiation takes the molecule from the bottom level of S_0 to one of the S_1 levels, where nonradiative decay quickly brings it to the bottom level of S_1. The latter serves as the upper laser level, the lower laser level being one of the vibrational-rotational levels of S_0.

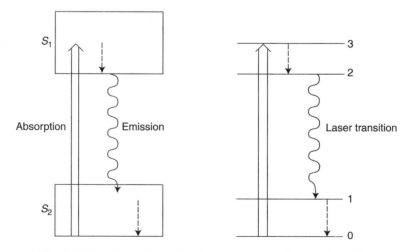

Figure 11.14 Approximate four-level picture of a dye laser transition.

Unfortunately, things are not as simple as the four-level picture suggests. For although $S \longleftrightarrow T$ transitions are radiatively forbidden, they may occur nonradiatively in collisions between molecules. This is called *intersystem crossing*.[10] The nonradiative, intersystem decay of S_1 into T_1, and T_1 into S_0, is indicated in Fig. 11.13.

Intersystem crossing has some undesirable ramifications for the pumping of dye lasers. The decay $S_1 \rightarrow T_1$, for instance, obviously reduces the number of molecules in the upper laser level, and in effect reduces the number of laser-active molecules. Since triplet–triplet transitions are strongly allowed, furthermore, the optical pumping process $S_0 \rightarrow S_1$ also induces absorptive transitions $T_1 \rightarrow T_2$, which have frequencies in the same region as $S_0 \rightarrow S_1$ transitions.

However, it is possible to get around the problems caused by intersystem crossing. By dissolving oxygen in the dye solution, for example, the decay rate for the intersystem crossing $T_1 \rightarrow S_0$ can be enhanced by orders of magnitude (to about $10^7\,\text{s}^{-1}$). Continuous-wave dye laser oscillation can then be achieved if the $T_1 \rightarrow S_0$ decay is faster than the $S_1 \rightarrow T_1$ decay. Otherwise the laser can only be operated in a pulsed mode in which the excitation pulse duration is short compared to the $S_1 \rightarrow T_1$ decay. In the latter case, loosely speaking, lasing occurs before S_1 has a chance to be significantly depleted by intersystem crossing.

Because of their large oscillator strengths, the $S_0 \rightarrow S_1$ transitions have small spontaneous emission lifetimes, typically in the nanosecond range. The upper laser level in particular decays quickly, and a flashlamp used to optically pump a dye laser must have fast-rising, high-intensity output in order to produce a significant population inversion. In this regard the requirements on the flashlamps are more stringent than in the case of ruby or Nd : YAG and Nd : glass lasers.

Dye lasers are also frequently optically pumped with the radiation from another laser. Pulsed N_2 lasers in the ultraviolet (337 nm) are particularly useful for pumping pulsed dye lasers. Continuous-wave dye lasers are frequently pumped by the blue-green radiation of an argon ion laser.

• One very important characteristic of laser dye molecules is that their emission spectra are shifted in wavelength from their absorption spectra. This fortunate circumstance prevents the laser radiation from being strongly absorbed by the dye itself. We can understand this characteristic based on the *Franck-Condon principle* and the fact that the vibrational relaxation associated with any electronic state is very rapid.

The Franck–Condon principle is basically just the statement that electron motion in molecules is very rapid compared to the vibrations of the individual atoms. This means that electronic transitions occur very quickly, with practically no adjustment of the interatomic coordinate R. We indicate in Fig. 11.15 an absorptive transition between the two lowest electronic states of a molecule. The transition is shown to proceed vertically (without change of R). It starts near the bottom of the potential curve of the lower electronic state because in thermal equilibrium most molecules will be in the lower vibrational states of the ground electronic level. Following the absorption, the molecule is in a vibrational state of the upper electronic level. The vibrational motion of the molecule then changes R. The vibrational state also relaxes quickly to $v = 0$, and so the eventual downward electronic transition (due to spontaneous emission) proceeds along a different vertical line. The emission wavelength is thus longer than the absorption wavelength. In Fig. 11.16 we

---

[10]There are in fact also radiative contributions to intersystem crossing because only dipole transitions are *strictly* forbidden. Higher-order multipole transitions are much less likely than allowed (dipole) transitions and therefore have much smaller transition rates.

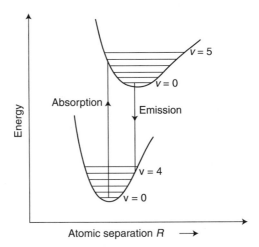

Figure 11.15   Illustration of the Franck–Condon principle. After absorption R changes due to the molecular vibration, so that the emission occurs at a different R, and therefore a different wavelength, than the absorption.

show the singlet-state absorption and emission spectra of a solution of rhodamine 6G, the most commonly used dye molecule. The emission spectrum is shifted to longer wavelengths as expected from the Franck–Condon pinciple. The shift between the peaks of the two curves is called the *Stokes shift*.   •

As noted earlier, the most important feature of dye lasers is their tunability. This tunability is a consequence of the broad emission curve of a dye molecule (Fig. 11.16), which allows dye laser radiation to extend over a broad band of wavelengths, typically 1–6 nm wide. Tuning within this band is accomplished by discriminating against most

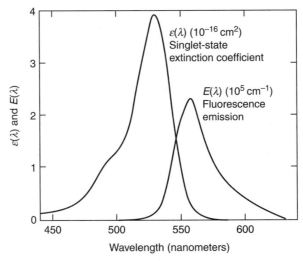

Figure 11.16   Absorption and emission spectra of the dye molecule rhodamine 6G in ethanol (10^{-4} molar solution). [From B. B. Snavely, *Proceedings of the IEEE* **57**, 1374 (1969).]

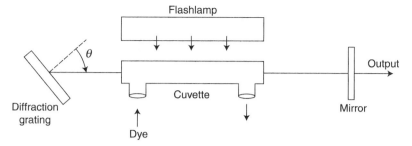

Figure 11.17 Littrow arrangement for wavelength tuning of a dye laser. Tuning is achieved by rotation of the grating.

of the wavelengths, i.e., by making the cavity loss larger than the gain for most wavelengths. The most common way of doing this makes use of the *Littrow arrangement* sketched in Fig. 11.17. In the arrangement shown one of the cavity mirrors is replaced by a diffraction grating, which reflects radiation of wavelength λ only in those directions satisfying the Bragg condition [Eq. (6.A.1)]

$$2d \sin \theta = m\lambda, \qquad m = 1, 2, \ldots, \qquad (11.11.1)$$

where d is the spacing between lines of the grating. Wavelengths not satisfying this condition are not fed back along the cavity axis and consequently have large losses. Thus, the bandwidth of the laser radiation is greatly reduced (typically to around 0.1 nm or less), and the tuning is accomplished by rotating the grating.

11.12   OPTICALLY PUMPED SOLID-STATE LASERS

A typical design for the optical pumping of a solid-state laser rod is shown in Fig. 11.18. The flashlamp typically contains Xe at a pressure of about a hundred or several hundred Torr. An electric discharge in the Xe gas produces an intense burst of spontaneous

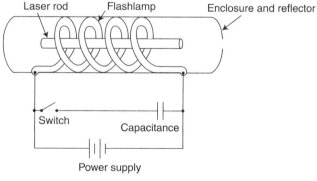

Figure 11.18 Helical flashlamp arrangement for the optical pumping of a solid-state laser. The mirrors may be external to the laser rod or may be silver coatings applied directly onto the ends of the rod.

emission lying in a visible spectral range that is absorbed by the laser rod. The flashlamp has a helical shape and wraps around the laser rod in order to expose as much of the rod as possible to the lamp's radiation.

Another flashlamp pumping configuration employs a flashlamp having the same (linear) cylindrical shape as the rod, and the lamp and rod are placed along the focal axes of an elliptical reflecting tube. This permits the focusing of a large portion of the lamp's output onto the laser rod.

Laser action was first obtained in 1960 by T. H. Maiman, using a ruby rod in the helical lamp configuration of Fig. 11.18. Maiman's original laser produced millisecond pulses of energy $\lesssim 1$ J, each pulse itself consisting of random microsecond pulses. We will consider the example of ruby laser a bit further here, as it illustrates some of the concepts involved in the optical pumping of other solid-state lasers.

Figure 11.19 shows the relevant energy levels of the Cr^{3+} ion in ruby. The levels labeled $^4F_1$ and $^4F_2$ (a conventional spectroscopic notation) are broad *bands* of energy, as indicated in the figure. The level labeled ^2E actually consists of two separate levels, labeled $2\bar{A}$ and $\bar{E}$ in conventional notation. These two levels are separated by about 29 cm^{-1}, or 8.7×10^{11} Hz, and it is the lower one, $\bar{E}$, that serves as the upper laser level in the ruby laser. The ground level is labeled $^4A_2$, and it serves as the lower laser level of the 694.3-nm laser transition.

The levels $^4F_1$ and $^4F_2$ each decay very rapidly into the ^2E level, the decay rate being about 10^7 s^{-1}. The decay process is not spontaneous emission but rather a nonradiative decay in which the energy lost by the chromium ions is converted to thermal energy (heat) of the crystal lattice. The upper laser level, on the other hand, is metastable. That is, it has a long lifetime, about 3 ms, for (spontaneous emission) decay to the ground level with the emission of 694.3-nm photons.

The absorption of radiation of wavelength around 400 or 570 nm will populate the $^4F_1$ or $^4F_2$ levels, respectively. Figure 11.20 shows the absorption coefficient versus wavelength in the visible region for ruby. Flashlamps used to pump ruby lasers should obviously emit radiation at those wavelengths where ruby is strongly absorbing. The rapid decay of the pumped bands into the upper laser level, with the ground level itself acting as the lower laser level, means that ruby is approximately a three-level laser system.

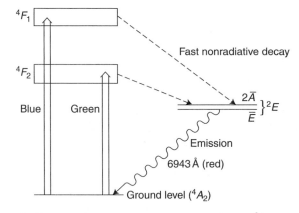

Figure 11.19   Simplified energy-level diagram of the Cr^{3+} ion in ruby.

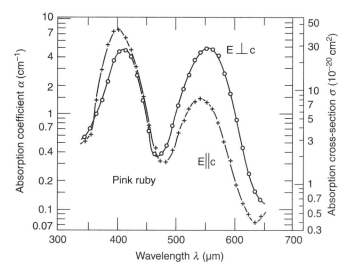

Figure 11.20 Absorption coefficient of ruby. [From D. C. Cronemeyer, *Journal of the Optical Society of America* **56**, 1703 (1966).]

In Section 4.10 we estimated a pumping power density of about 1 kW/cm^3 for laser oscillation in ruby. For a 1-ms excitation pulse this translates to a pumping energy density on the order of 1 J/cm^3. Since not all of the electrical energy delivered to the flashlamp is converted to radiation, and not all of the radiation is actually absorbed by the ruby rod, an electrical power on the order of megawatts must be delivered to the flashlamp in ruby lasers.

The very broad absorption coefficient of ruby (Fig. 11.20) and other solid-state or liquid laser materials is, of course, advantageous. It is well matched by the broad emission bandwidth of a "conventional" (i.e., nonlaser) light source such as a flashlamp, which is much like a blackbody radiator. A medium with very sharp absorption lines would be much more difficult to pump optically with a conventional broadband light source. In a Doppler-broadened gaseous absorber, for instance, an absorption linewidth is approximately v/c times the transition frequency, where v is the average atomic velocity and c is the speed of light (Problem 11.1). This v/c ratio is typically about 10^{-5} or 10^{-6}, which explains why optical pumping is seldom used for gas lasers. In solid-state or liquid media, however, the absorption linewidth is more like 10^{-2}–10^{-4} times the transition frequency.

A much more common solid-state system is the 1.06-μm Nd:YAG laser. The relevant energy levels of the Nd^{3+} ions in the YAG crystal lattice are shown in Fig. 11.21. Again we label the levels according to a conventional spectroscopic notation. In this case we have a good approximation to a four-level pumping scheme. As in ruby, the pump bands are broad and decay rapidly into the upper laser level. Because it is a four-level system, however, and furthermore has a much larger stimulated emission than ruby, the YAG laser has much lower pumping requirements (Section 4.10).

The Nd:glass laser is similar to Nd:YAG, except that the Nd^{3+} ions are present as impurities in glass rather than YAG. Unlike Nd:YAG, it is almost always used in the pulsed mode of excitation because the low thermal conductivity of glass makes it too difficult to cool efficiently under continuous excitation. The Nd:glass laser is

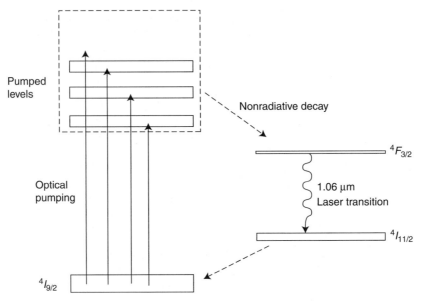

Figure 11.21   Simplified energy-level diagram for the Nd^{3+} ion in a YAG crystal.

especially useful for mode-locked operation because of its large gain linewidth ($\approx 3 \times 10^{12}$ Hz = 3 terahertz = 3 THz).

Solid-state lasers are much more efficiently pumped with diode lasers. Diode-pumped solid-state (DPSS) lasers were demonstrated in the 1960s, but it was only after substantial improvements in the performance of laser diodes (Chapter 15) that their advantages over flashlamp-pumped lasers were realized. "Wall-plug" efficiencies (output laser power divided by electrical power consumed by the laser diodes) of 25% or more are obtained with DPSS lasers, compared to $\sim 1\%$ in the case of flashlamp pumping; the efficiency with which diode laser power is absorbed by the laser crystal can be 80% or more, and the optical efficiency (output DPSS laser power divided by diode laser power) can exceed 50%. Such efficiencies result when the laser diode wavelength can be matched to a single absorption band of the lasing medium. A flashlamp emits over a broad spectrum including all wavelengths in the visible, and consequently there is much more wasted (and often deleterious) heat generated than in DPSS lasers. The operating lifetime of the diodes can be tens of thousands of hours compared to the much more frequent (and sudden) burnout of flashlamps and, because they do not require bulky power supplies, DPSS lasers are "all solid-state" and much more compact than lamp-pumped lasers.

Laser diodes with wavelengths around 810 nm are commonly used to pump 1064-nm Nd : YAG or other crystalline lasing media with similar absorption bands. For example, a commercial green laser pointer powered by two AAA batteries contains an 808-nm laser diode that pumps a Nd : YVO_4 (neodymium-doped yttrium orthovanadate) laser crystal that emits 1064 nm radiation, which is frequency doubled by a small intracavity KTP crystal (potassium titanium oxide phosphate, $KTiOPO_4$) to generate a few milliwatts of green (532 nm) light in a 1.1-mm-diameter output beam.

In *end-pumped* DPSS lasers, diode laser radiation is injected through an end facet of the laser crystal, so that the pump radiation propagates along the long axis of the crystal rod or slab. This produces high optical efficiency and also excellent beam quality characterized by small (<5) values of M^2 (Section 7.13), especially when the diode output is suitably shaped with fibers or other optical components to match the transverse mode profile of the laser resonator. End pumping is often used in small vanadate lasers to take advantage of the high optical efficiency and excellent lasing characteristics of Nd : YVO$_4$ while avoiding the difficulty found in growing large crystals of this material.

Because diode lasers are limited to typically a few watts of power, they are combined in bars or arrays in DPSS lasers. A diode bar typically contains $\sim$50 independent, edge-emitting diode lasers mounted on a chip, with a total output power $\sim$50 W. In high-power DPSS lasers in the kilowatt range, diode bars are stacked in two-dimensional arrays of $\sim$10 bars.

End pumping is normally limited in the amount of power it can deliver to the laser crystal. This is because the end facets of the crystal are usually only a few millimeters across, and focusing of high-power radiation onto them causes local heating that can result among other things in refractive-index variations in the crystal and therefore poor laser beam quality. An obvious solution to this problem is to spread the pump radiation over a broader area. This can be done by end-pumping a slab with wider end facets or, more commonly, by *side pumping* of the crystal to achieve pumping over a broader volume, analogous to the use of a helical flashlamp (Fig. 11.18). Although side-pumped DPSS lasers can produce cw output powers in the kilowatt range, and peak powers in the megawatt range with pulsed lasing, the beam quality is often poor. It should also be noted that the wavelengths available with DPSS lasers are limited by those available from laser diodes. The most common DPSS laser wavelengths at present are 1064 nm and the 532-nm radiation obtained by second-harmonic generation.

As mentioned in Section 11.8, the scaling of solid-state lasers to higher powers is often limited by heating of the laser rod. Besides the possibility of thermal damage, heat deposition can result in transverse variations of the refractive index; the resulting "thermal lensing" can seriously degrade the laser beam quality. A type of DPSS laser that largely eliminates this problem is the *thin-disk laser* illustrated in Fig. 11.22. Two advantageous features of the thin-disk design are apparent. First, the

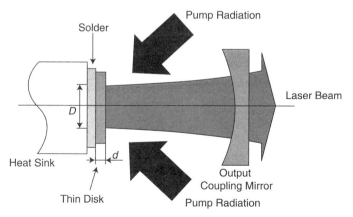

Figure 11.22 A thin-disk laser. The gain medium is a thin disk pumped by laser diode radiation. The back face of the disk reflects the pump and laser radiation and is attached to a heat sink. The front face is antireflection coated for pump and laser radiation. [From A. Giesen, *Laser Technik Journal*, June, No. 2, 42 (2005).]

surface-to-volume ratio of the disk is much larger than for a conventional laser crystal rod, and heat is efficiently dissipated in the water-cooled "heat sink." Second, for uniform pumping over the surface of the disk, there is little thermal lensing because temperature gradients are mainly in the longitudinal direction.

The disk typically has a thickness ~ 100–$200\ \mu m$ and a diameter of several millimeters. Additional optical components are used so that the pump radiation makes many passes through the disk, and it is possible in this way for the disk to absorb $\sim 90\%$ of the power from a diode stack. The output power can be increased simply by increasing the pump area. Several kilowatts of (multimode) continuous-wave power have been demonstrated with Yb : YAG thin-disk lasers. Excellent beam quality ($M^2 \cong 1.1$) and optical efficiencies $\sim 50\%$ have been realized with disk lasers at output powers of several hundred watts.

The first *tunable* solid-state laser to be marketed was the alexandrite laser in which, as in the ruby laser, the lasing species is Cr^{3+}. In the case of alexandrite the chromium ion substitutes for the Al^{3+} in chrysoberyl ($BeAl_2O_4$). Alexandrite has excellent thermal, mechanical, and optical properties that, in addition to tunability and a broad absorption bandwidth that allows relatively efficient flashlamp pumping or diode-laser pumping, make it especially attractive for high average-power operation.

A simplified energy-level diagram for Cr^{3+} in alexandrite is shown in Fig. 11.23. The relevant spectroscopy differs from that for the ruby laser (Fig. 11.19) in two important respects. First, the fast nonradiative decay of the pump bands labeled $^4T_1$ and $^4T_2$ results in the population of a "storage level" ^2E that, at temperatures not substantially below room temperature, results in thermal population of low-lying states in the $^4T_2$ manifold; any of these can serve as the upper laser level. Second, the lower laser level is not simply

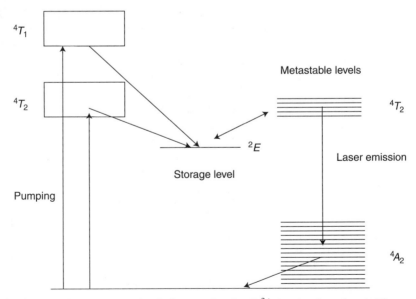

Figure 11.23 Simplified energy-level diagram for the Cr^{3+} ion in chrysoberyl. The separation ΔE between the storage level (radiative lifetime ~ 2 ms) and the "metastable" level (radiative lifetime $\sim 6\ \mu s$) is about $800\ cm^{-1}$, small enough that the metastable level can be thermally populated after the storage level is populated by the fast nonradiative decay from the pumped levels.

the electronic ground level $^4A_2$ but rather any of a broad manifold of vibrational states of this electronic level. The vibronic manifolds indicated in Fig. 11.23 make the alexandrite laser tunable in the near-infrared range $\sim$700–810 nm; the laser wavelength is selected by discriminating against all wavelengths but the one desired. At low temperatures laser emisison near 680 nm can occur directly from the storage level to the bottom of the $^4A_2$ manifold.

An unusual characteristic of alexandrite is that the gain increases with temperature up to about 225°C, at which point it is $\sim$4 times the gain at room temperature. This results from the increased thermal population of the upper laser level with temperature, that is, a larger Boltzmann factor $\exp(-\Delta E/k_B T)$. At higher temperatures the gain decreases due to increasing fluorescence from excited states and other effects. One such effect is *excited-state absorption* in which absorption occurs on transitions from excited states to higher excited states.

A much wider range of tunability in the near infrared is realized with Ti : sapphire lasers in which the lasing species is Ti^{3+} in Al_2O_3. Ti : sapphire has absorption bands in the 400–600 nm range, where high-power diode lasers are not available. It is usually pumped with 532-nm radiation from frequency-doubled Nd : YAG, or by an argon ion laser at 514.5 nm, and is well described by the four-level laser model with an absorption cross section $\sim$6.5 $\times$ 10^{-20} cm^2 at these pump wavelengths. The unique feature of Ti : sapphire is its extremely broad gain bandwidth, $\sim$100 THz, with gain possible at wavelengths between 650 and 1180 nm. This is about 20 times the gain bandwidth of the rhodamine 6G dye laser and more than 30 times that of Nd : YAG; it is in fact the broadest gain bandwidth of any existing laser. It also has a high thermal conductivity and other advantageous material properties.

Ti : sapphire has a peak stimulated emission cross section of about 3×10^{-19} cm^2 at 800 nm, an upper-level lifetime of about 3.2 μs, and a large quantum efficiency ($\sim$80%). It has a large saturation intensity, $\sim$200 kW/cm^2 compared to $\sim$3 kW/cm^2 in Nd : YAG, for example (Problem 11.8). For purposes of pulse amplification it also has the advantage of a high saturation fluence, $\phi_{sat} \sim 1$ J/cm^2 at 800 nm (see Section 6.12 and Problem 11.8). The relatively short upper-level lifetime makes flashlamp pumping inefficient.

Ti : sapphire lasers, both continuous-wave and pulsed, have replaced dye lasers in various applications requiring tunability in the near infrared. As with other tunable lasers, tuning of Ti : sapphire lasers may be accomplished in a variety of ways, for example, with diffraction gratings or prisms in a Littrow configuration (Fig. 11.17). For continuously tunable lasers *birefringent filters* (also called *Lyot filters*) have become the preferred wavelength selectors. These are basically thin, birefringent quartz plates placed inside the laser cavity at the Brewster angle to the laser beam. Rotation about the axis normal to the surface of the plate produces a rotation of the laser polarization except for the wavelength at which the net polarization rotation is zero after a round trip in the cavity. This is the selected wavelength; other wavelengths suffer reflection losses. Since there are no coatings or significant reflection losses at the selected wavelength, birefringent filters introduce less loss than gratings or prisms, and the bandpass can be narrowed using stacked filters. Birefringent filters can be used for cw as well as pulsed operation, but tuning for pulses in the femtosecond range (Section 11.13) requires filters of larger bandwidth. Wavelength tuning of femtosecond pulses from commercial Ti : sapphire lasers is accomplished with a motor-driven, movable slit that selects a particular wavelength dispersed by a prism. Such systems often come with several interchangeable mirrors allowing high reflectivities at different wavelengths.

11.13   ULTRASHORT, SUPERINTENSE PULSES

In Chapter 6 we discussed methods of producing very short, intense laser pulses. In particular, mode locking allows the generation of pulses much shorter than the time it takes for light to make a round trip inside the laser cavity. The large gain bandwidths of dye lasers allow the generation of mode-locked picosecond pulses; in 1981 a continuous train of $\lesssim 100$-fs pulses from a colliding-pulse dye laser was reported (Section 6.14). As discussed in Section 8.4, pulse compression techniques led a few years later to dye laser pulses as short as 6 fs. The development around the same time of Ti : sapphire and other solid-state lasers with huge gain bandwidths has resulted more recently in much more compact (and reliable) sources of laser pulses on the order of a few femtoseconds, i.e., on the order of a few optical periods or less. In this section we survey some of the principles underlying these developments.

Passive mode locking is generally preferable to active mode locking, primarily because there is no need to match the cavity length L and therefore the longitudinal mode spacing $c/2L$ to the frequency of a modulator. As discussed in Section 6.11, it is generally desirable for passive mode locking with a saturable absorber that the absorber have a short relaxation time. But absorbers with short relaxation times tend to have large saturation intensities and large small-signal absorption; this allows the gain in the laser to reach a high value while the loss is too large to allow lasing. Then, once the gain is sufficient to overcome the loss, the laser intensity grows rapidly and saturates the absorber, producing an intense, Q-switched mode-locked train that depletes the gain to a value below the threshold for laser oscillation. In other words, the large small-signal absorption and saturation intensity prevent the generation of a continuous mode-locked train. Q switching can be avoided in some laser designs: The passively mode-locked colliding-pulse laser, for instance, avoids rapid gain depletion by increasing the degree of saturation as the counterpropagating pulses overlap.

A practically instantaneous, passive method for producing mode-locked trains of ultrashort pulses is based on the "Kerr nonlinearity" responsible for self-focusing. This Kerr lens mode locking (KLM), which was discovered accidentally, follows from the fact that the central, most intense part of the intracavity laser beam undergoes the most self-focusing, while the least intense parts are too weak for the focusing effect of the Kerr (n_2) nonlinearity [Eq. (10.3.8)] to overcome diffraction (Fig. 11.24).

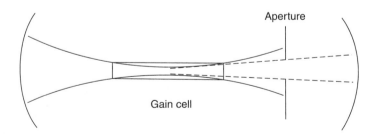

Figure 11.24   Kerr lens mode locking. The low-intensity "wings" of the beam intensity are blocked by an aperture and experience large loss, whereas the high-intensity, central portion of the beam undergoes self-focusing that keeps it confined near the optic axis so that it can pass through the aperture. There is therefore an intensity-dependent loss that results in (passive) mode locking.

The intensity-dependent loss therefore acts in a way similar to a saturable absorber to produce passive mode locking, an obvious difference being that now it is the gain cell itself that does the mode locking. The principal advantage of KLM is the fast response time: The nonlinear refractive index and self-focusing arise from light-scattering processes and adjust practically instantaneously to changes in the intensity. The response times of saturable absorbers, by contrast, are governed by the relatively long relaxation and recovery times of the absorber. Because Ti : sapphire and other laser rods are typically only ~ 2 cm long in order to keep group velocity dispersion (GVD) relatively small, the self-focusing effect is rather weak ($n_{2I} \sim 3 \times 10^{-16}$ cm^2/W for sapphire), and consequently KLM is typically not "self-starting" and must be initiated by some external means such as acousto-optic modulation. Alternatively, an additional cell containing a material with a larger nonlinear index than the gain cell can be inserted in the laser cavity.

The short lengths of the laser rods, and the addition of intracavity prisms to control GVD, allow the generation of pulses with durations, energies, powers, and repetition rates ~ 10 fs, 5 nJ, 1 MW, and 100 MHz, respectively, with Kerr lens mode locking. Figure 11.25 shows the essential features of the design of a KLM Ti : sapphire laser. Comparable laser outputs result when chirped mirrors (Section 8.4) are used for GVD compensation instead of intracavity prisms.

- The measurement of pulse durations less than about 100 fs cannot be done directly by conventional means since the resolving times of oscilloscopes and photodiodes are ~ 100 ps and ~ 1 ps, respectively. The fastest "direct" way of measuring the temporal variation of a pulse's intensity is to use a *streak camera* in which the pulse is mechanically or electronically deflected onto a detector, forming a streak of light from which a temporal intensity profile can be deduced. But even the fastest streak cameras have resolution times ~ 100 fs.

Ultrashort pulse durations have for many years been inferred from intensity correlation functions. Such a correlation function appears, for instance, when a pump pulse in second-harmonic generation is split by a 50/50 beam splitter, and the two resulting pulses with intensity $I(t)$ are recombined in a nonlinear crystal after one pulse is delayed by a time τ with respect to the other.

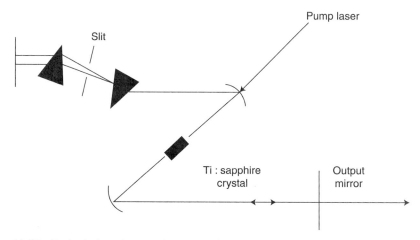

Figure 11.25   Basic design of a Kerr lens mode-locked laser with intracavity prisms for compensation of group velocity dispersion.

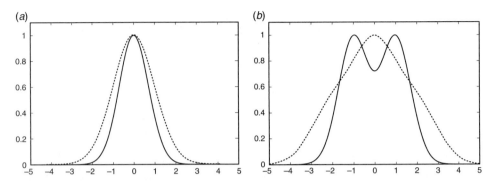

Figure 11.26   (*a*) Gaussian pulse (solid curve) and its time-averaged autocorrelation function (11.13.1) (hatched curve), both normalized. (*b*) As in (*a*) but for a more complicated pulse shape.

The second-harmonic intensity $I_{SHG}(t)$ in this case is proportional to $I(t)I(t - \tau)$ if pump depletion is negligible. For pulses too short to be resolved, the response of a detector to the second-harmonic intensity will be a time-averaged intensity autocorrelation function proportional to

$$G(t, \tau) = \int_{-\infty}^{\infty} I(t)I(t - \tau)\,dt. \tag{11.13.1}$$

Often $G(t, \tau)$ can provide a rough estimate of the pulse shape and duration, as shown in Fig. 11.26*a* for the case of a Gaussian pulse. However, it does not reproduce very well more complicated pulse shapes, as can be seen from the example of Fig. 11.26*b*, nor does it provide any phase information, such as whether a pulse is chirped.

A far more accurate technique that provides intensity as well as phase information is *frequency-resolved optical gating* (FROG). Consider again the example of second-harmonic generation. The second-harmonic electric field at time t is proportional to $\mathcal{E}(t)\mathcal{E}(t - \tau)$, where τ is again the delay time between the two pump electric fields, and the spectrum of this field is proportional to

$$S(\omega, \tau) = \left| \int_{-\infty}^{\infty} \mathcal{E}(t)\mathcal{E}(t - \tau)e^{-i\omega t}\,dt \right|^2. \tag{11.13.2}$$

Unlike (11.13.1), which is defined only in the time domain, the "FROG trace" $S(\omega, \tau)$ depends on both the frequency ω and the time τ. The field $\mathcal{E}(t)$ in (11.13.2) is "gated" by a delayed replica of itself, $\mathcal{E}(t - \tau)$. $S(\omega, \tau)$ is a time-dependent spectrum measured for varying delay times τ, and from this measurement Eq. (11.13.2) can be "inverted" to obtain the complex field envelope $\mathcal{E}(t)$. In other words, a measurement of $S(\omega, \tau)$ provides information about pulse intensity as well as phase.

Various algorithms are available for determining $\mathcal{E}(t)$ from the measured $S(\omega, \tau)$, typically an $N \times N$ set of real numbers with $N \sim 100$. A standard algorithm involves an initial guess for $\mathcal{E}(t)$ and therefore $\mathcal{E}(t)\mathcal{E}(t - \tau)$ and

$$s(\omega, \tau) = \int_{-\infty}^{\infty} \mathcal{E}(t)\mathcal{E}(t - \tau)e^{-i\omega t}\,dt. \tag{11.13.3}$$

The measured FROG trace $S(\omega, \tau)$ is then used to replace $|s(\omega, \tau)|$ by $\sqrt{S(\omega, \tau)}$, and numerical inversion of (11.13.3) with a Fast Fourier Transform then gives a new estimate for

$\mathcal{E}(t)\mathcal{E}(t - \tau)$, which is used in (11.13.2) to obtain an estimate for $S(\omega, \tau)$. The difference between this computed $S(\omega, \tau)$ and the measured FROG trace, characterized typically in a root-mean-square sense, represents an "error" used to obtain a better estimate for $\mathcal{E}(t)$. Iterations involving Fourier transformations between the time and frequency domains are continued until "convergence" is obtained to (presumably) the best estimate of the complex field envelope $\mathcal{E}(t)$ that is consistent with the measured FROG trace.

Like the intensity autocorrelation method, FROG can be used to measure a single pulse by arranging for different parts of the pulse to have different time delays τ. The details of various implementations of FROG, as well as numerical inversion algorithms, are described in more specialized literature.[11]   •

For some purposes it is desirable to generate even more powerful ultrashort pulses than those available from mode-locked lasers. Direct amplification, for example, the amplification of pulses from a mode-locked Ti : sapphire laser by a Ti : sapphire amplifier, is most efficient when the pulse fluence is comparable to or larger than the saturation fluence ϕ_{sat} (Section 6.12 and Problem 11.8). For a 10-fs pulse this implies intensities $\lesssim 10^{15}$ W/cm^2 in the case of Ti : sapphire at 800 nm ($\phi_{sat} \sim 1$ J/cm^2), and a problem with direct amplification is evident: The powerful pulses desired will strongly self-focus and cause optical damage to the Ti : sapphire rods.

The most powerful ultrashort pulses are generated not by direct amplification but by *chirped pulse amplification* (CPA). The idea is to stretch the pulses from a mode-locked laser, amplify them, and then recompress them after amplification. The stretching of the pulse durations by factors typically $\sim 10^3 - 10^4$ reduces their intensities by comparable factors, putting them below the amplifier damage threshold and also avoiding possible nonlinear effects that might cause spatial and temporal pulse reshaping or distortion; *this is the key feature of CPA*. Several amplifiers can be used, beginning with a multipass "preamplifier" after the pulse stretcher. The preamplifier increases pulse energies from the nanojoule range to $\sim 1 - 10$ mJ, which represents most of the overall amplification. It serves to boost the pulse energy to a level where efficient energy extraction occurs in subsequent amplifiers.

The pulse stretching in CPA is just the reverse of the pulse compression described in Sections 8.4 and 8.5. Consider, for example, a pulse that has been chirped by self-phase modulation (SPM). The instantaneous frequency (10.4.8) due to SPM increases with the time $t - z/c$: $\omega_{inst}(t) < \omega$ in the "front" part of the pulse ($t < z/c$), $\omega_{inst}(t) = \omega$ at the peak of the pulse ($t = z/c$), and $\omega_{inst}(t) > \omega$ in the "back" part of the pulse ($t > z/c$). In other words, SPM causes the instantaneous frequency to increase from the front part of the pulse to the back, a *positive* chirp. Thus, if a pulse is propagated through a nonlinear medium that causes chirping by SPM, it can be *stretched* by subsequent propagation in a linear medium with positive GVD, wherein the lower frequencies propagate faster than the higher frequencies: The front part of the pulse moves farther ahead as the back part falls farther behind.

Such pulse stretching is done with fibers, but for femtosecond pulses there are complications due to high-order dispersion, and stretching is more often accomplished with grating pairs. Figure 11.27 shows how the insertion of a "telescope" between two gratings can convert the negative GVD of a grating pair to a positive GVD; pulse compression can subsequently be done with a another (negative-GVD) grating pair.

---

[11]See, for instance, R. Trebino, *Frequency-Resolved Optical Gating: The Measurement of Ultrashort Laser Pulses*, Kluwer Academic, Norwell, MA, 2000.

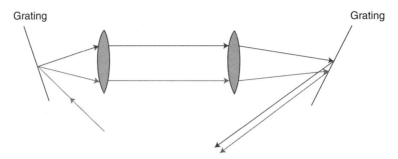

Figure 11.27   Pulse stretching by insertion of a "telescope" between a pair of gratings. The lenses and gratings cause lower frequencies to have smaller delay times than higher frequencies, and a pulse is stretched in time.

The chirped pulse compression methods described in Sections 8.4 and 8.5 are employed in CPA following stretching and amplification. In this case negative GVD produced by a grating pair causes the higher frequency components of a positively chirped pulse to propagate faster than the lower frequencies. The back of the pulse then "catches up" with the front of the pulse, resulting in its temporal compression.

Methods have been developed not only to compress or stretch femtosecond pulses but also to generate almost arbitrary pulse shapes. Basically, such methods spatially separate different frequency components of an input pulse and then modify in a prescribed way the amplitudes and relative phases of the spatially separated components. The different components are then recombined into a pulse having a spectrum that corresponds to the desired temporal shape.[12]

The inventions of Q switching and mode locking in the 1960s resulted later in "tabletop" lasers[13] with peak pulse powers in the gigawatt range and focused intensities $\sim 10^{15}$ W/cm^2; higher powers and intensities were prevented primarily by optical damage thresholds. With the invention of chirped pulse amplification in the mid-1980s, and the advent of media with huge gain bandwidths (Ti : sapphire in particular), the situation changed dramatically, and by the late 1990s peak powers in the terawatt range and focused intensities $\sim 10^{21}$ W/cm^2 were realized with tabletop systems. Ti : sapphire systems also increased repetition rates of earlier picosecond and femtosecond lasers so that average powers increased from ~ 10 mW to ~ 10 W.

To appreciate the magnitude of the pulse powers that have been obtained by chirped pulse amplification, recall that the Coulomb electric field that binds the valence electrons in atoms and molecules is $\sim e/4\pi\epsilon_0 a_0^2$, where e is the electron charge and a_0 the Bohr radius. This electric field is about 5×10^9 V/cm. The intensity of a monochromatic plane wave with the same electric field strength is about 3×10^{16} W/cm^2, much smaller than the intensities readily achievable by chirped pulse amplification (Problem 11.9). This

---

[12]For a review of femtosecond pulse shaping see, for instance, A. M. Weiner, *Review of Scientific Instruments* **71**, 1929 (2000).

[13]By "tabletop" lasers we mean systems that occupy an optical bench in a typical university laboratory. This excludes, for example, large laser systems designed with the goal of realizing in a controlled manner the extremely high temperatures and pressures required for nuclear fusion; they involve many beam lines and very large buildings. CPA laser peak powers in excess of a petawatt (10^{15} W) have been demonstrated at such facilities.

means among other things that traditional perturbative approaches to the theory of the interaction of light with atoms and molecules are inapplicable at the intensity levels made possible by CPA—the applied field is no longer weak compared to the binding field. At these intensity levels new phenomena appear such as the high-order harmonic generation mentioned at the end of Chapter 10; some such effects are considered in Section 14.7.

Once phenomena such as self-focusing, group velocity dispersion, and chirping are understood, the essential physics of chirped pulse amplification can be said to be conceptually straightforward. Quantitative analyses of CPA systems, however, are another matter; they involve the determination of the angular GVD of prisms and gratings, nonlinear spatial and temporal effects in pulse propagation, higher-order dispersion effects giving rise to nonlinear chirping and thereby affecting GVD compensation, amplified spontaneous emission, and *many* other factors. Any detailed analysis of the "front-end" Kerr lens mode-locked laser is in itself a nontrivial undertaking requiring numerical modeling.

One of the main factors that limit the intensities possible with CPA systems is optical damage by the stretched pulses to the amplifiers. For nanosecond pulses the fluence ϕ_d at which optical damage can be expected in Ti : sapphire, for instance, is about $20 \, \text{J/cm}^2$. Increasing pulse energies and intensities much beyond those achievable with table-top systems requires much larger beam areas than are possible with amplifying crystals of diameter ~ 1 cm. Beam areas $\sim 10 \, \text{m}^2$, or even larger areas conceivable with a large matrix of Ti : sapphire rods, have been considered for the extension of CPA techniques to exawatt (10^{18} W) and zetawatt (10^{21} W) powers and intensities $\sim 10^{28} \, \text{W/cm}^2$.[14]

11.14 FIBER AMPLIFIERS AND LASERS

In Section 8.7 we noted two developments that led to widespread use of optical fibers in communications: low-loss fibers and progress in diode laser technology. Research and development in these areas resulted in 1988 in the first transatlantic cable employing glass fibers. Attenuation in the fiber was so low that regenerators along the cable were spaced between about 40 and 70 km apart: At each regenerator (or "repeater") an optical signal from a modulated laser diode, weakened after propagation, was converted to an electric current that was then amplified electronically and used to regenerate the optical signal by driving another laser diode or by modulating its output.

Another major breakthrough occurred in the mid-1980s, when erbium-doped fibers serving as *all-optical* amplifiers were developed. In the early 1990s it was demonstrated that this amplification technique could increase the information capacity of fiber cables by a factor of about 100 compared to electronic amplification, and by 1996 both transatlantic and transpacific all-optical cables were installed. Some basic ideas behind optical fiber communications are discussed in Chapter 15. Here we begin by focusing on some characteristics of erbium-doped fiber amplifiers (EDFAs).[15]

[14]T. Tajima and G. Mourou, *Physical Review Special Topics—Accelerators and Beams* **5**, 031301 (2002).
[15]See, for instance, E. Desurvire, *Erbium-Doped Fiber Amplifiers. Principles and Applications*, Wiley, Hoboken, NJ, 2002.

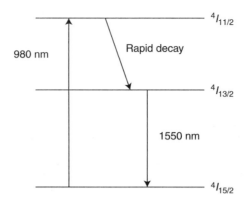

Figure 11.28   Three-level model for amplification at 1.55 μm in Er : glass.

The amplifying species in EDFAs is the Er^{3+} ion, just as another trivalent rare-earth ion, Nd^{3+}, is the lasing species in Nd : glass lasers. The advantage of erbium-doped fibers for telecommunications is the fact that they are efficient amplifiers within a pronounced transmission window of silica glass at 1.55 μm (Section 8.7). As in the case of Nd^{3+}, the amplifying transitions of interest in Er^{3+} occur between energy levels with the same total orbital angular momentum and spin quantum numbers, $L = 6$ and $S = \frac{3}{2}$, respectively, but with different values of J, that is, between levels labeled in spectroscopic notation as ^4I_J, where J is the total angular momentum quantum number (Section 3.13). [Recall that in this notation $L = 6$ is indicated by the letter I in the S, P, D, F, ... labeling of angular orbital momenta $L = 0, 1, 2, 3, \ldots$, respectively.] The possible values of J range in integral increments from $6 - \frac{3}{2} = \frac{9}{2}$ to $6 + \frac{3}{2} = \frac{15}{2}$ in the case of Er^{3+}. In the crudest approximation, amplification in Er : glass is described by the three-level system shown in Fig. 11.28: Pumping with diode laser radiation at 980 nm results in gain at 1.55 μm. Amplification at 1.55 μm can also be obtained by pumping with radiation at 1480 nm; in this case the upper level 3 is a Stark-split sublevel of $^4I_{13/2}$. The three-level model grossly oversimplifies the spectroscopy of Er : glass, just as the four-level laser model oversimplifies the spectroscopy of Nd : glass (Section 4.10).[16] It nevertheless provides a relatively simple and surprisingly accurate description of various important characteristics of EDFAs, and as such it has been used frequently in optical communications research.

The distribution of charge in the glass host produces an electric field, and this electric field acting on the erbium ions causes a splitting of their energy levels by the Stark effect; a level with total angular momentum quantum number J is split into $J + \frac{1}{2}$ sublevels that have a Boltzmann population distribution at the temperature T of the glass. The amplifier is described approximately by the three-level model (Fig. 11.28) with rates and cross sections involving averages over those for the transitions between the different sublevels of the levels 1, 2, and 3. Because of the spatial variation of the electric field in the crystal, different Er^{3+} ions have different Stark splittings, resulting in an inhomogeneous line broadening. The amplifying transition is also (homogeneously) broadened by

---

[16]At very low temperatures ($T \sim 77$K) only the lowest Stark-split sublevels have any significant population, and Er : glass in that case is approximately described by the four-level model for gain.

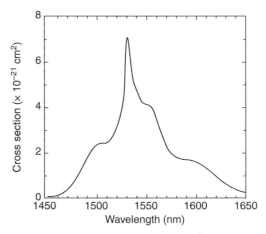

Figure 11.29   Measured stimulated emission cross section of Er^{3+} in a fluorophosphate glass at room temperature. [From W. J. Miniscalco and R. S. Quimby, *Optics Letters* **16**, 258 (1991).]

spontaneous emission. The absorption and emission spectra of the amplifying transition are therefore determined by overlapping Lorentzian lineshapes and depart significantly from a pure Lorentzian (or Gaussian) form; they are also temperature dependent. Figure 11.29 shows a measured stimulated emission cross section $\sigma(\lambda)$ for Er : glass. Note the large peak cross section, $\approx 7 \times 10^{-21}$ cm^2. The radiative lifetime of the upper level is about 10 ms, and the gain bandwidth is nominally taken to be about 30 nm, or 4 THz.

The gain coefficient for EDFAs exhibits the behavior expected from formula (4.12.3) for a three-level laser in which the pumping rate P is proportional to the pump laser power. At low pump power levels (a few milliwatts) the gain coefficient increases linearly with pump power, but then begins to level off and increase more slowly as the pump power is increased. Similarly, the gain coefficient saturates with increasing signal power in approximately the manner expected from (4.12.3). As we now show, the three-level amplifier model leads to a useful formula that, among other things, accounts for the saturation and threshold characteristics of EDFAs.

Recall Eqs. (4.7.4) for the population densities N_1 and N_2 in the three-level model. In terms of the field intensity I_s and the stimulated emission and absorption cross sections, σ_s^e and σ_s^a, respectively, we can write the equation for N_2 as[17]

$$\frac{dN_2}{dt} = -\Gamma_{12}N_2 + PN_1 - \frac{1}{h\nu_s}(\sigma_s^e N_2 - \sigma_s^a N_1)I_s, \qquad (11.14.1)$$

where the subscript s denotes the "signal" field that undergoes amplification. In the case of optical pumping with a field of intensity I_p, similarly,

$$PN_1 = \frac{\sigma_p^a}{h\nu_p}I_p, \qquad (11.14.2)$$

---

[17]In the case of degenerate atomic levels, for example, $\sigma_s^a = (g_2/g_1)\sigma_s^e$, and $\sigma_s^e N_2 - \sigma_s^a N_1$ takes the familiar form $\sigma_s^e[N_2 - (g_2/g_1)N_1]$.

with $\sigma_p^a \equiv \sigma^a(\nu_p)$ the absorption cross section at the pump frequency ν_p. We will generalize to allow the possibility that the pump can also stimulate transitions from N_2 to N_1, and replace (11.14.2) by

$$P(N_1 - N_2) = \frac{1}{h\nu_p}(\sigma_p^a N_1 - \sigma_p^e N_2)I_p, \tag{11.14.3}$$

where $\sigma_p^e = \sigma^e(\nu_p)$ is the stimulated emission cross section at the frequency ν_p; this form assumes that the level 3 that is directly[pumped decays rapidly to level 2, as is characteristic of the three-level model. The pump, in order to populate N_2 rather than deplete it, will generally be at a frequency for which σ_p^e is relatively small, but including σ_p^e will allow us to write our equations in a form that is more symmetrical in the different fields. For example, the generalization of (11.14.1) is

$$\frac{dN_2}{dt} = -\Gamma_{21}N_2 - \sum_j \frac{I_j}{h\nu_j}(\sigma_j^e N_2 - \sigma_j^a N_1). \tag{11.14.4}$$

In our model thus far, j here is summed over a signal field and a pump field, but it should be clear that this equation holds for any number of signal and pump fields with intensities I_j and frequencies ν_j; this is important because EDFAs for optical communications must amplify more than one input signal. The steady-state value $\overline{N}_2$ of the upper-level population is obtained as usual by putting $dN_2/dt = 0$:

$$\overline{N}_2 = -\frac{1}{\Gamma_{21}}\sum_j \frac{\Gamma_j I_j}{h\nu_j}[(\sigma_j^a + \sigma_j^e)\overline{N}_2 - \sigma_j^a N_T]. \tag{11.14.5}$$

We have used the fact that $N_1 + N_2 = \overline{N}_1 + \overline{N}_2 = N_T$ in the three-level model [Eq. (4.7.5)]. Because the transverse cross-sectional area of the field of frequency ν_j might be larger than the cross-sectional area of the active region, we have also introduced the *confinement factor* Γ_j (<1), which in the simplest description is just the ratio of the cross-sectional area of the active region to a cross-sectional area characterizing the field intensity I_j.

In the plane-wave approximation,

$$u_k \frac{\partial I_k}{\partial z} + \frac{1}{v_g}\frac{\partial I_k}{\partial t} = (\sigma_k^e N_2 - \sigma_k^a N_1)I_k, \tag{11.14.6}$$

where v_g is the group velocity and $u_k = +1$ for propagation in the forward ($+z$) direction and $u_k = -1$ for propagation in the backward ($-z$) direction. We again introduce the confinement factor and write the steady-state version of this equation in terms of the power $P_k = I_k S$ where S is the cross-sectional area of the active (gain) volume of the amplifier:

$$u_k \frac{dP_k}{dz} = \Gamma_k(\sigma_k^e \overline{N}_2 - \sigma_k^a \overline{N}_1)P_k = \Gamma_k[(\sigma_k^a + \sigma_k^e)\overline{N}_2 - \sigma_k^a N_T]P_k. \tag{11.14.7}$$

There is no competition for the letter P in the rest of this section, so we are departing from our usual practice of writing Pwr for power. From this expression we see that (11.14.5) is equivalent to

$$\overline{N}_2 = -\frac{1}{\Gamma_{21}S} \sum_j \frac{u_j}{h\nu_j} \frac{dP_j}{dz}, \tag{11.14.8}$$

and this in turn allows us to write (11.14.7) in the form

$$\frac{u_k}{P_k} \frac{dP_k}{dz} = -\alpha_k - \frac{h\nu_k}{P_k^{\text{sat}}} \sum_j \frac{u_j}{h\nu_j} \frac{dP_j}{dz}, \tag{11.14.9}$$

with

$$\alpha_k = \Gamma_k \sigma_k^a N_T \tag{11.14.10}$$

and

$$P_k^{\text{sat}} = \frac{\Gamma_{21}Sh\nu_k}{\Gamma_k(\sigma_k^a + \sigma_k^e)}. \tag{11.14.11}$$

P_k^{sat} is the saturation power for the field of frequency ν_k.[18]
 Now let us multiply both sides of (11.14.9) by u_k, use the fact that $u_k^2 = 1$, and integrate both sides of the resulting equation from $z = 0$ to $z = L$:

$$P_k(L) = P_k(0) \exp\left\{-u_k \alpha_k L - u_k \frac{h\nu_k}{P_k^{\text{sat}}} \sum_j \frac{u_j}{h\nu_j} [P_j(L) - P_j(0)]\right\}. \tag{11.14.12}$$

For forward-propagating fields the input and output ports for the amplifier are at $z = 0$ and $z = L$, respectively, whereas for backward-propagating fields the input and output ports are defined by $z = L$ and $z = 0$, respectively. So we define

$$P_i^{\text{in}} = P_i(0), \qquad P_i^{\text{out}} = P_i(L) \quad \text{for } u_k = +1, \tag{11.14.13a}$$

$$P_i^{\text{out}} = P_i(0), \qquad P_i^{\text{in}} = P_i(L) \quad \text{for } u_k = -1, \tag{11.14.13b}$$

in terms of which (11.14.12) is

$$P_k^{\text{out}} = P_k^{\text{in}} \exp\left\{-\alpha_k L + \frac{h\nu_k}{P_k^{\text{sat}}} \sum_j \frac{1}{h\nu_j} [P_j^{\text{in}} - P_k^{\text{out}}]\right\}. \tag{11.14.14}$$

---

[18]Note that for $\Gamma_k = 1$, $\Gamma_{21} = A_{21}$, and $\sigma_k^a + \sigma_k^e = 2\sigma_k^e$, P_k^{sat} is just the area S times the saturation intensity I_ν^{sat} defined by (4.11.2).

This relation among the different input and output powers in the amplifier may be simplified by defining

$$P_i^{\text{in,out}} = \frac{P_i^{\text{in,out}}}{h\nu_i}, \tag{11.14.15a}$$

$$P_i^{\text{sat}} = \frac{P_i^{\text{sat}}}{h\nu_i}. \tag{11.14.15b}$$

The P's represent photon fluxes rather than powers, and we use them to write the principal result of our analysis as

$$P_k^{\text{out}} = P_k^{\text{in}} \exp\left\{-\alpha_k L + \frac{P^{\text{in}} - P^{\text{out}}}{P_k^{\text{sat}}}\right\}, \tag{11.14.16}$$

where

$$P^{\text{in,out}} \equiv \sum_j P_j^{\text{in,out}}. \tag{11.14.17}$$

Summing both sides of (11.14.16) over all N beams, we obtain

$$P^{\text{out}} = \sum_{k=1}^{N} P_k^{\text{out}} = \sum_{k=1}^{N} a_k \exp\left\{-\frac{P^{\text{out}}}{P_k^{\text{sat}}}\right\}, \tag{11.14.18}$$

with

$$a_k \equiv P_k^{\text{in}} \exp\left\{-\alpha_k L + \frac{P^{\text{in}}}{P_k^{\text{sat}}}\right\}. \tag{11.14.19}$$

The amplifier in the three-level model is therefore characterized by two parameters for each frequency ν_k: the (small-signal) attenuation coefficient α_k and the saturation photon flux P_k^{sat}, both of which can be determined by measuring the transmission of light at frequency ν_k by the (unpumped) fiber. Then, given the N input photon fluxes P_k^{in}, we can solve (11.14.18) numerically for P^{out}, which can then be used in (11.14.16) to solve numerically for the output photon fluxes P_k^{out} of the N individual beams (Problem 11.10). Note that (11.14.16) *is valid regardless of the propagation directions (forward or backward) of the individual beams.* This fact, together of course with its analytical form, makes it much more useful than numerical solutions of the coupled differential equations for the N intensities.

Figure 11.30 shows comparisons of experimental data for the gain of an Er: glass fiber amplifier with theoretical predictions based on Eq. (11.14.16), in this case for a single pump field at 1480 nm and a single signal field at 1550 nm. The gain is expressed in decibels: $G(\text{dB}) = 10 \log_{10}[P_s^{\text{out}}/P_s^{\text{in}}]$, while the pump power is expressed in dBm (decibel-mW):

$$P \ (\text{dBm}) = 10 \log_{10}[P(\text{mW})], \qquad P(\text{mW}) = \text{power in mW}. \tag{11.14.20}$$

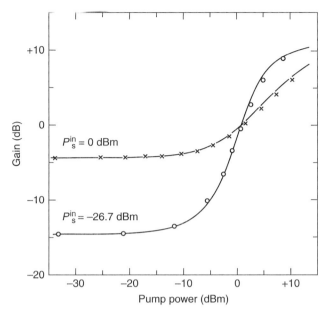

Figure 11.30 Gain at 1550 nm vs. laser pump power at 1480 nm for two different input signal powers: 0 dBm = 1 mW (crosses) and -26.7 dBm = 2.14 μW (circles). The solid curves are theoretical predictions based on Eq. (11.14.16) and transmission data for the parameters α_s, α_p, $\mathcal{P}_s^{\text{sat}}$, and $\mathcal{P}_p^{\text{sat}}$ for the fiber without pumping. [From A. A. M. Saleh, R. M. Jopson, J. D. Evankow, and J. Aspell, *IEEE Photonics Technology Letters* **2**, 714 (1990).]

Consider the case of a single signal beam and a single pump beam. The gain factors for the signal and pump are, respectively,

$$G_s = \frac{\mathcal{P}_s^{\text{out}}}{\mathcal{P}_s^{\text{in}}} = \exp\left\{ -\alpha_s L + \frac{\mathcal{P}^{\text{in}} - \mathcal{P}^{\text{out}}}{\mathcal{P}_s^{\text{sat}}} \right\}, \qquad (11.14.21a)$$

$$G_p = \frac{\mathcal{P}_p^{\text{out}}}{\mathcal{P}_p^{\text{in}}} = \exp\left\{ -\alpha_p L + \frac{\mathcal{P}^{\text{in}} - \mathcal{P}^{\text{out}}}{\mathcal{P}_p^{\text{sat}}} \right\}. \qquad (11.14.21b)$$

Because any gain in our three-level model can result only from excitation of level 3 and subsequently level 2, we cannot have $G_s > 1$ and $G_p > 1$; this would be inconsistent with the conservation of energy. If we have gain ($G_s > 1$) for the signal field, the pump must be attenuated ($G_p < 1$). We have treated the signal and pump fields in a symmetrical way in our equations, but they are distinguished physically by which is amplified and which is attenuated.

Suppose, for example, that we wish to determine the threshold pump power $(\mathcal{P}_p^{\text{in}})_{\text{th}}$ for amplification of a (co-propagating or counterpropagating) signal (Problem 11.11). This is the pump power for which $G_s = 1$, in which case the loss at the signal frequency is just compensated by the gain. In this case Eq. (11.14.21a) implies that

$\mathcal{P}^{\text{in}} - \mathcal{P}^{\text{out}} = \alpha_s L \mathcal{P}_s^{\text{sat}}$, or

$$(\mathcal{P}_p^{\text{in}})_{\text{th}} - \mathcal{P}_p^{\text{out}} = \alpha_s L \mathcal{P}_s^{\text{sat}}, \tag{11.14.22}$$

since $\mathcal{P}_s^{\text{in}} - \mathcal{P}_s^{\text{out}} = 0$ at threshold. Together with Eq. (11.14.21b), this gives

$$(\mathcal{P}_p^{\text{in}})_{\text{th}} = \frac{\alpha_s L \mathcal{P}_s^{\text{sat}}}{1 - \exp\left\{ -\alpha_p L + \dfrac{\mathcal{P}^{\text{in}} - \mathcal{P}^{\text{out}}}{\mathcal{P}_p^{\text{sat}}} \right\}}. \tag{11.14.23}$$

But (11.14.21a) with $G_s = 1$ implies

$$\exp\left\{ \frac{\mathcal{P}^{\text{in}} - \mathcal{P}^{\text{out}}}{\mathcal{P}_p^{\text{sat}}} \right\} = \exp\left\{ \frac{\alpha_s L \mathcal{P}_s^{\text{sat}}}{\mathcal{P}_p^{\text{sat}}} \right\}, \tag{11.14.24}$$

so that (Problem 11.11)

$$(\mathcal{P}_p^{\text{in}})_{\text{th}} = \frac{\alpha_s L \mathcal{P}_s^{\text{sat}}}{1 - \exp\left\{ \left[\alpha_s \mathcal{P}_s^{\text{sat}} / \mathcal{P}_p^{\text{sat}} - \alpha_p \right] L \right\}}. \tag{11.14.25}$$

Note that this threshold pump power for signal amplification is independent of the input signal power.

In addition to the simplified energy-level model for Er^{3+}, this EDFA model assumes, among other things, homogeneous broadening and plane-wave propagation of the signal and pump fields guided by the fiber. A very important omission in the model is amplified spontaneous emission (ASE, Section 6.13), which can result in gain saturation if it becomes sufficiently intense. In optical communications employing EDFAs the ASE co-propagating with the signal can also degrade the signal-to-noise ratio at the receiver. The omission of ASE for the modeling of gain and saturation characteristics can be expected to be valid if the ASE power remains small compared to the saturation power of the amplifier (Problem 11.12). The three-level model can be extended to include ASE, and various refinements of it are found to have generally good predictive value.[19]

Glass fibers doped with rare earths (erbium, neodymium, ytterbium, ...) and pumped with diode lasers can, of course, also serve as laser ocillators when end reflectors are added to form a resonator. In fact, fibers were among the first laser media. The first fiber laser was demonstrated in 1961, and following that work with 3-inch-long cladded Nd : glass rods, a 1-m fiber laser was demonstrated in which a 10-μm Nd : glass core was clad with 1-mm-diameter lower-index glass. The fiber was wound into a helix for efficient pumping by a flashlamp, and gains of approximately 50 dB were obtained.

One advantage of single-mode fiber lasers is their excellent beam quality ($M^2 \sim 1$), a consequence of the fact that the transverse spatial profile of the field is determined by the

[19]In the fiber literature this three-level model is usually referred to as a *two*-level model because the equations are written explicitly only for the populations of the levels 1 and 2 of the amplifying transition, and not for the level 3 that is pumped but decays rapidly to the level 2. See, for instance, *Desurvire* (footnote 15).

refractive indices and radii of the fiber core and cladding and is insensitive to mechanical or temperature fluctuations. The broad gain bandwidth allows wavelength tunability, while the high efficiency and broad absorption bandwidth allow pumping with low-power diode lasers that do not have to emit at a particular wavelength and do not have to be temperature stabilized. The bending and coiling of a fiber (Section 8.7) eliminates alignment considerations that might complicate other laser systems in some applications. While the most important application of fiber amplifiers and lasers has undoubtedly been in communications (Chapter 15), recent progress toward higher output powers will likely result in the replacement of some existing high-power lasers by much more compact and economical fiber lasers; for some applications, such as laser marking and bar-coding, fiber lasers are already being used and marketed.

In the single-mode fiber amplifiers and lasers considered thus far, the doped, active core is typically about 6–10 μm in diameter, and efficient coupling of the pump radiation into such a tightly confined core requires the spatial coherence of a single-mode pump (Chapter 13). Single-mode diode lasers are generally limited in power to a few watts, which therefore limits the output of a conventional single-mode fiber laser to a few watts. Significant progress in high-power fiber lasers occurred with the development of *double-clad fibers* (Fig. 11.31) in which the active core is surrounded by a lower-index, inner cladding (~100 μm), which in turn is surrounded by a lower-index, thinner outer cladding. The inner cladding has a much larger diameter than the core and has a

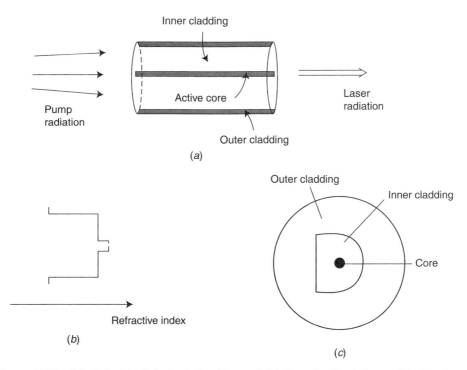

Figure 11.31 (*a*) A double-clad step-index fiber and (*b*) the refractive index profile. The inner cladding diameter is typically a few hundred times larger than the active core diameter, which is typically ≲10 μm. (*c*) A D-shaped inner cladding cross section that results in a greater filling factor than is obtained with a cylindrically symmetric double-clad fiber in which both the inner cladding and core cross sections are circular.

larger numerical aperture (Section 8.6). The basic idea here is that the spatially incoherent, divergent radiation from a high-power diode bar or stack can be injected into the large numerical-aperture double-clad fiber and be absorbed by the active core as it is guided down the length of the fiber. The absorption efficiency is defined by

$$\eta = \frac{P(0) - P(L)}{P(0)} = 1 - e^{-[F(S_{core}/S_{total})\alpha_p L]},$$ (11.14.26)

where $P(0)$ is the injected pump power, $P(L)$ is the remaining pump power after propagation over the length L of the fiber, α_p is again the core absorption coefficient at the pump wavelength, S_{core} is the cross-sectional area of the core, and S_{total} is the cross-sectional area of the inner cladding plus that of the core. F (<1) is a filling factor evaluated numerically based on Maxwell's equations and the appropriate boundary conditions. Such computations reveal that the modes of the cylindrically symmetric double-clad fiber we have implicitly assumed are such that the filling factor is in fact very small. Double-clad fibers are therefore designed to break the cylindrical symmetry; the D-shaped inner cladding shown in Fig. 11.31c is one of various designs that lead to filling factors ~ 0.5. With such designs the advantage of double-clad fibers is realized: Equation (11.14.26) implies that nearly all the pump power can be absorbed by the active core of a double-clad fiber if, for instance, the fiber is long enough. Since the inner cladding cross section is much larger than the core cross section, the absorbed power $\approx P(0)$ can be made much larger than that possible with single-clad fibers, and consequently much higher output laser powers are possible with double-clad fibers.

Because of their large surface-to-volume ratio, high-power fiber lasers do not suffer from heat dissipation problems that beset other lasers at high powers, and they are simply air cooled or, if necessary at high power levels, water cooled. The power scaling of fiber lasers and amplifiers is limited primarily by ASE, which limits the population inversion, and nonlinear optical effects including stimulated Raman scattering (SRS) and stimulated Brillouin scattering (SBS), which convert radiation in the core into radiation at non-lasing frequencies or into vibrational modes of the glass that can result in damage or fracture of the fiber. SRS and SBS are relatively weak in glass fibers, but become significant in fiber lasers because of the long path length: The low attenuation coefficient allows tens of meters of lasing fiber to be wrapped around a spool. Self-phase modulation is another nonlinear effect that can reduce peak powers in pulsed fiber lasers.

One way to reduce nonlinearities is to make the core diameter d_{core} larger. As Eq. (11.14.26) indicates, this can allow an increase in the absorption efficiency without increasing the length L of the fiber. It also reduces the threshold powers for SRS and SBS since these processes are intensity dependent: Increasing d_{core} increases the mode diameter of the lowest-order mode and therefore decreases the intensity for a given power. Simply increasing the core diameter, however, presents a problem when we recall that the condition for a single mode in a step-index fiber is [Eq. (8.7.1)]

$$\frac{\pi d_{core}}{\lambda} NA = \frac{\pi d_{core}}{\lambda} \sqrt{n_1^2 - n_2^2} < 2.405.$$ (11.14.27)

So if we increase d_{core} we must decrease the numerical aperture (NA) proportionately in order to prevent multimode oscillation that would generally render the beam quality of the laser unacceptable in many applications. But, as discussed in Section 8.7, decreasing NA could result in large bending losses.

An elegant way around this multimode problem with larger cores is to coil the fiber around a spool of appropriate radius. Recall from the discussion in Section 8.7 that higher-order modes suffer greater bending losses than the lowest-order mode. The bending radius can be chosen such that the power loss for the lowest-order mode is sufficiently small, while the loss in higher-order modes is sufficiently large, that only the lowest-order mode lases. This approach allows higher-power single-mode lasing not only because nonlinear effects are weakened or prevented but also because ASE in the core is reduced. This reduction is a consequence of the fact that spontaneous emission goes into all allowed modes of the field, and therefore only a relatively small portion of the total spontaneous emission undergoes amplification in the active core.

Single-mode high-power fiber lasers are sometimes said to be brightness upconverters in that they convert the spatially incoherent, low-brightness radiation from bars or stacks of laser diodes into spatially coherent, high-brightness radiation. There has been rapid progress in the development of high-power fiber lasers. Yb : glass fibers are especially useful as high-power gain media at ≈ 1.1 μm because of the large quantum efficiency ($\cong 90\%$) of ytterbium, which allows optical efficiencies $>80\%$ and little "thermal loading" of the gain medium by the pump. High-power fiber sources, like those based on other high-power laser systems, often consist of a laser followed by one or more amplifiers, the so-called master-oscillator-power-amplifier (MOPA) configuration. MOPAs have the advantage that the power in an amplifier is limited to the final output power, whereas that inside a laser is usually much larger than the output power (Chapter 5). Therefore, large output powers can be realized without excessively large and possibly damaging intracavity powers. MOPAs can also be advantageous when it is difficult to control the tunability or noise of the output of a high-power oscillator, and they can amplify weak modulated pulses with high fidelity.

• The end reflectors forming a fiber laser resonator can be dielectric mirrors butted onto the ends of the fiber or, more commonly in commercial fiber lasers, fiber Bragg gratings. A fiber Bragg grating (FBG) is essentially just a short (~ 1 cm) section of fiber with a refractive index that varies periodically with distance z along the fiber, for example, $n(z) = n + \Delta n(z)$ with

$$\Delta n(z) = a \sin \frac{2\pi z}{d_g} \tag{11.14.28}$$

and $|\Delta n| \ll n$. Light of wavelength λ incident on the FBG along the z direction is transmitted unless[20]

$$\lambda = 2nd_g, \tag{11.14.29}$$

in which case it is reflected. That is, the FBG serves as a reflector for light having a wavelength equal to twice the period d_g of the index grating times the refractive index of the fiber material.

The "photosensitivity" effect by which an index grating with a period on the order of an optical wavelength can be "written" into a fiber was discovered in the late 1970s. It was found that when an intense beam from an argon ion laser was injected into a germanosilicate fiber, the beam was increasingly reflected until, after a few minutes, it was almost totally reflected. This was interpreted as the result of an index grating formed by the standing-wave interference pattern

---

[20]See Eq. (6.A.3) with $\theta = \pi/2$ and $\lambda_s = d_g$.

formed by two counterpropagating fields.[21] How a spatial intensity variation translates into an index variation is not completely understood, but it is believed to result from photoionization resulting in a change in the absorption spectrum of the glass. The spatial variation of the absorption coefficient implies, from the Kramers–Kronig relation (Section 3.15), a corresponding spatial variation of the refractive index.

The period d_g here is obviously fixed by the wavelength of the laser that writes the grating. In the late 1980s, however, it was realized that side illumination of a doped fiber with two beams would allow the grating period to be "tuned" by varying the angle between the two beams. Consider, for example, two fields with the same amplitude and with wave vectors $\mathbf{k}_1$ and $\mathbf{k}_2$, $|\mathbf{k}_1| = |\mathbf{k}_2| = k = 2\pi/\lambda_w$. The interference of the two fields produces a total field proportional to $\exp(\mathbf{k}_1 \cdot \mathbf{r}) + \exp(i\mathbf{k}_2 \cdot \mathbf{r})$ and therefore an intensity having an interference term proportional to $\cos(\mathbf{k}_1 - \mathbf{k}_2) \cdot \mathbf{r}$. This interference term has a period $P = 2\pi/|\mathbf{k}_1 - \mathbf{k}_2|$ along the "z" direction defined by $\mathbf{k}_1 - \mathbf{k}_2$. Now

$$|\mathbf{k}_1 - \mathbf{k}_2| = [k_1^2 + k_2^2 - 2\mathbf{k}_1 \cdot \mathbf{k}_2]^{1/2} = k[2 - 2\cos\theta]^{1/2} = k\left[4\sin^2\left(\frac{\theta}{2}\right)\right]^{1/2}$$

$$= 2k\sin\left(\frac{\theta}{2}\right), \tag{11.14.30}$$

where θ is the angle between $\mathbf{k}_1$ and $\mathbf{k}_2$. Therefore

$$P = \frac{2\pi}{2k\sin(\theta/2)} = \frac{\lambda_w}{2\sin(\theta/2)}. \tag{11.14.31}$$

The period of an index grating written with two interfering beams of wavelength λ_w can, therefore, be varied from $\lambda_w/2$ to ∞, depending on the angle between the beams. In fact, this is the basic idea behind the manufacture of FBGs. The source of the writing beams is usually an ultraviolet laser (e.g., KrF at 248 nm) that illuminates a "phase mask" consisting of periodic corrugations such that two interfering beams at the desired angle are formed by diffraction. Fiber Bragg gratings have other applications besides their use as reflectors in fiber lasers.   •

A major recent development in optical fiber technology is the *photonic crystal fiber*. A *photonic crystal* is a material with a refractive index that varies periodically in one, two, or three dimensions. A fiber Bragg grating may be regarded as a one-dimensional photonic crystal; more complex photonic crystals can completely reflect light of a given wavelength and polarization for essentially any angle of incidence. The concept, if not the word, is an old one. Maxwell's equations applied to structures that are periodic on the scale of a wavelength have solutions in which only certain wavelengths are allowed, just as the Schrödinger equation for an electron in a crystal restricts the energies to certain bands separated by band gaps (Chapter 2). In the case of photonic crystals, forbidden gaps are called *photonic band gaps*. Photonic band-gap materials have been of particular interest for applications in which it is desirable to inhibit spontaneous emission: An

---

[21]Lord Rayleigh (1887) recognized the possibility of such an effect. In a paper dealing in part with "the propagation of waves through a medium endowed with a periodic structure," he cited experiments by Becquerel on the reflection of light from metals coated with a film and suggested an explanation for the observed strong reflection: "The various parts of the film ... may be conceived to be subjected, during exposure, to *stationary* luminous waves of nearly definite wave-length, the effect of which might be to impress upon the substance a periodic structure recurring at intervals equal to *half* the wave-length of the light ... In this way the operation of any kind of light would be to produce just such a modification of the film as would cause it to reflect copiously that particular kind of light."

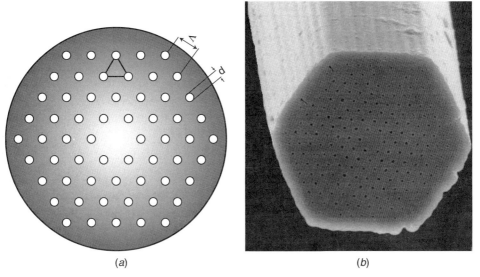

(a)   (b)

Figure 11.32 (*a*) Schematic illustration of the cross section of one type of photonic crystal fiber. Note the missing hole at the center. (*b*) Scanning electron microscope image of the PCF of this type used in early experiments demonstrating the modal properties of such a structure. [From P. Russell, *Science* **299**, 358 (2003) and J. C. Knight, *Nature* **424**, 847 (2003).]

excited atom whose transition frequency lies within a photonic band gap does not radiate.[22]

In a photonic crystal fiber (PCF) the "crystal" structure involves a regular pattern of thin air channels parallel to the fiber axis (Fig. 11.32). Unlike a step-index fiber, there is no high-index core that is materially different from the cladding. It is not obvious that such a structure can guide light: Is wave-guiding impossible without a high-index core, or will guiding occur because the solid core is surrounded by a medium of lower *average* (or "effective") refractive index? For such a complicated geometry, this question can be confidently answered only by numerical solutions of Maxwell's equations, or, better yet, by experiments. The first such experiments in the late 1990s revealed not only that single-mode guided beams were possible in PCFs, but more surprisingly that single-mode guiding occurred for a wide range (337–1550 nm) of wavelengths. The latter property, dubbed "endlessly single mode," is in marked contrast to a step-index fiber that, according to Eq. (11.14.27), becomes multimode when the wavelength λ is smaller than a certain value depending on the core diameter and the numerical aperture of the fiber.

In numerical computations of the modal properties of PCFs of the type shown in Fig. 11.32, it is found that only a single guided mode occurs if

$$\frac{d}{\Lambda} \lesssim 0.41, \tag{11.14.32}$$

where d and Λ are, respectively, the hole diameter and the center-to-center hole spacing (or "pitch") indicated in Fig. 11.32a. This condition is sometimes loosely interpreted by

---

[22]Recall also the remarks in Section 3.11 concerning the inhibition of spontaneous emission in cavity QED.

imagining that, since the air holes have a smaller refractive index than the surrounding silica, and therefore that total internal reflection can occur at their interface with the surrounding glass, the air holes define a sort of "sieve." The lowest-order mode has a diameter $\approx 2\Lambda$ and therefore cannot "slip through" the narrow gaps between the holes; it must be confined to the central core (missing-hole) region. Higher-order modes, however, have transverse spatial variations on a smaller scale, are able to slip through the gaps, and therefore are not "trapped" or confined unless the hole spacing is sufficiently small. The fact that (11.14.32) is insensitive to wavelength, or to d and Λ independently, means that single modes with large diameters are possible over a very wide range of wavelengths as long as a PCF like that in Fig. 11.32 is designed to satisfy (11.14.32). For the PCF of Fig. 11.32b, $d \sim 300$ nm, $\Lambda = 2.3$ µm, and $d/\Lambda = 0.13$.

Figure 11.33 shows results of numerical computations on modal characteristics of such a "missing-hole" PCF. For d/Λ satisfying (11.14.32) the PCF is "endlessly single-mode" with modes confined near the central core, whereas in the parameter region below the curve the PCF is multimode. In the parameter region above the curve the PCF is single mode, but the mode diameters are comparable to the diameter of the fiber, as opposed to modes satisfying (11.14.32), which are bell-shaped with mode diameter $\approx 2\Lambda$.

As discussed in Section 8.7, conventional fibers have unpredictable output polarizations because of random structural variations; for this reason amplifiers and other components of telecommunication systems must be polarization insensitive. PCFs with a twofold symmetry as opposed to the full circular cross-sectional symmetry of a conventional fiber, however, are found to be strongly birefringent and therefore are not susceptible to unpredictable polarization due to random variations in structure or temperature. Twofold symmetry can be realized, for instance, by using a regular pattern of holes with

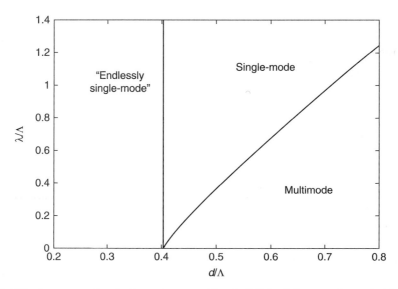

Figure 11.33 Parameters for single-mode vs. multimode PCFs of the type shown in Fig. 11.32. The curve is obtained from the formula $\lambda/\Lambda = 2.80[d/\Lambda - 0.406]^{0.86}$ obtained by B. T. Kuhlmey, R. C. McPhedran, and C. M. de Sterke, *Optics Letters* **27**, 1684 (2002), as a fit to results of numerical simulations.

two different diameters such that the pattern is unchanged under rotations by π about the fiber axis.

In Section 8.4 we alluded to the deleterious effect of group velocity dispersion in fiber communications. PCFs offer another advantage over conventional fibers in this regard: d and Λ and other design parameters can be selected such that the D parameter characterizing group velocity dispersion is very small. Numerical analyses of PCFs yield an effective refractive index from which D can be computed, and a PCF can then be designed to make D small. Experiments have demonstrated "ultra-flattened dispersion" in PCFs: $D \sim 0 \pm 1$ ps/(km-nm) over a wavelength range of hundreds of nanometers, compared to $D \sim 17$ ps/(km-nm) at 1.55 μm in conventional silica–glass fibers. PCFs in fiber communications could eliminate the need for the dispersion compensation techniques employed with step-index fibers (Chapter 15).

Hollow-core PCFs are especially attractive for guiding high-power radiation. In these fibers there is a hollow central core surrounded by a regular pattern of air channels with diameters smaller than the core. The confinement of an approximately Gaussian mode in the core is the result of a photonic band gap of the surrounding array of air channels. Modes within the band gap are confined to the hollow core and as such are far less susceptible to nonlinear effects associated with propagation in glass.

Research in photonic crystal fibers has expanded dramatically in recent years.[23] One of the important questions being addressed in connection with the possible replacement of step-index fibers by PCFs is whether the low attenuation of step-index telecom fibers ($\sim$0.2 dB/km, as mentioned in Section 8.7) can be achieved with PCFs. In addition to absorption and Rayleigh scattering, and losses due to bending for bending radii smaller than a few millimeters, PCFs can suffer losses due to scattering from surface structure and roughness at the glass–air interfaces; control of these losses involves details of the fabrication of these fibers that are beyond our scope here. Attenuations as low as $\sim$0.6 dB/km have been obtained with a solid-core PCF, while $\sim$1.2 dB/km is the lowest attenuation reported for a hollow-core PCF at the present time. Hollow-core fibers appear theoretically to be most promising in this regard due to the simple fact that the light propagates mainly within the hollow core.

Photonic crystal fiber lasers made by doping the central core region with rare-earth ions offer additional advantages besides the compactness, high gain, and broad gain bandwidth associated with other fiber lasers. In particular, mode diameters $\sim$10 times those of step-index fibers are possible, leading to substantially reduced nonlinearities and much higher powers. The advantages of double cladding—large numerical apertures and efficient coupling of pump power into the doped core—can also be realized with PCFs in which air channels surrounding the doped core act to guide the laser radiation while an additional air cladding confines the pump radiation (Fig. 11.34). The air cladding replaces the low-index, solid outer cladding of a conventional double-clad fiber. Such a design results in an inner cladding with high numerical aperture (NA $\sim$ 0.6) and therefore efficient coupling of pump power into the fiber. It also allows larger single-mode core diameters and smaller inner cladding diameters, and therefore shorter fiber lengths and reduced nonlinearities [Eq. (11.14.26)]. These designs are therefore of great interest for power scaling of fiber lasers. For the PCF of Fig. 11.34, single-mode output of 120 W was obtained with a 48-cm fiber having

---

[23]The reviews cited in Fig. 11.32 include pictures of various PCF designs.

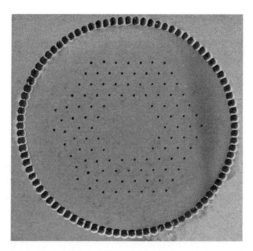

Figure 11.34 Cross section of a double-clad PCF laser. [From J. Limpert, F. Roser, T. Schreiber, and A. Tunnermann, *IEEE Journal of Selected Topics in Quantum Electronics* **12**, 233 (2006).]

core and inner cladding diameters of ~ 180 and 60 μm, respectively. It should also be noted that pumping of more conventional double-clad Yb fiber lasers with high-power diode stacks has yielded nearly single-mode output powers ~ 1 kW and good beam quality ($M^2 \approx 1.4$). At the present time the primary limitation to high-power scaling of fiber lasers appears to be the limited pump powers available from laser diodes. High-average-power (~ 100 W) pulsed fiber lasers employing chirped pulsed amplification and other techniques are also under development.[24]

- The GVD and SPM of optical fibers—that is, their dispersive and nonlinear properties—make them convenient for compression of the output pulses from mode-locked lasers. For wavelengths below about 1.3 μm the GVD (and therefore the chirp) is positive, and a pulse of sufficiently short duration and sufficiently high intensity will therefore be temporally and spectrally broadened by the combined effects of GVD and SPM in a fiber. It can then be compressed by employing the negative GVD of prisms and gratings.

  For wavelengths greater than 1.3 μm the negative GVD (and therefore negative chirp) of glass fibers allows pulse compression by SPM and GVD in a single fiber. This relatively simple technique requires, of course, that a pulse be sufficiently intense and sufficiently short for SPM and GVD, respectively, to be effective. When these conditions are met, a pulse can propagate without any change in its duration as the effects of SPM and negative GVD are effectively balanced. This temporal "soliton" effect has already been mentioned in Section 10.4. A rough estimate of the peak pulse intensity I_0 required for a pulse of duration τ_p to propagate as a soliton can be obtained by equating the characteristic length scale L_{GVD} for group velocity dispersion [Eq. (8.4.8)] to the characteristic length scale

$$L_{\mathrm{SPM}} = \frac{c}{\omega n_{2I} I_0} \tag{11.14.33}$$

for self-phase modulation. The peak "soliton power" obtained in this way is (Problem 11.13)

$$P \sim \frac{c|\beta|}{\omega n_{2I} \tau_p^2} = \frac{\lambda^3 |D| A_{\mathrm{eff}}}{4\pi^2 c n_{2I} \tau_p^2}, \tag{11.14.34}$$

---

[24]See the article cited in Fig. 11.34.

where A_{eff} is an effective cross-sectional area such that $P = I_0 A_{\text{eff}}$, and β and D are the GVD parameters defined by Eqs. (8.4.2) and (8.4.23), respectively. This estimate provides an order-of-magnitude approximation to the pulse power required for propagation without change in the pulse duration in a medium with SPM and negative GVD. It predicts that only modest pulse powers are required for soliton propagation in optical fibers, in agreement with observations (Problem 11.13).

Equation (11.14.34) implies that the soliton pulse energy times the pulse duration ($\propto P \times \tau_p \times \tau_p$) is proportional to the dispersion parameter $|D|$. This observation is the basis for another way in which pulses can be compressed: Decreasing $|D|$ along the length of a fiber, while keeping the pulse energy constant, results in a decrease in τ_p with propagation. Alternatively, in a fiber amplifier the pulse energy increases with propagation while the dispersion parameter is approximately constant, resulting again in pulse compression. Of course, such pulse compression only occurs if the pulse injected into the fiber has sufficient power for it to form a soliton, and the soliton must be stable with respect to the variations along the fiber. In fact, one of the most important characteristics of solitons is their stability: Provided that variations in the medium are not too strong, a soliton will "adiabatically" adjust to them such that the opposing causes of broadening and compression remain balanced and (11.14.34) remains applicable. •

11.15 REMARKS

The development of various kinds of lasers has been a largely evolutionary process involving the efforts of large numbers of people throughout the world. By any reasonable measure this development has been remarkable. Even the most optimistic early expectations for the development and application of lasers turned out to be too conservative.

Different lasers are very similar in their basic operating principles; the concepts of gain, threshold, and feedback are central to an understanding of any laser, regardless of the physical or chemical processes by which gain is established. Beyond that, however, it is obvious that different lasers differ greatly in their output power, tunability, complexity, size, cost, and other characteristics.

Laser output powers, for example, range from milliwatts in the case of laser pointers, for example, to the petawatt levels realized with chirped pulse amplification. Powers a thousand times greater than those generated by a large electrical power plant (~ 1 GW) are routinely produced with tabletop terawatt lasers, and it is not unusual for lasers to generate powers in excess of the total power consumption in the United States (~ 10 TW). Laser powers $\sim 10^{20}$ W appear to be feasible in more "futuristic" systems. By comparison, the total solar power received by the Earth is about $\sim 10^{17}$ W, and the power produced in the largest nuclear explosion to date (about 50 Megatons) was about $(50\,\text{Mt})(4 \times 10^{15}\,\text{J/Mt})/(40\,\text{ns}) \approx 5 \times 10^{24}$ W = 5 yottawatts.

The short pulses possible with lasers are the shortest man-made "events" ever created, and with them it has become possible, for instance, to follow chemical reactions in real time.[25] Attosecond (10^{-18} s) pulses have been measured in nonlinear processes made possible by ultrashort, superintense laser pulses (Section 14.7); an attosecond is to a second what a second is to the age of the universe! At the other extreme, laser technology has led to extremely stable, continuous-wave radiation that approaches an idealized monochromatic wave in its temporal characteristics.

---

[25]See Section 14.7. For a review of the development of ultrashort laser pulses and a historical perspective see N. Bloembergen, *Reviews of Modern Physics* **71**, S283 (1999).

Laser wavelengths span a large range of wavelengths from the far infrared to the ultra-violet, and many other wavelengths can be obtained by nonlinear frequency mixing processes such as second- and third-harmonic generation. Some of the most widely used lasers in the visible and near infrared are tunable over tens or even hundreds of nanometers. The wide gain bandwidths of lasers such as Ti : sapphire allow not only wide tunability but also the generation of ultrashort pulses by mode locking.

In this chapter we have considered only lasers based on stimulated emission involving at least one bound energy level. We have not considered *free-electron lasers* in which coherent radiation is generated with beams of unbound electrons in magnetic fields. Another active area of research not discussed here is the generation of coherent soft X-ray radiation. One approach to X-ray lasers employs a plasma column generated by high-power radiation, resulting under certain conditions in population inversion by electron-impact excitation. Coherent soft X-ray radiation has also been produced in high-order harmonic generation, and work is in progress with this approach to develop tabletop coherent X-ray sources.

Another important topic not considered here is that of *astrophysical masers*. These are galactic and intergalactic sources of microwave radiation produced by amplified spontaneous emission on rotational transitions of molecules including OH, H_2O, and HCN. An astrophysical maser was first observed at a wavelength of 18 cm in 1965; the "beaming," high intensity, polarization, and other properties of the observed radiation were so unusual that it was initially referred to as "mysterium." Within a few years it was concluded that this radiation must be the result of "maser" action on rotational transitions of OH molecules excited by radiation from a nearby star. In 1969 such radiation was observed at 1.35 μm and was attributed to a well-known rotational transition of H_2O. Many such astrophysical masers powered by energy from different astrophysical objects, including black holes, have been identified; they exist in part because collisional deexcitation of excited states is very slow at interstellar cloud densities. While their actual physical diameters can be on the order of the Earth–sun separation or more, these distant "masers" are regarded by astrophysicists as "compact" because of their small angular diameters, which have enabled high-resolution observations of the properties of black holes, for instance. We turn our attention in the Chapters 14 and 15 to some (mostly) more down-to-earth applications of stimulated emission.

APPENDIX: GAIN OR ABSORPTION COEFFICIENT FOR VIBRATIONAL-ROTATIONAL TRANSITIONS

The gain coefficient is given by the formula (Table 4.1)

$$g(v) = \frac{\lambda^2 A}{8\pi} \left(N_2 - \frac{g_2}{g_1} N_1 \right) S(v) \qquad (11.A.1)$$

if we take the refractive index $n \cong 1$. Let $N_2 = N(v_2, J_2)$ be the number density of molecules in vibrational level v_2 and rotational level J_2, the upper level of a molecular vibrational-rotational transition. Similarly, let $N_1 = N(v_1, J_1)$ be the number density of the lower level of the transition. Since the degeneracy of an energy level with rotational quantum number J is $2J+1$, we have

$$g(\nu) = \frac{\lambda^2 A}{8\pi}\left[N(v_2, J_2) - \frac{2J_2 + 1}{2J_1 + 1}N(v_1, J_1)\right]S(\nu). \qquad (11.A.2)$$

As noted in Chapter 2, rotational-level spacings in molecules are much smaller than vibrational-level spacings: The former lie in the microwave region, whereas the latter are in the infrared. In many molecules the separation of adjacent rotational energy levels is small compared to $k_B T$. Then the spacing between rotational levels is small compared to the average kinetic energy of a molecule. In this case we might expect that, as a result of collisions, the rotations of the molecule will be in thermal equilibrium at the gas (translational) temperature T. Then the rotational levels will be distributed according to the Boltzmann distribution:

$$N(v_2, J_2) = N(v_2)\frac{g_{J_2}}{Z}\exp\left(-\frac{E_{J_2}}{k_B T}\right), \qquad (11.A.3a)$$

$$N(v_1, J_1) = N(v_1)\frac{g_{J_1}}{Z}\exp\left(-\frac{E_{J_1}}{k_B T}\right). \qquad (11.A.3b)$$

Here $N(v_2)$ and $N(v_1)$ are the total vibrational population densities regardless of rotation, i.e.,

$$N(v_{2, 1}) = \sum_{J=0}^{\infty} N(v_{2,1}, J). \qquad (11.A.4)$$

$g_{J2} = 2J_2 + 1$ and $g_{J1} = 2J_1 + 1$ are the rotational-level degeneracies, and Z is the rotational partition function:

$$Z = \sum_{J=0}^{\infty} g_J e^{-E_J/k_B T}, \qquad (11.A.5)$$

where (Section 2.4)

$$E_J = hcB_e J(J + 1). \qquad (11.A.6)$$

For CO_2, $B_e \cong 0.39\ \text{cm}^{-1}$, and it is easily checked that $hcB_e \ll k_B T$, and therefore $E_J \ll k_B T$ except for very low temperatures or states with very large J. Partly for this reason we can expect that (11.A.3), the assumption of *rotational thermal equilibrium*, to be an excellent approximation for CO_2 and many other molecules. Using (11.A.6) in (11.A.5), we calculate

$$Z = \sum_{J=0}^{\infty} (2J + 1)e^{-(hcB_e/k_B T)J(J+1)} = \sum_{J=0}^{\infty} (2J + 1)e^{-xJ(J+1)}$$

$$\approx \int_0^{\infty} e^{-xy}\,dy = \frac{1}{x} = \frac{k_B T}{hcB_e} \qquad (x \ll 1). \qquad (11.A.7)$$

The replacement of the sum by the integral is accurate if $x \ll 1$; then we may replace $J(J + 1)$ by the continuous variable y. Note that $dy/dJ = 2J + 1$, which is used in writing the second line of (11.A.7). [The reader may check numerically the validity of (11.A.7).] Thus, using (11.A.3)–(11.A.7) in (11.A.2), we write the gain coefficient

$$g(v) = \frac{\lambda^2 A}{8\pi} \frac{\overline{B}_e}{T} (2J_2 + 1)[N(v_2)e^{-\overline{B}_e J_2(J_2+1)/T} - N(v_1)e^{-\overline{B}_e J_1(J_1+1)/T}]S(v), \qquad (11.A.8)$$

where $\overline{B}_e \equiv hcB_e/k_B$ is the rotational constant expressed in degrees Kelvin. For CO_2, $\overline{B}_e \cong 0.565K$.

The selection rule for allowed rotational transitions is $\Delta J = 0, \pm 1$; transitions $J - 1 \to J, J + 1 \to J$, and $J \to J$ are referred to as $P(J)$ branch, $R(J)$ branch, and $Q(J)$ branch transitions, respectively. Consider for definiteness a $P(J)$ branch, in which case (11.A.8) becomes

$$g(v) = \frac{\lambda^2 A}{8\pi} \frac{\overline{B}_e}{T} (2J_2 + 1)e^{-\overline{B}_e J_2(J_2+1)/T}[N(v_2) - N(v_1)e^{-2\overline{B}_e(J_2+1)/T}]S(v) \qquad (11.A.9)$$

for the gain on the vibrational-rotational transition $(v_2, J_2) \to (v_1, J_2 + 1)$. Expressions like this are found frequently in laser physics and spectroscopy. Note that only the vibrational-level-densities appear; the rotational-level densities are "frozen," so to speak, at the Boltzmann distribution (11.A.3). In a rate-equation model for gain or absorption on a vibrational-rotational transition, therefore, rate equations are necessary only for the vibrational populations rather than the myriad vibrational-rotational populations. The assumption of rotational thermal equilibrium leads therefore to an enormous simplification.

Equation (11.A.9) shows that there can be a positive gain even if $N(v_2) < N(v_1)$, provided that

$$N(v_2) > N(v_1)e^{-2\overline{B}_e(J_2+1)/T}. \qquad (11.A.10)$$

In other words, a population inversion can be achieved on a (P-branch) vibrational-rotational transition without having an inversion on the total vibrational populations. In this case gain is sometimes said to be due to a *partial population inversion*.

In the case of CO_2, Eq. (11.A.9) requires a slight modification, for it turns out that the (001) and (100) vibrational levels have associated with them only odd and even rotational quantum numbers, respectively.[26] The rotational partition function (11.A.7) is

---

[26]This is a consequence of the Pauli principle: The ^{16}O nuclei in CO_2 are bosons (their nuclear spin is 0), and so the complete wave function of the CO_2 molecule must not change sign when these two nuclei are interchanged. This is true for the nuclear spin part of the wave function and also for the electronic part of the ground electronic state of CO_2, and therefore the combined vibrational-rotational part of the wave function must not change sign when the two ^{16}O nuclei are interchanged. Under this interchange a state with rotational quantum number J changes sign if J is odd but not if J is even, a consequence of the fact that the dependence on the rotation angle θ of the wave function of a linear rotor is characterized by the associated Legendre function $P_J^M(\theta)$, which has the property $P_J^M(\theta + \pi) = (-1)^J P_J^M(\theta)$. The antisymmetric vibrational states change sign under the interchange whereas the symmetric vibrational states do not. Therefore, in order for the complete CO_2 wave function to be symmetric under interchange of the bosonic ^{16}O nuclei, only odd (even) values of J can occur for antisymmetric (symmetric) vibrational states. This is one of the many remarkable consequences of the Pauli principle for *identical* particles. In the case of the isotopic molecule $^{18}OC^{16}O$, odd and even values of J are observed to occur for both symmetric and antisymmetric vibrational states.

therefore effectively halved, and consequently the gain coefficient (11.A.9) is multiplied by 2. For the $(001) \rightarrow (100)$ vibrational band of CO_2, $A \approx \frac{1}{5}$ s^{-1}. Using $\lambda = 10.6$ μm, $\overline{B}_e = 0.565$K, and $J_2 = 19$, therefore, we obtain

$$g(10.6 \, \mu\text{m}) = 6.4 \times 10^{-10}[N(001) - 0.93N(100)]S(\nu) \text{ cm}^2\text{-s}^{-1} \qquad (11.\text{A}.11)$$

for the $P(20)$ branch, where the vibrational population densities $N(001)$ and $N(100)$ have units of cm^{-3} and the lineshape function $S(\nu)$ is expressed in seconds. These quantities will depend on factors such as pressure, temperature, and gas mix.

For definiteness, let us focus our attention on a situation in which the parameters are known reasonably well, namely, CO_2 in the Earth's atmosphere. For $T = 293$K and $P = 1$ atm, the total number density found from (3.8.20) is 2.5×10^{19} cm^{-3}. Assuming an atmospheric concentration of 0.033% CO_2, therefore, and a Boltzmann distribution of CO_2 vibrational levels, we estimate that at sea level the absorption coefficient for 10.6 μm radiation due to CO_2 is (Problem 11.14)

$$a(\nu) = -g(\nu) \approx 5.4 \times 10^3 S(\nu) \text{ cm}^{-1}\text{-s}^{-1}. \qquad (11.\text{A}.12)$$

It remains to estimate the lineshape factor $S(\nu)$.

At atmospheric pressures the 10.6-μm absorption line of CO_2 is collision broadened. At line center, therefore, $S(\nu) = 1/(\pi\delta\nu_0)$, with the collision linewidth $\delta\nu_0$ given by (3.8.12). Since N_2 and O_2 are the most frequent collision partners of CO_2 in the atmosphere, we can approximate $\delta\nu_0$ by including only contributions from N_2 and O_2. We will assume the following cross sections (Problem 11.14):

$$\sigma(CO_2, O_2) = 9.5 \times 10^{-15} \text{ cm}^2, \qquad (11.\text{A}.13\text{a})$$

$$\sigma(CO_2, N_2) = 1.2 \times 10^{-14} \text{ cm}^2, \qquad (11.\text{A}.13\text{b})$$

so that

$$\delta\nu_0 = 2.7 \times 10^9 \text{ s}^{-1} \qquad (11.\text{A}.14)$$

for the atmospheric concentrations ≈ 0.78 and 0.21 of N_2 and O_2, respectively, and therefore (Problem 11.14)

$$S(\nu) = \frac{1}{\pi\delta\nu_0} = 1.2 \times 10^{-10} \text{ s}^{-1}. \qquad (11.\text{A}.15)$$

It follows that

$$a(\nu) = 6.5 \times 10^{-7} \text{ cm}^{-1}, \qquad (11.\text{A}.16)$$

which is in good agreement with measurements of the absorption coefficient at 10.6 μm due to atmospheric carbon dioxide at low altitudes.[27] The attenuation coefficient for 10.6-μm radiation in the Earth's atmosphere also has a significant contribution from water vapor[27] and particulate matter ("aerosols").

---

[27]See, for instance, A. D. Wood, M. Camac, and E. T. Gerry, *Applied Optics* **10**, 1877 (1971).

PROBLEMS

11.1. **(a)** In the Introduction we noted that a transition between two nondegenerate states is always absorbing in thermal equilibrium, that is, the gain coefficient is negative. Show that the same conclusion holds regardless of the level degeneracies.

   (b) Show that the absorption linewidth of a Doppler-broadened transition is typically 10^{-5}–10^{-6} times the transition frequency.

   (c) Use numerical estimates to show that optical pumping of a gaseous laser medium with a blackbody source of radiation will usually not be very practical.

11.2. Consider an elastic collision between an electron and an atom. Show that the kinetic energy exchanged is relatively small.

11.3. For the excitation transfer process indicated in Fig. 11.1b, write rate equations like (11.3.3) for the populations N_A, N_B, and N_{B^*}.

11.4. According to our discussion in Section 11.4, the higher gain of the 3.39-μm transition in the He–Ne laser would ordinarily preclude lasing at 632.8 nm, unless the 3.39-μm line is deliberately suppressed. Why is this so?

11.5. In Section 11.5 we replaced the Boltzmann factor $e^{-\Delta E/k_B T}$ by unity, where ΔE is the energy difference between $He(2^1 S)$ and $Ne(3s_2)$. Was this a reasonable approximation for our purposes?

11.6. In Section 11.5 we mentioned that White and Gordon deduced relative level populations by monitoring intensities of spontaneously emitted "sidelight." Show that a knowledge of the Einstein A coefficient of a transition, combined with a frequency filter and an absolute intensity measurement, allows an absolute measurement of the upper-level population. [For instance, White and Gordon reported a $He(2^1 S)$ population of about 2.5×10^{11} cm^{-3} under lasing conditions.] Does radiative trapping affect such a measurement?

11.7. Consider the equation

$$\frac{dI}{dz} = \frac{g_0 I}{1 + I/I^{\text{sat}}}$$

describing the propagation of the intensity of a plane wave in an amplifier with small-signal gain g_0 and saturation intensity I^{sat}. Show that the intensity after a propagation distance z satisfies the transcendental equation

$$x = G_0 e^{-\alpha(x-1)},$$

where $x = I/I_0$, $G_0 = e^{g_0 z}$, $\alpha = I_0/I^{\text{sat}}$, and $I_0 = I(z = 0)$ is the intensity input to the amplifier. Make a plot of (a) x vs. G_0 for a range of values of α and (b) x vs. α for a range of values of G_0. (You can either solve numerically the transcendental equation for x or numerically integrate the differential equation for I.)

11.8. **(a)** Using the four-level model, estimate the saturation intensities of Nd : YAG and Ti : sapphire lasers.

(b) Estimate the saturation fluence at 800 nm of a Ti : sapphire amplifier.

(c) Using the Frantz–Nodvik model (Section 6.12), derive a formula relating the input and output pulse fluences, both expressed in units of the saturation fluence of the amplifier. Is there an optimal input fluence for efficient energy extraction from the amplifier?

11.9. **(a)** Calculate the Coulomb electric field between the electron and the nucleus in the hydrogen atom, using the Bohr model for the ground level of the electron.

(b) Calculate the intensity of a monochromatic plane wave with the electric field strength found in part (a), and compare it with the intensity 10^{21} W/cm^2 that can be obtained by chirped pulse amplification.

(c) Calculate the radiation pressure exerted by a field with the intensity 10^{21} W/cm^2, assuming the field is normally incident on a mirror. Compare it with estimates for the pressure at the center of the sun, which is believed to be around 3×10^{11} atm.

11.10. From curve fits to measured attenuation data in the experiments cited in Fig. 11.30 it was deduced that the absorption coefficients at 1550 and 1480 nm in the fiber were 0.876 and 0.792 m^{-1}, respectively, and the corresponding saturation powers were $h\nu_p \mathcal{P}_p^{\text{sat}} = 0.549$ mW and $h\nu_s \mathcal{P}_s^{\text{sat}} = 0.279$ mW. Using these data, and assuming a diode laser pump power of 5 mW (7 dBm), plot the gain G in decibels of the signal field for fiber lengths up to 10 m in length. Is there an optimal fiber length for maximal gain? If so, what is the physical basis for such an optimal length, that is, why doesn't the gain increase monotonically with the length of fiber?

11.11. **(a)** Using the three-level model and the data provided in the preceding problem, calculate the threshold diode laser pump power for amplification of 1550 nm radiation with a 3.87-m erbium-doped fiber. Compare your result to the experimentally determined value of 1.2 dBm.

(b) Show that $G_p < 1$ guarantees that the denominator in (11.14.25) is always positive.

11.12. **(a)** Amplified spontaneous emission in an amplifier can be expected to have a negligible effect on the amplification of an input signal if the ASE noise power $(\text{Pwr})_N$ (Section 6.13) is small compared to the gain saturation power. Using the data provided in Problem 11.11, estimate the amplifier gain G (dB) at which ASE might become significant.

(b) Even for high-gain amplifiers ASE can be expected to have a negligible effect on the amplification of an input signal if the input signal power is large compared to the effective noise input for ASE (Section 6.13). Assuming an EDFA gain bandwidth of 4 THz, and a spontaneous emission factor ~ 1, estimate the input signal power that should be exceeded if ASE is not to affect its amplification. [Note: In the paper cited in Fig. 11.30 it is stated that: "As a conservative guideline, [Eq. (11.14.16)] is valid for amplifiers with gains less than ~ 20 dB or for amplifiers with input signal powers greater than ~ -20 dBm."]

11.13. (a) Derive expression (11.14.33) for the propagation distance for which self-phase modulation can be expected to play a significant role in the propagation of a pulse with peak intensity I_0.

(b) Using the approximate formula (11.14.34), estimate the peak pulse power required for soliton propagation of a 1.55-μm, 10-ps pulse in an optical fiber with a dispersion parameter $D \approx 16 \, \text{ps}/(\text{km-nm})$ (Problem 8.5), a nonlinear refractive index coefficient $n_{2I} \approx 3.2 \times 10^{-20} \, \text{m}^2/\text{W}$, and an effective mode area $A_{\text{eff}} = \pi (2.5 \, \mu\text{m})^2$. (Answer: about 30 mW.) What is the peak intensity and the energy contained in a 10-ps Gaussian pulse with this effective cross-sectional area and peak power?

11.14. (a) Verify expression (11.A.12) for the absorption coefficient due to CO_2 of 10.6 μm radiation in the atmosphere.

(b) Using the collision cross sections (11.A.13), estimate the absorption linewidth of the 10.6-μm transition of CO_2 in the Earth's atmosphere. [For a discussion of these cross sections see T. W. Meyer, C. K. Rhodes, and H. A. Haus, *Physical Review* A**12**, 1993 (1975).]

(c) Verify (11.A.15) and (11.A.16).

12 PHOTONS

12.1 WHAT IS A PHOTON?

We have made frequent reference to photons of light. For some purposes it is convenient to think in terms of photons, whereas for others the wave description of light is indispensable. In this chapter we will discuss more carefully this wave–particle duality of light. We will also introduce the theory behind measurements that count photons and some aspects of the photon detectors used in these measurements.

The question whether light consists of particles or waves is, of course, a very old one. Newton, around 1700, thought it consisted of particles. Later, as a result of careful experimentation on the interference and diffraction of light, Newton's particles of light were abandoned in favor of a wave picture championed by Thomas Young and others. The wave theory was brilliantly formulated mathematically by Maxwell at roughly the time of the American Civil War, and the experiments of Heinrich Hertz (1887) gave convincing support for Maxwell's theory. Hertz confirmed that accelerated charges radiate electromagnetic waves, and that these waves have the same characteristics as visible light (Section 8.2).

By the 1890s Maxwell's theory, together with oscillator models for the response of matter to light, was used by Lorentz and others to explain, semiquantitatively, nearly all the known optical phenomena (Chapter 3). Many scientists held that Newton's theories of mechanics and gravitation, together with Maxwell's theory of electromagnetism, contained the fundamental laws of the universe, and that in principle everything about nature might one day be understood in terms of them. However, as is well known to every student of modern physics, there were some experimental results that did not fit into the picture. One was the spectrum of blackbody radiation, which led to the light-quantum hypotheses[1] of Planck (1900) and Einstein (1905). These ideas were used by Bohr in his theory of the hydrogen atom (1913), and a decade later led to quantum mechanics and the view that nature is fundamentally statistical rather than deterministic. According to quantum mechanics the wave and particle views of light are both oversimplifications; radiation and matter have both wave and particle attributes, or a "wave–particle duality."

What, then, is a photon? By considering a few examples, we will try to explain the essence of the answer given by quantum mechanics. The answer is subtle and also

---

[1]The term *photon* was coined in 1926 by the chemist G. N. Lewis. Before 1926 physicists spoke of "quanta" of light.

Laser Physics. By Peter W. Milonni and Joseph H. Eberly
Copyright © 2010 John Wiley & Sons, Inc.

very beautiful. The examples described in the next few sections deal with photon polarization, photon-induced recoil, and photon interference.

12.2 PHOTON POLARIZATION: ALL OR NOTHING

Consider the experiment shown in Fig. 12.1. A plane electromagnetic wave of frequency ν is incident on a polarizer oriented so that it passes radiation with polarization along the x direction but absorbs radiation with the orthogonal, y polarization. The incident light is assumed to be polarized at an angle θ with respect to the x direction (i.e., the electric field vector at any point on the wave oscillates along a line at an angle θ with respect to the x axis). According to the classical law of Malus, a fraction $\cos^2 \theta$ of the incident intensity will pass through the sheet. This is the prediction of classical physics.

Now suppose that the incident light is reduced in intensity so that only a single photon (of energy $h\nu$) is incident upon the polarizer. In this case the law of Malus appears to fail: The incident photon is not split by the polarizer. Instead experiments find that the entire photon either passes through the polarizer or is absorbed: It is "all or nothing" when a single photon is incident on the polarizer. The energy $h\nu$ is evidently an indivisible unit of energy for radiation of frequency ν.

If we repeat this one-photon experiment many times, always with the same source and arrangement, we find that sometimes the photon passes through the sheet and sometimes it does not. If θ is zero, however, we find that the incident photon always passes through the sheet, whereas if $\theta = 90°$, no photon ever passes through. Repetition of the experiment many times restores Malus' law, but now as the statement that $\cos^2 \theta$ is the *probability* that a photon polarized at an angle θ to the polarizer axis will pass through.

The situation here is akin to coin flipping. When we say that the probability of getting either heads or tails is $1/2$, we mean that in a large number of tosses heads and tails will

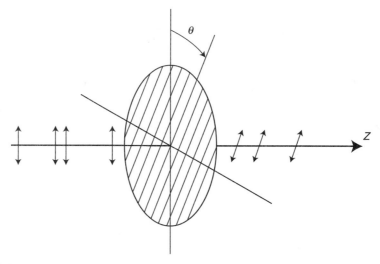

Figure 12.1 A plane monochromatic wave represented by vertically polarized photons is incident on a polarizing sheet. The transmitted field is linearly polarized at the angle θ of the polarizing axis of the sheet.

turn up approximately the same number of times. In any one toss of the coin, however, we get either heads or tails, just as our one-photon experiment records either an entire photon or nothing. Quantum mechanics asserts that the statistical aspect of our one-photon experiment is a fundamental characteristic of nature.

Suppose the incident light in our experiment has an energy corresponding to n photons. Since each photon has the probability $\cos^2 \theta$ of passing through the sheet, we expect on the average that $n \cos^2 \theta$ photons will be transmitted. As n increases, the deviation of the number of transmitted photons from the average $n \cos^2 \theta$ becomes smaller relative to the average number. It becomes increasingly more accurate to say that a fraction $\cos^2 \theta$ of the incident intensity is passed by the filter. In other words, *we approach the classical form of the law of Malus when the number of incident photons is large.* Most optical experiments involve enormous numbers of photons (Problem 12.1); in such experiments with polarized light the "all or nothing" nature of photon polarization may be ignored for all practical purposes.

Our discussion has assumed an ideal photon source, one that produces a single photon on demand. We have also assumed an ideal photon counter. Such an ideal device would count every incident photon, and it would give no spurious counts when there are no incident photons. Furthermore, it would respond immediately to the incident signal. Needless to say, there is no such perfect detector. Nevertheless, as discussed in Section 12.7, available detectors come fairly close to the ideal.

12.3 FAILURES OF CLASSICAL THEORY

A photon of frequency ν in free space carries not only an energy $h\nu$ but also a linear momentum of magnitude

$$p = \frac{h\nu}{c}. \tag{12.3.1}$$

In other words, the linear momentum of radiation of frequency ν is quantized in indivisible units of magnitude $h\nu/c$. Conservation of linear momentum demands that an atom that undergoes spontaneous emission must recoil with a linear momentum of magnitude (12.3.1). According to quantum mechanics, we cannot predict exactly in which direction the photon will be emitted, and therefore we cannot predict the direction of atomic recoil (Fig. 12.2).

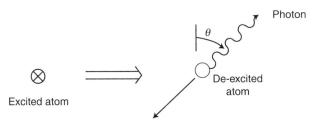

Figure 12.2 Conservation of linear momentum implies that an atom recoils when it undergoes spontaneous emission. The direction of photon emission (and atomic recoil) is not predictable.

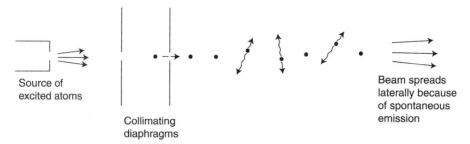

Source of
excited atoms

Collimating
diaphragms

Beam spreads
laterally because
of spontaneous
emission

Figure 12.3 A well-collimated atomic beam of excited atoms will spread laterally because of the recoil associated with spontaneous emission.

Experiments with beams of excited atoms confirm the recoil associated with spontaneous emission (Fig. 12.3). In fact, the recoil of a spontaneously emitting atom was inferred by O. R. Frisch in 1933 and has in more recent years been confirmed with greater accuracy.

It is not surprising that an atom recoils when it emits a photon. It is like the recoil a person feels upon firing a rifle, the person and the bullet corresponding to the atom and the photon, respectively. However, the recoil of a spontaneously emitting atom is not accounted for in the classical wave theory of radiation. Figure 12.4 shows why. Classical theory treats spontaneous emission as a smooth process, with radiation being continuously emitted more or less in all directions, according to the radiation pattern of the source. Classical radiation from an isolated system such as an atom is also characterized by inversion symmetry. That is, the intensity of spontaneous radiation emitted in the x direction is equal to the intensity emitted in the $-x$ direction, and so on for every direction. It is obvious that an emitter of this type suffers equal and opposite recoil forces along every axis and does not recoil at all. So classical wave theory predicts no recoil in spontaneous emission, in contradiction to results of experiment.

Stimulated emission and absorption also impart recoil momentum to an atom (Fig. 12.5) and the direction of atomic recoil follows exactly the direction of propagation of the incident radiation. It is interesting that the Doppler effect in the emission or absorption of radiation by a moving atom may be understood as a consequence of the

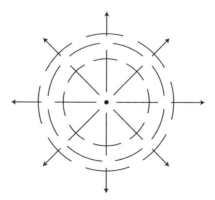

Figure 12.4 A source emitting a spherical wave cannot recoil because the spherical symmetry of the wave prevents it from carrying any linear momentum from the source.

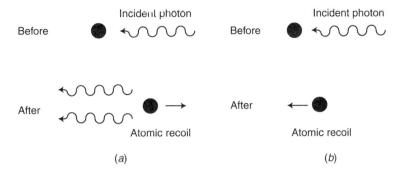

Figure 12.5 Atomic recoil associated with (*a*) stimulated emission and (*b*) absorption.

fact that photons carry energy $h\nu$ and linear momentum $h\nu/c$ (Problem 14.9). Atomic recoil in absorption and emission is the *sine qua non* of the Doppler cooling discussed in Section 14.4.

• The classic manifestation of the expression for photon momentum given in (12.3.1) is the Compton effect, as discussed in textbooks on modern physics. A. H. Compton's experiments were reported in 1923. Einstein had already inferred in 1917 that photon momentum causes an atom to recoil when it undergoes spontaneous emission. He did this by a careful analysis of thermal equilibrium between radiation and matter, showing that the Planck distribution *requires* atomic recoil in spontaneous emission (Section 14.4).

Regarding Einstein's inference in 1917 that photon momentum causes an atom to recoil when it undergoes spontaneous emission, it is interesting to remember that in his special theory of relativity he had even earlier (in 1905) proposed the equation

$$E = \sqrt{m^2c^4 + p^2c^2} \tag{12.3.2}$$

connecting the energy E and linear momentum p of a body of rest mass m. For a photon, $m = 0$ and $E = h\nu$, so (12.3.1) then follows from Einstein's 1905 relation. Nevertheless, the equation $p = h\nu/c$ for a photon was not stated by Einstein (or anyone else) before his study of 1917 on "The Quantum Theory of Radiation."   •

The photoelectric effect is often cited as evidence for the quantum nature of light. It is convincing evidence, but the evidence must be interpreted carefully because most of the basic features of the photoelectric effect can be explained without invoking photons or the quantum theory of radiation. This can be appreciated by recalling equation (3.4.33) for the rate of absorption by an atom in a field described by *classical* electromagnetic theory. This equation also applies when the electron makes a transition from a bound to a free state, as in the photoelectric effect. Then the equation predicts that electrons should be ejected as soon as the field is incident on the photoemissive surface, as is observed to occur. The observed proportionality of the photoelectron ejection rate to the radiation intensity is also contained in equation (3.4.33).

Finally there is Einstein's famous 1905 formula $K.E. = h\nu - W$, relating the kinetic energy of the ejected electron to photon energy $h\nu$, relative to the work function W of the photoemissive surface. This same feature is predicted by the absorption rate (3.4.33) obtained without invoking photons. Thus all of these features of the photoelectric effect, which are frequently said to require that light be described in terms of photons,

are explained by the transition theory of Chapter 3, which uses only an unquantized description of the radiation field together with quantum theory for the atom.[2]

However, a remaining key feature cannot be explained by this theory. Classical electromagnetic radiation can deliver energy to a target atom only as fast as its Poynting flux allows. This implies that a very weak incident field, requiring let us say 1 microsecond to deliver $h\nu$ of energy to a photoemissive surface, could never eject an electron with kinetic energy equal to $h\nu - W$ in a time shorter than 1 microsecond. But experiments have repeatedly shown that photoelectrons can be ejected with no delay at all, even for such weak radiation. This can be explained easily by a quantum mechanical interpretation of a weak radiation field, viz., as one in which photons arrive very rarely but randomly, and therefore occasionally even much earlier than a classical field can permit. This is the key experimental observation that shows why the particle rather than the wave aspect of photons is needed to fully understand the photoelectric effect.

Consider another situation involving an atom and a single photon. Take a single excited atom A and two identical photon detectors B and C, as illustrated in Fig. 12.6a. We imagine that B and C are perfect detectors in that each registers a count if and only if a photon of radiation is incident upon it. When A undergoes spontaneous emission, it need not register a count in either B or C, but there is some probability that a count is registered in B or C. We pose the following question: Is there any possibility that a coincidence occurs, that is, that the emitted radiation from A can be detected at both B and C?

The answer given by the quantum theory of radiation is clear: No. For A emits a single photon, and a photon can trigger the emission of only one photoelectron. Therefore, the radiation from A can register a count at either B or C, or neither, but never at both B and C.

Now think of the radiation from A as a classical electromagnetic wave. As such it has nonzero values over a certain region of space (and time). In particular, as illustrated in Fig. 12.6b, the field propagates outward from A and eventually reaches both B and C, so that there are measurable electric and magnetic fields at both B and C. And these

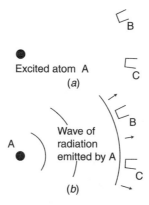

Figure 12.6 (a) An excited atom and two photodetectors. Can radiation emitted spontaneously by A be detected by both detectors? (b) According to classical radiation theory the field emitted by A will reach both B and C.

[2]For further discussion, see, for instance, M. O. Scully and M. Sargent III, *Physics Today*, March, 38 (1972).

fields can, according to classical electromagnetic theory, trigger photoelectric counts at both B and C. Thus, there is a clear disagreement here between the classical and quantum theories of radiation.

Experiment supports the quantum theory of radiation: The radiation emitted in a single atomic transition cannot be split into more than one unit or excite more than one detector.[3]

12.4 WAVE INTERFERENCE AND PHOTONS

Photons are energy quanta of the electromagnetic field, carrying energy $h\nu$ and linear momentum $h\nu/c$. They are often pictured as particles, like bullets or billiard balls, but having no mass and moving at the speed of light. This particle picture certainly helps us to understand atomic recoil in spontaneous emission, and a phenomenon such as Compton scattering, but what have these effects to do with the vast body of evidence for the wave nature of light? To see how quantum mechanics reconciles the wave and particle aspects of light, we will consider the example of the Young two-slit experiment (see Section 13.3), assuming that the slits are illuminated by a plane monochromatic wave. As illustrated in Fig. 12.7, the slits in this case produce a well-known interference pattern on a screen behind the slits.

The location of the intensity maxima is easily found from the condition for constructive interference of the two waves emerging from the slits: their path difference $s_2 - s_1$ must be an integral number of wavelengths. This gives the condition

$$y_n^{(\text{max})} = n\frac{\lambda D}{d}, \qquad n = 0, \pm 1, \pm 2, \dots \qquad (12.4.1)$$

for the intensity maxima on the observation screen, assuming that the two slits subtend a small angle at points of observation (Section 13.3). When the path difference $s_2 - s_1$

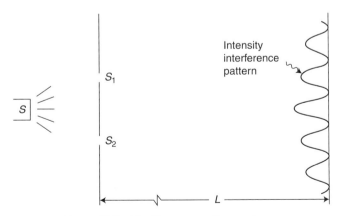

Figure 12.7   The Young two-slit experiment.

[3]J. F. Clauser, *Physical Review D* **9**, 853 (1974).

is an odd integral number of half wavelengths, however, there is complete destructive interference:

$$y_n^{(\min)} = \left(n + \frac{1}{2}\right) \frac{\lambda D}{d}, \quad n = 0, \pm 1, \pm 2, \ldots. \tag{12.4.2}$$

For optical wavelengths, the slit separation d in a demonstration of the Young experiment is typically on the order of a fraction of a millimeter, while D might be about a meter.

•   When Thomas Young (1773–1829) published his results in 1802, he encountered harsh criticism from the proponents of Newton's particle theory of light. Eventually, however, it was recognized by nearly all scientists that Young's experiment had put an end to Newton's particle theory. Young went on to other endeavors, and his accomplishments were remarkable. For instance, his interest in Egyptology led him to make a major contribution toward deciphering the Rosetta stone. His theory of color vision is still widely cited, as is his work on elasticity. He also made pioneering contributions to the studies of sound, tides, and the human voice, among other things. His work on the wave theory of light was done shortly after he began a medical practice in London.

Young's experiment did not actually require two slits. A narrow sunbeam entering a room through a small hole was instead split in two by a card of thickness (approximately 1 mm) slightly less than the diameter of the sunbeam, so that a "beam" passed along each side of the card. Young remarked that his experiment could be done "whenever the sun shines, and without any other apparatus than is at hand to everyone."   •

Imagine now that the Young experiment is performed with a single photon of light. Can we obtain Young's interference pattern in this case? According to quantum theory, the intensity interference pattern indicated in Fig. 12.7 represents only the relative *probability distribution* for detecting the photon somewhere on the observation screen. That is, a single photon by itself does not produce an interference pattern. Instead, there is a relatively high probability of detecting the photon at points satisfying the constructive interference condition (12.4.1), and the photon will never be found at points satisfying (12.4.2) because at such points the probability is exactly zero.

In other words, the quantum theory of radiation says that the classical interference pattern is correct, but not in the sense of classical theory. The strictly classical (wave) interpretation of the interference pattern is that the entire pattern is observed regardless of the intensity of the light incident on the screen containing the slits. However, the quantum interpretation is that, with light so dim that only a single photon is involved, we do not in fact observe a pattern but only a single photon at some point on the classically calculated pattern. To summarize: *The wave and particle aspects of two-slit interference are reconciled by associating a particle (photon) probability distribution function with the classical (wave) intensity pattern.*

At the risk of belaboring the point, we emphasize that the interference pattern *applies* to a single photon (in the probabilistic sense) but would not be *revealed* by a single photon. When the Young experiment is performed with a light beam containing a very large number of photons, even points of low probability (or, classically speaking, low intensity) receive some photons because of the large numbers involved. In this case we observe the complete interference pattern, just as if we perform the single-photon experiment a very large number of times. In nearly all interference experiments, of course, an extremely large number of photons is involved, as in all of classical optics

(Problem 12.1). Under such circumstances we observe entire intensity patterns, exactly as predicted by classical wave theory. This is why rather delicate experiments are necessary to see departures from classical wave theory. Because a succession of identical independent single-photon experiments eventually builds up the same pattern associated with a many-photons-at-once experiment, the entire pattern must be "known" to each single photon. It is conventional, therefore, to say in this context, following P. A. M. Dirac, that a photon interferes only with itself.

In quantum mechanics both radiation and massive particles (electrons, photons, neutrons, etc.) display a wave–particle duality. A particle with linear momentum of magnitude p has a de Broglie wavelength $\lambda = h/p$ associated with it. For a photon the de Broglie wavelength is $\lambda = h/(h\nu/c) = c/\nu = \lambda$, just the wavelength of a wave of frequency ν. The wave fields in this case satisfy Maxwell's equations. For material particles the wave fields satisfy Schrödinger's equation.

The association of a particle probability function with a wave interference pattern correctly describes many phenomena that cannot be explained with wave or particle concepts alone. Wave and particle attributes are interwoven in the microscopic world, both being oversimplifications derived from our experience with macroscopic phenomena.

- The first observation of an entire interference pattern with light so dim that only one photon at a time could be detected was made by G. I. Taylor in 1909. At the suggestion of J. J. Thomson, Taylor sought to determine whether diffraction of light is affected if no more than one "light quantum" at a time is incident on a photographic plate. Taylor observed the diffraction (shadow) pattern of a needle using light from a flame illuminating a slit and attenuated by smoked glass. The experiments were carried out at the house outside London of the 24-year-old Taylor's parents; the exposure time was about 3 months, during part of which Taylor was away on a yacht. He found the same intensity pattern regardless of the light intensity; this came as a surprise to Thomson, who expected that statistical fluctuations associated with single-quantum events would diminish the visibility of interference fringes. In other words, Taylor (indirectly) observed the buildup of an interference pattern from repeated single-photon experiments.[4] In more recent times this effect has been demonstrated more directly in two-slit interference with photomultipliers for photon detection.[5] •

12.5 PHOTON COUNTING

Imagine an ideal detector in which every incident photon ejects a photoelectron. By counting the number of photoelectrons, we are in effect counting the number of incident photons. Of course, real experiments are much more involved than this idealization would suggest, but let us suppose nevertheless that we can count photons in this manner.

Consider the following experiment. We have a shutter that we can open and close instantaneously. We place the shutter between a photon detector and an incident quasimonochromatic beam of light. We open the shutter at some time t and close it at a time $t + T$, and record the number of photons that were counted while the shutter was open. Next we open the shutter for another time interval T, and again take note

[4]For an entertaining discussion ("Take a Photon . . . ") of this and related phenomena see O. R. Frisch, *Contemporary Physics* **7**, 45 (1965).
[5]S. Parker, *American Journal of Physics* **39**, 420 (1971); **40**, 1003 (1972).

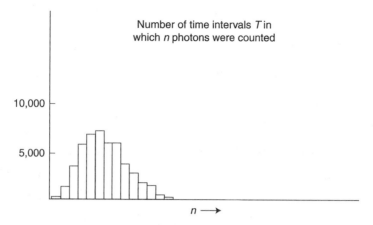

Figure 12.8   A hypothetical histogram of photon counts.

of the number of photons counted in that "run." We repeat this procedure several thousand times and make a histogram of the number of photons counted in a time interval T. Our histogram might look like that shown in Fig. 12.8.

What we would like is a theory of such a photon count distribution. Based on the scatter in the number of photon counts from one run to the next, we base our theory on probabilities. That is, we only try to calculate the probability $P_n(T)$ of counting n photons in a given time interval T.

Consider a very short subinterval Δt within the time interval T. We assume that $p(t)$ Δt, the probability of ejecting one electron in the time interval Δt, is given by

$$p(t)\,\Delta t = \alpha I(t)\,\Delta t, \tag{12.5.1}$$

where $I(t)$ is the incident intensity, averaged over a few optical periods, and α is a constant depending on the details of the photon detector (density of atoms, size of exposed surface, etc.). The time interval Δt is taken to be so short that the probability of ejecting more than one electron during this time is completely negligible. Δt is not an intrinsic parameter of our experiment but merely a theoretical construct, and so we can make it as small as we please.

Let $P_n(t)$ be the probability of counting n photons during a time t ($0 \le t \le T$) during which the shutter is open. It is convenient to consider $P_n(t + \Delta t)$, where Δt is defined above. There are two mutually exclusive ways of getting n photons in the time interval $t + \Delta t$: We can get $n - 1$ photons in the time interval t and 1 more in the interval from t to $t + \Delta t$, or n photons in the time t and none in the interval from t to $t + \Delta t$. The probability of the first way is

$$P_{n-1}(t)p(t)\,\Delta t = (\text{probability of } n - 1 \text{ photons in time } t)$$

$$\times (\text{probability of 1 photon in } \Delta t). \tag{12.5.2}$$

These probabilities multiply for the same reason that the probability of getting two heads in two successive flips of a coin is given by the product $(1/2)\,(1/2) = 1/4$. Similarly the probability of the second alternative is

$$P_n(t)[1 - p(t)\,\Delta t] = \text{(probability of } n \text{ photons in time } t)$$
$$\times \text{(probability of no photon in } \Delta t). \qquad (12.5.3)$$

Since each alternative leads to the same end result—n photons counted in a time interval $t + \Delta t$—we have

$$P_n(t + \Delta t) = P_{n-1}(t)p(t)\,\Delta t + P_n(t)[1 - p(t)\,\Delta t]. \qquad (12.5.4)$$

Here we add the probabilities for the two possible alternatives because they are mutually exclusive. The probability of getting one head and one tail in two flips of a coin, for instance, is the probability $(1/2)(1/2)$ of getting a head and then a tail, plus the probability $(1/2)(1/2)$ of getting a tail followed by a head.

Rearranging (12.5.4), we have

$$\frac{P_n(t + \Delta t) - P_n(t)}{\Delta t} = [P_{n-1}(t) - P_n(t)]p(t). \qquad (12.5.5)$$

Since Δt is at our disposal, let us make it so small that the left side becomes the derivative of $P_n(t)$:

$$\frac{P_n(t + \Delta t) - P_n(t)}{\Delta t} \cong \lim_{\Delta t \to 0} \frac{P_n(t + \Delta t) - P_n(t)}{\Delta t} = \frac{dP_n}{dt}. \qquad (12.5.6)$$

Therefore (12.5.5) becomes a differential equation for $P_n(t)$:

$$\frac{dP_n}{dt} = p(t)[P_{n-1}(t) - P_n(t)], \qquad (12.5.7)$$

or, using (12.5.1),

$$\frac{dP_n}{dt} = \alpha I(t)[P_{n-1}(t) - P_n(t)]. \qquad (12.5.8)$$

To compare the theory leading to (12.5.8) with the results of our photon counting experiment, we must solve (12.5.8) for the probability $P_n(T)$ of counting n photons in any one of our counting intervals of duration T. It is shown below that the desired solution is

$$P_n(T) = \frac{[X(T)]^n}{n!} e^{-X(T)}, \qquad (12.5.9)$$

where

$$X(T) = \alpha \int_0^T I(t)\,dt. \qquad (12.5.10)$$

Equation (12.5.9) is correct as far as it goes, but there is a modification of our theory to be made before we can meaningfully compare it to experiment. In an actual experiment we are dealing with a large number of time intervals T, each starting at a different time t. For a time interval from t to $t + T$ rather than from 0 to T we must replace (12.5.10) by

$$X(t, T) = \alpha \int_t^{t+T} I(t') \, dt'. \tag{12.5.11}$$

It is convenient to write this as

$$X(t, T) = \alpha T \bar{I}(t, T), \tag{12.5.12}$$

where

$$\bar{I}(t, T) = \frac{1}{T} \int_t^{t+T} I(t') \, dt' \tag{12.5.13}$$

is the average incident intensity during the time interval from t to $t + T$. Then (12.5.9) becomes

$$P_n(t, T) = \frac{1}{n!} [X(t, T)]^n \exp[-X(t, T)]. \tag{12.5.14}$$

Finally, to compare with experiment we must average (12.5.14) over all the "starting times" t. That is, the theoretical photon counting probability distribution is

$$P_n(T) = \langle P_n(t, T) \rangle = \left\langle \frac{1}{n!} [\alpha T \bar{I}(t, T)]^n \exp[-\alpha T \bar{I}(t, T)] \right\rangle, \tag{12.5.15}$$

where $\langle \cdots \rangle$ denotes an average over t. Equation (12.5.15), which was first derived by L. Mandel in 1958, is used frequently in the analysis of photon counting experiments. In the following section we will consider an especially important example of the use of this formula.

- To verify (12.5.9), consider the function

$$P_n(t) = \frac{1}{n!} X(t)^n e^{-X(t)}. \tag{12.5.16}$$

The derivative of this function is

$$\begin{aligned}
\frac{dP_n}{dt} &= \frac{1}{n!} X(t)^n \frac{d}{dt} e^{-X(t)} + \frac{1}{n!} e^{-X(t)} \frac{d}{dt} X(t)^n \\
&= \frac{1}{n!} X(t)^n \left(-\frac{dX}{dt} e^{-X(t)} \right) + \frac{1}{n!} e^{-X(t)} \left[nX(t)^{n-1} \frac{dX}{dt} \right] \\
&= \frac{dX}{dt} \left\{ \frac{[X(t)]^{n-1}}{(n-1)!} e^{-X(t)} - \frac{[X(t)]^n}{n!} e^{-X(t)} \right\} \\
&= \frac{dX}{dt} [P_{n-1}(t) - P_n(t)].
\end{aligned} \tag{12.5.17}$$

From (12.5.10) it follows that

$$\frac{dX(t)}{dt} = \alpha \frac{d}{dt} \int_0^t I(t')\, dt' = \alpha I(t), \tag{12.5.18}$$

and therefore (12.5.17) is just (12.5.8). In other words, the function (12.5.16) is a solution of the differential equation (12.5.8).

We want a solution of (12.5.8) satisfying $P_n(0) = 0$ because the probability of counting any photons at exactly the time $t = 0$, when the shutter has suddenly been opened, is zero. The function (12.5.16) satisfies this condition because $X(0) = 0$. Therefore (12.5.16) with $t = T$, i.e., (12.5.9), is the desired solution for $P_n(t = T)$. ●

12.6 THE POISSON DISTRIBUTION

Suppose the intensity $I(t)$ of the incident beam in our photon-counting experiment is constant:

$$I(t) = I = \text{const.} \tag{12.6.1}$$

Then (12.5.13) becomes simply

$$\bar{I}(t,\, T) = \frac{1}{T} \int_t^{t+T} I(t')\, dt' = \frac{1}{T} I \int_t^{t+T} dt' = I. \tag{12.6.2}$$

In this case the starting-time average in (12.5.15) is superfluous. Using (12.6.2) in (12.5.15), we obtain the photon-counting probability distribution

$$P_n(T) = \frac{(\alpha I T)^n}{n!} e^{-\alpha I T}, \tag{12.6.3}$$

or

$$P_n(T) = \frac{(\bar{n})^n}{n!} e^{-\bar{n}}, \tag{12.6.4}$$

where

$$\bar{n} = \alpha I T. \tag{12.6.5}$$

The probability distribution (12.6.4) is called the *Poisson distribution*. Note that it is properly normalized to unity, as any valid probability distribution must be. That is, the sum of the probabilities of all possible outcomes is equal to one:

$$\sum_{n=0}^{\infty} P_n = \sum_{n=0}^{\infty} \frac{(\bar{n})^n}{n!} e^{-\bar{n}} = e^{-\bar{n}} \left(\sum_{n=0}^{\infty} \frac{(\bar{n})^n}{n!} \right) = 1, \tag{12.6.6}$$

since the sum in parentheses is just the power series for the function $e^{\bar{n}}$.

One important quantity we can calculate, given $P_n(T)$, is the average number of photons counted in a time interval T, denoted $\langle n(T) \rangle$:

$$\langle n(T) \rangle = \sum_{n=1}^{\infty} n P_n(T) = \sum_{n=1}^{\infty} n \frac{(\bar{n})^n}{n!} e^{-\bar{n}} = \bar{n} e^{-\bar{n}} \sum_{n=1}^{\infty} \frac{(\bar{n})^{n-1}}{(n-1)!}$$

$$= \bar{n} e^{-\bar{n}} \sum_{n=0}^{\infty} \frac{(\bar{n})^n}{n!} = \bar{n} e^{-\bar{n}} e^{\bar{n}} = \bar{n} = \alpha I T. \tag{12.6.7}$$

This is a reasonable result: The average number of photons counted in a time T is equal to the rate of ejection of photoelectrons, αI, times the time T.

We can also calculate the average of the square of the number of photons counted in a time interval T (Problem 12.4):

$$\langle n(T)^2 \rangle = \sum_{n=0}^{\infty} n^2 P_n(T) = \bar{n}^2 + \bar{n}. \tag{12.6.8}$$

The quantity

$$\langle \Delta n(T)^2 \rangle = \left\langle [n(T) - \langle n(T) \rangle]^2 \right\rangle, \tag{12.6.9}$$

i.e., the average of the square of the deviation $n(T) - \langle n(T) \rangle$ of $n(T)$ from its average value $\langle n(T) \rangle$, is the mean-square deviation of $n(T)$ from its average. It gives a measure of the "spread" of $n(T)$ values about the average. From the definition (12.6.9) it follows that

$$\langle \Delta n(T)^2 \rangle = \left\langle n(T)^2 - 2n(T)\langle n(T) \rangle + \langle n(T) \rangle^2 \right\rangle = \langle n(T)^2 \rangle - 2\langle n(T) \rangle^2 + \langle n(T) \rangle^2$$

$$= \langle n(T)^2 \rangle - \langle n(T) \rangle^2, \tag{12.6.10}$$

since $\langle\langle \ldots \rangle\rangle$ is the same as $\langle \ldots \rangle$. From (12.6.7) and (12.6.8), therefore,

$$\langle \Delta n(T)^2 \rangle = \bar{n} \tag{12.6.11}$$

for the Poisson distribution. Similarly, the root-mean-square (rms) deviation, $\Delta n(T)_{\text{rms}}$, for the Poisson distribution is

$$\Delta n(T)_{\text{rms}} = \sqrt{\bar{n}}. \tag{12.6.12}$$

In Fig. 12.9 we plot the Poisson distribution for various values or $\bar{n}$. As $\bar{n}$ increases, the rms deviation in $\bar{n}$ increases as $\sqrt{\bar{n}}$, but the relative rms deviation, $\Delta n_{\text{rms}}/\bar{n}$, decreases.

Actually (12.5.15) reduces to a Poisson distribution not only when $I(t)$ is constant, but more generally whenever the *mean* intensity $\bar{I}(t, T)$ given by (12.5.13) is independent of t. In this more general case the average over t in (12.5.15) is again superfluous because everything to be averaged over t is in fact independent of t. We obtain the Poisson distribution (12.6.4) with the average photon count

$$\bar{n} = \alpha \int_{t}^{t+T} I(t')\, dt' = \alpha \int_{0}^{T} I(t')\, dt' \tag{12.6.13}$$

in a time interval T.

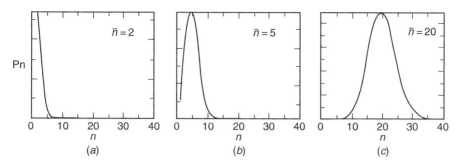

Figure 12.9   The Poisson distribution (12.6.4) for $\bar{n} = 2$, $\bar{n} = 5$, and $\bar{n} = 20$.

One important example of a field having Poisson photon counting statistics is the idealized monochromatic plane wave:

$$E(z, t) = E_0 \cos \omega\left(t - \frac{z}{c}\right), \qquad (12.6.14)$$

in which case the cycle-averaged intensity $I(t) = (c\epsilon_0/2)E_0^2$ is obviously independent of t [and therefore so is $\bar{I}(t, T)$]. In Chapter 13 we will see that laser light and ordinary light give rise to different photon statistics, the Poisson distribution applying to laser radiation.

● The Poisson distribution appears in many contexts in science and engineering. One famous application is to radioactive decay: If a sample emits α particles, say, at a rate of r particles per second, then the average number of α particles emitted in a time interval T is $\bar{n} = rT$, and the probability that n particles are counted in any time interval T is given by (12.6.4). Numerous other applications are discussed in textbooks on probability and statistics.   ●

12.7 PHOTON DETECTORS

Radiation detectors are of two basic types: There are *thermal detectors*, which respond to the heating of some part of the detector by incident light, and *photon detectors*, which measure directly the rate of absorption of photons. Examples of thermal detectors are the Golay cell, which detects the expansion of a gas heated by the absorption of light, the bolometer, which responds to a change in electrical resistance of a material that is heated by absorption, and the calorimeter, which measures the temperature rise of an absorber. In this section we will describe some general aspects of *photon detectors*, which count photons (or, more precisely, photoelectrons) without first converting light to heat. The operating principle of photon detectors is the photoelectric effect: Incident photons release electrons from bound states, leading to an electric current and an output signal in the form of a current or a voltage drop across a resistor in series with the detector.

Photon counting has become an indispensable tool in many applications requiring detection of very weak light. Such applications arise in astronomy, medical diagnostics, telecommunications, and biology, to name a few. Detection of molecular fluorescence by photon counting, for instance, plays an important role in DNA sequencing.

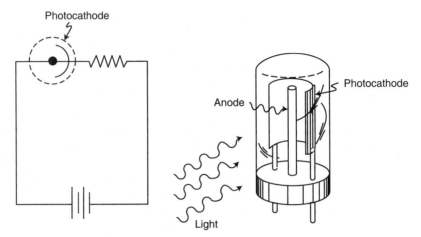

Figure 12.10   A vacuum photodiode acts as a current source in an electric circuit. The measured current in the circuit is proportional to the rate of absorption of incident photons.

A *phototube* is a photon detector whose principle of operation is based on the "external" photoelectric effect in which electrons are released from the surface of a material. Its simplest form is the *vacuum photodiode*, an evacuated tube containing a photoemissive surface called the photocathode, and an anode that collects electrons (because of its higher voltage) emitted from the photocathode (Fig. 12.10). The photoelectric ejection of electrons by incident radiation thus gives rise to an electric current in a circuit containing the phototube. It is the electric current that is measured, but since this current is proportional to the number of incident photons, the phototube responds to the rate of absorption of incident photons, that is, it is a "photon detector."

Phototubes are most effective in the ultraviolet, visible, and near-infrared portions of the electromagnetic spectrum. At lower frequencies the incident photons are not energetic enough to eject electrons from photoemissive surfaces, i.e., to overcome the "work function," the energy required to release an electron. At higher frequencies there is another technical difficulty, namely the absorption of the incident radiation by the window of the phototube; this difficulty may be alleviated by coating the window with a phosphor that emits at a lower frequency than it absorbs.

The sensitivity of a phototube may be increased by filling it with a low-pressure gas. An electron ejected from the photocathode can ionize the atoms of the gas, producing more electrons (called "secondary electrons"), which can themselves collide with atoms to produce more electrons. This electron avalanche process results in a greater current, and thus makes the phototube more sensitive to low-level radiation. A vacuum photodiode may have a response time, determined by the transit time of the photoelectrons to the anode, as low as 10 ns; the addition of gas to increase the sensitivity, however, tends to increase this response time by orders of magnitude.

Short response times, and the high sensitivity required for single-photon detection, are achieved with *photomultipliers*. A photomultiplier tube (PMT) is basically a phototube with a series of additional anodes called dynodes, each at a higher voltage (typically ~ 100 V) than the preceding one. The dynodes are coated with a material that loses secondary electrons when it is struck by incident electrons. An electron ejected from the photocathode by incident radiation is electrostatically focused to the first dynode, generating secondary electrons. These are focused to the next dynode, producing more electrons, and the process continues with a total of perhaps 10 dynodes in all. There is thus a multiplication of the number of electrons released at the photocathode, or equivalently an amplification of the current. The electron multiplication factor is typically on the order of 10^6–10^7. This results in high sensitivity combined with short response times. A photomultiplier is thus capable of a prompt response to a single photon, but not all incident photons eject electrons from the photocathode, so photomultipliers typically operate with quantum efficiencies (the ratio of the average number of electrons ejected from the photocathode to the average number of incident photons) below 30% in the visible and below 10% in the near-infrared. Moreover, not every electron released at the photocathode makes its way to the first dynode; the collection efficiency, or the probability that an electron released at the photocathode arrives at the first dynode, depends on the PMT geometry and other factors, and is typically >75%. The "detection efficiency" is defined as the product of the quantum and collection efficiencies.

Because the anode current increases very rapidly with the voltage, the output signal of a PMT is very sensitive to voltage fluctuations, so that the high-voltage power supply must be very stable.

The detection accuracy of a PMT is limited by *dark-current noise*, or anode dark current, the detector current that appears even in the absence of incident radiation as a consequence of the thermionic emission of electrons mainly from the photocathode but also from the dynode surfaces. Because these surfaces by design have small work functions W, there is thermionic current $i(T)$ even at room temperature. The temperature dependence of this current is described by the *Richardson–Dushman equation*:

$$i(T) = CST^2 e^{-W/k_B T}, \qquad (12.7.1)$$

where C is a constant that depends on the photocathode material, S is the photocathode surface area, and k_B is Boltzmann's constant. Clearly, the dark current is reduced dramatically as the temperature T is lowered. For a work function of 1.5 eV, for instance, $i(260\text{K}) \sim 10^{-4} i(300\text{K})$ (Problem 12.5). For photon-counting applications PMTs are often equipped with thermoelectric (Peltier) coolers in order to minimize dark counts (from perhaps 10^6 counts per second to ~ 10 or less, although these numbers depend strongly on the type of photocathode). The PMT sensitivity in its response to radiation typically improves as well with cooling. Response curves and dark counts per second at various temperatures are among the specifications provided by PMT vendors.

Figure 12.11 indicates how light incident on the photocathode leads to a stream of electrons that are multiplied by the dynode chain to produce current pulses at the anode. For light of sufficient intensity the pulses indicated in Fig. 12.11b overlap to such an extent that there is a (fluctuating) current at the anode, as indicated in Fig. 12.11c. The PMT in this case is said to operate in the *analog mode*. In this case the photons are too closely spaced, loosely speaking, to be counted individually.

Figure 12.12 shows what happens when the incident light intensity is sufficiently low that the current pulses at the anode do not overlap but are well-separated compared to both their duration and the resolution time of the PMT. The PMT in this case is said to operate in the digital or *photon-counting mode*.

Photomultipliers (and other useful photon detectors) have the important property of *linearity* over a wide range of incident intensities: A plot of the output signal versus intensity is essentially a straight line. Beyond some power the PMT response saturates,

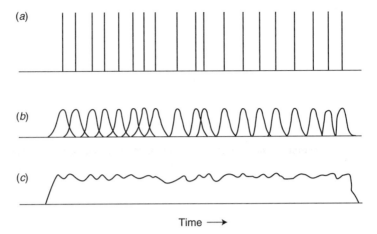

Figure 12.11 Generation of a PMT output signal. (*a*) Electrons released at photocathode; (*b*) electric current pulses at the anode; and (*c*) PMT output signal.

that is, it no longer increases with increasing intensity, and beyond some maximum power specified by the manufacturer the PMT could be damaged.

For the detection of pulsed light it is important that a photodetector be able to respond to rapid variations of the incident intensity. Figure 12.13 indicates two important times that affect the temporal response of a PMT. One of these is the electron transit time from the photocathode to the anode or, more precisely, the time it takes for the output pulse at the anode to reach its peak value after the photocathode is irradiated by a very short pulse of light. This time, together with the time constants of the electric circuitry used to convert the anode output to a meter reading, obviously affects the response time of the PMT to incident radiation. Because of the statistical nature of the electron multiplication process, there is a spread, or *dispersion*, in the electron transit time that affects the resolution time, or the accuracy with which light can be time resolved by the PMT. This determines in part the other time indicated in Fig. 12.13, namely the *rise time*. PMT rise times are typically about a few nanoseconds, whereas electron transit times are typically on the order of tens of nanoseconds. The rise time affects the maximum count rate of a PMT, which is typically $\sim 10^6 - 10^8$ s^{-1}, or $1 - 100$ MHz.

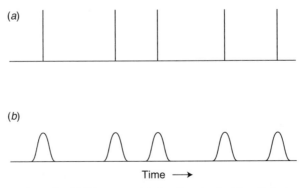

Figure 12.12 Generation of a PMT output signal when the incident light intensity is very low. (*a*) Electrons released at photocathode; (*b*) PMT output signal.

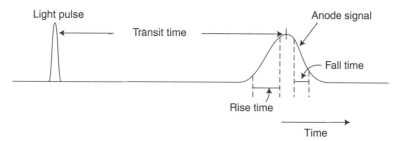

Figure 12.13 The electron transit time and the rise time at the anode of a PMT. The rise time is defined conventionally as the time it takes for the anode signal to increase from 10% of its peak value to 90% of its peak value. The fall time, which generally differs from the rise time, is defined similarly.

• The electron transit time dispersion, together with the electron multiplication factor, can be used to estimate the PMT output signal in the case of a single incident photon releasing a single electron from the photocathode. Assuming an electron multiplication factor of 5×10^6, the charge at the anode is $(5 \times 10^6)(1.6 \times 10^{-19} \text{ C}) = 8 \times 10^{-13} \text{ C}$. If the electron transit time dispersion is 5 ns, the peak current at the anode will be about $(8 \times 10^{-13} \text{ C})/(5 \text{ ns}) \sim 10^{-4} \text{ A}$.

The input voltage to the electric circuitry that converts the anode signal to a meter reading of "photon counts" is determined by the anode current and the load resistance R in series with the PMT. A given small current will produce a larger voltage as R is increased, but an increase in R will also increase the rise time because of the larger RC time constant. Too large a value of R can also affect the potential difference between the anode and the nearest dynode, leading to a non-linear variation with incident light intensity of the PMT output signal. These and other factors dictate the selection of the load resistance. A typical value for R in PMTs used for photon counting is 50 Ω. For this value the voltage in our example is $\sim(10^{-4} \text{ A})(50 \ \Omega) = 5 \text{ mV}$.   •

Dark current, background radiation, and other sources of noise tend to produce photo-electron counts that are spread over a wide range, whereas the signal counts tend to lie in a fairly narrow range of values. An important part of PMT circuitry is a pulse height *discriminator* that accepts only pulses within a preset range, so that noise pulses outside the range of signal pulse heights are eliminated. This discrimination produces a substantially larger signal-to-noise ratio than that obtained by simply counting all pulses.

The *internal* photoelectric effect is the basis for another type of photon detector in which an incident photon is not sufficiently energetic to free an electron from the surface of a material, but nevertheless gives the electron more energy inside the material. *Photoconductivity* occurs when an electron is promoted to the conduction band of a semiconductor by the absorption of a photon of energy $h\nu$ greater than the band gap. That is, shining light on a photoconductive material produces electron–hole pairs (Section 2.7). An applied voltage causes the electrons and holes to drift in opposite directions, producing a current and a voltage drop across a load resistor.

The electrons typically move faster than the holes, and therefore their transit time across the photoconductor is less than that of the holes, and it can also be less than the electron–hole recombination time. When an electron reaches the anode, the external circuit replaces it with another electron, maintaining the continuity of the current. This continues until recombination occurs. If the electron transit time τ_e is shorter than the recombination time τ_r, therefore, each electron–hole pair produced by the absorption of a photon adds a charge Ge to the circuit, where the effective gain $G = \tau_r/\tau_e$ can be as high as 10^6.

Common photoconductive materials for photon detectors are lead sulfide and cadmium sulfide. The latter is used in the visible (e.g., in camera light meters), whereas the former has a good response for wavelengths as high as 3 or 4 μm in the infrared. Photoconductors have longer response times—typically greater than a microsecond—than vacuum photodiodes or photomultipliers.

Faster response times are possible with *photodiodes* in which the internal photoelectric effect occurs at a *pn* junction (Section 2.7). The junction is reverse biased in order to make electrons flow to the *n* side and holes to the *p* side, thereby generating a current from the absorbed light. Since recombination does not occur in the depletion layer, there is no gain in the photodiode. Silicon is the most commonly used photodiode material in the visible and near-infrared spectral range between 400 and 1100 nm. The *responsivity* $\mathcal{R}$, defined as the ratio i/P, where i is the electric current and P is the incident light power, peaks for silicon photodiodes at about 0.7 A/W at 700 nm. They are used in commercial laser power meters, some of which can measure powers ~ 1 nW.

The PIN photodiode results from the insertion of a layer of an "intrinsic" semiconductor (undoped or slightly doped) between the p and n layers. The i region effectively widens the depletion layer so that more light is absorbed, and therefore more electron–hole pairs are generated. Response times ~ 10 ps are possible with PIN photodiode detectors but, because they do not have gain, they require external electronic amplifiers if their sensitivity to incident light is to be increased. Their compactness and low cost make PIN photodiodes especially useful in telecommunications and medical instrumentation.

Greater sensitivity is realized with the *avalanche photodiode* (APD), a reverse-biased photodiode that is the semiconductor analog of a photomultiplier tube. In the APD a strong ($\sim 10^5$ V/cm) reverse-bias electric field causes the charge carriers to accelerate to energies large enough to cause impact ionization of the crystal lattice. The bias voltage is chosen to be slightly below the breakdown voltage V_{br} at which impact ionization due to thermally generated carriers can occur without any incident light. The additional charge carriers produced in the ionization then cause further ionization, and there results a cascade or "avalanche" multiplication of the current. In other words the APD, unlike the photodiode, exhibits gain. The gain is very sensitive to the value of the bias voltage, and consequently a very stable power supply is required to maintain a constant gain.

Typical APD gains are ~ 100, too small for single-photon detection. For this purpose APDs are operated in the *Geiger mode* in which the bias voltage is slightly *above* the breakdown voltage V_{br}. Biasing above breakdown can be done for a short time before impact ionization and avalanche breakdown occur; the generation of an electron–hole pair by the absorption of a photon then causes a large current pulse. Gains approaching those of PMTs are achieved. Obviously, the bias voltage must be reduced and the current pulse dissipated before another photon can be counted. This is done with a "quench circuit" that reduces and resets the bias voltage following the detection of a current pulse. An important consideration in Geiger-mode APDs is the fact that charges can be trapped around defects in the semiconductor and released at a later time, causing spurious current pulses, or "after-pulsing," as well as dark counts.

Unlike a photodiode or an "analog-mode" APD in which the bias voltage is below V_{br} and the output signal is a current that varies linearly with light intensity, the output of a Geiger-mode APD is a current pulse. A change in light intensity results in a change in the rate at which current pulses are produced. In this sense an APD operated in the Geiger mode is analogous to the Geiger counter used in nuclear physics. Silicon, germanium, and InGaAs APDs are used for low-level light detection in the range of 400–1100,

800–1600, and 900–1700 nm, respectively. Compared with PMTs they have higher quantum efficiencies (>70% at the peak of their responsivity curve, around 700 nm for silicon), greater linearity, lower noise levels, and are, of course, more compact.

Avalanche photodiodes are less economical than PIN photodiodes, and the sensitivity of their multiplication factors to temperature variations requires additional electronics for temperature control. Their fast response times and sensitivity to very weak optical fields, however, make them attractive in applications such as high-bandwidth ($\gtrsim 1$ Gbits/s) fiber-optic communications (Section 15.6). PIN diodes have response times comparable to those of APDs and do not require the large bias voltages used in APDs.

Photodetection is never free of noise. There are many sources of noise, and the subject is too broad to treat in any detail here.[6] One important concept for photon-counting applications is *shot noise*, or noise associated with the discreteness of a quantity whose measurement is described by a Poisson distribution. The term is said to derive from the audible noise produced by gun shot fired on a target. An example of shot noise is the "photon noise" arising from the fact that the photon number of a field incident on a detector is not fixed at a certain value but fluctuates. This translates into a random fluctuation of the photoelectron count. If the incident photons produce an average of N_p photoelectron counts in a certain time interval τ, and if the photon distribution is Poissonian, then the rms deviation of the photoelectron count due to photon noise is $\sqrt{N_p}$.[7]

The discreteness of the electrons constituting dark current leads again to shot noise: If N_d is the number of dark counts in the time interval τ, then the rms deviation of the photoelectron count due to dark current in the interval τ is $\sqrt{N_d}$.

Another source of shot noise in photon counting is background radiation arising, for instance, from thermal sources. Background noise is especially important in the middle and far infrared, where there is substantial thermal radiation at room temperature (Problem 3.1). An average number N_b of counts produced by background light in the time interval τ gives an rms deviation $\sqrt{N_b}$ in the photoelectron count if we again assume Poisson statistics.

The process of absorbing a photon and creating a photoelectron or an electron–hole pair is, like the spontaneous emission of a photon, probabilistic and therefore also a source of noise. If an incident photon creates a photoelectron (or electron–hole pair) with probability η, and the incident photon flux is described by a Poisson distribution with mean photon number $\bar{n}$ in the time interval τ, then the rms deviation of the photoelectron number in the time interval τ is $\eta\bar{n}$.

We can derive an expression for the rms fluctuation of photocurrent due to shot noise in a PMT or PIN photodiode in the following way, assuming that the current fluctuations may be described in terms of electron pulses of duration τ and fluctuating electron number N. The mean current during each pulse in this simplified model is $\langle i \rangle = \langle N \rangle e/\tau$, and the variance is

$$\langle (i - \langle i \rangle)^2 \rangle = \langle \Delta i^2 \rangle = \left(\frac{e^2}{\tau^2} \right) \langle \Delta N^2 \rangle = \left(\frac{e^2}{\tau^2} \right) \langle N \rangle = \frac{e}{\tau} \langle i \rangle, \qquad (12.7.2)$$

---

[6]Various vendors provide excellent descriptions of operating and noise characteristics of photodetectors. Their websites are among the best sources for up-to-date information.

[7]This applies to any radiation that is well described by classical electromagnetic theory. It is possible to produce nonclassical *squeezed* states of light that exhibit photon noise below the shot level associated with Poisson statistics.

where we have invoked Poisson statistics in writing $\langle \Delta N^2 \rangle = \langle N \rangle$. The time interval τ may be identified with the inverse of the frequency bandwidth B. This is basically the shortest response time of the photodetector and is determined by charge diffusion and transit times as well as the RC time constant of the circuit. Taking $\tau = (2B)^{-1}$ in our simplistic model gives a well-known result of more rigorous analyses:

$$\Delta i_{\text{rms}} = \sqrt{\langle \Delta i^2 \rangle} = \sqrt{2e\langle i \rangle B} \quad \text{(shot noise)}. \tag{12.7.3}$$

This implies an electrical noise power $\Delta i_{\text{rms}}^2 R_L = 2e\langle i \rangle B R_L$, where R_L is the load resistance in the circuit. The mean current $\langle i \rangle$ here is the sum of the signal current i_S (proportional to the incident optical power) and the dark current i_D.

In the case of an APD with a gain or multiplication factor M (the average number of secondary charge carriers generated for each light-generated carrier), the signal current, for example, is $i_S = M\eta e(\text{Pwr}/h\nu)$, compared to $i_S = \eta e(\text{Pwr}/h\nu)$ for a PIN photodiode. The mean-square shot noise current would similarly be predicted to increase by a factor M^2 compared to that of a PIN photodiode. However, *gain always amplifies noise as well as signals*, and as a consequence of gain in an APD the mean-square shot noise current is actually increased by a factor $M^2 F(M)$, where $F(M)$ is the *excess noise factor*. The effect of $M^2 F(M)$ may be approximated as a multiplication of the mean-square shot noise current by a factor M^n, where n is typically between 2 and 3. M can be quite large, typically ~ 10 and ~ 100, respectively, in germanium and silicon photodiodes.

Johnson noise, first identified by J. B. Johnson in the 1920s, refers to the thermal fluctuations of charge carriers in any resistor at a finite temperature T; it is essentially a consequence of the Brownian motion of electrons in the resistor. The mean-square voltage noise in the frequency band $[\nu, \nu + B]$ for a resistance R_L at temperature T is, for $k_B T \gg h\nu$,

$$\langle V_N^2 \rangle = 4k_B T R_L B, \tag{12.7.4}$$

where k_B is Boltzmann's constant. The corresponding rms current fluctuation is

$$\Delta i_{\text{rms}} = \sqrt{\frac{\langle V_N^2 \rangle}{R_L^2}} = \sqrt{\frac{4k_B T}{R_L} B} \quad \text{(thermal noise)}. \tag{12.7.5}$$

This implies an electrical noise power $\Delta i_{\text{rms}}^2 R_L = 4k_B T B$, independent of R_L. These results for Johnson noise are applicable to APDs as well as PIN photodiodes.[8]

Although we are dealing here with ensemble averages, and assuming implicitly that the random processes are ergodic, it should of course be remembered that the noise currents associated with shot noise or Johnson noise in a physical system vary randomly in time. Fourier analysis of their time variations allows averages such as Δi_{rms} to be expressed as integrals of "power spectra" over frequency. The power spectra are generally relatively constant with frequency (in which case one refers to "white noise"), and therefore a measurement of Δi_{rms}^2, for example, will produce a value proportional to the

---

[8]Note the similarity to Eq. (6.13.20). Formulas for Johnson noise may be derived along essentially the same lines as our derivation of (6.13.20), and different derivations may be found, for instance, in textbooks on statistical physics.

frequency bandwidth involved in the measurement. This is why the average noise powers for both shot noise and Johnson noise are proportional to B.

Taking both shot noise and thermal noise into account, we have a signal-to-noise ratio for power (Pwr $= i^2R$):

$$\text{SNR} = \frac{i_S^2 R_L}{2e(i_S + i_D)BR_L + 4k_B TB} \tag{12.7.6}$$

for a PIN photodiode. In terms of the incident optical power Pwr and the responsivity $\mathcal{R}$, the signal current is $i_S = \eta e(\text{Pwr}/h\nu) = \mathcal{R}\text{Pwr}$ and the signal-to-noise ratio is

$$\text{SNR} = \frac{\mathcal{R}^2 R_L(\text{Pwr})^2}{2e\mathcal{R}BR_L(\text{Pwr}) + 2ei_D BR_L + 4k_B TB}, \tag{12.7.7}$$

The dependence of SNR on optical power obviously depends on the relative magnitudes of the signal-current shot noise, the dark-current shot noise, and the thermal noise. Equation (12.7.7) suggests that the signal-to-noise ratio can be increased by increasing the load resistance R_L. However, increasing R_L increases the response time $R_L C$, where C is the capacitance (typically in the picofarad range) characterizing the depletion layer, so that R_L cannot be increased without reducing the speed of response to a time-varying intensity.

For an APD (12.7.6) and (12.7.7) are modified simply by multiplying the numerators by M^2 and the shot-noise terms in the denominators by $M^2 F(M)$:

$$\text{SNR} = \frac{M^2 \mathcal{R}^2 R_L(\text{Pwr})^2}{2M^2 F(M)e\mathcal{R}BR_L(\text{Pwr}) + 2eM^2 F(M)i_D BR_L + 4k_B TB} \quad \text{(APD)}. \tag{12.7.8}$$

Because of the excess noise factor $F(M)$, the signal-to-noise ratio of an APD is usually less than that of a PIN photodiode when shot noise is dominant, e.g., when the signal is large. When thermal noise is dominant, however, the factor M^2 in the numerator implies that an APD will have a much greater SNR. Of course, the shot-noise terms in the denominator of (12.7.8) increase with increasing M, and in practice APDs are designed to have multiplication factors that optimize the signal-to-noise ratio.

If, as is often the case, the thermal noise is dominant, the signal-to-noise ratio is proportional to the square of the optical power:

$$\text{SNR} \cong \left(\frac{M^2 \mathcal{R}^2 R_L}{4k_B TB} \right)(\text{Pwr})^2, \tag{12.7.9}$$

with $M = 1$ for a PIN photodiode. Photodetection circuitry generally includes amplifiers and, as remarked earlier, noise as well as signals are amplified. The amplified noise can be accounted for by replacing the temperature T in the formulas for Johnson noise by an effective temperature $T_{\text{eff}} = F_n T$, where F_n is typically $\sim 2-6$.

The *minimum detectable signal* $(\text{Pwr})_{\min}$ is defined as the mean signal for which the SNR is unity, and (12.7.9) implies that $(\text{Pwr})_{\min} \propto \sqrt{B}$ when thermal noise dominates, or whenever the signal-current shot noise is negligible compared to either dark-current shot noise or thermal noise. If signal-current shot noise is much larger than dark-current

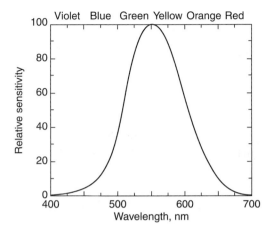

Figure 12.14 Relative sensitivity of the human eye as a function of wavelength for a "standard observer." (From D. Halliday and R. Resnick, *Physics*, Wiley, New York, 1978.)

shot noise and thermal noise, on the other hand, (12.7.8) implies that $(\text{Pwr})_{min} \propto B$. The *noise equivalent power* (NEP) of a photodetector is defined as the optical power needed to give SNR $= 1$ when $B = 1$ Hz.

• The human eye is a surprisingly good photon detector. The detector units on the retina are of two types, called rods and cones because of their shapes. The response of the rods and cones is different at different wavelengths. The cones are responsible for color vision and have a peak sensitivity at about 560 nm, while the rods are most sensitive at wavelengths around 510 nm.

The rods and cones also have different saturation properties. The rods saturate more easily and become relatively ineffective at high light levels, but are more sensitive at very low light levels. A comparatively high light level (and thus a strong contribution from the cone response) is necessary for color vision. The cones are more sensitive to longer wavelengths than the rods, as evidenced by the difficulty we have seeing red through dark-adapted eyes (Fig. 12.14). Dark-adapted eyes operate almost exclusively by rod response and are almost color blind.

The sensitivity of the dark-adapted eye was studied in the classic experiments of M. H. Pirenne and his colleagues in the early 1940s.[9] They found that the typical human retina can respond to < 10 photons around 510 nm, with an efficiency of about 60%. The overall detection efficiency of the eye is actually much less than this because 80–90% of the light incident upon the eye is lost before it can be absorbed by rod cells. (Of course, many animals have enormously more efficient and sensitive eyes than we do.)

The experiments revealed that a rod cell in the human retina can absorb a single photon. Observers were asked whether they saw a light flash when their dark-adapted eyes were exposed to a very weak pulse of light. From a series of such trials a "probability of seeing" was calculated. Reasoning that the mean number $\bar{n}$ of photons arriving at the retina was proportional to the light intensity I incident on the eye, and assuming Poisson statistics for n, Pirenne et al. assumed that the theoretical probability of seeing when at least θ photons arrive at the retina is

$$P_{\text{seeing}}(\theta) = \sum_{n=\theta}^{\infty} P(n) = \sum_{n=\theta}^{\infty} \frac{\bar{n}^n}{n!} e^{-\bar{n}}, \qquad (12.7.10)$$

[9] See F. Rieke and D. A. Baylor, *Reviews of Modern Physics* **70**, 1027 (1998), and references therein.

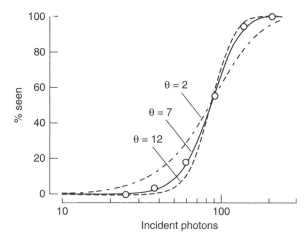

Figure 12.15 Theoretical probability of seeing [Eq. (12.7.11)] for different values of the threshold photon number θ, compared with the experimental seeing frequency for a single observer. Since the shapes of the different curves depend primarily on θ, each curve was shifted along the log I axis (see text). [From F. Rieke and D. A. Baylor, *Review Modern Physics* **70**, 1027 (1998).]

or, since $\bar{n} = \alpha I$, where α is a constant,

$$P_{\text{seeing}}(\theta) = \sum_{n=\theta}^{\infty} \frac{(\alpha I)^n}{n!} e^{-\alpha I}. \tag{12.7.11}$$

The (unknown) constant α depends among other things on the observer's age. However, when $P_{\text{seeing}}(\theta)$ is plotted vs. the logarithm of I, the *shape* of the curve depends on log I while different values of α result only in a translation of the curve along the x axis. Therefore, a comparison of Eq. (12.7.11) with the "experimentally" determined probability of seeing θ or more photons can be made without having to know the value of α. Figure 12.15 shows such a comparison. From such data Pirenne et al. concluded that the dark-adapted eye can detect 5–7 photons. Moreover, because the light arriving at the retina is spread over an area covering many rods, it follows that *a single rod cell can detect a single photon.*

The time constant of the human eye is about 0.1 s, and this is associated with the "persistence of vision" that makes movies possible: We do not notice the changing of frames if the rate is faster than about 15 frames per second. Moreover, the eye does not "integrate" a signal much beyond about 0.1 s. The experimental results of Pirenne and collaborators, for instance, were obtained with millisecond pulses of light. If the same total electromagnetic energy were incident upon the eye over a longer time, say 1 s, there would be no response because too few photons would be available over the 0.1-s integration period. ●

12.8 REMARKS

The photon concept is an integral part of the modern theory of light. In this chapter we have described, conceptually, some effects that cannot be understood without knowing that there are indivisible units (photons) of radiation energy and momentum. However, the wave concept is an equally important part of the modern theory of light. For practical applications it is sometimes more convenient to think in terms of light waves and sometimes in terms of photons. Because they involve enormously large numbers of photons, where the classical wave theory is quite adequate, most practical properties of lasers can be explained by classical wave theory together with the quantum theory of atoms, molecules, and solids.

We also described some general aspects of photon detection and counting, and have discussed the extent to which they can be interpreted using classical wave theory. As discussed in the following chapter, nonclassical properties of light are revealed only by optical measurements involving field correlation functions higher than "first order."

While there is no experiment that contradicts any predictions of the quantum theory of radiation, it continues to fascinate and not infrequently to surprise many scientists. Readers who find the photon concept difficult to comprehend—as well as those who find it easy—might do well to remember something Einstein said near the end of his life: "All these fifty years of conscious brooding have brought me no closer to the answer to the question 'what are light quanta?' Nowadays every rascal thinks he knows it, but he is mistaken."

PROBLEMS

12.1. (a) Estimate the average number of photons per second per square centimeter reaching Earth's surface from the sun.

(b) A laser pointer puts out 1 mW at a wavelength of 532 nm. How many photons are emitted per second?

(c) Suppose you are seated 10 m from a screen onto which a speaker shines his laser pointer. Assume that the screen scatters 20% of the radiation incident upon it, and that your eye's pupil has a diameter of 5 mm. What is the rate at which photons from the laser pointer enter your eye?

12.2. (a) Two polarizers are placed one over the other with their polarization axes orthogonal. Is there a nonzero probability that any photon will pass through both sheets?

(b) A third polarizer is inserted between the two polarizers in part (a), so that its axis makes an angle of $45°$ with respect to each of the other two. Is it possible for any photon to pass through all three sheets?

12.3. At points of constructive interference in the Young two-slit experiment, the intensity is twice the intensity calculated by adding the intensities associated with each individual slit. This does not violate the principle of conservation of energy. Indicate why not (a) qualitatively and (b) semiquantitatively.

12.4. Note that the Poisson distribution (12.6.4) has the property

$$\bar{n}\frac{\partial}{\partial\bar{n}}[e^{\bar{n}}P_n(T)] = nP_n(T)e^{\bar{n}}.$$

(a) Show therefore that $\langle n\rangle = \sum_{n=0}^{\infty} nP_n(T) = \bar{n}e^{-\bar{n}}\partial(e^{\bar{n}})/\partial\bar{n}$.

(b) Find a similar formula for $n^k P_n(T)e^{\bar{n}}$, and use it to verify (12.6.8) for $\langle n^2\rangle$ and to compute the analogous result for $\langle n^3\rangle$.

- (c) Assuming a Poissonian photon-counting distribution (12.6.3), what is the probability that at least one photon will be counted in a time interval T?
- (d) Show that the probability distribution for the time interval τ between two successive photon counts is $p(\tau) = \alpha I \exp(-\alpha I \tau)$.

12.5. (a) Assuming a PMT photocathode surface with a work function of 1.5 eV, estimate the reduction in dark current as the temperature is lowered from 300 to 260K.

- (b) Assuming a dark current of 10^4 electrons/s at room temperature, estimate the minimal power of the incident light signal required for the photoelectric cathode current to exceed this value, assuming the PMT has a quantum efficiency of 30%. What is the corresponding minimal power required at $T = 260$K?

12.6. (a) A direct current of 0.5 μA is displayed on an oscilloscope with bandwidth 1 MHz. Calculate the rms shot-noise current.

- (b) Consider a 50-Ω resistor at room temperature. Calculate the rms thermal noise current, the rms thermal noise voltage, and the thermal noise power in a bandwidth of 1 MHz.
- (c) A PIN photodiode has a responsivity of 1 A/W. Calculate the rms shot-noise current when light of power 1 μW is incident on the detector, assuming a detection bandwidth of 10 MHz.

13 COHERENCE

13.1 INTRODUCTION

Laser radiation can be both quantitatively and qualitatively different from ordinary radiation like that from the sun or a fluorescent lamp. There are the obvious differences, such as the very bright and nearly monochromatic nature of laser light, and its propagation as directed beams, but there are also subtle differences that distinguish laser radiation in other ways. For instance, if the light from the sun were filtered in such a way that only a single, quasi-monochromatic and unidirectional component of it remained, it could still be distinguished from laser radiation.

Of course, the obvious differences are very important, and some were reviewed briefly in Chapter 1. In many applications, for instance, it is only *brightness* that is needed. For this reason we will begin in the following section with a discussion of this aspect of laser radiation. The remainder of the chapter deals with *coherence* properties of radiation. After a careful consideration of the concept of coherence we can begin to appreciate the fundamental differences between lasers and other light sources.

13.2 BRIGHTNESS

Consider a thermal source of radiation. The radiation inside a blackbody cavity has a spectral energy density $\rho(\nu)$ given by (3.6.1). If we divide the frequency band into small finite elements $\delta\nu$, the intensity of radiation "at" frequency ν emitted by a blackbody is $I_\nu = \int_{\delta\nu} I(\nu)\,d\nu \approx I(\nu)\,\delta\nu$, or

$$I_\nu = \frac{1}{4}c\rho(\nu)\,\delta\nu = \frac{(2\pi h\nu^3/c^2)\,\delta\nu}{e^{h\nu/k_BT} - 1}.\qquad(13.2.1)$$

As discussed in Section 3.6, there are two reasons for the factor $\frac{1}{4}$, both stemming from the fact that an ideal blackbody emits radiation isotropically.

As an example of a thermal source of radiation, consider the sun. For wavelengths between about 10^2 and 10^7 nm, the solar spectrum is approximately that of a blackbody at $T \approx 6000K$ (Section 3.6). At the He–Ne laser wavelength $\lambda = 632.8$ nm, the intensity

Laser Physics. By Peter W. Milonni and Joseph H. Eberly
Copyright © 2010 John Wiley & Sons, Inc.

given by (13.2.1) for such a blackbody is

$$I_\nu = (1.14 \times 10^{-7} \text{ J/m}^2)\, \delta\nu. \qquad (13.2.2)$$

If we take $\delta\nu = 100$ MHz, then $I_\nu = 1.14 \text{ mW/cm}^2$.

Now a 632.8-nm He–Ne laser might have an output power Pwr $= 1$ mW and a Gaussian beam spot size $w_0 = 1$ mm. The peak intensity at the waist of the (lowest order) Gaussian beam in this case is (Problem 7.3) $I_{max} = 2\text{Pwr}/\pi w_0^2 = 64 \text{ mW/cm}^2$. The spectral width of the laser may be larger than the value $\delta\nu = 100$ MHz used in our example above, or it could be much smaller. The point is that even the low output powers of He–Ne lasers give higher radiation intensities, within a narrow spectral range, than the sun at its surface.

The total intensity radiated by a blackbody is obtained by integrating (13.2.1) over all frequencies, which gives the Stefan law (3.6.24). The total intensity of radiation from the sun is found to be (Problem 13.1)

$$I = 6.4 \times 10^7 \text{ J/m}^2/\text{s} = 6.4 \text{ kW/cm}^2 \quad \text{(sun)}. \qquad (13.2.3)$$

This intensity is beyond the range of (unfocused) He–Ne lasers, but it is easily exceeded by higher-power lasers, both cw and pulsed. Furthermore the laser radiation is all concentrated within a narrow frequency range, whereas (13.2.3) is a sum over all frequencies, distributed according to the Planck law.

• One convenient way to measure these differences in brightness between laser radiation and thermal radiation is to recall (Section 3.6) that the average number of photons per mode in a thermal (blackbody) field is $(e^{h\nu/k_B T} - 1)^{-1}$. For the solar temperature $T = 5800$K, this number (defined as the photon "degeneracy factor" in Section 13.12) is about 0.02. For a laser, on the other hand, this number can be enormously greater, as we will see in Section 13.12. The very first gas laser constructed, for instance, had a photon degeneracy factor of about 10^{12}. •

For many purposes, power and intensity are not adequate measures of "brightness." Instead, we define the *brightness* (or *radiance*) of a source as the emitted power per unit area per unit solid angle. This concept of brightness is useful in practical applications, especially when the radiation from a source is to be focused by a lens to increase its intensity. A fundamental theorem in optics states that *the brightness of a source is an invariant quantity*, unchangeable by a lens or any other passive optical system.[1] That is, the intensity of a light beam can be increased by focusing, but the brightness cannot.

An important aspect of brightness is that it is inversely proportional to the solid angle. The solid angle subtended by a beam is proportional to the square of the divergence angle θ (Fig. 13.1). For a Gaussian beam, for instance, the solid angle is

$$\Omega = \pi\theta^2 = \frac{\lambda^2}{\pi w_0^2} \qquad (13.2.4)$$

and is thus inversely proportional to the beam area (πw_0^2). Since brightness is power per unit area per unit solid angle, it is clear that the brightness of a Gaussian beam does not change as it propagates. Furthermore, since a Gaussian beam remains Gaussian under

---

[1]This is true provided the refractive indices of the object and image spaces are the same—a minor technical point that will not concern us.

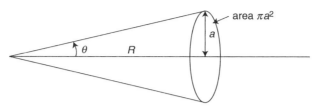

Figure 13.1   The solid angle Ω associated with a divergence angle θ is $\Omega \equiv \pi a^2 / R^2 = \pi \theta^2$ if $\theta = \tan^{-1}(a/R) \approx a/R$, i.e., if θ is small.

focusing by a lens, it is also clear that the brightness of a Gaussian beam cannot be changed by focusing it down to a smaller, more intense spot.

Laser beams have a high degree of directionality, that is, their divergence angles are small. Therefore their solid angles of divergence are also small, and consequently the brightness of a laser beam is very high. To see this, consider again the peak intensity of a (lowest-order) Gaussian beam at its waist (Problem 7.3):

$$I_{\max} = \frac{2\text{Pwr}}{\pi w_0^2},\tag{13.2.5}$$

where Pwr is the total power transported by the beam. From (13.2.4) it follows that the corresponding brightness is

$$B = \frac{I_{\max}}{\Omega} = \frac{2\text{Pwr}}{\lambda^2}.\tag{13.2.6}$$

For a He–Ne laser with Pwr $= 1\,$mW and $\lambda = 632.8\,$nm this brightness is 5×10^5 W/cm^2-sr. This is a modest brightness for lasers. A Q-switched laser, for instance, might have a brightness of 10^{12} W/cm^2-sr, and brightnesses many orders of magnitude larger than this are possible. Conventional light sources, even very powerful ones, have much lower brightnesses because their radiation lacks the directionality of laser beams. For example, the sun has a brightness of about $130\,$W/cm^2-sr at its surface; this is hundreds of times smaller than the brightness of He–Ne lasers, and trillions of times smaller than the brightnesses possible with high-power lasers.

In many applications laser radiation is focused to produce an intense spot. Brightness is extremely important in this respect because *the intensity that can be obtained in the focal plane of a lens is proportional to the brightness of the beam.*

Consider the focusing of a Gaussian beam. Without focusing, the peak intensity at the beam waist is given by (13.2.6). With a lens of focal length f we can focus the beam to a spot size (Problem 13.2)

$$w_f = \frac{\lambda f}{\pi w_0}\tag{13.2.7}$$

in the focal plane of the lens. The focused beam is still Gaussian, and so its peak intensity is given by (13.2.5) with w_0 replaced by w_f:

$$I_{\max}(f) = \frac{2\text{Pwr}}{\pi w_f^2} = \frac{2\text{Pwr}}{f^2}\frac{\pi w_0^2}{\lambda^2} = \frac{2\text{Pwr}}{f^2 \Omega},\tag{13.2.8}$$

where Ω is the solid-angle divergence of the unfocused beam. This result indicates that, for a beam of given area, the intensity that can be obtained by focusing is directly proportional to the beam brightness. It shows explicitly why the small divergence (i.e., directionality) of laser beams is so important for obtaining high intensities by beam focusing.

Consider again the example of a He–Ne laser with Pwr $= 1 \, \text{mW}$, $\lambda = 632.8 \, \text{nm}$, $w_0 = 1 \, \text{mm}$. From Table 7.1 we compute a divergence angle $\theta = 2 \times 10^{-4}$ rad, and therefore $\Omega = \pi \theta^2 = 1.3 \times 10^{-7}$ sr. Equation (13.2.8) gives $I_{\text{max}}(f) = 15 \, \text{kW/cm}^2$ for the peak intensity of a beam focused with a lens of focal length $f = 1 \, \text{cm}$. A laser with the same λ and w_0, but a power of 1 W, gives $I_{\text{max}}(f) = 15 \, \text{MW/cm}^2$. Such estimates of focal-spot intensities apply to the ideal case of perfect beam quality ($M^2 = 1$) and are therefore, as discussed below, somewhat high; but they serve to indicate the sorts of intensities possible even with lasers with modest output powers. The large intensities are a consequence of the low divergence angles (high brightness) of laser beams. Divergence angles of many lasers vary from a few tenths of millirads to 10 mrad; an ordinary flashlight, by contrast, might be characterized by a divergence angle $\sim 10°$. It is the brightness of laser beams, together of course with the powers achievable, that makes them useful in applications such as drilling and welding. It is even easy to vaporize metal surfaces with laser radiation, so that in laser welding special care must be taken to avoid vaporization.

An ordinary lamp emitting a power P in all directions may be shown to give an intensity

$$I \approx \frac{\text{Pwr}}{f^2} \qquad (13.2.9)$$

in the focal plane of a lens. This differs from (13.2.8) by the absence of Ω in the denominator, which is very small for a laser beam. Consequently, lamps would have to emit tens or hundreds of thousands of watts to match the intensities achievable by focusing low-power milliwatt He–Ne lasers.

13.3 THE COHERENCE OF LIGHT

The essential features of light coherence are displayed in the Young two-slit interference experiment as shown in Fig. 13.2. Light from a source S is incident upon a screen containing two narrow slits, S_1 and S_2. At a second screen, a distance L away from the first screen, we observe the intensity distribution of the light emerging from the two slits. For some sources we see interference fringes on the second screen, i.e., the intensity is not simply the sum of the intensities associated with each slit. In such cases we say that the radiation has a certain *coherence*, i.e., the ability to form fringes.

An elementary approach to the explanation of the interference fringes is to assume both that the source emits monochromatic radiation, and that the slit separation d is much smaller than the screen separation L. At a point P on the observation screen the path difference from the two slits is (Fig. 13.3)

$$l = d \sin \theta \approx d \frac{Y}{L} \qquad (13.3.1)$$

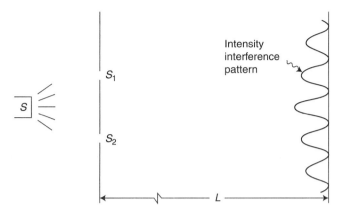

Figure 13.2 The Young two-slit interference experiment.

if the two slits subtend a small angle θ at P. Since $\omega/c = 2\pi/\lambda$, the intensity has its maxima at points P for which this optical path difference is an integral number of wavelengths, i.e., $\omega(l_2 - l_1)/c = n \times 2\pi$. This condition gives

$$Y_n^{(\text{max})} = n\frac{\lambda L}{d}, \quad n = 0, 1, 2, \ldots \tag{13.3.2a}$$

for the location of the intensity maxima on the observation screen. There is destructive interference of the light from the two slits if the path difference is an odd integral number of half wavelengths, i.e., $\omega(l_2 - l_1)/c = (n + \frac{1}{2}) \times 2\pi$. At the points where y has one of the values

$$Y_n^{(\text{min})} = \left(n + \frac{1}{2}\right)\frac{\lambda L}{d}, \quad n = 0, 1, 2, \ldots, \tag{13.3.2b}$$

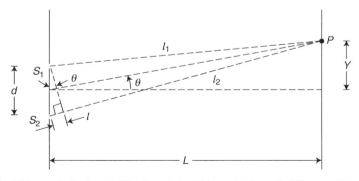

Figure 13.3 If the angle θ subtended by S_1 and S_2 at P is small, the path difference l is approximately dY/L.

therefore, there are minima in the intensity pattern on the observation screen. For optical wavelengths the separation

$$\Delta Y = \lambda L/d \qquad (13.3.3)$$

between intensity maxima (minima) on the observation screen is quite small even for very closely spaced slits. For example, for $\lambda = 600$ nm, $d = 0.1$ mm, and $L = 20$ cm, we compute $\Delta Y = 1.2$ mm.

However, a real experiment will lack some or all of the ideal features just employed, and so a real interference pattern will have features not predicted by our discussion so far. For example, in a real experiment performed with an ordinary source of light, such as a mercury arc lamp, or a sodium lamp, or some other nonlaser source of light, the maxima and minima are accurately located by Eqs. (13.3.2a), but as we increase d the interference pattern becomes less sharp, and beyond a certain value of d the interference fringes disappear altogether. A similar effect is observed when we hold d fixed and bring the observation screen closer to the slits. Then the interference fringes fade out and the intensity at any point P on the observation screen becomes simply the sum of the intensities associated with the two slits individually.

In addition, a real source of radiation will not be perfectly monochromatic. If the source emits radiation with a spread $\delta\lambda$ in wavelengths about λ, interference maxima of one wavelength may coincide with minima of another, causing the interference pattern to be washed out. The difference in fringe separation for two wavelengths separated by $\delta\lambda$ is, from (13.3.3), $\delta(\Delta Y) = L\delta\lambda/d$. The washing out of the interference pattern when the source is not perfectly monochromatic is therefore minimal if

$$\frac{\delta(\Delta Y)}{\Delta Y} = \frac{L\delta\lambda/d}{L\lambda/d} = \frac{\delta\lambda}{\lambda} = \frac{\delta\nu}{\nu} \ll 1. \qquad (13.3.4)$$

Radiation with bandwidth $\delta\nu \ll \nu$ is said to be *quasi-monochromatic*.

Because interference fringes are never ideally sharp, it is convenient to have a way to characterize the sharpness of the interference fringes. The quantity defined for this purpose is called *visibility*, defined by

$$V = \frac{I_{\max} - I_{\min}}{I_{\max} + I_{\min}}. \qquad (13.3.5)$$

Here $I_{\max}$ and $I_{\min}$ are respectively the maximum and minimum intensities on the observation screen. If at any point there is complete destructive interference, so that $I_{\min} = 0$, it follows from (13.3.5) that the visibility $V = 1$, its maximum possible value. If, on the other hand, there is no interference pattern, in which case the fields from the slits add *incoherently*, then $I_{\min} = I_{\max}$ and therefore $V = 0$. The visibility thus provides a quantitative measure of the sharpness of interference fringes. In this way, visibility is a measure of the *coherence* of light.

Our example of a real experiment indicates that coherence depends on the experimental situation (e.g., on the slit separation d, distance L, and bandwidth $\delta\lambda$). If the light source in our experiment were a laser instead of a thermal source, we would typically find high fringe visibilities for considerably larger slit separations. We say, somewhat loosely, that the laser radiation is much more coherent than thermal radiation.

In order to understand the results of the two-slit experiment for different sources, we must next consider more carefully what it is that the experiment actually measures. This will also lead us to a deeper appreciation of the concept of coherence.

13.4 THE MUTUAL COHERENCE FUNCTION

Suppose a quasi-monochromatic field is incident upon the screen containing the two slits in Fig. 13.3, and that the field is uniform over each slit. Suppose that the width of each slit is a. In order to have any interference pattern on the observation screen, the slits must cause enough diffraction that the field from each slit has a transverse spread much larger than ΔY [Eq. (13.3.3)]. Since a diffraction angle $\theta \sim \lambda/a$ is associated with a slit of width a (recall Section 7.11), we require that

$$\theta L = \frac{\lambda L}{a} \gg \Delta Y = \frac{\lambda L}{d}, \tag{13.4.1}$$

or $a \ll d$. Provided the slits are much narrower than their separation, therefore, the fields from the slits will be diffracted enough to produce interference fringes on the screen. We will assume this condition is satisfied.

If we consider a quasi-monochromatic field rather than a perfectly monochromatic one, it is no longer appropriate to write

$$E(\mathbf{r}, t) = \mathcal{E}(\mathbf{r})e^{-i\omega t} \tag{13.4.2}$$

for the complex electric field [cf. Eq. (7.4.2)]. If we have a field with N distinct frequency components, we might write instead

$$E(\mathbf{r}, t) = \sum_m \mathcal{E}_m(\mathbf{r})e^{-i\omega_m t}. \tag{13.4.3}$$

In the case $N = 1$, we recover (13.4.2) with $\mathcal{E}(\mathbf{r}) = \mathcal{E}_1(\mathbf{r})$. More generally, we may be dealing with a field having a continuous distribution of frequencies, in which case we replace (13.4.3) by the so-called *analytic signal*[2]

$$E(\mathbf{r}, t) = \int_0^\infty \tilde{\mathcal{E}}(\mathbf{r}, \omega)e^{-i\omega t}\, d\omega, \tag{13.4.4}$$

where $\tilde{\mathcal{E}}(\mathbf{r}, \omega)$ is the Fourier transform of the real physical field:

$$\tilde{\mathcal{E}}(\mathbf{r}, \omega) = \frac{1}{2\pi}\int_{-\infty}^\infty [E(\mathbf{r}, t) + E^*(\mathbf{r}, t)]e^{i\omega t}\, dt. \tag{13.4.5}$$

In practice, instead of (13.4.3) or (13.4.4), it is often more useful to write

$$E(\mathbf{r}, t) = \mathcal{E}(\mathbf{r}, t)e^{-i\omega t}, \tag{13.4.6}$$

where the complex amplitude $\mathcal{E}(\mathbf{r}, t)$ varies very slowly in time.

---

[2]In optical coherence theory, $\frac{1}{2}E(\mathbf{r}, t)$ is called the *analytic signal* associated with the field $\mathrm{Re}[E(\mathbf{r}, t)]$. It is usually denoted by $V(\mathbf{r}, t)$.

All of formulas (13.4.3)–(13.4.6) indicate possible generalizations of (13.4.2) and they all indicate that the complex field E is to be associated with the *positive frequency part* of the real field. This is seen quite clearly in (13.4.4), where the frequency integral is over the range 0 to ∞. [It is unfortunately firmly conventional that the *positive* frequency part of the field is the one that goes with the exponential $e^{-i\omega t}$ having the *negative* sign.]

We will consistently use (13.4.6) to denote a quasi-monochromatic field. Given the identity

$$\text{Re}[E(\mathbf{r}, t)] = \frac{1}{2}[E(\mathbf{r}, t) + E^*(\mathbf{r}, t)], \qquad (13.4.7)$$

we can then write the intensity in terms of the positive and negative frequency parts of the field as follows:

$$I(\mathbf{r}, t) = c\epsilon_0[\text{Re } E(\mathbf{r}, t)]^2 = \frac{c\epsilon_0}{4}[E^2(\mathbf{r}, t) + E^{*2}(\mathbf{r}, t) + 2E^*(\mathbf{r}, t)E(\mathbf{r}, t)]. \qquad (13.4.8)$$

In the special case of a purely monochromatic field, for instance, $E(\mathbf{r}, t)$ is given by (13.4.2), so that

$$I(\mathbf{r}, t) = \frac{c\epsilon_0}{4}[\mathcal{E}^2(\mathbf{r})e^{-2i\omega t} + \mathcal{E}^{*2}(\mathbf{r})e^{2i\omega t} + 2|\mathcal{E}(\mathbf{r})|^2]. \qquad (13.4.9)$$

For optical fields, ω is on the order of 10^{15} sec^{-1}, which means that the first two terms in (13.4.9) oscillate sinusoidally with a period far shorter than the resolving time of any available detector. That is, no available detector will be able to follow the rapid oscillations at frequency 2ω in (13.4.9). What is measured is an average of (13.4.9) over many optical periods. Since the first two terms of (13.4.9) average to zero over an optical period, we ignore them and write the measurable intensity as

$$I(\mathbf{r}, t) = \frac{c\epsilon_0}{2}|\mathcal{E}(\mathbf{r})|^2. \qquad (13.4.10)$$

In the quasi-monochromatic case, the first two terms in (13.4.8) execute almost purely sinusoidal oscillations which likewise average to zero over any realistic measurement time. In other words, in general we can write the measurable intensity as

$$I(\mathbf{r}, t) = \frac{c\epsilon_0}{2}|E(\mathbf{r}, t)|^2. \qquad (13.4.11)$$

In the two-slit interference experiment, the field $E(P, t)$ at the point P on the observation screen is the sum of the fields diffracted by the slits. The field $E(S_1, t)$ will give rise to the field

$$E_1(P, t) = K_1 E\left(S_1, t - \frac{l_1}{c}\right) \qquad (13.4.12)$$

at the point P. The *retardation time* l_1/c is just the time it takes for light to propagate from S_1 to P (Fig. 13.3). K_1 is a function of the distance l_1 and of other geometrical details of the particular experimental arrangement. It may be derived from diffraction theory, but for our purposes its precise form is unnecessary; it will be convenient, however, to know that K_1 (and K_2 below) is a pure dimensionless imaginary number, i.e., $K = -K_1^*$. This property of K_1 and K_2 is related to the factor of i used in our statement of Huygens' principle, Eq. (7.10.1). In any case (13.4.12) has an intuitively reasonable form. It says simply that the field $E_1(P, t)$ at time t is due to diffraction of the field that was incident on S_1 at the earlier time $t - l_1/c$.

The field $E(P, t)$ at P due to both slits is then

$$E(P, t) = E_1(P, t) + E_2(P, t) = K_1 E\left(S_1, t - \frac{l_1}{c}\right) + K_2 E\left(S_2, t - \frac{l_2}{c}\right). \quad (13.4.13)$$

Using this result in (13.4.11), we have

$$I(P, t) = \frac{c\epsilon_0}{2}\left[\left|K_1 E\left(S_1, t - \frac{l_1}{c}\right)\right|^2 + \left|K_2 E\left(S_2, t - \frac{l_2}{c}\right)\right|^2\right.$$
$$\left. + c\epsilon_0|K_1 K_2|\mathrm{Re}\left[E^*\left(S_1, t - \frac{l_1}{c}\right)E\left(S_2, t - \frac{l_2}{c}\right)\right]\right] \quad (13.4.14)$$

for the intensity at point P on the observation screen, averaged over a few optical periods.

In addition to the fluctuations of intensity due to the regular but very rapid time variation of the factors $e^{\pm 2i\omega t}$ that we have discarded, every light field is subject to small irregular fluctuations arising from a variety of causes. One fundamental source of such fluctuations is the (necessarily random) spontaneous emission component of every light beam. Other less fundamental but generally much more important sources include fluctuations in the atmosphere and mechanical vibrations of optical elements in the path of the beam.

To account for the influence of these fluctuations in a detailed way would be impossible. Fortunately, it is satisfactory to assume that these fluctuations can be treated in an average sense. Recall that we adopted such a point of view in treating the effect of collisions on a Lorentz oscillator [Section 3.8] when we assumed that, on average, an oscillator has zero displacement and velocity immediately after every collision. On this basis we developed equations of motion for the average oscillator rather than trying to account for the details of the collisional history of an individual atom.

For the same reasons we will now assume that the average light field is representative of the collection of all possible light fields compatible with the fluctuations mentioned. This imaginary collection of light fields can be termed a statistical *ensemble* of light fields, and we expect that observable properties of light fields can be associated with averages over this collection, so-called ensemble averages. We will denote an ensemble average by angular brackets, so the average intensity will be written $\langle I \rangle$. Thus, following (13.4.11), we have

$$\langle I(P, t)\rangle = \frac{c\epsilon_0}{2}\langle E(P, t)E^*(P, t)\rangle \quad (13.4.15)$$

and so on. Note that the ensemble average is not the same as a time average. In particular, the ensemble average may be time-dependent, and a time average by definition could not be. Of course, there can arise situations in which a light signal is detected (by counting photons, say) over a long time period $2T$, not instantaneously at time t. In this case the time-dependent ensemble average must be further time-averaged in order to correspond to the measuring process and we can add a bar above the symbol to denote this:

$$\langle \bar{I}(P, t) \rangle = \frac{1}{2T} \int_{-T}^{T} \langle I(P, t + t') \rangle \, dt'. \tag{13.4.16}$$

Upon averaging both sides of (13.4.14) over the field fluctuations, we obtain

$$\langle I(P, t) \rangle = \frac{c\epsilon_0}{2} |K_1|^2 \left\langle \left| E\left(S_1, t - \frac{l_1}{c}\right) \right|^2 \right\rangle + \frac{c\epsilon_0}{2} |K_2|^2 \left\langle \left| E\left(S_2, t - \frac{l_2}{c}\right) \right|^2 \right\rangle$$

$$+ c\epsilon_0 |K_1 K_2| \mathrm{Re}\left\langle E^*\left(S_1, t - \frac{l_1}{c}\right) E\left(S_2, t - \frac{l_2}{c}\right) \right\rangle. \tag{13.4.17}$$

The function

$$\Gamma(\mathbf{r}_1, t_1; \mathbf{r}_2, t_2) = \langle E^*(\mathbf{r}_1, t_1) E(\mathbf{r}_2, t_2) \rangle \tag{13.4.18}$$

appearing in (13.4.17) is called the *mutual coherence function* of the fields at $\mathbf{r}_1, t_1$ and $\mathbf{r}_2, t_2$. It is also called the two-point function or autocorrelation function of the electric field. In terms of the mutual coherence function, we may write (13.4.17) as

$$\langle I(P, t) \rangle = \langle I_1(P, t) \rangle + \langle I_2(P, t) \rangle$$

$$+ c\epsilon_0 |K_1 K_2| \mathrm{Re}\left[\Gamma\left(S_1, t - \frac{l_1}{c}; S_2, t - \frac{l_2}{c}\right) \right], \tag{13.4.19}$$

where

$$\langle I_i(P, t) \rangle = \frac{c\epsilon_0}{2} |K_i|^2 \left\langle \left| E\left(S_i, t - \frac{l_i}{c}\right) \right|^2 \right\rangle = |K_i|^2 \left\langle I\left(S_i, t - \frac{l_i}{c}\right) \right\rangle$$

$$(i = 1, 2), \tag{13.4.20}$$

is the intensity that would be measured at P if slit S_i were acting alone, i.e., if the other slit were closed. The intensity (13.4.19) is not just the sum of the intensities I_1 and I_2 associated with each slit alone, unless the mutual coherence function vanishes. We see, therefore, that the mutual coherence function is intimately connected with the ability of the fields to produce interference fringes, i.e., to act coherently.

The definition of the mutual coherence function is the principal result of this section. In the following sections we will use this important quantity to discuss the concepts of spatial coherence and temporal coherence.

13.5   COMPLEX DEGREE OF COHERENCE

We will find it convenient to have a normalized form of the mutual coherence function, which we can connect to measures of coherence such as visibility. The term for the standard normalized measure is the *complex degree of coherence*.

We will begin by considering sources of radiation that have reached a more or less "steady-state" operation after they have been turned on. The radiation from such sources may be assumed in most instances to have a property called *stationarity*.

A stationary field has the property that the mutual coherence function $\Gamma(\mathbf{r}_1, t_1; \mathbf{r}_2; t_2)$ depends on t_1 and t_2 only through the difference $t_2 - t_1$. This is sometimes expressed by saying that the mutual coherence function for a stationary field is independent of the origin of time. For instance, for a stationary field, for any time increment s,

$$\Gamma(\mathbf{r}_1, t_1; \mathbf{r}_2, t_2) = \Gamma(\mathbf{r}_1, t_1 + s; \mathbf{r}_2; t_2 + s), \tag{13.5.1}$$

and so we can shorten the notation and write

$$\Gamma(\mathbf{r}_1, t_1; \mathbf{r}_2, t_2) = \Gamma(\mathbf{r}_1, \mathbf{r}_2, t_2 - t_1) = \Gamma(\mathbf{r}_1, \mathbf{r}_2, \tau) \tag{13.5.2}$$

for stationary fields, where $\tau = t_2 - t_1$.

Note that, since

$$\langle I(P, t) \rangle = \frac{c\epsilon_0}{2} \langle E^*(P, t)E(P, t) \rangle = \frac{c\epsilon_0}{2} \Gamma(P, t; P, t), \tag{13.5.3}$$

the time independence of the measured intensity of a stationary field follows from the property (13.5.1) of the mutual coherence function, i.e.,

$$\langle I(P, t) \rangle = \langle I(P) \rangle = \frac{c\epsilon_0}{2} \Gamma(P, P, 0) \tag{13.5.4}$$

in the notation (13.5.2). Equation (13.5.1) may therefore be considered as the defining characteristic of a stationary field.

A trivial example of a stationary field is a perfectly monochromatic field. To see this, use (13.4.2) in (13.4.18):

$$\Gamma(\mathbf{r}_1, t_1; \mathbf{r}_2; t_2) = \langle E^*(\mathbf{r}_1, t_1)E(\mathbf{r}_2, t_2) \rangle = \langle \mathcal{E}^*(\mathbf{r}_1)\mathcal{E}(\mathbf{r}_2)e^{-i\omega(t_2-t_1)} \rangle$$

$$= \langle \mathcal{E}^*(\mathbf{r}_1)\mathcal{E}(\mathbf{r}_2) \rangle e^{-i\omega\tau}. \tag{13.5.5}$$

Thus the mutual coherence function depends on t_1 and t_2 only through the difference $t_2 - t_1 = \tau$, and so a monochromatic field is stationary.

In the two-slit experiment, the intensity (13.4.19) becomes

$$\langle I(P, t) \rangle = I_1 + I_2 + c\epsilon_0 |K_1 K_2| \mathrm{Re} \left[\Gamma \left(S_1, S_2, \frac{l}{c} \right) \right], \quad l = l_2 - l_1, \tag{13.5.6}$$

when a stationary field is incident on the slits. Here

$$I_i = \langle I_i(P) \rangle = |K_i|^2 \langle I(S_i) \rangle \tag{13.5.7}$$

is the intensity associated with slit S_i alone, and is independent of time because the source is stationary. The intensity (13.5.6) is also time-independent.

In the case of a monochromatic field, the mutual coherence function is given by (13.5.5), and so

$$\langle I(P) \rangle = I_1 + I_2 + c\epsilon_0 |K_1 K_2| \text{Re} \left[\mathcal{E}^*(S_1)\mathcal{E}(S_2)e^{-i\omega l/c} \right]. \tag{13.5.8}$$

Suppose that $\mathcal{E}(S_1) = \mathcal{E}_0$ and $\mathcal{E}(S_2) = \mathcal{E}_0 e^{-i\Phi}$. This means that the fields at the slits differ only by a constant phase term. Then

$$c\epsilon_0 |K_1 K_2| \text{Re} \left[\mathcal{E}^*(S_1)\mathcal{E}(S_2)e^{-i\omega l/c} \right] = c\epsilon_0 |K_1 K_2 \mathcal{E}_0^2| \cos\left(\frac{\omega l}{c} + \Phi \right)$$

$$= 2\left(\frac{c\epsilon_0}{2} |K_1 \mathcal{E}_0|^2 \frac{c\epsilon_0}{2} |K_1 \mathcal{E}_0|^2 \right)^{1/2} \cos\left(\frac{2\pi l}{\lambda} + \Phi \right)$$

$$= 2(I_1 I_2)^{1/2} \cos\left(\frac{2\pi l}{\lambda} + \Phi \right) \tag{13.5.9}$$

and therefore

$$\langle I(P) \rangle = I_1 + I_2 + 2(I_1 I_2)^{1/2} \cos\left(\frac{2\pi l}{\lambda} + \Phi \right). \tag{13.5.10}$$

If $\Phi = 0$, there is constructive interference at point P if $\cos(2\pi l/\lambda) = 1$, i.e., if the path difference l in Fig. 13.3 is an integral number of wavelengths. Similarly, there is destructive interference at points on the observation screen where l is an odd integral number of half-wavelengths. We have merely justified our assumptions leading to (13.3.2a) and (13.3.2b).

If $\Phi \neq 0$, (13.5.10) implies there is constructive interference at points P where $2\pi l/\lambda + \Phi = 2\pi n$, or

$$l = n\lambda + \lambda(\Phi/2\pi), \quad n = 1, 2, \ldots . \tag{13.5.11}$$

If the two slits subtend a small angle at P, then (13.5.11) leads to

$$Y_n^{(\text{max})} = n\frac{\lambda L}{d} + \frac{\Phi}{2\pi}\frac{\lambda L}{d}, \quad n = 0, 1, 2, \ldots, \tag{13.5.12a}$$

or

$$Y_n^{(\text{min})} = \left(n + \frac{1}{2} \right)\frac{\lambda L}{d} + \frac{\Phi}{2\pi}\frac{\lambda L}{d}, \quad n = 0, 1, 2, \ldots \tag{13.5.12b}$$

in place of (13.3.2a) and (13.3.2b). These equations indicate that a phase difference Φ has the effect of shifting the positions of the intensity maxima and minima by the amount $\Phi\lambda L/2\pi d$. The overall interference pattern, though shifted upwards or downwards (depending on the sign of Φ), is otherwise basically unchanged. In particular, the separation Δ between intensity maxima (and minima) is the same as the value (13.3.3) for the case $\Phi = 0$.

The intensity on the observation screen in the two-slit experiment is given by (13.5.6) whenever the field is stationary. It is convenient to write this as

$$\langle I(P) \rangle = I_1 + I_2 + 2\sqrt{I_1 I_2}\, \mathrm{Re}[\gamma(S_1, S_2, l/c)], \tag{13.5.13}$$

where the dimensionless complex number γ is defined by

$$\gamma(S_1, S_2, l/c) = \frac{(c\epsilon_0/2)|K_1 K_2|\Gamma(S_1, S_2, l/c)}{\sqrt{I_1 I_2}} = \frac{(c\epsilon_0/2)|K_1 K_2|\Gamma(S_1, S_2, l/c)}{\sqrt{|K_1|^2\langle I(S_1)\rangle \cdot |K_2|^2\langle I(S_2)\rangle}}$$

$$= \frac{(c\epsilon_0/2)\Gamma(S_1, S_2, l/c)}{\sqrt{\langle I(S_1)\rangle\langle I(S_2)\rangle}}, \tag{13.5.14}$$

or in general

$$\gamma(\mathbf{r}_1, \mathbf{r}_2, \tau) = \frac{\Gamma(\mathbf{r}_1, \mathbf{r}_1, \tau)}{\sqrt{\Gamma(\mathbf{r}_1, \mathbf{r}_2, 0)\Gamma(\mathbf{r}_2, \mathbf{r}_2, 0)}}. \tag{13.5.15}$$

$\gamma(\mathbf{r}_1, \mathbf{r}_2, \tau)$ is called the *complex degree of coherence*. As we will now show, it is intimately related to the fringe visibility V defined by Eq. (13.3.5).

13.6 QUASI-MONOCHROMATIC FIELDS AND VISIBILITY

In the case of a purely monochromatic field, where $\Gamma(\mathbf{r}_1, \mathbf{r}_2, \tau)$ is given by (13.5.5), the complex degree of coherence is simply

$$\gamma(\mathbf{r}_1, \mathbf{r}_2, \tau) = e^{-i\omega\tau}. \tag{13.6.1}$$

For quasi-monochromatic light, we may assume that

$$\gamma = |\gamma(\mathbf{r}_1, \mathbf{r}_2, \tau)|e^{-i\Phi}e^{-i\omega\tau}, \tag{13.6.2}$$

where ω is the central frequency and $|\gamma(\mathbf{r}_1, \mathbf{r}_2, \tau)|$ is a slowly varying function of τ compared with $e^{-i\omega\tau}$. In this case, (13.5.13) becomes

$$\langle I(P) \rangle = I_1 + I_2 + 2\sqrt{I_1 I_2}\left|\gamma\left(S_1, S_2, \frac{l}{c}\right)\right|\cos\left(\frac{2\pi l}{\lambda} + \Phi\right). \tag{13.6.3}$$

Consider a region around P much larger than a wavelength. Over this region the factor $\cos(2\pi l/\lambda)$ varies rapidly between -1 and $+1$ as l is varied, whereas the (slowly varying) factor $|\gamma(S_1, S_2, l/c)|$ is practically unchanged. Thus the maximum and minimum intensities in the neighborhood of P are

$$I_{max} = I_1 + I_2 + 2\sqrt{I_1 I_2}|\gamma| \quad \text{and} \quad I_{min} = I_1 + I_2 - 2\sqrt{I_1 I_2}|\gamma|, \tag{13.6.4}$$

and so the fringe visibility (13.3.5) is

$$V = \frac{2\sqrt{I_1 I_2}|\gamma|}{I_1 + I_2}. \tag{13.6.5}$$

The modulus γ of the complex degree of coherence is thus a direct measure of the fringe visibility.

In the special case $I_1 = I_2$, the visibility and the modulus of the complex degree of coherence are identical:

$$V = |\gamma|. \tag{13.6.6}$$

From the definition of the visibility, it is clear that $0 \leq V \leq 1$ in general, and therefore also that[3]

$$0 \leq |\gamma| \leq 1. \tag{13.6.7}$$

When $|\gamma| = 1$ or 0, we have complete coherence or incoherence, respectively. When $0 < |\gamma| < 1$, the light is said to be *partially coherent*.

In many cases, it is useful to have a specific model of Γ that exhibits the commonly observed property that Γ decreases as the separation $\tau = |t_2 - t_1|$ increases. For one such model of a quasi-monochromatic laser field with frequency ω_L,

$$\Gamma = \langle \mathcal{E}_0^*(\mathbf{r}_1)\mathcal{E}_0(\mathbf{r}_2)\rangle e^{-|\tau/\tau_{\rm coh}|}e^{-i\omega_L\tau}, \tag{13.6.8}$$

where $\tau_{\rm coh}$ is called the field's correlation time. Obviously, the field is poorly correlated with itself (Γ is small) over time displacements greater than $|t_2 - t_1| \approx \tau_{\rm coh}$. According to the Wiener–Khintchine theorem, the (stationary) field's spectrum $S(\omega)$ can be defined as the Fourier transform of its autocorrelation function, so that in this case we can easily determine

$$S(\omega) = 2\pi\langle \mathcal{E}_0^*(\mathbf{r}_1)\mathcal{E}_0(\mathbf{r}_2)\rangle \frac{1/\pi\tau_{\rm coh}}{(\omega - \omega_L)^2 + (1/\tau_{\rm coh})^2}. \tag{13.6.9}$$

The laser spectrum is thus predicted to be a smoothly peaked function centered at ω_L, with a half-width $1/\tau_{\rm coh}$. The spectral lineshape of most lasers is *not* Lorentzian, but the other features of this model are satisfactory, especially the identification of the spectral linewidth with the inverse coherence time of the light field (recall Section 1.2).

It is important to note that for a monochromatic field

$$|\gamma| = 1 \quad \text{(monochromatic field)}, \tag{13.6.10}$$

which follows trivially from (13.6.1). Therefore, the idealized monochromatic field always gives the maximum possible fringe visibility. However, nonmonochromatic radiation can also satisfy the condition $|\gamma| = 1$ for complete coherence.

The mutual coherence function $\Gamma(\mathbf{r}_1, \mathbf{r}_2, \tau)$ is the average of the product of $E^*(\mathbf{r}_1, t)$ and $E(\mathbf{r}_2, t + \tau)$. During the measurement time over which the average is taken, the fields at

---

[3]This general property of γ may be derived from the definition (13.5.14) and the Schwarz inequality.

$\mathbf{r}_1$, t and $\mathbf{r}_2$, $t + \tau$ may undergo rapid fluctuations. This is to be expected, because the fields are due to a large number of individual radiators (atoms and molecules) that themselves fluctuate due to collisions, thermal motion, etc. The functions Γ and γ characterize the degree to which the fields at $\mathbf{r}_1$, t and $\mathbf{r}_2$, $t + \tau$ are correlated, or able to produce interference fringes, in spite of these fluctuations.

The total electric field is always the sum of the fields from all the individual sources. The intensity is proportional to the square of the total field, and therefore has contributions arising from the interference of the fields from different sources. If we had a hypothetical detector that could respond *instantaneously* to the field fluctuations, we would always measure interference fringes. Usually the interference terms fluctuate too rapidly to be observed in a realistic measurement time. In other words, whether the fields at $\mathbf{r}_1$, t and $\mathbf{r}_2$, $t + \tau$ exhibit any mutual coherence depends not only on the intrinsic properties of the fields, or their sources, but also on what we measure.

13.7 SPATIAL COHERENCE OF LIGHT FROM ORDINARY SOURCES

We have seen two examples in which a phase shift Φ appears in an expression for registered intensity in (13.5.10) and (13.6.3). We can think of an ordinary light source as one that introduces a different Φ from each of its infinitely many infinitesimal radiating units. Such a source is completely incoherent in this sense, that it has no phase regularity at all over its surface. It should be obvious that such a source cannot give rise to a field having an intensity function with regularities such as interference fringes that can be seen on an observing screen. Nevertheless, such a field can be coherent in a more subtle way that we will now discuss. The essential element, as we will see, is the presence of a substantial distance between the source and points on the observation screen. Then two different points ($\mathbf{r}_1$ and $\mathbf{r}_2$) on the screen will each be receiving more or less the same jumble of randomly phased fields from the source, and will therefore have nearly the same character. This makes the received fields at points $\mathbf{r}_1$ and $\mathbf{r}_2$ mutually coherent, although not individually coherent in the sense of contributing to a fringe pattern.

While the detected intensity in the scenario just described will be unpatterned on the observation screen, we recognize that it is a single-point measure, point by point on the screen. The mutual coherence function, on the other hand, deals with fields at two different spatial points ($\mathbf{r}_1$ and $\mathbf{r}_2$) *and* at two different times (t_1 and t_2). To isolate the spatial characteristics of the mutual coherence function, we take $t_1 = t_2 = t$; then $\Gamma(\mathbf{r}_1, t; \mathbf{r}_2, t)$ determines the mutual coherence of the fields at two different points in space at the same time. In this case we speak of *spatial coherence*. For stationary sources, spatial coherence is characterized by the mutual coherence function $\Gamma(\mathbf{r}_1, \mathbf{r}_2, 0)$ and the complex degree of coherence $\gamma(\mathbf{r}_1, \mathbf{r}_2, 0)$. This is the case of most practical interest, and so we will henceforth always assume stationarity.

To test experimentally for the spatial coherence of the light on the screen, we can place pinholes at P_1 and P_2, and detect the mutual coherence at those points indirectly, by determining whether or not light transmitted through the pinholes can create fringes when falling on a second farther observation screen. The light phases at P_1 and P_2 will not be constant, but the phase *difference* can be constant, and in that case the farther screen will show an interference pattern, just as in the two-slit experiment with a phase difference Φ between the fields at the slits [Eqs. (13.5.12)]. If the field phases at P_1 and P_2 are fluctuating and not sticking together, or cohering, the fields they transmit will not be able to produce fringes, and we label the fields as incoherent.

It may seem surprising at first that an ordinary light source consisting of myriads of individual, independent radiators, can emit light with any spatial coherence at all. However, ordinary sources can (and frequently do) produce spatially coherent fields, fields which give rise to interference fringes when used to illuminate the slits in a two-slit experiment, for instance. We will now explain how an ordinary incoherent source can emit spatially coherent radiation. As a result of our discussion, we will be able to understand the experimentally observed decrease of the fringe visibility in the Young setup as slit separation d is increased, as described in Section 13.3.

It is worth emphasizing that for a single polarization of monochromatic radiation $|\gamma(\mathbf{r}_1, \mathbf{r}_2, 0)| = 1$; [recall Eq. (13.6.1)]. Needless to say, monochromaticity is an idealization that cannot be attained in the real world. However, we can produce quasi-monochromatic radiation quite readily, and so we will focus our attention on this more realistic case. We will denote by λ the central wavelength of our quasi-monochromatic field. The spread in wavelengths, $\delta\lambda$, is very small compared to λ for such a field [Eq. (13.3.4)].

We note first that a *point source* of radiation, one with dimensions much smaller than a wavelength, will always produce spatially coherent radiation. For even though the radiated field may vary quite erratically in its amplitude and phase, as a result of fluctuations in the source, every point on the wavefront has the *same* variation, that dictated by the single point source (Fig. 13.4). Thus the variations are perfectly correlated across any wavefront, and the emitted field is spatially coherent. The real question, therefore, is how an actual *extended* source, comprising many independently fluctuating point sources, can produce spatially coherent radiation.

Consider the case of two independent point sources, one on the axis in a two-slit experiment and the other a distance ρ off axis (Fig. 13.5). The source on the axis is equidistant from the two slits and, because it is a point source, produces spatially coherent radiation. It therefore produces fringes on the observation screen with perfect visibility ($V = 1$); the positions of intensity maxima and minima are given by (13.3.2a) and (13.3.2b) for $L \gg d$.

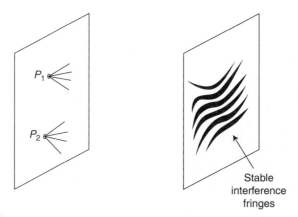

Figure 13.4   Two points P_1 and P_2 transmitting light from an equidistant common point source have exactly the same amplitude and phase variations, these being determined only by the fluctuations in the source.

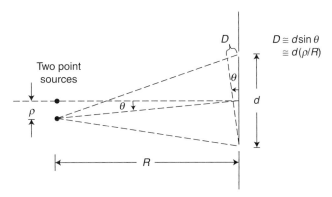

Figure 13.5   Two point sources in a two-slit experiment, one equidistant from the slits, the other displaced by ρ from the axis of equidistance. For the second source the distances to the slits differ by D. (The two angles labeled by θ are equal because their sides are perpendicular.)

The second point source is not equidistant from the two slits. The difference in path length is $D = \rho d / R$ (Fig. 13.5), corresponding to a phase difference

$$\Phi = \frac{2\pi D}{\lambda} - \frac{2\pi(\rho d)}{\lambda R}. \tag{13.7.1}$$

This point source therefore produces an interference pattern with intensity maxima and minima given by (13.5.12). In other words, the interference pattern associated with the second source is shifted a distance

$$\frac{\Phi}{2\pi} \frac{\lambda L}{d} = \frac{\rho L}{R} \tag{13.7.2}$$

from the interference pattern of the source on axis.

If the two point sources are completely independent, their fields fluctuate independently, and do not interfere for any measurable time interval. However, an interference pattern may still be observed, for the fringes associated with one (point) source may practically coincide with the fringes of the other. This happens if the displacement (13.7.2) of their interference patterns is small compared to the fringe spacing ΔY [Eq. (13.3.3)] of the interference pattern associated with each individual source, i.e., if

$$\frac{\rho L}{R} < \Delta Y = \frac{\lambda L}{d}. \tag{13.7.3}$$

Equation (13.7.3) says that, using the two point sources of Fig. 13.5, there will be interference fringes in the two-slit experiment if the slit separation d is small enough, namely if

$$d < \frac{\lambda}{\rho} R. \tag{13.7.4}$$

The factor λ / ρ is approximately the diffraction angle for light of wavelength λ incident upon an aperture of radius ρ (Section 7.11). This connection with diffraction theory is the essence of the *van Cittert–Zernike theorem*, which relates the mutual coherence

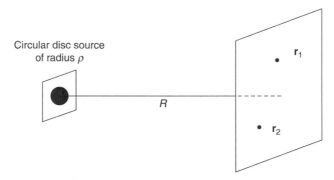

Circular disc source
of radius ρ

R

$\mathbf{r}_1$

$\mathbf{r}_2$

Figure 13.6   The van Cittert–Zernike theorem gives $|\gamma(\mathbf{r}_1, \mathbf{r}_2, 0)|$ in a plane a distance R from an ordinary (nonlaser) source of quasi-monochromatic radiation. We consider the case of a circular disk source of radius ρ, for which $|\gamma(\mathbf{r}_1, \mathbf{r}_2, 0)|$ is given by (13.7.5).

function of the field from an ordinary source to the diffraction pattern for an aperture of the same dimensions as the source. We will discuss one important example.

Consider a plane circular disk source of radius ρ. The van Cittert–Zernike theorem gives a simple expression in this case for the degree of spatial coherence $|\gamma(\mathbf{r}_1, \mathbf{r}_2, 0)|$ in a plane parallel to the source and a distance R from it (Fig. 13.6) in terms of the first-order Bessel function $J_1(x)$:

$$|\gamma(\mathbf{r}_1, \mathbf{r}_2, 0)| = \left| \frac{2J_1(x)}{x} \right|, \tag{13.7.5}$$

where

$$x = \frac{2\pi\rho}{\lambda R} |\mathbf{r}_1 - \mathbf{r}_2| = \frac{2\pi\rho d}{\lambda R}, \tag{13.7.6}$$

and it is assumed that ρ and d are much smaller than R. Comparing this with our results in Section 7.11, we see that $|\gamma(\mathbf{r}_1, \mathbf{r}_2, 0)|^2$ is the Airy pattern associated with Fraunhofer diffraction by a uniformly illuminated circular *aperture* of radius ρ. Equation (13.7.5) gives the degree of coherence of the radiation at a distance R from the source. Since $2J_1(x)/x \approx 0.88$ for $x = 1$, the radiation has a degree of spatial coherence $|\gamma(\mathbf{r}_1, \mathbf{r}_2, 0)| \geq 88\%$ if $x \leq 1$, i.e., if

$$\frac{2\pi\rho d}{\lambda R} \leq 1, \tag{13.7.7}$$

or $d \leq (1/2\pi)(\lambda R/\rho)$. In other words, the radiation has a high degree of spatial coherence over a circular area of diameter

$$d_{\text{coh}} = \frac{1}{2\pi} \frac{\lambda R}{\rho} \approx 0.16 \frac{\lambda R}{\rho}. \tag{13.7.8}$$

This result of the van Cittert–Zernike theorem supports our intuitive argument leading to (13.7.4).

Now we can understand the results of the two-slit experiment when an ordinary source is used (Section 13.3). The light from such a source has spatial coherence over a limited area ($\approx \pi d_{\mathrm{coh}}^2/4$) on the screen containing the slits. As the slit separation is increased, the degree of coherence of the fields at the two slits decreases. For slit separations large compared to d_{coh}, the fringe visibility approaches zero. The fringe visibility also decreases if the slit separation is kept constant and R is decreased, i.e., if the source is brought closer to the slits. Actually, if the slit separation is increased so that the fringe visibility falls from near unity to zero, a continued increase in the slit separation causes the visibility to increase and then decrease again repeatedly, but with very small secondary maxima. This is simply a reflection of the oscillatory behavior of the Bessel function $J_1(x)$ in (13.7.5) (cf. Fig. 6.13). A similar result is obtained if the slit separation is fixed while R is decreased.

The spatial filtering discussed in Section 7.13 is a way to obtain a spatially coherent beam of large area from an ordinary source of radiation (see Fig. 7.33). The radiation from the source is focused to a spot near a pinhole of radius a. The pinhole, which acts as a plane circular source, is in the focal plane of a second lens of focal length f. The beam emerging from the second lens is spatially coherent over a circle of diameter given by (13.7.8) with $\rho = a$ and $R = f$:

$$d_{\mathrm{coh}} = 0.16 \frac{\lambda f}{a}. \tag{13.7.9}$$

A small pinhole therefore gives rise to a large spatially coherent beam.

• Consider as an example the light from the sun. As an approximation let us treat the sun as a disk source of mean wavelength 550 nm. Over what linear dimensions on Earth is the light from the sun spatially coherent? The answer is given by (13.7.8) with $\rho = 6.96 \times 10^8$ m, the radius of the sun, and $R = 1.5 \times 10^{11}$ m, the mean distance of the Earth from the sun:

$$d_{\mathrm{coh}} = \frac{(0.16)(550 \times 10^{-9}\,\mathrm{m})(1.5 \times 10^{11}\,\mathrm{m})}{6.96 \times 10^8\,\mathrm{m}} = 0.02\,\mathrm{mm}. \tag{13.7.10}$$

We could use the arrangement of Fig. 7.33 to achieve a larger region of spatial coherence, but the larger beam area would result in a diminution of intensity. Lasers, on the other hand, can give large coherence areas *and* high intensity.

We can write (13.7.8) in the form

$$d_{\mathrm{coh}} = 0.16 \frac{\lambda}{\theta}, \tag{13.7.11}$$

where $\theta = \rho/R$ is the angle subtended by the source at the observation plane. The sun, for instance, subtends an angle $\theta \approx 4.6 \times 10^{-3}$ rad at the Earth.

The smaller the angle θ subtended by the source, the greater the diameter d_{coh} over which its radiation is spatially coherent. The star Betelguese, for instance, subtends an angle $\approx 2 \times 10^{-7}$ rad at the Earth. For $\lambda = 550$ nm, therefore, (13.7.8) gives $d_{\mathrm{coh}} \approx 0.8$ m, In other words, stellar radiation is spatially coherent over fairly large areas at the Earth's surface. There are techniques that take advantage of this to measure the angular diameters of stars.

The planets in our solar system typically subtend angles several orders of magnitude larger than those of stars. Their radiation (i.e., the solar radiation they scatter to the Earth) is therefore spatially coherent over much smaller distances at the Earth. This is partly responsible for the fact that stars twinkle, whereas planets normally do not. The twinkling is an interference effect arising from refractive-index fluctuations in the Earth's atmosphere. Similarly, distant streetlights appear to twinkle, whereas closer ones do not. Such effects were noted by Aristotle (384–322 B.C.), even before streetlights were common in Athens.   •

13.8 SPATIAL COHERENCE OF LASER RADIATION

Laser radiation can have a high degree of spatial coherence. This in itself is not remarkable, for we have seen that spatially coherent beams of light can be obtained from the radiation of ordinary sources. What is unique about lasers is that they can combine spatial coherence with high intensity, or at least an intensity high enough to be useful in applications. It is this property of lasers that makes them so useful for holography, for instance.

A laser oscillating on a single transverse mode has perfect spatial coherence.[4] A two-slit experiment will show interference fringes of high visibility. For a Gaussian mode, for instance, sharp fringes are observed even if the slit separation is considerably larger than the spot size w of the laser beam.

It is worth noting that this spatial coherence has nothing to do with stimulated emission *per se*. The spatial coherence of a laser oscillating on a single transverse mode is a consequence of the fact that the field is a mode of a resonator. As a result, the field values at any two points across the wavefront are perfectly correlated, i.e., in step with one another. For instance, even an emitter operating below the threshold for laser oscillation exhibits perfect spatial coherence if the radiation is associated with a single transverse mode.

However, *a laser operating on more than one transverse mode does not have perfect spatial coherence*. In particular, a laser operating on many transverse modes has spatial coherence properties much like those of ordinary sources of radiation, where the van Cittert–Zernike theorem is applicable. This is why single-mode operation is so important in holography, for instance.

To get an intuitive picture of the reason that oscillation on more than one transverse mode reduces spatial coherence, recall that different transverse modes have different field distributions, as in the case of a lowest-order Gaussian beam compared with a higher-order one. It can thus be imagined that the different modes are being excited by quite different groups of active atoms, and are therefore associated with completely independent sources. This brings us close to our picture of an ordinary source of radiation.

Figure 13.7 shows results of an experiment to determine the spatial coherence of a 632.8 nm He–Ne laser. The degree of spatial coherence was determined by a two-slit arrangement with aperture spacings from 2 to 20 mm. When the laser was oscillating on a single transverse mode, the result was $|\gamma(\mathbf{r}_1, \mathbf{r}_2, 0)| \approx 1$. However, when the resonator

---

[4]This is true even if the laser is oscillating on more than one longitudinal mode. See Section 13.11. It should be noted, however, that this conclusion is based on scalar wave theory in which polarization effects are ignored. If the resonator is such that the polarization varies azimuthally across the output beam, for instance, the complex degree of coherence of the laser radiation can be 1 for some pairs of points and 0 for others. See D. P. Brown, A. K. Spilman, T. G. Brown, R. Borghi, S. N. Volkov, and E. Wolf, *Optics Communications* **281**, 5287 (2008).

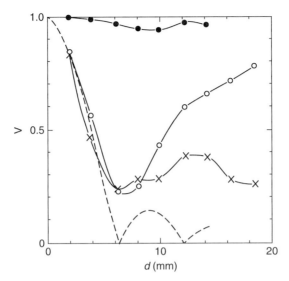

Figure 13.7 $V = |\gamma(\mathbf{r}_1, \mathbf{r}_2, 0)|$ for a 632.8 nm He–Ne laser, as a function of $d = |\mathbf{r}_1 - \mathbf{r}_2|$. Observations of single-mode (closed circles), double-mode (open circles), and multimode (crosses) types of operation are shown along with the calculated visibility function of a thermal source (dashed curve). From M. Young and P. L. Drewes, *Optics Communications* **2**, 253 (1970).

was adjusted so that two transverse modes oscillated, $|\gamma(\mathbf{r}_1, \mathbf{r}_2, 0)|$ dropped dramatically with increasing $|\mathbf{r}_1 - \mathbf{r}_2|$. In fact, $|\gamma|$ for the case of only two transverse modes already approaches that for an ordinary incoherent (e.g., thermal) source. As the number of oscillating transverse modes increases further, $|\gamma(\mathbf{r}_1, \mathbf{r}_2, 0)|$ comes close to the functional form (13.7.5) for a thermal source.

These experimental results, and others like them, show how crucial oscillation is on a single transverse mode for the spatial coherence of laser radiation. Unfortunately, the restriction to a single transverse mode often reduces the total output power of the laser. For when several modes oscillate, each with their different field distributions, the overall mode volume covers a greater portion of the available gain medium.

●   One result of the spatial coherence of a laser beam is the *speckle effect* that is observed when an expanded laser beam shines on a surface with fine-scale irregularities (e.g., a "diffuse" surface like a wall). The reflected light has a speckled appearance, consisting of irregularly-shaped but sharply-defined bright and dark areas. The bright and dark areas are associated, respectively, with constructive and destructive interference of the light from the various surface scattering elements (Fig. 13.8). Because the surface has more or less random irregularities, the speckle pattern itself appears random and irregular. Laser speckle is a consequence of spatial coherence: if the radiation incident on the scattering surface were not spatially coherent, the uncorrelated fluctuations in the field at nearby points on the surface would wash out the interference pattern.

When we view an object illuminated by laser light, we often find it difficult to focus on it. This is because our eyes involuntarily try to focus on the speckle. This cannot be done, because the speckle pattern is not "on" the object or any other plane in space. Indeed, we see the interference pattern even if we focus our eyes on a plane *between* the object and ourselves.

If a near-sighted observer moves his head from side to side, the speckle pattern appears to move in the opposite direction, whereas a far-sighted person will see it moving in the same direction he is moving his head. This effect has a simple explanation (Problem 13.3).   ●

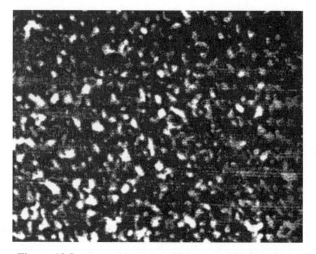

Figure 13.8   A speckle pattern [Courtesy of R. W. Boyd].

13.9 DIFFRACTION OF LASER RADIATION

The high degree of directionality of laser beams, and therefore their high brightness, is intimately related to their spatial coherence. Our treatment of diffraction in Chapter 7 assumed perfect spatial coherence: we dealt with time-independent field amplitudes and phases, so there were no fluctuations of these quantities to be averaged. For a spatially coherent beam propagating in free space, the divergence angle obeys the relation

$$\theta \sim \frac{\lambda}{D}, \tag{13.9.1}$$

where D is the beam diameter. The precise value of the divergence angle depends on the intensity distribution across the beam. For a Gaussian beam $\theta = (2/\pi)(\lambda/D)$, where $D = 2w_0$.

If the beam has only partial spatial coherence, the Huygens wavelets from different points on the beam do not all add up coherently. Imagine, for instance, that the beam is spatially coherent only over distances $d < D$ across the beam. In this case the divergence angle is

$$\theta \sim \frac{\lambda}{d} > \frac{\lambda}{D}. \tag{13.9.2}$$

In other words, if the beam is not spatially coherent, the divergence angle is greater than in the spatially coherent case with the same intensity distribution. In particular, the divergence angle associated with a laser operating on more than one transverse mode will generally be greater than that for the single-mode case.

The divergence angle of a laser beam can be reduced simply by increasing the beam diameter. This may be done, for instance, by letting the beam pass backwards through a

Keplerian telescope (Fig. 7.24). The divergence angle is inversely proportional to the beam diameter, and so the angle θ_f after passage through the telescope is related to the initial angle θ_i by

$$\frac{\theta_f}{\theta_i} = \frac{D_i}{D_f} = \frac{1}{M_T}, \tag{13.9.3}$$

where M_T is the magnification of the telescope. Low divergence angles are obiously essential in such applications as alignment or surveying, where a laser beam is used as a straight line.

Because of diffraction, a laser beam cannot be focused with a lens to the geometrical point predicted by ray optics (but see footnote 6 in Sec. 7.11). The beam divergence is minimized, however, if the beam is spatially coherent. In this case, because diffraction sets the ultimate lower limit on the beam spread, we say we have reached the *diffraction limit*. Realization of the diffraction limit in practice requires that aberrations and other defects in components such as lenses and mirrors are negligible. Unfortunately, the term "diffraction limit" is used without general agreement on its precise meaning; frequently its intended meaning has to be understood from the context in which it is used.

13.10 COHERENCE AND THE MICHELSON INTERFEROMETER

If we take $\mathbf{r}_1 = \mathbf{r}_2 = \mathbf{r}$ in the definition (13.4.18), then $\Gamma(\mathbf{r}_1, t_1; \mathbf{r}_2, t_2)$ determines the mutual coherence of the fields at the same point in space but at two different times. In this case we speak of *temporal coherence*. We are interested in stationary fields, for which $\Gamma(\mathbf{r}, t_1; \mathbf{r}, t_2) = \Gamma(\mathbf{r}, \mathbf{r}, \tau)$ depends on t_1 and t_2 only through the difference $\tau = t_2 - t_1$.

The significance of temporal coherence can be illustrated by considering as an example the Michelson interferometer shown in Fig. 13.9. The incident beam is split by a 50:50 beam splitter (BS) into two beams of equal intensity. One of these beams is reflected off mirror M_1 and makes its way to BS again, where part of it is transmitted. Similarly, the other beam reflects off mirror M_2 and propagates back to BS, where part of it is reflected. At a point such as P in Fig. 13.9 there is thus a superposition of two fields. Because each of these fields is twice incident on BS, it has an intensity one-fourth the intensity (at R, say) originally incident upon the interferometer; the field amplitude,

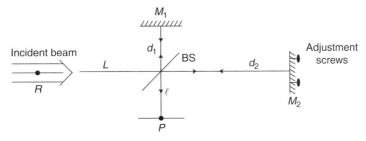

Figure 13.9 Basic setup for a Michelson interferometer.

therefore, has been cut in half. So the total field at P at time t is

$$E(P,\ t) = \frac{1}{2}E\left(R,\ t - \frac{l_1}{c}\right) + \frac{1}{2}E\left(R,\ t - \frac{l_2}{c}\right), \tag{13.10.1}$$

where

$$l_1 = L + 2d_1 + l, \qquad l_2 = L + 2d_2 + l. \tag{13.10.2}$$

In writing (13.10.1) we are assuming that the field amplitudes are reduced only by the beam splitter. The first term is the field at P resulting from propagation via the upper arm of the interferometer in Fig. 13.9. Except for the factor $\frac{1}{2}$, this field is the same as that at R at the earlier time $t - l_1/c$, where l_1/c is the time it takes light to propagate from R to P via the upper arm. The second term has the same interpretation, except that it arises because of the second arm of the interferometer.

The intensity measured at P is

$$\langle I(P,\ t)\rangle = \frac{c\epsilon_0}{2}\langle|E(P,\ t)|^2\rangle = \frac{c\epsilon_0}{8}\left|E\left(R,\ t - \frac{l_1}{c}\right)\right|^2 + \frac{c\epsilon_0}{8}\left|E\left(R,\ t - \frac{l_2}{c}\right)\right|^2$$

$$+ \frac{c\epsilon_0}{4}\operatorname{Re}\left[E^*\left(R,\ t - \frac{l_1}{c}\right)E\left(R,\ t - \frac{l_2}{c}\right)\right]. \tag{13.10.3}$$

For stationary fields, every term in this equation is independent of t, and furthermore the mutual coherence function appearing on the right depends only on the path difference

$$\left(t - \frac{l_2}{c}\right) - \left(t - \frac{l_1}{c}\right) = \frac{l_1 - l_2}{c} = 2\frac{d_1 - d_2}{c} = \tau. \tag{13.10.4}$$

The measured intensity at P for a stationary field is thus

$$\langle I(P)\rangle = \tfrac{1}{4}[\langle I(R)\rangle + \langle I(R)\rangle + c\epsilon_0\operatorname{Re}\Gamma(R,\ R,\ \tau)]$$
$$= \tfrac{1}{2}\langle I(R)\rangle[1 + \operatorname{Re}\gamma(R,\ R,\ \tau)], \tag{13.10.5}$$

where the complex degree of coherence is defined by (13.6.3):

$$\gamma(R,\ R,\ \tau) = \frac{(c\epsilon_0/2)\Gamma(R,\ R,\ \tau)}{\sqrt{\langle I(R)\rangle\langle I(R)\rangle}} = \frac{(c\epsilon_0/2)\Gamma(R,\ R,\ \tau)}{\langle I(R)\rangle}. \tag{13.10.6}$$

In the case of perfectly monochromatic light, for which γ is given by (13.6.4), we have from (13.10.5),

$$\langle I(P)\rangle = \frac{1}{2}\langle I(R)\rangle(1 + \cos\omega\tau) = \langle I(R)\rangle\cos^2\frac{1}{2}\omega\tau = \langle I(R)\rangle\cos^2\left[\frac{\omega}{c}(d_1 - d_2)\right]$$

$$= \langle I(R)\rangle\cos^2\left[\frac{2\pi}{\lambda}(d_1 - d_2)\right]. \tag{13.10.7}$$

There is therefore constructive interference at P when

$$|d_1 - d_2| = n\lambda, \qquad n = 0, 1, 2, \ldots, \qquad (13.10.8a)$$

and destructive interference when

$$|d_1 - d_2| = (n + \tfrac{1}{2})\lambda, \qquad n = 0, 1, 2, \ldots, \qquad (13.10.8b)$$

just as we should have expected. As the arm separation $|d_1 - d_2|$ is varied, there is a sequence of alternately bright and dark spots at P.

For quasi-monochromatic light, where γ is given by (13.6.5), we obtain from (13.10.5) the intensity

$$\langle I(P) \rangle = \frac{1}{2} \langle I(R) \rangle \left[1 + |\gamma(R, R, \tau)| \cos\left(\frac{2\pi}{\lambda}(d_1 - d_2) \right) \right]. \qquad (13.10.9)$$

The visibility in this case is

$$V = \frac{\langle I(P) \rangle_{\max} - \langle I(P) \rangle_{\min}}{\langle I(P) \rangle_{\max} + \langle I(P) \rangle_{\min}} = |\gamma(R, R, \tau)|. \qquad (13.10.10)$$

The Michelson interferometer thus provides a way of measuring temporal coherence, just as the Young two-slit experiment may be used to measure spatial coherence (Problem 13.4).

• The Michelson interferometer was invented by Albert A. Michelson, who began his study of optics as a student at the U.S. Naval Academy. There he was considered below average in seamanship, but he excelled in science. His best-known work involved the use of his interferometer to test for the motion of the Earth through the "ether." The null result of the Michelson–Morley experiment in 1887 led eventually to the abandonment of the ether concept.

The Michelson interferometer can be used to determine the wavelength of quasi-monochromatic radiation (Problem 13.5), and in this role it has for a long time been a very useful spectroscopic tool. •

13.11   TEMPORAL COHERENCE

It is found experimentally that the visibility (13.10.10) decreases with increasing τ. Furthermore the visibility decreases more rapidly for larger bandwidths $\delta\nu$ of the (quasi-monochromatic) radiation. In other words, the more nearly monochromatic the radiation, the greater its temporal coherence.

To understand this, suppose we have radiation of spectral width $\delta\lambda$ incident upon a Michelson interferometer. Then the total intensity at P is the sum of contributions like (13.10.7) if we add intensities of different frequency components (Problem 13.6). Since each wavelength component of the incident radiation is associated with a different pattern of bright and dark spots as $|d_1 - d_2|$ is varied, the pattern will be smeared out if $\delta\lambda$ is large enough. We will assume for simplicity that the intensity is constant for wavelengths between $\lambda - \tfrac{1}{2}\delta\lambda$ and $\lambda + \tfrac{1}{2}\delta\lambda$, and zero outside this range, as shown in

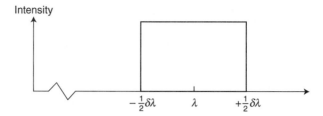

Figure 13.10   Hypothetical intensity distribution as a function of wavelength.

Fig. 13.10. Then we expect that the interference pattern is smeared out if $|d_1 - d_2|$ is large enough that the largest wavelength $\lambda + \frac{1}{2}\delta\lambda$ corresponds to an intensity maximum, whereas the smallest wavelength $\lambda - \frac{1}{2}\delta\lambda$ corresponds to an intensity minimum (or vice versa). From (13.10.8) we have therefore the two conditions

$$|d_1 - d_2| = n(\lambda + \tfrac{1}{2}\delta\lambda), \tag{13.11.1a}$$
$$|d_1 - d_2| = (n + \tfrac{1}{2})(\lambda - \tfrac{1}{2}\delta\lambda) \tag{13.11.1b}$$

in this case, or

$$\frac{|d_1 - d_2|}{\lambda + \frac{1}{2}\delta\lambda} = n, \tag{13.11.2a}$$

$$\frac{|d_1 - d_2|}{\lambda - \frac{1}{2}\delta\lambda} = n + \frac{1}{2}, \tag{13.11.2b}$$

where n is an integer. Subtraction of the first of these equations from the second yields

$$|d_1 - d_2|\left(\frac{1}{\lambda - \frac{1}{2}\delta\lambda} - \frac{1}{\lambda + \frac{1}{2}\delta\lambda}\right) = \frac{1}{2}. \tag{13.11.3}$$

Since $\delta\lambda \ll \lambda$, we can combine the two fractions to obtain

$$|d_1 - d_2|\frac{\delta\lambda}{\lambda^2} = \frac{1}{2}, \tag{13.11.4}$$

where we have dropped $(\delta\lambda)^2/4$ in the denominator compared to λ^2. Thus, we find

$$|d_1 - d_2| = c\tau = \frac{\lambda^2}{2\delta\lambda}. \tag{13.11.5}$$

Since $\lambda = c/v$, it follows that differential increments of wavelength and frequency are related by

$$\left|\frac{d\lambda}{dv}\right| = \frac{c}{v^2} = \frac{\lambda}{v}, \tag{13.11.6}$$

and so we have, for sufficiently small finite positive increments,

$$\frac{\delta\lambda}{\delta\nu} = \frac{\lambda}{\nu},$$ (13.11.7)

or

$$\frac{\delta\lambda}{\lambda} = \frac{\delta\nu}{\nu}.$$ (13.11.8)

Using this relation in (13.11.5), we obtain

$$c\tau = \frac{\lambda}{2}\frac{\lambda}{\delta\lambda} = \frac{\lambda\nu}{2\delta\nu} = \frac{c}{2\delta\nu},$$ (13.11.9)

that is,

$$\tau = \frac{|d_1 - d_2|}{c} = \frac{1}{2\delta\nu}.$$ (13.11.10)

Equation (13.11.10) gives the value of the interferometer path separation $|d_1 - d_2|$, or time difference τ, at which we expect the interference pattern to be smeared out. For separations larger than τ the visibility should be very small or zero. In agreement with experiment, τ decreases with increasing bandwidth $\delta\nu$. The intensity distribution shown in Fig. 13.10 was chosen for convenience, so the relation (13.11.10) derived from it cannot be regarded as fundamental. Also, as in the case of spatial coherence, there is some arbitrariness to the boundary between coherence and incoherence. Instead of (13.11.10) it is conventional to define

$$\tau_{\text{coh}} = \frac{1}{2\pi\delta\nu}$$ (13.11.11)

as the *coherence time* of quasi-monochromatic radiation of bandwidth $\delta\nu$. The distance $c\tau_{\text{coh}}$ is called the *coherence length*. If a beam is divided into two parts, the coherence length is the path difference beyond which there will be very little interference (or fringe visibility) when the two fields are superposed. Note that the coherence length arises from temporal coherence and is thus unrelated to the coherence area of Section 13.7, which is a measure of spatial coherence.

A good nonlaser source of "monochromatic" radiation might have a bandwidth $\delta\nu = 100$ MHz. This translates into a coherence time

$$\tau_{\text{coh}} = \frac{1}{(2\pi)(10^8 \text{ s}^{-1})} = 1.6 \text{ ns}$$ (13.11.12)

and a coherence length

$$c\tau_{\text{coh}} = 48 \text{ cm}.$$ (13.11.13)

More typical of such sources are coherence times and lengths on the order of 10^{-10} s and a few centimeters, respectively (Problem 13.7). With such sources the path separation (e.g., in a Michelson interferometer) must be less than a centimeter or two if interference fringes are to be observed.

A laser operating on a single transverse mode will have perfect spatial coherence, whereas its temporal coherence will be determined by the bandwidth of the output radiation. If it is operating on a single longitudinal mode, $\delta\nu$ is often so small that the coherence length is practically infinite for many purposes.

A laser operating on more than one longitudinal mode, however, can have a much larger bandwidth, and therefore a much smaller coherence length, than in the single-mode case. Many He–Ne lasers, for instance, operate on two longitudinal modes separated in frequency by $c/2L$. In this case $\delta\nu \sim c/2L$, and therefore $\tau_{coh} \sim L/\pi c$, so $c\tau_{coh} \sim L/\pi$. The coherence length of the laser radiation in this case is less than the length of the laser itself.

If many longitudinal modes are lasing, the laser may emit radiation over virtually the entire gain bandwidth. If we have a gain bandwidth of 30 GHz, for example, the coherence length is on the order of $c/(30 \text{ GHz}) = 1$ cm if lasing occurs over the entire gain bandwidth and on many modes. A laser operating on several longitudinal and transverse modes can therefore resemble a thermal source in both its temporal and spatial coherence properties. It remains true, of course, that the laser can emit more power than one can ever hope to obtain from a conventional source of radiation.

In the special case of mode-locked lasers, where many longitudinal modes oscillate in phase, the output is a train of *phase-locked* pulses and the spectrum is a frequency comb (Section 14.7). The coherence length is determined by the duration of the individual pulses, and since they can be extremely short, the coherence length can be very small. For pulses in the femtosecond range, coherence lengths are measured in microns; this makes them useful in optical coherence tomography (Section 14.7).

- The quantity $|\gamma(\mathbf{r}, \mathbf{r}, \tau)|$ is related to the Fourier transform of the spectral lineshape function. For a Lorentzian lineshape of HWHM $\delta\nu$, for instance, it may be shown that [cf. (13.6.8)]

$$|\gamma(\mathbf{r}, \mathbf{r}, \tau)| = e^{-2\pi\delta\nu\tau} = e^{-\tau/\tau_{coh}}, \qquad (13.11.14)$$

where τ_{coh} is given by (13.11.11). In this case, therefore, $c\tau_{coh}$ is just the value of the path separation $|d_2 - d_1|$ at which the visibility drops to e^{-1}, which incidentally illustrates again that there is no sharp boundary between temporal coherence and incoherence, just as there is no sharp boundary between spatial coherence and incoherence.

In the Young two-slit experiment the modulus $|\gamma(S_1, S_2, l/c)|$ appearing in (13.6.3) may be replaced by $|\gamma(S_1, S_2, 0)|$ if $l/c \ll \tau$, that is, if the path difference l for the two slits is small compared with the coherence length of the radiation. This condition is frequently well satisfied in practice, even for a "monochromatic" thermal source or a laser operating on many longitudinal modes. In this case the Young experiment gives us a direct measure of spatial coherence, as we assumed in our discussion. This allows us to speak separately of "spatial" and "temporal" coherence.

13.12 THE PHOTON DEGENERACY FACTOR

Consider a thermal source of quasi-monochromatic radiation of bandwidth $\delta\nu$. The radiation from this source is spatially coherent over an area

$$A_{coh} \sim d_{coh}^2 \sim \frac{\lambda^2 R^2}{\rho^2} \sim \frac{\lambda^2 R^2}{S} \qquad (13.12.1)$$

at a distance R from the source, with S the source area. A_{coh} is called the *coherence area*. The product of the coherence area and the coherence length, $c\tau_{\text{coh}} \sim c/\delta v$, defines the *coherence volume*:

$$V_{\text{coh}} = A_{\text{coh}} \times c\tau_{\text{coh}} = \frac{c\lambda^2 R^2}{S\delta v}. \tag{13.12.2}$$

In this section we will ignore geometrical details involving factors of π, 2, 2π, etc. We will only concern ourselves with general orders of magnitude that are independent of source shape—square or circular or whatever.

Let us now think in terms of photons and consider the number of photons crossing the coherence area in a coherence time. We denote this dimensionless number by δ:

$$\delta = \Phi A_{\text{coh}} \tau_{\text{coh}}, \tag{13.12.3}$$

where Φ is the photon flux, i.e., the number of photons crossing a unit area per unit time. For blackbody radiation $\rho(v)\delta v$ is the energy per unit volume in the frequency interval from v to $v + \delta v$, and so $\rho(v)\delta v/hv$ is the number of photons per unit volume in this interval; $\rho(v)$ is the spectral energy density (3.6.1). The photon flux from a blackbody source of surface area S is therefore

$$\Phi_v \sim c \frac{\rho(v)\delta v}{hv} \frac{S}{4\pi R^2} \tag{13.12.4}$$

at a distance R from the source. The factor $S/4\pi R^2$, the ratio between the emitting surface area and the surface area of a sphere of radius R, takes account of the inverse square law for photon flux. That is, $R^2 \Phi_v$ must be a constant, independent of R. From (13.12.3) and (13.12.4), therefore,

$$\delta \sim \left(\frac{c\rho(v)\delta vS}{4\pi hvR^2}\right)\left(\frac{\lambda^2 R^2}{S}\right)\left(\frac{1}{\delta v}\right) = \frac{c\rho(v)\lambda^2}{4\pi hv} = \frac{c^3}{4\pi hv^3}\rho(v). \tag{13.12.5}$$

δ is the number of photons crossing the coherence area A_{coh} during a coherence time τ_{coh}. Blackbody radiation is unpolarized. The number of photons *of a particular polarization* crossing A_{coh} in a time τ_{coh} is therefore half the value (13.12.5):

$$\delta = \frac{c^3}{8\pi hv^3}\rho(v) = \frac{1}{e^{hv/k_B T} - 1}, \tag{13.12.6}$$

the last step following from the Planck law (3.6.1).

From Eq. (3.6.20) we recognize δ in (13.12.6) as the average number of thermal photons in a mode of frequency v. In other words, *the average number of photons crossing an area equal to A_{coh} in a time equal to τ_{coh} is equal to the average number of photons per mode.* This number δ is called the *photon degeneracy factor*, or simply the *degeneracy parameter*. It represents a "degeneracy" in the sense that the δ photons are not distinguished from each other by spatial or temporal labels. For thermal radiation δ is much smaller than one, being typically 10^{-2} or 10^{-3} (recall the black-dot passage in Section 13.2).

It is clear that the degeneracy parameter is equal to the average number of photons in a volume V_{coh}: we simply write (13.12.3) as

$$\delta = \frac{\Phi}{c} A_{coh}(c\tau_{coh}) = \frac{\Phi}{c} V_{coh} = NV_{coh}, \qquad (13.12.7)$$

where N is the density of photons, related to the photon flux Φ by the equation $\Phi = cN$.

How many photons per mode are in a *laser* beam? Our discussion of cavity modes in Chapter 7 assumed a perfectly monochromatic field. Such a field has perfect spatial and temporal coherence; the coherence volume is infinite in this hypothetical limit. Now a single-mode laser is spatially coherent over the entire beam area A, but its temporal coherence is limited by the frequency bandwidth $\delta\nu$. To estimate δ in this case let us assume, as we have proven for thermal radiation, that δ is equal to the number of photons crossing an area equal to A_{coh} in a time τ_{coh}:

$$\delta = \Phi A_{coh}\tau_{coh} \sim \frac{I}{h\nu} A \frac{1}{\delta\nu} = \frac{IA}{hc}\frac{\lambda}{\delta\nu} = \frac{Pwr}{hc}\frac{\lambda}{\delta\nu}, \qquad (13.12.8)$$

where I is the intensity and $Pwr = IA$ the beam power.

As an example, consider a single-mode, 632.8-nm He–Ne laser emitting a power $Pwr = 1$ mW with a bandwidth $\delta\nu = 500$ Hz:

$$\delta = \frac{(10^{-3}\,\text{J s}^{-1})(632.8 \times 10^{-9}\,\text{m})}{(6.625 \times 10^{-34}\,\text{J s})(3 \times 10^{8}\,\text{m s}^{-1})(500\,\text{s}^{-1})} = 6.4 \times 10^{12}. \qquad (13.12.9)$$

This is fantastically larger than the photon degeneracy factor of blackbody radiation. Furthermore, it is not difficult to exceed (13.12.9) by orders of magnitude in well-stabilized single-mode lasers. And so, although ordinary thermal sources can emit radiation as spatially and temporally coherent as laser radiation, they could produce such huge numbers of degenerate photons only with source temperatures above about 10^{16}K, temperatures unknown in the universe.

- The coherence volume has an interesting interpretation in terms of the Heisenberg uncertainty principle. According to this principle of quantum theory, there is a fundamental limitation on the accuracy to which the position and momentum of a particle can be simultaneously measured. The uncertainties Δx, Δy, and Δz in the position coordinates of the particle are related to the uncertainties Δp_x, Δp_y, and Δp_z in the momentum components by (still ignoring factors of 2π)

$$\Delta x\,\Delta p_x \geq h, \qquad \Delta y\,\Delta p_y \geq h, \qquad \Delta z\,\Delta p_z \geq h. \qquad (13.12.10)$$

The uncertainty $\Delta V = \Delta x\,\Delta y\,\Delta z$ of the volume within which the particle can be localized in a measurement is therefore

$$\Delta V \sim \frac{h^3}{\Delta p_x\,\Delta p_y\,\Delta p_z}. \qquad (13.12.11)$$

Let us apply (13.12.11) to the case of a photon of light of frequency ν and bandwidth $\delta\nu$ propagating in the z direction. Since the momentum of a photon is $p = h/\lambda = h\nu/c$, we have

$$\Delta p_z = \frac{h}{c}\delta\nu. \qquad (13.12.12)$$

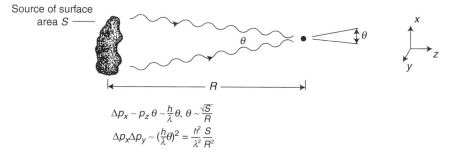

$$\Delta p_x \sim p_z \, \theta \sim \frac{h}{\lambda} \, \theta, \; \theta \sim \frac{\sqrt{S}}{R}$$

$$\Delta p_x \Delta p_y \sim \left(\frac{h}{\lambda} \theta\right)^2 = \frac{h^2}{\lambda^2} \frac{S}{R^2}$$

Figure 13.11 Photons from a source of area S. The uncertainties in the momentum components p_x and p_y are directly proportional to the angle θ subtended by the source.

At a distance R from the source of surface area S, the uncertainties in Δp_x and Δp_y are (Fig. 13.11)

$$\Delta p_x \sim \frac{h}{\lambda}\left(\frac{S}{R^2}\right)^{1/2}. \tag{13.12.13}$$

Combining (13.12.11)–(13.12.13), we obtain

$$\Delta V \sim \frac{c\lambda^2 R^2}{S\delta\nu}. \tag{13.12.14}$$

We recognize the right-hand side as the coherence volume V_{coh} given by (13.12.2).

We can summarize this result as follows. Photons emitted from a given source have linear momenta that can only be determined within a certain tolerance owing to the finite area of the source and the finite bandwidth of the radiation. Associated with this uncertainty in momenta is an uncertainty in the volume within which any measurement can locate the photon. This volume is equal to the coherence volume of the radiation. This connection between the Heisenberg uncertainty principle and coherence, or the ability to produce interference effects, is a general feature of quantum theory. ●

13.13 ORDERS OF COHERENCE

The spatial and temporal coherence properties of radiation determine the degree to which the fields at two points in space, and at two points in time, are found to interfere. We have seen that thermal sources, for instance, emit radiation that can be made to have the same degree of spatial and temporal coherence as laser light, even if the laser is well stabilized and oscillating on a single cavity mode. Of course, the use of pinholes, lenses, and wavelength filters to increase the coherence of radiation from a nonlaser source can severely reduce the light intensity far below that available from lasers.

Aside from this difference in photon number, we might ask whether there is any difference *in principle* between coherent thermal radiation and coherent laser radiation: If they have the same central frequency, the same bandwidth, the same degree of spatial coherence, and the same intensity, can we distinguish thermal radiation from laser radiation?

It seems clear that we cannot distinguish them by measuring the location and visibility of interference fringes in a Young or Michelson experiment. In fact no experiment that divides a wavefront into two parts and then measures in some way the mutual coherence function can show any difference.

However, it turns out that we can distinguish between the two radiation fields if we undertake more sophisticated experiments. The mutual coherence function only characterizes *first-order coherence*. In general, we can define *nth-order coherence functions*. A second-order coherence function, for instance, is

$$\Gamma^{(2)}(\mathbf{r}_1, t_1; \mathbf{r}_2, t_2 \mid \mathbf{r}_3, t_3; \mathbf{r}_4, t_4) = \langle E^*(\mathbf{r}_1, t_1)E^*(\mathbf{r}_2, t_2)E(\mathbf{r}_3, t_3)E(\mathbf{r}_4, t_4)\rangle. \quad (13.13.1)$$

It depends on *four* space–time points. A general *n*th-order coherence function will likewise be a function of 2*n* space–time variables. Experiments that measure such higher-order coherence functions, and the corresponding higher-order complex degrees of coherence, can distinguish laser radiation from thermal radiation, even if the two fields are identical in their spatial and temporal coherence functions.

The spatial and temporal coherence already described in this chapter depend on the mutual coherence function Γ, which is a measure of first-order coherence. In other words, *ordinary spatial and temporal coherence are only manifestations of first-order coherence*, the ability of radiation to produce interference effects at two space–time points $(\mathbf{r}_1, t_1)$ and $(\mathbf{r}_2, t_2)$.

In the great majority of laser applications only first-order coherence properties, like directionality and quasi-monochromaticity, are important. In some applications not even these coherence properties, but only the high intensity of laser radiation, really matter. We will, therefore, not devote much space to higher-order coherence or the types of experiments that measure higher-order coherence properties of radiation. In Section 13.15, however, we will discuss an important example of a second-order coherence function in connection with photon bunching.

It should be mentioned that in laser applications the word "coherence" is used mainly in situations involving only first-order coherence. For instance, a beam may be said to be "coherent" if it produces interference fringes in a Michelson interferometer with a certain path separation, or if it gives interference fringes in a two-slit experiment with a certain slit separation.

13.14 PHOTON STATISTICS OF LASERS AND THERMAL SOURCES

We have introduced some of the main ideas of optical coherence theory in terms of the classical wave theory of light. The particle aspect of light was used only in Section 13.12, where we found it convenient to think in terms of photons in order to elucidate the meaning of the volume of coherence. There we found that lasers and thermal sources are drastically different in terms of the number of photons they can put into a single field mode.

As discussed in Chapter 12, it is possible to count photons (or rather, to count, for instance, the photoelectrons ejected from a photoemissive surface by the absorption of photons). Such experiments very clearly reveal that the difference between lasers and thermal sources is not merely a matter of how many photons can be generated.

They show that lasers are different from conventional sources in a *fundamental* way because their higher-order coherence functions are different.

In Section 12.5 we considered a typical photon-counting experiment, which records the number of photoelectrons ejected by a light beam during a time interval T. By repeating the experiment a large number of times, the probability distribution $P_n(T)$ for the number n of photoelectrons counted during a time T can be determined, and under ideal circumstances we may say that this is the probability distribution for photons counted in a time T. Using again our ensemble-average notation, we found that, if the intensity

$$\langle \bar{I}(t,\, T) \rangle = \frac{1}{T} \int_t^{t+T} \langle I(t') \rangle \, dt' \tag{13.14.1}$$

averaged over the counting time interval from t to $t + T$ is a constant, independent of t, then $P_n(T)$ is the Poisson distribution:

$$P_n(T) = \frac{\bar{n}^n}{n!} e^{-\bar{n}}, \tag{13.14.2}$$

where $\bar{n}$ is the average number of photons counted in a time interval T.

Equation (13.14.2) may be expected to apply to a cw laser beam. It also applies to a beam of thermal radiation, provided the intensity (13.14.1) is independent of t. This will be true if the counting interval T is large enough, because fluctuations in the thermal-field intensity will be averaged out. In fact, T is "large enough" if it is large compared to the coherence time $\tau_{\text{coh}} = \frac{1}{2}\pi\delta\nu$ of the (quasi-monochromatic) thermal radiation.

Suppose, however, that T is small compared to the coherence time of a thermal source. Then the fluctuations in the intensity during a time T are not averaged out, and the Poisson distribution (13.14.2) for the photon counts does not apply. To obtain $P_n(T)$ in this case, let us consider a single-mode thermal field. Such a field is in thermal equilibrium at some temperature T_e, and the probability that there are exactly n photons in the field is given by the Boltzmann law:

$$p(n) = \frac{e^{-E_n/k_B T_e}}{\sum_{m=0}^{\infty} e^{-E_m/k_B T_e}} = \frac{e^{-nh\nu/k_B T_e}}{\sum_{m=0}^{\infty} e^{-mh\nu/k_B T_e}}, \tag{13.14.3}$$

since $E = nh\nu$ is the energy of a state with n photons of frequency ν. The series can be summed easily:

$$\sum_{m=0}^{\infty} e^{-mh\nu/k_B T_e} = \sum_{m=0}^{\infty} x^m = (1-x)^{-1}, \qquad x = e^{-h\nu/k_B T_e}. \tag{13.14.4}$$

Therefore,

$$p(n) = (1-x)x^n = (1 - e^{-h\nu/k_B T_e})e^{-nh\nu/k_B T_e}. \tag{13.14.5}$$

This implies the familiar result for the average photon number $\bar{n}$:

$$\bar{n} = \sum_{n=0}^{\infty} np(n) = (1-x) \sum_{n=0}^{\infty} nx^n = x(1-x)^{-1} = (x^{-1} - 1)^{-1}$$

$$= (e^{h\nu/k_B T_e} - 1)^{-1}. \tag{13.14.6}$$

Using this result in (13.14.5), we may write the probability distribution $p(n)$ in the form

$$p(n) = \frac{\bar{n}^n}{(\bar{n} + 1)^{n+1}}, \tag{13.14.7}$$

which is the *Bose–Einstein distribution*.

- It is instructive to obtain (13.14.7) in a different way, using the fact that the intensity of light from a thermal source of mean intensity $\langle I \rangle$ has the probability distribution

$$P(I) = \frac{1}{\langle I \rangle} e^{-I/\langle I \rangle}, \tag{13.14.8}$$

as shown in the following section. For this purpose we first recall the formula (12.5.14) for the photon counting distribution, assuming a stationary (e.g., thermal) light source in which the intensity given by (12.5.13) is independent of t. Then (12.5.14) reduces to

$$P_n(T) = \frac{1}{n!} [\alpha T \bar{I}(T)]^n e^{-\alpha T \bar{I}(T)}. \tag{13.14.9}$$

now suppose that $\bar{I}$ undergoes statistical fluctuations about a mean value $\langle I \rangle$, and that its variations are described by the probability distribution (13.14.8). In this case the photon-counting distribution is not simply (13.14.9); we must instead take into account the statistical distribution of $\bar{I}(T)$ and replace (13.14.9) by the ensemble average:

$$P_n = \int_0^{\infty} \frac{1}{\langle I \rangle} e^{-I/\langle I \rangle} \frac{1}{n!} [\alpha T I]^n e^{-\alpha T I} \, dI = \frac{[\alpha T \langle I \rangle]^n}{[1 + \alpha T \langle I \rangle]^{n+1}} = \frac{\bar{n}^n}{[\bar{n} + 1]^{n+1}}. \tag{13.14.10}$$

The reader may wish to think about the extent (if any) to which this derivation of the Bose–Einstein distribution formula requires quantum theory. ●

Returning now to our photon counting with $T \ll \tau_{\text{coh}}$, we observe that for a single mode $\Delta \nu \to 0$ and therefore $\tau_{\text{coh}} \to \infty$. We can therefore assume $T \ll \tau_{\text{coh}}$, which leads us to suspect that $p(n)$ in (13.14.7) gives the photon-counting probability distribution for a thermal field when the counting interval T is small compared to the coherence time τ_{coh}. This suspicion can be justified by a more rigorous approach, but we will just assume it is true: The probability of counting n photons of a thermal field during a time interval $T \ll \tau_{\text{coh}}$ is

$$P_n = \frac{\bar{n}}{(\bar{n} + 1)^{n+1}} \quad \text{(thermal field)}, \tag{13.14.11}$$

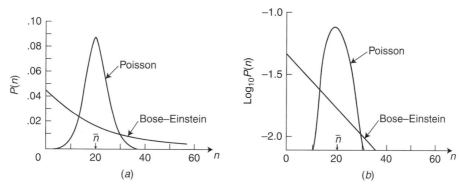

Figure 13.12   Poisson and Bose–Einstein distributions on linear (*a*) and semilog (*b*) plots for $\bar{n} = 20$.

where $\bar{n}$ is the average number of photons counted in a time interval T, as opposed to

$$P_n = \frac{\bar{n}^n e^{-\bar{n}}}{n!} \quad \text{(laser)} \tag{13.14.12}$$

for a laser beam. Figure 13.12 compares the probability distributions (13.14.11) and (13.14.12) for the case $\bar{n} = 20$. It is clear that a laser and a thermal source, even though they may give the same average number $\bar{n}$ of photons during a counting interval, will nevertheless have completely different photon statistics. The two fields may have the same average intensity, the same frequency and bandwidth, and the same first-order spatial and temporal coherence properties, but they are nevertheless different.

Photon-counting experiments have accurately confirmed the predictions (13.14.11) and (13.14.12). They have also shown, again in agreement with theory, that the light from a laser below the threshold for oscillation has the photon statistics of a thermal field. In other words, a laser below threshold is characterized by the photon-counting probability distribution (13.14.11), whereas above threshold it is characterized by the Poisson distribution (13.14.12).

Since

$$\ln P_n = \ln \frac{\bar{n}^n}{(\bar{n} + 1)^{n+1}} = n \ln \frac{\bar{n}}{\bar{n} + 1} - \ln (\bar{n} + 1), \tag{13.14.13}$$

for the Bose–Einstein distribution, a plot of $\ln P_n$ vs. n is a straight line. Figure 13.13 shows experimental results for a 632.8-nm He–Ne laser, both below and above the threshold for oscillation. Below threshold the observed distribution is given accurately by the straight line (13.14.13), whereas above threshold good agreement is found with the Poisson distribution.

In Section 5.14 we found that a remarkable transition occurs as a laser goes from below threshold to above threshold: the average photon number changes rather abruptly by orders of magnitude. Now we see that the threshold point is also the boundary between two completely different types of photon statistics.

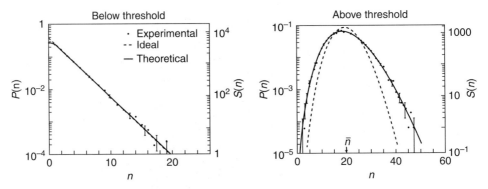

Figure 13.13   Poisson statistics of a He–Ne laser above and below threshold. [From C. Freed and H. A. Haus, *IEEE Journal of Quantum Electronics* **QE-2**, 190 (1966).]

The coherence time τ_{coh} of thermal sources is usually so small that as a practical matter the counting interval $T \gg \tau_{\mathrm{coh}}$ in a photon-counting experiment. As discussed following Eq. (13.14.2), the photon-counting distribution in this case will be Poissonian. However, it is possible to produce "pseudothermal" light such that $T \ll \tau_{\mathrm{coh}}$. In the case of a thermal source the radiation consists of a superposition of a large number of fields with randomly varying amplitudes and phases (Section 13.15). This suggests that if laser radiation is incident on a large collection of randomly varying scattering centers, the total scattered field will have the statistical features characteristic of thermal radiation. Such pseudothermal radiation has been realized, for instance, by scattering laser light off small ($<1\ \mu$m) plastic balls undergoing Brownian motion in water. The coherence time of the scattered radiation in this case is ~ 100 ms. Pseudothermal radiation has also been produced using a rotating ground-glass plate. Because of surface irregularities the radiation transmitted by the plate, rotated about an axis parallel to an incident laser beam, is a superposition of fields from a large number of independent scattering centers, which is approximately a superposition of fields with randomly varying amplitudes and phases. The coherence time of the transmitted (pseudothermal) light depends on the rotation speed of the glass plate, and can easily be made much larger than a counting time T (typically 1–10 ms) for which a meaningful number of photons can be counted. The photon-counting distribution of pseudothermal light obtained by such methods has been found to be accurately described by the Bose–Einstein distribution.[5]

In Section 12.6 we noted that the mean-square deviation of the photon number is given by

$$\langle \Delta n^2 \rangle = \langle (n - \overline{n})^2 \rangle = \overline{n} \quad \text{(Poisson distribution)} \qquad (13.14.14)$$

for the Poisson distribution. For the Bose–Einstein distribution it is not difficult to show that a different result holds:

$$\langle \Delta n^2 \rangle = \overline{n}^2 + \overline{n} \quad \text{(Bose–Einstein distribution).} \qquad (13.14.15)$$

[5]See, for instance, P. Koczyk, P. Wiewiór, and C. Radzewicz, *American Journal of Physics* **64**, 240 (1996) and references therein.

Thus, whereas the relative rms deviation for the Poisson distribution is

$$\frac{(\Delta n)_{\text{rms}}}{\bar{n}} = \frac{1}{\sqrt{\bar{n}}} \quad \text{(Poisson distribution)}, \tag{13.14.16}$$

we have instead

$$\frac{(\Delta n)_{\text{rms}}}{\bar{n}} = \sqrt{1 + \frac{1}{\bar{n}}} \quad \text{(Bose–Einstein distribution)}, \tag{13.14.17}$$

for the Bose–Einstein distribution. Whereas the relative rms deviation from the mean for the Poisson distribution decreases toward zero with increasing $\bar{n}$, the corresponding deviation for the Bose–Einstein distribution approaches unity. Thus, the fluctuations in the photon number for a thermal field can be much more pronounced than in the case of a coherent laser field.

•   Equation (13.14.15) is sometimes referred to as the *Einstein fluctuation formula.* Using Eq. (13.14.6) for $\bar{n}$, and the Planck formula

$$\rho(\nu) = \frac{8\pi h\nu^3}{c^3} \bar{n} \tag{13.14.18}$$

for the spectral energy density of thermal radiation, we obtain

$$\langle \Delta n^2 \rangle = \bar{n}^2 + \bar{n} = \frac{c^3}{8\pi h\nu^3} \left[\frac{c^3}{8\pi h\nu^3} \rho^2(\nu) + \rho(\nu) \right], \tag{13.14.19}$$

The mean-square deviation from the average energy for thermal radiation of frequency ν is

$$\langle \Delta E_\nu^2 \rangle = (h\nu)^2 \langle \Delta n^2 \rangle [8\pi\nu^2 V \, d\nu/c^3], \tag{13.14.20}$$

where the factor in brackets is the number of field modes in the volume V and in the frequency interval $[\nu, \nu + d\nu]$, as derived in Section 3.12. Thus, by using (13.14.19) in (13.14.20), we find

$$\langle \Delta E_\nu^2 \rangle = \left[h\nu\rho(\nu) + \frac{c^3}{8\pi\nu^2} \rho^2(\nu) \right] V \, d\nu, \tag{13.14.21}$$

which is the form of the fluctuation formula for thermal radiation found by Einstein in 1909.

The Einstein fluctuation formula is important historically as the first indication of the wave–particle duality of light. The first term inside the brackets in (13.14.21) can be derived from the classical Poisson statistics of distinguishable particles, whereas the second term follows from a purely wave approach to thermal radiation, as was shown by Lorentz. Thus, the fluctuation formula (13.14.21) has both particle and wave contributions. Einstein showed that this formula is a direct consequence of the Planck law for $\rho(\nu)$.   •

It should be clear by now that the Poisson distribution plays a central role in the theory of photon counting. The Bose–Einstein distribution, similarly, is important as the fundamental photon distribution of thermal radiation. But these are not the only photon-counting distributions that can be measured. More generally, if the mean intensity $\bar{I}(T)$ has a probability distribution $P(I)$, then the measured photon counting distribution is given by the formula [cf. (12.5.15)]

$$P_n = \int_0^\infty P(I) \frac{1}{n!} [\alpha I T]^n e^{-\alpha I T} \, dI. \qquad (13.14.22)$$

The Poisson factor in this expression accounts for the probability distribution of photon counts that appears even if the mean intensity $\bar{I}(T)$ is constant. The distribution function $P(I)$ accounts for the fact that $\bar{I}(T)$ may itself fluctuate. Equation (13.14.10) is an

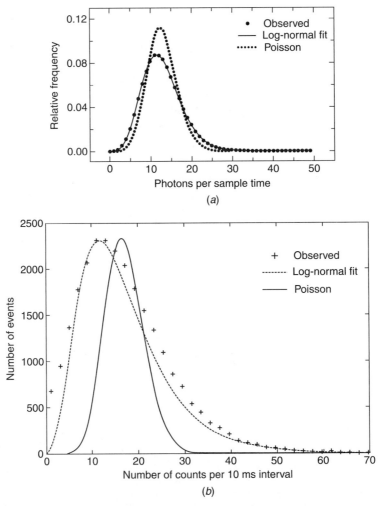

Figure 13.14   (*a*) Measured photon counting distribution of filtered starlight, compared with theory [Eq. (13.14.22)] assuming the log-normal distribution with $\sigma_I^2 = 0.058$. [From D. Dravins, L. Lindegren, E. Mezey, and A. T. Young, *Publications of the Astronomical Society of the Pacific* **109**, 173 (1997).] (*b*) Measured photon counting distribution for laser radiation propagated over a 10-km horizontal path in air, compared with theory assuming a log-normal distribution with $\sigma_I^2 = 0.094$. [From P. W. Milonni, J. H. Carter, C. G. Peterson, and R. J. Hughes, *Journal of Optics B: Quantum and Semiclassical Optics* **6**, S742 (2004).]

example of (13.14.22) for the special case $P(I) = (1/\langle I \rangle)\exp(-I/\langle I \rangle)$. Another example is provided by the propagation of radiation in Earth's atmosphere. In this case the fluctuations (scintillations) of the intensity are described approximately by the log-normal distribution, i.e., $P(I)$ is given by Eq. (8.13.7). Figure 13.14a shows a measured photon-counting distribution for filtered starlight at 550 nm for a counting time interval $T = 200\,\mu s$; a good fit to the data is obtained for an assumed intensity variance $\sigma_I^2 = 0.058$ (Section 8.13). Similarly, Fig. 13.14b shows the measured photon-counting distribution for $T = 10$ ms after laser radiation of very low intensity has propagated over a 10-km horizontal path in air. A good fit to the data is obtained when $P(I)$ is assumed to be a log-normal distribution with $\sigma_I^2 = 0.094$.

13.15   BROWN–TWISS CORRELATIONS

We have emphasized that first-order coherence does not in general distinguish between lasers and ordinary, thermal sources of radiation, even though the two have measurably different photon statistics. Expressions (13.14.14) and (13.14.15) for the variance $\langle \Delta n^2 \rangle$ suggest that lasers and ordinary sources may be distinguished by their *second*-order coherence properties, and this is indeed the case. We will now describe an experiment concerned with second-order coherence properties of radiation.

The experiment is sketched in Fig. 13.15. Quasi-monochromatic radiation is incident upon a $50:50$ beam splitter, so that half the original intensity is directed to a photomultiplier tube PM1, the other half to a second photomultiplier PM2. The responses of the photomultipliers are used to determine the average of the product of $I_1(t)$ and $I_2(t+\tau)$, where I_1 and I_2 are the intensities incident on PM1 and PM2, respectively. We denote this average by $C(\tau)$:

$$C(\tau) = \langle I_1(t)I_2(t+\tau) \rangle. \tag{13.15.1}$$

The instrumentation performing this correlation is simply called the "correlator" in Fig. 13.15. We assume the field has the property of stationarity, so that $\langle I_2(t+\tau) \rangle = \langle I_1(t) \rangle \equiv I$ for a $50:50$ beam splitter. As usual the intensities $I_1(t)$ and $I_2(t)$ are

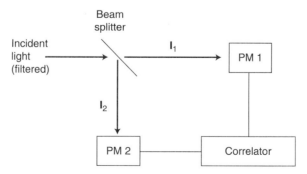

Figure 13.15   Experimental setup to determine second-order coherence properties of light by measuring $C(\tau)$ [Eq. (13.15.1)].

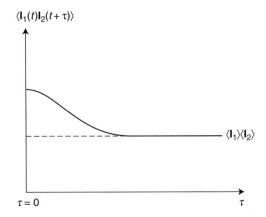

Figure 13.16   Photon bunching, or Hanbury Brown–Twiss intensity correlation near $\tau = 0$.

themselves averages over a few optical periods. Note that, from definition (13.15.1), we have

$$C(\tau) = \left(\frac{c\epsilon_0}{2}\right)^2 \langle E^*(\mathbf{r}_1, t)E^*(\mathbf{r}_2, t + \tau)E(\mathbf{r}_2, t + \tau)E(\mathbf{r}_1, t)\rangle. \qquad (13.15.2)$$

That is, $C(\tau)$ is a second-order coherence function. Here $\mathbf{r}_1$ and $\mathbf{r}_2$ refer to points on PM1 and PM2, respectively. Since spatial variations of the field will not play an important role in our discussion, we will ignore them.

The second-order coherence function $C(\tau)$ was first measured by R. H. Brown and R. Q. Twiss in the 1950s. Brown and Twiss used radiation from a mercury arc lamp and obtained a result like that shown in Fig. 13.16. The most significant feature of Fig. 13.16 is the positive correlation for small values of the delay time τ: for small τ, $\langle I_1(t)I_2(t+\tau)\rangle$ is *larger* than $\langle I_1(t)\rangle\langle I_2(t + \tau)\rangle = I^2$. This *Brown–Twiss effect*[6] is explainable in either classical or quantum mechanical terms. We begin with a quantum mechanical description, focusing our attention on the correlation with zero time delay, the quantity $C(0) = \langle I_1(t)I_2(t)\rangle$.

We will assume again that quantum mechanical expectation values are directly applicable to the interpretation of a photon-counting experiment when the counting time is short compared with the coherence time. The first question we must then address is that of associating a quantum mechanical expectation value with $C(0)$. Now a measurement of the intensity of a quasi-monochromatic field records a quantity proportional to $\langle n \rangle = \bar{n}$, the expectation value of the photon number, and so we might suspect that an experiment such as that of Fig. 13.15 measures a quantity proportional to $\langle n^2 \rangle$. In fact, such an experiment measures not $\langle n^2 \rangle$ but $\langle n^2 \rangle - \langle n \rangle$. To appreciate why this is so, suppose the field incident on the beam splitter contains one and only one photon, i.e., it is in a stationary state of energy $h\nu$. In this case $\langle n^2 \rangle = 1$. But the experiment of Fig. 13.15 must give $C(0) = 0$ in this case because the beam splitter cannot split the incident photon. Thus, $C(0)$ cannot be proportional to $\langle n^2 \rangle$. On the other hand

[6]Brown's given name Robert is sufficiently common that he deliberately emphasized his middle name Hanbury by spelling it out in signing his articles. The erroneous impression is widespread that his family name is actually Hanbury Brown, and the Brown–Twiss effect is almost universally referred to as the Hanbury Brown–Twiss effect, a usage we adopt hereafter.

$\langle n^2 \rangle - \langle n \rangle = 0$ in this example, so that $\langle n^2 \rangle - \langle n \rangle$ is at least a plausible possibility for the quantum expectation value given by $C(0)$. We will simply outline a derivation of this result. A proof is given in the black-dot section below.

In an experiment like that indicated in Fig. 13.15, in which PM1 and PM2 can simultaneously record photon counts, we are interested, according to (13.15.2), in the average value of a product of four fields. If the incident field is in a stationary state of n photons, this expectation value turns out to be proportional to $n(n-1)$. This result may be understood as follows. The probability of counting a photon is proportional to n, the number of photons in the field. Thus, the probability of then counting a second photon is proportional to $n-1$, the number of photons remaining after the first is removed by the detection process. The probability of counting the two photons is then proportional to the average of the product of n and $n-1$, that is, to the expectation value $\langle n(n-1) \rangle = \langle n^2 \rangle - \langle n \rangle$.

Since $\langle \Delta n^2 \rangle = \langle n^2 \rangle - \langle n \rangle^2$, it follows that

$$C(0) \propto \langle n^2 \rangle - \langle n \rangle = \langle \Delta n^2 \rangle + \langle n \rangle^2 - \langle n \rangle = \langle \Delta n^2 \rangle + \bar{n}^2 - \bar{n}. \qquad (13.15.3)$$

For a thermal source $\langle \Delta n^2 \rangle$ is given by (13.14.15), and so

$$C(0) = \bar{n}^2 + \bar{n} + \bar{n}^2 - \bar{n} = 2\bar{n}^2. \qquad (13.15.4)$$

Now $C(0) = \langle I_1(t)I_2(t) \rangle$ and $\langle I_1(t) \rangle = \langle I_2(t) \rangle = I \propto \bar{n}$, and therefore (13.15.4) implies that

$$\frac{\langle I_1(t)I_2(t) \rangle}{\langle I_1(t) \rangle \langle I_2(t) \rangle} = 2 \quad \text{(thermal radiation)}. \qquad (13.15.5)$$

According to this result, a thermal field exhibits excess intensity fluctuations (i.e., the ratio is greater than 1) for $\tau = 0$, as shown in Fig. 13.16. This is the Hanbury Brown–Twiss effect for thermal radiation. It is seen from our derivation to be a consequence of the Bose–Einstein statistics of thermal radiation.

This Hanbury Brown–Twiss correlation is often called *photon bunching*, for it indicates a tendency for photons to arrive simultaneously at PM1 and PM2, i.e., to bunch together. This bunching also occurs if PM1 and PM2 are close together or coincident, in which case we may say that *for thermal radiation there is a statistical tendency for photons to arrive in pairs*.

Suppose, however, that the incident light is from a laser. In this case the photon statistics is Poissonian, and $\langle \Delta n^2 \rangle$ is given by (13.14.14). From (13.15.3) we then have

$$C(0) \propto \bar{n} + \bar{n}^2 - \bar{n} = \bar{n}^2, \qquad (13.15.6)$$

or

$$\frac{\langle I_1(t)I_2(t) \rangle}{\langle I_1(t) \rangle \langle I_2(t) \rangle} = 1 \quad \text{(laser)}. \qquad (13.15.7)$$

Thus, a field described by Poisson photon statistics shows no Hanbury Brown–Twiss correlations. There are no excess fluctuations. The photon arrivals are statistically independent. This distinction between laser radiation and thermal radiation has been accurately confirmed experimentally.

• In more advanced analyses it is convenient to define the non-Hermitian operators $\mathbf{a}$ and $\mathbf{a}^\dagger$ from the coordinate $\mathbf{q}$ and momentum $\mathbf{p}$:

$$\mathbf{a} = (2\hbar\omega)^{-1/2}(\mathbf{p} - i\omega\mathbf{q}), \qquad \mathbf{a}^\dagger = (2\hbar\omega)^{-1/2}(\mathbf{p} + i\omega\mathbf{q}) \qquad (13.15.8)$$

for a harmonic oscillator of angular frequency ω and unit mass; in the quantum theory of radiation an electromagnetic field mode of frequency ω corresponds exactly to such an oscillator. The operators $\mathbf{a}$ and $\mathbf{a}^\dagger$ satisfy the commutation relation

$$\mathbf{a}\mathbf{a}^\dagger - \mathbf{a}^\dagger\mathbf{a} = 1, \qquad (13.15.9)$$

as is easily verified from definitions (13.15.8) and the commutation relation $\mathbf{q}\mathbf{p} - \mathbf{p}\mathbf{q} = i\hbar$. When the operator $\mathbf{a}$ acts on a stationary state $|n\rangle$ of oscillator energy $n\hbar\omega$, it converts it to a state $|n-1\rangle$; $\mathbf{a}$ is therefore called a *lowering* (or annihilation) operator. Similarly $\mathbf{a}^\dagger$ acting on the state $|n\rangle$ produces the state $|n+1\rangle$, and so $\mathbf{a}^\dagger$ is called a raising (or creation) operator. In the quantum theory of radiation these operators are called photon annihilation and creation operators, and it may be shown that they represent the quantized complex field amplitudes E and E^*, respectively. Thus, we have

$$C(0) \propto \langle \mathbf{a}^\dagger\mathbf{a}^\dagger\mathbf{a}\mathbf{a} \rangle \qquad (13.15.10)$$

for a single-mode field. Using (13.15.9), we have

$$\mathbf{a}^\dagger\mathbf{a}^\dagger\mathbf{a}\mathbf{a} = \mathbf{a}^\dagger(\mathbf{a}\mathbf{a}^\dagger - 1)\mathbf{a} = \mathbf{a}^\dagger\mathbf{a}\mathbf{a}^\dagger\mathbf{a} - \mathbf{a}^\dagger\mathbf{a}. \qquad (13.15.11)$$

Now from (13.15.8) it follows that

$$\hbar\omega\mathbf{a}^\dagger\mathbf{a} = \tfrac{1}{2}(\mathbf{p}^2 + \omega^2\mathbf{q}^2) - \tfrac{1}{2} = \mathbf{H} - \tfrac{1}{2}\hbar\omega, \qquad (13.15.12)$$

where $\mathbf{H}$ is the Hamiltonian operator for the mode "oscillator," with eigenvalues $(n + \tfrac{1}{2})\hbar\omega$. It follows that $\mathbf{a}^\dagger\mathbf{a}$ is the operator associated with the photon number n, and so from (13.15.10) and (13.15.11) we have

$$C(0) \propto \langle n^2 \rangle - \langle n \rangle, \qquad (13.15.13)$$

which is the result used above.

The key step in the derivation of (13.15.13) is the identification (13.15.10) of $C(0)$ with $\langle \mathbf{a}^\dagger\mathbf{a}^\dagger \mathbf{a}\mathbf{a} \rangle$. In particular, the ordering of the $\mathbf{a}$, $\mathbf{a}^\dagger$ operators is absolutely crucial. This ordering, in which $\mathbf{a}^\dagger$'s appear to the left of $\mathbf{a}$'s, is called *normal ordering*. For a discussion of the physical motivation for normal ordering, we refer the reader to more advanced treatises.[7]   •

The term "photon bunching" for the Hanbury Brown–Twiss effect might convey the impression that the effect cannot be understood in classical terms. However, it is possible, and instructive, to explain the effect without invoking photons and the quantum theory of radiation. For this purpose we consider a model of a thermal source in which there are N atoms, atom j radiating a field whose time dependence is given by

$$E_j(t) = \mathcal{E}_0 e^{-i(\omega t + \phi_j)}, \qquad (13.15.14)$$

---

[7]See, for instance, L. Mandel and E. Wolf, *Optical Coherence and Quantum Optics*, Cambridge University Press, Cambridge, UK, 1995.

where the phase ϕ_j varies randomly in time due to collisions among the atoms. The total complex field is

$$E(t) = \mathcal{E}_0 e^{-i\omega t} \sum_{j=1}^{N} e^{-i\phi_j}, \tag{13.15.15}$$

and the total intensity is [recall (13.4.11)]

$$I(t) = \frac{1}{2} c\epsilon_0 |\mathcal{E}_0|^2 \sum_{i=1}^{N} \sum_{j=1}^{N} e^{i(\phi_i - \phi_j)} \tag{13.15.16}$$

when we average over a few periods of time $2\pi/\omega$. Now if the randomly varying phases ϕ_i are statistically independent, the average of $e^{i(\phi_i - \phi_j)}$ is zero unless $i = j$, in which case it is unity [recall Problem 6.5]. Thus, the average value of $I(t)$ is

$$\langle I(t) \rangle = \frac{1}{2} c\epsilon_0 |\mathcal{E}_0|^2 \sum_{j=1}^{N} (1) = \frac{1}{2} c\epsilon_0 N |\mathcal{E}_0|^2. \tag{13.15.17}$$

Note that this is a classical average over the random phases, not a quantum mechanical expectation value.

Similarly, the average square of the intensity corresponding to (13.15.16) is proportional to

$$\langle I^2(t) \rangle = \left(\frac{1}{2} c\epsilon_0 \right)^2 |\mathcal{E}_0|^4 \sum_i \sum_j \sum_l \sum_m \langle e^{i(\phi_i - \phi_j + \phi_l - \phi_m)} \rangle. \tag{13.15.18}$$

Because the ϕ_j's are independent, the summand above is nonzero only when $i = j, l = m$ or when $i = m, j = l$. Thus,

$$\langle I^2(t) \rangle = \left(\frac{1}{2} c\epsilon_0 \right)^2 |\mathcal{E}_0|^4 \left(\sum_{i=1}^{N} \sum_{l=1}^{N} (1) + \sum_{i=1}^{N} \sum_{j=1}^{N} (1) \right) = \left(\frac{1}{2} c\epsilon_0 \right)^2 |\mathcal{E}_0|^4 (N^2 + N^2)$$

$$= 2 \langle I(t) \rangle^2. \tag{13.15.19}$$

The factor of 2 leads to the result (13.15.5), i.e., to the photon bunching or Hanbury Brown–Twiss intensity correlation. Obviously, we have here only a very crude model of a thermal source. The point is that the uncorrelated emissions of rapidly fluctuating fields lead naturally to the positive intensity correlations found in the Hanbury Brown–Twiss experiment.

If the atoms of a source all act coherently to produce a total complex field with a single phase $\phi(t)$,

$$E(t) = N\mathcal{E}_0 e^{-i(\omega t + \phi)}, \tag{13.15.20}$$

and therefore a total (cycle-averaged) intensity

$$I = \frac{c\epsilon_0}{2} |N\mathcal{E}_0|^2, \tag{13.15.21}$$

then obviously (13.15.7) is satisfied (i.e., there are no Hanbury Brown–Twiss correlations) simply because I is constant. Note that this absence of Hanbury Brown–
Twiss correlations applies even if the phase ϕ in (13.15.20) is randomly varying,
because (13.15.21) is valid regardless of the variations of the single phase ϕ. Thus,
we have a simple, classical explanation for the absence of Hanbury Brown–Twiss correlations in an ideal laser beam.

● These results can be generalized as follows. Instead of (13.15.15) let us write

$$E(t) = e^{-i\omega t} \sum_{j=1}^{N} A_j e^{i\phi_j} = S e^{-i\omega t}, \tag{13.15.22}$$

where now we allow for both the (real) amplitude A_j and the phase ϕ_j of the jth source to vary. The
complex quantity $S = X + iY$, where the real quantities X and Y are given by

$$X = \sum_{j=1}^{N} A_j \cos \phi_j, \qquad Y = \sum_{j=1}^{N} A_j \sin \phi_j. \tag{13.15.23}$$

We assume the random variables A_j and ϕ_j are uncorrelated, and that the phases ϕ_j are uniformly
and independently distributed over the interval $[0, 2\pi]$, so that $\langle \cos \phi_j \rangle = \langle \sin \phi_j \rangle = \langle \cos \phi_j \sin \phi_j \rangle = 0$ and $\langle \cos \phi_i \cos \phi_j \rangle = \langle \sin \phi_i \sin \phi_j \rangle = \frac{1}{2}$ if $i = j$ and 0 otherwise. Then

$$\langle X \rangle = \sum_{j=1}^{N} \langle A_j \cos \phi_j \rangle = \sum_{j=1}^{N} \langle A_j \rangle \langle \cos \phi_j \rangle = 0 \tag{13.15.24}$$

and

$$\langle X^2 \rangle = \sum_{i=1}^{N} \sum_{j=1}^{N} \langle A_i A_j \cos \phi_i \cos \phi_j \rangle = \sum_{i=1}^{N} \sum_{j=1}^{N} \langle A_i A_j \rangle \langle \cos \phi_i \cos \phi_j \rangle = \frac{1}{2} \sum_{j=1}^{N} \langle A_j^2 \rangle \tag{13.15.25}$$

As in Sections 8.11 and 8.13, we invoke now the central limit theorem of probability theory. In
the case $\langle X \rangle = 0$, the central limit theorem requires that the probability distribution of X is

$$P_X(X) = \frac{1}{\sigma\sqrt{2\pi}} e^{-X^2/2\sigma^2}, \qquad \sigma^2 = \langle X^2 \rangle. \tag{13.15.26}$$

The probability distribution $P_Y(Y)$ is likewise Gaussian, and it is easy to see that $\langle Y^2 \rangle = \langle X^2 \rangle = \sigma^2$.
Furthermore X and Y are independent random variables, and so the joint probability distribution

$$P_{XY}(X, Y) = P_X(X)P_Y(Y) = \frac{1}{2\pi\sigma^2} e^{-(X^2+Y^2)/2\sigma^2}. \tag{13.15.27}$$

We can write the probability distribution $P_{R\theta}(R, \theta)$ for the polar "coordinates" R, θ, defined by writing $X = R\cos\theta$, $Y = R\sin\theta$, by noting that

$$P_{R\theta}(R, \theta)\,dR\,d\theta = P_{XY}(X, Y)\,dX\,dY = P_{XY}(X, Y)R\,dR\,d\theta, \qquad (13.15.28)$$

and therefore

$$P_{R\theta}(R, \theta) = RP_{XY}(X, Y) = \frac{R}{2\pi\sigma^2}e^{-R^2/2\sigma^2}. \qquad (13.15.29)$$

The probability distribution for R is obtained by integrating this distribution over all possible values of θ:[8]

$$P_R(R) = \int_0^{2\pi} P_{R\theta}(R, \theta)\,d\theta = \frac{R}{\sigma^2}e^{-R^2/2\sigma^2}. \qquad (13.15.30)$$

The intensity is $I = |S|^2 = X^2 + Y^2 = R^2$. Using the fact that $P(I)\,dI = P(I)\,dR^2 = 2RP(I)$ $dR = P_R(R)\,dR$, we have $P(I) = (1/2\sigma^2)e^{-I/2\sigma^2}$ or, since $\langle X^2 \rangle + \langle Y^2 \rangle = \langle I \rangle = 2\sigma^2$,

$$P(I) - \frac{1}{\langle I \rangle}e^{-I/\langle I \rangle} \qquad (13.15.31)$$

for the probability distribution of the intensity of radiation from a large number of independent radiators. It follows in particular that

$$\langle I^n \rangle = \int_0^\infty I^n \frac{1}{\langle I \rangle}e^{-I/\langle I \rangle}\,dI = n!\langle I \rangle^n, \qquad (13.15.32)$$

of which (13.15.19) is a special case.   ●

Although classical explanations are possible for many aspects of the coherence properties of optical fields, quantum theory allows for a far richer variety of photon statistics. We have considered only two examples, namely the Poisson statistics of laser radiation and the Bose–Einstein statistics of thermal radiation. The latter exhibits photon bunching; the former does not. To underscore the fact that other kinds of photon statistics can arise, we consider briefly the possibility of *photon antibunching*, that is, reduced intensity fluctuations (Fig. 13.17), or the opposite of the Hanbury Brown–Twiss correlation. In this case there is a tendency for photons not to arrive in pairs, or even to arrive randomly, but to arrive "well spaced." In particular, there is zero probability of detecting a second photon immediately after the detection of a first.

Photon antibunching has been observed in the resonance fluorescence of an atom prepared as a two-state system by pumping with circularly polarized light (Section 14.3). The two-state atom is driven up and down between the two states by a strong resonant field. In addition to the stimulated transitions, the atom can spontaneously emit a photon when it is in the upper state. After such a spontaneous emission of a photon, however, the atom cannot emit a second photon until it has been pumped by the resonant field back

---

[8]This distribution was first obtained—without using the central limit theorem—by Lord Rayleigh (1887), and is called the *Rayleigh distribution*.

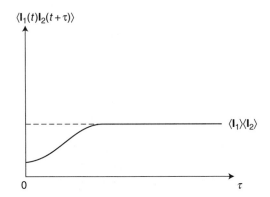

Figure 13.17   Photon antibunching near $\tau = 0$. [See H. J. Kimble, M. Dagenais, and L. Mandel, *Physical Review Letters* **39**, 691 (1977)].

into the upper state. Thus, the fluorescence (spontaneous emission) radiation displays a reduced intensity correlation, i.e., photon antibunching.

A field has first-order coherence if its complex degree of coherence $\gamma(\mathbf{r}_1, \mathbf{r}_2, \tau)$ has modulus unity. This occurs if and only if the mutual coherence function (13.4.18) factors:

$$\Gamma(\mathbf{r}_1, \mathbf{r}_2, \tau) = \langle \mathcal{E}^*(\mathbf{r}_1, t)\mathcal{E}(\mathbf{r}_2, t + \tau) \rangle = \langle \mathcal{E}^*(\mathbf{r}_1, t)\rangle\langle\mathcal{E}(\mathbf{r}_2, t + \tau) \rangle. \qquad (13.15.33)$$

This factorization condition for first-order coherence follows easily from the definition (13.6.3) of the complex degree of coherence. Similarly, a field is said to have *second-order coherence* if (13.15.7) is satisfied; this condition is equivalent to a factorization of the second-order coherence function of the field. In general, a field is said to have nth-order coherence if its nth-order coherence function factors. This definition of coherence leads to an elegant formulation of the quantum theory of coherence.[9]

PROBLEMS

13.1. **(a)** Assume that the sun is a 5800-K blackbody radiator. Calculate the intensity of radiation from the sun.

   (b) What is the intensity of solar radiation at the Earth's surface? (Answer: about 1.4 kW/m^2.)

13.2. In Section 7.6 we obtained the result (7.6.20) for the new waist of a focused Gaussian beam, assuming that the lens intercepts the beam at its waist. Show that (13.2.7) gives the spot size in the focal plane of the lens.

13.3. Explain why a nearsighted person sees a speckle pattern moving in the direction opposite to the head motion. (Recall that the brain inverts the image on the retina.) What would a farsighted person see? A person with no visual disorder? [See

---

[9]For a more extensive introduction to the subject see E. Wolf, *Introduction to the Theory of Coherence and Polarization of Light*, Cambridge University Press, Cambridge, 2007.

N. Mohon and A. Rodemann, *Applied Optics* **12**, 783 (1973).] If you wear corrective lenses, remember to remove them when you try this experiment. Explain why this effect of speckle is similar to the following one you can do while sitting at your desk. Focus your eyes on a distant object outside a window, and note the apparent motion of an object on the window sill as you move your head up and down or side to side. Then focus on the object on the sill, and note the apparent motion of the object outside when you move your head.

13.4. In an actual Michelson interferometer using an extended source of light (see Fig. 13.9), we observe circular interference fringes when viewing M_1 through BS if M_1 and M_2 are perpendicular (one of the mirrors of the interferometer has tilting screws so that the mirror orientation is adjustable). Explain the appearance of circular fringes. Do you expect to see circular fringes if the mirrors are not perpendicular?

13.5. Explain how a Michelson interferometer can be used to determine the wavelength of quasi-monochromatic radiation.

13.6. Why do we normally add intensities of different frequency components of radiation to obtain the total measured intensity?

13.7. **(a)** Estimate the bandwidth and coherence time of white light, assuming it comprises wavelengths from 400 to 700 nm.

   (b) Show that the coherence length of white light is on the order of the wavelength.

   (c) Is is possible to observe "white-light fringes"? [See A. Michelson, *Light Waves and Their Uses*, University of Chicago Press, Chicago, 1906.]

13.8. Suppose a mercury arc lamp, emitting 546.1-nm radiation with a bandwidth of 1 GHz, is placed behind a 0.1-mm-diameter circular aperture in an opaque screen. Beyond the screen is a second one with two narrow slits in it, and interference fringes are observed on a third screen a distance 3 m from the second. Calculate the slit separation for which the interference fringes first disappear.

13.9. A piece of transparent ground glass is placed before the two slits of a Young interference experiment and rotated rapidly. It is found that spatially coherent radiation does not produce any interference fringes in this modified two-slit experiment. Explain this observation. [See W. Martienssen and E. Spiller, *American Journal of Physics* **32**, 919 (1964).]

14 SOME APPLICATIONS OF LASERS

In the early years of their development, lasers were regarded by skeptics as "a solution looking for a problem." More and more "problems" were found, and lasers have become an important part of the science and technology of our time, with applications ranging from medical to military. Lasers have been used in distance and velocity measurements, holography, printers, bar coding, CD players, surgery, and many other areas for many years now, and such "everyday" applications will not be touched upon here. We will instead consider, in addition to some aspects of the medical applications of lasers, just a few examples of the importance of lasers in research and emerging technology. The special role of diode and fiber lasers in telecommunications is the main subject of the next chapter.

14.1 LIDAR

Light detection and ranging (lidar) dates back to the 1930s, but because of lasers it has become one of the primary tools in atmospheric and environmental research. There are several types of lidar, all involving a transmitter of laser radiation and a receiver for the detection and analysis of backscattered light (Fig. 14.1). Before describing some lidar techniques and the information they provide, we will derive an equation for the number of photons counted at the receiver. Our analysis will apply directly to the most common type of lidar system, that in which the transmitter and receiver are located at essentially the same place; the laser beam in this case is typically sent through the receiver telescope. This is called the "monostatic," as opposed to "bistatic," configuration.

Consider first the case in which the backscattered light is resonance fluorescence, that is, spontaneous emission from atoms (or molecules) excited by laser radiation. Suppose that a laser beam propagates vertically from ground level to an altitude z and excites atoms within a column between altitudes z and $z + \Delta z$ ($\Delta z \ll z$). The radiated energy in time Δt from these atoms is

$$\Delta E_{\text{atoms}}(z, t) = h\nu A_{21} \, \Delta t \int_{z}^{z+\Delta z} dz' \, N(z') \int_{-\infty}^{\infty} dx \int_{-\infty}^{\infty} dy \, p(x, y, z', t), \qquad (14.1.1)$$

Laser Physics. By Peter W. Milonni and Joseph H. Eberly
Copyright © 2010 John Wiley & Sons, Inc.

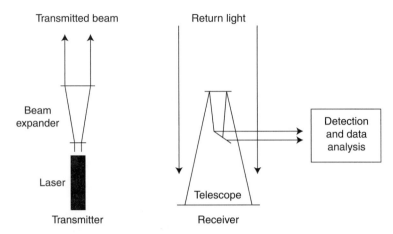

Figure 14.1 Basic elements of a standard lidar system. A beam expander is usually used at the transmitter in order to reduce the divergence of the laser beam before it propagates into the atmosphere. The receiver includes a wavelength filter, a photodetector, and computers and electronics for data acquisition and analysis.

where $N(z)$ is the number density of atoms at altitude z, ν is the transition frequency (assumed to be equal to the laser frequency), A_{21} is the spontaneous emission rate, and $p(x, y, z, t)$ is the excited-state probability at time t for an atom at a point x, y, z. If the excited atoms radiate isotropically, the power incident on a receiver of area A_r is simply

$$\mathrm{Pwr}(z, t) = T_0(z)\frac{A_r}{4\pi z^2}\Delta E_{\mathrm{atoms}}(z, t)$$

$$= T_0(z)\frac{A_r}{4\pi z^2}h\nu A_{21}\int_z^{z+\Delta z} dz'\, N(z')\int_{-\infty}^{\infty} dx\int_{-\infty}^{\infty} dy\, p(x, y, z', t), \quad (14.1.2)$$

where $T_0(z)$ is the atmospheric transmission coefficient at the transition frequency for propagation to ground from altitude z (or from ground to z), and is assumed not to vary significantly between z and $z + \Delta z$, and we continue to write Pwr for power, to distinguish it easily from P's denoting pumping rate and pressure and probability, etc.

We will assume that the excited-state probability $p(x, y, z, t)$ changes predominantly by spontaneous emission; this is a reasonable assumption at high altitudes, where molecular densities are sufficiently small that collisional deexcitation is negligible. Suppose that the laser radiation is in the form of a pulse of duration τ_p long compared to the radiative lifetime $1/A_{21}$. Then, for times t much longer than the radiative lifetime (Problem 14.1),

$$p(x, y, z, t) \cong \frac{\sigma(z, \nu)}{A_{21}h\nu}I(x, y, z, t), \quad (14.1.3)$$

where $\sigma(z, \nu)$ is the absorption cross section at z at the laser frequency ν, and so

$$\mathrm{Pwr}(z, t) = T_0(z)\frac{A_r}{4\pi z^2}\int_z^{z+\Delta z} dz'\, N(z')\sigma(z', \nu)\int_{-\infty}^{\infty} dx\int_{-\infty}^{\infty} dy\, I(x, y, z', t) \quad (14.1.4)$$

in this approximation.

Now consider the number of photons counted in a time interval from $2z/c$ to $2z/c+\tau_d$ by a detector at the receiver. $2z/c$ is the time it takes for the laser radiation to propagate to altitude z, plus the time for photons radiated at z to propagate back to ground.[1] The number of photons counted from the "bin" $[z, z + \Delta z]$ in this time interval is

$$\mathcal{N}(z) = \eta \times \frac{1}{h\nu} \times \int_{2z/c}^{2z/c+\tau_d} dt\, \text{Pwr}(z, t)$$

$$= \eta T_0(z)\frac{A_r}{4\pi z^2}\frac{1}{h\nu} \int_z^{z+\Delta z} dz'\, N(z')\sigma(z', \nu) \int_{-\infty}^{\infty} dx \int_{-\infty}^{\infty} dy \int_{2z/c}^{2z/c+\tau_d} dt\, I(x, y, z', t),$$

$$(14.1.5)$$

where $\eta\, (\leq 1)$ is the photon counting efficiency at the receiver. If τ_d is much larger than the pulse duration, the integral of $I(x, y, z', t)$ over x, y, and t in (14.1.5) is just the pulse energy $E(z')$ at z'. Since $E(z') = T_0(z')E_p \cong T_0(z)E_p$, where E_p is the energy of the pulse leaving the transmitter, we write (14.1.5) as

$$\mathcal{N}(z) = \eta T_0^2(z)\frac{A_r}{4\pi z^2}\frac{E_p}{h\nu} \int_z^{z+\Delta z} dz'\, N(z')\sigma(z', \nu). \qquad (14.1.6)$$

The column length Δz from which photons can reach the detector during the time interval $[2z/c,\ 2z/c + \tau_d]$ is determined by $2\Delta z/c = \tau_d$, or $\Delta z = c\tau_p/2$. We have assumed that this *range bin length* is made small enough that $N(z')$ and $\sigma(z', \nu)$ do not vary significantly within it. Then we obtain from (14.1.6) the *lidar equation*

$$\mathcal{N}(z) = \eta T_0^2(z)\frac{A_r}{4\pi z^2}\frac{E_p}{h\nu} N(z)\sigma(z, \nu)\frac{c\tau_d}{2}, \qquad (14.1.7)$$

for the number of photon counts at the receiver from a range bin of length $\Delta z = c\tau_d/2$ at altitude z. There is clearly a trade-off between the range bin length and the strength of the photon counting signal: The smaller the range bin, the greater the "range resolution" but the smaller the number of photons backscattered from it.

Suppose some atmospheric constituent is probed as a function of altitude and the detection electronics is designed to "reset to zero" at the end of each counting interval, after which it begins counting photons from the next range bin. This *range gating* continues for some number of range bins, and photon counts per range bin are accumulated with multiple laser pulses fired at some repetition rate R_{rep} over some "integration time" τ. The number of photons accumulated from a range bin of length Δz over the integration time τ is obtained by multiplying (14.1.7) by $R_{\text{rep}}\tau$. Then the lidar equation for the number of photon counts from a range bin of length Δz at altitude z takes the form

$$\mathcal{N}(z) = \eta T_0^2(z)\frac{(\text{Pwr})_L \tau A_r}{h\nu\, z^2} N(z)\sigma^{\text{B}}(z, \nu)\, \Delta z, \qquad (14.1.8)$$

where $(\text{Pwr})_L = E_p R_{\text{rep}}$ is the laser power. We have assumed that the scattering (or resonance fluorescence) is isotropic, i.e., that it is the same over all 4π steradians about the scatterer; in this case the cross section for scattering in the backward direction

---

[1]Whether we use here the speed of light in vacuum (c) or the phase velocity or, more appropriately, the group velocity, is inconsequential because the atmosphere is so weakly dispersive.

is $\sigma(z, \nu)/4\pi$, where $\sigma(z, \nu)$ is the total (over all angles) scattering cross section. But in general the scattering (or resonance fluorescence) is not isotropic, and to indicate this we have written $\sigma^B(z, \nu)$ in (14.1.8) instead of $\sigma(z, \nu)/4\pi$, where $\sigma^B(z, \nu)$ is the "differential" cross section for backscattering, and reduces to $\sigma(z, \nu)/4\pi$ if the scattering happens to be isotropic.

Various expressions relating to (14.1.7) and (14.1.8) are found in the lidar literature. For example, the average power at the receiver implied by (14.1.7) is

$$\text{Pwr}(z) = \frac{h\nu\mathcal{N}(z)}{\eta\tau_d} = T_0^2(z)\frac{A_r}{z^2}E_P N(z)\sigma^B(z, \nu)\frac{c}{2}$$
$$= (\text{Pwr})_0 T_0^2(z)\frac{A_r}{z^2}N(z)\sigma^B(z, \nu)\frac{c\tau_p}{2}, \qquad (14.1.9)$$

where $(\text{Pwr})_0$ is the single-pulse average power defined as E_p/τ_p. Similarly, if we take the time τ_d to be some arbitrary number of single pulse durations we infer from (14.1.7) that the photon count obtained from a single laser pulse is

$$\mathcal{N}(z) = \eta T_0^2(z)\frac{A_r}{z^2}\frac{E_p}{h\nu}N(z)\sigma^B(z, \nu)\frac{c\tau_p}{2}. \qquad (14.1.10)$$

What is called "the lidar equation" in the literature may refer to any of expressions (14.1.7)–(14.1.10) or to some other variant of (14.1.7).

In practice the number of "background counts" $\mathcal{N}_B$, which includes detector dark counts (Section 12.7) and counts from any background light sources (e.g., the sun), must be added to the right-hand side of whatever form of the lidar equation is used.

If laser pulses propagate from ground at a zenith angle ψ, where $\psi = 0$ defines vertical propagation, then the distance of atoms at altitude z from the receiver is $R = z/\cos \psi$, and this distance replaces z in the lidar equation. Similarly the lidar equation is often written with $T_0^2(R)$ expressed as an integral over the path to R:

$$T_0^2(R) = \exp\left[-2\int_0^R a(R')\,dR'\right], \qquad (14.1.11)$$

where $a(R')$ is the atmospheric attenuation coefficient at propagation distance R' at the wavelength of interest. For instance, $a(R')$ due to Rayleigh scattering will obviously decrease with altitude as the density of scatterers decreases.

It has been assumed that the backscattered photons come from resonance fluorescence, but the lidar equation is in fact more broadly applicable, as its form suggests. If the backscattered photons are due to scattering by air molecules (predominantly N_2 and O_2), the cross section $\sigma^B(z, \nu)$ in the lidar equation is the Rayleigh backscattering cross section, which is

$$\sigma_R^B(z, \nu) = \frac{\pi^2[n^2(z, \lambda) - 1]^2}{N^2(z)\lambda^4} \quad \begin{array}{l}\text{(Rayleigh backscattering cross section} \\ \text{for wavelength } \lambda = c/\nu), \end{array} \quad (14.1.12)$$

as shown below. The laser radiation is assumed to be narrowband and at the same frequency as the scattered radiation. Since Rayleigh scattering does not involve a change in molecular energy levels, it is often called *elastic scattering* in the lidar literature. This terminology applies also to the more general case of *Mie scattering* in which the scatterers are not necessarily small compared to the wavelength. The lidar equation is based on the approximation that multiple scattering is negligible; this is the approximation that each molecule scatters only the laser radiation and not the radiation scattered in its direction by another molecule.

We have implicitly assumed in deriving the lidar equation that the field of view of the receiving optics is larger than the laser beam divergence. It should also be noted that, although we have referred to photon counts at the detector, as is appropriate when return signals are measured with photomultiplier tubes or avalanche photodiodes in the Geiger mode, strong return signals are usually recorded in the analog mode and then converted to digital form (Section 12.7).

● The electric field from an electric dipole oscillating along a direction $\hat{\boldsymbol{\varepsilon}}$ is proportional to $(\hat{\boldsymbol{\varepsilon}} \cdot \hat{\mathbf{r}})\hat{\mathbf{r}} - \hat{\boldsymbol{\varepsilon}}$ in the radiation zone, where $\hat{\mathbf{r}}$ is the unit vector pointing from the dipole to the point of observation [Eq. (8.9.13)]. The scattered power in the direction $\hat{\mathbf{r}}$ is therefore proportional to

$$[(\hat{\boldsymbol{\varepsilon}} \cdot \hat{\mathbf{r}})\hat{\mathbf{r}} - \hat{\boldsymbol{\varepsilon}}]^2 = 1 - (\hat{\boldsymbol{\varepsilon}} \cdot \hat{\mathbf{r}})^2. \tag{14.1.13}$$

For a dipole induced by an incident field, $\hat{\boldsymbol{\varepsilon}}$ is the polarization unit vector of the incident field. Let θ be the scattering angle, that is, the angle between the propagation direction $\hat{\mathbf{z}}$ of the incident field and the direction $\hat{\mathbf{r}}$ in which the scattered field is observed. We write

$$\hat{\mathbf{r}} = \hat{\mathbf{x}} \sin\theta \cos\phi + \hat{\mathbf{y}} \sin\theta \sin\phi + \hat{\mathbf{z}} \cos\theta, \tag{14.1.14}$$

where $\hat{\mathbf{x}}$ and $\hat{\mathbf{y}}$ are orthogonal unit vectors in the plane perpendicular to $\hat{\mathbf{z}}$ and ϕ is the angle between $\hat{\mathbf{x}}$ and the projection of $\hat{\mathbf{r}}$ onto the xy plane. Then, for incident radiation polarized along $\hat{\mathbf{x}}$ or $\hat{\mathbf{y}}$, respectively, we have

$$1 - (\hat{\boldsymbol{\varepsilon}} \cdot \hat{\mathbf{r}})^2 = 1 - \sin^2\theta \cos^2\phi \quad (x \text{ polarization}), \tag{14.1.15a}$$

$$= 1 - \sin^2\theta \sin^2\phi \quad (y \text{ polarization}). \tag{14.1.15b}$$

In the case of unpolarized incident radiation the angular dependence of the scattered power can be obtained simply by taking the average of (14.1.15a) and (14.1.15b), since the x and y components of the incident field have the same (average) intensity:

$$\tfrac{1}{2}(1 - \sin^2\theta \cos^2\phi) + \tfrac{1}{2}(1 - \sin^2\theta \sin^2\phi) = 1 - \tfrac{1}{2}\sin^2\theta = \tfrac{1}{2}(1 + \cos^2\theta). \tag{14.1.16}$$

The same result is obtained for incident light that is circularly polarized because the x and y field components again have equal intensities. The integral of (14.1.16) over all solid angles Ω is

$$\int_{4\pi} d\Omega \, \tfrac{1}{2}(1 + \cos^2\theta) = \int_0^{2\pi} d\phi \int_0^{\pi} d\theta \sin\theta \, \tfrac{1}{2}(1 + \cos^2\theta) = \frac{8\pi}{3}, \tag{14.1.17}$$

and so we define the *differential scattering cross section* $d\sigma/d\Omega$ such that $\int_0^{2\pi} d\phi \int_0^{\pi} d\theta \sin\theta (d\sigma/d\Omega) = \sigma_R$, where σ_R is the *total* cross section for Rayleigh scattering defined by (8.9.5):

$$\frac{d\sigma}{d\Omega} = \sigma_R \times \frac{3}{8\pi} \frac{1}{2}(1 + \cos^2\theta) = \frac{\pi^2[n^2(z, \lambda) - 1]^2}{\lambda^4 N^2(z)} \frac{1}{2}(1 + \cos^2\theta), \tag{14.1.18}$$

for $n(\lambda) \cong 1$. Similary, for incident light linearly polarized in the x direction,

$$\frac{d\sigma}{d\Omega} = \frac{\pi^2[n^2(z, \lambda) - 1]^2}{\lambda^4 N^2(z)}(1 - \sin^2 \theta \cos^2 \phi), \tag{14.1.19}$$

and for y-polarized light $\cos^2 \phi$ is replaced by $\sin^2\phi$ in this formula. Therefore, the differential cross section for backscattering ($\theta = \pi$) is

$$\sigma_R^B(z, \nu) \equiv \frac{d\sigma}{d\Omega}(\theta = \pi) = \frac{\pi^2[n^2(z, \lambda) - 1]^2}{N^2(z)\lambda^4} \tag{14.1.20}$$

for *any* polarization of incident light.

Using (14.1.20) and the formula (8.10.5) for the refractive index of air, we obtain for the quantity $N(z)\sigma_R^B(z, \nu)$ in the lidar equation for Rayleigh scattering the approximation (Problem 14.2)

$$N(z)\sigma_R^B(z, \nu) \cong \frac{3.3 \times 10^4}{\lambda^4} \frac{P(z)}{T(z)} \text{ m}^{-1}, \tag{14.1.21}$$

where $P(z)$ and $T(z)$ are, respectively, the atmospheric pressure (millibars) and temperature (Kelvin) at altitude z and λ is the wavelength in nanometers.[2]

Rayleigh lidars typically probe the atmosphere at altitudes above about 30 km, where scattering from aerosols is negligible.[3] They have been used to infer atmospheric temperature distributions at such altitudes. To get a rough idea of the number of photon counts predicted by the lidar equation, consider a fairly typical sort of Rayleigh lidar system in which 532-nm, 1-J laser pulses at a repetition rate of 30 Hz propagate vertically to 30 km, and backscattered photons are returned from a 150-m range bin ($\tau_d = 1 \, \mu$s) to a 50-cm-diameter receiver aperture over an integration time of 1 min. At $z = 30$ km the atmospheric pressure and temperature are approximately 12 mbar and 250K, respectively; then, from (14.1.8) and (14.1.21) (Problem 14.2),

$$\mathcal{N}(z) \sim 3.2 \times 10^6 \times \eta T_0^2(z) \tag{14.1.22}$$

from a 150-m range bin at 30 km. A reasonable estimate of $T_0^2(z)$ at 30 km is about 0.6 for visible wavelengths far from any absorption resonances of air; the attenuation in this case is due predominantly to Rayleigh scattering from air molecules and to scattering from aerosols. Assuming $\eta \sim 0.5$ for the detection efficiency, we obtain $\mathcal{N}(z) \sim 10^6$ photon counts. Note that the power $h\nu\mathcal{N}(z)/(\eta\tau)$ backscattered to the receiver during the integration time in this example is about 10^{-14} W, compared to an average laser power of $(30 \text{ Hz})(1 \text{ J}) = 30$ W.   ●

The contribution of Rayleigh scattering by air molecules to the backscattered signal can be calculated, since the backscattering cross section and the attenuation coefficient for Rayleigh scattering are known as a function of pressure and temperature. The spectral width of the backscattered light depends on the velocity distribution of the air molecules. Since any aerosol particles present are much more massive and therefore have much smaller thermal velocities than air molecules, the spectral width of backscattered light from them is much narrower than that for Rayleigh scattering by air molecules. Narrow-bandwidth filters can be used to distinguish the contributions of air and aerosol

---

[2]Since $\sigma_R^B(z, \nu)$ is a *differential* cross section, $N(z)\sigma_R^B(z, \nu)$, strictly speaking, has dimensions of (length)$^{-1}$ (steradian)$^{-1}$ rather than (length)$^{-1}$. This distinction does not affect computations of backscattered photon numbers.

[3]In the lidar literature Rayleigh scattering refers specifically to scattering by molecules rather than by aerosol particles, which may or may not be small compared to the wavelength.

particles to the elastic backscatter, and therefore to determine distributions of aerosol concentrations. Similarly, whereas the absolute Doppler shift due to line-of-sight wind is the same for light scattered by aerosols and by air molecules, the *relative* Doppler shift for aerosols is much larger; in other words, the ratio of the Doppler shift resulting from being carried along by wind to the average thermal velocity is larger for aerosols. The large relative frequency shift allows the Doppler shift to be accurately measured as a separation between two narrow "spikes" in the spectrum of the backscattered light. This allows accurate profiling of atmospheric winds and is the basis for the atmospheric laser Doppler instrument (ALADIN) planned for operation aboard the European Space Agency's ADM-Aeolus "wind watch" satellite. This lidar system will operate at the 355-nm wavelength of frequency-tripled Nd : YAG laser radiation. It is expected to provide 100 wind profiles per hour with a range resolution of 1 km for altitudes up to 30 km.

Elastic-backscatter lidars can detect the presence of large, nonspherical particles (e.g., ice crystals or soot). For spherical particles the backscattered light has the same polarization as the laser radiation. Nonspherical particles, however, result in a "depolarization" of backscattered light; analysis of this depolarization can therefore determine whether such particles are present. Polarization-sensitive elastic-backscatter lidar has been useful, for instance, in studies of the distribution of ice crystals in cirrus clouds and of aerosols in dust layers.

Resonance fluorescence lidar studies of sodium and other mesospheric metallic species (e.g., potassium and iron) believed to result from meteoric ablation have also been important in atmospheric research. Sodium is found in a layer about 30 km wide centered at about 95 km.[4] It has been especially useful in resonance fluorescence lidar because of its strong D line transitions and its *relatively* large density (only a few thousand atoms per cubic centimeter!) in the mesosphere.

To be specific, let us consider, as in Section 3.13, the $3S_{1/2}(F = 2) \leftrightarrow 3P_{3/2}$ D_2 transition. We can estimate as follows the number of photon counts $\mathcal{N}$ that can be obtained from the full width of the sodium layer with laser radiation at this transition frequency. Assuming n_b range bins of length Δz [$\sim 30\,\mathrm{km}/n_b$] in the lidar equation (14.1.8), we write

$$\mathcal{N} \sim \sum_{m=1}^{n_b} \mathcal{N}(z + m\,\Delta z) \sim \eta T_0^2 \frac{(\mathrm{Pwr})_L \tau\, A_r}{h\nu} \frac{A_r}{z_{\mathrm{eff}}^2}\, \sigma_{\mathrm{Na}}^B \sum_{m=1}^{n_b} N(z + m\,\Delta z)\,\Delta z$$

$$= \eta T_0^2 \frac{(\mathrm{Pwr})_L \tau\, A_r}{h\nu} \frac{A_r}{z_{\mathrm{eff}}^2}\, \sigma_{\mathrm{Na}}^B C_s, \tag{14.1.23}$$

where C_s (m^{-2}) is the sodium *column density*. We have replaced $T_0(z)$ and z within the sodium layer by effective values T_0 and z_{eff}, and σ_{Na}^B is the backscattering cross section for sodium resonance fluorescence, which for our rough estimates is assumed to be isotropic; thus we take

$$\sigma_{\mathrm{Na}}^B = \frac{1}{4\pi}(9.3 \times 10^{-16}\,\mathrm{m}^2), \tag{14.1.24}$$

---

[4]The sodium layer is seen as a faint, thin yellow arc in a photograph taken by an astronaut during the September, 1992 flight of the space shuttle *Endeavour*. This photograph is reproduced in W. J. Wild and R. Q. Fugate, *Sky and Telescope* **87**, 10 (1994).

where 9.3×10^{-16} m^2 is the total (absorption) cross section calculated in Section 3.13 [Eq. (3.13.10)] for $T = 200$K, which is approximately the temperature in the mesosphere. Then, from (14.1.23) with $z_{\text{eff}} = 95$ km, we obtain the number of photons per unit area and per unit time arriving at the receiver during the integration time τ:

$$\frac{\mathcal{N}}{\eta A_r \tau} \sim 2.4 \times 10^{-8} T_0^2 (\text{Pwr})_L (\text{W}) C_s (\text{m}^{-2}) \text{ photons/m}^2/\text{s}. \qquad (14.1.25)$$

The sodium column density C_s varies with location, time of year, and even time of day, typical values being $3 - 6 \times 10^{13}$ m^{-2}, corresponding to peak sodium densities of around a few thousand atoms per cubic centimeter. Using $C_s = 4 \times 10^{13}$ m^{-2} and $T_0^2 \sim 0.5$, we estimate that $\mathcal{N}/(\eta A_r \tau) \sim 5 \times 10^5 \, T_0^2$ photons/m^2/s per watt of laser power arrive at the detector. This is much smaller than the photon flux from low-altitude Rayleigh backscattering, which can be eliminated from the measured sodium resonance fluorescence signal by range gating [for pulses shorter than about $(95 \text{ km})/c \sim 300 \, \mu\text{s}$] such that photon counting at the receiver is begun only after a sufficiently long time following the launch of a laser pulse.

We have made assumptions about column density and temperature in order to estimate photon returns, but, of course, one application of lidar is to *determine* such quantities from measured photon counts. The temperature profile of the mesosphere, for example, has been measured by sodium resonance fluorescence lidar based on the temperature dependence of the absorption cross section shown in Fig. 3.20. The maximum and minimum backscattering cross sections occur at frequencies we denote by ν_{max} and ν_{min}, respectively, and the ratio of these cross sections depends on the temperature. The ratio of range-gated photon return signals taken at the two laser frequencies ν_{max} and ν_{min}, for example, can therefore be compared with the theoretical ratio (Fig. 3.20) to infer the temperature as a function of altitude in the mesosphere. Figure 14.2 shows results of

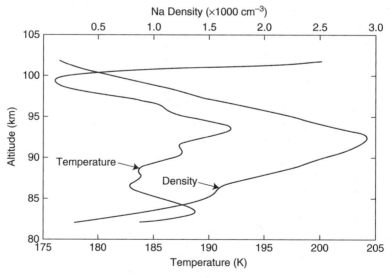

Figure 14.2 Temperature and sodium density in the mesosphere over Ft. Collins, Colorado, obtained from sodium resonance fluorescence lidar. [From R. E. Bills, C. S. Gardner, and C.-Y. She, *Optical Engineering* **30**, 13 (1991).]

such measurements for the temperature as well as the sodium density. The relatively high-resolution ($\Delta z = 1$ km) data were obtained with a cw dye laser tuned alternately between two frequencies.

● The sodium density can be obtained by normalizing the sodium photocounts to the Rayleigh photocounts from aerosol-free altitudes ($\gtrsim 35$ km) at which the Rayleigh scattering is due almost entirely to air molecules; this has the advantage of effectively eliminating uncertainties in the atmospheric transmission coefficient $T_0(z)$. Thus, from Eq. (14.1.8) it follows that the sodium density $N_{Na}(z)$ at altitude z can be expressed as (Problem 14.3)

$$N_{Na}(z) = \left[\frac{N_R(z_R)\sigma_R^B(z_R, \nu)}{\sigma_{Na}^B(z, \nu)} \right] \frac{z^2}{z_R^2} \frac{\mathcal{N}_{Na}(z)}{\mathcal{N}_R(z_R)}, \tag{14.1.26}$$

Here z_R and z are the altitudes for which the Rayleigh and sodium photon counts $\mathcal{N}_R(z_R)$ and $\mathcal{N}_{Na}(z)$, respectively, are taken, and $N_R(z_R)$ is the density of molecular (Rayleigh) scatterers at z_R. The factor in brackets on the right-hand side is known theoretically as a function of the laser frequency [cf. (14.1.21)], z^2/z_R^2 is determined by the altitudes probed, and the ratio of sodium and Rayleigh photocounts is measured to infer $N_{Na}(z)$.

In deriving the lidar equation, we used the proportionality (14.1.3) between the excited-state population and the intensity; this ignores the possibility of saturation or coherent excitation effects that depend on the temporal variations of the electric field. In most lidar systems saturation or coherent excitation effects are of no concern, and the forms of the lidar equation we have written are quite adequate. It should be remembered, however, that backscattering cross sections generally depend on the polarization of the laser. In narrowband lidars, moreover, it is often necesary to account for the laser linewidth via a convolution of the frequency-dependent cross section with the laser spectrum. ●

In *differential absorption lidar* (DIAL) the wavelength dependence of the transmission coefficient T_0 is used to determine the densities of absorbing species. Consider the lidar equation for the power at the receiver [cf. (14.1.9)] when the laser is at wavelength λ:

$$\text{Pwr}(R, \lambda) = (\text{Pwr})_0(\lambda) \frac{A_r}{R^2} N(R) \sigma^B(R, \lambda) \Delta R e^{-2 \int_0^R dR' a(R', \lambda)}, \tag{14.1.27}$$

where we have used (14.1.11) and have written R instead of z to allow for arbitrary zenith angles; ΔR is the range bin length, $(\text{Pwr})_0(\lambda)$ is the laser power at $R = 0$ at wavelength λ, and $a(R', \lambda)$ is the atmospheric attenuation coefficient at wavelength λ and range R'. The basic idea of DIAL is to measure photon return signals at two wavelengths λ_{on} and λ_{off}, where λ_{on} is a wavelength that is "on" resonance with an absorption line of the atmospheric molecule of interest and λ_{off} is "off" resonance. In other words, the molecule of interest absorbs at wavelength λ_{on} but not at λ_{off}. From (14.1.27) we obtain the ratio

$$\frac{\text{Pwr}(R, \lambda_{on})}{\text{Pwr}(R, \lambda_{off})} = \exp\left(-2 \int_0^R dR' [a(R', \lambda_{on}) - a(R', \lambda_{off})] \right). \tag{14.1.28}$$

We are assuming that the laser powers as well as the range bin lengths are the same at the two wavelengths, and that the two backscattering cross sections due, for instance, to Rayleigh scattering, are also approximately the same.

The attenuation at the two wavelengths is attributable to scattering from air molecules and aerosols as well as to absorption by any molecules with absorption lines near these wavelengths. Suppose, however, that only a particular molecular species contributes significantly to the integral over R of the *difference* $a(R, \lambda_{on}) - a(R, \lambda_{off})$; this, of course, is the species probed by DIAL with (slightly) different laser wavelengths λ_{on} and λ_{off}. Then we can ignore all contributions to (14.1.28) except that from this particular molecular species. Since

$$a(R', \lambda_{on}) - a(R', \lambda_{off}) = N_a(R')[\sigma_a(R', \lambda_{on}) - \sigma_a(R', \lambda_{off})], \qquad (14.1.29)$$

where N_a and σ_a are the number density and absorption cross section, respectively, of this particular molecule, it follows from (14.1.28) that

$$\frac{\text{Pwr}(R, \lambda_{on})}{\text{Pwr}(R, \lambda_{off})} = \exp\left[-2 \int_0^R dR' \, N_a(R') \, \Delta\sigma_a(R')\right], \qquad (14.1.30)$$

where we define the "differential absorption cross section"

$$\Delta\sigma_a(R') = \sigma_a(R', \lambda_{on}) - \sigma_a(R', \lambda_{off}). \qquad (14.1.31)$$

Differentiation of (14.1.30) yields

$$N_a(R) = \frac{-1}{2\Delta\sigma_a(R)} \frac{d}{dR}\left[\ln \frac{\text{Pwr}(R, \lambda_{on})}{\text{Pwr}(R, \lambda_{off})}\right], \qquad (14.1.32)$$

or, since the return signals are, of course, recorded in discrete range bins ΔR,

$$N_a(R) \cong \frac{1}{2\Delta\sigma_a(R)\,\Delta R} \ln\left[\frac{\text{Pwr}(R + \Delta R, \lambda_{off})}{\text{Pwr}(R, \lambda_{off})} \frac{\text{Pwr}(R, \lambda_{on})}{\text{Pwr}(R + \Delta R, \lambda_{on})}\right]. \qquad (14.1.33)$$

Sometimes DIAL is said to be (approximately) "self-calibrating": by measuring the return signals at the two wavelengths λ_{on} and λ_{off}, the number density $N_a(R)$ in different range bins is determined independently of the area of the receiving aperture, the photon-counting efficiency, the absolute laser power, the density and cross sections of the back-scatterers, and the atmospheric transmission coefficient $T_0(R)$. It is, of course, essential that the differential absorption cross section $\Delta\sigma_a(R)$ of the molecule of interest be known from theory or from laboratory experiments. DIAL has played a very important role in the determination of absolute concentrations of ozone, water vapor, carbon dioxide, and other environmentally important absorbers. It has also been used for temperature profiling of the atmosphere based on the known concentration of O_2 and the temperature dependence of its differential absorption cross section.

• Lidar signals are often very weak, and it is sometimes essential to detect them in such a way as to discriminate against background noise. This is done by *heterodyne detection*, as opposed to the "direct detection" we have presumed. In heterodyne detection the return light is superposed with light from a "local oscillator," a cw laser at frequency ω_{LO}. Let $E_S \cos(\omega_S + \phi_S)t$ and $E_{LO} \cos(\omega_{LO}t + \phi_{LO})$ denote the electric fields of the signal and local oscillator, respectively. The electric current i_D of a photodetector will be proportional to the total intensity $\varepsilon_0 c[E_S \cos(\omega_S t + \phi_S) + E_{LO} \cos(\omega_{LO}t + \phi_{LO})]^2$, or

$$i_D \propto \tfrac{1}{2}E_S^2 + \tfrac{1}{2}E_{LO}^2 + E_S E_{LO} \cos\left[(\omega_S - \omega_{LO})t + (\phi_S - \phi_{LO})\right] \qquad (14.1.34)$$

when terms at frequencies $2\omega_S$, $2\omega_{LO}$, and $\omega_S + \omega_{LO}$ are dropped under the assumption that these frequencies are so large that the detector can only respond to their time average, which is zero. In addition to a dc component, the current at the detector oscillates at the difference (beat) frequency $\omega_S - \omega_{LO}$, typically at radio frequencies, and with an amplitude proportional to $\sqrt{(\mathrm{Pwr})_S (\mathrm{Pwr})_{LO}}$, where $(\mathrm{Pwr})_S$ and $(\mathrm{Pwr})_{LO}$ are the powers of the signal and local oscillator. The principal advantage of heterodyning is the greater signal-to-noise ratio achievable with a sufficiently strong local oscillator field.

Heterodyne detection is "coherent" in that it responds to the signal field rather than directly to the intensity; it records information about the phase (and polarization) as well as the power of the signal. Lidars employing heterodyne detection are referred to as coherent DIAL, coherent Doppler, and so forth. Heterodyne detection requires a highly stable pulsed laser and local oscillator as well as sufficiently fast detectors; while offering better noise performance than direct detection, it is more sensitive to phase perturbations and misalignments. Because of the speckle effect due to atmospheric turbulence (Sections 8.11 and 13.8), there are shot-to-shot fluctuations in the return signal from each range bin, and a time averaging must be performed to smooth out these fluctuations. The heterodyne signal decreases for ranges R greater than the coherence length of the source (Section 13.11). •

Other lidar techniques are based on Raman scattering, an *inelastic* scattering process involving a change in the vibrational-rotational state of a molecule. The scattered radiation is shifted in frequency from the incident laser radiation by the change in the vibrational-rotational energy. Since the energy levels are distributed according to a Boltzmann distribution at the ambient temperature, Raman lidars have been used for temperature profiling of the atmosphere. Unlike resonance fluorescence lidar or DIAL, Raman lidars do not require specific laser wavelengths that match an absorption line of an atmospheric constituent; Raman cross sections are approximately proportional to λ^{-4}, as in Rayleigh scattering, so that the laser wavelengths used (often between about 320 and 550 nm) are typically smaller than in other lidars. The cross sections tend to be small, rendering Raman lidars most useful for probing constituents with relatively high concentrations. One of their principal applications is to measurements of water vapor concentrations in the troposphere, the lowest and densest layer of Earth's atmosphere containing nearly all its water vapor.[5]

The implementations and results of lidar are far too diverse to describe in any detail here. Mobile lidar systems housed in trucks and airplanes are used to monitor concentrations in air of many constituents including ozone, carbon dioxide, methane, water vapor, industrial emissions, and pollutants. Airborne lasers in the visible and ultraviolet (e.g., frequency-doubled Nd : YAG laser radiation at 532 nm) are used in fluorescence lidars in which chlorophyll, for example, absorbs at 532 nm and fluoresces at 685 and 740 nm; the strength of the radiation at these wavelengths provides information about forest ecosystems and environmental variations. Fluorescence submarine lidars have been developed for the detection of accidental or illegal chemical discharges in seawater. All such lidars employ some variant of the basic lidar equation and software for the "inversion" of the lidar equation to retrieve the desired information from measured return signals.

The first spaceborne lidar—the Lidar-in-Space Technology Experiment (LITE)—was launched in September, 1994 as the primary payload aboard the U.S. space shuttle

---

[5]The troposphere extends from ground to ~ 8 km at the poles and to ~ 18 km at the equator. It is where weather "happens."

Discovery. The transmitter was a Nd : YAG laser with frequency doublers and triplers generating 532 and 355 nm wavelengths at a 10-Hz pulse repetition rate. The receiving telescope had a 1-m diameter, and photomultipliers (for 532 and 355 nm) and an avalanche photodiode (for 1064 nm) were used for the detection of return signals; the vertical range resolution was about 15 m. This experimental system, the data from which were validated by six lidar-equipped aircraft as well as more than 90 ground-based systems in 20 countries, demonstrated that return signals in space could be obtained from ground or close to ground, and it succeeded in identifying storm systems, dust layers, and complex cloud structures. Earth-orbiting lidar systems under development will take comprehensive data from clouds and aerosols in order to obtain information needed for climate modeling.

Lasers have long been used for distance and velocity measurements. Range finders employ photodetection and timing electronics to measure the time T it takes for a light pulse to reach an object and reflect back to the transmitter: The distance to the object is $d = cT/2$. The advantages of laser pulses over radio-frequency systems (radar) are their very short durations, directionality, and their intensities that result in much stronger echo signals. On July 20, 1969, Apollo 11 astronauts placed an 18-inch-square reflector on the moon, and return pulses from a ruby laser at the Lick Observatory in California were first detected on August 1. Distances between points on Earth and the moon were determined to an accuracy of a few inches. The U.S. Laser Geodynamics Satellite (LAGEOS) launched in 1976 is a sphere of mass 407 kg and diameter 60 cm covered with 426 retro-reflectors for satellite laser ranging, similar in principle to lunar laser ranging. Laser tracking of the position of LAGEOS in its orbital altitude $\sim$5900 km revealed small variations in the length of the Earth day. A second LAGEOS satellite was launched in 1992 to provide more data relevant to seismic activity. Both satellites remain in service as of this writing.

Laser ranging can also be done interferometrically. In a Michelson interferometer (Section 13.10), for example, the fringe maxima and minima are interchanged when the arm separation is changed by $\lambda/2$: The magnitude of this change in the arm separation can be determined in terms of the wavelength λ by counting the number of fringe shifts as the change occurs. This technique is used routinely in length calibrations and machining applications.

Lasers are also used to measure velocities based on the Doppler effect. Laser velocimeters, some employing heterodyne detection, allow accurate measurements of velocities of aircraft and fluid flows, for example. Velocities can also be measured by *time of flight*: The time T for a laser pulse to propagate to a target and back implies the distance $d = cT/2$ to the object, and the slope of the plot of distance vs. time is the velocity. This is how police lidar guns work. They typically employ GaAs diode lasers at an eye-safe wavelength (904 nm) and average power ($\sim$50 μW). Since $\sim$100 pulses are used to compute the velocity, a typical pulse repetition rate of 300 Hz implies that only $\sim$0.3 s is needed for a velocity determination—not much time for a speeding driver to react and slow down.

14.2 ADAPTIVE OPTICS FOR ASTRONOMY

In Chapter 8 we introduced concepts such as the refractive index structure constant C_n^2, the coherence diameter r_0, and the seeing angle θ that are especially important in the theory of the propagation of light in the turbulent atmosphere. We showed that even under good seeing conditions a large ground-based telescope, while collecting more light and allowing the observation of far fainter objects, will have no greater image resolution than a much smaller telescope of comparable optical quality. Were it possible to eliminate the effects of atmospheric turbulence, the image-resolving capability of ground-based telescopes could exceed that of a space telescope (Section 8.11). Substantial enhancement of the imaging performance of ground-based telescopes is in fact possible with *adaptive optics*.

The degradation of image quality caused by atmospheric turbulence arises primarily from phase distortions across an incoming wavefront; recall the discussion in Section 8.11, where we showed that relatively small phase variations can substantially degrade image quality. The basic idea of adaptive optics is to measure the phase distortions, that is, the variations in local time lags across a wavefront, and to correct for them by advancing the phase at points where the atmosphere has retarded it and retarding it at points where the atmosphere has advanced it. The result, ideally, is a "corrected" wavefront in which transverse phase variations have been removed and the image of an object in a telescope is free of the blurring effect of the phase-distorting atmosphere. The advancement and retardation of the local phases are done with a deformable mirror surface, or "rubber mirror," as indicated in Fig. 14.3. The phase distortions across an incident wavefront—or more precisely the transverse phase *gradients*—are measured by a wavefront sensor and used by a wavefront reconstructor to compute approximations to the actual phase gradients. These are converted to voltages that drive an array of mechanical actuators (or "pistons") that adjust the shape of the deformable mirror surface so that the phase variations across an incoming wavefront are (ideally) absent in the reflected wavefront.

The most common type of wavefront sensor in adaptive optics is the *Shack–Hartmann sensor*, an array of lenslets or "subapertures" (Fig. 14.4). Each lenslet produces a spot on a detector array in the focal plane, and the displacement of the spot from its local null position is a measure of the local phase gradient—and therefore the local ray propagation direction—of the incoming wavefront at that subaperture. (The null positions of the focal spots can be defined using an undistorted incoming wavefront

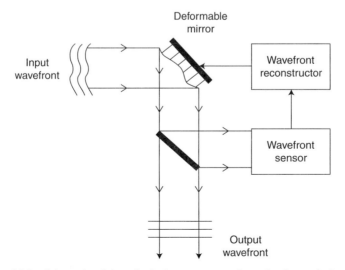

Figure 14.3 Schematic of the principal components of an adaptive optical system.

Input wavefront

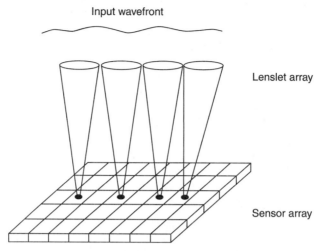

Lenslet array

Sensor array

Figure 14.4   Shack–Hartmann sensor consisting of an array of small lenslets (subapertures), each of which focuses a small portion of an incoming wavefront onto a spot on a detector array in the focal plane.

from a laser.) The Shack–Hartmann sensor thus transforms the local phase gradients of the incoming wavefront into a matrix of focal-spot displacements. The spots have a diffracted-limited form if the lenslet diameters are smaller than the atmospheric coherent diameter r_0 (Section 8.11). The detection array used to determine the phase gradient associated with each subaperture is usually a "quad cell," which consists of four detectors (e.g., avalanche photodiodes) forming the quadrants of a square: The different photon counts in the four quadrants provide the measure of the local phase gradient.

The performance of the adaptive optical system depends, among other things, on how accurately the local phase gradients can be measured. Measurement errors arise from uncertainties in the centroid positions of the focal spots in the Shack–Hartmann sensor. These uncertainties are due to photon number fluctuations of the incoming wavefront and to detector readout noise: Both types of noise diminish as the average number of incoming photons per subaperture increases.

The wavefront "reconstruction" process involves algorithms and software for estimating the actual phase profile from the measured matrix of phase gradients. This reconstruction provides the information used to adjust the deformable mirror and retard or advance the phase in such a way as to obtain the flat, undistorted output wavefront indicated in Fig. 14.3.

Deformable mirrors are either segmented or continuous. The segmented type consists of closely spaced planar mirror segments, each of which can be independently positioned, whereas the continuous type uses a single reflecting sheet (Fig. 14.5). Continuous deformable mirrors require more sophisticated control algorithms but avoid alignment and edge diffraction complications that arise with segmented mirrors.

When employed at a telescope the adaptive optical system of Fig. 14.3 operates as follows. Light that has been phase distorted by atmospheric turbulence enters the telescope and reflects off the deformable mirror. Part of the reflected light is incident on the wavefront sensor. The sensor measures the phase "errors" (deviations of focal spots from null positions), which would vanish if the deformable mirror were such that the reflected

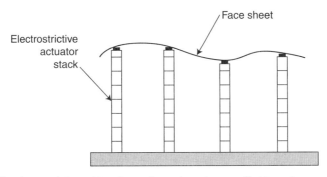

Figure 14.5 Continuous deformable mirror whose shape is controlled by voltages applied to a stack of electrostatic actuators (pistons). Deflections $\sim \pm 1\,\mu m$ are typically obtained with drive voltages $\sim \pm 100$ V.

wavefront had no phase distortions. The wavefront reconstructor then converts the sensor output into voltages that drive the "pistons" of the deformable mirror to change its shape in such a way as to reduce the phase errors measured by the sensor. In other words, the adaptive optical system functions as a servo-control loop—a very complicated one because of the large number of elements (pistons) that must be simultaneously controlled. Moreover, it must correct for phase errors on a time scale short compared to the ~ 10 ms in which phase distortions of incoming light change because of atmospheric fluctuations (Section 8.11). Deformable mirrors typically have much faster response times, so the rapidity of atmospheric fluctuations is not problemmatic.

Adaptive optical systems also employ a fast steering or "tip-tilt" mirror to remove the overall phase slope or tilt (Section 8.11) of the wavefront; tilt causes the image wander or "dancing" that is most responsible for the blurring of images obtained with long exposure times (Section 8.12). A tip-tilt sensor controls the steering mirror to stabilize the image for correction by the deformable mirror of "higher-order" phase distortions. Tilt typically accounts for more than 80% of the total power in wavefront distortion, but it can be followed and removed by the steering mirror with the light from stars as dim as magnitude 17–19; stars of the required magnitude cover about 60% of the sky.[6]

- A highly simplified model shows how wavefront sensing errors decrease with increasing light intensity. Consider instead of a quad cell a two-segment detector geometry that divides the image plane in the neighborhood of a focal spot from a subaperture into the regions $x < 0$ and $x > 0$. Let N_L and N_R be the number of photons counted in the regions $x < 0$ and $x > 0$, respectively, in a given time interval. Since the direction of ray propagation at the subaperture is perpendicular to the local wavefront, a phase tilt implies $N_R \neq N_L$: The approximately linear phase difference across the small subaperture results in a measured phase gradient

$$\Phi \propto \frac{N_R - N_L}{N_R + N_L}, \tag{14.2.1}$$

[6]It should be noted that wavefront tilt cannot be compensated using (single-wavelength) artificial laser guide stars because of "reciprocity:" The downward-propagating light from the guide star follows the same path as the upward-propagating laser beam, and so the guide star image will not appear to wander in the focal plane of the telescope.

and fluctuations ΔN_R and ΔN_L in N_R and N_L will cause a fluctuation

$$\Delta \Phi \propto \frac{(N_R + N_L)(\Delta N_R - \Delta N_L) - (N_R - N_L)(\Delta N_R + \Delta N_L)}{(N_R + N_L)^2}$$

$$= \frac{N_L \Delta N_R - N_R \Delta N_L}{(N_R + N_L)^2} \tag{14.2.2}$$

in the measured phase. Assuming $\langle \Delta N_R \rangle = \langle \Delta N_L \rangle = \langle \Delta N_R \Delta N_L \rangle = 0$, $\langle N_R \rangle = \langle N_L \rangle \equiv N/2$, and Poisson statistics for N_R and N_L [$\langle \Delta N_R^2 \rangle = \langle N_R \rangle$, $\langle \Delta N_L^2 \rangle = \langle N_L \rangle$], we obtain a mean-square fluctuation

$$\langle \Delta \Phi^2 \rangle \propto 1/N \tag{14.2.3}$$

for the measured phase, where N is the average photon count, or, more precisely, the average number of photoelectrons collected per subaperture. Wavefront sensing therefore increases in accuracy with increasing light intensity.   •

    Astronomical sources are too faint for the wavefront sensor to make accurate enough phase measurements for correction of higher-order distortions. Adaptive optical systems for astronomy therefore use the brighter light from a reference "guide star." Ideally the light from the guide star is distorted in the same way by atmospheric turbulence as the light from an astronomical object of interest, so that removal of the phase distortions of the light from the guide star will result in an improved image of the object. In order for the light from the guide star and the object to be phase distorted in the same way by atmospheric turbulence, the distortions should derive from the same "patch" of sky. The angle θ subtended by this patch at the receiver should be no larger than roughly the coherence diameter r_0 divided by the effective height (~ 10 km) of the atmosphere, in this case, $\theta \lesssim 20$ cm/10 km $= 4$ arcsec $= 19$ μrad for $r_0 = 20$ cm in the visible. This maximum allowable angular separation between the object and the guide star is called the *isoplanatic angle*. Unfortunately, there is a dearth of sufficiently bright stars that can be used as guide stars—it is estimated, for instance, that only about 0.1% of the sky has stars sufficiently bright to serve as guide stars in the visible.

    However, lasers allow *artificial* guide stars to be created anywhere in the sky. For example, the Rayleigh-backscattered light from a laser beam propagating up into the atmosphere can serve as a guide star. The resonance fluorescence from laser-excited mesospheric sodium atoms provides another type of artificial guide star. Mesospheric sodium has a major advantage over Rayleigh scattering as a guide star: It is at an altitude (~ 95 km) much higher than can be probed with sufficient photon returns by Rayleigh scattering. The light from it therefore suffers phase distortions attributable to essentially the full height of the atmosphere, and so an adaptive optical system employing it as a guide star will "correct" the wavefront from an astronomical object at "infinity." Figure 14.6 shows how an adaptive optical system and an artificial guide star can be employed at a ground-based telescope. Experiments have demonstrated the feasibility of adaptive optics using artificial as well as natural guide stars (Figs. 14.7 and 14.8).[7] We will confine the remainder of our discussion mainly to the sodium guide star.

---

[7]For a comprehensive review of optical technology for improving the resolution of large ground-based telescopes see, for instance, M. C. Roggemann, B. M. Welsh, and R. Q. Fugate, *Reviews of Modern Physics* **69**, 437 (1997).

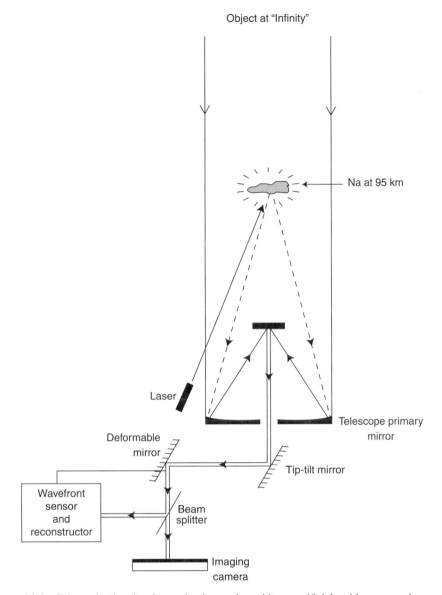

Figure 14.6 Schematic showing how adaptive optics with an artificial guide star can be used to improve the resolution of a telescope.

It is estimated that photon returns from the sodium layer of $\sim 10^6$ photons/m^2/s are required for adaptive optics as described above to result in a Strehl ratio of $\exp(-1.0) = 0.37$, a nominal value associated with generally adequate imaging.[8] Calculations of photon returns from the sodium guide star are the same as for the sodium resonance

[8]Estimates of required photon numbers for desired Strehl ratios, and comparisons of theoretical and experimental photon returns for various laser pulse formats, are discussed, for instance, in P. W. Milonni, R. Q. Fugate, and J. M. Telle, *Journal of the Optical Society of America A* **15**, 217 (1998).

Figure 14.7   The 3.5-m telescope facility at the Starfire Optical Range in New Mexico with a 589-nm beam propagating to the mesosphere to produce a sodium guide star. (Photo courtesy of J. M. Telle.)

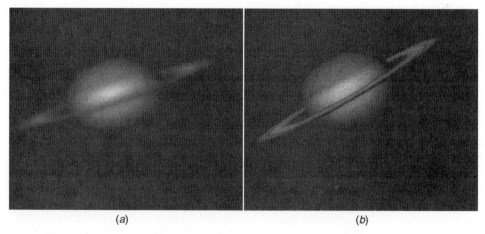

(a)   (b)

Figure 14.8   One of the early demonstrations of adaptive optics in astronomy (July, 1994). Images of Saturn and one of its moons (Titan) obtained (a) without and (b) with adaptive optics using a Rayleigh guide star at 656.4 nm with the 1.5-m telescope at the Starfire Optical Range in New Mexico. (Photo courtesy of R. Q. Fugate.)

fluorescence lidar and are generally consistent with observations. For example, formula (14.1.25) predicts a return photon flux of 1.6×10^6 photons/m^2/s per watt of cw laser power at 589 nm, compared with a reported 1.9×10^6 photons/m^2/s per watt when $T_0^2 C_s = 6.7 \times 10^{13}$ m^{-2}.[9] The reported data were actually the average of photon returns obtained with linearly and circularly polarized excitation, which gave 1.5×10^6 photons/m^2/s/W and 2.2×10^6 photons/m^2/s/W, respectively. For reasons discussed in the following section, circular polarization generally results in larger photon return fluxes. With a 30-W, circularly polarized transmitted beam, for example, the photon return was approximately 70×10^6 photons/m^2/s, whereas formula (14.1.25), which is applicable for laser intensities sufficiently small that the polarization effects discussed in the following section are insignificant, predicts a return of 52×10^6 photons/m^2/s for 30 W of transmitted power.

•   The intensity of light from an artificial guide star can be characterized by the "apparent magnitude" m used to quantify the visually perceived brightness of stars. For reasons grounded more in tradition than logic, the apparent magnitude is defined as

$$m = -[19 + 2.5 \log I], \tag{14.2.4}$$

where I is the intensity in W/m^2. The smaller m, the brighter the object. For the sun, the brightest object in the sky, $I \cong 1.4$ kW/m^2 at Earth's surface, and so $m \cong -26.9$. For the full moon, $m = -12.6$, while for Sirius (the brightest star aside from the sun), $m = -1.5$. The faintest stars observable with the naked eye have apparent magnitudes $m \cong 6$, whereas the Hubble space telescope can detect objects as faint as magnitude 30 in the visible. A sodium guide star yielding 70×10^6 photons/m^2/s at ground level has an apparent magnitude of about 7.6.   •

Quasi-monochromatic irradiation of mesospheric sodium excites only a narrow velocity group within the approximately 3-GHz-wide Doppler absorption profile (Fig. 3.20). By appropriate phase modulation the spectrum of laser radiation can be made to cover essentially the entire Doppler absorption profile. Various pulse-train excitation schemes have been employed, including trains of subnanosecond pulses producing the coherent excitation effects described in Chapter 9.

Saturation of the sodium transitions occurs at sufficiently high laser powers. If, for example, the D$_2$ line is excited by phase-modulated light whose spectrum is relatively flat over the absorption cross section, then formula (4.11.2) with the D$_2$ radiative rate $A_{21} = 1/(16 \text{ ns})$ and an "average" cross section $\bar{\sigma} \sim 2 \times 10^{-16}$ m^2 at 200K (Fig. 3.20) implies $I^{\text{sat}} \sim 5$ W/cm^2 for the saturation intensity. Suppose, for example, that such a phase-modulated pulse with a duration long compared to the radiative lifetime saturates the D$_2$ transition such that the excited-state probability for an atom with coordinates (x, y) transverse to the propagation direction has the quasi-steady-state value

$$\bar{p}(x, y) = \frac{\frac{1}{2} I(x, y)/I^{\text{sat}}}{1 + I(x, y)/I^{\text{sat}}} \tag{14.2.5}$$

[9]J. Telle, J. Drummond, C. Denman, P. Hillman, G. Moore, S. Novotny, and R. Fugate, *Proceedings of the SPIE (International Society for Optical Engineering)* **6215**, 62150K-1 (2006).

during the pulse. Suppose also that the average intensity at the mesosphere may be approximated by the Gaussian form[10]

$$I(x, y) = I_0 e^{-2(x^2+y^2)/w^2} = I_0 e^{-2r^2/w^2} \tag{14.2.6}$$

(w is typically ~ 1 m). Then the rate at which photons are backscattered from such a pulse to a receiving aperture of area A_r at ground is

$$
\begin{aligned}
\mathcal{R} &= \frac{T_0 C_s}{4\pi z^2} A_r \int_{-\infty}^{\infty} dx \int_{-\infty}^{\infty} dy \frac{A_{21} \times \frac{1}{2} I(x, y)/I^{\text{sat}}}{1 + I(x, y)/I^{\text{sat}}} \\
&= \frac{T_0 C_s}{4\pi z^2} A_r (2\pi) \int_0^{\infty} dr\, r \frac{A_{21} \times \frac{1}{2} I_0 e^{-2r^2/w^2}/I^{\text{sat}}}{1 + I_0 e^{-2r^2/w^2}/I^{\text{sat}}} \\
&= \frac{T_0 C_s}{4\pi z^2} A_r \frac{\pi w^2}{2} \ln\left(1 + \frac{I_0}{I^{\text{sat}}}\right),
\end{aligned}
\tag{14.2.7}
$$

aside from a possible factor accounting for anisotropy of the backscatter. Such an approximate functional dependence of $\mathcal{R}$ on the peak intensity I_0 provides a good fit to observed photon returns from "long," phase-modulated laser pulses.[8]

Work on sodium guide star includes research into methods of generating high-power radiation at 589 nm. Chains of dye lasers have been used to generate trains of pulses of sufficient intensity to strongly saturate the mesospheric sodium D_2 line, but these systems are large and complicated and the overall efficiency is low. Sum-frequency (three-wave mixing) generation using Nd : YAG laser transitions at 1.064 and 1.319 μm ($1/1064 + 1/1319 = 1/589$) appears especially promising for laser guide stars,[9] and fiber lasers capable of yielding sufficiently high powers at 589 nm are also an attractive possibility.

In Section 8.11 we introduced the seeing angle as a measure of angular resolution, and estimated that under good seeing conditions (characterized by a coherence diameter $r_0 = 10$ cm in the visible) the angular resolution of a 10-m telescope is about 1 arcsec compared to its theoretical diffraction-limited value of 0.01 arcsec. In other words, compensation for atmospheric turbulence could ideally result in an improvement in angular resolution by a factor of 100! Aside from far more detailed imaging, such an improvement in angular resolution would have important benefits for ground-based astronomical spectroscopy: It would allow the use of smaller spectroscopic slits and therefore a reduction of background radiation in the spectroscopy of very faint objects.

• Adaptive optics was first proposed in 1953 by the astronomer H. W. Babcock, who envisioned a deformable mirror based on an oil film whose thickness over a mirror surface could be controlled electrostatically. Similar ideas were advanced by V. P. Linnik. These concepts were impossible to implement without the computers and image-monitoring devices that came much later.

The concept of an artificial guide star, like much of adaptive optics, originated in classified military research on satellite surveillance and other applications. One type of guide star or "beacon" used in satellite surveillance is a "glint" of sunlight reflected by a satellite. The use

---

[10]The laser radiation propagating to the mesosphere is of course subject to speckle and other effects of turbulence discussed in Chapter 8.

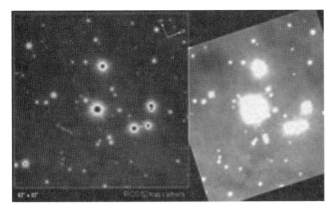

Figure 14.9 One of the early demonstrations of the use of the sodium guide star for adaptive optics on a large telescope (October, 2006). Images of the Trapezium region of the Orion Nebula with (left) and without (right) adaptive optics. (© Subaru Telescope, National Astronomical Observatory of Japan.)

of Rayleigh-scattered light as a beacon for adaptive optics was first suggested in the late 1970s. In 1982 mesospheric sodium was proposed as a beacon for adaptive optics in a classified report;[11] in 1985 it was suggested as a guide star for adaptive optics on telescopes,[12] and the first experimental studies in that direction were reported in 1987.[13] In 1991 most of the military research on adaptive optics and laser guide stars in the United States was declassified.   •

Adaptive optics with a sodium guide star is not a perfect solution to the problem of imaging through atmospheric turbulence. For one thing, *focus anisoplanatism*—the fact that the light from the guide star is a spherical wave rather than the plane wave from an astronomical object at "infinity"—implies that the guide star light does not "sample" exactly the same part of the turbulent atmosphere as the light from an object at infinity. And as already mentioned, a natural guide star is still required in order to compensate for the overall phase tilt caused by the atmosphere. But the results thus far have been impressive, and development of adaptive optical telescope systems with the sodium guide star is proceeding rapidly. Figure 14.9 shows an example of the image improvement obtained with the 8.2-m Subaru telescope at Mauna Kea, Hawaii, one of a growing number of very large (6–10 m) ground-based telescopes employing adaptive optics with sodium guide stars. The adaptive optics system used a 188-subaperture Shack–Hartmann wavefront sensor, about 2 m wide, and a 13-cm-diameter deformable mirror with 188 actuators. The 589-nm beam was obtained by sum-frequency generation with 1064- and 1319-nm Nd : YAG laser radiation; the 589-nm radiation was sent by a photonic crystal fiber from the room housing the lasers and the nonlinear optics to the launch telescope. The diffraction-limited angular resolution at the 2.2-μm imaging wavelength is $\sim 2.2 \times 10^{-6}/(8.2) = 0.27$ μrad = 0.06 arcsec, compared to a 0.6-arcsec resolution without any adaptive optics. With adaptive optics and the sodium guide star the angular resolution was nearly diffraction-limited.

[11]W. Happer, G. J. MacDonald, C. E. Max, and F. J. Dyson, *Journal of the Optical Society of America B* **11**, 263 (1994).
[12]R. Foy and A. Labeyrie, *Astronomy and Astrophysics* **152**, L29 (1985).
[13]L. A. Thompson and C. S. Gardner, *Nature* **328**, 229 (1987).

14.3 OPTICAL PUMPING AND SPIN-POLARIZED ATOMS

Aside from occasional allusions to the possibility of using polarized light to preferentially populate particular magnetic substates (cf. Sections 3.7 and 4.11), we have presumed that degenerate states differing only in the magnetic quantum number m are equally populated. In the calculation of the absorption cross section for the sodium D_2 line in Section 3.13, for example, the three magnetic substates of the $3S_{1/2}(F = 1)$ level and the five magnetic substates of the $3S_{1/2}(F = 2)$ level were assigned equal populations, as must be the case for degenerate states in thermal equilibrium. The process by which a departure from a thermal distribution occurs due to irradiation with light, as in the creation of population inversion in dye or solid-state lasers, for example, is called *optical pumping*. The term often refers specifically to the redistribution of hyperfine levels and magnetic substates by a resonant atom–field interaction, and it is in this sense that we will use it here. Optical pumping as such was well understood by the late 1940s but, as with so many other aspects of spectroscopy, it came into very widespread use primarily because of the quasi-monochromaticity, directionality, tunability, and intensity of light made possible by the laser.

Let us briefly review the physical significance of the magnetic substates. An atom has a magnetic dipole moment proportional to its angular momentum $\mathbf{F}$, and for a level with total angular momentum quantum number F there are $2F + 1$ degenerate substates corresponding to "magnetic" quantum numbers $m = -F, -F + 1, \ldots, -1, 0, 1, \ldots, F - 1, F$. (Recall the simplified discussion of the hydrogen atom in Section 2.2, or Fig. 3.19 for the hyperfine structure of the sodium D_2 line.) An atom in a state with magnetic quantum number m will be found in a measurement to have a component of angular momentum $m\hbar$ along *any* chosen "quantization axis," which is usually called the z axis. We are denoting the total angular momentum by $\mathbf{F}$, the standard notation for the sum of the electron orbital angular momentum $\mathbf{L}$, the electron spin angular momentum $\mathbf{S}$, and the nuclear spin angular momentum $\mathbf{I}$. Atoms with an odd isotope number (the sum of the number of protons and the number of neutrons in the nucleus) have a net nuclear spin, hyperfine structure, and a total angular momentum $\mathbf{F} = \mathbf{L} + \mathbf{S} + \mathbf{I}$. For atoms with even isotope numbers the net nuclear spin $\mathbf{I} = 0$ and the total angular momentum is $\mathbf{L} + \mathbf{S}$; in this case $\mathbf{F}$ in the discussion to follow, unless otherwise noted, is actually $\mathbf{L} + \mathbf{S}$ and is conventionally denoted by $\mathbf{J}$.

In the presence of a weak magnetic field $\mathbf{B} = B_z\hat{z}$ the degeneracy of the different magnetic substates is removed by the Zeeman effect: The state F, m is shifted in energy by an amount proportional to $\mathbf{F} \cdot \mathbf{B} = mB_z$:

$$\Delta E_{F,m} = \mu_B g_F m B_z, \tag{14.3.1}$$

where $\mu_B = e\hbar/2m_e = 9.274 \times 10^{-24}$ joule/tesla is the *Bohr magneton* (e and m_e are the electron charge and mass, respectively) and g_F is a so-called Landé g factor. For example, the Zeeman shift of a substate m of a hyperfine level with total angular momentum quantum number F (Section 3.13) is given approximately by (14.3.1) with

$$g_F \cong g_J \frac{F(F + 1) + J(J + 1) - I(I + 1)}{2F(F + 1)},$$

$$g_J \cong 1 + \frac{J(J + 1) + S(S + 1) - L(L + 1)}{2J(J + 1)}. \tag{14.3.2}$$

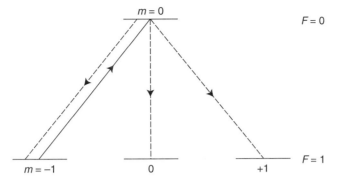

Figure 14.10   Optical pumping with σ_+ light of an atomic transition with lower- and upper-level angular momentum quantum numbers 1 and 0, respectively. The σ_+ light induces a transition indicated by the solid arrow, whereas spontaneous emission can occur on the transitions shown by the dashed arrows.

For sodium, $I = \frac{3}{2}$, $S = \frac{1}{2}$, $g_{J=3/2}(3P_{3/2}) = \frac{4}{3}$, and $g_{J=1/2}(3S_{1/2}) = 2$. Then, for example, $g_{F=1}(3S_{1/2}) = -\frac{1}{2}$ and $g_{F=1}(3P_{3/2}) = \frac{2}{3}$.

Now the fact that light carries intrinsic angular momentum implies that there can be a change Δm in the magnetic quantum number when light is absorbed or emitted. Circularly polarized photons carry angular momentum ± 1 (in units of $\hbar$) directed along the propagation direction $\hat{z}$, and the *selection rule* for allowed transitions is $\Delta m = \pm 1$ when the quantization axis is taken to be the $\hat{z}$ direction. For linearly polarized photons the selection rule is $\Delta m = 0$ with the quantization axis along the polarization direction, as discussed below. Based on these selection rules, it is easy to see how polarized light can be used to preferentially populate a particular magnetic substate, or in other words to "align" (or "spin-polarize") an atom so that along a particular direction its magnetic dipole moment has only one possible value. We will illustrate this with a few examples.

The solid arrow in Fig. 14.10 shows the $\Delta m = +1$ transition allowed when light with σ_+ circular polarization is resonant with an atomic transition having a lower level with three degenerate magnetic substates ($m = -1, 0, 1$) and an upper level with a single magnetic substate ($m = 0$).[14] The only allowed absorptive transition that can be induced by σ_+ light is between the lower state with $m = -1$ and the upper state with $m = 0$.[15] Spontaneous emission from the upper states, however, is constrained only by the selection rule $\Delta m = 0, \pm 1$, since we cannot associate any special direction or quantization axis with it. The upper state $m = 0$ can therefore decay spontaneously into any of the three lower states in Fig. 14.10. When a spontaneous transition occurs from the upper state to either the $m = 0$ or $m = 1$ lower state, the atom remains in that lower state

[14]σ_+ and σ_- circular polarizations are defined with respect to the *transitions* they can produce rather than with respect to left- or right-hand circular polarization of the light itself. Thus, for example, σ_+ light propagating in the $+z$ direction (the direction of the quantization axis) is said in optics to be left-hand circularly polarized, as is σ_- light propagating in the $-z$ direction. σ_+ photons have angular momentum $\pm \hbar$ along their direction of propagation and therefore can induce $\Delta m = \pm 1$ absorptive transitions. It is best in this context to avoid referring to "left" or "right" circular polarizations.
[15]Stimulated emission from the upper $m = 0$ state to the lower $m = -1$ state is also allowed, but it does not change the fact that the atom ends up eventually with zero population in the $m = -1$ state.

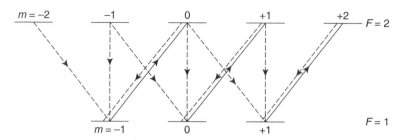

Figure 14.11   Optical pumping with σ_+ light of an atomic transition with lower- and upper-level angular momentum quantum numbers 1 and 2, respectively. σ_+ light induces transitions shown by solid arrows, whereas spontaneous emission occurs on transitions indicated by dashed arrows.

because there are no allowed transitions from it that can be induced by the applied σ_+ light. It is obvious then that, after a sufficient number of absorption and emission transitions, the atom will end up with zero probability of being in the $m = -1$ state, and therefore will cease to absorb any σ_+ light. In fact it is easy to see that any transition with a lower level having a greater (or equal) number of magnetic substates than the upper level will eventually cease to absorb σ_+ light; the same conclusion holds for applied σ_- light.

The same sort of thing happens when resonant linearly polarized light (labeled by π instead of σ) is incident on the atom with the upper and lower levels of Fig. 14.10. Because of the $\Delta m = 0$ selection rule for induced transitions in this case, while again $\Delta m = 0, \pm 1$ for spontaneous emission transitions, the atom after a few induced and spontaneous transitions will have zero probability of being in the $m = 0$ lower state and will not absorb the applied π light. The conclusion that π light will cease to absorb is also reached (of course!) if we choose to take the direction of light propagation rather than the direction of polarization as the quantization axis and regard the π-polarized light as a superposition of σ_+ and σ_- light [cf. (14.3.4)]. In terms of this quantization axis, with the magnetic quantum number denoted m', absorptive transitions occur only from the $m' = \pm 1$ lower states, and after a spontaneous transition from the upper state to the $m' = 0$ lower state the atom can no longer absorb.[16]

Consider next an example in which the lower level of a transition has a smaller number of magnetic substates than the upper level. Figure 14.11 shows allowed induced and spontaneous transitions when resonant σ_+ light is incident on an atom with lower and upper levels having three and five degenerate magnetic substates, respectively. It is clear that in this case the atom will eventually have zero probability of being in any state other than the lower $m = 1$ state or the upper $m = 2$ state. After it is fully "pumped" by σ_+ light it can only make spontaneous and stimulated transitions between these two states—*it is a two-state atom.* The circularly polarized light has "spin-polarized" the atom in the sense that its component of angular momentum along the direction $\hat{z}$ of field propagation (the quantization axis) is either $\hbar$ (when it is in the lower, $m = 1$ state) or $2\hbar$ (when it is in the upper, $m = 2$ state); if the σ_+ light is shut

[16]The m' states can be expressed as a linear combination of the m states, and vice versa. In general magnetic substates corresponding to two different quantization axes making an angle θ are related by rotation matrices $d_{mm'}^{(F)}(\theta)$. See, for instance, A. R. Edmonds, *Angular Momentum in Quantum Mechanics*, Princeton University Press, Princeton, NJ, 1996.

off, the atom will stay in the lower $m = 1$ state after making a spontaneous transition to that state. With σ_- light, similarly, a two-state atom is created with lower and upper states $m = -1$ and $m = -2$, respectively.

- We will derive the *general* selection rule $\Delta m = 0, \pm 1$ for electric dipole transitions using the example of a one-electron, spinless atom and the fact that a transition from a state 1 of energy E_1 to a state 2 of energy $E_2 > E_1$ is governed by the matrix element [see Eq. (9.3.16) and Problem 14.6]

$$\mathbf{x}_{21} \cdot \hat{\boldsymbol{\varepsilon}} = \int \phi_2^*(\mathbf{x})(\mathbf{x} \cdot \hat{\boldsymbol{\varepsilon}})\phi_1(\mathbf{x}) \, d^3x. \tag{14.3.3}$$

The polarization unit vector $\hat{\boldsymbol{\varepsilon}}$ will in general have components in all three directions $\hat{x}$, $\hat{y}$, and $\hat{z}$ defined by some Cartesian coordinate system. Consider polarization in the xy plane. It is convenient in this case to combine the x and y components and define σ_+ and σ_- circularly polarized light with complex unit polarization vectors (Problem 3.4)

$$\hat{\boldsymbol{\varepsilon}}_\pm = \frac{1}{\sqrt{2}}(\hat{x} \pm i\hat{y}), \tag{14.3.4}$$

where the orthogonal unit vectors $\hat{x}$ and $\hat{y}$ are perpendicular to the light propagation direction $\hat{z}$. To evaluate (14.3.3) we express the hydrogen wave functions in terms of spherical coordinates (r, θ, ϕ) defined with respect to the same (x, y, z) coordinate system, i.e., $x = r \sin \theta \cos \phi$, $y = r \sin \theta \sin \phi$, and $z = r \cos \theta$. Now all we need to know to derive the selection rule for the magnetic quantum number is that, for a state of our hydrogen atom with principal, orbital angular momentum, and magnetic quantum numbers n, ℓ, m, respectively, the dependence of the wave function $\phi_{n\ell m}(r, \theta, \phi)$ on the azimuthal angle ϕ is described entirely by a factor $\exp(im\phi)$. Thus, for an electric dipole transition from a state with quantum numbers n, ℓ, m to a state with quantum numbers n', ℓ', m' the matrix element (14.3.3) is proportional to

$$\int_0^{2\pi} e^{-im'\phi} e^{\pm i\phi} e^{im\phi} \, d\phi = \int_0^{2\pi} e^{-i(m'-m\mp 1)\phi} \, d\phi, \tag{14.3.5}$$

since $\mathbf{x} \cdot \hat{\boldsymbol{\varepsilon}}_\pm = (x\hat{x} + y\hat{y} + z\hat{z}) \cdot \hat{\boldsymbol{\varepsilon}}_\pm = (1/\sqrt{2})(x \pm iy) = (1/\sqrt{2})(r \sin \theta \cos \phi \pm ir \sin \theta \sin \phi) = (1/\sqrt{2})r \sin \theta \exp(\pm i\phi)$. m and m' have only integer values, and consequently the integral (14.3.5) is zero unless the exponent in the integrand is zero, i.e., unless the selection rule $m' = m + 1$ (for σ_+ light) or $m' = m - 1$ (for σ_- light) is satisfied.

The matrix element for the stimulated emission transition from the excited state 2 with magnetic quantum number m' to the lower energy state 1 with magnetic quantum number m is just the complex conjugate of (14.3.3) (Problem 14.6). In this case the same selection rule applies ($\Delta m = \pm 1$) but the magnetic quantum number *decreases* when the transition is induced by σ_+ light and *increases* when the transition is induced by σ_- light.

For light polarization parallel to the z axis we have $\mathbf{x} \cdot \hat{\boldsymbol{\varepsilon}} = \mathbf{x} \cdot \hat{z} = z = r \cos \theta$, and the matrix element (14.3.3) is proportional to

$$\int_0^{2\pi} e^{-im'\phi} e^{im\phi} \, d\phi = \int_0^{2\pi} e^{-i(m'-m)\phi} \, d\phi, \tag{14.3.6}$$

which vanishes unless $m = m'$. In other words, for linearly polarized light we have the selection rule $\Delta m = 0$ for allowed transitions. In this case the quantization axis (z) has been chosen to be along the direction of field polarization, whereas for circular polarization we took it to be the direction of field propagation.

Selection rules for the total angular momentum quantum number are derived in textbooks on quantum mechanics or atomic physics: $\Delta F = 0, \pm 1$, but $F = 0 \leftrightarrow F' = 0$ transitions are forbidden, that is, their electric dipole transition matrix elements are zero. If F is the total angular momentum quantum number in the case of nonzero nuclear spin ($\mathbf{F} = \mathbf{J} + \mathbf{I}$, the sum of electronic and nuclear angular momenta), we also have the selection rules $\Delta J = 0, \pm 1$ ($J = 0 \leftrightarrow J' = 0$ forbidden) and $\Delta m_J = 0, \pm 1$.

There are also selection rules for magnetic dipole and electric quadrupole transitions, which have much smaller transition rates than electric dipole transitions (typically about 10^5 and 10^8 times smaller, respectively). The selection rules for magnetic dipole transitions are $\Delta F = 0, \pm 1$ and $\Delta m = 0, \pm 1$, whereas for electric quadrupole transitions $\Delta F = 0, \pm 1, \pm 2$ and $\Delta m = 0, \pm 1, \pm 2$. In all cases $F = 0 \leftrightarrow F' = 0$ transitions are forbidden.   •

Let us return to the example of the sodium D_2 line. In Section 3.13 we calculated the absorption cross section shown in Fig. 3.20 under the assumption that the eight magnetic substates of the two $3S_{1/2}$ hyperfine levels were equally populated, as is approximately the case in thermal equilibrium at temperatures for which $k_B T$ is large compared to the 1.77 GHz $3S_{1/2}(F = 1) \leftrightarrow 3S_{1/2}(F = 2)$ hyperfine splitting. Suppose we irradiate the sodium atom with 589-nm σ_+ light that can induce transitions out of both the $3S_{1/2}$ hyperfine levels. From Fig. 3.19 and the selection rules $\Delta F = 0, \pm 1$ and $\Delta m = 1$ it can be seen that, for irradiation times long compared to the 16-ns radiative lifetime of the excited states, the sodium atom will be a two-state atom: only the $3S_{1/2}(F = 2, m = 2)$ and $3P_{3/2}(F = 3, m = 3)$ states will have nonzero occupation probabilities. The optical pumping to this two-state system occurs in ~ 20 excitation and decay transitions, after which the absorption cross section will have only a single peak instead of the two appearing in Fig. 3.20. As discussed below, this results in stronger absorption than in the case of the unpolarized D_2 line. If the σ_+ light is shut off after the two-state atom has been realized, the sodium atom after spontaneous emission from the $3P_{3/2}(F = 3, m = 3)$ state will be in the single, spin-polarized state $3S_{1/2}(F = 2, m = 2)$. In this state the electron with spin z-component m_S and the nucleus with spin z-component m_I have their spins aligned: $m = m_F = m_L + m_S + m_I = 0 + \frac{1}{2} + \frac{3}{2} = 2$.

Suppose instead that we irradiate the sodium atoms with narrowband, linearly polarized (π) radiation that can only induce transitions from the $3S_{1/2}(F = 2)$ level. Because the atoms excited to the $3P_{3/2}(F = 1, 2, 3)$ levels can undergo spontaneous transitions to $3S_{1/2}(F = 1, m = \pm 1)$, the sodium atoms in this case will become *transparent* to the π light after several excitation and decay transitions.

Absorption can increase or decrease in strength, therefore, when the incident light causes different, degenerate magnetic substates to have different populations. Such optical pumping occurs in the absence of collisions and magnetic fields that tend to redistribute the m-state populations. In collisions of sodium atoms, for example, the valence electrons can be exchanged; these *spin-exchange* collisions can quickly "thermalize" the populations of the magnetic substates. Spin-relaxation collisions can have large collision cross sections ($\sim 10^{-14}$ cm^2 for alkali atoms) and can prevent or rapidly destroy spin polarization. In a dilute vapor with relatively infrequent atom–atom collisions, atom–wall collisions can likewise prevent spin polarization.

Spin relaxation rates are greatly reduced in experiments with atomic beams, where collisions are effectively avoided. They are also substantially reduced by the use of "buffer" gases, especially inert gases, that cause very little spin relaxation in collisions with polarized atoms. Buffer gases at pressures of typically a few Torr are used to slow down the diffusion of polarized atoms to the cell walls and to maintain their polarization for times on the order of seconds or longer in the case of alkali atoms. Coating the cell walls with paraffins and other materials that do not cause significant spin relaxation is another frequently used method for preserving spin polarization.

Magnetic fields, including Earth's magnetic field, can also induce transitions between different magnetic substates F, m and F, m'. Such *static* **B** fields cannot induce transitions between states of different energy, and, in particular, cannot cause transitions between states of different total angular momentum quantum number F. The frequency with which a static **B** field induces oscillations between different m states is on the order of $\mu_B B / 2\pi\hbar$, as shown below. For Earth's magnetic field, for example, $B \approx 0.5\,\text{G}$ (gauss) $= 0.5 \times 10^{-4}$ T (tesla), and so the time scale for spin depolarization due to the geomagnetic field is typically on the order of a microsecond. Optical pumping experiments often employ a magnetic field applied along the direction of spin polarization; since the energy of the magnetic dipole moment **M** of the atom in a magnetic field **B** is $-\mathbf{M} \cdot \mathbf{B}$, the field serves to maintain the spin polarization in the presence of any stray, weak magnetic fields $\mathbf{B}_s$ for which $\mathbf{M} \times \mathbf{B}_s \neq 0$.

• For an example of how a magnetic field can affect optically pumped atoms, consider the experiment sketched in Fig. 14.12. Light propagating in the y direction and linearly polarized along x is resonant with an atomic transition that we assume for simplicity has a lower level with $F = 0$ and an upper level with $F = 1$. There is also a static **B** field along the z direction, which we will take to be the quantization axis.

For the incident electric field we write

$$\mathbf{E}(y, t) = \hat{x} E_0 e^{-i(\omega t - ky)} = \frac{1}{2}[(\hat{x} + i\hat{y}) + (\hat{x} - i\hat{y})]E_0 e^{-i(\omega t - ky)}$$

$$= \frac{1}{\sqrt{2}}[\hat{\boldsymbol{\varepsilon}}_+ + \hat{\boldsymbol{\varepsilon}}_-]E_0 e^{-i(\omega t - ky)}, \tag{14.3.7}$$

where as always it is implicit that we are to take the real part. The $\hat{\boldsymbol{\varepsilon}}_+$ and $\hat{\boldsymbol{\varepsilon}}_-$ components result in nonvanishing probability amplitudes for the $F = 1, m = 1$ and $F = 1, m = -1$ atomic states, respectively. We denote the stationary-state wave functions (eigenfunctions) of these two states by $\Phi_1(\mathbf{x})$ and $\Phi_{-1}(\mathbf{x})$, respectively, and similarly let $\Phi_0(\mathbf{x})$ denote the wave function for the

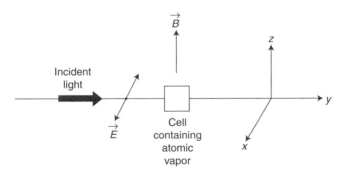

Figure 14.12 Light polarized along x propagates in the y direction and excites atoms in a dilute vapor. In the absence of a magnetic field the atoms do not radiate any light in the x direction. With a magnetic field **B** in the z direction, however, resonance fluorescence polarized along y is observed in the x direction. This is the Hanle effect.

$F = 0$, $m = 0$ lower state of the (one-electron) atom. Then for the wave function at time t for an atom in the field (14.3.7) we write

$$\psi(\mathbf{x}, t) = a_0(t)\Phi_0(\mathbf{x}) + a_1(t)\Phi_1(\mathbf{x}) + a_{-1}(t)\Phi_{-1}(\mathbf{x}). \tag{14.3.8}$$

Let us calculate the expectation value $\mathbf{p}(t)$ of the electric dipole moment of an atom described by this wave function:

$$\mathbf{p}(t) = e \int d^3x \, \psi^*(\mathbf{x}, t)\mathbf{x}\psi(\mathbf{x}, t) = e\mathbf{x}_{0,1}a_0^*(t)a_1(t) + e\mathbf{x}_{0,-1}a_0^*(t)a_{-1}(t) + \text{c.c.}, \tag{14.3.9}$$

since the only nonvanishing dipole matrix elements are between the state $F = 0$, $m = 0$ and the states $F = 1$, $m = \pm 1$. In terms of spherical coordinates r, θ, ϕ, $\Phi_0(\mathbf{x}) = \Phi_0(r, \theta)$ and $\Phi_{\pm 1}(\mathbf{x}) = \Phi(r, \theta)e^{\pm i\phi}$, so that for the x, y, and z components of $\mathbf{x}_{0,\pm 1}$ we obtain

$$x_{0,\pm 1} = \int d^3x \, \Phi_0^*(r, \theta)x\Phi(r, \theta)e^{\pm i\phi}$$

$$= \int_0^\infty dr \, r \int_0^\pi d\theta \sin\theta \, \Phi_0^*(r, \theta)r \sin\theta \, \Phi(r, \theta) \int_0^{2\pi} d\phi \cos\phi \, e^{\pm i\phi}$$

$$\equiv \mu \int_0^{2\pi} d\phi \cos\phi \, e^{\pm i\phi} = \pi\mu, \tag{14.3.10a}$$

$$y_{0,\pm 1} = \mu \int_0^{2\pi} d\phi \sin\phi \, e^{\pm i\phi} = \pm i\pi\mu, \tag{14.3.10b}$$

$$z_{0,\pm 1} = \int_0^\infty dr \, r \int_0^\pi d\theta \sin\theta \, \Phi_0^*(r, \theta)r \cos\theta \, \Phi(r, \theta) \int_0^{2\pi} d\phi e^{\pm i\phi} = 0, \tag{14.3.10c}$$

and therefore $\mathbf{x}_{0,\pm 1} = x_{0,\pm 1}\hat{x} + y_{0,\pm 1}\hat{y} + z_{0,\pm 1}\hat{z} = \pi\mu(\hat{x} \pm i\hat{y})$ and

$$\mathbf{p}(t) = \pi e\mu[a_0^*(t)a_1(t)(\hat{x} + i\hat{y}) + a_0^*(t)a_{-1}(t)(\hat{x} - i\hat{y})] + \text{c.c.} \tag{14.3.11}$$

We will assume now that the light incident on an atom in Fig. 14.12 is in the form of a short pulse that excites the atom at time $t = t_0$, after which the state probability amplitudes evolve according to the time-dependent Schroödinger equation [cf. (9.2.2)]

$$\dot{a}_0 = 0, \tag{14.3.12a}$$

$$\dot{a}_1 = -i(\omega_0 + \omega_L)a_1, \tag{14.3.12b}$$

$$\dot{a}_{-1} = -i(\omega_0 - \omega_L)a_{-1}, \tag{14.3.12c}$$

where $\hbar\omega_L = \mu_B g_1 B$ and $-\hbar\omega_L$ are the Zeeman shifts of the $F = 1$, $m = 1$ and $F = 1$, $m = -1$ states, respectively [Eq. (14.3.1)] and ω_0 is the $F = 0 \leftrightarrow 1$ transition frequency of the unperturbed atom. Then

$$a_0(t) = c_0, \tag{14.3.13a}$$

$$a_1(t) = c_1 e^{-i(\omega_0 + \omega_L)(t - t_0)}, \tag{14.3.13b}$$

$$a_{-1}(t) = c_{-1} e^{-i(\omega_0 - \omega_L)(t - t_0)}, \tag{14.3.13c}$$

where c_0, c_1, and c_{-1} are the probability amplitudes at $t = t_0$. Since the linearly polarized pulse is a superposition of $\hat{\varepsilon}_+$ and $\hat{\varepsilon}_-$ components of equal amplitude, we can assume that

$c_0^* c_1 = c_0^* c_{-1} \equiv \rho$; we also assume that the phases of the probability amplitudes are chosen to make ρ real. Then it follows from (14.3.11) and some simple algebra that

$$\mathbf{p}(t) = 4\pi e \mu \rho \cos \omega_0(t - t_0)\{\hat{x}\cos[\omega_L(t - t_0)] + \hat{y}\sin[\omega_L(t - t_0)]\}, \tag{14.3.14}$$

where $\omega_L = \mu_B g_1 B/\hbar$ is the *Larmor frequency*.

This result suggests—correctly—that when $\mathbf{B} = 0$ ($\omega_L = 0$) the resonance fluorescence from the atoms in the experiment of Fig. 14.12 is always linearly polarized along the same axis (x) as the polarization of the incident field: The atomic dipole oscillates along the x axis and therefore does not radiate in the $\pm x$ directions. When $\mathbf{B} \neq 0$, however, the dipole has a component along y, produces resonance fluorescence polarized along y, and therefore *does* radiate in the $\pm x$ directions. This effect of a magnetic field on resonance fluorescence is called the *Hanle effect*.

The Hanle effect is easily explained with the classical electron oscillator model (Chapter 3) when we include the force $e\mathbf{v} \times \mathbf{B}$ in the equation of motion for the electron displacement $\mathbf{x}$. In our example the electric field polarized along x causes the electron to oscillate with a component of velocity $\mathbf{v}$ along x, so that a magnetic field along z results in a force $e\mathbf{v} \times \mathbf{B}$ along y. The oscillating electric dipole moment of the atom therefore acquires a y component that results in y-polarized radiation in the $\pm x$ directions.

We have not accounted for the damping of the expectation value $\mathbf{p}(t)$ of the atomic dipole moment as a result of the spontaneous emission from the excited states. To do so requires only that we replace the density matrix element ρ in (14.3.14) by $\rho \exp(-\gamma t/2)$, where γ is the spontaneous emission rate, which is the same for the degenerate magnetic substates of a given energy level:

$$\mathbf{p}(t) = 4\pi e \mu \rho \cos \omega_0(t - t_0)e^{-\gamma(t-t_0)/2}\{\hat{x}\cos[\omega_L(t - t_0] + \hat{y}\sin[\omega_L(t - t_0)]\}. \tag{14.3.15}$$

If a polarizer (or "analyzer") on the x axis is oriented to fully transmit radiation polarized along a direction $\hat{a}$ making an angle φ with respect to $\hat{z}$, the time-averaged intensity $I(t)$ of the resonance fluorescence measured by a detector behind the polarizer will be proportional to the time average of $[\mathbf{p}(t) \cdot \hat{a}]^2$, or (Problem 14.7)

$$I(t) \propto e^{-\gamma(t-t_0)} \cos^2\{\omega_L(t - t_0) - \varphi\}. \tag{14.3.16}$$

This is the basis of a useful technique for measuring excited-state lifetimes $(1/\gamma)$. In fact "the Hanle effect has been developed into one of the most reliable methods for measuring the lifetimes of excited levels of atoms and molecules."[17] This is explored further in Problem 14.7.

Two points about our simplified approach to the Hanle effect are noteworthy. First, $\rho \neq 0$ means that the Hanle effect involves off-diagonal coherence of the atomic density matrix (Chapter 9); this explains why it attracted interest during the development of quantum theory in the 1920s. Second, as suggested by the classical oscillator model, expressions like (14.3.15) for the Hanle effect, with relatively small modifications, describe more general cases than the simplest one we have considered of a transition between levels with angular momentum quantum numbers 0 and 1. ●

We have considered specifically some examples involving polarized light, but in fact a redistribution of magnetic substates occurs also for *unpolarized* light, provided it is anisotropic (e.g., unidirectional). As long as the light is not isotropic the spherical symmetry of an unperturbed atom is "broken" and different magnetic substates defined with respect to some quantization axis interact differently with different field polarization

[17]A. Corney, *Atomic and Laser Spectroscopy*, Oxford University Press, Oxford, 2006, p. 478. Chapter 15 of this book is an extensive treatment of the Hanle effect.

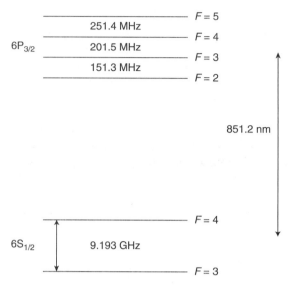

Figure 14.13 Simplified energy-level diagram of ^{133}Cs, showing the hyperfine *clock transition* at 9.193 GHz. The nuclear and electronic angular momenta ($I = \frac{7}{2}$ and $J = \frac{1}{2}$, respectively) result in a splitting of the 6S$_{1/2}$ ground level into two hyperfine levels with $F = 3$ ($= \frac{7}{2} - \frac{1}{2}$) and $F = 4$ ($= \frac{7}{2} + \frac{1}{2}$).

components, regardless of whether the field has a definite polarization. Polarized emission lines from the solar corona, for example, are observed and explained as a consequence of the interplay of the sun's magnetic field and optical pumping with the (unpolarized) radiation from the photosphere.[18]

Optical pumping has been a very useful tool in basic atomic physics, especially in experimental studies of hyperfine structure and nuclear magnetic moments, atomic collisions, and other aspects of spectroscopy. In the remainder of this section we briefly discuss an application of optical pumping in atomic frequency standards. The following section describes another important application.

• The idea that ground- and excited-state populations of atomic energy levels can deviate substantially from a thermal distribution as a result of resonant atom–field interactions, and in particular that atoms can become spin polarized, is attributed to Alfred Kastler, who was awarded the 1966 Nobel Prize in Physics for his research on optical pumping. Kastler's work included the prediction and observation of the polarization of the "twilight glow" resulting from the excitation of the mesospheric sodium D lines by sunlight. The earliest observations of spin polarization of atomic ground levels by optical pumping were reported by J. Brossel, Kastler, and J. Winter in 1952 and by W. B. Hawkins and R. H. Dicke in 1953. •

Since 1967 the second in the International System of Units (SI) has been defined as "the duration of 9,192,631,770 periods of the radiation corresponding to the transition between the two hyperfine levels of the ground state of the cesium-133 atom" (Fig. 14.13). Time and frequency standards have evolved over many years and are an

---

[18]J. Trujillo Bueno, E. Landi Degl'Innocenti, M. Collados, L. Merenda, and R. Manso Sainz, *Nature* **415**, 403 (2002).

integral part of such applications as satellite communications and especially the U.S. Global Positioning System (GPS), which would not exist without them; every GPS satellite is equipped with an atomic clock. They are also required for basic scientific experiments aimed at determining whether various fundamental "constants" actually vary in time. The details of the design and implementation of these systems are complex, but the basic operating principles are relatively simple. In the most widely used type of atomic clock the frequency of radiation near an atomic absorption line is varied and controlled to lock it to the peak absorption frequency ν_0. Then a number $\mathcal{N}$ of oscillation periods $1/\nu_0$ of the radiation is counted to determine a "standard" time interval; thus the time taken to count $\mathcal{N} = 9,192,631,770$ cycles of radiation at the $6S_{1/2}(F = 3) \rightarrow 6S_{1/2}(F = 4)$ transition of cesium is, by definition, a second. Until recently it has been necessary in atomic clocks to employ microwave frequencies, which are small enough to allow accurate electronic counting of cycles (Section 14.7).

Cesium has been used in atomic clocks since the 1950s, and hundreds of commercial time and frequency standards based on its "clock transition" are currently in operation. Like all atomic transition frequencies, the cesium clock frequency is fundamentally the same everywhere aside from generally calculable shifts due to electric, magnetic, and gravitational fields. Cesium, while hardly the only atom used for atomic clocks, has a vapor pressure that allows relatively intense atomic beams to be produced, a large mass resulting in small thermal velocities and Doppler shifts and, in common with all alkali atoms, only two lower hyperfine levels (Fig. 14.13). Other advantages of cesium include its relatively large clock transition frequency, so that a large Q factor $\nu_0/\Delta\nu_0$ is obtained for a given linewidth $\Delta\nu_0$, and the fact that this transition is only weakly affected by small electric fields that may be present.

In a type of atomic clock that served as the primary frequency standard from the late 1960s until about 1990, a beam of cesium atoms, all in the (approximately equally populated) $F = 3$ and $F = 4$ hyperfine levels, pass through a strong and spatially inhomogeneous "Stern–Gerlach" magnetic field in region A that deflects atoms in different hyperfine states (and therefore with different magnetic moments) by different amounts (Fig. 14.14) and in one of two directions determined by the "polarity" (the sign of the magnetic quantum number m) with respect to the field.[19] A second inhomogeneous magnetic field in region B is designed to deflect atoms further, such that, absent anything else, no atoms would be "focused" onto a hot-wire atom detector; such a detector causes ionization of atoms incident upon it and thereby an electric signal proportional to the number of incident atoms. However, if an atom undergoes a transition between the $F = 3$ and $F = 4$ levels in a region C between A and B, the deflection caused by magnet B reverses that caused by A, so that *those atoms that have undergone a transition* are focused onto the atom detector. In other words, magnet A selects atoms in certain magnetic substates, while magnet B is used for the detection of atoms that have made a transition. Transitions are effected in region C by a 3.26-cm microwave field that is resonant with the $F = 3$, $m = 0 \leftrightarrow F = 4$, $m = 0$ transition in the presence of a weak, (nearly) uniform magnetic field (the "C field") that Zeeman splits the different magnetic substates; the $m = 0$ substates are chosen because they have no (first-order) Zeeman shifts [Eq. (14.3.1)]. Region C is surrounded by a high-permeability material in order to shield it from Earth's magnetic field. The C field, perpendicular to the atomic

---

[19]Because the atoms have a distribution of velocities, it is generally not possible by this method to put only atoms in one particular magnetic substate in region C of Fig. 14.14.

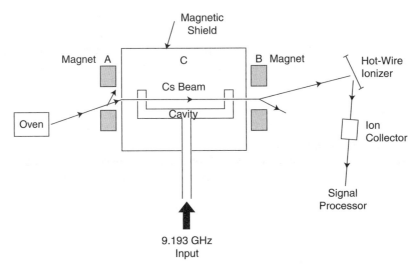

Figure 14.14 Schematic of a cesium atomic clock. The atomic beam is split by a state-selecting magnetic field A into two beams. Atoms selected to enter region C are irradiated with microwave radiation resonant with the clock transition indicated in Fig. 14.13. Those that have undergone the clock transition are focused by the magnetic field B onto an atom detector, the response of which is maximized when the microwave frequency is matched to the clock transition. Feedback circuitry keeps the microwave frequency locked to this value.

beam, acts to Zeeman shift $m \neq 0$ states in the presence of residual stray magnetic fields away from the $F = 3$, $m = 0 \leftrightarrow F = 4$, $m = 0$ resonance, which has only a very small, second-order ($\propto B^2$) Zeeman shift.

The narrow linewidth of the clock transition results in a sharp resonance frequency ν_0, the frequency at which the applied microwave field produces the largest signal from the atom detector. A feedback loop keeps the field locked to this frequency, and the field cycles provide the periodic "clicks" used to keep time. The locked microwave frequency is downconverted electronically and used in a servo loop to lock a quartz oscillator at a frequency of 5 MHz; this is used to generate a signal consisting of one pulse per second, the "output" of the atomic clock.

In this scheme most of the atoms in the atomic beam—about $\frac{15}{16}$ of them if only the $6S_{1/2}(F = 4, m = 0)$ state, say, were selected—are rejected, resulting in a smaller signal-to-noise ratio than would otherwise be possible. State *preparation* by optical pumping, however, has increased the accuracy of atomic frequency standards. For example, all the atoms entering region C can be prepared in the $6S_{1/2}(F = 3)$ level by irradiating them with laser radiation tuned to the $6S_{1/2}(F = 4) \leftrightarrow 6P_{3/2}(F = 3)$ transition, the upper state of which has a radiative lifetime of about 30 ns. Spontaneous transitions occur to both $6S_{1/2}(F = 4)$ and $6S_{1/2}(F = 3)$, and transitions out of the latter level do not occur because they are too detuned from the laser radiation; therefore all the atoms are pumped into that level. A second laser can be used to optically pump all the atoms into the $6S_{1/2}(F = 3, m = 0)$ state, thus preparing them for the $6S_{1/2}(F = 3, m = 0) \leftrightarrow 6S_{1/2}(F = 4, m = 0)$ clock transition in region C.

Optical pumping can also be used for state *detection*. For example, a laser tuned to the $6S_{1/2}(F = 4) \leftrightarrow 6P_{3/2}(F = 5)$ transition results in allowed spontaneous transitions from $6P_{3/2}(F = 5)$ to $6S_{1/2}(F = 4)$, so that the detection of photons radiated in these transitions indicates that atoms in region C have undergone the clock transition. In other words, in optical pumping schemes the state selection magnets A and B are replaced by lasers, the atom detector is replaced by a photodetector, and signal-to-noise ratios are increased because atoms are *prepared* in desired states rather than just selected from an ensemble of atoms distributed over many states. Optical pumping is employed in the cesium frequency standard NIST-7 of the U.S. National Institute of Standards and Technology (NIST). This served as the primary frequency standard in the United States from 1993 to 1999 but has been replaced in this role by the NIST-F1 standard employing laser cooling (Section 14.4).

The accuracy $\Delta \nu_0$ with which the cesium resonance frequency can be determined is limited by the transit time t_d over which the atoms interact with the microwave field; the fractional width $\Delta \nu_0 / \nu_0$ is inversely proportional to the "interrogation time" t_d, i.e., $\Delta \nu_0 / \nu_0 \propto 1/\nu_0 t_d$ (Section 9.11). It is therefore advantageous to make t_d as large as possible. Cesium, in addition to the fact that it has the largest vapor pressure and the largest ground-level hyperfine splitting of any alkali, has the advantage of a large mass and therefore small thermal velocities (typically ~ 250 m/s). Relatively long interrogation times (~ 0.004 s for a path of length $L \sim 1$ m) are therefore obtained with cesium. However, it is impractical to produce magnetic fields that are sufficiently homogeneous over such lengths; small field inhomogeneities give rise to line-broadening effects. For this reason atomic frequency standards, which require extremely narrow resonance lines, employ a U-shaped microwave cavity (Fig. 14.14) that takes advantage of the Ramsey method of separated oscillatory fields. As discussed in Section 9.11, this avoids line broadening due to field inhomogeneities while allowing large transit times and the very sharp resonance of the central Ramsey fringe.

Inaccuracies $\Delta \nu_0 / \nu_0$ as small as about 5×10^{-15} have characterized cesium frequency standards of the type just described. This corresponds to a clock accuracy of about 1 s in 6 million years. The atomic fountain clocks described in the following section are more accurate by about an order of magnitude. These measurements of the cesium clock frequency are probably the most accurate measurements ever made of any physical quantity.

The cesium atomic clock is *passive*, as opposed to *active* frequency standards based on masers. The hydrogen maser frequency standard operating on the 21-cm hyperfine transition $1S_{1/2}(F = 1, \ m = 0) \rightarrow 1S_{1/2}(F = 0, \ m = 0)$ of atomic hydrogen, for example, operates on the basis of a quartz oscillator locked to the frequency of the hydrogen maser radiation.

The technical details involved in the operation of atomic clocks and their application to the GPS, for example, are complex and too far removed from laser physics to delve into here. The interested reader can easily find more information on the websites of companies that sell atomic clocks and of national laboratories, such as NIST, that maintain "primary" time and frequency standards.

In Section 3.7 we wrote rate equations for the occupation probabilities of degenerate upper states and degenerate lower states of transitions between two energy levels. We briefly discussed conditions under which the degenerate states of equal energy could be assumed to be equally populated, including the case where the atom is irradiated with isotropic rather than unidirectional radiation, or where collisions act to equalize the degenerate-state populations, or where the light intensity is too small to produce significant optical pumping. Under such conditions degeneracy results simply in factors like the ratio g_2/g_1 appearing in formulas for the small-signal gain or absorption coefficient. This is the case for most naturally occurring optical phenomena and may therefore be called the case of "natural excitation." In this respect a remark from a classic work on atomic spectroscopy is revealing: "If the excitation occurs in some definitely non-isotropic way, as by absorption from a unidirectional beam of light . . . , large departures from natural excitation may be produced. The study of such effects raises a whole complex of problems somewhat removed from the main body of spectroscopy."[20] Whether *laser* spectroscopy is considered to be "somewhat detached" from the main body of spectroscopy is only a matter of viewpoint, of course. However, the reader should be aware that formulas for absorption, emission, and dispersion found in pre-laser reference material may not be directly applicable to atoms prepared by laser radiation, which is definitely not isotropic!

• It is often necessary in detailed computations to know how electric dipole transition matrix elements depend on the quantum numbers F, m and F', m' of the two states of an atomic transition. Formulas expressing this dependence may be found in the book by Condon and Shortley[20] and other monographs.[21] Here we present a few pertinent formulas.

The transition electric dipole moment between states with angular momentum quantum numbers F, m and F', m' is nonvanishing only if $q = m' - m$ is 0 or ± 1, as discussed earlier. This dipole matrix element is denoted $\langle F'm'|d_q|Fm \rangle$ in the "bra-ket" notation of quantum theory. According to the *Wigner–Eckart theorem*, it has the form

$$\langle F'm'|d_q|Fm \rangle = (-1)^{F'-m'} \begin{pmatrix} F' & 1 & F \\ -m' & q & m \end{pmatrix} \langle F'\|d\|F \rangle, \qquad (14.3.17)$$

where $\langle F'\|d\|F \rangle$ is the *reduced matrix element* and the quantity in large parentheses is the *3j symbol*, numerical values for which can be found in books[21] or on the Web.

The reduced matrix element is independent of m and m'. In the case of hyperfine transitions it depends not only on F and F' but also on the electron angular momentum quantum numbers J and J' of the two states and on the nuclear spin angular momentum quantum number I:

$$\langle F'\|d\|F \rangle = (-1)^F \sqrt{(2F+1)(2F'+1)} \begin{Bmatrix} J' & I & F' \\ F & 1 & J \end{Bmatrix} \langle J'\|d\|J \rangle. \qquad (14.3.18)$$

Here $\langle J'\|d\|J \rangle$ is a further "reduced" matrix element, and the quantity in curly brackets is the *6j symbol* that, like the 3j symbol, is tabulated in various places.[21] The numerical value of $\langle J'\|d\|J \rangle$ follows from the formula

$$\frac{1}{\tau_{\text{rad}}} = \frac{4\bar\omega^3}{3\hbar c^3} \frac{1}{2J'+1} |\langle J'\|d\|J \rangle|^2 \qquad (14.3.19)$$

for the radiative lifetime τ_{rad} of the upper state with quantum number J'; the value of τ_{rad} is usually provided by experiment. Here, owing to the smallness of the hyperfine splittings, $\bar\omega$ is simply

---

[20]E. U. Condon and G. H. Shortley, *The Theory of Atomic Spectra*, Cambridge University Press, London, 1959, p. 97.

[21]See, for instance, R. D. Cowan, *The Theory of Atomic Structure and Spectra*, University of California Press, Berkeley, CA, 1981. Different conventions, which are of no physical consequence when followed consistently, are used in the definitions of the 3j and 6j coefficients. For example, the reduced matrix element $\langle J'\|d\|J \rangle$ is sometimes defined such that a factor $2J + 1$ appears on the right-hand side of Eq. (14.3.19). We follow the conventions of *Cowan*, which is consistent with *Condon and Shortley*. We also write the electric dipole operator, which we have usually denoted by μ, as d here in order to conform to a conventional notation used in defining reduced matrix elements and related quantities.

an average transition frequency, for example, $\bar{\omega} = 2\pi c/\lambda$ with $\lambda = 589.0$ nm in the case of the sodium D_2 transitions. The spontaneous emission rate for the particular transition F', $m' \to F, m$ is

$$A(F', m' \to F, m) = \frac{4\bar{\omega}^3}{3\hbar c^3} |\langle F'm'|d_q|Fm\rangle|^2, \tag{14.3.20}$$

and the total rate of spontaneous emission from the state F', m' may be shown from the properties of the $3j$ and $6j$ symbols to be just $1/\tau_{\text{rad}}$:

$$\sum_{F,m} A(F', m' \to F, m) = \frac{1}{\tau_{\text{rad}}}. \tag{14.3.21}$$

So all the hyperfine states of the excited energy level have the same radiative lifetime, as implied by (14.3.19).

The Rabi frequency for the F, $m \leftrightarrow F'$, m' transition in the case of a linearly polarized field $\mathbf{E}(t) = E_0\hat{x}\cos(\omega t + \phi)$ is

$$\chi_{F',m';F,m} = \langle F'm'|d_{q=0}|Fm\rangle \frac{E_0}{\hbar}, \tag{14.3.22}$$

whereas for a circularly polarized field $\mathbf{E}(t) = (1/\sqrt{2})E_0[\hat{x}\cos(\omega t + \phi) \pm \hat{y}\sin(\omega t + \phi)]$,

$$\chi_{F',m';F,m} = \langle F'm'|d_{q=\pm 1}|Fm\rangle \frac{E_0}{\hbar}. \tag{14.3.23}$$

●

14.4 LASER COOLING

Atoms recoil when they emit or absorb light, as required by conservation of linear momentum and the fact that photons carry linear momentum. Einstein (1909) inferred from his analysis of thermal radiation that atoms recoil not only when they absorb radiation but also when they undergo spontaneous or stimulated emission: They must do so if their average kinetic energy as they absorb and emit thermal radiation at temperature T is to be equal to the value $\frac{3}{2}k_B T$ required by the equipartition theorem of statistical mechanics. (k_B is Boltzmann's constant.) As discussed in Chapter 12, the recoil of an atom in spontaneous emission in particular provides strong evidence for the validity of the photon concept. In fact, the Doppler effect can be understood simply as a consequence of the fact that a photon of frequency ν absorbed (or emitted) by a moving atom carries linear momentum $h\nu/c$ (Problem 14.9).

The recoil forces exerted on atoms by resonant laser radiation are used to slow atoms to very small velocities and thereby to cool gases to extremely low temperatures. In this section we discuss some of the basic physics of laser cooling.

Since a photon carries a linear momentum $h\nu/c = \hbar\omega/c = \hbar k$, the force (= rate of change of linear momentum) on an atom is $F = \hbar k R_{\text{abs}}$, where R_{abs} is the rate at which photons are absorbed. In the case of a two-state atom,

$$R_{\text{abs}} = \left(\frac{dP_2}{dt}\right)_{\text{abs}}. \tag{14.4.1}$$

$(dP_2/dt)_{\text{abs}}$, the rate at which the upper-state probability P_2 changes due to absorption, may be obtained from the optical Bloch equations (9.5.1). Since $P_2 = \rho_{22} = \frac{1}{2}(\rho_{22} - \rho_{11}) + \frac{1}{2}(\rho_{22} + \rho_{11}) = \frac{1}{2}(w + 1)$,

$$\left(\frac{dP_2}{dt}\right)_{\text{abs}} = \frac{1}{2}\left(\frac{dw}{dt}\right)_{\text{abs}} = -\frac{1}{2}\chi v, \qquad (14.4.2)$$

which in steady state is obtained as usual by equating the derivatives in (9.5.1) to zero. The resulting expression for the absorption rate in a monochromatic field has a familar form:

$$R_{\text{abs}} = \left(\frac{dP_2}{dt}\right)_{\text{abs}} = \frac{\frac{1}{2}\chi^2/\beta}{1 + \Delta^2/\beta^2 + \chi^2/\beta A_{21}} = \frac{\frac{1}{2}A_{21}I/I^{\text{sat}}}{1 + \Delta^2/\beta^2 + I/I^{\text{sat}}}, \qquad (14.4.3)$$

where we have used the fact that the intensity I is proportional to χ^2 to write R_{abs} in terms of the line-center saturation intensity I^{sat}. The force on the atom is therefore

$$F = \frac{\frac{1}{2}A_{21}I/I^{\text{sat}}}{1 + \Delta^2/\beta^2 + I/I^{\text{sat}}}\hbar k. \qquad (14.4.4)$$

In the case under consideration of radiative broadening, $\beta = A_{21}/2$ (Section 9.4).

Consider as an example a sodium atom that has been optically pumped by circularly polarized light into the state $3S_{1/2}(F = 2, m = 2)$. For $I = I^{\text{sat}} = 6.3$ mW/cm² the acceleration is found from (14.4.4) to be

$$a = 4.6 \times 10^5 \text{ m/s}^2 \qquad (14.4.5)$$

for laser radiation at resonance ($\Delta = 0$) with the atom (Problem 14.8). This is 5×10^4 times the acceleration due to gravity and indicates that resonant laser radiation of very modest intensity can significantly affect how atoms move about. It should be noted, however, that for this simplified calculation we have ignored the Doppler shift $kv = \omega v/c$, which contributes to an atom–field detuning.

The force (14.4.4) will obviously slow down an atom moving oppositely to the propagation direction of laser radiation, but eventually it will make the atom turn around and speed up. Suppose, however, that there are *two* laser beams, propagating in opposite directions. Assuming that both lasers have the same intensity and the same detuning Δ from resonance with a stationary atom, the total force they exert on an atom with velocity v is

$$F = \frac{\frac{1}{2}A_{21}I/I^{\text{sat}}}{1 + (\Delta + kv)^2/\beta^2 + I/I^{\text{sat}}}\hbar k - \frac{\frac{1}{2}A_{21}I/I^{\text{sat}}}{1 + (\Delta - kv)^2/\beta^2 + I/I^{\text{sat}}}\hbar k. \qquad (14.4.6)$$

The first term is the force exerted by the laser beam with propagation direction parallel to that of the atom's velocity v, so that it exerts a force in the same direction and is resonant with the atomic transition when $\omega = \omega_0 + kv$. (Recall that $\Delta = \omega_0 - \omega$, the detuning of the field frequency ω from the atomic transition frequency ω_0.) The second term is the force exerted by the laser with propagation direction opposed to the atom's velocity, so that it is at resonance when $\omega = \omega_0 - kv$. For velocities such that the Doppler shift kv is small in magnitude compared to β and $|\Delta|$, a binomial expansion of (14.4.6), together with the fact that $\beta = A_{21}/2$, gives

$$F \cong -\kappa v, \qquad \kappa = \frac{4(I/I^{\text{sat}})\hbar k^2 \Delta/\beta}{(1 + \Delta^2/\beta^2 + I/I^{\text{sat}})^2} \qquad (14.4.7)$$

for the total force. This force acts to "damp" the atom's velocity if $\kappa > 0$, i.e., if $\Delta > 0$, which means that the laser frequency ω is smaller than the atomic transition frequency ω_0. This has a simple interpretation. The laser beam propagating in the direction opposite to the atom velocity and exerting a retarding force on it is seen by the atom to be Doppler shifted closer to resonance, since its frequency in the laboratory frame is below the atomic resonance frequency; the counterpropagating beam exerting a force acting to increase the atom's velocity, however, is seen by the atom to be Doppler shifted further away from resonance. The net effect, called *Doppler cooling*, is therefore to slow down the atom.

The net force (14.4.6) is zero if $\Delta = 0$ and acts to increase rather than decrease an atom's velocity if the lasers are tuned *above* the atomic resonance ($\Delta < 0$). Figure 14.15 plots F vs. v for $\Delta = \beta$ and $I = 0.1 I^{\text{sat}}$. It is seen from this figure and Eq. (14.4.6) that the force is always in a direction opposite to v for $\Delta > 0$, but is very small for atomic velocities much greater in magnitude than $\Delta/k = \lambda\Delta/2\pi$. For our sodium example with $\Delta = \beta$, $\lambda\Delta/2\pi = \lambda A_{21}/4\pi = 3$ m/s: An atom's velocity must already be relatively small in order to slow it further by Doppler cooling. $|kv| < \beta$ defines a "velocity capture range," i.e., the velocities for which atoms are significantly slowed by Doppler cooling. The cooling of "warmer" atoms is discussed below.

The force (14.4.7) implies that an atom in the field of two counterpropagating laser beams tuned below resonance will eventually come to rest. But our analysis thus far

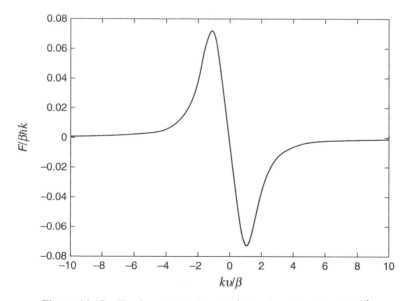

Figure 14.15 The force (14.4.6) vs. kv/β for $\Delta = \beta$ and $I = 0.1 I^{\text{sat}}$.

has ignored the fact that the absorption process itself, while producing no net recoil when $v \cong 0$ [Eq. (14.4.7)], causes the mean-square momentum of the atom to increase: The atom can recoil with momentum $\pm \hbar k$, depending from which laser it absorbs a photon. Thus, for small v, the rate at which the mean-square momentum of an atom increases due to absorption is $2R_{abs}$, where R_{abs} is given by (14.4.3) (for $v \cong 0$) and the factor 2 comes from the fact that we have two laser beams of equal intensity. In addition, the atom can be driven from the ground state back to the ground state by resonance fluorescence and in so doing recoil with momentum $\hbar k$. This recoil, again, has no preferred direction and is therefore zero *on average*, but, like absorption, it causes the mean-square momentum of the atom to grow at the rate $2R_{abs}(\hbar k)^2$. The increase of the average kinetic energy E of an atom of mass M due to absorption followed by emission is therefore

$$\left(\frac{dE}{dt}\right)_{\text{heating}} = \frac{d}{dt}\frac{\langle p^2 \rangle}{2M} = \frac{4R_{abs}(\hbar k)^2}{2M}, \tag{14.4.8}$$

and this prevents an atom from coming to a complete stop.

The cooling force (14.4.6) causes the average kinetic energy to decrease at the rate

$$\left(\frac{dE}{dt}\right)_{\text{cooling}} = \langle Fv \rangle = -\kappa \langle v^2 \rangle = -\frac{2\kappa}{M}\left(\frac{1}{2}M\langle v^2 \rangle\right) = -\frac{2\kappa}{M}E. \tag{14.4.9}$$

Setting the sum of (14.4.8) and (14.4.9) to zero, we obtain the equilibrium kinetic energy $E = (R_{abs}/\kappa)(\hbar k)^2$, or, from (14.4.3) and (14.4.7),

$$E \approx \frac{\hbar \beta^2}{4\Delta}\left(1 + \frac{\Delta^2}{\beta^2}\right). \tag{14.4.10}$$

We are assuming for simplicity that $I/I^{\text{sat}} \ll 1$, that is, that we are in the linear absorption regime. This is consistent with our neglect of stimulated emission in deriving (14.4.8).

According to the equipartition theorem of statistical mechanics, in thermal equilibrium at temperature T the average kinetic energy $E = \left(\frac{1}{2}\right)k_B T$ for motion along one axis, that axis here being defined by the two counterpropagating laser beams. Then (14.4.10) implies the equilibrium temperature

$$T \approx \frac{\hbar \beta^2}{2\Delta k_B}\left(1 + \frac{\Delta^2}{\beta^2}\right) \tag{14.4.11}$$

for a gas in the field of two counterpropagating laser beams tuned below resonance. $dT/d\Delta = 0$ when $\Delta = \beta$, and for this value of the detuning the temperature is minimized:

$$T \approx \frac{\hbar \beta}{k_B}. \tag{14.4.12}$$

The smallest possible temperature achievable by Doppler cooling is therefore estimated to be

$$T_{\min}^D = \frac{\hbar\beta}{k_B} = \frac{\hbar A_{21}}{2k_B}. \qquad (14.4.13)$$

For the sodium D_2 line, for example, $T_{\min}^D \approx 240\,\mu K$, corresponding to an rms velocity $\approx 0.3\,m/s$. Note also that the time scale for Doppler cooling in this case, which according to Eq. (14.4.9) is $\sim M/2\kappa$, is about $16\,\mu s$ for $I/I^{\text{sat}} = 0.1$. In other words, extremely low temperatures can be achieved in very short times by Doppler cooling with very modest laser intensities.

Two counterpropagating laser beams slow the atoms' velocities only along one axis. Doppler cooling in all three dimensions of space is realized using three pairs of counterpropagating beams, each with detuning $\Delta > 0$, along three orthogonal axes (x, y, and z). An atom moving in any direction then experiences "viscosity" similar to that of a particle in a fluid, while the random "kicks" (recoils) it gets in absorption and emission are analogous to the thermal fluctuations resulting from collisions of a particle with the molecules of a fluid. Because of these analogies, a vapor undergoing three-dimensional Doppler cooling has come to be referred to as *optical molasses*.

Optical molasses was first observed in 1985. The temperature of Doppler-cooled sodium vapor was inferred from the time taken for atoms to leave the confinement region after the lasers were all turned off, and the result was consistent with the expected $T_{\min}^D \approx 240\,\mu K$. Further experimentation during the next few years at Bell Laboratories and NIST, however, resulted in temperatures $\approx 40\,\mu K$, well below the theoretical minimum for Doppler cooling.

The explanation of this sub-Doppler cooling invokes effects not included in the derivation of (14.4.13), beginning with the fact that the atoms undergo optical pumping in a field having a *spatially varying polarization*. We follow here a simplified model that brings out the essential features,[22] assuming two counterpropagating plane waves with orthogonal linear polarizations. At a point z along the axis of propagation we write the total electric field as (the real part of)

$$\mathbf{E}(z, t) = \frac{1}{\sqrt{2}} E_0(\hat{x}e^{ikz} + \hat{y}e^{-ikz})e^{-i\omega t}, \qquad (14.4.14)$$

or

$$\mathbf{E}(z, t) = \frac{1}{\sqrt{2}} E_0(\hat{x} + \hat{y}e^{-2ikz})e^{-i(\omega t - kz)} = \hat{\boldsymbol{\varepsilon}}(z)E_0 e^{-i(\omega t - kz)}, \qquad (14.4.15)$$

where

$$\hat{\boldsymbol{\varepsilon}}(z) = \frac{1}{\sqrt{2}}(\hat{x} + \hat{y}e^{-2ikz}) \qquad (14.4.16)$$

[22]J. Dalibard and C. Cohen-Tannoudji, *Journal of the Optical Society of America B* **6**, 2023 (1989).

is the complex unit polarization vector, defined with respect to propagation in the positive z direction. Now

$$\hat{\boldsymbol{\epsilon}}(z) \rightarrow \frac{1}{\sqrt{2}}(\hat{x} + \hat{y}) \qquad \text{for } kz = \frac{2\pi z}{\lambda} = 0,$$

$$\rightarrow \frac{1}{\sqrt{2}}(\hat{x} - i\hat{y}) = \hat{\boldsymbol{\epsilon}}_- \quad \text{for } kz = \frac{\pi}{4},$$

$$\rightarrow \frac{1}{\sqrt{2}}(\hat{x} - \hat{y}) \qquad \text{for } kz = \frac{\pi}{2},$$

$$\rightarrow \frac{1}{\sqrt{2}}(\hat{x} + i\hat{y}) = \hat{\boldsymbol{\epsilon}}_+ \quad \text{for } kz = \frac{3\pi}{4},$$

$$\rightarrow \frac{1}{\sqrt{2}}(\hat{x} + \hat{y}) \qquad \text{for } kz = \pi, \tag{14.4.17}$$

and so on: As z changes in steps of $\lambda/8$, the field polarization changes from linear (along $\hat{x} + \hat{y}$) to σ_- circular to linear (along $\hat{x} - \hat{y}$) to σ_+ circular to linear (along $\hat{x} + \hat{y}$) ..., and so forth.[23] Because the transition matrix elements between magnetic substates are different for these different field polarizations, the populations of different magnetic substates of an atom will vary as it moves along z. This occurs whenever the two counter-propagating fields have different polarizations and the total field therefore has a polarization gradient.

There is another effect that was neglected in our discussion of Doppler cooling: Energy levels of atoms are shifted in an electric field. This is an electric analog of the Zeeman shifts in a magnetic field. As is the case with Zeeman shifts, these *Stark shifts* (or "light shifts") are generally different for different magnetic substates. Light shifts of ground magnetic states play a key role in sub-Doppler cooling; in the simplified model followed here it is assumed that the ground level has an orbital angular momentum quantum number $J = \frac{1}{2}$ and therefore only two magnetic substates, $m = \pm \frac{1}{2}$. For the excited level it is assumed that $J = \frac{3}{2}$ and therefore $m = -\frac{3}{2}, -\frac{1}{2}, \frac{1}{2}, \frac{3}{2}$ (Fig. 14.16a). We will assume laser intensities small enough that excited-state populations are negligible and focus attention on the $m = \pm \frac{1}{2}$ ground states. As shown below, the light shifts of these two states at any point z in the field (14.4.14) are

$$\Delta E(m = -\tfrac{1}{2}) = -2V - V \sin 2kz, \tag{14.4.18a}$$

$$\Delta E(m = +\tfrac{1}{2}) = -2V + V \sin 2kz. \tag{14.4.18b}$$

For our purposes it will not be necessary to have a numerical value for V, which is positive for the case of interest in which the laser frequency ω is less than the atomic transition frequency ω_0.

---

[23]Note that any phase difference between the two counterpropagating waves can be discarded simply by defining the $z = 0$ origin appropriately.

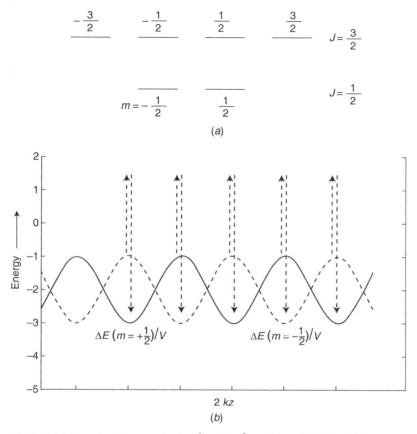

Figure 14.16   (a) Magnetic substates of a $J = \frac{1}{2} \leftrightarrow J = \frac{3}{2}$ transition. (b) Light shifts (14.4.18) of the $m = \pm\frac{1}{2}$ states of the $J = \frac{1}{2}$ ground level. The dashed arrows indicate transitions between the light-shifted $J = \frac{1}{2}$, $m = \pm\frac{1}{2}$ ground states and the $J = \frac{3}{2}$ level at points for which $\sin 2kz = 1$.

Optical pumping by the field (14.4.14) results in transfer of population between the ground $m = \pm\frac{1}{2}$ states. Writing (14.4.16) in the form

$$\hat{\boldsymbol{\varepsilon}}(z) = \frac{1}{\sqrt{2}} \left[\frac{1}{\sqrt{2}}(\hat{\boldsymbol{\varepsilon}}_+ + \hat{\boldsymbol{\varepsilon}}_-) - \frac{i}{\sqrt{2}}(\hat{\boldsymbol{\varepsilon}}_+ - \hat{\boldsymbol{\varepsilon}}_-)e^{-2ikz} \right]$$

$$= \frac{1}{2}\hat{\boldsymbol{\varepsilon}}_+(1 - ie^{-2ikz}) + \frac{1}{2}\hat{\boldsymbol{\varepsilon}}_-(1 + ie^{-2ikz}), \tag{14.4.19}$$

where $\hat{\boldsymbol{\varepsilon}}_\pm$ are the complex unit polarization vectors for $\sigma_\pm$ light [Eq. (14.3.4)], we see that the intensity of σ_+ light at any point z is proportional to $(\frac{1}{4})|1 - i\exp(-2ikz)|^2 = (\frac{1}{2})(1 - \sin 2kz)$, while the intensity of σ_- light is proportional to $(\frac{1}{4})|1 + i\exp(-2ikz)|^2 = (\frac{1}{2})(1 + \sin 2kz)$:

$$I(\sigma_\pm) = \frac{1}{2}I_0(1 \mp \sin 2kz), \tag{14.4.20}$$

where I_0 is the maximum intensity. The intensities $I(\sigma_\pm)$ determine the time scale τ_p for population transfer by optical pumping between the $m = \pm\frac{1}{2}$ ground states. For σ_+ light, for instance, absorptive transitions are allowed from the $m = -\frac{1}{2}$ ground state to the excited $m = +\frac{1}{2}$ state and from the $m = +\frac{1}{2}$ ground state to the $m = +\frac{3}{2}$ excited state. As discussed in the preceding section, all the population in steady state will be in the $m = +\frac{1}{2}$ ground state if the intensity is small enough that saturation of the $m = +\frac{1}{2} \leftrightarrow m = +\frac{3}{2}$ transition is negligible, even if all the population initially resides in the $m = -\frac{1}{2}$ ground state. In other words, σ_+ light transfers population from the $m = -\frac{1}{2}$ ground state to the $m = +\frac{1}{2}$ ground state. Similarly, σ_- light transfers population from the $m = +\frac{1}{2}$ ground state to the $m = -\frac{1}{2}$ ground state. Depending on z, either transfer rate can be dominant since the relative strengths of the σ_+ and σ_- components vary with z according to (14.4.20).

We can now understand, qualitatively, how the polarization gradient and optical pumping cause atoms to lose kinetic energy. Consider an atom at a point z such that the field polarization is σ_-. At such a point the atom will be in the state $J = \frac{1}{2}$, $m = -\frac{1}{2}$ in steady state as a result of optical pumping. Now suppose the atom is moving to the right with a velocity v such that $v\tau_p \approx \lambda/4$, so that after the time τ_p it will be at a point where the field polarization is σ_+ (Fig. 14.16). In other words, the atom will have "climbed the potential hill" from a trough to a peak in the light-shifted energy. At the peak the light polarization is σ_+ and the population transfer rate from $m = -\frac{1}{2}$ to $m = +\frac{1}{2}$ is greatest. The population transfer is not instantaneous, and if we imagine that the atom stays in the $m = -\frac{1}{2}$ state as it climbs the potential hill, and is then optically pumped at the top of the hill into the $m = +\frac{1}{2}$ state at the bottom, we can see from Fig. 14.16b that it emits a larger photon frequency than it absorbs and therefore that its kinetic energy decreases. After the absorption and emission the atom again starts climbing the potential hill. In reality, of course, the absorption, emission, and optical pumping are statistical processes. But it should be clear that, because the population transfers between the $m = \pm\frac{1}{2}$ states are not instantaneous but require a finite time $\sim\tau_p$, there is a net tendency for kinetic energy to be converted to potential energy and therefore for the atoms to be slowed. Regardless of whether an atom is moving to the right or to the left, it finds itself continually climbing a potential hill and losing kinetic energy in this so-called *Sisyphus cooling.*[24]

- The light shifts can be derived by first considering the work involved in inducing an electric dipole moment $\mathbf{p} = \alpha\mathbf{E}$. Since the potential energy of an electric dipole moment $\mathbf{p}$ in an incremental electric field $d\mathbf{E}$ is $-\mathbf{p} \cdot d\mathbf{E}$, this work is

$$W = -\int_0^{\mathbf{E}} \mathbf{p} \cdot d\mathbf{E} = -\alpha \int_0^{\mathbf{E}} \mathbf{E} \cdot d\mathbf{E} = -\frac{1}{2}\alpha\mathbf{E}^2. \tag{14.4.21}$$

The factor $\frac{1}{2}$ appears because the dipole is *induced* by the field. In the case of a monochromatic field $\mathbf{E}_0 \cos \omega t$, the cycle-averaged energy

$$\Delta E_i = -\frac{1}{4}\alpha_i(\omega)\mathbf{E}_0^2 \tag{14.4.22}$$

---

[24]"The gods had condemned Sisyphus to ceaselessly rolling a rock to the top of a mountain, whence the stone would fall back of its own weight."—Albert Camus, *The Myth of Sisyphus and Other Essays*, Vintage Books, New York, 1955, p. 119.

is the "light shift" of an atom in a state with polarizability $\alpha_i(\omega)$. The polarizability is in general complex, and $\alpha_i(\omega)$ here is implicitly assumed to be its real part.

The nearly resonant atom–field interaction is characterized by a detuning $\Delta = \omega_0 - \omega$, in which case the polarizability of the lower state of the transition can be inferred from Eq. (9.6.30) and the formula $n(\omega) = 1 + N\alpha(\omega)/2\varepsilon_0$ for the refractive index $n(\omega)$ of a dilute gas of N atoms per unit volume [cf. Section 3.15]:

$$\alpha_i(\omega) = \frac{|\mu_{ji}|^2}{\hbar} \frac{\Delta}{\Delta^2 + \beta^2} \tag{14.4.23}$$

for the lower state.[25] In the more general case in which the state i can make allowed transitions to more than one other state, this generalizes to

$$\alpha_i(\omega) = \frac{\Delta}{\hbar(\Delta^2 + \beta^2)} \sum_j |\mu_{ji}|^2, \tag{14.4.24}$$

and the light shift (14.4.22) becomes

$$\Delta E_i = -\frac{\Delta}{4\hbar(\Delta^2 + \beta^2)} \mathbf{E}_0^2 \sum_j |\mu_{ji}|^2 \equiv -CI \sum_j |\mu_{ji}|^2, \tag{14.4.25}$$

where I is the intensity and C is positive for $\Delta > 0$ (i.e., for the field tuned below resonance). It is assumed that the states j all have the same energy, so that a single detuning Δ appears in this equation, and that all the transitions $i \leftrightarrow j$ are characterized by the same homogeneous linewidth. This simplification applies in fact to the example of interest here.

Consider now the two ground states of Fig. 14.16. For the $J = \frac{1}{2}$, $m = -\frac{1}{2}$ state,

$$\Delta E(m = -\tfrac{1}{2}) = -CI(\sigma_+)\left|\langle \tfrac{3}{2} \tfrac{1}{2}|d_1|\tfrac{1}{2} - \tfrac{1}{2}\rangle\right|^2 - CI(\sigma_-)\left|\langle \tfrac{3}{2} - \tfrac{3}{2}|d_{-1}|\tfrac{1}{2} - \tfrac{1}{2}\rangle\right|^2, \tag{14.4.26}$$

since it has an allowed transition to $J = \frac{3}{2}$, $m = \frac{1}{2}$ for σ_+ light and an allowed transition to $J = \frac{3}{2}$, $m = -\frac{3}{2}$ for σ_- light. We use here the notation of Eq. (14.3.17) for the electric dipole matrix elements. Since we are ignoring any hyperfine structure for our model atom, the matrix elements $\langle J'm'|d_q|Jm\rangle$ (with $q = m' - m$) are given simply by (14.3.17) with $F' = J'$ and $F = J$. Then

$$\left|\langle \tfrac{3}{2} \tfrac{1}{2}|d_1|\tfrac{1}{2} - \tfrac{1}{2}\rangle\right|^2 = \begin{pmatrix} \tfrac{3}{2} & 1 & \tfrac{1}{2} \\ -\tfrac{1}{2} & 1 & -\tfrac{1}{2} \end{pmatrix}^2 \left|\langle \tfrac{3}{2}\|d\|\tfrac{1}{2}\rangle\right|^2 = \tfrac{1}{12}\left|\langle \tfrac{3}{2}\|d\|\tfrac{1}{2}\rangle\right|^2 \tag{14.4.27}$$

and

$$\left|\langle \tfrac{3}{2} - \tfrac{3}{2}|d_{-1}|\tfrac{1}{2} - \tfrac{1}{2}\rangle\right|^2 = \begin{pmatrix} \tfrac{3}{2} & 1 & \tfrac{1}{2} \\ \tfrac{3}{2} & -1 & -\tfrac{1}{2} \end{pmatrix}^2 \left|\langle \tfrac{3}{2}\|d\|\tfrac{1}{2}\rangle\right|^2 = \tfrac{1}{4}\left|\langle \tfrac{3}{2}\|d\|\tfrac{1}{2}\rangle\right|^2. \tag{14.4.28}$$

Therefore

$$\Delta E(m = -\tfrac{1}{2}) = -C\left[\tfrac{1}{12}I(\sigma_+) + \tfrac{1}{4}I(\sigma_-)\right]\left|\langle \tfrac{3}{2}\|d\|\tfrac{1}{2}\rangle\right|^2, \tag{14.4.29}$$

[25] Note from (9.6.30) that the polarizability of the upper state j is $\alpha_j(\omega) = -\alpha_i(\omega)$. Here we are concerned only with ground-state light shifts and therefore ground-state polarizabilities.

and, from Eq. (14.4.20),

$$\Delta E(m = -\tfrac{1}{2}) = -\tfrac{1}{12}CI_0\left|\langle\tfrac{3}{2}||d||\tfrac{1}{2}\rangle\right|^2(2 + \sin 2kz) \equiv -V(2 + \sin 2kz) \tag{14.4.30}$$

for the light shift of an atom in state $J = \tfrac{1}{2}$, $m = -\tfrac{1}{2}$ at the point z in the field (14.4.14). Similarly,

$$\begin{aligned}
\Delta E(m = +\tfrac{1}{2}) &= -CI(\sigma_+)\left|\langle\tfrac{3}{2}\,\tfrac{3}{2}|d_1|\tfrac{1}{2}\,\tfrac{1}{2}\rangle\right|^2 - CI(\sigma_-)\left|\langle\tfrac{3}{2}-\tfrac{1}{2}|d_{-1}|\tfrac{1}{2}\,\tfrac{1}{2}\rangle\right|^2 \\
&= -C\left[\tfrac{1}{4}I(\sigma_+) + \tfrac{1}{12}I(\sigma_-)\right]\left|\langle\tfrac{3}{2}||d||\tfrac{1}{2}\rangle\right|^2 \\
&= -V(2 - \sin 2kz).
\end{aligned} \tag{14.4.31}$$

Since V and the optical pumping time τ_p are directly proportional and inversely proportional, respectively, to the intensity, the rate $1/\tau_{\text{cool}} = E^{-1}(dE/dt)_{\text{cool}} \propto V\tau_p$ for Sisyphus cooling is independent of intensity, in contrast to the rate (14.4.9) for Doppler cooling, which is proportional to intensity. As in the case of Doppler cooling, the heating rate is proportional to the intensity. Since the equilibrium temperature is proportional to the heating rate times τ_{cool}, it is proportional to intensity for Sisyphus cooling. The velocity capture range is also intensity dependent, again in contrast to Doppler cooling: Our simplified discussion shows that the optimal situation for Sisyphus cooling occurs when $v\tau_p \sim \lambda$, that is, when an atom with velocity v moves a distance $\sim\lambda$ during an optical pumping time τ_p, so that $kv \sim 1/\tau_p$, implying a velocity capture range inversely proportional to τ_p and therefore directly proportional to the laser intensity.

At low laser intensities the cooling in optical molasses is observed to be less effective in the presence of magnetic fields, which cause transitions among magnetic substates and thereby weakens the Sisyphus effect. If the laser intensity is sufficiently large, however, the cooling becomes less sensitive to magnetic fields because the light shifts and optical pumping rates become larger. Detailed analyses of Sisyphus cooling are found to be consistent with such experimental observations as well as with the extremely low temperatures—as small as a few microkelvins—that have been realized in optical molasses.

The temperatures obtained by Sisyphus cooling can be lowered by lowering the ground-state light shift, for example, by decreasing the laser intensity or by increasing the detuning. But there are limits even to Sisyphus cooling. The smallest average kinetic energy for atoms absorbing and emitting photons is that associated with the recoil of a nearly stationary atom when it absorbs or emits a *single photon*. The mean-square recoil momentum, $(\hbar\omega_0/c)^2$, implies a minimum average kinetic energy $(\hbar\omega_0/c)^2/2M = h^2/2M\lambda^2$ for an atom with mass M and transition wavelength λ, and therefore a temperature

$$T_{\text{recoil}} = \frac{h^2}{Mk_B\lambda^2}. \tag{14.4.32}$$

For $M = 23$ amu and $\lambda = 589$ nm (sodium), $T_{\text{recoil}} = 2.4$ μK; for $M = 133$ amu and $\lambda = 852$ nm (cesium), $T_{\text{recoil}} = 0.2$ μK. Temperatures several times larger than T_{recoil} have been obtained by Sisyphus cooling. By employing quantum interference effects similar to those used to inhibit absorption in electromagnetically induced transparency (Section 9.10), it has been possible to cool atoms to temperatures well below the recoil limit.

• The force on an electric dipole moment **p** in an electromagnetic field is given by the "Lorentz force,"

$$\mathbf{F} = (\mathbf{p} \cdot \nabla)\mathbf{E} + \frac{\partial \mathbf{p}}{\partial t} \times \mathbf{B}. \tag{14.4.33}$$

To obtain the force on an electric dipole in the electromagnetic field

$$\mathbf{E}(\mathbf{r}, t) = \tfrac{1}{2}\mathbf{E}_0(\mathbf{r})e^{-i\omega t} + \tfrac{1}{2}\mathbf{E}_0^*(\mathbf{r})e^{i\omega t}, \qquad \mathbf{B}(\mathbf{r}, t) = \tfrac{1}{2}\mathbf{B}_0(\mathbf{r})e^{-i\omega t} + \tfrac{1}{2}\mathbf{B}_0^*(\mathbf{r})e^{i\omega t}, \tag{14.4.34}$$

we write

$$\mathbf{p}(\mathbf{r}, t) = \tfrac{1}{2}\mathbf{p}_0(\mathbf{r})e^{-i\omega t} + \tfrac{1}{2}\mathbf{p}_0^*(\mathbf{r})e^{i\omega t}, \tag{14.4.35}$$

with $\mathbf{p}_0(\mathbf{r}) = \alpha(\omega)\mathbf{E}_0(\mathbf{r})$ and $\alpha(\omega) = \alpha_R(\omega) + i\alpha_I(\omega)$ the complex polarizability. It follows from the Maxwell equation $\nabla \times \mathbf{E} = -\partial\mathbf{B}/\partial t$ that $\mathbf{B}_0 = -(i/\omega)\nabla \times \mathbf{E}_0$, and from these expressions that the cycle-averaged z component of the force (14.4.33) is

$$F_z(\mathbf{r}) = \frac{1}{4}\alpha_R(\omega)\frac{\partial}{\partial z}|\mathbf{E}_0(\mathbf{r})|^2$$
$$- \frac{1}{2}\alpha_I(\omega)\mathrm{Im}\left[E_{0x}(\mathbf{r})\frac{\partial E_{0x}^*}{\partial z} + E_{0y}(\mathbf{r})\frac{\partial E_{0y}^*}{\partial z} + E_{0z}(\mathbf{r})\frac{\partial E_{0z}^*}{\partial z}\right]. \tag{14.4.36}$$

The x and y components of the force are obtained by replacing $\partial/\partial z$ with $\partial/\partial x$ and $\partial/\partial y$, respectively. The formula (14.4.36) also gives the dipole force on a small dielectric sphere (Section 14.5).[26]

Consider as an example a two-state atom, for which the complex, near-resonance polarizability can be inferred from (9.6.18), for instance:

$$\alpha(\omega) = \frac{|\mu|^2}{\hbar}\frac{1}{\Delta - i\beta} \qquad (\Delta = \omega_0 - \omega). \tag{14.4.37}$$

The first term on the right-hand side of (14.4.36) is therefore the z component of

$$\mathbf{F}_{\mathrm{dipole}} = -\nabla[U(\mathbf{r})], \tag{14.4.38}$$

where

$$U(\mathbf{r}) = -\frac{|\mu|^2}{4\hbar}\frac{\Delta}{\Delta^2 + \beta^2}|E_0|^2, \tag{14.4.39}$$

which is just the light shift (14.4.22). In other words, the "dipole force" $\mathbf{F}_{\mathrm{dipole}}$ is the force resulting from a gradient of the light-shift energy; this is the basis for the *optical lattices* discussed in the following section.

To interpret the other part of the force in (14.4.36), let us assume for simplicity the plane wave

$$\mathbf{E}_0(\mathbf{r}) = \hat{x}E_0 e^{ikz}. \tag{14.4.40}$$

---

[26]See, for instance, P. C. Chaumet and M. Nieto-Vesperinas, *Optics Letters* **25**, 1065 (2000).

In this case the part of the total force that is proportional to the imaginary part of the polarizability in (14.4.36) is

$$\mathbf{F}_{\text{recoil}} = \frac{1}{2} \frac{|\mu|^2}{\hbar} \frac{\beta}{\Delta^2 + \beta^2} E_0^2 k \hat{z} = \hbar k \hat{z} \frac{\frac{1}{2} \chi^2 \beta}{\Delta^2 + \beta^2} = \hbar k \hat{z} R_{\text{abs}}, \qquad (14.4.41)$$

for a two-state atom, R_{abs} being just the absorption rate [cf. Eq. (14.4.3)] when the intensity is well below I^{sat}, as assumed here. $\mathbf{F}_{\text{recoil}}$ is just the recoil force (14.4.4) exerted on the atom.

The dipole force arises from spatial variations in the Stark shift $U(\mathbf{r})$ of an atom, whereas the recoil force is a direct consequence of absorption (and, more generally, emission). The recoil force has its maximum value at resonance ($\Delta = 0$) and decreases as $1/\Delta^2$ when the field is far off resonance, whereas the dipole force vanishes at resonance and decreases more slowly, as $1/|\Delta|$, far off resonance.[27] The recoil force saturates with intensity according to the formula (14.4.3). The potential energy (14.4.39) determining the dipole force via (14.4.38) at large intensities takes the form (Problem 14.10)

$$U(\mathbf{r}) = -\frac{1}{2} \hbar \Delta \ln \left[1 + \frac{I}{I_{\nu}^{\text{sat}}} \frac{\beta^2}{\Delta^2 + \beta^2} \right]. \qquad (14.4.42)$$

We have assumed in our discussion of laser cooling that only the recoil force affects the motion of an atom. This assumption is justified if the field is well described as a plane wave ($\nabla E_0 = 0$). In general, however, the dipole force must be taken into account and, as discussed in the following section, it can significantly affect the motion of atoms in optical fields.

We mentioned at the beginning of this section that Einstein invoked recoil in absorption and emission in his treatment of thermal radiation. Assuming an isotropic and unpolarized field with spectral density $\rho(\omega)$, Einstein obtained a cooling rate of the form (14.4.9) with "friction" coefficient

$$\kappa = \left(\frac{\hbar \omega}{c^2} \right) (P_1 - P_2) B \left[\rho(\omega) - \frac{\omega}{3} \frac{d\rho}{d\omega} \right]. \qquad (14.4.43)$$

P_1 and P_2 are the lower- and upper-state probabilities of the two-state transition of frequency ω and B is the Einstein coefficient for absorption in a broadband field (Section 3.6). The retarding force $F = -\kappa v$ arises from Doppler shifts and "aberration" in the broadband field. The cooling rate in thermal equilibrium is balanced by the recoil heating rate, which Einstein calculated to be

$$\left(\frac{dE}{dt} \right)_{\text{heating}} = \frac{1}{3} \left(\frac{\hbar \omega}{c} \right)^2 P_1 B \rho(\omega). \qquad (14.4.44)$$

From this result, together with (14.4.9), (14.4.43), and $E = (\frac{1}{2}) k_B T$, therefore, the sum of the heating and cooling rates is zero when $\rho(\omega)$ satisfies

$$\rho - \frac{\omega}{3} \frac{d\rho}{d\omega} = \frac{\hbar \omega}{3 k_B T} \frac{P_1}{P_1 - P_2} \rho \qquad (14.4.45)$$

with $P_2/P_1 = \exp(-\hbar \omega / k_B T)$ in thermal equilibrium. The solution for $\rho(\omega)$ with "initial condition" $\rho(0) = 0$ is exactly the Planck spectrum.   `   ●

---

[27] In terminology that seems to generate more confusion than insight, the recoil and dipole forces are sometimes attributed to "real" and "virtual" transitions, respectively.

The lowest temperatures obtained by laser cooling are realized with trapped atoms (Section 14.5). In order for atoms to be captured by a trap, their velocities must be small, much smaller than the typical mean velocity of atoms emerging from an oven, which is ~ 800 m/s for sodium atoms at a vapor temperature $T = 600$K. Before atoms can be trapped and cooled to temperatures in the microkelvin range and below, therefore, they must first be cooled to temperatures well below their oven vapor temperature. This is done with a counterpropagating laser beam with frequency ω tuned below the atomic resonance frequency ω_0, so that an atom with an initial velocity v experiences the largest retarding force when $\omega_0 = \omega + kv$, that is, when it is at resonance with the Doppler-shifted field frequency. The atom moves increasingly out of resonance, however, as its velocity decreases, and the degree of slowing it experiences is substantially reduced unless ω or ω_0 can be varied to maintain resonance. The field frequency can be changed by chirping, but the most commonly employed method of maintaining resonance is to change the atomic resonance frequency with a magnetic field that varies along the atomic beam. In this *Zeeman slowing* the magnetic field along the atomic beam is varied by varying the winding of a solenoid, as indicated in Fig. 14.17.

● To get an idea of the sort of numbers involved in Zeeman slowing, let us assume that atoms start out with a velocity $v_i = 800$ m/s and are brought to a complete stop ($v_f = 0$) after a distance z. Using $a = 4.6 \times 10^5$ m/s^2 [Eq. (14.4.5)] for the deceleration of sodium atoms in a field of intensity $I = I^{\text{sat}} = 6.3$ mW/cm^2, we calculate that the time taken for the atoms to be stopped is $t = (v_i - v_f)/a = 1.7$ ms, and the distance over which this occurs is $z = \frac{1}{2}at^2 = 70$ cm.

The Doppler shift at a distance z from the trap (Fig. 14.17) is $kv(z) = \omega v(z)/c$. To estimate the magnetic field required to compensate for this Doppler shift, we assume that the Landé g factors of the upper and lower states of the atomic transition differ by a factor ~ 1—a reasonable approximation—so that the Zeeman shift of the transition frequency is $\Delta\omega_0 \sim \mu_B B_z/\hbar$ for σ_+ light ($\Delta m = +1$). The magnetic field required to keep an atom in resonance with the field as it slows down is therefore

$$B_z(z) = \frac{\hbar\omega}{\mu_B c}v(z) \sim \frac{\hbar\omega_0}{\mu_B c}at = \frac{\hbar\omega_0}{\mu_B c}\sqrt{2az}. \tag{14.4.46}$$

For the sodium transition at 589 nm we obtain, assuming again the accelaration $a = 4.6 \times 10^5$ m/s^2,

$$B_z(z) \sim 120\sqrt{z} \text{ gauss}, \tag{14.4.47}$$

where z (in centimeters) is the distance from the trap and we have used the fact that a tesla is 10^4 gauss. ●

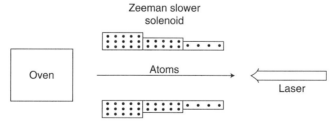

Figure 14.17 A Zeeman slower. Atoms are slowed by the force of a counterpropagating laser beam and are kept near resonance with the Doppler-shifted laser field by the Zeeman shift of their transition frequency in a magnetic field that varies along the direction of the atomic beam. The laser and magnetic field parameters are chosen such that the atoms are slowed enough to allow them to be captured in a trap.

Laser cooling is used to slow cesium atoms and thereby to obtain the extremely small transit-time line-broadening characteristic of atomic fountain clocks that, as noted earlier (Section 14.3), are about 10 times more accurate than atomic beam clocks. Some features of a fountain clock are indicated in Fig. 14.18. Cesium atoms in a trap (Section 14.5) are cooled to microkelvin temperatures at the intersection of six laser beams, as discussed above, and then "launched" upward. The launching can be done by the force on the atoms of a single laser beam, but this results in heating of the atoms as they absorb and emit photons and suffer recoil kicks. The atoms are instead launched by up-shifting the frequency of the upward-propagating laser beam and down-shifting the frequency of the downward-propagating laser beam to produce a *moving standing wave*. If the up-shifted and down-shifted frequencies are, respectively, $v + \Delta v$ and $v - \Delta v$, this standing wave moves upward with a velocity $v = \lambda \, \Delta v$ (Problem 14.11), and the atoms move along with the standing wave without any average recoil velocity: The Doppler effect causes the atoms to see the same frequency v for both fields, resulting in optical molasses in the moving frame. The atoms continue moving upward at a few meters/ second after the lasers are turned off (Problem 14.11), and, after being optically pumped into one of the states of the clock transition, they enter the microwave cavity, which serves as the first field for the Ramsey method of separated oscillatory fields. The atoms pass through the cavity, reach their apogee, and then fall by gravity and experience the second Ramsey field as they pass through the microwave cavity a second time. For a total atom path length of $\sim$30 cm the Ramsey interrogation time $T \sim 1$ s and therefore the Ramsey fringes are extremely sharp (Section 9.11). The detection of atoms that have made the clock transition is done by irradiating the atoms that fall through the microwave cavity with laser radiation and counting resonance fluorescence photons, and the peak of the atomic resonance is determined by the frequency of the microwave field that produces the largest fluorescence signal. The essential difference from the older atomic beam clock, then, is the narrower resonance of the cesium clock transition resulting from the long interrogation time made possible by laser cooling.

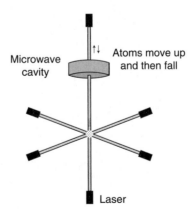

Figure 14.18   An atomic fountain in which atoms make two passes through a microwave cavity, corresponding in effect to the Ramsey method of separated oscillatory fields. The long Ramsey interrogation time (Section 9.11) is the primary reason for the high degree of accuracy of atomic fountain clocks.

14.5 TRAPPING ATOMS WITH LASERS AND MAGNETIC FIELDS

Laser cooling as such does not *trap* atoms since atoms can diffuse out of the laser fields. For trapping we require forces acting to spatially confine atoms. The most ubiquitous atom trap is the *magneto-optical trap* (MOT) employing both lasers and magnetic fields. Before discussing the operating principles of a MOT we will consider briefly the trapping of atoms by a static magnetic field, specifically the *quadrupole* magnetic field

$$\mathbf{B} = \mathcal{B}[x\hat{x} + y\hat{y} - 2z\hat{z}], \tag{14.5.1}$$

where $\mathcal{B}$ is the magnetic field gradient along the x and y axes; the gradient of $\mathbf{B}$ along the z axis is $-2\mathcal{B}$, as required by the Maxwell equation $\nabla \cdot \mathbf{B} = 0$. The field (14.5.1) is produced by a pair of Helmholtz coils (Fig. 14.19) when the coil separation is 1.25 times the coil radius.

The Zeeman-shifted energy of an atom in a state i in a magnetic field of magnitude B has the form [Eq. (14.3.1)]

$$E_i(B) = E_i(B = 0) - \mu_i B, \tag{14.5.2}$$

where μ_i ($\propto m\mu_B$) is the magnetic dipole moment of state i along the direction of the magnetic field. A state with $\mu_i > 0$ therefore has its lowest Zeeman-shifted energy at points in space where the magnetic field is strongest, and an atom in such a state will experience a force $-\nabla(-\mu_i B)$ acting to move it to a point where the magnetic field is higher. A state with $\mu_i > 0$ is said to be a *high-field seeker*. Similarly a state with $\mu_i < 0$ is a *low-field seeker*.[28] The potential energy function $-\mu_i B \sim \mu_B B = 9.274 \times 10^{-24}B$ J, or $0.67B$ K when expressed as temperature in degrees Kelvin (K). Thus, for the magnetic fields typical of laboratory experiments ($B \ll 1$ tesla), the "depth" of a magnetic trap is much less than 1K; atoms must be cooled to such temperatures before they can be magnetically trapped. Trapping is accomplished much more effectively by combining magnetic fields with lasers, as we now discuss.

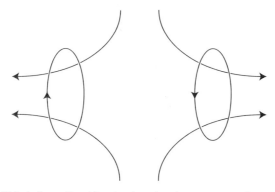

Figure 14.19   Two Helmholtz coils with opposing electric currents produce a quadrupole magnetic field.

---

[28]In regions free of electric currents there are no local maxima of $|\mathbf{B}|$, and consequently only low-field seekers can be trapped by magnetic fields alone. This is discussed by W. H. Wing, *Progress in Quantum Electronics* **8**, 181 (1984).

Equation (14.5.2) implies that atomic transition frequencies in a magnetic field such as (14.5.1) are position-dependent. The force exerted by a laser field on an atom therefore depends on the position of the atom, and it is this position-dependent force that confines atoms in a magneto-optical trap. Unlike a purely magnetic trap, a MOT acts not only to trap atoms but also to cool them with counterpropagating laser beams as discussed in the preceding section. To understand this in more detail we consider a model atom with a $J = 0$ ground level and a $J = 1$ upper level. The atom is assumed to be on the z axis, where the magnetic field (14.5.1) is $\mathbf{B} = -2\mathcal{B}z\hat{z}$. The frequencies of the $\Delta m = 0$ and $\Delta m = \pm 1$ transitions are

$$\omega_0(J=0, m=0 \rightarrow J=1, m=0) = \omega_0,$$

$$\omega_0(J=0, m=0 \rightarrow J=1, m=+1) = \omega_0 - \frac{\mu_B g \mathcal{B}z}{\hbar} = \omega_0 + \frac{2\mu_B g \mathcal{B}z}{\hbar} = \omega_0 + \Delta\omega_0,$$

$$\omega_0(J=0, m=0 \rightarrow J=1, m=-1) = \omega_0 + \frac{\mu_B g \mathcal{B}z}{\hbar} = \omega_0 - \frac{2\mu_B g \mathcal{B}z}{\hbar} = \omega_0 - \Delta\omega_0.$$

$$(14.5.3)$$

where $\Delta\omega_0 = 2\mu_B g \mathcal{B}z/\hbar$ and ω_0 is the transition frequency in the absence of the magnetic field. The recoil force exerted on the atom by $\sigma_\pm$ light of frequency ω is given by Eq. (14.4.4) with $\Delta = \omega_0 - \omega \pm \Delta\omega_0$:

$$F(\sigma_\pm) = \frac{\frac{1}{2}A_{21}I/I^{\text{sat}}}{1 + (1 \pm \Delta\omega_0/\beta)^2} \hbar k \cong \frac{\frac{1}{4}A_{21}I/I^{\text{sat}}}{1 \pm \Delta\omega_0/\beta} \hbar k \qquad (14.5.4)$$

for $\Delta\omega_0 \ll \beta$ and $I \ll I^{\text{sat}}$. To simplify the algebra we have set $\omega_0 - \omega = \beta = A_{21}/2$, and we are ignoring for the moment any motion of the atom. Under these assumptions the average recoil force when the atom is irradiated with σ_+ light propagating in the $+z$ direction and σ_- light propagating in the $-z$ direction (Fig. 14.20) is

$$F(\sigma_+) - F(\sigma_-) \cong \left[\frac{\frac{1}{4}A_{21}I}{I^{\text{sat}}} \right] \left[\frac{1}{1 + \Delta\omega_0/\beta} - \frac{1}{1 - \Delta\omega_0/\beta} \right] \hbar k \cong -\left[\frac{\frac{1}{2}A_{21}I}{I^{\text{sat}}} \right] \hbar k \frac{\Delta\omega_0}{\beta}$$

$$= -\left[\frac{2I}{I^{\text{sat}}} \frac{\mu_B g \mathcal{B}}{\hbar} \hbar k \right] z \equiv -k_s z \qquad (14.5.5)$$

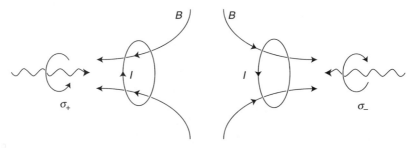

Figure 14.20   One-dimensional model of a magneto-optical trap (MOT). After W. D. Phillips in *Fundamental Systems in Quantum Optics*, eds. J. Dalibard, J.-M. Raimond, and J. Zinn-Justin (North-Holland, Amsterdam, 1992), p. 165.

when the two fields have the same intensity I. The combined effects of the magnetic field and the counterpropagating laser fields produce a restoring force with "spring constant" k_s. The approximation (14.5.5) is valid when the magnetic field and the distance from the center of the trap ($z = 0$) are sufficiently small.

In addition to the trapping effect in a MOT there is cooling by the counterpropagating laser beams: The MOT produces a trapped optical molasses. Let us continue with our one-dimensional model and include the effects of both the trapping and Doppler cooling forces on the motion of an atom of mass M. The equation of motion for the coordinate z of the atom is $M\, d^2z/dt^2 = -k_s z - av$, or

$$\frac{d^2z}{dt^2} + \gamma_t \frac{dz}{dt} + \omega_t^2 z = 0. \tag{14.5.6}$$

Here the oscillation frequency ω_t for an atom in the MOT is given by

$$\omega_t^2 = \frac{k_s}{M} = \frac{2I\mu_B g \mathcal{B} k}{M I^{\text{sat}}} \tag{14.5.7}$$

and the damping constant $\gamma_t = \kappa/M$, where κ is defined by (14.4.7). As in (14.5.4) we simplify some algebra by taking $\Delta = \omega_0 - \omega = \beta$. Then, if I/I^{sat} is small,

$$\gamma_t \cong \frac{I\hbar k^2}{M I^{\text{sat}}}. \tag{14.5.8}$$

Equation (14.5.6) is the familiar equation for a damped harmonic oscillator and expresses the fact that atoms in a MOT are both trapped and cooled. For $M = 23$ amu, $\lambda = 589$ nm, and $g \sim 1$, we estimate $\omega_t/2\pi \sim 2$ kHz and $\gamma_t \sim 10^5$ s^{-1} for $I/I^{\text{sat}} = 0.3$ and $\mathcal{B} = 10$ G/cm, as is fairly typical of MOTs. The atom oscillations are typically "overdamped" ($\gamma_t > \omega_t$), as this example suggests. In an actual MOT the oscillation frequencies are different along the x, y, and z axes but have similar magnitudes.

The force (14.5.5) exerted on atoms by the counterpropagating laser beams in a MOT is typically considerably larger than the magnetic force $\nabla(\mu_B g B) = \mu_B g \mathcal{B}$ for atoms at a distance of a wavelength or so from the center of the trap (Problem 14.12). Typically, 10^{18}–10^{21} atoms/m^3 are trapped, and the size of the trapped "cloud" can be inferred from the temperature T: According to the equipartition theorem, the average potential energy $\frac{1}{2}k_s z^2$ and the average kinetic energy $\frac{1}{2}Mv^2$ are equal to $\frac{1}{2}k_B T$ in thermal equilibrium, so that the cloud size is estimated to be

$$z_{\text{cloud}} \sim \sqrt{k_B T/k_s}, \tag{14.5.9}$$

which is typically $\sim 10^{-2}$ cm for temperatures $T \sim 200$ μK obtained by Doppler cooling (Problem 14.12). The temperature T can be inferred by turning off the trap and determining the velocities of escaping atoms by their time of flight from the trap.

These results of our highly simplified, one-dimensional model for a MOT are in reasonable accord with those of three-dimensional numerical analyses that account for hyperfine structure and all the magnetic substates of real atoms. When such analyses include polarization gradients and other effects, they become quite complicated, and in fact the physics of atom cooling and trapping is a subject to which entire books are devoted.[29] Magnetic traps have been developed over many years, especially in connection with the confinement of hot plasmas in nuclear fusion research, and there is an extensive literature on the subject.

Techniques for the cooling and trapping of *ions* differ in several respects from those we have described for neutral atoms. In this case the electric charge of the particles allows them to be efficiently cooled and trapped with static electric and magnetic fields. In a *Penning trap*, for example, ions are confined in the x and y directions by a magnetic field along z while a quadrupole electric field confines them in the z direction. Cooling can be done by applying a laser beam along some other direction; the laser field cools ions moving toward it and pushes them back to the center of the trap where they exchange energy in collisions with hotter ions, resulting in a cooling of the ion "cloud." Ion cooling with a single laser beam is done similarly in a *Paul trap*, which confines ions with oscillating electric fields. The repulsive forces between ions make it difficult or impossible to realize Bose–Einstein condensation and, since we highlight that topic in the following section, we will not consider ion traps any further.

● We have described the motion of atoms in a MOT using elementary classical mechanics rather than quantum mechanics. Classical theory is a good approximation provided the atoms are not too cold and therefore their de Broglie wavelengths are not too large. We discuss this point a bit more in the following section.

According to the formula (2.3.8) for a particle in a harmonic-oscillator potential, the motion of an atom in a MOT is characterized by the quantized energies $E_n = \hbar\omega_t(n + \frac{1}{2})$, $n = 0, 1, 2, 3, \ldots$, and so we can expect classical theory to be accurate if $\hbar\omega_t/k_B T$ is small. But at sufficiently low temperatures quantum effects become important, and in particular the zero-point energy $\frac{1}{2}\hbar\omega_t$ puts a lower limit on the temperature to which atoms can be cooled in a MOT:

$$T_{\min} = \frac{\hbar\omega_t}{k_B}. \tag{14.5.10}$$

This limiting temperature is very small (Problem 14.12), but it has been closely approached in experiments in which a single beryllium ion is confined in a Paul trap with $\omega_t \sim 100\,\text{MHz}$.   ●

There is a great deal more to be said about laser cooling and trapping of atoms than is warranted here.[29] In the remainder of this section we will touch on two applications of the dipole force (14.4.38).

The light shifts (14.4.18) we calculated in connection with Sisyphus cooling imply that the model atoms in their $m = \pm\frac{1}{2}$ ground states experience periodic potentials in the counterpropagating laser fields; the period of these potentials is half the wavelength of the presumed plane wave. The dipole force given by Eq. (14.4.38),

$$\mathbf{F}_{\text{dipole}} = -\nabla[U(\mathbf{r})] = \nabla\left[\tfrac{1}{4}\alpha_R(\omega)\mathbf{E}_0^2(\mathbf{r})\right] \tag{14.5.11}$$

when we take $\mathbf{E}_0$ to be real, implies that an atom finds itself in a periodic potential $U(\mathbf{r})$ whenever the light intensity is periodic, for example, when the atom is in a standing-wave field. The intersection of two plane waves at an angle θ results in a periodic

[29]See, for instance, H. J. Metcalf and P. van der Straten, *Laser Cooling and Trapping*, Springer, New York, 1999. A tutorial on the design and operation of an "inexpensive" MOT is given by C. Wieman, G. Flowers, and S. Gilbert, *American Journal of Physics* **63**, 317 (1995).

potential with period $\lambda/[2\sin(\theta/2)]$ for a polarizable particle [cf. Eq. (11.14.31)], and, more generally, pairs of interfering fields can produce a great variety of two- and three-dimensional potentials with different periodicities along different directions. An atom in such an "egg-crate" potential, called an *optical lattice*, behaves very much like an electron in a crystal, where the allowed energies are restricted to certain bands depending on the crystal structure (Chapter 2). The light shifts can be comparable to or greater than the average kinetic energy of atoms loaded into the lattice from a MOT. By varying the laser intensity or detuning, the atoms can be tightly or weakly bound by the potential wells of the "egg crate"; they can be confined—"held in midair" by laser beams—in potential wells comparable in their spatial extent to an optical wavelength. If the field frequency is such that $\alpha_R(\omega) > 0$, the dipole force moves atoms toward maxima of the electric field; if $\alpha_R(\omega) < 0$, it moves atoms toward field minima. Either way, the atoms can be trapped by the optical lattice, and the trapping potentials are increased or decreased as the field intensity is increased or decreased. Thus it is possible with optical lattices to obtain a "phase-space density" of atoms large enough to realize "all-optical" Bose–Einstein condensation (Section 14.6).

Absorption and recoil in optical lattices can be minimized by detuning the lasers far from atomic resonances. The peak values of the light-shift potentials are usually expressed in units of the atomic recoil energy (Problem 14.14). Another attractive feature of optical lattices is that the interaction strengths of atoms at different lattice sites can be controlled by designing the lattice appropriately, and relaxation processes can be effectively eliminated. Fundamental quantum phenomena predicted for electrons in crystal lattices but difficult to observe because of electron–electron and electron–photon interactions can be probed very "cleanly" with atoms stored in optical lattices. Transferring atoms from a Bose–Einstein condensate in a magnetic trap to an optical lattice has led to some remarkable experimental studies, one of which is described in the following section.

We have restricted our discussion of laser trapping thus far to atoms, but lasers are also used to trap many other particles including, for example, DNA molecules, biological cells, and small dielectric particles. The physical basis for this trapping is again the dipole force (14.5.11). Consider a nonabsorbing dielectric sphere with radius a smaller than the wavelength of incident light. With n the refractive index of the dielectric material, and n_b the refractive index of the medium in which the sphere is placed, the polarizability α_R of the sphere is given by electromagnetic theory as

$$\alpha_R = 4\pi\epsilon_0 n_b^2 \left(\frac{n^2 - n_b^2}{n^2 + 2n_b^2}\right) a^3. \tag{14.5.12}$$

(α_R, n, and n_b are all evaluated at the frequency of the incident light.) It follows that the dipole force (14.5.11) is in the direction of increasing electric field if $n > n_b$, and in the direction of decreasing electric field if $n < n_b$. A sphere immersed in water and irradiated with a Gaussian beam at an optical wavelength will experience a radial force toward the axis of the beam if its refractive index n is greater than about 1.33. However, there is also a force of radiation pressure in the direction of propagation of the beam (Problem 14.15):

$$\mathbf{F}_{\text{rad}} = \frac{n_b^2}{4\pi\epsilon_0}\frac{\omega^4}{3c^3}\alpha_R^2(\omega)\mathbf{E}_0^2 = \frac{8\pi}{3c}n_b^5\left(\frac{\omega}{c}\right)^4\left(\frac{n^2 - n_b^2}{n^2 + 2n_b^2}\right)^2 a^6 I. \tag{14.5.13}$$

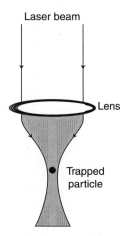

Figure 14.21   "Optical tweezer" used in microscopy. A particle is trapped where the field intensity is greatest.

In order to trap a particle with $n > n_b$ both radially and axially with a focused Gaussian beam, for example, the axial gradient of the field in the direction of the focal spot must be large enough that the dipole force (14.5.11) acting to push the particle toward the focus exceeds the force $\mathbf{F}_{rad}$ acting to push it away. This condition is not difficult to satisfy. Figure 14.21 shows a commonly used setup for trapping a dielectric particle with a single laser beam, an example of an *optical tweezer*. (The terminology comes from the use of laser beams to noninvasively "take hold" of a particle and move it from one place to another.) Diode-pumped Nd : YAG lasers are commonly used in biological applications of laser tweezers; the 1064-nm wavelength is not absorbed by water and does not damage biological samples. The frequency-doubled Nd : YAG radiation at 532 nm, by contrast, is strongly absorbed by many samples and can serve as an "optical scissors" used in conjunction with an optical tweezer. Optical tweezer technology and its applications are advancing rapidly and, as with other applications of lasers, we cannot go into it in any depth without straying into topics quite distinct from laser physics as such.[30]

14.6   BOSE–EINSTEIN CONDENSATION

The classical mechanics of an object with linear momentum p is valid if the de Broglie wavelength $\lambda_{deB} = h/p$ is sufficiently small, just as ray optics well describes some aspects of the propagation of light when the wavelength λ [equal to the photon de Broglie wavelength $h/(h\nu/c)$] is small compared to apertures and other things affecting the propagation. For an ideal gas with N atoms per unit volume at temperature T, for example, the wave nature of the atoms is not expected to play any role unless their de Broglie wavelengths are comparable to the interatomic spacing $\sim N^{-1/3}$. Using

---

[30]A large list of papers on optical tweezers, with succinct commentaries, has been compiled by M. J. Lang and S. M. Block, *American Journal of Physics* **71**, 201 (2003).

$p^2/2M = \frac{3}{2}k_BT$ to calculate λ_{deB} for atoms of mass M, we estimate therefore that the temperature must be less than

$$T_c \sim \frac{1}{3}\frac{h^2}{Mk_B}N^{2/3} \qquad (14.6.1)$$

if the wave nature of the atoms is to be important. A gas below this critial temperature might be expected to behave as a single "matter wave" rather than as a collection of independently moving atoms, much as laser light is better described as a wave rather than as a collection of independently propagating photons. The possibility that particles of *matter* can behave *collectively* as a wave exhibiting interference and diffraction effects follows from a prediction in 1924–1925 by Einstein, who showed that, at sufficiently low temperatures, *noninteracting* particles can "condense" into a single quantum state of zero velocity. This is called *Bose–Einstein condensation* (BEC).

The critical temperature T_c for BEC is generally extremely small. Consider, for example, a gas of ^{87}Rb atoms at a density $N = 2.5 \times 10^{18}$ m^{-3}, about 10^{-7} times the density of air at standard temperature and pressure (STP). Using (14.6.1) we calculate $T_c \sim 130$ nK, much smaller even than the sort of temperatures quoted in our discussions of Doppler and Sisyphus cooling. Nevertheless, as discussed below, BEC has in fact been realized in atomic clouds, including ^{87}Rb at the density 2.5×10^{18} atoms/m^3.

• Einstein deduced the condensation effect along the lines of the following simplified argument. Consider the number of possible quantum states $d\mathcal{N}_E$ of a particle with energy in the interval $[E, E+dE]$ in an ideal gas of volume V. This is analogous to the number (3.12.12) of field modes; in the case here of particles of mass M, all occupying the same internal quantum state (e.g., the same electronic state of an atom),

$$d\mathcal{N}_E = \frac{1}{(2\pi)^3}V\,d^3k = \frac{V}{(2\pi)^3}4\pi k^2\,dk. \qquad (14.6.2)$$

This is derived in the same fashion as (3.12.12), except that (i) no factor of 2 associated with photon polarization appears, and (ii) k here is related to the energy E by $E = p^2/2M = \hbar^2 k^2/2M$, i.e., by relating a particle's momentum p to its de Broglie wavelength ($p = 2\pi\hbar/\lambda_{\mathrm{deB}} = \hbar k$). It follows that $d\mathcal{N}_E = \rho(E)\,dE$, where the "density of states" is

$$\rho(E) = \frac{VM^{3/2}}{\pi^2\hbar^3\sqrt{2}}E^{1/2}. \qquad (14.6.3)$$

If the atoms are bosons, the number of particles $\mathcal{N}$ at temperature T is

$$\mathcal{N} = \mathcal{N}_0(T) + \int_0^\infty \frac{\rho(E)\,dE}{z(T)e^{E/k_BT} - 1}. \qquad (14.6.4)$$

The first and second terms on the right-hand side are the numbers of particles with energy $E = 0$ and $E > 0$, respectively. Because of $z(T)$, whose functional form we will not require, the denominator in the second term differs from the $(e^{E/k_BT} - 1)^{-1}$ familiar from the case of thermal photons. $z(T)$ appears because, unlike the number of photons in thermal equilibrium, the total number of particles (atoms) of an ideal gas in thermal equilibrium is a conserved quantity; $z(T)$ is *determined* by this number. Since $[z(T)e^{E/k_BT} - 1]^{-1}$ represents the average number of particles in a particular state of energy $E > 0$, and as such must be positive, we must have $z(T) \geq 1$. It then follows from (14.6.3) that the integrand of the second term in (14.6.4) vanishes when $E = 0$: as stated, the integral is the density of particles with energy E *greater than 0*.

The Bose–Einstein "condensation" of particles into the state with $E = 0$ occurs only at very low temperatures. It can be seen from (14.6.4) that, as the temperature decreases, $z(T)$ must also decrease in order to keep the total number of particles fixed. As $T \to 0$, $z(T) \to 1$, its smallest possible value and the value that maximizes the number of particles with $E > 0$. Assuming $z(T_c) \cong 1$, we define the critical temperature T_c for BEC as the temperature at which $E > 0$ for *all* the particles:

$$\mathcal{N} = \int_0^\infty \frac{\rho(E)\, dE}{e^{E/k_B T_c} - 1} = \frac{V M^{3/2}}{\pi^2 \hbar^3 \sqrt{2}} \int_0^\infty \frac{\sqrt{E}\, dE}{e^{E/k_B T} - 1} = 2.612 V \left(\frac{2\pi M k_B T_c}{h^2} \right)^{3/2}, \tag{14.6.5}$$

or

$$T_c = 0.084 \frac{h^2}{M k_B} N^{2/3} \qquad (N = \mathcal{N}/V). \tag{14.6.6}$$

Below this temperature *some* of the particles will be in the condensed phase $[\mathcal{N}_0(T) > 0]$. T_c as defined here is about $\frac{1}{4}$ the cruder estimate (14.6.1). At temperatures $T < T_c$,

$$\mathcal{N} = \mathcal{N}_0(T) + \int_0^\infty \frac{\rho(E)\, dE}{e^{E/k_B T} - 1} = \mathcal{N}_0(T) + 2.612 \left(\frac{2\pi M k_B T}{h^2} \right)^{3/2}$$

$$= \mathcal{N}_0(T) + 2.612 \left(\frac{2\pi M k_B T_c}{h^2} \right)^{3/2} \left(\frac{T}{T_c} \right)^{3/2} = \mathcal{N}_0(T) + \mathcal{N} \left(\frac{T}{T_c} \right)^{3/2}. \tag{14.6.7}$$

The fraction of particles in the condensed state is therefore

$$\frac{\mathcal{N}_0(T)}{\mathcal{N}} = 1 - \left(\frac{T}{T_c} \right)^{3/2}. \tag{14.6.8}$$

The length

$$\lambda_T = \left(\frac{h^2}{2\pi M k_B T} \right)^{1/2} \tag{14.6.9}$$

is called the *thermal de Broglie wavelength*, and $N\lambda_T^3$ is called the *phase-space density*. From (14.6.7) it follows that the condition $T < T_c$ for BEC can be expressed in terms of the phase-space density:

$$N\lambda_T^3 > 2.612. \tag{14.6.10}$$

Usually $z(T)$ is written as $e^{-\mu/k_B T}$, where μ, the *chemical potential*, is a thermodynamic quantity defined as the change in the energy of a system when a particle is added while the volume and entropy are kept constant. In the case of a BEC at a fixed temperature, the number of particles with energy $E > 0$ is constant [cf. Eq. (14.6.4)], so that any added particles must become part of the zero-velocity condensate. In other words, the chemical potential is nearly zero and therefore $z(T) \cong 1$ for a BEC, as we have assumed in our derivation of T_c. In Einstein's work a parameter related to $z(T)$ appeared via a "Lagrange multiplier" used in imposing the constraint that the particle number is constant. Einstein suggested that condensation might be observed with a gas of electrons—this was before it was understood that particles are either bosons or fermions, that no two fermions can occupy the same quantum state, and therefore that the condensation effect cannot occur with (unpaired) electrons or any other fermions.   ●

Before laser cooling and trapping, experimental studies of Bose–Einstein condensates were limited primarily to liquid ^{4}He. At the required temperatures most substances solidify, and in those that do not the atoms interact too strongly to approximate an ideal gas. Liquid ^{4}He, the superfluidity of which has traditionally been associated with Bose–Einstein condensation, is the major exception. The small mass and therefore large zero-point energy of helium atoms prevents solidification at atmospheric pressures, and below a critical temperature of about 3K, some fraction of the atoms go into a superfluid state in which the liquid has zero viscosity (Problem 14.16). At lower temperatures nearly all the atoms are in the superfluid component. While many studies strongly support the conclusion that superfluid ^{4}He is a Bose–Einstein condensate, the density is sufficiently large that interatomic interactions are significant, as evidenced by the mere fact that low-temperature ^{4}He is a liquid. One of the most remarkable features of Einstein's prediction, after all, is that condensation can occur in the absence of any particle interactions.

In dilute atomic clouds at temperatures below that required for Bose–Einstein condensation, the average distance between atoms is fairly large, typically $\sim 10^2$ nm. At such separations, interactions resulting in collisions that are deleterious to BEC are weak if the atoms are spin polarized, as discussed below. However, these weak interactions actually turn out to be essential for meeting the major experimental challenge, which has been to realize the exceedingly small critical temperatures that, because of the low densities of dilute gases, are much smaller even than those obtained by Sisyphus cooling, as noted earlier. If we assume a density of 10^{18} atoms/m^3 in an atom trap, the critical temperature predicted by Eq. (14.6.6) is about 20 nK for ^{87}Rb, for example, whereas the recoil limit is about 400 nK. Actually, as also discussed below, critical temperatures for trapped atoms can be considerably larger than that given by (14.6.6), but they are still much smaller than temperatures reached by laser cooling.

The temperatures required for BEC in atomic clouds are realized by *evaporative cooling*. In this technique the six cooling laser fields (Section 14.4) are turned off and the low-field-seeking atoms are magnetically trapped; the lasers are turned off in order to avoid photon scattering processes that limit the achievable densities and temperatures. The magnetic trapping field is then slowly reduced, so that the (Zeeman energy) depth of the magnetic trap is lowered, allowing atoms with sufficient kinetic energy to escape the trap while cooler atoms remain. An rf magnetic field can also be applied to induce transitions between magnetic substates such that atoms with the largest kinetic energies are changed from low-field seekers to (untrapped) high-field seekers. The effects of the reduced trapping field and the rf field are then to allow hotter atoms to be removed from the cloud while cooler ones remain, analogous to the cooling of a cup of coffee by evaporation; note that atom interactions in the form of elastic collisions are essential for evaporative cooling, that is, for keeping the cloud thermalized as the evaporation proceeds. Evaporative cooling from a laser-cooled temperature of ~ 10 μK to ~ 200 nK, sufficient for the formation of a spin-polarized condensate of about 2000 ^{87}Rb atoms, was first achieved in 1995; a zero-velocity BEC component of the cloud was verified by a time-of-flight measurement of the atoms' velocity distribution when the cloud was allowed to freely expand. Following those experiments, BEC has been demonstrated by similar methods in different gases. The basic technique has usually been to use Zeeman slowing to cool an atomic beam from an oven, capture and laser-cool the slowed atoms in a MOT, turn off the lasers and magnetically trap the atoms, and then evaporatively cool the atoms (typically a few million of them) to below the critical BEC temperature.[31]

A serious problem that had to be overcome in the first experiments concerned the trapping in the static magnetic quadrupole field after the lasers were turned off. An atom moving in such a field experiences a time-dependent field that can cause it to make a transition from a low-field-seeking magnetic substate to an untrapped, high-field-seeking state if this time variation is sufficiently rapid, that is, if the magnetic field the atom sees has a Fourier frequency component near the frequency for a transition between two Zeeman-shifted magnetic substates. Near the center of the trap the magnetic field and therefore the Zeeman splittings are small enough for even a slow atom to experience a time-dependent field that causes its magnetic moment to flip into an untrapped state: The quadrupole trap has in effect a "hole" near its center that prevents atoms from being trapped for very long. In one of the original experiments this problem was solved by applying an oscillatory magnetic field such that the total (time-averaged) magnetic field did not vanish anywhere in the trap and therefore there was no "hole." In another a "blue-tuned" laser at a frequency giving a negative atomic polarizability was applied such that atoms were repelled from the hole. Experiments that followed have employed different magnetic field configurations to circumvent the problem.

- To derive the critical temperature T_c for atoms in a magnetic trap with restoring forces along the x, y, and z axes and atom oscillation frequencies ω_x, ω_y, and ω_z along these axes, we first recall that the quantized energy levels for the atomic motion in the trap are

$$E = \left(n_x + \tfrac{1}{2}\right)\hbar\omega_x + \left(n_y + \tfrac{1}{2}\right)\hbar\omega_y + \left(n_z + \tfrac{1}{2}\right)\hbar\omega_z \qquad (n_x, n_y, n_z = 0, 1, 2, \ldots). \qquad (14.6.11)$$

Consider the number $\mathcal{N}_E$ of possible states with energies less than E. For cases of interest in the trapping of neutral atoms, the energies may be assumed to be much greater than the zero-point energies $\tfrac{1}{2}\hbar\omega_x$, $\tfrac{1}{2}\hbar\omega_y$, and $\tfrac{1}{2}\hbar\omega_z$. Then for the calculation of $\mathcal{N}_E$ the integers n_x, n_y, and n_z may be replaced by continuous variables $E_x/\hbar\omega_x$, $E_y/\hbar\omega_y$, and $E_z/\hbar\omega_z$:

$$\mathcal{N}_E = \frac{1}{\hbar^3\,\omega_x\omega_y\omega_z} \int_0^E dE_x \int_0^{E-E_x} dE_y \int_0^{E-E_x-E_y} dE_z = \frac{E^3}{6\hbar^3\,\omega_x\omega_y\omega_z}. \qquad (14.6.12)$$

The number of states with energy between E and $E + dE$ is therefore

$$d\mathcal{N}_E = \frac{E^2\,dE}{2\hbar^3\omega_x\omega_y\omega_z} = \rho(E)\,dE, \qquad (14.6.13)$$

which defines the density of states $\rho(E)$. Proceeding now as in (14.6.5)–(14.6.8), using this density of states instead of the density of states (14.6.3) for untrapped particles, we obtain the critical temperature

$$T_c = 0.94\frac{\hbar(\omega_x\omega_y\omega_z)^{1/2}}{k_B}\mathcal{N}^{1/3} \qquad (14.6.14)$$

---

[31]For references to the original work and more detailed discussions see, for instance, C. J. Pethick and H. Smith, *Bose–Einstein Condensation in Dilute Gases*, Cambridge University Press, Cambridge, 2004.

and the condensate fraction

$$\frac{\mathcal{N}_0(T)}{\mathcal{N}} = 1 - \left(\frac{T}{T_c}\right)^3. \tag{14.6.15}$$

Unlike the critical temperature (14.6.6) for free particles, (14.6.14) depends on the number of particles rather than the density. In the original ^{87}Rb experiments the critical temperature was about 170 nK.

Another important consideration in the physics of cold atomic gases is spin polarization. BECs in magnetic traps normally consist of spin-polarized atoms, e.g., alkali atoms in which the valence electrons have "spin up." The atom interaction involving spin-up (or spin-down) electrons differs from the interaction when the spins are opposite because the Pauli principle allows two electrons in different spin states to occupy the same atomic orbital, as occurs in covalent bonding. If the two electrons have the same spin, however, the Pauli principle forbids them from sharing the same orbital, and the atom–atom interaction in this case is consequently weaker. Of particular interest for BEC are the so-called *doubly polarized* and *maximally stretched* states of the electronic ground state; spin-depolarizing collisions of atoms in these states have very small cross sections and the atoms are low-field seekers. In the doubly polarized state the spin components of the electron (m_J) and the nucleus (m_I) are aligned and have their largest allowed values, for example, $m_I = \frac{3}{2}$ and $m_J = \frac{1}{2}$ for Na or ^{87}Rb (nuclear spin $I = \frac{3}{2}$); the quantization axis is in the direction of the magnetic field. Thus, for Na or ^{87}Rb, both of which have $F = 1$ and $F = 2$ hyperfine ground levels, the doubly polarized state has the quantum numbers $F = 2$, $m_F = 2$. In the maximally stretched state, $F = I - \frac{1}{2}$ and $m_F = -(I - \frac{1}{2})$, for example, $F = 1$, $m_F = -1$ for Na or ^{87}Rb. Collisions of atoms in these states are primarily the "good" collisions necessary to keep the gas thermalized during evaporative cooling, as opposed to the "bad" collisions that act to quench the low-field-seeking trapped states. •

The technique of evaporative cooling was developed about a decade prior to its playing an essential role in the first observations of BEC in dilute gases. It was conceived for the cooling of magnetically trapped hydrogen, which was considered at the time the most promising candidate for BEC of an atomic gas; this consideration stemmed from the fact that the interactions of spin-polarized hydrogen atoms are very weak, preventing the formation of hydrogen molecules. The longest-wavelength transition from the ground state of hydrogen is the Lyman-α line at 122 nm, too short for laser cooling with available lasers. This short wavelength and the small mass of the H atom also imply too large a recoil temperature ($T_{\text{recoil}} = 1.3$ mK) for BEC (Problem 14.16). BEC in magnetically trapped hydrogen was achieved in 1998 by cryogenic cooling followed by evaporative cooling. The BEC transition was observed at $T_c = 50$ μK at a density of 1.8×10^{20} atoms/m^3. The presence of a condensate was inferred from a large shift in the absorption frequency of the two-photon 1S → 2S transition (Problem 14.9).

Atoms in a Bose–Einstein condensate can be "loaded" into an optical lattice from a magnetic trap by applying off-resonance laser beams to produce the desired lattice spacings and light-shift potential depths. With the magnetic trapping fields turned off, the atoms are left trapped solely by the laser fields forming the lattice. The potential depth is usually sufficiently large that, according to classical mechanics, the atoms cannot cross the potential energy barrier between sites and move about the lattice. However, according to quantum theory there is a nonzero probability that they can "tunnel" across the barriers, and in the case of a Bose–Einstein condensate this can result in an atomic matter wave whose phase is approximately constant while the number of atoms at each lattice site fluctuates. This is analogous to a coherent light wave in which the phase but not the number of photons is fixed (Chapter 12): The smaller the fluctuations in the phase, the greater the fluctuations in the photon number, and vice versa. In the case of a fixed number of atoms at each site of an optical lattice, the phase of the matter wave is not well defined and, by analogy to a similar situation for electrons in a crystal, the atoms are said to form a *Mott insulator*. In this case, no interference effects involving a coherent matter wave are observed when the gas is released from the lattice. If, on the other hand, the phase of the matter wave is well defined while the number of atoms at each site fluctuates, matter–wave interference effects can be observed. By changing the potential depth of an optical lattice it has been possible to observe the *Mott transition* between a Mott insulator and the ("superfluid") case of coherent matter waves exhibiting interference. It has also been demonstrated that evaporative cooling leading to a Bose–Einstein condensate can be done with atoms trapped in an optical lattice, without any magnetic trapping field. In this case the evaporative cooling results from lowering the power of the laser beams defining the lattice.

Two-slit interference and other effects of coherent matter waves have been observed in experiments with optical lattices and other cold-atom systems, and the new field of *atom optics* is based on the wave properties of cold atoms. Atomic interferometers employing atom–wave analogs of mirrors, gratings, and other optical elements will very likely be used in applications such as the detection of rotations and gravitational gradients.[32]

Bose–Einstein condensates have been produced by trapping atoms with magnetic fields from planar wire structures fabricated lithographically on a substrate. The small scale ($\sim 0.1 - 10$ μm) of these structures allows strong magnetic field gradients to be produced with small currents (< 1 A) and low power dissipation, and the wire patterns can be designed to form, for instance, magnetic traps similar to the Helmholz coil quadrupole trap described earlier. The tight trapping results in rapid cooling and the formation of a BEC in a second or less. The motion and positions of atoms trapped a few microns above the surface of these structures are more controllable than in the case of MOTs in which the trapping fields are produced by lasers and coils outside the atomic cloud. For example, it has been demonstrated that a BEC cloud can be moved about in a prescribed fashion on such an *atom chip*. Integrated atom chips including tiny lasers and readout electronics might eventually be the basis of miniaturized sensors, atomic clocks, and other devices.

• Atoms in which the numbers of electrons, protons, and neutrons are even (odd) integers are bosons (fermions). Neutral atoms (number of electrons equal to the number of protons) are therefore bosons or fermions depending on whether the number of neutrons is even or odd, respectively. Thus, ^{7}Li (atomic number $Z = 3$, neutron number $N = 4$) is a boson, whereas ^{6}Li ($Z = 3$, $N = 3$) is a fermion. It is a remarkable consequence of quantum statistics that a Bose–Einstein condensate can be made with ^{7}Li but not with ^{6}Li, which differs from ^{7}Li only in having one less neutron in the nucleus.

While fermionic atoms cannot form BECs, they can be cooled and trapped with lasers and magnetic fields. Techniques have been developed to control with applied fields the interatomic forces in cold fermion gases in such a way that pairs of atoms behave compositely as bosons, somewhat similar to the way coupled pairs of electrons behave as bosons in a superconductor. •

---

[32]See, for instance, M. A. Kasevich, *Science* **298**, 1363 (2002) for an overview of matter–wave experiments and applications of atom interferometry.

14.7 APPLICATIONS OF ULTRASHORT PULSES

Ultrashort laser pulses make possible an astounding degree of time resolution, and the extremely high intensities that often characterize them are of interest for basic research as well as for a growing number of applications. In this section brief introductions are given to three new fields created by ultrashort laser pulses.

Time Resolution of Atomic and Molecular Processes

In the 1870s the photographer Eadweard Muybridge carried out experiments on the Palo Alto, California, farm of Leland Stanford to answer the question whether all four hooves of a horse are off the ground at any point in its trot. Using a series of cameras whose shutters were triggered by strings placed along a track, and later a periodic mechanical triggering, Muybridge eventually achieved millisecond time resolution—probably the record at the time—and found that the answer to the question was Yes.

Great progress in the resolution of short-duration events followed advances in electronics and other areas. But before femtosecond lasers[33] there was no way of following in real time such things as the vibrations of atoms in molecules or the detailed time evolution of a chemical reaction. Chemical reactions at room temperatures typically involve atomic or molecular velocities $\sim 10^3$ m/s and distances $\sim 10^{-10}$ m, implying a time scale $\sim (10^{-10} \text{ m})/(10^3 \text{ m/s}) = 100$ fs. Observation of the evolution of a reaction therefore requires a time resolution of about $1/10 \times 100$ fs $= 10$ fs, about 1000 times shorter than that possible with the fastest available electronics.

However, resolution on femtosecond time scales is still possible. We saw an example of this in Section 11.13, where we described how the durations of femtosecond laser pulses can be inferred from measurements of a field autocorrelation function. The "event" of a laser pulse is recorded via an interference of the pulse with itself: The pulse is used to measure its own time evolution. But how can femtosecond pulses be used to time-resolve processes occurring in and among atoms and molecules? To illustrate how this is done, we will focus on the vibrations of a diatomic molecule.

Recall that the potential energy function $V(R)$ for the bonding of the atoms of a diatomic molecule is well approximated near its minimum by a parabola, that is, by a harmonic-oscillator potential [cf. Fig. 2.6]. The vibrational energy levels are therefore approximately those of a harmonic oscillator [Eq. (2.3.8)]; anharmonic effects are accounted for by formula (2.3.12), where the frequency ω_e characterizes the molecule and the vibrational mode. To the extent that the vibrations are harmonic, we can picture a diatomic molecule as two atoms whose separation R oscillates sinusoidally in time with a period $1/\omega_e$ that is typically $\sim 10^{-13}$ s.

Figure 14.22 shows potential energy functions $V_0(R)$, $V_1(R)$, and $V_2(R)$ for three electronic levels 0, 1, and 2 of a hypothetical diatomic molecule. A field of frequency ν_{pump} can induce transitions between vibrational states of electronic levels 0 and 1, and a field of frequency ν_{probe} can induce transitions between vibrational states of electronic levels 1 and 2, as indicated. In accordance with the Franck–Condon principle (Section 11.11), transitions are indicated by vertical lines, i.e., they occur without a change in the

---

[33]"Femtosecond" here refers to pulse durations between roughly 1 fs and 0.1 ps $= 100$ fs.

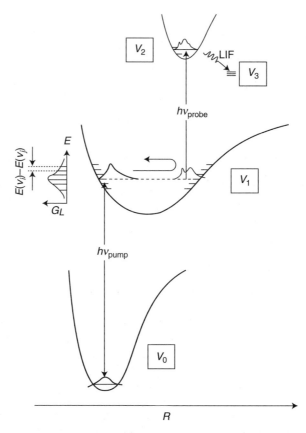

Figure 14.22 Time resolution of molecular vibrations using femtosecond pump and probe pulses. The pump pulse excites a molecule to an electronic level 1 with potential energy function $V_1(R)$, and after some time delay τ the probe pulse is applied to excite the molecule to a level 2 that decays by fluorescence to level 3. During the time τ the interatomic separation R changes, so that the strength of the fluorescence signal depends on τ. The variation with τ of this signal therefore reflects the variation with τ of the distance between the atoms. [From M. Gruebele and A. H. Zewail, *Journal of Chemical Physics* **98**, 883 (1993).]

interatomic separation R. Excitation to the electronic level 2 by absorption of radiation at ν_{pump} followed by absorption of radiation at ν_{probe} can be monitored by detection of the fluorescence accompanying transitions from 2 to some other level 3 ("laser-induced fluorescence," indicated by LIF in Fig. 14.22).

Suppose a laser pulse at frequency ν_{probe} is applied after some time τ following a pulse at frequency ν_{pump}. During this time the interatomic separation R of a vibrating molecule in electronic level 1 will have changed. If the durations of the pump and probe pulses are short compared to the vibrational period, the variation with τ of the LIF from level 2 signifies the variation of R with τ. Oscillations of the interatomic separation R are reflected in oscillations in the observed LIF signal. If, as estimated above, the period of the molecular vibrations in electronic level 1 is 100 fs, these vibrations can be observed in "real time" using pump and probe pulses of duration less that 100 fs. This is the basic idea behind time-resolved studies of atomic and

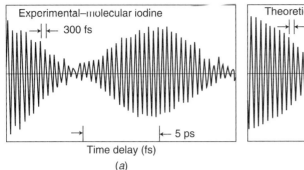
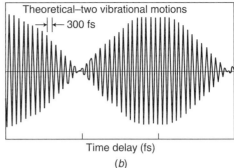

Figure 14.23 (*a*) Femtosecond pump-probe data showing vibrations of molecular iodine with an oscillation period of 300 fs. The much longer oscillation period of 10 ps is due to the anharmonicity of the vibrations. (*b*) Calculated signal with the two principal oscillation periods seen in (*a*). [From M. Dantus, R. M. Bowman, and A. H. Zewail, *Nature* **343**, 737 (1990).]

molecular processes with femtosecond lasers: Pump pulses set the "zero" of time from which temporal variations are determined with time-delayed probe pulses. The time delay τ between the pump and probe pulses can be varied by changing the difference ΔL in their path lengths: $\tau = \Delta L/c$. A path difference of a micron, for example, corresponds to a time delay of 3.33 fs.

Figure 14.23 shows the first reported time resolution of molecular vibrations. The molecule in the experiments was room temperature I_2, for which the vibrational period of the electronic level "2" is known from spectroscopic studies to be ≈ 300 fs. The pump and probe wavelengths were 620 and 310 nm, respectively. The vibrational period is clearly seen in the data. The amplitude of the LIF signal is modulated at a longer period (10 ps) as a consequence of anharmonicity. On a much longer time scale a period ~ 600 ps is observed as a consequence of the *rotation* of the I_2 molecule.

In our simplified discussion we have imagined that the interatomic separation R follows a classical trajectory, when in fact it must be described statistically using a quantum-mechanical wave function $\Psi(R, t)$: The probability that R is between R and $R + dR$ ($dR \ll R$) at time t is $|\psi(R, t)|^2 \, dR$. The lowest energy ($v = 0$) vibrational stationary state (or eigenstate), for example, is described by a time-independent wave function whose squared modulus in the harmonic-oscillator approximation is a Gaussian function of R. (Recall the remarks near the end of Section 11.10.) This is indicated for the electronic level with potential curve $V_0(R)$ in Fig. 14.22. The bandwidth of the pump pulse results in nonvanishing occupation probabilities for a range of vibrational eigenstates of the electronic level 1, as also indicated in Fig. 14.22. To simplify matters, let us suppose that only two eigenstates, with vibrational quantum numbers v and $v + 1$ and wave functions $\phi_v(R)$ and $\phi_{v+1}(R)$, have significant excitation probabilities, so that the wave function describing the vibrational state of the molecule in electronic level 1 at the time $t = 0$ immediately after irradiation by the pump pulse is

$$\psi(R, 0) = a_v(0)\phi_v(R) + a_{v+1}(0)\phi_{v+1}(R). \tag{14.7.1}$$

The probability amplitudes satisfy the time-dependent Schrödinger equations

$$i\hbar\dot{a}_v(t) = E_v a_v(t) \qquad \text{and} \qquad i\hbar\dot{a}_{v+1}(t) = E_{v+1}a_{v+1}(t) \qquad (14.7.2)$$

in the absence of any perturbation of the molecule [cf. Eq. (3.A.7)], so that $a_v(t) = a_v(0)\exp(-iE_v t/\hbar)$, $a_{v+1}(t) = a_{v+1}(0)\exp(-iE_{v+1}t/\hbar)$, and the vibrational wave function is

$$\psi(R,\,t) = a_v(0)e^{-iE_v t/\hbar}\phi_v(R) + a_{v+1}e^{-iE_{v+1}t/\hbar}\phi_{v+1}(R) \qquad (14.7.3)$$

at a time t after irradiation of the molecule by the pump pulse and before irradiation by the probe pulse. The probability distribution for the interatomic separation R is therefore

$$|\psi(R,\,t)|^2 = |a_v(0)|^2|\phi_v(R)|^2 + |a_{v+1}(0)|^2|\phi_{v+1}(0)|^2$$
$$+ 2\text{Re}\left[a_{v+1}^*(0)a_v(0)\phi_{v+1}^*(R)\phi_v(R)e^{i[E_{v+1}-E_v]t/\hbar}\right]. \qquad (14.7.4)$$

This oscillates in time at the frequency $(E_{v+1} - E_v)/h$ which, in the harmonic-oscillator approximation, is just $c\omega_e$ [Eq. (2.3.12)]. This oscillatory behavior of the probability distribution for R is indicated for the potential curve $V_1(R)$ in Fig. 14.22.

Now consider the probability amplitude for the transition to a vibrational eigenstate v' of electronic level 2 when the probe pulse is applied at a time τ after the pump pulse. Assuming it is small enough for lowest-order perturbation theory to be accurate, this amplitude is proportional to

$$\int_{-\infty}^{\infty}\phi_{v'}^*(R)D\psi(R,\,\tau)\,dR = a_v(0)e^{-iE_v\tau/\hbar}\int_{-\infty}^{\infty}\phi_{v'}^*(R)D\phi_v(R)\,dR$$
$$+ a_{v+1}(0)e^{-iE_{v+1}\tau/\hbar}\int_{-\infty}^{\infty}\phi_{v'}^*(R)D\phi_{v+1}(R)\,dR$$
$$\equiv D_{v',v}a_v(0)e^{-iE_v\tau/\hbar} + D_{v',v+1}a_{v+1}(0)e^{-iE_{v+1}\tau/\hbar}, \qquad (14.7.5)$$

where $\phi_{v'}(R)$ is the wave function for the vibrational eigenstate v' for the potential energy curve $V_2(R)$ and D is an electric dipole moment operator that depends on the polarization of the probe field. The probability amplitude will of course also depend on the electric field of the probe; this dependence can be ignored for our purpose here, which is simply to note that the LIF signal intensity of interest, which is proportional to $|\int_{-\infty}^{\infty}\phi_{v'}^*(R)D\psi(R,\,\tau)\,dR|^2$ for a pump-probe delay τ, will, like (14.7.4), oscillate at the frequency $(E_{v+1} - E_v)/h \cong c\omega_e$. Because the vibrational states are approximately equally spaced, this conclusion holds also if more than two vibrational states of the electronic level 1 are populated by the pump pulse. The (anharmonic) deviations from equal spacings result in additional frequency components in the LIF signal.

Quantum mechanics therefore predicts the same sort of LIF signal intensity expected from a classical trajectory picture of molecular vibrations: The signal oscillates with the pump-probe delay τ at principally the vibrational frequency $c\omega_e$, with additional frequency components due to anharmonicity. The wave function (14.7.3) depends on the probability amplitudes [$a_v(t)$ and $a_{v+1}(t)$] of the two states described by $\phi_v(R)$ and $\phi_{v+1}(R)$ and not just their occupation probabilities [$|a_v(t)|^2$ and $|a_{v+1}(t)|^2$]. In other words, it is a coherent superposition of two states, just as the wave function (9.3.3) is a coherent superposition of the states of an atom modeled as a two-state system. In general, the broad bandwidth of a femtosecond pulse, together with the approximately equal spacing of vibrational levels, results in a so-called *coherent wave packet*:

$$\psi(R, t) = \sum_{v=0}^{\infty} a_v(t)\phi_v(R). \tag{14.7.6}$$

The number of vibrational levels with significant probability amplitudes $a_v(t)$ is typically $\sim 5-10$ in studies of the type considered here. Of course, a more realistic theory requires that rotational energy levels be included. Since rotational periods are typically $\sim 10^3$ times larger than vibrational periods, an effect of rotations is to introduce periodicities on a much longer time scale than is shown in Fig. 14.23.

Fourier analysis of time series such as the data plotted in Fig. 14.23 can be used to extract spectroscopic information about molecular vibrations and rotations, i.e., to determine vibrational and rotational constants. This *femtosecond wave packet spectroscopy* can be performed on molecules in cells as well as in beams, since collision times are so much longer than the pulse durations and delay times (Problem 14.17).

Femtosecond lasers have made it possible to follow the movements of atoms in chemical processes, creating the field of *femtochemistry*. For example, the breaking of the chemical bond between two atoms of a diatomic molecule has been time-resolved by the femtosecond pump-probe technique. A pump pulse creates a coherent wave packet that oscillates within the potential energy well characterizing the bond. If ν_{pump} is sufficiently large, the outer wings of the wave packet can escape the well, that is, the molecule can dissociate. If the time-delayed probe pulse is at the frequency for which there is absorption when the molecules are closest together, for instance, an LIF signal oscillates as the molecule vibrates, but the oscillation is damped because of dissociation. Fluorescence from a dissociation product excited by a laser at the appropriate wavelength can also track the dissociation in real time.

The femtosecond pump-probe method has been employed in studies of subpicosecond processes in liquids and solids as well as in gases. It has spawned another field, *femtobiology*, based on time-resolved studies of phenomena such as vision and photosynthesis at the molecular level.

With the advent of attosecond pulses (see below) it has become possible to observe phenomena on a time scale characterizing the motion of atomic electrons.[34] Because electrons have a much smaller mass than atoms, their movements can be expected, generally speaking, to be faster and therefore to be significant on much shorter time scales; the relevant time scale for an atomic electron can be roughly estimated using the Bohr model, according to which the orbital period of the electron in the ground state of hydrogen is about 150 attoseconds (as). With pump-probe techniques it is even possible to follow the oscillations in time of the electric field of a femtosecond optical pulse. The details of such measurements are complicated, but the basic concept can

---

[34]See, for instance, R. Kienberger, E. Goulielmakis, M. Uiberacker, A. Baltuska, V. Yakovlev, F. Bammer, A. Scrinzi, Th. Westerwalbesloh, U. Kleineberg, U. Heinzmann, M. Drescher, and F. Krausz, *Nature* **427**, 817 (2004).

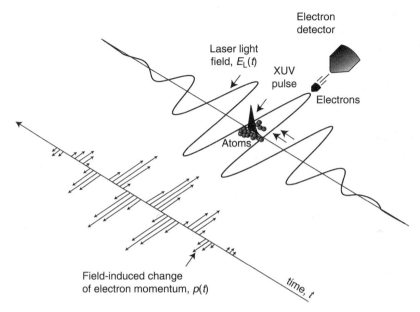

Figure 14.24   Attosecond burst of radiation (XUV pulse) photoionizes an atom, and a time-delayed femtosecond pulse then imparts a momentum impulse to the freed electron, affecting its time of flight to a detector. By repeating the TOF measurements with identical pulses and different time delays, the oscillations of the electric field of the femtosecond pulse can be determined. [From E. Goulielmakis, M. Uiberacker, R. Kienberger, A. Baltuska, V. Yakovlev, A. Scrinzi, Th. Westerwalbesloh, U. Kleineberg, U. Heinzmann, M. Drescher, and F. Krausz, *Science* **305**, 1267 (2004).]

be simply illustrated as in Fig. 14.24. The arrows in the lower left part of the figure indicate the electric field $E(t)$ of a few-cycle, linearly polarized femtosecond pulse whose oscillations are to be measured. This field is applied at a time t after the time $t = 0$ at which an extreme-ultraviolet (XUV) attosecond "burst" of radiation photoionizes an atom to produce a free electron. The field $E(t)$ does not cause any significant ionization of the atom, but it does impart to the electron a momentum impulse

$$\Delta p(t) = e \int_{t}^{\infty} E(t') \, dt'. \tag{14.7.7}$$

Depending on the sign of this impulse along the direction pointing to the electron detector in Fig. 14.24, the time of flight of the electron is either increased or decreased; in other words, time-of-flight (TOF) measurements on the electrons provide information about the *amplitude and phase* of $E(t)$, and repeated TOF measurements with identical pulses and different time delays then allow the temporal variation of $E(t)$ to be mapped out (Fig. 14.25). Thus, for two different times differing by a time δt much smaller than the period of the femtosecond field oscillations,

$$\Delta p\left(t - \frac{1}{2}\delta t\right) - \Delta p\left(t + \frac{1}{2}\delta t\right) = e \int_{t - \frac{1}{2}\delta t}^{t + \frac{1}{2}\delta t} E(t') \, dt' \cong e\,\delta t E(t), \tag{14.7.8}$$

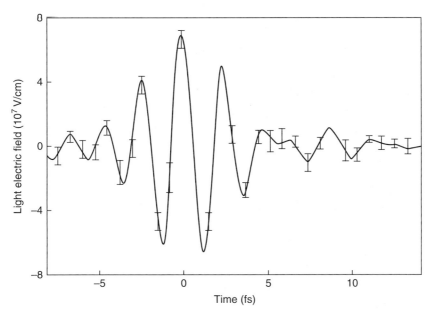

Figure 14.25 Electric field oscillations of a few-cycle femtosecond pulse, determined from electron time-of-flight data. [From E. Goulielmakis, M. Uiberacker, A. Baltuska, V. Yakovlev, F. Bammer, A. Scrinzi, Th. Westerwalbesloh, U. Kleineberg, U. Heinzmann, M. Drescher, and F. Krausz, *Science* **305**, 1267 (2004).]

or

$$E(t) \cong \frac{\Delta p\left(t - \frac{1}{2}\delta t\right) - \Delta p\left(t + \frac{1}{2}\delta t\right)}{e\,\delta t}, \qquad (14.7.9)$$

and $E(t)$ can be determined by TOF measurements in which t is varied over the duration of the femtosecond pulse.

• Among the highly nontrivial details ignored in our discussion is the need for precise timing between the attosecond and femtosecond pulses. For the data shown in Fig. 14.25, the ionizing pulses were 250-as bursts of 13.4-nm (93-eV) XUV radiation produced via harmonic generation with 0.5-mJ, 750-nm, ~5-fs Ti : sapphire laser pulses incident on a neon target (see below). The attosecond and femtosecond pulses then propagated along the same direction to a second neon target where the photoionization occurs. This second neon target, a <50-μm jet, was at the focus of a specially manufactured spherical mirror, such that the focused femtosecond pulse incident on the jet had a diameter >60 μm. Because of their much longer wavelengths, the femtosecond pulses diverged more in their propagation to the mirror, spreading over a diameter on the mirror of about 25 mm compared to a much smaller spot for the attosecond pulses. Piezoelectric adjustment of the movable central part of the mirror with nanometer precision allowed the time delay between the femtosecond and attosecond pulses incident on the second neon target to be controlled with attosecond precision [(1 nm)/c = 3.3 as].   •

High-Harmonic Generation

In Chapter 10 we explained various nonlinear optical phenomena by expanding the polarization **P** in powers of the electric field strength. As remarked at the end of that chapter, this perturbative approach cannot be usefully applied when, for instance, the radiation is so intense that many harmonics of the incident radiation are generated. In this regime of "extreme nonlinear optics" the field so strongly affects an atom that it is anything but a small perturbation.

The electric fields resulting from mode locking and chirped pulse amplification can approach or exceed the fields acting on atomic electrons (Section 11.13). For the hydrogen atom, for example, the electric field at the electron is about 5×10^9 V/cm; a laser with this electric field strength would have an intensity of about 3×10^{16} W/cm^2, which is smaller than the intensities obtainable by mode-locking and CPA techniques (Problem 11.9). Similarly the energies of photons that can be created by harmonic generation can exceed the binding energies of atomic electrons, for example, the 13.6-eV ionization energy of the hydrogen atom. For this reason photons created in high-harmonic generation are often characterized in terms of energies in electron volts rather than their associated wavelengths or frequencies: near-IR to optical wavelengths correspond to $\sim 1-2$ eV ($\sim 1200-600$ nm), XUV wavelengths to $\sim 10-100$ eV ($\sim 120-12$ nm), and soft X rays are characterized by photon energies up to ~ 1 keV (~ 1 nm). For a Ti:sapphire laser at a wavelength of 800 nm, by comparison, the photon energy is about 1.5 eV.

It is not surprising that an extremely intense field can generate high-harmonic radiation, given that the response of matter to high-intensity radiation is nonlinear (Section 10.1 and Problem 10.1). The electric field from an atom is proportional to $d^2\mathbf{p}/dt^2$ in the radiation zone, where $\mathbf{p}(t)$ is the induced electric dipole moment. Integrating the equations determining $\mathbf{p}(t)$ in the model of a two-state atom in a field of frequency ω, for example, and then computing the Fourier transform $S(\Omega)$ of the field radiated by $\mathbf{p}(t)$, one finds for sufficiently high intensities that $S(\Omega)$ has peaks at harmonics $\Omega = N\omega$ for a large set of odd integers N. By performing such an exercise, one can also demonstrate the inefficacy of an approach based on a perturbative expansion of $\mathbf{p}(t)$ in powers of the electric field.

High-harmonic generation (HHG) has most often been achieved by focusing laser pulses onto a small cell or jet of atoms with a high ionization potential, i.e., an inert gas. Figure 14.26 shows an HHG spectrum measured in one of the earliest such experiments. The basic features of this harmonic distribution have been amply confirmed in many later experiments: There is a steep decline in the intensity of the first few harmonics, followed by a plateau of roughly comparable intensities and then a sharp cutoff beyond which no higher harmonics are observed. Note also that only odd harmonics appear, as expected from inversion symmetry (Chapter 10).

The development of Ti:sapphire systems yielding extremely intense pulses has greatly extended the range of HHG photon energies, so much so that compact femtosecond laser sources are now used to generate coherent (laserlike) XUV and soft X-ray radiation for various applications. The photon energies are limited by the cutoff feature of HHG spectra, e.g., $N = 33$ in Fig. 14.26. Experimental results have been consistent with the following "universal" cutoff, i.e., the largest photon energy achievable by HHG in a gas of atoms with ionization potential I_p:

$$E_{\max} = I_p + 3.17 U_p, \qquad (14.7.10)$$

where U_p is the *ponderomotive energy*, defined as $e^2 E_0^2 / 4m\omega^2$, where E_0 is the electric field strength of the laser radiation at the fundamental (angular) frequency ω and e and m

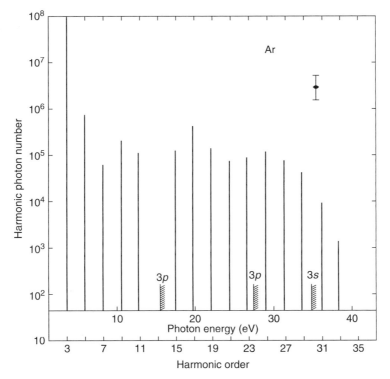

Figure 14.26 Distribution of harmonics in an early experiment on high-harmonic generation. The harmonics were generated when 3×10^{13} W/cm^2 mode-locked Nd:YAG pulses were focused onto a 15-Torr, $\sim$1-mm argon jet. $3s$, $3p$, and $3p^+$ indicate ionization energies for levels of Ar and Ar$^+$. [From X. F. Li, A. L'Huillier, M. Ferray, L. A. Lompré, and G. Mainfray, *Physical Review A* **39**, 5751 (1989).]

are the electron charge and mass, respectively. U_p is the cycle-averaged "quiver energy" of an electron in an electric field $\mathbf{E}_0 \cos \omega t$:

$$m\ddot{\mathbf{x}} = e\mathbf{E}_0 \cos \omega t, \qquad v(t) = \dot{\mathbf{x}}(t) = \frac{eE_0}{m\omega} \sin \omega t, \qquad \frac{1}{2} mv^2(t) = \frac{e^2 E_0^2}{2m\omega^2} \sin^2 \omega t,$$

$$U_p \equiv \frac{e^2 E_0^2}{4m\omega^2} = 0.93 \times 10^{-13} I\lambda^2 \text{ eV}, \tag{14.7.11}$$

where I (W/cm^2) is the intensity and λ (μm) is the wavelength of the presumed monochromatic field. The relation (14.7.10) has a surprisingly simple explanation, as follows.[35]

At the high intensities needed for HHG we can expect ionization. It is not necessary for our purposes to delve into the quantum theory of the ionization process; in fact we will describe the freed electron using classical mechanics, assuming that at the instant t_i at which ionization occurs, the electron is momentarily at rest at the position $x = 0$

[35]K. C. Kulander, K. J. Schafer, and J. L. Krause, in *Super-Intense Laser-Atom Physics*, eds. B. Piraux, A. L'Huillier, and K. Rząṣewski (Plenum, New York, 1993), p. 106; P. B. Corkum, *Physical Review Letters* **71**, 1994 (1993).

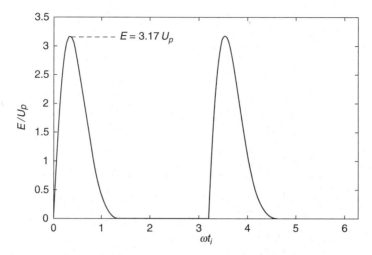

Figure 14.27   Numerical results from Eqs. (14.7.12) for the kinetic energy of an electron at the moment it returns to the parent ion after ionization at time t_i. The largest kinetic energy the electron can have when it hits the ion is computed to be $3.17U_p$.

of the ion in a field $E_0 \cos \omega t$ linearly polarized along the x direction. Ignoring any effect of the parent ion, we describe the electron's motion at times $t > t_i$ by the classical Newton equation, $m\ddot{x} = eE_0 \cos \omega t$, which gives

$$v(t, t_i) = \dot{x}(t, t_i) = \frac{eE_0}{m\omega} (\sin \omega t - \sin \omega t_i), \tag{14.7.12a}$$

$$x(t, t_i) = -\frac{eE_0}{m\omega^2} [\cos \omega t - \cos \omega t_i + \omega(t - t_i) \sin \omega t_i] \tag{14.7.12b}$$

for the chosen initial conditions. Equation (14.7.12b) can be solved for the time $t_f (>t_i)$ at which an electron freed from the atom at time t_i and "quivering" in the field can collide with the ion, that is, the time t_f for which $x(t_f, t_i) = 0$. The velocity at which the electron strikes the ion in this classical model is $v(t_f, t_i)$, and the kinetic energy is $E(t_f, t_i) = \frac{1}{2}mv^2(t_f, t_i)$. Thus, for any particular ionization time t_i we can calculate the time t_f at which the freed electron can "recollide" with the ion, and from that the electron's kinetic energy $E(t_f, t_i)$ at the moment of impact (Problem 14.19). Figure 14.27 shows numerical results for $E(t_f, t_i)$ obtained in this way for values of ωt_i between 0 and 2π. For ωt_i such that $E(t_f, t_i) = 0$ in the figure the freed electron never returns to the ion.[36]

The largest possible value of $E(t_f, t_i)$ is $3.17U_p$, which is found to occur at $\omega t_i \cong 0.30$ rad $= 17°$; the corresponding value of t_f is given by $\omega t_f \cong 4.45$ rad. Thus, $t_f - t_i = (4.45 - 0.30)/\omega = 1.8$ fs for an 800-nm Ti : sapphire pulse, compared with the 2.7-fs period of a monochromatic wave at 800 nm. In reality, of course, the electric field is not periodic, and during a few-cycle pulse, for instance, the electron can return to the ion a few times at most. The main point for our purposes is that *in an initial recollision the electron's kinetic energy cannot exceed* $3.17U_p$. For a gas of atoms in which these ionizations and recollisions occur, the kinetic energies vary over the distribution

---

[36]We are ignoring any motion of the electron transverse to the direction of field polarization. Such motion will reduce the number of electrons that actually recollide with their parent ions.

of ionization times, but the kinetic energy with which any electron can hit its parent ion cannot exceed $E_{max} = 3.17U_p$.

We have avoided any analysis of the ionization process, and likewise we will not analyze in any detail what happens when the electron collides with the ion. It can scatter off the ion and in so doing absorb energy from the field, or it can lose energy by releasing another electron from the ion by collisional ionization. It can also lose energy by radiation since together with the ion it forms a dipole driven by the laser field; based on the remarks above, any emitted photons must be at odd harmonics of the field. The classical model, of course, can only be taken so far, but it suggests that the largest possible energy that the electron could lose whenever it recollides with the ion is $E_{max} + I_p$, this being the largest possible energy lost when the electron is recaptured by the ion and falls into the ground atomic level of energy $-I_p$ from which it was freed in the first place. Equating this largest possible energy loss to the largest possible energy of a radiated photon, we arrive at the HHG cutoff relation (14.7.10). Electrons recolliding with their ions with kinetic energies smaller than E_{max} result in radiation at lower harmonics.

Further support for this model of high-harmonic generation may be found in its predictions regarding the laser polarization, which we have assumed to be linear. If we write the Newton equation of motion for an ionized electron in a circularly polarized field, we find that the electron never returns to the ion, in which case there should be no HHG. In fact it is found experimentally that high-harmonic intensities decrease very rapidly with increasing polarization ellipticity of the laser field, which is 0 for linear polarization and 1 for circular.

• The model just described is called the *three-step model*: (1) an atom is ionized by the field, (2) the freed electron is driven back into the ion by the field, and (3) an electron returning to the ion emits a high-harmonic photon. The ionization is assumed to occur by a tunneling process that is best understood in the limiting case of a *static* electric field $\mathcal{E}$. In this case an electron in the hydrogen atom, for example, has a total potential energy $-e^2/r + e\mathcal{E}z$ if the applied static field is along the z direction. Thus, the applied field effectively lowers the "Coulomb barrier" confining the electron, allowing it to escape (ionize) by quantum mechanical tunneling; the stronger the applied field, the more the barrier is lowered and the greater is the rate of tunneling.

Although the applied field in the three-step model for HHG is certainly not static, the assumption that ionization occurs by tunneling appears to be a good approximation if the field frequency is not too large. This condition may be expressed in terms of the *Keldysh parameter* defined as

$$\gamma = \sqrt{\frac{I_p}{2U_p}} = \frac{\omega}{\omega_t}, \qquad (14.7.13)$$

where $\omega_t = eE_0/\sqrt{2mI_p}$. γ delineates approximately between ionization by tunneling and by a multiphoton ionization process, tunneling dominating if $\gamma < 1$. The assumption of tunneling ionization in the three-step model is implicit in the initial condition that the electron velocity is zero at time t_i, since in the case of tunneling ionization, as opposed to multiphoton ionization, the initial kinetic energy of the electron can be assumed to be small. In the case of tunneling, furthermore, the ionization typically occurs in a short time compared to the field period, and the amplitude of the electron quiver is larger than an atomic dimension (Problem 14.19). For the ground state of the hydrogen atom, for example, the rate of tunneling ionization in the static-field approximation is given by quantum theory as

$$R_{ion}(t) = 4\omega_{at}\frac{E_{at}}{E(t)}\exp\left[-\frac{2}{3}\frac{E_{at}}{E(t)}\right], \qquad (14.7.14)$$

where $\omega_{at} = (1/4\pi\epsilon_0)^2 me^4/\hbar^3 = 4.1 \times 10^{16}$ s^{-1} and $E_{at} = (1/4\pi\epsilon_0)e/a_0^2 = 5 \times 10^9$ V/cm is the electric field on the electron ($a_0 =$ Bohr radius). Note the *exponential sensitivity* of the tunneling rate on the electric field strength $E(t)$. For a multiphoton ionization process involving the absorption of q photons, by contrast, the ionization rate is approximately proportional to E_0^{2q}.

For a linearly polarized field $E(t) = E_0 \cos \omega t$ the cycle-averaged tunneling ionization rate is

$$\bar{R}_{ion} = \frac{1}{\pi}\int_{-\pi/2}^{\pi/2} R_{ion}(t)d(\omega t) = \frac{4\omega_{at}}{\pi}\frac{1}{F}\int_{-\pi/2}^{\pi/2}\frac{1}{\cos x}e^{-2/(3F\cos x)}\,dx, \tag{14.7.15}$$

where $F \equiv E_0/E_{at}$ (>0). Since the integrand is strongly peaked around $x = 0$, we make the approximation $1/\cos x \cong 1/(1 - x^2/2) \cong 1 + x^2/2$ in the exponential and replace $\cos x$ by 1 in the prefactor:

$$\bar{R}_{ion} \approx \frac{4\omega_{at}}{\pi}\frac{1}{F}\int_{-\pi/2}^{\pi/2} e^{-(2/3F)(1+x^2/2)}\,dx = \frac{4\omega_{at}}{\pi}\frac{2}{F}e^{-2/3F}\int_0^{\pi/2} e^{-x^2/3F}\,dx$$

$$= \frac{4\omega_{at}}{\pi}\frac{2}{F}e^{-2/3F}\sqrt{3F}\int_0^{\pi/(2\sqrt{3F})} e^{-u^2}\,du \cong \frac{4\omega_{at}}{\pi}\frac{2}{F}e^{-2/3F}\sqrt{3F}\int_0^{\infty} e^{-u^2}\,du$$

$$= 4\omega_{at}\sqrt{\frac{3}{\pi F}}e^{-2/3F} = \sqrt{\frac{3F}{\pi}}R_{ion}^{static}, \tag{14.7.16}$$

where $R_{ion}^{static} = (4\omega_{at}/F)\exp(-2/3F)$ is the ionization rate in a static field of strength E_0. We have assumed that $F \ll 1$, as is the case in most situations of interest, allowing us to replace the upper integration limit by ∞ in the second line.

For circularly polarized light the magnitude of the electric field is independent of time, and the tunneling ionization rate is the same as that for a static field.

The calculation of the HHG spectrum and cutoff in the three-step model is based on the expression [cf. Eq. (3.A.13)]

$$\mathbf{p}(t) = \int d^3x\,\psi^*(\mathbf{x}, t)e\mathbf{x}\psi(\mathbf{x}, t) \tag{14.7.17}$$

for the expectation value of the electric dipole moment when an atomic electron is described by a wave function $\psi(\mathbf{x}, t)$. The electron in a strong field can either remain in the initial atomic ground state with wave function $\psi_g(\mathbf{x}, t)$, or it can be in a continuum state with wave function $\psi_c(\mathbf{x}, t)$ if ionization occurs. If it is assumed that the occupation probabilities of excited bound states are negligible, $\psi(\mathbf{x}, t)$ in Eq. (14.7.17) takes the form $\psi(\mathbf{x}, t) = a_g\psi_g(\mathbf{x}, t) + a_c\psi_c(\mathbf{x}, t)$, where $|a_g|^2$ and $|a_c|^2$ are, respectively, the probabilities that the electron is in the ground state or a continuum state. Assuming furthermore that the ionization probability is small, or in other words that the ground state is negligibly depleted, the dipole moment (14.7.17) is determined approximately by the real part of $\int d^3x\,\psi_g^*(\mathbf{x}, t)e\mathbf{x}\psi_c(\mathbf{x}, t)$, and this expression can be evaluated "semiclassically" using results of the classical model of electron–ion recollisions.[35]   •

According to the three-step model, electrons ionized by the laser field at different times recollide after different times with their parent ions, and their different kinetic energies upon recollision result in a distribution of radiated photon energies over the odd harmonics of the laser. The HHG radiation is assumed to be sufficiently weak that it does not produce any significant ionization. Each electron that does recollide with its parent ion does so by first being pulled away from the ion by the field and

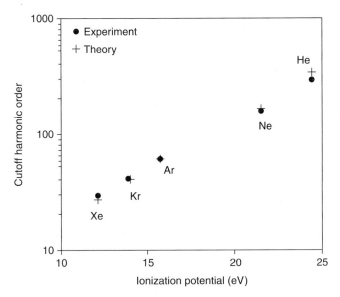

Figure 14.28 Experiment vs. theory for the cutoff HHG photon energies in inert gases. Note the logarithmic scale. [From Z. Chang, A. Rundquist, H. Wang, M. M. Murnane, and H. C. Kapteyn, *Physical Review Letters* **79**, 2967 (1997).]

then being forced back into the ion, the process occurring over a half-cycle of the field. After each half-cycle of the laser field, therefore, there is a burst of radiation whose spectrum consists of all the harmonics up to the cutoff photon energy, and whose duration is much shorter than a field cycle.

The cutoff relation (14.7.10) is an approximate, theoretical upper limit to the HHG photon energy, given the ionization potential and the laser intensity and frequency. In practice, the observed cutoff energy is generally smaller than that predicted by (14.7.10) for several reasons.[37] One is that at sufficiently high intensities the ground state is fully depleted and all the atoms are ionized by the leading edge of the laser pulse; the intensity I appearing in (14.7.11) is then replaced by a smaller "saturation" intensity I_s that can be estimated from quantum theoretical calculations of ionization rates. Figure 14.28 shows a comparison of experimentally observed HHG photon cutoff energies to the theoretical cutoff given by (14.7.10) with calculated values of I_s used for I in the expression for the ponderomotive energy. In these experiments 800-nm, 26-fs pulses from a Ti : sapphire laser system were focused onto gas jets from 1-mm-diameter nozzles, such that the gas pressures were ~ 8 Torr in the ~ 100-μm focal region. The peak intensity of the 20-mJ laser pulses was $\sim 6 \times 10^{15}$ W/cm^2.

The three-step model applies to a single atom, and as such does not account for propagation effects such as phase mismatching. Furthermore, the intensity used to estimate the cutoff energy might well be higher than that in the medium under conditions of significant ionization, resulting in a smaller than calculated cutoff energy: The refractive index of a plasma, given approximately by Eq. (3.14.13), decreases with electron density, which is largest where the field intensity is greatest and causes the most ionization.

---

[37]See, for instance, K. Miyazaki and H. Takada, *Physical Review A* **52**, 3007 (1995).

Therefore, a beam with a Gaussian transverse spatial profile, for example, will experience a smaller refractive index near the beam axis and can undergo self-defocusing as it ionizes atoms in its path. The good agreement between the experimental results shown in Fig. 14.28 with the predictions based on the single-atom cutoff relation (14.7.10) was attributed in part to the low gas density; at higher densities, where propagation effects are more significant, the cutoff energies were smaller, as expected.

Thus far, we have considered HHG mainly in terms of the frequencies generated, and have not addressed the question of the temporal nature of the HHG pulses aside from concluding from the three-step model that their duration should be extremely short. Their short wavelength, broad spectral width, and low intensities make them unamenable to the standard autocorrelation techniques described in Section 11.13, and studies of their temporal characteristics became possible only relatively recently. It is now well established that HHG produces attosecond pulse trains, and, as discussed below, can even produce the *single* attosecond pulses desired for time-resolved studies of subfemtosecond phenomena [cf. Figs. 14.23 and 14.24].

The very broad frequency spectrum associated with high-harmonic generation immediately suggests the possibility of producing attosecond pulses. Recall from our discussion of mode locking in Section 6.8 that if N phase-locked fields with equal amplitudes and with angular frequencies separated by Δ are superposed, the resulting field is a train of pulses separated in time by $T = 2\pi/\Delta$, each pulse having a duration $\tau = T/N$. In HHG, where only odd harmonics of the fundamental (laser) frequency ω appear, we have frequencies separated by 2ω, suggesting that the generation of N harmonics might result in a pulse train with $T = \pi/\omega$ and $\tau = \pi/N\omega$. For example, 100 harmonics of 880-nm radiation might form a train of 15-as pulses separated by 1.5 fs. However, the equal spacing of the high harmonics is by no means sufficient to make an attosecond pulse: The harmonics must also be phase-locked.

Quantum mechanical calculations reveal that the harmonics generated by a single atom are not phase-locked. It is found that, in the language of the semiclassical three-step model, the main contribution to each harmonic comes from two electron trajectories that correspond to different ionization and recollision times but that recollide with the ion with the same kinetic energy. Based on their differences in ionization and recollision times, these two types of trajectory are referred to as "long" and "short." The two types of trajectory give rise to single-atom HHG emission in the form of a pulse train in which there are mainly *two* pulses per half-cycle ($T = \pi/\omega$) of the laser field. However, nonlinear propagation effects in a gas act in such a way that the harmonics are generated by either the long or the short trajectories, depending on the focusing geometry, *and the harmonics are phase-locked*. The HHG emission is then in the form of a train of extremely short pulses, *one* per half-cycle.[38] Experimental results are consistent with these predictions. It has been found, for instance, that the 5 odd harmonics 11–19 generated by 40-fs Ti : sapphire laser pulses ($T = \pi/\omega = 1.35$ fs) in argon are phase-locked and can combine to produce trains of 250-as pulses separated by 1.35 fs.[39]

Pump-probe time resolution of subfemtosecond phenomena, as discussed in the preceding subsection, generally requires *single* attosecond pulses rather than pulse trains. *Pulse pickers* that select a single pulse (or a sequence of pulses to produce a train with

[38]P. Antoine, A. L'Huillier, and M. Lewenstein, *Physical Review Letters* **77**, 1234 (1996).

[39]P. M. Paul, E. S. Toma, P. Breger, G. Mullot, F. Auge, P. Balcou, H. G. Muller, and P. Agostini, *Science* **292**, 1689 (2001).

a smaller repetition rate) from a mode-locked train of picosecond or femtosecond pulses typically employ a Pockels cell between two crossed polarizers, much like a light modulator (Fig. 5.19). The repetition rate of attosecond pulses in HHG, however, is far too fast for any electronics that can switch a Pockels cell. Single attosecond pulses have been produced by a *polarization gating* technique that relies on the strong dependence of HHG on the laser ellipticity, as noted earlier; the idea, basically, is to form a laser pulse with a time-dependent ellipticity, such that HHG can occur only within a narrow time window. Here we will briefly describe single attosecond pulse generation in the particular case of a few-cycle laser pulse. Based on the three-step model in which electrons emit short bursts of radiation only every half-cycle of the field, a few-cycle laser pulse can be expected to produce at most only a few pulses of high-harmonic radiation.

For the electric field of a linearly polarized laser pulse used to generate high harmonics we write

$$E(t) = \mathcal{E}(t)\cos(\omega t + \phi). \tag{14.7.18}$$

Usually, we can make the assumption that the envelope $\mathcal{E}(t)$ is slowly varying in time compared to the carrier wave $\cos(\omega t + \phi)$. Since $\omega \approx 10^{15}\,\mathrm{s}^{-1}$ at optical frequencies, this is a valid approximation for pulses in which $\mathcal{E}(t)$ varies negligibly on a femtosecond time scale, for example, for picosecond pulses; for such pulses the constant phase ϕ in (14.7.18) is unimportant. We used this approximation in Chapter 9, for instance, to formulate the theory of resonant pulse propagation based on the Maxwell–Bloch equations. For pulses lasting only a few cycles of the carrier wave, this approximation of a slowly varying envelope breaks down. Moreover, the phase ϕ for such pulses plays an important role in HHG, which depends sensitively on how the electric field varies over the duration of the laser pulse. In particular, different half-cycles of the field have different electric field amplitudes (Fig. 14.29) and will therefore lead to different peak (cutoff) photon energies.

Consider the example of a Gaussian pulse envelope:

$$E(t) = \mathcal{E}_0 e^{-t^2/\tau^2} \cos(\omega t + \phi), \tag{14.7.19}$$

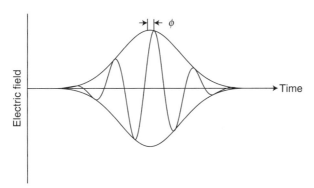

Figure 14.29 Carrier-envelope phase ϕ of a few-cycle pulse gives the phase difference between the peak of the pulse envelope and the nearest peak of the carrier wave.

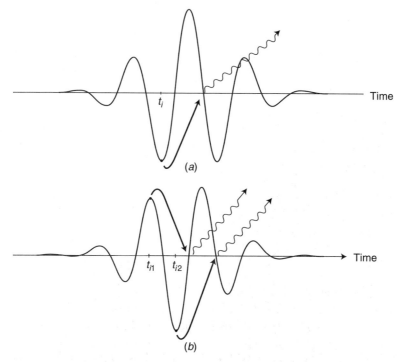

Figure 14.30   Dependence of peak-energy high-harmonic generation on the carrier-envelope phase of a few-cycle pulse. The peak pulse intensity is large enough to give rise to tunneling ionization and therefore bursts of high-harmonic radiation at times near that at which the peak of the pulse envelope occurs. In (a) the CEP is zero. Atoms undergo tunneling ionization at a time t_i, and an electron freed at this time, as indicated by the solid arrow, recollides with the parent ion half a field cycle later, emitting a burst of radiation according to the three-step model. In (b) the CEP is nonzero, and tunneling ionization of atoms is indicated at the two times t_{i1} and t_{i2}. The radiation that occurs half a field cycle after ionization now results in two bursts of radiation. In each case the electron "trajectories" shown are those for which the electron recolliding with the ion experiences the strongest field half-cycle, resulting in bursts of radiation at the peak photon energy.

for a pulse duration amounting to just a few cycles (Fig. 14.29). The pulse envelope peaks at $t = 0$, which coincides with the peak modulus of the electric field only if $\phi = 0$, where ϕ, the *carrier-envelope phase* (CEP), specifies the difference in time from the peak of the envelope to the nearest peak of the carrier. As illustrated in Fig. 14.30, a few-cycle laser pulse with zero carrier-envelope phase results in a single burst of radiation at the peak photon energy, whereas a nonvanishing CEP can produce two. Techniques have been developed to measure and control the carrier-envelope phase of few-cycle laser pulses and thereby to reliably produce single attosecond pulses by high-harmonic generation.[40]

As discussed in Chapter 10, harmonic generation is generally very inefficient without phase matching. Gaseous media are nearly always used for high-harmonic generation

---

[40]See C. A. Haworth, L. E. Chipperfield, J. S. Robinson, P. L. Knight, J. P. Marangos, and J. W. G. Tisch, *Nature Physics* **3**, 52 (2007) and references therein.

in order to minimize absorption, and the ionization that accompanies the process results in large phase mismatches between the fundamental and the co-propagating harmonics due to the plasma dispersion. Since gases are not birefringent, they do not permit angle phase matching based on birefringence. It has been demonstrated, however, that HHG conversion efficiencies can be subtantially increased by a quasi-phase-matching in which counterpropagating Ti : sapphire pulses in a waveguide containing an inert gas modulate the field in such a way that phase matching is effectively realized periodically along the waveguide.[41] Enhancement of HHG conversion efficiencies will have important consequences for attophysics and in a broad range of applications requiring coherent XUV and soft X-ray sources.

Frequency Combs and Optical Frequency Metrology

Throughout its history optical spectroscopy has been based on measurements of wavelengths rather than frequencies. Wavefront distortions result in different path lengths for different parts of a wave, limiting the most precise measurements to relative accuracies ($\Delta\lambda/\lambda$) of about 10^{-10}. Measurements of *frequencies* could be far more accurate since, as noted in Section 6.8, atomic clocks allow measurements of time to relative accuracies on the order of 10^{-15}. While frequency counters allow measurements of microwave frequencies as high as a few tens of gigahertz, no available electronics can count the $\sim 10^{15}$ cycles per second of *optical* waves. This problem was circumvented in the late 1990s with the use of *frequency combs*, allowing measurements of absolute optical frequencies (*optical frequency metrology*). By an "absolute" frequency (or an absolute measurement of frequency) we mean one that involves the second as *defined* by the cesium clock transition (Section 14.3); absolute frequencies are measured by referencing them to the oscillations of this transition. Since optical frequencies are $\sim 10^5$ times larger than that of the clock transition, measurements of absolute optical frequencies require that a very large frequency gap must somehow be bridged from the microwave to the optical.

Recall that the output of a mode-locked laser consists of a continuous, periodic train of pulses (Figs. 6.9 and 6.10). The spectrum of the pulse train is a *comb* of frequencies, such that the (angular-frequency) spacing ω_r between the "teeth" of the comb is equal to $2\pi/T$, where T is the spacing in time between pulses [Fig. 6.8 and Eq. (6.8.5b)]. ω_r depends on the length of the laser cavity and is in the radio-frequency range, typically 10^7–$10^9\,\text{s}^{-1}$. The frequency combs used in optical frequency metrology are obtained from mode-locked lasers; for reasons discussed below, the comb frequencies are

$$\omega_n = n\omega_r + \omega_0, \qquad (14.7.20)$$

where ω_0, like ω_r, is very much smaller than an optical frequency; n is a positive integer, typically $\sim 10^6$ (Section 6.8). This formula, as well as an expression for the "offset" frequency ω_0, will be derived below. We will also describe how ω_0 can be measured.

A frequency comb acts as a "ruler" for the measurement of optical frequencies. To determine the absolute frequency ω_L of a single-mode laser, for example, a measurement

---

[41]X. Zhang, A. L. Lytle, T. Popmintchev, X. B. Zhou, H. C. Kapteyn, M. M. Murnane, and O. Cohen, *Nature Physics* **3**, 270 (2007).

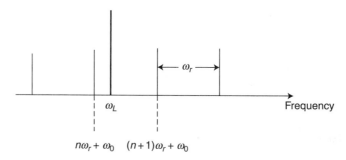

Figure 14.31   Frequency comb spectrum and position of optical frequency ω_L with respect to the closest comb frequency $n\omega_r + \omega_0$. Heterodyning results in beat notes determined by $\omega_L - n\omega_r - \omega_0$.

can be made of the signal detected by a photodetector when the single-mode laser and the mode-locked laser fields are superposed (heterodyned). As can be seen from Fig. 14.31, the rf signal will peak not only at the comb frequencies but also at beat frequencies determined by the difference $\omega_L - n\omega_r - \omega_0$ between the laser frequency and the nearest comb frequency (Problem 14.21). Since ω_r and ω_0 are known, n can be determined if ω_L is known beforehand (e.g., by having measured the wavelength with a wavemeter) to within $\pm \omega_r/4$; this yields ω_L. If such information about ω_L is not known beforehand, n can be determined by varying the comb spacing ω_r of the mode-locked laser. The frequency comb "ruler" can be applied in other ways. For example, the single-mode laser radiation can be frequency doubled, and the difference frequencies $\omega_1 = \omega_L - n\omega_r - \omega_0$ and $\omega_2 = 2\omega_L - 2n\omega_r - \omega_0$ between ω_L and ω_{2L} and the nth and $2n$th comb lines can be measured by heterodyning. Again n can be determined if ω_L is known beforehand to within $\pm \omega_r/4$, or by "dithering" ω_r if it is not. Then ω_L is given in terms of the known quantities ω_r, ω_1, ω_2 and n by the relation $\omega_2 - \omega_1 = \omega_L - n\omega_r$. The measured frequencies are "absolute" when they are referenced to a frequency standard, for example, when cycles per second are counted with a cesium atomic clock. It is the fact that n is very large ($\sim 10^6$) that allows the "bridging of the gap" between rf and optical frequencies.

Measurements of absolute optical transition frequencies have been made by locking the frequency of a laser to a transition (Section 5.13) and then measuring the absolute frequency of the laser by the frequency comb technique. Such methods have been used to measure the 1S–2S transition of the hydrogen atom, for example, to a few parts in 10^{14}. These ultra-high-precision techniques are valuable not only for spectroscopy and for applications including atomic clocks and navigation, but also for basic research in quantum electrodynamics and in determining whether the fundamental constants of nature might actually change over time.

To explain the form (14.7.20) of the frequency comb of a mode-locked laser, consider the electric field $E(t)$ from the laser incident on a photodiode; $E(t)$ is a sum of pulses separated by a time $T = 1/\omega_r$. Assuming the pulses are identical and described by an electric field $\mathcal{E}(t) \exp(-i\omega_c t)$, where $\mathcal{E}(t)$ is a pulse envelope and ω_c a carrier frequency, we write

$$E(t) = \sum_m \mathcal{E}(t - mT)e^{-i\omega_c(t - mT)}, \qquad (14.7.21)$$

where the summation is over a large number of integers m. The Fourier transform of this field is

$$
\begin{aligned}
\tilde{E}(\omega) &= \frac{1}{2\pi}\int_{-\infty}^{\infty} E(t)e^{i\omega t}\,dt = \frac{1}{2\pi}\int_{-\infty}^{\infty} dt \sum_{m} \mathcal{E}(t - mT)e^{-i\omega_c(t-mT)}e^{i\omega t} \\
&= \frac{1}{2\pi}\sum_{m}\int_{-\infty}^{\infty} dt\, \mathcal{E}(t)e^{-i\omega_c t}e^{i\omega(t+mT)} \\
&= \frac{1}{2\pi}\sum_{m} e^{im\omega T}\int_{-\infty}^{\infty} dt\, \mathcal{E}(t)e^{i(\omega-\omega_c)t} \\
&= \tilde{\mathcal{E}}(\omega - \omega_c)\sum_{m} e^{im\omega T},
\end{aligned}
\tag{14.7.22}
$$

where $\tilde{\mathcal{E}}(\omega)$ is the Fourier transform of the envelope of a single pulse. We have already encountered the sum in the last line in Section 6.7. For the large number of summands appropriate for a mode-locked laser, it consists of a series of peaks separated by $2\pi/T$, i.e., a series of peaks at $\omega_n = 2\pi n/T$, where n is an integer and T is the pulse spacing. So (14.7.22) merely tells us what we already know from Chapter 6 about mode-locked pulse trains.

The assumption of identical pulses in a mode-locked train, however, is not valid when the pulses are very short and therefore have a large bandwidth causing significant dispersion in the laser medium and optical elements. In this case the pulse envelope will propagate within the laser cavity at a group velocity v_g while the carrier wave at frequency ω_c propagates at the phase velocity v_p (Sections 8.3 and 8.4). This results in a carrier-envelope phase difference that we denote by $\Delta\phi$. Now based on the "bouncing-ball" picture of the intracavity field of a mode-locked laser (Section 6.8), we infer that each successive pulse from the laser has an additional carrier-envelope phase difference $\Delta\phi$ over that of the preceding pulse, so aside from some overall phase we can assign a carrier-envelope phase $m\Delta\phi$ to the mth pulse of the mode-locked train. We therefore replace (14.7.21) by

$$
E(t) = \sum_{m} \mathcal{E}(t - mT)e^{-i\omega_c(t-mT)}e^{im\Delta\phi}
\tag{14.7.23}
$$

and (14.7.22) by

$$
\begin{aligned}
\tilde{E}(\omega) &= \frac{1}{2\pi}\int_{-\infty}^{\infty} E(t)e^{i\omega t}\,dt = \frac{1}{2\pi}\int_{-\infty}^{\infty} dt \sum_{m} \mathcal{E}(t - mT)e^{-i\omega_c(t-mT)}e^{im\Delta\phi}e^{i\omega t} \\
&= \frac{1}{2\pi}\sum_{m} e^{im(\omega T+\Delta\phi)}\int_{-\infty}^{\infty} dt\, \mathcal{E}(t)e^{i(\omega-\omega_c)t} \\
&= \tilde{\mathcal{E}}(\omega - \omega_c)\sum_{m} e^{im(\omega T+\Delta\phi)}.
\end{aligned}
\tag{14.7.24}
$$

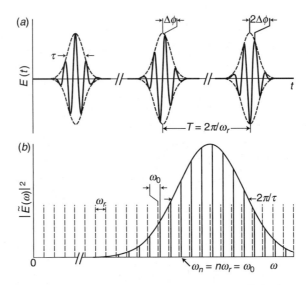

Figure 14.32   A mode-locked pulse train in (*a*) the time domain and (*b*) the frequency domain. Pulse-to-pulse carrier-envelope phase shifts $\Delta\phi$ in the time domain translate in the frequency domain to an offset ω_0 from integral multiples of the pulse repetition frequency ω_r. [After S. T. Cundiff and J. Ye, *Reviews of Modern Physics* **75**, 325 (2003).]

The only difference from (14.7.22) is that the peaks in the spectrum now occur at $\omega_n T + \Delta\phi = 2\pi n$:

$$\omega_n = n\omega_r - \frac{\Delta\phi}{T} \equiv n\omega_r + \omega_0, \tag{14.7.25}$$

with $\omega_0 = -\Delta\phi/T = -(\Delta\phi/2\pi)\omega_r$. Equation (14.7.24) describes the pulse train in the frequency domain, as opposed to the time-domain expression (14.7.23) (Fig. 14.32).

To use the frequency comb as a "ruler" to measure optical frequencies, it is necessary of course to have accurate values of ω_r and ω_0. Whereas ω_r can be measured from the beat signal when the pulse train is incident on a photodiode, the measurement of the frequency offset ω_0 is not so straightforward. A common way of doing so is to frequency-double the nth comb line (frequency ω_n) and measure the beat signal between its second harmonic ($2\omega_n$) and the $2n$th comb line (ω_{2n}):

$$2\omega_n - \omega_{2n} = 2(n\omega_r + \omega_0) - (2n\omega_r + \omega_0) = \omega_0. \tag{14.7.26}$$

Our simplified discussion ignores some critical issues in the actual implementation of frequency combs.[42] We have implicitly assumed in writing (14.7.26), for instance, that the frequency comb from a mode-locked laser covers at least an octave of frequencies, that is, that the highest frequencies in the comb are twice as large as the lowest. Although a full octave is not strictly necessary, it is highly advantageous because it permits straightforward determinations, as described, of both rf frequencies ω_r and ω_0; once they are measured and stabilized, optical frequency metrology becomes a going concern.

---

[42]For a comprehensive discussion see, for instance, S. T. Cundiff and J. Ye, *Reviews of Modern Physics* **75**, 325 (2003) and the many references therein.

An unchirped, transform-limited Gaussian laser pulse of duration τ has a FWHM spectral width $\Delta\Omega = (4 \ln 2)/\tau$ [Eq. (8.5.22)]. For this width to span an octave around the central frequency ω_c, we require $\Delta\Omega = \omega_c$, or $\tau = (4 \ln 2)/\omega_c = (4 \ln 2)/2\pi$ cycles, i.e., the pulse would have to have a duration of roughly a single cycle or less. Because of its extremely broad spectrum, the primary generator of frequency combs in optical frequency metrology is the Ti:sapphire laser with Kerr lens mode locking and compensation for group velocity dispersion (Fig. 11.25), but even the radiation from this system does not usually span an octave. For this reason octave-spanning frequency combs have been produced by spectrally broadening mode-locked pulses using self-phase modulation in fibers.

Recall from Section 10.4 that a nonlinear refractive index causes spectral broadening due to "self-chirping" or self-phase modulation. Group velocity dispersion of a short pulse causes it to broaden in time, reducing the peak intensity and therefore the self-phase modulation (Section 8.4). It has been found, however, that when a femtosecond pulse propagates in a photonic crystal fiber (Section 11.14) in which the silica core is surrounded by a particular arrangement of air holes, there is nearly zero group velocity dispersion near the 800-nm central wavelength of Ti:sapphire. This allows mode-locked pulses from a Ti:sapphire laser to propagate with little temporal broadening while maintaining the high intensities needed for substantial self-phase modulation and spectral broadening; more than an octave around 800 nm has been obtained in this way.[43] This is an example of *supercontinuum generation* in which a very broad ("white light") spectrum is created from the much narrower spectrum of a laser. Self-focusing and other nonlinear phenomena in addition to self-phase modulation play a role in determining the white-light pulses generated by a mode-locked pulse train, and it is not at all obvious that successive pulses are strongly correlated in phase, as they must be if they are to serve as a broadband frequency comb (Problem 14.21). Interference experiments have demonstrated nevertheless that the supercontinuum white-light pulses generated by mode-locked laser pulses of interest for optical frequency metrology can in fact be phase-locked. In other words, supercontinuum generation can be used to convert a phase-locked laser pulse train into an *octave-spanning* phase-locked train.

A crucial factor for optical frequency metrology is the stabilization of the frequencies ω_r and ω_0 against unpredictable fluctuations in the laser cavity length. A small fraction of the mode-locked Ti:sapphire laser output, typically consisting of $\sim$30-fs pulses, can be used to measure ω_r (or a high harmonic of ω_r) with a photodiode and using a feedback loop to adjust the cavity length and lock ω_r to a frequency referenced to an atomic clock or a GPS-controlled quartz oscillator. For some purposes, such as measuring the difference in frequency between two lasers, locking of ω_r alone suffices, whereas for absolute optical frequency measurements it is also necessary to lock ω_0. For this purpose the pulse train is injected into the photonic crystal fiber to generate an octave-spanning frequency comb, and part of the output from the fiber is used to measure ω_0 as described above [Eq. (14.7.26)]. One way of varying ω_0 in order to lock it is to swivel the laser feedback mirror of the layout shown in Fig. 11.25. Since the spectrum of the laser radiation with intracavity prisms is not uniform over the mirror, a swivel introduces a frequency-dependent phase shift and results in a group delay and a change in $\Delta\phi$ and therefore ω_0.

[43]D. J. Jones, S. A. Diddams, J. K. Ranka, A. Stentz, R. S. Windeler, J. L. Hall, and S. T. Cundiff, *Science* **288**, 635 (2000).

Locking ω_r and ω_0 at particular frequencies allows the synthesis of a comb of optical frequencies referenced to an atomic clock. *Optical synthesizers* providing $\sim$500,000 absolute frequencies covering most of the optical and near-infrared spectrum, and containing the femtosecond laser and all the other instrumentation needed for comb generation in a unit occupying $\sim$1 m^2 on an optical bench, are now commercially available for absolute frequency measurements, high-precision spectroscopy, precise distance measurements, and other applications. It has been said that having such a synthesizer is like having $\sim$500,000 extremely stable and precisely tuned lasers all at once. Optical synthesizers will likely be used to make optical clocks by locking a laser to the extremely narrow resonance of a cold atom. Because of the much shorter periods of optical cycles compared to microwaves, optical clocks will in turn result in even more precise timekeeping than is now possible.

• Prior to frequency comb generators absolute optical frequencies were measured by constructing long, complex "harmonic chains" to multiply the cesium clock frequency by harmonic generation in crystals, electronic frequency mixing, and other techniques. Because of their complexity and the large amount of laboratory space required, and because they could only be used to measure a single optical frequency, only a few such harmonic chains were developed. The frequency comb technique has revolutionized optical frequency metrology, and high-harmonic generation might extend the range of frequency metrology to XUV and soft X-ray frequencies.

Knowing that the spectrum of a mode-locked laser consists of a comb of frequencies, one might wonder in hindsight why the idea of a frequency comb as a ruler for optical frequency measurements was not put into practice until relatively recently. One reason, it seems, is that it was not realized just how uniform the frequency combs are; frequency metrology requires that the lines be "exactly" equally spaced [cf. Eq. (14.7.20)]. One might reasonably expect that dispersion effects in the laser would cause the mode spacing to vary slightly across the comb. Experiments in the late 1990s, however, revealed—surprisingly—that the mode spacing in a Kerr lens mode-locked laser with dispersion compensation (Fig. 11.25) is uniform and equal to the pulse repetition frequency to at least one part in 10^{17}, *even after the spectrum of the pulse train is broadened by propagation in a fiber!* Part of the 2005 Nobel Prize in Physics was awarded to J. L. Hall and T. W. Hänsch "for their contributions to the development of laser-based precision spectroscopy, including the optical frequency comb technique." •

14.8 LASERS IN MEDICINE

Lasers continue to play a large role in medicine. Our brief overview will focus on some qualitative aspects of laser–tissue interactions and on the growing importance of laser science in medical imaging.

The first medical application of lasers was in ophthalmology, just a few years after the first demonstrations of laser oscillation in the early 1960s. It was known for centuries that visual loss can result from prolonged, direct viewing of the sun. This occurs due to a burning and consequent scarring of the macula, the central part of the retina responsible for acute vision (Fig. 14.33). It was suspected for some time that a localized burning and scarring might actually be useful for the treatment of certain visual disorders, and in the late 1940s G. Meyer-Schwickerath demonstrated that burns produced by white-light sources such as the sun and xenon lamps could be used to connect the retina to its substratum tissue. The brightness of lasers made it immediately apparent that they could be

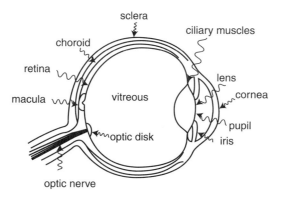

Figure 14.33 The human eye.

used in ophthalmology. In particular, it was apparent that lasers could be used for photocoagulation of small areas of ocular tissue, and that this could be done with very short exposure times. Furthermore the monochromaticity of laser radiation allows selective coagulation of a particular tissue, since the photocoagulation process is initiated by wavelength-dependent absorption of light. Laser therapy has not only replaced some older surgical procedures for visual disorders but has also been used for conditions where effective treatments were previously unavailable.

•   For safety's sake it must be remembered that even a low-power laser can damage the retina. To estimate the intensity focused on the retina when the eye is in the direct path of a laser beam, assume that the lens of the eye has a focal length of 2.3 cm and that the pupil diameter is 2 mm. The diameter of the spot on the retina is $f\theta$, where θ is the divergence angle of the laser beam. [See Eq. (13.2.7) for this formula in the special case of a Gaussian beam.] The fraction of the laser power entering the eye is equal to the square of the pupil diameter d divided by the square of the laser beam diameter D. The intensity of the focused radiation at the retina is therefore

$$I = \mathrm{Pwr}\left(\frac{d}{D}\right)^2 \frac{1}{\pi(f\theta)^2/4} = \frac{4(\mathrm{Pwr})d^2}{\pi D^2 f^2 \theta^2}. \tag{14.8.1}$$

Consider a green laser pointer with Pwr = 1 mW, beam diameter $D = 1.5$ mm, and divergence angle $\theta = 1.4$ mrad. In this example $I \sim 200$ W/cm^2. This estimate assumes perfect transmittivity of the eye (not too bad an approximation in the visible) and direct viewing with the eye held fixed and not blinking. But it illustrates the important point: Even very low-power lasers are potentially hazardous.[44] A 100-W lightbulb would not create nearly such a hazard; it is the small divergence angle θ of a laser that makes the focused intensity so large.   •

Laser therapies may be broadly classified as photoabsorptive or photodisruptive. Photoabsorption results in electronic and vibrational excitation, breaking of molecular bonds, and a rise in temperature. Large biological molecules, such as proteins, can undergo conformational changes when the temperature is increased. The result is a *thermal denaturation* in which certain biological functions are lost or impaired due,

---

[44]The retinal damage threshold in the case of continuous illumination can be as low as 2–3 W/cm^2. [A. M. Clarke, W. T. Ham, Jr., W. J. Geeraets, R. C. Williams, and H. A. Meuller, *Archives of Environmental Health* **18**, 424 (1969).]

for instance, to changes in cell membranes. Such thermal denaturation is responsible for effects such as inflammation and coagulation of tissue. Through a normal reparative response it leads to a scar, which can serve to connect tissues such as the sensory and pigmented layers of the retina. In other words, coagulation can "weld" disconnected tissue back together. Coagulation can also occlude (close) and destroy blood vessels; this is used in the treatment of diabetic retinopathy, in which fragile blood vessels may appear on the retina. These abnormal vessels tend to break, and the hemorrhaging can cause loss of vision. Since the pigmented cells behind the nerve-containing outer layer of the retina absorb strongly in the blue-green, whereas the neural retina, lens, cornea, and vitreous do not, the green argon ion laser is the primary tool for this procedure.

Photocoagulation begins with absorption of light, the primary absorbers of light in the human eye being melanin, hemoglobin, and xanthophyll. The ocular medium itself is transparent between about 380 and 1400 nm. At shorter wavelengths the lens and cornea are absorbing, and at longer wavelengths the primary absorber is water. Lasers allow a selective energy treatment in that the total energy of irradiation during the exposure can be accurately controlled. This allows a trained ophthalmologist to preselect a certain energy that, based on experience, will produce a minimal degree of coagulation. The energy can then be increased gradually until a desired degree of coagulation has occurred, with minimal damage or side effects.

Exposure time is of course another very important consideration. For a given total energy, the temperature rise of the irradiated tissue increases with decreasing pulse duration, since there is less time for thermal diffusion to the surrounding tissue. With short pulses, therefore, the temperature rise can be quite large and can lead to vaporization at the irradiated spot; the clinical manifestation of this vaporization is the appearance of gas bubbles near the target. The vaporization can also generate pressure waves strong enough to damage eye tissue.

In contrast to photocoagulation, photodisruptive laser therapies are nonthermal. Photodisruption is initiated by ionization from the intense heat produced by a pulsed laser at the target. The resulting plasma absorbs energy from the laser and becomes very hot, expands rapidly, and produces a shock wave that can blast a hole in an ocular membrane. Such a hole can be several times larger than the waist of the laser beam at the focal point. In laser iridectomy for the treatment of glaucoma, a hole is made in the iris to relieve the elevated intraocular pressure.

A common application of laser photodisruption of tissue is in treating the clouding of the posterior capsule membrane behind the lens, an occasional complication of cataract surgery. In this outpatient treatment, called posterior capsulotomy, a pulsed Nd : YAG photodisruptor tears a hole in the clouded membrane, opening a path for clear vision. This is done without damage to the retina by focusing the laser to a spot just behind the lens. The divergence of the beam beyond the focus then results in a reduced intensity at the retina. The attenuation of the beam by the plasma created also helps to reduce the intensity at the retina.

Laser in situ keratomileusis (LASIK), introduced in the mid-1990s, has become the most common surgical procedure for refractive correction. After a metal blade is used to make a flap in the outer layer of the cornea, an ArF excimer laser (193 nm, $\sim$10-ns pulses, $\sim$100-Hz pulse repetition rate) penetrates the inner part of the cornea and breaks molecular bonds, allowing cells to escape in the form of a tiny "mushroom cloud." Following this ablative sculpting with the excimer laser, the flap is closed and serves as a natural "bandage." Most complications from LASIK are flap-related; in a recent innovation the flap is made with a mode-locked, femtosecond diode-pumped Nd : YAG laser at 1.053 μm. The positions and focal depths of successive pulses on the cornea are computer controlled, the pulses creating, by ionization and the formation of an expanding plasma that results in bubbles in the corneal tissue, thousands of holes ("cavitation bubbles") of diameter $\sim 2-3$ μm. The pattern of holes defines the flap and its thickness ($\sim 10^2$ μm) and "hinge," and the flap is then lifted prior to the application of the sculpting excimer laser. The flap is created in a matter of seconds, and with the programmed application of the femtosecond laser is done much more precisely than is possible with a hand-held blade. The low-energy femtosecond pulses have no effect on the intracorneal tissue until they reach their programmed depth; they do not cause any temperature rise or shock waves.

For precise corneal sculpting, information about the aberrations in the patient's eye must be input to the software controlling the excimer laser pulses. This information is obtained by wavefront sensing with an "aberrometer," which is usually a Shack–Hartmann sensor as described in Section 14.2. An eye-safe laser enters the eye and the aberrations of the originally "flat" wavefront are measured after it propagates through the cornea, vitreous, and lens and then reflects off the retina and makes a return pass through the eye. As in adaptive optics, the displacements of spots focused onto a CCD sensor by the lenslet array correspond to local phase gradients of the wavefront incident on the array, which in turn are determined by the refractive imperfections in the entire optical system of the eye (Fig. 14.34). The measured spot displacements and a computer are used to control the excimer laser pulses that sculpt the cornea to correct for these imperfections.

- There is still much to be learned about laser–tissue interactions. Given that living tissue is about 80% water by weight, it might be expected, for example, that the effects of the UV excimer laser pulses used in LASIK on living tissue can be understood in large part from how they affect water. But there are important differences in the way these pulses interact with water and with living tissue. One is that tissue elasticity limits the growth of cavitation bubbles, so that they

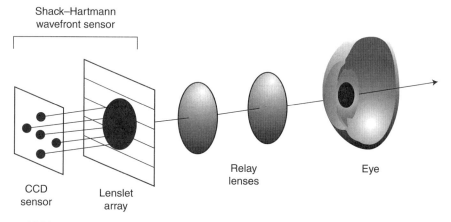

Figure 14.34 An "aberrometer" for determining the aberrations of an eye using a Shack–Hartmann wavefront sensor. The displacements of spots on the CCD array are a measure of the aberrations and are used to control an excimer laser for the corneal sculpting needed for refractive correction. [After L. N. Thibos, *Journal of Refractive Surgery* **16**, S563 (2000).]

are much smaller than in water; the photodisrupted regions of living tissues can be as small as a few hundred nanometers. This is beneficial in that it allows greater cutting precision than might be expected based on experiments with water.

As already noted, photodisruptive laser therapies rely on ionization and plasma formation. As discussed in Section 10.10, this optical breakdown is an avalanche process that proceeds from the presence or creation of a few electrons. In the optical breakdown of pure water these "seed" electrons are created by multiphoton ionization, which requires high intensities. In many biological tissues, however, there are large concentrations of a molecule (NADH) that is ionized by low-intensity UV radiation, yielding the seed electrons necessary for plasma formation; such tissues can be cut with lower laser intensities than tissues lacking this molecule.[45]   •

The same properties of lasers—brightness, directionality, and monochromaticity—used in eye surgery are employed in many other surgical procedures. Apart from opthalmology, laser surgery appears to be most well established in dermatology and gynecology, but it is now commonly used in many other specialties. One of its advantages is that blood loss is greatly reduced compared to surgeries with knives. Together with magnification, this greatly facilitates the removal of tiny structures. In laryngeal surgery, for example, a laser and an endoscope make possible a degree of precision difficult to obtain by other procedures. In cancer surgery with lasers there is less danger than with cold-knife surgery of dislodging cancer cells and worsening the malignancy.

The most ubiquitous laser in nonophthalmic surgery has been the CO_2 laser. This is not surprising, given that water strongly absorbs at 10.6 μm; the absorption coefficient of living tissue is roughly 200 cm^{-1} at 10.6 μm. Thus, CO_2 laser radiation interacts strongly with living tissue, and a focused CO_2 laser beam can produce explosive boiling of tissue fluids, resulting in an incision as the focal point is moved. Hand-held CO_2 surgical lasers, which have for many years been commercially available, are used, for instance, to remove arterial plaque during open-heart surgery.

Photodynamic therapy, which was invented in the early 1900s, came into widespread use in the mid-1980s and is now performed mainly with red diode lasers or diode laser arrays. This procedure begins with the injection into the bloodstream of a "precursor" that results in a photosensitizer, a substance that absorbs light of a particular wavelength. The targeted tissue is irradiated for several minutes with red laser light or LED radiation transported by a fiber. When the photosensitizer molecules absorb the light, they make a singlet–singlet transition and then undergo intersystem crossing to an excited, metastable triplet state (Section 11.11). From this state the excitation energy can be transferred to oxygen molecules in tissues, which in the process are excited from the ground triplet state to an excited singlet state. Singlet oxygen happens to be very reactive and, depending on the photosensitizer used, will react with and destroy undesirable (e.g., cancerous) cells. Photodynamic therapy is usually confined to about a 1-cm layer of tissue, the absorption depth of the light, and is therefore most effective for the treatment of small tumors on the skin or on internal organs accessed with optical fibers.

While the CO_2, argon ion, Nd : YAG, and excimer lasers remain the primary medical lasers, it is expected that diode-pumped fiber lasers, particularly because of their compactness, will in the near future take on increased importance as commercial development proceeds. Similarly, femtosecond lasers, including femtosecond fiber lasers, will

[45]M. S. Hutson and X. Ma, *Physical Review Letters* **99**, 158104 (2007).

likely become more widely used in medicine, having already become effective and commercially viable tools in ophthalmology.

Laser physics and optical science more generally have spawned many medical diagnostic tools such as fiber endoscopes and other instruments that continue to be developed and improved. We will conclude this section with brief descriptions of two relatively new imaging technologies. In each case we focus on the basic principles and refer the reader to the specialized literature for details of clinical implementations.

- The imaging technique discussed below, based on spin-polarized gases, is basically a new way of magnetic resonance imaging (MRI). It is perhaps useful therefore to review briefly some basic principles of MRI.

As noted in Section 14.3, nuclei with an odd number of nucleons have spin. The nucleus of interest for MRI is the hydrogen nucleus (proton)—hydrogen atoms make up about 60% of the human body by number and about 10% by mass. The proton has spin $\frac{1}{2}$, and so in a static magnetic field $\mathbf{B}_0 = B_0\hat{z}$ it has two allowed energy levels, E_+ and E_-, corresponding to two different spin orientations ("spin up" and "spin down") with respect to $\hat{z}$, and the transition frequency between these two levels is [Eq. (14.3.1)]

$$\omega_p = \frac{E_+ - E_-}{\hbar} = \left[\left(\tfrac{1}{2}\mu_N g B_0\right) - \left(-\tfrac{1}{2}\mu_N g B_0\right)\right]/\hbar = \frac{\mu_N g B_0}{\hbar} \equiv \gamma B_0. \tag{14.8.2}$$

μ_N and g are, respectively, the nuclear Bohr magneton and the proton g factor, and $\gamma \cong 2\pi \times 42.576$ MHz/tesla is the *gyromagnetic ratio* of the proton; γ relates the magnetic dipole moment $\mathbf{m}$ and the angular momentum (spin) vector $\mathbf{s}$: $\mathbf{m} = \gamma\,\mathbf{s}$, with $|\mathbf{s}| = \hbar/2$ for spin $\frac{1}{2}$. An extremely strong field $\mathbf{B}_0$, as large as 1.5 T or more generated with liquid-helium-cooled superconducting coils, is used in MRI; for $B_0 = 1.5$ T, $\omega_p = 63.86$ s^{-1}. The torque on a magnetic dipole moment $\mathbf{m}$ in a magnetic field $\mathbf{B}$ is $\mathbf{m} \times \mathbf{B}$, which gives the rate of change of the angular momentum $\mathbf{s}$, so that

$$\frac{d\mathbf{m}}{dt} = \gamma\,\mathbf{m} \times \mathbf{B}. \tag{14.8.3}$$

This equation says that the magnetic moment $\mathbf{m}$ precesses about $\mathbf{B}_0$, and that the precessional frequency is $\omega_p = \gamma|\mathbf{B}|$.

In a sample exposed to a magnetic field $B_0\hat{z}$ and in thermal equilibrium at temperature T, there will be a net bulk magnetization $M_0\hat{z}$, i.e., an imbalance in the densities of spin-up and spin-down protons:

$$\begin{aligned}
M_0 &= P_D\gamma\frac{\hbar}{2} \times \frac{e^{-E_-/k_BT} - e^{-E_+/k_BT}}{e^{-E_-/k_BT} + e^{-E_+/k_BT}} = P_D\gamma\frac{\hbar}{2}\tanh\left(\frac{E_+ - E_-}{2k_BT}\right) \\
&= P_D\gamma\frac{\hbar}{2}\tanh\left(\frac{\hbar\gamma B_0}{2k_BT}\right) \\
&\cong P_D\frac{\hbar^2\gamma^2 B_0}{2k_BT}
\end{aligned} \tag{14.8.4}$$

for $\hbar\gamma B_0/2k_BT \ll 1$, where P_D is the proton density. For $T = 300$K and $B_0 = 1.5$ T, $\hbar\gamma B_0/2k_BT = 5.1 \times 10^{-6}$, implying that the thermal magnetization in MRI is very small.

In MRI an rf field propagating perpendicular to the direction $\hat{z}$ of the static field, and having a frequency close to ω_p, induces transitions between the levels E_+ and E_- via the interaction with the proton spins of its time-dependent magnetic field $\mathbf{B}_1$. According to (14.8.3) such a field will result in a component of the magnetization perpendicular to the axial direction $\hat{z}$, i.e., it causes the magnetization $\mathbf{M}_0$ to tilt away from the $\hat{z}$ direction by some tipping angle Θ and therefore to precess about the strong static field $B_0\hat{z}$. This in turn results in a time-dependent magnetic flux perpendicular to $\hat{z}$ that, from Faraday's law of induction ($\nabla \times \mathbf{E} = -\partial \mathbf{B}/\partial t$), generates a voltage in an induction "pick-up" coil. This voltage is the MRI signal.

Considered as a two-state system, the proton can be described by Bloch equations; in fact, as noted in Section 9.5, the Bloch equations were first derived in the context of nuclear magnetic resonance (NMR) theory. When relaxation processes are included, the NMR Bloch equations take the form (9.5.3), where T_1 and T_2' are "longitudinal" and "transverse" relaxation times for the damping of the components of the magnetization parallel and perpendicular to $\mathbf{B}_0$, $\Delta = \omega_p - \omega$, the Rabi frequency $\chi = \gamma B_1$, and $w_0 = M_0$. The "area" of the pulse B_1, i.e., the integral over time of the Rabi frequency, Eq. (9.5.21), defines the tipping angle Θ. A pulse with $\Theta = \pi$, for example, will cause a magnetic moment to make a 180° flip.

After the magnetization is "tipped" by the rf pulse, it relaxes back to its equilibrium value M_0. The transverse (x, y) components, proportional to M_0, decay exponentially at a rate determined by T_2' as well as by an "inhomogeneous broadening" time T_2^* due to spatial inhomogeneities in $\mathbf{B}_0$ and therefore different precessional frequencies of different spins; as was noted in Section 9.11 in connection with the Ramsey method of separated fields, no one has solved the problem of making a magnetic field that is uniform over a large region of space (in the case of MRI, a region very roughly the size of a human body). The decay of the transverse components of the magnetization results in an exponentially damped voltage signal in a pick-up coil, a so-called *free induction decay* signal. The longitudinal (z) component of the magnetization increases after a tipping pulse at the rate $1/T_1$ ($<1/T_2'$): $M(t) = M_0[1 - \exp(-t/T_1)]$ [see Eq. (9.5.3c) with $\chi = 0$, $w = M$, and $w_0 = M_0$].

Spins in different tissues have different resonance frequencies ω_p, relaxation times T_1, T_2', proton densities P_D, and magnetizations M_0 in a magnetic field. These are determined, for instance, by different amounts of water (long relaxation times, ~ 1 s) and fat (short relaxation times, $\sim 10^2$ ms). Thus, the gray matter of the brain has relaxation times $T_1 \sim 760$ ms and $T_2' \sim 77$ ms, whereas for the white matter $T_1 \sim 510$ ms and $T_2' \sim 67$ ms.[46] They are distinguished by the different signals produced in a pick-up coil following the tipping rf pulses. Various "pulse sequences" with different pulse areas, repetition rates, and other characteristics have been devised to produce MRI signals; a commonly used pulse sequence for eliminating T_2^* effects is the *spin echo* sequence discovered by E. L. Hahn in the early 1950s.[47]

In MRI the voltage signals are converted to a spatial map (image) of proton densities using magnetic field gradients. A "slice" of the sample is selected by a magnetic field gradient G_z along the z direction. Different slices of the sample along z have different precession frequencies $\omega_p = (\gamma B_0 + \gamma G_z z)$, and only one slice will be resonant with the rf field and produce a signal. The thickness of the slice is determined by the bandwidth of the rf pulse, and can be changed by changing G_z. This provides one (z) dimension of localization within the sample. A magnetic field gradient G_x orthogonal to z performs *frequency encoding* by causing different spins within a slice to have slightly different precession frequencies: a spin with spatial coordinate x along this gradient has a precession frequency $\omega_p(x) = \gamma(B_0 + G_x x)$ and therefore oscillates as $\exp[-i\gamma(B_0 + G_x x)t]$. Aside from the rapidly varying factor $\exp(-i\gamma B_0 t)$, therefore, the magnetization of a

---

[46]J. L. Prince and J. Links, *Medical Imaging Signals and Systems* (Prentice Hall, Upper Saddle River, NJ, 2005).

[47]A conceptually similar phenomenon—*the photon echo*—occurs for atomic transitions. See, for instance, L. Allen and J. H. Eberly, *Optical Resonance and Two-Level Atoms*, (Dover, New York, 1987), Chapter 9.

volume element at x oscillates as $\exp(-i\gamma G_x xt)$, and the signal at a pick-up coil from all the spins along x is

$$S(t) = \kappa \int_{-\infty}^{\infty} M_0(x) e^{-i\gamma G_x xt} \, dx, \tag{14.8.5}$$

where κ is a calibration constant. Defining $k_x = \gamma G_x t$, we replace (14.8.5) by

$$s(k_x) = \int_{-\infty}^{\infty} M_0(x) e^{-ik_x x} \, dx. \tag{14.8.6}$$

Inversion of this Fourier transform gives

$$M_0(x) = \frac{1}{2\pi} \int_{-\infty}^{\infty} s(k_x) e^{ik_x x} \, dk_x. \tag{14.8.7}$$

In other words, frequency encoding during the signal readout allows a computation of a one-dimensional map $[M_0(x)]$ of the magnetization in the selected slice of the sample. To obtain a two-dimensional image of each slice, a third magnetic field gradient G_y along the y direction is applied prior to the frequency conversion and signal readout. G_y has the same effect on the spins as G_x, but it is turned off prior to the signal readout. This leaves spins at different points along the y direction with different phases, similar to the way frequency encoding imparts different frequencies. The effect of this *phase encoding* is to replace (14.8.7) by a two-dimensional Fourier transform of the form

$$M_0(x, y) = \left(\frac{1}{2\pi}\right)^2 \int_{-\infty}^{\infty} \int_{-\infty}^{\infty} s(k_x, k_y) e^{i(k_x x + k_y y)} \, dk_x \, dk_y \tag{14.8.8}$$

giving the two-dimensional map $M_0(x, y)$ for the selected slice: 256 frequency-encoding sequences, each involving 256 phase-encoding steps, produce a 256×256 pixel array that is converted to a gray-scale image on a film. •

Very detailed images of soft tissues are obtained by magnetic resonance imaging. For porous tissues such as the lung and colon, however, the air-filled spaces do not have enough protons to generate a meaningful magnetization signal. Larger magnetizations could be realized by increasing the magnetic field strength B_0 or by lowering the temperature [Eq. (14.8.4)]. But current MRI scanners already use extremely strong magnetic fields, and increasing B_0 further would increase the complexity and the cost beyond their already high levels. And numerical estimates based on (14.8.4) indicate that the temperatures required would freeze patients to death—not a viable option. A different approach, based on the injection into porous tissue of a spin-polarized gas, was proposed in 1994.[48] The idea is to spin-polarize a gas by optical pumping (Section 14.3) and then inject it into a porous tissue; for the lungs, the patient simply inhales the spin-polarized gas, whose atoms have their spins and therefore magnetic dipole moments aligned. The degree of magnetization per particle is enormously larger than that implied by the thermal-equilibrium expression (14.8.4) for the protons in tissues imaged by standard MRI; this compensates for the low density of the gas, so that MRI as described above

[48]M. S. Albert, G. D. Cates, B. Driehuys, W. Happer, B. Saam, C. S. Springer, Jr., and A. Wishnia, *Nature* **370**, 199 (1994).

can be done with the "breathable magnets" of the spin-polarized gas. For lung imaging, the patient inhales the spin-polarized gas before entering the MRI chamber.

Only certain atoms are suitable as breathable magnets. The atoms should have a nuclear spin in order to have a nuclear magnetic moment, and this eliminates all atoms with an even number of nucleons (Section 14.3). While alkali atoms are easy to spin-polarize with resonant laser radiation, the fact that they react violently with water or oxygen obviously rules them out. The spin-polarized atoms should have no toxic effects and should have long spin relaxation times so that their magnetization is sufficiently long-lived. They should not interact to form molecules that result in strong depolarization of the nuclear spins. Such considerations point to noble gases with nuclear spin $I = \frac{1}{2}$, and ^{3}He and ^{129}Xe in particular. Aside from the "Donald Duck voice" effect of helium, there are no adverse effects associated with it. Xenon, likewise, has for many years been used as an anesthetic and poses no known health risks, and moreover the small dosages of spin-polarized ^{129}Xe that are sufficient for MRI reduce the anesthetic effects. ^{129}Xe has some advantages over ^{3}He in that it is more readily and inexpensively available. Unlike ^{3}He, it occurs naturally at a concentration $\sim 10^{-7}$ in the atmosphere and is a by-product, along with the more abundant ^{131}Xe, of the distillation of air for the commercial production of O_2 and N_2.

Optical pumping of ground-state noble-gas atoms is generally impractical; for He, for example, the transition from the ground level to the first excited level has a wavelength of 58 nm and cannot be accessed with available lasers. Helium has been spin-polarized by exciting it to a metastable excited level in an electric discharge and then optically pumping it with a laser at 1083 nm. Both ^{3}He and ^{129}Xe are spin-polarized in large concentrations by spin-exchange collisions (Section 14.3) with optically pumped alkali atoms. The most common way of doing this is to optically pump Rb atoms with 795 nm, circularly polarized laser radiation, usually obtained with a high-power (~ 100 W) diode laser array. The electronic spin polarization of the Rb atoms is then transferred to the ^{3}He or ^{129}Xe nuclei in collisions involving a magnetic hyperfine interaction. Since the hyperfine interactions are weak, the exchange probability in a collision is small, and it can take seconds or minutes for spin polarization of ^{129}Xe nuclei by this method, and hours to do so for ^{3}He. Spin relaxation of ^{129}Xe is sufficiently slow that the degree of polarization is limited by the Rb polarization. On the other hand, ^{3}He has a larger magnetic moment than ^{129}Xe (about three times larger), and thus far greater degrees of spin polarization have been realized with it (~ 20–50% compared to ~ 10–30% for ^{129}Xe). ^{3}He is found to produce sharper MRI images and has been used in most lung imaging studies. ^{129}Xe has one extremely attractive property in contrast to ^{3}He: It dissolves in blood and in many tissues. The circulation time of the blood in a human body is ~ 15 s, whereas ^{129}Xe in the blood maintains its polarization for tens of seconds, enough time for it to be used for brain imaging, for instance.

Unlike conventional MRI, where between rf sampling pulses the magnetization $M_0 \hat{z}$ is restored by relaxation to thermal equilibrium in a time T_1 typically ~ 1 s, in MRI with a spin-polarized gas the magnetization is reduced with each successive sampling pulse and can be restored only by adding more polarized gas. The pulse sequencing is also different from conventional MRI; for example, $\pi/2$ pulses are not used in the spin-polarized case because such a pulse would tip the Bloch vector into the xy plane and thereby destroy the polarization with a single pulse. Typically ~ 100 pulses are used to produce an image, each pulse having the same small area. For pulses with an area of 0.20 rad (11.5°), for example, the magnetization after 100 sampling pulses is reduced by a factor $1 - [\cos(0.20)]^{100} = 0.87$, assuming that no other processes act to reduce the magnetization during the pulse sequence. The fact that 100 rf pulses can be applied without having to wait for the magnetization to be restored by slow relaxation processes is important for lung imaging, for example, where a patient cannot be expected to hold his breath for too long.

Another attractive feature of MRI with spin-polarized gases is that they do not require the huge magnetic fields of conventional MRI systems. In the latter the magnetization is proportional to B_0 [Eq. (14.8.4)] and the voltage signal in a pick-up coil is, by Faraday's law of induction, proportional to the rate of change of the magnetization, that is, to $\omega_p B_0 = \gamma B_0^2$; a large B_0 is required to generate a large enough signal in the presence of noise associated with amplifiers and resistances in the system, since the magnetization is entirely thermal in origin and therefore small. When the magnetization is due to a spin-polarized gas, however, the factor B_0 on the right-hand side of (14.8.4) does not enter into the determination of the signal since the magnetization of the gas is not thermally equilibrated; the signal is then proportional to B_0, which only determines ω_p, not the magnetization. It has been demonstrated that magnetic fields ~ 20 G can be used for MRI with spin-polarized gases, so that the cost and other disadvantages of extremely strong magnetic fields are avoidable.

Spin polarization of ^{3}He gas persists for several days in glass containers, and frozen ^{129}Xe maintains its polarization for comparable durations. Compact systems producing spin-polarized gases for clinical lung imaging have been developed. A flask containing ^{3}He, N_2, and a small amount of Rb is heated to vaporize the Rb. Circularly polarized laser radiation optically pumps the Rb vapor and results after a few hours in the polarization of most of the ^{3}He. The flask is then cooled to condense out the Rb, and the spin-polarized ^{3}He gas is drawn into a plastic bag for later inhalation by a patient.

In the first MRI images demonstrated with spin-polarized ^{129}Xe gas, a xenon density of $\sim 1.2 \times 10^{19}$ cm^{-3} filled the excised lung of a mouse,[48] compared to the $\sim 5 \times 10^{22}$ cm^{-3} proton concentrations of tissues imaged in conventional MRI. Since this demonstration many other studies have been reported, and this new approach to MRI imaging appears to be progressing rapidly toward widespread clinical use.[49]

Another novel imaging method is based on the connection between the bandwidth of light and interference. As discussed in Section 13.11, the coherence length of light decreases with increasing bandwidth; the larger the bandwidth, the smaller the arm separation has to be in a Michelson interferometer, for instance, in order to observe interference fringes. Femtosecond laser pulses, for example, have coherence lengths on the order of microns, so that no interference fringes are observed in a Michelson interferometer if the arm separation exceeds, say, a few microns (Section 13.11). This is the basis for *optical coherence tomography* (OCT), and the basic idea behind it is sketched in Fig. 14.35.[50]

[49]Reviews with many references to research papers in both the physics and medical literature have been published by T. Chupp and S. Swanson, *Advances in Atomic, Molecular, and Optical Physics* **45**, 41 (2001) and S. J. Kadlecek, K. Emami, M. C. Fischer, M. Ishii, Y. Jiangsheng, J. M. Woodburn, M. NikKhah, V. Vahdat, D. A. Lipson, J. E. Baumgardner, and R. R. Rizi, *Progress in Nuclear Magnetic Spectroscopy* **47**, 187 (2005).
[50]D. Huang, E. A Swanson, C. P Lin, J. S Schuman, W. G Stinson, W. Chang, M. R Hee, T. Flotte, K. Gregory, C. A Puliafito, and J.G. Fujimoto, *Science* **254**, 1178 (1991). A review of optical coherence tomography "from bench to bedside" is given by A. M. Zysk, F. T Nguyen, A. L. Oldenburg, D. L. Marks, and S. A. Boppart, *Journal of Biomedical Optics* **12**, 051403 (2007).

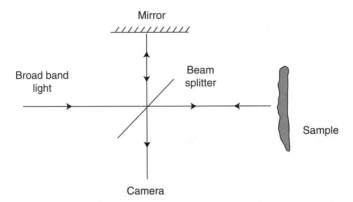

Figure 14.35 Simplified illustration of optical coherence tomography. Light from a broadband source (short coherence length) enters a Michelson interferometer, one arm of which contains a sample S (e.g., a retina or a fingertip). Interference fringes are observed only for path separations, determined by the topography of S, that differ by less than the coherence length. This allows depth resolution of the sample, while the camera provides two-dimensional profiling, so that three-dimensional imaging of S with micron-scale resolution is achieved if the coherence length of the light source is on the order of a micron.

The broadband source in OCT is typically "superfluorescent" (mirrorless) diode laser radiation transported by a fiber, but femtosecond lasers and supercontinuum radiation in fibers are also used in order to obtain extremely small coherence lengths. Compared to imaging techniques such as MRI and X-ray computed tomography (CT), OCT has a much smaller imaging depth (typically a few millimeters) but offers greater resolution of fine structure. Because the light source is nonionizing, it allows long exposure times that are not possible with X rays. Commercially available OCT systems are compact and less complex than MRIs or CTs, and their clinical utility has been demonstrated in oncology, opthalmology, dermatology, and other areas.

14.9 REMARKS

Lasers have made possible the most accurate determination of several physical quantities, the smallest measured time intervals, the most accurate clocks, the lowest temperatures ever realized, and some of the highest powers ever generated. They have been used to determine distances between Earth and the moon to an accuracy of a few centimeters and to enable ground-based telescopes to produce images comparable in quality to those of a space telescope; to determine concentrations of atmospheric constituents in the atmosphere as well as temperature, density, and wind profiles; to produce ultracold gases and a state of matter in which large numbers of atoms are described by a single quantum state and in which the interference of atom waves can be observed and put to practical use; to time-resolve chemical and biological processes; to determine our grocery bills, print what we read, record our music, and cut and weld materials used in many other aspects of everyday life. Their importance in medicine continues to grow and they are an integral part of modern communications and the Internet. The list of applications has no foreseeable end.

The operating principle of every laser is, of course, stimulated emission of radiation. As discussed in Section 3.6, it was Einstein who first realized that matter can radiate by stimulated emission as well as by spontaneous emission. But no one at the time seems to have imagined the possibility of radiation amplifiers and oscillators, much less their widespread use decades later. While some applications could be confidently predicted when the first lasers were constructed in the early 1960s, many others were not foreseen by even the most imaginative scientists and engineers. The reader might find it interesting to think about which laser applications were the most unpredictable, and to speculate on directions future applications might take.

PROBLEMS

14.1. (a) Using Eqs. (3.7.5) for the case in which the level populations of a two-level atom change by absorption and stimulated and spontaneous emission in a narrowband field, show that the upper-state probability $p(t)$ changes according to the equation

$$\dot{p}(t) = -A_{21}p(t) + \frac{\sigma(\nu)}{h\nu}I(t)$$

if the lower-state probability $\cong 1$.

(b) Assuming that the laser pulse duration is long compared to the radiative lifetime, derive Eq. (14.1.3).

14.2. (a) Verify the approximation (14.1.21) for Rayleigh backscattering at frequencies roughly in the visible range.

(b) Verify Eq. (14.1.22) for the number of backscattered photons for the assumed lidar system parameters.

14.3. Show that the sodium density at different altitudes in the mesosphere can be determined from the ratio of sodium and Rayleigh photocounts as in Eq. (14.1.26). What assumptions are implicit in this equation?

14.4. (a) Estimate the maximum photon flux (number of photons per unit area per unit time) at ground level that can be obtained when the 589-nm D_2 line of mesospheric sodium is uniformly irradiated with resonant radiation having a 1-m^2 spot size at the mesosphere.

(b) What would be the apparent magnitude of a guide star that produces this photon flux?

(c) Approximately what fraction of quasi-monochromatic 589-nm laser radiation propagated from ground is absorbed by the mesospheric sodium layer? (Assume that the laser power is sufficiently low that saturation effects are negligible.)

(d) Why should the saturation formula (14.2.5) be applicable? Shouldn't the absorption or gain coefficient for a Doppler-broadened transition saturate according to the formula (4.14.7)?

14.5. (a) The reflector placed on the moon by Apollo 11 astronauts consists of 100 corner cubes, each about 1.5 inches across. Estimate the diameter at Earth of the return ruby laser pulses. (A *corner cube* consists of three reflecting edges intersecting at right angles. Any incident ray is reflected parallel to itself.)

(b) The laser pulses from a police lidar gun have a beam divergence angle of 3 mrad. Estimate the diameter of these pulses at a distance of 300 m from the policeman. Can you think of ways to thwart this type of "radar"?

14.6. Show that the matrix element for an electric dipole transition in which light is absorbed and an atom goes from a state 1 with magnetic quantum number m to a state 2 with magnetic quantum number m' is given by (14.3.3), and that the reverse, stimulated emission transition is given by the complex conjugate of (14.3.3).

14.7. (a) Show that a detector placed along the x axis, and responding to an average over times long compared to ω_0^{-1} of the power radiated by the electric dipole (14.3.15), will record an intensity with the time dependence (14.3.16) if it is behind a polarizer oriented at an angle φ with respect to the z axis.

(b) Suppose that the atoms are excited by linearly polarized light as in Fig. 14.12 and that the detector records the time-integrated intensity

$$\mathcal{J} \propto \int_{-\infty}^{t} dt_0\, I(t)$$

with $t \gg 1/\gamma$. Show that $\mathcal{J}$ is proportional to

$$\frac{1}{2\gamma} + \frac{\gamma \cos 2\varphi}{\gamma^2 + 4\omega_L^2} + \frac{2\omega_L \sin 2\varphi}{\gamma^2 + 4\omega_L^2}.$$

Plot this signal vs. ω_L for (i) $\varphi = 0$, (ii) $\varphi = \pi/4$, (iii) $\varphi = \pi/2$, and (iv) $\varphi = 3\pi/4$. What experimental situation does this signal describe? Compare your results with the corresponding experimental results shown in Fig. 14.36.

14.8. (a) Show that optical pumping with circularly polarized light results in complete transparency in the case of the sodium D_1 line in the absence of any spin relaxation effects.

(b) Calculate the electric dipole matrix element $\langle F'm'|d_{q=1}|Fm\rangle$ for the transition $3P_{3/2}$ ($F = 3$, $m = 3$) $\rightarrow$ $3S_{1/2}(F = 2, M = 2)$ of sodium. (You will have to look up the values of the *3j* and *6j* symbols for the set of quantum numbers $F' = 3$, $m' = 3$, $F = 2$, $m = 2$, $J' = \frac{3}{2}$, $J = \frac{1}{2}$, $I = \frac{3}{2}$.)

(c) Confirm that the saturation intensity for this transition in the case of pure radiative broadening is 6.3 mW/cm^2, as stated in Section 4.11.

(d) Verify Eq. (14.4.5).

14.9. (a) An atom has a transition frequency $\nu_0 = (E_2 - E_1)/h$, where 1 and 2 refer to the ground level and first excited level, respectively, and it is moving with velocity v away from a source of radiation of frequency ν. Using conservation of energy and linear momentum, and assuming that line broadening is

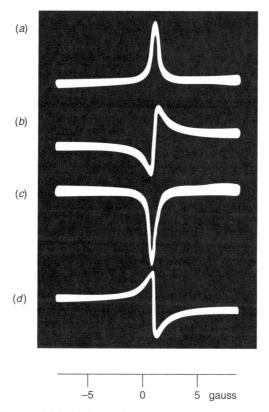

Figure 14.36 Time-integrated Hanle signals from mercury atoms for polarizer angles (*a*) $\varphi = 0$, (*b*) $\varphi = \pi/4$, (*c*) $\varphi = \pi/2$, and (*d*) $\varphi = 3\pi/4$. (See Problem 14.7.) [From B. P. Kibble and G. W. Series, *Proceedings of the Physical Society of London* **78**, 70 (1961).]

negligible, show that the atom will absorb a photon of frequency ν provided that

$$\nu = \nu_0\left(1 + \frac{v}{c}\right).$$

Assume that $v \ll c$, so that relativistic effects may be ignored.

(b) Show that there is a *recoil shift*

$$\nu - \nu_0 = \frac{h\nu_0^2}{2Mc^2}$$

in the absorption frequency, where M is the atomic mass.

(c) Derive an expression for the recoil shift in the case of *two-photon* absorption. Calculate the recoil shift in the case of the 1S–2S two-photon transition in atomic hydrogen, and compare your answer with the measured recoil shift of 6.7 MHz reported by D. G. Fried, T. C. Killian, L. Willmann, D. Landhuis, S. C. Moss, D. Kleppner, and T. J. Greytak, *Physical Review Letters* **81**, 3811 (1998).

(d) Estimate the critical temperature for Bose–Einstein condensation of atomic hydrogen at a density of 1.8×10^{20} atoms/m^3.

14.10. Derive the formula (14.4.42) for the potential energy function for the dipole force. [Note: You might use the fact that the dipole force in (14.4.36) applies even under conditions where the polarizability can be saturated. You could also proceed as in the derivation of the light shift (14.4.22).]

14.11. (a) Show that the superposition of an upward-propagating laser beam of frequency $\nu + \Delta\nu$ and a downward-propagating laser beam of frequency $\nu - \Delta\nu$ results in a "moving standing wave" with upward velocity $\lambda\Delta\nu$ ($\lambda = c/\nu$).

   (b) Estimate the length traveled and the time taken for a cesium atom in an atomic fountain to reach its apogee and return to its original position after being launched by a moving standing wave with $\lambda = 852$ nm (Fig. 14.18) and $\Delta\nu = 1.6$ MHz.

14.12. (a) Compare the optical trapping force to the magnetic force acting on atoms in a MOT, and show that the optical force is dominant for atoms at a distance of a wavelength or more from the center of the trap.

   (b) Use Eq. (14.5.9) to estimate the size of the cloud of sodium atoms trapped in a MOT.

   (c) Estimate the zero-point energy of a sodium atom in a magnetic trap with $\mathcal{B} = 10$ G/cm.

14.13. A sodium atom in a MOT with a magnetic field gradient of 30 G/cm and counter-propagating laser fields of intensity 1 mW/cm^2, detuned from the D$_2$ transition by 30 MHz, is displaced from the center of the trap. Estimate the time it takes for the atom to move to the center of the trap.

14.14. Suppose that sodium atoms are in an optical lattice formed by a laser of intensity 4 mW/cm^2 detuned by 60 GHz from the D$_2$ line. Show that the peak value of the light-shift potential energy function is about 10 times the sodium recoil energy. [J. H. Denschlag, J. E. Simsarian, H. Häffner, C. McKenzie, A. Browaeys, D. Cho, K. Helmerson, S. L. Rolston, and W. D. Phillips, *Journal of Physics B: Atomic, Molecular and Optical Physics* **35**, 3095 (2002).]

14.15. (a) Derive the expression (14.5.13) for the (cycle-averaged) force of radiation pressure on a dielectric sphere with a radius small compared to the radiation wavelength. (Hint: Show that this force is equal to $n_b P/c$, where P is the Rayleigh-scattered power.)

   (b) A focused laser beam is to be used to balance the force of gravity on a latex sphere of diameter 5 μm. Assuming that latex has a density of 1.05 g/cm^3 and a refractive index of 1.6 at the laser wavelength, estimate the laser intensity and power required when the sphere is to be held in air, assuming a Gaussian laser beam focused to a spot of diameter 10 μm. Repeat these estimates when the latex sphere is in water.

14.16. (a) Estimate the critical temperature for Bose–Einstein condensation of liquid ^{4}He, for which the density is 2.2×10^{28} atoms/m^3.

(b) Calculate the recoil temperature T_{recoil} for the laser cooling of ^{87}Rb atoms at the 780 nm transition.

(c) Calculate the recoil temperature for hydrogen atoms at the 122-nm Lyman-α transition, and compare it with the critical temperature obtained in Problem 14.9 for BEC in hydrogen.

14.17. (a) Assuming a wave function of the form (14.7.6), and probability amplitudes $a_v(t) = a_v(0) \exp(-iE_v t/\hbar)$, derive an expression for the expectation value of the interatomic separation R. Compare your formula to that obtained if there is no "off-diagonal coherence," that is, if collisions or other "decoherence" effects cause the quantities $a_v^*(t)a_{v'}(t)$ to decay rapidly in time for $v \neq v'$.

(b) Explain how signals such as that shown in Fig. 14.23 can be used to obtain values for vibrational and rotational constants of a molecule.

(c) Our simplified analysis of the time resolution of molecular vibrations with femtosecond pump and probe pulses in Section 14.7 assumes a single molecule. Under what conditions can we apply such a single-molecule theory to an ensemble of molecules in order to interpret experimental data such as shown in Fig. 14.23?

14.18. (a) Use Eq. (3.14.6) and the formula $n^2(\omega) - 1 = (N/\varepsilon_0)\alpha(\omega)$ for the refractive index to obtain

$$\alpha_i(\omega) = \frac{e^2}{m} \sum_j \frac{f_j}{\omega_j^2 - \omega^2}$$

for the polarizability of an atom in state i, where the summmation is over all states j connected to state i by oscillator strengths f_j and transition frequencies ω_j.

(b) What modifications reduce this expression to the approximation (14.4.23)?

(c) Show that the light shift obtained using the polarizability derived in part (a) reduces to the ponderomotive energy U_p [Eq. (14.7.11)] in the limit in which the field frequency ω is much greater than any of the transition frequencies ω_j.

14.19. (a) Show that the times t_f at which the electron returns to the ion in the model presented in Section 14.7 are given by solutions of the equation

$$\cos\theta_f - \cos\theta_i + (\theta_f - \theta_i)\sin\theta_i = 0,$$

where $\theta_f = \omega t_f$ and $\theta_i = \omega t_i$, t_i being a time at which ionization occurs. Show also that the electron's kinetic energy when it reaches the ion is $E(t_f, t_i) = 2U_p(\sin\theta_f - \sin\theta_i)^2$, where U_p is the ponderomotive energy.

(b) Solve numerically for $E(t_f, t_i)$ for θ_i between 0 and $\pi/2$ and verify the results plotted in Fig. 14.27. In particular, verify that the maximum value of $E(t_f, t_i)$ is $3.17U_p$.

(c) What is the probability that an electron will recollide with its parent during the first cycle of the laser field following ionization?

(d) How are the predictions of the model changed when the laser radiation is *circularly* polarized?

(e) Using the Bohr model, compare the amplitude of the electron's quivering in the field to the size of an atom when the Keldysh parameter γ given by (14.7.13) is less than 1.

14.20. Estimate the tunneling ionization rate for a hydrogen atom in (a) a linearly polarized field of intensity 3×10^{16} W/cm^2 and (b) a circularly polarized field of the same intensity.

14.21. (a) Describe the spectrum obtained by the heterodyning of a single-mode laser with a mode-locked laser, assuming the two fields have frequencies as indicated in Fig. 14.31.

   (b) Explain why there must be phase coherence among pulses equally spaced in time if the pulse train is to serve as a comb for optical frequency metrology.

15 DIODE LASERS AND OPTICAL COMMUNICATIONS

15.1 INTRODUCTION

Any means of communication, whether smoke signals or telephone, involves a transmitter and a receiver of information. In optical communications an electrical signal is converted to a modulated light wave, which then transmits the signal to a receiver. At the receiver the light signal is converted back to an electrical signal, where it might, for example, be "read" as a television picture.

A monochromatic wave would not convey any information.[1] To transmit information with an electromagnetic wave, we have to turn it on and off or, as in AM or FM radio, modulate it in some way. The rate at which information can be transmitted obviously depends on the rate at which modulations are impressed on the wave. This modulation rate must be slow compared to the carrier frequency that might, for example, be a broadcast frequency to which your radio is tuned. The higher the carrier frequency, therefore, the higher the rate at which information can be conveyed; for this reason the great potential of *optical* communication was recognized for a long time.

Optical communication became practical with the advent of lasers. But major hurdles stood in the way of long-distance optical communication by laser radiation in air: The atmosphere introduces signal distortion due to turbulence (Chapter 8), and even under the most favorable weather and seeing conditions only direct line-of-sight communication (e.g., satellite-to-satellite) would be possible if the atmosphere were the "transmission channel."

These obstacles were overcome around 1970 with the development of low-loss optical fibers that could serve as transmission channels at near-infrared wavelengths (Section 8.7). At the same time light-emitting diodes and room-temperature laser sources suitable for efficient coupling of light into fibers were undergoing rapid development. By 1977 optical communication systems based on these developments were commercially available, and in 1988 the first transatlantic undersea optical cable was installed. Optical networks employing lasers and fibers have in large part been responsible for the expansion of the Internet and its ever-increasing economic and societal impact.

---

[1]Recall earlier remarks concerning radio wave modulation (Section 6.10) and group velocities exceeding the speed of light in vacuum (Section 8.3).

Laser Physics. By Peter W. Milonni and Joseph H. Eberly
Copyright © 2010 John Wiley & Sons, Inc.

This chapter focuses on some basic physics of diode lasers and their role in fiber-optic communications. We also introduce some elementary concepts of *information theory* in order to better understand the relation between bandwidth and the rate of information transmission in the inevitable presence of noise.

15.2   DIODE LASERS

In Section 2.9 we discussed light-emitting diodes (LEDs) in which radiation results from the radiative recombination of electrons and holes at a *pn* junction. If there is a large enough density of electrons and holes in a biased junction of doped *p* and *n* materials, this radiation can *stimulate* the recombination of electrons and holes, and lasing results if the amplification of radiation by stimulated emission exceeds the loss. This is the operating principle of semiconductor diode lasers. Diode lasers are ubiquitous in bar-code readers, laser pointers, printers, and compact disc players, for example; we will focus primarily on their properties of interest for fiber-optic communications. In addition to their wavelengths, efficiencies, and extremely small size, the most distinctive feature of semiconductor diode lasers for optical communications is the degree to which their output can be modulated by varying the applied current. In this section we introduce some general features of diode lasers, including their gain characteristics that follow from the energy bands and the Fermi–Dirac distribution of electrons in semiconductors.

To get some idea of the sort of parameters that characterize diode lasers, consider first the geometry shown in Fig. 15.1. The gain medium consists of a *pn* junction with an active region of width *d*. In this region an "injection" current results in a sufficient number of electron–hole pairs to produce gain; *d* may be estimated from the diffusion length of the charge carriers and is very small, typically ~ 1 µm in a homojunction laser. The width *D*, which is discussed further below, indicates the size of the transverse mode volume of the radiation field; whereas in most lasers $D < d$ (Chapter 7), the reverse can be true in diode lasers. Needless to say, neither *d* nor *D* are sharply defined.

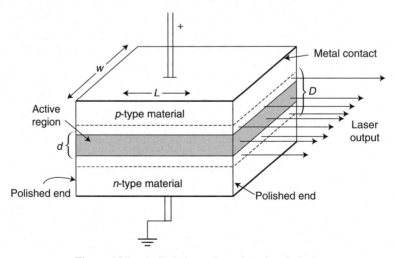

Figure 15.1   A diode laser (homojunction design).

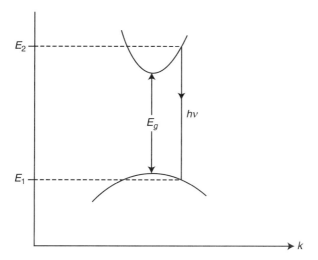

Figure 15.2 $E-k$ curves for the conduction and valence bands of a direct-band-gap semiconductor. A transition from an energy level E_2 to an energy level E_1 is accompanied by the emission of a photon of frequency $\nu = (E_2 - E_1)/h$.

The threshold gain for a diode laser may be expressed in terms of a threshold injection current. The familiar expression (3.12.11) for the gain coefficient,[2]

$$g(\nu) = \frac{\lambda^2 A_{21}}{8\pi n^2}\left(N_2 - \frac{g_2}{g_1}N_1\right)S(\nu), \qquad (15.2.1)$$

was derived assuming a transition between discrete bound states of energy E_2 and E_1 ($E_2 - E_1 = h\nu$) characterized by a lineshape function $S(\nu)$. In diode lasers amplification occurs between states within a range of energies E_2 and E_1 (cf. Fig. 15.2), and the spontaneous radiative transitions are electron–hole recombinations ocurring at a rate we will denote by $1/\tau_R$; τ_R is typically ~ 3 ns for (direct-band-gap) semiconductor gain media. Thus, we replace (15.2.1) by

$$g(\nu) = \frac{\lambda^2}{8\pi n^2 \tau_R}\left[\rho_c(E_2)\frac{dE_2}{d\nu} - \rho_v(E_1)\frac{dE_1}{d\nu}\right]. \qquad (15.2.2)$$

$\rho_c(E_2)\,dE_2$ is the volume density of states in the energy interval $(E_2, E_2 + dE_2)$ of the conduction band, and $\rho_v(E_1)\,dE_1$ is the density of states in the interval $(E_1, E_1 + dE_1)$ of the valence band; we are ignoring for the moment the fact that different states in these energy intervals have different occupation probabilities. The factors $dE_2/d\nu$ and $dE_1/d\nu$ appear because $\rho_c(E_2)$ and $\rho_v(E_1)$ are defined with respect to energy increments, whereas $S(\nu)$ is defined with respect to frequency increments $[\int_0^\infty S(\nu)\,d\nu = 1]$. Besides replacing A_{21} in (15.2.1) by $1/\tau_R$, we have dropped the degeneracy ratio g_2/g_1; the only degeneracy of interest here will be that associated with electron spin, and will be

---

[2]Recall that λ throughout this book is the wavelength in vacuum ($\lambda = c/\nu$), and that A_{21} is the spontaneous emission rate *in the medium* and is itself proportional to n if local field corrections are negligible (Section 3.12).

dealt with later. Equation (15.2.2) gives the gain coefficient for the transition of frequency $v = (E_2 - E_1)/h$ between two particular energy levels E_2 and E_1 within continuous distributions of upper- and lower-level energies (Fig. 15.2).

The threshold gain coefficient is given by Eq. (4.3.13):

$$g_t = a - \frac{1}{2L} \ln r_1 r_2, \tag{15.2.3}$$

where L is the length of the gain region, r_1 and r_2 are the mirror reflectivities, and a is the cavity loss per unit length due to intracavity field attenuation effects other than transmission through the mirrors. The threshold condition for laser oscillation at frequency v, therefore, takes the form

$$\left[\rho_c(E_2) \frac{dE_2}{dv} - \rho_v(E_1) \frac{dE_1}{dv} \right]_t = \frac{8\pi n^2 \tau_R}{\lambda^2} \left(a - \frac{1}{2L} \ln r_1 r_2 \right). \tag{15.2.4}$$

To simplify as much as possible, let us assume that $\rho_c(E_2) \, dE_2/dv - \rho_v(E_1) \, dE_1/dv \approx \rho_c(E_2) \, dE_2/dv \approx h\rho_c(E_2)$ and that a current injects electrons into the conduction band of the active (gain) region with energies in a range $\sim \Delta E_2$ about E_2. Then the density of conduction-band electrons is $\sim \rho_c(E_2) \, \Delta E_2$ under the assumption that every state in the energy interval $(E_2, E_2 + \Delta E_2)$ of the conduction band is occupied by an electron. We denote by J the current density (flow of charge per unit area per unit time) and write J/ed for the electron injection rate per unit volume. Electrons are also being lost at a total rate R_e due to radiative as well as nonradiative recombination processes, and in steady state the injection rate equals the loss rate: $J/ed = R_e \rho_c(E_2) \, \Delta E_2$. To simplify still further we will assume that the "internal quantum efficiency" $\eta = 1/(R_e \tau_R)$, the fraction of injected electrons undergoing radiative recombination, is close to unity— not a bad approximation for diode lasers. With all these assumptions and approximations we can write the threshold condition in terms of a threshold current density J_t:

$$J_t = \frac{8\pi n^2}{\lambda^2} \frac{\Delta E_2}{h} (ed) \left(a - \frac{1}{2L} \ln r_1 r_2 \right). \tag{15.2.5}$$

This expression does not account for the fact that the mode volume is wider than the width d of the active region. The greater density of charge carriers in the active region results in a greater refractive index than in the surrounding medium, and therefore some confinement of radiation to the active region. This is called *index guiding*; there is also *gain guiding* due just to the fact that there is gain in the active region but not in the surrounding material. But if these "wave-guiding" effects are weak, the radiation has a mode volume of width $D > d$. The effective gain coefficient is then smaller than (15.2.2) by a factor $\approx d/D$, and the actual threshold current density is a factor D/d larger than (15.2.5). For our purposes of making some rough estimates for homojunction lasers, we will assume $D \approx d$.

Diode lasers usually do not require mirrors for feedback. This is because the refractive index n is large enough for significant Fresnel reflection at the semiconductor–air interfaces. From the Fresnel formulas for normal incidence we have the reflection coefficient [Eq. (5.A.6)]

$$r = \left(\frac{n-1}{n+1}\right)^2 \tag{15.2.6}$$

if we approximate the refractive index of air by unity. For GaAs $n \approx 3.6$ and therefore $r \approx (2.6/4.6)^2 \approx 0.32$ for the reflection coefficient of the laser "mirrors." By polishing the two opposite ends of the diode, and leaving the remaining sides rough (so that they are poor specular reflectors), laser oscillation is favored along the axis joining the polished ends (Fig. 15.1). Sometimes coatings are applied to the polished ends to increase their reflectivities.

GaAs, with a band gap such that $\lambda \approx 860$ nm, is a diode laser medium for which experiments and numerical models suggest the very rough estimate $\Delta E_2/h \approx 10^{13}$ Hz. With this estimate and the fairly typical values $d = 2$ μm, $L = 500$ μm, and $a = 10$ cm^{-1}, we calculate from (15.2.5) that $J_t \approx 500$ A/cm^2. For a junction area $lw = 500 \times 100$ μm^2 (Fig. 15.1), this implies a threshold current $J_t Lw \approx 250$ mA.

Note that the gain region is less than a millimeter across in any direction in our example: Diode lasers are tiny! The fact that the internal quantum efficiency is close to unity suggests that diode lasers are also potentially very efficient, and in fact they are among the most efficient of all lasers, with typical overall efficiencies (laser output power divided by input electrical power) ~ 30–40% or more.

Our estimate of 500 A/cm^2 for the threshold current density of a GaAs laser is in order-of-magnitude agreement with measurements for low-temperature GaAs lasers. However, the considerations leading up to this result were obviously highly simplified, and it turns out that, for all but the very lowest operating temperatures, our simplistic theory predicts much too small a value for the threshold current density of homojunction GaAs lasers.

At room temperature, current densities more like 500,000 A/cm^2 are needed to reach threshold. However, current densities are dramatically smaller for structures employing heterojunctions, stripes, quantum wells, and other concepts.

Gain Coefficient

Before turning attention to these other structures, we will derive an equation for the gain coefficient of a diode laser, starting from the fact that the $E - k$ curves (Chapter 2) of the conduction and valence bands are approximately parabolic, as shown in Fig. 15.2. Recall that the $E - k$ relation for a *free* electron of mass m is $E = \hbar^2 k^2/2m$ [Eq. (2.A.4)]. For an electron of energy E_2 in the conduction band ($E_2 > E_c$) of a semiconductor, similarly,

$$E_2 \cong E_c + \frac{\hbar^2 k^2}{2m_c}, \tag{15.2.7a}$$

and likewise, for a hole of energy E_1 in the valence band ($E_1 < E_v$),

$$E_1 \cong E_v - \frac{\hbar^2 k^2}{2m_v}; \tag{15.2.7b}$$

m_c and m_v are the effective masses of conduction-band electrons and valence-band holes, respectively (Chapter 2); for GaAs, for example, $m_c \cong 0.07m$ and $m_v \cong 0.50m$. A radiative recombination transition from a level E_2 of the conduction band to a level E_1 of the valence band occurs with the emission of a photon of frequency ν given by

$$hv = E_2 - E_1 = E_c - E_v + \frac{\hbar^2 k^2}{2}\left(\frac{1}{m_c} + \frac{1}{m_v}\right) = E_g + \frac{\hbar^2 k^2}{2m_r}, \tag{15.2.8}$$

where $E_g = E_c - E_v$ is the gap energy ($\cong 1.44$ eV for GaAs) and $m_r = m_c m_v/(m_c + m_v)$ is the reduced electron–hole effective mass. It follows from (15.2.7) and (15.2.8) that

$$E_2 = E_c + \frac{m_r}{m_c}(hv - E_g), \tag{15.2.9a}$$

$$E_1 = E_2 - hv = E_v - \frac{m_r}{m_v}(hv - E_g). \tag{15.2.9b}$$

The density of states $\rho_c(E)$ can be derived in essentially the same way as Eq. (14.6.3), with a few modifications. First, (14.6.3) applies to particles in a particular spin state, and must be multiplied by 2 to allow for the two possible spin states ("up" and "down") of the electrons in a semiconductor. Second, Eq. (14.6.3) was derived under the assumption that $E = \hbar^2 k^2/2M$, that is, for free particles of mass M, whereas (15.2.7a) is appropriate for an electron in the conduction band of a semiconductor. These considerations lead to the density of states (Problem 15.1)

$$\rho_c(E) = \frac{\sqrt{2}m_c^{3/2}}{\pi^2 \hbar^3}(E - E_c)^{1/2} \qquad (E > E_c). \tag{15.2.10}$$

For the valence band, similarly, the density of states is

$$\rho_v(E) = \frac{\sqrt{2}m_v^{3/2}}{\pi^2 \hbar^3}(E_v - E)^{1/2} \qquad (E < E_v). \tag{15.2.11}$$

One more fact of utmost importance must be accounted for: No two electrons can occupy the same state, i.e., electrons are fermions. States of electrons in a solid are defined by the **k** vector and the spin quantum number, and the Pauli exclusion principle forbids electrons from all "piling up" into a state of lowest energy. Instead they fill up allowed energy states in accordance with the exclusion principle, one electron per state, up to some maximum energy E_f, the *Fermi energy*. For any finite temperature T, some of the electrons reach energies larger than E_f; the average number of electrons with energy E is given by the *Fermi–Dirac distribution function*:

$$F(E, E_f) = \frac{1}{e^{(E-E_f)/k_B T} + 1}. \tag{15.2.12}$$

At $T = 0$, $F(E, E_f)$ is 1 for $E < E_f$ and 0 for $E > E_f$, whereas for $T > 0$, $F(E, E_f) > 0$ for energies $E > E_f$. E_f plays the same role as the chemical potential μ of Section 14.6, where we introduced the Bose–Einstein occupancy factor $[e^{(E-\mu)/k_BT} - 1]^{-1}$ as opposed to the occupancy factor (15.2.12) for fermions: It is determined by the condition that the total number of particles is fixed. Thus, if N_c is the density of conduction-band electrons,

$$N_c = \frac{\sqrt{2}m_c^{3/2}}{\pi^2\hbar^3} \int_{E_c}^{\infty} \frac{(E - E_c)^{1/2}\, dE}{e^{(E-E_{fc})/k_BT} + 1}, \tag{15.2.13}$$

where E_{fc} is a Fermi energy characterizing the energy distribution of conduction-band electrons. The density N_v of holes in the valence band can similarly be related to a Fermi energy E_{fv} for valence-band electrons: The occupancy factor for a hole of energy E is equal to the *nonoccupancy* factor $1 - F(E, E_{fv}) = 1 - [e^{(E-E_{fv})/k_BT} + 1]^{-1}$ for an *electron* of energy E in the valence band, and therefore

$$
\begin{aligned}
N_v &= \frac{\sqrt{2}m_v^{3/2}}{\pi^2\hbar^3} \int_{-\infty}^{E_v} (E_v - E)^{1/2} \left(1 - \frac{1}{e^{(E-E_{fv})/k_BT} + 1} \right) dE \\
&= \frac{\sqrt{2}m_v^{3/2}}{\pi^2\hbar^3} \int_{-\infty}^{E_v} \frac{(E_v - E)^{1/2}\, dE}{e^{(E_{fv}-E)/k_BT} + 1}.
\end{aligned}
\tag{15.2.14}
$$

Taking into account the Fermi–Dirac distribution, we replace (15.2.10) by[3]

$$\rho_c(E) = \frac{\sqrt{2}m_c^{3/2}}{\pi^2\hbar^3} (E - E_c)^{1/2} \frac{1}{e^{(E-E_{fc})/k_BT} + 1} \qquad (E > E_c). \tag{15.2.15}$$

The Fermi–Dirac distribution applies at thermal equilibrium, in which case there can be no gain. However, in writing (15.2.15) we do not require that the conduction-band electrons and valence-band holes are thermally equilibrated *with each other*. Provided that relaxation processes, mainly electron–electron and electron–phonon collisions, are fast enough, and much faster than interband transitions, it is reasonable to assume that the conduction-band electrons and the valence-band holes are described approximately by thermal distributions. In fact, the intraband relaxation times are typically on the order of picoseconds, whereas the interband transition times are on the order of nanoseconds. Under such circumstances we can define the separate, "quasi-Fermi levels" E_{fc} and E_{fv}, whereas in global thermal equilibrium there is only a single Fermi level $E_f = E_{fc} = E_{fv}$.

In the same way we account for the Fermi–Dirac distribution by replacing the valence-band density of states (15.2.11) by

$$\rho_v(E) = \frac{\sqrt{2}m_v^{3/2}}{\pi^2\hbar^3} (E_v - E)^{1/2} \frac{1}{e^{(E-E_{fv})/k_BT} + 1} \qquad (E < E_v). \tag{15.2.16}$$

---

[3]This now represents a density of electrons rather than a density of *states*, i.e., $\rho_c(E)\, dE$ is the average number of electrons per unit volume in the energy interval $[E, E + dE]$.

The expressions (15.2.2), (15.2.9), (15.2.15), and (15.2.16) then imply the gain coefficient[4] (Problem 15.1)

$$g(\nu) = C(h\nu - E_g)^{1/2} \left[\frac{1}{e^{(E_2 - E_{fc})/k_B T} + 1} - \frac{1}{e^{(E_1 - E_{fv})/k_B T} + 1} \right]$$

$$(h\nu = E_2 - E_1 > E_g), \quad C \equiv \sqrt{2} m_r^{3/2} \lambda^2 / 4\pi^2 \hbar^2 n^2 \tau_R. \qquad (15.2.17)$$

• The factor in brackets in (15.2.17) can be arrived at equivalently as follows. The probability of a downward (stimulated-emisson) electron–hole recombination transition from level E_2 in the conduction band to level E_1 in the valence band is proportional to the occupancy factor $F(E_2, E_{fc})$ for an electron of energy E_2 in the conduction band, times the occupancy factor $1 - F(E_1, E_{fv})$ that there is a hole (i.e., that there is no electron) of energy E_1 in the valence band. The probability of an upward (absorptive) electron–hole recombination transition from level E_1 in the valence band to level E_2 in the conduction band is proportional to the occupancy factor $F(E_1, E_{fv})$ that there is an electron of energy E_1 in the valence band, times the nonoccupancy factor $1 - F(E_2, E_{fc})$ that there is no electron of energy E_2 in the conduction band. The gain coefficient is therefore proportional to

$$F(E_2, E_{fc})[1 - F(E_1, E_{fv})] - F(E_1, E_{fv})[1 - F(E_2, E_{fc})]$$
$$= F(E_2, E_{fc}) - F(E_1, E_{fv})$$
$$= \left[\frac{1}{e^{(E_2 - E_{fc})/k_B T} + 1} - \frac{1}{e^{(E_1 - E_{fv})/k_B T} + 1} \right], \qquad (15.2.18)$$

as in (15.2.17).   •

The gain coefficient (15.2.17) vanishes if $h\nu < E_g$, obviously because a photon at such a frequency cannot satisfy the energy conservation condition $h\nu = E_2 - E_1$ (Fig. 15.2). It follows furthermore from (15.2.17) that (positive) gain requires $E_2 - E_{fc} < E_1 - E_{fv}$, or $E_2 - E_1 = h\nu < E_{fc} - E_{fv}$. In other words, the gain coefficient (15.2.17) is positive only for frequencies ν satisfying

$$E_g < h\nu < E_{fc} - E_{fv}. \qquad (15.2.19)$$

$g(\nu)$ depends on the Fermi energies E_{fc} and E_{fv}, and these energies depend on the electron and hole concentrations in the rather complicated way expressed by (15.2.13) and (15.2.14). Equation (15.2.13), for instance, implies that E_{fc} increases with N_c. This can be seen from the fact that the denominator in the integrand of (15.2.13) decreases with increasing E_{fc}, so that, with all other parameters held fixed, $dN_c/dE_{fc} > 0$ and therefore $dE_{fc}/dN_c > 0$. (See also Problem 15.1 for the special case $T = 0$.) Since $g(\nu)$ increases with E_{fc}, it increases with N_c, which in turn implies that it should increase with current density. Closer examination, taking account of the dependence of $g(\nu)$ on E_{fv} and N_v, confirms that this is indeed the case, and moreover that the gain *bandwidth* as well as the peak gain should increase with current density (cf. Fig. 15.3). This property of diode lasers is somewhat unusual: In most other types of laser the gain bandwidth is

[4]More detailed discussions of the assumptions leading to (15.2.17) may be found, for instance, in M. Shur, *Physics of Semiconductor Devices*, Prentice-Hall, Englewood Cliffs, NJ, 1990, or B. E. A. Saleh and M. C. Teich, *Fundamentals of Photonics*, Wiley, New York, 2007.

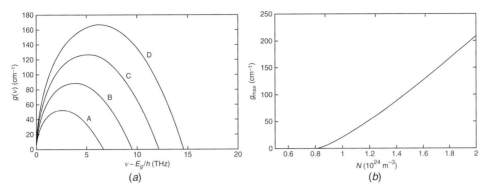

Figure 15.3 (a) Computed gain coefficients $g(\nu) > 0$ vs. $\nu - E_g/h$ for $T = 300$K. Results are shown for four different values of N: (A) 1.2×10^{24}, (B) 1.4×10^{24}, (C) 1.6×10^{24}, and (D) 1.8×10^{24} m^{-3}. The frequencies $\nu - E_g/h$ are in units of THz (10^{12} Hz). (b) Computed peak values of the gain coefficient vs. N.

approximately independent of the pumping strength, at least insofar as power broadening and other effects that might affect the lineshape function $S(\nu)$ can be ignored.

The conduction-band electron density $N_c = \bar{N}_c + N$, where N is the density due to the applied current and $\bar{N}_c$ is the density in the absence of the current. Similarly the valence-band hole density $N_v = \bar{N}_v + N$, where $\bar{N}_v$ is the hole density without the current. We can take the electron and hole densities due to the current to be the same (N), since injection of an electron into the conduction band leaves a hole in the valence band. Equations (15.2.13) and (15.2.14) determining E_{fc} and E_{fv} are then [see the note in Problem 15.2]

$$\bar{N}_c + N = \frac{\sqrt{2}m_c^{3/2}}{\pi^2\hbar^3} \int_{E_c}^{\infty} \frac{(E - E_c)^{1/2}\,dE}{e^{(E - E_{fc})/k_BT} + 1}, \tag{15.2.20a}$$

$$\bar{N}_v + N = \frac{\sqrt{2}m_v^{3/2}}{\pi^2\hbar^3} \int_{-\infty}^{E_v} \frac{(E_v - E)^{1/2}\,dE}{e^{(E_{fv} - E)/k_BT} + 1}. \tag{15.2.20b}$$

A semiconductor is characterized among other things by an energy gap E_g, carrier charge densities $\bar{N}_c$ and $\bar{N}_v$, and effective masses m_c and m_v that depend on the doping and the energy bands ($E - k$ curves) of the material. Given these quantities and the injected electron density N, the energies $E_c - E_{fc}$ and $E_v - E_{fv}$ are determined for any temperature T by Eqs. (15.2.20). Once these energies are known, Eqs. (15.2.9) can be used to compute the gain coefficient $g(\nu)$ [Eq. (15.2.17)] for any frequency ν. Figure 15.3 shows results of such computations for an example in which $E_g = 0.96$ eV, $m_c = 0.059m$, $m_v = 0.44m$, $n = 3.5$, $\tau_R = 2.5$ ns, and $\bar{N}_c = \bar{N}_v = 2 \times 10^{23}$ m^{-3}. We have plotted $g(\nu)$ vs. ν for four different values of N, assuming $T = 300$K.

Figure 15.3 brings out some important points. First, as noted above, the gain bandwidth increases with N and therefore with current density. The gain bandwidth for the largest value of N (curve D) is 14.7 THz. The computed values of $E_{fc} - E_c$ and $E_{fv} - E_v$ for this N are, respectively, 0.092 eV and 0.031 eV, and the gain bandwidth is that expected from Eq. (15.2.19): $(0.092 - 0.031 = 0.061)$ eV, or $\Delta\nu = 14.7$ THz.

Second, the frequency at which the gain is largest *at fixed temperature* is seen to increase with N and therefore with current. Both of these features are observed with diode lasers; the peak gain coefficient at fixed temperature shifts in frequency by typically a few megahertz per microampere change in injection current. Another difference between diode lasers and most other types of laser is evident in Fig. 15.3, namely that diode lasers can have huge gain coefficients. These large gains are necessary for lasing, given the large threshold gains (15.2.3) implied by the short gain lengths L and the low facet reflectivities. Of course, the threshold gains can be reduced with reflective coatings, but other losses, including the spreading of light outside the gain region and the nonspecular light scattering at layer interfaces, often remain relatively large.[5]

In Fig. 15.3 we also plot the peak gain coefficients g_{max}. The smallest value of N that results in a positive gain is computed to be $8.1 \times 10^{23}\,\mathrm{m}^{-3}$. If we argue as before that N and the current density J are related by $J = (ed/\tau_R)N$, and assume $d = 2\,\mu\mathrm{m}$, then the current density needed for a positive gain in this example is

$$\frac{(1.6 \times 10^{-19}\,\mathrm{C})(2 \times 10^{-6}\,\mathrm{m})(8.1 \times 10^{23}\,\mathrm{m}^{-3})}{2.5 \times 10^{-9}\,\mathrm{s}} = 1.0 \times 10^{4}\,\mathrm{A/cm}^2. \qquad (15.2.21)$$

Note that we have assumed an internal quantum efficiency $\eta = 1$. If the radiative recombination rate divided by the total recombination rate is $\eta = 0.90$, for example, then this estimate for the current density must be multiplied by $1/\eta = 1.1$.

For N greater than about $1.2 \times 10^{24}\,\mathrm{m}^{-3}$ the peak gains plotted in Fig. 15.3 satisfy

$$g_{max} \cong \sigma[N - N_{tr}] \qquad (15.2.22)$$

with $\sigma = 1.9 \times 10^{-20}\,\mathrm{m}^2$ and $N_{tr} = 9.5 \times 10^{23}\,\mathrm{m}^{-3}$. N_{tr} is the value of N needed for "transparency," i.e., for the gain to have a positive value instead of a negative value corresponding to absorption. Writing $J = edR_eN$ and $J_0 = edR_e\Delta N_{tr}$, we obtain a gain–current relation of the form frequently used to relate gain to current for diode lasers:

$$g_{max} \approx \beta(J - J_0) \qquad \left(\beta \equiv \frac{\sigma\eta\tau_R}{ed}\right). \qquad (15.2.23)$$

We can define a threshold current density J_t by equating (15.2.23) to the threshold gain coefficient (15.2.3):

$$\frac{J_t}{d} \approx \frac{e}{\tau_R}\left(\frac{g_t}{\sigma} + N_{tr}\right) \qquad (15.2.24)$$

for $\eta \approx 1$. Using the parameters assumed in obtaining Fig. 15.3, and $a = 10\,\mathrm{cm}^{-1}$ and $L = 500\,\mu\mathrm{m}$ as in our earlier estimate for the threshold gain of a diode laser,

---

[5]Manufacturers of diode lasers generally consider information about the actual design and preparation of the facets to be proprietary.

we calculate

$$\frac{J_t}{d} \approx 7.2 \times 10^3 \, \text{A}/(\text{cm}^2\text{-}\mu\text{m}), \qquad (15.2.25)$$

which is comparable to measured values. The threshold current is $I_t = $ (junction area) $\times J_t = LwJ_t$.

Steady-state laser oscillation requires that electrons be injected into the conduction band at a rate of at least I_t/e. With an injection current $I > I_t$ the laser power radiated by stimulated radiative recombination of electrons and holes is just $\eta(I - I_t)/e$ times the energy $h\nu$ of each photon emitted (Problem 15.3):

$$P = \eta\left(\frac{I - I_t}{e}\right)h\nu. \qquad (15.2.26)$$

(In this chapter no confusion can arise so we use simply P instead of Pwr to designate power.) In the uniform-field approximation the fraction of this power emerging as *output* laser radiation is the output coupling loss coefficient divided by the total loss coefficient:

$$f = \frac{-\frac{1}{2L}\ln(r_1 r_2)}{a - \frac{1}{2L}\ln(r_1 r_2)}, \qquad (15.2.27)$$

and therefore the steady-state output power is approximately

$$P_{\text{out}} = \eta\left(\frac{I - I_t}{e}\right)h\nu \frac{-\frac{1}{2L}\ln(r_1 r_2)}{a - \frac{1}{2L}\ln(r_1 r_2)}. \qquad (15.2.28)$$

Consider as an example a laser with a threshold current density given by (15.2.25). For a heterojunction laser with $L = 200 \, \mu\text{m}$, $w = 10 \, \mu\text{m}$, and $d = 0.1 \, \mu\text{m}$ (see below), $I_t = 14$ mA. For $\eta = 0.9$, $a = 10 \, \text{cm}^{-1}$, and $r_1 = r_2 = (n - 1)^2/(n + 1)^2 = (2.5/4.5)^2 = 0.31$,

$$P_{\text{out}} \approx 0.7(I - 14) \, \text{mW} \qquad (I \text{ in mA}) \qquad (15.2.29)$$

for $\lambda = 1.3 \, \mu\text{m}$. Currents I of a few tens of milliamps are therefore predicted to result in output powers on the order of a few milliwatts, as is typical of room-temperature heterojunction diode lasers with the parameters we have assumed.

The model leading to the gain coefficient (15.2.17) thus provides reasonably good qualitative and (semi)quantitative agreement with measurements made with diode lasers. A more "first-principles" computation of the gain coefficient and the gain–current relation would require energy-band computations for any given semiconductor material, doping, temperature, etc. (recall the Kronig–Penney model of Chapter 2 for the illustrative example of a one-dimensional crystal), whereas we have simply *assumed* values for the band gap and effective masses, for example, that would be deduced from such computations (or from experiments).

• The model used to derive (15.2.17) is an approximation to a more general theory. One approximation implicit in (15.2.17) is that the **k** vector of an electron does not change in an interband transition, allowing us to consider only vertical ("momentum-conserving") transitions as shown in Fig. 15.2 (Section 2.9). This approximation is implicit in Eqs. (15.2.7), where it is assumed that $\mathbf{k}$ is the same for both energy levels E_1 and E_2. We have also assumed that the line-shape function $S_{ep}(\nu)$ for the homogeneous broadening due to electron–electron and electron–phonon collisions has a very narrow width $\Delta\nu_c$ near $\nu = (E_2 - E_1)/h$ compared to the width of $g(\nu)$ determined by (15.2.17). As already noted, collision times are roughly on the order of picoseconds. Then $\Delta\nu_c \approx (10^{12}/2\pi)$ Hz, which is small compared to the width of the gain curve shown in Fig. 15.3a, for instance.

Similar approximations can be used to model the emission characteristics of a light-emitting diode. In this case the quantity of interest is not a gain or absorption coefficient but a spontaneous emission rate. To obtain the spectrum of spontaneously emitted light when a current is applied to an LED, for example, we first use (15.2.18) to write the gain coefficient (15.2.17) in the form

$$g(\nu) = C(h\nu - E_g)^{1/2}F(E_2, E_{fc})[1 - F(E_1, E_{fv})]$$
$$- C(h\nu - E_g)^{1/2}F(E_1, E_{fv})[1 - F(E_2, E_{fc})]. \qquad (15.2.30)$$

This is the difference of two terms, the first of which we identify with stimulated emission from level E_2 and the second with absorption from level E_1. Comparison with (15.2.1) or (15.2.2) shows that the spectrum of spontaneous emission is proportional to the first term:

$$R_{sp}(\nu) \propto (h\nu - E_g)^{1/2}F(E_2, E_{fc})[1 - F(E_1, E_{fv})]$$
$$= (h\nu - E_g)^{1/2}\left[\frac{1}{e^{(E_2 - E_{fc})/k_BT} + 1}\right]\left[1 - \frac{1}{e^{(E_1 - E_{fv})/k_BT} + 1}\right]. \qquad (15.2.31)$$

In both (15.2.30) and (15.2.31), of course, $h\nu > E_g$. We ignore a factor $\lambda^2 = c^2/\nu^2$, which is nearly constant over the effective width of the function $R_{sp}(\nu)$. In Fig. 15.4 we plot (15.2.31), computed in the same way as the gain coefficient by first determining the quasi-Fermi levels. We have used the same parameters as in Fig. 15.3 and assumed a value of N too small for positive gain.

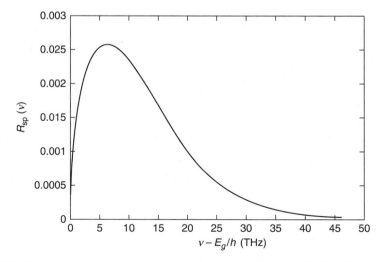

Figure 15.4 Computed spontaneous emission spectrum (arbitrary units) for the parameters of Fig. 15.3 but with $N = 2 \times 10^{23}$ m^{-3}.

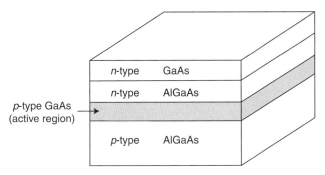

Figure 15.5   A double-heterojunction diode laser.

A more detailed theory of gain, absorption, and spontaneous emission in semiconductor devices includes not only "nonvertical" transitions (cf. Section 2.9) and their matrix elements but also effects such as a reduction of the energy gap E_g due to the injected charge carriers. Thus, for example, the range of frequencies for which $g(\nu) > 0$ is found to extend to energies $h\nu$ below the constant energy gap that we have assumed in obtaining the results shown in Fig. 15.3. The basic features of the variation of gain with injected charge carriers shown in Fig. 15.3, however, are not substantially altered.[6]   •

Most applications of diode lasers call for room-temperature operation. The large current densities required for the homojunction design of Fig. 15.1, the simplest type of diode laser, then pose serious problems. This difficulty was overcome in the 1970s with the development of heterojunction diode lasers. Figure 15.5 illustrates a "double" heterojunction design, which employs not only n- and p-type GaAs layers, but also n- and p-layers of an AlAs–GaAs alloy, denoted AlGaAs. Because of the band-gap differences at the two GaAs–AlGaAs junctions, there is a greater confinement of electrons and holes in the active region (p-type GaAs in the example shown). This comes about because the band-gap differences act in effect as potential energy barriers for the electrons and holes, preventing them from diffusing out of the active GaAs layer. Furthermore the greater refractive index of GaAs compared to the AlGaAs compound helps to confine the radiation to the active region by index guiding, thus reducing the width d appearing in (15.2.24) and therefore the threshold current density. In addition, the loss coefficient a is smaller than in the homojunction case, simply because any radiation that "spills over" into the AlGaAs layers finds itself in a nonresonant, nonabsorbing medium (because of the large band gap compared to the radiation frequency). For these reasons, cw heterojunction lasers can operate at room temperatures with current densities typically ~ 1000 A/cm^2.

Threshold currents are reduced still further by confining the current across the active region to a narrow stripe, as shown in Fig. 15.6. This may be done as shown by building high-resistance regions into the diode. For a double heterojunction laser with a gain region of length $L = 1$ mm and width $w = 10\,\mu$m, the threshold current is $\sim (10^3$ A/cm$^2)$ $(10^{-1}$ cm$)$ $(10^{-3}$ cm$)$ $\sim 10^2$ mA, and a typical output power for such a device ranges from a few milliwatts to a few tens of milliwatts [cf. (15.2.29)]. Another

---

[6]In order to compare with computations including nonvertical transitions we chose the parameters for these figures in accordance with those assumed by N. K. Dutta, *Journal of Applied Physics* **51**, 6095 (1980) for a In$_{0.72}$Ga$_{0.28}$As$_{0.6}$P$_{0.4}$ laser. Results of that work corresponding to our Fig. 15.3 are reproduced in Fig. 16.2–3 of Saleh and Teich, *op. cit.*, see footnote 4.

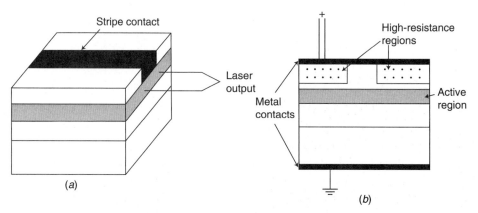

Figure 15.6   (*a*) Diode laser with a stripe contact to confine the current to a small part of the active layer. (*b*) Stripe contact obtained by building high-resistance regions into a laser diode.

advantage of the stripe geometry is that the cross-sectional area of the output radiation is reduced, making it easier to couple the radiation into an optical fiber. Stripe lasers also tend to have a more stable output.

Quantum Wells

As noted, the energy gaps between valence and conduction bands are different in the GaAs and AlGaAs materials of Fig. 15.5 (Section 2.9), and this results in a potential energy barrier for electrons in the conduction band and for holes in the valence band. The electrons and holes in the gain region are therefore in potential energy wells, or *quantum wells*. In the simplest approximation these quantum wells are modeled as infinite square wells for which the allowed electron energies in the conduction band are (Problem 2.7)

$$E_{cq} = E_c + \frac{q^2 \pi^2 \hbar^2}{2m_c d^2} \qquad (q = 1, 2, 3, \ldots), \tag{15.2.32}$$

and similarly the allowed hole energies in the valence band are

$$E_{vq} = E_v - \frac{q^2 \pi^2 \hbar^2}{2m_v d^2} \qquad (q = 1, 2, 3, \ldots). \tag{15.2.33}$$

If d is very small, these energy levels cannot be well approximated by continua; they have only certain allowed values, and radiative transitions between these levels in *quantum well lasers* occur at wavelengths that depend on the thickness d of the active region. The laser frequency for oscillation between the lowest energy level of the conduction band and the highest level of the valence band, for example, is

$$\nu = \frac{1}{h} \left[E_g + \frac{\pi^2 \hbar^2}{2d^2} \left(\frac{1}{m_c} + \frac{1}{m_v} \right) \right] = \frac{1}{h} \left(E_g + \frac{\pi^2 \hbar^2}{2m_r d^2} \right). \tag{15.2.34}$$

The components k_x, k_y of the electron **k** vectors parallel to the two planes separated by d in a quantum well laser take on a continuous range of values. When all three components of **k** take on a continuous range of values, the number of electron states in a volume V and in a differential volume d^3k of k space is $2V\,dk_x\,dk_y\,dk_z/(2\pi)^3$, where the factor 2 accounts for the two possible states of electron spin for each **k**. For a quantum well, however, there are $2dk_x\,dk_y/(2\pi)^2 = 2(2\pi k_\parallel\,dk_\parallel)/(2\pi)^2 = (k_\parallel/\pi)\,dk_\parallel$ states per unit area associated with the "parallel" components of **k** ($k_\parallel^2 = k_x^2 + k_y^2$), and so the number of states per unit *volume* in the interval $[k_\parallel, k_\parallel + dk_\parallel]$ is $(1/d) \times (k_\parallel/\pi)\,dk_\parallel \equiv \tilde{\rho}(k_\parallel)\,dk_\parallel$. The $E - k$ relation (15.2.7a) is similarly modified when the electrons are confined in a quantum well:

$$E = E_c + E_q + \frac{\hbar^2 k_\parallel^2}{2m_c} \qquad (q = 1, 2, 3, \ldots), \qquad (15.2.35)$$

where E_q is defined by (15.2.32). Using the relation $\rho_q(E) = \tilde{\rho}(k_\parallel)(dk_\parallel/dE)$, we obtain

$$\rho_{cq}(E) = \frac{m_c}{\pi\hbar^2 d} \qquad (E > E_c + E_q) \qquad (15.2.36)$$

for the density of states of conduction-band electrons with quantum number q; if $E < E_c + E_q$, $\rho_{cq}(E) = 0$. A similar expression for the density of states for the valence band, $\rho_{vq}(E)$, is obtained by replacing m_c by m_v. The gain coefficient for allowed transitions, which conserve both the **k** vector and the quantum number q, may be derived (and computed) along the same lines as (15.2.17). Allowing for all possible quantum numbers q for the conduction-band electrons and valence-band holes, we obtain

$$g(v) = \frac{\lambda^2}{8\pi n^2 \tau_R} \frac{2m_r}{\hbar d} \left[\frac{1}{e^{(E_2 - E_{fc})/k_B T} + 1} - \frac{1}{e^{(E_1 - E_{fv})/k_B T} + 1} \right]$$

$$\times \sum_{q=1}^{\infty} \theta\left(hv - E_g - \frac{q^2 \pi^2 \hbar^2}{2m_r d^2} \right). \qquad (15.2.37)$$

We have introduced the unit step function (or "Heaviside function") $\theta(x)$, defined such that $\theta(x)$ is 1 for $x > 0$ and 0 for $x < 0$. According to Eq. (15.2.37) the dependence on frequency of the gain coefficient of a quantum well laser has a "staircase" behavior, exhibiting step-like increases at frequencies $v = (E_g + q^2 \pi^2 \hbar^2/2m_r d^2)/h$, $q = 1, 2, 3, \ldots$; as the injection current is increased, gain first appears at the frequency (15.2.34). As in bulk semiconductor lasers, the dependence of the quasi-Fermi levels on the charge carrier densities results in higher gain and larger gain bandwidth as the injection current is increased. Injection currents are often such that only the $q = 1$ "step" satisfies the condition $E_g < hv < E_{fc} - E_{fv}$ for positive gain [Eq. (15.2.19)].

Quantum well lasers differ from older types of double-heterojunction lasers in that the thickness d of the active region is much smaller. This gives them the important advantage of larger gain per injected electron, and consequently higher efficiencies and much smaller threshold currents than bulk semiconductor lasers. The higher efficiency is related to the "staircase" nature of the effective density of states in (15.2.37), which results in a greater fraction of electrons in lasing energy states. The smaller number of charge carriers required to realize a particular level of gain also results in weaker internal losses than bulk semiconductor lasers; threshold current densities $<100 \, \text{A/cm}^2$ are possible. Typical threshold currents and output powers are $\lesssim 1 \, \text{mA}$ and $\lesssim 100 \, \text{mW}$, the output powers of individual quantum well lasers being limited by optical damage to the small-area facets. The d dependence of the electron and hole energies also allows the flexibility of selecting different laser wavelengths by fabricating lasers with different thicknesses d of the active region. Quantum well lasers tend to be less sensitive to temperature and, for reasons discussed in Section 15.3, to allow faster modulation rates. They have replaced bulk double-heterojunction lasers in many applications.

The active layer width d can be as small as 10 nm or less in a quantum well laser, while the field mode width D might be ~ 10 times or more larger. The effective gain coefficient, or "modal gain coefficient," is therefore smaller than g by a factor $\sim d/D$. Low-refractive-index *confinement layers* are usually added to either side of the quantum well, resulting in greater confinement of the field within the well. Another way to increase the modal gain coefficient is to have multiple quantum wells in the gain region. The modal gain coefficient of a *multiquantum well laser* is then approximately the sum of the gains of the individual wells. In addition, the current within each well can be below a level that would saturate the gain, so that the total modal gain for a given current is maximized.

Equations (15.2.32) and (15.2.33) apply when the electrons and holes are confined along one direction, perpendicular to the junction layers. Different material structures can also result in confinement in two or three dimensions, in which case the potential energy wells are referred to as *quantum wires* and *quantum dots*, respectively. Such structures are fabricated by molecular beam epitaxy or chemical vapor deposition, and at this writing quantum wire and quantum dot lasers are being actively researched and developed.

● Lasing can also occur on the long-wavelength ($\sim 5 - 10 \, \mu\text{m}$) *intraband* transitions between discrete energy levels of a quantum well rather than on the interband transitions involving electron–hole recombination. In a *quantum cascade laser* a voltage is applied across a series of alternating high- and low-band-gap materials, such that an electron in any one of ~ 20 wells finds itself in a potential well. In the applied electric field an electron can tunnel from one quantum well to the next, and in so doing it loses energy in *discrete* steps as it jumps from an energy level in one well to a different energy level of the adjacent well. This results in the emission of a large number of photons as each electron "cascades" down the series of quantum wells. Quantum cascade lasers are most useful for the generation of long-wavelength radiation that would be difficult to obtain with conventional (interband) diode lasers because of the requirement of a small band gap and therefore substantial numbers of thermally excited electrons and holes.   ●

Mode Properties

In optical communication systems employing wavelength-division multiplexing (Section 15.6), different wavelengths are separately modulated and transmitted through a single fiber in order to transmit different signals simultaneously. This requires that the different wavelengths be sufficiently far apart that interference among the different signals is negligible. The gain bandwidths of diode lasers (cf. Fig. 15.3) allow laser oscillation on several longitudinal modes over bandwidths $\sim 2 \, \text{nm}$ at low powers ($\sim 1 \, \text{mW}$), while the useful bandwidth of a fiber might be ~ 10 times greater; this would allow only ~ 10 "channels" to be transmitted by the fiber. Increasing the number of possible channels therefore requires that the laser linewidth be reduced. Linewidth reduction also limits group velocity dispersion and "mode-hopping" effects in which fluctuations in temperature or injection currents cause fluctuations in the laser wavelength.

An obvious way to achieve single-longitudinal-mode oscillation is to reduce the length L of a diode laser so that the mode spacing ($\cong c/2nL$) is sufficiently large that only a single frequency within the gain bandwidth can lase (Section 5.10). Even for L as small as 100 μm, however, the mode spacing of edge-emitting diode lasers is typically small compared to the very large gain bandwidth (Problem 15.4). More generally, as with all laser media with predominantly homogeneous broadening, gain "clamping" tends to favor single-mode oscillation, but spatial hole burning, which is especially significant because of the short cavity length and therefore the small number of spatial field cycles in the gain medium, often results in oscillation on several longitudinal modes (Section 5.10). As the injection current and power level are raised, the temperature rises, which results in an increase in the output wavelength, typically by ~ 0.3 nm/K. This often causes the wavelength of a free-running diode laser to jump to a larger value, drifting gradually to higher values between jumps.

A diode laser can be tuned to a particular frequency by forming an "external cavity" using a diffraction grating at some distance from the diode; this is basically just the Littrow method used for dye lasers and other tunable lasers (Sections 11.11 and 11.12). Rotation of the grating selects a particular lasing frequency, other frequencies within the gain bandwidth being prohibited from lasing by their smaller feedback and larger loss. In optical communication systems, however, it is far more practical to use diode lasers with built-in Bragg gratings. In a *distributed Bragg reflector* (DBR) design, for example, Bragg gratings are fabricated on both sides of the gain medium, and act in effect as end mirrors that reflect in only a narrow band of frequencies.

Wavelength tuning is commonly done by fabricating the laser such that there is a periodic corrugation forming a grating along the length of active layer. If d_g is the spatial period of the grating and n is the refractive index, the lasing wavelength will be that satisfying the Bragg condition (11.14.29): $\lambda = 2nd_g$ or, more generally, $\lambda = 2nd_g/m$ for diffraction order m ($= 1, 2, 3, \ldots$). In such a *distributed feedback* (DFB) *laser* the reflection and feedback occur continuously throughout the gain medium rather than at end mirrors. The lasing wavelength can be further controlled by varying the injection current, which changes the density of charge carriers and therefore the refractive index n. DFB lasers have much greater wavelength stability than most ordinary lasers with end mirrors.

The small size of diode lasers has some important consequences when we compare their operating characteristics to other types of laser. For example, unlike most other lasers, the fundamental laser linewidth due to spontaneous emission noise is much larger in diode lasers precisely because of their small size [Eq. (5.11.12)], and furthermore this linewidth is enhanced by effects that can usually be ignored in other lasers [cf. Eq. (5.11.16) and Section 15.4]. Mode locking, which requires a modulation frequency equal to the mode spacing, provides another example. Mode locking of a diode laser by modulation of the injection current would require a modulation frequency ≈ 100 GHz for a cavity length $L \approx 500$ μm (Problem 15.4); such a fast modulation of an electric current cannot be done with conventional electronics. Mode locking of diode lasers can instead be done by the external-cavity technique—placing a mirror at some distance away from a facet of the diode that has been antireflection coated. This reduces the

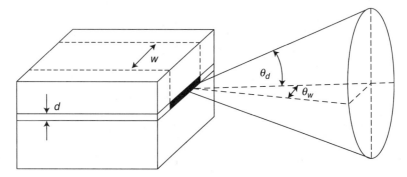

Figure 15.7   Angular distribution of radiation from a diode laser of wavelength λ. $\theta_d \sim \lambda/d$, $\theta_w \sim \lambda/w$.

required modulation frequency to approximately $c/2L_{\text{ext}}$, where L_{ext} is the length of the extended cavity.

An immediate consequence of the small size of a diode laser is the large divergence angle of the emitted radiation. This results simply from the small "aperture" of a diode laser. Since radiation of wavelength λ has a divergence angle $\sim \lambda/D$ due to diffraction when it passes through an aperture of diameter D [Eq. (7.11.10)], a gain region of thickness d and width w in a diode laser implies two divergence angles, $\theta_d \sim \lambda/d$ and $\theta_w \sim \lambda/w$, as shown in Fig. 15.7. For example, for $\lambda = 860\,\text{nm}$, $d = 2\,\mu\text{m}$, and $w = 10\,\mu\text{m}$, $\theta_d \sim 25°$ and $\theta_w \sim 5°$. The "fanning out" of diode laser radiation is therefore much more pronounced than in other types of laser, and moreover the radiation pattern is asymmetrical due to the significantly different values of d and w.

Thus far, we have considered only *edge-emitting* diode lasers in which the laser radiation is in a direction approximately parallel to the active layer (cf. Fig. 15.1). Of course, there is also gain in directions perpendicular to the active layer, allowing the possibility of "surface emission." Since the width d of the active quantum well layer is much smaller than its length L, the gain-length product gd for surface emission is much smaller than it is for edge emission. To overcome reflection losses, g must be larger than the threshold value $-(1/2d)\ln(r_1 r_1)$ in order to have a surface emission laser; in other words, the product $r_1 r_2$ of the reflectivities must be greater than $\exp(-2gd)$. If $g = 200\,\text{cm}^{-1}$ and $d = 1\,\mu\text{m}$, for example, this requires $r_1 r_2 > 0.98$, which cannot be realized in practice by simple Fresnel reflection. In vertical cavity surface-emitting lasers (VCSELs) large reflectivities are obtained with Bragg gratings consisting of embedded layers of semiconductors with different energy band gaps and alternatingly high and low refractive indices, as shown in Fig. 15.8. Layer materials, thicknesses, and spacings can be chosen such that reflectivities >0.99 are obtained at a chosen wavelength, and consequently threshold currents are small ($<1\,\text{mA}$), which allows fast modulation of the laser radiation (Section 15.3) in optical communications. Surface emission results in a larger output aperture and therefore a smaller beam divergence angle (typically $5°$–$10°$) than edge emission. Furthermore the radiation pattern is symmetrical (and nonastigmatic), making it easier to collimate the output beam with a simple lens and inject it into a fiber. The mode spacing $\approx c/2nd$, compared to $\approx c/2nL$ for an edge emitter (Problem 15.4), is large compared to the gain bandwidth and therefore results in single-mode oscillation. Many commercial VCSELs are based on GaAs/AlGaAs and emit in a

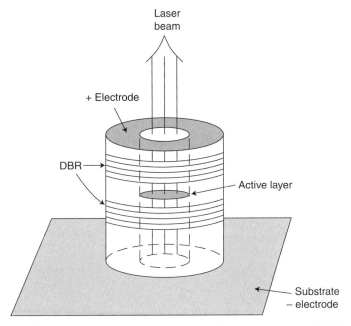

Laser
beam

+ Electrode

DBR

Active layer

Substrate
− electrode

Figure 15.8   Schematic of a vertical cavity surface-emitting laser (VCSEL).

range ~850 ± 100 nm, although VCSELs at other wavelengths, including the visible, are not unusual. VCSELS based on GaInNAs or InAlGaAsP emit at longer wavelengths, including the 1.3- and 1.55-μm wavelengths for optical communications (Section 15.6). Laser mice employing VCSELS are now common; they are able to track surface variations much more accurately than LED optical mice, and their low-power consumption makes them especially attractive for battery-powered operation.

A large number of VCSELs can be "grown" together on a semiconductor wafer, making the testing and production process more economical than in the case of edge emitters. Two-dimensional arrays of thousands of VCSELs emitting ~10^2 W can be fabricated such that different individual lasers can be separately controlled and can have different wavelengths determined by the widths of the active layers. Laser printers employing VCSEL arrays have recently been marketed.

- The diode laser bars and stacks presently used to pump high-power solid-state or fiber lasers and amplifiers (Sections 11.12 and 11.14) or in other applications consist of edge-emitting lasers (Fig. 15.9), each bar emitting ~50 W or more. The use of bars instead of a wide single active layer avoids the problem of amplified spontaneous emission parallel to the active layer, where the gain-length product would be large. Since the lasers operate independently, the angular distribution of the radiation has the same properties as that from each laser. In particular, the beam quality associated with the angle θ_w in Fig. 15.7 tends to be low because of the (relatively) large width w, along which the active medium might have significant nonuniformity. For coupling into a fiber the asymmetric output beam and the large divergence angle associated with the angle θ_d in Fig. 15.7 require a high numerical-aperture, aspherical lens. Beam-shaping optical components can also be used to "symmetrize" the output beam so that it has an approximately circular waist. For some beam delivery applications different lasers in a bar are coupled to different fibers, which are then bundled.

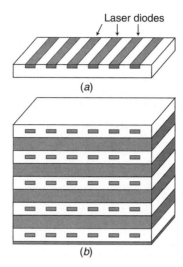

Figure 15.9   (*a*) A diode-laser bar, consisting perhaps of ∼50 individual edge-emitting diode lasers, each with an active layer of length $L \sim 1$ mm and a width $w \sim 0.1$ mm for a total width ∼10 mm. (*b*) A diode stack of bars.

A diode stack typically consists of perhaps 20 bars for a total power of a few kilowatts or more. Both bars and stacks require heat sinks to keep the temperature constant, which is necessary for stable and efficient operation. Bars are cooled by mounting the semiconductor chip on a heat sink that might use flowing water in microchannels. In the case of stacks there are heat sinks between the bars, limiting how closely the bars can be stacked.   ●

15.3   MODULATION OF DIODE LASERS

Laser radiation can be modulated using, for example, a Pockels cell and polarizers (Section 5.12). A great advantage of diode lasers in optical communications is that they can be modulated *directly* just by modulating the injection current. Equally important is the fact that modulation can be done at a very high rate, allowing a high rate of transmission of information. In this section we present a simplified model to explain the most important characteristics of this modulation.

Our discussion of gain in diode lasers in the preceding section focused on steady-state laser operation. To treat a time-dependent injection current, we will resort to a phenomenological rate equation model for the densities of photons and injected electrons, assuming that neither depends significantly on position within the active layer and that lasing occurs primarily at a single frequency. While this model is simplistic in several respects, it expresses some generic features of the interaction of light with semiconductor gain media, and as such has been found useful as a starting point for the analysis of time-dependent phenomena in various types of diode laser. We denote the photon density by Q and write for it the familiar rate equation

$$\frac{dQ}{dt} = GQ - \frac{Q}{\tau_p},$$ (15.3.1)

where G is the (temporal) gain coefficient at the frequency of interest and $1/\tau_p$ is the rate at which photons are lost to output coupling, scattering, etc. For the density N of electrons we write the rate equation

$$\frac{dN}{dt} = \frac{1}{V}\frac{I}{e} - \frac{N}{\tau_R} - GQ. \tag{15.3.2}$$

The first term on the right-hand side is the rate at which N increases due to an injection current I in the active region of volume V, and the second is the rate at which it decreases due to electron–hole recombination at the rate $1/\tau_R$. The last term is the rate of change of N due to the interaction of radiation with the electrons and holes. The form of this term ensures that, if there were no injection current or recombination, the sum of the total number of "excitations" is constant, i.e., $d(N + Q) = 0$. For G we assume the form suggested by the numerical results presented in the preceding section: $G = v_g g = v_g \sigma(N - N_{tr}) \equiv C_s(N - N_{tr})$. Then

$$\frac{dQ}{dt} = C_s(N - N_{tr})Q - \frac{Q}{\tau_p} \tag{15.3.3a}$$

and

$$\frac{dN}{dt} = \frac{1}{V}\frac{I}{e} - \frac{N}{\tau_R} - C_s(N - N_{tr})Q. \tag{15.3.3b}$$

The steady-state values of N and Q for a constant current I are found as usual by setting $dN/dt = dQ/dt = 0$:

$$\bar{Q} = \frac{\tau_p I}{Ve} - \left(\frac{\tau_p}{\tau_R}N_{tr} + \frac{1}{C_s\tau_R}\right), \tag{15.3.4a}$$

$$\bar{N} = N_{tr} + \frac{1}{C_s\tau_p}. \tag{15.3.4b}$$

Now suppose that the current I is not constant but has a small time-dependent modulation component $I_m(t)$: $I(t) = \bar{I} + I_m(t)$. This results in time-dependent perturbations $Q_m(t)$ and $N_m(t)$ of Q and N: $Q(t) = \bar{Q} + Q_m(t)$ and $N(t) = \bar{N} + N_m(t)$. Assuming the magnitudes of $I_m(t)$, $Q_m(t)$, and $N_m(t)$ are small compared to their respective steady-state values $\bar{I}$, $\bar{Q}$, and $\bar{N}$, we obtain from (15.3.3) approximate equations for the modulated parts of Q and N:

$$\frac{dQ_m}{dt} = C_s\bar{Q}N_m, \tag{15.3.5a}$$

$$\frac{dN_m}{dt} = \frac{1}{V}\frac{I_m}{e} - \left(\frac{1}{\tau_R} + C_s\bar{Q}\right)N_m - \frac{1}{\tau_p}Q_m. \tag{15.3.5b}$$

Suppose furthermore that the current modulation is periodic, with frequency ω_m:

$$I(t) = \bar{I} + I_m(t) = \bar{I} + \tilde{I}_m e^{-i\omega_m t}, \tag{15.3.6}$$

with $\tilde{I}_m$ a constant modulation amplitude. Using this expression in (15.3.5), and writing similarly

$$Q(t) = \overline{Q} + \tilde{Q}_m e^{-i\omega_m t} \qquad \text{and} \qquad N(t) = \overline{N} + \tilde{N}_m e^{-i\omega_m t}, \tag{15.3.7}$$

we obtain

$$i\omega_m \tilde{Q}_m + C_s \overline{Q} \tilde{N}_m = 0, \tag{15.3.8a}$$

$$\frac{1}{\tau_p} \tilde{Q}_m - \left(i\omega_m - \frac{1}{\tau_R} - C_s \overline{Q} \right) \tilde{N}_m = \frac{1}{V} \frac{\tilde{I}_m}{e}. \tag{15.3.8b}$$

We are interested in particular in how the photon density is modulated by the modulation of the injection current, and solve (15.3.8) for $\tilde{Q}_m$:

$$Q(t) = \overline{Q} + \tilde{Q}_m e^{-i\omega_m t}$$

$$= \overline{Q} - \left[\frac{C_s/eV}{\omega_m^2 + i(1/\tau_R + C_s\overline{Q})\omega_m - C_s\overline{Q}/\tau_p} \right] \tilde{I}_m \overline{Q} e^{-i\omega_m t}. \tag{15.3.9}$$

The modulus of the factor in brackets has its maximum value at the current modulation frequency

$$(\omega_m)_{\text{max}} = \sqrt{\frac{C_s \overline{Q}}{\tau_p} - \frac{1}{2} \left(\frac{1}{\tau_R} + C_s \overline{Q} \right)^2}. \tag{15.3.10}$$

The photon lifetime τ_p is typically much smaller than the recombination lifetime τ_R. For example, even if we ignore all losses except for output coupling and assume end facets with reflectivities $r_1 = r_2 = 0.99$, $1/\tau_p \cong -(c/2nL)\ln(r_1 r_2) = 1.7 \times 10^9 \text{ s}^{-1}$, or $\tau_p = 0.6$ ns for $L = 500$ μm, whereas τ_R is usually $\sim$3–4 ns, as noted earlier. Since $C_s \overline{Q}$ is also typically on the order of nanoseconds, we can approximate (15.3.10) by

$$(\omega_m)_{\text{max}} = \sqrt{C_s \overline{Q}/\tau_p} = \sqrt{v_g \sigma \overline{Q}/\tau_p}. \tag{15.3.11}$$

This result shows why diode lasers can be modulated at high rates: They have large values of the cross section σ, or in other words a large gain per injected electron, and usually a small photon lifetime τ_p. If we relate $\overline{Q}$ to the laser output power P_{out} using the output coupling fraction f defined by (15.2.27), we can approximate (15.3.11) by (Problem 15.5)

$$(\omega_m)_{\text{max}} = \sqrt{\frac{v_g \sigma P_{\text{out}}}{f h \nu V}} \simeq \sqrt{\frac{\sigma \lambda P_{\text{out}}}{n h f V}} \tag{15.3.12}$$

if we approximate the group velocity v_g by the phase velocity c/n.

Let us return to the example in the preceding section, where we obtained $\sigma = 1.9 \times 10^{-20} \text{ m}^2$ [Eq. (15.2.22)]. Assuming $L = 250$ μm, $w = 3$ μm, $d = 0.1$ μm, $\lambda = 1.3$ μm, Fresnel reflection coefficients $r_1 = r_2 = r = 0.31$ (for $n = 3.5$), $a = 10 \text{ cm}^{-1}$, and an output power of 5 mW, we calculate $f = 0.8$ and

$$\nu_{\max} = \frac{(\omega_m)_{\max}}{2\pi} \sim 5 \text{ GHz.} \tag{15.3.13}$$

Figure 15.10 plots the modulus of the factor in brackets in (15.3.9) for the parameters assumed in obtaining the estimate (15.3.13). The predicted response is seen to be approximately Lorentzian. The important conclusion of this analysis is that diode laser intensities can be modulated at high frequencies, and over a fairly large bandwidth, by direct modulation of injection currents.

However, our simplified model does not place any upper limit on the modulation frequency. According to Eq. (15.3.12), $(\omega_m)_{\max}$ is linearly proportional to the square root of the laser power. It suggests, for instance, that modulation frequencies much greater than (15.3.13) could be obtained by increasing the laser power as much as possible as long as other effects such as damage to the laser facets do not come into play. In reality, in contrast to the nearly Lorentzian response shown in Fig. 15.10, there is a rapid rolloff in the modulated intensity when the injection current is modulated at very high frequencies ω_m. In particular, the monotonic increase in $(\omega_m)_{\max}$ with photon density predicted by (15.3.11) breaks down when, for instance, two-photon absorption becomes significant, or when $\overline{Q}$ becomes sufficiently large to cause a rise in the temperature T appearing in the gain coefficient (15.2.17). As T increases, the energy distributions of electrons and holes broaden, and the densities of electrons and holes undergoing radiative recombination at the lasing frequency therefore decrease, so that the gain decreases with increasing Q. To account for such *gain suppression* effects, let us first note from (15.3.4b) that $G = C_s(N - N_{\text{tr}}) = C_s(N - \overline{N}) + G_t$, where $G_t = 1/\tau_p$ is the value of G needed to reach the laser threshold. Now consider again a (small) periodic modulation of the injection current as in (15.3.6)–(15.3.8), this time accounting phenomenologically for the

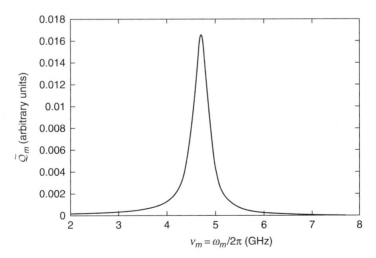

Figure 15.10 Response of the modulated intensity to the injected current modulation frequency, as predicted by small-signal analysis of the rate equation model [Eqs. (15.3.1) and (15.3.2)]. The modulus of the factor in brackets in (15.3.9) is plotted in scaled units. The parameters assumed are those used in obtaining (15.3.13).

decrease of G with Q by writing

$$\delta G \cong C_s \tilde{N}_m + \frac{\partial G}{\partial Q} Q_m \equiv C_s \tilde{N}_m - r_g Q_m \qquad (15.3.14)$$

for the change in G, due to the current modulation, in the small-signal analysis of Eq. (15.3.3). This results in the replacement of Eqs. (15.3.8) by (Problem 15.5)

$$(i\omega_m - r_g \overline{Q})\tilde{Q}_m + C_s \overline{Q}\tilde{N}_m = 0, \qquad (15.3.15a)$$

$$\left(\frac{1}{\tau_p} - r_g \overline{Q}\right)\tilde{Q}_m - \left(i\omega_m - \frac{1}{\tau_R} - C_s \overline{Q}\right)\tilde{N}_m = \frac{1}{V}\frac{\tilde{I}_m}{e}, \qquad (15.3.15b)$$

which, of course, reduce to Eqs. (15.3.8) when $r_g = 0$. The solution of these equations for $\tilde{Q}$ implies that the modulation of Q is strongest when (Problem 15.5)

$$\omega_m = (\omega_m)_{max} \cong \sqrt{\frac{C_s \overline{Q}}{\tau_p} - \frac{1}{2} r_g^2 \overline{Q}^2}, \qquad (15.3.16)$$

if, in addition to the assumptions made in writing (15.3.11), we assume that $r_g \overline{Q} \gg 1/\tau_R$ and $r_g \gg C_s$; for present purposes we simply assert that this is generally a good approximation. The obvious difference between this result and (15.3.11) is that $(\omega_m)_{max}$ does not increase monotonically with $\overline{Q}$. Equation (15.3.16) implies that $(\omega_m)_{max}$ is largest when $\overline{Q} = C_s/r_g^2 \tau_p$, and that at this photon density

$$(\omega_m)_{max} = \frac{1}{\sqrt{2}} \frac{C_s}{r_g \tau_p}. \qquad (15.3.17)$$

Experimental data suggest $r_g \sim 3 \times 10^{-12}$ m^3/s as a rough but reasonable estimate for our purposes.[7] Then, for the same laser parameters used to obtain (15.3.13), we estimate from (15.3.17) that

$$\nu_{max} = \frac{(\omega_m)_{max}}{2\pi} \sim 30 \text{ GHz}. \qquad (15.3.18)$$

Because of gain suppression effects, current modulation frequencies much larger than this in our example would not produce much laser intensity modulation.

From Eq. (15.3.15a) we have $\tilde{N}_m = -(C_s \overline{Q})^{-1}(i\omega_m - r_g \overline{Q})\tilde{Q}_m$, or, in terms of $N_m(t)$ and $Q_m(t)$,

$$N_m(t) = \frac{1}{C_s}\left[\frac{1}{\overline{Q}}\frac{dQ_m}{dt} + r_g Q_m(t)\right]. \qquad (15.3.19)$$

---

[7]L. A. Coldren and S. W. Corzine, *Diode Lasers and Photonic Integrated Circuits*, Wiley, New York, 1995. See also A. Yariv, *Optical Electronics in Modern Communications*, 5th ed., Oxford University Press, New York, 1997, Chapter 15.

This follows from our assumption of a periodic modulation ($i\omega_m \mathcal{Q}_m = -d\mathcal{Q}_m/dt$), but the linear relation between $\tilde{N}_m$ and $\tilde{\mathcal{Q}}_m$, and the fact that $\mathcal{Q}_m(t)$ can be expressed as a (Fourier) superposition of monochromatic components, gives (15.3.19) a general validity under the assumption used in deriving it—that $\tilde{N}_m$ and $\tilde{\mathcal{Q}}_m$ are small compared to $\bar{N}$ and $\bar{\mathcal{Q}}$, respectively. In other words, any modulation of the injection current results in a modulation of the density N of injected electrons:

$$N(t) = \bar{N} + N_m(t) = \bar{N} + \frac{1}{C_s}\left[\frac{1}{\bar{\mathcal{Q}}}\frac{d\mathcal{Q}_m}{dt} + r_g \mathcal{Q}_m(t)\right] \equiv \bar{N} + \Delta N(t). \qquad (15.3.20)$$

In addition to the obvious modulation of the gain that this implies, there is also a modulation of the refractive index. Recall that, in terms of the complex refractive index $n = n_R + in_I$, a plane wave experiences amplification (or attenuation) and a phase shift determined by n_R and n_I, respectively:

$$e^{ikz} = e^{i\omega(n_R + in_I)z/c} = e^{\beta z/2}e^{i\omega n_R z/c} \qquad (\beta = -2\omega n_I/c). \qquad (15.3.21)$$

Identifying the amplification rate $G = (c/n_R)\beta = -2\omega n_I/n_R$, and using again the relation $G = C_s(N - N_{tr})$, we can relate n_I to the density N:

$$n_I = -\frac{C_s n_R}{2\omega}(N - N_{tr}). \qquad (15.3.22)$$

Assuming that the relative change in n_R is small compared to that of $N(t)$, therefore, we can relate the change in the imaginary part of the refractive index to the change in the electron density $\Delta N(t)$:

$$\Delta n_I(t) = -\frac{C_s n_R}{2\omega}\Delta N(t). \qquad (15.3.23)$$

The Kramers–Kronig relation between n_R and n_I (Section 3.15) implies that there must also be a change $\Delta n_R(t)$ in n_R. It is conventional in diode laser theory to relate $\Delta n_R(t)$ to $\Delta n_I(t)$ as follows:

$$\Delta n_R(t) = \alpha \, \Delta n_I(t). \qquad (15.3.24)$$

For reasons discussed in the following section, α is called the *linewidth enhancement factor*.

Recall that the longitudinal mode frequencies ν of a laser resonator are approximately inversely proportional to the (real part of the) refractive index $n_R(\nu)$. [See, e.g., Eq. (5.9.2) with $l = L$, the case of interest here.] Thus, $\Delta n_R(t)/n_R = -\Delta\nu(t)/\nu$ and, from (15.3.24), (15.3.23), and (15.3.20),

$$\Delta\nu(t) = \frac{\alpha C_s}{4\pi}\Delta N(t) = \frac{\alpha}{4\pi}\left[\frac{1}{\bar{\mathcal{Q}}}\frac{d\mathcal{Q}_m}{dt} + r_g \mathcal{Q}_m(t)\right]. \qquad (15.3.25)$$

This shows that *modulation of the laser intensity by modulation of the injection current results in a frequency chirp*. This is a very important consideration in fiber-optic communication systems because a frequency chirp implies a spectral broadening of a light pulse, and group velocity dispersion in a fiber causes a pulse temporal broadening proportional to the spectral width of the pulse (Section 8.4) and the propagation distance. Equation (15.3.25) implies that too rapid a modulation of the intensity might result in large temporal broadening of pulses and therefore a deleterious pulse overlap in a long optical fiber (or any other medium with group velocity dispersion). In other words, long-distance, high-speed communication of information might require that a diode laser be externally modulated rather than "internally" modulated by the injection current.

15.4   NOISE CHARACTERISTICS OF DIODE LASERS

Like all lasers, diode lasers exhibit unavoidable, random fluctuations in the phase and the intensity of their radiation. In Section 5.11 we derived the (Schawlow–Townes) linewidth arising from spontaneous emission noise, and remarked that for diode lasers, in contrast to most other lasers, this fundamental source of noise typically dominates "technical noise." We noted furthermore that for diode lasers, again in contrast to most other lasers, corrections to the Schawlow–Townes formula due to "excess spontaneous emission noise," and especially to a coupling of the laser phase and intensity, can be significant. We now discuss in greater detail these two corrections to the Schawlow–Townes formula. Aside from the appearance of the dimensionless quantities K and α, the analyses below may be viewed as two derivations of the fundamental laser linewidth in addition to the one presented in Section 5.11.

Excess Spontaneous Emission Noise: The K Parameter

When the mirrors of a laser are not highly reflecting, the expression for the Schawlow–Townes linewidth must be corrected to include the K parameter associated with "excess spontaneous emission noise," as mentioned in Section 5.11. The K parameter for a Fabry-Pérot resonator with mirror reflectivities r_1 and r_2 is defined by Eq. (5.11.15). For the reflectivities $r_1 = r_2 = 0.3$ corresponding to Fresnel reflection with $n = 3.5$ [Eq. (15.2.6)], $K = 1.13$, while for $r_1 = r_2 = 0.05$, $K = 2.01$. This correction to the Schawlow–Townes linewidth is therefore modest for most situations of practical interest, but it can be significant for very lossy cavities. Since a full derivation of the K parameter is rather complicated, we will only present a plausibility argument for formula (5.11.15).

We start from the rate equation for the photon number q in a gain medium described simply by the occupation densities N_2 and N_1 of the upper and lower levels, respectively, of the lasing transition:[8]

$$\frac{dq}{dt} = c\sigma(N_2 - N_1)q + c\sigma N_2 - \gamma q. \tag{15.4.1}$$

---

[8]In order to compare with the expression for the Schawlow–Townes linewidth in Section 5.11, we approximate the refractive index by 1 in our discussion here of the K parameter. The effect of the refractive index on the Schawlow–Townes linewidth is contained in the expression (15.4.35).

The second term on the right accounts for spontaneous emission into the lasing mode, and as such is equal to the stimulated emission rate from the upper level N_2 when the photon number $q = 1$ (Section 3.7). In steady-state oscillation ($dq/dt = 0$),

$$cg = \frac{\gamma}{1 + n_{sp}/q} \cong \gamma - \gamma\frac{n_{sp}}{q} \equiv \gamma - \xi, \tag{15.4.2}$$

where $g = \sigma(N_2 - N_1)$ is the steady-state gain coefficient and we again define the "spontaneous emission factor" $n_{sp} = N_2/(N_2 - N_1)$. The steady-state gain is *less than* the loss $\gamma = -(c/2L)\ln(r_1r_2)$: Spontaneous emission, represented by ξ, puts photons into the lasing mode and therefore reduces the gain required for a steady-state photon number q. We have generally assumed that the number of photons due to spontaneous emission is much smaller than the number due to stimulated emission and taken $cg = \gamma$, the gain = loss condition for laser oscillation. This is a superb approximation for calculating laser intensity and power, but it would imply a laser linewidth of zero.

Let us first show how (15.4.2) leads to the Schawlow–Townes linewidth. The linewidth $\Delta\nu$ of a quasi-monochromatic field of central frequency ν is inversely proportional to the coherence time τ_{coh} [Eq. (13.11.11)] determined by the mutual coherence function $\Gamma(\mathbf{r}_1, t; \mathbf{r}_2, t + \tau)$ for $\mathbf{r}_1 = \mathbf{r}_2$. [Recall, e.g., Eq. (13.5.5).] Since this mutual coherence function, which we denote simply by $\Gamma(\tau)$, is proportional to the ensemble average $\langle E^*(t)E(t + \tau)\rangle$ of the product of the complex conjugate of the electric field at a time t and the electric field at a time $t + \tau$, it is seen from (15.4.1) that (for a stationary field) it satisfies (Problem 15.6)

$$\frac{d\Gamma}{d\tau} = -i\omega\Gamma + \frac{1}{2}(cg - \gamma)\Gamma = -i\omega\Gamma - \frac{1}{2}\xi\Gamma \qquad (\omega = 2\pi\nu). \tag{15.4.3}$$

We do not include any contribution corresponding to the second term on the right-hand side of (15.4.1). This is because of the random nature of spontaneous emission; in a more rigorous, fully quantum mechanical approach, this term makes no net contribution to the ensemble average of the mutual coherence function, basically just because the electric field produced by spontaneous emission is zero *on average*. The factor $\frac{1}{2}$ in (15.4.3) is just a consequence of the fact that q is proportional to the field intensity. Thus,

$$\Gamma(\tau) = \Gamma(0)e^{-i\omega\tau}e^{-\xi\tau/2} = \Gamma(0)e^{-i\omega\tau}e^{-\tau/\tau_{coh}}. \tag{15.4.4}$$

The linewidth of quasi-monochromatic radiation with this mutual coherence function is proportional to $1/\tau_{coh} = \xi/2 = \gamma n_{sp}/2q$. More precisely the laser spectrum, defined by the Fourier transform of (15.4.4) [cf. Eq. (13.5.5)], is a Lorentzian lineshape function centered at ν and having a full width at half-maximum

$$\Delta\nu = \frac{1}{\pi\tau_{coh}} = \frac{\gamma n_{sp}}{2\pi q}, \tag{15.4.5}$$

which is equivalent to the expression (5.11.13) for the Schawlow–Townes linewidth (Problem 15.6).

This shows again that the Schawlow–Townes linewidth is a consequence of spontaneous emission. It was assumed in its derivation, however, that the intracavity laser field is uniform and characterized completely by a photon number q, whereas we know from Section 5.5 that there are substantial spatial variations of the field when the output coupling is large. Large output coupling, furthermore, implies large loss and therefore that a high gain is needed to support laser oscillation. Since spontaneous emission noise is amplified in a gain medium (Section 6.13), we must take this into account in calculating the linewidth of a laser with large output coupling, for example, a Fabry-Pérot laser resonator with small mirror reflectivities. In other words, for lasers with large output coupling we must deal with two effects that were ignored in our derivation of the Schawlow–Townes linewidth: (i) spatial variations of the intracavity field and (ii) the amplification of spontaneous emission.

It will be convenient for our purposes to address the problem using the concept of "effective noise." As discussed in Section 6.13, in the quantum theory of radiation a field mode of frequency v has a zero-point energy $\frac{1}{2}hv$; we can imagine an infinite set of plane-wave modes, each undergoing quantum fluctuations with mean energy $\frac{1}{2}hv$ per mode. Let us assume that this effective noise radiation can be reflected, transmitted, and amplified in just the same way as "ordinary" radiation, and consider a laser resonator with mirror power reflectivities r_i and transmissivities $t_i = 1 - r_i$ $(i = 1, 2)$ (Fig. 15.11). Thus, for example, noise radiation can enter the resonator at $z = 0$, and a part of it exits at $z = L$ after amplification, as indicated in Fig. 15.11a. It can also be reflected at $z = L$ and, after another pass through the gain medium, emerge as a left-going wave at

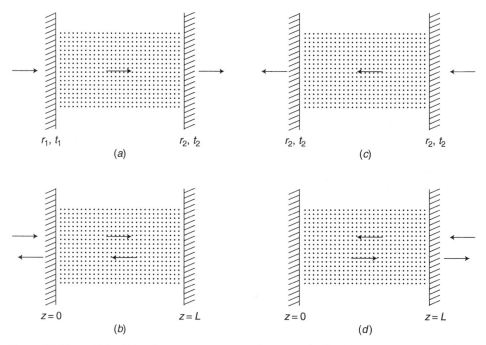

Figure 15.11   A Fabry-Pérot laser resonator with mirror power reflectivities r_1, r_2 and transmissivities t_1, t_2. The gain medium produces a single-pass power amplification $\mathcal{G}^2 = \exp{(gL)}$. The four contributions to the amplification of "effective input noise" are indicated.

$z = 0$, as shown in Fig. 15.11b. Noise radiation entering the resonator at $z = L$ can similarly follow one of the two paths shown in Figs. 15.11c and 15.11d.

How does the gain medium affect the effective noise intensity? Consider, for example, the effect of path (b). The modification of the noise intensity for this path derives from (i) the transmissivity t_1; (ii) the gain $\mathcal{G}$ resulting from propagation from $z = 0$ to $z = L$; (iii) the reflection coefficient r_2 at $z = L$; (iv) further gain $\mathcal{G}$ on going from $z = L$ to $z = 0$; and (v) the transmissivity t_1. Thus, the noise radiation propagating leftward in the region $z < 0$ is modified in intensity by the product of the factors involved in processes (i)–(v): $t_1 \mathcal{G} r_2 \mathcal{G} t_1$. The path shown in Fig. 15.11c similarly adds an amplified component $t_2 \mathcal{G} t_1$ to the noise radiation propagating leftward in the region $z < 0$, so that the intensity of this radiation is amplified altogether by the factor $t_1 \mathcal{G} r_2 \mathcal{G} t_1 + t_2 \mathcal{G} t_1$. Similarly, the noise intensity propagating rightward in the region $z > L$ is amplified by the factor $t_1 \mathcal{G} t_2 + t_2 \mathcal{G} r_1 \mathcal{G} t_2$ due to the two paths shown in Figs. 15.11a and 15.11d. The noise intensity (at the laser frequency ν) outside the resonator is therefore modified by a total factor $t_1 \mathcal{G} r_2 \mathcal{G} t_1 + t_2 \mathcal{G} t_1 + t_1 \mathcal{G} t_2 + t_2 \mathcal{G} r_1 \mathcal{G} t_2$ compared to what it would be in free space. Using the threshold (steady-state) condition $\mathcal{G}^2 r_1 r_2 = 1$ for laser oscillation, we find by simple algebra that

$$t_1 \mathcal{G} r_2 \mathcal{G} t_1 + t_2 \mathcal{G} t_1 + t_1 \mathcal{G} t_2 + t_2 \mathcal{G} r_1 \mathcal{G} t_2 = \left(\frac{t_1}{\sqrt{r_1}} + \frac{t_2}{\sqrt{r_2}} \right)^2. \tag{15.4.6}$$

The four paths shown in Fig. 15.11 are the only ones that modify the noise intensity. Suppose, for example, that we include a path in which noise radiation is incident from the left at $z = 0$, passes through the gain medium, reflects off the mirror at $z = L$, makes another pass through the gain medium, reflects off the mirror at $z = 0$, passes again through the gain medium, and then exits the resonator through the mirror at $z = L$. The intensity of noise radiation propagating through the mirror at $z = L$ is therefore modified compared to its free-space level by the factor $t_1 \mathcal{G} r_2 \mathcal{G} r_1 \mathcal{G} t_2$; using again the threshold condition $r_1 r_2 \mathcal{G} \mathcal{G} = 1$, we see that this factor is just $t_1 \mathcal{G} t_2$, the amplification factor already accounted for by the path shown in Fig. 15.11. Similar considerations for other paths involving multiple passes through the resonator should convince the reader that the four paths shown in Fig. 15.11 account for all the amplification of noise radiation.

It is less clear without a more rigorous and detailed analysis than is appropriate here that the factor (15.4.6) associated with the amplification of noise radiation also accounts for the amplification of spontaneous emission noise, although the discussion in Section 6.13 certainly lends support to such a surmise. The laser linewidth calculated on the basis of the amplification of "effective input noise" is found to be

$$\Delta \nu = \frac{n_{sp} h \nu}{2 \pi P_{out}} \left(\frac{c}{2L} \right)^2 \left(\frac{t_1}{\sqrt{r_1}} + \frac{t_2}{\sqrt{r_2}} \right)^2, \tag{15.4.7}$$

which is to be compared to the Schawlow–Townes linewidth defined by (5.11.12) or (5.11.13) with $l = L$:

$$\Delta \nu_{ST} = \frac{n_{sp} h \nu}{2 \pi P_{out}} \left[\frac{c}{2L} \ln (r_1 r_2) \right]^2. \tag{15.4.8}$$

After some straightforward algebra using $t_i = 1 - r_i$, $i = 1, 2$, we obtain

$$\Delta\nu = \left[\frac{\frac{t_1}{\sqrt{r_1}} + \frac{t_2}{\sqrt{r_2}}}{\ln(r_1 r_2)}\right]^2 \Delta\nu_{ST} = \left[\frac{(\sqrt{r_1} + \sqrt{r_2})(1 - \sqrt{r_1 r_2})}{\sqrt{r_1 r_2}\ln(r_1 r_2)}\right]^2 \Delta\nu_{ST} = K\Delta\nu_{ST}, \quad (15.4.9)$$

which is just Eq. (5.11.14).[9]

We have thus been led to the correction of the Schawlow–Townes formula by the K parameter by taking account of the *amplification* of spontaneous emission noise in a laser. This amplification always occurs but is significant only when the output coupling is large (i.e., when r_1 and r_2 are small in Fabry-Pérot resonators) and, consequently, the gain required for laser oscillation is large; then the "excess spontaneous emission noise" characterized by K can be measurably greater than unity. If, on the other hand, r_1 and r_2 are close to unity, then $K \cong 1$.

The K parameter—often referred to as the *Petermann factor* in the literature—has for many years been treated in different ways and with considerable discussion as to its proper interpretation.[10] It is worth noting that the K parameter does not describe an increased *rate* of spontaneous emission; the rate of spontaneous emission is given by the second term on the right-hand side of (15.4.1). Rather, it describes the *amplification* of spontaneous emission.

We have restricted our discussion to Fabry-Pérot resonators, but any laser resonator is characterized by a K parameter that can act as a significant device parameter whenever transmission or diffractive losses are large. Not surprisingly, large K-factor enhancements of the Schawlow–Townes linewidth—as large as ~ 200 or more—have been measured with unstable-resonator lasers. From a mathematical perspective the K parameter is associated with the nonorthogonality of the modes of a lossy resonator: The cavity modes are eigenfunctions of a *non*-Hermitian operator and are not orthogonal.

Let us write Eq. (15.4.2) in the form

$$q = \frac{cgn_{sp}}{\gamma - cg} = \frac{gn_{sp}}{g_t - g}, \quad (15.4.10)$$

where again g_t is the threshold gain coefficient and g in the correct version of this formula is the *modal* gain coefficient introduced earlier. When the amplification of spontaneous emission is accounted for, the right-hand side of this equation, which as we have seen is associated with spontaneous emission noise, is multiplied by K:

$$q = K\frac{gn_{sp}}{g_t - g}. \quad (15.4.11)$$

This formula in one form or other is often cited to explain why gain-guided diode lasers tend to oscillate on many longitudinal modes, whereas index-guided lasers typically

[9]Experimental results consistent with formula (15.4.9) have been reported by W. A. Hamel and J. P. Woerdman, *Physical Review Letters* **64**, 1506 (1990).

[10]A more rigorous analysis and review of this and other aspects of "quantum noise" in semiconductor laser physics is given by C. H. Henry and R. F. Kazarinov, *Reviews of Modern Physics* **68**, 801 (1996). This work also presents formulas for the K parameter that generalize the expression used here for longitudinal modes of Fabry-Pérot resonators.

oscillate mainly on just one or two modes. It implies that, for a fixed laser power (or photon number q), a larger value of K requires a larger value of the difference $g_t - g$ between loss and gain. This in turn implies that the *relative* differences between $g_t - g$ for different modes decrease. Since very small values of this relative difference can significantly affect mode discrimination, therefore, a larger K implies a smaller degree of mode discrimination. In gain-guided lasers there is greater loss and a larger K parameter than in index-guided lasers due to the fact that there is less lateral confinement of the field. Gain guiding consequently allows more longitudinal modes to lase than index guiding. It was in this context that the K parameter was first investigated in the late 1970s and early 1980s. The expression we have obtained for the K parameter [Eq. (15.4.9)] applies to index-guided lasers, but the physical interpretation of the K parameter in terms of amplified spontaneous emission noise applies also to gain-guided lasers.

Phase-Intensity Coupling: The α Parameter

We consider next the correction to the Schawlow–Townes formula associated with the linewidth enhancement factor α defined by Eq. (15.3.24). To simplify as much as possible we assume in the wave equation (8.2.13) an electric field and a polarization density of the form

$$E(z, t) = \hat{x}\mathcal{E}(t)e^{-i(\omega t - kz)} \qquad \text{and} \qquad P(z, t) = \hat{x}\mathcal{P}(t)e^{-i(\omega t - kz)}. \qquad (15.4.12)$$

If the complex amplitudes $\mathcal{E}(t)$ and $\mathcal{P}(t)$ vary slowly compared to $\exp(-i\omega t)$, we can drop their second derivatives with respect to time in the wave equation and write (Problem 15.5)

$$\left(\frac{\omega^2}{c^2} - k^2\right)\mathcal{E} + 2i\frac{\omega}{c^2}\dot{\mathcal{E}} = -\frac{\omega}{\epsilon_0 c^2}(\omega\mathcal{P} + 2i\dot{\mathcal{P}}) \qquad (15.4.13)$$

in the approximation of slowly varying envelopes (cf. Section 9.6). Following a procedure similar to that in Section 8.3, we next write

$$\mathcal{E}(t) = \int_{-\infty}^{\infty} d\Delta\tilde{\mathcal{E}}(\omega + \Delta)e^{-i\Delta t}, \qquad (15.4.14a)$$

$$\mathcal{P}(t) = \int_{-\infty}^{\infty} d\Delta\tilde{\mathcal{P}}(\omega + \Delta)e^{-i\Delta t} = \int_{-\infty}^{\infty} d\Delta\chi(\omega + \Delta)\tilde{\mathcal{E}}(\omega + \Delta)e^{-i\Delta t}, \qquad (15.4.14b)$$

and make a first-order Taylor series approximation to the susceptibility $\chi(\omega)$ about the carrier frequency ω:

$$\begin{aligned}
\mathcal{P}(t) &= \epsilon_0 \int_{-\infty}^{\infty} d\Delta[\chi(\omega) + \chi'(\omega)\Delta]\tilde{\mathcal{E}}(\omega + \Delta)e^{-i\Delta t} \\
&= \epsilon_0\chi(\omega)\mathcal{E}(t) + i\epsilon_0\chi'(\omega)\frac{d}{dt}\int_{-\infty}^{\infty} d\Delta\tilde{\mathcal{E}}(\omega + \Delta)e^{-i\Delta t} \\
&= \epsilon_0\chi(\omega)\mathcal{E}(t) + i\epsilon_0\chi'(\omega)\frac{d\mathcal{E}}{dt}, \qquad (15.4.15)
\end{aligned}$$

where $\chi(\omega) = n^2(\omega) - 1$ is the (linear) electric susceptibility and $\chi' \equiv d\chi/d\omega$. Then, in terms of the refractive index $n(\omega)$ and the group velocity v_g (Section 8.3), (15.4.13) becomes (Problem 15.5)

$$\frac{d\mathcal{E}}{dt} = \frac{iv_gc}{2n\omega}\left(n^2\frac{\omega^2}{c^2} - k^2\right)\mathcal{E}. \tag{15.4.16}$$

An injection current will change the refractive index of the active medium. We denote by $n_0(\omega)$ the index in the absence of a current and by $\Delta n(\omega) = \Delta n_R(\omega) + i\Delta n_I(\omega)$ the change in the index caused by an injection current, and use $n = n_0 + \Delta n_R + i\Delta n_I$ in (15.4.16). Taking n_0 to be real and $k = n_0\omega/c$, and assuming that Δn_R and Δn_I are small enough that Δn_R^2, Δn_I^2, and $\Delta n_R\Delta n_I$ can be ignored, we replace (15.4.16) by

$$\dot{\mathcal{E}} \equiv \frac{d\mathcal{E}}{dt} = \frac{iv_g\omega}{c}(\Delta n_R + i\,\Delta n_I)\mathcal{E} = -\frac{v_g\omega}{c}\Delta n_I(1 - i\alpha)\mathcal{E}, \tag{15.4.17}$$

where α is defined by (15.3.24).

The imaginary part of the index change, Δn_I, causes changes in the real part of the field amplitude $\mathcal{E}$, i.e., it is associated with gain (G) and loss (γ) in the medium:

$$-\frac{v_g\omega}{c}\Delta n_I = \frac{1}{2}(G - \gamma), \tag{15.4.18}$$

and (15.4.17) becomes

$$\dot{\mathcal{E}} = \tfrac{1}{2}(G - \gamma)(1 - i\alpha)\mathcal{E}, \tag{15.4.19}$$

or, in terms of the real field amplitude $|\mathcal{E}|$ and phase ϕ ($\mathcal{E} = |\mathcal{E}|e^{i\phi}$),

$$\frac{d}{dt}|\mathcal{E}|^2 = \mathcal{E}\dot{\mathcal{E}}^* + \dot{\mathcal{E}}\mathcal{E}^* = (G - \gamma)|\mathcal{E}|^2, \tag{15.4.20a}$$

$$\dot{\phi} = -\frac{\alpha}{2}(G - \gamma). \tag{15.4.20b}$$

Since $|\mathcal{E}|^2$ is proportional to the field intensity and therefore to the number of photons q in our model, we can write these equations equivalently as

$$\dot{q} = (G - \gamma)q, \tag{15.4.21a}$$

$$\dot{\phi} = -\frac{\alpha}{2}(G - \gamma) = -\frac{\alpha}{2q}\dot{q} = -\frac{\alpha}{2}\frac{d}{dt}(\ln q). \tag{15.4.21b}$$

This exhibits the coupling between the field phase and intensity, the strength of which depends on the α parameter.

Equations (15.4.21) do not include effects of spontaneous emission. To calculate the principal quantity of interest here—the mean-square phase fluctuation—we will follow

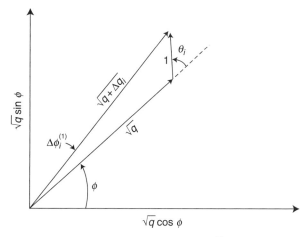

Figure 15.12 A complex field represented as a "phasor" $\sqrt{q}e^{i\phi}$. The ith spontaneous emission of a photon produces a rotation of the phasor by θ_i as shown. [Adapted from C. H. Henry, *IEEE Journal of Quantum Electronics* **QE-18**, 259 (1982).]

an intuitive approach for the effect on the phase of spontaneous emission.[11] The basic idea behind this approach is indicated in Fig. 15.12. A change $\Delta\phi_i^{(1)}$ in the phase ϕ due to the ith spontaneous emission event can be read off directly from the change in the phasor shown in Fig. 15.12:

$$\Delta\phi_i^{(1)} \cong \frac{1}{\sqrt{q}}\sin\theta_i. \tag{15.4.22}$$

There is another contribution to the change in ϕ in a spontaneous emission event, this one due to the phase-intensity coupling described by (15.4.21b); integrating both sides of that equation from a time before the ith spontaneous emission event occurs to a time immediately after, we obtain

$$\Delta\phi_i^{(2)} = -\frac{\alpha}{2}[\ln(q+\Delta q_i) - \ln(q)] \cong -\frac{\alpha}{2q}\Delta q_i \tag{15.4.23}$$

if the change Δq_i in the length of the phasor of Fig. 15.12 is small compared to the length q of the phasor before spontaneous emission. Δq_i can be deduced with reference to Fig. 15.12 and the law of cosines: $q + \Delta q_i = q + 1 + 2\sqrt{q}\cos\theta_i$, or $\Delta q_i = 1 + 2\sqrt{q}\cos\theta_i$. Therefore

$$\Delta\phi_i^{(2)} \cong -\frac{\alpha}{2q}(1 + 2\sqrt{q}\cos\theta_i) = -\frac{\alpha}{2q} - \frac{\alpha}{\sqrt{q}}\cos\theta_i. \tag{15.4.24}$$

Then for the change in ϕ in the ith spontaneous emission event we write

$$\Delta\phi_i = \phi_i^{(1)} + \Delta\phi_i^{(2)} = -\frac{\alpha}{2q} + \frac{1}{\sqrt{q}}(\sin\theta_i - \alpha\cos\theta_i), \tag{15.4.25}$$

[11]C. H. Henry, *IEEE Journal of Quantum Electronics* **QE-18**, 259 (1982).

and the total phase change after $\mathcal{N}$ spontaneous emission events is

$$\Delta\phi = \sum_{i=1}^{\mathcal{N}} \Delta\phi_i = \frac{1}{\sqrt{q}} \sum_{i=1}^{\mathcal{N}} (\sin\theta_i - \alpha\cos\theta_i). \qquad (15.4.26)$$

We have dropped a constant phase $-(\alpha/2q)\mathcal{N}$, as we are only interested in phase *fluctuations*. The random nature of spontaneous emission implies that the angles θ_i are independent random variables with average values $\langle\theta_i\rangle = 0$ and $\langle\sin\theta_i\rangle = \langle\cos\theta_i\rangle = 0$, $\langle\sin^2\theta_i\rangle = \langle\cos^2\theta_i\rangle = \frac{1}{2}$. Then

$$\langle\Delta\phi^2\rangle = \frac{1}{q} \sum_{i=1}^{\mathcal{N}} (\langle\sin^2\theta_i\rangle + \alpha^2\langle\cos^2\theta_i\rangle) = \frac{1}{q} \sum_{i=1}^{\mathcal{N}} \left(\frac{1}{2} + \frac{1}{2}\alpha^2\right)$$

$$= \frac{\mathcal{N}}{2q}(1 + \alpha^2). \qquad (15.4.27)$$

For the (average) number of spontaneous emission events in a time τ we assume $\mathcal{N} = C\tau$, where C is the rate of spontaneous emission into the lasing mode (Section 5.11). Then

$$\tau^{-1}\langle\Delta\phi^2\rangle = \frac{C}{2q}(1 + \alpha^2). \qquad (15.4.28)$$

As in the preceding subsection we calculate the linewidth $\Delta\nu$ using the mutual coherence function $\Gamma(\tau)$. Aside from irrelevant factors, we may define this mutual coherence function as the ensemble average (Problem 15.6)

$$\Gamma(\tau) = qe^{-i\omega\tau}\langle e^{i\Delta\phi}\rangle \qquad (15.4.29)$$

under the assumptions that the photon number q is constant in steady-state laser oscillation and that the laser field is stationary (Section 13.5). According to (15.4.26) and (15.4.27) we can regard $\Delta\phi$ as a sum of a large number of independent random variables having zero mean and variance $\sigma^2 = \langle\Delta\phi^2\rangle = C\tau(1 + \alpha^2)/2q$. We will therefore assume, based on the central limit theorem, that the phase excursion $\Delta\phi$ of the laser field has a Gaussian distribution with zero mean [cf. Eq. (13.15.26)]:

$$P(\Delta\phi) = \frac{1}{\sigma\sqrt{2\pi}} e^{-\Delta\phi^2/2\sigma^2}, \qquad (15.4.30)$$

and therefore

$$\Gamma(\tau) = qe^{-i\omega\tau} \int_{-\infty}^{\infty} d(\Delta\phi)e^{i\Delta\phi}P(\Delta\phi) = e^{-i\omega\tau}\frac{q}{\sigma\sqrt{2\pi}} \int_{-\infty}^{\infty} d(\Delta\phi)e^{i\Delta\phi}e^{-\Delta\phi^2/2\sigma^2}$$

$$= e^{-i\omega\tau}e^{-\sigma^2/2} = e^{-i\omega\tau}e^{-\tau/\tau_{\text{coh}}}, \qquad (15.4.31)$$

where $\tau_{\text{coh}}^{-1} \equiv C(1 + \alpha^2)/4q$. Thus, $\Delta\nu \propto C(1 + \alpha^2)/4q$. More precisely, the laser spectrum, defined by the Fourier transform of (15.4.31), is a Lorentzian lineshape function centered at $\nu = \omega/2\pi$ with width (FWHM)

$$\Delta\nu = \frac{1}{\pi\tau_{\text{coh}}} = \frac{C}{4\pi q}(1 + \alpha^2). \tag{15.4.32}$$

Except for a factor $\frac{1}{2}$ discussed below, this reduces when $\alpha = 0$ to the formula (5.11.6), which leads in turn to the expression (5.11.13) for the Schawlow–Townes linewidth $\Delta\nu_{\text{ST}}$. That is, the phase-intensity coupling characterized by the parameter α increases the Schawlow–Townes linewidth by the factor $1 + \alpha^2$. In deriving (5.11.13) we ignored internal cavity losses as well as dispersion in the gain medium, but these effects are easily accounted for as follows. The output power P_0 *per facet* of a diode laser is related to the cavity photon number q by the formula

$$P_0 = \frac{1}{2}qh\nu\left(-\frac{v_g}{2L}\ln r^2\right) = \frac{1}{2}qh\nu\left(-\frac{v_g}{L}\ln r\right) \equiv \frac{1}{2}qh\nu v_g a_m \tag{15.4.33}$$

if both facets are assumed to have the same reflectivity r. Similarly, using relations similar to those employed in Section 5.11, we can write the spontaneous emission rate C into the lasing mode as (Problem 15.5)

$$C = v_g n_{\text{sp}} g_t = v_g n_{\text{sp}}(a + a_m), \tag{15.4.34}$$

in which a is the loss coefficient in the active medium [Eq. (15.2.3)]. Equations (15.4.32)–(15.4.34) allow us to relate $\Delta\nu$ to the output power per facet and other quantities characterizing a particular laser:[12]

$$\Delta\nu = \frac{h\nu v_g^2 n_{\text{sp}} g_t a_m}{8\pi P_0}(1 + \alpha^2) \equiv (1 + \alpha^2)\Delta\nu'_{\text{ST}}, \tag{15.4.35}$$

$\Delta\nu'_{\text{ST}}$ is (correctly) half the Schawlow–Townes linewidth modified to allow an internal loss $a \neq 0$ and to include dispersion via a group velocity $v_g \neq c$.

Before discussing some details concerning the derivation of (15.4.35), let us consider a numerical example. Figure 15.13 shows experimental data, obtained with a scanning Fabry-Pérot interferometer, for the linewidth of a heterojunction diode laser as a function of the power P_0. The measured linewidth confirms the $1/P_0$ dependence predicted by (15.4.35). The parameters entering (15.4.35) for these measurements were:[11] $a = 45 \text{ cm}^{-1}$, $a_m = 39 \text{ cm}^{-1}$, $g_t = a + a_m = 84 \text{ cm}^{-1}$, $h\nu = 1.5 \text{ eV}$, $n_{\text{sp}} \cong 2.6$, and $v_g = c/4.33$.[13] With these numbers we calculate from Eq. (15.4.35) the laser

---

[12]This is equivalent to formula (26) of Henry, *op. cit.*, footnote 11.

[13]n_p, which can be deduced from the ratio of spontaneous emission and gain spectra, was determined in different experiments cited by Henry, *op. cit.*, as was the group velocity v_g, which can be inferred from a measured longitudinal mode spacing $v_g/2L$ (Problem 15.4).

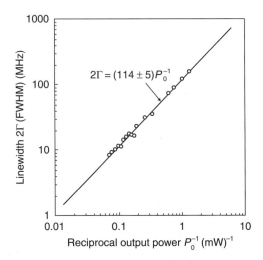

Figure 15.13   Measured linewidth of a single-mode diode laser as a function of power P_0. [From M. W. Fleming and A. Mooradian, *Applied Physics Letters* **38**, 511 (1981).]

linewidth

$$\Delta \nu \cong \frac{3.9}{P_0(\text{mW})} (1 + \alpha^2) \text{ MHz.} \tag{15.4.36}$$

The experimental data of Fig. 15.13 are accurately fit by the formula $\Delta \nu = (114 \pm 5)/P_0(\text{mW})$, implying that $\alpha \cong 5.3$. In other experiments the parameter α for a heterojunction laser was inferred by a measurement of the change with carrier density of the gain (and therefore Δn_I).[11] The Kramers–Kronig relation between the real (Δn_R) and imaginary (Δn_I) parts of Δn yielded the estimate $\alpha = \Delta n_R / \Delta n_I \approx 6$, consistent with the value needed to bring (15.4.36) into agreement with the data shown in Fig. 15.13.[14]

Other experiments have confirmed that the fundamental linewidth of diode lasers is much larger than in other lasers, where it is usually so small as to be practically irrelevant; recall our estimate of $\Delta \nu$ for a He–Ne laser following Eq. (5.11.13). It has also been confirmed that the Schawlow–Townes linewidth $\Delta \nu_{\text{ST}}$ underestimates the observed linewidths by factors $1 + \alpha^2 \approx 30$.

• The difference between (5.11.6) and (15.4.32) with $\alpha = 0$ is related to the fact that both amplitude and phase fluctuations contribute in general to the laser linewidth. In the near-threshold, linear regime of laser oscillation the amplitude and phase fluctuations contribute equally to the linewidth. In the above-threshold, nonlinear regime, however, the amplitude is stabilized by gain clamping and only phase fluctuations contribute to the linewidth. Thus, the linewidth (15.4.32) is half that given by (5.11.6), which is expressed in a form that includes contributions from both amplitude and phase fluctuations.

The phase excursion $\Delta \phi$ in (15.4.31) takes on all values between $-\infty$ and ∞, but of course it should only be necessary to work with phases between 0 and 2π. If $\Delta \phi$'s in a range small

---

[14]A different procedure for measuring α is described, for instance, in K.-G. Gan and J. E. Bowers, *IEEE Photonics Technology Letters* **16**, 1256 (2004).

compared to 2π make the dominant contribution to (15.4.31), then it is irrelevant as a practical matter whether the integration extends from $-\infty$ to ∞ or from 0 to 2π. In any event, a more rigorous approach to the "phase diffusion" responsible for the laser linewidth fully validates (15.4.31).

Our derivation of (15.4.35) assumes a Fabry-Pérot cavity, but similar expressions apply in other configurations. For a distributed feedback laser, for example, (15.4.35) can be applied provided that we use the appropriate output coupling loss for a_m. Except for diode lasers, the α parameter is typically very small, and the factor $1 + \alpha^2$ makes a very small contribution to the Schawlow–Townes linewidth, which is already extremely small for most (nonsemiconductor) lasers. As noted in Section 5.11, in most lasers the primary causes of the laser linewidth are "technical" factors such as mirror vibrations, temperature fluctuations, etc. The appearance of the α parameter in expressions such as (15.3.25) indicates that it plays a role not only in the laser linewidth but also in various other effects that depend on phase variations and fluctuations.[15]   •

The derivation leading to (15.4.35) ignored the amplification of spontaneous emission noise by the gain medium. It should be clear from our interpretation of the K and α parameters that the enhancement of the linewidth by *both* "excess spontaneous emission noise" and phase-intensity coupling is accounted for when (15.4.35) is replaced by

$$\Delta\nu = K(1 + \alpha^2)\,\Delta\nu'_{\text{ST}}. \tag{15.4.37}$$

Intensity Noise

Spontaneous emission causes laser radiation to have random fluctuations in *power* as well as phase. Thus, the output power of a single-mode laser is described by

$$P(t) = P_{\text{out}} + \Delta P(t). \tag{15.4.38}$$

P_{out} is the average output power and $\Delta P(t)$ is the fluctuation at time t of the power from its average value. As usual $\langle X \rangle$ denotes the average over the ensemble of all possible values of X. We will assume that the "random process" $\Delta P(t)$ is stationary and ergodic, so that $\langle \Delta P(t)\,\Delta P(t + \tau)\rangle$ is independent of t and ensemble averages are equal to time averages (Section 8.11).

A conventional measure of power fluctuations is the *relative intensity noise* (RIN)

$$\text{RIN}(\omega) = \frac{S_P(\omega)}{P_{\text{out}}^2}, \tag{15.4.39}$$

where the spectral power density $S_P(\omega)$ is defined in this context as the Fourier transform of the autocorrelation function $\langle \Delta P(t)\Delta P(t + \tau)\rangle$:

$$S_P(\omega) = \int_{-\infty}^{\infty} \langle \Delta P(t)\,\Delta P(t + \tau)\rangle e^{-i\omega\tau}\,d\tau. \tag{15.4.40}$$

---

[15]A review of early work leading to our present understanding of the α parameter is given by M. Osiński and J. Buus, *IEEE Journal of Quantum Electronics* **QE-23**, 9 (1987).

From the inverse Fourier transform of (15.4.40) we have[16]

$$\langle \Delta P(t)\, \Delta P(t+\tau)\rangle = \frac{1}{2\pi}\int_{-\infty}^{\infty} S_P(\omega)e^{i\omega\tau}\, d\omega, \tag{15.4.41}$$

and in particular the relative mean-square power fluctuation is

$$\frac{\langle \Delta P^2\rangle}{P_{\text{out}}^2} = \frac{\langle [P(t)-P_{\text{out}}]^2\rangle}{P_{\text{out}}^2} = \frac{\langle \Delta P(t)\,\Delta P(t)\rangle}{P_{\text{out}}^2} = \frac{1}{2\pi P_{\text{out}}^2}\int_{-\infty}^{\infty} S_P(\omega)\, d\omega$$

$$= \frac{1}{2\pi}\int_{-\infty}^{\infty} \text{RIN}(\omega)\, d\omega, \tag{15.4.42}$$

from which we can define a laser signal-to-noise ratio:

$$\text{SNR}_L = \sqrt{\frac{P_{\text{out}}^2}{\langle \Delta P^2\rangle}} = \frac{\sqrt{2\pi}}{\sqrt{\int_{-\infty}^{\infty}\text{RIN}(\omega)\, d\omega}}. \tag{15.4.43}$$

A model for $\text{RIN}(\omega)$ and its dependence on the device parameters of a diode laser starts from the coupled equations for the photon density Q, the carrier density N, and the phase ϕ:

$$\frac{dQ}{dt} = (G-\gamma)Q, \tag{15.4.44a}$$

$$\frac{dN}{dt} = \frac{1}{V}\frac{I}{e} - \frac{N}{\tau_R} - GQ, \tag{15.4.44b}$$

$$\frac{d\phi}{dt} = -\frac{\alpha}{2}(G-\gamma), \tag{15.4.44c}$$

where $G = v_g\sigma(N-N_{\text{tr}}) = C_s(N-N_{\text{tr}})$ is the gain factor. The first two equations are equivalent to Eqs. (15.3.3a) and (15.3.3b), respectively, with $\gamma = 1/\tau_p$, and the third equation is identical to (15.4.21b). Since Q is proportional to the output power P,

$$\text{RIN}(\omega) = \frac{S_Q(\omega)}{\bar{Q}^2}, \tag{15.4.45}$$

with

$$S_Q(\omega) = \int_{-\infty}^{\infty} \langle \Delta Q(t)\, \Delta Q(t+\tau)\rangle e^{-i\omega\tau}\, d\tau. \tag{15.4.46}$$

Suppose first that Q and N are perturbed from their steady-state values $\bar{Q}$ and $\bar{N}$ by δQ and δN, with $|\delta Q| \ll \bar{Q}$ and $|\delta N| \ll \bar{N}$. In a linear stability analysis of the coupled equations (15.4.44a) and (15.4.44b) we obtain equations for δQ and δN by keeping

---

[16]The fact that the spectral density and the autocorrelation function are Fourier transforms of each other is the Wiener–Khintchine theorem, which we invoked in Section 13.5 to relate the spectrum of a stationary field to its autocorrelation function.

only terms linear in the perturbations. The resulting equations are identical in form to Eqs. (15.3.5) with $I_m = 0$, since here we are assuming a constant current I:

$$\delta \dot{Q} = C_s \overline{Q} \delta N, \tag{15.4.47a}$$

$$\delta \dot{N} = -\left(C_s \overline{Q} + \frac{1}{\tau_R} \right) \delta N - \gamma \, \delta Q. \tag{15.4.47b}$$

Writing $\delta Q = \delta Q_0 e^{\Omega t}$ and $\delta N = \delta N_0 e^{\Omega t}$, we solve the resulting two linear algebraic equations for $\Omega = \Omega_R + i\Omega_I$ and find that the real part of Ω (Ω_R) is negative, i.e., that the steady-state solutions $\overline{Q}$ and $\overline{N}$ of Eqs. (15.4.44a) and (15.4.44b) are stable (Problem 15.7). The imaginary part of Ω (Ω_I) is the frequency of the oscillations that are damped at the exponential rate Ω_R; this frequency is found to be

$$\Omega_I = \sqrt{C_s \gamma \overline{Q}} = \sqrt{C_s \overline{Q}/\tau_p} \tag{15.4.48}$$

in the same approximation used in writing (15.3.11). In other words, from the discussion following (15.3.11), we can expect typical relaxation oscillation frequencies (Section 6.3) of diode lasers to be on the order of a few gigahertz.

Equations (15.4.44) do not include effects of spontaneous emission. As we saw in our discussion of the α parameter, spontaneous emission gives rise to mean-square fluctuations in the phase ϕ. It also results in mean-square fluctuations in Q and N. Because of these fluctuations relaxation oscillations are not exponentially damped but rather are "driven" to persist by random spontaneous emission events. Including effects of spontaneous emission noise in Eqs. (15.4.44), we can calculate $\langle \Delta Q(t) \, \Delta Q(t + \tau) \rangle$ and therefore its Fourier transform $S_Q(\omega)$ and the relative intensity noise, RIN(ω).[17] Figure 15.14 shows results of the calculation for a typical, strongly index-guided 1.3-μm InGaAs laser. RIN(ω) is strongest near the relaxation oscillation frequency (15.4.48) and decreases with increasing laser power, as would be expected if the intensity noise is due primarily to spontaneous emission rather than to "technical noise" or to noise associated with electron–hole recombination. These results are consistent with determinations of RIN(ω) made by measuring the laser power with a fast photodetector and $S_P(\omega)$ with an electronic spectrum analyzer. For diode laser powers of a few milliwatts the signal-to-noise ratio is typically found both theoretically and experimentally to be roughly 20 dB,[17] i.e.,

$$\frac{\sqrt{\langle \Delta P^2 \rangle}}{P_{\text{out}}} = \frac{1}{\text{SNR}_{\text{L}}} \sim \frac{1}{10^2}. \tag{15.4.49}$$

For multimode lasers the relative intensity noise defined in terms of the total power over all modes is found to vary roughly in the same way as the single-mode results shown in Fig. 15.14, although the power in some modes can exhibit much larger relative intensity noise than the total power.

---

[17]For details of the calculation see, for instance, G. P. Agrawal and N. K. Dutta, *Semiconductor Lasers*, 2nd ed., Van Nostrand Reinhold, New York, 1993, Chapter 6.

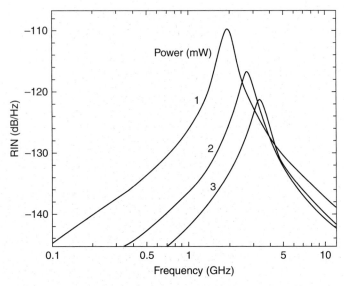

Figure 15.14   Relative intensity noise RIN(ω) calculated for three different power levels for a 1.3-μm InGaAs laser. (Adapted from Fig. 6.10 of G. P. Agrawal and N. K. Dutta, *Semiconductor Lasers*, 2nd ed., Van Nostrand Reinhold, New York, 1993.)

Intensity noise is an important consideration in optical communications because it is a source of fluctuations in the current

$$i(t) = \eta \frac{eP(t)}{h\nu} \qquad (15.4.50)$$

measured by a detector. Here η is the quantum efficiency of the detector and $P(t)$ is the incident power of light of frequency ν. In the simplest approximation, ignoring complications associated with photoelectron counting statistics, it follows from (15.4.50) that

$$\langle \Delta i^2 \rangle = \eta^2 \left(\frac{e}{h\nu}\right)^2 \langle \Delta P^2 \rangle = \eta^2 \left(\frac{e}{h\nu}\right)^2 \frac{P_{out}^2}{2\pi} \int_{-\infty}^{\infty} \text{RIN}(\omega)\, d\omega = \eta^2 \left(\frac{e}{h\nu}\right)^2 \frac{P_{out}^2}{\text{SNR}_L^2}, \quad (15.4.51)$$

or

$$\frac{\sqrt{\langle \Delta i^2 \rangle}}{\langle i \rangle} \sim \frac{1}{\text{SNL}_L}. \qquad (15.4.52)$$

In Section 15.6 we discuss the implications of intensity noise and other sources of noise for the design of optical fiber communication links.

15.5   INFORMATION AND NOISE

All modern methods of telecommunication involve electrical signals. In telephony, for example, sound waves modulate electric currents at the transmitter, and currents at the receiver generate modulated sound waves. The speed and accuracy with which the audio signal is reconstructed at the receiver depends on the physical properties of the transmitting and receiving systems as well as the transmission channel (e.g., copper wires). The transmitter, the channel, and the receiver all introduce noise that prevents the original signal from being communicated with perfect fidelity. In fiber-optic communications electrical signals are converted to optical signals that are transmitted by a fiber and then converted back to an electrical signal at the receiver. Noise appears in the electronics, the transmitting diode laser or LED, the fiber, and in the detection of the optical signal. Attenuation in the fiber limits the distance over which a signal can be transmitted without amplification along the way, which itself introduces noise. In this section we review some basic concepts relating to the transmission of information in the presence of noise, beginning with the quantification of information in terms of binary digits.

Imagine some process that always produces one of two possible outcomes, like heads or tails of a coin flip. We can assign to each outcome a label such as "yes" or "no," or "heads" or "tails," or simply one of the binary digits 0 and 1. Each outcome then provides one binary digit, or *bit* of information. Similarly, if there are four possible outcomes, two bits of information specify each outcome. To see this, imagine the four possible outcomes are represented by four slots in a line. Then one bit of information is required to specify whether the outcome belongs to, say, one of the left pair of slots or one of the right pair, and one more bit of information then fully specifies the outcome. In general, if each possible, equally likely outcome of some process is represented by a string of N 0's and 1's, the *information* associated with each outcome is defined as

$$I = \log_2 N \text{ bits.} \tag{15.5.1}$$

Information as such may or may not relate well to colloquial concepts of information. Some source may generate numbers or symbols that have no "meaning" or "value," but we nevertheless *define* its information content by (15.5.1). On the other hand, everyone would agree that *Hamlet* and *Macbeth* each contain a great deal of "information" in the everyday sense of the term, but if a source of some sort generates either the entire text of *Hamlet* or the entire text of *Macbeth* with equal likelihood, the reception of *Macbeth*, say, provides a mere 1 bit ($\log_2 2$) of information as defined by Eq. (15.5.1).

Consider, for example, the outdated but instructive example of a teletype system in which five binary digits are represented by holes punched in a transmission tape. We might label a punched hole by 1 and the absence of a hole by 0. There are $2^5 = 32$ possible sequences of 0's and 1's, enough to allow for the 26 letters of the English alphabet plus some control signals such as start and stop. Each sequence of five 0's and 1's (e.g., 00101) carries $I = \log_2 2^5 = 5$ bits of information.

In communications we are especially concerned with the *rate* of information transfer. If the transmission speed in our teletype example is 60 five-letter words per minute, then the information transmission rate is (60×5 letters/minute $\times$ 5 bits/letter), or 25 bits/s (bps).

Consider as another example a black-and-white television signal in which the information is contained in small picture elements (pixels) of different light intensity. If we assume the eye can distinguish 10 intensity gradations, and that each pixel has an intensity corresponding to one of these gradations, then each pixel contains $I = \log_2 10 = 3.32$ bits of information. Assuming 30 frames/s, 525 lines/frame, and 500 pixels/line, we have the information rate $(30)(525)(500)(3.32) = 2.61 \times 10^7$ bits/s $= 26.1$ Mbps.

- The characterization of information by bits and bytes [1 byte (B) = 8 bits] in computers and digital electronics ultimately derives, of course, from the (millions of) transistors acting as on/off (or 0/1) switches in integrated circuits. Examples: (i) a computer with a 32-bit CPU can access $2^{32} = 4.3 \times 10^9$ different memory addresses at a time; (ii) the text of this book uses about 2.2 MB of memory on the authors' computers; (iii) a CD stores about 700 MB, whereas a DVD game disc might store 8.5–50 GB; (iv) a 56k dial-up modem ("modem" being a contraction of "modulator-demodulator"), which modulates and demodulates analog signals from a telephone line to encode and decode digital information in terms of 0's and 1's, transfers data at a rate of 56 kilobits per second (Kbps); (v) a cable modem for Internet access might transfer data at a "bandwidth" of a megabit per second (Mbps) or more; (vi) modems that modulate and demodulate light transmitted by undersea optical fibers have bandwidths of about a gigabit per second (Gbps); and (vii) the human genome is estimated to contain about 800 MB of information.

In the American Standard Code for Information Interchange (ASCII) each character in a text is represented by a byte, a packet of 8 bits; this allows $2^8 = 256$ different characters to be represented. For example, a file consisting of the sentence "Lasers are based on the principle of stimulated emission of radiation." has a size of 71 bytes (59 letters, 11 spaces, and a period). Each character in ASCII is represented by a particular string of eight 0's and 1's. The ASCII space character, for example, corresponds in binary notation to $00100000 = 0(2^7) + (0)2^6 + 1(2^5) + 0(2^4) + 0(2^3) + 0(2^2) + 0(2^1) + 0(2^0) = 32$ in decimal notation.[18]

The terms "megabyte" and "gigabyte" often refer to $2^{20} = 1024^2$ and $2^{30} = 1024^3$ bytes and not to 1000^2 and 1000^3 bytes, respectively, in characterizing file sizes and random-access memory (RAM) as opposed to hard disk and flash memories. In network applications a megabyte per second (MBps) and a gigabyte per second (GBps) refer to the "decimal" definitions—10^6 and 10^9 bytes per second, respectively. •

In digital modulation systems information is conveyed by a sequence of pulses corresponding to 0's and 1's depending on whether their amplitudes are less than or greater than some particular value. A major advantage of digital systems, in addition to the fact that digitized information can be stored and retrieved electronically, is their accuracy, as indicated in Fig. 15.15. A signal in the form of a stream of 0's and 1's (Fig. 15.15*a*) is shown degraded by noise as in Fig. 15.15*b*. In Fig. 15.15*c*, however, the original signal is accurately recovered by letting any measurement (e.g., of current or voltage) above a certain "decision level" define a 1 and any measurement below that level a 0.

The first step, conceptually, in a pulse code modulation (PCM) scheme (Fig. 15.15) is to digitize an analog signal, which might, for instance, be a modulated current generated by voice, music, or video. By the nature of practical electronics it will have a finite bandwidth and a highest frequency. Digitization is done by *sampling* (or integrating) the continuous analog signal at different times. For illustrative purposes let us consider a "toy model" in which the analog signal to be digitized is a current that varies in time as in Fig. 15.16*a* with a noise level of 0.1 units, that is, the peak current shown would be measured as 0.8 ± 0.05 units (e.g., mA). We divide the range of currents into 8 bins (0.0–0.1, 0.1–0.2, etc.), which we label 0, 1, 2, 3, 4, 5, 6, 7, or, in binary notation, (000), (001), (010), (011), (100), (101), (110), (111). Each sampled current then represents $\log_2 8 = 3$ bits of information.

The question immediately arises: How well does the digitized signal represent the original analog signal? In other words, what *sampling rate* is required to accurately represent the continuous analog signal by a stream of bits? The answer is provided by the *Nyquist sampling theorem*: The sampling rate must be at least twice the largest frequency

---

[18]The Microsoft Windows calculator accessory in "scientific mode," for example, allows direct conversion between binary and decimal representations.

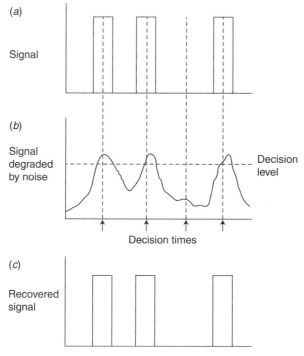

(a)

Signal

(b)

Signal
degraded
by noise

Decision
level

Decision times

(c)

Recovered
signal

Figure 15.15   A digital signal (a), degraded by noise (b), can be exactly reconstituted (c) even in the presence of substantial noise.

contained in the Fourier decomposition of the analog signal. Equivalently, the sampling rate must be at least twice the *bandwidth* of the analog signal.

•   This can be understood as follows. Let us represent the analog signal $A(t)$ of duration T as a Fourier series:

$$A(t) = \sum_j \left(a_j \cos \frac{2\pi jt}{T} + b_j \sin \frac{2\pi jt}{T} \right). \tag{15.5.2}$$

We assume for simplicity, but with no real loss of generality, that the lowest frequency appearing in the Fourier decomposition (15.5.2) is 0, i.e., that $A(t)$ has a "dc" component. If the largest frequency component is ν_{max}, corresponding to a maximum j given by $j_{max}/T = \nu_{max}$, or $j_{max} = \nu_{max}T$, then the Fourier decomposition (15.5.2) involves a set of j_{max} frequency components. Sampling $A(t)$ at any particular time provides one condition the $2j_{max}$ numbers a_j and b_j must satisfy; to obtain $2j_{max}$ linear equations, enough to solve for the $2j_{max}$ "unknowns" a_j and b_j, we must sample $A(t)$ at $2j_{max}$ different times, corresponding to a sampling rate $2j_{max}/T = 2\nu_{max}$. ν_{max} gives the range of frequencies in the Fourier decomposition of $A(t)$, i.e., it is the bandwidth of the analog signal. With this sampling rate the analog signal can be optimally synthesized from its digital representation using (15.5.2). In the simple example shown in Fig. 15.16 the sampling is done at equally spaced time intervals, which is the preferred sampling in practice.   •

Let us return now to our toy model in which each sampled value of the analog signal represents 3 bits of information. If the sampling rate is $2B$, where B is the bandwidth of the analog signal, and if the sampled values are transmitted by

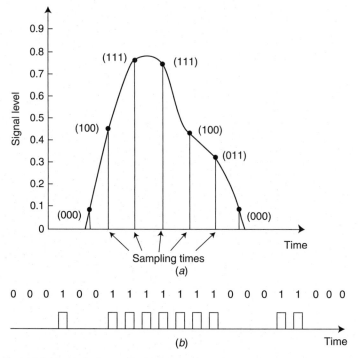

Figure 15.16 (*a*) Digitization of a continuous analog signal. The currents in this case are divided into 8 bins and the sampled values are labeled by the bin numbers in binary notation. (*b*) Transmission of the signal as a bit stream, with pulses shown as idealized square waves.

stringing them together in a bit stream as indicated in Fig. 15.16*b*, then the information transmission rate is (3 bits/sample)(2*B* samples/s) = 6*B* bps. In a more realistic scenario we might represent the current values by, say, $16 = 2^4$ bins and the bandwidth might be something like 3 kHz for a voice signal (roughly the bandwidth of sound waves in ordinary speech), in which case the transmission rate would be 4 bits × 6 kHz = 24 kbps.

In addition to allowing more accurate communication of information in the presence of noise (Fig. 15.15), pulse code modulation has the advantage that digital receivers can more precisely "decide" whether a pulse is present or not than an analog receiver can determine a particular signal level within a continuum; these are extremely important advantages. But there is, as always, a trade-off: More bandwidth is required. For an analog signal of bandwidth *B* we must sample at the rate 2*B*, and to transmit *N*-bit digitized signal levels we therefore require a bandwidth 2*BN*, *N* times the bandwidth of the analog signal. As noted at the beginning of this chapter, higher bandwidths are possible with higher carrier-wave frequencies, which explains the dominance of optical systems in contemporary telecommunications.

Consider now a generalization of our toy model in which an analog signal varies in strength from 0 to *s*, and that in the presence of noise of average strength *n* we digitize it into *s*/*n* "levels." Including the decision level 0, then, we have a total of 1 + *s*/*n* different levels for each sampled value of the signal. If, based on the Nyquist theorem, we sample $\mathcal{N} = 2BT$ values of the signal in a time *T*, where *B* is the signal bandwidth, then the total number of possible combinations of sampled values is

$$\left(1 + \frac{s}{n}\right)^{\mathcal{N}} = \left(1 + \frac{s}{n}\right)^{2BT}. \tag{15.5.3}$$

In terms of the signal and noise *powers*, which we take to be proportional to $S_p = s^2$ and $N_p = n^2$, respectively, this can be written

$$\left(1 + \frac{S_p^{1/2}}{N_p^{1/2}}\right)^{2BT} \cong \left(1 + \frac{S_p}{N_p}\right)^{BT}, \tag{15.5.4}$$

for $S_p \gg N_p$, which corresponds to

$$\log_2\left(1 + \frac{S_p}{N_p}\right)^{BT} = BT \log_2\left(1 + \frac{S_p}{N_p}\right) \quad \text{bits}, \tag{15.5.5}$$

or a bit rate

$$C = B \, \log_2\left(1 + \frac{S_p}{N_p}\right) \quad \text{bps}. \tag{15.5.6}$$

In information theory a rigorous derivation of this formula for the *capacity* of a noisy channel is obtained under the reasonable assumptions that the noise in the transmission channel has zero mean and is additive, Gaussian, and "white" in the sense that its power spectrum is approximately independent of frequency. C as given by (15.5.6) is shown to be the maximum rate at which information can be transmitted over a channel of bandwidth B when the average received signal and noise powers are, respectively, S_p and N_p.[19]

• Bandwidth in the case of analog systems refers to the range of frequencies contained in the Fourier decomposition of a signal and is expressed in hertz. This bandwidth might be defined, for instance, as the difference between the maximum and minimum frequencies, or the full-width-at-half-maximum width of the Fourier spectrum, etc. In digital systems and computer technology, however, "bandwidth" generally refers to a transmission rate, expressed in bits/second. Equation (15.5.6) provides a relation between these different meanings of bandwidth. As we have defined it, B (in hertz) is simply the largest frequency needed for an accurate Fourier synthesis of an analog signal, assuming that the Fourier spectrum has a dc component (i.e., a minimum frequency of zero). On the other hand, C is a bit rate, expressed in bits/second. Equation (15.5.6) implies that this "bandwidth" in bits/second is numerically larger than the bandwidth in hertz if the signal-to-noise ratio $S_p/N_p > 1$. For instance, a bandwidth of 10 MHz might in a particular system result in a bit rate of 50 Mbps, corresponding to "5 bits per hertz." •

The definition (15.5.1) assumes equally likely outcomes. More generally, if a given outcome or "symbol" in a finite stream of symbols occurs with probability P_i, the

---

[19]Equation (15.5.6) for the channel capacity is often called Shannon's formula, after Claude E. Shannon, who is generally regarded as the principal originator of information theory. One of the best semipopular expositions of information theory remains that by J. R. Pierce, *Symbols, Signals and Noise*, Harper & and Row, New York, 1961.

information obtained from an observation of it is defined as

$$I_i = \log_2 \frac{1}{P_i} \text{ bits.} \qquad (15.5.7)$$

If a particular outcome is certain (probability 1), then obviously no information is conveyed when it occurs: $I_i = 0$. It is useful to define the average information per symbol, or *entropy*,[20]

$$H = \sum_{i=1}^{N} P_i \log_2 \frac{1}{P_i} = - \sum_{i=1}^{N} P_i \log_2 P_i, \qquad (15.5.8)$$

where P_i is the probability of the ith outcome. H is a measure of "surprise": If a highly improbable ($P_i \ll 1$) event occurs, this occurrence is "surprising" and—by definition—carries a lot of information. If, on the other hand, an event is highly predictable, we do not get much information when we observe it. If a particular outcome is absolutely certain, its occurence provides no information whatsoever ($H = 0$) that was not known before its observation.

The relation (15.5.6) shows that the rate of transmission of information is limited by the bandwidth of the communication system. Its importance lies in part in a fundamental theorem of information theory to the effect that, if R is the rate at which a source generates entropy H, and C is the capacity of a noisy transmission channel, it is possible to transmit the source output over the channel *with arbitrarily small error* provided that $R < C$. This leads to *source compression* techniques for deliberately reducing the information content of the source (by, e.g., using a coarser signal digitization) in order to transmit information more reliably (and more quickly) over a noisy channel. It would take us well beyond our scope here to discuss these matters further, but we note that readers who have downloaded music files on the Internet have benefited from source compression algorithms. In any event, it must be emphasized that no transmission channel is free of noise. This might, for instance, be thermal (Johnson) noise in conducting wires or amplified spontaneous emission noise in fiber amplifiers. Moreover, while the formula (15.5.6) for the channel capacity assumes among other things that the noise is independent of bandwidth, it will be recalled that both Johnson noise and amplified spontaneous noise are in fact proportional to the bandwidth. Writing $N_p = N_0 B$, where N_0 is the noise power per unit bandwidth, we can replace (15.5.6) by

$$C = B \log_2 \left(1 + \frac{S_p}{N_0 B} \right) \text{ bps.} \qquad (15.5.9)$$

This shows that the channel capacity cannot be increased indefinitely by increasing the bandwidth; subject to the assumptions about the transmission channel under which (15.5.6) is valid, the maximum channel capacity for a given received signal power S_p

---

[20]This terminology arose from the similarity of (15.5.8) to entropy as defined in statistical mechanics. E. T. Jaynes and others have interpreted thermodynamic entropy as an example of a more general informational entropy.

and a noise power N_0 per unit bandwidth is

$$C_{\max} = \lim_{B \to \infty} \left[B \log_2 \left(1 + \frac{S_p}{N_0 B} \right) \right] = \log_2 \left[\lim_{B \to \infty} \left(1 + \frac{S_p}{N_0 B} \right)^B \right] = \log_2 e^{S_p/N_0}$$

$$= \frac{S_p}{N_0 \ln 2}. \tag{15.5.10}$$

Noise causes bit errors—a 1 pulse to be interpreted as a 0 or a 0 as a 1. As a general rule, an acceptable *bit error rate* (BER) in telecommunications is about 10^{-9}, that is, one out of every 10^9 received bits is in error.[21] Assuming the noise that causes these errors has a Gaussian distribution with zero mean, consistent with the central limit theorem for noise arising from many small, additive contributions, we write [cf. (13.15.26)]

$$P(i_N) = \frac{1}{\sigma \sqrt{2\pi}} e^{-i_N^2/2\sigma^2} \tag{15.5.11}$$

for the probability distribution of the noise current i_N at the receiver. $\sigma^2 = \langle i_N^2 \rangle$ is the mean-square noise current, which will have contributions from shot noise, intensity noise of the transmitting laser, Johnson noise, etc., but for now we leave it unspecified; a numerical example is given in the following section. We suppose that a 1 pulse is associated with a signal current i_S and a 0 pulse with a zero current. Then a 0 pulse is recorded as a 1 pulse if the noise current exceeds a threshold value of, say, $i_S/2$ for recording a 1. In this case the current $i_S + i_N$ exceeds the threshold for recording a 1 pulse even if $i_S = 0$. Similarly, the noise current can cause a 1 pulse to be recorded as a 0 if $i_N < -i_S/2$, in which case $i_S + i_N$ is below the threshold $i_S/2$ even if $i_S = 1$. The bit error rate is therefore half the probability that $i_N > i_S/2$ plus half the probability that $i_N < i_S/2$, since on average half the pulses are 1's and half are 0's:

$$\text{BER} = \frac{1}{2} \int_{i_S/2}^{\infty} P(i_N)\, di_N + \frac{1}{2} \int_{-\infty}^{i_S/2} P(i_N)\, di_N = \frac{1}{\sigma \sqrt{2\pi}} \int_{i_S/2}^{\infty} e^{-i_N^2/2\sigma^2}\, di_N, \tag{15.5.12}$$

since $P(i_N)$ is symmetric about $i_N = 0$. A simple change of variables allows us to write this as

$$\text{BER} = \frac{1}{\sqrt{\pi}} \int_{i_S/(2\sqrt{2}\sigma)}^{\infty} e^{-x^2}\, dx = \frac{1}{2} \text{erfc}\left(\frac{i_S}{2\sqrt{2}\sigma} \right) = \frac{1}{2} \text{erfc}\left(\frac{i_S}{2\sqrt{2\langle i_N^2 \rangle}} \right), \tag{15.5.13}$$

where erfc is the complementary error function defined by (3.10.8). In Fig. 15.17 we plot BER as a function of $i_S^2/\langle i_N^2 \rangle$; i_S^2 and $\langle i_N^2 \rangle$ are proportional to the received signal and noise powers, respectively. From this plot it follows that a bit error rate BER $\lesssim 10^{-9}$ requires a signal-to-noise power ratio $i_S^2/\langle i_N^2 \rangle \gtrsim 140$ (21.5 dB).

---

[21] In some applications much smaller bit error rates are required (and realized). Note that the bit error "rate" is actually a probability, not a rate as such; the more appropriate term "bit error *ratio*" is sometimes used.

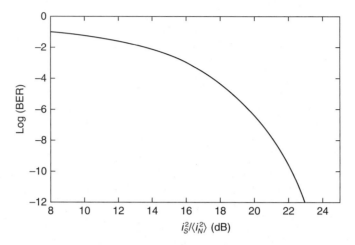

Figure 15.17   Bit error rate (BER) vs. $i_S^2/\langle i_N^2 \rangle$ (in dB) calculated from Eq. (15.5.13). (Adapted from A. Yariv, *Optical Electronics in Modern Communications*, 5th ed., Oxford University Press, New York, 1997, Fig. 10–20.)

In *error correction coding* a bit sequence is replaced by many different longer sequences with "redundancy," each long sequence being identified with the same original sequence. Bit errors in the original sequence are corrected when the original bit sequence is recovered from the longer sequences.[22] Error correction coding is widely used in computer technology. In fiber-optic communication it reduces the signal-to-noise ratio needed, for instance, to transmit a signal over some distance without amplification. Together with dispersion management and wavelength-division multiplexing (see below), it has dramatically increased the bit rates of long-haul fiber-optic communication systems.

15.6   OPTICAL COMMUNICATIONS

A voice signal with a bandwidth of 3 kHz can be transmitted over large distances as a modulated current on a telephone line. However, simultaneous transmission of, say, 200 such voice signals without interference requires a bandwidth $\sim 2 \times 3\,\text{kHz} \times 200 \sim 1\,\text{MHz}$. At these high frequencies resistance increases because of the "skin effect" (currents become tightly confined to the surface of a wire) and signal power is also dissipated as radiation. Coaxial and twisted-pair cabling allow bit rates of hundreds of megabits per second, but high-frequency ohmic losses limit transmission distances to $\sim 1\,\text{km}$ between repeaters. Wireless systems usually permit larger distances between repeaters (e.g., cell phone towers), but their bit rates are ultimately limited to $\sim 1\%$ of their microwave carrier frequencies, or a few gigabits per second. The capacity of a communication channel is often characterized by its *bit rate–distance product*, the product of the bit rate times the distance between repeaters. The primary advantage of fiber-optic communication lies in bandwidth and bit rate–distance product: The high carrier frequencies of optical waves, together with the large transmission bandwidths and

[22]See, for instance, Pierce, *op. cit.*, footnote 19.

low attenuation coefficients of fibers, allow data to be transmitted at very high bit rates over large distances. And because signals in optical communications are transmitted by light rather than electric currents, they are not distorted by external electromagnetic fields.

The components of a fiber-optic communication system are shown schematically in Fig. 15.18. A bit stream in the form of electric current pulses modulates the light from a diode laser to produce a stream of "0" and "1" optical pulses that are coupled into an optical fiber; the modulation can be done either directly by varying the injection current for the laser or by external (electro-optical) modulation of the laser output. After transmission through the fiber—which might also include a series of fiber amplifiers in the case of a "wide area network" (WAN) involving distances of hundreds of kilometers or more—the light pulses are coupled out of the fiber and onto a detector, whose output is demodulated to produce the electrical output signal. In this section we consider some basic power, noise, and dispersion characteristics of the generic fiber-optic communication system of Fig. 15.18.

Optical fibers were discussed in Chapter 8, where we noted that the glass fibers used in communications are characterized by attenuation factors of only ~ 0.2 dB/km at 1.55 μm; at this wavelength the power loss in propagating through a kilometer of fiber is only about 4%. Diode lasers emitting at this and other wavelengths where fibers have very small attenuation have been developed specifically as transmitters in fiber-optic communications (Section 15.2) and, fortuitously, erbium-doped fiber amplifiers (EDFAs) can amplify radiation over a wide range (~ 4 THz) at these wavelengths (Section 11.14) and serve as "all-optical" repeaters. For short transmission distances repeaters are not necessary, whereas for "long-haul" transmission they are essential. Fiber lengths and bit rates in optical communications are determined not only by attenuation and gain, but also by noise and by the dispersion effects discussed in Chapter 8. We will first consider the *power budget* of a fiber link.

The optical power P at the receiver is determined by the laser power P_L input to the fiber, attenuation in the fiber, and any amplifiers in the link:

$$P = P_L e^{-a_0 L} G A_s^{-1}, \tag{15.6.1}$$

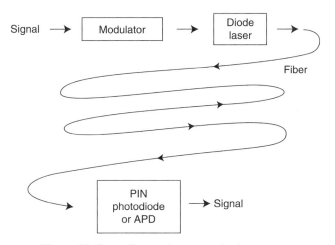

Figure 15.18 A fiber-optic communication system.

where a_0 is the fiber attenuation coefficient, L is the fiber length, G is the total power amplification factor due to any and all amplifiers, and we have included a factor A_s^{-1} (<1) to account for loss at points where two fibers are connected (spliced) or where light is coupled into or out of the fiber at the transmitter and receiver. Since decibels are simply additive, it is convenient to express powers in dBm (Section 11.14) and gains and losses in dB. Taking the logarithm (base 10) of both sides of (15.6.1) and then multiplying through by 10, we obtain

$$\mathcal{P} = \mathcal{P}_L - \mathcal{A} + \mathcal{G} - \mathcal{A}_s, \tag{15.6.2}$$

for the power at the receiver in dBm. $\mathcal{P}_L$ (dBm) is the laser power and $\mathcal{A} = 10\log(e^{-a_0 L})$, $\mathcal{G} = 10\log G$, and $\mathcal{A}_s = 10\log A_s$ are all expressed in dB. If $\mathcal{P}_{\min}$ (dBm) is the minimum power detectable at the receiver, or the *receiver sensitivity*, then the power budget for the fiber link is expressed as

$$\mathcal{P}_L - \mathcal{A} + \mathcal{G} - \mathcal{A}_s = \mathcal{P}_{\min} + \mathcal{M}, \tag{15.6.3}$$

where $\mathcal{M}$ (dB) is a "system margin" included to ensure that the power at the receiver will be large enough to exceed the smallest detectable power and to guard against any losses that might develop in system components over time. Equation (15.6.3) can be used to calculate, for instance, the maximum fiber length for $\mathcal{G} = 0$, that is, the maximum distance between repeaters, for a specified system margin $\mathcal{M}$ when $\mathcal{P}_L$, a_0, $\mathcal{A}_s$, and $\mathcal{P}_{\min}$ are known.

Such a calculation depends not only on the attenuation and gain properties of the fiber link but also on the bit rate. The dependence on the bit rate comes from the dependence of the minimum detectable power $\mathcal{P}_{\min}$ on the bandwidth B. As discussed in Section 12.7, $\mathcal{P}_{\min}$ is approximately proportional to either B or $\sqrt{B}$, depending on the dominant source of noise at the receiver. In any case, the fact that $\mathcal{P}_{\min}$ increases with bandwidth implies from (15.6.3) that, other things being equal, the transmitted laser power $\mathcal{P}_L$ or the gain $\mathcal{G}$ in the fiber link must increase with bandwidth and therefore with bit rate in order to maintain a specified system margin $\mathcal{M}$.

• There is a "quantum limit" to $\mathcal{P}_{\min}$ that depends on the acceptable bit error rate. If the average number of photons counted in a "1" light pulse is $\bar{n}$, the probability of detecting n photons when a "1" pulse is incident on the detector is

$$P_n = \frac{(\bar{n})^n e^{-\bar{n}}}{n!}. \tag{15.6.4}$$

We assume there are no photons in a "0" pulse. The probability that a 1 pulse is incorrectly identified as a 0 pulse, i.e., the probability that no photons are detected when a 1 pulse is incident on the detector, is

$$P_{n=0} = \frac{(\bar{n})^0 e^{-\bar{n}}}{0!} = e^{-\bar{n}}. \tag{15.6.5}$$

Assuming an equal number of 0 and 1 pulses, therefore, the bit error rate is

$$\text{BER} = \tfrac{1}{2} e^{-\bar{n}}, \tag{15.6.6}$$

and, if we require a bit error rate of 10^{-9} (Section 15.5), the minimum required value $\bar{n}_{min}$ of $\bar{n}$ is given by $\frac{1}{2}e^{-\bar{n}_{min}} = 10^{-9}$, or $\bar{n}_{min} = 20$. Assuming again that there are no photons in 0 light pulses, and that 0 and 1 pulses are equally likely on average, we require an average of $n_{min} = \frac{1}{2}\bar{n}_{min} = 10$ photons per pulse to realize a bit error rate of 10^{-9}. For light of frequency ν this implies a minimum average optical power $\mathcal{P}^q_{min}$ equal to $10h\nu$ times the bit rate. For light of wavelength 1.55 μm and a bit rate of 100 Mbs, therefore, $\mathcal{P}^q_{min} = 1.3 \times 10^{-10}$ W, or -68.9 dBm. More generally, for a bit rate b_r (bps) and a minimum of n_{min} photons per bit on average needed to achieve a specified bit error rate, the average power at the receiver must exceed the quantum limit

$$\mathcal{P}^q_{min} = n_{min}h\nu b_r. \tag{15.6.7}$$

The minimum detectable power $\mathcal{P}_{min}$ of an actual receiver is usually much larger than the quantum limit for bit error rates of practical interest. A PIN receiver, for example, might have a minimum detectable power corresponding to 3000 photons at 1.55 μm and 100 Mbs, or $\mathcal{P}_{min} = 3.8 \times 10^{-8}$ W $= -44.1$ dBm.   •

Suppose, for example, that we have a single-mode fiber into which is coupled $\mathcal{P}_L = 1$ mW (0 dBm) of 1.55-μm diode laser radiation, and that along the fiber there are 10 splices, each with an attenuation $\mathcal{A}_s = 0.1$ dB, as is typical. Suppose further that the receiver has a minimum detectable power $\mathcal{P}_{min} = 10^{-8}$ W (-50 dBm) for the particular bit rate of interest and that a (typical) system margin $\mathcal{M} = 6$ dB is imposed. What is the fiber length L for which we can meet the power budget (15.6.3) without any amplifiers? Equation (15.6.3) with $\mathcal{G} = 0$ (no amplifiers) implies

$$\mathcal{A} = 10\log\left(e^{a_0 L}\right) = 4.34a_0 L = \mathcal{P}_L - \mathcal{A}_s - \mathcal{P}_{min} - \mathcal{M}, \tag{15.6.8}$$

or

$$L = \frac{1}{4.34a_0}[0 - 10 \times 0.1 - (-50) - 6]. \tag{15.6.9}$$

Since $a_0 = 4.96 \times 10^{-2}$ km^{-1} for $\mathcal{A} = 0.2$ dB/km, we calculate $L = 200$ km as the maximum fiber length for which the system margin can be met without amplifiers or regenerators.

Such estimates do not address the question of whether the BER is acceptable. Figure 15.17 gives the BER as a function of the signal-to-noise ratio $i_S^2/\langle i_N^2 \rangle$. If we require a BER of 10^{-10}, for example, we must have $i_S^2/\langle i_N^2 \rangle > 22$ dB. The signal current $i_S = \eta e(P/h\nu)$, where P is the optical power at the receiver and η is the probability that an incident photon creates an electron–hole pair in a PIN or APD detector (Section 12.7). $P = P_L e^{-a_0 L}$, where L is the length of fiber traversed from the laser transmitter to the receiver, so we can write the signal current and its square at the receiver as

$$i_S = \frac{\eta e P_L}{h\nu}e^{-a_0 L}, \qquad i_S^2 = \left(\frac{\eta e P_L}{h\nu}\right)^2 e^{-2a_0 L}. \tag{15.6.10}$$

We are assuming for simplification here that splicing and connection losses are negligible.

The noise current has several contributions. In Section 12.7 we discussed signal-current shot noise, dark-current shot noise, and Johnson noise. Their total contribution to the mean-square current noise is [Eqs. (12.7.3) and (12.7.5)]

$$\langle i_N^2 \rangle = \Delta i_{\text{rms}}^2 = 2e(i_S + i_D)B + \frac{4k_BT}{R_L}B$$

$$= 2e\left(\frac{\eta e P_L}{h\nu}\right)Be^{-a_0L} + 2ei_DB + \frac{4k_BT}{R_L}B \quad \text{(shot noise and Johnson noise).} \quad (15.6.11)$$

In (15.6.10) and (15.6.11) P_L is the average laser power. As discussed in Section 15.4, the laser intensity undergoes fluctuations due primarily to spontaneous emission in the laser medium, and the intensity noise results in photocurrent noise when the laser radiation is incident on a detector. For the mean-square current noise in a bandwidth B we have [Eq. (15.4.51)]

$$\langle i_N^2 \rangle = \left(\frac{\eta e P_L e^{-a_0L}}{h\nu}\right)^2 (\text{RIN})B \quad \text{(laser intensity noise).} \quad (15.6.12)$$

Here RIN is taken to be the relative intensity noise at the center of the frequency band of width B. Assuming that dark-current shot noise is negligible compared to the other sources of current fluctuations, we take the total mean-square current noise to be

$$\langle i_N^2 \rangle = \left(\frac{2\eta e^2 P_L}{h\nu}\right)Be^{-a_0L} + \frac{4k_BT}{R_L}B + \left(\frac{\eta e P_L e^{-a_0L}}{h\nu}\right)^2(\text{RIN})B, \quad (15.6.13)$$

which implies the signal-to-noise ratio

$$\frac{i_S^2}{\langle i_N^2 \rangle} = \frac{(\eta e P_L/h\nu)^2 e^{-2a_0L}}{(2\eta e^2 P_L/h\nu)Be^{-a_0L} + (4k_BT/R_L)B + (\eta e P_L e^{-a_0L}/h\nu)^2(\text{RIN})B}. \quad (15.6.14)$$

If we require a BER of 10^{-10} or less, this ratio must exceed 22 dB, or 158. This condition depends on the carrier frequency, average power, and intensity noise of the transmitting laser, the detector efficiency, load resistance, and bandwidth, and the fiber attenuation and length. It can be used to determine, for example, the length of fiber needed to realize a specified bit error rate without repeaters.

It is seen from (15.6.14) that the signal-to-noise ratio at the receiver decreases with increasing fiber length. In other words, in order to realize a particular bit error rate with the laser power and other parameters fixed, the fiber length must be kept below a certain value in the absence of repeaters; subject to the assumptions and approximations used in obtaining (15.6.14), this value determines the spacing of the repeaters (fiber amplifiers) between the transmitter and the receiver. Equation (15.6.14) shows that at sufficiently large distances the dominant source of noise is thermal:

$$\frac{i_S^2}{\langle i_N^2 \rangle} \cong \left(\frac{\eta e P_L}{h\nu}\right)^2 \frac{R_L}{4k_BTB}e^{-2a_0L}. \quad (15.6.15)$$

For a rough estimate let us assume the (somewhat) typical values $P_L = 1$ mW, $\eta = 0.5$, $R_L = 50\,\Omega$, $B = 200$ MHz (corresponding to a bit rate of, say, 400 Mbps), and $a_0 = 5 \times 10^{-2}$ km^{-1} at the wavelength $\lambda = c/\nu = 1.55$ μm. Recall from Section 12.7 that the effect of noise due to electronic amplification of the current at the receiver is to replace the physical temperature T in (15.6.15) by an effective value $T_{\text{eff}} > T$; we will take $T_{\text{eff}} = 3(293\text{K})$, consistent with the remarks in Section 12.7. With these assumptions, the maximum fiber length L that allows a signal-to-noise ratio $i_S^2/\langle i_N^2 \rangle > 158$, and therefore a bit error rate $< 10^{-10}$, is 95 km.

The formulas for power budgeting and signal-to-noise ratio show that the performance of a fiber-optic communication system should improve with increasing laser power P_L. However, laser powers can only be increased so far before nonlinear-optical effects in the fiber, in particular stimulated Brillouin scattering, begin to have deleterious effects. The threshold power for the onset of SBS at wavelengths near 1.5 μm in fibers can be as small as a few milliwatts, although it is considerably larger for the relatively large spectral widths typical of pulses used in communications. At much larger powers stimulated Raman scattering (SRS) can similarly cause loss of laser power in a fiber as radiation at other wavelengths is generated.

A basic and extremely important consideration in fiber-optic communication is fiber dispersion (Chapter 8). There is first of all material dispersion, the dependence of the fiber refractive index on frequency; this results in group velocity dispersion and pulse broadening that, unless reduced in some way, can seriously limit bit rates (Section 8.4). Intermodal dispersion can also cause pulse broadening and overlap to a degree that increases with the length and numerical aperture of a multimode fiber but, as noted in Section 8.6, it is absent in single-mode fibers. Multimode fibers are used only for short-distance communications, typically within or between buildings. Whereas LEDs can be used as transmitters in such applications, their nondirectional emission results in inefficient coupling of radiation into single-mode fibers; diode lasers are therefore required for long-haul communications.

In addition to material dispersion, single-mode fibers also exhibit polarization mode dispersion arising from the birefringence of an optical fiber: Two orthogonal polarizations have different refractive indices and therefore different group velocities (Section 8.7). Because of the unpredictable nature of the output polarization, it is important that the optical elements of a fiber-optic communication system be insensitive to polarization. Even if the light injected into a fiber is strongly polarized, temperature variations in the fiber can result in effectively random polarization of the light exiting the fiber, and different frequency components will generally have different polarizations. These effects can also cause pulses to overlap, but they are usually weak except at very high bit rates ($\gtrsim 10$ Gbps), and telecommunication fiber cables are now manufactured such that polarization mode dispersion is fairly small even at these rates.

Material dispersion is of course most pronounced for the short pulses used in high-bit-rate systems. In Section 8.4 we estimated the spreading in time of a pulse as a function of the propagation distance and the dispersion parameter of the fiber [cf. Eqs. (8.4.21)–(8.4.23)]. We found, for example, that for a standard telecom fiber an initial 10-ps pulse at a 1.55-μm carrier wavelength will have a duration approximately three times larger after propagating 15 km in the fiber. The maximum bit rate for a 100-km distance is similarly calculated to be $\sim 1/(200\text{ ps}) = 5$ Gbps, compared to a theoretical bit rate of 100 Gbps without group velocity dispersion (Problem 15.11). Amplifiers overcome the problem of attenuation in long-distance fiber-optic communications, but they do not correct for group velocity dispersion.

As discussed in Section 8.4, the effect of group velocity dispersion is reduced (or "compensated") when there are two propagation paths with opposite dispersion parameters (D). We showed how GVD compensation for ultrashort pulse generation is done with prisms and gratings. In long-haul and high-bit-rate fiber-optic communication systems dispersion management is essential. It is accomplished by introducing dispersion compensation modules (DCM) consisting, for example, of a fiber Bragg grating or a loop of "dispersion-shifted fiber" with a suitable dispersion parameter and length. Insertion losses associated with the DCMs must be included in detailed power budget calculations. Standard telecom fibers have minimum dispersion at 1.3 μm and minimum attenuation at 1.55 μm. InGaAsP diode lasers can operate at either wavelength, the latter used most often for long-haul communications of hundreds of kilometers or more. Diode laser transmitters presently operate at a few milliwatts of output power and can transmit bit streams at about 40 Gbps.

The development in the late 1980s of erbium-doped fiber amplifiers (EDFAs) for amplification of wavelengths near 1.55 μm was crucial to the advance of long-haul fiber-optic communications. Direct optical amplification can, of course, also be done with current-driven diode laser amplifers, but EDFAs have several major advantages over this method as well as techniques involving electronic amplifiers. Being fibers, they are easily connected to the transmission fiber, and are pumped by diode lasers very much like the diode laser transmitters. They have a large amplification bandwidth (∼4 THz), their response is approximately linear, they have better noise characteristics than diode laser amplifiers, and the amplification is polarization-independent. Their importance for long-haul optical communication is comparable to that of vacuum-tube amplifiers for the development of long-distance telephone communication around 1915.[23] As noted in Section 8.7, the low attenuation of telecom fibers allows distances ∼70 km or more between repeaters in undersea fiber cables.

Different signals can, of course, be transmitted simultaneously over separate fibers in a cable. *Time-division multiplexing* (TDM) allows transmission of different signals over a *single* fiber. A time-division multiplexer "interleaves" the 0's and 1's of N different bit streams so that, if the time separation between bits for each stream is T, the new stream has a bit separation of T/N (Fig. 15.19) and therefore the channel capacity is increased from $1/T$ to N/T. So-called T1 and T3 lines for telephone and Internet connections, for example, have standardized data rates of 1.544 and 44.736 Mbps, respectively, and allow correspondingly $N = 24$ and 672 separate voice channels, for example, if we assume

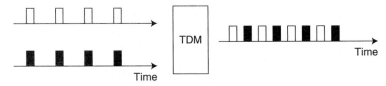

Figure 15.19   Time-division multiplexing illustrated for an example with two separate bit streams.

[23]An important difference between long-distance telephone service before and after fiber-optic communications is in the cost per distance. The largest profits of telephone companies used to come from long-distance calls, whereas fiber-optic systems—research and development on which was funded in large part from those profits—allow telephone service at a cost that is practically independent of distance.

64 kbps per channel. The standardized optical network rates designated OC-192 and OC-768 correspond, respectively, to data rates of 10 and 40 Gbps, the latter allowing over 500,000 telephone conversations to take place simultaneously over a single line. TDM bit rates are limited by practical electronic response times to a few tens of gigabits per second.

Fiber channel capacity is greatly increased by *wavelength-division multiplexing* (WDM). In this approach N different wavelengths or "channels" from N diode lasers, or perhaps a tunable laser that can be rapidly switched among N different output wavelengths, are separately modulated to produce N different bit streams that propagate together along the fiber. At the end of the transmission channel the different wavelength components are separated and each enters one of N different receivers. This allows tens of bit streams, or channels, each at ~ 10 Gbps, to be transmitted by a single fiber. The standardized central channel frequencies agreed upon by the International Telecommunication Union are integral multiples of 50 GHz, so that EDFAs with bandwidths of 4 THz allow the simultaneous amplification of 4 THz$/$50 GHz $= 80$ different channels. For this purpose the difference between the central channel frequencies must be very small—on the order of 50 GHz—and in this case the technique is referred to as *dense wavelength-division multiplexing* (DWDM). Because of the small channel separation, DWDM generally requires transmitting lasers with temperature control in order to keep the channel frequencies approximately constant. If transmission of only a few channels is sufficient, the channel separation can be much larger and no temperature control is required; in this case the multiplexing is called *coarse wavelength-division multiplexing* (CWDM). The bit rates possible with DWDM are truly enormous—in the range of terabits$/$second (Tbps). This compares to the bit rates of a few gigabits per second cited earlier for wireless, microwave communications. In other words, the large bandwidths of optical fibers, and in particular the large-gain bandwidths of erbium-doped fiber amplifiers, allow bit rates ~ 1000 times greater than what is possible with wireless communication systems.

- The first transatlantic fiber-optic cable, installed in 1988, operated at 1.3 μm at a bit rate of 2×280 Mbps on two single-mode fibers with attenuation of 0.4 dB$/$km. This was followed in 1991 by a transatlantic cable with twice the bit rate on fiber with an attenuation of 0.2 dB$/$km at 1.55 μm and repeater spacings of about 120 km. Several other transoceanic fiber cables were installed in the following few years. In 1995 EDFAs were used for the first time in transatlantic cables; they were spaced by 45 km between the United States and France and by 75 km in the section between France and the United Kingdom. In 1999 a transatlantic fiber-optic cable with a bit rate of 5 Gbps was upgraded to 20 Gbps by adding WDM components on both sides of the Atlantic. The oceans are now laced with dozens of fiber-optic cables; maps of cables connecting the continents may be found on the Web. DWDM and other developments have made Tbps transmission rates practical. There is an ever-increasing demand for bandwidth, and more cables are presently being installed or planned with bit rates ~ 10 Tbps.

Cable television companies began replacing copper cables with fibers in the 1990s, enabling them also to extend services to include Internet access. Long-distance telephone companies have been able to avoid the cost of laying new underground fiber by using WDM to increase bandwidths. In fact the cost of laying new fiber is the main reason that network connections to homes still largely employ copper wires rather than optical fibers.   •

In this brief introduction we have only considered the most important and common type of fiber-optic communication system and have not discussed other systems or components that are being actively researched and developed. The latter include *coherent optical communication systems* in which the frequency or phase of light is modulated rather than the intensity, and detection at the receiver involves homodyne or heterodyne techniques. In such systems the phase noise and linewidth of the transmitting laser take on added importance in determining the receiver sensitivity.

In *soliton communication systems* the light pulses propagating in a fiber are the shape- and duration-preserving solitons discussed in Section 9.9 in the context of self-induced transparency and in Section 11.14 in the case of pulse propagation in fibers. In a fiber a temporal soliton can result from the combined effects of group velocity dispersion and self-phase modulation, and we showed that the required powers for such solition formation are rather modest (Problem 11.13). The promise of soliton communication lies mainly in the fact that pulse spreading does not occur and dispersion management is unnecessary.

The design of a fiber-optic communication system obviously involves many different factors and components. In addressing only some of the most basic optical physics in the current technology we have ignored consideration of the coupling of the laser pulses into the fiber; the electronic components used for sampling, digitization, and decision levels for the input electrical signal to be transmitted; internal or external modulation of the laser transmitter; detection circuitry including electronic amplifiers; noise associated with amplified spontaneous emission in any erbium-doped fiber amplifiers in the system; and optical instrumentation for wavelength multiplexing and demultiplexing. Nor have we discussed the possible use and advantages of transmission in photonic crystal fibers, for instance. The reader can undoubtedly think of other important elements and possibilities involved in the design of a fiber-optic communication system.[24]

PROBLEMS

15.1. (a) Derive the density of states (15.2.10).

(b) Verify Eq. (15.2.17) for the gain coefficient.

(c) Consider a doped semiconductor with N_c conduction-band electrons per unit volume and N_v valence-band holes per unit volume at $T = 0$. Show that the quasi-Fermi levels are given by

$$E_{fc} = E_c + (3\pi^2)^{2/3} \frac{\hbar^2}{2m_c} N_c^{2/3} \quad \text{and} \quad E_{fv} = E_v - (3\pi^2)^{2/3} \frac{\hbar^2}{2m_v} N_v^{2/3}.$$

15.2. Using the results quoted in the text for the quasi-Fermi levels E_{fc} and E_{fv} corresponding to curve D of Fig. 15.3, plot the Fermi–Dirac occupancy factor as a function of energy $E' = E_2 - E_c$ in electron volts for conduction-band electrons. Also plot the occupancy factor for valence-band holes as a function of energy $E'' = E_v - E_1$. [Note: The calculation of $E_c - E_{fc}$ and $E_v - E_{fv}$ using Eqs. (15.2.20) can be done by simple iteration. For example, Eq. (15.2.20a)

---

[24]See, for instance, E. Desurvire, *Erbium-Doped Fiber Amplifiers. Principles and Applications*, Wiley, Hoboken, NJ, 2002; G. P. Agrawal, *Lightwave Technology: Components and Devices*, Wiley, New York, 2004; G. P. Agrawal, *Lightwave Technology: Telecommunication Systems*, Wiley, New York, 2005.

can be cast in the form

$$x = \frac{N_0}{N + \bar{N}_c} \int_0^\infty \frac{y^{1/2} dy}{e^y + 1/x},$$

with $x = \exp[(E_c - E_{fc})/k_B T]$ and $N_0 = \sqrt{2} m_c^{3/2} (k_B T)^{3/2} / \pi^2 \hbar^3$, and this equation can be solved iteratively for x starting from some initial "guess" on the right-hand side.

15.3. Based on the "gain clamping" of a laser in steady-state oscillation, justify the expression (15.2.26) for the output power of a diode laser.

15.4. Consider a diode laser with a frequency-dependent refractive index n in the active region of length L. Show that the longitudinal mode spacing is

$$\Delta \nu = \frac{c}{2L[n + \nu \, dn/d\nu]} = \frac{v_g}{2L},$$

where v_g is the group velocity and n and $dn/d\nu$ are evaluated at the laser frequency. [Hint: Start from the condition $\exp(2ikL) = 1$, or $2nL = m\lambda$, where m is a very large integer. This implies $\lambda + m \, \Delta\lambda = 2L \, \Delta n$ with $\Delta\lambda$ and Δn the changes in the mode wavelength and the refractive index, respectively, when m changes by 1.] Under what conditions does this equation apply to *any* laser? Show that, in terms of wavelength $\lambda = c/\nu$, the mode spacing is

$$\Delta\lambda = \frac{\lambda^2/2L}{n - \lambda \, dn/d\lambda}.$$

15.5. **(a)** Derive Eqs. (15.4.13) and (15.4.16).

(b) The phase change (15.4.24) was obtained using the expression $\Delta q_i = 1 + 2\sqrt{q} \cos \theta_i$ for the change in the photon number due to a single spontaneous emission event. The average change in the photon number is $\langle \Delta q_i \rangle = 1$, since $\langle \cos \theta_i \rangle = 0$, but the mean square, $\langle \Delta q_i^2 \rangle$, is large, since $q \gg 1$. Explain how $\langle \Delta q_i^2 \rangle$ can be much greater than 1 even though it results from the spontaneous emission of a single photon.

(c) Derive Eq. (15.4.34).

(d) Why is the linewidth of a diode laser so large compared to other lasers?

(e) In (15.4.17) and (15.4.18) Δn_I characterizes gain and loss *in the active medium*, whereas in (15.4.34) the loss includes that due to *output coupling from the facets* as well as internal cavity loss. Does this apparent inconsistency affect the validity of (15.4.37)?

15.6. **(a)** Derive Eq. (15.4.3) for the mutual coherence function.

(b) Verify that (15.4.5) is equivalent to the Schawlow–Townes linewidth defined by expression (5.11.13).

(c) Justify expression (15.4.29) for the mutual coherence function.

15.7. **(a)** Show that the steady-state solutions of Eqs. (15.4.44a) and (15.4.44b) are stable against small perturbations.

(b) Show that the angular frequency of relaxation oscillations is given approximately by (15.4.48).

15.8. A pulse code modulation system has a root-mean-square voltage noise equal to 5% of the threshold voltage above which a pulse is identified as a "1." Estimate the bit error rate.

15.9. Estimate the bit error rate for a signal-to-noise ratio $i_S^2/\langle i_N^2 \rangle$ equal to (i) 10 dB, (ii) 15 dB, (iii) 20 dB, and (iv) 25 dB.

15.10. Consider a long communications channel, such that the signal power S_p at the receiver has been so attenuated that it is much smaller than the receiver noise power N_p, which we assume to be due entirely to thermal fluctuations at temperature T. Show that the channel capacity $C \cong 1.44 S_p / k_B T$.

15.11. Consider a telecom fiber with group velocity dispersion characterized by $\beta = 22$ ps^2/km at 1550 nm [Eq. (8.4.2)]. A 1550-nm Gaussian pulse with initial duration $\tau_p = 10$ ps is injected into this fiber. Calculate the pulse duration after a propagation distance of (i) 1 km, (ii) 10 km, and (iii) 100 km.

16 NUMERICAL METHODS FOR DIFFERENTIAL EQUATIONS

The main text in Chapters 1–15 deals almost everywhere either with analytic solutions to oversimplified equations or the derivation of equations that are more realistic but do not have analytic solutions. In this chapter we address the need to use numerical methods to attack almost all of the problems that arise in practical situations related to laser physics. For some readers the present chapter will seem unnecessary in view of the availability of various packaged numerical equation solvers in common use. Nevertheless a brief discussion of the solution of differential equations of the type appearing frequently in laser physics seems not altogether superfluous. Packaged programs are not always the most efficient way for beginners to obtain rapid and accurate solutions to ordinary and partial differential equations, especially multi-dimensional versions of them. Together with freely available compilers for FORTRAN and other languages, the programs below can be used by anyone with a personal computer to solve a rather wide variety of equations encountered in laser physics (and other areas). In the three sections below, called Appendices, we list explicit FORTRAN programs, the programming language with which we happen to be most familiar. The choice of a particular language is becoming less and less restrictive with the increasing availability of free software for conversion between FORTRAN and C++ and other languages.

16.A FORTRAN PROGRAM FOR ORDINARY DIFFERENTIAL EQUATIONS

The Runge–Kutta algorithm is an easily implemented and frequently used method for the numerical integration of ordinary differential equations. For the first-order differential equation

$$\frac{dy}{dx} = f(x, y), \tag{16.A.1}$$

the fourth-order Runge–Kutta algorithm for $y(x + h)$ is

$$y(x + h) = y(x) + \tfrac{1}{6}(k_1 + 2k_2 + 2k_3 + k_4), \tag{16.A.2}$$

Laser Physics. By Peter W. Milonni and Joseph H. Eberly
Copyright © 2010 John Wiley & Sons, Inc.

where

$$k_1 = hf(x, y),$$

$$k_2 = hf\left(x + \frac{h}{2}, y + \frac{k_1}{2}\right),$$

$$k_3 = hf\left(x + \frac{h}{2}, y + \frac{k_2}{2}\right),$$

(16.A.3)

$$k_4 = hf(x + h, y + k_3).$$

The form of (16.A.2) and (16.A.3) is such as to duplicate the Taylor series for $y(x + h)$ up to fourth order in the step size h. The method is easily extended to systems of equations, as in the example below. The Runge–Kutta method is described in detail in many textbooks on mathematical methods and numerical analysis. For the reader's convenience we list here a FORTRAN program for the Runge–Kutta integration of the two coupled equations (6.4.3):

```
        program main
        dimension y(2),dy(2),w(2,5)
        t=0.
        y(1)  = 1.e-3
        y(2)  = 2.0
        dt = .01
        nstep = 1000
        do 1 n = 1, nstep
        call rung(2,y,dy,t,dt,w)
   1    continue
        print 100,t,y(1),y(2)
 100    format(3e16.7)
        stop
        end
        subroutine deriv (t,y,dy)
        dimension y(2),dy(2)
        dy(1)=(y(2)-1.)*y(1)
        dy(2)=-y(1)*y(2)
        return
        end
        subroutine rung(n,y,dy,t,dt,w)
c       n is the number of equations
        dimension y(n),dy(n),w(n,5)
        do 10 j= 1,n
  10    w(j,1) = y(j)
        call deriv(t,y,dy)
        do 20 j=1,n
```

```
20      w(j,2)=dy(j)*dt
        z=t+.5*dt
        do 40 i=2,3
        do 30 j=1,n
30      y(j)=w(j,1)+w(j,i)/2.
        call deriv(z,y,dy)
        do 40 j= 1,n
40      w(j,i+1)=dy(j)*dt
        z=t+dt
        do 50 j=1,n
50      y(j)=w(j,1)+w(j,4)
        call deriv(z,y,dy)
        do 60 j= 1,n
60      w(j,5)=dy(j)*dt
        do 70 j=1,n
70      y(j)=w(j,1)+(w(j,2)+2.*(w(j,3)+w(j,4))+w(j,5))/6.
        t=t+dt
        return
        end
```

The dependent variables are stored in the array y. In this case [Eqs. (6.4.3)] $y(1) = x$ and $y(2) = y$. t is the independent variable (τ), and the step size dt [denoted by h in (16.A.3)] is taken to be 0.01. In general dt should be taken small enough to give an accurate solution of the equations, but not so small that the program is unnecessarily time-consuming. The accuracy of the solution can always be checked by halving the step size and noting whether there is a significant change in the computed solution.

nstep is the number of integration steps. Each call to rung moves time forward by dt. In the way our program is set up, therefore, the final values of y after the last call to rung are $y(1) = x(t_{max})$ and $y(2) = y(t_{max})$, where $t_{max} = $ nstep*dt. In general, the intermediate values of y can be stored in arrays for plotting purposes.

Subroutine deriv simply defines the derivatives. In our example $dy(1) = dx/d\tau$ and $dy(2) = dy/d\tau$. Subroutine rung does a fourth-order Runge–Kutta integration, using the derivatives defined in deriv. n is the number of (first-order) simultaneous differential equations to be solved; in our example n = 2. w is a work array that must be dimensioned n by 5. Only the main and deriv routines in our example depend on the specific problem, the rung subroutine being a "canned" routine, is usable as it stands in every problem. It may be worth noting that rung can also be used to solve higher-order differential equations of mixed order. For example, to solve the system

$$\frac{d^2x}{dt^2} + x = 0, \qquad x(0) = \left(\frac{dx}{dt}\right)_{t=0} = 1 \qquad (16.A.4)$$

using rung, we let $y(1) = x$, $y(2) = dx/dt$. Then (16.A.4) is equivalent to the two first-order equations

$$dy(1) = y(2), \qquad dy(2) = -y(1) \qquad (16.A.5)$$

with initial values $y(1) = y(2) = 1$.

`rung` can be used as it stands to solve complex as well as real systems of equations. One simply writes the equations of `deriv` in complex form and declares the appropriate variables complex. The program above is set up specifically to solve an autonomous set of equations of the type shown in Eqs. (6.4.3), where the right sides do not depend explicitly on the independent variable, but it is easily modified to solve non-autonomous systems.

16.B   FORTRAN PROGRAM FOR PLANE-WAVE PROPAGATION

A detailed account of propagation of laser radiation generally requires numerical solutions of coupled partial differential equations. We will describe two computational techniques for obtaining solutions of typical forms of such equations.

In the plane-wave approximation the evolution of the field with distance of propagation z and time t typically involves the partial derivatives $\partial/\partial z$ and $\partial/\partial t$ in the combination $\partial/\partial z + (1/c)\partial/\partial t$, as in Eqs. (9.6.18) and (9.6.21), for example. It is convenient to introduce new independent variables $\eta = z$ and $\tau = t - z/c$ to convert this combination to the single derivative $\partial/\partial\eta$; this procedure was followed in Section 8.4, where the group velocity v_g rather than the vacuum speed of light c appeared in the propagation equation of interest. In numerical computations it is usually also convenient to scale both dependent and independent variables and work with dimensionless quantities. How the scaling is done will depend, of course, on the particular set of equations to be solved.

We will consider a relatively simple but illustrative example of the propagation of intensity in a medium assumed to consist of N two-state atoms per unit volume. The coupled atom-field equations in our example are

$$\frac{\partial I}{\partial z} + \frac{1}{c}\frac{\partial I}{\partial t} = \sigma N I w, \tag{16.B.1a}$$

$$\frac{\partial w}{\partial t} = -A_{21}(w + 1) - \frac{2\sigma}{\hbar\omega} I w. \tag{16.B.1b}$$

These equations follow from (9.6.17) and (9.6.21) when we assume $\Gamma_1 = \Gamma_2 = 0$ and again define $w = (N_2 - N_1)/N$ and the stimulated emission cross section $\sigma = \omega|\mu|^2\beta/[\epsilon_0\hbar c(\Delta^2 + \beta^2)]$ for a homogeneously broadened transition. (If we assume that the intensity is in the form of a pulse of sufficiently short duration that the first term on the right-hand side of (16.B.1b) can also be dropped, then our model reduces to the "Frantz–Nodvik model" of Section 6.12. The reader can use the analytical results obtained there as a check of the FORTRAN program listed below.) It is straightforward to apply the numerical approach we now describe to more general situations in which, for example, it is the propagation of the complex electric field envelope $\mathcal{E}(z, t)$ rather than the intensity that is of interest, or when the field is coupled to more than two atomic states or when the right-hand side of (16.B.1a) is replaced by some prescribed function of the intensity or of z and t.

In our example it is convenient to define the new, dimensionless independent variables $\eta = \sigma N z$ and $\tau = A_{21}(t - z/c)$, in terms of which our model equations become

$$\frac{\partial I}{\partial \eta} = Iw, \tag{16.B.2a}$$

$$\frac{\partial w}{\partial \tau} = -(w + 1) - \frac{2\sigma}{\hbar \omega A_{21}} Iw. \tag{16.B.2b}$$

We will also scale the intensity by defining $\tilde{I} = (2\sigma/\hbar \omega A_{21})I$:

$$\frac{\partial \tilde{I}}{\partial \eta} = \tilde{I}w, \tag{16.B.3a}$$

$$\frac{\partial w}{\partial \tau} = -(w + 1) - \tilde{I}w. \tag{16.B.3b}$$

In this form all variables are dimensionless. We will assume a temporally Gaussian pulse of duration τ_p incident on the medium at $z = 0$: $I(z = 0, t) = I_0 e^{-(t-t_0)^2/\tau_p^2}$, or

$$\tilde{I}(\eta = 0, \tau) = \tilde{I}_0 e^{-(\tau - \tau_0)^2/\tilde{\tau}_p^2}, \tag{16.B.4}$$

where $\tilde{I}_0 = (2\sigma/\hbar \omega A_{21})I_0$ and $\tilde{\tau}_p = A_{21}\tau_p$ is the pulse duration in units of the radiative lifetime $1/A_{21}$. We also assume that all the atoms are initially excited ($w = 1$).
    Equation (16.B.2a) implies

$$\tilde{I}(\eta + \Delta\eta, \tau) \cong [1 + w(\eta, \tau)\Delta\eta]\tilde{I}(\eta, \tau) \tag{16.B.5}$$

for small $\Delta\eta$. We can use this first-order (in $\Delta\eta$) forward stepping in η to obtain $\tilde{I}(\eta, \tau)$ for all values of τ by integrating (16.B.2b) over τ for each step in η. The FORTRAN program planewave implements this scheme for the numerical solution of equations (16.B.2). The output arrays xarray and yy at each step in η are the temporal profiles of $\tilde{I}$ and w at η, and can be plotted versus the time array xx.

```
      program planewave
c           this program is used to solve equations (16.B.2)
c           elen is the length of the medium
c           tlen is the total time interval, chosen here to
            be ten times the pulse duration
c           ndt is the number of time steps, each of size
c           dt=tlen/ndt
c           nde is the number of steps in eta, each of size
c           de=elen/nde
c           pulsdur is the pulse duration
c           xipeak is the peak intensity of the assumed
c           temporally gaussian incident pulse
c           elen, tlen, pulsdur, and xipeak are all scaled
```

```
c          and dimensionless
           common/field/xint
           common/fieldd/ndt,pulseo(500),pulsdur,xipeak,tlen
           dimension y(1),dy(1),w(1,5),xarray(500),
           >xx(500),yy(500)

           elen=10. ; ndt=500 ; nde=100
           pulsdur=0.10 ; tlen=10.*pulsdur ; xipeak=1.0
           de=elen/nde; dt=tlen/ndt
c          define the incident intensity temporal profile
c          and store in array pulseo
           call spulse

           do 1 it=1,ndt
           xarray(it)=pulseo(it)
     1     continue
           do 2 iz=1,nde
           y(1)=1.0
           tin=0.0
           do 3 it=1,ndt
           xint=xarray(it)
           xarray(it)=(1.0+y(1)*de)*xint
           call rung(1,y,dy,tin,dt,w)
c          rung is the fourth-order runge-kutta integrator
c          (Appendix 16.A)
           xx(it)=tin
           yy(it)=y(1)
     3     continue
c            xarray(it) and y(1) are respectively the
c            temporal profiles of the pulse intensity
c            and the population difference w at distance
c            eta=iz*de into the medium
c            xx(it) is the array of time increments
     2     continue
           stop
           end


           subroutine spulse
c          this subroutine defines the temporal profile of
c          the incident pulse
           common/fieldd/ndt,pulseo(500),pulsdur,xipeak,tlen
           cent=0.5*tlen
           dt=tlen/ndt
           do 1 i=1,ndt
           t=(i*dt-cent)/pulsdur
           pulseo(i)=xipeak*exp(-t*t)
     1     continue
           return
           end
```

```
subroutine deriv(x,y,dy)
common/field/xint
dimension y(1),dy(1)
dy(1)=-(y(1)+1.)-xint*y(1)
return
end
```

16.C FORTRAN PROGRAM FOR PARAXIAL PROPAGATION

When the plane-wave approximation is not made, the propagation equations obviously become more complicated. We now describe a computational approach for problems of this sort, using the example of second-harmonic generation when the paraxial wave approximation is made for pump and second-harmonic pulses and when the two pulses have different group velocities and group velocity dispersion. Including diffraction and temporal variations, we replace Eqs. (10.7.5) by

$$\frac{1}{2ik_\omega}\left(\frac{\partial^2 \mathcal{E}_\omega}{\partial x^2}+\frac{\partial^2 \mathcal{E}_\omega}{\partial y^2}\right)+\frac{\partial \mathcal{E}_\omega}{\partial z}+\frac{1}{v_\omega}\frac{\partial \mathcal{E}_\omega}{\partial t}+\frac{i}{2}\beta_\omega\frac{\partial^2 \mathcal{E}_\omega}{\partial t^2}=i\omega\sqrt{\frac{\mu_0}{\epsilon_\omega}}\,\bar{d}\mathcal{E}_\omega^*\mathcal{E}_{2\omega}, \quad (16.C.1a)$$

$$\frac{1}{2ik_{2\omega}}\left(\frac{\partial^2 \mathcal{E}_{2\omega}}{\partial x^2}+\frac{\partial^2 \mathcal{E}_{2\omega}}{\partial y^2}\right)+\frac{\partial \mathcal{E}_{2\omega}}{\partial z}+\frac{1}{v_{2\omega}}\frac{\partial \mathcal{E}_{2\omega}}{\partial t}+\frac{i}{2}\beta_{2\omega}\frac{\partial^2 \mathcal{E}_{2\omega}}{\partial t^2}=i\omega\sqrt{\frac{\mu_0}{\epsilon_{2\omega}}}\,\bar{d}\mathcal{E}_\omega^2. \quad (16.C.1b)$$

We assume perfect phase matching and ignore "walkoff" associated with birefringence (Section 8.8), which is often a small effect; phase matching and walkoff are easily included in the numerical approach we now describe. Equations (16.C.1) can be simplified slightly by making a transformation similar to that made in replacing (16.B.1) by (16.B.2); in this case we define $\eta = z$ and $\xi = t - z/v_\omega$ and replace (16.C.1) by

$$\frac{1}{2ik_\omega}\left(\frac{\partial^2 \mathcal{E}_\omega}{\partial x^2}+\frac{\partial^2 \mathcal{E}_\omega}{\partial y^2}\right)+\frac{\partial \mathcal{E}_\omega}{\partial \eta}+\frac{i}{2}\beta_\omega\frac{\partial^2 \mathcal{E}_\omega}{\partial \xi^2}=i\omega\sqrt{\frac{\mu_0}{\epsilon_\omega}}\,\bar{d}\mathcal{E}_\omega^*\mathcal{E}_{2\omega}, \quad (16.C.2a)$$

$$\frac{1}{2ik_{2\omega}}\left(\frac{\partial^2 \mathcal{E}_{2\omega}}{\partial x^2}+\frac{\partial^2 \mathcal{E}_{2\omega}}{\partial y^2}\right)+\frac{\partial \mathcal{E}_{2\omega}}{\partial \eta}+\left(\frac{1}{v_{2\omega}}-\frac{1}{v_\omega}\right)\frac{\partial \mathcal{E}_{2\omega}}{\partial \xi}+\frac{i}{2}\beta_{2\omega}\frac{\partial^2 \mathcal{E}_{2\omega}}{\partial \xi^2}=i\omega\sqrt{\frac{\mu_0}{\epsilon_{2\omega}}}\,\bar{d}\mathcal{E}_\omega^2.$$

$$(16.C.2b)$$

We define dimensionless independent variables $Z = \eta/L$, $X = x/a$, $Y = y/a$, and $T = \xi/\tau$, the lengths L and a and the time τ at this point being arbitrary. We also define dimensionless dependent variables F and H by writing

$$\mathcal{E}_\omega = \sqrt{\frac{2I_0}{n_\omega\epsilon_0 c}}F, \qquad \mathcal{E}_{2\omega} = \sqrt{\frac{2I_0}{n_{2\omega}\epsilon_0 c}}H. \quad (16.C.3)$$

$|F|^2$ and $|H|^2$ are, respectively, the fundamental intensity and the second-harmonic intensity in units of the (arbitrary) intensity I_0. These scalings transform

Eqs. (16.C.2) to

$$\frac{\partial F}{\partial Z} - \frac{i\lambda L}{4\pi a^2 n_\omega}\left(\frac{\partial^2 F}{\partial X^2} + \frac{\partial^2 F}{\partial Y^2}\right) + \frac{i\beta_\omega L}{2\tau^2}\frac{\partial^2 F}{\partial T^2} = iCF^*H, \qquad (16.C.4a)$$

$$\frac{\partial H}{\partial Z} - \frac{i\lambda L}{8\pi a^2 n_{2\omega}}\left(\frac{\partial^2 H}{\partial X^2} + \frac{\partial^2 H}{\partial Y^2}\right) + \frac{L}{\tau}\left(\frac{1}{v_{2\omega}} - \frac{1}{v_\omega}\right)\frac{\partial H}{\partial T} + \frac{i\beta_{2\omega}L}{2\tau^2}\frac{\partial^2 H}{\partial T^2} = iCF^2. \quad (16.C.4b)$$

Here $\lambda = 2\pi c/\omega$ is the fundamental wavelength and

$$C = \bar{\omega}\bar{d}L\sqrt{\frac{2I_0}{n_\omega^2 n_{2\omega}(\epsilon_0 c)^3}} \qquad (16.C.5)$$

is a dimensionless constant.

A simplified "split-step" approximation to the solution of Eqs. (16.C.4) is obtained as follows. Given $F(X, Y, Z = 0, T)$ and $H(X, Y, Z = 0, T)$ at the input plane $(Z = 0)$ of the medium, we first solve the differential equations

$$\frac{\partial F}{\partial Z} = iCF^*H, \qquad (16.C.6a)$$

$$\frac{\partial H}{\partial Z} = iCF^2 \qquad (16.C.6b)$$

over a step ΔZ, keeping fixed the X, Y, and T dependence of F and H. Then use the resulting F and H as initial values to obtain approximations to $F(X, Y, \Delta Z, T)$ and $H(X, Y, \Delta Z, T)$ by solving the equations

$$\frac{\partial F}{\partial Z} - \frac{i\lambda L}{4\pi a^2 n_\omega}\left(\frac{\partial^2 F}{\partial X^2} + \frac{\partial^2 F}{\partial Y^2}\right) + \frac{i\beta_\omega L}{2\tau^2}\frac{\partial^2 F}{\partial T^2} = 0, \qquad (16.C.7a)$$

$$\frac{\partial H}{\partial Z} - \frac{i\lambda L}{8\pi a^2 n_{2\omega}}\left(\frac{\partial^2 H}{\partial X^2} + \frac{\partial^2 H}{\partial Y^2}\right) + \frac{L}{\tau}\left(\frac{1}{v_{2\omega}} - \frac{1}{v_\omega}\right)\frac{\partial H}{\partial T} + \frac{i\beta_{2\omega}L}{2\tau^2}\frac{\partial^2 H}{\partial T^2} = 0 \quad (16.C.7b)$$

for the entire range of values of X, Y, and T of interest. These approximations are then used as initial conditions to solve again (16.C.6) and (16.C.7), giving us an approximation to $F(X, Y, 2\Delta Z, T)$ and $H(X, Y, 2\Delta Z, T)$. This procedure of solving (16.C.6) and (16.C.7) separately (in "split steps") and sequentially is continued for as many steps as are necessary to advance to $Z = Z_{\text{final}}$.

The solution of the differential equations for the "diffractionless" step [Eqs. (16.C.6) in our example] must, in general, be done numerically using, for instance, a Runge–Kutta integrator. Equations of the form (16.C.7), being linear, can be solved using Fourier transforms; what makes the Fourier transform approach especially efficient is the Fast Fourier Transform (FFT) algorithm for the numerical evaluation of Fourier transforms.

We define a Fourier transform $\tilde{F}(\alpha, \beta, Z, \gamma)$ by writing

$$F(X, Y, Z, T) = \int_{-\infty}^{\infty} d\alpha \int_{-\infty}^{\infty} d\beta \int_{-\infty}^{\infty} d\gamma \tilde{F}(\alpha, \beta, Z, \gamma) e^{-2\pi i [\alpha X + \beta Y + \gamma T]}, \qquad (16.C.8)$$

or equivalently

$$\tilde{F}(\alpha, \beta, Z, \gamma) = \int_{-\infty}^{\infty} dX \int_{-\infty}^{\infty} dY \int_{-\infty}^{\infty} dT \, F(X, Y, Z, T) e^{2\pi i [\alpha X + \beta Y + \gamma T]}. \qquad (16.C.9)$$

In "frequency space" Eq. (16.C.7a), for example, is simply

$$\frac{\partial}{\partial Z} \tilde{F}(\alpha, \beta, Z, \gamma) = -iV(\alpha, \beta, \gamma) \tilde{F}(\alpha, \beta, Z, \gamma), \qquad (16.C.10)$$

or

$$\tilde{F}(\alpha, \beta, Z + \Delta Z, \gamma) = \tilde{F}(\alpha, \beta, Z, \gamma) e^{-iV(\alpha, \beta, \gamma)\Delta Z}, \qquad (16.C.11)$$

with

$$V(\alpha, \beta, \gamma) \equiv \frac{\pi \lambda L}{a^2 n_\omega} (\alpha^2 + \beta^2) - \frac{2\pi^2 \beta_\omega L}{\tau^2} \gamma^2. \qquad (16.C.12)$$

X, Y, and T are replaced in computations by discrete variables $n_X \Delta X$, $n_Y \Delta Y$, and $n_T \Delta T$, and $F(X, Y, Z, T)$ is defined on a lattice:

$$F(X, Y, Z, T) \quad \longrightarrow \quad F(n_X \Delta X, n_Y \Delta Y, Z, n_T \Delta T) \equiv \mathcal{F}(n_X, n_Y, n_T, Z), \qquad (16.C.13)$$

where $n_X = 0, 1, \ldots, N_X - 1$, $n_Y = 0, 1, \ldots, N_Y - 1$, and $n_T = 0, 1, \ldots, N_T - 1$. It will be convenient to take the integers N_X, N_Y, and N_T to be even.

The Fourier transform of F is also discretized in computations:

$$\tilde{F}(\alpha, \beta, Z, \gamma) \quad \longrightarrow \quad \tilde{F}(\alpha_\ell, \beta_m, Z, \gamma_n) \equiv \tilde{\mathcal{F}}(\alpha_\ell, \beta_m, \gamma_n, Z), \qquad (16.C.14)$$

and we write

$$\tilde{\mathcal{F}}(\alpha_\ell, \beta_m, \gamma_n, Z) = \sum_{n_X=0}^{N_X-1} \sum_{n_Y=0}^{N_Y-1} \sum_{n_T=0}^{N_T-1} \mathcal{F}(n_X, n_Y, n_T, Z) e^{2\pi i [n_X \alpha_\ell \Delta X + n_Y \beta_m \Delta Y + n_T \gamma_n \Delta T]}.$$

$$(16.C.15)$$

Note that α_ℓ, β_m, and γ_n are not restricted to positive values.

The smaller we choose ΔX, for instance, the larger are the frequencies α_ℓ that we can "resolve" in the X variations of $\mathcal{F}$. In fact, there is a maximum frequency that can be resolved for a given ΔX, the *Nyquist critical frequency* given by $\alpha_{max} = 1/(2\Delta X)$. Therefore, we take the frequencies α_ℓ in (16.C.15) to range from $-1/(2\Delta X)$ to $-1/(2\Delta X)$. Furthermore, since n_X takes on N_X different values, we can obtain from the equations (16.C.15) only N_X frequencies α_ℓ. So we take

$$\alpha_\ell = \frac{\ell}{N_X \Delta X}, \qquad \ell = -\frac{N_X}{2}, \ldots, \frac{N_X}{2}, \qquad (16.C.16a)$$

and similarly

$$\beta_m = \frac{m}{N_Y \Delta Y}, \quad m = -\frac{N_Y}{2}, \ldots, \frac{N_Y}{2}, \quad \gamma_n = \frac{n}{N_T \Delta T}, \quad n = -\frac{N_T}{2}, \ldots, \frac{N_T}{2}, \qquad (16.C.16b)$$

and replace (16.C.15) with

$$\tilde{\mathcal{F}}(\ell, m, n, Z) = \sum_{n_X=0}^{N_X-1} \sum_{n_Y=0}^{N_Y-1} \sum_{n_T=0}^{N_T-1} \mathcal{F}(n_X, n_Y, n_T, Z) e^{2\pi i[n_X \ell/N_X + n_Y m/N_Y + n_T n/N_T]}. \quad (16.C.17)$$

α_ℓ, for instance, appears from (16.C.16a) to have $N_X + 1$ values, which contradicts our assertion that there are at most N_X values. But, from (16.C.17), $\tilde{\mathcal{F}}(-\ell, m, n, Z) = \tilde{\mathcal{F}}(N_X - \ell, m, n, Z)$, and in particular $\tilde{\mathcal{F}}(-N_X/2, m, n, Z) = \tilde{\mathcal{F}}(N_X/2, m, n, Z)$, so there are, in fact, N_X independent values of α_ℓ. It is conventional to take the integer ℓ to vary from 0 to $N_X - 1$, in which case the values $1 \leq \ell \leq N_X/2 - 1$ correspond to positive frequencies, $\ell = 0$ to zero frequency, and $N_X/2 + 1 \leq \ell \leq N_X - 1$ to negative frequencies. Using the same convention for labeling the frequencies β_m and γ_n, we write the inverse Fourier transform of (16.C.17) as

$$\mathcal{F}(n_X, n_Y, n_T, Z) = N_X N_Y N_T \sum_{\ell=0}^{N_X-1} \sum_{m=0}^{N_Y-1} \sum_{m=0}^{N_T-1} \tilde{\mathcal{F}}(\ell, m, n, Z) e^{-2\pi i[n_X \ell/N_X + n_Y m/N_Y + n_T n/N_T]}. \quad (16.C.18)$$

The factor $N_X N_Y N_T$ ensures that (16.C.17) follows from (16.C.18) and vice versa.

The numerical solution of Eq. (16.C.7a) for each propagation step ΔZ therefore proceeds by computing the Fourier transform (16.C.17), evaluating (for each ℓ, m, and n)

$$\tilde{\mathcal{F}}(\ell, m, n, Z + \Delta Z) = \tilde{\mathcal{F}}(\ell, m, n, Z) e^{-iV(\ell,m,n)\Delta Z} \qquad (16.C.19)$$

with

$$V(\ell, m, n) = \frac{\pi \lambda L}{a^2 n_\omega}\left[\left(\frac{\ell}{N_X \Delta X}\right)^2 + \left(\frac{m}{N_Y \Delta Y}\right)^2\right] - \frac{2\pi^2 \beta_\omega L}{\tau^2}\left(\frac{n}{N_T \Delta T}\right)^2, \qquad (16.C.20)$$

and then performing the inverse Fourier transform (16.C.18) for each n_X, n_Y, and n_T. Equation (16.C.7b) is, of course, solved in the same fashion at each propagation step.

In our example we assume $\mathcal{E}_{2\omega}(x, y, z = 0, t) = 0$ and initial Gaussian profiles in space and time for $\mathcal{E}_\omega$:

$$\mathcal{E}_\omega(x, y, z, t) = \mathcal{E}_\omega^{(0)} e^{-(x^2+y^2)/w_0^2} e^{-t^2/\tau_p^2}. \qquad (16.C.21)$$

In terms of F and the scaled variables X, Y, and T, this translates to

$$F(X, Y, Z = 0, T) = F_0 e^{-(X^2+Y^2)} e^{-T^2} \qquad (16.C.22)$$

when we choose $a = w_0$ and $\tau = \tau_p$. The definitions (16.C.3) imply that $|F_0|^2$ is the peak intensity of the input pump field and, as already noted, $|F(X, Y, Z, T)|^2$ and $|H(X, Y, Z, T)|^2$ are, respectively, the intensities, in units of I_0, of the fundamental and second-harmonic fields at X, Y, Z, T. We assume the fundamental to have a wavelength $\lambda = 1064$ nm and to be incident on a crystal for which $\bar{d} = 3.4 \times 10^{-24}$. For type I phase matching with an ordinary pump and an extraordinary second harmonic we take $n_\omega = 1.494$, $n_{2\omega} = 1.471$, $v_\omega = 1.966 \times 10^8$ m/s, $v_{2\omega} = 1.941 \times 10^8$ m/s, $\beta_\omega = 7.411 \times 10^{-21}$ s^2/m, and $\beta_{2\omega} = 1.870 \times 10^{-21}$ s^2/m; these happen to be appropriate values for the crystal KDP. If we choose $I_0 = 1$ GW/cm^2 and $L = 1$ cm, then $C = 1.09$. In the program `paraxial` listed below we let $F_0 = 1.0$, $w_0 = 0.1$ cm, $\tau_p = 1$ ns, and the propagation distance (`zlength`) $= 1$ cm. We assume there is no input second-harmonic field: $H = 0$ at $Z = 0$.

Different FFT programs are available for the evaluation of discrete Fourier transforms. `paraxial` calls the readily available and well-documented subroutine `fourn`[1] and puts the F and H arrays (`dataf` and `datah`) in the "normal FORTRAN form" required there.

FFT routines such as `fourn` require the number of elements in the arrays to be a power of 2. `paraxial` as listed below assumes $N_X = N_Y = 128$ ($= 2^7$) and $N_T = 64$. The X, Y, and T grid sizes (`xsize`, `ysize`, and `tsize`) are all taken to be 5.0, and the dimensionless propagation length Z goes from 0 to `zlen` $= 1$ in `lz` $= 10$ steps. Of course, all these can be changed. In particular, `xsize`, `ysize`, and `tsize` should always be made large enough that F and H in our example remain negligibly small at the grid boundaries.

`paraxial` can be checked and applied in different ways. For example, if we turn off the coupling between F and H by setting $C = 0$ (or by just "commenting out" the call to subroutine `medium`) we can check that the results obtained with `paraxial` conform with known analytical results for Gaussian beam propagation.

The subroutine `plotfile` is used for plotting the input pump intensity at $Z = 0$ versus X for $Y = T = 0$, and the output second-harmonic intensity versus Y for $X = T = 0$ and versus T for $X = Y = 0$. `plotfile` stores the data in the file `pl` to be read or plotted. We use `plotfile` for X, Y, and T variations in order to indicate how these variations are retrieved using the normal Fortran ordering of the F and H files.

---

[1]W. H. Press, S. A. Teukolsky, W. T. Vetterling, and B. P. Flannery, *Numerical Recipes* in FORTRAN 77, second edition, Cambridge University Press, New York, 1986.

The propagation length (1 cm) in `paraxial` is such that neither diffraction nor group velocity dispersion are significant; the much longer crystal lengths required for these effects to be important in our example would be unrealistic. However, small-scale phase variations across the input pump beam, or much shorter pulse durations, could introduce substantial spatial and temporal variations of the fields, as the reader may verify in "numerical experiments."

`paraxial` employs the simplest implementation of the FFT-based split-step method and as such is accurate to first order in ΔZ; it is easy to modify the program, using a "symmetrized" split-step procedure,[2] so that it is accurate to second order. To reduce the computation times we have used in `medium` a second-order Runge–Kutta algorithm rather than the fourth-order version. For many problems of interest `paraxial` requires computational times on the order of a few minutes or less on typical laptop computers.

```
        program paraxial
c       this program solves time-dependent wave equations including
c       diffraction
c       this version solves the equations (16.C.2)
c       lx, ly, lt are the numbers of grid points in x, y, and t,
c       respectively
c       lz is the number of propagation steps in the z direction
        common/fort/np,nhalf
        common/propmed/C,npp
        complex xi,f(128,128),h(128,128)
        dimension xx(128),xintf(128),xinth(128),frqx(128),
      >frqy(128),frqt(128)
        dimension nn(3),dataf(4194304),datah(4194304)
        dimension cf(4194304),sf(4194304),ch(4194304),sh(4194304)
c       note: 4194304=2(128)(128)(128)
        open(unit=2,file="pl")

        tpi=6.28318530717959 ; pi=.5*tpi ; xi=cmplx(0.0,1.0)
        zlength=1.0e-2 ; wspot=1.0e-3 ; xlam=1.064e-6 ; taup=10.0e-9
        lx=128 ; ly=lx ; lt=64 ; lz=10
        xsize=5.0 ; ysize=5.0 ; tsize=5.0 ; zlen=1.0

c       initialize data for grid size, etc.
        lx2=lx/2 ; ly2=ly/2 ; lt2=lt/2 ; lxx=2*lx ; lxx1=lxx-1 ; lz1=lz-1
        dx=xsize/lx ; dy=ysize/ly ; dt=tsize/lt ; dz=zlen/lz
        nxy2=lx*ly ; nxy=2*nxy2 ; nbig=nxy*lt ; nhalf=nbig/2 ; np=nbig-1
        npp=np; nn(1)=lx ; nn(2)=ly ; nn(3)=lt

        C=1.09 ; f0=1.0 ; xnf=1.494 ; xnh=1.471

c       define group velocities and group velocity coefficients of f and h
c       fields
```

---

[2]See, for instance, J. A. Fleck, J. R. Morris, and M. D. Feit, *Applied Physics* **10**, 129 (1976) and more recent literature on split-step methods.

```
      groupf=1.966e8
      grouph=1.941e8
      betaf=7.411e-21
      betah=1.870e-21

      diffrf=pi*xlam*zlength/wspot**2*dz/xnf
      ff=2*pi**2*betaf*zlength/taup**2*dz
      diffrh=pi*0.5*xlam*zlength/wspot**2*dz/xnh
      grh=tpi*(zlength/taup)*(1./grouph-1./groupf)*dz
      hh=2*pi**2*betah*zlength/taup**2*dz

      z=0.0
c     define initial (x,y) field distribution at z=0,
c     assumed gaussian
      f0=1.
      do 5 i=1,lx
      x=(i-lx2)*dx
      do 6 j=1,ly
      y=(j-ly2)*dy
      f(i,j)=f0*exp(-(x**2+y**2))
      h(i,j)=0.0
    6 continue
    5 continue
c     define spatial frequencies for x,y grids, in normal fortran form
c     assumed in fourn
      call freq(frqx,lx,dx)
      call freq(frqy,ly,dy)
c     define the temporal frequencies for t grid
      call freq(frqt,lt,dt)

c     put arrays into normal fortran form
      do 100 m=1,lt
      t=(m-lt2)*dt
      fm=exp(-t**2)
c     fm defines the temporal profile of the pulse intensity
c     we are assuming here a gaussian temporal profile.
      k=0
      jp=1
      do 10 j=1,nxy,2
      k=k+1
      if(k.gt.lx) then
      k=1
      jp=jp+1
      endif
      jj=(m-1)*nxy+j
c     also put the appropriate sines and cosines in the form appropriate
c     to the normal fortran form assumed for the data arrays
      vf=diffrf*(frqx(k)**2+frqy(jp)**2)-ff*frqt(m)**2
```

```
       vh=diffrh*(frqx(k)**2+frqy(jp)**2)-grh*frqt(m)-hh*frqt(m)**2
       cf(jj)=cos(vf)
       sf(jj)=sin(vf)
       ch(jj)=cos(vh)
       sh(jj)=sin(vh)
       dataf(jj)=real(f(k,jp))*fm
       dataf(jj+1)=aimag(f(k,jp))*fm
       datah(jj)=0.0
       datah(jj+1)=0.0
  10   continue
 100   continue

c      define initial x intensity profile along a slice with y=t=0
       i=0
       do 800 j=1,lxx1,2
       k=nhalf+nxy2+j
       i=i+1
       xx(i)=i*dx
       xintf(i)=dataf(k)**2+dataf(k+1)**2
 800   continue
       call plotfile(2,lx,xx,xintf)

       do 20 nnn=1,lz
       if(nnn.eq.1) go to 21
c      go to real spacetime domain
       call fourn(dataf,nn,-1)
       call fourn(datah,nn,-1)
  21   continue
c      modify field by effect of the medium
       call medium(dataf,datah,dz)
c      go to (spatial and temporal) frequency space
       call fourn(dataf,nn,1)
       call fourn(datah,nn,1)
c      free propagation by dz in frequency space
       do 320 j=1,np,2
       datfj=dataf(j)*cf(j)+dataf(j+1)*sf(j)
       datfj1=dataf(j+1)*cf(j)-dataf(j)*sf(j)
       dataf(j)=datfj
       dataf(j+1)=datfj1
       dathj=datah(j)*ch(j)+datah(j+1)*sh(j)
       dathj1=datah(j+1)*ch(j)-datah(j)*sh(j)
       datah(j)=dathj
       datah(j+1)=dathj1
 320   continue
  20   continue
c      back to real spacetime
       call fourn(dataf,nn,-1)
       call fourn(datah,nn,-1)
```

```
c       define y intensity profile of second harmonic along x=t=0
c       note: t=0 refers to temporal center of the input pulse
        do 81 j=1,ly
        k=nhalf+lx+2*(j-1)*lx
        xx(j)=j*dy
        xinth(j)=datah(k)**2+datah(k+1)**2
   81   continue
        call plotfile(2,lx,xx,xintf)

c       define t intensity profile of second harmonic along x=y=0
        do 84 j=1,lt
        k=nxy2+lx+(j-1)*nxy
        xx(j)=j*dt
        xinth(j)=datah(k)**2+datah(k+1)**2
   84   continue
        call plotfile(2,lt,xx,xinth)
        stop
        end


        subroutine fourn(data,nn,isign)
c       this FFT subroutine is described and listed in Press et al.[1]
c       use common/fort/np,nhalf in fourn


        subroutine medium(dataf,datah,dz)
        common/propmed/C,npp
c       this routine calculates the change in the field over a step deltaz
c       due to the medium
c       a second-order runge-kutta algorithm is used
        dimension dataf(1),datah(1)
        do 1 j=1,npp,2
        fr=datah(j) ; fi=datah(j+1)
        hr=datah(j) ; hi=datah(j+1)
        frk1=dz*C*(fi*hr-fr*hi) ; fik1=dz*C*(fr*hr+fi*hi)
        hrk1=-2.*dz*C*fr*fi ; hik1=dz*C*(fr*fr-fi*fi)
        fri=fr+frk1 ; fii=fi+fik1
        hri=hr+hrk1 ; hii=hi+hik1
        frk2=dz*C*(fii*hri-fri*hii) ; fik2=dz*C*(fri*hri+fii*hii)
        hrk2=-2.*dz*C*fri*fii ; hik2=dz*C*(fri*fri-fii*fii)
        fr=fr+.5*(frk1+frk2) ; fi=fi+.5*(fik1+fik2)
        hr=hr+.5*(hrk1+hrk2) ; hi=hi+.5*(hik1+hik2)
        dataf(j)=fr ; dataf(j+1)=fi
        datah(j)=hr ; datah(j+1)=hi
    1   continue
        return
        end


        subroutine plotfile(lu,n,x,y)
        dimension x(1),y(1)
```

```
      do 100 k=1,n
      write(lu,1500) x(k),y(k)
 100  continue
1500  format(1p2e15.7)
      return
      end

      subroutine freq(frq,lx,delta)
      dimension frq(1)
      lx2=lx/2
      lxx=lx2+1
      lxx1=lxx+1
      xnd=1./(lx*delta)
      do 1 i=1,lx2
      frq(i)=(i-1)*xnd
  1   continue
      frq(lxx)=1./(2.*delta)
      do 2 i=lxx1,lx
      frq(i)=-(lx-i+1)*xnd
  2   continue
      return
      end
```

INDEX